Aquatic Plants

of Northern and Central Europe
including Britain and Ireland

Published by Princeton University Press
41 William Street, Princeton, New Jersey 08540
99 Banbury Road, Oxford OX2 6JX

press.princeton.edu

Library of Congress Control Number 2023938119

ISBN 978-0-691-25101-1
ISBN (e-book) 978-0-691-25102-8

British Library Cataloging-in-Publication Data is available

Editorial: Robert Kirk and Megan Mendonça
Production Editorial: Jill Harris
Jacket Design: Wanda España
Production: Steven Sears
Publicity: William Pagdatoon and Caitlyn Robson
Copyeditor: Jennifer McClain and Jens Christian Schou

Jacket image: Front: *Groenlandia densa* by Klaus van de Weyer. Back: (from left to right) photography by Klaus van de Weyer, illustration by Jens Christian Schou

Layout and drawings: Jens Christian Schou

This book has been composed in Calibri

The publisher would like to acknowledge the authors of this volume for providing the print-ready files from which this book was printed.

Printed on acid-free paper. ∞

Printed in Italy

10 9 8 7 6 5 4 3 2 1

Jens Christian Schou, Bjarne Moeslund, Klaus van de Weyer, Richard V. Lansdown,
Gerhard Wiegleb, Peter Holm, Lars Baastrup-Spohr and Kaj Sand-Jensen

Aquatic Plants

of Northern and Central Europe including Britain and Ireland

AAGE V. JENSEN NATURFOND

WILDGuides

PRINCETON UNIVERSITY PRESS

The authors

Jens Christian Schou (born 1954) is a self-made botanist, with particular interests in Nordic botany. Jens was a schoolteacher for 36 years, but during a holiday in 2001-2003 was employed as a botanical draftsman for the project "Bestimmen lehrnen online" at the German Ministry of Research and Education. Jens is author and illustrator of identification works on Cyperaceae, Rubus, Hieracium, Apiaceae, Poaceae, Juncaceae and aquatic plants, lead author of the comprehensive work *Danmarks Vandplanter*, and author a large number of articles on Danish botany. He contributes as an artist for *Flora Ecuador* and is referee of aquatic plants for the Atlas Flora Danica project. Jens draws and paints plants and animals to illustrate books and scientific publications, posters and exhibitions. Some of his works as a nature photographer can be found at www.biopix.com.

Bjarne Moeslund (born 1954) obtained an MSc in biology from Aarhus University in 1981 and since then has worked as an environmental consultant, first in Bio/consult A/S, then in Orbicon A/S and currently in WSP Denmark A/S. His work has spanned a very wide range of projects with a focus on monitoring and assessment of the habitat quality as well as restoration and management of aquatic and wetland habitats. He is author/co-author of numerous technical reports as well as several scientific articles. In 1990 he was in charge of the preparation of the first Danish guide to aquatic plants, and in 1997 was co-author of *Danmarks Vandplanter*. Bjarne has presented numerous training courses on the identification of aquatic plants and on river management; he has also contributed to several technical guidelines on monitoring and assessment of aquatic habitats.

Klaus van de Weyer (born 1961) is a freshwater ecologist and botanist. He has worked on macrophytes for 35 years and is a recognised expert in his field. He is a member of several German botanical committees including CEN/DIN and ICPR. Klaus is founder and managing director of the environmental consultancy lanaplan, specialising in freshwater macrophytes. He has designed different assessment systems for macrophytes in riverine and lacustrine ecosystems under the Water Frame Work Directive in Germany. Klaus has also published several papers concerning macrophyte identification, ecology, distribution and their threat status in Germany (www.lanaplan.de). He is a scuba diver and underwater photographer. Klaus has provided training courses on macrophyte ecology and identification since 2000.

Richard Lansdown (born 1962) has worked for more than 40 years in wetland conservation, specialising in the survey and identification of aquatic plants. He has surveyed macrophytes in rivers for more than 20 years throughout Britain and Ireland and has studied the ecology and identification of aquatic plants throughout the world. He is chair of the IUCN SSC Freshwater Plant Specialist Group and the Natural England Aquatic Plant Taxon Group and serves as referee for *Callitriche*, *Ranunculus* subgenus *Batrachium*, *Sparganium*, *Zannichellia* and the Lemnaceae. Richard has run courses on aquatic plant identification for more than 15 years in the UK and France.

Gerhard Wiegleb (born 1948) holds a PhD in biology from the University of Göttingen and is currently professor emeritus of general ecology at Brandenburg University of Technology in Cottbus. He has been working on macrophyte ecology since 1972, with the aim of developing valid assessment systems for lowland rivers based on macrophyte presence and abundance. He has published more than 300 papers and book articles on freshwater ecology, restoration ecology, conservation biology, biodiversity ethics, environmental law, and aquatic plant taxonomy, including publication of both global and regional accounts of the genus *Potamogeton* sensu lato. A worldwide revision of *Ranunculus* section *Batrachium* became his major field of interest in 2012, leading to publication of a monographic account to the section in 2017.

Peter Holm (born 1953) is a schoolteacher and a self-taught expert in identification of macro invertebrates and aquatic plants in rivers, streams and lakes. He was employed at the environmental office in Aarhus County at the Danish Ministry of Environment and at a private company between 1986 and 2014. He carried out water quality assessments for more than 35 years and contributed to implementation of the Water Framework Directive in Denmark, including extensive surveys both for preparation of river basin management plans and to inform wetland restoration. Peter has run training courses on macro invertebrates and aquatic plants since 1990.

Lars Baastrup-Spohr (born 1982) is an assistant professor of Freshwater Ecology at the University of Copenhagen. He has a master's degree and PhD in biology from the Freshwater Biology Laboratory, University of Copenhagen. His research focuses on aquatic plant distribution, biodiversity and historical development, as well as on the effect of restoration of lakes and streams on aquatic plants. Lars teaches courses in freshwater ecology and biodiversity.

Kaj Sand-Jensen (born 1950) is professor of Freshwater Ecology at the University of Copenhagen and formerly professor in plant ecology and physiology at Aarhus University in the 1980s. Kaj is an author of numerous scientific articles and several books on biodiversity, ecosystem ecology and plant ecophysiology in streams, lakes and coastal marine waters. He is an engaged naturalist who has published books and popular science articles on Danish nature; he is also a member of the Danish Natural Science Academy and the Danish Royal Scientific Society.

Contents

Preface

We are extremely fond of aquatic plants. As naturalists, we find them beautiful and fascinating. As taxonomists, we strive to identify the species in the many different life forms in which they occur. As scientists, we reveal how their distribution can tell us about past and present environmental conditions in aquatic habitats. Enjoying the beauty of species and using them in science requires proper identification. Identifying aquatic vascular plants has been challenging in the past due to lack of well-illustrated identification keys, a frequent lack of flowering parts and the wide variation in vegetative traits.

In this book we offer the illustrated identification keys and the descriptions and illustrations needed to identify the more than 400 aquatic, vascular plant species, subspecies and hybrids found in Northern and Central Europe from Germany to Greenland, from Estonia to Britain and Ireland, and from Poland to the Netherlands. We also offer a brief introduction to the history of aquatic plant research, species distribution patterns and the type of habitats for aquatic plants within our target area. Practical recommendations for the study of aquatic plants are also included.

Five of us have a background in publishing the book on Danish aquatic plants (*Danmarks Vandplanter* 2017). When Klaus van de Weyer, Gerhard Wiegleb (Germany) and Richard Lansdown (Britain), agreed to be co-authors, we were able to expand our analyses beyond the Nordic borders. We felt confident that we had an expert group of field botanists and scientists who knew all the aquatic species well throughout the geographic region with 17 individual areas or countries.

Aage V. Jensen's Nature Foundation has supported us with grants for travel expenses and for illustrations with line drawings and photographs. We thank Karl Georg Bernhardt, Alexander A. Bobrov, Victor Chepinoga, Sven Dahlke, Martin Dančák, Niels Faurholdt, Claudia Ferguson-Smyth, Thomas Franke, Michael Hassler, David Hlisnikovský, Mogens Holmen, Klaus Høiland, Jan-Thomas Johansson, Uwe Koenzen, Birgit Knudsen, Sebastian Meis, Inge Nagstrup, Jana Navrátilová, Henrik Ærenlund Pedersen, Rikke Persson, Frank Pätzold, Heidi Rauers, Irene Strang, Wim van der Ven, Filip Verloove, Åke Widgren, Vilhelm Wrigstedt who all have made supplementary photos available to us. John Bruinsma, Laura Grīnberga, Ari Hyvärinen, Zdenek Kaplan, Helle Mäemets, Jan Prančl, Zofija Sinkevičienė, Rune Svensson, Jari Särkkä and Benjamin Øllgaard have all helped us in various ways during the work.

Kevin Murphy, University of Glasgow, improved and corrected the English text. Retired publisher Axel Kielland connected us with publisher Robert Kirk of Princeton University Press, who offered to publish and distribute the book. We are delighted with the final book and look forward to seeing it widely used by naturalists and professionals assessing aquatic habitats and wetlands throughout and outside Northern and Central Europe, Britain and Ireland.

Introduction

Travelling down the river

In pouring rain, on a summer's day in 2012, we sailed the lowermost 25 km of the River Gudenå from Ulstrup to the mouth of the river in Randers Fjord with the towering skipper, Bjarne Moeslund, standing in the stern of the wide-bellied inflatable boat. As we progressed, raindrops on our faces were replaced by an inner excitement over the botanical miracles that were revealed in the stream below us. The River Gudenå had certainly not become barren.

We discovered 30 m long stands and up to 5 m long shoots of a *Ranunculus baudotii* hybrid as well as equally large *Potamogeton* species and hybrids that rose from the river bottom 2–3 m below us. Along the banks, the vegetation formed an almost impenetrable mass all the way to the surface. In the middle of the stream there was open water above the vegetation that swayed back and forth carried by the large swirls in the fast current. If we came too close to the bank to examine an interesting plant stand that Bjarne had spotted from his elevated position in the stern, we got stuck and had to push ourselves free and clean the entangled plant stems off the propeller.

Bjarne, Peter Holm and Jens Chr. Schou had previously walked and sailed part or the entire length of the 176 km River Gudenå, the longest in Denmark, but they had rarely seen such dense aquatic vegetation. The other members of the excursion - Lars Baastrup-Spohr, Zdeněk Kaplan, Jan Prančl and Kaj Sand-Jensen – were on this river boat trip for the first time. We were all there due to an enquiry from the two Czechs, who wanted to check whether Johannes Baagøe's 17 hybrids of *Potamogeton*, collected from mainly Jutland streams more than 100 years earlier, were still growing there. The hybrids had been identified based on their anatomy and morphology and specimens are preserved in Baagøe's special herbarium of the Botanical Museum in Copenhagen. Zdeněk Kaplan had studied the famous herbarium and wanted to apply genetic analysis to either confirm or dismiss the existence of the allegedly numerous hybrids. J. Hagström had accepted their validity in his monography (1916), while later scholars had reduced them to form variations within species. Here we were, 117 years after J. Baagøe's and C. Raunkiær's first botanical expedition to the River Gudenå and other localities in Jutland in the summer of 1895. We soon realised that we had arrived in the Garden of Eden and found a plant richness that we had never experienced before. However, which species grew here and whether Baagøe's species and hybrids were still present were the two questions we wanted to answer.

Gigantic specimens of *Potamogeton lucens* and *P. praelongus* with leaves 20–30 cm long were quite frequent. We also found *P. perfoliatus* and three hybrids involving those three species: *P.*

River Gudenå, Denmark in pouring rain. Photo JCS.

×*cognatus (P. perfoliatus × praelongus)*, *P.* ×*salicifolius (P. lucens × perfoliatus)* and *P.* ×*undulatus (P. crispus × praelongus)*. *Potamogeton lucens* reflected the light from its lattice of veins in the fresh green leaves completely devoid of the greyish layer of marl seen on individuals from lakes. The hybrid *P.* ×*cognatus* was earlier known from four locations in Denmark, but presently it is only found in the lower part of the River Gudenå along which we sailed. We noted the marked white veins on a background of the dark green leaf lamina – for sure a very beautiful specimen. Several of the participants had previously only seen small individuals of *Potamogeton compressus* characterised by its long strap-formed leaves, resembling *Zostera* leaves, hence the synonym *P. zosterifolius*. Nowadays, it is a rare species in much of Europe, but here it was present as well-grown individuals with the characteristic reddish tinge and in decently large stands.

The four fully equipped botanists participating in the first aquatic expedition to streams and lakes in Jutland in 1895. From the left F. Kølpin Ravn, school teacher Hansen, C. Raunkiær and J. Baagøe with pipe and plant specimen box. Mindeblad fra Expeditionen til de jyske søer og åer, 1895.

We did not re-discover three *Potamogeton* hybrids that Baagøe and Raunkiær had noted in the River Gudenå in 1895. One example: The hybrid *P.* ×*fluitans (P. natans × lucens)* used to be very common in extensive stands in the river, but although both parent species still grow here, the hybrid was not to be found. Across Denmark, the hybrid was previously known from about 30 localities, while today only 8 localities are left. A similar decline has also taken place for 16 *Potamogeton* species, among the country's 20 species, that Baagøe and Raunkiær recorded along 13 stretches in different streams during their expedition to Jutland in 1895. When Riis and Sand-Jensen (2001) re-visited the same stretches 100 years later, they could only find 7 species; the most robust and pollution-tolerant ones, whereas the large slowly growing species, such as *P. lucens* and *P. praelongus*, and species that were already rare in 1895 (e.g., *P. rutilus*) had vanished over the past 100 years due to channelization, overzealous weed cutting and dredging, as well as pollution of upstream lakes leading to outlet of water turbid with phytoplankton. While the mean richness was 6.0 *Potamogeton* species per stretch in 1895, the number had dropped to 2.8 species 100 years later.

Even though we did not re-discover all Baagøe's species and hybrids in the lower River Gudenå, the experience was still very pleasant compared with the country's other streams because it testified to a substantial vitality among stream plants, once the worst negative impacts have been removed. In the River Gudenå, weed cutting had become less intense and 40 years of effort to reduce nutrients from waste water to lakes upstream in the Gudenå system had been effective. Simultaneously, establishment of the mussel *Dreissena polymorpha* in the lakes and the river itself had contributed to remove phytoplankton and make the water clearer and suitable for submerged plants.

When we pulled the boat ashore, the rain had subsided and we all had broad smiles on our faces. The two Czech researchers had never seen such an impressive plant growth in their homeland rivers and the locals could only applaud. The total number of aquatic species was well above 30. When Zdeněk's genetic analyses of the hybrids and species of *Potamogeton* collected became available, as a result of the visit to River Gudenå and the other streams in Jutland, the conclusion was that Baagøe's hybrids were supported by the genetics (Kaplan & Fehrer 2013). There were some surprises regarding parent species. A broad-leaved *Stuckenia* that we used to call *S. pectinata* var. *interrupta* turned out to be a hybrid *S.* ×*bottnica* with the widely distributed *S. pectinata* as its one parent, while the other was *S. vaginata*. The latter species has never been recorded in Denmark; the closest locality is in the Gulf of Bothnia. *Stuckenia* ×*bottnica* in Denmark can be regarded as a relic. Its existence testifies that many hybrids possess a high vitality and that cross-fertilisation is not a prerequisite for long-term survival of an aquatic taxon. Plants can live well solely by vegetative growth and dispersal. Just think about the immigrant *Elodea canadensis* that became naturalised and now is one of the most common species in European lakes and rivers solely on the basis of vegetative growth and dispersal. Pollination and seed set are not easy among submerged plants in lakes, not to mention those submerged in swift flow in the middle of a river.

Potamogeton ×undulatus in the River Gudenå, Denmark. Photo KvdW.

One might think that everybody was happy with the improved nature qualities of the River Gudenå, including the luxuriant flora. Many years of effort and economic investments since 1970 to improve the water quality and ecological quality of lakes and streams in its catchment have, finally, been successful. However, this is not necessarily the view of all local people. Over the last decades, higher precipitation – remember the heavy downpour on our river journey – has increased the need for higher downstream flow through the river, while the improved growth of stream plants tends to slow down the flow and elevate the water level, hence increasing the risk of flooding of low-lying soils and properties built too close to the river. Thus, a local political majority has initiated extra weed cuttings in conflict with the legal directive of one annual weed cut for the river. This ancient battle in many cultivated lowlands of Europe between the capacity of streams and rivers to divert water from their catchment as opposed to their environmental and ecological qualities has, once again, become current due to the prospect of higher flow and water levels forecasted as a result of climate change. The River Gudenå has become a fierce battlefield despite the fact that it is Denmark's and one of Europe's most precious rivers for aquatic plants and invertebrates and one of the few rivers that has carried much of the nineteenth century's nature quality into the twenty-first century.

For the Danish participants, the 2012 excursion to the River Gudenå and streams in Jutland had another consequence in addition to improved identification of certain species and hybrids of *Callitriche* and *Potamogeton*. We discovered that species concepts of the Czechs and the Danish were not always in line. In one case, subsequent genetic analysis revealed that one species discussed was in fact a hybrid. The Jutland trip and later trips through Denmark and Sweden stimulated us to publish a large, comprehensive book on all known Danish aquatic plants (Schou et al. 2017). With identification keys, descriptions, drawings and photos of so many Danish species available, we also had a fair coverage of the species in Central and Northern Europe, including the Baltic nations, Britain and the Republic of Ireland.

At the same time, we realised that Klaus van de Weyer and Gerhard Wiegleb in Germany had been working for some time with new plant guides, better *Ranunculus* taxonomy and the idea of revising *Süsswasserflora von Mitteleuropa*. It was a perfect opportunity to team up with Klaus and Gerhard to ensure expertise from Central Europe as well as to include expertise from Britain. British plant specialist Richard Lansdown offered to join us, while plant ecologist Kevin Murphy in Scotland agreed to help with terminology and language. On this basis, the team decided to initiate the work to publish *Aquatic Plants of Northern and Central Europe including Britain and Ireland* to stimulate interest in aquatic plants and hopefully strengthen both their protection and their application in assessments of environmental and ecological quality. Thus, the journey down the river Gudenå initiated a long and historical (see below) journey to produce this English guide to aquatic plants.

Our goal with *Aquatic Plants of Northern and Central Europe including Britain and Ireland*, covering a geographic area stretching from Germany to Greenland, and Estonia to the Republic of Ireland, is to provide a guide book for accurate identification of all aquatic plants occurring within the area, through high-quality identification keys and illustrations as well as descriptions of the species, supported by global and regional maps of species distribution. Finally, we describe the typical habitats and selected aspects of their biology. We knew in advance that good identification keys to aquatic plants were lacking in several countries and that different species perceptions existed between Scandinavia and Central Europe, for instance within the genus *Ranunculus*. We wanted to eliminate these limitations with this book for all of the 16 nations included: Belgium, Czech Republic, Denmark and Greenland, Estonia, Finland, Germany, Latvia, Lithuania, Luxembourg, Netherlands, Norway, Iceland, Poland, Sweden, the Republic of Ireland and Britain.

Potamogeton ×cognatus in the River Gudenå, Denmark. Photo JCS.

From left to right: Bjarne Moeslund, John Bruinsma, Ulrike Hamann and snorkeling Klaus van de Weyer visiting the River Gudenå on an extremely hot summer's day in 2019. Photo JCS.

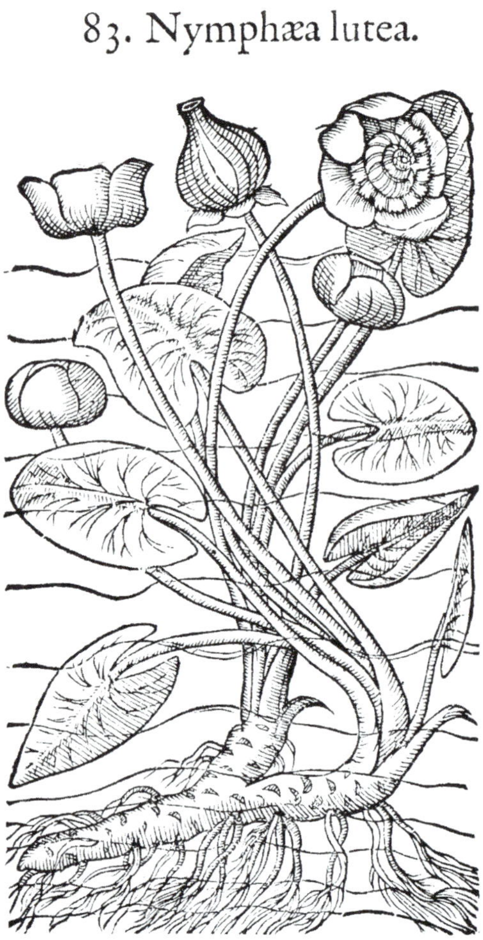

Woodcut of *Nuphar lutea* in Simon Paulli, 1648: Flora Danica Det er: Dansk Urtebog.

Carl von Linné (1707–1788). Swedish botanist, zoologist and taxonomist, who formalised the modern system of naming species. He studied in Uppsala and in the Netherlands. He was sent to different regions of Sweden to evaluate the economic potentials, while also collecting new species, like his later disciples, visiting remote parts of the world. Many species carry his author name, e.g., *Homo sapiens* L. Oil painting by J. H. Scheffel, 1739.

Travelling through time

Studies of aquatic plants have long been part of general studies of plants. Quite naturally, there has been a particular focus on the use of plants in medicine and as cultivars in fields and gardens. Systematic European studies of plants for medical use began in Italy more than 1000 years ago and gradually spread to Northern Europe, primarily via the monasteries, over the next 200 years. Herbals, books written by authors with a mixed education and practical training – as plant collectors, botanists, pharmacists and medical doctors – were popular during medieval times and were occasionally published in the national language (as opposed to Latin) to promote their use by ordinary people who could not afford to consult a doctor, but had to manage with those herbs that grew wild in their vicinity or were available in herb gardens. Some herbals were quite extensive, with drawings and text descriptions of the medical uses of several hundred species; among them several amphibious and a few truly aquatic species (e.g., *Nuphar lutea* and *Lemna minor*).

Evaluation of agricultural and industrial production capacity and the possibility of improvements took place throughout the eighteenth Century, often by means of royal envoys, for example, Carl von Linné's journey to Lapland and other regions of Sweden. Linné included the results of meticulously detailed botanical surveys; while visiting the calcareous islands, Öland and Gotland in the Baltic Sea in 1741, he described 116 plant species new to Sweden. On his visit to Västergötland, he noted in his diary the long, floating leaves of *Sparganium gramineum* in lake vegetation and their ability to dampen the waves as does floating *Sargassum* in the Sargasso Sea. The King in Denmark-Norway was also open to the idea of combining economics and botany and set up a royal society for agricultural advice with the aim of improving crop yields and soil fertility. In the mid-1800s the scientific background became professional through establishment of agricultural universities in most countries, whose main goals were to optimise plant cultivation and livestock farming.

General studies of plants, terrestrial as well as aquatic, have passed through several historical phases in Central and Northern Europe. They began with intricate descriptions of the species, often under different names in the 1500–1600s. Binomial naming and unique description of species were standardised in the Linnéan tradition from about 1753 onwards. This was followed by species identification via keys in early national excursion floras, among the first of these being Bentham's British flora from 1858. Studies of aquatic

George Christian Oeder (1728-1791). German-Danish botanist, medical doctor and economist (agrarian reformist). From 1753 he led the publication of the monumental botanical plate work, *Flora Danica*, which was planned to include all plants in the Danish-Norwegian Kingdom, including Schleswig-Holstein, Oldenburg-Delmenhorst, Greenland, Iceland and the Faroe Island. He travelled in Norway in 1758-1760. Illustration from the Flora Danica project, 1791.

An exquisite drawing of *Heloscadium inundatum* from *Flora Danica*, 1763. During a period of 123 years, from 1761 to 1883, 3240 copper engraved, folio-sized plates were produced. Among these were 28 plates showing *Potamogeton* spp., 9 *Batrachium* spp., 4 *Callitriche* spp. and 8 *Sparganium* spp. More Flora Danica illustrations on p. 177 and 281.

plant anatomy, morphology and biology were initiated in 1880–1930, at the same time as the first ecological studies commenced in relation to plant distribution in lakes and streams. Advanced studies of aquatic plant eco-physiology developed further in the 1930s.

All these forementioned disciplines remain alive and active. Improved excursion floras with descriptions and illustrations of the species are a precondition to ensure reliable species identification and subsequent mapping of species distribution, as well as documenting the arrival of new species and population trends of established species. Publication of *Danske Vandplanter* (Moeslund et al. 1990) with good illustrations and determination keys significantly boosted knowledge of aquatic plants, in general, and their application in regional and national nature and environmental assessment, planning and management, in particular. Hopefully *Aquatic Plants of Northern and Central Europe including Britain and Ireland* can fulfil the same goal.

Technological advancements have also improved our knowledge of aquatic plants. Scuba diving has enabled the discovery of rare species growing submerged in very small stands and identification of the ultimate lower depth boundaries of plant growth. Better physio-chemical methods, combined with introduction of computers and statistical analyses, have advanced studies of plant ecology, physiology and evolution. Recent genetic analyses have shown the validity of many hybrids and resulted in the dismissal of some species and the discovery of new species, but in particular have led to changes in the taxonomic affiliation at genus and family levels, while also improving knowledge of aquatic plant phylogeny.

In the following chapters we offer a short historical overview of aquatic plant studies in Central and Northern Europe until 1950–1960. Early historical research on aquatic plants had a very strong foundation within this region, leading to our own interest in the subject. Later studies became extremely numerous and expanded to most of the world, but their description is beyond the scope of our book.

Species, identification and taxonomy

Local floras existed in Central and Northern Europe in the 1500–1600s. The system for Carl von Linné nomenclature and classification was published in the mid-1700s, establishing a solid foundation for taxonomy and nomenclature, including many of the aquatic plant genera and species recognised today. Linné included aquatic plants,

Fig. 38. ***Potamogeton gramineus*** (c. $\frac{2}{3}$). Dannelsen af overvintrende Skudkjæder fra de øverste Bladaxler paa et med Flydeblade udstyret Skud, *I; a, b, c, d, e, f,* Blade paa Skud *I;* fra Blad ***a*'s** Axel en Skudkjæde af 5 Skudgererationer: ***a*₁**, ***a*₂**, *a*₃, ***a*₄**, ***a*₅**; ligesaa fra Blad *b***'s** Axel; ogsaa i Blad *c***'s** Axel sees et Skud. (T. af I. R.)

Christen Raunkiær (1860-1938). Professor of botany in Copenhagen. Renowned for his work on the biology of monocotyledons (1890), including aquatic species. His ecological work is on: 1. the relative frequency of plant life forms in the flora and the correspondence to the climate of different world regions, and 2. the common pattern of the relative species abundance in plant communities (few abundant species and many infrequent species).
Left: Portrait of Christen Raunkiær 1896. https://www.wikiwand.com. Public domain.
Right: Illustration from Raunkiær 1890.

such as 12 *Potamogeton* species, in *Species Plantarum* (1753), and 9 *Potamogeton* species in Sweden in the second edition of *Flora Svecica* (1755). More aquatic plants were included in S. Liljeblad's *Swedish Flora* (1792–1816). *Flora Danica,* a comprehensive, illustrated work covering the entire Danish-Norwegian-Icelandic kingdom, was initiated in 1761 by the German-born G. Oeder and continued by seven Danish botanists until its completion in 1883. It is famous for its 3240 high-quality, hand-drawn colour illustrations. Among the species treated are 195 aquatic species which also are included here, e.g., drawings of *Ceratophyllum demersum, Apium inundatum, Lobelia dortmanna, Pilularia globulifera* and *Ranunculus aquatilis.* While these illustrations are beautiful and attractive, however, they are of limited value for the purpose of accurate species identification.

Samuel Liljeblad (1761-1815). Swedish botanist, economist, apostle of Linné. Liljeblad was appointed professor and later Rector of Uppsala University. He made excursions throughout Sweden and into Finland, publishing a flora of Sweden in several editions and in Swedish.
Painting by unknown artist. Sok.riksarkivet.se

The species-rich aquatic genus *Potamogeton,* with many well-founded hybrids, has been a particular research object in Europe through the years as emphasised in the opening description of the journey down the River Gudenå. C. Raunkiær, the later world-famous plant ecologist, included 20 species and 8 hybrids in the still valid determination keys in his Danish excursion flora from 1890. Overall interest in *Potamogeton* species and hybrids was extensive around the year 1900 with comprehensive monographic studies published in the United Kingdom (A. Fryer) and Sweden (J. Hagström). The genus is richest in species in the Northern Hemisphere; P. Graebner in 1907 proposed 87 species, 47 hybrids and numerous varieties globally. A global evaluation reduced the numbers to 69 unique species and 40 confirmed hybrids (Wiegleb & Kaplan 1998). World Flora Online presently accepts 90 *Potamogeton* species worldwide (Murphy et al. 2019).

Recent advances in methods for genetic research have enabled a better understanding of the taxonomic status of and relationships between some aquatic plants. DNA analysis has led to the bringing together of some taxa formerly recognised as distinct species within a single species concept, while others formerly recognised as single species have been shown to involve more than one distinct taxon. Higher-level changes have also been made, with the families Callitrichaceae and Hippuridaceae now considered to fall within the Plantaginaceae (Karlsson 2012). In this guide we follow the taxonomic arrangement of the Angiosperm Phylogeny Group (APG) (Karlsson 2012).

Anatomy and morphology

Studies on the anatomy and morphology of water plants had a strong foundation in Germany with K. von Glöbel in the 1890s and H. Glück in the early 1900s. However, by 1870 E. Askenasy had already characterised the dissected aerial leaves and the laminar floating leaves of *Ranunculus aquatilis*, both leaf types having stomata and irregularly arranged epidermis cells, in contrast to the finely dissected underwater leaves, lacking stomata and with epidermis cells arranged in regular rows. In Denmark, C. Raunkiær published in 1896 a comprehensive study of the natural history of the monocotyledons. He included entirely new aspects, such as winter buds of species and the relationship between air-filled spaces in *Carex* fruits and their floating abilities and growth habitats occupied along the

Agnes Arber (1879-1960). English plant anatomist-morphologist, historian of botany and philosopher of biology. She is known for her botanical work and books on monocotyledons, including grasses, and for the book *Water Plants* (1920). She was the third woman to be elected as a fellow of the English Royal Society.
Left: Illustration from Agnes Arber 1920. *Hydrocharis morsus ranae*. Right: Photo of Agnes Arber, 1911, taken by Edward Alexander Newell Arber (1870–1918). Public domain.

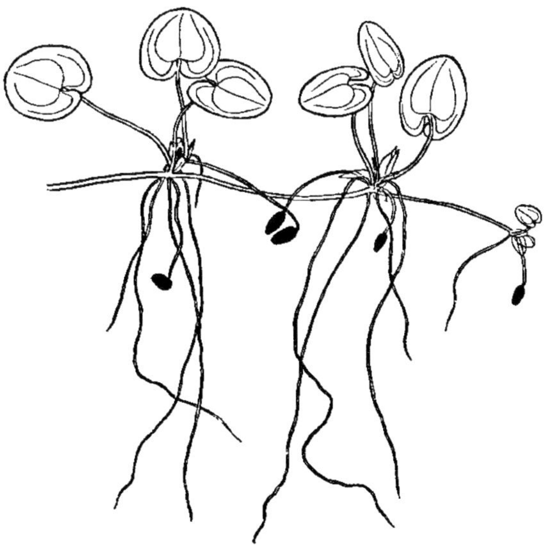

land-water gradient. In England, the grass specialist Agnes Arber 1920 conducted comprehensive anatomical-morphological studies on aquatic plants, and much later C. D. Sculthorpe (1967) published a global overview of the biology of aquatic plants.

These botanists were all highly interested in anatomical-morphological adaptations to plant life under water and on damp soil. They presented examples of leaf dimorphy, such as the following: Underwater leaves can be thin, finely dissected, with a thin cuticle and prominent air-filled spaces, but with no cell differentiation of the mesophyll and no stomata. In contrast, aerial leaves can be thicker, with entire edges, have differentiated mesophyll in palisade and spongy tissues with little air spaces, thicker cuticle with wax impregnation and many stomata. A. Allsopp related in 1954–1965 the morphogenesis to environmental conditions and cellular contents of sugar and phytohormones. Since 2000, experimental studies on phytohormones have revealed the variable impact on leaf dimorphy of aquatic and amphibious plants and the adaptations of different leaf forms to light intensity, supply and evaporation of water as well as supply of CO_2 and bicarbonate.

Among amphibious plants, the vascular tissues are better developed and reinforced with lignin in individuals on land, while air spaces in the cortex are more prominent in individuals under water. Hugo Glück showed in experiments that many aquatic plants can grow on land; nowadays this observation is used in practice to multiply aquarium plants in CO_2-enriched air.

Fritz Gessner's experiments in 1937 showed that finely dissected water leaves in *Proserpinaca palustris* obtained twice the photosynthetic rate relative to dry weight compared to undivided leaves, presumably due to the large surface area relative to volume in the *Myriophyllum*-like leaves. Much later eco-physiological studies have related photosynthetic rates to the full suite of anatomical-morphological differences between leaves and different environmental conditions regarding CO_2 and bicarbonate concentrations and flow velocities. Later studies have also revealed gas transport between leaves and roots and its positive relationship to the magnitude and continuity of air spaces. Likewise, water transport from roots to leaves depends on the continuity of the vascular transport system and the existence of open water pores on the leaf surfaces, allowing unimpeded water transport through the plant.

In comparison with lakes, studies in streams are much fewer. In Central and Northern Europe, many species often grow in both environments, but the abundance of species and the main environmental drivers are very different between lakes and streams. Studies on stream plants' anatomy and morphology focus to a greater extent on relations to flow velocity and sediment type than on relations to light intensity and water chemistry, which are the most prominent themes in studies on lake plants.

Ecology

Danish botanist Eugen Warming embraced many subjects such as taxonomy, anatomy, morphology and ecology, and he wrote international textbooks on plant biology, taxonomy and ecology. Today he is known as the founder of plant ecology with the publication of the first and highly influential ecological textbook from 1895, translated into German in 1896, shortly after to Polish and Russian and in revised edition to English in 1909. Warming had an in-depth knowledge of arctic-alpine, temperate and tropical vegetation. As a young student, he studied for three years in Brazil and later edited the first comprehensive flora of Brazil. He was a specialist on the peculiar anatomy-morphology of the aquatic family Podostemaceae. In his ecological work, Warming described zonation and succession of the vegetation in sand dunes facing the North Sea, including the morphological growth response of the species to wind and sand deposition, which

Eugenius Warming (1841-1920). Professor of botany in Stockholm and Copenhagen. Warming published the first and highly influential textbook in plant ecology (1895). He was active in many botanical disciplines: anatomy-morphology of Podostemaceae, Brazil's flora, textbooks on general and systematic botany as well as the vegetation of Denmark, Greenland, Iceland and the Faroe Islands. Royal Library, Copenhagen

inspired H. Cowles in his zonation studies in the dunes of Lake Michigan, regarded as the start of ecology in the USA. Many aspects of early limnological botany have been summarised by G. E. Hutchinson (1975).

The first careful maps of swamp, floating and submerged vegetation in lakes were constructed by Boye Petersen in Lake Furesø and neighbouring lakes in 1911–1913 and by W. H. Pearsall in the English Lake District in 1917–1923. Petersen focused on the importance of wind exposure for swamp and floating species and on water depth and water clarity for the composition and depth penetration of submerged species. Pearsall was interested in the importance of organic and mineral composition of the sediment for species distribution; this aspect was later examined experimentally by R. Misra. Pure organic mud is unsuitable for root anchorage, and submerged aquatic species grow poorly on mud compared with mixtures, including fine-grained minerals such as clay. The distribution of submerged species between lakes in relation to water pH was established in an early extensive Danish study by J. Iversen in 1929, and plant distribution in relation to the combination of pH and calcium in the lake water was established in several Swedish studies by G. Lohammer, A. Almestrand and A. Lundh between 1938 and 1965. Iversen went a step further in 1946 and evaluated the combined influence of water alkalinity and nutrients for submerged plant distribution. Alkalinity is closely related to calcium, pH and bicarbonate. Bicarbonate's importance for the global distribution of submerged plants was only finally established in 2019 (Iversen et al. 2019).

Both Boye Petersen's and Pearsall's early studies showed the positive relationship between lower depth boundaries of submerged plants and water clarity. Studies on Lake Furesø and neighbouring lakes in 1950 documented the marked upward migration of lower depth boundaries due to eutrophication and blooms of shading phytoplankton over 40 years. Studies of 27 Finnish lakes by L. Maristo in 1941 showed a remarkable linear relationship between the lower depth boundaries of submerged plants and the Secchi transparency. While aquatic mosses in most cases formed the lower depth boundary, reaching about 6 m in Finnish humic, softwater lakes, charophytes and angiosperms formed a common lower boundary of 8 m in the alkaline Lake Furesø in 1911. Lake eutrophication already appeared then to become a national and global threat to the distribution and survival of submerged species.

Life forms can be described in broad categories in relation to water level and sediment type as suggested by C. Raunkiær in 1904 and later by many others, including the Dutch C. den Hartog and S. Segal in 1964. Alternatively, species of widely different taxonomy can be grouped morphologically according to the same growth form as adaptation to life in a specific environment. An example is the isoetid growth form of *Isoetes lacustris*, *Littorella uniflora* and *Lobelia dortmanna* with stiff leaves in a rosette arising from a short stem. The short stem and leaves with well-developed air spaces is an adaptation to CO_2 supply via the roots from nutrient-poor sediment in softwater lakes. This coupling between anatomy-morphology and eco-physiological functions is examined today both in the laboratory and in the field by refined physio-chemical microanalyses of oxygen, CO_2, pH and Fe in leaves, roots and sediments.

Eco-physiology

Physiology and its coupling to ecology had a strong experimental tradition in Germany from the late 1800s. K. Arens in the 1930s studied the polarised *Potamogeton* leaves having a high pH and calcium-carbonate precipitation on the upper surface and low pH on the lower leaf surface. This inspired Danish E. Steemann-Nielsen to his quantitative studies of relative CO_2 and bicarbonate supply to photosynthesis of different plant types in the 1940s. Later, details on the mechanism of bicarbonate use by Dutch H. Prins and the existence or non-existence of bicarbonate use in many species by Scottish D. Spence were continued by several younger researchers. While eco-physiology had a strong platform in the above-mentioned countries, it was virtually absent in other parts of Northern Europe. It is remarkable that the strong German focus on aquatic plants' use of and adaptation to light and inorganic carbon from K. Arens, F. Gessner and F. Ruttner between 1930 and 1960 disappeared with them and was not taken up again until after 2010.

Research traditions

Traditions of aquatic plant research can persist across centuries, but can also arise, become amplified and disappear after some years. Charophytes are not included in this book, but the study of them has had a strong and lasting tradition at several universities in Poland. In contrast, Gessner's eco-physiological research on aquatic plants, summarised in two books (1955 and 1959), and S. Thunmark's and G. Lohammer's long-term studies on aquatic plants within and among Swedish lakes were not continued to the same extent after them, while E. Steemann Nielsen's influence persists among researchers at several Danish limnological laboratories. In the Netherlands, C. den Hartog, as professor in hydrobotany and through his initiative to establish the journal *Aquatic Botany* with a focus on freshwater and marine plants, has had a lasting influence on a new generation of successful researchers with variable biological, ecological and physiological focus on aquatic plants. In Norway, B. Rørslett and co-workers established a highly interesting focus on the biodiversity of aquatic plants, which was later expanded by Estonian, Finnish and Scottish researchers to include more countries and ultimately the entire world. Those recent biodiversity analyses are dependent on good floras to enable accurate identification of species across countries and to be accompanied by distribution maps of species. At all times, persistence and ability to synthesise the best from various disciplines have been important for scientific progress and for the key role of hydro-botany in scientific knowledge and in practical management of lakes and streams. Nonetheless, it all starts with knowing the species.

The author team, except Kaj Sand-Jensen, at an excursion to streams and lakes in Jutland in 2022.
From the left Richard Lansdown, Bjarne Moeslund, Jens Christian Schou, Klaus van de Weyer, Gerhard Wiegleb, Lars Baastrup-Spohr and Peter Holm.
Photo John Bruinsma.

Species distribution and diversity in the study area

Alisma wahlenbergii is endemic to the Baltic Sea and adjacent lakes in Finland, Sweden and western Russia. Hailuoto, Finland. Photo KvdW.

Our study area includes 17 geographic regions (countries or territories), from Germany and the Czech Republic in the south to Iceland and Greenland in the north, and from Estonia, Latvia and Lithuania in the east to Britain and Ireland in the west. We are familiar with the aquatic habitats and plant species, and their appearance, within this large geographic area and present photographs to illustrate the species and their environment.

In total we examine 318 aquatic species within the whole target geographic region. The aquatic species comprise permanently submerged, floating-leaved and amphibious species. The classification of species into these life forms is not easy, and many show more than one life form. Some submerged species may often, other species rarely, form floating leaves or even grow on wet substrate exposed to air and thereby act as amphibious. Likewise, some of the amphibious species commonly grow under water (e.g., *Veronica anagallis-aquatica*), while others rather rarely occur submersed (e.g., *Ranunculus flammula*). Among the 318 species, we classify 27% as primarily submerged species, 18% as floating-leaved species, and 55% as amphibious species.

The total number of species depends on how many of the wetland species are regarded as amphibious species and hence become included in the aquatic species pool. The perception of amphibious versus terrestrial species may well differ between different studies. We do not include species whose identification was uncertain in parts of the geographic area, and we omit hybrids from the total given above, although they are well accepted and described in the main text. A few subspecies, regarded as true species in some countries, are also included.

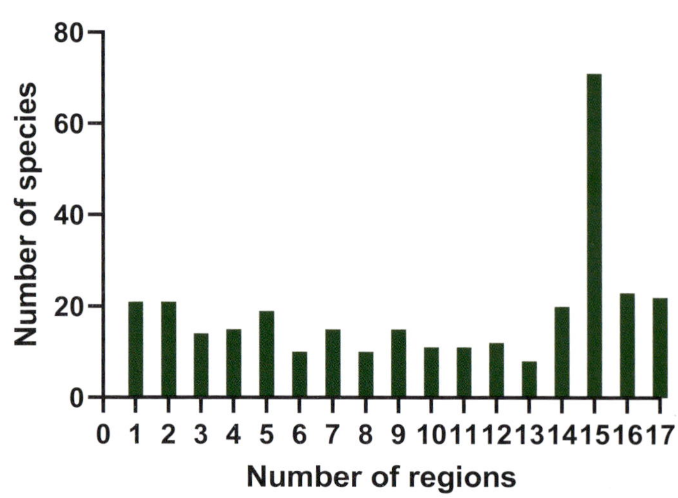

Figure 1. Number of species present in only 1 or up to all 17 geographic regions. The highest column, e.g., shows that 71 species were present in 15 of the 17 regions.

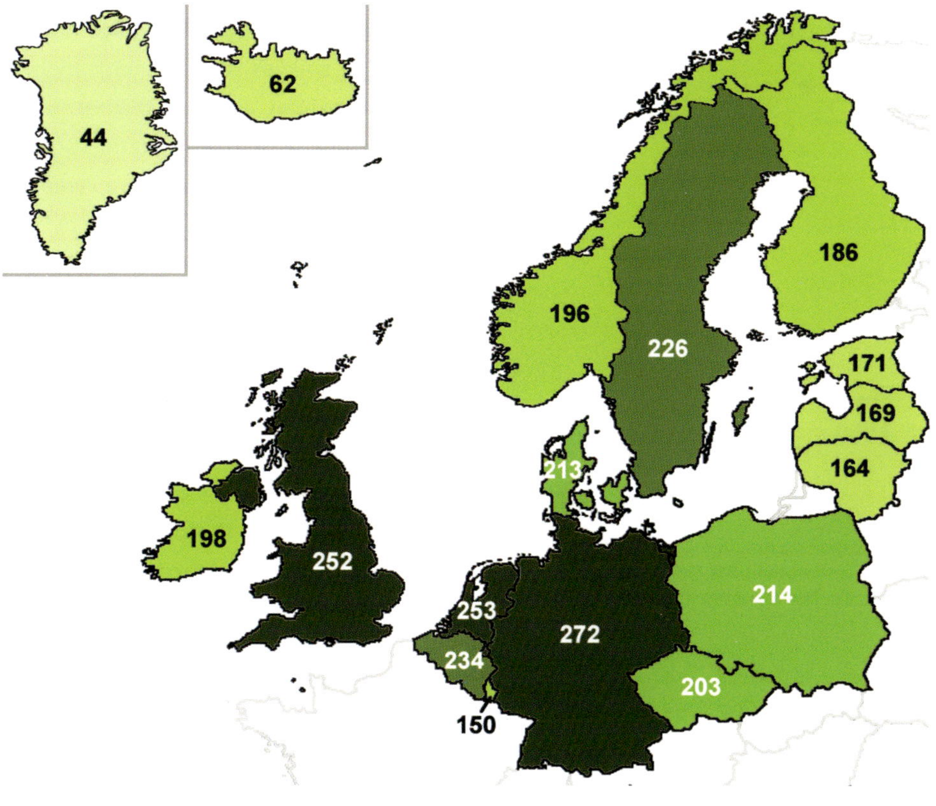

Figure 2. Number of species in the different areas and countries within the study area.

Most species are widely distributed, as 33% of them are recorded in 15-17 of the 17 geographic regions and only 17.6% are restricted to 1-3 regions (Figure 1). Twenty-two of the analysed species are present within all 17 geographic areas, and among these most widespread species are *Potamogeton alpinus, P. gramineus, P. natans* and *P. perfoliatus*. Seven additional *Potamogeton* species grow in 15-16 of the geographic areas, mainly missing from Greenland and Iceland. Other examples of species present in all areas include *Catabrosa aquatica, Callitriche brutia, C. palustris, Limosella aquatica*, and *Myriophyllum alterniflorum*.

The most narrowly distributed species, present in only one region, comprised 21 species (6.6% of all 318 species). They grow primarily in Britain (e.g., *Potamogeton epihydrus* and *Zannichellia obtusifolia*) or in Germany (e.g., *Najas gracillima, Vallisneria australis* and *Zannichellia peltata*). Most of them are alien species that have dispersed in different ways (e.g., aquarium facilities). None of the 21 narrowly distributed species are common within the geographic region where they have been recorded and none of them are resident species that historically had a wider distribution.

Despite a large part of species showing broad distributional ranges, a number of species exhibit more restricted geographical occurrences. Fifty-three of the species show a southwestern distribution including Denmark, and Germany and westwards. The pool of northern species is rather limited; 12 species only occur in the northernmost countries of our focus area (Greenland, Iceland, Norway, Finland and Sweden). Remarkably, there is only one species, *Eleocharis carniolica*, showing an eastern distribution from the mentioned southwestern or northern regions. Furthermore, *Nymphaea tetragona* and *Najas tenuissima* have a northeastern distribution in Finland and the Baltic countries. However, this does not change the overall picture that the vast majority of the more narrowly distributed species are found in the southwestern parts of the study area and elevate the total species richness there.

Species richness

After having summarised all species distributions, we can document the species richness and a large species similarity in species composition across the entire region. Apart from Greenland and Iceland, both with very low numbers of species (44 and 62) due to their location in the arctic zone to the far north, the 15 remaining countries vary relatively little in total species richness (150-272; Figure 2). The number of amphibious species is consistently higher than the sum of submerged and floating-leaved species throughout the entire geographic area. In the south-north gradient from Germany to Greenland, the percentage of amphibious species of all aquatic species is 56%-58% in Germany, Denmark, Norway and Iceland, and it is slightly higher at 64% in Greenland, but the number of species is low in Greenland and the difference is very small.

The relatively high richness of aquatic species in our focus area is noteworthy. In global analyses, total species number of aquatic species (approximately 3500 species) comprises about 1% of global species richness of aquatic plus terrestrial vascular species (roughly 350,000 species; Murphy et al. 2019). Within our focus area, the percentage of aquatic vascular species relative to all vascular species is about 10%. For example, in Denmark we record 213 aquatic species, which is 10.8% of the 1970 species of all indigenous and resident alien species (Hartvig 2015). About the same percentage of aquatic vascular plants is present in Germany (8.2%), Greenland (8.8%) and the other geographic regions. If we only count submerged and floating-leaved species, their contribution to the total species pool remains relatively high.

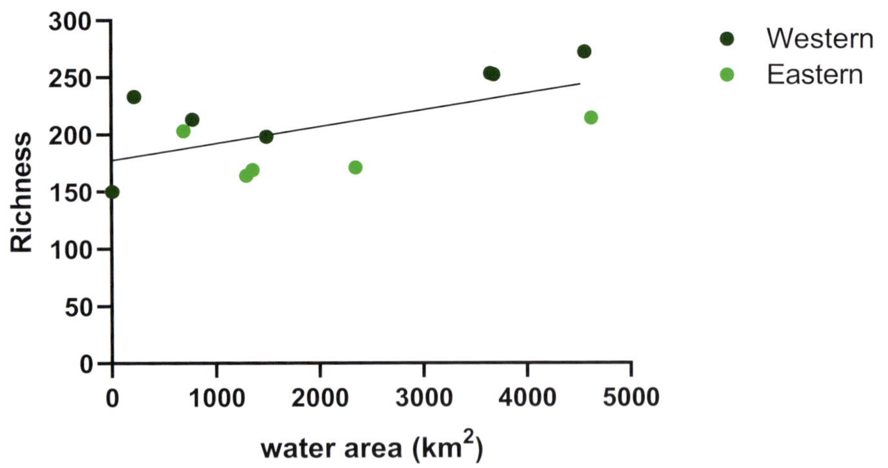

Figure 3. Species richness versus combined area of lakes, ponds and streams in regions located to the west or to the east. Western regions: Ireland, Britain, Belgium, Luxembourg, the Netherlands, Denmark; eastern regions: Estonia, Latvia, Lithuania, Poland, Czech Republic. The regions to the north are not included due to their generally lower species richness.

Many classic biodiversity studies report increasing species richness with higher available habitat area and higher habitat heterogeneity. The aquatic species richness increased linearly and weakly with the combined surface area of freshwater lakes, ponds and streams in the separate regions (without including the more northern Greenland, Iceland, Sweden, Norway and Finland). The regions located to the west generally had a higher species richness for the same surface area than the regions located to the east (Figure 3). The positive, though small, influence of surface area on species richness also was apparent from the presence of 150 species in Luxembourg compared to 233 species in Belgium with much larger freshwater areas.

Their omnipresence may have caused the historic and contemporary strong interest in aquatic plants among naturalists and researchers across the entire geographic area. Moreover, it is a strong argument for the relevance of this book; it treats a large and widespread species group and offers keys, illustrative drawings and photos, which ensure accurate species identification.

The book also includes all or the majority of aquatic species in European countries close by. The aquatic flora of Austria, Switzerland, Slovenia, Slovakia, Belarus, Leningrad area and France north of the Loire is completely covered by the book. Even Croatia, Serbia and Romania up to the Carpathians are floristically similar.

The strongest gradient in species richness within the large geographic area is from south to north following the markedly falling temperatures from temperate Germany (272 species) through Denmark (213) and Norway (196) to arctic Iceland (62) and Greenland (44; Figure 2). A weaker east-west gradient of species richness is from Estonia, Latvia and Lithuania (164-171) to Britain and Ireland (198-252). Likewise, in the east-west gradient located further to the south there are fewer species in the large country of Poland (214) than in the much smaller country of the Netherlands (253) located to the west. Both these patterns most likely result from the high prevalence of species with a western geographical distribution.

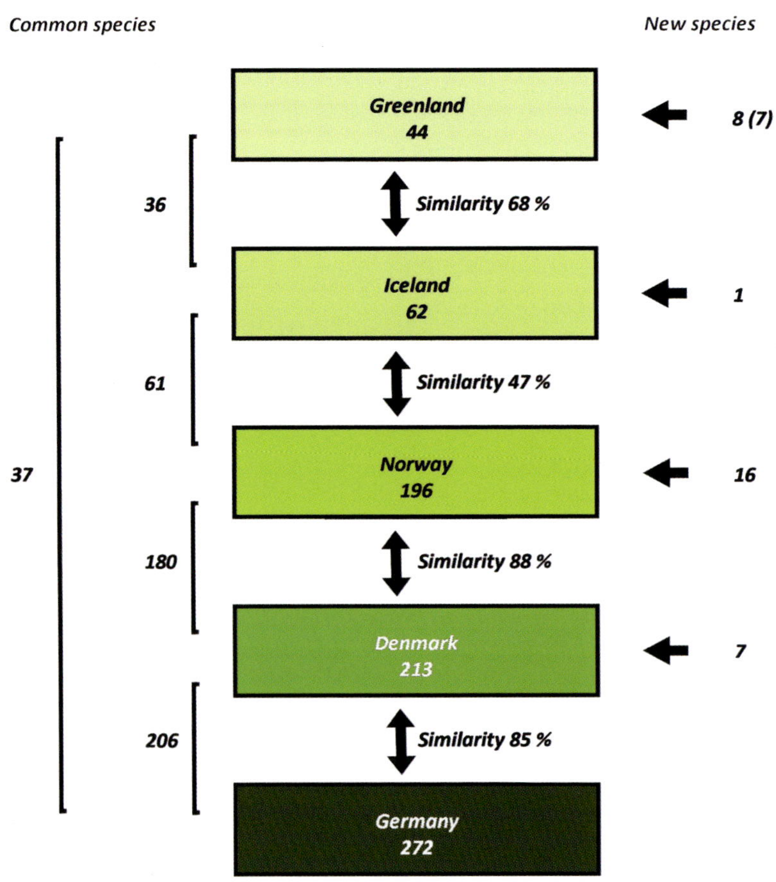

Figure 4. Diagram showing species richness in boxes along the south-north gradient from Germany to Greenland. Numbers of shared species between neighbouring areas are shown to the left, while numbers of new incoming species are given to the right. Similarity of species composition (Sørensen's similarity index) between the regions is shown between the boxes.

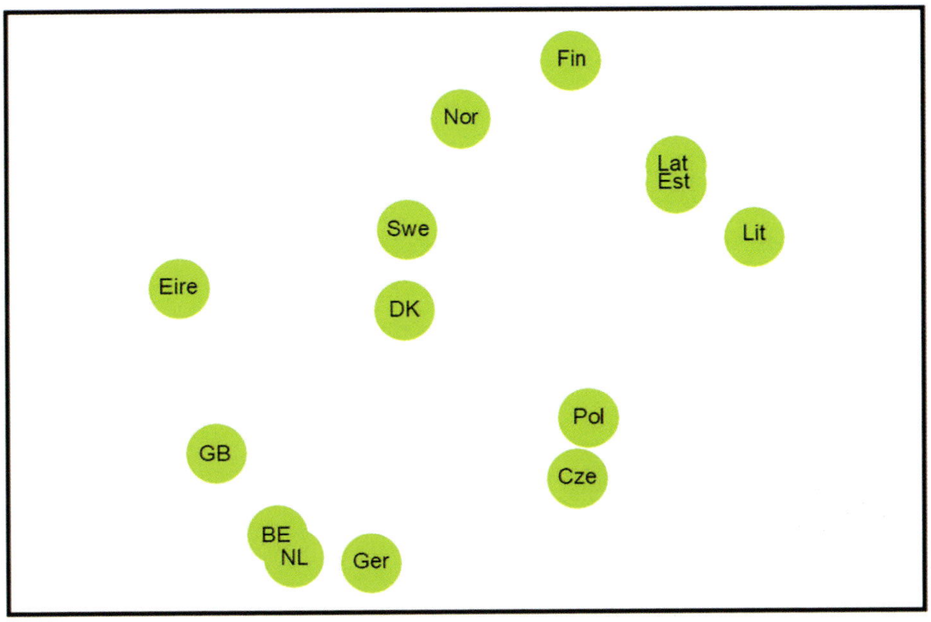

Figure 5. Similarity of species composition based on pairwise comparison of Sørensen's similarity index between 14 regions closely resembles the geographic location (ordination plot). The species-poor Greenland, Iceland and Luxembourg were omitted.

Similarity of species composition

The most pronounced gradient in species richness runs south-north from 272 species in Germany, over 213 in Denmark and 196 in Norway, to only 62 in Iceland and 44 species in Greenland (Figures 3 and 4). It is characteristic that most species in each region along this route are also present in the nearest region located to the south and present in Germany as well. Among the 44 aquatic species in Greenland, 36 species grow in Iceland, 37 species grow in Germany and only 2 species are unique to Greenland (i.e., *Potamogeton groenlandicus* and *Callitriche heterophylla*). This means that the decrease in similarity of species composition from south to north is mainly due to loss of species along the route and not to replacement of species. Only 7 species in Greenland, or 16% of all species, have been replaced along the route.

It is common to compare similarity in species composition between two areas by using Sørensen's similarity index. It is simply

$$\frac{2c}{a + b}$$

where c is the number of common species for two areas, while a and b are their respective species numbers. If no species are in common, the index is 0, while the index is 1 when all species are in common.

In the south-north gradient from Germany to Greenland, the similarity index is high between Germany and Denmark (0.85) and Denmark and Norway (0.88), while it is markedly lower between Norway and Iceland (0.47) and between Iceland and Greenland (0.68; Figure 2). Between Germany and Greenland, being located in different world ecozones, the similarity index is only 0.23. It is important to note that the low similarity indices are not due to substantial replacement of species, but mainly due to loss of species along the south-north route.

After omitting arctic Greenland and Iceland, and the small Luxembourg, all three of low species richness, we made a pairwise analysis of Sørensen's similarity in species composition among all 14 remaining regions. The result is a multidimensional figure which, when projected onto a plane, generates a pattern closely resembling the geographic location of the regions (Figure 5). Thus, regions located far from each other both north to south and east to west have a lower similarity in species composition than neighbouring regions. This result, combined with falling species richness from west to east (Figures 2 and 4), could indicate that there has been dispersal limitation of species from west to east following the last Ice Age (Weischel's glaciation), when areas to the west (i.e., southern Britain, southern Ireland, Belgium, the Netherlands, most of Germany, southwest Denmark) remained unglaciated, while the other Nordic countries, the Baltic states and most of Poland were fully glaciated. The low similarity between countries with similar environmental conditions and size, such as the Baltic countries and Denmark, support the idea that geographical gradients rather than environmental differences are shaping the current biodiversity patterns of aquatic plants in Northwest Europe.

Habitats

Rheocrene springs emerge horizontally from one or more sources at the foot of hills and pass over a bed of gravel and stones through numerous braided stream channels with swift current. The channels are commonly bordered by dense stands of helophytes, such as *Berula erecta*, *Cardamine amara*, *Veronica anagallis-aquatica* and *Epilobium* spp., while the submerged vegetation is sparse and dominated by mosses, liverworts and algae. With almost constant water temperature, year-round springs are cool habitats in summer and warm habitats in winter. The latter condition allows many plants to overwinter with fresh green leaves on short shoots, making springs appear green even in cold winters. Kovads Bæk, Denmark. Photo PH.

Aquatic habitats display high spatial variation, and the same applies to the plant species and the composition of communities occupying these habitats. The experience of meeting new species and new combinations of species is thrilling when moving from one place to another within a single ecosystem and between neighbouring ecosystems, or when travelling across extensive environmental gradients in the landscape.

Down the stream from the spring to the lower course, aquatic plants are adapted to extensive spatial changes in site dimension, current velocity, temperature, bottom substrate and dissolved inorganic carbon and nutrients in the water, to mention just a few conditions that strongly affect the species. Aquatic mosses and submerged forms of amphibious species may grow in the temperature- and flow-constant spring, while species of *Callitriche*, *Ranunculus* and *Potamogeton* gradually appear when the stream becomes wider, deeper, more nutrient rich and warmer.

Naturalists may enjoy the shifts in species composition and distribution and wonder, perhaps even investigate, which factors cause these shifts. Because many environmental parameters change in accordance with location in the stream network or with depth in lakes, it remains a challenge to isolate the main factors that are responsible for the spatial changes of species. Factors that are perhaps little considered may turn out to be important. Perhaps management of the stream by weed cutting and dredging benefited one species (e.g., *Sparganium emersum*), but harmed others, or historical factors such as dispersal and colonisation of a foreign species almost two centuries ago (e.g., *Elodea canadensis*) may prove to be more relevant than any contemporary factor. Thus, we must be ready for surprises regarding the species we find, and the reasons why they are here. Fortunately, this is what makes both natural history and science attractive: the species occurrences and the ecological perspectives are constantly changing and stimulate asking and answering new questions. Likely, aquatic ecosystems and their species distribution have undergone more extensive changes than most other natural ecosystems.

Moving from lakes located in well-leached sandy soils to lakes in nutrient-rich clayish soils, the availability of inorganic nutrients and hydrogen carbonate (bicarbonate) changes steeply and leads to development of plant communities with virtually no species in common. Thus, small species with rosette growth form, such as *Lobelia dortmanna, Littorella uniflora* and the like, may dominate in the nutrient-poor, low-alkaline lakes with mineral sediment, while tall species of *Potamogeton* may dominate in the nutrient-rich, high-alkaline lakes. Analyses show that the ability to use sediment CO_2 among rosette species and bicarbonate in the water among tall *Potamogeton* species is essential to account for this distinct difference in distribution patterns. Here, we apparently have found a factor – usable source of inorganic carbon for plant photosynthesis and growth – which at least in part explains the distribution of aquatic plants.

Other important environmental conditions obviously exist. No plants grow in the dark in very deep or turbid water; few plants can withstand fierce physical disturbance on exposed shores of large lakes; and most freshwater species cannot grow in saline water above a threshold of 3‰-5‰ NaCl. However, most changes of species distribution and plant community composition are gradual and also strongly influenced by biotic conditions, such as competition or facilitation among species that we still know relatively little about. However, the main habitat types and associated species are recognisable, and we describe them here.

High-altitude flush, Corrie Fee, Scotland. Photo RVL.

Springs

Flowing waters in the form of streams and rivers are fed by both outflowing groundwater and surface runoff from rainfall. The outflow of groundwater may be very diffuse and concealed below the surface of running as well as still waters, or it may take place in the form of distinct natural formations called springs. They occur when water pressure within the aquifer causes a natural flow of groundwater onto the earth's surface.

In contrast to the simple definition, natural springs occur in a number of main types, each spanning a wide range of sizes, surroundings and light regimes as well as hydrochemical and hydromorphological conditions. These parameters define the suitability of each spring as a growing place for aquatic plants. Thus, some springs are without vegetation or harbour only a few species, while others support a number of species covered by this study, commonly in addition to mosses, liverworts and algae (not covered by this study).

Springs are among the most endangered running water habitats. Existing springs often suffer from disturbed hydrology as well as deteriorated water chemistry due to the land use in their catchments.

Limocrene spring. The groundwater emerges vertically from one or more point sources at the bottom of a deep pool. Despite the constantly clear water, the number of aquatic plant species is commonly very low. In this spring pool the only two species are *Berula erecta* and *Hippuris vulgaris*, both permanently submerged and green all year. Store Blåkilde, Denmark. Photo BM.

Limocrene spring in Endla Nature Reserve, Estonia. Photo PH.

River Gauja, Latvia. The river is characterised by patch-like assemblages of plant communities in sheltered stream sections forming narrow belts of emerged plants, such as *Phalaris arundinacea, Glyceria fluitans, Butomus umbellatus, Equisetum fluviatilis, Mentha aquatica* and *Sparganium erectum*. The submerged vegetation is dominated by *Stuckenia pectinata, Potamogeton perfoliatus, P. natans, Elodea canadensis* and *Ranunculus trichophyllus*. Photo PH.

Running waters (watercourses)

Running waters (watercourses) are pathways in the hydrological cycle transporting the surplus precipitation, including upwelling groundwater, from land to sea embedded in distinct landscape structures, such as catchments, floodplains and watercourses themselves. Each reach of a watercourse is part of a hydrological continuum that spans from source to sea and is characterised by the unidirectional flow that causes a continuous and irreversible loss of sediments and minerals from land to sea. All running waters are fundamentally different from each other, both at the scale of the whole river and the river reach. Watercourses are thus highly individualistic study objects for the ecologist.

The differences at the largest scale are due to the climate, geology and relief of the catchments. Macroclimatic factors decide the precipitation regime and the valley formation. Geological parameters such as rock type (hard vs. soft; calcareous vs. acid) influence the geomorphological, hydraulic and hydrochemical processes. The relief of the landscape (lowland vs. upland or high mountain) influences bed load, discharge and current velocity in a complex way. At the local scale (the reach), many parameters define the structure of the habitats for macrophytic aquatic plants. These are width, depth, slope, current velocity, discharge, groundwater inflow, light regime, water turbidity and colour, sediment type, bank morphology, oxygen content and hydrochemistry. All these factors have in most North European watercourses been severely altered during the past centuries by human interference.

The general importance of light means that both small and medium-sized streams surrounded by trees or running through dense forest are commonly without or with only sparse and scattered aquatic vegetation, despite otherwise ideal habitat conditions.

Although the effective number of different running water types is theoretically infinite, we have selected a few examples to illustrate the diversity of running waters in the geographic region covered by this study. We use the size (width of the streambed) of watercourses as the main parameter for grouping. In our region width varies from less than a metre to more than 100 metres, and depth varies from a few centimetres to more than 10 metres.

The size differences among running waters are clearly reflected in all languages of the region. In English "brook" is a commonly used term for smaller watercourses, "stream" for midsized ones, and "river" for the larger ones. There are no naturally given threshold values for the three size categories, but for practical reasons such values have been established in the European hydrological, geomorphological and ecological literature, with 2, 5, 10 and 25 m being the most commonly used thresholds.

Running waters mostly share their aquatic plant flora with still waters. There are only a few species among vascular plants of our region, which have clear habitat preferences for running waters (so-called rheophytes) and grow exclusively in streams and rivers. Examples are *Ranunculus fluitans*, some of the hybrid species derived from that species (*R. penicillatus*, *R. pseudofluitans*) and *Oenanthe fluviatilis*. On the other hand, many species typically growing in still waters are less frequent or absent in running waters. This is true for some free-floating genera, such as *Azolla* and *Utricularia*. Running waters harbour on the other hand more aquatic forms of amphibious and wetland species, especially where they have strong seasonal water level fluctuations.

Brooks and small streams (width mostly < 2 m)

Narrow running waters are commonly bordered by a riparian vegetation. In the case of brooks and small streams it makes an additional difference for the occurrence of aquatic plants whether the water is flowing in a stream bed lying high in the terrain or in a streambed that has been dug deep below the surrounding terrain. In the former, the light at the bottom of the stream bed may still be sufficient for aquatic plant growth, while in the latter the riparian vegetation will outshadow all or most growth of aquatic vegetation. Since small streams as a group have undergone very comprehensive channelisation and dredging in lowland regions, mainly due to the requirement for efficient drainage of farmland, most of them are poor in aquatic plants even in the open landscape.

Small natural streams flowing high in the terrain receiving enough light can harbour several submerged species such as *Callitriche* spp. and *Ranunculus* spp., aquatic forms of helophytic species such as *Berula erecta*, *Veronica anagallis-aquatica*, *Epilobium* spp., *Myosotis* spp. and small not very competitive species such as *Montia fontana* and *Ranunculus hederaceus*, in addition to amphibious species such as *Glyceria* spp. and *Catabrosa aquatica*. Spring brook, Gettrup near Hobro, Denmark. Photo JCS.

Dense stands of *Erythranthe peregrina* and *E.* ×*robertsii* lining an upland stream in the Scottish borders, Glengonnar Water, Dumfries and Galloway, Scotland. Photo RVL.

Among the running waters in the lowland part of the target region, the brooks and small streams are the group that has undergone by far the most comprehensive physical and morphological changes in the form of channelisation and dredging. This has caused very significant deteriorations of the habitat quality for aquatic plants. The photo shows a less shaded small stream with lush stands of *Callitriche platycarpa*, along the banks surrounded by *Berula erecta* and scattered *Mentha aquatica*. Lundbæk, Jutland, Denmark. Photo BM.

Medium-sized streams (width mostly between 2 and 10 m)

In medium-sized streams, shading from the riparian vegetation becomes less and less important with increasing width for the occurrence of aquatic plants within the streambed. While at the same time the depth of streams is not critically limiting the amount of light at the bottom, medium-sized streams are potential habitats for both well-developed and species-rich aquatic vegetation. Medium-sized streams originally constitute the major part of the macrophyte area in streams in the natural landscape. The removal of forests alongside the watercourses in the cultural landscape has amplified this effect by improvement of the light regime. Whether the light potential can be exploited depends, among other things, on the nature of the riverbed. Soft, in particular sandy soils and moderate water velocities in lowland watercourses provide the best basis, while gravelly or rocky streambeds and high water velocities may exclude or strongly limit the presence of aquatic vegetation (except bryophytes), although riffles and runs with gravelly substrate may harbour very well developed stands of flow-tolerant species, e.g., *Ranunculus* spp.

In terms of aquatic plants, medium-sized streams are generally among the most species-rich ones. All species which can grow in small streams and larger streams (or small upper courses and wide lower courses) can grow here as well. Generally, the number of aquatic species increases with increasing size of the riverbed within this size class, which may be an effect of increasing habitat diversity with increasing width. However, anthropogenic impact such as channelization, dredging and pollution can neutralise this general pattern, causing both quantitative and qualitative, mainly negative deviations.

Large streams (width between 10 and 25 m)

In large streams light can become a limiting factor for the occurrence of aquatic plants due to either depth or turbidity. In upland streams varying

Medium-sized stream with clear water in a wide and shallow streambed which despite channelisation and repeated weed cutting is harbouring a very well developed and well-structured submerged vegetation where *Ranunculus aquatilis*, *Berula erecta* and *Callitriche hamulata* form a mosaic of species and growth forms. Note the total absence of emergent helophytes. River Kastbjerg Å at Norup Denmark. Photo BM.

Mountainous streams and streams in rocky terrain display high physical habitat variation, but due to a rocky or stony riverbed without or with only very little soft sediments, generally high water velocities and occasional floodings, such streams commonly harbour none or only very few of the species covered by this study. Medium-sized and fast-flowing streams in volcanic terrain with no submerged vegetation as contrasted to a very species rich and lush riparian vegetation. Iceland. Photo BM.

Medium-sized, fast-flowing lowland stream in rocky terrain with only a sparse submerged vegetation of *Myriophyllum alterniflorum*. Gårdshult Naturreservat, Halmstad Kommun, Sweden. Photo BM.

discharge also plays an important role. If light conditions are good, large streams may harbour a remarkably well developed and species-rich aquatic vegetation comprising many submerged species as well as many helophytes and their submerged forms. Among the submerged plants the genus *Potamogeton* (including *Stuckenia*) is commonly represented by several species and often hybrids, and many species develop exceedingly long shoots swaying in the current.

River Gudenå between Silkeborg and Randers, Denmark, is an example of a large and deep stream that due to its clear water is remarkably rich in aquatic plants. The photo shows a very large stand of *Stuckenia pectinata*, which in other parts of the stream is replaced by similarly large stands of *Potamogeton lucens, P. perfoliatus* or the hybrid *P. ×undulatus*. Photo BM.

Rivers (width mostly > 25 metres)

As the size and depth increase and the watercourses become categorised as rivers, the habitat quality for aquatic plants commonly decreases due to insufficient light at the bottom. With respect to both the submerged and emergent vegetation, rivers have many similarities with lakes. In addition to the effect of depth itself, in rivers the light factor is influenced by both natural and anthropogenically induced turbidity. Light conditions therefore mostly limit the potential habitat for aquatic plants to narrow bands along the banks and at the margins of islands within the riverbed. Here emergent but flood-resistant plants are favoured in relation to submerged plants, which are negatively affected both by insufficient light and mechanical disturbance. Emergent plants can easily recolonise the riverbed from the banks after destruction by flood or heavy bed loads.

Extensive beds of *Ranunculus fluitans* in the River Creuse, near Ruffec, Parc Naturel Régional de la Brenne, Indre, France. Photo RVL.

Ditches and canals

In addition to channelised and regulated natural running waters, a large number of man-made ditches and canals have also been constructed for diversion of drainage water from farmland and urban areas, as irrigation ditches, for water supply for fishponds, for water diversion to and from watermills, for navigation and for many other purposes, many of which are to a large extent still fulfilled in the modern cultural landscape. Despite their origin and repeated dredging, ditches and canals can harbour well-developed aquatic vegetation comprising submerged, free-floating and emergent plants. This re-

Ditch with a very well developed and species rich aquatic vegetation. The ditch is dredged at annual intervals to maintain the flow capacity. This "intermediate disturbance" allows the occurrence of a wide variety of submerged aquatic plants, such as *Potamogeton acutifolius*, *P. ×sparganifolius*.
Vlijmen, the Netherlands. Photo KvdW.

Lake Biale in Wigierski National Park, Poland, is an example of a large, clear-water lake. Photo PH.

quires however that water quality is not deteriorated by eutrophication, acidification, leakage of ochre after groundwater lowering and inorganic pollution. Also intensive dredging, weed cutting and flow regulation below ponds or drinking water reservoirs may have a negative influence.

Still waters

Still waters constitute a diverse group of aquatic habitats in terms of relevant ecological parameters. The lake itself is a holistic unit characterised by size, depth, shape of shoreline, morphometry and sediment structure. It is embedded into a basic hydrological and geomorphological setting, comprising the surrounding landscape (catchment), its geology, geochemistry and climate. Directly important for plant growth are chemical factors such as pH, inorganic carbon (alkalinity) and nutrients as well as physical factors such as temperature and not least light. Anthropogenic effects strongly modify the natural settings in the cultural landscape, especially eutrophication, acidification and hydrological management.

The most important parameters for macrophyte growth are alkalinity and light availability due to their direct relation to physiological processes such as gas exchange, plant nutrition and photosynthesis. While alkalinity, or rather the availability of inorganic carbon, may vary among seasons, light availability is mainly related to water depth. In eutrophic lakes the light regime is negatively influenced by turbidity and plankton development. In addition, nutrients such as phosphorus and nitrogen can become limiting for aquatic plants' growth, directly or indirectly through their effect on water turbidity. Thus, in terms of habitat characteristics the range of still water types is in fact almost infinite within the geographical range of this book. We here give some examples to illustrate the diversity of still waters.

The geographic area covered by this study contains a number of very big and mostly deep lakes. From a distance these lakes seem devoid of aquatic plants, which however occur in a relatively narrow band along the shores, only rarely to depths in excess of 10 metres. Lake Constance on the border between Germany, Switzerland, and Austria. Inserted: *Potamogeton friesii*. Photos KvdW.

Large lakes

The number of species present in any habitat most commonly increases by increasing size. Thus, large lakes are most usuallly richer in aquatic plant species than small lakes. This size effect is mostly caused by the higher habitat diversity of large lakes. Although large lakes are less numerous than small lakes, they may harbour a large proportion of the aquatic flora of a region.

In large deep lakes the aquatic flora is mostly restricted to the littoral zone due to restrictions of light availability for photosynthesis. In deep natural lakes with clean clear water the aquatic vegetation may extend to a depth of 10–15 meters. Mostly, the aquatic vegetation is restricted to depths of less than 5 metres due to turbidity and the shading effect of phytoplankton. Towards the shoreline submerged plants are furthermore often limited by competition with emergent helophytes and floating-leaf plants, although some submerged species benefit from the shelter against waves provided by floating-leaved plants and helophytes. Commonly, we find a distinct zonation of species and growth form types, comprising a reed belt (with dominant helophytes), a zone dominated by floating-leaved plants and a zone dominated by submerged plants.

Large shallow lakes up to a depth of 5 m can be potentially colonised by all major growth form types due to the availability of light down to the bottom. Shallow lakes with clear water often exhibit widespread, diverse and dense stands of plants arranged in a mosaic pattern rather than as a distinct zonation. In more turbid shallow lakes the submerged vegetation may be quantitatively well developed due to the ability of many species (such as *Potamogeton*, *Stuckenia*, and *Myriophyllum*) to develop long, branched and canopy-forming shoots. This way they can overcome the lack of light at the bottom by developing the main part of their photosynthetically active biomass to the upper part of the water column. Plants without this ability to compensate for the lack of light at the bottom are to large

Large deep clear water Lake Nors Sø, NW Jutland, Denmark. Below the surface a well-developed and species-rich underwater vegetation is found, comprising a wide range of species and growth forms. When large deep lakes become polluted by nutrients, the submerged vegetation becomes more and more restricted to the shallow parts of the littoral zone which are at the same time occupied by the dense reed stands. Eutrophication and subsequently increased turbidity therefore constitute a severe threat to the occurrence of submerged plants in the deeper water. Photo JCS, photo of diver Uwe Koenzen.

Shallow, calcareous lake bordered by *Cladium mariscus* and *Schoenus* spp. and with a sparse and open submerged vegetation of *Najas marina* and *Chara* spp. Norrsund, Fårö, Gotland, Sweden. Photo JCS.

Nutrient-poor lakes with low alkalinity and pH below 7 have become increasingly rare in the intensively cultivated and eutrophicated landscapes within the geographical area covered by this book. The characteristic vegetation of such lakes, some of which are both large and deep while others are smaller and shallower, comprises only a few specialised species, such as *Lobelia dortmanna*, *Littorella uniflora*, *Isoetes* spp., *Juncus bulbosus* and *Subularia aquatica*. Lake Södra Vixen, Sweden. Photo JCS.

extent excluded from shallow turbid lakes or become restricted to narrow belts along the shoreline. The closer they grow to the shoreline, the more they compete with emergent helophytes. Helophytes can gain a competitive advantage over true hydrophytic plants by transporting nutrients and oxygen from semiterrestrial stands to deep water areas by means of stolons and rhizomes.

Small lakes, ponds and dams

The countries covered by this study use different threshold values to distinguish between large and small lakes. Regardless of this, the number of small lakes exceeds the number of large lakes almost everywhere, commonly by several orders of magnitude. This group of still waters shows an enormous diversity and variation in terms of all ecological parameters relating to aquatic plant growth. It is far beyond the scope of this book to describe and illustrate the full range of small still waters. We only give a few examples to illustrate the diversity and variation within this group.

The individual small lake, pond or dam is often colonised by only a limited number of species. Due to their sheer numbers and, additionally, their diversity, small lakes represent a significant proportion of the total range of habitats for aquatic plants. Furthermore, several species are particularly or exclusively associated with small still waters or only with some special types of small waters. Among the latter are temporary waters, as both temporary water filling and desiccation are preconditions for the occurrence of some specialised species.

Many small still waters are natural, but in cultural landscapes the natural ones are exceeded in number by waters that have emerged as a result of human activity. Some may have been created more or less unintentionally by sand, gravel, marl, or peat excavation, while others were created by damming of running waters for fisheries, mining, irrigation or storm water retention.

In recent years many additional small

A small deep lochan typical of the Outer Hebrides supporting species such as *Nymphaea alba* subsp. *occidentalis*, *Potamogeton epihydrus* and *Sparganium angustifolium*. Near Askenish, South Uist, Scotland. Photo RVL.

Due to a rocky bottom, often in combination with humic and low-alkaline water, smaller mountainous lakes and ponds commonly offer limited conditions for occurrence of rooted macrophytes, in this example *Isoetes echinospora* and *Sparganium* sp. Lifjell at Blefjell, Norway. Photo BM.

Fishponds are artificial lakes made for fish farming (usually carps and pikes). They are often situated in connection to monasteries or castles. The flat, wet and muddy banks are habitat for *Cyperus fuscus*, and *Eleocharis, Schoenoplectus, Elatine* and *Callitriche* species. Hausdülmen, Germany. Photo KvdW.

waters were created in the context of wetland restoration projects or as compensation measures in the context of environmental impact assessment.

Many small waters are subjected to a strong influence of nutrients from the surrounding cultural landscapes and have thereby experienced severe deterioration of habitat quality. In less intensively cultivated landscapes and especially in near-natural areas, many small still waters have maintained good habitat quality. Despite their location and technical function, even storm water retention ponds in urban areas may exhibit a rich aquatic flora and vegetation.

Coastal waters

Most of the species covered by this book are obligate freshwater species. However, a small number of species extend from freshwater habitats into brackish coastal water habitats, such as ditches and canals in coastal marshes. coastal lagoons, estuaries, fjords and especially the northern part of the Baltic Sea. Only a small group of species is largely restricted to brackish water, e.g., *Ruppia* spp., *Zannichellia* spp. and *Eleocharis parvula*, requiring a salinity of between c. 5‰ and 20‰. Around the lower threshold value, the highest number of freshwater species is found in coastal waters. Most common are *Stuckenia pectinata*, *Ranunculus baudotii*, *R. circinatus* and *Myriophyllum spicatum*. A very small number of species, e.g., *Zostera* spp., are obligate saltwater species requiring a salinity of more than 5‰–6‰. At higher salinities (> 20‰ freshwater species occur only exceptionally, while brackish water species can survive with low population densities.

Rain retention basin with large formations of *Ludwigia palustris*. The changing climate with higher precipitation and heavy rain events is causing the public authorities to establish more rain water retention ponds, protecting urban areas and arable fields against flooding and protecting running waters against hydraulic overload and nutrient load. Despite their origin, function, hydraulic pressure and nutrient load, these man-made ponds often become rapidly invaded by aquatic plants. Steinhorster Becken, Germany. Photo KvdW.

The Gulf of Bothnia is fed by several large rivers, and the salinity is as low as 2‰ and drops to almost zero close to river inlets. In the photo the water level is at about 1 m below normal, and most plants are beached on the mud and extend into shallow water. The swards include a wide range of taxa, such as *Alisma wahlenbergii, Callitriche hermaphroditica, Elatine hydropiper, Eleocharis acicularis, Isoetes echinospora, Myriophyllum verticillatum, Potamogeton friesii, P. gramineus, P. perfoliatus, P. pusillus, P. ×nitens, Ranunculus confervoides, Sagittaria natans, Stuckenia filiformis, S. pectinata, S. ×suecica* and *Subularia aquatica.* Varjakansaari, Finland. Photo RVL.

Small lakes between the inland ice and the fjord close to Kangerlussuaq, Greenland. The arctic regions generally have few aquatic plants, and from Greenland only 44 species are known, half of which are helophytes. Photo Jan-Thomas Johansson.

Basic guidelines for sampling and identification of aquatic plants

A common garden rake with close-fitting tines is the tool for sampling under numerous conditions. The garden rake is also useful as a walking stick in difficult terrain. Being applicable only in shallow water, the rake allows targeted sampling of entire plants, ensuring sampling of all potentially necessary plant characters. Equipped with distance markings, the shaft enables easy measurement of water depth while sampling. In addition to sampling, the rake also serves a safety measure when wading shallow waters. Photo BM.

Based on our own experience, we provide some basic and general guidelines for sampling, tools, handling and identification of aquatic plants.

However, these basic guidelines do not replace, but only supplement, the specific technical instructions and protocols applicable to aquatic vegetation surveys and monitoring under national monitoring programs, laws and international directives.

Sampling tools

Sampling of plants in aquatic habitats may present several challenges and difficulties that in most cases require the use of tools:

A common garden rake with a 1.5–2 m long shaft with 5 or 10 cm markings on the shaft to allow measurement of depth is the standard tool for sampling in shallow waters.

A common garden rake with a 2–4 m long telescopic shaft with 5 or 10 cm markings on the shaft to allow measurement of depth is the standard tool for sampling in medium-deep waters.

A grapnel, which is a specially designed rake head mounted on a long line, is a tool for sampling waters that cannot be approached by wading or by boat.

A water scope enables below-surface inspection of submerged plants.

Polarizing sunglasses reduce surface reflections and facilitate below-surface inspection of submerged plants from above, mainly under sunny conditions.

In addition to these tools, there is a general need for personal equipment such as rubber boots, waders, a life jacket and in some situations also some kind of boat.

A garden rake with a long telescopic shaft increases the range of sampling. Due to the extended range, sampling with the telescopic rake is more often in the blind, making targeted sampling of entire plants more difficult or less likely. Nevertheless, the telescopic rake commonly gives surprisingly positive results when used at depths otherwise inaccessible. Photo left Heidi Rauers, photo right PH.

The grapnel comes in many different designs – some factory made, others custom- or homemade. A weird-looking, however quite efficient homemade grapnel is illustrated here - a rake head mounted on a rack from a dishwasher to weigh down the rake head and keep it in position while it is being pulled across the bottom surface. The grapnel can be thrown from the shore or used to retrieve plants from deeper water while surveying by boat. Although a long line extends the range of the grapnel beyond that of the shafted rakes, the grapnel is the least suitable of the rakes for sampling targeted and entire plants. Photos KvdW and JCS.

The water scope is a useful tool that enables in situ inspection of submerged vegetation without being disturbed by the reflections in the water surface. In addition, the use of a water scope often enables a more targeted sampling of specimens when using a rake. However, the use of it requires clear water. Photo KvdW.

Requiring both special equipment and skills diving is an exclusive method used when surveying and sampling aquatic vegetation in deep waters according to specific programs and protocols. Photo Heidi Rauers.

Sampling strategy

The wide range of aquatic habitats as well as the internal variation within aquatic habitats commonly require different surveying and sampling strategies in order to obtain the full picture of the species composition and distribution.

Detailed topographical maps and aerial photos as well as bathymetric and geological maps are useful tools in both planning and conducting surveys and sampling. If available, previous recordings may be very useful.

Aquatic vegetation in shallow waters can commonly be surveyed and sampled on foot using a rake and a water scope, sometimes supplemented by a grapnel. For personal safety it is recommended to wear a life jacket when wearing waders. Photo JCS.

Surveying and sampling in deeper waters – still waters as well as running waters – require the use of some form of a boat. The sampling tools are telescopic rake, grapnel and water scope. For personal security, a life jacket is standard equipment when surveying and sampling from a boat. Photo JCS.

Snorkelling and scuba diving are the best methods for surveying, sampling and photographing aquatic vegetation in deep waters, and sometimes also in shallow waters. However, it is a method only for the few, as it requires special equipment and diving training as well as special tools for underwater sampling and registration. Furthermore, safety regulations normally require the diver to be accompanied by a diver assistant. Photo RVL.

Storage in the field

While surveying aquatic habitats, there may often be a need to store collected plants temporarily for later determination or conservation.

Plastic bags and buckets may be needed for general use, along with cooler bins to keep sampled material cool and fresh until final handling.

A spacious plastic bag with a little water or a piece of wet paper or cloth is in most cases suitable and sufficient for temporary transportation and storage of aquatic plants. Leave some air in the bags before sealing off in order to avoid squeezing and breaking the plants, thereby losing their spatial appearance. Place a label in the bag with adequate information, preferably written on waterproof paper with a lead pencil. All bags should be labelled on the outside with a waterproof pen. Photo Sebastian Meis.

Identification

This book enables identification of the aquatic vascular plant species present within in the specified study area. Although many characters can be seen by the naked eye, a good handheld lens is a necessary tool for examination of species characters, and in some cases, even a microscope is needed.

A handheld lens with 10x or (even better) 15–20x magnification, a stereo microscope with 50–100x magnification and a microscope with +100x magnification are all useful to have on hand.

Handheld lens. Photo JCS.

Time and cultivation

Surveying aquatic habitats at any time within the growth season or even all year long, will to a large extent enable correct determination of many species. However, the identification of several species requires presence of flowers and/or mature fruits or sporangia. If these characters are not present at the time of sampling, plants may be brought home for cultivation until flowering and/or fruiting.

Alternatively, the habitats must be visited two or more times during the season or the year. The latter approach often applies to scientific studies.

Potamogeton coloratus in culture. Photo BM.

Documentation

In addition to sampling, identification and preparation of voucher collections, it is valuable to have a detailed and correct documentation of the findings:

Photos of habitat – overviews as well as details.

Photos of sampled plants – full figure as well as sections and details.

Site information – topographical name, exact location (GPS), location details (depth, sediment type, etc.).

Date and information about the collector.

When surveying and sampling according to specific programs and protocols, further requirements may apply.

A good overview photo as well as a more detailed photo of the plant(s) in the habitat are useful supplements to the sampled material itself. Include notes as to where photos are taken. Some cameras are waterproof and allow underwater photography. Also, ordinary cameras equipped with a polarizing filter make it possible to take good photos of underwater plants. Photos JCS.

For many species photos cannot replace collected material as documentation, but photos can be a very valuable supplement, e.g., for documentation of colours which cannot be seen on dried material. Photos of sampled plants should ideally be taken with a white background, e.g., in a large white tray. Photo JCS.

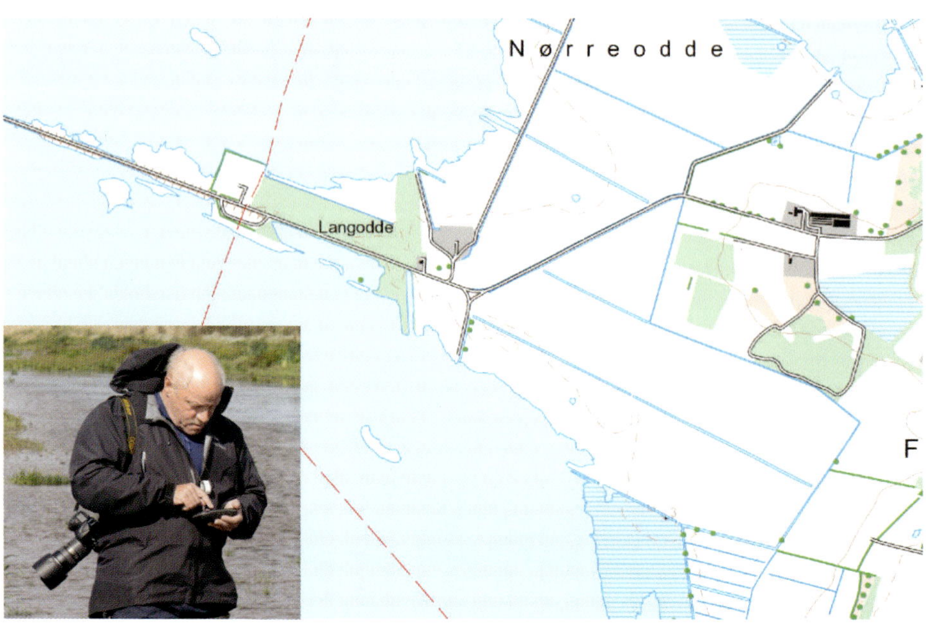

In addition to map information, a modern handheld GPS unit is very useful for accurate positioning of findings and for later retrieval of the finding place, respectively distribution of data on finding sites and maps. Photo JCS.

Preservation

The sampled plants may be preserved either as documentation for the findings, as contributions to a herbarium or as reference material (voucher collection). To ensure the value and future applicability, it is important to preserve adequately and store carefully labelled material.

Preservation of plants and parts of plants in +70% ethanol removes all the green colour but leaves the hard structures as well as the spatial appearance intact. Photos JCS.

Dry pressing of plants has for hundreds of years been the predominant method of preserving plants. Herbaria around the world contain numerous old, but still usable pressed specimens. Plants for pressing can in many cases be arranged directly on the dry herbarium sheet, but in the case of soft and limb plants, a special technique must be used: Fill a large tray with a few centimetres of clean water. Place a sheet of paper at the bottom of the tray. Place the plant material – one species at a time - in the tray and organise it over the submerged paper as one would do on dry paper. Then slowly pull the paper and the overlying plant material horizontally out of the tray while adjusting continuously the position of the plant material if necessary. Remove excess water by placing the paper on a newspaper or a piece of cloth. Then place a clean sheet of herbarium paper or better a sheet of blotting paper on top of the plant material and place this "sandwich" between sheets of newspaper. Now the sample is ready for pressing. Place the label from the sampling site next to the specimen or write a new label. Photos PH.

Herbarium sheet from 1901 with a well-organised specimen and a label with all necessary and relevant information about the find. The necessary information could be identification (valid name + author), place (exact location, country, province, county or municipality, field number in case of sampling campaigns, GPS coordinates, short habitat description, date, collector(s), collector number, identifier (if another person).

Main identification key

Determination of plants by use of this key should be based on more or less complete plants.

1 Plants free floating, not rooted in the sediment (***1***). In case of fragments, try both options
Subkey A

1 Plants rooted in the sediment 2

2 All leaves arising from the base of the plant, or from nodes of rooted stolons (***2***)
Subkey B

2 At least some leaves arising from an above-ground stem - this also includes leaves represented by sheaths on the lower part of stem (***7***) 3

3 Floating leaves with swollen petioles, gathered in an apical rosette (***3***)
240 *Trapa natans*

3 Floating leaves (if present) without swollen petioles. Habit different from (***3***) 4

4 Leaves or branches in whorls - at least sometimes more than two leaves arising at a node (***4***)
Subkey C

4 Leaves or branches not in whorls - at most 2 leaves arising at a node 5

5 Leaves opposite (***5***) - a few alternate leaves may be present at base of stem
Subkey D

5 Leaves alternate (***6***) - a few opposite leaves may be present in the apical part of stem 6

6 At least some leaves distinctly petiolate
Subkey E

6 Leaves sessile or represented by sheaths without or with much reduced lamina (***6***, ***7***)
Subkey F

1

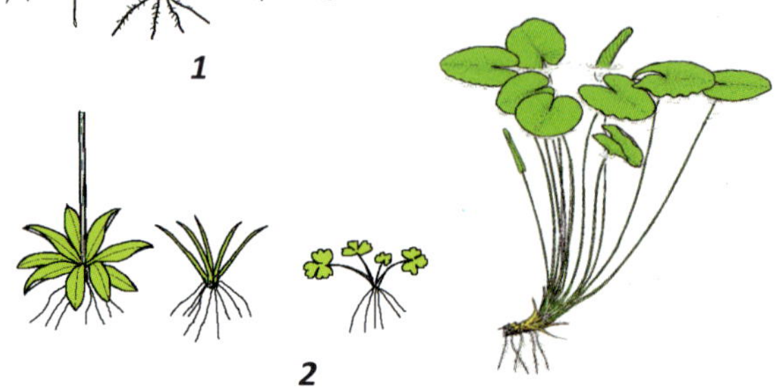

2

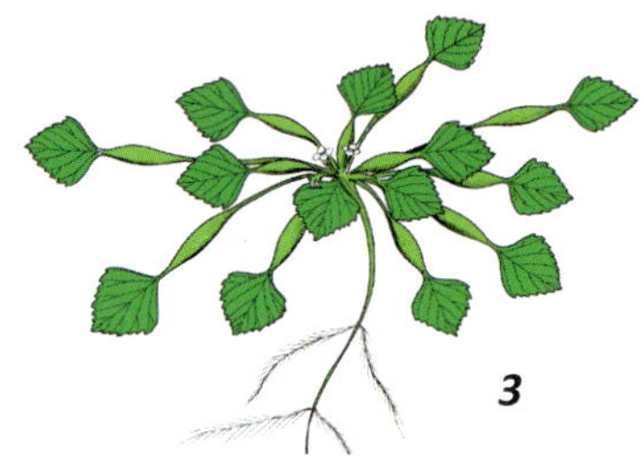

3

4

5

6

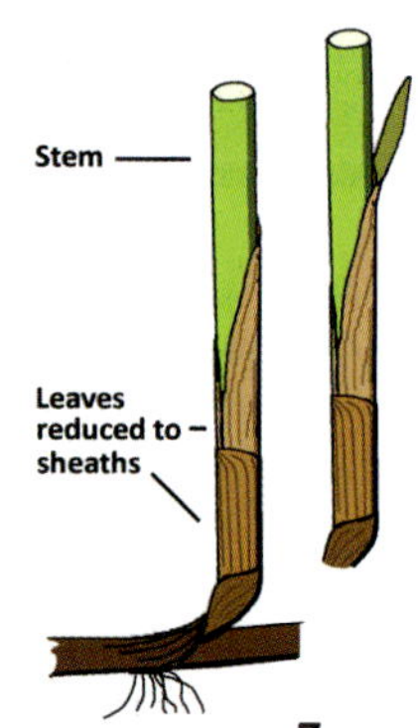

7

Subkey A

Free-floating plants not rooted in the sediment.
Note: Fragments of plants which would normally be rooted may be found detached and free floating; these can easily be mistaken for free-floating plants. If a satisfactory match is not reached using subkey A, please try using the other subkeys.

1 Plant less than 2 cm long, not differentiated into stem and leaves (***1***)
(*Lemna trisulca* (***1a***) may be interpreted as having stem and leaves, but should be keyed out here.)
 Lemna, Spirodela, Wolffia - use key p. 88
1 Plant more than 2 cm, differentiated into leaves and stem (the stem may be very short and hidden by the leaves (***6, 7***)) 2

2 Leaves in whorls (***2***, ***3***, ***4a***, ***4b***) 3
2 Leaves not in whorls 5

3 Leaves with trap-like structures (***2***)
 259 *Aldrovanda vesiculosa*
3 Leaves without trap-like structures (***3***) 4

4 Leaves 1–2 times bifurcate, distinctly dentate (***4a***). Fruits with or without two spines up to 6 mm long at the base
 197 *Ceratophyllum demersum*
4 Leaves 3–4 times bifurcate, more or less inconspicuously dentate (***4b***). Fruits smooth or warty sometimes with scattered minute emarginate spines
 199 *Ceratophyllum submersum*

5 Leaflets finely divided with more or less capillary segments (usually with some bladder-like traps (***5***))
 Utricularia - use key p. 658
5 Leaves not divided into capillary segments, traps lacking 6

6 Leaves tapering gradually from near base, sharply toothed (***6***)
 52 *Stratiotes aloides*
6 Leaves not tapering gradually from base and not sharply toothed 7

7 Leaves all arising from the same point 8
7 Leaves or leaflets arising from a stem 10

8 Leaves sessile, plicate (***7***)
 14 *Pistia stratiotes*
8 Leaves petiolate 9

9 Petioles not swollen, 1–2 mm in diameter, solid (***8***)
 51 *Hydrocharis morsus-ranae*
9 Petioles highly swollen, spongy (***9***)
 16 *Eichhornia crassipes*

10 Leaves linear
 163 *Isolepis fluitans*
10 Leaves not linear 11

11 Leaves scale-like, densely overlapping along a stem (***10***)
 365-366 *Azolla*
11 Leaves broadly rounded and more or less opposite (***11***)
 363-364 *Salvinia*

Subkey B

All leaves arising from the base of the plant. Plants rooted in the sediment.
Note: Couplet 2 requires a complete plant - if this is not an option, try both options for a satisfactory match.

1 At least some leaves distinctly petiolate ... 2
1 All leaves without distinct petiole ... 27

2 Plants with stolons or creeping rhizome ... 3
2 Plants without stolons or creeping rhizome ... 18

3 Lamina 3- or 4-foliolate (***1***, ***1a***) ... 4
3 Lamina not 3- or 4-foliolate ... 5

4 Leaves with 3 equal, 4–7 cm long, obovate to oval leaflets (***1a***)
335 *Menyanthes trifoliata*
4 Leaves with 4 equal, 0.6–3 cm long, obovate leaflets (***1***)
362 *Marsilea quadrifolia*

5 Leaves with a deep basal incision (***2***, ***3***, ***4***, ***17***, ***18***) ... 6
5 Leaves without a deep basal incision ... 12

6 Leaves sagittate - at least some of them (***2***)
42-48 *Sagittaria* - use key p. 128
6 Leaves not sagittate ... 7

7 Leaves lanceolate to almost linear (***8a***)
42-48 *Sagittaria* - use key p. 128
7 Leaves circular, ovate or reniform ... 8

8 Leaves reniform, lobed (***4***)
345 *Hydrocotyle ranunculoides*
8 Leaves almost circular to ovate, not lobed (***3***) ... 9

9 Lower side of leaf with dense covering of wart-like papillae (***8***). Leaf-margin wavy, scalloped
336 *Nymphoides peltata*
9 Lower side of leaf without papillae. Leaf-margin entire ... 10

10 Leaves with 2 curved primary veins on each side of the midrib converging at the leaf tip
51 *Hydrocharis morsus-ranae*
10 Lateral veins radiating out from a single point or pinnately arranged along the central vein (***17***,***18***) ... 11

11 Leaf veins forming a reticulum towards the leaf margin (***17***)
1-3 *Nymphaea* - use key p. 62
11 Leaf veins forked towards the leaf margin (***18***)
4-7 *Nuphar* - use key p. 63

12 Leaves peltate (***5***)
344-347 *Hydrocotyle* - use key p. 689
12 Leaves not peltate ... 13

13 Leaf sheaths less than 1 cm wide. Slender plants 14
13 Leaf sheaths 1.5–6 cm wide. Robust plants 15

14 Leaves with evident transverse veins (***6***)
38 *Baldellia repens*
14 Leaves without obvious transverse veins (***7***)
303 *Limosella aquatica*

15 Petiole about as long as lamina or longer. Lamina lanceolate to ovate or linear (***9***) 16
15 Petiole much shorter than lamina. Lamina 50–135 cm long, 30–70 cm wide, obovate to spathulate (***12***) 17

12

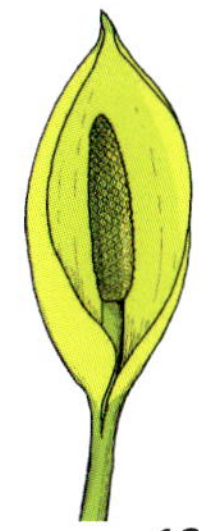
13

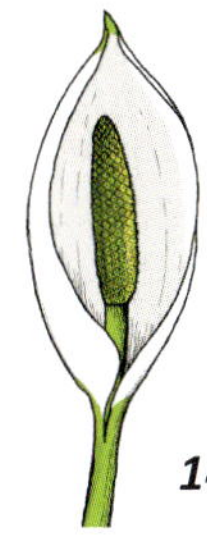
14

15

16 Inflorescence racemose with flowers in whorls (***11***)
42-48 *Sagittaria* - use key p. 128
16 Inflorescence a yellow, cylindrical spadix (***10***)
15 *Orontium aquaticum*

17 Spathe yellow (***13***)
12 *Lysichiton americanus*
17 Spathe white (***14***)
13 *Lysichiton camtschatcensis*
(Intermediate plants may be hybrids; please see page 85.)

18 Leaves deeply 3-lobed (***15***)
218 *Ranunculus sceleratus*
18 Leaves not lobed 19

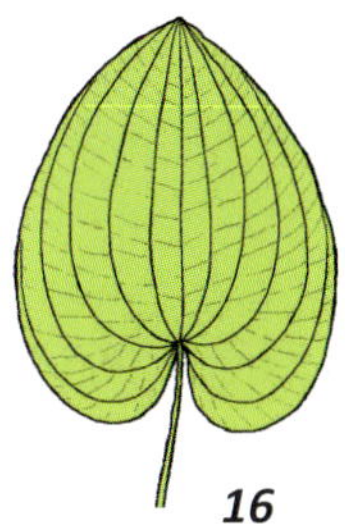
16

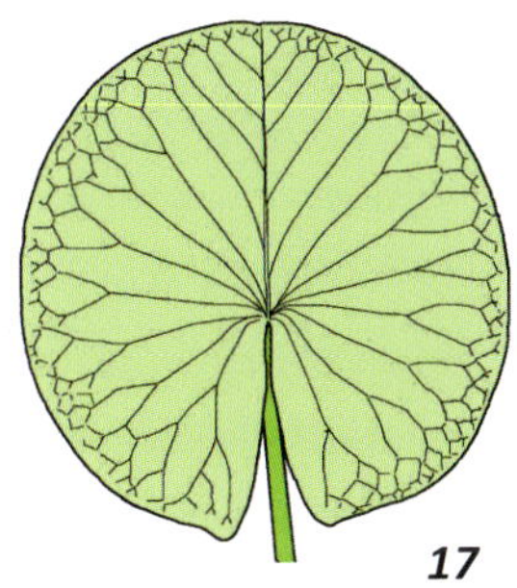
17

18

19 Leaves with a deep basal incision (***16***, ***17***, ***18***) 20
19 Leaves without a deep basal incision 22

19

20

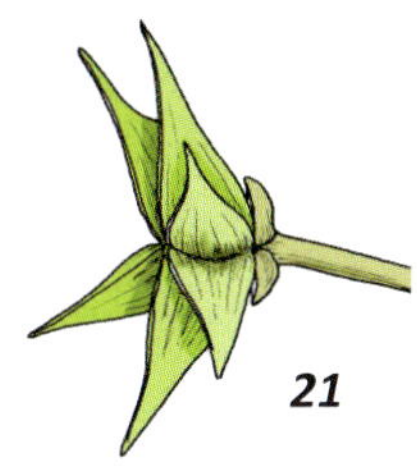
21

20 Lateral veins arched, parallel to the midvein and not branched (***16***)
39 *Caldesia parnassifolia*
20 Lateral veins radiating out from a single point or pinnately arranged along the central vein (***17***,***18***) 21

21 Leaf veins forming a reticulum towards the leaf margin (***17***)
1-3 *Nymphaea* - use key p. 62
21 Leaf veins forked towards the leaf margin (***18***)
4-7 *Nuphar* - use key p. 63

22 Inflorescence forked. Rhizome black
49 *Aponogeton distachyos*
22 Inflorescence not forked. Rhizome not black 23

22

23

24

25

23 Carpels forming a globose head (***19***)
37 *Baldellia ranunculoides*
23 Carpels in a single whorl (***20***, ***21***) 24

24 Carpels with long straight beak and spreading like a star (***21***)
40 *Damasonium alisma*
24 Carpels with short ± curved beak and not spreading like a star (***20***) 25

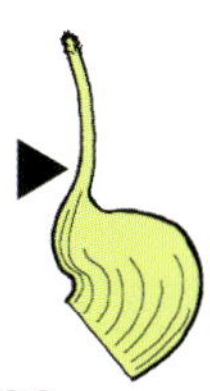
26

27

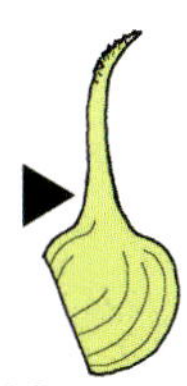
28

29

25 Styles 0.7–1.4 mm long, not hooked (***22***). Stamens with broad bases (***23***). Anthers 0.6–1.3 mm long 26
25 Styles 0.3–0.7 mm long, hooked (***24***). Stamens with narrow bases (***25***). Anthers 0.3–0.7 mm long
35 *Alisma gramineum*

26 Style fine and of even width from base to tip (***26***). Petals irregularly toothed (***27***). Stomata on the upper side of leaves 39–51 µm long

33 *Alisma plantago-aquatica*

26 Style broad at the base, narrower towards the apex (***28***). Petals with a prominent middle tooth (***29***). Stomata on the upper side of leaves 59–69 µm long.

34 *Alisma lanceolatum*

(Intermediate plants may be hybrids; see page 111.)

27 At least some leaves flaccid, ribbon-like, submerged 28

27 No leaves flaccid, not ribbon-like, submerged or not 35

28 Leaves faintly denticulate towards apex (***30, 30a***) 29

28 Leaves not denticulate towards apex 30

29 Ribbon-like, submerged leaves 2-4 mm wide

145-146 *Schoenoplectus*

29 Ribbon-like, submerged leaves more than 4 mm wide

64 *Vallisneria spiralis*

64a *Vallisneria australis*

30 Leaves seen in backlight with a pattern of almost quadrangular air chambers (***31***)

114-122 *Sparganium* - use key p. 278

30 Leaves seen in backlight with a pattern of oblong rectangular air chambers (***32, 33***) 31

31 Leaves with very prominent transverse secondary veins and 3–5 primary veins (***33***)**.** Leaves exude white latex when damaged

42-48 *Sagittaria* - use key p. 128

31 Leaves without or with weak transverse secondary veins and 3–14 primary veins (***32***). Leaves do not exude white latex when damaged 32

32 Leaves with 5–14 primary veins and weak transverse secondary veins (***32***)

145-146 *Schoenoplectus*

32 Leaves with central vein and marginal veins, without transverse secondary veins (***34***) 33

33 Plant with creeping stems rooting at nodes

41 *Luronium natans*

33 Plant without creeping stems 34

34 Leaves 3–15 mm wide

33 *Alisma plantago-aquatica*

34 *Alisma lanceolata*

35 *A. gramineum*

34 Leaves 1–3 mm wide

36 *Alisma wahlenbergii*

35 Plants with stolons or creeping rhizome 36

35 Plants without stolons or creeping rhizome 44

36 Leaf margin distinctly dentate (***35***)

52 *Stratiotes aloides*

36 Leaf margin not dentate 37

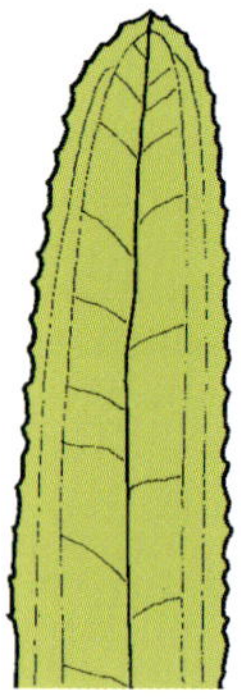

30

30a

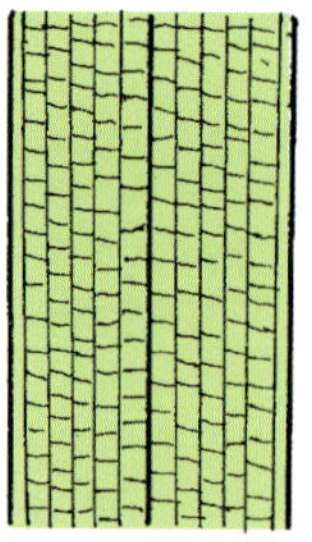

31

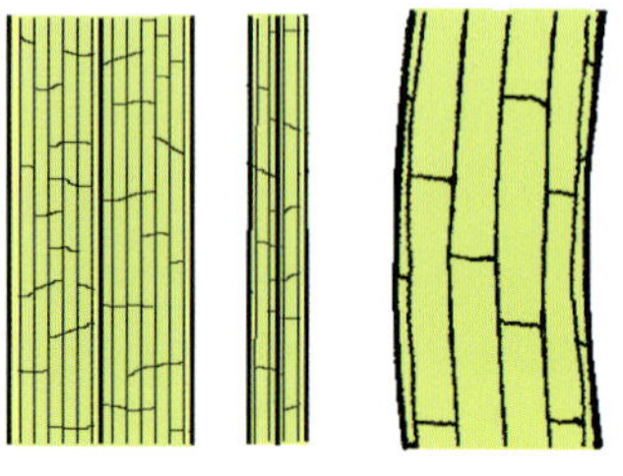

32

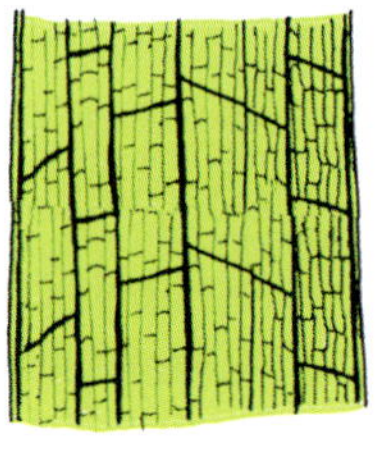

33

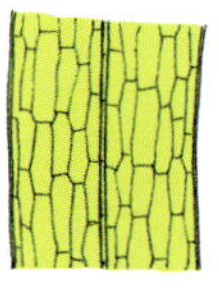

34

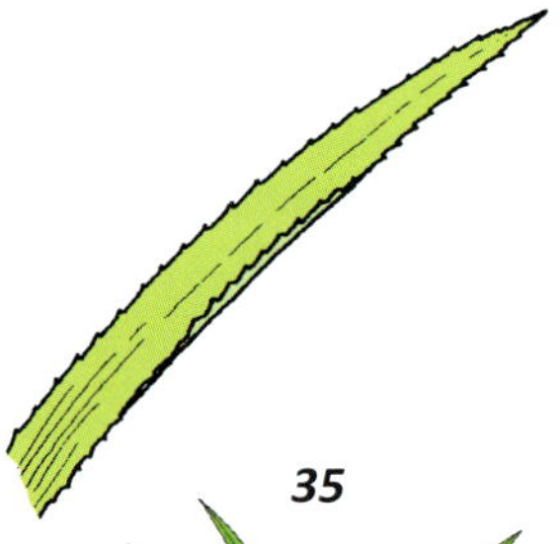

35

35a

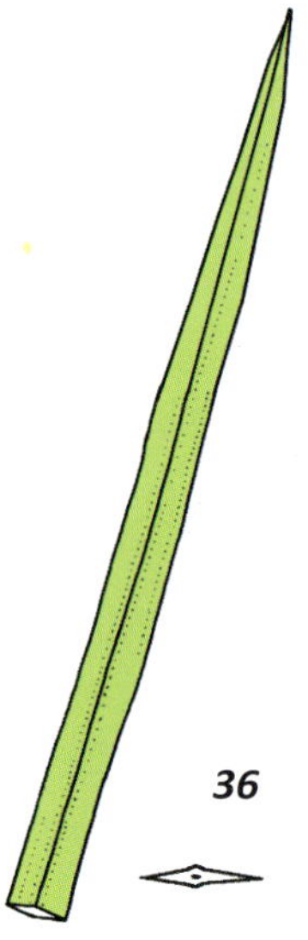

36

37

37 Leaves in cross section with 4 to 20 almost parallel air chambers (***35a***)
126-132 *Typha* - use key p. 280

37 Leaves in cross section without such air chambers 38

38 Leaves flat to slightly rhombic in cross section, 1–3 cm wide (***36***) 39

38 Leaves terete to semi-terete and 0.1–0.6 cm wide, or triangular in cross section and 0.3–1 cm wide (***37***) 40

38

39

39 Leaves grey green to glaucous, without an aromatic smell when crushed
113 *Iris pseudacorus*

39 Leaves yellow green, with an aromatic smell when crushed
9 *Acorus calamus*

40 Leaves triangular in cross section (***37***)
50 *Butomus umbellatus*

40 Leaves not triangular in cross section 41

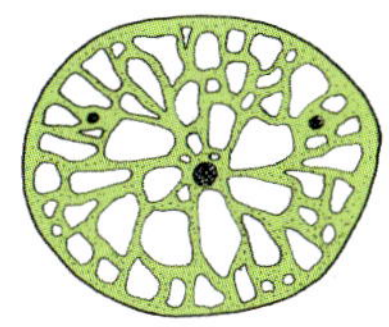

40

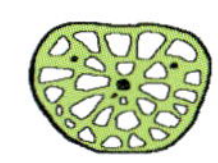

41

41 With a smell of cumin/coriander when crushed. Leaves with flat upper side and convex lower side (***38***)
38 *Baldellia repens*

41 Without noticeable smell when crushed. Leaves terete or ovate in cross section (***40***, ***41***) 42

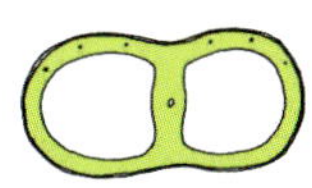

42

43

42 Leaves solitary or two together from a horizontal creeping stem. Young leaves unfurling from coiled apex (***39***)
361 *Pilularia globulifera*

42 Leaves several together forming rosette. Leaves not coiled when young 43

43 Submerged leaves 3–5 mm in diameter, terete (***40***). Leaves on emergent plants 2–3 mm wide, ovate in cross section when young with somewhat hairy margins
296 *Littorella uniflora*

43 Submerged leaves 1–2 mm in diameter, terete. Leaves on emergent plants 1–2 mm wide, slightly flattened (***41***), margins glabrous even when young
304 *Limosella australis*

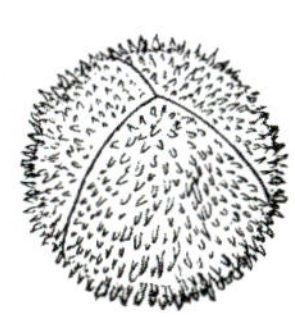

44

45

44 Leaves with 2 or 4 large air channels - visible in cross section (***42, 43***) 45

44 Leaves without conspicuous air channels 47

45 Leaves with 2 large air channels (***42***)
334 *Lobelia dortmanna*

45 Leaves with 4 large air channels (***43***) 46

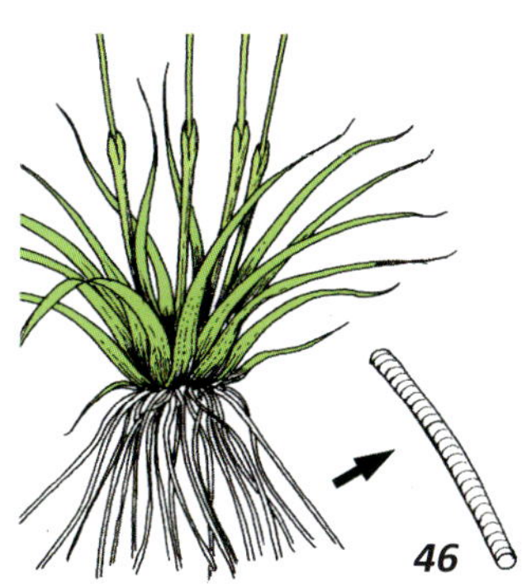

46

46 Leaves yellow green, long tapering, rather flaccid. Megaspores white, covered with fine spikes (***44***)
359 *Isoetes echinospora*

46 Leaves dark green, abruptly tapering, stiff. Megaspores grey, with wart- or crest-like formations (***45***)
360 *Isoetes lacustris*

(Intermediate plants may be hybrids; - see page 715.)

47 Roots white, appearing articulated (***46***)

133 *Eriocaulon aquaticum*

47 Roots white, not appearing articulated 48

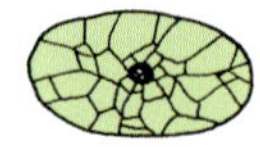

47

48 Leaves with a flat upper side and an arched lower side (***38***). With a smell of cumin/coriander when crushed

37 *Baldellia ranunculoides*

48 Leaves flattish elliptic in cross section (***47***). Without noticeable smell when crushed

254 *Subularia aquatica*

Subkey C

Leaves or branches in whorls - at least sometimes more than two leaves arising at a node.

Note: Charophytes are often mistaken for aquatic vascular plants - they are not treated in this book.

1 Most leaves divided (some undivided floating leaves may also be present) 2
1 Leaves not divided 5

2 Leaves petiolate, fan shaped (***1***)
8 *Cabomba caroliniana*
2 Leaves sessile, not fan shaped 3

3 Leaves pinnately divided (***2***)
Myriophyllum - use key p. 516
3 Leaves dichotomously divided (***3***) 4

4 Leaves 1–2 times bifurcate, distinctly dentate (***4***). Fruits with or without two spines up to 6 mm long at base
197 *Ceratophyllum demersum*
4 Leaves 3–4 times bifurcate, more or less inconspicuously dentate (***5***).
Fruits smooth or warty sometimes with scattered minute emarginate spines
199 *Ceratophyllum submersum*

5 Stem enclosed by a sheath above each stem node. Often with whorls of 3-dimensional branches arising at the nodes (***6***)
367 *Equisetum* species
5 Stem not enclosed by a sheath above each stem node 6

6 Leaves pinnately veined 7
6 Leaves with a single central vein 9

7 Leaves 1–3.5 cm long and up to 5 mm wide 21
7 Leaves more than 3.5 cm long and wider than 5 mm 8

8 Leaf margin entire. Flowers yellow, 5-merous
267 *Lysimachia vulgaris*
8 Leaf margin dentate, crenate or undulate. Flowers blue, lilac or white, 4-merous
Veronica species - see **subkey D**, couplet 30

9 Leaves with sheaths (***7***)
59-63 *Najas* - use key p. 156
9 Leaves without sheaths 10

10 Leaf margin without teeth (check by use of a hand lens - especially towards the leaf tip) 11
10 Leaf margin fine to coarsely toothed or with small, retrorse prickles (***8-10***, ***24-27***) 15

1

2

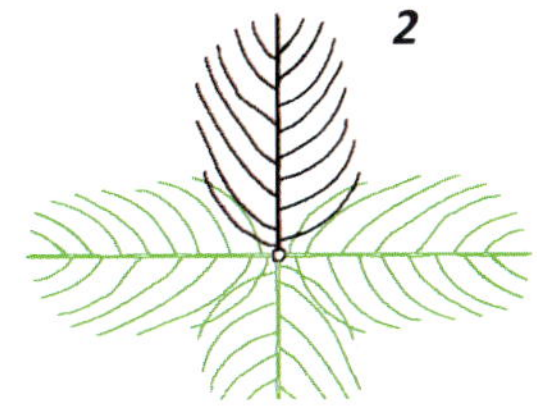

3

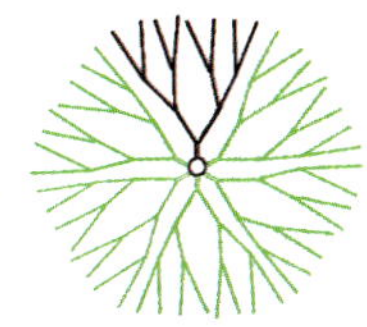

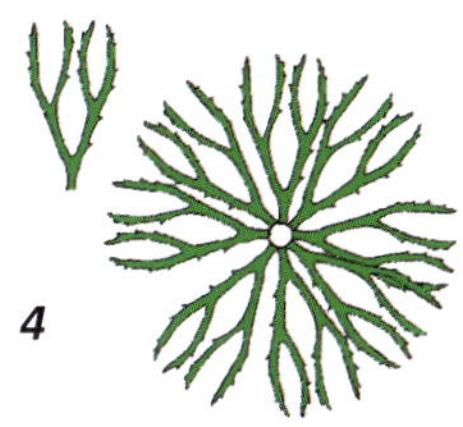

4

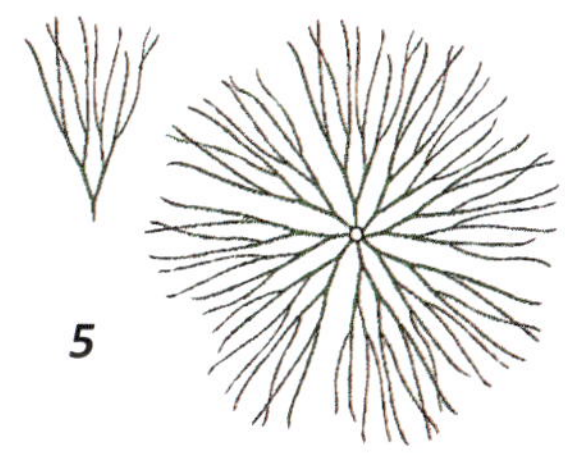

5

6

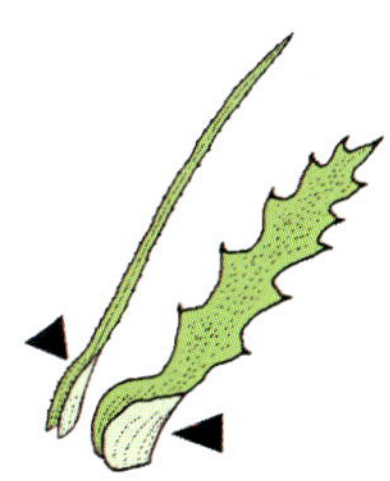

7

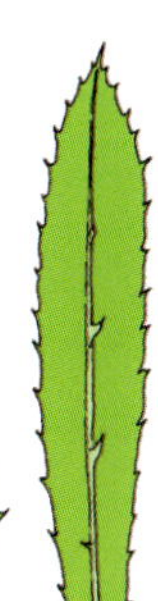

8 ***9*** ***10***

11 Leaves with an apical hydatode (***11***).
Plant without rhizome
233 *Elatine alsinastrum*
11 Leaves without an apical hydatode.
Plant with rhizome 12

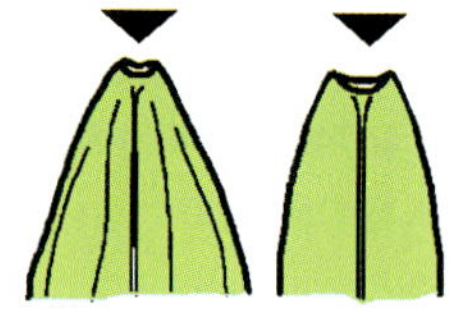
11

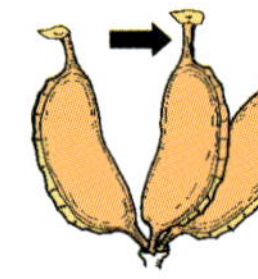
12

12 Leaves narrowly linear, with a stipule-like
ligule (***13***). Fruits with a beak (***12***)
104-107 *Zannichellia* - use key p. 260
12 Leaves tapering or oblong ovate to spathulate,
without ligule. Fruits without a beak 13

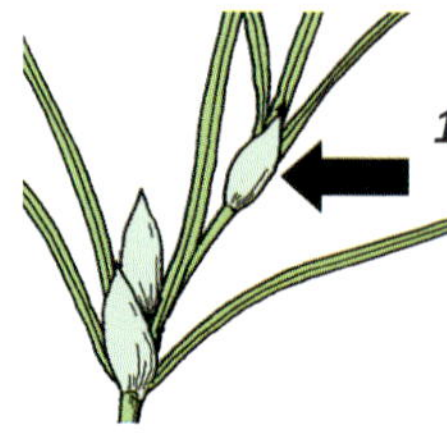
13

13 Leaves in whorls of 8–12(–16) (***14***). Stamens with
filaments longer than the mature anther (***15***)
293 *Hippuris vulgaris*
13 Leaves in whorls of 4–7 (rarely only 3). Stamens
with filaments shorter than the mature
anther (***17***) 14

14

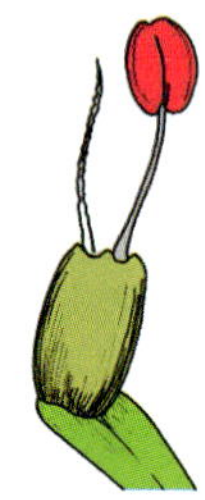
15

14 Leaves usually in whorls of 4 (3 or 5) (***16***)
294 *Hippuris tetraphylla*
14 Leaves usually in whorls of 6 (5 or 7) (***18***)
295 *Hippuris lanceolata*

15 Leaves fine to coarsely toothed, hydromorphic.
Stem terete 16
15 Leaves with small retrorse spines, xeromorphic.
Stem angled 21

16

17

18

16 Leaves sharp, almost spiny dentate (***10***), mostly
in whorls of 5–8. Stipules present (***9***)
57 *Hydrilla verticillata*
16 Leaves finely toothed (***9***, ***21-22***) mostly in whorls
of 3–5 or spirally arranged but pseudo-whorled
above. Stipules absent 17

17 Leaves in whorls of 3 18
17 Leaves in whorls of 4–5 or spirally arranged
but pseudo-whorled distally 20

18 Leaves obtuse or subacute (***19***), 0.8–2.3 mm
wide measured 0.5 mm below apex (***21***)
54 *Elodea canadensis*
18 Leaves acute, 0.2–0.7 mm wide measured
0.5 mm below apex (***22***) 19

19

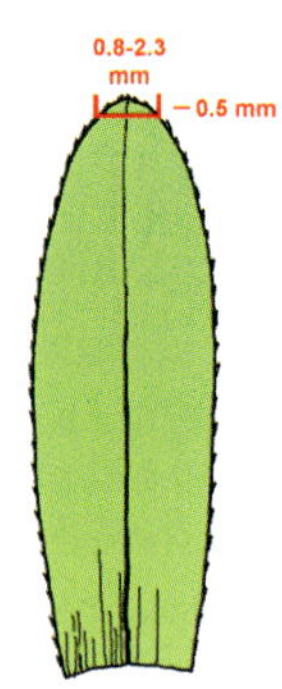

21

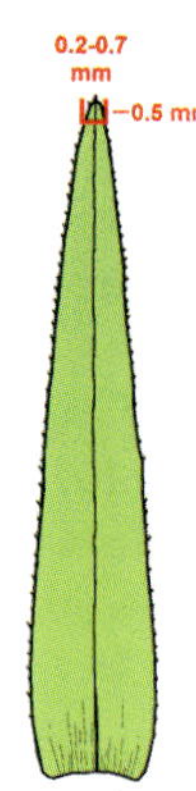

22

19 Leaves with 0.06–0.09(–0.10) mm long marginal
teeth. Root tips white to greenish, when fresh.
Leaves often strongly recurved (***20***)
55 *Elodea nuttallii*
19 Leaves with (0.08–)0.11–0.14 mm long marginal
teeth. Root tips red, when fresh. Leaves not
strongly recurved
56 *Elodea callitrichoides*

20

20 Leaves in whorls of 4–5
53 *Egeria densa*
20 Leaves spirally arranged (***23***) but pseudo-whorled
distally
58 *Lagarosiphon major*

23

21 Stem angles very rough of backwardly directed spines. Leaves apiculate or mucronate at apex (***27***)
273 *Galium uliginosum*

21 Stem angles smooth or slightly rough. Leaves obtuse to acute (***24-26***) 22

22 Leaves linear oblong to obovate, 1.5–3(–5) mm wide (***24-25***). Pedicels scarcely divaricate at fruiting
270 *Galium palustre*

22 Leaves linear, up to 1 mm wide (***26***). Pedicels strongly divaricate at fruiting
272 *Galium constrictum*

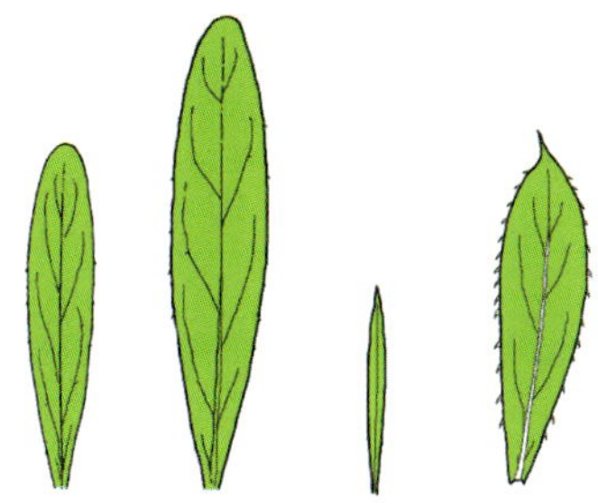

24 ***25*** ***26*** ***27***

Subkey D

Leaves opposite, but a few alternate leaves may be present at base of stem.

1 At least some of the leaves divided 2
1 All leaves undivided 3

2 Divided leaves trifoliate
339-341 *Bidens* - use key p. 680
2 Divided leaves fan-like with fine segments (***1***)
8 *Cabomba caroliniana*

1

3 Leaves with only a single central vein (examine the leaf in backlight and by use of a hand lens) 4
3 Leaves not with a central vein only 13

4 Leaves with sheaths 5
4 Leaves without sheaths 7

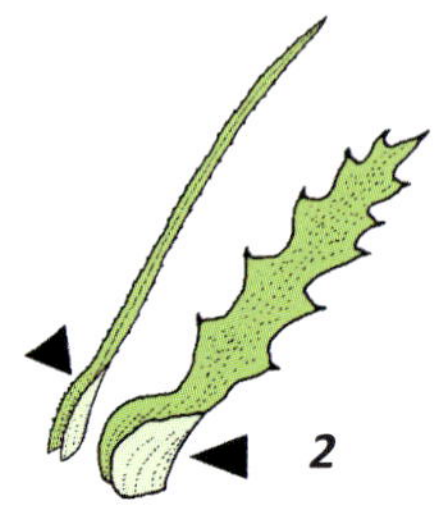

2

5 Leaf margin fine to coarsely toothed from base to apex (***2***)
59-63 *Najas* - use key p. 156
5 Leaf margin entire except for the finely denticulated apical part 6

6 Leaf apex obtuse, symmetrical (***3***). Peduncles 5–30 cm long, coiled when fruiting
108 *Ruppia cirrhosa*
6 Leaf apex acute, asymmetrical (***4***). Peduncles 1–2.6 cm long, not coiled when fruiting
109 *Ruppia maritima*

3

4

7 Leaves with an apical notch (***5***)
281-292 *Callitriche* - use key p. 600
7 Leaves without an apical notch 8

8 Leaves spathulate to obovate 9
8 Leaves narrowly linear to lanceolate 10

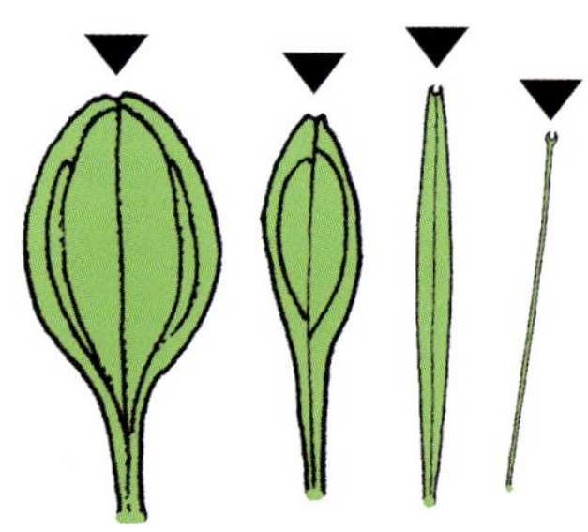

5

9 Stem rooting at nodes
234-237 *Elatine* - use key p. 533
9 Stem not rooting at nodes
261-263 *Montia* spp.

10 Fruits with a beak (***6***). Leaves not fleshy, narrowly linear, with a stipule-like ligule (***7***)
104-107 *Zannichellia* - use key p. 260
10 Fruits without a beak. Leaves fleshy, linear to lanceolate, without ligule (***8***) 11

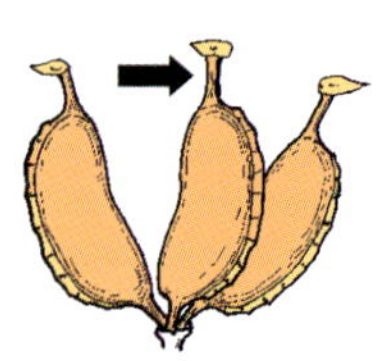

6

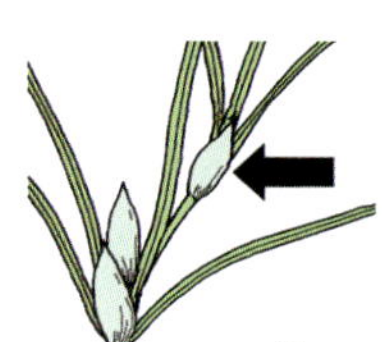

7

11 Flowers sessile (***9***). Leaves 3–6 mm long, 1–1.5 mm wide
224 *Crassula aquatica*
11 Flowers pedicellate (***10***). Leaves 4–20 mm long, 1.0–2.0 mm wide 12

8

9

12 Pedicels not longer than the neighbouring leaves (***10***)
225 *Crassula helmsii*
12 Pedicels longer than the neighbouring leaves
225a *Crassula peduncularis*

10

13 Stem terete 14
13 Stem bluntly to sharply angular or grooved 34

14 Leaves sessile 15
14 At least some leaves petiolate 26

15 Leaves 1.2–3.5 cm wide 16
15 Leaves less than 1.2 cm wide 20

16 Leaf margin serrate or lobed 17
16 Leaf margin entire. Flowers yellow 19

17 Leaves serrate 18
17 Leaves lobed or sinuate (***11***)
343 *Shinnersia rivularis*

11

12

18 Leaves coarsely serrate (***12***). Flowers yellow, many together in capitula
337 *Bidens cernua*
18 Leaves finely serrate. Flowers pink to rose purple or white solitary
241 *Epilobium hirsutum*

19 Leaves with orange, glandular dots above only (use hand lens). Flowers in dense semi-globular racemes from the axils of the leaves on the middle part of the stem (***13***)
266 *Lysimachia thyrsiflora*
19 Leaves with orange, glandular dots on both sides (use hand lens). Flowers in large, terminal, open clusters from leaf axils and top of stem
267 *Lysimachia vulgaris*

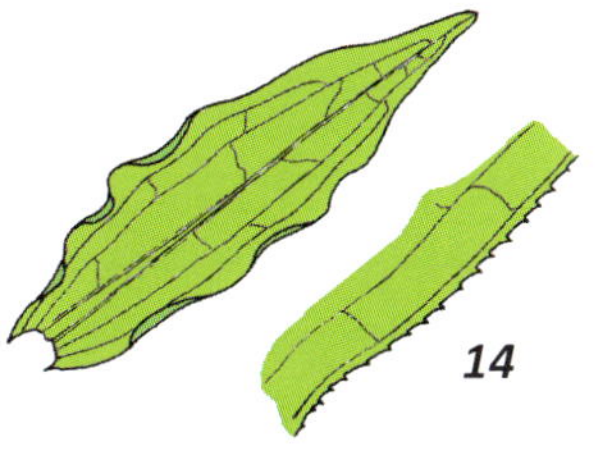

14

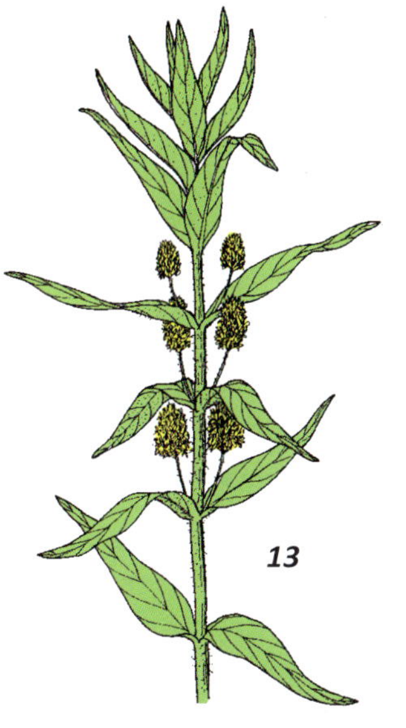

13

20 Leaf margin finely denticulate (use hand lens) (***14***)
65 *Groenlandia densa*
20 Leaf margin entire 21

21 Stem rooting at nodes 22
21 Stem not rooting at nodes 24

22 Flowers without petals and sepals. Leaves with an apical notch (***5***)
281-292 *Callitriche* - use key p. 600
22 Flowers with reddish or yellow petals and green sepals. Leaves without an apical notch, but eventually with a distinct hydathode 23

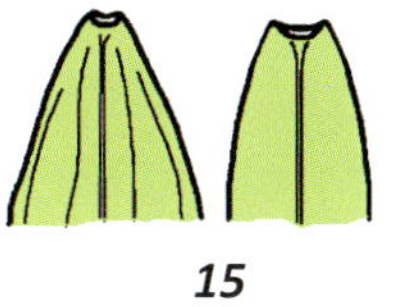

15

23 Leaves less than 4 mm wide. Leaf apex with a distinct hydathode (***15***)
234-237 *Elatine* - use key p. 533
23 Leaves more than 5 mm wide. Leaf apex without a distinct hydathode (***16***)
238 *Hypericum elodes*

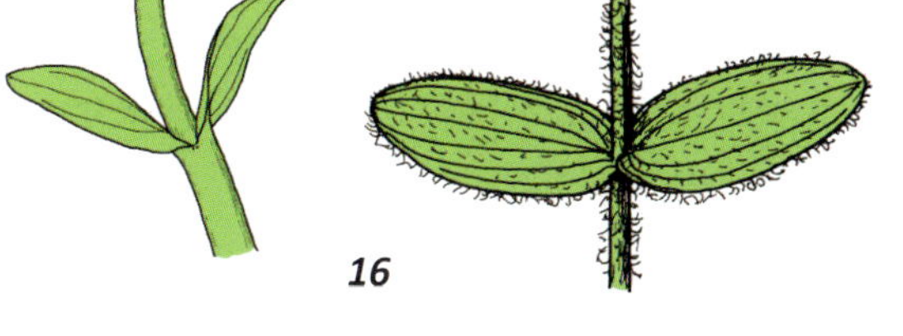

16

24 Upper side of leaves with numerous orange dots easily seen in backlight by use of a hand lens
266 *Lysimachia thyrsiflora*
24 Upper side of leaves without orange dots 25

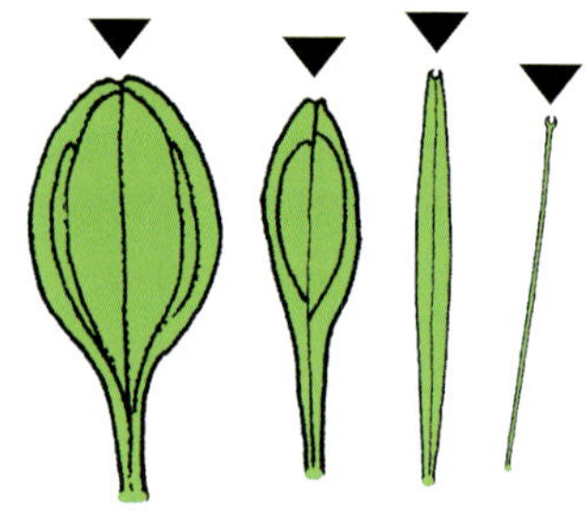
17a

25 Leaves thin, with an apical notch (***17a***). Plant submerged to partly floating or emerged and laying flat to the ground. Flowers without petals
281-292 *Callitriche* - use key p. 601
25 Leaves thick, without an apical notch. Plant emergent. Flowers with white petals
261 *Montia fontana*

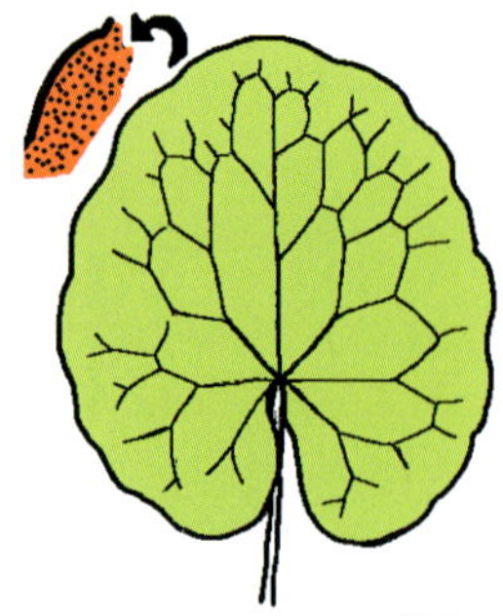
17

26 Lamina with a deep basal incision (***17***)
336 *Nymphoides peltata*
26 Lamina without a deep basal incision 27

27 Leaf margin more or less prominently toothed 28
27 Leaf margin entire 30

28 Leaf base with auricles (***11***)
343 *Shinnersia rivularis*
28 Leaf base without auricles 29

29 Flowers numerous in capitula, yellowish
339-341 *Bidens* - use key p. 680
29 Flowers solitary in leaf axils, yellow to orange blotched, red or lavender blue
314-326 *Erythranthe/Mimulus* - use key p. 648

30 Flowers without petals and sepals. Leaves with an apical notch (***5***)
281-292 *Callitriche* - use key p. 600
30 Flowers with petals and sepals. Leaves without an apical notch 31

31 Lamina 3–4(–6) cm long and 1.5–2.0(–3.5) cm wide. Flowers in racemes.
301 *Veronica beccabunga*
31 Lamina less than 3 cm long. Flowers single or few together 32

32 Stem rooting at nodes 33
32 Stem not rooting at nodes
261 *Montia fontana*

33 Leaves spathulate to narrowly lanceolate. Flowers 3- to 4-merous
234-237 *Elatine* - use key p. 533
33 Leaves ovate to suborbicular. Flowers 5-merous (***18***)
265 *Lysimachia tenella*

18

34 All leaves sessile 35
34 At least some leaves petiolate 40

35 Leaves 2–6 mm long. Calyx shining white to pinkish, with small purple dots (***19***)
260 *Illecebrum verticillatum*
35 Leaves more than 10 mm long. Calyx green 36

19

36 Leaves fused at base by a narrow ciliate flange of tissue (***20***). Capsule narrow, up to 9 mm long
313 *Hygrophila polysperma*

36 Leaves clasping the stem and without cilia
Veronica spp. - inflorescence needed 37

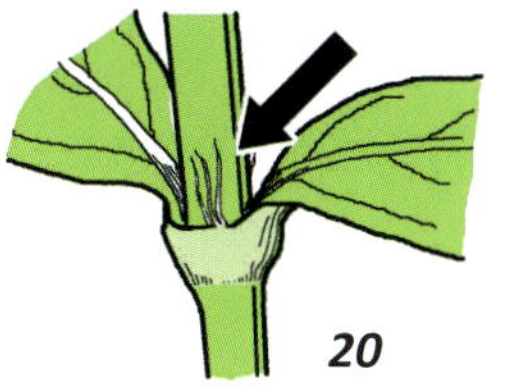

20

37 Leaf margin with small, widely spaced, perpendicularly protruding teeth (***21***). With one raceme per node (***21a***). Capsule opening into 2 valves
302 *Veronica scutellata*

37 Leaf margin not with small, widely spaced, perpendicularly protruding teeth. With two racemes per node (***22***). Capsule opening into 4 valves 38

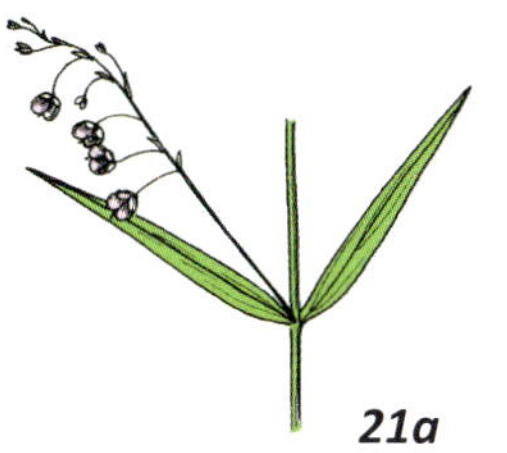

21a

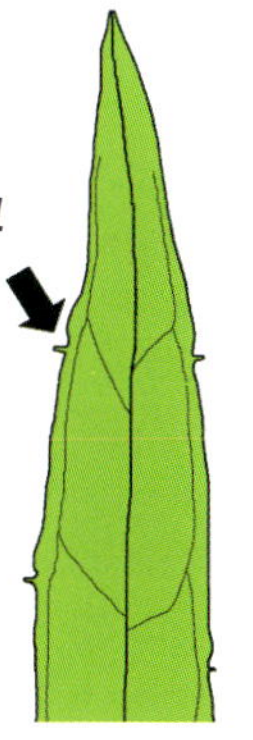

21

38 Capsule ovate to elliptic, distinctly longer than wide (***23***)
299 *Veronica anagalloides*

38 Capsule orbicular, at most slightly longer than wide 39

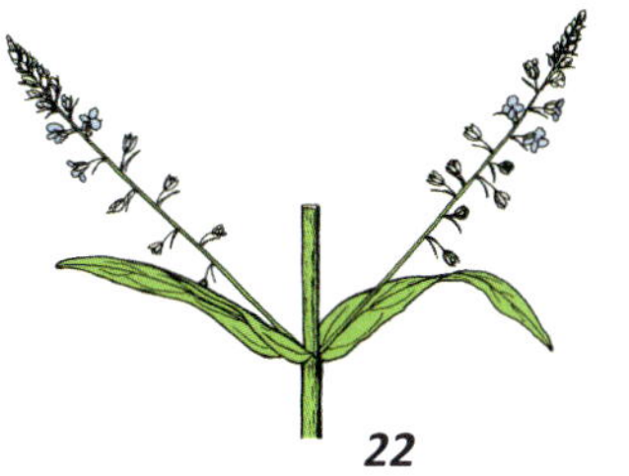

22

39 Pedicels in the fruiting stage protruding at an angle of 90° (***24***). Flowers predominantly pinkish (rarely white), with the veins not extended almost to the edge of the petals
300 *Veronica catenata*

39 Pedicels in the fruiting stage protruding at an angle of 45° (***25***). Flowers predominantly lilac blue (rarely white), with the veins extended almost to the edge of the petals
297 *Veronica anagallis-aquatica*
Vigorous, but almost completely sterile plants; possibly the hybrid **298 *Veronica ×lackschewitzii***

23

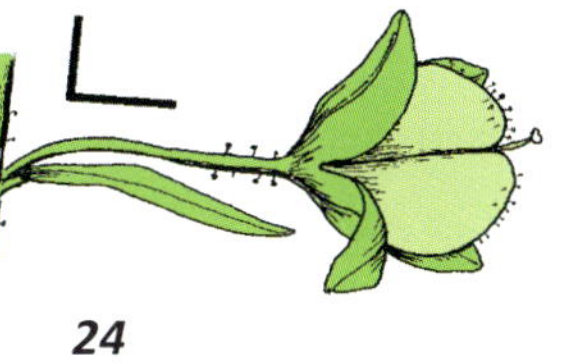

24

40 Stem rooting at nodes 41

40 Stem not rooting at nodes 48

41 Leaf margin serrate. Corolla dark to bright blue, rarely white or pinkish
301 *Veronica beccabunga*

41 Leaf margin entire. Corolla yellow or pink, in some species rudimentary or falling early 42

42 Leaves fused at base by a narrow ciliate flange of tissue (***20***)
313 *Hygrophila polysperma*

42 Leaves without cilia at base 43

25

43 Leaves rounded; not acute at apex 44

43 Leaves acute at apex 45

44 Leaves almost circular, shortly petiolate. Upper side of leaf with numerous, submerged, dark, glandular dots (***26***). Flowers large, yellow, 5-merous
268 *Lysimachia nummularia*

44 Leaves more or less widely spathulate. Central vein usually red coloured. Upper side of leaf without glandular dots (***27***). Flowers small, pink, 6-merous.
239 *Lythrum portula*

26

27

45 Flowers 5-merous. Petals pink (***18***)
265 *Lysimachia tenella*
45 Flowers 4-merous. Petals yellow, up to 3–5 mm long or absent 46

46 Petals absent (***28***)
242 *Ludwigia palustris*
46 Petals present, small (***29***) 47

47 Petals 3–5 mm long
244 *Ludwigia repens*
47 Petals 0.5 mm long
243 *Ludwigia ×kentiana*

48 Leaves when crushed releasing a distinct odour of mint. Flowers several together in axillary whorls or in a terminal spike or head
305-312 *Mentha* - use key p. 640
48 Leaves when crushed not releasing a distinct odour of mint. Flowers solitary or in racemes from leaf axils 49

49 Leaf margins entire. Leaves fused at base by a narrow ciliate flange of tissue (***20***)
313 *Hygrophila polysperma*
49 Leaf margins toothed. Leaf bases without cilia 50

50 Flowers solitary in leaf axils. Petals yellow to orange blotched, red or lavender blue, forming a zygomorphic corolla
314-326 *Erythranthe/Mimulus* - use key p. 648
50 Flowers in racemes from leaf axils. Petals blue or white 51

51 Leaves all shortly petiolate, ovate to oblong, fleshy (***30***)
301 *Veronica beccabunga*
51 Leaves sessile in the upper part of the plant, ovate lanceolate to lanceolate, not fleshy (***31***)
297 *Veronica anagallis-aquatica*

28

29

30

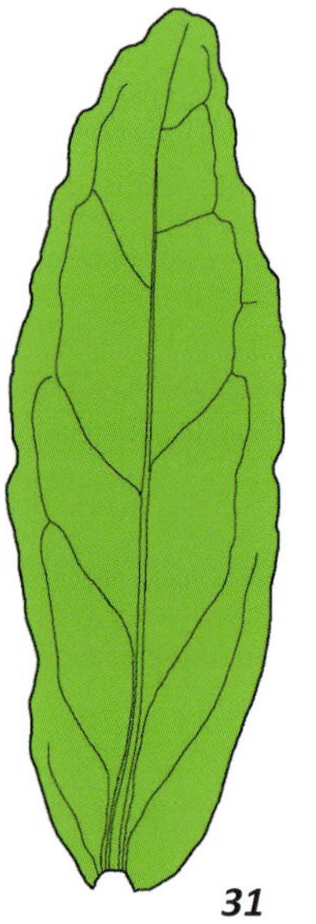

31

Subkey E

Leaves alternate, but a few opposite leaves may be present in the apical part of stem. At least some leaves distinctly petiolate.

1	Leaves divided	2
1	Leaves not divided	14
2	Leaves pinnately divided	3
2	Leaves not pinnately divided	12
3	With aerial leaves only or submerged leaves almost like aerial leaves	4
3	With submerged leaves only or aerial leaves very different from submerged leaves	10
4	Stem with numerous hairs and glandular hairs. Leaves with terminal and 2(–3) pairs of leaflets (***1***) **232 *Comarum palustre***	
4	Stem glabrous. Leaves with terminal and 3 or more pairs of leaflets	5
5	Leaf sheath enclosing the stem (***2***). Flowers in capitula (***3***) **342 *Cotula coronopifolia***	
5	Leaf sheath not enclosing the stem. Flowers not in capitula	6
6	Flowers in umbels **348-358 Apiaceae** - use key p. 692	
6	Flowers in racemes	7
7	Flowers white	8
7	Flowers yellow **251-253 *Rorippa*** species and hybrids - please read the descriptions	
8	Anthers violet. Stem solid **247 *Cardamine amara***	
8	Anthers yellow. Stem hollow	9
9	Seeds with a fine structure of 100–150(–190) polygonal depressions - 11–20 across the broadest width (***4***). Siliquae usually with one row of seeds in each loculi (***5***). Terminal leaflet tapered at base (***6***) **248 *Nasturtium microphyllum***	
9	Seeds with a coarse structure of 25–50(–70) polygonal depressions - 6–12 across the broadest width (***7***). Siliquae usually with two rows of seeds in each loculi (***8***). Terminal leaflet rounded to slightly cordate at base (***9***) **250 *Nasturtium officinale*** The hybrid between the two *Nasturtium* species, **249 *Nasturtium ×sterile***, is intermediate between the parental species, has usually sterile pollen and poor seeds.	

10 All leaves submerged, uniform, 1-pinnate (***10***). Only inflorescence raised above the water surface
264 *Hottonia palustris*
10 With aerial and submerged leaves. The latter of variable shape and incision (***11***) 11

10

11

11 Flowers in umbels
348-358 Apiaceae - use key p. 692
11 Flowers in racemes
251 *Rorippa amphibia*

12 Leaves with one large and two small leaflets (***12***)
274 *Solanum dulcamara*
12 Leaves not with one large and two small leaflets 13

12

13

13 Leaves with three equal, 4–7 cm long obovate to oval leaflets (***13***)
335 *Menyanthes trifoliata*
13 Leaves 3–5 lobular or divided into capillary segments (***14***). Many species with both laminar and capillary leaves
202-223 Ranunculaceae - use key p. 452

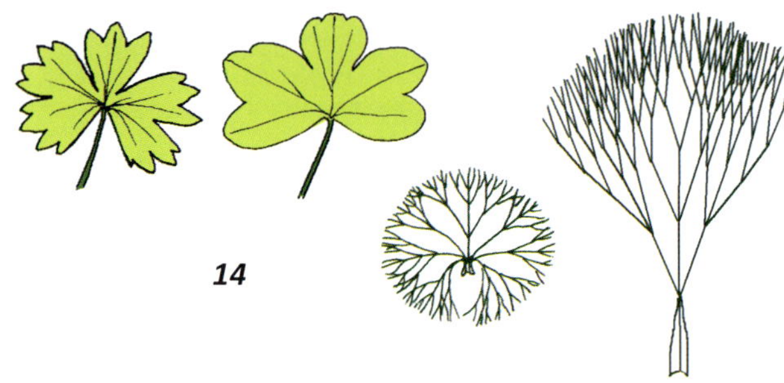
14

14 Leaves all linear or reduced to sheaths at base of stem 15
14 All or some leaves not linear or reduced to sheaths 24

15 Plant with aerial or emergent leaves
Grass-like plants - use key p. 318
15 Plant submerged or floating. Without leaves above water 16

16 Leaves with sheaths or reduced to sheaths 17
16 Leaves without sheaths 23

17 All leaves reduced to sheaths
Grass-like plants - use key p. 318
17 At least some leaves not reduced to sheaths 17

18 Leaves with free ligules (***15***)
99-103 *Stuckenia* - use key p. 246
18 Leaves without free ligules 19

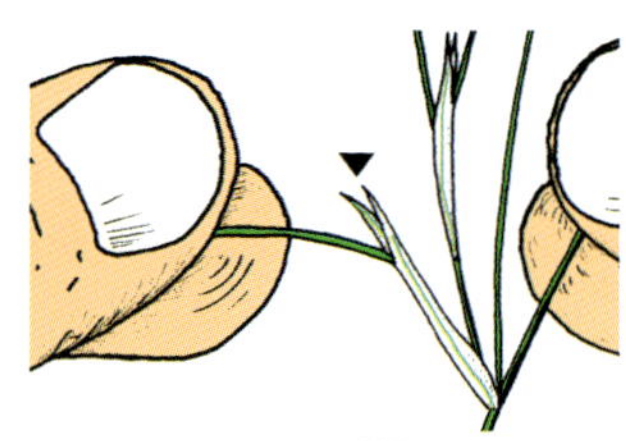
15

19 Leaves flat 20
19 Leaves terete
Juncus - use key p. 319

20 Plants in salt- or brackish water 21
20 Plants in freshwater 22

21 Leaves dark green, becoming almost black when dried. Perennial plants with creeping rhizomes
110-112 *Zostera*
21 Leaves whitish green, not becoming almost black when dried. Annual plant
342 *Cotula coronopifolia*

22 Leaves flat, not fleshy. Perennial plants with creeping rhizomes
145-151 *Schoenoplectus* - use key p. 337
22 Leaves flat and fleshy. Annual plant
342 *Cotula coronopifolia*

23 Leaves densely set along the stem and curved backwards (***16***)
58 *Lagarosiphon major*
23 Leaves not densely set and not curved backwards
66-97 *Potamogeton* - use key p. 170

24 Stipules forming a membranous ochrea surrounding the stem (***17***) 25
24 Stipules if present not forming an ochrea 27

25 Tepals 5 without swollen tubercles (***18***). Inflorescence spike-like 26
25 Tepals 6 in 2 whorls of 3, at least some of the inner with swollen tubercles (***19***). Inflorescence paniculate or racemose with flowers in whorls
257-258 *Rumex* spp.

26 Perennial. Inflorescence dense, erect. Ochreas without bristles at the margins
255 *Persicaria amphibia*
26 Annual. Inflorescence open, slender. Ochreas with bristles at the margins
256 *Persicaria hydropiper*

27 Base of lamina cordate 28
27 Base of lamina rounded or tapered into the petiole 32

28 Lower side of leaf with dense covering of wart-like papillae (***20***)
336 *Nymphoides peltata*
28 Lower side of leaf without papillae 29

29 Leaf veins palmate, reticulate (***21***)
200-201 *Caltha palustris*
29 Leaf veins pinnate or arched and parallel 30

30 Leaves submerged and flaccid
219 *Ranunculus lingua*
30 Leaves aerial and firm 31

31 Inflorescence a yellow-green spadix with an inside white spathe (***22***)
11 *Calla palustris*
31 Inflorescence a spike of bluish-violet or (rarely) white flowers
17 *Pontederia cordata*

32 Lamina 50–135 cm long and 30–70 cm wide. Inflorescence a spadix of yellowish to green flowers and with a large yellow or white spathe (***23-24***) 33
32 Lamina smaller. Inflorescence different 34

16

17

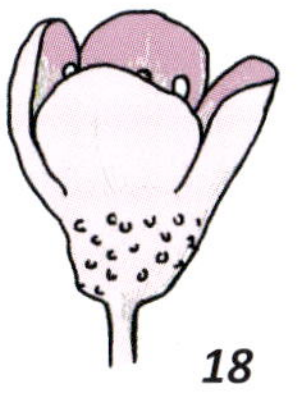
18

19

20
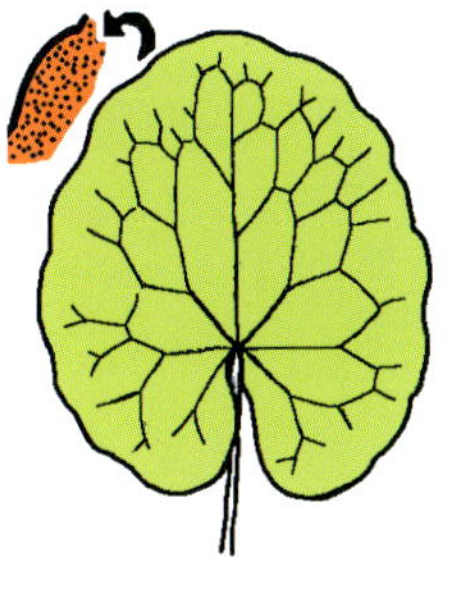

21

22

33 Spathe yellow (***23***)

12 *Lysichiton americanus*

33 Spathe white (***24***)

13 *Lysichiton camtschatcensis*

23

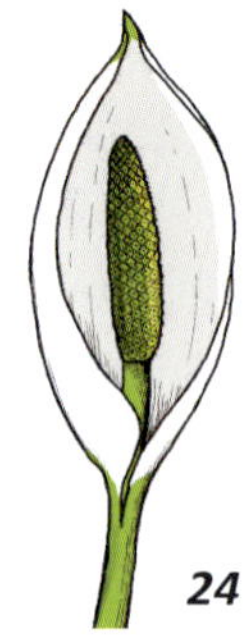

24

34 Leaf margin serrate

251 *Rorippa amphibia*

34 Leaf margin entire 35

35 Leaf base with stipules (look carefully) 36

35 Leaf base without stipules 38

36 Leaf veins pinnate 37

36 Leaf veins parallel

66-97 *Potamogeton* - use key p. 170

37 Stipules obtuse to rounded (***25***). Flowers with sepals up to 10 mm long and petals 7–17 mm

246 *Ludwigia peploides*

37 Stipules triangular (***26***). Flowers with sepals up to 18 mm long and petals 12–25 mm long

245 *Ludwigia grandiflora*

25

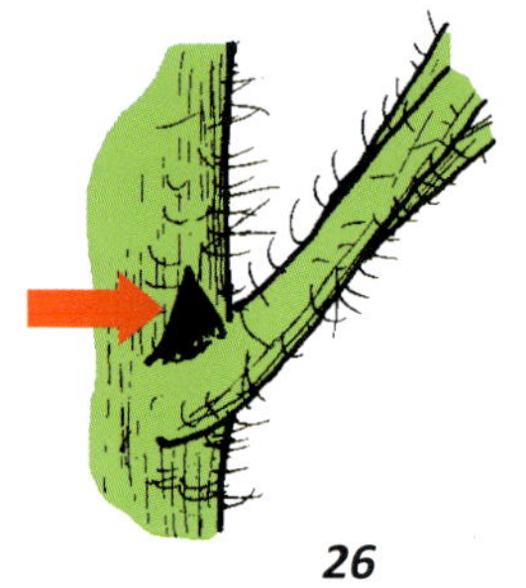

26

38 Plant sprawling and with a woody base

274 *Solanum dulcamara*

38 Plant erect or ascending, herbaceous 39

39 Leaves hairy

275-280 *Myosotis* - use key p. 596

39 Leaves glabrous 40

40 Leaves broadly spathulate (***27***)

269 *Samolus valerandi*

40 Leaves lanceolate to ovate or narrowly spathulate (***28***)

219-223 Ranunculaceae - use key p. 452

27

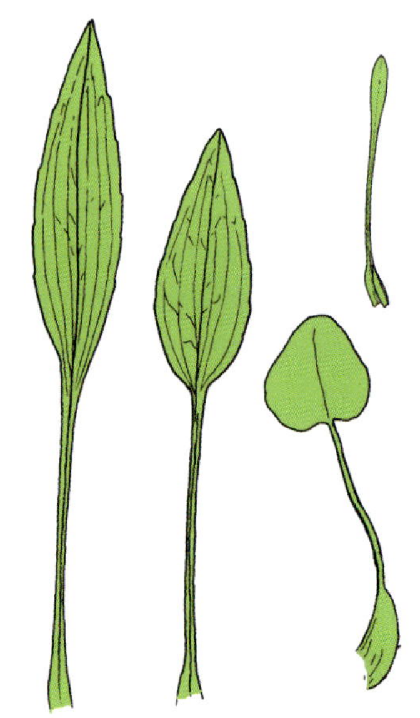

28

Subkey F

Leaves alternate, but a few opposite leaves may be present in the apical part of stem. Leaves sessile or represented by sheaths without or with much reduced lamina.

1 Leaves divided 2
1 Leaves not divided 3

2 Leaves pinnately divided (***1***)
264 *Hottonia palustris*
2 Leaves not pinnately divided
202-223 Ranunculaceae - use key p. 452

1

3 Leaves with ligule (look carefully - especially in the apical part of the stem) 4
3 Leaves without ligule 9

4 Leaves without leaf sheath
66-97 *Potamogeton* - use key p. 170
4 Leaves with leaf sheath 5

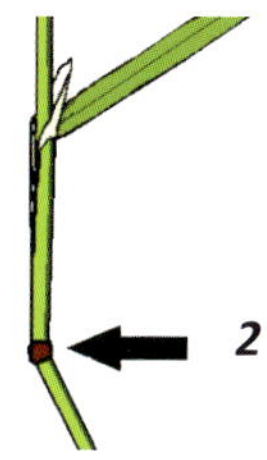

2

5 Stem with conspicuous, often thickened nodes (***2***)
180-196 Poaceae - use key p. 412
5 Stem without conspicuous or thickened nodes 6

6 Leaf sheaths closed, tubular. Flowers not 4-merous 7
6 Leaf sheaths open with overlapping edges or tubular at base for 2–3 mm or more. Flowers 4-merous, perianth of 4 sepals
98-103 *Stuckenia* - use key p. 246

7 Leaves submerged or aerial, not ribbon-like
138-179 Cyperaceae - use key p. 328
7 Leaves submerged, ribbon-like 8

8 Leaves with primary veins and weaker transverse secondary veins -
see **subkey B**, couplet 27
8 Leaves with primary veins but without transverse secondary veins (marine species) 22

2a

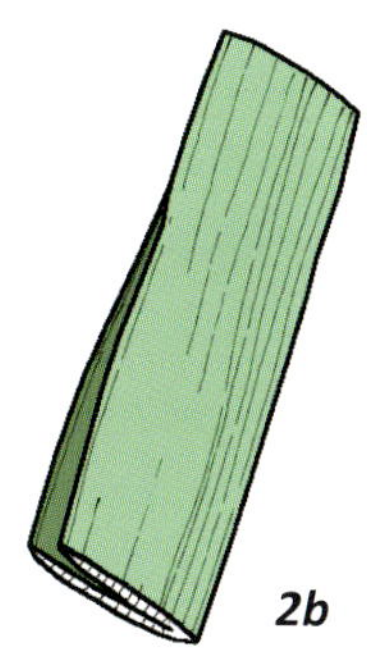

2b

9 Leaves only present as sheaths with or without a very much reduced lamina
138-179 *Cyperaceae* - use key p. 328
9 Leaves with flat or terete lamina 10

10 Leaves in cross section with 4 to 30 almost parallel air chambers (***2a***) 11
10 Leaves in cross section without such air chambers 12

11 Leaf base folded (***2b***) and partly enclosing the leaf above it, fan-shaped arranged
113 *Iris pseudacorus*
11 Leaf base not folded, not fan-shaped arranged
126-132 *Typha* - use key p. 280

12 Plants from freshwater habitats 13
12 Plants from brackish or saline habitats 20

13 Leaves terete 14
13 Leaves flat at least on the upper side 16

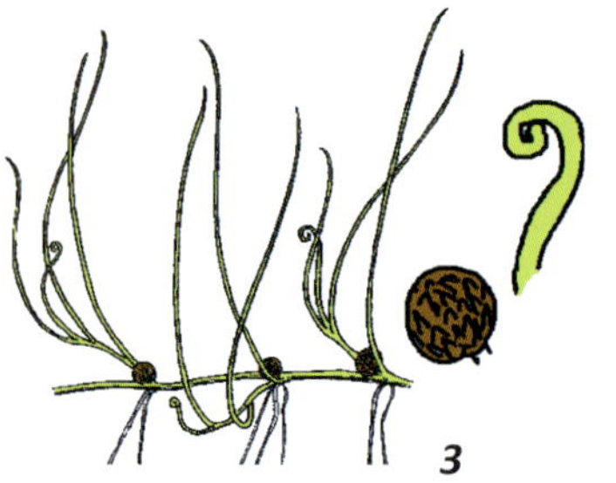

3

14 With brownish, globular sporangia at leaf bases. Leaves coiled when young (***3***)
361 *Pilularia globulifera*
14 Without sporangia. Leaves not coiled when young 15

15 Stem erect or ascending. Leaves with air channels (use a hand lens) and slightly septate
137 *Juncus bulbosus*
15 Stem creeping. Leaves without air channels and not septate
221 *Ranunculus reptans*

16 Leaves appressed hairy, scattered on the stem
275-280 *Myostis* - use key p. 596
16 Leaves glabrous 17

4

17 Leaves densely spirally arranged along the stem (***4***)
58 *Lagarosiphon major*
17 Leaves not spirally arranged along the stem 18

18 Leaves more than 2 mm wide 19
18 Leaves up to 1 mm wide
163 *Isolepis fluitans*

19 Leaves glaucous or grey green, fan-shaped arranged
113 *Iris pseudacorus*
19 Leaves yellow green to green, not fan-shaped arranged
114-122 *Sparganium* - please use key p. 278

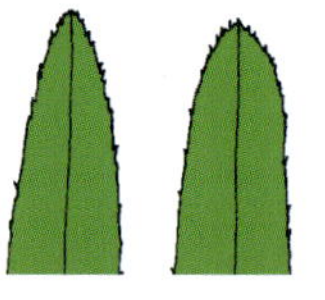

5

6

20 Leaf tip finely dentate 21
20 Leaf tip rounded or with a notch 22

21 Leaf tip rounded, symmetrical with many regularly set teeth (***5***). Peduncle slender, 5 to 30 cm long, spirally coiled after fertilisation
108 *Ruppia cirrhosa*
21 Leaf tip acute, asymmetrical with many irregularly set teeth (***6***). Peduncle 1–2 cm long, not spirally coiled
109 *Ruppia maritima*

7

8

22 Leaves 1-veined. Drupes dark brown, without ribs
112 *Zostera noltii*
22 Leaves 3–9-veined. Drupes light brown, with longitudinal ribs 23

23 Stigmas about twice as long as the style (***7***)
110 *Zostera marina*
23 Stigmas about as long as the style (***8***)
111 *Zostera angustifolia*

Descriptions and maps

The species treated in this guide were selected based on the definition of aquatic plants below, which follows the European Standard "EN 15460:2007 – Water quality – Guidance standard for the surveying of macrophytes in lakes".

1. Full accounts of hydrophytes (aquatic vascular plants, defined as an "aquatic plant that is usually rooted under water with floating or submerged leaves, or totally free floating at the water surface").
2. Full accounts of helophytes (defined as a "plant that is normally rooted under water with emergent shoots, typically growing in marginal or marshy areas") where we consider that these only occur where surface water persists for a period of months.
3. Other selected helophytes where these may be confused with taxa included under 1 and 2, providing enough information to inform identification.
4. Other taxa which are capable of growing at least with their roots submerged for extended periods and which therefore will occur together with taxa covered by 1 and 2.

Except where specified, nomenclature follows Plants of the World Online (POWO), Royal Botanic Gardens, Kew.

The descriptions are presented in a standardised format to facilitate comparisons between species.

Indication of size, e.g., (3–)5–7(–10) mm, should be read: Normally 5–7 mm, rarely down to 3 mm or up to 10 mm.

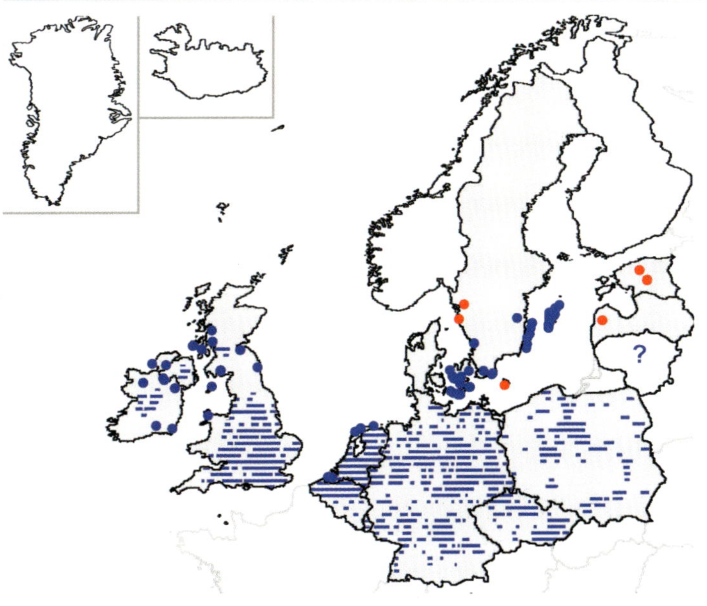

Map showing the distribution of species within the region. The grey background indicates countries in which the species occurs. A question mark is used where information is lacking or unreliable. Abundance is shown either by full shading where the species is common or dashes showing scattered occurrence. Dots indicate isolated occurrences. An explanation of red dots is given below the map when these are used.

Distribution information has been derived from the major floras, with additional information from specific sources which are listed in the literature section at the end of the book.

The world map is based on the UN geographical subregions, but with additional divisions in North and South America, Siberia and Indonesia. The islands of Greenland, Iceland, Madagascar and Japan are also separated.

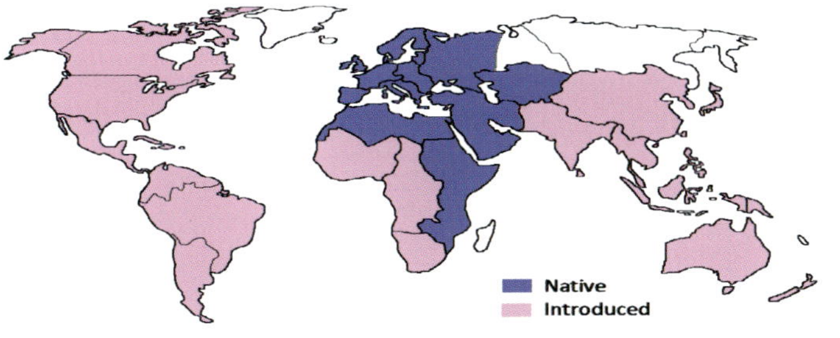

The different shading indicates native or introduced status as shown. Global distribution information follows Plants of the World Online (powo.science.kew.org), supplemented by other sources which are listed in the literature section.

Nymphaeaceae

Key to *Nymphaea*

1 Petals white, pink or red, occasionally with a yellow tinge at base 2
1 Petals yellowish 8

2 Leaf lobes rounded. Flowers floating on the water. Upper side of leaves green or with reddish spots 3
2 Leaf lobes with an apical projection. Flowers more or less raised above the water. Upper side of leaves green 7

3 Flowers white or rarely with faint pink streaks. Upper side of leaves green. Receptacle rounded or square at base, recessed or not around the petiole 4
3 Flowers pink or red. Upper side of leaves often speckled with green and coppery red. Receptacle square at base and recessed around the petiole (***1***)
    N. ×laydekeri Lat.-Marl. p.70.

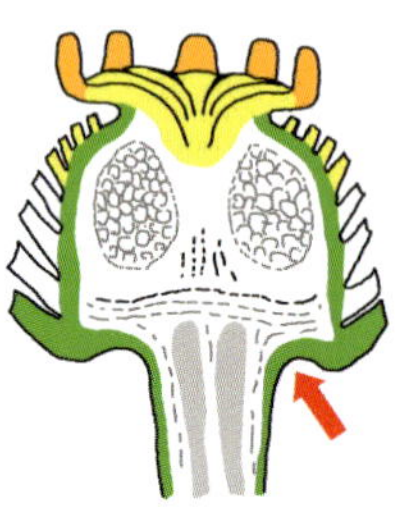

Receptacle at base with a recession around the petiole.

1

4 Receptacle square at base with ridges from the corners towards the pedicel (***2***). Leaf blades less than 10 cm wide, dark red to brownish on the underside. Flowers small with petals less than 35 mm long, white or rarely with light pink streaks
    3 *N. tetragona*
4 Receptacle rounded or square at base without ridges from the corners towards the pedicel. Leaf blades usually more than 10 cm wide, green or reddish brown on the underside in with age. Flowers normally larger, white, pink or red 5

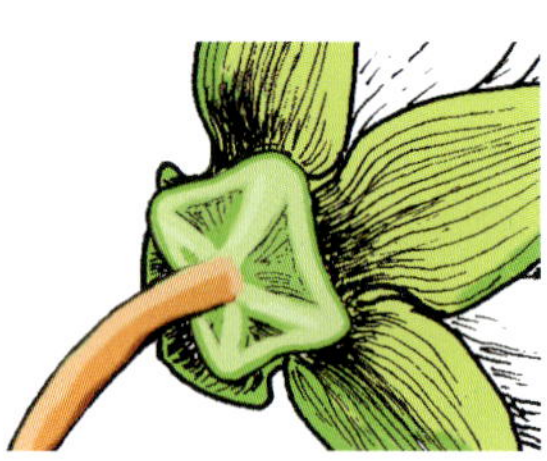

2

5 Filaments of the inner stamens linear, less than 2 mm wide, usually narrower than the anthers (***3***). Stigmatic disc yellow, 13–25 mm wide, with (10–)14–24 rays. Receptacle rounded at base (***4***) 6
5 Filaments of the inner stamens lanceolate, more than 2 mm wide and almost twice as wide as the anthers (***5***). Stigmatic disc yellow to red, 7–16 mm wide, with 6–12(–15) rays. Receptacle square at base (***6***)
    2 *N. candida*

3

4

5

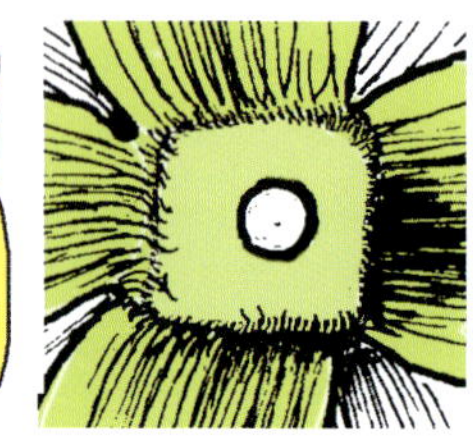

6

6 Stamens borne almost to top of the ovary (***7***). Flowers 9–20 cm across, opening wide. Stigmatic disc with more than 14 rays
    1 *N. alba* subsp. ***alba***
6 Stamens not borne on upper part of the ovary (***8***). Flowers 5–12 cm across, never opening wide. Stigmatic disc with fewer than 16 rays
    1a *N. alba* L. subsp. ***occidentalis***

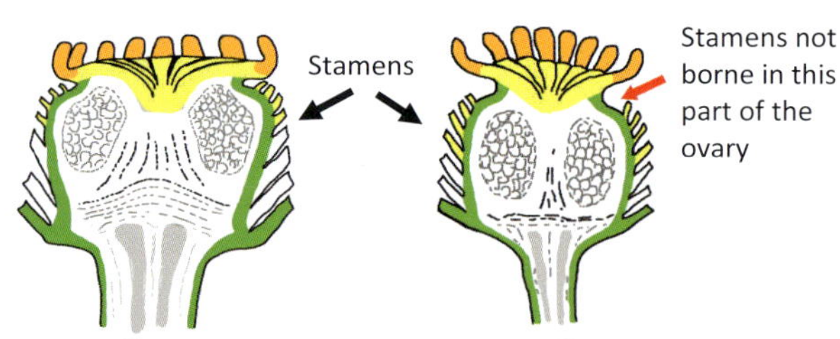

7 ***8***

Cross section of fruit. Stamens, petals and sepals cut off.

7 Flowers white. Pollen viable and uniform
    N. odorata p.71.
7 Flowers mostly rose red. Pollen partly aborted and malformed
    N. ×marliacea p.70.

8 Flowers less than 5 cm wide. Leaf blade less than 10 cm wide
    N. mexicana* × *N. tetragona p.71.
8 Flowers more than 5 cm wide. Leaf blade more than 10 cm wide
    N. ×thiona p.71. or ***N. ×marliacea*** p.70.

Key to *Nuphar*

1 Anthers shorter than filaments.
Fruit shaped like an urn with a narrow, elongate neck (***1***) 2

1 Anthers as long as or longer than filaments.
Fruit ovoid obovate, without a narrow elongate neck (***2***) or decaying, not developing 5

1

2

2 Stigmatic disc entire or crenate, yellow (***3, 4***).
Petioles trigonous to flattened on the adaxial side in cross section 3

2 Stigmatic disc deeply lobed, yellow or red (***5***).
Petioles circular or elliptic to flattened in cross section 4

3

4

5

3 Stigmatic disc entire or with slightly waved margin, 7–19 mm wide (***3***).
Fruit 2.6–4.5 cm long and 1.9–3.4 cm wide.
Leaf blades 16–30 cm long
4 *N. lutea*

3 Stigmatic disc with undulate to crenate margin, 5–9 mm wide (***4***).
Fruit 1.7–3.0 cm long and 1.0–2.0 cm wide.
Leaf blades 9–20 cm long
5a *N. ×spenneriana*

4 Leaf blades 12–35 cm long, 1.5–2.5 times longer than wide. Flowers 2–3.5 cm wide.
Petioles 3–14 mm wide, circular in cross section (***6***)
7 *N. japonica*

4 Leaf blades 4–15 cm long, less than 1.5 times longer than wide. Flowers 1.0–2.5 cm wide.
Petioles 1–5 mm wide, elliptic in cross section to flattened (***7***)
5 *N. pumila*

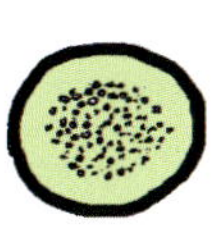

6

7

5 Leaf blades circular to broadly ovate, 1–2.5 times longer than wide (***8***). Sepals and petals yellow, rarely reddish or peach coloured 6

5 Leaf blades elliptic to lanceolate, 3–7 times longer than wide (***9***). Sepals and petals yellow
N. sagittifolia (Walter) Pursh
(seldom cultivated but so far not naturalised in the area)

8

9

6 Anthers dehiscing, pollen yellowish white in two lines, contrasting with the colour of the anther.
Leaves persisting at least into September
6 *N. advena*
(occasionally cultivated but so far not naturalised in the area)

6 Anthers not dehiscing, of uniform colour, reduced to two raised lines at the end of the filament.
Leaves typically degrading in July
6a *N. ×porphyranthera*

Nymphaeaceae

1 *Nymphaea alba* L. subsp. *alba*

DK: Hvid Nøkkerose **N:** Kvit nykkerose **S:** Sydnäckros **FIN:** Valkolumme **IS:** -
GB: White Water-Lily **NL:** Witte waterlelie **F:** Nénuphar blanc
D: Weiβe Seerose **CZ:** Leknín bílý **PL:** Grzybienie białe
EST: Valge vesiroos **LV:** Baltā ūdensroze **LT:** Paprastoji vandens lelija **Ru:** Кувшинка белая

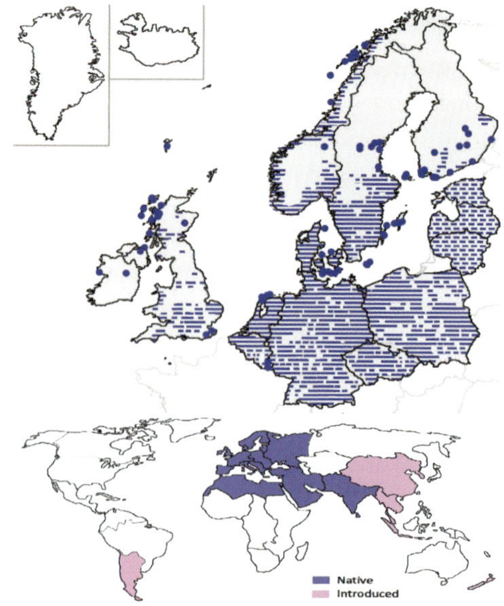

Nymphaea alba in still water in a canal near Hadsund, Jutland, Denmark. Photo JCS.

Perennial hydrophyte. Rhizome up to 1 m long and 10 cm wide, horizontal to vertical, mostly unbranched, covered in round scars from broken-off petioles and stipules. The scars with markings from 4 air canals. Roots brownish.

Stem: Absent.

Leaves: Floating, spirally arranged from the rhizome. Lamina 15–35 cm long, 9–34 cm wide, almost orbicular, flat, coriaceous, firm, at base with a narrow, deep incision. Upper surface dark green to purplish, somewhat shiny, with furrows along the midrib and main veins. Lower surface mostly purplish. Veins more or less straight, outer veins forming a net pattern. The main veins of the basal lobes are almost straight in the basal third. Petiole terete, 4–11 mm wide, with 4 main (or 2 in very small plants) and several minor air canals, with a single 8–9 cm long stipule at base. Small, flaccid, triangular or oval, submerged leaves only in young plants.

Flowers: 6–12(–20) cm in diameter, spreading. Receptacle 1.1–2.3 cm wide, base rounded, without a rim. Sepals 4(–5) more or less equalling the petals, 3–4 times as long as the diameter of the receptacle. Green on the abaxial side, whitish on the adaxial. Petals 15–25(–40), white or rarely reddish, the outermost 8 cm long, the inner gradually shorter, lanceolate, spirally arranged. Stamens borne almost to the top of the ovary, the outer ones with petaloid filaments. Filaments of the inner stamens linear, less than 2 mm wide, usually narrower than the anthers. Anthers sulphur yellow. Pollen grains papillose. Stigmatic disc flat or slightly concave, with 13–16(–20) brownish rays.

Fruits: 2–4.5 cm long, obovate, densely covered with scars from fallen anthers almost to the stigmatic disc. Seeds 2–3.5 mm long and 1–2 mm wide, ovate, blackish brown, up to 1700 in each fruit.

Flowering: June-August.

Biology: Flowers close at night and in overcast conditions. Self-pollinated or pollinated by small beetles. After flowering the peduncle coils and pulls the fruit down below the surface. The mature fruit opens below water and releases the seeds in a spongy, air filled tissue lifting them to the surface, where they are dispersed by water movement. Overwinters as a leafless rhizome.

Habitats: Lakes, ponds, streams and brackish bays. Prefers a soft substrate. Grows to 3 m depth. Sensitive to waves.

Distribution within the region: Native, but often introduced even in natural water bodies. Common in most of the area except northern Fennoscandia and parts of the British Isles.

Characteristics and similar species: *Nymphaea alba* is recognised by the almost orbicular, floating leaves and white flowers with rounded receptacle. Small specimens may be confused with **1a *N. alba*** subsp. ***occidentalis***.

2 *N. candida* has a square base to the receptacle and inner stamens with lanceolate filaments broader than the anther.

3 *N. tetragona* differs from *N. alba* by the square base of the receptacle and sepals more or less 1.5 times as long as the diameter of the receptacle.

Nymphaea alba leaf. Bornholm, Denmark. Photo JCS.

Identification of the numerous introduced and more or less naturalised cultivars and garden hybrids is extremely complicated and relies on the characters presented in the key.

Species of **5-7 *Nuphar*** differ by broadly ovate to long sagittate leaves with green lower surface and side veins not reticulate but dichotomously divided towards the margin. Most of the *Nuphar* species form flaccid, light green submerged leaves.

Hybrids: *Nymphaea alba* and **2 *N. candida*** hybridise (***N. ×borealis*** Camus) and backcross frequently where the two species meet and are intermediate between the parents. Hybrids have partly undeveloped pollen and poor seed germination.

See also **3 *N. tetragona*.**

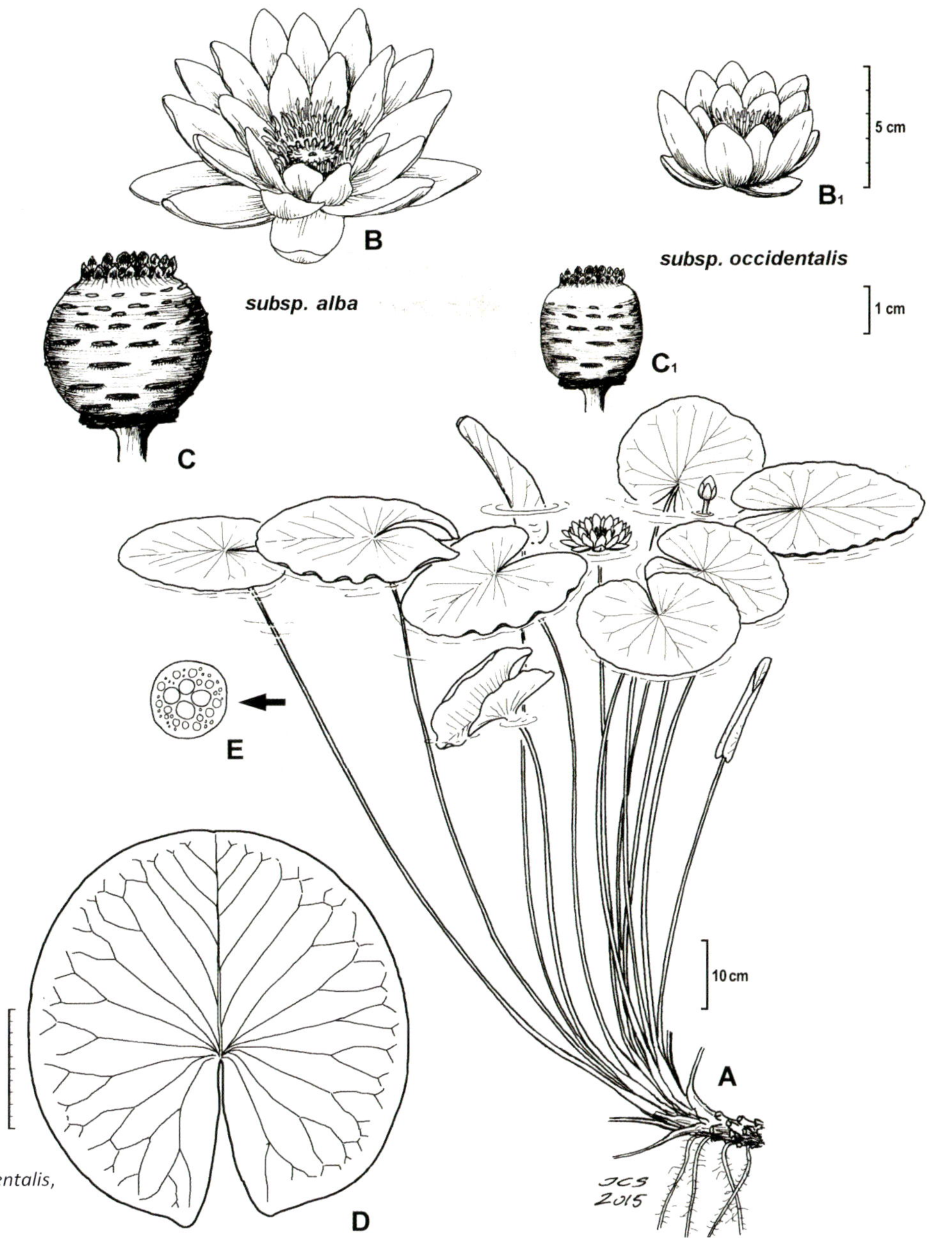

Nymphaea alba
A habit **B** flower subsp. *alba*, **B1** flower subsp. *occidentalis*, **C** fruit subsp. *alba*, **C1** fruit subsp. *occidentalis*, **D** leaf subsp. *alba*, **E** cross section of petiole.

Nymphaeaceae

1a *Nymphaea alba* L. subsp. *occidentalis* (Ostenf.) Hyl.

Syn.: *N. alba* L. var. *minor* DC.
DK: Vestlig Nøkkerose **N:** Smånykkerose **S:** Atlantnäckros **FIN:** - **IS:** -
GB: - **NL:** - **F:** Nénuphar de l'ouest
D: - **CZ:** - **PL:** -
EST: - **LV:** - **LT:** - **Ru:** -

Nymphaea alba subsp. *occidentalis*.
V-Jutland, Denmark. Photo JCS.

Differs from subsp. ***alba*** in smaller flowers, 5–8(–12) cm in diameter and never fully opened. Leaves 9–13 cm long. Stigmatic disc with 8–15 rays and without stamens on the upper part of the ovary (see p. 62, figure 8 and figure C1 above).

Habitats: Small oligotrophic water bodies.

Distribution within the region:
Native. Ireland, Britain, Denmark, Norway, Sweden and Germany?

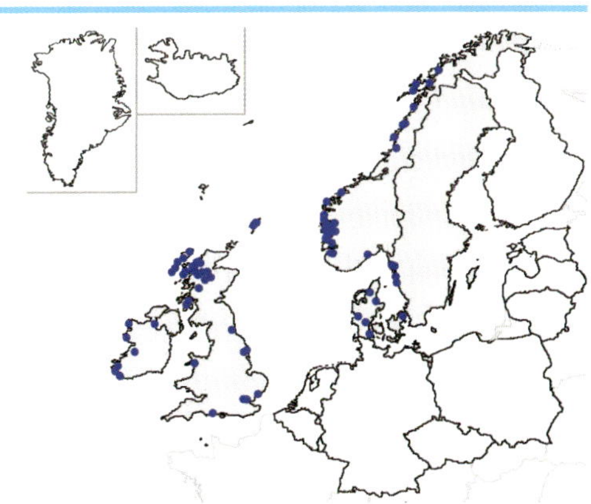

The taxonomic status of subsp. *occidentalis* is unclear, and it is connected to subsp. *alba* by a range of morphological intermediates. It is considered a variety of *N. alba* by some authors.

Nymphaeaceae

2 *Nymphaea candida* C.Presl & J.Presl

DK: Liden Nøkkerose **N:** Kantnykkerose **S:** Nordnäckros **FIN:** Pohjanlumme **IS:** -
GB: Shining water-lily **NL:** Noordelijke waterlelie **F:** Nénuphar blanc boreal
D: Kleine Seerose **CZ:** Leknín bělostný **PL:** Grzybienie północne
EST: Väike vesiroos **LV:** Sniegbaltā ūdensroze **LT:** Mažažiedė vendens lelija **Ru:** Кувшинка чистобелая

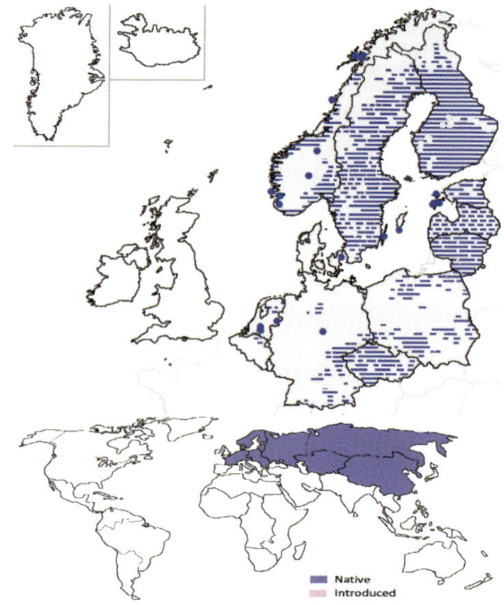

Nymphaea candida. Skärhamn, Bohuslän, Sweden. Photo Jan-Thomas Johansson.

Perennial hydrophyte. Rhizome up to 35 cm long and 9 cm wide, vertical to ascending, unbranched, covered in round scars from broken-off petioles and stipules. The scars with markings from 4 air canals. Roots brownish.

Stem: Absent.

Leaves: Floating, spirally arranged from the rhizome. Lamina 12–27 cm long, 10–26 cm wide, ovate to suborbicular, flat, coriaceous, firm, with a narrow, deep incision at the base. Upper surface dark green to purple tinged, somewhat shiny, with a furrow along the midrib and main veins. Lower surface mostly purplish. Veins more or less straight, reticulate toward the margin. The main vein of the basal lobes is though more or less evenly curved for its entire length. Petiole terete, 2.5–8.5 mm wide, with 4 major (2 in very small plants) and several minor air canals, with a single c. 4 cm long stipule at the base. Small, flaccid, triangular or oval, submerged leaves only on young plants.

Flowers: 6–12(–20) cm in diameter, cup shaped. Receptacle 0.9–2.3 mm wide, square at base and with a rim formed by the sepal bases.

Sepals 4 more or less equalling the petals, 3–4 times the diameter of the receptacle. Green on the abaxial side, whitish on the adaxial. Petals 15–20, white or slightly reddish, lanceolate, the outermost 3–7 cm long, the inner gradually shorter, spirally arranged.

Stamens borne almost to the top of the ovary, the outer ones with petaloid filaments. Filaments of the inner stamens lanceolate, more than 2 mm wide and almost twice as wide as the anthers. Anthers sulphurous yellow. Pollen grains papillose. Stigmatic disc more or less concave, with (6–)8–12(–15) brownish rays.

Fruits: 2–4 cm long, ovoid, densely covered with scars from fallen anthers almost to the stigmatic disc. Seeds 3–4 (–5) mm long and 1–2 mm wide, ovate, black or brownish, 200–700 in each fruit.

Flowering: June-August.

Biology: Flowers close at night and in overcast conditions. Self-pollinated or pollinated by small beetles. Overwintering by the leafless rhizome.

Flowers of *N. candida*. Note the square base of the receptacle. Vederslövssjön, Vaxjö, Sweden. Photo JCS.

Nymphaea candida. Vederslövssjön, Vaxjö, Sweden. Photo JCS.

Basal lobes of leaves. In *N. alba* (left) the main vein is almost straight on the basal third, but in *N. candida* (right) it is almost evenly curved for its entire length. Photos KvdW.

After flowering the peduncle coils and pulls the fruit below the surface. The mature fruit opens under water and releases the seeds in a spongy, air-filled tissue lifting them to the surface, where they are dispersed by water movement.

Habitats: Lakes, ponds, ditches, slow-flowing streams and brackish bays. Prefers soft substrates. Grows to 2–3 m depth. Sensitive to waves.

Distribution within the region:
Native, but sometimes introduced even in natural water bodies. Common in the Baltic, most of Fennoscandia except Denmark (introduced), rather rare to rare in the Netherlands, Germany, Poland and Czech Republic.

Characteristics and similar species:
Very similar to **1 *N. alba*** in which the base of the receptacle is rounded, inner stamens with linear filaments as wide as the anther, main vein of the basal lobes almost straight at the basal third. **3 *N. tetragona*** also has a receptacle square at base, but sepals only about 1.5 times as long as the diameter of the receptacle.

Identification of the numerous introduced and more or less naturalised cultivars and garden hybrids is extremely complicated and relies on the characters presented in the key.

5-7 *Nuphar* species differ by having broadly ovate to long sagittate leaves with green lower surface and side veins not reticulate but dichotomously divided towards the margin. Most of the *Nuphar* species form flaccid, light green submerged leaves.

Hybrids: See 1 *N. alba* and 3 *N. tetragona*.

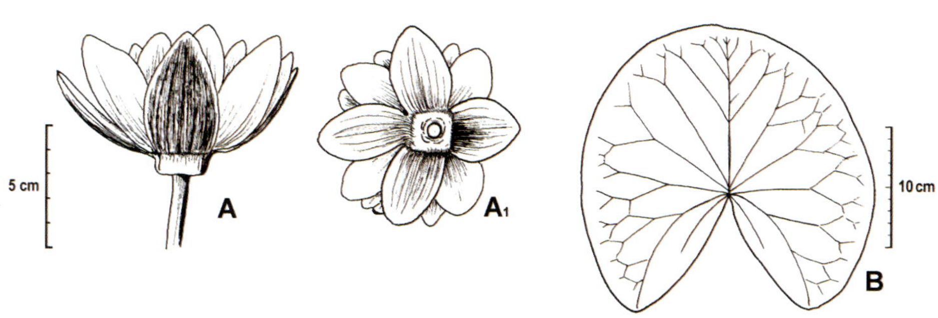

Nymphaea candida
A Flower side view, **A1** flower from below, **B** leaf.

Nymphaeaceae

3 *Nymphaea tetragona* Georgi

Syn.: *N. pygmaea* W.T.Aiton; *Castalia tetragona* (Georgi) G.Lawson
DK: Dværg-Nøkkerose **N:** Finsk nykkerose **S:** Finsk näckros **FIN:** Suomenlumme **IS:** -
GB: Pygmy water-lily **NL:** - **F:** Nénuphar tetragona
D: Zwergseerose **CZ:** - **PL:** -
EST: Kandiline vesiroos **LV:** Mazā ūdensroze **LT:** Mažoji vandens lelija **Ru:** Кувшинка четырёхгранная

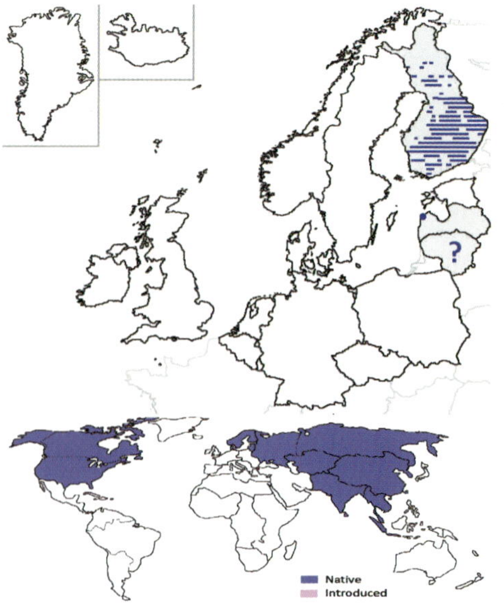

Nymphaea tetragona. Kermesaari, Finland. Photo Jan-Thomas Johansson.

Perennial hydrophyte. Rhizome 5–12 cm long and 1.5–3.5 cm wide, vertical to ascending, unbranched, covered in remnants of petioles and stipules. Scars with markings from 2 air canals. Roots brownish.

Stem: Absent.

Leaves: Floating, spirally arranged from the rhizome. Lamina 3.5–10(–15) cm long, 2.5–8(–10) cm wide, ovate, glabrous, rather thin, and often somewhat undulate, with a narrow, deep incision at base. Upper surface green to dark green, somewhat shiny, with faint furrows along the midrib and main veins. Lower surface green sometimes with reddish-brown or purple tones. Petiole slightly flattened, 2–3.5 mm wide, with 2 main and several minor air canals, at base with a single 1–2 cm long stipule. Small, flaccid, triangular or oval, submerged leaves only produced by young plants.

Flowers: Floating, 3–5(–7.5) cm in diameter. Receptacle 1.3–2.1 cm wide, square at base with a distinct rim formed by the sepal bases and with ridges from the corners towards the petiole. Sepals 4, persistent, more or less equalling the petals, 1.7–3.0 cm long, 1.2–2.1 cm wide, about 1.5 times as long as the diameter of the receptacle. Green on the abaxial side, whitish on the adaxial. Petals 8–12(–15), white, lanceolate, 1.7–3.0 cm long, the outer in whorls of 4, the inner spirally arranged, more or less appressed to each other forming a cup. Stamens 30–70, filaments widest above middle, longer than anthers, the outer ones without transitions to the petals. Anthers yellowish orange, 1.5–3 mm long. Pollen grains more or less smooth. Stigmatic disc purple or orange to yellow, more or less flat with (5–)6–10 rays.

Fruits: c. 2.5 cm long, conical, enclosed in the persistent sepals appears almost pyramidal. Seeds 2–3.5mm × 1.5–2 mm, ovate, greenish brown, 200–300 per fruit.

Flowering: June-September.

Biology: Flowers close at night and in overcast conditions. Pollination by small beetles or self-pollination. Overwintering as a leafless rhizome. After flowering the peduncle coils and pulls the fruit below the surface. The mature fruit opens under water and releases the seeds in a spongy, air-filled tissue lifting them to the surface, where they are dispersed by water movement.

Nymphaea tetragona. Ilsalmi, Finland. Photo JCS.

Nymphaea tetragona. Flower. Kermesaari, Finland. Photo Jan-Thomas Johansson.

Nymphaea tetragona. Receptacle square at base, with a marked rim and ridges from the corners towards the petiole. Ilsalmi, Finland. Photo JCS.

Habitats: In still water in humic lakes, ponds, forest swamps and slow-flowing streams. Prefers soft substrates. Grows to 2–3 m depth. Sensitive to wave action.

Distribution within the region:
Native in Finland and Latvia.

Characteristics and similar species:
Nymphaea tetragona can be recognised by its small flowers and leaves, square base of the receptacle with a distinct rim and ridges from the corners towards the petiole and the sepals about 1.5 times as long as the diameter of the receptacle.

2 *Nymphaea candida* is sometimes rather similar, but has sepals 3–4 times as long as the diameter of the receptacle and stamens with a gradual transition outwards to the petals.

5-7 Nuphar species differ in their broadly ovate to long sagittate leaves with green lower surface and side veins not reticulate but dichotomously divided towards the margin. Most *Nuphar* species form flaccid, light green submerged leaves.

Hybrids: *Nymphaea alba x tetragona*, and *N. candida x tetragona* are known from Finland. The hybrids have a mixture of flower and leaf characters from the parental species, and pollen only partly developed.

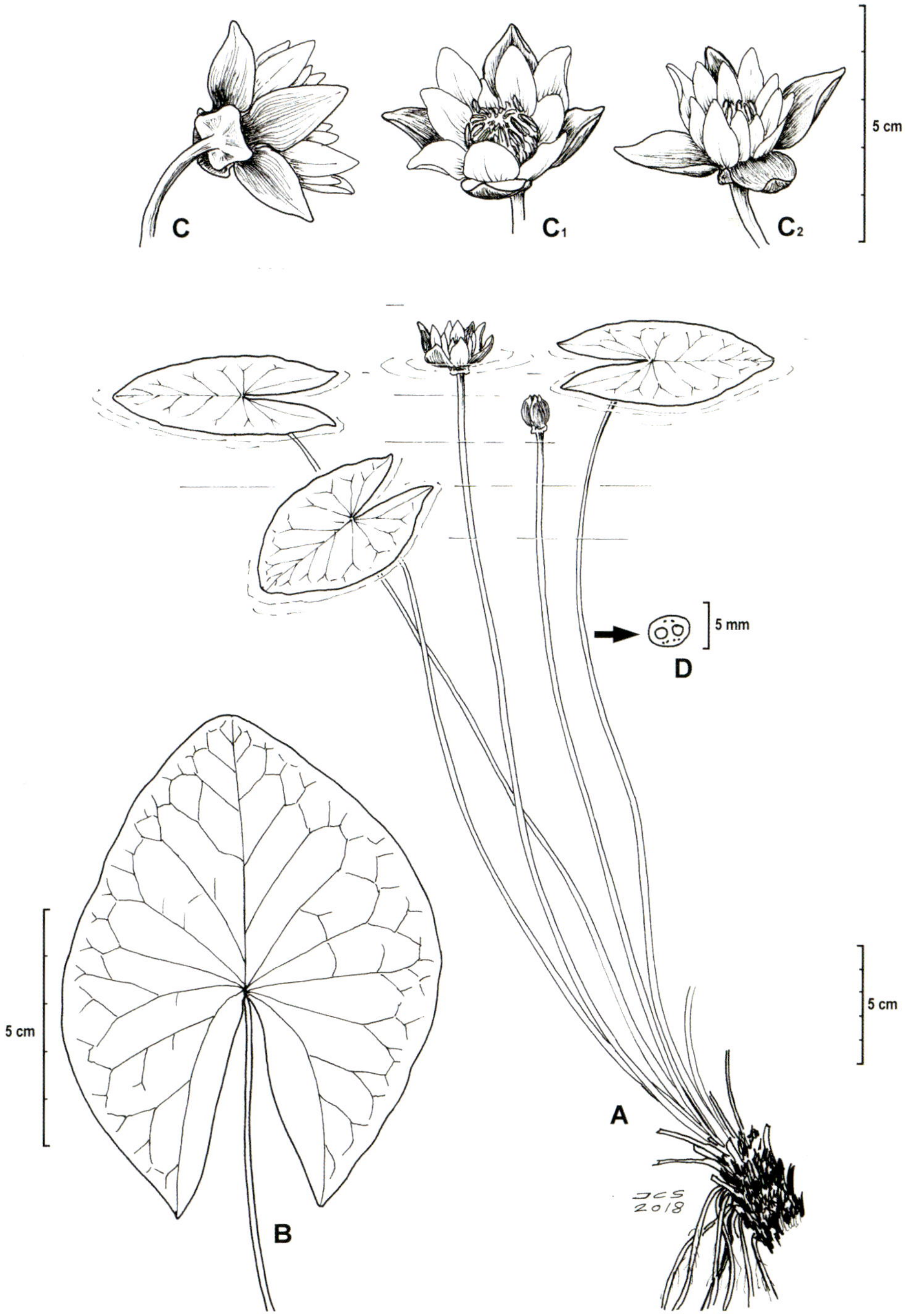

Nymphaea tetragona
A habit **B** leaf, **C** flower – side view, **C1, C2** open flower, **D** cross section petiole.

Introduced *Nymphaea* species and hybrids

Correct identification of the numerous hybrids and cultivars is highly problematic and a subject more suited to horticultural specialists than botanists. The key and these pages are intended as a guide in an attempt to help botanists put names to some of the more frequently encountered hybrids and cultivars.

The natural aquatic vegetation of village ponds and lakes is often augmented with exotic-looking *Nymphaea* hybrids, such as this red-flowered *N.* ×*marliacea*. Foulum near Viborg, Denmark. Photo JCS.

Nymphaea ×*marliacea*. Graamyr, Gudhjem, Bornholm, Denmark. Photo JCS.

Nymphaea* ×*marliacea Lat.-Marl. is one of the most common hybrids cultivated and introduced to natural water bodies. The flowers are large and usually pink to red or white with pink tones, floating or somewhat raised above the water. The parents are usually *N. alba* and the North American *N. odorata* Aiton.

Nymphaea* ×*laydekeri Lat.-Marl. is another red cultivar, but with small flowers and leaves. The parents are *N. alba* and *N. tetragona*.

Nymphaea alba f. ***rosea*** is similar but has rounded leaf lobes.

Nymphaea ×thiona in a moat at Castle Dyck, North Rhine-Westphalia, Germany. Photo KvdW.

The yellow-flowered hybrids have *N. mexicana* Zucc. from the southeastern USA and Mexico as one of the parents.

Nymphaea mexicana* × 3 *N. tetragona (traded as *×helvola*) has flowers less than 5 cm wide and leaf blades less than 10 cm wide.

Nymphaea ×thiona D.B.Ward has larger flowers and leaves. It is the result of the crossing of *N. mexicana* x *odorata*.

Nymphaea odorata Aiton is white flowered and in many ways similar to *N. alba*, but the leaf lobes have an apical projection.

Nymphaea ×thiona in a moat at Castle Dyck, North Rhine-Westphalia, Germany. Photo KvdW.

Nymphaeaceae

4 *Nuphar lutea* (L.) Sm.

DK: Gul Åkande **N:** Gul Nykkeros **S:** Gul Näckros **FIN:** Isoulpukka **IS:** -
GB: Yellow Water-Lily **NL:** Gele plomp **F:** Jaunet d'eau, Nenuphar jaune
D: Gelbe Teichrose **CZ:** Stulík žlutý **PL:** Grążel żółty
EST: Kollane vesikupp **LV:** Dzeltenā lēpe **LT:** Paprastoji lūgnė **Ru:** Кубышка жёлтая

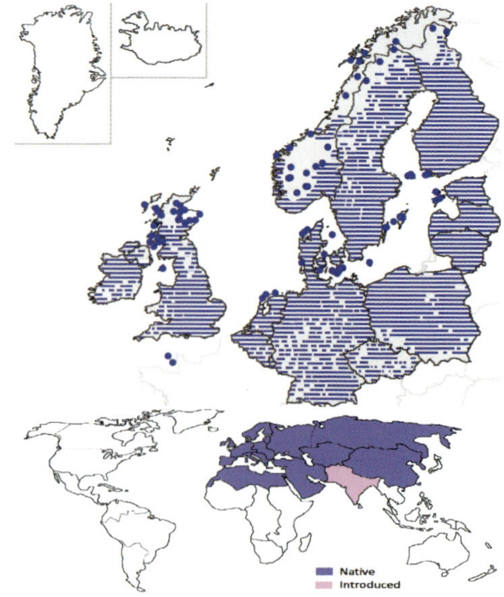

Flowering *N. lutea* in slow-flowing water. River Gudenå, Randers, Denmark. Photo JCS.

Perennial hydrophyte. Rhizome 50–250 cm long and 3–8 cm wide, horizontally creeping, branched, covered in numerous triangular to diamond-shaped scars from broken-off petioles. Scars with markings from 9 or more air canals. Roots brownish.

Stem: Absent.

Leaves: Floating or sometimes emergent, spirally arranged from the rhizome. Lamina 12–40 cm long, 9–30 cm wide, broadly ovate, pale green, coriaceous, somewhat dull, glabrous on both sides even when young, with a narrow, deep incision at base. Main veins several times dichotomously divided towards the margin, but not forming a net pattern. Petioles rather stiff, bluntly trigonous in cross section and with numerous air canals. Submerged leaves almost always present, up to 30 cm wide, wider than long, thin, semi-transparent, wavy, yellowish green to brownish, with 10–15 cm long petioles.

Flowers: 4–6 cm in diameter. Sepals 5–6, 2–3.5 cm long, broadly obovate, yellow, almost all alike. Petals 15–20, yellow, spathulate, c. 10 mm long, longer than the numerous stamens. Anthers 3–7 mm long, 3–4 times longer than wide. Stigmatic disc 8–12 mm in diameter, more or less orbicular with entire margin, slightly concave, yellow or sometimes purplish. The 15–20 brownish rays do not reach the margin.

Fruits: 3.5–6 cm long, urnshaped with a slightly elongated, usually straight neck. The fruit contains 100–400 seeds. Seeds c. 5 mm long, 3 mm wide, ovate, yellowish brown, shiny.

Flowering: June-August.

Biology: Unlike *Nymphaea* species, the fruit ripens floating on the surface. It splits and divides into several oblong white, spongy lamellae, seeds which are released sink, while those contained within the lamellae float. In fast-flowing water the plants have submerged leaves only. The petals have nectaries which attract flies for pollination. Vegetative propagation occurs by rhizome fragments.

Habitats: In both nutrient-rich and nutrient-poor waters to a depth of 4–5 metres, usually forming large pure stands. Lakes, ponds, streams, brackish bays, often on soft substrate.

Nuphar lutea with submerged leaves. Lake Wolfskuhle, Rheinberg, Germany. Photo KvdW.

Distribution within the region:
Native. Common more or less throughout; absent in Greenland and Iceland.

Characteristics and similar species:
Nymphaea lutea can be recognised by 5 sepals, anthers shorter than filaments, fruit urnshaped with a narrow, elongated neck and entire stigmatic disc with 15–20 rays, not reaching the margin.

5 ***N. pumila*** has smaller leaves with silky hairs on the lower side and a deeply lobed stigmatic disc with 8–11 rays.

6 ***N. advena*** has anthers longer than filaments, different shape of fruit and erect leaves. **1-3** ***Nymphaea*** species have more or less orbicular leaves with veins forming a reticulate pattern towards the margin.

336 ***Nymphoides peltata*** has smaller leaves with veins forming a reticulate pattern, a papillose underside and undulate margin.

Hybrids:
N. ×spenneriana Gaudin
(*N. lutea x pumila*). See **5** ***N. pumila.***

N. ×porphyranthera Lansdown & Ruhsam *(N. advena x lutea).*
See **6** ***N. advena.***

Nuphar lutea. Rhizome with scars from old leaves. Soendersund, Nees, Denmark. Photo JCS.

The orbicular stigmatic disc on fruit and flower of *N. lutea*. The stamens are recurved and the petals fairly visible. Fruit River Gudenå, flower Tversted Lakes, Denmark. Photo JCS.

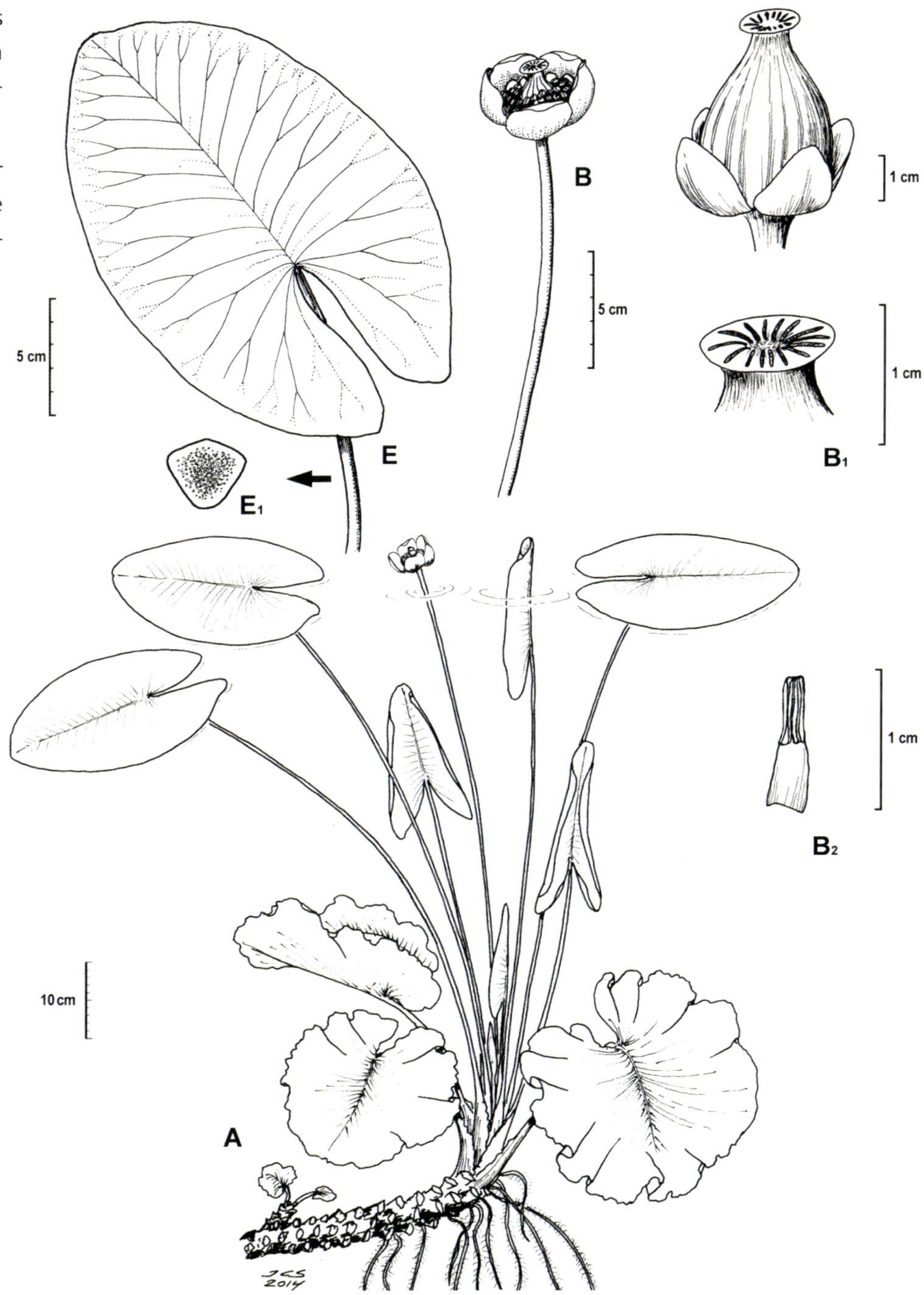

Nuphar lutea
A habit with floating and submerged leaves. **B** flower, **B1** fruit and detail of stigmatic disc, **B2** stamen, **E** floating leaf, **E1** cross section of petiole.

Nymphaeaceae

5 *Nuphar pumila* (Timm) DC.

DK: Liden Åkande **N:** Soleinykkeros **S:** Dvärgnäckros **FIN:** Konnanpulpukka **IS:** -
GB: Least Water-Lily **NL:** Kleine plomp **F:** Nenuphar nain
D: Zwerg-Teichrose **CZ:** Stulík malý **PL:** Grążel drobny
EST: Väike vesikupp **LV:** Mazā lēpe **LT:** Mažažiede lūgnė **Ru:** Клубышка малая

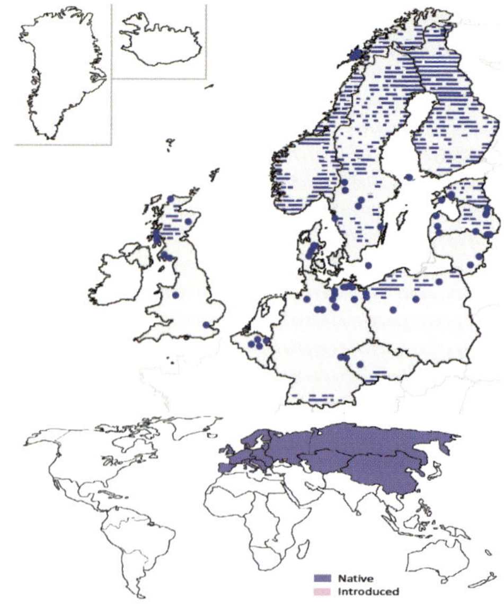

Nuphar pumila. Rødesø, Salten, Jutland, Denmark. Photo JCS.

Perennial hydrophyte. Rhizome 20–70 cm long and 1–3 cm wide, creeping, branched, covered in numerous lens-shaped to ovate scars from broken-off petioles. Roots brownish.

Leaves: Floating, spirally arranged from the rhizome. Lamina 5–17 cm long, 4–13 cm wide, elliptic to ovate, green to purple, rather thin, somewhat dull, glabrous on the upper side, finely silky haired on the lower side at least when young, with a narrow, deep incision at the base. Main veins several times dichotomously divided towards the margin, but not reticulate. Petioles compressed, lens shaped in cross section and with numerous air canals. Submerged leaves not always present, 6–12 cm wide, wider than long, thin, semi-transparent, wavy, greenish to purplish brown, with 10–20 cm long petioles.

Flowers: Small, 1.5–3.5 cm in diameter. Sepals (4–)5(–7), 1.2–2 cm long, ovate, adaxially yellow, abaxially green sometimes with yellow margins, more or less uneven. Petals 8–13, yellow, 4–5 mm long, almost as long as the 50 stamens. Anthers 1–2 mm long, only slightly longer than wide. Stigmatic disc 6–8 mm in diameter, star shaped, almost flat, yellow, the 8–11 brownish rays reaching the margin.

Fruits: 2–3 cm long, urn-shaped with a slightly elongated, slightly crooked neck. The fruit contains more than 70 seeds. Seeds 3.5–4 mm long, 2 mm wide, ovate, pale brown.

Flowering: June-August.

Biology: The petals with nectaries attract flies for pollination. Seed dispersal - see **4 *N. lutea***.

Habitats: In nutrient-poor ponds and lakes, in sheltered areas with still water in rivers. Prefers soft substrates. Grows in water 0.5–3.5 m deep, but does not flower in water over 2 m deep.

Distribution within the region:
Native. Scattered to rare throughout, more frequent in the north.

Characteristics and similar species:
Smaller but otherwise rather similar to **4 *N. lutea.*** This latter with markedly larger leaves, glabrous on the underside, larger flowers, 4–6 cm in diameter, stigmatic disc orbicular with 15–20 rays.

336 *Nymphoides peltata* has very different flowers, leaves with reticulate venation, the lower side reddish papillose and without silky hairs.

Hybrids: 5a *Nuphar ×spenneriana* Gaudin (*N. lutea x pumila*) is intermediate between the parents in terms of habit. It is easiest recognised by the crenulate, not star-shaped, stigmatic

Nuphar pumila with floating and submerged leaves, flower and fruit. Inserted: Detail of the lower side of the lamina with silky hairs. Rødesø, Salten, Jutland, Denmark. Photo JCS.

Nuphar pumila. Typical fruit with star-shaped stigmatic disc and the neck slightly crooked. Flower with 5 sepals. Rødesø, Salten, Jutland, Denmark. Photo JCS.

disc with 9–15 rays. The underside of leaves has only colourless silky hairs near the margin. It is partly fertile (pollen fertility 15%) and probably able to backcross. It is rare and typically occurs where both parents are present, but may occur in the absence of either parent.

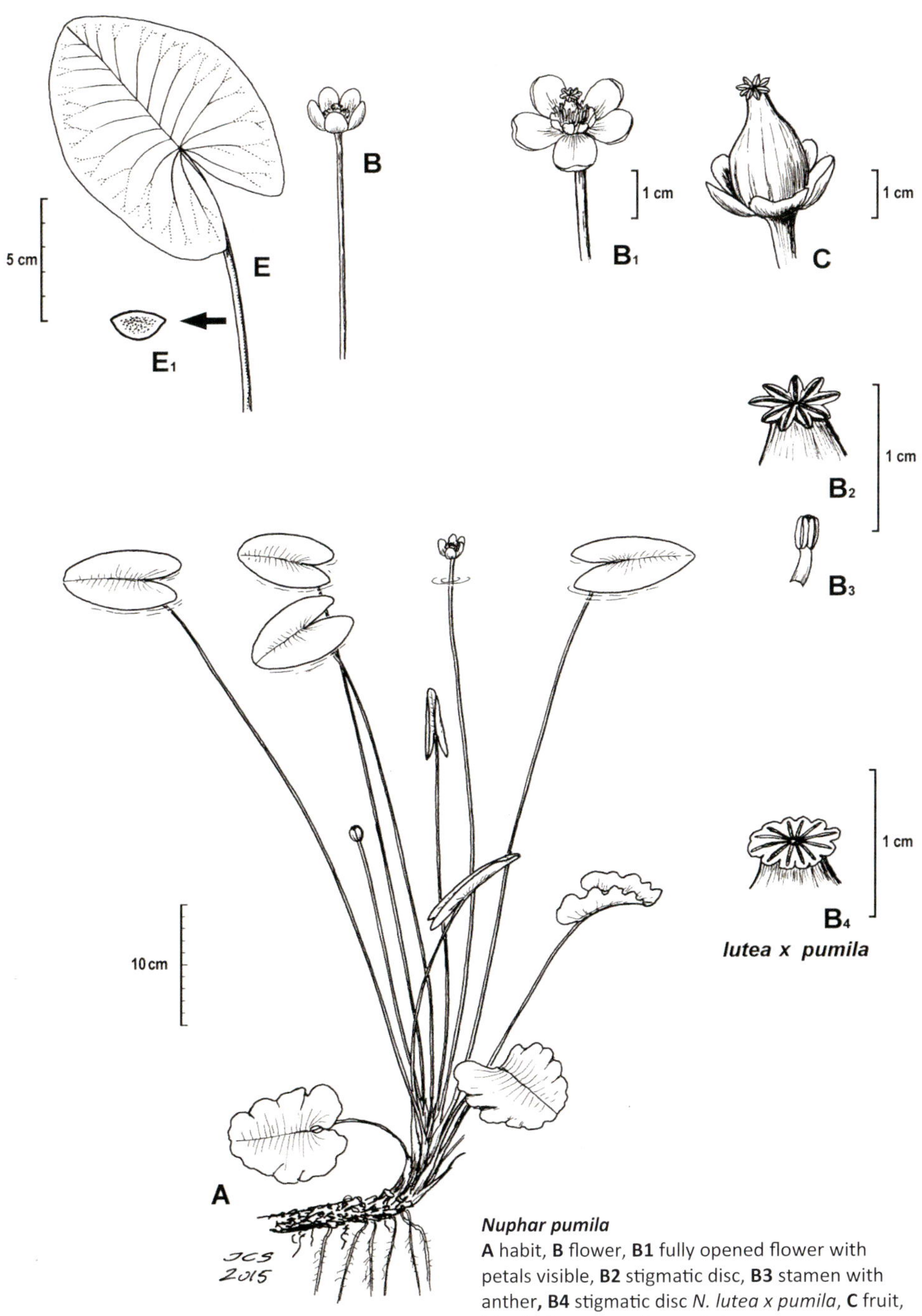

Nuphar pumila
A habit, **B** flower, **B1** fully opened flower with petals visible, **B2** stigmatic disc, **B3** stamen with anther, **B4** stigmatic disc *N. lutea x pumila*, **C** fruit, **E** leaf, **E1** cross section petiole.

Stigmatic discs from above. Top: *Nuphar pumila*, Middle *N. ×spenneriana*, Bottom *N. lutea*. Photos KvdW.

Nymphaeaceae

6 ***Nuphar advena*** (Aiton) Aiton f.

DK: - **N:** - **S:** - **FIN:** - **IS:** -
GB: Spatter-dock **NL:** Tuinplomp **F:** -
D: Amerikanische Teichrose **CZ:** - **PL:** -
EST: - **LV:** - **LT:** - **Ru:** Кубышка пришлая

Nuphar advena. Spottiswoode Loch, near Duns, Berwickshire, Scotland. Photo RVL.

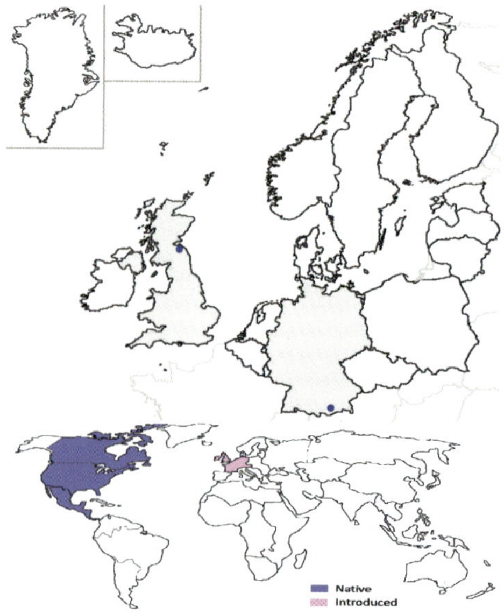

6a *N. ×porphyranthera* Lansdown & Ruhsam *(N. advena* x *lutea)* has anthers indehiscent and more or less the same length as filaments and most leaves erect, degrading early in the season.

Perennial hydrophyte. Rhizome 3–10 cm wide.

Leaves: Usually emergent and erect, sometimes floating. Lamina of emergent leaves 12–40 cm long, 10–31 cm wide, ovate oblong to suborbicular, green, glabrous to pubescent, with a narrow, deep incision at base. With 16–32 lateral veins on each side of the midrib. Petioles terete, 4–11 mm in diameter.

Flowers: 2–4.5 cm in diameter. Sepals usually 6, broadly ovate, adaxially yellow, rarely with reddish tones. Anthers 3–9 mm long. Stigmatic disc 9–26 mm in diameter, orbicular, entire, concave, yellow, the 10–24 rays terminating 1–3 mm from the margin.

Fruits: 2–5.5 cm long, ovoid, without an elongated neck. Fruit wall ribbed. Seeds 3–6.5 mm long, 3–5 mm wide.

Flowering: June-October.

Biology: See **4 *N. lutea***.

Habitats: Lakes, ponds, pools and slow-flowing rivers. It grows to a depth of 1.5 metres.

Distribution within the region:
Most records from Britain are errors for *N.* ×*porphyranthera,* and only a single population of *N. advena* has been confirmed in Scotland.

Characteristics and similar species:
Easily confused with **4 *N. lutea***, which differs in the petioles being bluntly trigonous in cross section, urn-shaped fruit with elongated neck and usually 5 sepals, the leaves normally floating.

7 *Nuphar japonica* is also quite similar. Its leaves are emergent, but the fruit is shaped like an urn with an elongated neck and a star-shaped stigmatic disc with 9–17 rays.

Nuphar ×porphyranthera. Left: Gravetye Manor, East Grinstead. Right: Oulton, Leeds, Britain. Photo RVL.

Nuphar advena. Spottiswoode Loch, near Duns, Berwickshire, Scotland. Photo RVL.

Nymphaeaceae

7 *Nuphar japonica* DC.

DK: Japansk Åkande **N:** - **S:** Japansk näckros **FIN:** - **IS:** -
GB: East Asian Water-Lily **NL:** - **F:** Nénuphar du Japon
D: - **CZ:** - **PL:** -
EST: - **LV:** - **LT:** - **Ru:** Кубышка японская

Perennial hydrophyte. Rhizome 1-3 cm wide.

Leaves: Emergent or floating. Lamina of 12–35 cm long, 6.5–18 cm wide, with a 2–10 cm deep basal incision, ovate to oblong ovate, green, glabrous to pubescent on the lower side. Petioles subterete, 3–10(–14) mm in diameter.

Flowers: 2–3.5 cm in diameter. Sepals usually 5, broadly ovate, yellow, rarely with reddish tones. Anthers 2.5–5 mm long. Stigmatic disc 5–7 mm in diameter, deeply lobed, yellow or rarely with reddish tones, the 9–17 rays terminating at the margin.

Fruits: 2–3.5 cm long, urn shaped, with a prominent neck. Fruit wall smooth to slightly furrowed. Seeds ovoid.

Flowering: June-September.

Distribution within the region:
Introduced and naturalised at Zealand, Denmark and in Bavaria, Germany.
Native in Japan.

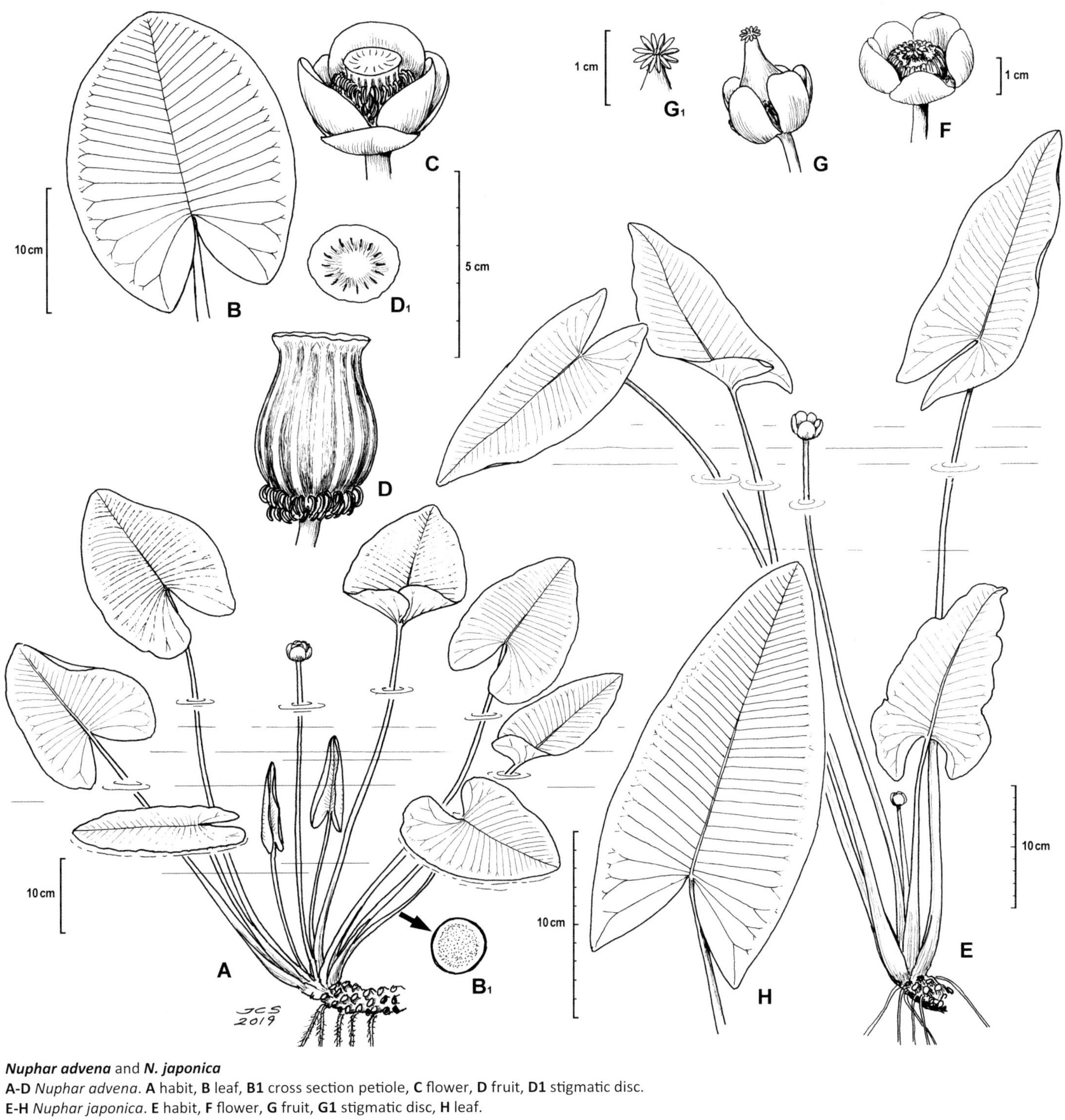

Nuphar advena **and** ***N. japonica***
A-D *Nuphar advena*. **A** habit, **B** leaf, **B1** cross section petiole, **C** flower, **D** fruit, **D1** stigmatic disc.
E-H *Nuphar japonica*. **E** habit, **F** flower, **G** fruit, **G1** stigmatic disc, **H** leaf.

Cabombaceae

8 *Cabomba caroliniana* A.Gray

DK: Grøn Cabomba **N:** - **S:** Kabomba **FIN:** Karkeaviuhkalehti **IS:** -
GB: Fanwort **NL:** Waterwaaier **F:** Cabomba de Caroline
D: Karolina-Haarnixe **CZ:** - **PL:** -
EST: - **LV:** - **LT:** - **Ru:** Кабомба каролинская

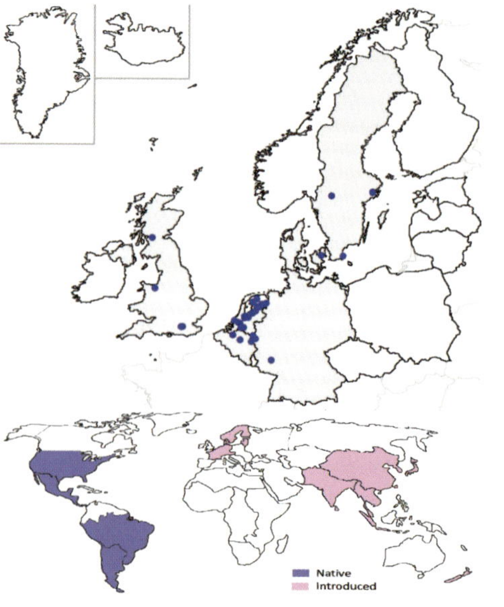

Cabomba caroliniana in a dense formation. At a distance it may look like *Ranunculus circinatus*. Hedeland, Roskilde, Denmark. Photo Henrik Ærenlund Pedersen.

Perennial hydrophyte. Rhizome creeping, slender. Rooting from lower stem-nodes and rhizome.

Stem: Up to 3–4(–10) m long, terete, greenish to brownish, with scattered white or brownish hairs, somewhat brittle.

Leaves: Floating leaves only present on flowering shoots, blades 0.5–2 cm long, 1–3 mm wide, attached at the centre, narrowly elliptic, entire or sometimes with a basal incision. Pedicels long. Submerged leaves opposite or in whorls of 3. Lamina up to 6 cm wide, green, circular or kidneyshaped in outline, several times divided into fine, flat branches, which together with the rather long pedicel gives the leaf a feathery, fan-like appearance. The submerged parts of the plant secrete a gelatinous, slimy mucous covering leaves and stem.

Flowers: 3-merous, 6–15 mm in diameter, raised above the surface. Sepals white or cream sometimes with a pinkish tinge at the tips. Petals similar to sepals, but with yellow, auriculate nectaries at base. 3 or 6 stamens and (2–)3 (–4) pistils.

Fruits: Aggregate fruit, each segment 4–7 mm long, containing 1–3, short papillose seeds.

Flowering: August-September.

Biology: *C. caroliniana* plants are able to occupy large water bodies rapidly by division of the stem and rhizome forming dense mats. Uprooted, floating plants can continue to grow for up to 2 months. In late autumn the stems become defoliated, brittle and rigid, fragments acting as turions remain green and enable the plant to overwinter. The flowers open late in the morning and close in the evening. Each flower opens for 2 days - the first day as female, the second as male releasing the pollen. Pollination is by small flies attracted by the nectaries. The fruit matures under water, and the seeds are released as the fruit decays.

Habitats: More or less nutrient rich, neutral to alkaline water. Small lakes, ponds, ditches, canals and rivers. It grows to a depth of 3 metres.

Distribution within the region: Introduced and naturalised as an aquarium escape in the Netherlands, Belgium, Germany, Denmark and Sweden, as well as a single site in Britain. Considered invasive outside its natural range, being able to outcompete native plants.

Cabomba caroliniana. Flowers and floating leaves. Hedeland, Roskilde, Denmark. Photo Henrik Ærenlund Pedersen

Cabomba caroliniana. Flower and submerged plants. Utrecht, the Netherlands. Photos KvdW.

Characteristics and similar species: *Cabomba caroliniana* differs from similar species in the region by the stalked, flat fan-shaped submerged leaves, which are opposite or in whorls of 3. The finely divided leaves and white flowers of *C. caroliniana* could be confused with **206-207 *Ranunculus*** spp., but *Ranunculus* have alternate leaves divided in threes. Species of **197-198 *Ceratophyllum*** and **226-231 *Myriophyllum*** differ in sessile or whorled leaves.

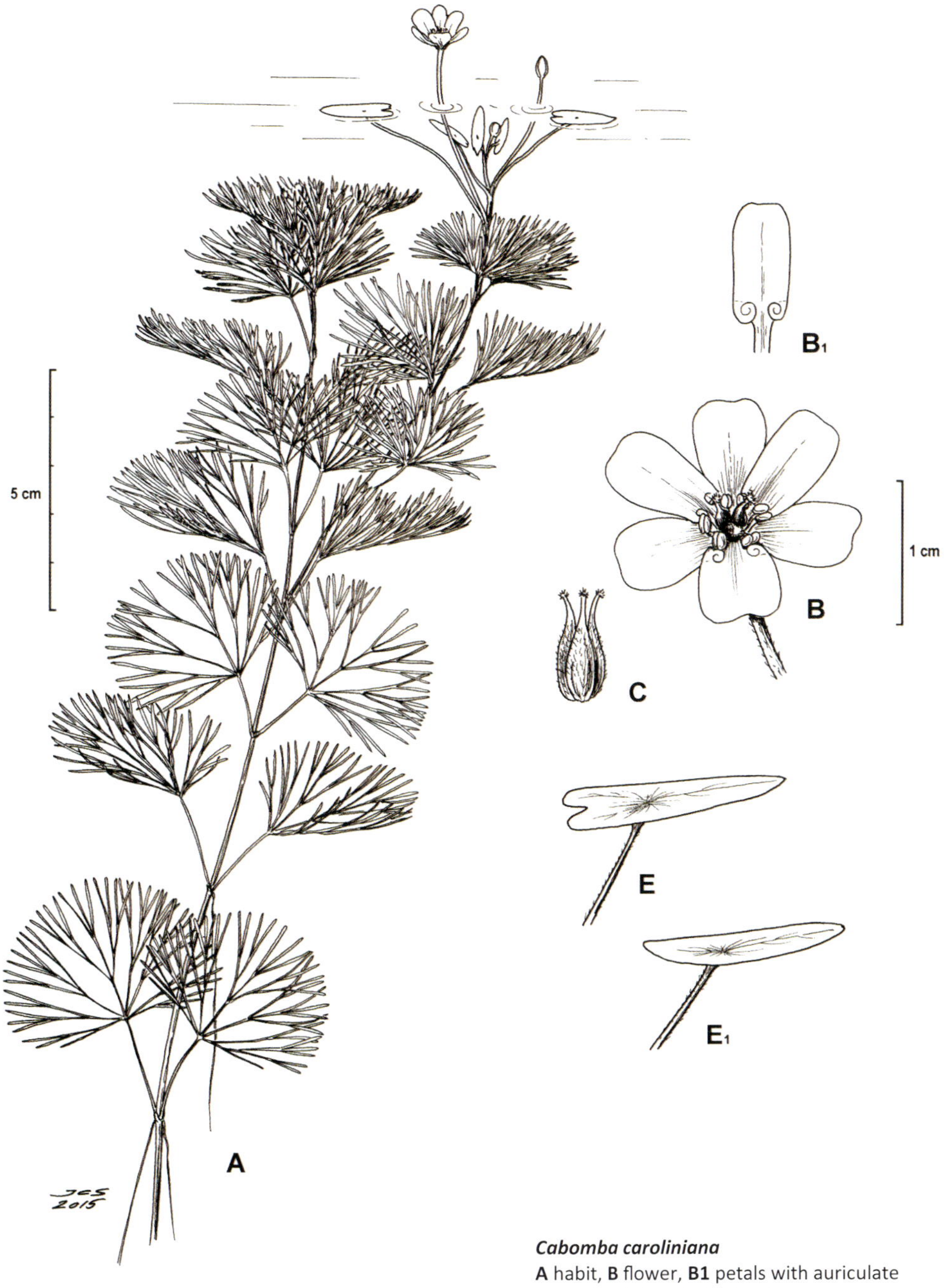

Cabomba caroliniana. Stem showing opposite leaves. Photo JCS.

Cabomba caroliniana
A habit, **B** flower, **B1** petals with auriculate nectaries, **C** fruit, **E, E1** floating leaves.

Acoraceae

9 *Acorus calamus* L.

DK: Kalmus **N:** Kalmusrot **S:** Kalmus **FIN:** Rohtokalmojuuri **IS:** -
GB: Sweet Flag **NL:** Echte kalmus **F:** Acore calame
D: Kalmus **CZ:** Puškvorec obecný **PL:** Tatarak zwyczajny
EST: Harilik kalmus **LV:** Smaržīgā kalme **LT:** Balinis ajeras **Ru:** Аир обыкновенный

Acorus calamus in shallow water in river Viskan, Halland, Sweden. Photo BM.

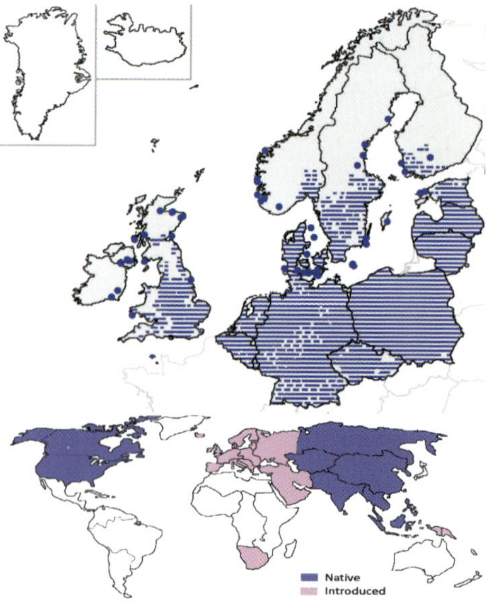

Perennial helophyte. Rhizome horizontal, 1–1.5(–3) cm wide, aromatic, rooting from the lower side.

Stem: 0.5–1.0 m long, 1–1.5 cm wide, flattened-triangular in cross section, bright green, erect. With a leaf-like extension at the top, which may be regarded as a spathe, giving the impression, that the inflorescence is lateral.

Leaves: 0.5–1.5 m long, 1–2.5 cm wide, distichous, ensiform, flattened-rhombic in cross section, laterally compressed with a sharp keel on the adaxial side of the distinct midrib and the margins fused into one on the abaxial side, somewhat shiny, bright green, often transversely wrinkled. Acute at the apex, sheathing at the often pink to winered base. As with all parts of the plant, with a marked aromatic smell similar to tangerines when crushed.

Inflorescence: Spadix 5–10 cm long and 1–1.5 cm wide, yellowish green and composed of nearly 700 small, spirally arranged flowers.

Flowers: About 3 mm across, with 6 free, obovate perianth leaves, 6 stamens, fused carpels and 3-locular ovary. The style is short and wide.

Fruits: Oblong to obovoid, reddish berries. European plants are sterile and do not produce fruits.

Flowering: June-July.

Biology: Worldwide, *Acorus calamus* is a cytologically and morphologically variable species with diploid, triploid and tetraploid populations. In Europe and eastern North America, the plants are triploids producing only abnormal, sterile pollen, and propagating only by division of the rhizome.

Habitats: In shallow, more or less nutrient-rich water close to the shores of rivers, streams, canals, ditches, lakes and ponds.

Distribution within the region:
Introduced as a medical plant and naturalised.

Characteristics and similar species:
When sterile it can be mistaken for **113 *Iris pseudacorus*, 118-122 *Sparganium*** spp. and **126-132 *Typha*** spp., but these lack the pleasant aromatic smell. *Iris* and *Acorus* leaves have an edge facing the stem or axis, while *Sparganium* and *Typha* leaves have a flat side facing the stem or axis. *Iris pseudacorus* has bluish-green leaves.

In autumn *A. calamus* starting to die back, changing colour to golden brown. It is often found in water bodies associated with medieval buildings, such as Hammershus Castle, Bornholm, Denmark. Even before the Reformation the bitter and aromatic rhizome was used in monasteries as a cure against digestive problems and other diseases. Photo JCS.

Acorus calamus. The tiny flowers are densely crowded on the spadix. Hammershus Castle, Denmark. Photos JCS.

Acorus calamus has transversely wrinkled leaves - this though can also be seen in *Iris pseudacorus*. Photo KvdW.

Acorus gramineus. Inflorescence and part of leaf.

10 *Acorus gramineus* Aiton

GB: Slender Sweet Flag **F:** Acore graminée

A. gramineus is much smaller than *A. calamus* in all dimensions, with leaves up to 50 cm long, but less than 10 mm wide and with an indistinct midrib. Spadix narrower, 3–5 mm wide and more or less erect. Planted as an ornamental and naturalised in scattered sites in Britain.

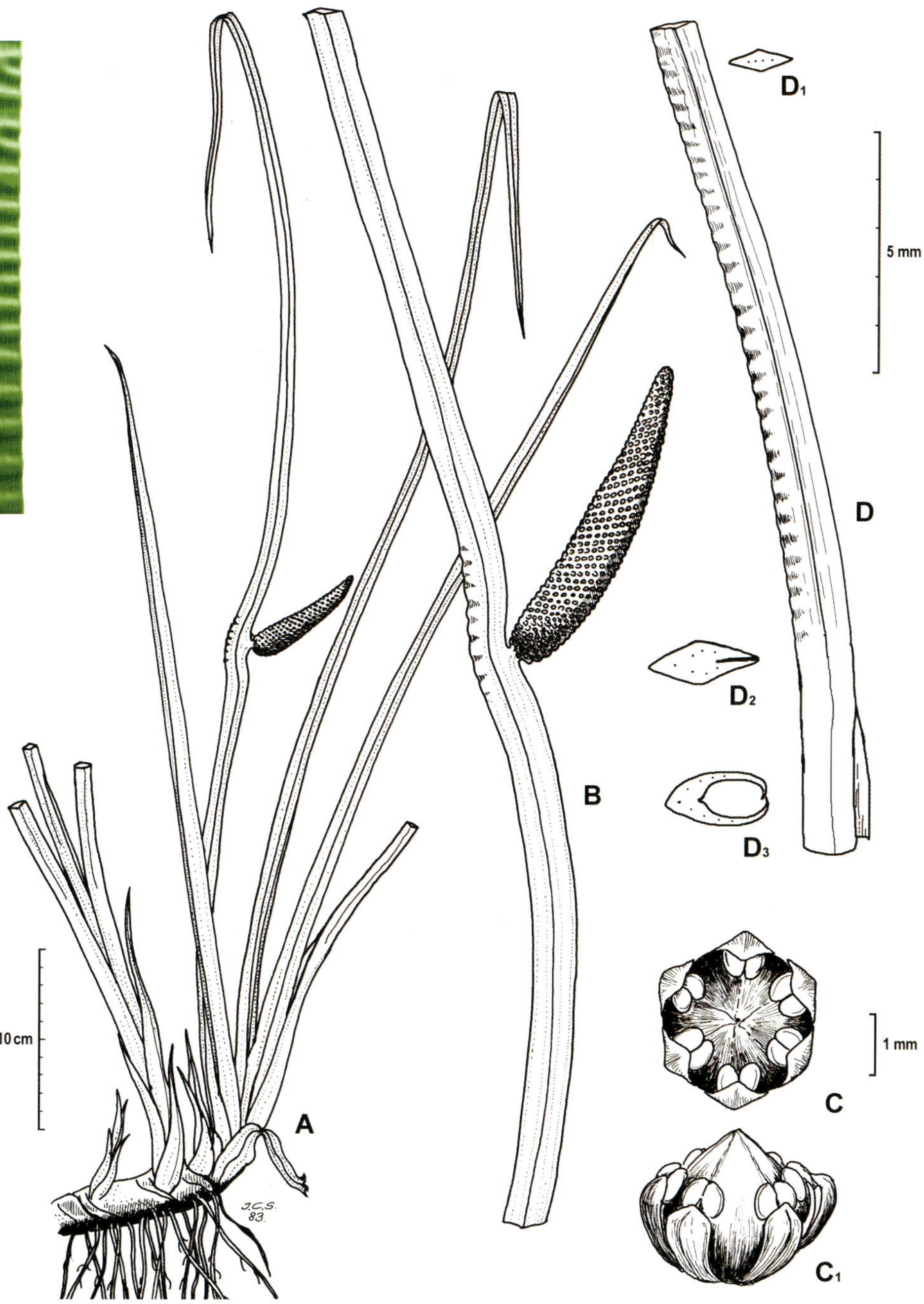

Acorus calamus
A habit, **B** stem with spadix and leaf like spathe, **C** flower top view, **C1** flower side view, **D** part of leaf. **D1** cross section of leaf sheath**, D2-D3** cross section of leaf - keel to the left and the two fused margins to the right.

Araceae

11 *Calla palustris* L.

DK: Kærmysse **N:** Myrkongle **S:** Missne **FIN:** Suovehka **IS:** -
GB: Bog Arum **NL:** Slangenwortel **F:** Calla des marais
D: Schlangenwurz **CZ:** D'áblík bahenní **PL:** Czermień błotna
EST: Soovõhk **LV:** Purva cūkausis **LT:** Pelkinis žinginys **Ru:** Белокрыльник болотный

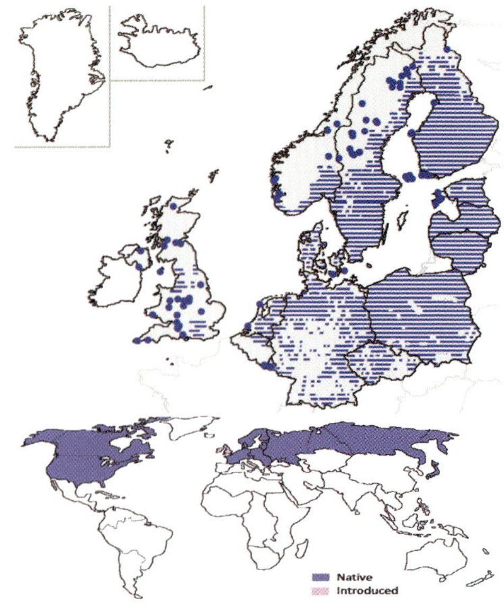

Calla palustris with *Lemna minor* in an overgrown moat. Fussingø, Jutland, Denmark. Photo JCS.

Perennial helophyte. Rhizome up to half a metre long and 3 cm in diameter, green or whitish. Roots 1–1.5 mm in diameter, whitish to pale brown.

Stem: Ascending, 1–1.5 cm in diameter, 15–25 cm long, terete.

Leaves: All basal, distichous, alternate. Lamina 4–15 cm long, broadly cordate, entire, somewhat shiny, green - lower side a bit paler than the upper side, veins rather inconspicuous, curved inwards and united close to the margins. Leaf tip with hydathodes. Petiole up to 30 cm long, sheathing at base and with a ligule 3–5 cm long, which soon splits in two.

Inflorescence: Spadix 2–4 cm long, 1.5–2 cm thick, cylindrical, yellowish green, consisting of about 80 small, spirally arranged flowers. Spathe 3–7 cm long, ovate, cuspidate at apex, flat, white inside and pale green outside, persisting in fruit.

Flowers: Bisexual, but uppermost flowers often male, without perianth. 6 or more white stamens, ovary 1-celled, stigma disc shaped.

Fruits: The ripe scarlet red berries are 5–8 mm wide, containing 6–10 cylindrical, brown seeds 4–5 mm long, with abundant endosperm.

Flowering: June-July.

Biology: Vegetative propagation by rhizome division, often forming dense stands in still water. Anchored in shallow water, the long, branched rhizomes often float and contribute to the formation of floating mats. Flowering starts from the base of the spadix and moves upwards as the spadix develops. The flowers are pollinated by insects. They are female at first, and the stamens develop only after the stigma has withered, thereby preventing self-pollination of individual flowers. It is typically abundantly fertile. The berries float, but are also dispersed by birds which eat them or by sticking to the legs of mammals and birds. The whole plant and especially the berries are very poisonous to humans.

Habitats: In *Alnus* swamps, peat bogs, ditches, in shallow water on muddy lake-shores and along slow-flowing streams.

Distribution within the region:
Native and locally common throughout much of the region, rare or absent from some areas. Introduced and very locally naturalised in Britain.

Even without flowers *C. palustris* is easily recognised by the somewhat shiny, cordate leaves. Småland, Sweden. Photo JCS.

Flowering *C. palustris* - note the white stamens and the already somewhat swollen fruits with withered stigmas. Fussingø, Jutland, Denmark. Photo JCS.

Nearly mature fruiting head of *C. palustris*. Later the berries turn scarlet red - very pretty but also very poisonous, as is the entire plant.
Fussingø, Denmark. Photo JCS.

Characteristics and similar species:

Calla palustris is easily recognised by the shiny, green, cordate leaves and in flower by the inflorescence and the white spathe, which is flat inside. *Calla palustris* can be distinguished from species with a similar leaf outline, such as **33 *Alisma plantago-aquatica*** by the leaf venation and from **335 *Menyanthes trifoliata*** by the trifoliate leaf blades of the latter.

Zantedeschia aethiopica (L.) Spreng. is native to southern Africa and very rarely found as a garden escape on the banks of reservoirs and ponds in the Netherlands and Britain. It is easily recognised by the white spathe up to 25 cm, yellow spadix and arrow-shaped, dark green leaves up to 45 cm long.

Zantedeschia aethiopica.
Stevensweert, Netherlands. Photo KvdW.

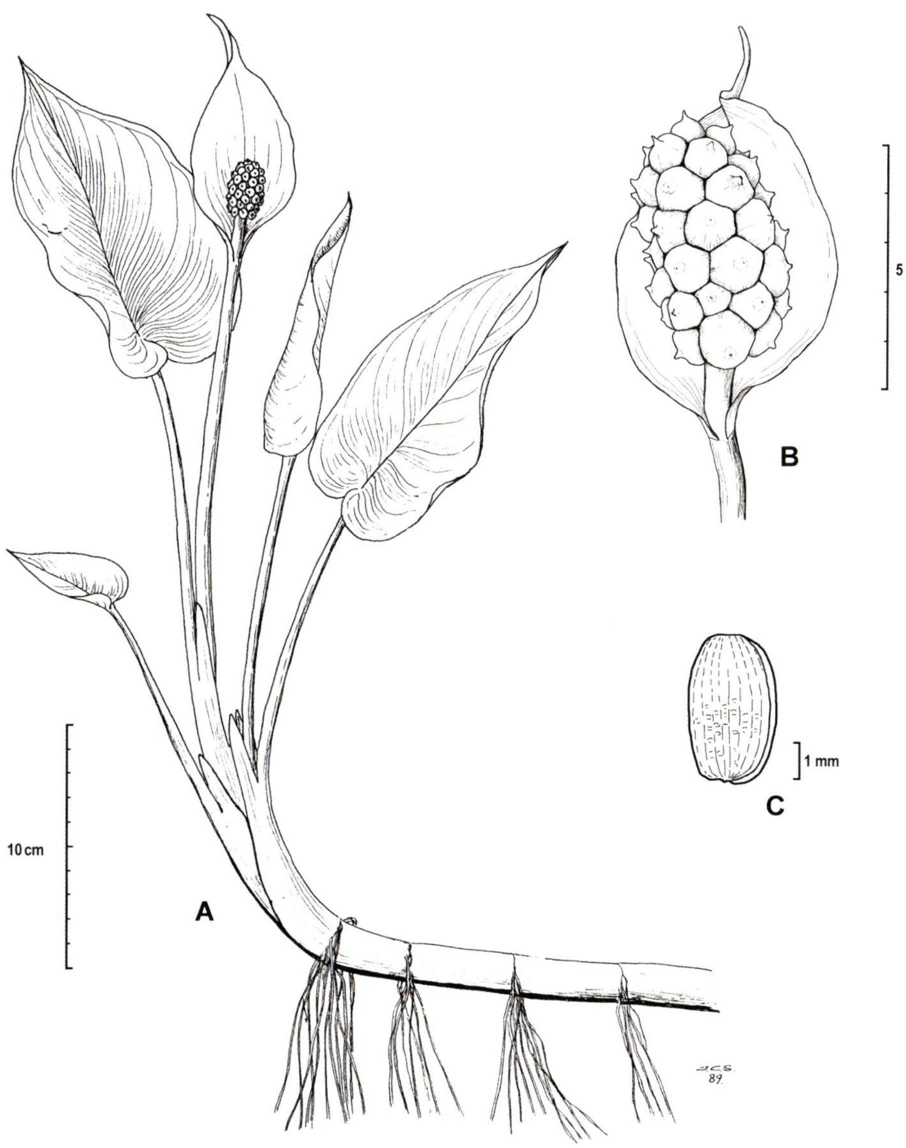

Calla palustris
A habit, **B** fruiting head, **C** seed.

Araceae

12 *Lysichiton americanus* Hultén & H.St.John

DK: Gul Kæmpekalla **N:** Skunk-kala **S:** Skunkkalla **FIN:** Keltamajavankalla **IS:** -
GB: American Skunk-Cabbage **NL:** Moeraslantaarn **F:**
D: Amerikanische Scheinkalla **CZ:** - **PL:** Tulejnik amerykanski
EST: Ameerika kevadvohk **LV:** - **LT:** - **Ru:** Лизихитон американский

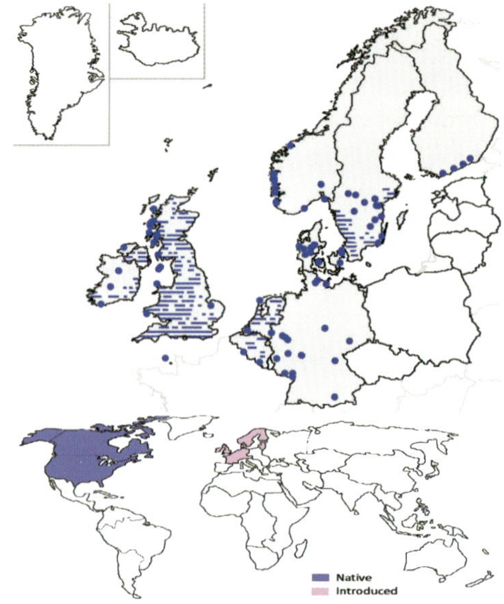

The large, obovate to elliptic leaves of *L. americanus* can reach a height of almost 1.5 m. Jardin du Conservatoire Botanique National de Brest, France. Photo RVL.

Perennial helophyte. Rhizome 30 cm or more long, 3–5 cm in diameter, horizontal. Roots white.

Stem: Short, vertical.

Leaves: 50–135 cm long, 30–70 cm wide, obovate to broadly spathulate or elliptic, eventually elongate, shiny, green. Midrib very robust. Petiole short, robust, broadly channelled.

Inflorescence: Spadix 4–14 cm long, cylindrical, with numerous, dense flowers. Spathe 30–40 cm long, with a sheathing basal part densely covering the stipe of the spadix, obtuse or cuspidate at apex, yellowish green or yellow, withering in fruit.

Flowers: Bisexual or the uppermost male, with 4(–6) greenish perianth segments, 4(–6) stamens and sessile stigma. Ovary (1–)2 celled.

Fruits: The ripe greenish to reddish berries are about 6 mm in size, broadly ovate and containing 1–2 greyish-brown or reddish-brown seeds.

Flowering: April-May.

Biology: Vegetative propagation by rhizome division and by seed. A single plant can cover an area of 1 m^2. Growth is slow, but in time (80 years or more) it is capable of forming large populations. The flowers appear early in the year and a little before the leaves. The unpleasant smell attracts different kinds of insects, such as flies and beetles for pollination, but self-pollination also occurs. The fruit set is apparently quite good.

Habitats: In nutrient-rich habitats. Marshes, fens, bogs, ponds, forest swamps, ditches, along stream and river banks and lake sides. Often associated with formal gardens.

Distribution within the region: Introduced to Britain at the beginning of the twentieth century as an ornamental garden plant. Since then it has

Cultivated *L. americanus*. Botanical Garden, Aarhus, Denmark. Photo JCS.

spread and become naturalised in moist habitats in several countries. It is regarded as invasive in Europe; it has proved difficult to control and spreads rapidly by seeds.

Characteristics and similar species: Even without flowers *Lysichiton* species are easy to recognise by the large, obovate to elliptical leaves and by the unpleasant smell.

Peltandra virginica is sometimes mistaken for *Lysichiton* spp., but differs in leaf shape - see p. 135.

Hybrids: The hybrid between the two *Lysichiton* species:

L. ×hortensis J.D.Arm. & B.W.Phillips is recognised by the colour of the spathe being pale yellow to cream. It is known from a single site in Britain and a few sites in Belgium.

Lysichiton americanus
A habit after flowering, **B** spadix and spathe, **C** flower - side view.

Araceae

13 *Lysichiton camtschatcensis* (L.) Schott

DK: Hvid Kæmpekalla **N:** - **S:** Vit skunkkalla **FIN:** **IS:** -
GB: White skunk cabbage **NL:** - **F:** Lysichite blanc
D: Weiße Scheinkalla **CZ:** - **PL:** Tulejnik kamczacki
EST: - **LV:** - **LT:** - **Ru:** Лизихитон камчатский

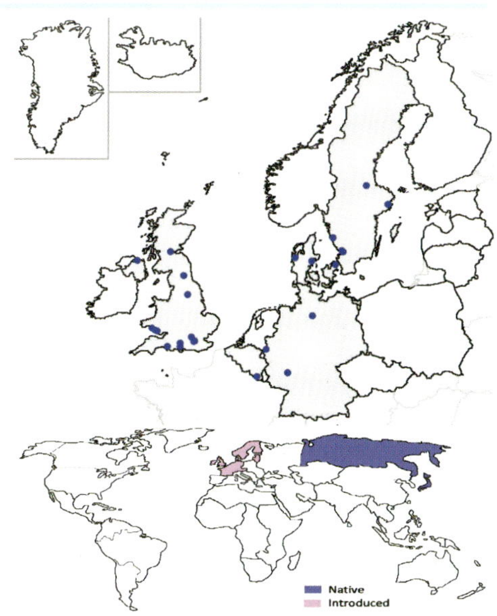

Cultivated *L. camtschatcensis*. Botanical Garden, Aarhus, Denmark. Photo JCS.

With the exception of the white spathe, similar to *L. americanus* both in habit and biology, as well as in preferred habitats.

Lysichiton camtschatcensis was introduced to Europe as a garden plant and rarely naturalized.

Araceae

14 *Pistia stratiotes* L.

DK: Muslingeblomst **N:** Muslingblom **S:** Musselblomma **FIN:** **IS:** -
GB: Water Cabbage **NL:** Mosselplant **F:** Laitue d'eau
D: Wassersalat **CZ:** Babelka řezanovitá **PL:** -
EST: - **LV:** - **LT:** - **Ru:** Пистия слоистая

Pistia stratiotes in cultivation in a garden centre, Britain. Photo RVL.

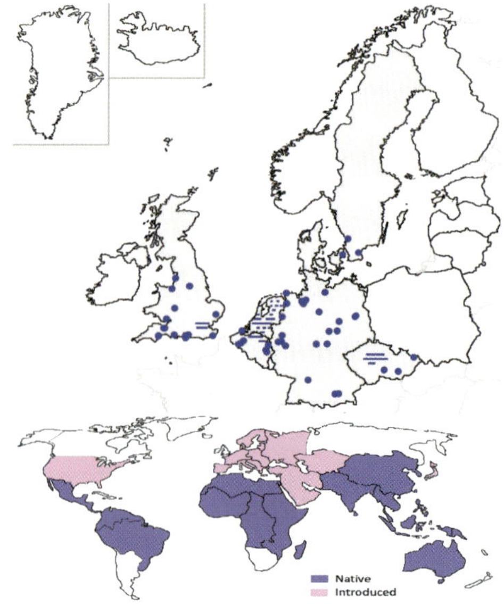

Annual or perennial, floating hydrophyte, with free roots hanging in the water.

Leaves: In a rosette, 3–15 cm long, green, pubescent, sessile, broadly spatulate to ovate, somewhat plicate, with 7–15 veins.

Inflorescence: Spadix short, covered by a whitish-green spathe and hidden between the leaves in the centre of the rosette.

Flowers: Unisexual, with 2–8 male flowers in top of the spadix and 1 female below.

Distribution within the region: Introduced and escaped from cultivation to natural habitats. It has apparently not overwintered in Britain.

Araceae

15 *Orontium aquaticum* L.

DK: Guldkolbe **N:** Gyllen klubb **S:** Guldkolv **FIN:** **IS:** -
GB: Golden-club **NL:** - **F:** Oronce
D: Gold-Keule **CZ:** - **PL:** -
EST: - **LV:** - **LT:** - **Ru:** -

Orontium aquaticum. New Forest, Britain. Photo RVL.

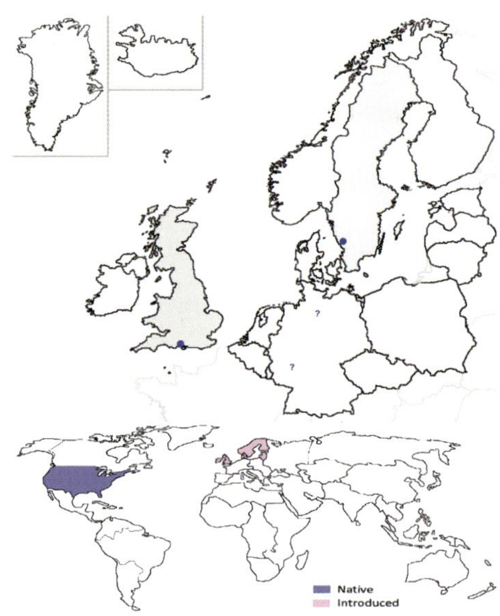

Perennial hydrophyte. Rhizome 1.5–3 cm thick.

Leaves: In a rosette, erect or floating, long petiolate. Lamina 30–40 cm long, somewhat obtuse, ovate, bluish green with a waxy, water-repellent surface.

Inflorescence: Spadix cylindrical, yellow. Stipe long, flaccid, brownish at the base becoming wider and white just below the inflorescence, with a short, greenish spathe at the base that soon withers away.

Distribution within the region: Introduced in Europe and locally naturalised in Britain, Sweden and Germany?

Pontederiaceae

16 *Eichhornia crassipes* (Mart.) Solms

DK: Vandhyacint **N:** Vannhyasint **S:** Vattenhyacint **FIN:** **IS:** -
GB: Water Hyacinth **NL:** Gewone waterhyacint **F:** Eichhornie commune
D: Dickstielige Wasserhyazinthe **CZ:** Tokozelka nadmutá **PL:** -
EST: Vesihüatsint **LV:** - **LT:** - **Ru:** Эйхорния толстоножковая

Eichhornia crassipes Amazonas, Brasilia. Photo BM. Detail of swollen leaf base, Photo JCS.

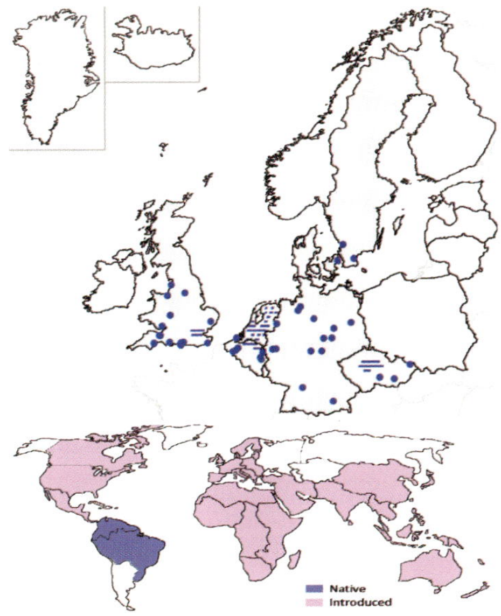

Perennial, free-floating hydrophyte.

Leaves: In a rosette. Lamina 4–15(–25) cm long, ovate with more or less cordate base. Petiole up to 30 cm long, highly swollen, spongy, air filled.

Inflorescence: Spike 2–16 cm long with 3–20 flowers, at base with 2 dissimilar spathes, the lower leaf-like.

Flowers: 4–6 cm across, lilac, more or less actinomorphic with yellow spot on the upper one of the 6 perianth leaves which are fused at the base. 6 stamens.

Distribution within the region:
Introduced and escaped from cultivation to natural habitats. It has apparently never over wintered in Britain.

Pontederiaceae

17 *Pontederia cordata* L.

DK: Hjerte-Vandkærte **N:** - **S:** Pontederia **FIN:** **IS:** -
GB: Pickerelweed **NL:** Moerashyacint **F:** Pontédéria à feuilles en coeur
D: Herzblättriges Hechtkraut **CZ:** Modráska srdčitá **PL:** -
EST: - **LV:** - **LT:** - **Ru:** Понтедерия сердцелистная

Pontederia cordata. Britain. Photos RVL.

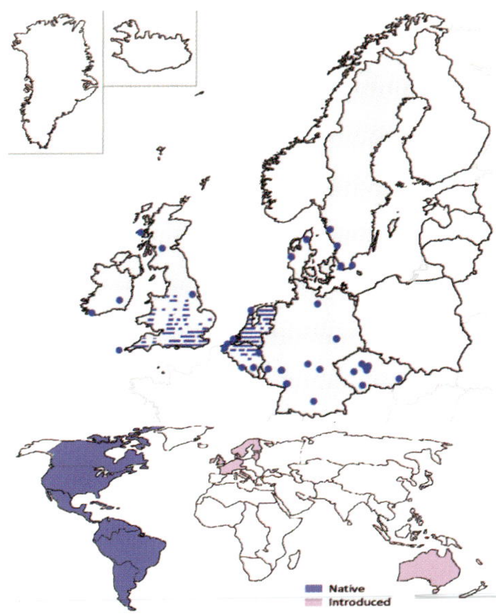

Perennial helophyte. Rhizome short, vertical, with roots in the substrate.

Stem: Erect, up to 120 cm tall.

Leaves: Upper stem leaves sessile, enclosing the stem, the other short or long petiolate, with elongated, cordate lamina. Submerged leaves in rosettes, linear.

Inflorescence: Spike 2–15 cm long.

Flowers: Bluish violet, funnel shaped with 6 petals, 5 oblong and alike, the latter slightly wider, forming an upward lip with 2 yellow spots.

Distribution within the region:
Introduced and escaped from cultivation to natural habitats, widely established and persisting in the wild in Britain and elsewhere.

Araceae

Key to *Lemna*, *Spirodela* and *Wollfia*

In *Lemna, Spirodela* and *Wolffia* the stem is transformed into a leaf-like structure - a frond, with flat branches, but without leaves. Root(s) are produced from the lower side of the frond. The plants float due to air-filled tissue (aerenchyma) below the epidermis. The frond often turns red (due to anthocyanin) on the upper surface and in some species also on the lower surface.

Propagation is mainly clonal by buds, but sexual reproduction also occasionally occurs in most of the species. Pollination of the very tiny flowers takes place just above the water surface with the help of wind or perhaps small insects. Overwintering by small, compact shoots with only a thin layer of aerenchyma, sinking to the bottom, as the mother plant withers. In spring the aerenchyma layer expands and the plant rises to the surface. Some species overwinter as specialised, starchy shoots (turions). Hybridisation occurs and can blur the boundaries between taxa.

---

1 Frond with 1–20 roots and two lateral reproductive pouches (***1***) 2

1 Frond without roots and with one terminal reproductive pouch (***2***). ***Wolffia*** 12

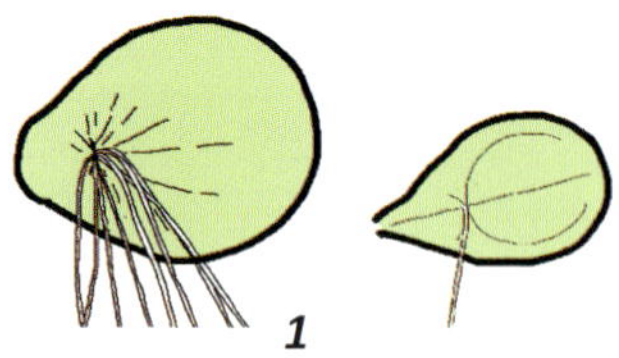

1

2

2 Frond with (1–)2–20 roots and (3–)5–16 veins (***3***). Lower side with a small, scale-like leaflet covering the base of the roots (***5-7***). ***Spirodela*** 3

2 Frond with 1 root and 1–5 veins (***4***). Lower side without a scale-like leaflet covering the base of the roots. ***Lemna*** 5

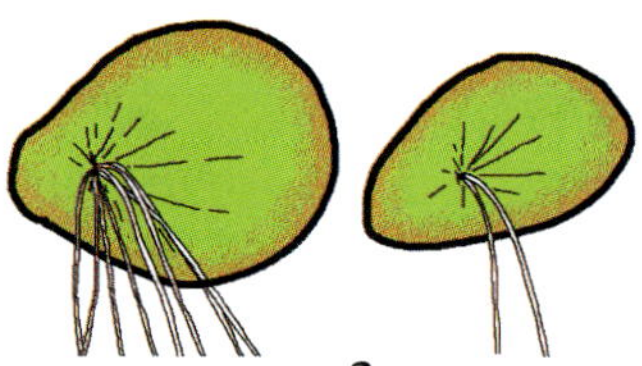

3

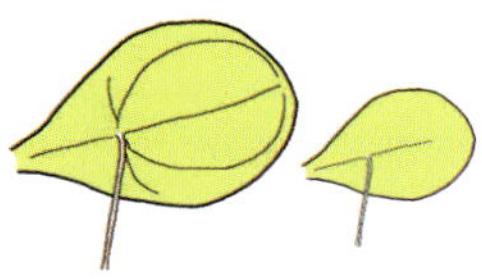

4

Spirodela

3 Frond with 7–20 roots and 7–16 veins, 1–1.5 times as long as wide. 1–5 of the roots perforating the scale-like leaflet, never all roots perforating (***5-7***) 4

3 Frond with (1–)2–7 roots and 3–7 veins, 1.5–2 times as long as wide. All roots perforating the scale-like leaflet (***6***)

27 *S. oligorrhiza*

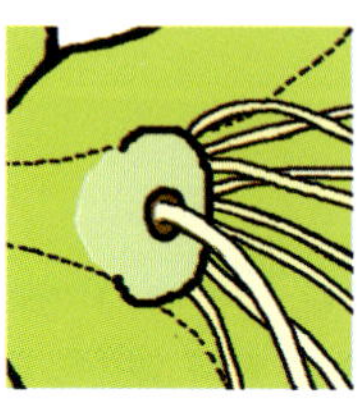

5

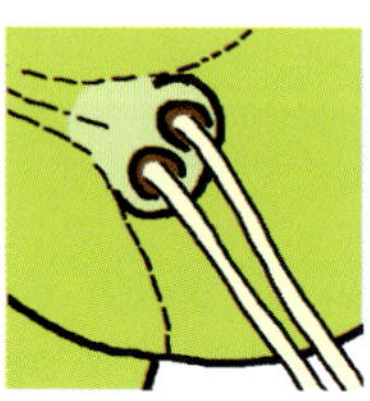

6

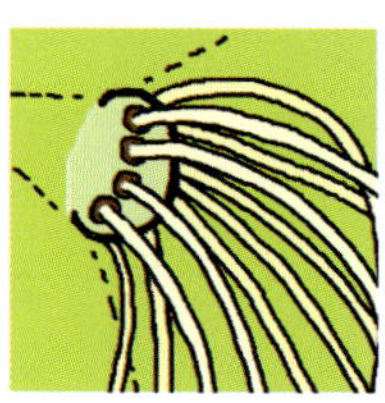

7

4 Only 1 of the roots perforating the scale-like leaflet (***5***)

26 *S. polyrhiza*

4 3–5 of the roots perforating the scale-like leaflet (***7***)

28 *S. punctata*

Lemna

5 Frond finely dentate, lanceolate, with a narrow stalk-like part at base (***8***). Frond suspended in the water column

25 *L. trisulca*

5 Frond with entire margins, ovate, without a narrow stalk-like part at base (***9***). Frond floating 6

8

9

6 Root sheath winged at the base (***10***). Frond never with red pigmentation

21 *L. aequinoctialis*

6 Root sheath not winged at the base (***11***). Frond sometimes with red pigmentation 7

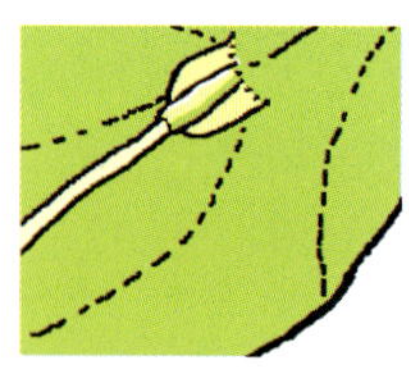

10

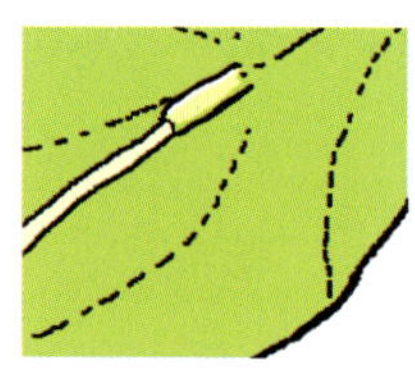

11

7 Fully developed frond with 3–5(–7) veins (***12***), not translucent 8

7 Fully developed frond with only 1 vein (***13***), somewhat translucent 11

12

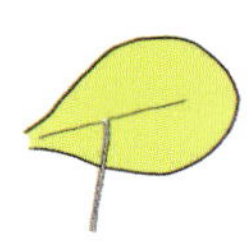

13

(Note: Young fronds of *Lemna gibba* and *minor* can have only 1 vein.)

8 Frond with a row of almost equal papillae on the upper side (***14***). With red colouration on the lower side, at least around the root base (***16***)

22 *L. turionifera*

8 Frond without papillae or with a row of unequal papillae - the one above the node and at apex bigger than the other (***15***). With or without red colouration on the lower side 9

14

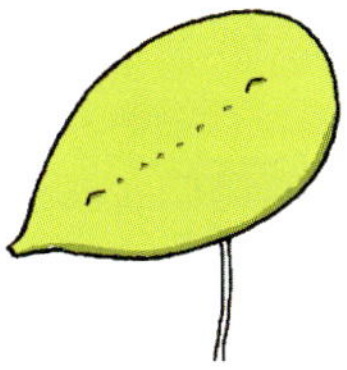

15

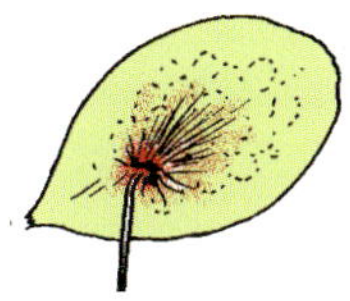

16

9 Lower side of frond with a tissue of 0.3–0.7 mm wide, aerial chambers, spongy and swollen or flat (***17***)

18 *L. gibba*

9 Lower side of frond with a tissue of less than 0.3 mm wide aerial chambers, flat (***18***) 10

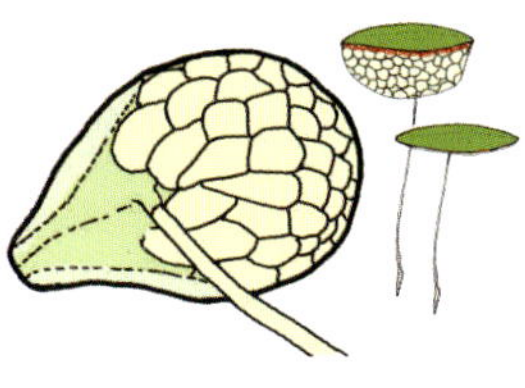

17

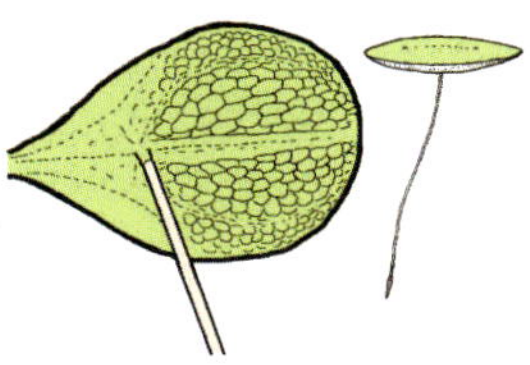

18

10 Lower side of frond with red colours (***16***)

23 *L. ×japonica*

10 Lower side of frond never with red colours (***18***)

19 *L. minor*

11 Vein as long as or shorter than the aerenchymatic part of the frond. Frond symmetrical, also at base, 1–1.6 times as long as wide (***19***)

20 *L. minuta*

11 Vein longer than the aerenchymatic part of the frond. Frond mostly asymmetrical, especially at base, 1.3–3 times as long as wide (***20***)

24 *L. valdiviana*

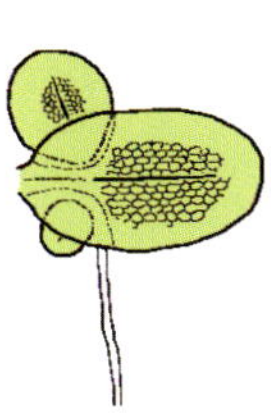

19

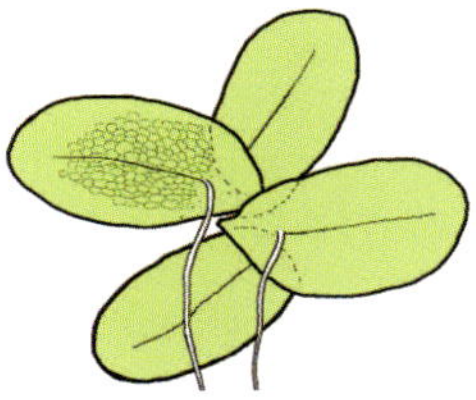

20

Wolffia

12 Frond boat-like, 2–3 times as deep as wide, widest at the water surface (***21, 22***)

30 *W. australiana*

12 Frond globoid to ovoid, 1–1.5 times as deep as wide, widest below the water surface (***24, 26***) 13

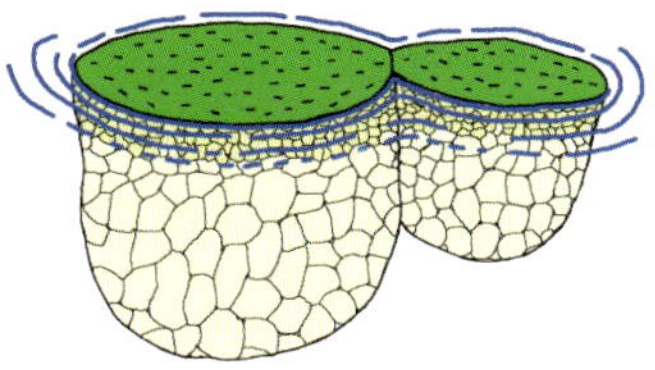

***21** side view*

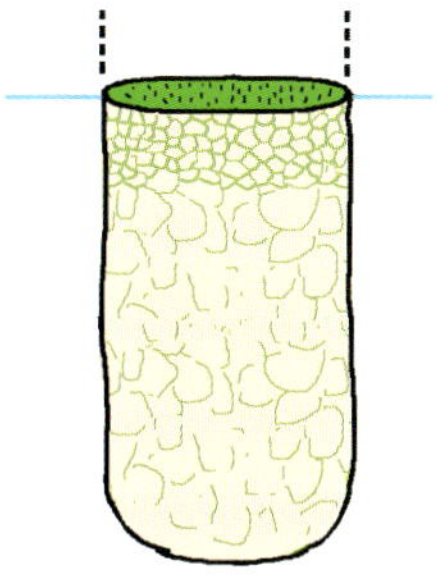

***22** front view*

13 Frond on the upper side shiny, dark green, with (20–)30–120 stomata (***23***), widest just below the water surface (***24***)

29 *W. arrhiza*

13 Frond on the upper side light green, with 1–15(–30) stomata (***25***), widest somewhat below the water surface (***26, 28***) 14

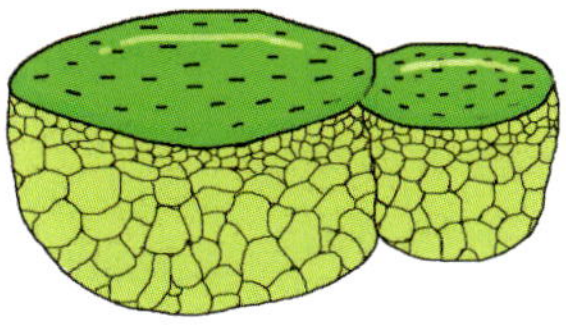

***23** side view*

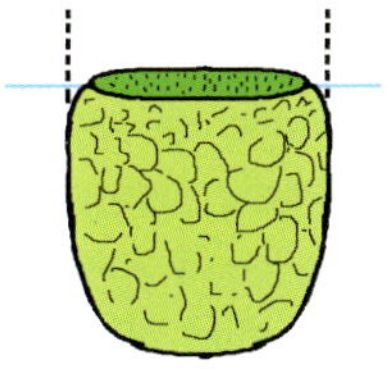

***24** front view*

14 Frond 1–1.3 times as long as wide, 0.4–1.2 mm wide (***25, 26***)

31 *W. columbiana*

14 Frond 1.3–1.7 times as long as wide, 0.3–0.6 mm wide (***27, 28***)

32 *W. globosa*

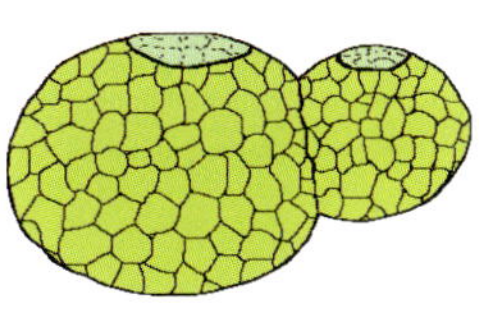

***25** side view*

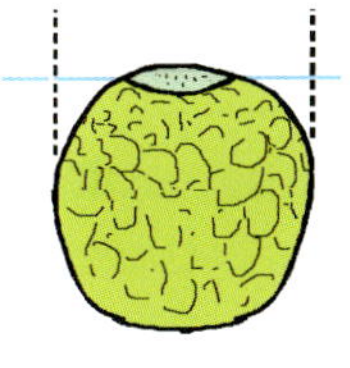

***26** front view*

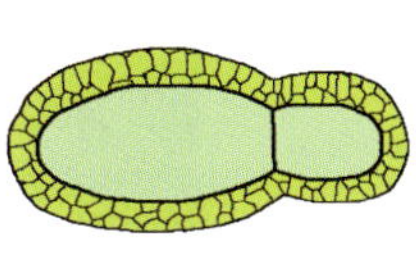

***27** top view*

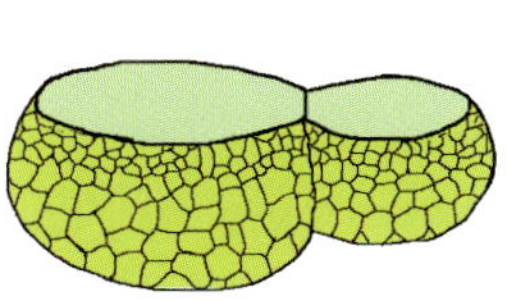

***28** side view*

Araceae

18 *Lemna gibba* L.

DK: Tyk Andemad **N:** Klumpandemat **S:** Kupandmat **FIN:** Kupulimaska **IS:** -
GB: Fat Duckweed **NL:** Bultkroos **F:** Lentille d'eau gibbeuse
D: Buckel-Wasserlinse **CZ:** Okřehek hrbatý **PL:** Rzęsa garbata
EST: Küürlemmel **LV:** Kuprainais ūdenszieds **LT:** Kuprotoji plūdena **Ru:** Ряска горбатая

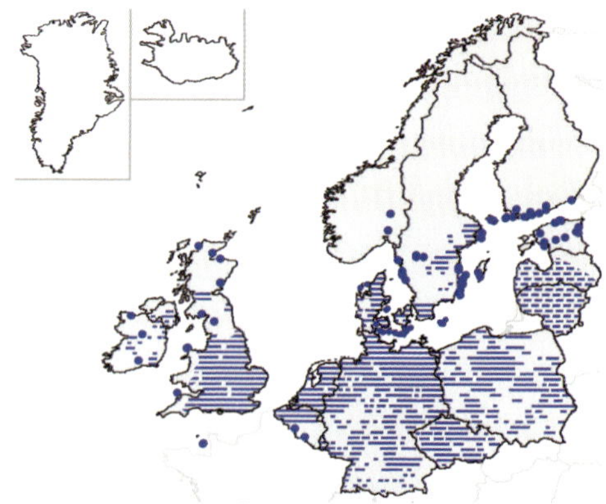

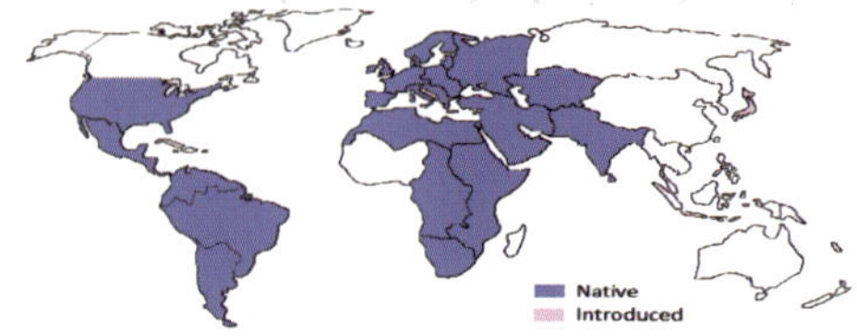

Lemna gibba - some green, some with reddish-brown pigmentation - mixed with *Spirodela polyrhiza*. River Gudenå, Randers, Denmark. Photo JCS.

Perennial, floating hydrophyte. With 1 root, up to 12 cm long from each frond. The root with an acute cap, 0.8–1.8 mm long.

Frond: Solitary or adhering in small groups, broadly ovate, (3–)3.5–6 mm long and up to 4 mm wide, with 3–5 veins. Upper side slightly arched, with or without scattered papillae, bright "apple" green, mottled yellow green often with reddish pigmentation, lower side often slightly reddish brown along the edge, but otherwise pale green, usually more or less flat, occasionally more or less vaulted and almost hemispherical in shape. Lower side, even on flat specimens, with a tissue of 0.3–0.7 mm wide aerial chambers, seen with a magnifying glass as a mesh of 10–20 cells.

Inflorescence: Monoecious. Much reduced and very tiny, enclosed within a membranous spathe. With only one female and one male flower borne from (1–)2 hollows on the margin of the frond. Flowering very rare.

Flowers: Male flower with 2 stamens. Female flower more or less bottle-shaped with a funnel-shaped stigma.

Fruits: Elliptical, 0.6–1 mm long, winged upwards, wings 0.1–0.2 mm wide. The fruit contains (1–)3–4(–5) ribbed seeds.

Flowering: June-July.

Biology: Propagation by budding from pouches in the margins of the frond, rarely by seed. *L. gibba* does not form turions, but overwinters as flat, starchy fronds floating or as buds in dead, sunken fronds.

Often in extensive, dense carpets, completely covering entire water bodies and with other species of *Lemna*. Dispersal by water, birds and human activities.

Habitats: In nutrient-rich, usually eutrophic, still or very slow-flowing water in ditches, canals, streams and ponds - e.g., typically in stock ponds and ditches in grassland.

Distribution within the region:
Native. Rather common throughout most of the area, but absent from Iceland and Greenland, as well as northern parts of Norway, Sweden and Finland.

Lemna gibba. The tissue on the lower side of the frond can be more or less inflated. River Noerreaa, Randers, Denmark. Photo JCS.

Typical *L. gibba* with inflated lower side and reddish pigmentation on the upper side. River Noerreaa, Randers, Denmark. Photo JCS.

Characteristics and similar species:
In its characteristic form, with inflated lower side, *L. gibba* is unmistakable, but the species is not always as thick as is stated in most floras. Flat specimens, as shown in the photo on p. 90 at the bottom right, can be mistaken for **22 *L. turionifera***, **23 *L. ×japonica*** or **19 *L. minor***, but should be recognised by the 10–20, 0.3–0.7 mm wide, transparent air spaces on the lower side. **26 *Spirodela polyrhiza*** and **27 *Spirodela oligorrhiza*** have more than one root from each frond.

The liverwort ***Ricciocarpos natans*** (L.) Corda has a dichotomously divided thallus - see photo p. 93 (bottom).

Slightly inflated form of *L. gibba* seen from below. Note the large air spaces. Photo KvdW.

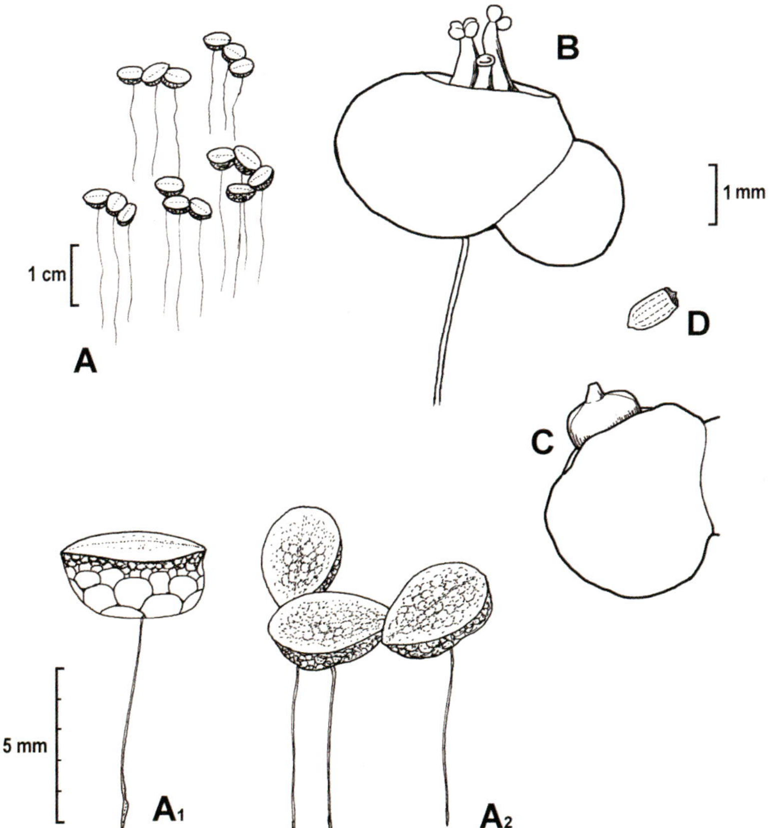

Lemna gibba
A, A1, A2 habit, **B** flowering plant, **C** fruiting plant, **D** seed.

Araceae

19 *Lemna minor* L.

DK: Liden Andemad **N:** Andemat **S:** Andmat **FIN:** Pikkulimaska **IS:** -
GB: Common Duckweed **NL:** Klein kroos **F:** Petite lentille-d'eau
D: Kleine Wasserlinse **CZ:** Okřehek menší **PL:** Rzęsa drobna
EST: Väike lemmel **LV:** Mazais ūdenszieds **LT:** Mažoji plūdena **Ru:** Ряска малая

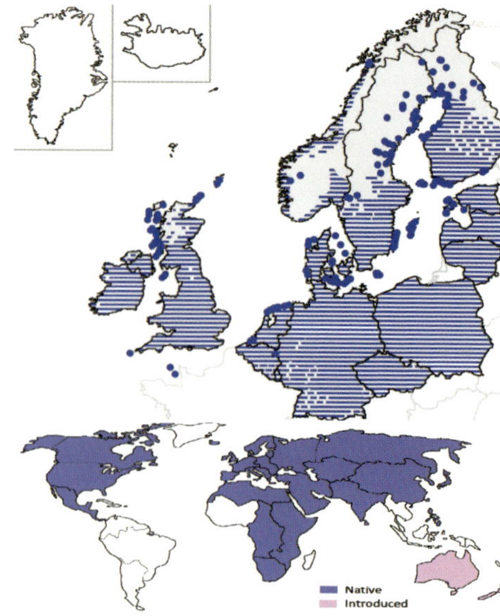

Flowering *L. minor*. One of the two stamens has already released its pollen. Hobro, Denmark. Photo JCS.

Perennial, floating hydrophyte. With 1 root, 1–4(–10) cm long from each frond. The root with an obtuse cap, 0.4–1.0 mm long.

Frond: Solitary or adhering in small groups, broadly ovate or elliptic, (2–)3–5 mm long and up to 3 mm wide, with 3–5 veins. Upper side slightly arched, light green or with reddish pigmentation, with or without a row of papillae of varying sizes, the most prominent over the node and at the apex. Lower side slightly arched, light green and never reddish. Lower side with a tissue of aerial chambers 0.10–0.25 mm wide - seen in a magnifying glass as a mesh of 40–60 cells.

Inflorescence: Monoecious. Much reduced and very tiny, enclosed within a membranous spathe. With only one female and one male flower borne from (1–)2 hollows on the margin of the frond. Flowering rather rare.

Flowers: Male flower with 2 stamens. Female flower more or less bottle shaped with a funnel-shaped stigma.

Fruits: Elliptical, 0.8–1.0 mm long, winged upwards, wings 0.05–0.10 mm wide. The fruit contains only 1 seed, which is ornamented with 12–15, longitudinal ribs.

Flowering: June-July.

Biology: Propagation by budding from pouches in the margins of the frond, sometimes by seed. *Lemna minor* does not form turions, but overwinters as small, starchy fronds. Often forming dense carpets and often with other species of *Lemna*. Dispersal by water and birds.

Habitats: In mesotrophic to eutrophic, still or very slow-flowing water in ditches, canals, streams and ponds.

Distribution within the region:
Native. Common throughout most of the area, but absent from Iceland and Greenland.

Characteristics and similar species:
Lemna minor is recognised by the flat frond with a light green lower side which never has red pigmentation and

Lemna minor seen in backlight from below. Note the 3-5 veins. Hobro, Denmark. Photo JCS.

Fruiting *L. minor*. Note the wing on the upper part of the fruit. Lower left: The mature seed released from the fruit sac. Hobro, Denmark. Photo JCS.

A variegated population with green and red pigmented plants of *L. minor*. The lower side of all plants remains green. Ny Frederikskog, SV-Jutland, Denmark. Photo JCS.

with a tissue of air chambers 0.10–0.25 mm wide - seen in a magnifying glass as a mesh of 40–60 cells. If a row of papillae is visible on the upper side, they are unequal in size - the terminal ones are larger.

20 ***Lemna minuta*** is generally smaller. Frond 0.8–3.0(–4.0) mm long, with a regular outline and a pale line along the centre of the upper side.

22 ***Lemna turionifera*** and **23** ***L. ×japonica*** have some reddish pigmentation on the lower side of the frond, at least around the root base, and a row of more or less equal papillae on the upper side.

18 ***Lemna gibba*** has different front outline and 10–20, 0.3–0.7 mm wide air cells on the lower side. Lower side sometimes swollen.

26 ***Spirodela polyrhiza*** has several roots from each frond.

The liverwort ***Ricciocarpos natans*** (L.) Corda has a fringe of purplish-black pendent scales and a dichotomously divided thallus - see photo (right).

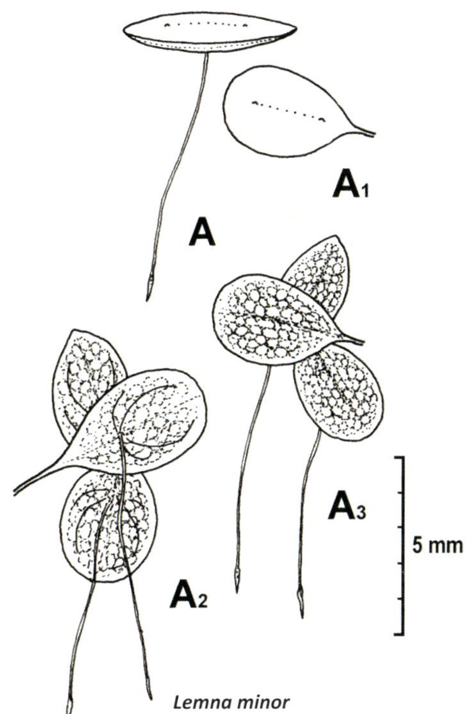

Lemna minor
A, A1 habit with row of papillae,
A2, A3 habit upper and lower side.

Lemna minor and the liverwort *Ricciocarpus natans* in shallow water at Djurle Vaatmark, Sweden. Because of its small size and floating thallus *R. natans* is sometimes mistaken for *Lemna* species, but its thallus is dichotomously divided. Photo JCS.

Araceae

20 *Lemna minuta* Kunth

Syn.: *L. minuscula* Herter nom. illegit.; *L. minima* Phil.
DK: Enstrenget Andemad **N:** - **S:** Kölandmat **FIN:** - **IS:** -
GB: Least Duckweed **NL:** Dwergkroos **F:** Lentille-d'eau minuscule
D: Zierliche Wasserlinse **CZ:** Okřehek nejmenší **PL:** -
EST: Pisi lemmel **LV:** - **LT:** - **Ru:** Ряска мелковатая

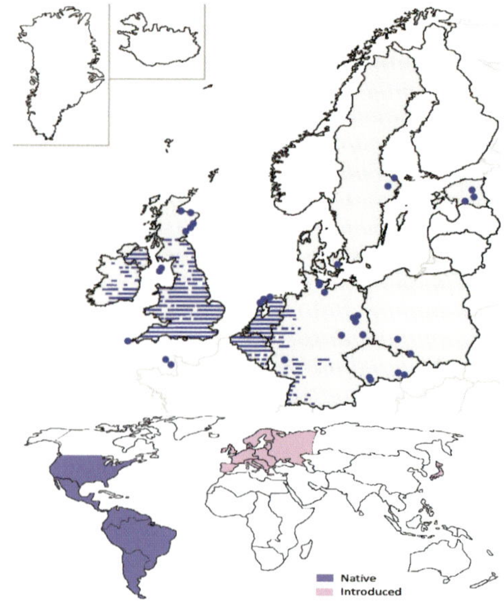

Lemna minuta and a few *L. minor*. Specimens of *L. minuta* growing in sunlit sites are generally smaller (1–2 mm long) than specimens from shady sites (up to 3–4 mm long). Germany. Photo KvdW.

Perennial, floating hydrophyte. With 1 root, up to 1.5 cm long from each frond. The root with a rounded or somewhat acute cap.

Frond: Solitary or adhering in small groups, obovate to elliptical, 0.8–4.0 mm long, 0.5–2.5 mm wide, 1–1.6 times as long as wide, flat to slightly thickened, somewhat translucent, pale green and never with reddish tones (anthocyanin), margins entire. With 1 often rather indistinct vein, not more than 2/3 of the distance from the root to the apex. Upper side often with a low ridge appearing as a pale line, with or without small papillae. Lower side with a tissue of aerial chambers much smaller than 0.3 mm.

Inflorescence: Monoecious. Much reduced and very tiny, enclosed within a membranous spathe. With only one female and one male flower borne from (1–)2 hollows on the margin of the frond. Flowering rather rare.

Flowering *L. minuta*. Photo Wim van der Ven.

Flowers: Male flower with 2 stamens. Female flower more or less bottle shaped with a funnel-shaped stigma.

Fruits: 0.6–1.0 mm long, unwinged. The fruit contains only 1 seed with 12–15 ribs.

Flowering: June-July.

Biology: Propagation by budding from pouches on the margins of the frond, sometimes by seed. *Lemna minuta* does not form turions, but overwinters as small, starchy fronds. Often forming extensive, dense carpets, completely covering entire water bodies and mixed with other species of *Lemna*. Dispersal by water, birds and human activities.

Habitats: In mesotrophic to slightly eutrophic, still or slow-flowing water in ditches, canals, streams and ponds.

Distribution within the region:
Alien. Introduced and naturalised in Europe since 1965. Ireland, Britain, Netherlands, Belgium, Luxembourg, Germany, Poland, the Czech Republic, Sweden, Denmark and Estonia.

Characteristics and similar species: *Lemna minuta* is easily confused with other species of *Lemna*, but is recognised by the regular frond outline, translucence and often the pale line along the centre on the upper side. The single vein is hard to discern even when backlit.

Lemna minuta. Upper side. The faint ridge on the upper side is visible, whereas the vein is seen only in backlight or on decaying material. Photo KvdW.

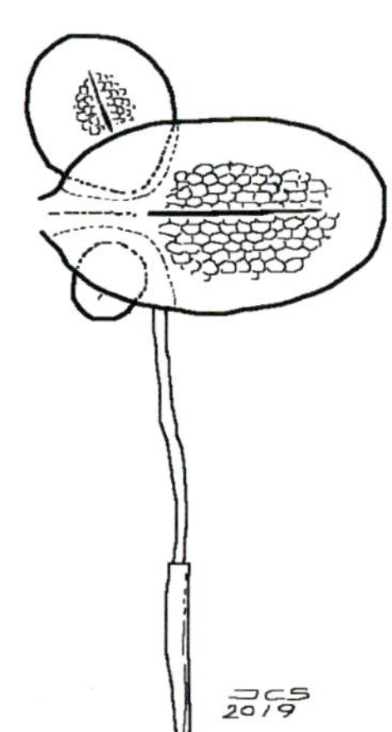

Lemna minuta habit. Note the short, single vein.

21 *Lemna aequinoctialis* Welw.

DK: - **N:** - **S:** - **FIN:** - **IS:** -
GB: Tropical Duckweed **NL:** - **F:** -
D: Weniggerippte Wasserlinse **CZ:** - **PL:** -
EST: - **LV:** - **LT:** - **Ru:** Ряска тропическая

Lemna aequinoctialis mixed with *Wolffia* cf. *globosa*. Note the long, narrow fronds. Botanical Garden Heidelberg, Germany. Photo KvdW.

Perennial, floating hydrophyte. With 1 root, up to 3 cm long from each frond. Root sheath winged at base, tip of cap acute.

Frond: Solitary or together in small groups, ovate, 1–6 mm long, 1–3 times as long as wide, flat, green and never with reddish tones (anthocyanin), margins entire. With 3 veins. Upper side with a row of papillae, the most prominent above the root attachment and near the apex. Lower side with a tissue of aerial chambers much smaller than 0.3 mm.

Lemna aequinoctialis. From above. Photo KvdW.
Side view. Photo Wim van der Ven.

Inflorescence: Monoecious. Much reduced and very tiny, enclosed within a membranous spathe. With only one female and one male flower borne from (1–)2 hollows on the margin of the frond. Flowering unknown in Europe.

Flowers: Male flower with 2 stamens. Female flower more or less bottle shaped with a funnel-shaped stigma.

Fruits: Elliptical, 0.5–0.8 mm long, without wings. The fruit contains only 1 seed with 8–26 ribs.

Flowering: June-July.

Biology: Propagation by budding from pouches in the margins of the frond, and by seed. *Lemna aequinoctialis* does not form turions, but overwinters as small, starchy fronds. Usually in large numbers. Dispersal by water, birds and human activities.

Habitats: In mesotrophic to eutrophic, still or very slow-flowing water in canals, streams and ponds.

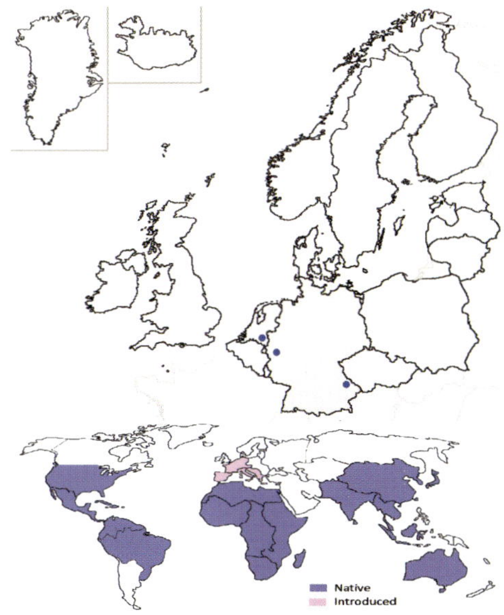

Distribution within the region:
Alien. Introduced and rare, locally naturalised in the Netherlands and Germany.

Characteristics and similar species:
Lemna aequinoctialis is recognised by the winged base of the root sheath.

Note: Two *Lemna* species have a winged root sheath, *L. aequinoctialis* and *L. perpusilla* Torr., which can only be distinguished by the number of ribs on the seeds. However, populations of *Lemna* with winged root sheaths in Europe have never been found flowering and so they have not been critically determined. They are referred to here as *L. aequinoctialis* as their discovery was published under that name.

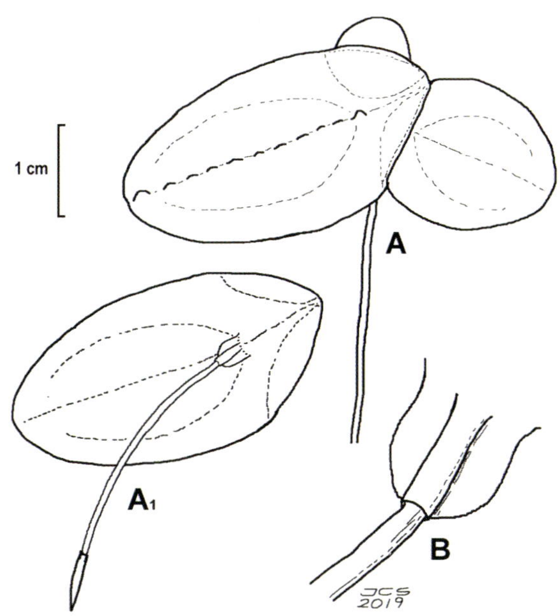

Lemna aequinoctialis.
A habit upper side, **A1** habit lower side,
B base of root sheath.

Araceae

22 *Lemna turionifera* Landolt

DK: Rød Andemad **N:** Strengandemat **S:** Röd andmat **FIN:** Itulimaska **IS:** -
GB: Red Duckweed **NL:** Knopkroos **F:** Lenticule à turion
D: Rote Wasserlinse **CZ:** Okřehek červený **PL:** Rzęsa turionowa
EST: Punatäpp-lemmel **LV:** Vairvasiņu ūdenszieds **LT:** - **Ru:** Ряска туриононосная

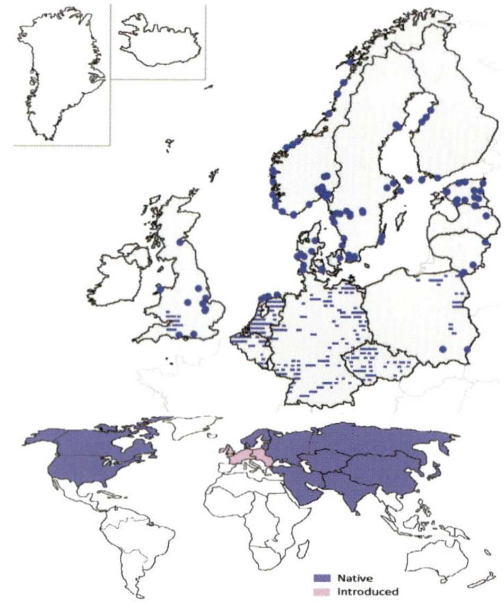

Lemna turionifera with few *Spirodela polyrhiza*. River Thames at Windsor, Britain. Photo RVL.

Perennial, floating hydrophyte. With 1 root, 1–4(–10) cm long from each frond. The root with an obtuse cap, 0.4–1.0 mm long.

Frond: Solitary or in small groups, elliptical to ovate, (1.5–)2–3(–4) mm long and up to 3 mm wide, with 3(–5) veins. Upper side slightly arched, olive green and often with reddish pigmentation, with a row of papillae of equal size. Lower side slightly arched, light green and always with some reddish pigmentation at least in an area around the root base. Lower side with a tissue of 0.1–0.3 mm wide aerial chambers, seen in a magnifying glass as a mesh of 40–60 cells.

Inflorescence: Monoecious. Much reduced and very tiny, enclosed within a membranous spathe. With only one female and one male flower borne from (1–)2 hollows on the margin of the frond. Flowering rare.

Flowers: Male flower with 2 stamens. Female flower more or less bottle shaped with a funnel-shaped stigma.

Fruits: Elliptical, 0.5–0.6 mm long, unwinged, containing only 1 seed. The seed remains in the fruit sac after maturing.

Flowering: June-July.

Biology: Propagation by budding from pouches at the margins of the frond, rarely by seed. *Lemna turionifera* is the only species within the genus to form turions. The turions are small, 0.8–1.6 mm long, elliptical, olive green to red brown, starchy, and without roots. They are heavier than water and sink to overwinter. Dispersal by water, birds and human activities.

Habitats: In mesotrophic to eutrophic, still or very slow-flowing water in ditches, canals, streams and ponds, also in backwaters of larger rivers.

Distribution within the region:
Native to temperate areas of Northern Europe. Presumed introduced to the Netherlands, Belgium, Germany, Poland and Czech Republic.

Characteristics and similar species:
Lemna turionifera resembles **19 *L. minor*** and **20 *L. minuta*,** but differs from these by the reddish pigmentation on the lower side and the row of papillae

Flowering *L. turionifera* in early July 2015. The distinct row of papillae more or less equal in size is visible with the naked eye, but are better seen by use of a magnifying glass and obliquely incident light. Ny Frederikskog, SW-Jutland, Denmark. Photo JCS.

Left: *Lemna turionifera* seen from below. Note the purplish-brown turion in the upper left corner and the reddish pigmentation spreading from the attachment point of the root. Right: *Lemna minor* seen from below. The lower side shows no reddish pigmentation at all. Photo JCS.

equal in size on the upper side. Late in the year the reddish pigmentation is sometimes missing and identification must then be based on the row of equal papillae and presence of turions.

18 ***L. gibba*** has a tissue of 10–20, 0.3–0.7 mm wide translucent cells on the lower side of the frond. The lower side rarely has reddish tones and if so the colouration spreads from the margin, not from the root base.

For separation from **23** ***L. ×japonica*** - see below. **26** ***Spirodela polyrhiza*** has several roots from each frond.

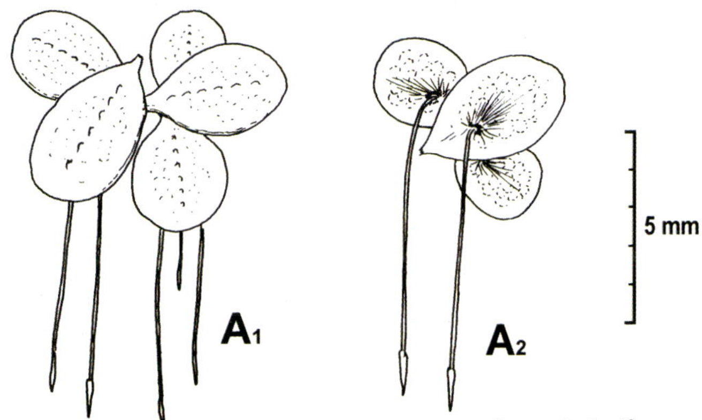

Lemna turionifera
A1, A2 habit upper and lower side.

Araceae

23 *Lemna ×japonica* Landolt

DK: Japansk Andemad **N:** Japanandemat **S:** Japansk andmat **FIN:** - **IS:** -
GB: - **NL:** - **F:** -
D: - **CZ:** - **PL:** -
EST: - **LV:** - **LT:** - **Ru:**

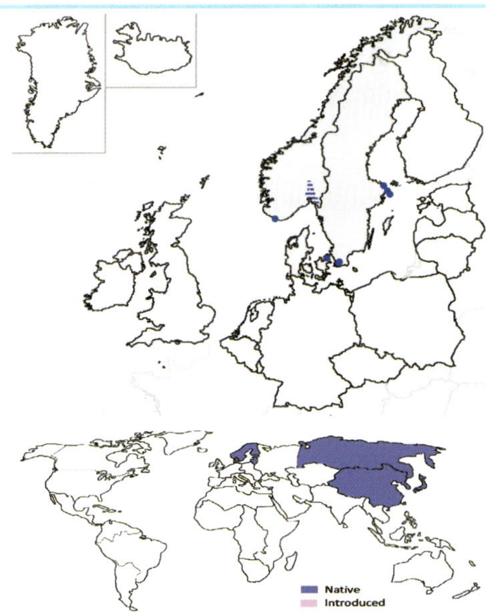

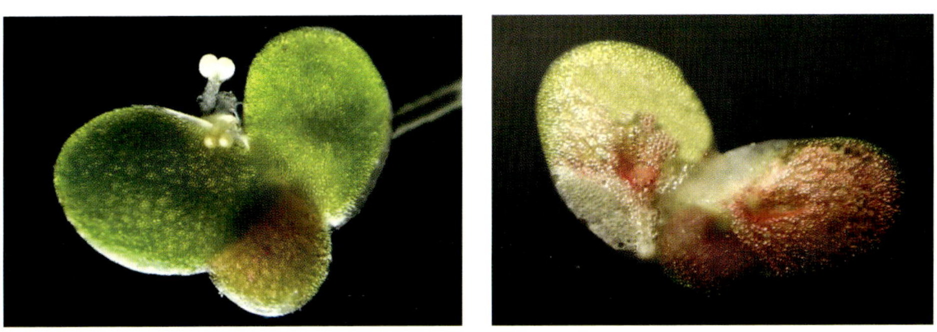

Lemna ×japonica. Left: Flowering frond, upper side. Right: Lower side of frond with red colours. Lake Skenkelsø, Sjælland, Denmark. Photos JCS.

Morphologically resembles a combination of **19** ***L. minor*** and **22** ***L. turionifera*** and is of hybrid origin. The upper side is similar to *L. minor*, with a number of small, sometimes indistinct papillae of different sizes - the largest at the apex and above the root base. The lower side resembles *L. turionifera* with varying amounts of reddish pigmentation.

Distribution within the region:
Presumed native in Norway, Sweden and Denmark.

Characteristics and similar species:
Lemna ×japonica differs from *L. turionifera* by the papillae on the upper side being more indistinct and of a different size. *Lemna minor* always lacks reddish pigmentation on the lower side.

Araceae

24 *Lemna valdiviana* Phil.

DK: - **N:** - **S:** - **FIN:** - **IS:** -
GB: Valdivia Duckweed **NL:** - **F:** -
D: - **CZ:** - **PL:** -
EST: - **LV:** - **LT:** - **Ru:** -

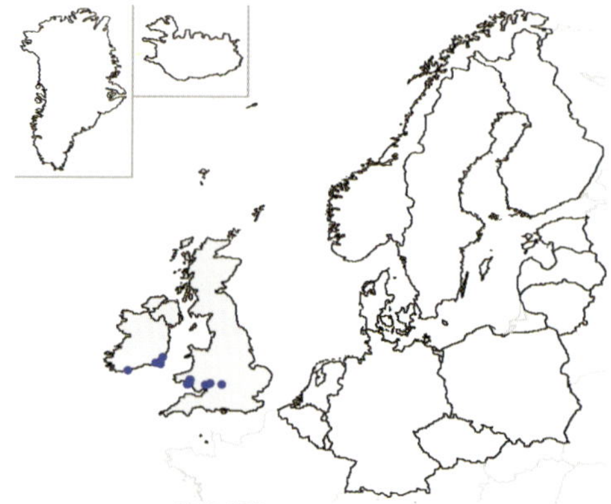

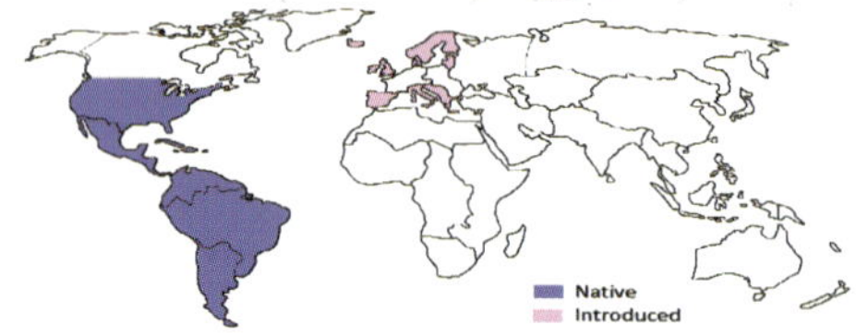

Lemna valdiviana. Ornamental pond, Cotswold Wildlife Park, Burford, Oxfordshire, Britain. Photo RVL.

Perennial, floating hydrophyte. With 1 root, up to 1.5 cm long from each frond. The root with a rounded or somewhat acute cap.

Frond: Usually adhering in groups of 4(–7), ovate to elliptical, usually asymmetrical, 1.7–2.6(–4.0) mm long, 0.6–1.0(–1.6) mm wide, 1.3–3 times as long as wide, very thin, translucent, pale green sometimes with some whitish cells, margins entire, often hyaline. With 1 vein, typically more than 3/4 of the distance from root base to the shoot apex, and as long as or extending beyond the tissue of air chambers on the lower side. Air chambers much smaller than 0.3 mm. Upper side with or without small papillae.

Inflorescence: Monoecious. Much reduced and very tiny, enclosed within a membranous spathe. With only one female and one male flower borne from (1–)2 hollows on the margin of the frond. Flowering very rare.

Flowers: Male flower with 2 stamens. Female flower more or less bottle shaped with a funnel-shaped stigma.

Fruits: 1.0–1.35 mm long, unwinged. The fruit contains only 1 seed with 15–29 ribs.

Flowering: June-July. To our knowledge flowering is not noticed in Europe.

Biology: Propagation by budding from pouches on the margins of the frond, rarely by seed. *Lemna valdiviana* does not form turions, but overwinters as starchy fronds. Often forming extensive, dense carpets, completely covering entire water bodies and mixed with other species of *Lemna.* The fronds float, sometimes just below the water-surface, particularly in winter. Dispersal by water, birds and human activities.

Habitats: In mesotrophic to eutrophic, still or slow-flowing water. In ornamental ponds in wildlife parks and botanical gardens, in flooded gravel pits and slow-flowing areas of streams and rivers.

Distribution within the region:
Alien. Introduced and naturalised in Britain and Ireland.

Lemna valdiviana. Cotswold Wildlife Park, Burford, Oxfordshire, Britain. Photo RVL.

Lemna valdiviana together with *L. minor* and a single frond of *Spirodela polyrrhiza*. Ornamental pond. Twigworth, Gloucestershire, Britain. Photo RVL.

Characteristics and similar species:
Lemna valdiviana is easily confused with **20 *Lemna minuta.*** The most reliable character for separation of the two species is the relative length of the vein in relation to the distance between the root base and the apex. While the vein is often obscure in *L. minuta*, it is usually more prominent in *L. valdiviana*. The shape of the frond, symmetrical or not, is another, less reliable character. This species is often first noticed because of the tendency of plants to adhere in groups of four.

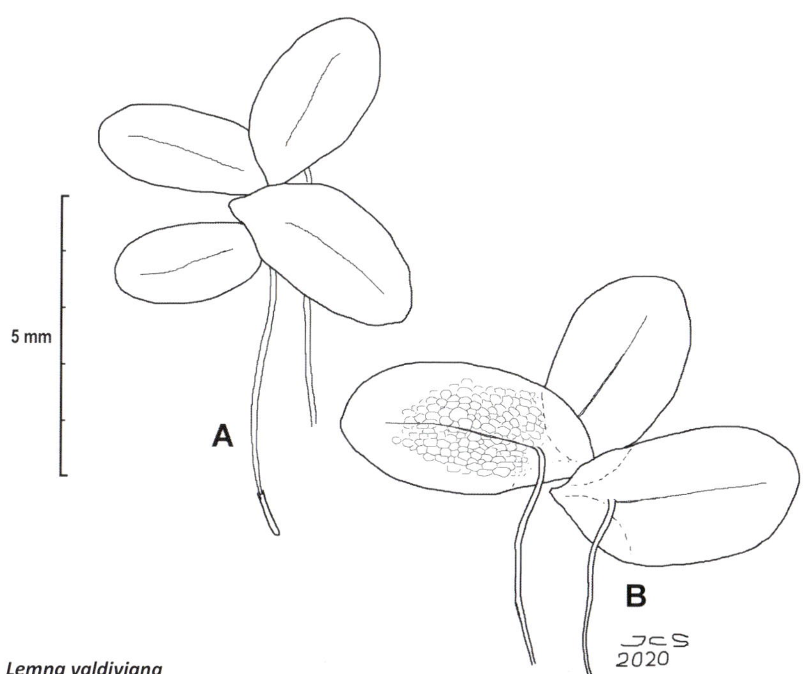

Lemna valdiviana
A habit upper side, **B** habit lower side.

Araceae

25 *Lemna trisulca* L.

DK: Kors-Andemad **N:** Krossandemat **S:** Korsandmat **FIN:** Ristilimaska **IS:** -
GB: Ivy-leaved Duckweed **NL:** Puntkroos **F:** Lentille d'eau à trois sillons
D: Dreifurchen-Wasserlinse **CZ:** Okřehek trojbrázdý **PL:** Rzęsa trójrowkowa
EST: Ristlemmel **LV:** Trejdaivu ūdenszieds **LT:** Trilypė plūdena **Ru:** Ряска тройчатая

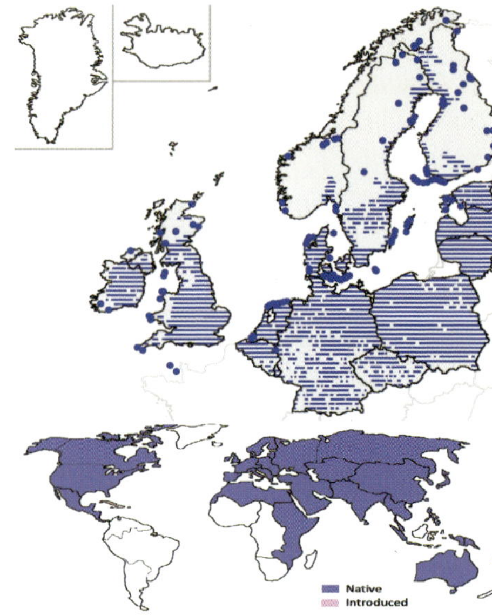

Submerged *L. trisulca* in a ditch south of Hoejer, SV-Jutland, Denmark. Photo JCS.

Perennial hydrophyte typically suspended beneath the water surface, usually only flowering shoots floating at the surface. With or without 1 root, 1–4 cm long from each frond. Root tip with an acute cap.

Frond: Vegetative shoots 3–50 together in cross-like, branched chains. Fronds elongate lanceolate, 2–8(–10) mm long and up to 4 mm wide, abruptly narrowed at base to a 2–20 mm long stalk, 1–3 veins, entire, or occasionally finely toothed towards the tip, transparent, light to grey green, upper side with fine papillae, sterile shoots without stomata.
Fertile shoots 1–3 together detached from the mother plant, not branching, ovate, shorter and thicker than sterile shoots, with serrate margin and shorter stipe, with stomata on the upper surface.

Inflorescence: Monoecious. Much reduced and very tiny, enclosed within a membranous spathe. With only one female and one male flower borne from hollows on the margins of specialised shoots floating on the water surface. Flowering rather rare, initiated by long days.

Flowers: Male flower with 2 stamens. Female flower more or less bottle shaped with a funnel-shaped stigma.

Fruits: Elliptical, 0.6–0.9 mm long, with 0.15 mm wide wings in the upper part. The fruit contains only 1 seed, about 1 mm long with 12–15 ribs.

Flowering: June-July.

Biology: Propagation by budding from pouches in the margins of the frond, rarely by seed. *Lemna trisulca* can occur as very large populations which may form mats at the surface. Plants sink to the bed in the winter, rising again in the spring. It does not form turions. Dispersal by water, birds and human activities.

Habitats: In mesotrophic to eutrophic, still or very slow-flowing water in ditches, canals, streams and ponds, also in lakes or in oligotrophic water.

Distribution within the region:
Native. Throughout most of the area, but scattered in the northern part and absent from Iceland and Greenland.

Characteristics and similar species: Apart from the very rare and introduced *Lemna valdiviana. L. trisulca* is

Submerged *L. trisulca* showing the characteristic cross-like branching from the two pouches on each side of the frond. Ny Frederikskog, Hoejer, SV-Jutland, Denmark. Photo JCS.

Flowering *L. trisulca*. The fronds are shorter and thicker than the sterile, submerged plants and characteristically curved to lift the flower free of the water. Pollen is released from one anther at a time and transported to the stigma by water movement. Photo KvdW.

the only *Lemna species* in the area which remains suspended in the water column.

The long-stalked lanceolate fronds are finely toothed towards the tip, and with its cross-like branching unlike other aquatic plants.

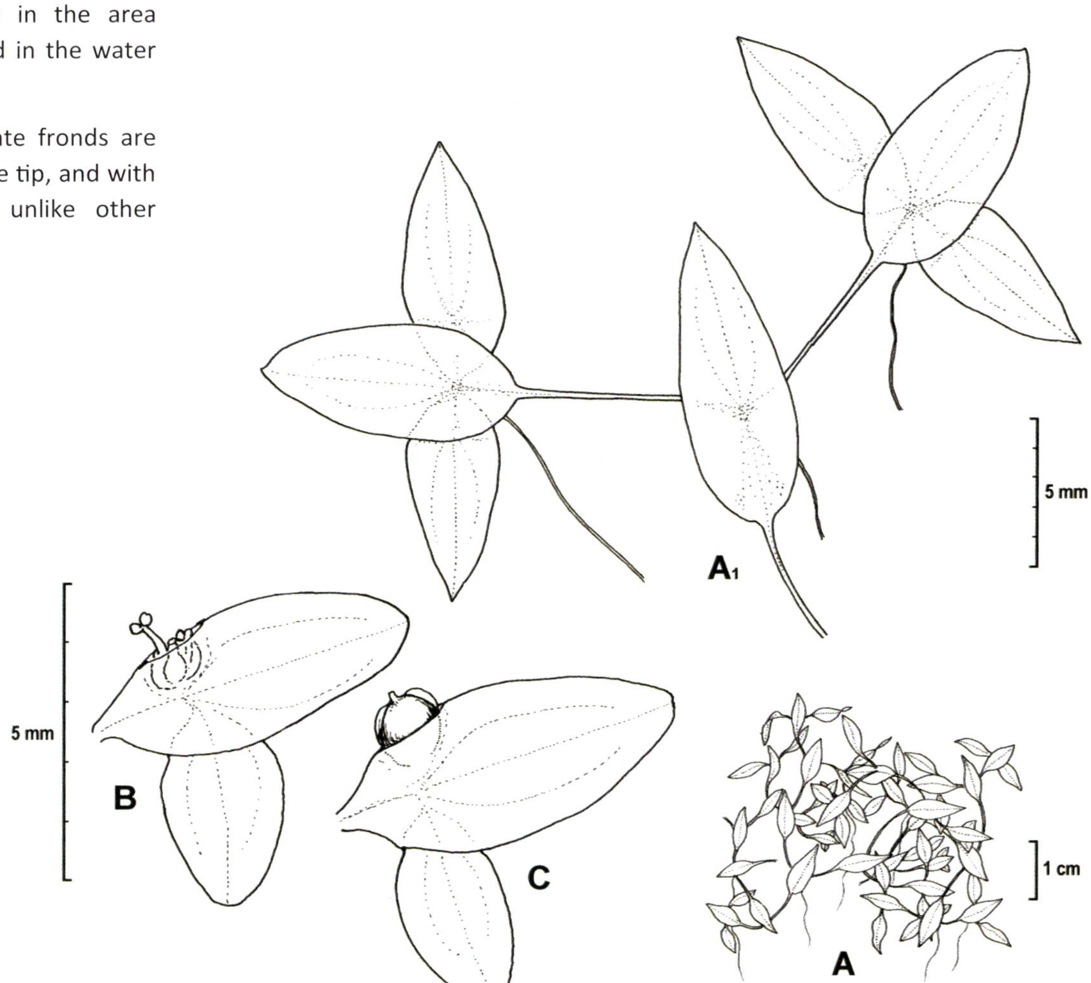

Lemna trisulca
A, A1 habit sterile plant, **B** flowering fertile shoot,
C flowering fertile shoot.

Araceae

26 *Spirodela polyrhiza* (L.) Schleid.

Syn.: *Lemna polyrhiza* L.

DK: Stor Andemad **N:** Stor andemat **S:** Stor andmat **FIN:** Isolimaska **IS:** -
GB: Greater Duckweed **NL:** Veelwortelig kroos **F:** Lentille d'eau à plusieurs racines
D: Vielwurzelige Teichlinse **CZ:** Závitka mnohokořenná **PL:** Spirodela wielokorzeniowa
EST: Hulgajuurine vesilääts **LV:** Parastā spirodela **LT:** Daugiašaknė maurė **Ru:** Многокоренник обыкновенный

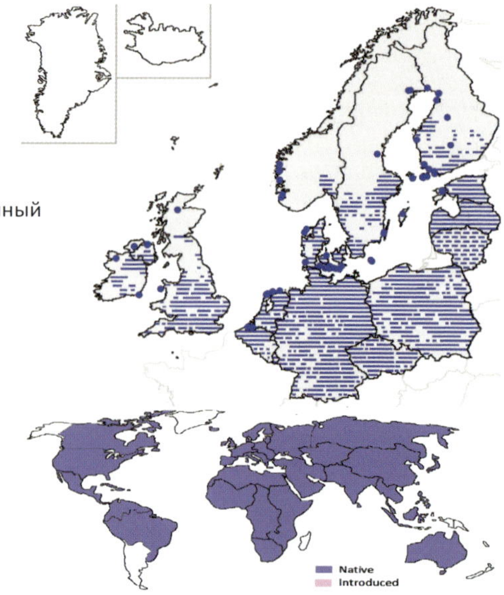

Spirodela polyrhiza. Lake Tange, Ans, Jutland, Denmark. Photo JCS.

Perennial, floating hydrophyte. With 5–15 roots from each frond, but only 1 of the roots perforating the scale-like leaflet (prophyllum), root cap acute.

Fronds: Solitary or 2–5 attached, almost circular to obovate, asymmetrical at base, (3–)5–8(–12) mm long and up to 8 mm wide, with (5–)7–12(–15) veins. Upper side slightly domed, often with a row of papillae along the mid-vein and eventually along the side veins, yellowish green to green, sometimes with reddish to purple pigmentation (anthocyanin). Lower side more or less flat, mostly with red to purple pigmentation or greenish.

Inflorescence: Monoecious. Much reduced and very tiny, enclosed within a membranous spathe. With only one female and one male flower borne from (1–)2 hollows on the margin of the frond. Flowering rare.

Flowers: Male flower with 2 stamens. Female flower more or less bottle shaped with a funnel-shaped stigma.

Fruits: Elliptical, 1.0–1.5 mm, with 0.15 mm wide wings in the upper part. The fruit contains 1(–2), 1.0–1.5 mm long seeds with 12–20 distinct ribs.

Flowering: June-July.

Biology: Propagation by budding from pouches in the margins of the frond, rarely by seeds. Usually in large numbers and often with species of *Lemna*. Late in the year flat, starchy, 1–3 mm long, reniform, brownish turions without roots are formed. They are heavier than water and sink to the bottom to overwinter. See photo on opposite page. Dispersal by water and birds.

Habitats: In nutrient-rich, typically eutrophic, still or very slow-flowing water in ditches, canals, ponds, lakes and streams.

Distribution within the region:
Native. Common throughout most of the area, scattered or absent in northern Norway, Sweden and Finland. Absent from Iceland and Greenland.

Characteristics and similar species:
Spirodela polyrhiza differs from **18-24 *Lemna*** species in size, number of veins and 5–15 roots from each frond.

28 *S. punctata* differs in having 3–5 roots perforating the scale-like leaflet.

27 *S. oligorrhiza* differs in having all roots penetrating the scale-like leaflet, narrower frond outline and by a row of papillae along the midvein on the upper side - the terminal papillae most prominent.

Spirodela polyrhiza - view from below. New shoots are formed from pouches at the margins of the mother plant. Under unfavourable conditions or late in the year, starchy, specialised shoots without roots (turions) are produced for overwintering. Fladbro, Randers, Denmark. Photo JCS.

Spirodela polyrhiza. Plants with purple pigmentation on the upper side. Randers, Denmark. Photo JCS.

Spirodela polyrhiza. Close up lower side of frond showing only 1 root perforating the scale-like leaflet. Balkaakra, Skaane, Sweden. Photo Jan-Thomas Johansson.

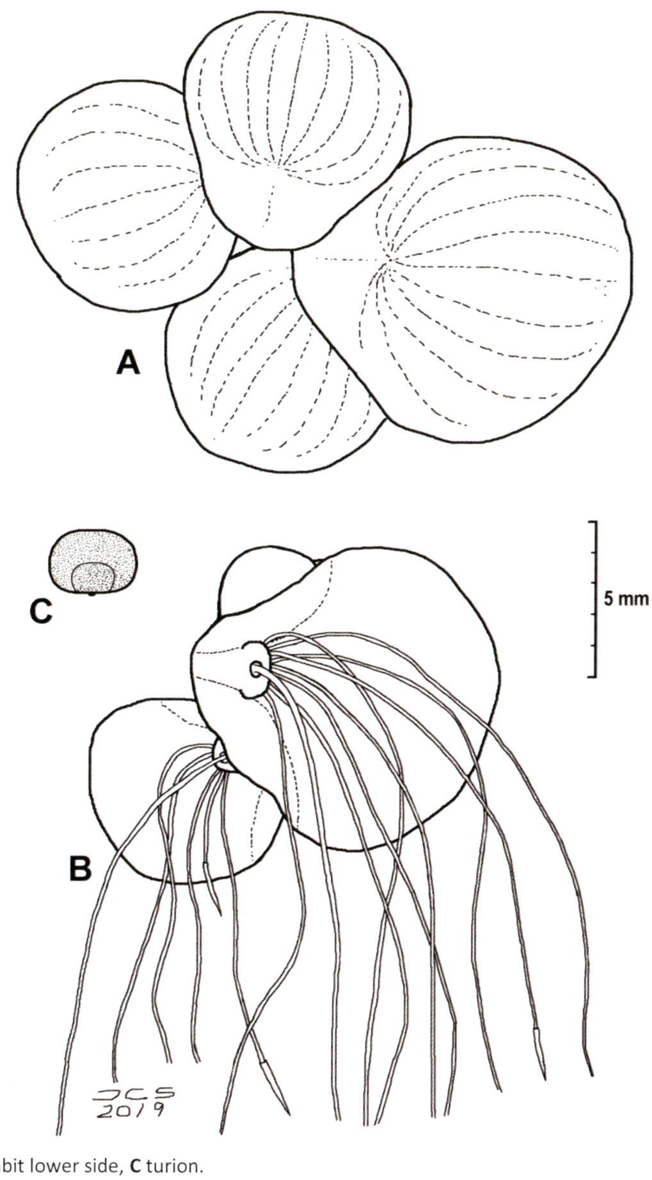

Spirodela polyrhiza
A habit upper side, **B** habit lower side, **C** turion.

Araceae

27 *Spirodela oligorrhiza* (Kurz) Hegelm.

Syn.: *Landoltia punctata* (G.Mey.) Les & D.J.Crawford; *S. punctata* auct. non (G.Mey.) C.H.Thomps.
DK: - **N:** - **S:** Prickandmat **FIN:** - **IS:** -
GB: - **NL:** Smal kroos **F:** -
D: Geringwurzelige Teichlinse **CZ:** - **PL:** -
EST: - **LV:** - **LT:** - **Ru:** -

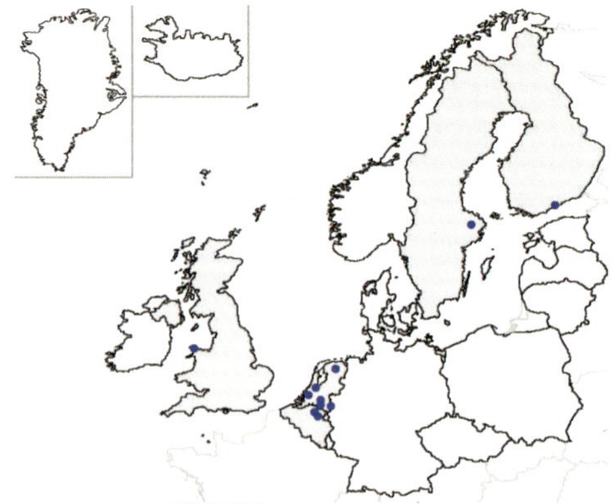

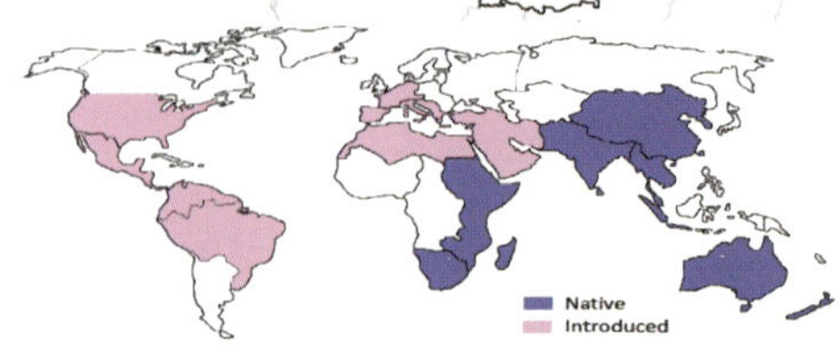

Spirodela oligorrhiza - upper and lower side. Photo Wim van der Ven.

Perennial, floating hydrophyte. With 2–7(–12) roots from each frond, all penetrating the scale-like leaflet (prophyllum), root cap acute.

Fronds: Solitary or 2–6 together in small groups, obovate to elliptical, 1.5–8.0 mm long, 1.5–2 times as long as wide, with 3–7 veins. Upper side slightly domed and always with a row of papillae along the mid-vein, green to olive-green, mostly with red-coloured margins. Lower side flat or somewhat domed, often with some red to purple pigmentation or greenish.

Inflorescence: Monoecious. Much reduced and very tiny, enclosed within a membranous spathe. With only one female and one male flower borne from (1–)2 hollows on the margin of the frond. Flowering rare.

Flowers: Male flower with 2 stamens. Female flower more or less bottle shaped with a funnel-shaped stigma.

Fruits: Elliptical, 0.8–1.0 mm, laterally winged. Seed with 10–15 distinct ribs.

Flowering: No information.

Biology: Propagation by budding from pouches in the margins of the frond, rarely by seed. *S. oligorrhiza* does not produce turions. Dispersal is most likely to be by water and birds.

Habitats: In nutrient-rich, usually eutrophic, still or very slow-flowing water. In Europe mainly as weed in containers in garden centres and ponds.

Distribution within the region:
Introduced and naturalised in Netherlands, Belgium, Sweden and Finland.

Characteristics and similar species:
Spirodela oligorrhiza differs from **18-24 *Lemna*** species in number of veins and 2–7 roots from each frond.

It is similar to **28 *S. punctata*** and **26 *S. polyrhiza*** but differs in having all roots perforating the scale-like leaflet, the frond outline and by the row of papillae along the midvein on the upper side - of which the terminal papule is most prominent.

Note: There has been considerable confusion over the correct names which should be applied to the two alien *Spirodela* species which have been recorded in Europe. The correct nomenclature was clarified by Ward (2011), including neotypification of the name *Spirodela punctata* (G.Mey.) C.H.Thomps. This account employs the names in the sense that they were used by Ward (2011).

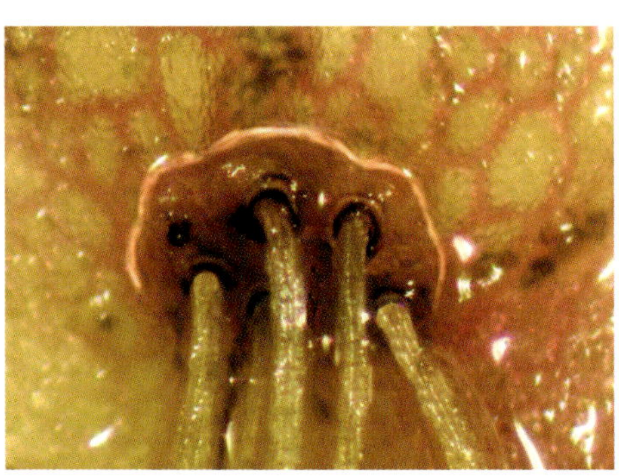

Spirodela oligorrhiza. Close-up of lower frond side showing all roots perforating the scale-like leaflet. Photo KvdW.

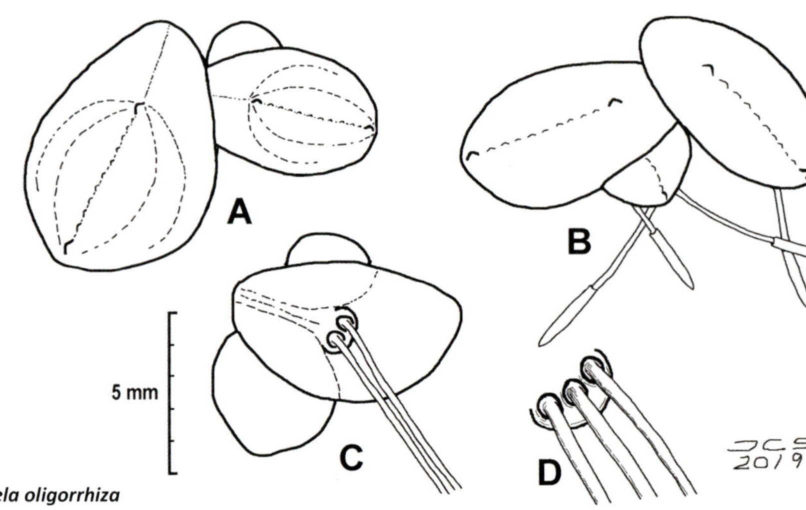

Spirodela oligorrhiza
A, B habit upper side, **C** habit lower side, **D** scale-like leaf perforated by all the roots.

Araceae

28 *Spirodela punctata* (G. Mey.) C.H.Thomps.

Syn.: *S. intermedia* W.Koch
DK: - **N:** - **S:** Mellanandmat **FIN:** - **IS:** -
GB: - **NL:** - **F:** -
D: - **CZ:** - **PL:** -
EST: - **LV:** - **LT:** - **Ru:** -

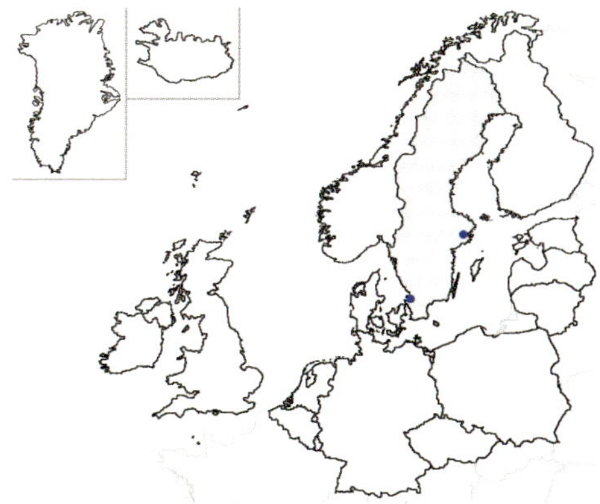

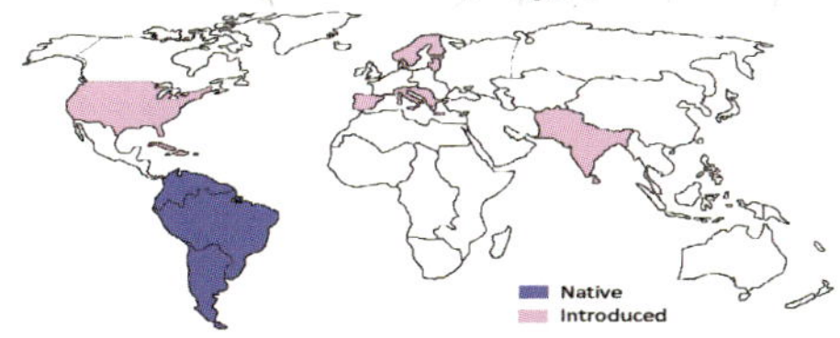

Spirodela punctata. Cultivated. Sweden. Photo JCS.

Spirodela punctata. Close-up of lower frond side showing some roots perforating the scale-like leaflet. Photo JCS.

Perennial, floating hydrophyte. With 6–20 roots, 0.7–3.5 cm long from each frond, 3–5 of them penetrating the scale-like leaflet (prophyllum), root cap 0.8–1.3 mm long, acute.

Fronds: Solitary or 2–7 together in small groups, subrotund to elliptical, asymmetrical at base, 3–12 mm long, 1–1.3 times as long as wide, with 3–7 veins. Upper side flat to slightly domed, with numerous small papillae, green to olive green, often with brownish-coloured margins. Lower side flat or somewhat domed, normally with brownish pigmentation.

Inflorescence: Monoecious. Much reduced and very tiny, enclosed within a membranous spathe. With one female and one male flower borne from (1–)2 hollows on the margin of the frond. Flowering rare.

Biology: Flowers or fruits not seen.

Habitats: In nutrient-rich, usually eutrophic, still or very slow-flowing water. Previous European finds are restricted to greenhouse tanks.

Distribution within the region:
Introduced in Sweden, but not naturalised.

Characteristics and similar species: Similar to **26 *S. polyrhiza*** but differs in having 3–5 roots penetrating the scale-like leaflet.

27 *S. oligorrhiza* is also somewhat similar, but differs in all roots penetrating the scale-like leaflet, narrower frond outline and by having a row of papillae along the midvein on the upper side - the terminal papule most prominent.

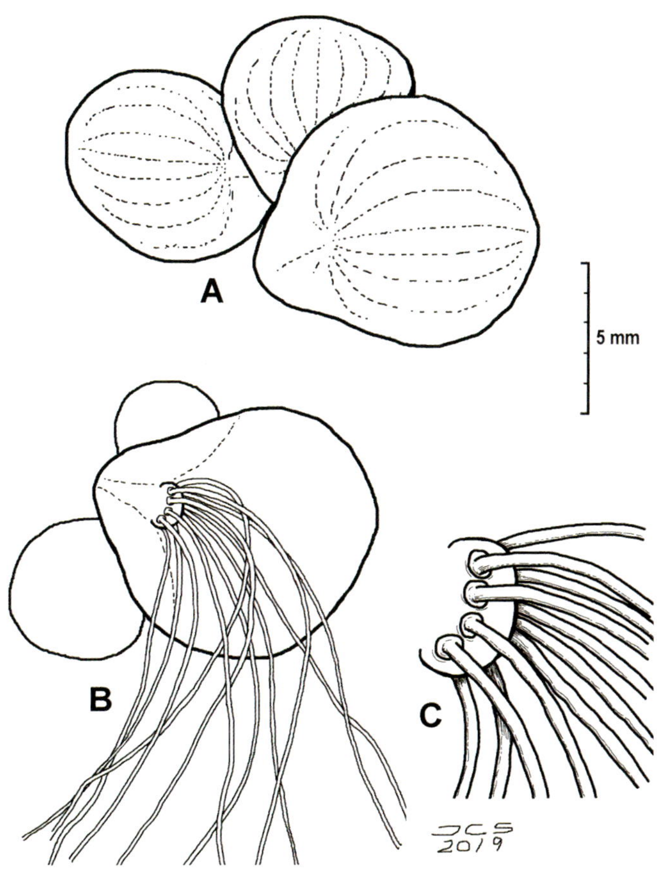

Spirodela punctata
A habit upper side, **B** habit lower side, **C** scale-like leaf perforated by some of the roots.

Araceae

29 ***Wolffia arrhiza*** (L.) Horkel ex Wimm.

DK: Dværgandemad **N:** - **S:** Dvärgandmat **FIN:** - **IS:** -
GB: Rootless Duckweed **NL:** Wortelloos kroos **F:** Wolffia sans racines
D: Wurzellose Zwergwasserlinse **CZ:** Drobnička bezkořenná **PL:** Wolffia bezkorzeniowa
EST: - **LV:** - **LT:** Bešaknė volfija **Ru:** Вольфия бескорневая

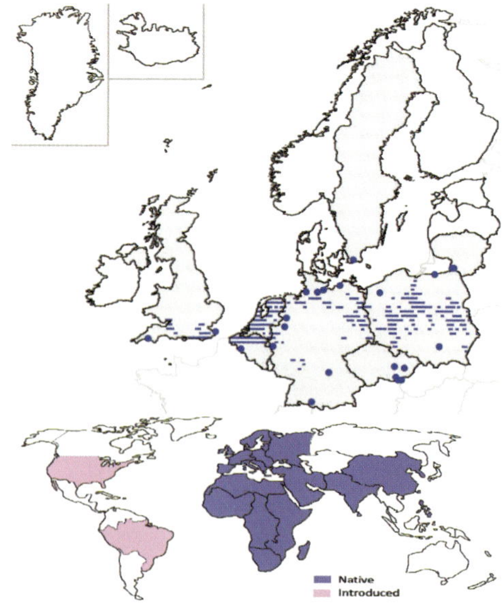

Wolffia arrhiza. At least 80% of the width is above the water surface. Photo Wim van der Ven.

Frond of *W. arrhiza* - side view. Photo Wim van der Ven.

Perennial, floating hydrophyte. Without roots.

Fronds: Solitary or 2-3 together in small groups, spherical to ellipsoid, without veins. Fronds (0.5-)0.7-1.3 mm long, (0.4-)0.5-1.0 mm wide, 1-1.3 times as long as wide, greatest width just below the water surface. At least 80% of the width above the surface (seen from above with a translucent edge on each side). Frond as deep as wide. Upper side flat to slightly domed, intensely green, shiny, with (20-)30-120 stomata. Lower side greenish, very swollen, the upper cells smaller than the cells at the bottom of the frond.

Inflorescence: Monoecious. Much reduced and very tiny, without a membranous spathe. With only one female and one male flower borne from a hollow on the upper side of the frond. Flowering rare.

Flowers: Male flower with 1 stamen. Female flower more or less bottle shaped with a disc-shaped stigma without pigment.

Fruits: One-seeded, globose.

Flowering: Very rare in Europe.

Biology: Propagation by budding from pouches in the margins of the frond, rarely by seed. Turions are not produced. Dispersal by water and birds.

Habitats: In eutrophic, still water in ditches, moats and ponds.

Distribution within the region:
Native. Locally common to scattered in the southern part of the region.

Characteristics and similar species: Easily mistaken for other ***Wolffia*** species - for identification please see the following species or use the key.

Wolffia arrhiza growing with *Spirodela polyrhiza* and *Lemna minor* in a moat at Stjärneholm ruined castle, Skåne, Sweden. Photo JCS.

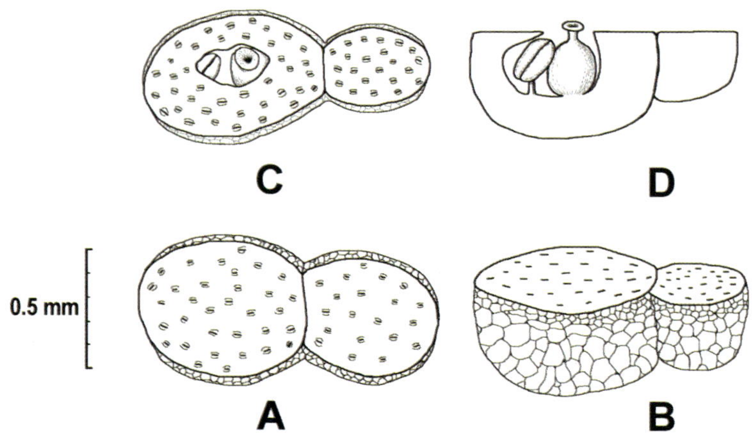

Wolffia arrhiza
A habit from above, **B** habit side view, **C** flowering plant from above, **D** flowering plant in cross section.

Araceae

30 ***Wolffia australiana*** (Benth.) Hartog & Plas

DK: - **N:** - **S:** -**FIN:** - **IS:** -
GB: - **NL:** Smalle Wolffia **F:** -
D: - **CZ:** - **PL:** -
EST: - **LV:** - **LT:** - **Ru:** -

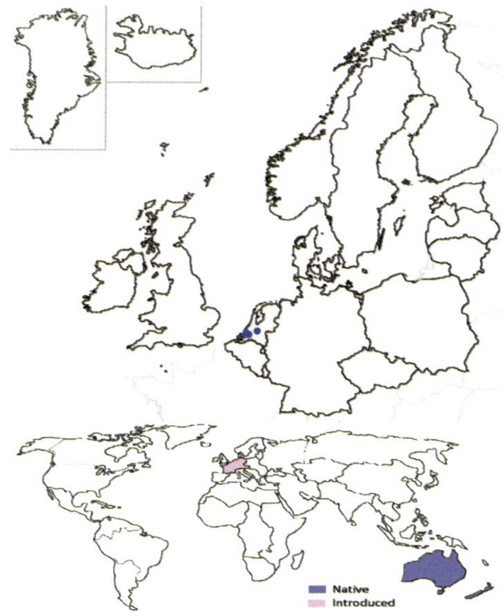

Wolffia australiana. Stomata visible as white dots. Photo Wim van der Ven.

Frond of *W. australiana* - side view. Photo Wim van der Ven.

Fronds: (0.5–)1.0–1.3(–1.5) mm long, (0.3–)0.5–0.7(–0.8) mm wide, 1.3–1.7 times as long as wide, greatest width at the water surface and therefore without a visible translucent edge on each side when viewed from above. Frond 2–3 times as deep as wide. Upper side flat to slightly domed, intensely green, shiny, with 60–120 stomata. Lower side almost colourless, very swollen, the upper cells smaller than the cells at the bottom of the frond.

Inflorescence: Monoecious. Much reduced and very tiny, without a membranous spathe. With only one female and one male flower borne from a hollow on the upper side of the frond. Flowering rare.

Flowers: Male flower with 1 stamen. Female flower with pigmented stigma.

Fruits: One-seeded, globose.

Flowering: Very rare in Europe.

Biology: Propagation by budding from pouches in the margins of the frond, rarely by seed. No turions. Dispersal by water and birds.

Habitats: In eutrophic, still water.

Distribution within the region: Introduced and naturalised in the Netherlands since 2014.

Characteristics and similar species: Differ from **29 *Wolffia arrhiza*** by the almost colourless lower side, 2–3 times as deep as wide.

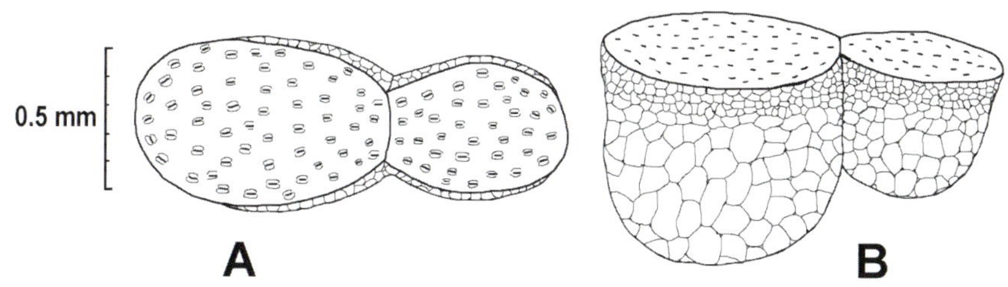

Wolffia australiana
A habit from above, **B** habit in side view.

Araceae

31 *Wolffia columbiana* Karsten

DK: - **N:** - **S:** -**FIN:** - **IS:** -
GB: - **NL:** Colombiaanse Wolffia **F:** -
D: Kolumbianische Zwergwasserlinse **CZ:** - **PL:** -
EST: - **LV:** - **LT:** - **Ru:**

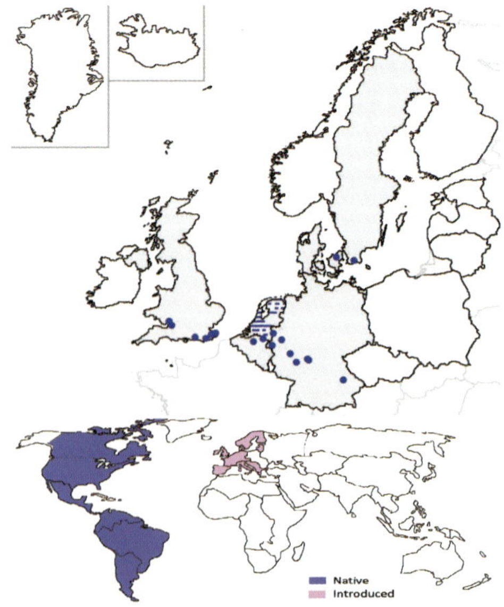

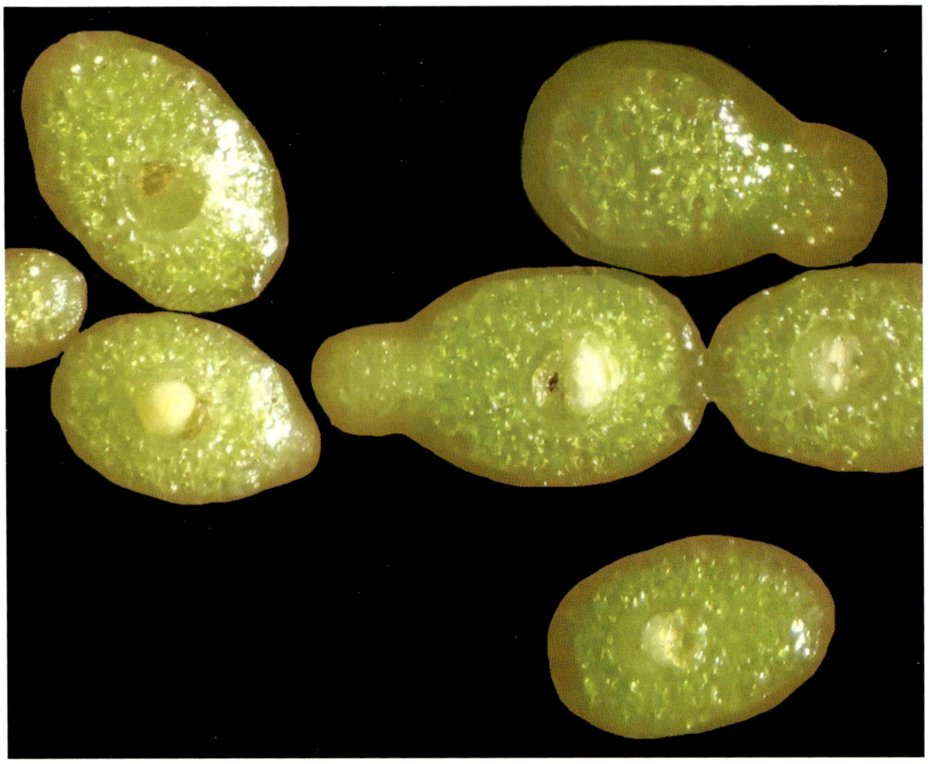

Flowering *W. columbiana*. Photo Wim van der Ven.

Frond of *W. columbiana* - side view. Photo JCS.

Fronds: (0.5–)0.7–1.2(–1.4) mm long, (0.5–)0.6–1.1(–1.2) mm wide, 1–1.3 times as long as wide, greatest width below the water surface. Seen from above, it appears that the emergent part has a broad translucent edge throughout. Frond 1–1.3 times as deep as wide. Upper side with a flat, pale green central area with 1–15(–30) stomata. Lower side transparent green, very swollen, the upper cells not smaller than the cells at the bottom of the frond.

Inflorescence: Monoecious. Much reduced and very tiny, without a membranous spathe. With only one female and one male flower borne from a hollow on the upper side of the frond. Flowering rare.

Flowers: Male flower with 1 stamen. Female flower with unpigmented stigma.

Fruits: One-seeded, globose.

Flowering: Very rare in Europe.

Biology: Propagation by budding from pouches in the margins of the frond, rarely by seed. Turions not produced. Dispersal by water and birds.

Habitats: In eutrophic, still water.

Distribution within the region: Introduced and naturalised in Britain, Denmark, Sweden, Netherlands, Belgium and Germany.

Characteristics and similar species: Differs from other ***Wolffia*** species in the area by the almost spherical fronds with only a small portion of the upper side above the water surface.

Wolffia columbiana. Sjælland, Denmark. Photo JCS.

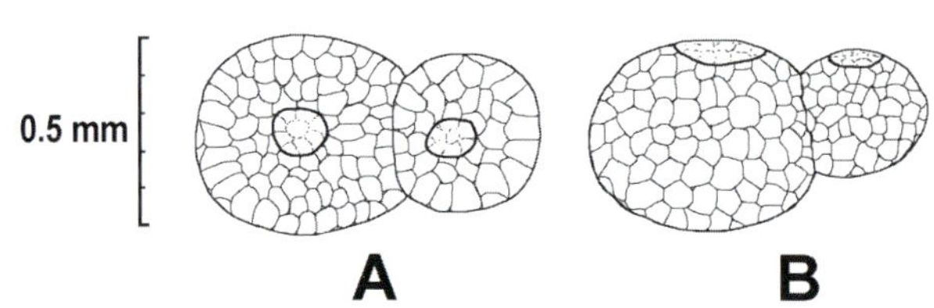

Wolffia columbiana
A habit from above, **B** habit in side view.

Araceae

32 *Wolffia globosa* (Roxb.) Hartog & Plas

DK: - **N:** - **S:** -**FIN:** - **IS:** -
GB: - **NL:** Kleine Wolffia **F:** -
D: - **CZ:** - **PL:** -
EST: - **LV:** - **LT:** - **Ru:**

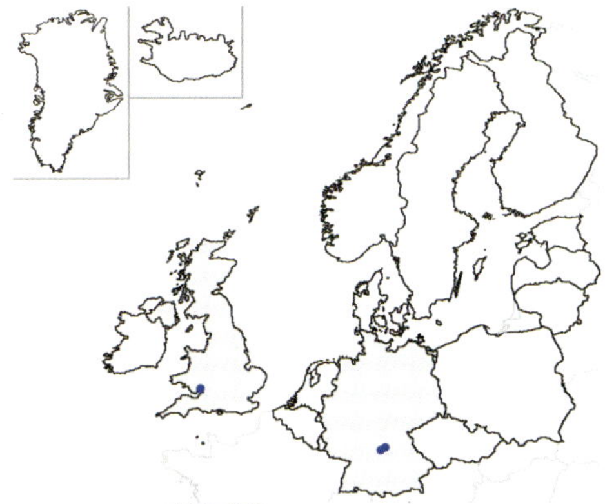

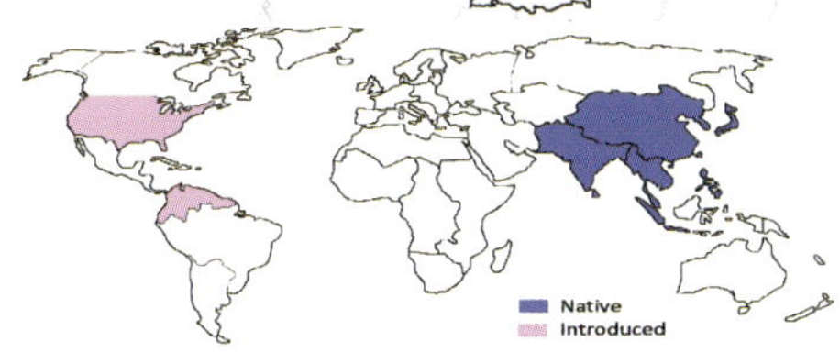

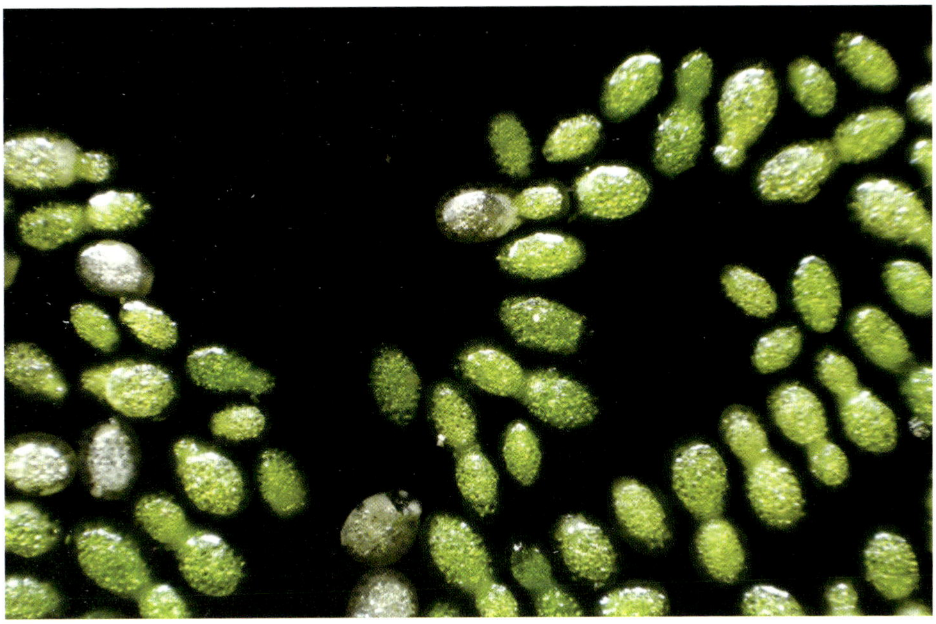

Wolffia globosa in culture. Photo Win van Der Ven.

Frond of *W. globosa* - side view.
Photo KvdW.

Fronds: (0.4–)0.5–0.8(–0.9) mm long, (0.3–)0.4–0.6 mm wide, 1.3–1.7 times as long as wide, with the greatest width just below the water surface. Seen from above, it appears that the emergent part has a translucent margin throughout. Frond 0.7–1.1 times as deep as wide. Upper side with a more or less flat, pale green central area with 8–25 (–35) stomata. Lower side transparent green, not swollen, the upper cells slightly smaller than the cells at the bottom of the frond.

Inflorescence: Monoecious. Much reduced and very tiny, without a membranous spathe. With only one female and one male flower borne from a hollow on the upper side of the frond. Flowering rare.

Flowers: Male flower with 1 stamen. Female flower with unpigmented stigma.

Fruits: One-seeded, about 0.30 mm long.

Flowering: No information.

Biology: Propagation by budding from pouches in the margins of the frond, rarely by seed. No turions. Dispersal by water and birds.

Habitats: In eutrophic, still water.

Distribution within the region: Introduced and naturalised in Britain and Germany.

Characteristics and similar species: Differs from other ***Wolffia*** species in the area by its small size.

31 *Wolffia columbiana* is of a similar colour, but is larger and wider.

Wolffia globosa is the smallest flowering plant on earth. Here in culture in a film roll box with a diameter of 31 mm. Photo Wim van der Ven.

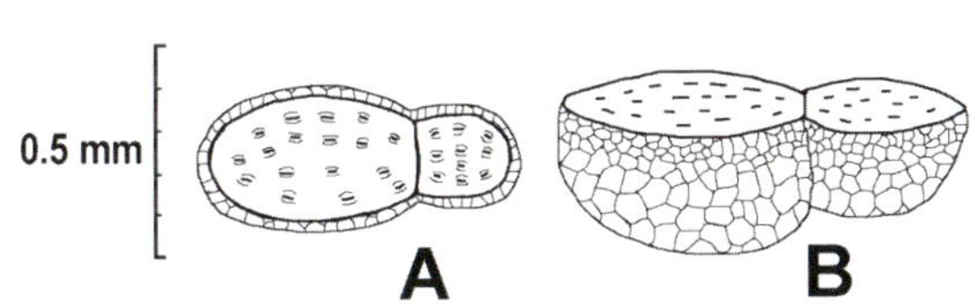

Wolffia globosa
A habit from above, **B** habit in side view.

Alismataceae

33 *Alisma plantago-aquatica* L.

DK: Vejbred-Skeblad **N:** Vassgro **S:** Svalting **FIN:** Ratamosarpio **IS:** -
GB: Common Water-plantain **NL:** Grote waterweegbree **F:** Alisma plantain
D: Gemeiner Froschlöffel **CZ:** Žabník jitrocelový **PL:** Żabieniec babka wodna
EST: Harilik konnarohi **LV:** Parastā cirvene **LT:** Gyslotinis dumblialaiškis **Ru:** Частуха болотная

Alisma plantago-aquatica. Onsild, Hobro, Denmark. Photo JCS.

Perennial helophyte with numerous short roots. Rhizome up to 5 cm long and 4 cm diameter, vertical, bulbous, without stolons.

Leaves: In rosettes from the upper part of the rhizome and of different shape depending on whether submerged, floating or emergent. Submerged leaves thin, greyish green to green, 10–20(–80) cm long, 3–10 mm wide, with 1–3 veins. Floating leaves with flaccid petiole up to 55 cm long, lamina 2–8 cm long, 0.5–3 cm wide, apex acute and base rounded to truncate. Emergent leaves greyish to dark green, petiole up to 35 cm long, lamina 4.5–20 cm long, 3–10 cm wide, ovate to ovate elliptic with 5–7 primary veins transversely connected by numerous secondary veins; apex acute and base cordate to rounded. Stomata on the upper side of the leaves 39–51 µm long. Plants growing in deep water may produce very long petiolate leaves which do not always reach the surface.

Inflorescence: 20–100 cm high, erect, paniculate, richly flowering in the upper half, distinctly longer than the leaves, often longer than wide, with whorled branches subtended by bracts.

Flowers: 3-merous. Sepals green with membranous margins. Petals 4–6 mm long, whitish to faint reddish violet toned, with a yellow base and somewhat uneven denticulate margin. 6 stamens very wide at base, filaments 1.5–3 mm long, anthers elliptical, 1–1.3 mm long, yellow green. Pollen grains on average 25–26 µm. Styles 0.8–1.4 mm long, fine and of even width from base to tip. Stigma finely papillose, short, measuring 1/6–1/8 of the style length.

Fruits: Achenes greyish yellow, obovate, 2–2.8 mm long, usually with only one groove on the back.

Flowering: June-September.

Biology: Flowers only open in daylight. Pollination especially by flies attracted to nectar produced at the base of stamens.

Habitats: In more or less nutrient-rich ponds, ditches, canals, slow-flowing streams, river backwaters and lakes.

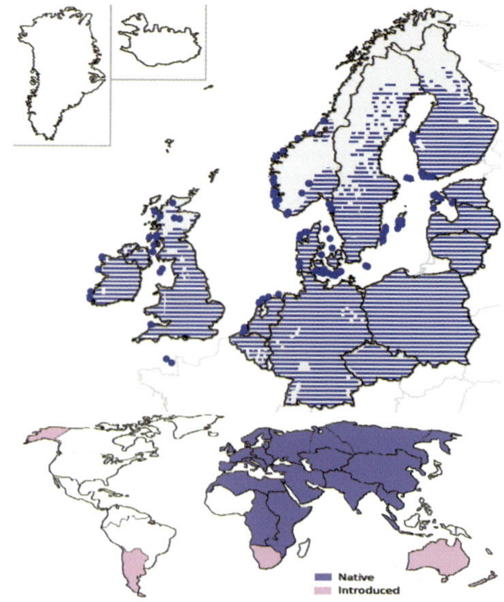

Distribution within the region:
Native. Widespread and abundant more or less throughout the region north to the northern coast of the Gulf of Bothnia.

Characteristics and similar species:
Sterile, submerged plants with ribbon-like leaves can be mistaken for species of **115-122** ***Sparganium***, **145-146** ***Schoenoplectus***, **45** ***Sagittaria*** and **41** ***Luronium natans*** (see these). *Alisma* species with only ribbon-shaped leaves cannot be identified to species. Small forms can be mistaken for **37-38** ***Baldellia*** (see these), and for **359** ***Isoëtes echinospora*** - the latter with leaves with 4 air channels (cross section).

Emergent forms are very similar to **34** ***Alisma lanceolatum***, which has styles broad at the base, narrower toward the apex, achenes often with 2 furrows on the back, more oblong anthers, petals with a prominent tooth in the middle, stomata on the upper side of leaves 59-69 µm long.

Alisma plantago-aquatica. Young plant with thin ribbon-like leaves and floating leaves. Lake Førby Sø, Thy, Denmark. Photo JCS.

39 ***Caldesia parnassifolia*** differs in its cordate leaf bases, turions and different fruits.

Small plants with only floating leaves, may resemble **220** ***Ranunculus flammula***, which has different venation and usually toothed leaf margins.

Hybrids: ***A. ×rhicnocarpum*** Schotsman (*A. plantago-aquatica × lanceolatum*) is largely sterile and intermediate between the parents. Identification based on leaf shape alone is not reliable. Rare and only with the parents.

The petals of *A. plantago-aquatica* are unevenly toothed, but without a prominent middle tooth. Nørre Onsild, Hobro, Denmark. Photo JCS.

Alisma plantago-aquatica. Close-up of flower showing styles and ovaries. Photo RVL.

Alisma plantago-aquatica

A habit emergent plant, **A1** young plant with ribbon-like and floating leaves, **B** flower, **B1** ovary with style, **B2** stamen with a wide base and almost circular anther, **C** achene in side view, **C1** achene with 1 furrow on the back, **E1** detail of ribbon-like leaf with midvein and air chambers.

Alismataceae

34 *Alisma lanceolatum* With.

DK: Lancet-Skeblad **N:** Gotlandsvassgro **S:** Gotlandssvalting **FIN:** - **IS:** -
GB: Narrow-leaved Water-plantain **NL:** Slanke waterweegbree **F:** Flûteau lancéolé
D: Lanzett-Froschlöffel **CZ:** Žabník kopinatý **PL:** Żabieniec lanzetowaty
EST: Süstlehine konnarohi **LV:** Šaurlapu cirvene **LT:** Lancetinis dumblialaiškis **Ru:** Частуха ланцетная

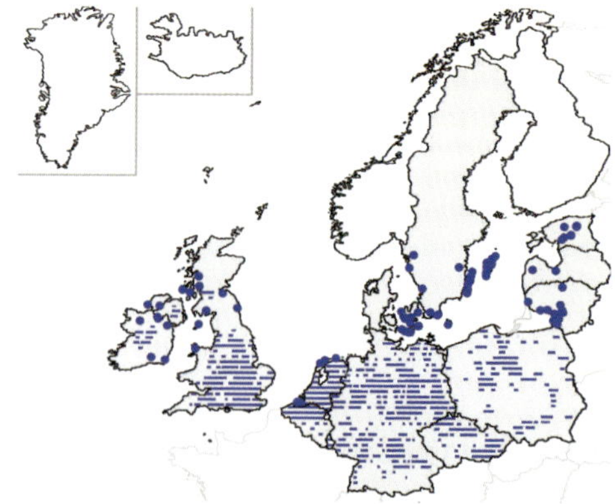

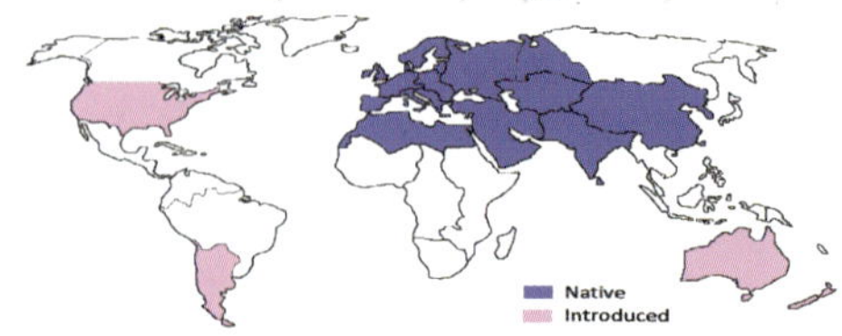

Alisma lanceolatum. Jordtorp, Öland, Sweden. Photo JCS.

Perennial helophyte with numerous short roots. Rhizome up to 5 cm long and 3 cm diameter, vertical, bulbous, without stolons.

Leaves: In rosettes from the upper part of the rhizome and of different shape depending on water depth. Submerged, ribbon-like leaves are thin, 10–80 cm long and 3–10 mm wide. Emergent or terrestrial leaves usually more or less erect, with lamina 4.5–20 cm long, up to 4.5 cm wide, narrowly ovate to elliptic lanceolate, greyish green to dark green, with (3–)5–7 veins, with acute apex and at base narrowed into the petiole, which may be up to 20 cm long. The veins are transversely connected with numerous secondary veins. Stomata on the upper side of the leaves 59–69 μm long. Floating leaves are rare, smaller and narrower than emerged leaves.

Inflorescence: 20–70 cm tall, erect, paniculate, richly flowering in the upper half, distinctly longer than the leaves, rarely longer than wide, with whorled branches subtended by bracts.

Flowers: 3-merous. Sepals green with membranous margins. Petals 4–6.5 mm long, very pale pink to faintly violet-toned, with a yellow base and somewhat uneven denticulate margin - the middle tooth often prominent. 6 stamens very wide at base, filaments 2.5 mm long, anthers oblong elliptical, 0.6–1.1 mm long, yellow green. Pollen grains on average 30–32 μm. Styles 0.7–1.1(–1.3) mm long, broad at the base, narrower toward the apex, somewhat curved. Stigma roughly papillose representing 1/3–1/5 of the style length.

Fruits: Achenes greyish yellow, obovate, 2–2.9 mm long, usually with 2 grooves on the back, less often with a single furrow.

Flowering: June-September.

Biology: Flowers only open in daylight. Pollination mainly by flies attracted to nectar produced at base of the stamens.

Habitats: On more or less calcareous soil in or by ponds, ditches, canals, lakes, slow-flowing streams and rivers.

Distribution within the region:
Native. Scattered to locally frequent in southern parts of the region, rare in eastern Denmark, southern Sweden and the Baltic states, absent from the north.

Characteristics and similar species:
Similar to **33** ***A. plantago-aquatica,*** which usually has wider leaves with truncate or rounded base, styles uni-

Alisma lanceolatum has faint violet-toned petals with a prominent middle tooth. Jordtorp, Öland, Sweden. Photo JCS.

Alisma lanceolatum. The anthers are oblong elliptical. Photo KvdW.

Alisma lanceolatum. Close-up of flower showing styles and ovaries. Photo RVL.

formly fine from the base, achenes with only 1 furrow on the back, circular to elliptic anthers and 39–51 µm long stomata on the upper side of the leaves.

35 *Alisma gramineum* differs in the anther less than 0.5 mm long, roundish, and styles shorter than the ovary when flowering and strongly recurved.

For other similar taxa see ***A. plantago-aquatica.***

Hybrids: See *A. plantago-aquatica.*

Form of *A. lanceolatum* with limp leaves, Congresbury Yeo, Britain. Photo RVL.

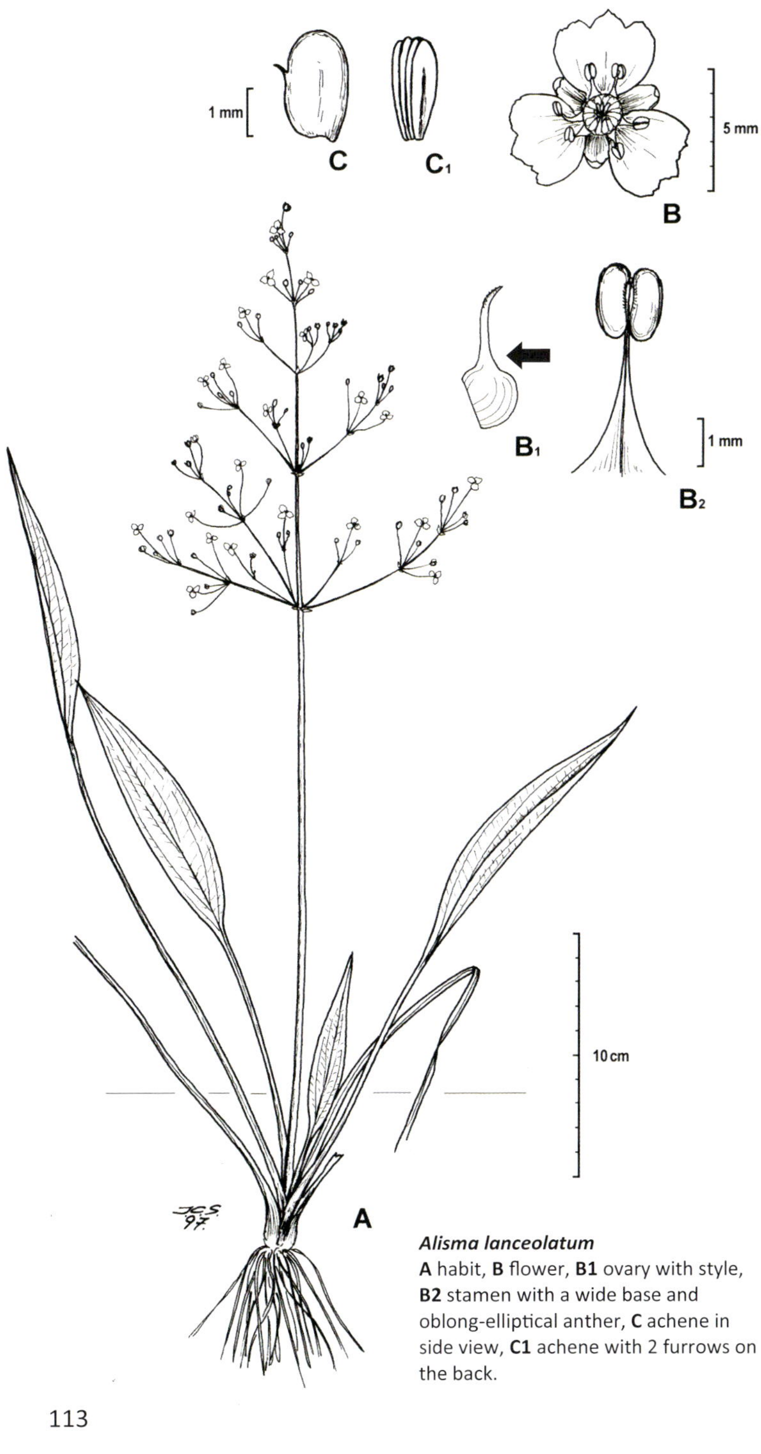

Alisma lanceolatum
A habit, **B** flower, **B1** ovary with style, **B2** stamen with a wide base and oblong-elliptical anther, **C** achene in side view, **C1** achene with 2 furrows on the back.

Alismataceae

35 *Alisma gramineum* Lej.

Syn.: *Alisma arcuatum* Michalet

DK: Kortskaftet Skeblad **N:** Grasvassgro **S:** Grässvalting **FIN:** - **IS:** -
GB: Ribbon-leaved Water-plantain **NL:** Smalbladige waterweegbree **F:** Alisma graminée
D: Gras-Froschlöffel **CZ:** Žabník trávolistý **PL:** Żabieniec trawolistny
EST: Väike konnarohi **LV:** Zālainā cirvene **LT:** Siauralapis dumblialaiškis **Ru:** Частýха злáковая

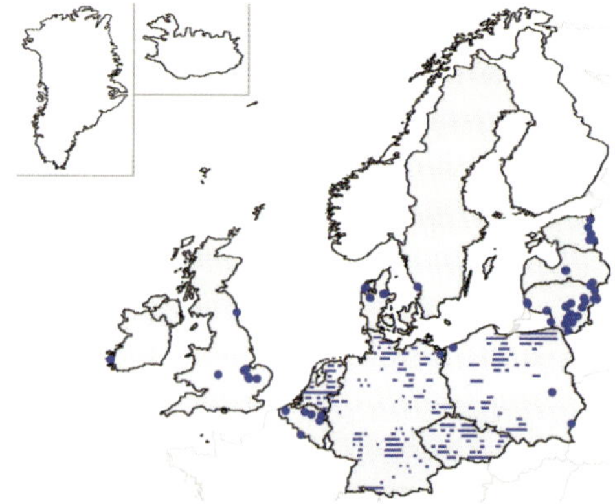

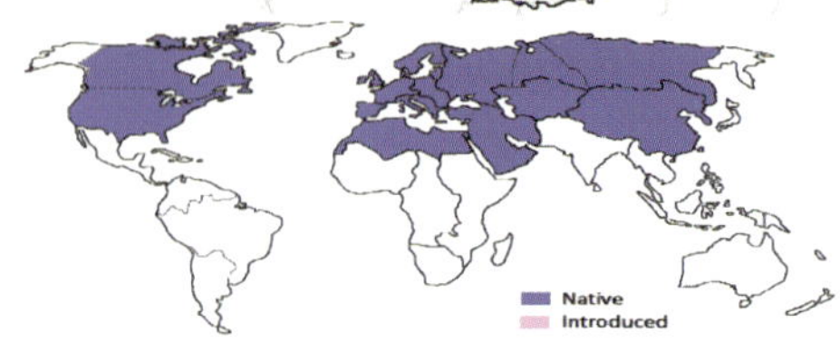

Flowering *A. gramineum* with both long, submerged or floating, ribbon-like leaves and erect leaves with lamina. Bislicher Insel, Germany. Photo KvdW.

Perennial helophyte with numerous short roots. Rhizome short up to 1.5 cm diameter, vertical, bulbous, without stolons.

Leaves: In rosettes from the upper part of the rhizome and of different shape depending on water depth. Submerged or floating leaves thin, up to 100 cm long and 3–15 mm wide, ribbon-like. Emerged or terrestrial leaves usually more or less erect, with lamina 2–6(–15) cm long, 1.5–4.5(–8) cm wide, narrowly ovate to elliptic lanceolate, green to dark green, with 3-5(-7) veins, with somewhat acute apex and narrowed base. The veins are transversely connected with numerous secondary veins.

Inflorescence: 20-200 cm high, erect, paniculate, richly flowering, longer than wide, with whorled branches subtended by 5-13 mm long bracts.

Flowers: 3-merous. Sepals 2-4 mm long, green with membranous margins. Petals 4-6 mm long, white to faintly violet toned, with a yellow base and uneven denticulate margin. 6 stamens narrow at base, filaments 1-1.5 mm long, anthers globular, 0.3-0.5 mm long, yellow green. Pollen grains on average 27 µm. Styles 0.3-0.7 mm long, curved like a hook.

Fruits: Achenes greyish yellow, obovate, 2-2.7 mm long, thick walled, usually with 2 grooves on the back.

Flowering: June-September.

Biology: Emergent plants with flowers open in daylight. Pollination especially by flies attracted to nectar produced at base of the stamens. Submerged plants have cleistogamous, self-pollinating flowers.

Habitats: In clean, naturally nutrient-rich, neutral to alkaline water. Lakes, slow-flowing rivers and weakly brackish fjords, as well as in canals, ditches and gravel pits.

Distribution within the region:
Native. Rare and declining in Ireland, Britain, Belgium, Denmark, Estonia and Latvia; extinct in Sweden. Scattered to locally frequent in the Netherlands, Germany, Czech Republic, Poland and Lithuania.

Characteristics and similar species:
Sterile, submerged plants with ribbon-like leaves may be mistaken for species

Submerged *A. gramineum* at 1 metre depth in Lake Fussing Sø, Jutland, Denmark. Photo JCS.

of **115-122** ***Sparganium***, **145-146** ***Schoenoplectus***, **45** ***Sagittaria sagittifolia*** and **41** ***Luronium natans***, but can be recognised by the venation and cell-pattern of the leaves - see these.

36 ***Alisma wahlenbergii*** (Holmb.) Juz. is usually totally submerged, with ribbon-like leaves 1-3 mm wide, and inflorescence with shorter bracts, 2-5 mm long.

Alisma gramineum is most easily distinguished from **34** ***A. lanceolatum*** and **33** ***A. plantago-aquatica*** by floral characters, particularly the shape of the styles, stamens and achenes; vegetative characters are less reliable.

Alisma gramineum. Emerged leaf. Lake Fussing Sø, East-Jutland, Denmark. Photo BM.

Alisma gramineum flower - note the hook-shaped styles. Photo KvdW.

Submerged *A. gramineum*. Toeppersee, Germany. Photo KvdW.

Alisma gramineum
A, A1, A2 habits - submerged, partial submerged plant, emergent, **B** flower, **B1** ovary and hooked style, **B2** stamen with narrow base and globular anther, **C** achene sideview, **C1** achene with 2 furrows on the back, **E1** section of ribbon-like leaf.

Alismataceae

36 *Alisma wahlenbergii* (Holmb.) Juz.

Syn.: *Alisma gramineum* Lej. subsp. *wahlenbergii* Holmb.
DK: Liden Skeblad **N:** - **S:** Småsvalting **FIN:** Upossarpio **IS:** -
GB: Baltic Water-plantain **NL:** - **F:** -
D: - **CZ:** - **PL:** -
EST: Põhja-konnarohi **LV:** - **LT:** - **Ru:** Частуха Валенберга

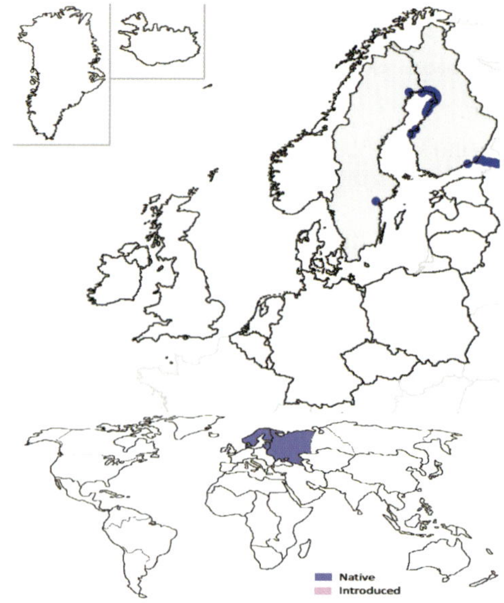

Short-lived perennial helophyte with numerous short roots. Rhizome small, less than 5 mm diameter, vertical, bulbous. Without stolons.

Leaves: Up to 15 in rosettes from the upper part of the rhizome and of different shape depending on water depth. Submerged plants develop ribbon-like, 10–40 cm long, 1–3 mm wide, flat, yellow-green, thin leaves with 1(–3) veins. Emergent plants have shortly petiolate lanceolate leaves, 3–8 mm long, 2–5 mm wide, with a distinct lamina.

Inflorescence: 4–15(–20) cm high, usually curved, paniculate, distinctly shorter than the leaves, branches in 2–3 whorls, subtended by bracts 2–5 mm long.

Flowers: 3-merous, usually cleistogamous. Open flowers are 5–7 mm in diameter but seldom seen in nature, and only on emergent inflorescences at low water levels. Sepals green with membranous margins. Petals slightly longer than sepals, 1.5–2.5 mm long, rather narrow, white. 6 stamens narrow at base, filaments 1 mm long, anthers globular, 0.4 mm long, yellowish green. Styles curved like a hook.

Fruits: Achenes 12–15, greyish yellow, obovate, 1.5–2 mm long, thin walled usually with 2 furrows on the back.

Flowering: July-August.

Biology: Flowers mainly under water, cleistogamous and self-pollinating. Fruits ripen in August-September and are spread by water and drifting ice.

Habitats: Lakes, lagoons and coastal inlets. In clear, naturally mesotrophic to eutrophic, fresh or brackish water (salinity of less than 3–4 ‰). It grows on sandy substrates, occasionally mixed with silt or clay, typically at depths of 5 to 45 cm, occasionally up to 1.5 metres or even to 4 metres in clear water.

Alisma wahlenbergii is a poor competitor and disappears when the vegetation gets too dense or tall as a result of increased eutrophication. It is often associated with small aquatic plants such as *Potamogeton filiformis*, *Elatine hydropiper* and *Zannichellia palustris*. Hailuoto, Finland. Photo KvdW.

Distribution within the region:
Native. Endemic to Finland, Sweden and western Russia; within the region it occurs in Lake Mälaren and in the northern part of the Gulf of Bothnia.

Characteristics and similar species:
Without inflorescence and with narrow, ribbon-like leaves in rosettes, *A. wahlenbergii* may be mistaken for small plants of **145-146** ***Schoenoplectus***, which have bluish-green to dark green leaves, or for young **50** ***Butomus umbellatus*** with leaves trigonous in cross section. **35** ***Alisma gramineum*** is quite similar but larger in all its parts - see this. **37-38** ***Baldellia*** has a strong smell when crushed and **359** ***Isoëtes echinospora*** has leaves with 4 air channels in cross section.

Alisma wahlenbergii in shallow water in the Gulf of Bothnia, Finland. Photo KvdW.

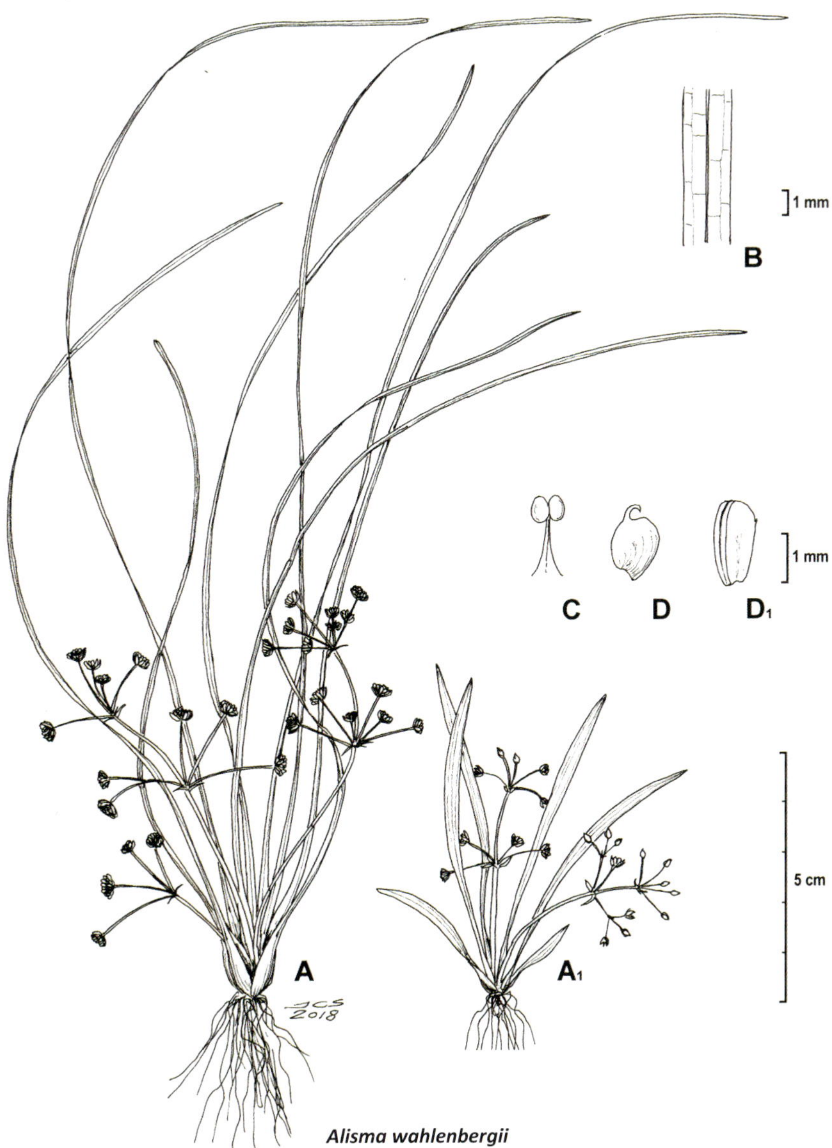

Alisma wahlenbergii
A habit submerged plant, **A1** habit emerged plant, **B** part of leaf, **C** stamen, **D** ovary and hooked style, **D1** achene.

Alismataceae

37 *Baldellia ranunculoides* (L.) Parl.

Syn.: *Echinodorus ranunculoides* (L.) Engelm.
DK: Almindelig Søpryd **N:** - **S:** Flocksvalting **FIN:** - **IS:** -
GB: Lesser Water-plantain **NL:** Stijve moerasweegbree **F:** Flûteau fausse-renoncule
D: Hahnenfußähnlicher Igelschlauch **CZ:** - **PL:** Żabieniec jaskrowata
EST: - **LV:** - **LT:** Vėdrynlapis šilininkas **Ru:** Балделлия лютиковая

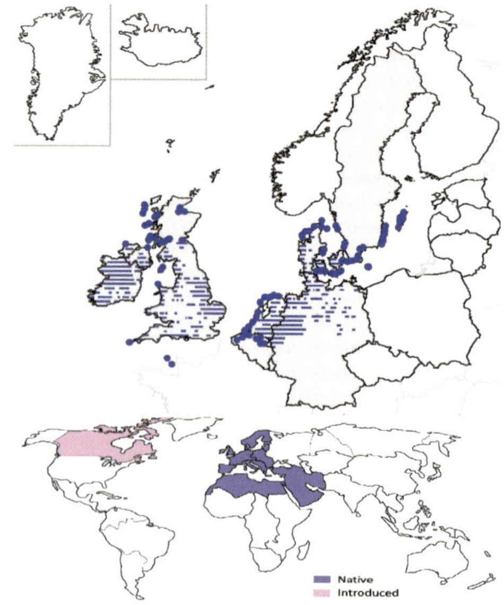

Partly submerged *B. ranunculoides* in a small lake in coastal heathland. Thy, Denmark. Photo JCS.

Perennial hydrophyte or helophyte without stolons. Rhizome short, 1–1.5 cm thick, vertical, densely covered in leaf sheaths - almost like an onion. Roots numerous, without root hairs.

Leaves: 5–25 in rosette. Primary leaves submerged, with flat upper side and arched lower side, 4–5 mm wide, almost transparent and linear with long acuminate, subulate apex, with membranous margin at base partially clasping the rhizome. Submerged plants may also have long-petiolate, floating or aerial leaves with 3–8 cm long, lanceolate to ovate lamina. Lamina with 3 veins connate at apex and several secondary veins. Terrestrial plants with aerial leaves only.

Inflorescence: Scape 5–20 cm long, up to 2.5 mm diameter, leafless, erect or prostrate, but very seldom rooting. Plants with 1–2 panicles, each carrying 1–2(–3) whorls of 3–11 flowers on long, straight pedicels.

Flowers: 3-merous, 13–18 mm in diameter. Sepals 2.5–3.6 mm long, green. Petals 6–8 mm wide, 7–11 mm long, usually not overlapping, white or faintly pink with a yellow base and dentate margin. 6 stamens with 0.8–1.1 mm long anthers.

Fruits: Achenes 26–38 together in spherical heads, 6–8 mm in diameter. Achene 2.0–2.7 mm long, ovoid, acute, curved, more or less glabrous, without papillae, with 3 prominent dorsal and 2 ventral ribs.

Flowering: June-August.

Biology: Only 1(–2) flowers in each whorl open on any given day. The flowers are visited by insects, but are capable of self pollination - flowers closing while pollen is still viable. Fruit set is typically very good. Vegetative propagation by budding from leaf axils.

Habitats: On sandy or muddy substrate in or by lakes, ponds, streams and ditches with clear, oligotrophic to mesotrophic, weakly acidic to alkaline or occasionally weakly brackish water.

Distribution within the region:
Native. Western Europe from southern Scandinavia and Britain to the Netherlands, Belgium and Germany.

Characteristics and similar species:
Both species of *Baldellia* emit a strong, sharp odour when crushing the leaves. This character is useful for separating them from similar taxa, such as delicate or submerged forms of **33-36 *Alisma*** spp. and **41 *Luronium natans***, or sub-

The smell of coriander or old apple skins by crushing the leaves and the missing stolons distinguishes *B. ranunculoides* from similar species. Submerged plants with triangular (left) or linear (right) submerged leaves. Photos RVL and JCS. Inserted: Plants with floating linear leaves. Photo KvdW.

Baldellia ranunculoides. 5-6 cm tall plant on dried-up bottom. The former submerged leaves are being replaced by aerial leaves with lamina. Öland, Sweden. Photo JCS.

Baldellia ranunculoides. Flower and fruit head. Thy, Denmark. Photo JCS.

merged plants with only linear leaves, such as **259 *Isoëtes echinospora*** (leaves with 4 air channels) and **254 *Subularia aquatic*** (annual with different-looking inflorescence).

38 *Baldellia repens* is closely related, but can be recognised by its clonal growth formed by rooting, prostrate inflorescences, larger flowers and smaller, papillose achenes.

Note: The newest theories suggest that *Baldellia* survived the latest ice age in two separate populations. The Italian-Balkan population became *B. ranunculoides* while the population on the Iberian Peninsula became *B. repens*. According to DNA-investigations the two species are clearly separated, but hybridise in the northern part of the range.

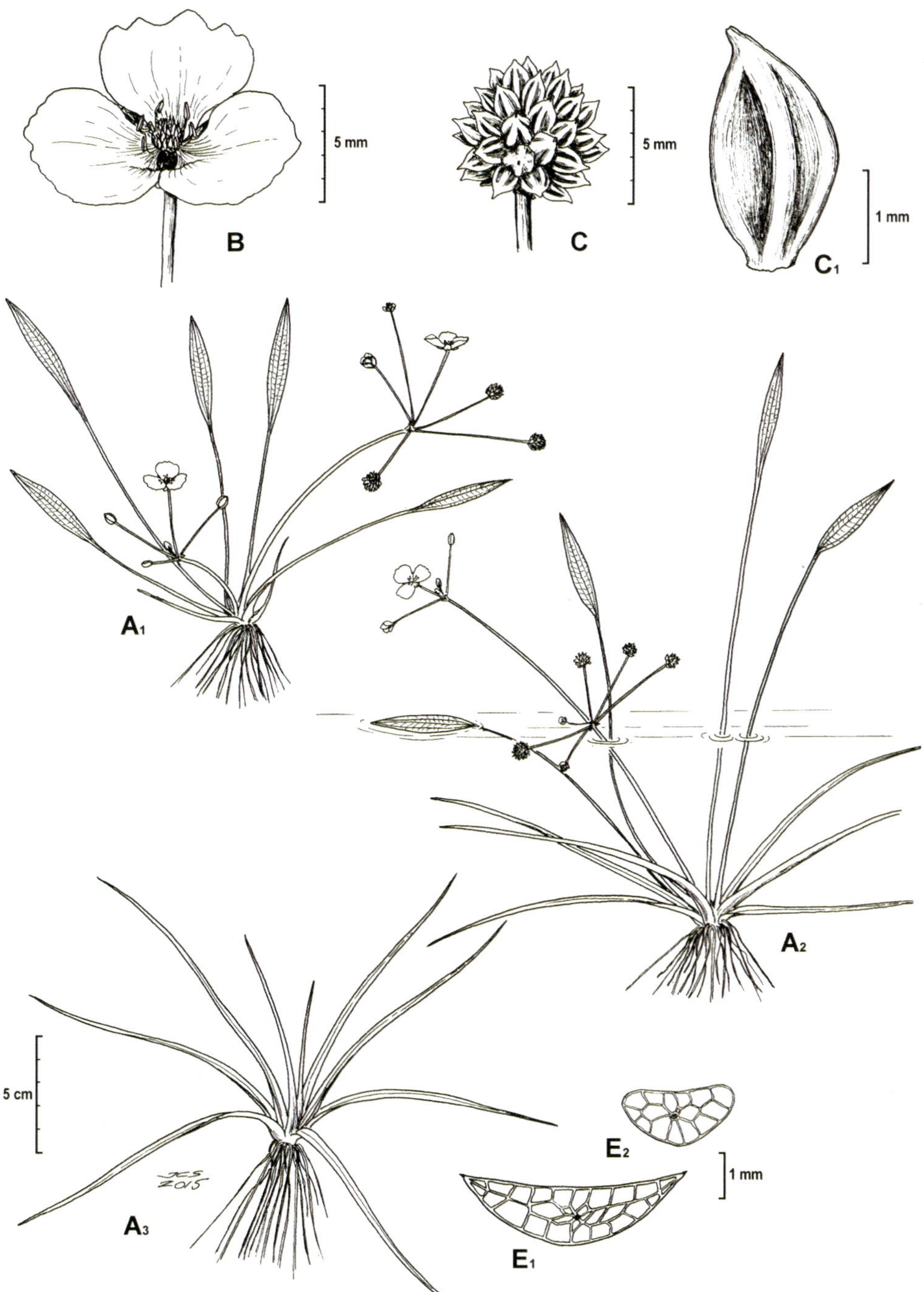

Baldellia ranunculoides
A1 habit terrestrial plant, **A2** habit partly submerged plant, **A3** habit submerged form, **B** flower, **C** fruiting head, **C1** achene, **E1**, **E2** cross section from base and middle section of submerged leaf.

Baldellia ranunculoides. Sections submerged, linear leaf. Photos KvdW.

Floating/aerial leaf.

Alismataceae

38 *Baldellia repens* (Lam.) Ooststr. ex Lawalrée

Syn.: *B. ranunculoides* (L.) Parl. subsp. *repens* A. & D.Löve; *Echinodorus ranunculoides* (L.) Engelm. var. *repens* (Lam.) Nyman

DK: Krybende Søpryd **N:** Soleigro **S:** Revsvalting **FIN:** - **IS:** -
GB: Creeping Water-plantain **NL:** Kruipende moerasweegbree **F:** Flûteau rampant
D: Kriechender Igelschlauch **CZ:** - **PL:** - **EST:** - **LV:** - **LT:** Vėdrynlapis šilininkas **Ru:** -

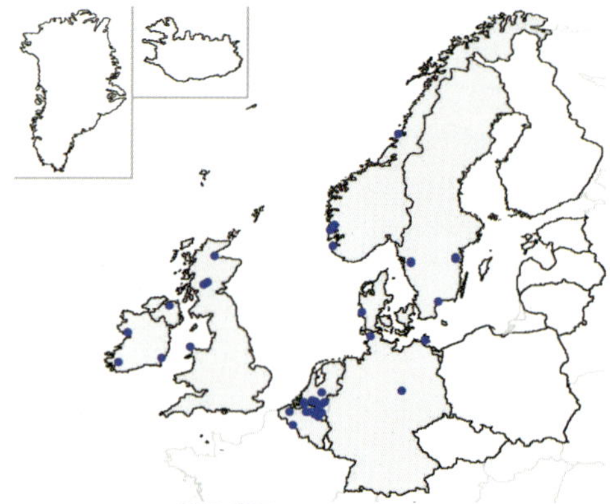

Baldellia repens growing in the drawdown zone of a large lake. Etang de Bellebouche, Parc Naturel Regional de la Brenne, France. Photo RVL.

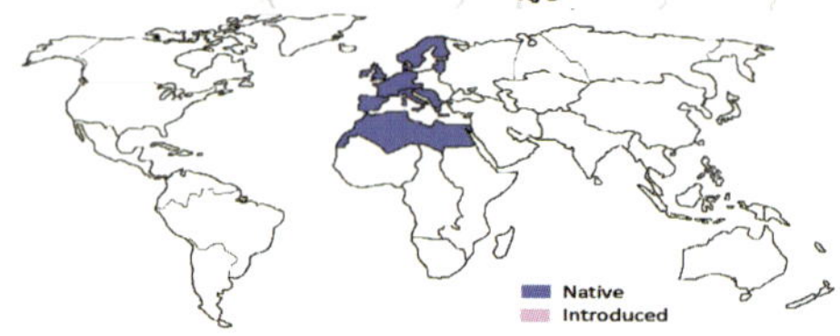

Perennial hydrophyte or helophyte with clonal growth by inflorescence stolons. Rhizome short, 1–1.5 cm diameter, vertical, densely covered in leaf sheaths - almost like an onion. Roots numerous, without root hairs.

Leaves: 5–25 in a rosette. Primary leaves submerged, with flat upper side and arched lower side, 4–5 mm wide, almost transparent and linear with long acuminate, subulate apex, with membranous margin at base partially clasping the rhizome. Submerged plants may also have long-petiolate, floating or aerial leaves with 3–8 cm long, lanceolate to ovate lamina. Lamina with 3 veins connate at apex and several secondary veins. Terrestrial plants only have aerial leaves.

Inflorescence: Scape about 1 mm diameter, prostrate or ascending, leafless, rooting and forming new plants upon contact with the substrate. Plants with 1–2 panicles, each carrying 1–2(–3) whorls of 2–5(–7) flowers on long, often somewhat curved pedicels.

Flowers: 3-merous. 15–22 mm in diameter. Sepals 3–4.5 mm long, green. Petals 9–13 mm wide and 9–13 mm

Baldellia repens. Beautifully developed, cushion-shaped clone in 10 cm deep water. Changing water level means that *B. repens* does not necessarily flower every year, but continues submerged vegetative growth. Lake Filsø, Denmark. Photo JCS.

Baldellia repens has fruits with fine papillae, as opposed to *B. ranunculus*, where the fruits are more or less smooth. Lake Filsø, Denmark. Photo BM.

Fruit heads from *B. repens* (left) and *B. ranunculoides* (right) - same scale. Photo JCS.

Baldellia repens has usually wide and overlapping petals. Lake Filsø, Denmark. Photo JCS.

long, usually overlapping, white or faintly pink with yellow base and more or less entire margin. 6 stamens with 1.1–1.5 mm long anthers.

Fruits: Achenes 12–24 together in spherical heads, 4–5 mm in diameter. Achene 1.8–2 mm long, ovoid, acute, curved, finely papillose, with 3 prominent dorsal and 2 ventral ribs.

Flowering: June-October.

Biology: Only 1(–2) flowers in each whorl open each day. *Baldellia repens* is said to be self-incompatible producing only sparse fruits, but indoor cultivation experiments with rich fruit set show that this is not true. Vegetative propagation by inflorescence stolons and to some degree by budding from leaf axils.

Habitats: On sandy or silty substrate in or by lakes and pools, with clear, oligotrophic to mesotrophic, weakly acidic water, also in streams to 1 metre depth.

Distribution within the region:
Native. Britain, southern Scandinavia, Netherlands, Belgium and Germany.

Characteristics and similar species:
See **37** ***Baldellia ranunculoides***.

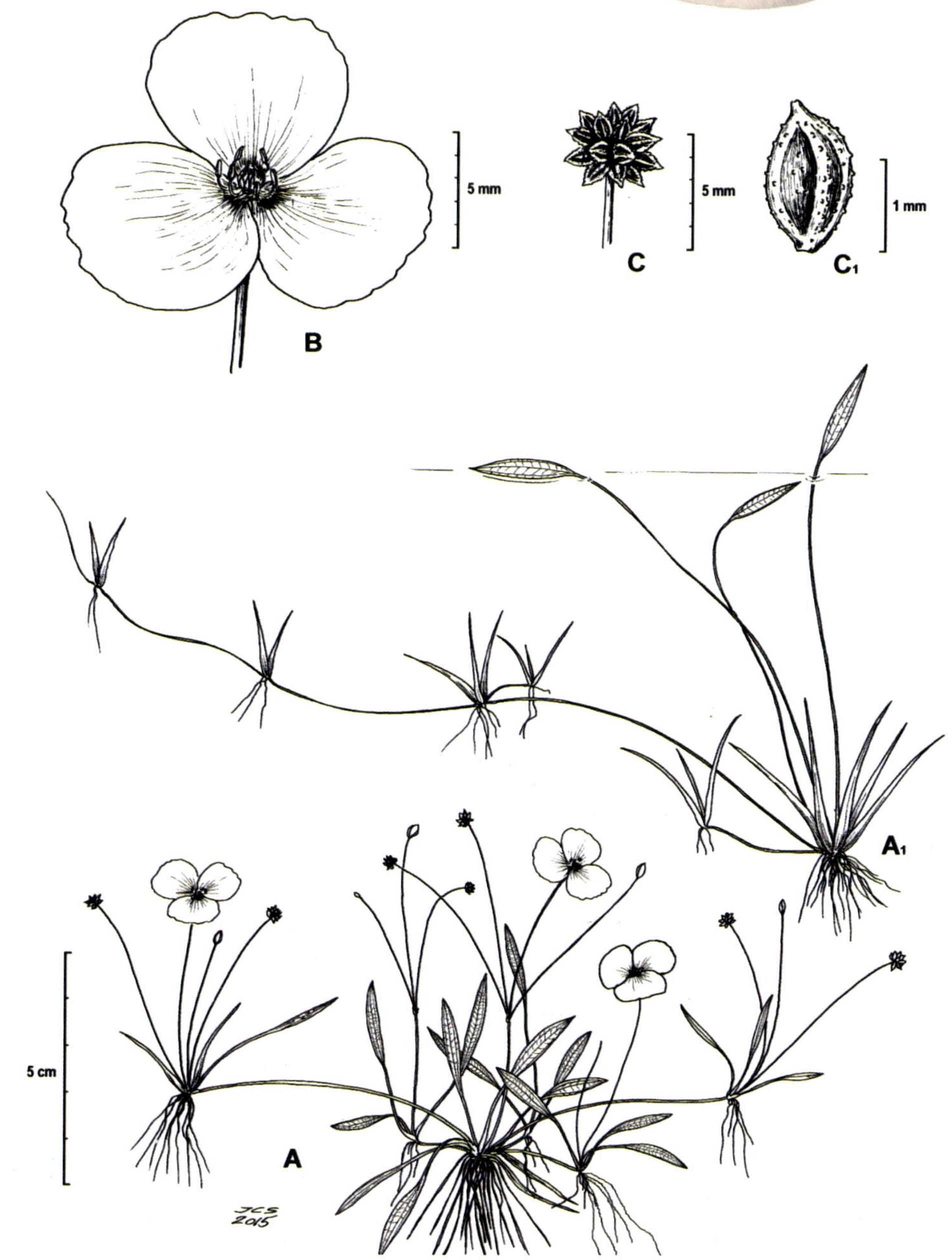

Baldellia repens
A Newly established landform, **A1** submerged form, **B** flower, **C** fruiting head, **C1** achene.

Alismataceae

39 *Caldesia parnassifolia* (L.) Parl.

Syn.: *Caldesia reniformis* (D.Don) Makino; *Alisma parnassifolium* Bassi ex L.
DK: - **N:** - **S:** - **FIN:** - **IS:** -
GB: Caldesia **NL:** - **F:** Caldésie à feuilles de parnassie
D: Herzblattfroschlöffel **CZ:** - **PL:** Kaldesia dziewięciornikowata
EST: - **LV:** Sirdslapu kaldēzija **LT:** Širdžialapė kaldezija **Ru:** Кальдезия белозоролистная

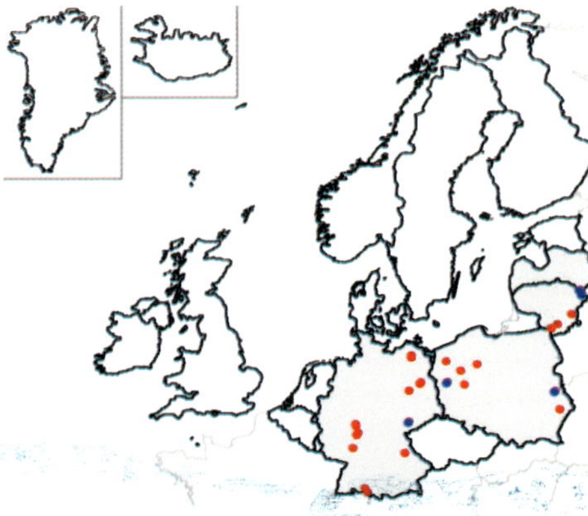

Red dots indicate extinct populations.

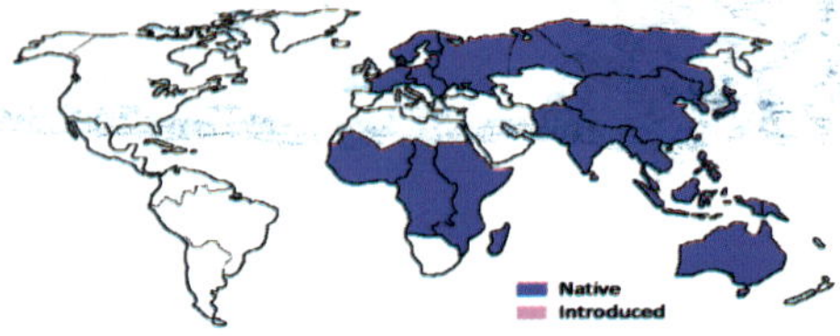

Flowering *C. parnassifolia*. Schwandorf, Bavaria, Germany. Photo KvdW.

Annual or perennial hydrophyte or helophyte with numerous roots. Rhizome short and thick, without stolons.

Leaves: In rosettes from the upper part of the rhizome. Submerged, ribbon like leaves are present only on young plants and soon decay. They are thin, 5–30 cm long and 2–5(–8) mm wide. Floating and aerial leaves with lamina 3–10 cm long, 2–6(–7) cm wide, ovate to broadly ovate, deeply cordate at base, obtuse or subobtuse at apex, green to dark green and often with purplish tones particularly on the lower side. With 5–15 parallel veins connected with numerous secondary veins. Petiole 5–100 cm long.

Inflorescence: 10–100 cm high distinctly longer than the leaves. Flowering part raised above the water 20–35 cm long, paniculate. Branches in whorls of three subtended by bracts. Lowest bracts oblong lanceolate, about 1 cm long, thickened. Flowers sometimes replaced by vegetative buds.

Flowers: 3-merous. Sepals spreading, 3–4 mm long, greenish, persistent. Petals c. 5 mm long, white, with entire or denticulate margin. 6 stamens, filaments 1.5–2 mm long, anthers elliptical, c. 1 mm long, yellow green. Carpels 6–10, ovate. Styles 1–2 mm long, slender, erect to slightly curved.

Fruits: Achenes yellowish green, obovoid, inflated, 3–4 mm long, 2–2.5 mm wide, with 3–5 longitudinal abaxial ridges.

Flowering: May-August.

Biology: Flowers open in the morning and each flower lasts about 8 hours. The flowering period for each plant is up to 80 days. The flowers are capable of self-pollination, but pollination by flies and bees results in higher seed set. Propagation by turions is of far greater importance than seeds. They are formed abundantly in the axils of bracts on submerged inflorescence-like shoots or in flowering inflorescences.

Habitats: On calcareous substrates in still, oligotrophic to mesotrophic, shallow water, typically at depths of 20–40 cm, but sometimes to 1 m. Ponds, ditches and small lakes.

Caldesia parnassifolia. Left: Infructescence. Right: Base of plant with numerous turions (white arrows). Etang Mouton, Parc Naturel Regional de la Brenne, France. Photos RVL.

Distribution within the region:
Native. Germany, Poland and Lithuania. Very rare and declining in Europe and threatened by habitat loss and pollution.

Characteristics and similar species:
Somewhat similar to **33 *A. plantago-aquatica,*** but differs in the ovate, deeply cordate lamina, the inflated achenes and the production of turions. Specimens with only floating leaves may be mistaken for **336 *Nymphoides peltata***, but differ in the venation of the leaves.

Caldesia parnassifolia. Floating leaves. Schwandorf, Bavaria, Germany. Photo KvdW.

Flower of *C. parnassifolia*. Del. JCS.

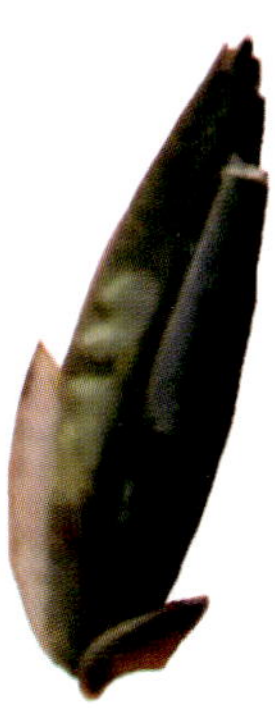

Turion. *C. parnassifolia*. Photo RVL.

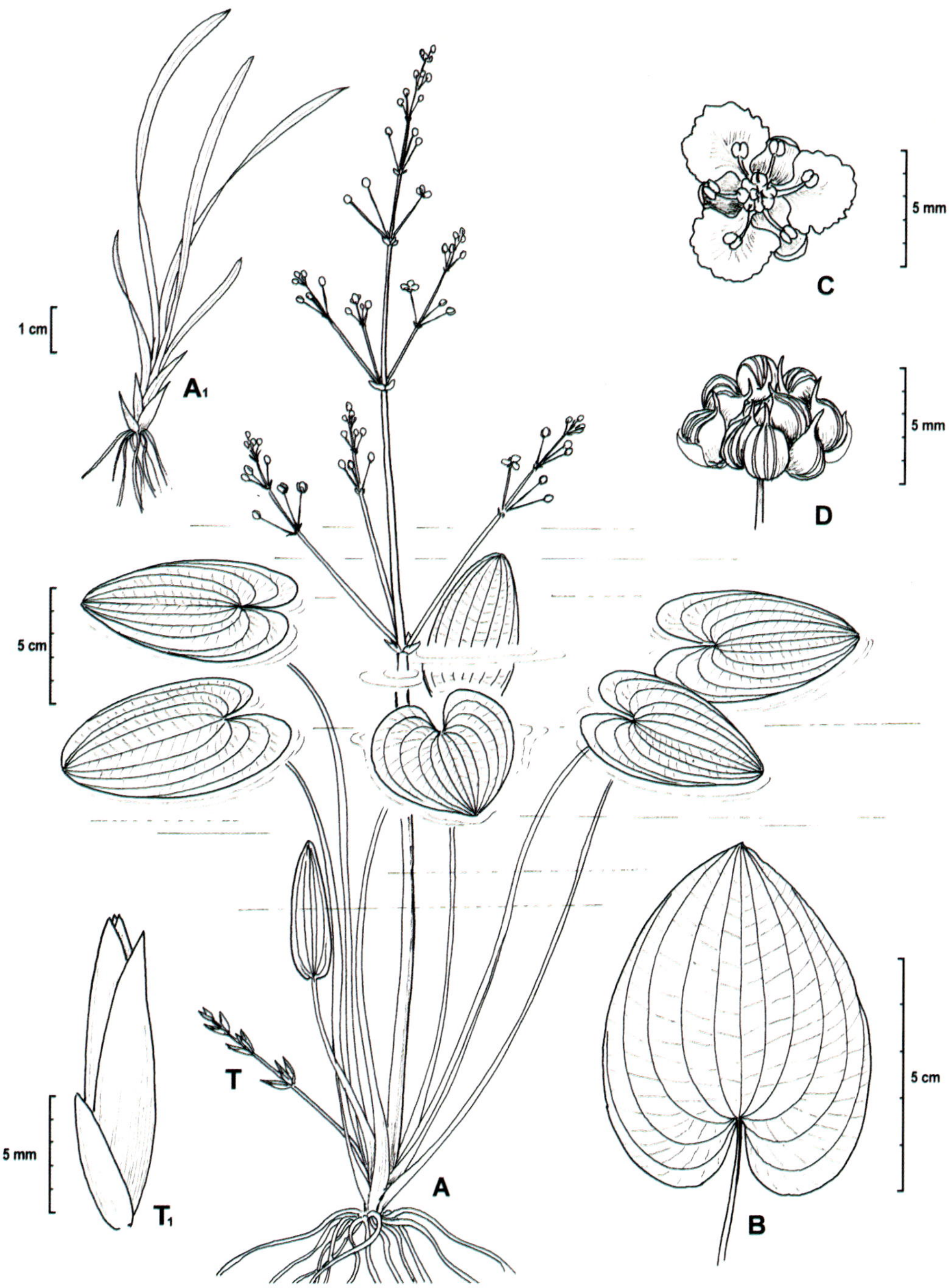

Caldesia parnassifolia
A habit, **A1** young plant developed from turion, **B** leaf, **C** flower, **D** infructescence, **T** shoot with turions, **T1** turion.

Alismataceae

40 *Damasonium alisma* Miller

Syn.: *D. stellatum* Persoon; *Alisma damasonium* L.
DK: - **N:** - **S: FIN:** - **IS:** -
GB: Starfruit **NL:** - **F:** Étoile d'eau
D: Sternfrucht **CZ:** - **PL:** -
EST: - **LV:** - **LT:** - **Ru:** Звездоплодник частуховидный

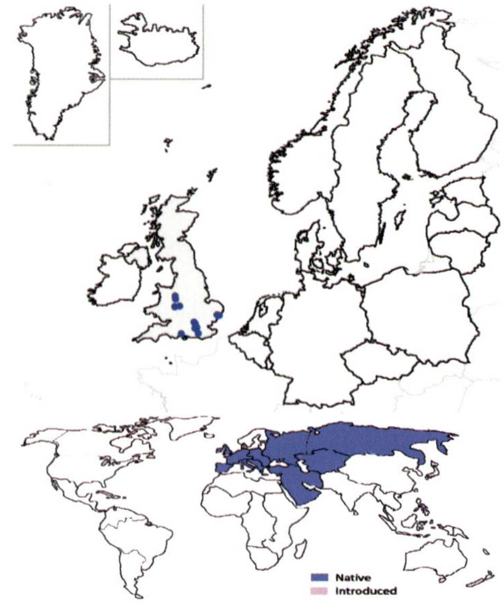

Damasonium alisma on temporarily dried-out, muddy bottom. Parc Naturel Regional de la Brenne, Indre, France. Photo RVL.

Annual or occasionally short-lived perennial hydrophyte or amphiphyte. Rhizome short, erect, without stolons. Roots numerous.

Leaves: In rosettes from the upper part of the rhizome. Submerged, thin, 5–20 cm long and 2–5 mm wide, ribbon-like leaves are present only on young or submerged plants, and soon decay when plants become exposed by falling water levels. Floating and emergent leaves with lamina 1.5–7 cm long, 0.5–3 cm wide, ovate to oblong, rounded to somewhat cordate at base, obtuse or subobtuse at apex, entire, yellowish green, firm to almost coriaceous, with 3 –5(–7) parallel veins connected by secondary veins. Petiole 10–30 cm long.

Inflorescence: 1–7 per plant, 15–60 cm tall, distinctly longer than the leaves, umbellate or racemose. Flowers in 1–3 simple whorls subtended by bracts. Lowest bracts oblong lanceolate, about 1 cm long.

Flowers: 3-merous. Sepals 3, 2.5–3.5 mm long, ovate, greenish with scarious margin, persistent. Petals 3–4 mm long, longer than sepals, oval, white to faint pinkish, with yellow spot at base, margin irregularly dentate. 6 stamens, filaments 1–1.5 mm long, anthers pale yellow. Carpels 6, each with 2 ovules. Styles terminal.

Fruits: Follicles united at base, spreading like a star, 5–14 mm long, laterally compressed, gradually narrowed into an elongated apical beak. Each follicle with 1 or 2 seeds, one larger than the other, 1.3–1.7 mm long.

Flowering: June-August.

Biology: *Damasonium alisma* flowers richly when growing in shallow water or on moist substrate. Flowers last only for a short while. They are pollinated by beetles and flies, but are also capable of self-pollination.
Germination depends on the site, as with most mud-plants, it can germinate in spring or autumn. If the autumn cohort survive the winter, then they are larger and have a competitive advantage over other plants. Seeds may remain dormant for years, building a seed bank of great importance for the survival of the species. Dormancy can be broken by disturbance to the seeds' environment, e.g., by cattle trampling and by spreading of seeds to new ponds on cattle hooves.

Habitats: *Damasonium alisma* typically grows in small ponds, on the drawdown

Damasonium alisma partly submerged. Inholms Claypits, Surrey, Britain. Photo RVL.

Damasonium alisma flowers richly to a depth of 1 metre. Inholms Claypits, Surrey, Britain. Photo RVL.

zones of lakes and in poached, muddy gateways in areas which are dry in summer. It has declined significantly throughout its range due to habitat degradation and the loss of small field ponds.

Distribution within the region: Native. *D. alisma* occurs in two areas, in western Europe and a separate population in Ukraine and southern Russia. It is probably extinct in Portugal and Italy and is known from two sites in England, but remains locally frequent in parts of France. In the east, it has been recorded from only three sites in the last 10 years.

Characteristics and similar species: *Damasonium alisma* is easily recognised by the fruits appearing like 6-rayed stars. Plants without flowers may be similar to other Alismataceae and confirmation is dependent on flower characters.

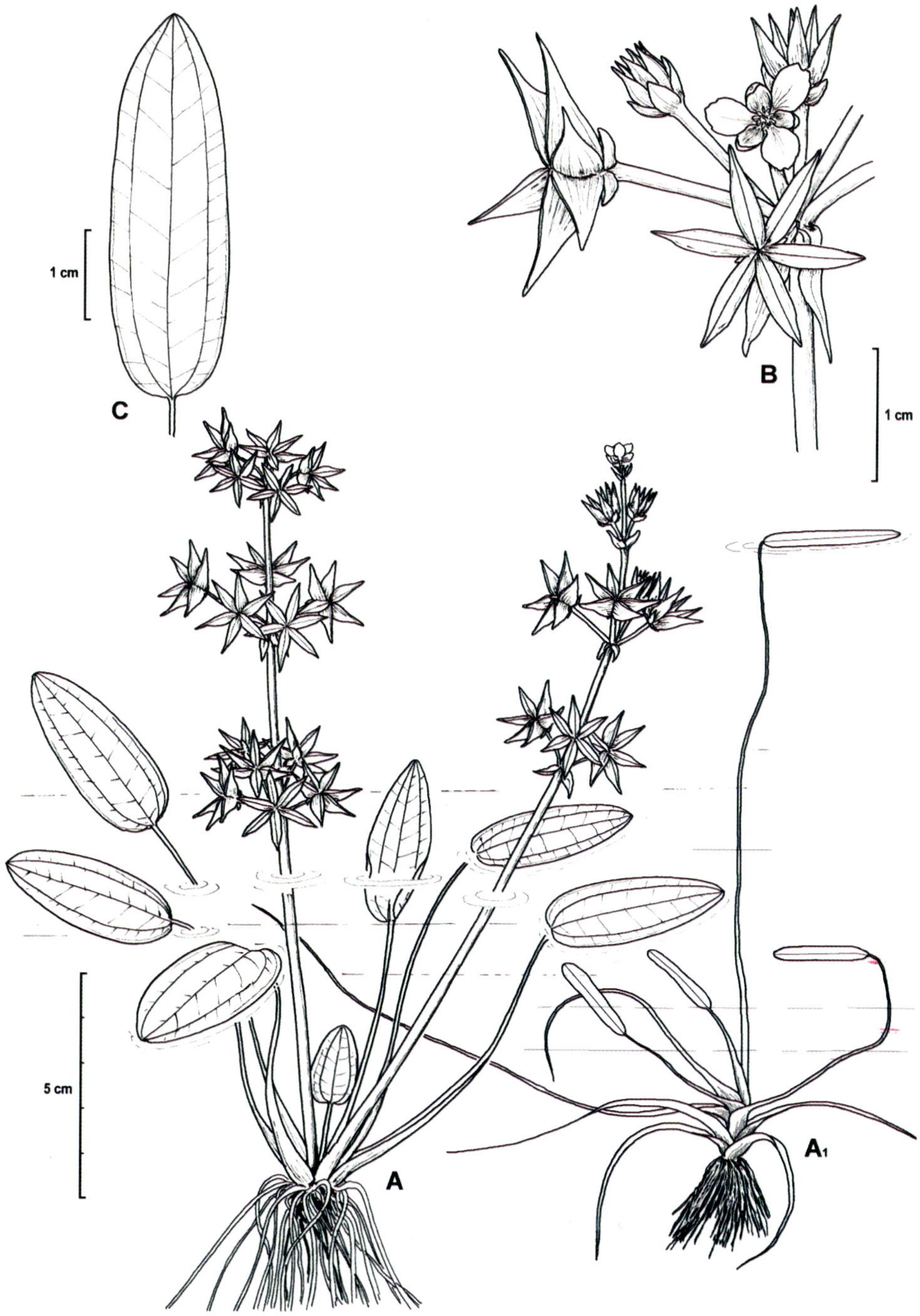

Damasonium alisma

A habit, **A1** habit young submerged plant, **B** part of inflorescence, **C** leaf.

Alismataceae

41 *Luronium natans* (L.) Raf.

Syn: *Alisma natans* L., *Elisma natans* (L.) Buchenau
DK: Vandranke **N:** Flytegro **S:** Flytsvalting **FIN:** - **IS:** -
GB: Floating Water-plantain **NL:** Drijvende waterweegbree **F:** Flûteau nageant
D: Froschkraut **CZ:** Žabníček vzplývavý **PL:** Elisma wodna
EST: - **LV:** - **LT:** Plūduriuojančioji šustinė **Ru:** Элизма плавающая

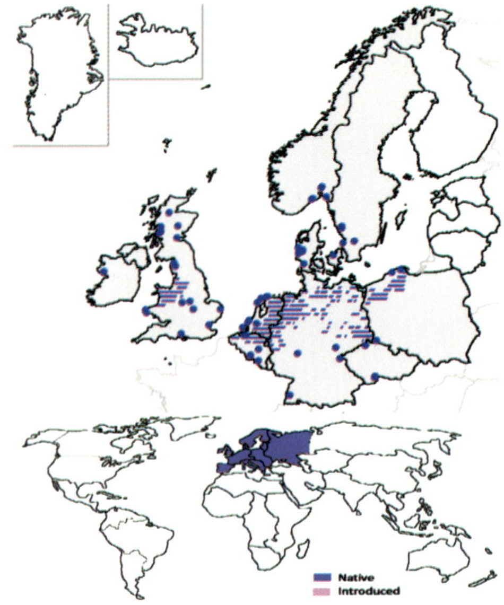

Flowering *L. natans*. Sydlige Parallelkanal, Skjern, Denmark. Photo JCS.

Perennial hydrophyte. Rhizome vertical, thin and soon withering from below. Roots white.

Stem: Completely submerged or rising through the water column bearing frequent plantlets. Flowering stems with 3-merous whorls of bracts subtending 2 flowers in the upper part and a new shoot, which can form new generations of flowers and shoots in a long chain.

Leaves: In rosettes. In still or slow-flowing water both ribbon-like submerged leaves and long-petiolate, floating leaves are developed. In fast-flowing water and in deep water only submerged leaves are developed. The submerged leaves are flaccid, linear, evenly tapered into an obtuse apex, up to 30 cm long, 3–7 mm wide, yellowish green with only a single vein. The middle section of the leaf has two layers of rectangular air chambers, but only one towards the apex. The floating leaves are long petiolate. Floating leaves and those of terrestrial plants have an ovate to elliptic lamina, rounded or cuneate at base, 2–3 cm long, 1–1.5 cm wide, green often with some reddish colouration on veins and the petiole, with 3 primary veins and transverse secondary veins.

Inflorescence: Floating, 10–40 cm long, umbellate or flowers in 2 or 3 whorls.

Flowers: 3-merous, monoecious, about 1.5 cm in diameter. Pedicels 5–10 cm long. Sepals 3–4 mm long, ovate, green. Petals up to 10 mm long, circular to broadly ovate, white with yellow base. Stamens 6, filaments 1–2 mm long, yellow green, anthers pale yellow. Carpels 6–9, forming a more or less globose head.

Fruits: Achenes 2.0–2.5 mm long, narrowly obovate, somewhat compressed, brown, with 12–15 longitudinal ribs.

Flowering: July- August.

Biology: The flowers are pollinated by small flies. After fertilisation, the pedicels bend and the fruits ripen under water. Flowers not reaching the surface remain closed and self-pollinate within an enclosed air bubble. *L. natans* overwinters as submerged plants with ribbon-like leaves. It is likely that vegetative propagation is the main method of reproduction and the only method in deeper water; many populations show very low fertility and appear to reproduce mainly by vegetative means. The procumbent stems, rooting at intervals, form dense populations of non-flowering, leafy shoots that may become detached and carried to new places by stock or human activity.

Typical formation of submerged *L. natans*. Photo KvdW.

Luronium natans. Flower and floating leaves. Sydlige Parallelkanal, Skjern, Denmark. Photos JCS.

Habitats: In clear, mostly acidic, oligotrophic to mesotrophic, slow-flowing water in ditches, canals, streams and rivers, also in still water in lakes and small seasonal or permanent ponds.

Distribution within the region: Native. Endemic to Western and Central Europe. Listed in the annex to the EU Habitat Directive.

Characteristics and similar species: *L. natans* differs from species of **66-71 *Potamogeton*** and **33-36 *Alisma*** by its small, elliptic, 3-veined floating leaves. When only submerged leaves are formed, it can be mistaken for *Alisma* spp., but these lack creeping stems with roots at intervals. Both **37-38 *Baldellia*** spp. have a strong smell when crushed.

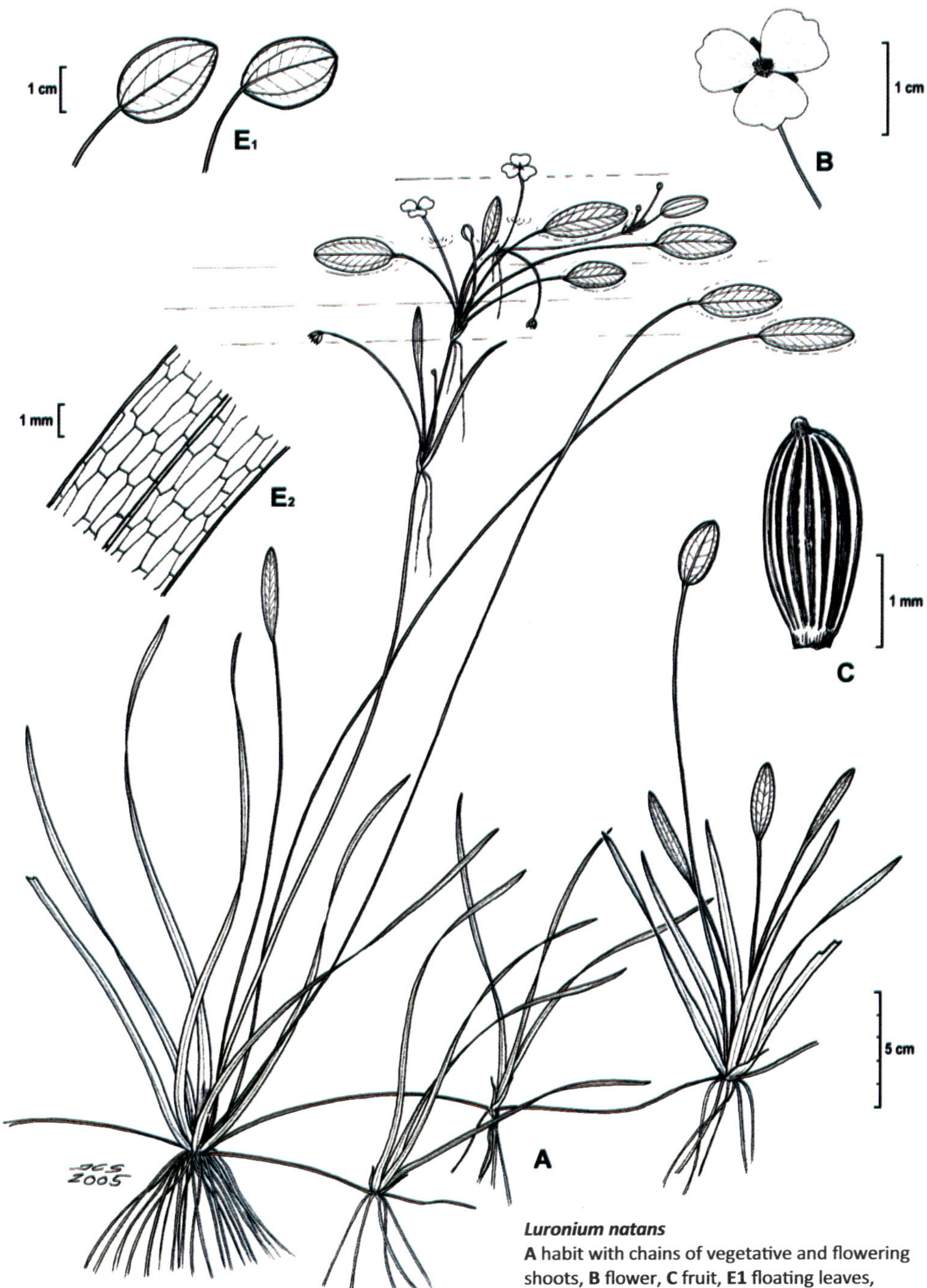

Luronium natans
A habit with chains of vegetative and flowering shoots, **B** flower, **C** fruit, **E1** floating leaves, **E2** section of submerged leaf.

Submerged *L. natans*. Sandpit, Handel, Netherlands. Photo KvdW.

Alismataceae

Key to *Sagittaria* L.

Determination of *Sagittaria* species without flowers and fruits is unreliable because of great plasticity in the size and shape of leaves. An increasing number of cultivated species of foreign origin are being found in Europe and causing considerable identification problems. The key and descriptions of such foreign species (*S. filiformis, S. latifolia, S. graminea* and *S. rigida*) presented here are based on American floras and should be used with caution. A modern revision of the entire genus is needed. Besides the species mentioned, *S. platyphylla* has been recorded from Italy.

1 Peduncle lax, generally submerged, more or less prostrate and supported by water with individual flowers floating or held erect on a delicate pedicel. Emergent leaves lacking except on terrestrial plants 2
1 Peduncle more or less erect, not supported by water. Emergent leaves usually present, with or without submerged and/or floating leaves 4

2 Anthers purple or with purplish tinge. If hastate leaves present, the lower lobes spreading
44 *S. ×lunulata*
2 Anthers yellow. If hastate leaves present, the lower lobes subparallel 3

3 Petals 1–6 mm long. Lamina of floating leaves 1.5–5 cm long. Peduncle delicate
42 *S. filiformis*
3 Petals 8–10 mm long. Lamina of floating leaves 3–15 cm long. Peduncle robust
43 *S. natans*

4 Stamens with glabrous filaments (***1***) 5
4 Stamens with pubescent filaments (***2***) 7

1

2

5 Petals with purple base. Achenes 4–6 mm long; anthers purple (exceptionally yellow)
45 *S. sagittifolia*
5 Petals entirely white. Achenes 2–3.5 mm long; anthers yellow 6

6 Petiole triangular in cross section. Fruiting pedicels spreading (obtuse angle to peduncle), Inflorescence with 2–8 whorls of flowers. Bracts connate much less than 1/4 total length
46 *S. latifolia*
6 Petiole 5-angled in cross section. Fruiting pedicels ascending (acute angle to peduncle). Inflorescence with 5–12 whorls of flowers. Bracts connate more than 1/4 total length
S. australis (J.G.Sm.) Small
(not treated in this book)

7 Female flowers sessile or subsessile. Bracts ovate
48 *S. rigida*
7 Female flowers pedicellate (pedicel more than 0.5 mm long). Bracts lanceolate 8

8. Emergent leaf blades less than 4 cm wide. Fruiting pedicel ascending
47 *S. graminea*
8 Emergent leaf blades up to 10 cm wide, rarely less than 4 cm wide. Fruiting pedicels recurved
S. platyphylla (Engelm.) J.G.Sm.
(not treated in this book)

Sagittaria filiformis. Shortheath, Britain. Photo RVL.

Alismataceae

42 ***Sagittaria filiformis*** J.G.Sm.

Syn.: *S. stagnorum* Small; *S. subulata* (L.) Buchenau var. *gracillima* (S.Watson) J.G.Sm.
DK: - **N:** - **S:** - **FIN:** - **IS:** -
GB: Narrow-leaved Arrowhead - **F:** -
D: - **CZ:** - **PL:** -
EST: - **LV:** - **LT:** - **Ru:**

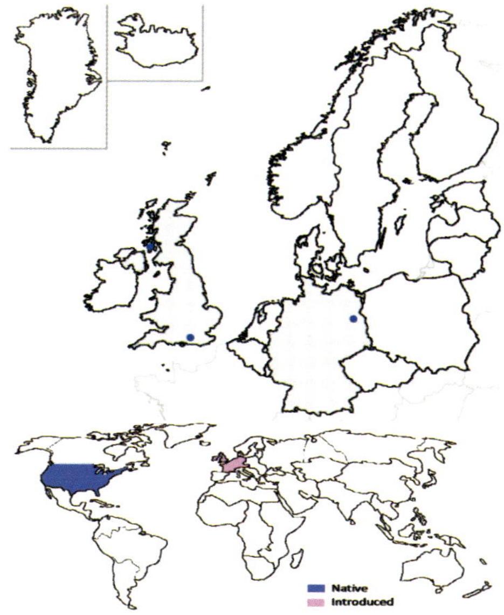

Perennial hydrophyte or amphiphyte. Rhizome short, vertical. Stolons slender, with apical turions, ovoid, slightly swollen and about 1.5 cm long. Roots whitish.

Leaves: In rosettes from the upper part of the rhizome and of different shape depending on water depth. Young and submerged plants develop ribbon-like leaves 1–15 mm wide and up to 2.5 metres long, with obtuse apex. Floating leaves with 15–50 mm long, elliptical to ovate oblong, entire lamina and long petioles. Emergent leaves normally absent.

Inflorescence: Monoecious. Scape up to 100 cm long, flexuous and slender, typically submerged with flowers floating or held above the water. Inflorescence racemose with 1–4(–10) whorls. Bracts connate for more than 1/4 of their length. Lower 1–2 whorls with female flowers and upper whorls with male flowers. Pedicels 1–6 cm long, rather slender.

Flowers: 3-merous. Sepals green, ovate, 3–4.5 mm long, those of the female flowers appressed in fruit. Petals 10–15 mm long, white. Male flowers with 7–9 stamens, filaments 1 mm long, glabrous, anthers yellow. Female flowers with numerous, short-styled carpels.

Fruits: Fruiting head globular, 7–10 mm in diameter. Achenes greyish, obovate, compressed, about 1.5–2.5 mm long, with lateral beak 1 mm long and crenate wings.

Flowering: June-October.

Biology: Pollination mainly by flies attracted to the large white flowers. The fruits are usually fertile. Overwinters and propagates by turions buried in the mud.

Habitats: Tidal rivers, slow flowing streams and ponds. The British population occurs in a lowland pond with acidic, shallow water.

Distribution within the region:
Introduced and naturalised in Hampshire, Britain, and at one location in Germany. Native to North America.

Characteristics and similar species:
S. filiformis can be distinguished from all other *Sagittaria* species by the delicate scape which is submerged or supported by the water column, with only individual flowers floating or emergent, combined with the yellow anthers, fruiting heads with reflexed pedicels and stamens with glabrous filaments, lack of emergent simple floating leaves and flattened submerged leaves.

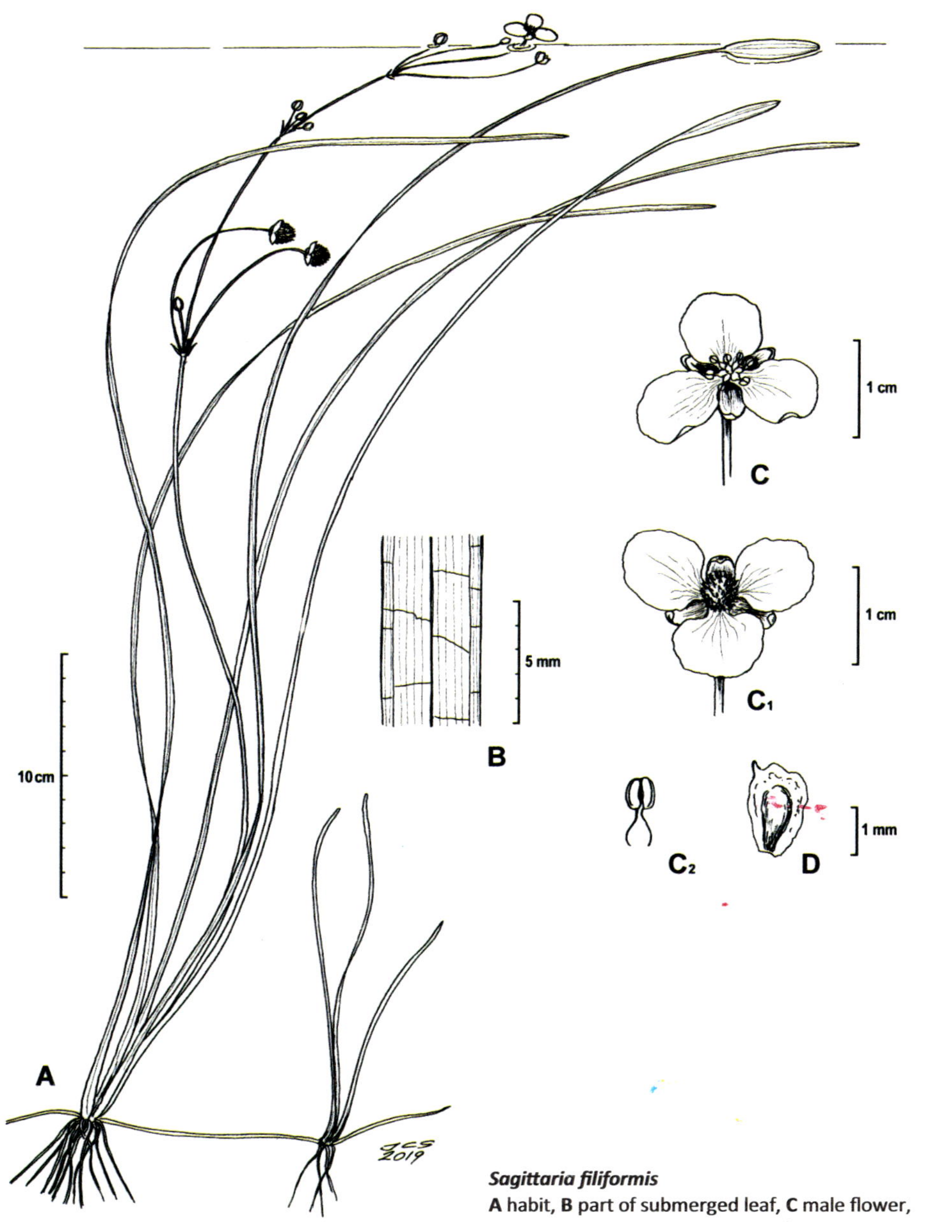

Sagittaria filiformis
A habit, **B** part of submerged leaf, **C** male flower, **C1** female flower, **C2** stamen, **D** achene.

Alismataceae

43 *Sagittaria natans* Pall.

DK: Svømmende Pilblad **N:** Buttpilblad **S:** Trubbpilblad **FIN:** Kelluskeiholehti **IS:** - **GB:** - **NL:** - **F:** - **D:** - **CZ:** - **PL:** - **EST:** - **LV:** - **LT:** - **Ru:** Стрелолист плавающий

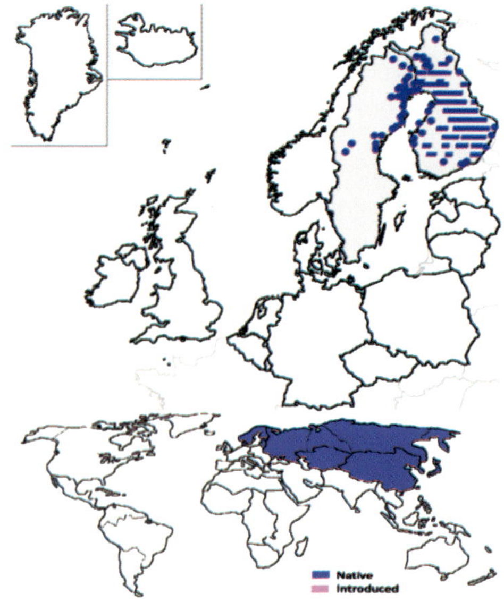

Flowering *S. natans*. Kilnäset, Sweden. Photo RVL.

Perennial hydrophyte or amphiphyte. Rhizome short, vertical. Stolons slender, 2–4 mm thick, not rooting, producing ovoid, slightly swollen apical turions, about 1.5 cm long late in the year. Roots whitish.

Leaves: In rosettes from the upper part of the rhizome and of different shape depending on water depth. Young and submerged plants develop thin, flaccid, light green to pale green ribbon-like leaves, 2–6 mm wide, with acuminate to narrowly obtuse, often mucronate apex. Floating leaves develop in still or slowly floating water and are always present on flowering plants. Lamina of floating leaves 3–15(–20) cm long, 0.3–3 cm wide, with 3–5 veins connected by secondary veins, green to brownish green sometimes with a purplish tinge, entire, oblong ovate to lanceolate, rounded at base - sometimes with small, more or less parallel basal lobes. Petioles slender, 20–70(–120) cm long, with 10–15 cm long sheaths at base. Emergent leaves absent.

Inflorescence: Monoecious. Scape 20–90 cm long, angular. Inflorescence racemose. Lower 1–2 whorls with female flowers and upper 1–4 whorls with male flowers. Pedicels 0.5–3 cm long, rather slender, bluntly trigonous or more or less circular in cross section.

Flowers: 3-merous. Sepals brownish, ovate, 5–7 mm long. Petals 6–10 mm long, white with a yellow base. Male flowers with numerous stamens with 1 mm long, glabrous filaments and yellow anthers. Female flowers with numerous, short-styled carpels.

Fruits: Fruiting head globular, 8–10 mm in diameter. Achenes greyish, obovate, compressed, 2.4–3.5 mm long, with an inconspicuous beak and membranous margins.

Flowering: July-August.

Biology: Plants growing at depths of 0.5 to 1.5 or even 2 metres will flower at the surface. Overwinters as turions. Pollination mainly by flies attracted by the large white flowers. The fruits are usually fertile, but vegetative propagation by turions seems to be of some importance too.

Habitats: In oligotrophic to moderately eutrophic, fresh or sometimes weakly brackish water to 2 metres depth.
Lakes, slow-flowing rivers, estuaries and protected sea creeks.

Distribution within the region:
Native. Within the region known only from Sweden and Finland.

Characteristics and similar species:
Easily mistaken for **44 *S. ×lunata***, which differ in having spreading, rather than subparallel, lower lobes of floating leaves. *S. natans* is typically highly fertile, whereas *S. ×lunata* is sterile. Submerged plants with ribbon-like leaves may be mistaken for species of **114-122 *Sparganium*** or **145-146 *Schoenoplectus***, but can be recognised by the prominent, oblique secondary veins connecting the primary veins.

Sagittaria natans with typical developed floating leaves. Kilnäset, Sweden. Photo KvdW.

Alismataceae

44 ***Sagittaria ×lunata*** C.D.Preston & Uotila *(S. natans x sagittifolia)*

Flowering *S. ×lunata*. Norrbotten, Laiva-vigen, Sweden. Photo Vilhelm Wrigstedt.

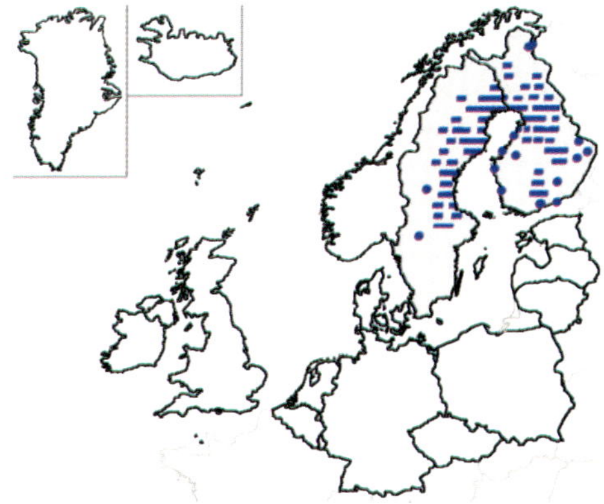

Distribution within the region:
Native. Sweden, Finland and the north-western part of Russia.

Sagittaria ×lunata is almost always sterile with empty fruits. Filaments of the male flowers are glabrous. In terms of habit it is more or less intermediate between the parents.

It differs from **45 *S. sagittifolia*** in the narrower and more acute submerged leaves, in the more elongate floating leaves and in the colour of the anthers, which are yellowish with a more or less prominent purple tinge.

Compared to **43 *S. natans*** the hybrid differs in the colour of the anthers, as well as in the floating leaves, which are more frequently lobed and usually with a broader sinus between the basal lobes; emergent leaves sometimes develop in shallow water.

Habitats: In oligotrophic to moderately eutrophic, fresh or sometimes weakly brackish water to 2 metres depth. Lakes, slow-flowing rivers, estuaries and sheltered coastal creeks.

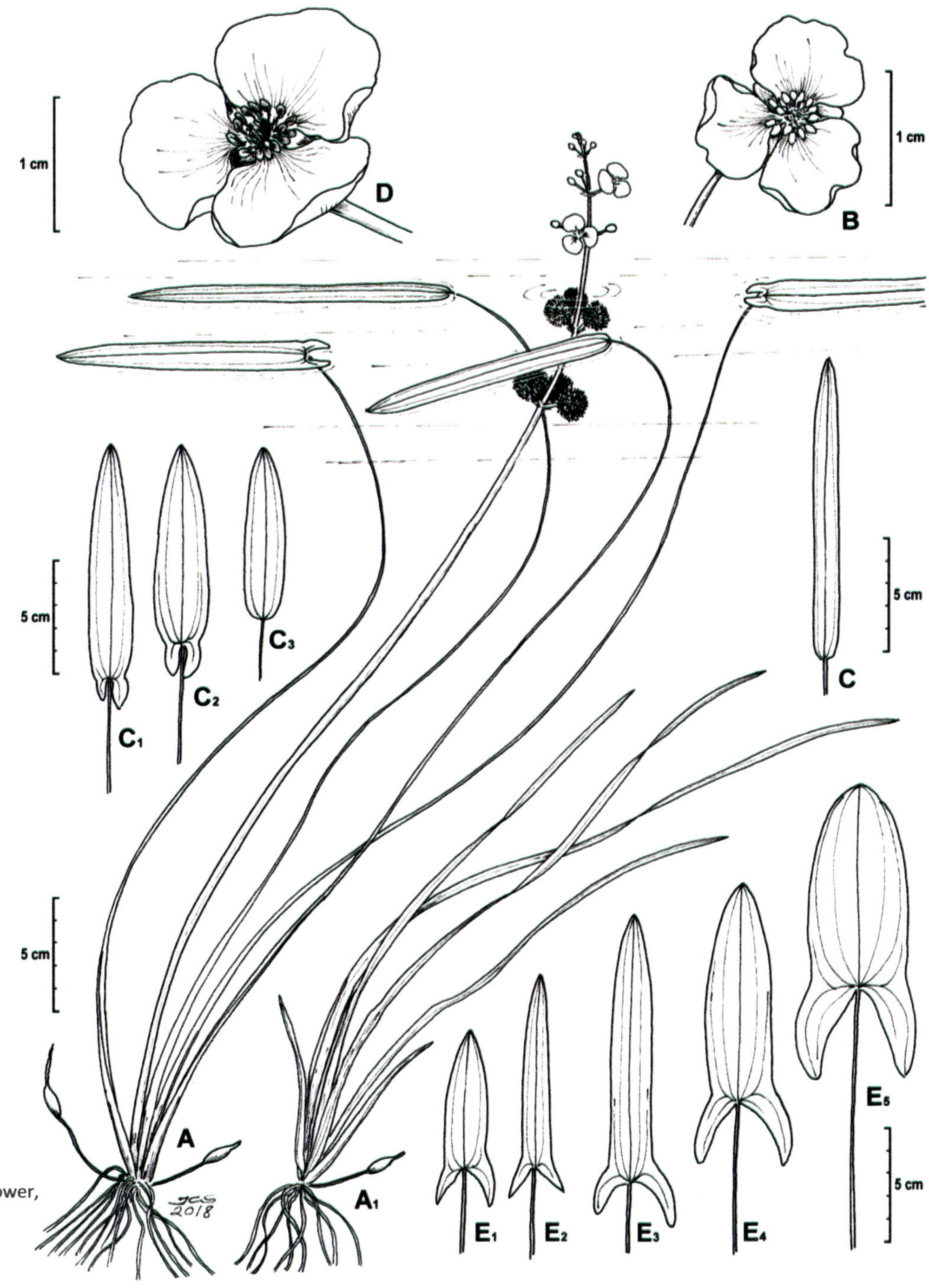

Sagittaria natans
A habit, **A1** habit submerged plant, **B** male flower, **C-C3** floating leaves.
Sagittaria ×lunata
D male flower, **E1-E5** floating leaves.

Alismataceae

45 *Sagittaria sagittifolia* L.

DK: Almindelig Pilblad **N:** Pilblad **S:** Vanlig Pilblad **FIN:** Pystykeiholehti **IS:** -
GB: Common Arrowhead **NL:** Gewoon pijlkruid **F:** Flèche-d'eau
D: Gewöhnliches Pfeilkraut **CZ:** Šípatka střelolistá **PL:** Strzałka wodna
EST: Jõgi-kõõlusleht **LV:** Parastā bultene **LT:** Strėlialapė papliauška **Ru:** Стрелолист обыкновенный

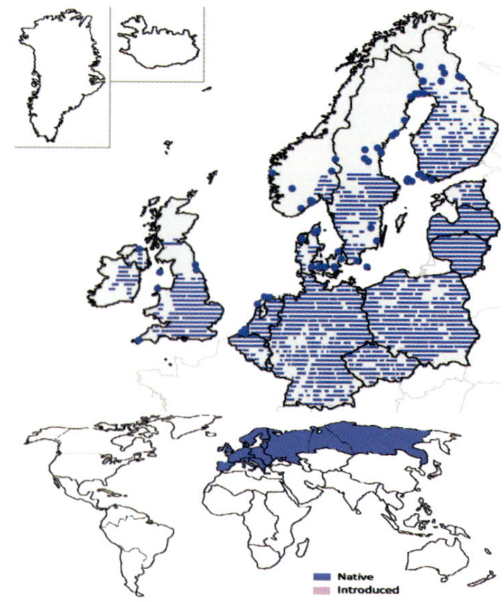

Flowering *S. sagittifolia* showing different leaf forms. Ny Frederikskog, Højer, Denmark. Photo JCS.

Perennial hydrophyte or amphiphyte. Rhizome short, vertical and swollen. Stolons slender, 3–7 mm diameter, not rooting, late in the year developing apical turions, which are up to 3 cm long and ovoid. Roots whitish.

Leaves: In rosettes from the upper part of the rhizome and of different shape depending on water depth and other factors such as flow. Young and submerged plants develop thin, flaccid, light green to pale green, ribbon-shaped, shortly acuminate leaves, 25–200 cm long, (5–)10–20(–30) mm wide. These usually have 5 veins and prominent, oblique secondary veins. In still or shallow water, floating and emergent leaves develop. These are longpetiolate with a light green to dark green lamina - the floating leaves often have a tinge of violet. Lamina and lobes with 3–5 veins connected by secondary veins. Lamina of emergent leaves 5–15 (–20) cm long of variable width and shape, from spoon- to arrow shaped with long basal lobes. Floating leaves similar to the emergent leaves, but wider and more rounded. Transitional leaves between submerged and floating leaves are common, and likewise between floating and emergent leaves.

Inflorescence: Monoecious. Scape 30–90 cm long, angular. Inflorescence racemose. Lower 1–3 whorls with female or rarely hermaphrodite flowers with short, thick pedicels. Upper 3–7 whorls with male flowers with slender pedicels up to 3 cm long. The female flowers have usually gone over when the male flowers open.

Flowers: 3-merous. Sepals green, ovate, 5–7 mm long. Petals 10–15 mm long and up to 20 mm wide, white with a purplish base. Male flowers with numerous stamens, glabrous filaments with 1 mm long and dark violet anthers. Female flowers with numerous, short-styled carpels forming a 4–6 mm wide, globular head surrounded by rudimentary, sterile stamens.

Fruits: Fruiting head globular, 1–1.5 mm in diameter. Achenes greyish yellow, obovate, compressed, 4–6 mm long, with a slightly curved, c. 1 mm long beak, and broad membranous margins.

Flowering: July-August.

Biology: Overwinters as leafless stolons and turions. Pollination mainly by flies attracted by the large white flowers.

Submerged leaves of *S. sagittifolia*. River Skjern Å, Denmark. Photo JCS.

Sagittaria sagittifolia. From the left: male flower, female flower and fruit head. The globular fruit heads resemble those of *Sparganium* spp., but are situated 2 or 3 in whorls. Flowers - River Nørreå, Randers, Denmark. Fruit head - canal at Filsø, SW-Jutland, Denmark. Photos JCS.

Fruit set is often poor and vegetative propagation by stolons, turions and detached shoots appears to be more frequent. The achenes float or may stick to birds and in this way spread to other water bodies.

Habitats: *Sagittaria sagittifolia* occurs in eutrophic, fresh or sometimes weakly brackish water of rivers, canals, ditches, lakes and ponds.

Distribution within the region:
Native. Common throughout much of the region, but absent from Iceland, Greenland, western Norway and the extreme north of Finland.

Characteristics and similar species:
Submerged plants with ribbon-like leaves may be mistaken for species of **115-122 *Sparganium*** or **145-146 *Schoenoplectus***, but can be recognised by the prominent, oblique secondary veins connecting the primary veins.

46 *Sagittaria latifolia* differs by usually larger emergent leaves, white petals without any purple, yellow anthers and 2.5–4 mm long achenes. Submerged plants cannot be separated with certainty from *S. sagittifolia*. **43 *S. natans*** has yellow anthers, 3–4 mm long achenes with an inconspicuous beak, and submerged leaves only 2–6 mm wide. **44 *S. ×lunata*** is more or less intermediate between *S. natans* and *S. sagittifolia* but rarely produces emergent leaves and the anthers are yellowish with a purple tinge.

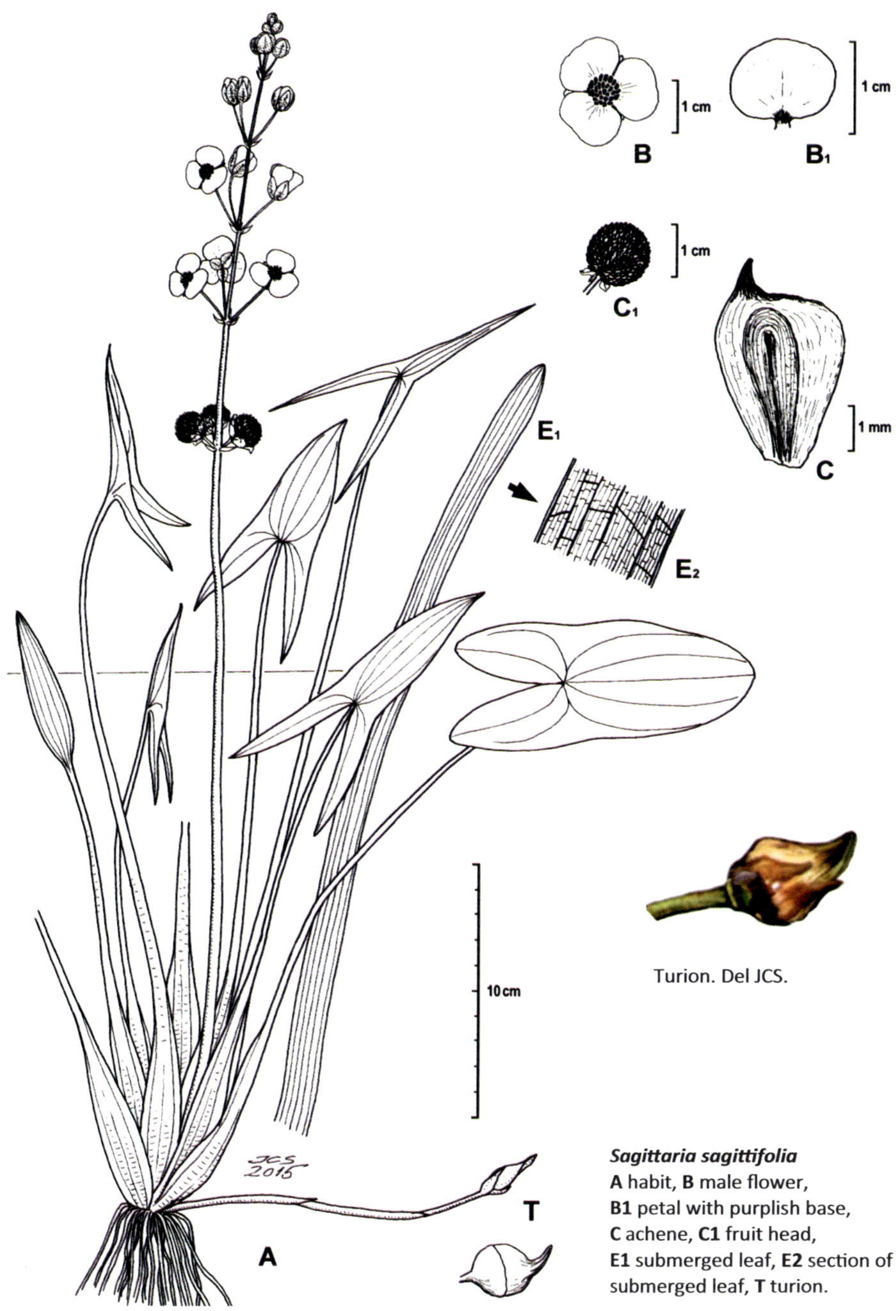

Turion. Del JCS.

Sagittaria sagittifolia
A habit, **B** male flower,
B1 petal with purplish base,
C achene, **C1** fruit head,
E1 submerged leaf, **E2** section of submerged leaf, **T** turion.

Alismataceae

46 *Sagittaria latifolia* L.

DK: Bredbladet Pilblad **N:** - **S:** Bredpilblad **FIN:** - **IS:** -
GB: Duck-potato **NL:** Breed pijlkruid **F:** Sagittaire à larges feuilles
D: Breitblättriges Pfeilkraut **CZ:** Šípatka širolistá **PL:** Strzałka szerokolistna
EST: - **LV:** - **LT:** - **Ru:** Стрелолист широколистный

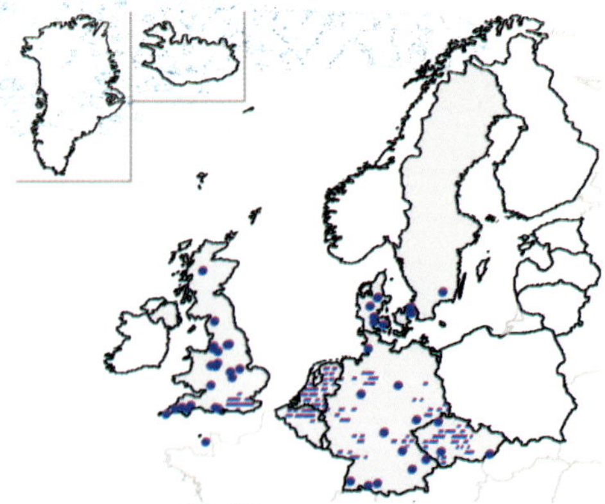

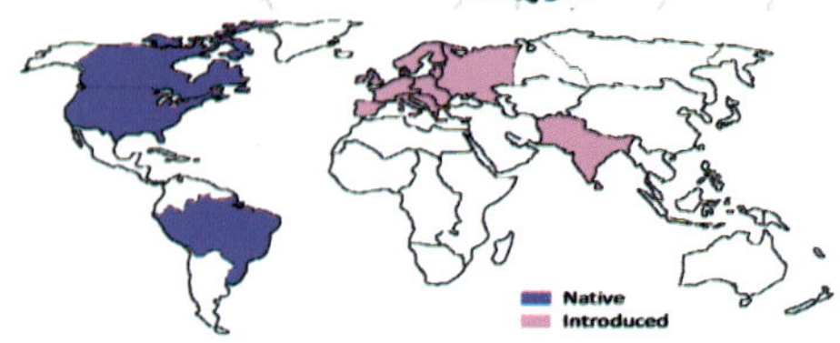

Sagittaria latifolia. Geldrop, Netherlands. Photo KvdW.

Perennial hydrophyte or amphiphyte. Rhizome short, vertical and swollen. Stolons slender, 1 cm diameter, not rooting, with ovoid, swollen apical turions up to 3 cm long late in the year. Roots whitish.

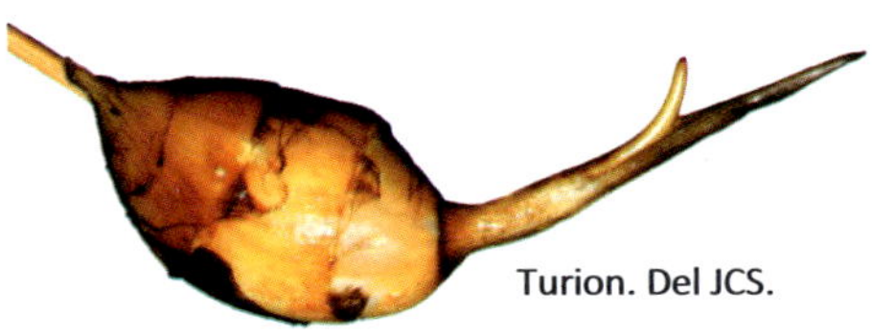

Turion. Del JCS.

Leaves: In rosettes from the upper part of the rhizome and of different shape depending on factors such as water depth and flow. Young and submerged plants develop thin, flaccid, light green to pale green ribbon-shaped, shortly acuminate leaves, up to 60 cm long. These usually have 5 primary veins and prominent, oblique secondary veins. In still or shallow water emergent or more rarely floating leaves develop. These are long petiolate with a light green to green lamina. Lamina and lobes with 3–7 veins connected by secondary veins.
Lamina of emerged leaves up to 35 cm long, 5–12 cm wide, variable in shape, from spathulate to broadly arrow shaped with long, broad basal lobes.

Inflorescence: Monoecious or dioecious. Scape 20–90(–150) cm long, angular. Inflorescence racemose, with flowers arranged in 2–8 whorls subtended by bracts. The lower 1–3 whorls female and the upper male or most often all whorls of the same sex. Rachis sometimes hairy.

Flowers: 3-merous. Sepals green or reddish, 7–11 mm long. Petals 15–20 mm long and up to 20 mm wide, white. Male flowers with numerous stamens with 1.5–2 mm long, yellow to yellowish-brown anthers, filaments glabrous. Female flowers with numerous, short-styled carpels.

Fruits: Fruiting head globular, 1–1.5 mm in diameter. Achenes greyish yellow, obovate, compressed, 2.3–3.5 mm long, with an almost perpendicular protruding beak c. 1 mm long and wide-winged margin.

Flowering: July-September.

Biology: Overwinters as leafless stolons and turions. Pollination mainly by flies attracted by the large white flowers. The fruit set is often poor and vegetative reproduction by stolons, turions and detached shoots appears to be more frequent. The achenes float or may stick to birds and be dispersed by them to other water bodies.

Habitats: In eutrophic water. Rivers, canals, ditches, lakes and ponds.

Distribution within the region:
Introduced to garden ponds as an ornamental plant and now spread to natural habitats. In several European countries regarded as invasive. Native to North America and northern South America.

Characteristics and similar species:
Submerged plants with ribbon-shaped leaves may be mistaken for species of **115-122** ***Sparganium*** or **145-146** ***Schoenoplectus***, but can be recognised by the prominent, oblique secondary veins connecting the primary veins. Plants with emergent leaves and flow-

Sagittaria latifolia. Cultivated. Aarhus, Denmark. Photo JCS.

Sagittaria latifolia. Male flower. Photo KvdW.

Female flower. Photo BM.

Emergent leaf. Photo JCS.

ers may be mistaken for **45 *S. sagittifolia***, but differ in the leaves which are typically larger, white petals lacking purple colouration and smaller achenes.

46a *Sagittaria montevidensis* Cham. & Schltdl. is rather similar in habit but with cream to white petals with a red-coloured spot at base. Native in America. Introduced in Sonderhausen, Germany.

Peltandra virginica (L.) Schott (Araceae) is sometimes found as a garden escape. It has very large, emergent leaves with almost similar leaf shape as *S. latifolia*, but different venation. Native in North America, naturalised in Denmark.

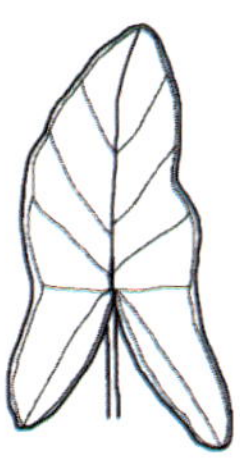

Peltandra virginica leaf.

Peltandra virginica. Arreskov Sø, Fyn, Denmark. Photo Rikke Persson.

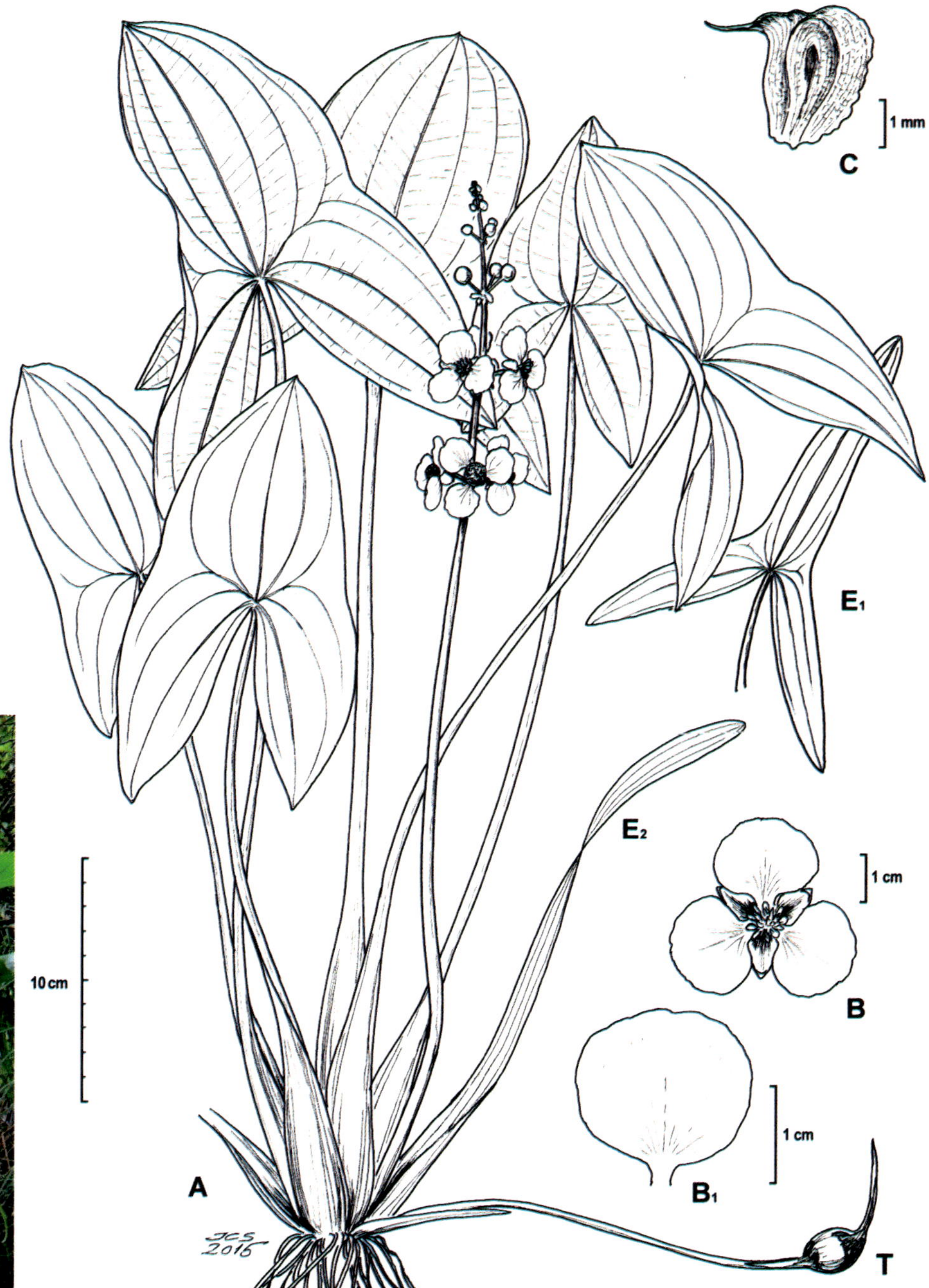

Sagittaria latifolia
A habit, **B** male flower, **B1** petal, **C** achene, **E1** emergent leaf with narrow lobes, **E2** submerged leaf, **T** turion.

Alismataceae

47 *Sagittaria graminea* Michx.

DK: - **N:** - **S:** - **FIN:** - **IS:** -
GB: Grass-leaved Arrowhead - **F:** -
D: - **CZ:** - **PL:** -
EST: - **LV:** - **LT:** - **Ru:** Стрелолист злаковый

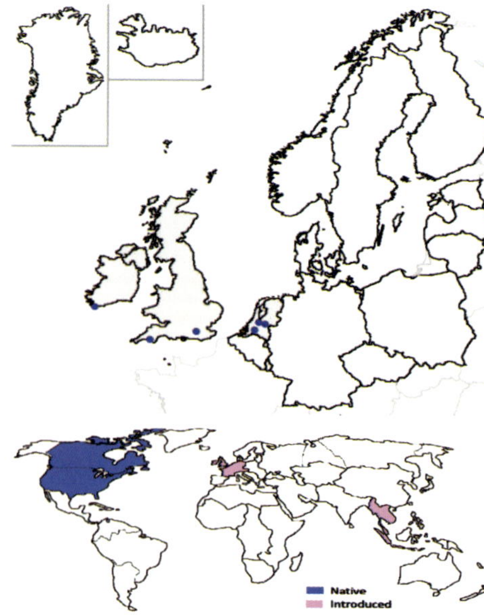

Perennial hydrophyte or amphiphyte. Rhizome short, stolons without turions.

Leaves: Submerged leaves phyllodial, flattened on the adaxial face, with a ridge on the abaxial face. Floating leaves oblong elliptic. Emergent leaves with lamina linear to oblanceolate and up to 4 cm wide.

Inflorescence: Monoecious. Scape 30–50 cm long, angular. Inflorescence emergent, racemose, with flowers in 2–12 whorls. Lower whorls with female or rarely hermaphrodite flowers with thick ascending pedicels. Upper whorls with male flowers on slender, erect pedicels. Bracts connate to approximately half their length.

Flowers: 3-merous. Sepals 3–6 mm long, reflexed in fruit. Petals 10–20 mm long and up to 20 mm wide, white or pink. Filaments with scale-like hairs.

Fruits: Fruiting head globular. Achenes obovate, 1.2–3.0 mm long, winged on the margins and with a lateral beak, 0.2 mm long.

Flowering: July-August.

Habitats: In eutrophic, fresh or sometimes weakly brackish water. Rivers, canals, ditches, lakes and ponds.

Distribution within the region:
Introduced and naturalised in Ireland and the Netherlands. Native to eastern North America.

Characteristics and similar species:
S. graminea can be distinguished from other *Sagittaria* species by the combination of simple emergent leaves, the erect inflorescence on which fruiting female flowers are on ascending pedicels which are more than 0.5 cm long and the stamens with yellow anthers and pubescent filament.

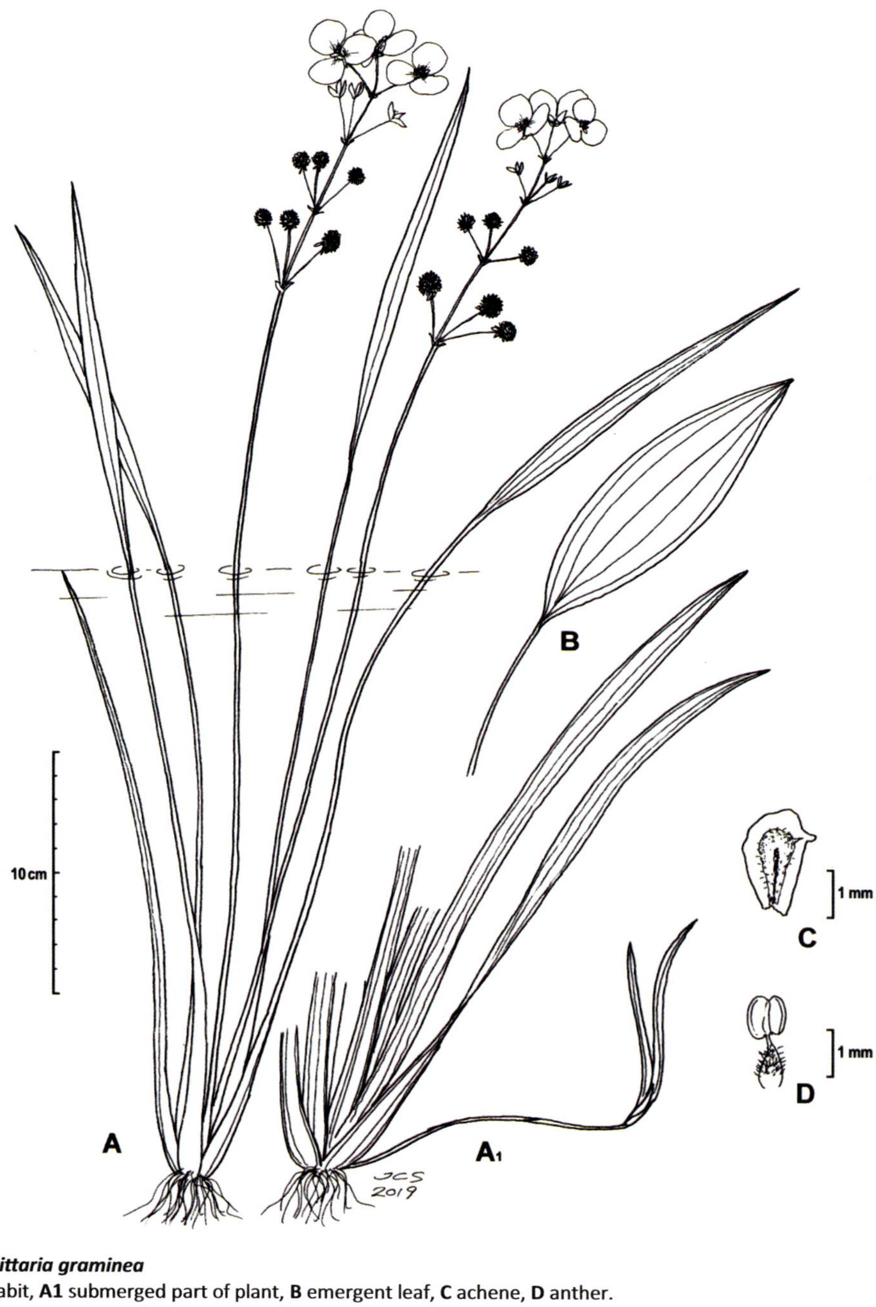

Sagittaria graminea
A habit, **A1** submerged part of plant, **B** emergent leaf, **C** achene, **D** anther.

Alismataceae

48 *Sagittaria rigida* Pursh

Syn.: *S. heterophylla* Pursh

DK: - **N:** - **S:** - **FIN:** - **IS:** - **GB:** Canadian Arrowhead - **F:** - **D:** - **CZ:** - **PL:** - **EST:** - **LV:** - **LT:** - **Ru:**

Sagittaria rigida. Exeter Canal, Devon, Britain. Photo RVL.

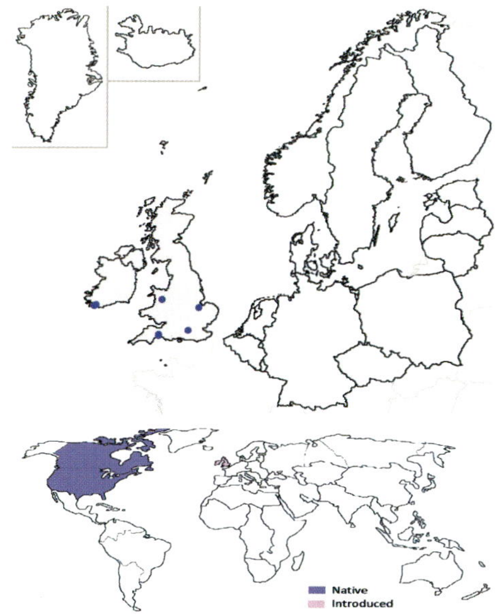

Perennial hydrophyte or amphiphyte. Rhizome short, stolons with apical turions.

Leaves: Submerged leaves phyllodial, flattened, 30–70 cm long, when reaching the surface sometimes widened and floating. Emergent leaves with linear to elliptic to broadly lanceolate lamina, 5–15 cm long and 0.5–12 cm wide, rarely with narrow basal lobes.

Inflorescence: Monoecious. Scape 15–100 cm long. Inflorescence emergent, racemose, with flowers in 2–8 whorls. Lower whorls with sessile female flowers - the pedicel sometimes to 0.5 cm long. Upper whorls with male flowers on very slender, 1.5–3 cm long pedicels. Bracts connate for more or less 1/4 of their length.

Flowers: 3-merous. Sepals 4–7 mm long, reflexed in fruit. Petals 10–15(–20) mm long and up to 20 mm wide, white. Filaments with scale-like hairs.

Fruits: Fruiting head globular, 1–1.7 cm in diameter. Achenes obovate, 2–3 mm long, winged on the margins and with a lateral beak 0.8–1.4 mm long.

Flowering: July-October.

Habitats: In eutrophic, fresh or brackish water. Rivers, canals, ditches, lakes and ponds.

Distribution within the region: Introduced and naturalised in Ireland and Britain. Native to eastern North America.

Characteristics and similar species: *Sagittaria rigida* can be recognised by the sessile or almost sessile female flowers, the hairy filaments and elliptic to broadly lanceolate lamina of the emergent leaves.

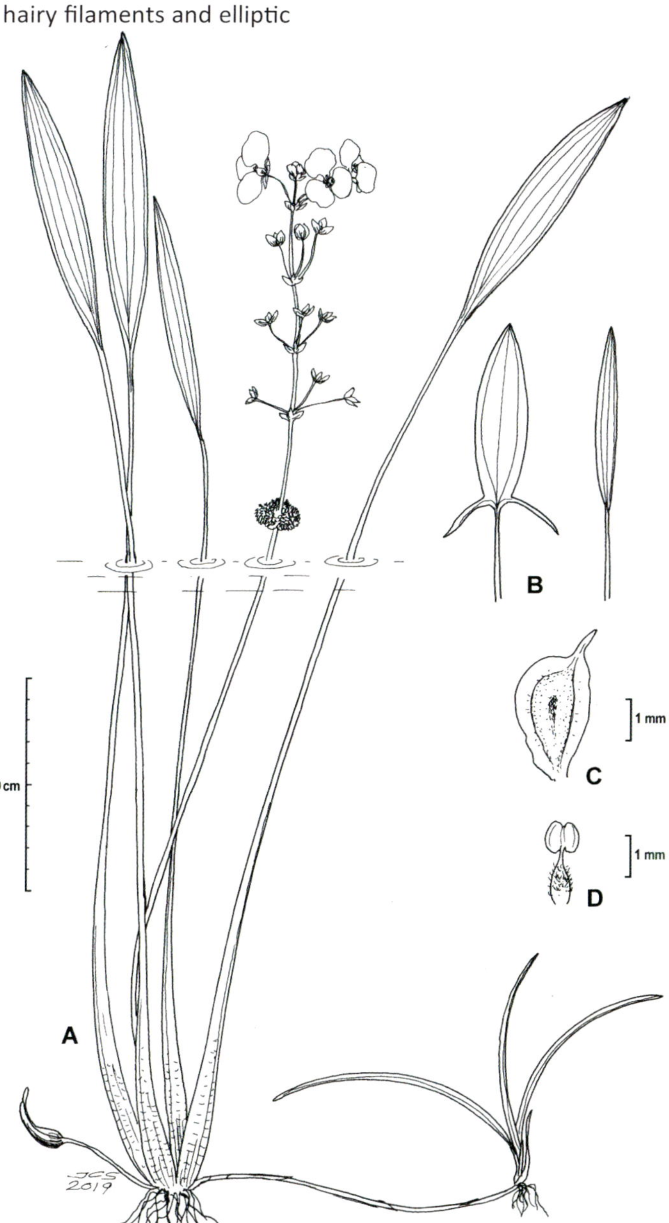

Sagittaria rigida **A** habit, **B** emergent leaves, **C** achene, **D** anther.

Aponogetonaceae

49 *Aponogeton distachyos* L.f.

DK: - **N:** - **S:** - **FIN:** - **IS:** -
GB: Cape Pondweed **NL:** - **F:** Aponogéton à deux épis
D: Wasserähre **CZ:** - **PL:** -
EST: - **LV:** - **LT:** - **Ru:** Апоногетон двуколосый

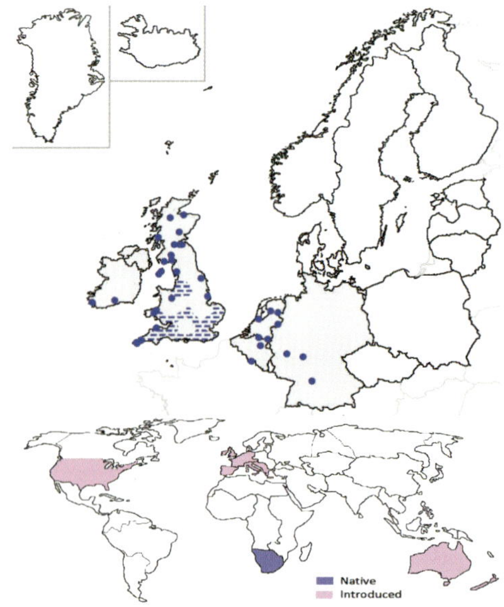

Aponogeton distachyos was introduced to Europe as an ornamental plant for garden ponds. Cultivated. Swansea, Britain. Photo RVL.

Perennial hydrophyte. Rhizome black, erect, up to 6 cm long and 3 cm in diameter. Without stolons. Roots numerous, white.

Leaves: In rosettes from the upper part of the rhizome. Leaves all floating with lamina 5–25 cm long, 1.5–8 cm wide, elliptic to oblong, rounded to attenuate at base, rounded to somewhat acute at apex, entire, light to dark green, often with red or brownish colouration, firm to almost coriaceous, with 7–9 parallel veins connected by secondary veins. Petiole 30–100 cm long, sheathing at base. The first leaves produced from seedlings are submerged and phyllodial, but laminar; floating leaves gradually develop as plants mature.

Inflorescence: 1–6 per plant. Peduncle up to 80 cm high, 3–6 mm wide, becoming markedly wider towards the spike and lifting it just above the surface of the water. Spike forked into two 2–6 cm long branches facing each other, with a deciduous spathe at the base. Each branch of the spike with 10–14 flowers arranged in two rows.

Flowers: Each flower subtended by 1 (–2), bract-like, ovate perianth segments, 10–15 mm long and 3.5–6 mm wide, white or pinkish, which at maturity become greenish and elongate to c. 3 cm. 6–20 stamens with 3–4.5 mm long, white filaments and dark purple to almost black anthers.
Carpels 2–6, up to 1.5 cm long, each with several ovules. Styles 1, short, terminal.

Fruits: Follicles in groups of 3(–9), free, up to 22 mm long and 6 mm in diameter, with a straight or curved terminal beak 5mm long. Each follicle contains (1–)2–4 (–6) flattened, ellipsoid seeds, 8–10 mm long.

Flowering: June-September.

Biology: The flowers have a sweet scent and are pollinated by bees and other insects, but the flowers are also capable of self-pollination. The seeds ripen under water and are able to float for a while.
In South Africa it is widely cultivated for its edible rhizome, buds and flowers.

Habitats: In shallow, slightly acidic water in sunlit ponds, lakes and canals.

The inflorescence of *A. distachyos* is forked. Kirstenbosch Botanical Garden, Cape Town, South Africa. Photo Jan-Thomas Johansson.

Aponogeton distachyos is easily recognised when flowering. Cultivated. Swansea, Britain. Photo RVL.

Distribution within the region:
Introduced and locally naturalised in Ireland, Britain, the Netherlands, Belgium and Germany. Native to South Africa.

Characteristics and similar species:
Aponogeton distachyos is recognised by the forked inflorescence and the shape and venation of the floating leaves. When not flowering it could be mistaken for some species of *Potamogeton* from which it can be distinguished by the leaves, which all arise from the rhizome rather than alternately from a stem. It could also be mistaken for some species of Alismataceae, such as *Damasonium alisma* or *Luronium natans*, from which it can be distinguished by dense venation of the leaves and black rhizome.

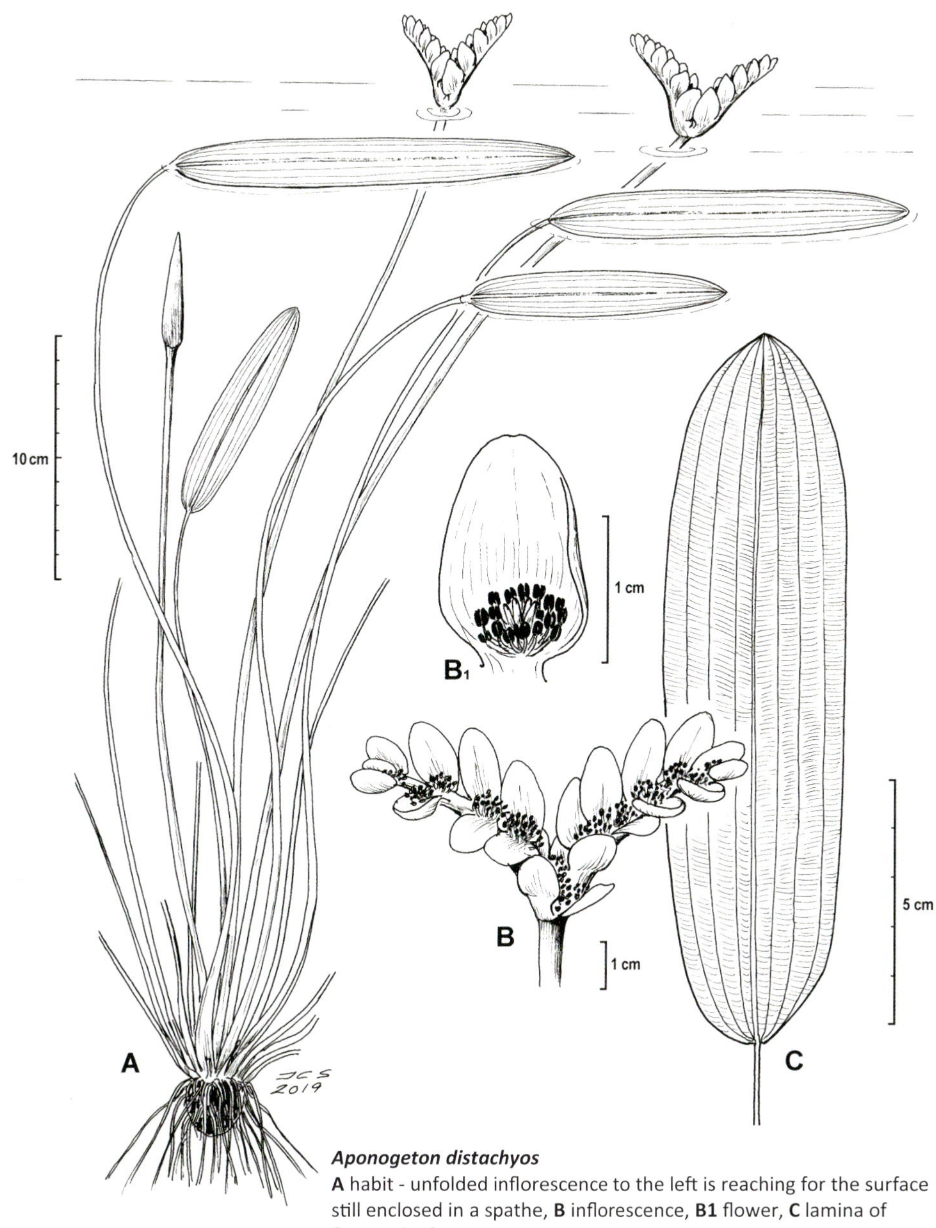

Aponogeton distachyos
A habit - unfolded inflorescence to the left is reaching for the surface still enclosed in a spathe, **B** inflorescence, **B1** flower, **C** lamina of floating leaf.

Butomaceae

50 *Butomus umbellatus* L.

DK: Brudelys **N:** Brudelys **S:** Blomvass **FIN:** Sarjarimpi **IS:** -
GB: Grass Rush **NL:** Zwanenbloem **F:** Butome en ombelle
D: Schwanenblume **CZ:** Šmel okoličnatý **PL:** łączeń baldaszkowy
EST: Harilik luigelill **LV:** Čemurainais puķumeldrs **LT:** Skėtinis bėžis **Ru:** Сусак зонтичный

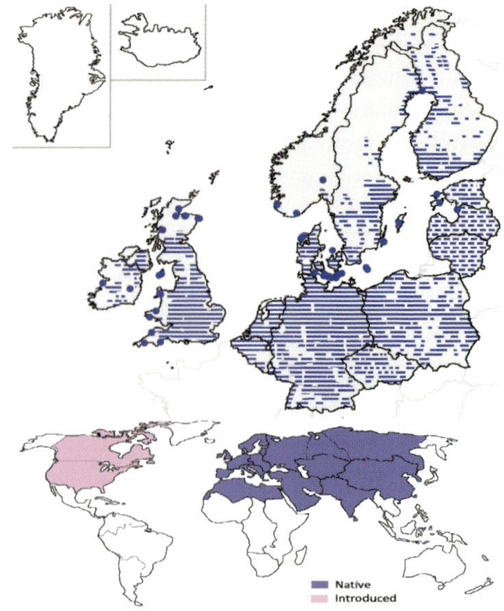

Butomus umbellatus and *Sagittaria sagittifolia* close to the shore of River Skjern Å, Denmark. Photo JCS.

Perennial hydrophyte or amphiphyte. Rhizome horizontal, up 2 cm in diameter, with short nodules and lateral buds. Roots numerous from the lower side of the rhizome.

Leaves: All basal, distichously arranged on the rhizome, 50–120 cm. They are dark green, sheathing and c. 1 cm wide at base, becoming triangular in cross section, usually flatter and often twisted toward the apex. Submerged leaves are similar to emergent leaves. In flowing water, where the species can form large populations, the leaves are normally submerged, with a flatter triangular cross section, but still rather stiff. Young submerged plants with a leaf width of 3–4 mm may feel rather soft in the water, but do not droop when lifted above the water.

Inflorescence: Scape, 50–150 cm high, terete, dark green to brownish, terminating in an apical flower and 3 helical constricted tassels, each subtended by a 2–3 cm long bract. All together the tassels form a 15–20 cm wide umbel-like inflorescence, with flowers opening one at a time over a long period.

Flowers: 3-merous. Pedicels 5–10 (–12) long, stiff. Sepals 10–12 mm long, obovate, brownish pink to dark purple with a pinkish to white margin, rounded at apex. Petals 10–15 mm long, white to pink with reddish to violet veins. Stamens 9, anthers reddish, pollen orange to brown, filaments 5–6 mm long, white - pink towards the base. Carpels 6, connate at base. Sepals and petals persisting in fruit.

Fruits: The 6 follicles are purplish brown, each of them containing numerous ovules. Seeds reddish brown, cylindrical, 1.5–1.8 mm long and 0.6 mm wide, with 8–10 prominent ribs.

Flowering: July-August.

Biology: The flowers are visited by flies, but fruiting is often sparse and propagation by detached, lateral buds from the rhizome appear to be more important for dispersal within water bodies. In late autumn the leaves die back and only the leafless rhizomes overwinter.

Submerged *B. umbellatus*. Inserted: A twisted section of a leaf and a cross section. Photos KvdW and JCS.

Butomus umbellatus. Left: Flower in male stage - the outer 6 anthers are open, while the inner 3 are closed. They will open when the stigmas are ready thereby making self-pollination possible. Aarhus Botanical Garden, Denmark. Right: Flower with mature follicles. Nørresø, Tønder, Denmark. Photos JCS.

Habitats: Typically in the margins of nutrient-rich streams, canals, ditches, lakes and ponds. It will also form large populations in the channels of slow-flowing rivers.

Distribution within the region: Native, but sometimes introduced even in natural water bodies. Common throughout, except northern Fennoscandia.

Characteristics and similar species: The large umbel-like inflorescence and the dark green leaves with triangular cross section make *Butomus* easy to recognise. When not flowering it is easily overlooked, particularly among stands of emergent monocots and is most easily spotted by the leaves, which taper gradually to the apex and often lean out from a single point.

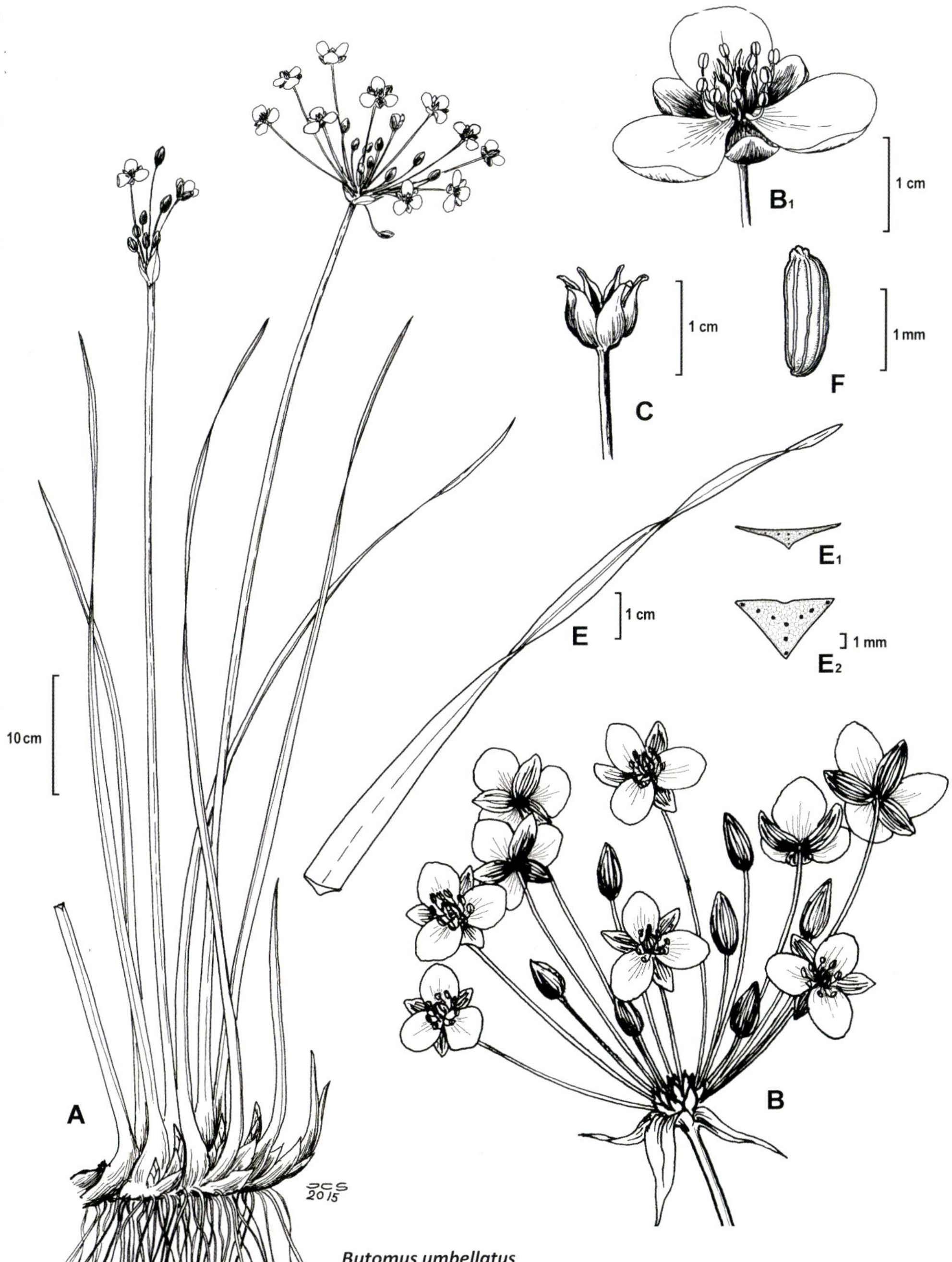

Butomus umbellatus
A habit, **B** inflorescence, **B1** flower, **C** 6 follicles connate at base, **E** leaf, **E1** cross section submerged leaf, **E2** cross section aerial leaf, **F** seed.

Hydrocharitaceae

51 *Hydrocharis morsus-ranae* L.

DK: Frøbid **N:** Froskebit **S:** Dyblad **FIN:** Kilpukka **IS:** -
GB: Frogbit **NL:** Kikkerbeet **F:** Hydrocharis des grenouilles
D: Froschbiss **CZ:** Voďanka žabí **PL:** Żabiściek pływający
EST: Konnakilbukas **LV:** Parastā mazlēpe **LT:** Plūduriuojantysis vandenplūkis **Ru:** Водокрáс лягушáчий

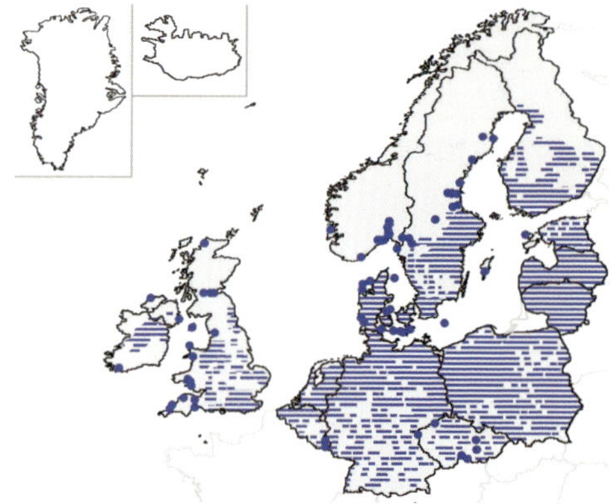

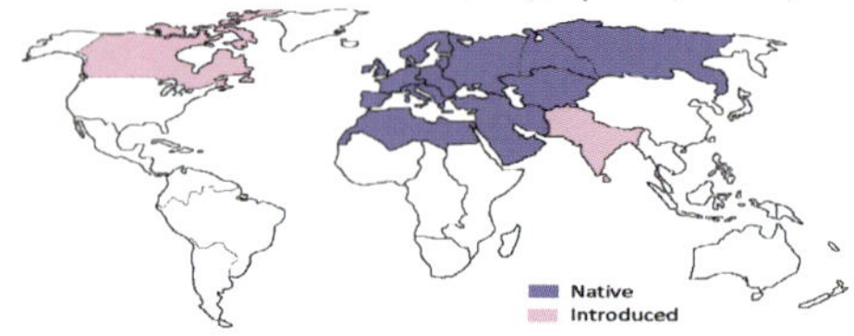

Flowering *H. morsus-ranae*. Ditch. Ny Frederikskog ved Højer, Jutland, Denmark. Photo JCS.

Perennial, free-floating hydrophyte. Stolons 5–20 cm long, 2 mm diameter, rooting from the nodes and terminating in a leaf rosette. Not rooted in substrate, initially green but soon turning white, unbranched, with numerous, long root hairs.

Stem: Leafy, vertical, about 1 cm long.

Leaves: 3–10, in a rosette from the stem. Petiole 5–10(–14) cm long, slender, widening at base and with 2 membranous stipules, protecting the leaf before it unfolds. Lamina floating, 2–7 cm long, circular to heart shaped, with a deep basal incision, the two lobes often overlapping. Upper surface shiny, bright green with a bronze tinge, lower surface often reddish or purplish. With 2 curved primary veins on each side of the midrib converging at the leaf tip, and many secondary cross-veins.

Inflorescence: Male and female flowers on separate plants or rarely with male and female flowers on the same plant but from different rosettes. Male flowers 2–4 together on pedicels c. 1 mm wide, 1–5 cm long from a 1.0–1.2 cm long spathe of 2 membranous bracts. Peduncle 2–5 cm long. Female flower solitary on a c. 2 mm wide, 3–8 cm long pedicel from a basal spathe of 1 bract. Peduncle absent.

Flowers: 3-merous, unisexual. Male flowers with 3 blunt pink to greenish sepals 4–5 mm long and 3 broadly obovate to orbicular 9–20 mm long petals, white with yellow at base. Stamens yellow, 9–12 in whorls of 3 - stamens on the outer whorl mostly sterile. Female flower similar to male flower, but petals often slightly smaller, 10–15 mm long. Stigmas 6 yellowish and divided for about half their length. 3 staminodes functioning as nectaries.

Fruits: The berry-like capsules are globose, green, often striate or tinged with purple and up to 10 mm in diameter. Seeds c. 1 mm long, elliptic, covered with hollow spiraliform tubercles.

Flowering: July-August.

Biology: *Hydrocharis* is pollinated by small flying insects. The seeds germinate on the water surface and resemble *Lemna* plants. Vegetative reproduction by detaching rosettes or by green turions 7–10 mm long, which develop and detach from the apex of stolons and sink to the substrate, where they remain dormant until spring.

Habitats: Shallow, still or slow-flowing water in nutrient-rich pools, lakes, ditches, canals, large streams and rivers.

Distribution within the region:
Native. Widespread throughout much of the region but scarce or absent from

The stoloniferous growth makes it possible for *H. morsus-ranae* to rapidly cover large water surfaces. Mature fruits are usually hidden beneath the leaves. Jutland, Denmark. Photo JCS.

Hydrocharis morsus-ranae. Left: Female flower with 6 bifid stigmas. Right: Male flower with yellow stamens. Ny Frederikskog, Højer, Jutland, Denmark. Photos JCS.

Hydrocharis morsus-ranae. Capsule. Cantley marshes, Norfolk, Britain. Photo RVL.

northern Scandinavia, Iceland and Greenland.

Characteristics and similar species: Easily recognised by its free-floating habit and by the leathery, circular to heart-shaped floating leaves with cordate base and membranous stipules. Without flowers it might be confused with **336 *Nymphoides peltata***, which has a repeatedly dichotomously divided reticulate pattern of veins on the leaves. Young plants of **1-3 *Nymphaea*** spp. and **4-6 *Nuphar*** spp. are rooted in the substrate and both with different leaf venation.

Limnobium laevigatum (Humb. & Bonpl. ex Willd.) Heine is cultivated as an aquarium plant and rarely found as a throw-out or escape. It differs from *Hydrocharis morsus-ranae* by a thick layer of air-filled spongy tissue on the lower side of the leaves.

Native in Central and South America.

Limnobium laevigatum. Staffanstorp, Skåne, Sweden. Photo JCS.

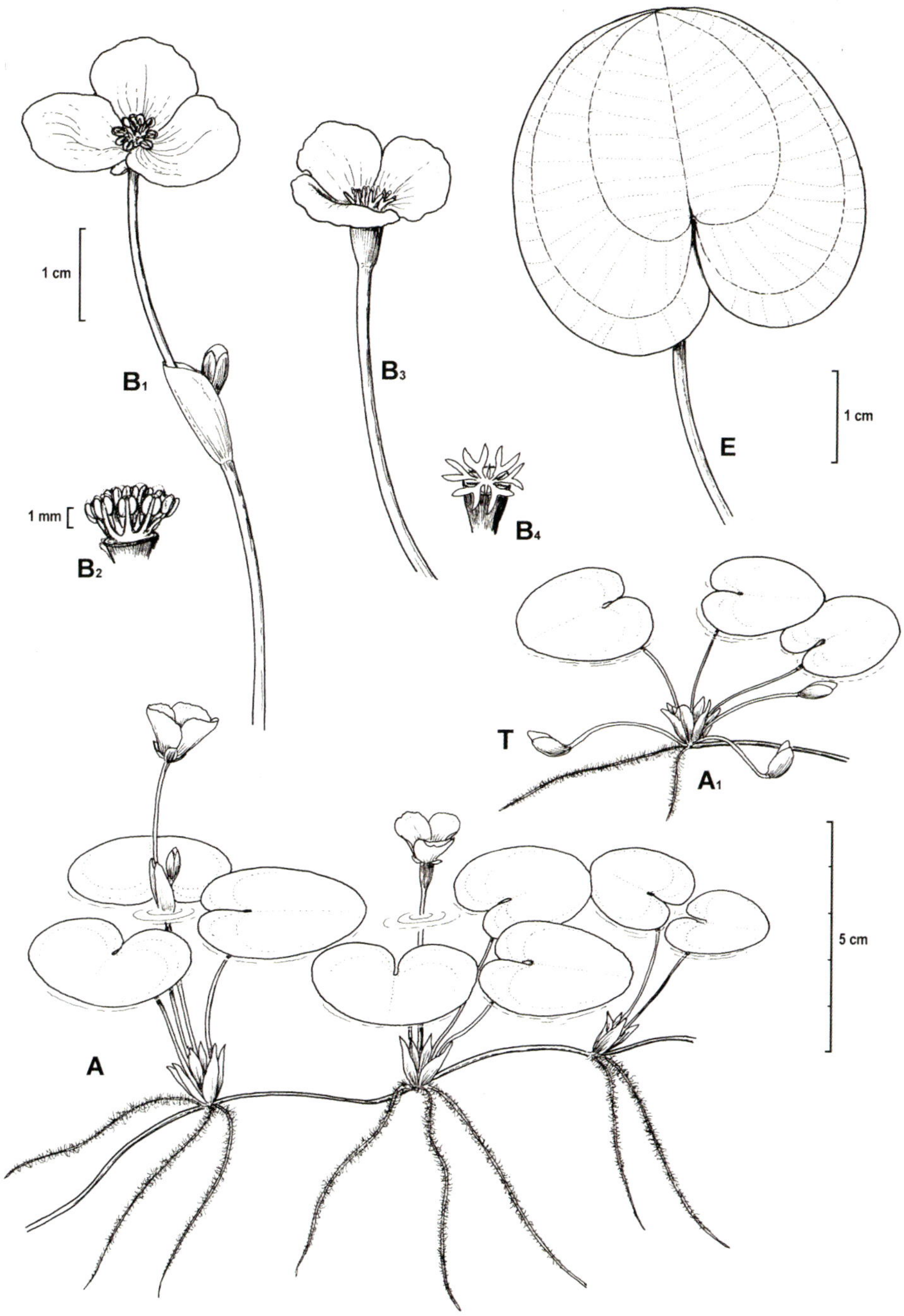

Hydrocharis morsus-ranae
A habit with male and female flowers, **A1** leaf rosette with 3 turions, **B1** male flowers, **B2** Male flower with stamens - perianth removed, **B3** female flower, **B4** female flower with 6 bifid stigmas - perianth removed, **E** floating leaf showing venation.

Hydrocharitaceae

52 *Stratiotes aloides* L.

DK: Krebseklo **N:** Vassaloe **S:** Vattenaloe **FIN:** Sahalehti **IS:** -
GB: Water Soldier **NL:** Krabbescheer **F:** Aloès d'eau
D: Krebsschere **CZ:** Řezan pilolistý **PL:** Osoka aloesowata
EST: Vesikarikas **LV:** Parastais elsis **LT:** Alavijinis aštrys **Ru:** Телорез обыкновенный

Flowering female *S. aloides*. Lilleborg, Almindingen, Bornholm. Photo JCS.

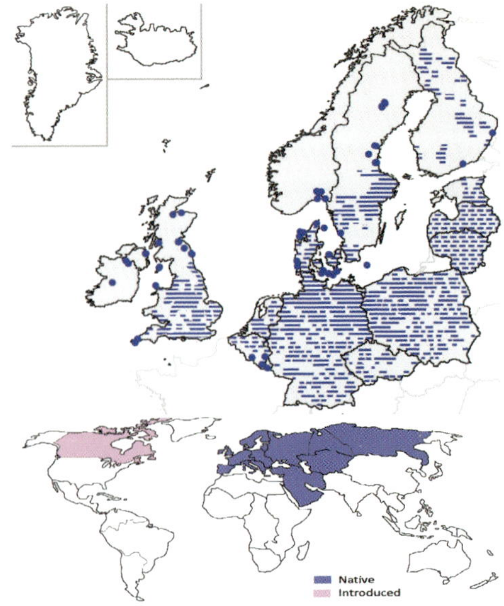

Perennial, free-floating hydrophyte, sometimes loosely anchored to the bottom by very long, unbranched roots.

Stem: Short and thick with up to 30 cm long, stolon-like branches carrying terminal rosettes.

Leaves: Spirally arranged in a funnel-shaped rosette from the upper part of the stem, with numerous longitudinal veins and cross-veins. Aerial leaves up to 40 cm long and 3 cm wide, dark green, stiff, linear, long acuminate, with sharply spinous-serrate margins. Submerged leaves up to 1 m long and about 1 cm wide, somewhat flaccid, light green. Lower leaves recurved and often with brownish-violet colours.

Inflorescence: Male and female flowers on separate plants. In both sexes, peduncle 10–30 cm long, about 1 cm wide, flat, with a claw-like spathe consisting of 2 free, overlapping, green 3-5 cm long bracts with a prominent, often spiny keel. Male flowers 2–3 on pedicels 3–10 cm long, each subtended by a bract inside the spathe. Female flower solitary, rarely 2, sessile.

Flower: 3-merous, unisexual. Sepals 3, greenish, ovate, c. 1–1.5 cm long. Petals 3, white - yellow at base, obovate, 2–2.5 cm long. Male flower with 12 or more yellowish stamens. Female flower with 6 yellowish, bifid stigmas and 15–30 thread-like staminodes functioning as nectaries.

Flowering: June-August.

Fruits: The berry-like, ovoid, hexagonal capsule contains 24 seeds. Seeds cylindrical, 6–11 mm long.

Biology: Roots and the oldest leaves die back in autumn, and only the youngest part of the plant remains, sinking to the bed where it overwinters. In April - May new roots develop and with sufficient light the plant begins photosynthesis, filling intercellular spaces in the leaves with gas, probably oxygen. The increased buoyancy lifts the plant to the surface to flower. Permanently submerged plants are known from oligo-mesotrophic lakes. The showy flowers smell like rotten meat and attract flies for pollination. Seeds are released from the ripe fruit in a gel-like mass and sink to the bottom. Vegetative propagation by stolon-like branches makes it possible for the plant rapidly to cover small waters and suppress all other aquatic plants. Populations covering entire ditches are increasingly common in Britain.

Habitats: Eutrophic ponds, pools, ditches and canals, oligo-mesotrophic lakes.

Distribution within the region:
Native. Throughout most of the region except Iceland and Greenland. In the northern part of its range and in the British Isles only female plants are pres-

In oligo-mesotrophic lakes *S. aloides* can be submerged for the whole year. Gültzsee, Germany. Photo KvdW.

Stratiotes aloides. Male plants. Fauler See, Germany. Photos KvdW.

Dense population of *S. aloides* almost covering the water surface. Cantley marshes, Norfolk, Britain. Photo RVL.

ent, and populations in the southern part of its range are predominantly male. Both sexes occur in the centre of its range.

Characteristics and similar species: Easily recognised by the stiff, linear sharply spinous-serrate leaves in floating or submerged rosettes and the claw-like spathes on top of the peduncles.

Stratiotes aloides. Female flower with 6 bifid stigmas surrounded by yellow staminodes functioning as nectaries. Hobro, Denmark. Photo JCS.

Stratiotes aloides. The "claw" - upper part of peduncle with a spathe of two green bracts, enclosing flowers or fruit. Hobro, Denmark. Photo JCS.

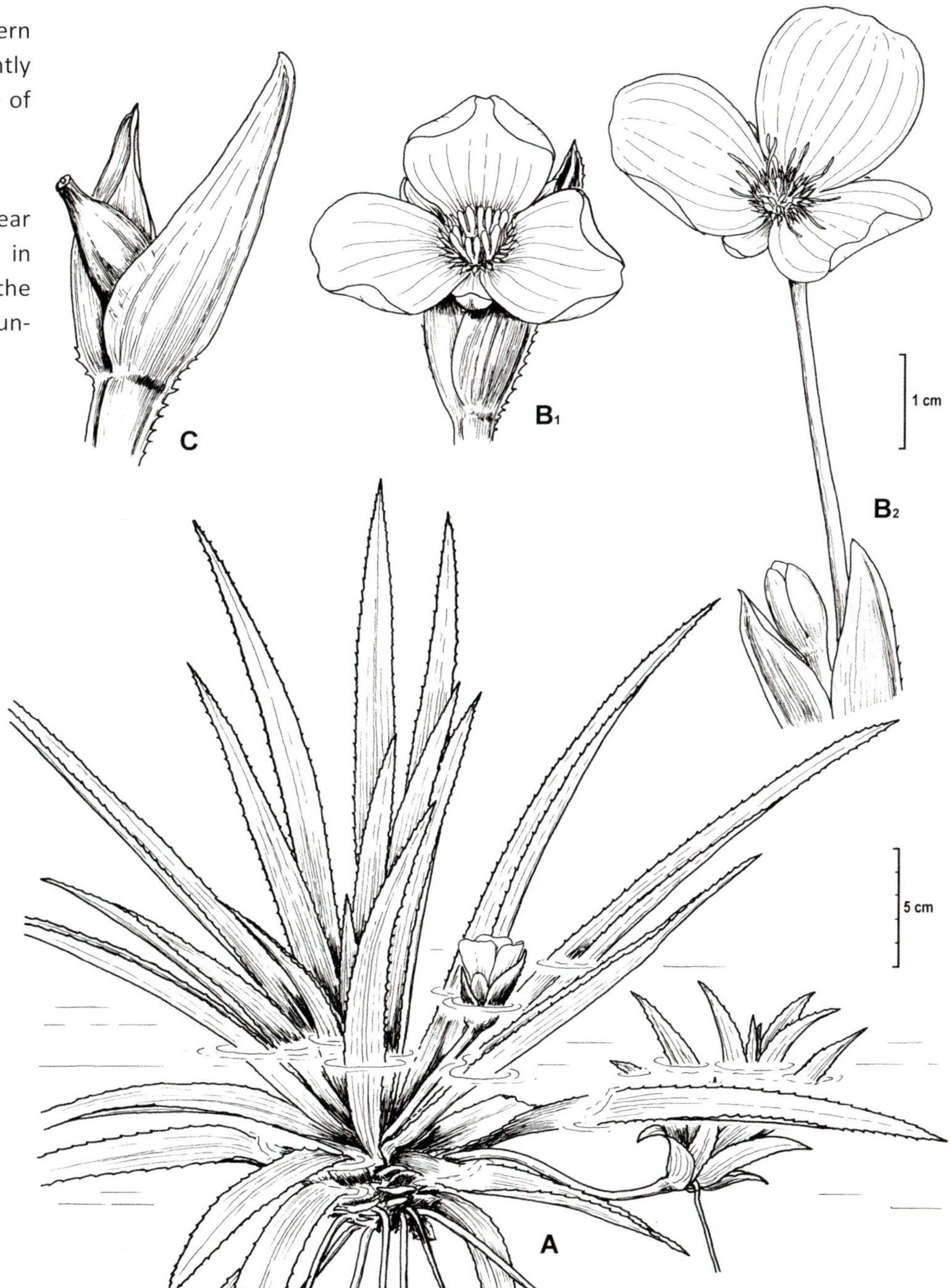

Stratiotes aloides
A habit, **B1** female flower, **B2** male flowers, **C** infructescence.

Hydrocharitaceae

53 *Egeria densa* Planch.

Syn.: *Anacharis densa* (Planch.) Victorin; *Elodea densa* Casp.
DK: - **N:** Brasiliansk vasspest **S:** Stor vattenpest **FIN:** Tiuhavesirutto **IS:** -
GB: Large-flowered Waterweed **NL:** Egeria **F:** Élodée dense
D: Dichtblättrige Wasserpest **CZ:** Morovinka hustolistá **PL:** -
EST: - **LV:** - **LT:** - **Ru:** Эгерия густая

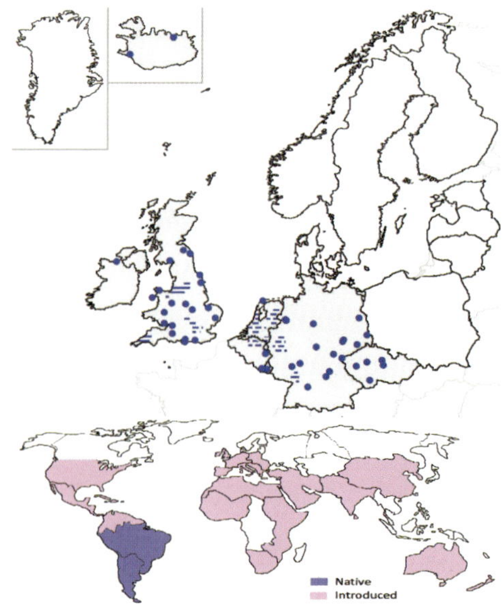

Flowering *E. densa*. River Erft, Germany. Photo KvdW.

Perennial, submerged hydrophyte. Free floating or attached to the bottom by adventitious, unbranched roots growing only from nodes with branches. Rhizome absent.

Stem: Usually up to 1(–3) metres long, 2–3 mm diameter, more or less branched, terete, with short internodes, yellowish green to light green.

Leaves: In whorls of 4(–5), 8 on double nodes. Leaves 15–30(–40) mm long and 1.5–5 mm wide, sessile, linear to narrowly oblong, acute at apex, terminating in a spine, 1-nerved, finely toothed, spreading, bright green. Scale-like and opposite on the lower part of the stem. Stipules absent.

Inflorescence: Male and female flowers on separate plants. The inflorescences are axillary spathes formed by two united bracts. Male spathe 7.5–12 mm long. The 2–4(–5) flowers are carried to the surface by a pedicel up to 19 cm long, less than 1 mm diameter.
Female spathe 8–14 mm long. The single female flower is carried to the surface by the elongated, pedicel-like perianth tube up to 6 cm long, which remains attached to the plant.

Flowers: 3-merous. Male flowers with 3 sepals, 2.5–4.5 mm long, green, streaked with violet, ovate, and 3 petals, 5–10(–12) mm long, white, obovate. Stamens 9(–10) yellow to orange, filaments papillose.

Female flowers with 3 sepals, 3–4 mm long, green, streaked with violet, ovate, spreading and 3 petals, 4.0–8.5 mm long, white, obovate. 3 styles, up to 4 mm long, white, deeply divided. Staminodes 3, 1–2.4 mm long, clavate, yellow.

Fruits: Capsule 1.2–1.5 cm long, 4–5 mm wide. Seeds elliptic, 5.5–7.2 mm long.

Flowering: Late in the year depending on the temperature.

Biology: The rather large, white flowers have a strong scent, secrete nectar and are pollinated by flying insects. All introduced plants are males; even in its native range male plants are more common than female plants and seed production is rare. Vegetative propaga-

Egeria densa. Hungary. Photo KvdW.

Egeria densa. Leaf tips. Photo KvdW.

Egeria densa. Shoot apex. Photo KvdW.

tion is by stem fragments rooting from nodes. Plants overwinter in mud as stem fragments.

Habitats: In warm, still or slow-flowing water. Rivers, canals, ditches and hot springs.

Distribution within the region:
Introduced and naturalised as an aquarium escape, often in warmer waters. Scattered in Iceland and from Ireland east to the Czech Republic. Only male plants have been found in Europe.

Characteristics and similar species:
Egeria densa can be recognised by the long, linear to narrowly oblong, finely toothed leaves densely clothing the stem, usually in whorls of 4, and by the 1.2–2 cm wide, white flowers.

54 ***Elodea canadensis*** has shorter leaves in whorls of 3, and much smaller, pinkish flowers.

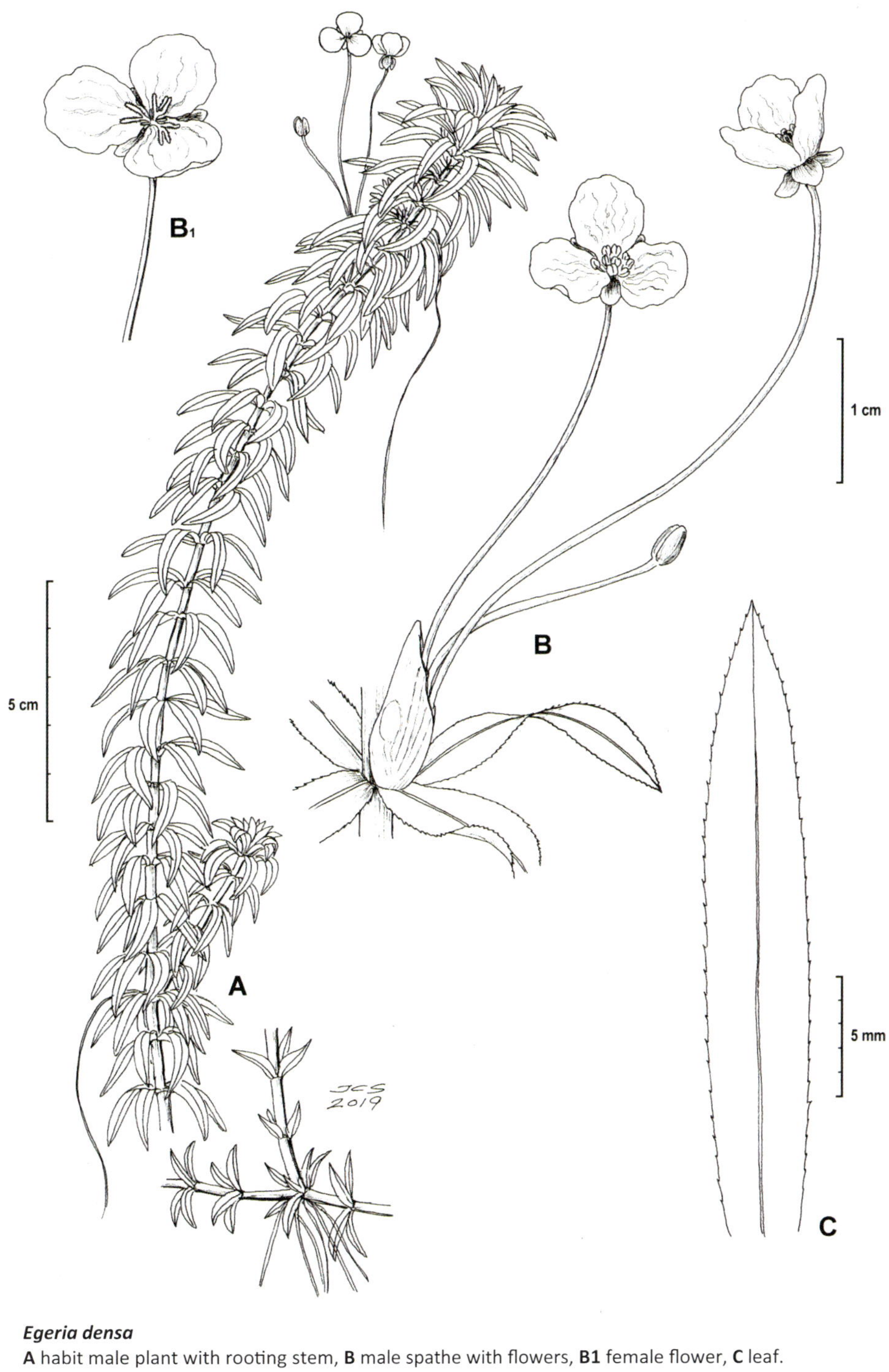

Egeria densa
A habit male plant with rooting stem, **B** male spathe with flowers, **B1** female flower, **C** leaf.

Hydrocharitaceae

54 *Elodea canadensis* L.C.Rich.

DK: Almindelig Vandpest **N:** Vasspest **S:** Vattenpest **FIN:** Kanadanvesirutto **IS:** -
GB: Canadian Waterweed **NL:** Brede waterpest **F:** Élodée du Canada
D: Kanadische Wasserpest **CZ:** Vodní mor kanadský **PL:** Moczarka kanadyjska
EST: Kanada vesikatk **LV:** Kanādas elodeja **LT:** Kanadinė elodėja **Ru:** Элодея канадская

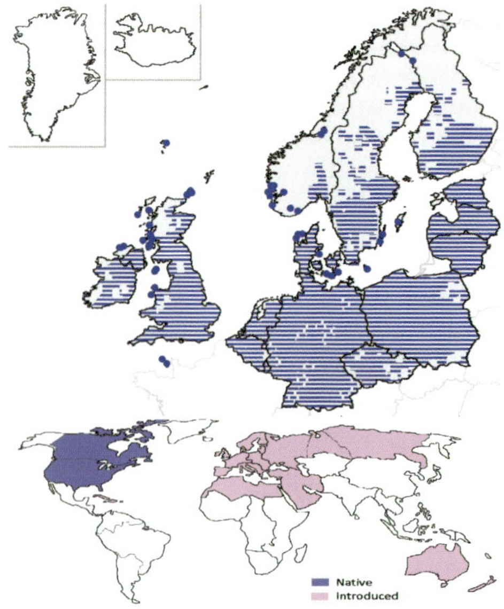

Flowering, female *E. canadensis.* Lake Filsoe, Jutland, Denmark. Photo JCS.

Perennial, submerged hydrophyte. Free floating or attached to the bottom by adventitious, unbranched roots growing only from nodes with branches. Root tips whitish to greenish when fresh.

Stem: Up to several metres long, more or less branched, terete, yellow green to whitish, sometimes with a reddish tinge, brittle and easily broken. With a reddish ring around the stem just below the leaf bases.

Leaves: In whorls of 3 (rarely 2 or 4–5). Leaves 5–14(–17) mm long and 1.5–5.5 mm wide (0.8–2.3 mm wide measured 0.5 mm from the apex), 2–5 times as long as wide, sessile, oblong ovate to lingulate, widest at or just below the middle, obtuse or subacute at apex, 1-nerved, finely toothed or entire, somewhat reflexed, dark green. Stipules 2, minute, almost circular.

Inflorescence: Male and female flowers on separate plants. The inflorescences are axillary spathes formed by two united bracts. Male spathe 6–12 mm long. The single flower is carried to the surface by a pedicel up to 15 cm long, thin, fragile and easily broken. Female spathe 8–18 mm long, tubular. The single female flower is carried to the surface by the strongly elongated, pedicel-like perianth tube up to 30 cm long, which remains attached to the plant.

Flowers: 3-merous. Male flowers with 3 violet-streaked green sepals, 3.0–4.5 mm long, 3 white petals 1.5–3.0 mm long and c. 9 stamens (7–18), all filaments proximally connate. Female flowers smaller than male flowers. With 3 erecto-patent, red-brownish sepals 2.5–3.5 mm long, with broad hyaline margins. The 3 petals are reflexed, 2–3 mm long and translucent white. With 3 mostly bifid, pinkish styles up to 4 mm long and 3 pale brownish staminodes.

Fruits: Capsule containing 1–5 seeds. Seeds fusiform, 4–6 mm long, without basal hairs.

Flowering: July-September.

Biology: The detached male flowers float on the surface and release pollen for aerial pollination of floating female flowers. Vegetative propagation is by rooting stem fragments. Wintergreen. Overwinters in mud as condensed,

The shoots of *E. canadensis* look almost like pipe cleaners. Germany. Photo KvdW.

Elodea canadensis. Female flower. River Skjern Å, Jutland, Denmark. Photo JCS.

Elodea canadensis. Shoot apex. Note the lingulate leaves in whorls of 3 and the red rings on the stem below the leaves. River Gudenå, Randers, Denmark. Photo JCS.

almost cone-like, swollen, short-leaved shoot apices produced in the late autumn.

Habitats: In almost all kinds of waters - ponds, ditches, canals, streams and lakes.

Distribution within the region:
Introduced and spread as an aquarium escape to waters more or less throughout the region, except Iceland and Greenland. Now declining in central Europe. With the exception of male plants reported from a single site in Scotland in 1879–1903 only female plants have been found in Europe. Native to North America.

Characteristics and similar species:
Elodea canadensis can be recognised by the obtuse or subacute leaves usually in whorls of 3 (rarely more, sometimes 2), 0.8–2.3 mm wide measured 0.5 mm below apex and the minutely serrulate leaf margins.

55 *E. nuttallii* and **56 *E. callitrichoides*** both have more acute, narrower leaves, 0.2–0.7 mm wide measured 0.5 mm below apex.

Compare with **53 *Egeria densa*, 57 *Hydrilla verticillata*** and **58 *Lagarosiphon major***.

Note: The "pedicel" of the female flower is actually a heavily elongated perianth tube.

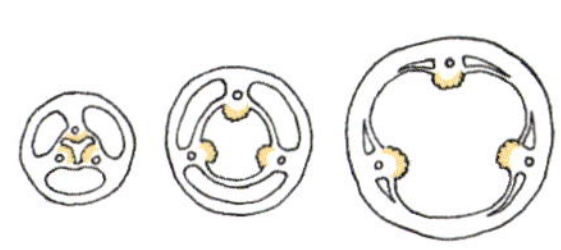

Cross sections of the "pedicel" show the 3 placentas continuing all the way down, attached to the inner surface. (Raunkiær 1890).

Elodea canadensis. Overwintering shoot. Smallhythe, Kent, Britain. Photo RVL.

Elodea canadensis. Leaf. Photo KvdW.

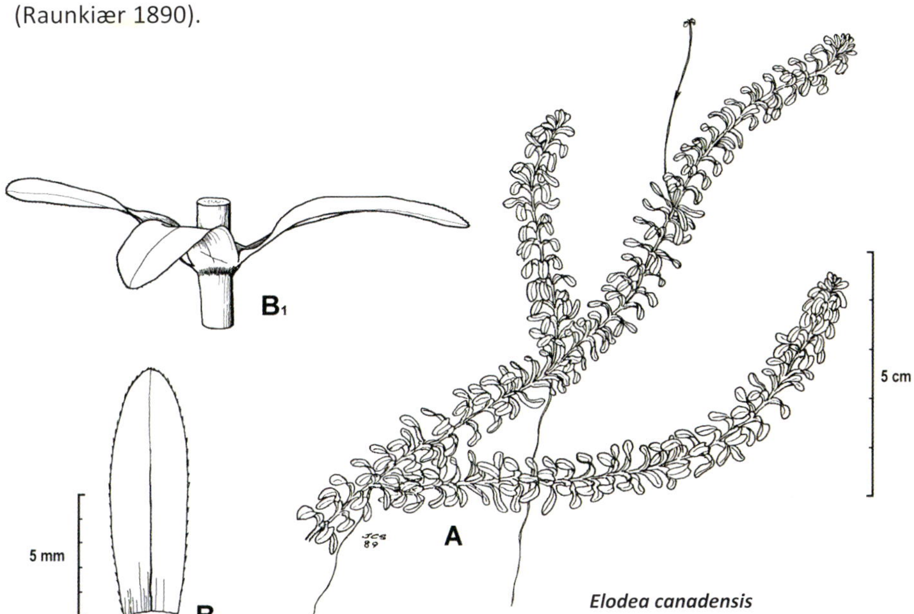

Elodea canadensis
A habit, **B** leaf, **B1** whorl of 3 leaves.

Hydrocharitaceae

55 *Elodea nuttallii* (Planch.) H.St.John

DK: Smalbladet Vandpest **N:** - **S:** Smal Vattenpest **FIN:** Kiehkuravesirutto **IS:** -
GB: Nuttall's Waterweed **NL:** Smalle waterpest **F:** Élodee de Nuttall
D: Nuttall-Wasserpest **CZ:** Vodní mor americký **PL:** -
EST: Väike vesikatk **LV:** - **LT:** - **Ru:**

Female *E. nuttallii* just before flowering. The strongly elongated pedicel-like hypanthium carries the flower to the surface. Ribe, Jutland, Denmark. Photo JCS.

Perennial, submerged hydrophyte. Free floating or attached to the bottom by adventitious, unbranched roots growing only from nodes with branches. Root tips whitish to greenish when fresh.

Stem: Up to several metres long, more or less branched, terete, yellow green, somewhat brittle and easily broken. With a reddish ring around the stem just below the leaf bases.

Leaves: In whorls of 3(–5). Leaves 6–35 mm long, 0.8–3.0 mm wide (measured 0.5 mm from the apex 0.2–0.7 mm wide), 3.5–10 times as long as wide, sessile, linear to lanceolate, widest near the base and gradually acuminate, acute at apex, 1-nerved, strongly recurved and often somewhat twisted, pale to dark green. Leaf margins with teeth 0.06–0.09(–0.10) mm long. Stipules 2, minute, elliptic.

Inflorescence: Male and female flowers on separate plants. The inflorescences are axillary spathes formed by two united bracts. Male spathe 2.2–4.0 mm long. The single male flower is detached as a bud and floats to the surface, where it opens and releases the pollen. Female spathe 8–14.5 mm long, tubular. The single female flower is carried to the surface by the strongly elongate, up to 10 cm long, pedicel-like perianth tube and remains attached to the plant.

Flowers: 3-merous. Male flowers about 4 mm in diameter, sepals greenish, 1.7–2.5 mm long and c. 1.5 mm wide, petals shorter, 0.5–1.5 mm long, translucent, white. Stamens 9 - inner 3 filaments connate, proximally forming a column. Female flowers 3.0–4.5 mm in diameter. Sepals and petals of equal length, 1.0–2.1 mm long, translucent,

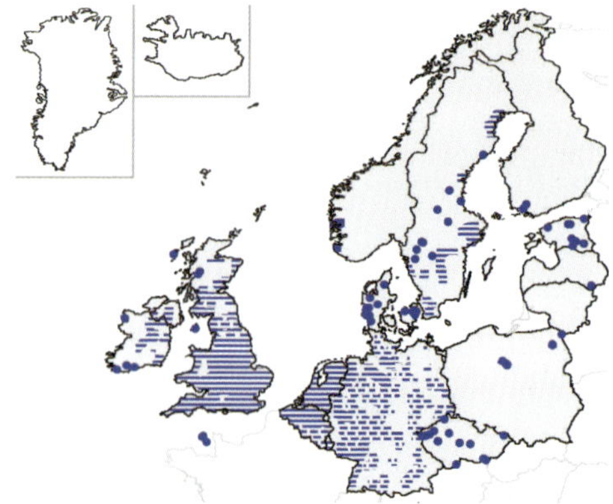

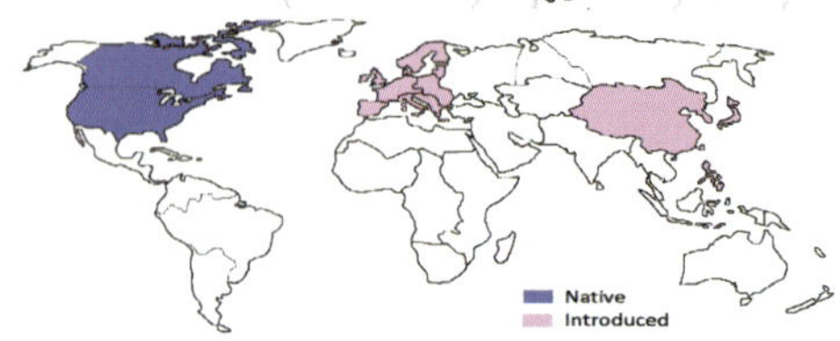

whitish. With 3 styles, up to 3.0 mm long, mostly bifid, pinkish and 3 light brownish staminodes.

Fruits: Capsule containing 1–5 seeds. Seeds fusiform, 4.0–4.6 mm long, basal hairs long.

Flowering: July-September.

Biology: Vegetative propagation by rooting stem fragments. *Elodea nuttallii* seems to overwinter by turions and prostrate shoots forming dense green mats on the bottom.

Habitats: In almost all kinds of non-acidic waters - ponds, ditches, canals, slow-flowing streams, rivers and lakes.

Distribution within the region:
Introduced and spreading throughout most of the area. Absent from Iceland and Greenland. Native to North America. Almost all plants outside its native

Elodea nuttallii. Female flower ready to unfold. Ribe, Jutland, Denmark. Photo JCS.

range are female. Male plants have been recorded from Germany.

Characteristics and similar species:
Elodea nuttallii looks similar to other species with leaves in whorls of 3–4.

54 ***Elodea canadensis*** has broader, lingulate leaves which are not twisted.

56 ***E. callitrichoides*** is almost identical to *E. nuttallii* - see below.

58 ***Lagarosiphon major*** has alternate leaves spirally arranged on the stem.

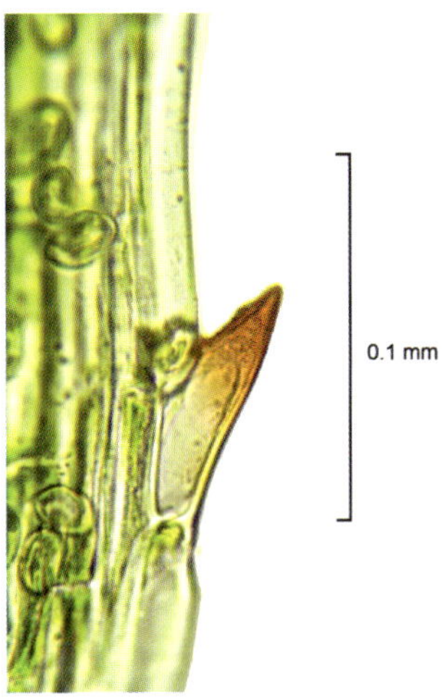

Elodea nuttallii. Leaf tooth. Photo KvdW.

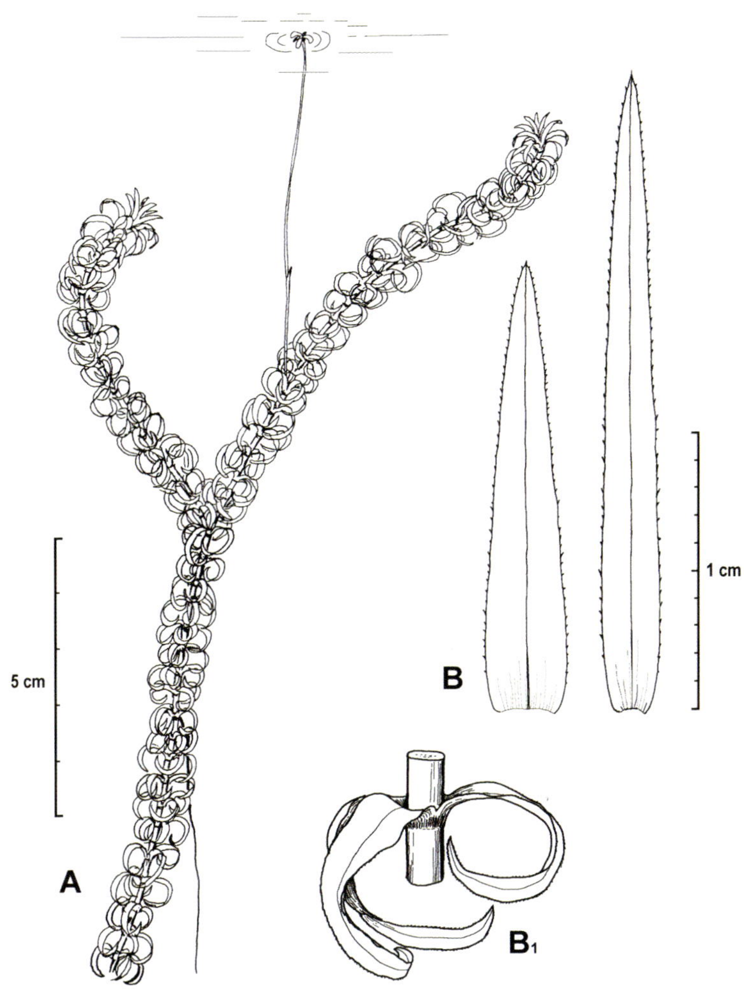

Elodea nuttallii
A habit flowering female plant, **B** leaves, **B1** whorl of 3 leaves.

Hydrocharitaceae

56 *Elodea callitrichoides* (Rich.) Casp.

DK: Argentinsk Vandpest **N:** - **S:** Argentinsk vattenpest **FIN:** Argentiinanvesirutto **IS:** -
GB: South American Waterweed **NL:** - **F:** -
D: Argentinische Wasserpest **CZ:** - **PL:** -
EST: - **LV:** - **LT:** - **Ru:**

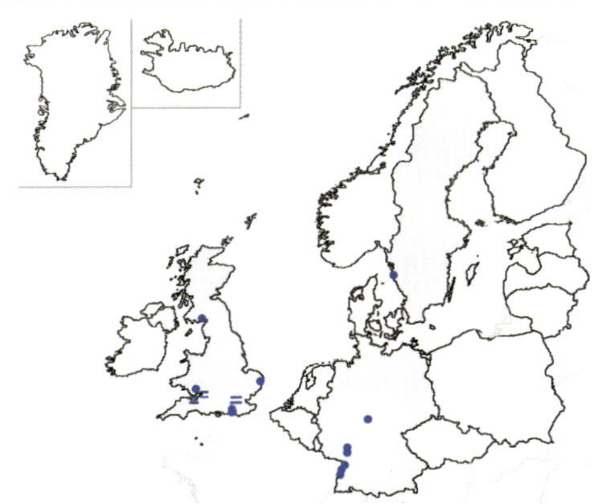

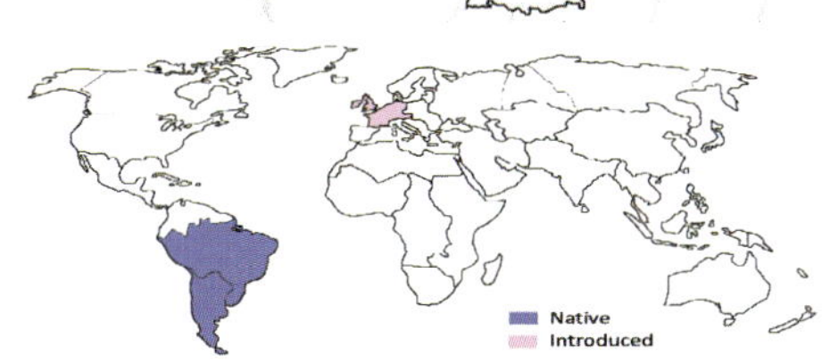

In many aspects similar to **55** ***E. nuttallii.*** It differs in the more flaccid and flatter spreading leaves that are not strongly recurved or twisted, and with marginal teeth (0.08–)0.11–0.14 mm long. Female flowers with a diameter of 5–8 mm, sepals 3.0–4.0 mm long. Male flowers larger, up to 10 mm in diameter, sepals 4.5–6.2 mm long. Root tips red when fresh.

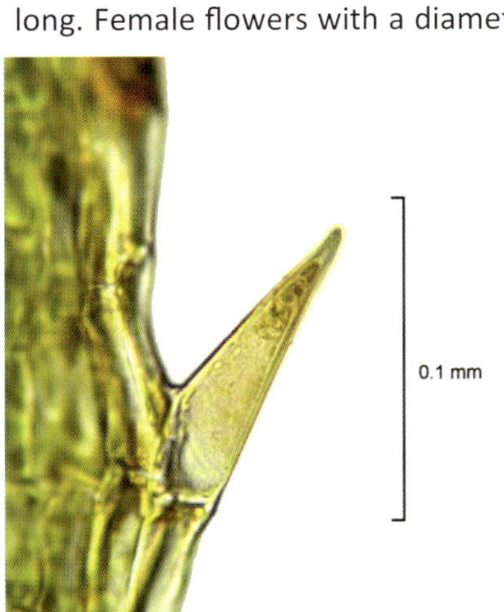

Elodea callitrichoides.
Leaf tooth. Photo KvdW.

Distribution within the region:
Introduced to Great Britain (where now extinct), Germany and Sweden. It may have been overlooked elsewhere. Native to South America. Only female plants are known in Europe.

Characteristics and similar species:
In addition to *E. nuttallii* it may be confused with **54** ***Elodea canadensis***, which has broader, lingulate leaves.

58 ***Lagarosiphon major*** has alternate leaves spirally arranged on the stem.

Herbarium specimen. Photo KvdW.

Hydrocharitaceae

57 ***Hydrilla verticillata*** (L.f.) Royle

DK: - **N:** - **S:** - **FIN:** - **IS:** -
GB: Esthwaite Waterweed **NL:** Hydrilla **F:** Hydrille de Lithuanie
D: Grundnessel **CZ:** - **PL:** -
EST: - **LV:** Mieturu hidrilla **LT:** Menturlapė ežerutė **Ru:** Гидрилла

Red dots indicate extinct populations.

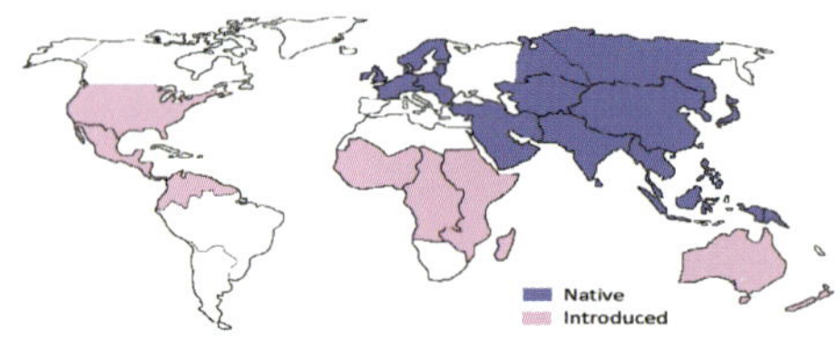

Shoots of *H. verticillata* can in a short period of time cover large areas of the water surface, limiting light availability for other aquatic plant species. Balaton region, Hungary. Photo KvdW.

Perennial or more rarely annual, submerged hydrophyte, forming dense mats, with shoots arising from a slender rhizome rooted in the substrate.

Stem: 20–100(–300) cm, 1–1.5 mm in diameter, branched, terete, yellow green to green.

Leaves: In whorls of 3–8. Leaves 10–20 (–40) mm long and 2–5 mm wide, sessile, linear to lanceolate, acute at apex, terminating in a spine, 1-nerved, sharply, almost spiny dentate and sometimes even with a few spines on the lower side of the prominent midrib. Stipules 2, c. 0.5 mm long, triangular, fringed with orange-brownish projections.

Inflorescence: Two biotypes are present. A type with male and female flowers on the same plant, and a type with male and female flowers on separate plants - sometimes with female and hermaphrodite plants. The inflorescences are axillary spathes formed by two united bracts. Male spathe c. 2.5 mm long, shortly pedicellate, almost globose and slightly spiny. The single male flower is detached as a bud and floats to the surface. Female spathe c. 5 mm long, tubular. The single female flower is carried to the surface by the elongate pedicel-like perianth tube, up to 10 cm long, and remains attached to the plant.

Flowers: 3-merous. Male flowers with 3 somewhat translucent, light greenish, streaked with pinkish, 1.2–3.0 mm long, ovate sepals. Petals shorter and narrower than sepals, translucent with pinkish streaks. Stamens 3, staminodes 3.

Female and hermaphrodite flowers with 3 oblong ovate, 1.5–4.0 mm long, translucent, slightly pinkish sepals and 3 similar, but narrower petals. Styles 3, up to 0.7 mm long, pinkish.

Fruits: Cylindrical, c. 7 mm long and 1.5 mm thick, containing 2–7 oblong-elliptic seeds.

Flowering: July-August.

Biology: The detached male flower opens on the surface and explosively releases the pollen for aerial pollination of the floating female flower. Vegetative propagation is by rooting stem fragments, by subterranean whitish turions up to 1.2 cm long, developed in the apex of stolons and by dark green turions up to 0.6 cm long produced from leaf axils and the apices of erect shoots. The turions remain viable for 4–7 years

Hydrilla verticillata. Dense shoots. Balaton region, Hungary. Photo KvdW.

and tolerate harsh winter conditions. A single plant is able to produce more than 5000 tubers/turions/m^2.

Habitats: In warm, oligo-mesotrophic to eutrophic, acidic or alkaline, still or slow-flowing, shallow water at depths of 0.5–2(–6) metres. In Asia it even occurs in brackish water at a salinity of 1‰–4‰. Ponds, lakes, canals, streams and rivers.

Distribution within the region:
Scattered populations are known in the region in Britain, Ireland, Germany, Poland, Lithuania, Latvia and Belarus, where it is presumed to be native.

Characteristics and similar species:
Hydrilla verticillata is a variable species both phenotypically and genetically. Despite this it can be recognised by its sharply dentate leaves in whorls of 3–8 and by the tuber- or potato-like turions terminal on the stolons.

It resembles **54-56** ***Elodea*** species and **53** ***Egeria densa,*** but these have minutely dentate or minutely spiny leaves, their basal scales without a fringe of brownish hairs and always without spines on the back of the midrib.

Note: The native European populations of *Hydrilla verticillata* may be taxonomically different from the invasive alien populations. They are typically much more delicate with longer internodes and fewer leaves at the node, thus more closely resembling **55** ***Elodea nuttallii***, from which they are initially best distinguished by the larger number of leaves at each node.

Hydrilla verticillata. Whorl of 8 leaves and leaf tip. Photos KvdW.

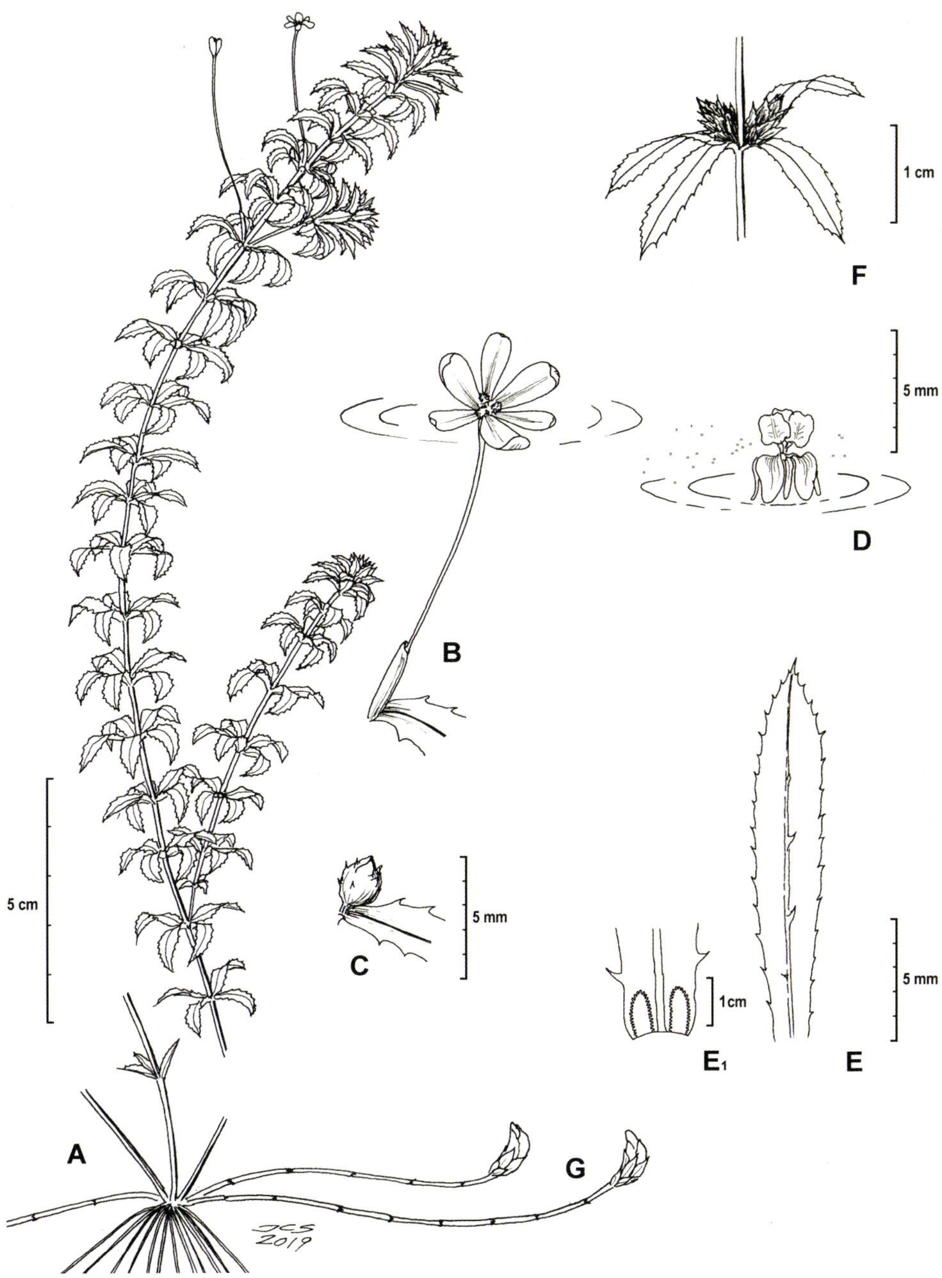

Hydrilla verticillata
A habit female plant, **B** female flower with spathe, **C** male spathe, **D** floating male flower releasing pollen, **E** lower side of leaf showing spines on the midrib, **E1** leaf base with basal scales, **F** turions in leaf axils, **G** subterranean turions terminal on stolons.

Hydrocharitaceae

58 *Lagarosiphon major* (Ridl.) Moss

DK: - **N:** - **S:** - **FIN:** Afrikanvesihäntä **IS:** -
GB: Curly Waterweed **NL:** Verspreidbladige waterpest **F:** Grand lagarosiphon
D: Grosse Wechselblatt-Wasserpest **CZ:** - **PL:** -
EST: - **LV:** - **LT:** - **Ru:**

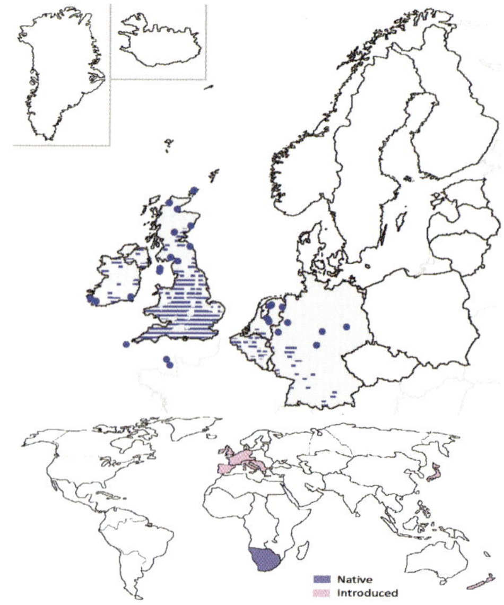

Lagarosiphon major. Mönchwaldsee, Hessen, Germany. Photo KvdW.

Perennial, submerged hydrophyte. Free floating or attached to the bottom by adventitious, unbranched roots growing from the nodes and internodes. Rhizome as thick as the stem.

Stem: Up to 5 metres long, 3–5 mm diameter, sparsely branched, terete, yellow to dark green, brittle, curving downward towards the apex.

Leaves: Crowded towards the apex, but lower down clearly alternate, spirally arranged and curved back towards the stem. 5–25 mm long, 2–3(–4) mm wide, sessile, stiff, gradually tapering to an acute apex, 1-nerved, dark green, minutely toothed - teeth usually one-celled. Stipules 2, minute.

Inflorescence: Male and female flowers on separate plants. The inflorescences are axillary spathes formed by two united bracts. Male spathe 3–5 mm long, ovoid, with 3–18 marginal teeth, containing up to 50 male flowers which detach as buds. Female spathe c. 3.5 mm long, ovoid, with 2–3 marginal teeth, subtending only one female flower.

Flowers: 3-merous. Male flowers with 6 pinkish perianth segments, 3 fertile stamens and 3 violet, thread-like, papillose staminodes.
Female flowers with 6 pinkish perianth segments and 3 bifid, violet, papillose styles. The female flowers are lifted to the surface by an elongated, pedicel-like hypanthium up to 15 cm long. Carpels 3.

Fruits: Capsule 4–5 mm long, ovate, containing 6–30 seeds.

Flowering: July-August.

Biology: Male and female flowers are produced under water, but float to the surface for aerial pollination - the female flower still attached to the plant by the elongated hypanthium. The numerous male flowers detach as independent buds from the male inflorescence and unfold like a bell with the perianth floating on the surface. The 3 fertile stamens are directed horizontally to meet the stigmas on the female flower, whereas the 3 sterile stamens are prolonged and erect, functioning as sails. The seeds mature under water. Vegetative propagation by rooting stem fragments. Overwintering as dor-

Lagarosiphon major. Mönchwaldsee, Hessen, Germany. Photo KvdW.

mant shoots and rhizomes or capable of remaining green throughout the winter beneath ice cover.

Habitats: Usually in eutrophic water in ponds, lakes, gravel pits, drainage ditches, canals, slow-flowing rivers, but also in oligotrophic water in quarries. In clear water it can grow to a depth of 10 m.

Distribution within the region:
Introduced and naturalised as an aquarium escape in Ireland, Britain, the Netherlands, Belgium and Germany. Only female plants are known outside its native range. Native in South Africa.

Characteristics and similar species:
Lagarosiphon major strongly resembles **53 *Egeria densa***, **54-56 *Elodea*** spp., and **57 *Hydrilla verticillata***, but differs in having alternate and recurved, spirally arranged leaves.

Lagarosiphon major. Mönchwaldsee, Hessen, Germany. Photo KvdW.

Lagarosiphon major. The spiral arrangement of the leaves are most distinct on the lower part of the stem. Photo KvdW.

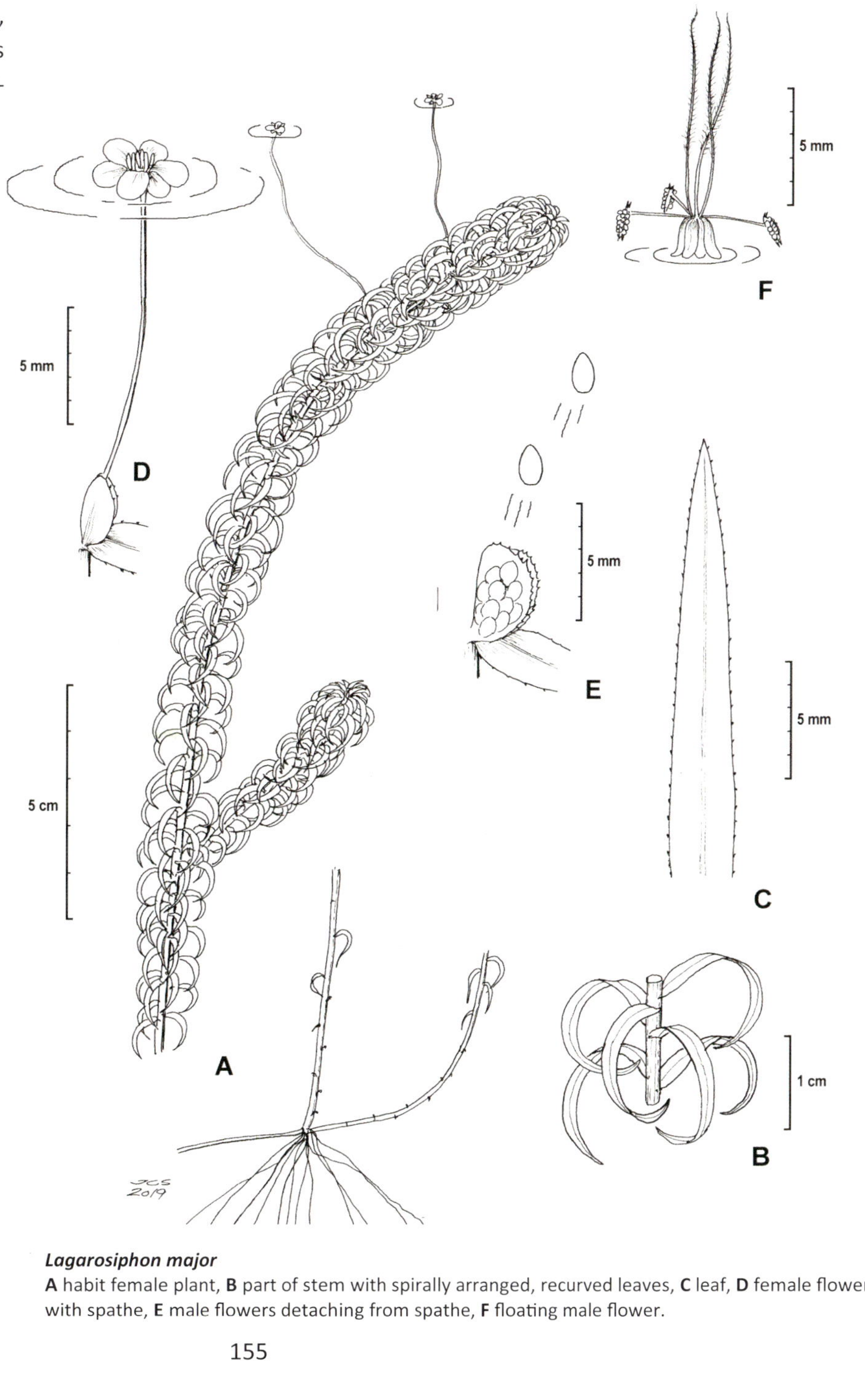

Lagarosiphon major
A habit female plant, **B** part of stem with spirally arranged, recurved leaves, **C** leaf, **D** female flower with spathe, **E** male flowers detaching from spathe, **F** floating male flower.

Hydrocharitaceae

Key to *Najas*

In *Najas* the leaves are all opposite, but on side shoots the basal pair is reduced to only one, giving the impression that the leaves are arranged in whorls of 3. The genus Najas comprises c. 37 species of which some are used as aquarium plants and some introduced and naturalised in parts of Europe.

1 Stem with spines (***1***) - at least in some parts - young plants sometimes without spines. Dioecious. Robust plants 2

1 Stem without spines (***2***). Monoecious. More or less slender plants 3

1

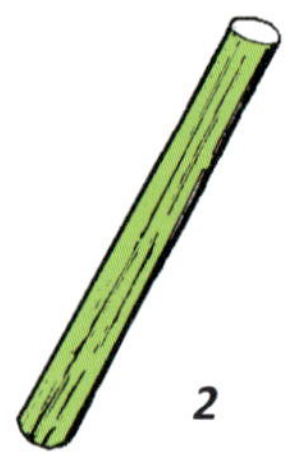

2

2 Seeds 3.0–4.2 mm long. Leaves 0.4–0.9(–1.5) mm wide (excluding teeth). Leaf sheath with 1–3(–4) teeth on each margin

59 *N. marina* var. ***intermedia***

2 Seeds (3.5–)4.2–6.5(–8.0) mm long. Leaves 0.7–1.9(–2.4) mm wide (excluding teeth). Leaf sheath with 0(–1) teeth on each margin

59 *N. marina* var. ***marina***

3 Leaf sheaths rounded, gradually narrowed into the lamina - without a shoulder (***3***). Leaves 0.5–0.7(–1.0) mm wide

61 *N. flexilis*

3 Leaf sheaths truncate - with a shoulder (***4***). Leaves 0.1–0.5 mm wide 4

3

4

4 Seed coat with areoles arranged in a ladder-like pattern, wider than long (***5***)

60 *N. minor*

4 Seed coat with areoles not arranged like a ladder - squarish or longer than wide (***6***) 5

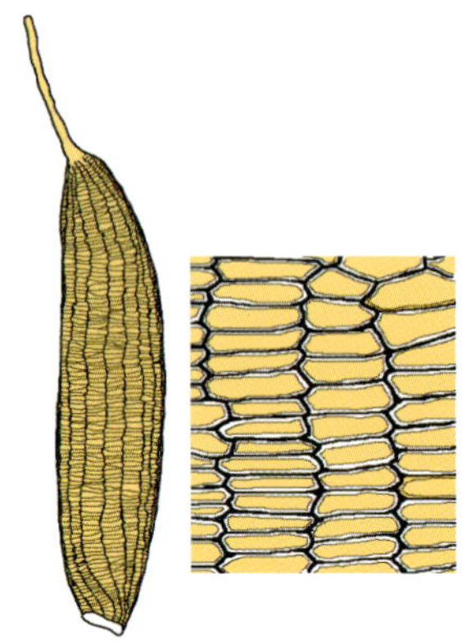

5

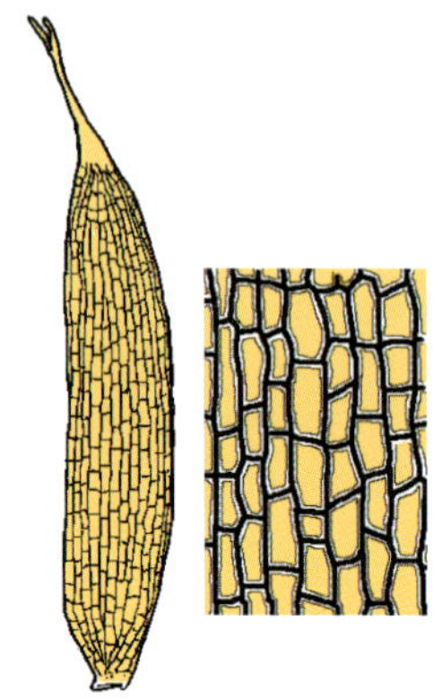

6

5 Seed surface smooth. Areoles 25–30(–35) in each longitudinal row

63 *N. tenuissima*

5 Seed surface pitted. Areoles 8–19 in each longitudinal row

62 *N. gracillima*

Najas cf. *guadelupensis*, Germany. Photo KvdW.

Najas guadalupensis (Sprengel) Magnus is known as an escape from horticulture in Denmark and probably from the same source in Germany. Native to North and South America. It differs from the similar **61 *N. flexilis*** in the seeds c. 2 mm long, dull, stigmas including styles 0.3–0.6 mm long, leaves 0.6–2.2 mm wide.

Hydrocharitaceae

59 *Najas marina* L.

Syn.: *N. major* All.; *N. polonica* Zalewski

DK: Stor Najade **N:** Stivt havfrugras **S:** Havsnajas **FIN:** Merinäkinruoho **IS:** -
GB: Holly-leaved Naiad **NL:** Groot nimfkruid **F:** Grande naïade
D: Großes Nixkraut **CZ:** Řečanka přímořská **PL:** Jezierza morska
EST: Meri-näkirohi **LV:** Jūras najāda **LT:** Didysis plukenis **Ru:** Наяда морская

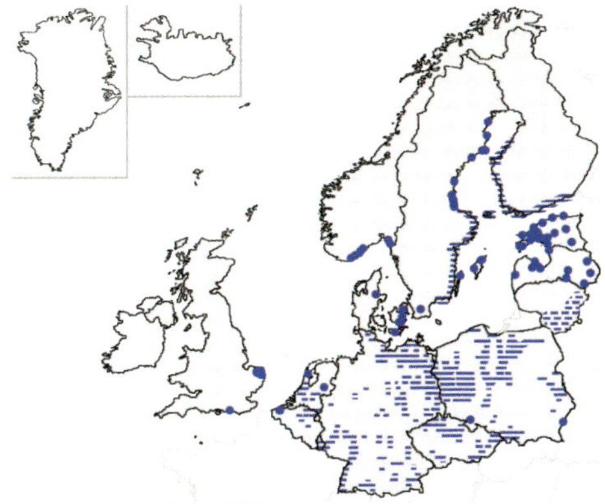

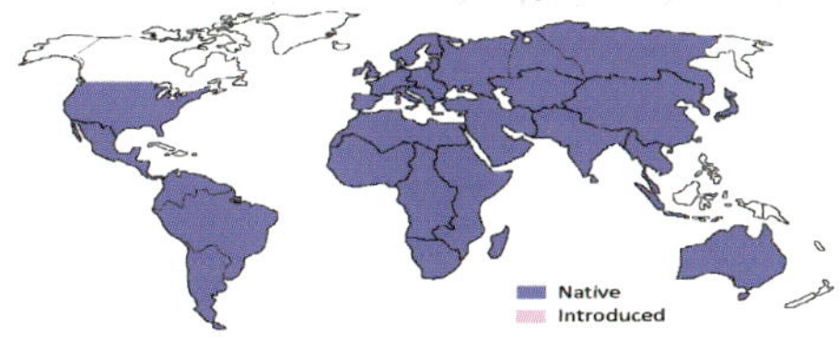

Najas marina var. *marina*. Brechtsee, Germany. Photo KvdW.

Annual, submerged hydrophyte. Roots white.

Stem: 5–50(–300) cm long, robust, stiff, somewhat fragile, with spines dichotomously divided at least in some parts, greenish or reddish to purplish brown.

Leaves: Opposite, but appearing to be in whorls of 3 (see note at top of p. 156), 1–4 cm long, 0.5–2.2 mm wide excluding teeth, linear, green to grey green or with reddish tones, without septa, stiff. Margins and sometimes the back of the midrib with scattered, forward- projecting, spine-like teeth. Leaf sheaths partly clasping the stem, with rounded shoulders, entire or with 1–3 small teeth, green or reddish, evenly narrowed into the lamina.

Inflorescence: Male and female flowers on separate plants. Flowers solitary in leaf axils.

Flowers: Male flower with only one anther enclosed in two thin scales, one of them fused with the anther. Female flower without scales, with a 3–8 mm long ovary and 2–3-lobed stigma.

Fruits: Seeds 3–8 mm long, up to 2.5 mm diameter, dull, dark brown, areoles uneven, rounded oblong and not arranged in rows.

Flowering: July-September.

Biology: Before pollination, the anther pushes through the enclosing scales and disperses a swarm of pollen, on which pollen tubes 2 mm long have already developed, for easier attachment to the stigmas on the female flowers. Seeds mature under water.

Habitats: Var. *marina* in shallow, eutrophic lakes. Var. *intermedia* in oligotrophic to mesotrophic lakes, estuaries and sheltered coastal creeks in slightly brackish water with a salinity up to 6 ‰.

Distribution within the region:

Native. Spreading in Germany and England, but rare and scattered in the south of the region, coastal further north.

Najas marina var. *marina*. Photo KvdW.

Najas marina var. *intermedia*. Young plant in shallow, calcareous water. Norrsund, Fårö, Gotland, Sweden. Photo JCS.

Najas marina var. *intermedia*. Lake Røgbølle, Lolland, Denmark. Photo JCS.

Najas marina L. is a variable species and often split into two varieties in the region, separated by leaf size and shape, with more varieties recognised outside the region. Most modern floras treat these as varieties or subspecies, but some authors treat them as separate species. The distinction between the two taxa is cause for much confusion, especially in specimens showing intermediate characters.

Najas marina L. var. ***marina***

Syn.: *N. marina* subsp. marina L.; *N. major* All.;

Leaves with only (3–)6–8(–10) teeth on each margin. Leaves (7–)20–50(–65) mm long and (2.1–)3–4(–5.7) mm wide (including teeth on both sides), 0.7–1.9 (–2.4) mm wide (excluding teeth). Leaf sheaths with rounded shoulders untoothed or rarely with one tooth.

Seeds (3.5–)4.2–6.5(–8.0) mm long.

Within the area known from Germany, Poland, Czech Republic and Lithuania.

Najas marina L. var. ***intermedia*** (Wolfg. ex Gorski) Rendle

Syn.: *N. marina* subsp. *intermedia* (Wolfg. ex Gorski) Casper; *N. intermedia* Wolfg. ex Gorski

Leaves with (2–)4–7(–10) teeth on each margin and often with additional teeth on the back of the midrib. Leaves (4–)8–20(–50) mm long and 1.1–3.3(–4) mm wide (including teeth on both sides), 0.4–1.5(–2) mm broad (excluding teeth). Leaf sheaths with somewhat angular shoulders with 1–3 (–4) teeth.

Seeds 3.0–4.2 mm long.

Known from all countries within the area except Greenland, Iceland and Ireland.

Characteristics and similar species:
Najas marina is characterised by the rather stiff leaves which are opposite or in whorls of 3 with prominent spiny teeth, the stem with spines and the solitary, sessile flowers in leaf axils. The plant is stiff and spiny to the touch, resembling **197 *Ceratophyllum demersum***, although the two species are structurally very different.

Other species of ***Najas*** differ in the stem lacking spines, narrower leaves and smaller fruits.

104-107 *Zannichellia* spp. have a creeping rhizome and leaf margins without teeth.

54-56 *Elodea* spp. have a denser habit, recurved leaves in whorls of 3 and very delicate serrulate leaf margins.

Note: According to Bräuchler (2015) the correct name at the species level for *Najas marina* L. var. *marina* is ***N. major***

Var. *marina* and var. *intermedia*. Photo KvdW.

Seeds. Var. *marina* and var. *intermedia*. Photo KvdW.

Najas marina var. *intermedia*. Left: Male flower enclosed in thin scales. Right: Female flower. Lake Røgbølle, Lolland, Denmark. Photos JCS.

All., and for *Najas marina* L. var. *intermedia* (Wolfg. ex Gorski) Rendle is ***Najas marina*** L.

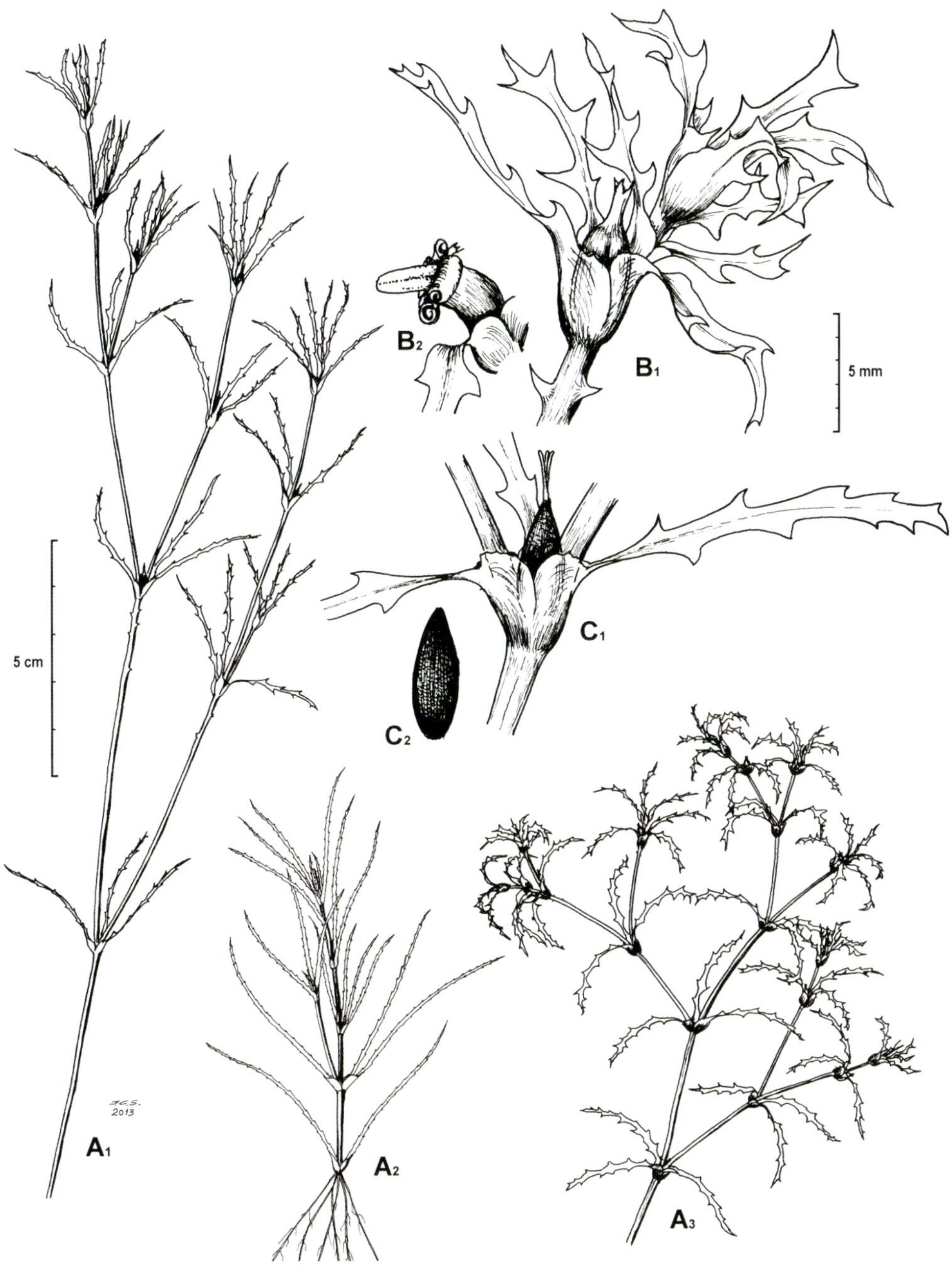

Najas marina var. ***intermedia***
A1 habit, **A2** habit young plant, **A3** habit, **B1** shoot tip with male flower enclosed in scales, **B2** anther pushing itself through the enclosing scales, **C1** female flower in leaf axil, **C2** seed.

Hydrocharitaceae

60 *Najas minor* All.

Syn.: *Caulinia minor* (All.) Coss. & Germ.
DK: - **N:** - **S:** - **FIN:** - **IS:** -
GB: Lesser Naiad **NL:** Klein nimfkruid **F:** Naïade mineure
D: Kleines Nixenkraut **CZ:** Řečanka menší **PL:** Jezierza mniejsza
EST: Väike näkirohi **LV:** Mazā najāda **LT:** Mažasis plukenis **Ru:** Наяда малая

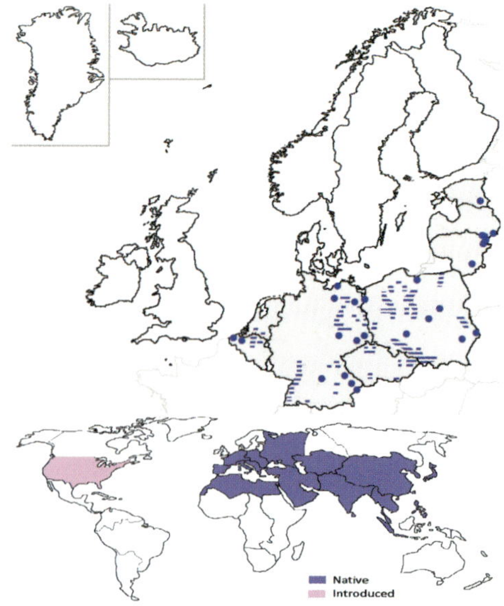

Najas minor often has a bushy or curly look because of the dense growth of recurved leaves. Tegeler See, Berlin, Germany. Photo KvdW.

Annual, submerged hydrophyte.

Stem: 5–25(–40) cm long, less than 0.5 mm wide, slender to quite robust, stiff, fragile and easily broken when old, without spines, dichotomously divided, dark green and highly branched, particularly toward the apex.

Leaves: Opposite, but appear to be in whorls of 3, (0.5–)1–2(–3) cm long, 0.4–0.5 mm wide excluding teeth, linear, green, septate, recurved on older plants, stiff. Margins serrulate, with (6–)10–18, c. 0.5 mm long, spiny teeth on each side. Leaf sheaths partly clasping the stem, truncate to auriculate, with 8–13 spines on each side, green.

Inflorescence: Male and female flowers on the same plant, but on different branches or nodes. Flowers solitary or a few together in leaf axils.

Flowers: Male flower with a single anther enclosed in two thin scales. Anther with only one pollen sac. Female flower without scale, with a 1.6–2.0 mm long ovary and 2-lobed stigma.

Fruits: Seeds 1.7–2.7 mm long, up to 0.7 mm wide, dull, dark brown, areoles broader than long, and ladder-like arranged in 12–18 in rows.

Flowering: June-September.

Biology: Before pollination, the anther emerges through the enclosing scales and disperses a swarm of pollen on which the pollen tubes are already developed and up to 2 mm long to facilitate attachment to the stigmas. Seeds mature under water.

Habitats: In still or slow-flowing, alkaline, meso- to eutrophic water. Lakes, ditches and canals.

Distribution within the region:
Native. Scattered throughout the south of the region, except Britain and Ireland, absent from the north.

Characteristics and similar species:
N. minor differs from other *Najas* species by the combination of the following characters: leaf sheaths truncate, leaves septate, seeds 1.7–2.7 mm long with areoles broader than long and arranged in 12–18 ladder-like rows, anther with only one pollen sac.

Najas minor. Tegeler See, Berlin, Germany. Photo KvdW.

Najas minor.
Female flowers/fruits in leaf axils.

Leaf sheaths and fruit.

Apical part of branch. Photos KvdW.

104-107 ***Zannichellia*** spp. has a creeping rhizome and untoothed leaf margins.

54-56 ***Elodea*** spp. has broader leaves, densely set in whorls of 3 along the whole stem, and different flowers.

Najas minor. Fruit. Photo JCS.

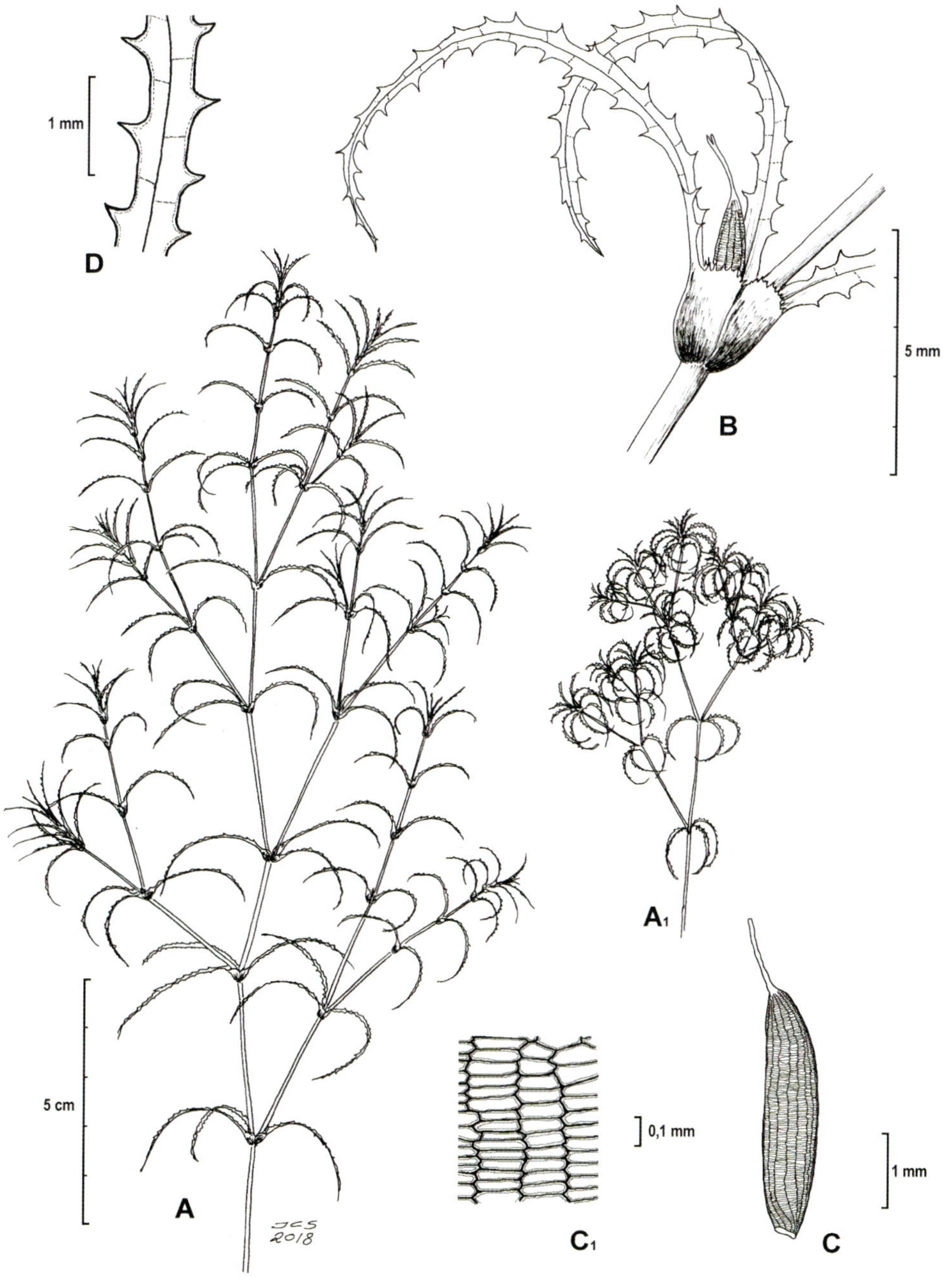

Najas minor
A, A1 habit, **B** female flower in leaf axil, **C** fruit, **C1** seed coat with ladder-like arranged areoles, **D** part of leaf showing septa.

Hydrocharitaceae

61 ***Najas flexilis*** (Willd.) Rostk. & W. L.E.Schmidt

Syn.: *Caulinia flexilis* Willd.
DK: Liden Najade **N:** Mjukt havfrugras **S:** Sjönajas **FIN:** Notkeanäkinruoho **IS:** -
GB: Slender Naiad **NL:** - **F:** Naïade flexible
D: Biegsames Nixkraut **CZ:** - **PL:** Jezierza giętka
EST: Nõtke näkirohi **LV:** Lokanā najāda **LT:** Lankstusis plukenis **Ru:** Наяда гибкая

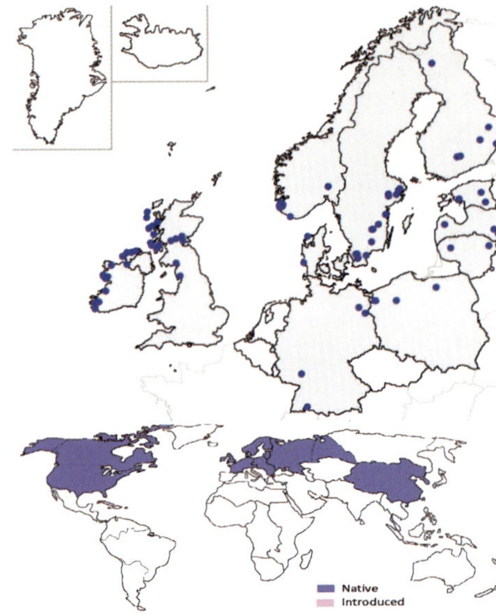

Najas flexilis and *Elodea canadensis* at 3 m depth. Södra Vixen, Sweden. Photo JCS.

Annual, submerged hydrophyte. Roots delicate, faintly reddish.

Stem: 3–25 cm long, 0.5–0.8 mm wide, terete, without teeth, brownish green, ascending or prostrate, slender, rather flaccid, flexible, but may become somewhat stiff and fragile especially at the end of the season, more or less branched.

Leaves: Opposite or appearing to be in whorls of 3 or more, 1–2 cm long and 0.5–0.7(–1.0) mm wide, linear, green to brownish green, without septa, acute. With 25–30 obliquely forward-pointing teeth on each margin, up to 0.1 mm long and 1-celled. Apex with a single tooth. The leaf sheath is approximately twice as wide as the lamina and similar with teeth on the margins, gradually narrowed into the lamina (without shoulder), more or less clasping the stem at the base.

Inflorescence: Male and female flowers on the same plant. Flowers solitary in leaf axils, but 1–3 flowers in a whorl of leaves.

Flowers: Male flower with a single anther enclosed in two thin scales, one of them fused to the anther. Female flowers without scales, with a 1.6–2.1 mm long ovary and 4-lobed stigma. Styles including stigmas 1–2.5 mm long.

Fruits: Seeds 2.2–3.5 mm long, up to 1 mm wide, shining dark brown to black, areoles uneven, squarish to hexagonal, in 30–40 lengthwise rows.

Flowering: July-September.

Biology: Before pollination, the anther emerges through the enclosing scales and disperses a swarm of pollen on which the pollen tubes are already developed and up to 2 mm long, to facilitate attachment to the stigmas. Seeds mature under water.

Habitats: In oligotrophic to moderately eutrophic lakes. Mostly at depths of 1.5–3.1 m, but sometimes in shallow water on sandy substrates overlayered by gyttja.

Distribution within the region: Native. Scattered and very local throughout the region.

Characteristics and similar species: *Najas flexilis* can be recognised by its flaccid appearance, the leaves which are up to 1 mm wide with small teeth on the margins and arranged in whorls of 3 or more, the leaf sheath which

Najas flexilis in shallow water with *Isoetes lacustris* and *Subularia*. Lough Mhaimin, Ireland. Photo KvdW.

Najas flexilis Female flower in leaf axil. Note the leaf margins with small teeth. Södra Vixen, Sweden. Photo JCS.

Najas flexilis. In Lake Nors, Denmark, the whole plant is typically covered with lime. Photo JCS.

Najas flexilis. Female flower. Photo KvdW.

Najas flexilis seed. Photo JCS.

narrows gradually into the lamina, without a shoulder. Both **63** ***N. tenuissima*** and **62** ***N. gracillima*** have narrower leaves (up to 0.5 mm) and leaf sheaths with a distinct shoulder. **59** ***N. marina*** and **60** ***N. minor*** are stiffer and robust. *N. marina* has spines on the stem and sometimes teeth on the midrib on the lower side of the leaves.

104-107 ***Zannichellia*** spp. have a creeping rhizome and leaves without teeth.

79-81 ***Potamogeton*** spp. have scattered leaves without marginal teeth.

Some ***Characeae*** may have a similar appearance - but are algae with each internode consisting of a single cell, often covered with a layer of long cells - the cortex and branches arising in whorls at the nodes.

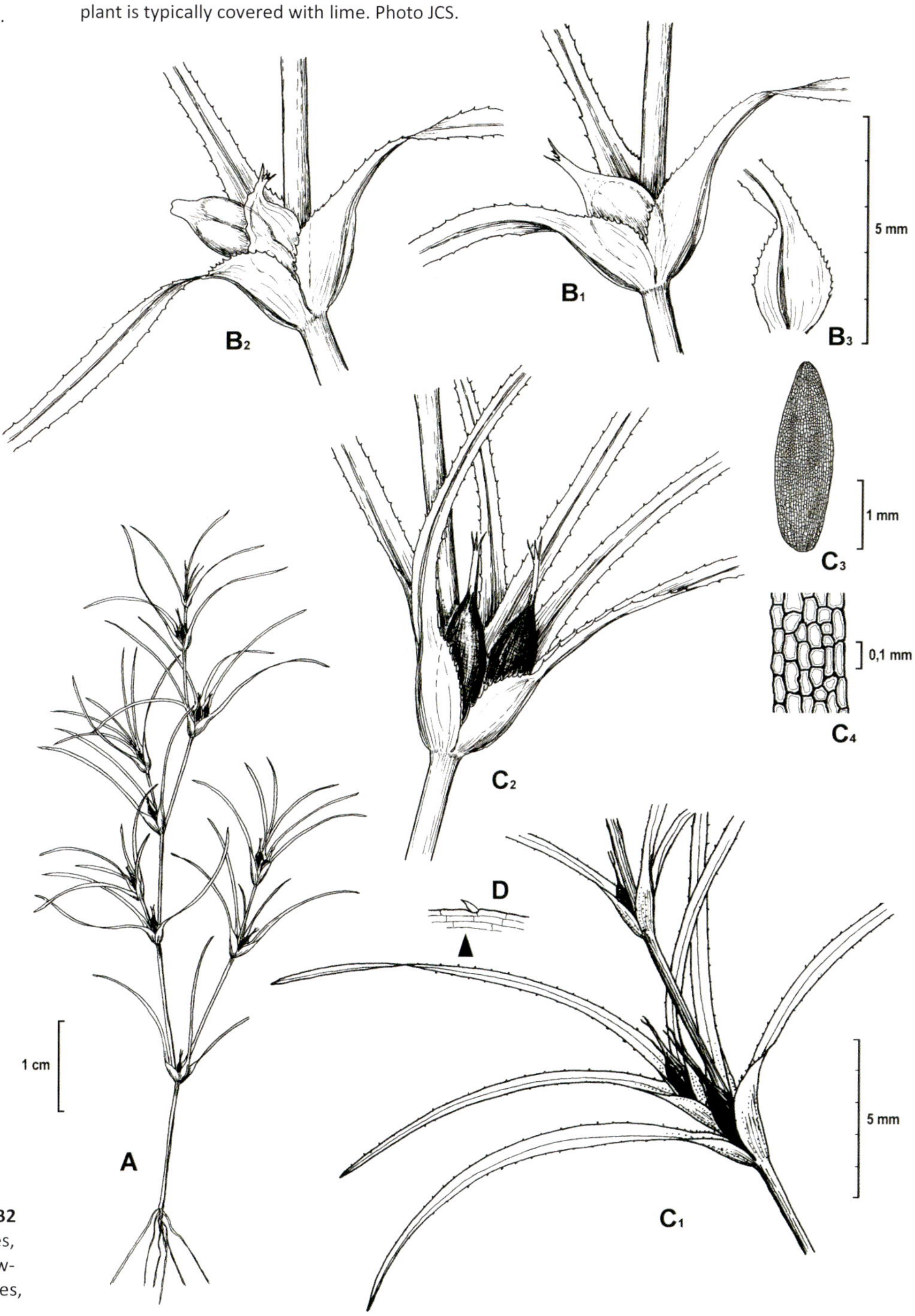

Najas flexilis
A habit, **B1** male flower enclosed in thin scales, **B2** anther pushing itself through the enclosing scales, **B3** leaf sheath, **C1, C2** stem part with female flowers in leaf axils, **C3** seed, **C4** seed coat with areoles, **D** leaf margin with tooth.

Hydrocharitaceae

62 ***Najas gracillima*** (Engelm.) Magnus

Syn.: *Caulinia amurensis* (Tzvelev) Tzvelev; *C. japonica* (Nakai) Nakai
DK: - **N:** - **S:** - **FIN:** - **IS:** -
GB: - **NL:** - **F:** -
D: Zierliches Nixenkraut **CZ:** - **PL:** -
EST: - **LV:** - **LT:** - **Ru:** Наяда изящнейшая

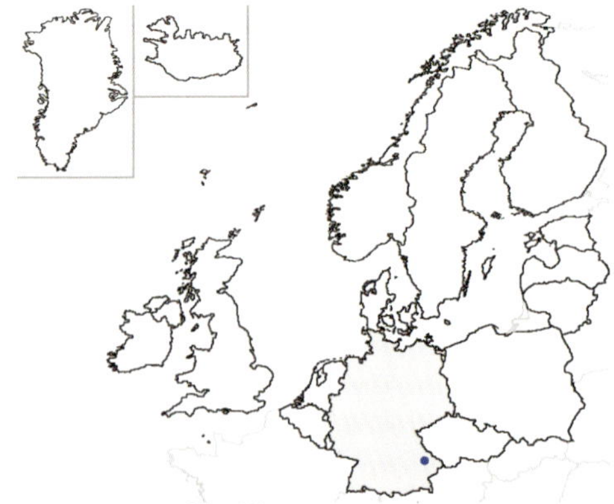

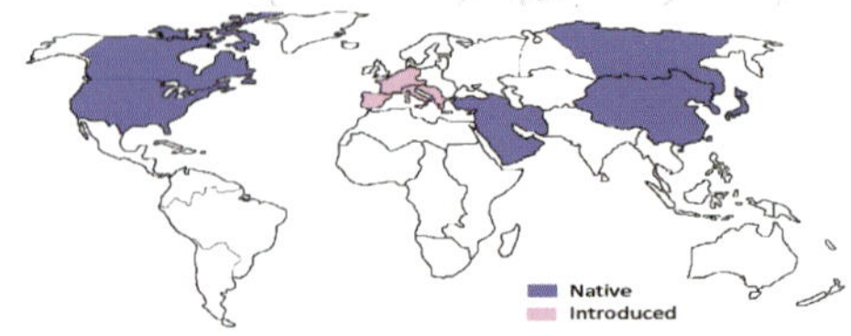

Najas gracillima densely covered with algae. Hafen Sand, Germany. Photo KvdW.

Annual, submerged hydrophyte. Roots white to brownish.

Stem: Green, 5–50 cm long, c. 0.5 mm wide, slender, without spines, dichotomously divided.

Leaves: Opposite, but appearing to be in whorls of 3, 0.9–2 cm long, 0.1–0.3 mm wide excluding teeth, linear, green, without septa, flaccid, acute. Margins serrulate, with spiny teeth 4–14, 0.05–0.1 mm long, on each side. Leaf sheaths green, partly clasping the stem, truncate to slightly auriculate, with 2–8 spines on each side.

Inflorescence: Male and female flowers on the same plant, but on different nodes, male flowers close to the top of the shoots. Flowers solitary or a few together in leaf axils.

Flowers: Male flower with a single anther enclosed by two thin scales. Anther with only one pollen sac. Female flower without scale, with a c. 0.6 mm long ovary and 2-3-lobed stigma.

Fruits: Seeds (1.8–)1.9–2.2(–2.7) mm long, 0.4–0.6 mm diameter, dull, yellow brownish. Areoles 8–19 in each longitudinal row, longer than broad and irregularly shaped.

Flowering: June-September.

Biology: Before pollination, the anther emerges through the enclosing scales and disperses a swarm of pollen on which the pollen tubes are already developed and up to 2 mm long to facilitate attachment to the stigmas. Seeds mature under water.

Habitats: In warm, still or slow-flowing, meso- to eutrophic water. Ponds, lakes, rice fields.

Distribution within the region:
Introduced and naturalised in Southern and Central Europe. Native to East Asia and North America.

Characteristics and similar species:
Najas gracillima differs from other *Najas* species by the combination of the following characters: Leaf sheaths truncate, leaves without septa, seeds 1.9–2.7 mm long with areoles irregularly shaped, longer than broad and arranged in more than 24 rows, anther with only one pollen sac.

104-107 ***Zannichellia*** spp. have a creeping rhizome and leaf margins without teeth.

54-56 ***Elodea*** spp. have broader leaves, densely set in whorls of 3 along the whole stem, and different flowers.

Najas gracillima, Hafen Sand, Germany. Photo KvdW.

Najas gracillima. Fruit in leaf axil. Photo KvdW.

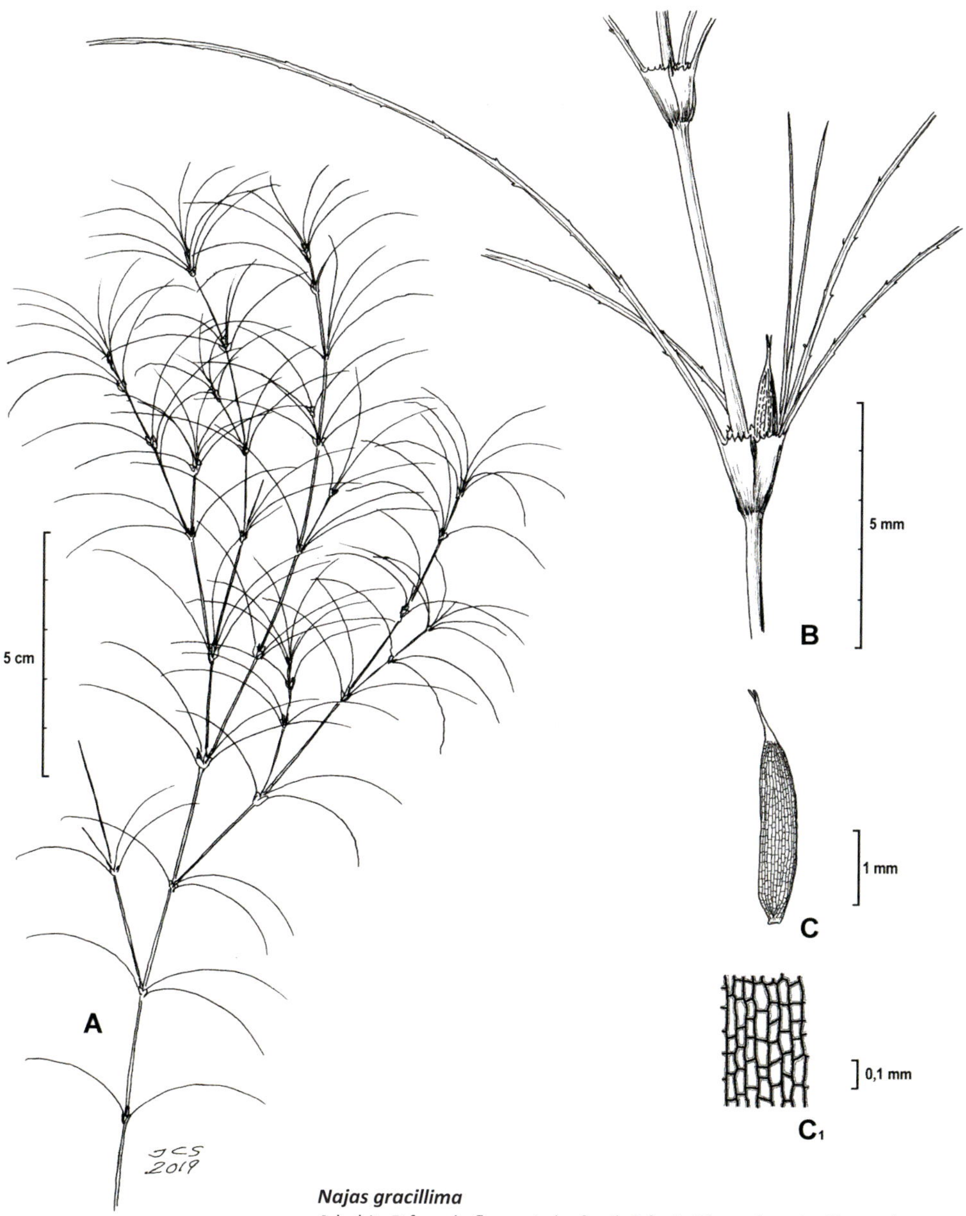

Najas gracillima
A habit, **B** female flower in leaf axil, **C** fruit, **C1** seed coat with areoles.

Hydrocharitaceae

63 *Najas tenuissima* (A.Braun) Magnus

Syn.: *Caulinia tenuissima* (A.Braun ex Magnus) Tzvelev
DK: Spæd Najade **N:** - **S:** Spädnajas **FIN:** Hentonäkinruoho **IS:** -
GB: - **NL:** - **F:** -
D: - **CZ:** - **PL:** -
EST: - **LV:** Smalkā najāda **LT:** - **Ru:** Наяда тончайшая

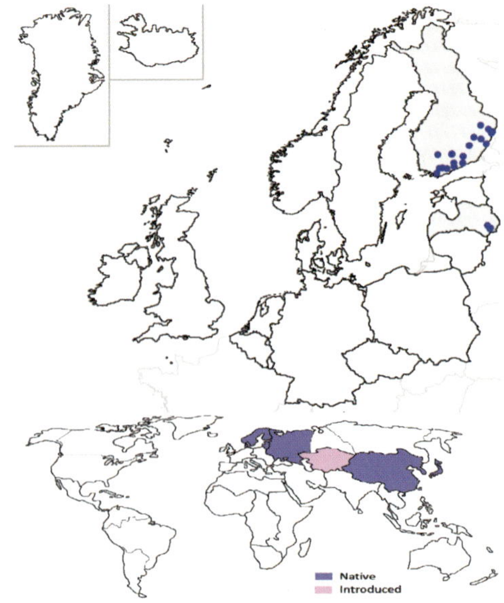

Najas tenuissima. Vyborg, Russia. Photo JCS.

Annual, submerged hydrophyte. Roots white to brownish.

Stem: 10–15(–20) cm long, c. 0.5 mm thick, slender, easily broken, without spines, dichotomously divided, green.

Leaves: Opposite, but appearing to be in whorls of 3, 0.7–2 cm long, 0.2–0.3 mm wide excluding teeth, linear, dark green, without septa, flaccid, acute. Margins serrulate, with 8–11 c. 0.1 mm long spiny teeth on each side. Leaf sheaths partly clasping the stem, truncate to shortly auriculate, with 4–6 spines on each side, green.

Inflorescence: Male and female flowers on the same plant, but on different nodes, male flowers close to the top of the shoots. Flowers solitary or few together in leaf axils.

Flowers: Male flower with a single anther enclosed in two thin scales. Anther with only one pollen sac. Female flower without scale, with a 0.7–1.3 mm long ovary and 2–3–lobed stigma.

Fruits: Seeds (2.2–)2.5–2.8(–3.0) mm long, 0.5–1.0 mm thick, dull, olive green to reddish. Areoles 25–30(–35) in each longitudinal row, longer than broad and irregularly shaped.

Flowering: June-September.

Biology: Before pollination, the anther emerges through the enclosing scales and disperses a swarm of pollen on which the pollen tubes are already developed and up to 2 mm long to facilitate attachment to the stigmas. Seeds mature under water and sink to the bottom. The seeds are rarely dispersed far from the parent plant, but may survive for some years in the seed bank.

Habitats: In clear, neutral, slightly eutrophic, groundwater-fed lakes. It prefers a light, open bottom with fine-grained minerals overlaid by neutral or slightly alkaline mud. In river estuaries in the Gulf of Finland.

Distribution within the region:
Native. Finland, Russia and Latvia. Declining because of eutrophication throughout its range.

Characteristics and similar species:
Najas tenuissima differs from other *Najas* species by the combination of the following characters: Leaf sheaths truncate to auriculate, leaves without septa, seeds (2.2–)2.5–2.8(–3.0) mm long with areoles irregularly shaped, longer than broad and arranged in more than 24 rows, anther with only one pollen sac.

104–107 ***Zannichellia*** spp. has a creeping rhizome and leaf margins without teeth.

54–56 ***Elodea*** spp. have shorter leaves relative to their width, in whorls of 3–5, more or less evenly spaced along the stem without a sheathing base.

Najas tenuissima. Vyborg, Russia. Photos JCS.

Najas tenuissima. Fruit almost smooth. Vyborg, Russia. Photo JCS.

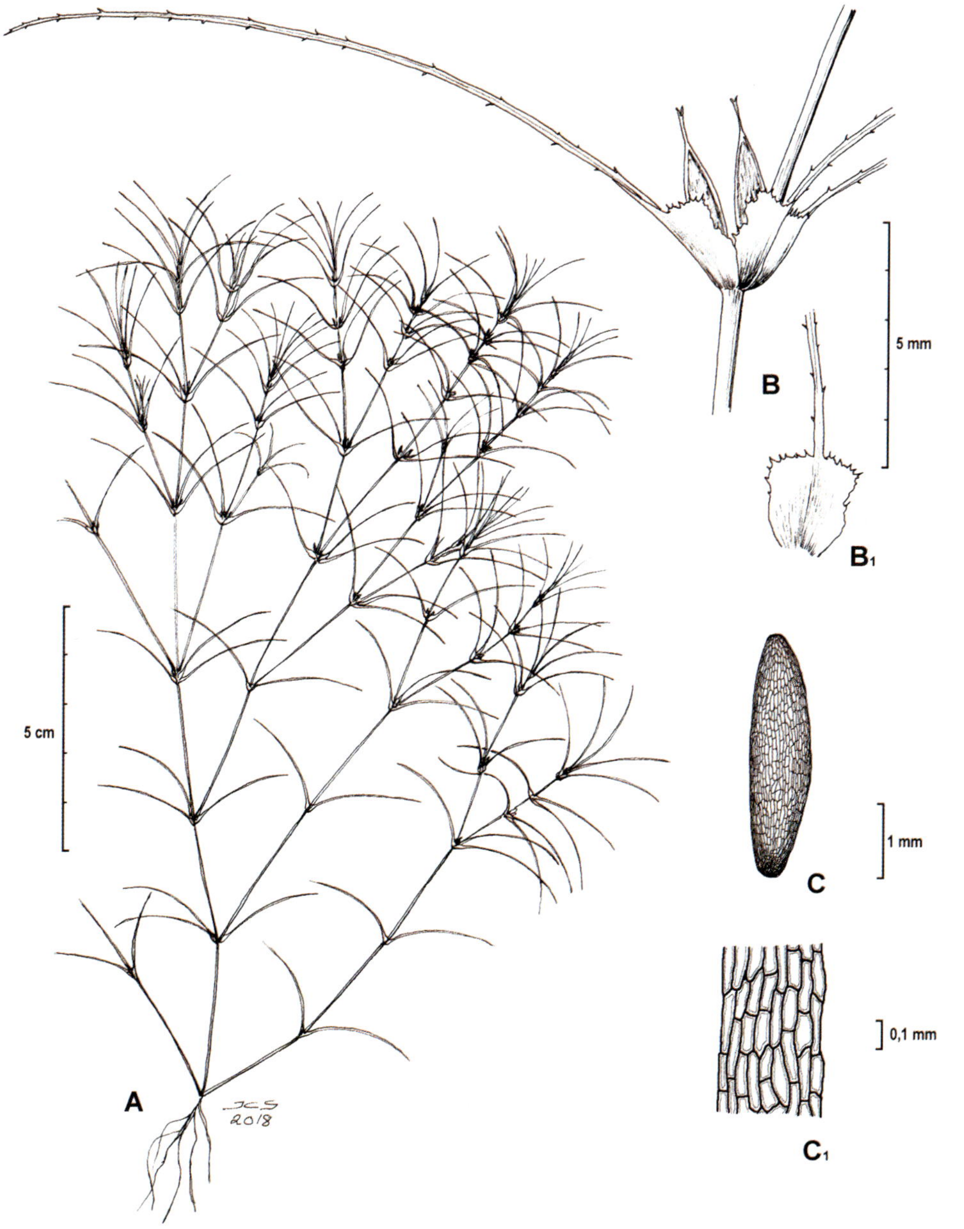

Nájas tenuissima
A habit, **B** female flowers in leaf axil, **B1** leaf sheath, **C** fruit, **C1** seed coat with areoles.

Hydrocharitaceae

64 *Vallisneria spiralis* L.

DK: - **N:** - **S:** - **FIN:** Kierrevallisneria **IS:** -
GB: Tapegrass **NL:** Vallisneria **F:** Vallisnérie en spirale
D: Gewöhnliche Wasserschraube **CZ:** Zákrutícha šroubovitá **PL:** Nurzaniec śrubowy
EST: - **LV:** - **LT:** - **Ru:** Валлиснерия спиральная

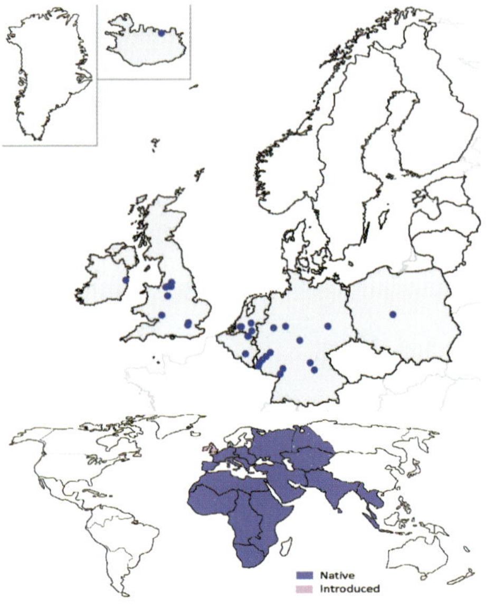

Vallisneria spiralis - female flowers stretch to reach the surface. Lago di Bolsena, Italy. Photo KvdW.

Perennial, submerged hydrophyte. Stolons up to 10 cm long, terete, coarse, whitish to brownish, often emerging from the sediment. Roots numerous, unbranched, white.

Stem: Very short, erect.

Leaves: All arising from basal rosettes. 5–100(–550) cm long, 0.5–1.0 cm wide ribbon-like, with 2 parallel veins and cross-veins on each side of the rather prominent midrib, translucent green with reddish-brown dots or streaks, flaccid, faintly denticulate towards the obtuse or rounded apex.

Inflorescence: Male and female flowers on separate plants. The inflorescences are axillary spathes formed by two united bracts. Male spathe ovoid, 6–9 mm long, containing numerous male flowers. Peduncle 2–7 cm. Female spathe tubular, 0.5–1.5 cm long, 2–3 mm thick, subtending a single female flower. Peduncle filiform, very long, spirally contracted after anthesis.

Flowers: Male flower c. 1 mm wide, with two stamens. Female flower 5–6 mm in diameter, sepals up to 4 mm long, pinkish white, petals very small. Ovary cylindrical, with many ovules. Styles 3, stigma bifid.

Fruits: Capsule up to 20 cm long, greenish yellow, with many seeds up to 2 mm long.

Flowering: June-October.

Biology: Male and female flowers are produced under water, but float to the surface for aerial pollination. The numerous male flowers detach as independent buds from the male inflorescence and open like a raft with only the perianth tips touching the surface. The female flowers remain attached to long thread-like peduncles and form a small depression on the water surface, which increases the likelihood of contact with the drifting male flowers. After pollination the spiral peduncle contracts, pulling the female flowers below the surface, where they ripen. Vegetative propagation by fragments and most importantly by stolons often results in dense monospecific stands.

The leaves of *V. spiralis* can be several metres long. Lago di Garda, Italy. Photo KvdW.

Vallisneria spiralis. Leaf tip with denticulate margins. Photo KvdW.

Vallisneria australis. Note the wide leaves - the single shot of *Lagarosiphon* in the centre is approximately 2 cm across. Mönchwaldsee, Frankfurt am Main, Germany. Photo KvdW.

Outside its native range, populations typically include only one sex - in Germany only male plants are known.

Habitats: In still or slow-flowing, warm, eutrophic water. Lakes, canals, rivers, hot springs.

Distribution within the region:
Native. Scattered in the southern part of the region, from Belgium east to Poland and in Iceland. Very locally naturalised in Britain.

Characteristics and similar species:
Vallisneria spiralis can be mistaken for other plants with long, submerged, ribbon-like leaves, such as *Sparganium* and *Sagittaria* species, but differs in the denticulate leaf margins and apex.

Vallisneria species are very popular as aquarium plants, and occasionally occur in the wild. The identification of such aliens can be very difficult, especially when the plants are sterile or young. The following alien species could occur:

64a *V. australis* S.W.L.Jacobs & Les has been detected in material from Belgium and Germany. It differs from **64 *V. spiralis*** in leaves 16–35 wide.

V. americana Michx., ***V. nana*** R.Br., ***V. neotropicalis*** Marie-Vict.

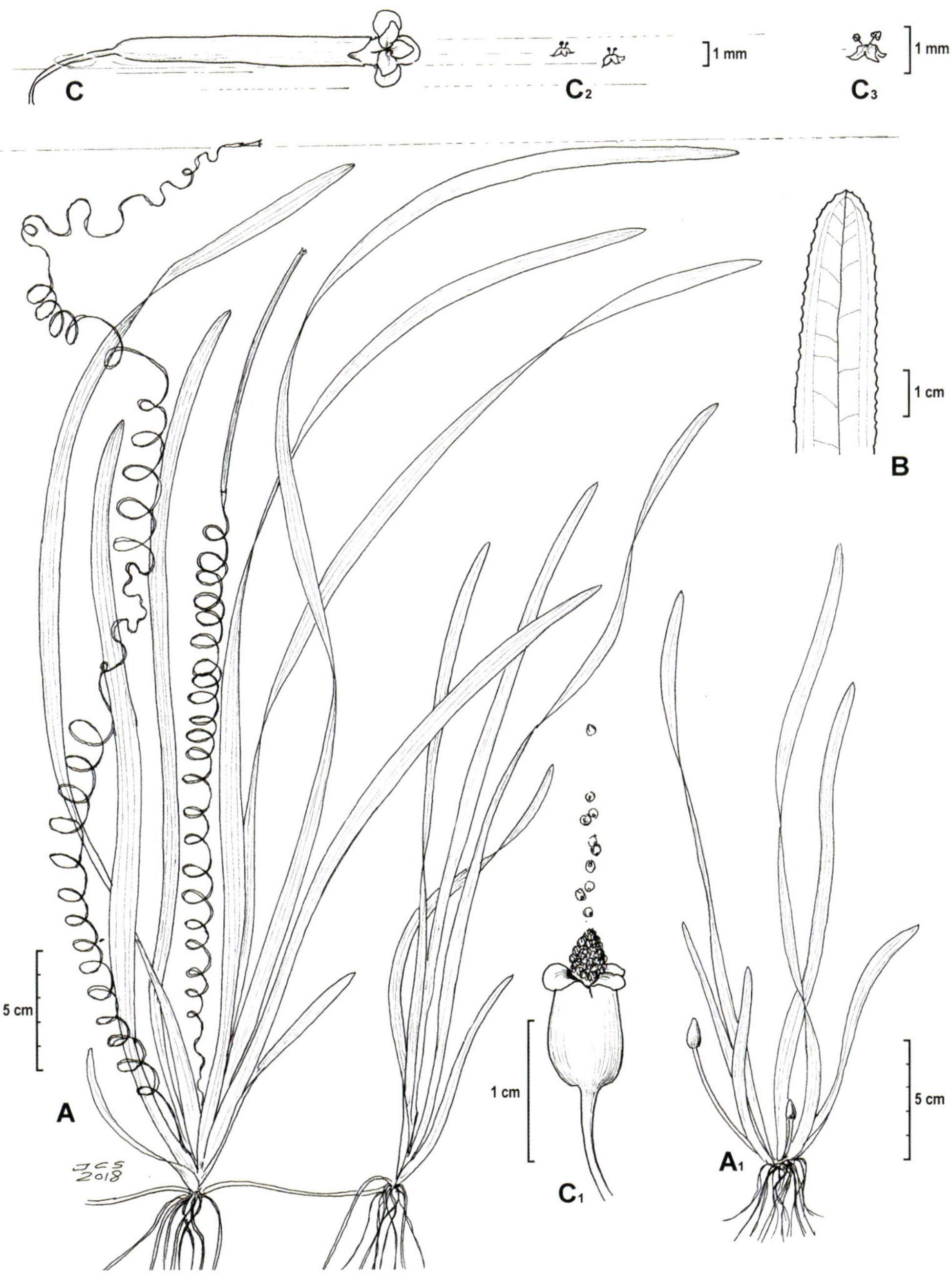

Vallisneria spiralis
A habit female plant, **A1** habit male plant, **B** leaf apex, **C** floating female flower, **C1** male spathe releasing flowers, **C2** floating male flowers, **C3** male flower.

Potamogetonaceae

Key to *Groenlandia*, *Potamogeton*, *Stuckenia* and *Zannichellia*

Species and most common hybrids

When collecting *Potamogeton* it is important that material selected is as representative of the population as possible. In several species and hybrids, the lower stem leaves differ from those higher up, and with shoot lengths of up to 5 metres, there is a risk that only the upper leaves will be collected. In most cases, *Potamogeton* species can reliably be identified using only a hand lens. In the case of some hybrids, microscopy of stem cross sections may be required - see page 220. In addition to the dichotomous key, a tabular key (p. 178-179) has been prepared for *Potamogeton* species and the most common hybrids.

Hybrids can be difficult to identify, and especially when it comes to hybrids between narrow-leaved *Potamogeton* species DNA sequencing may be required for confirmation. For that reason narrow-leaved *Potamogeton* hybrids are not included in the key. An introduction to *Potamogeton* hybrids, including a list of those known from the region, is presented on p. 221.

1 Leaf base enclosing stem - gently pull the leaf away from the stem (***1***)
Stuckenia - see p. 246

1 Leaf base not enclosing stem ... 2

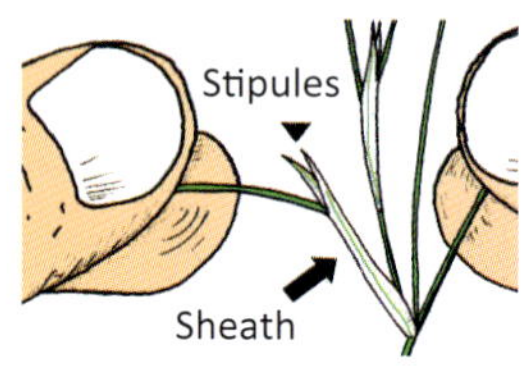

1

2 Leaves opposite or 3–4 at each node ... 3

2 Leaves alternate
Potamogeton ... 4

3 Leaves ovate oblong, margins finely denticulate (***2***)
65 *Groenlandia densa*

3 Leaves linear, margins entire
Zannichellia - see p. 260

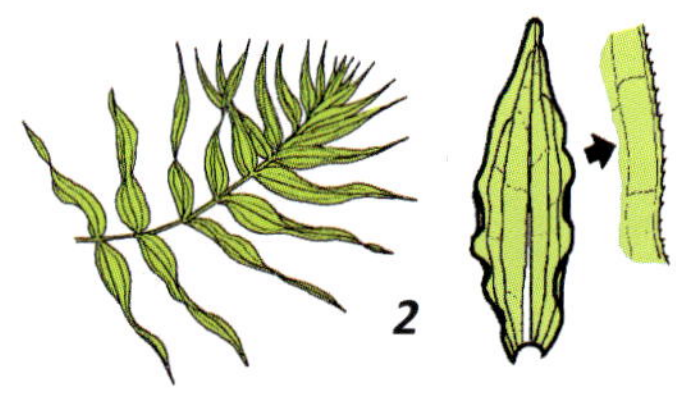

2

Potamogeton

4 All leaves linear to oblong with parallel margins sometimes undulate ... 5

4 At least some leaves lanceolate to elliptic with convex margins ... 20

5 Stem with a shallow groove on one or both of the broader sides (***3***). Leaves often undulate ... 6

5 Stem without grooves. Leaves not undulate ... 9

3

6 Leaf margin distinctly serrate (***4***) (visible to the naked eye)
75 *P. crispus*

6 Leaf margin obscurely serrate (use a hand lens) or entire ... 7
(In addition to the hybrids keyed out in couplet 7 and 8, juvenile plants of 75 *P. crispus* may also have entire leaf margins, especially during autumn and winter.)

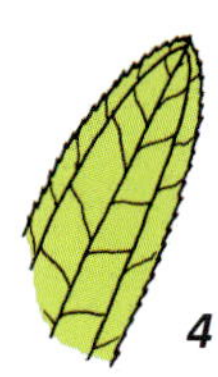

4

7 Leaves half-clasping the stem at base (***6***), 0.8–2.3 cm wide
93 *P.* ×*undulatus* (*P. crispus* × *praelongus*)

7 Leaves not clasping the stem, at most slightly auriculate (***5***) ... 8

8 Leaves 0.2–0.5 cm wide
94 *P.* ×*lintonii* (*P. crispus* × *friesii*)

8 Leaves 0.6–1.5 cm wide
91 *P.* ×*olivaceus* (*P. alpinus* × *crispus*)

5

6

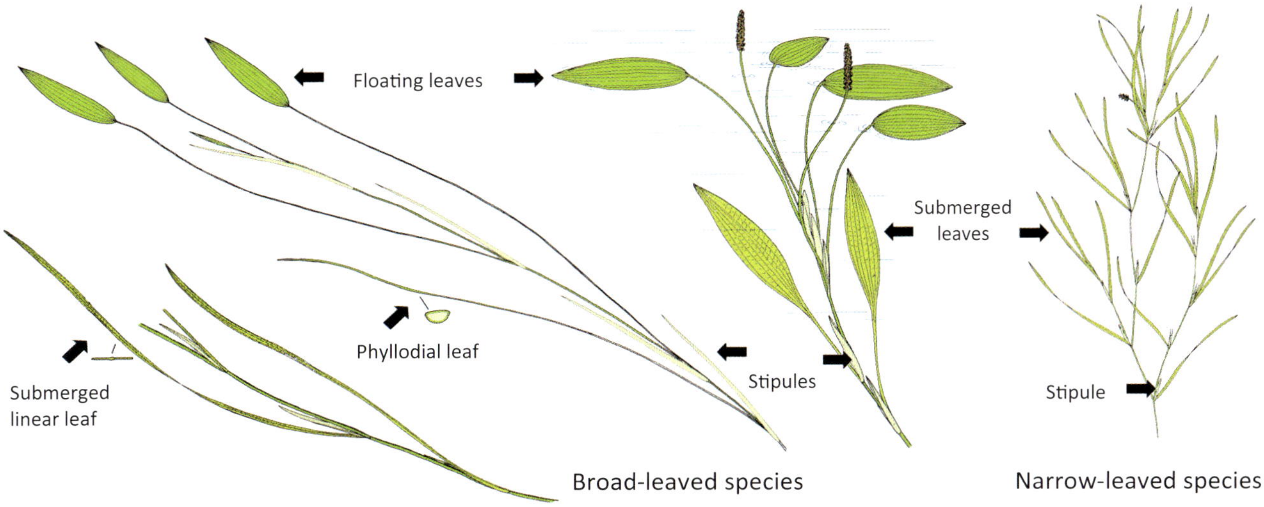

9 Leaves phyllodial, semi-terete, flat or shallowly concave on the adaxial side and convex on the abaxial side (***7***); veins 1–3, indistinct
66 *P. natans*

9 Leaves flat with a distinct midrib and lateral veins (***8***) 10

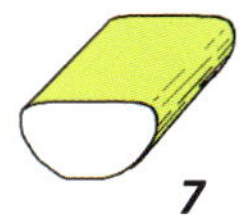
7

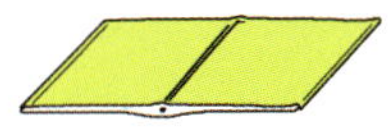
8

10 Rhizome present 11

10 Rhizome absent 12

11 Stem somewhat flat (***9***) or almost terete towards the apex
76 *P. epihydrus*

11 Stem terete (***10***)
88 *P. ×sparganiifolius* *(P. gramineus × natans)* or ***P. ×vepsicus*** *(P. natans × praelongus)*
(*P. ×sparganiifolius* and *P. ×vepsicus* cannot be separated without use of DNA sequencing, but see p. 226.)

9

10

12 Stem compressed to winged (***11***) 13

12 Stem terete (roll between your fingers) (***10***) 17

11

12

13 Leaves with several inconspicuous sclerenchymatous strands between the veins (use a hand lens against the light) (***13***). Stem flat or winged 14

13 Leaves without sclerenchymatous strands between the veins (use a hand lens against the light). Stem flat 15

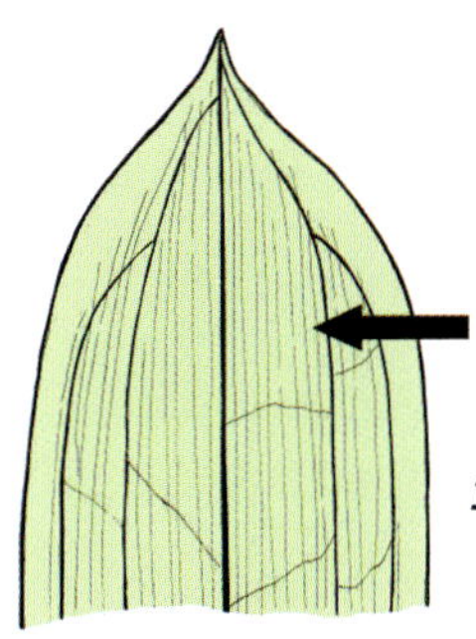
13

14 Leaves 3.5–8(–12) cm long, 1.5–4.0 mm wide; with 1 lateral vein on each side of the midrib. Stem flat, but not winged (***11***). Peduncle 0.5–2.0(–2.5) long, compressed, approximately 1–2 times the length of the inflorescence. Fruits with a tooth on the ventral side (***14***)
77 *P. acutifolius*
(The hybrid **97 *P. ×bambergensis*** may key out here - for separation see the text.)

14 Leaves 8.5–24 cm long, 3–6 mm wide; with 2 lateral veins on each side of the midrib. Stem winged (***12***). Peduncle 2.8–7.0 cm long, about 2–5 times the length of the inflorescence. Fruits without tooth (***15***)
78 *P. compressus*
(The hybrid **97 *P. ×bambergensis*** may key out here - for separation see the text.)

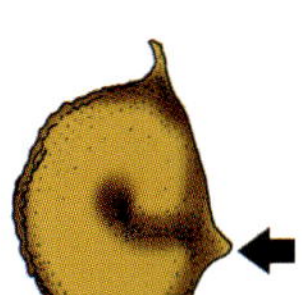
14

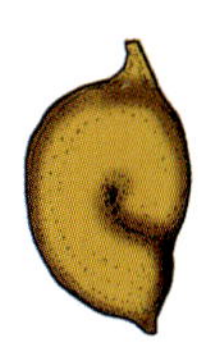
15

15 Leaves gradually tapering to a very sharp, pointed apex, 0.5–1.1 mm wide (***16***). Stipules with several distinct veins (prominent when dry) between the 2 primary veins
85 *P. rutilus*

15 Leaves obtuse to obtusely mucronate, 1.5–3.5 mm wide (***17,18***). Stipules with obscure to indistinct veins between the 2 primary veins — 16

16 Shoots particularly rich and densely branched in the upper part. Stipules open and convolute, rounded or more or less obtuse, not splitting into a V shape when decaying
79 *P. obtusifolius*

16 Shoots with branches more evenly distributed. Stipules connate at the base when young, with two very prominent veins, soon splitting into a V shape (***19***)
80 *P. friesii*

17 Leaves with a distinct band of pale tissue (lacunae) on each side of the midrib (***20***) and 2 well-developed glands at the nodes (***21***)
81 *P. berchtoldii*

17 Leaves without a distinct band of pale tissue (lacunae) on each side of the midrib (***22***) and without glands at the nodes (***23***) — 18

18 Stipules with several distinct veins between the 2 primary veins, tubular in the basal 2–4 mm, but soon splitting (***24***)
85 *P. rutilus*

18 Stipules with inconspicuous veins between the 2 primary veins, tubular or not in the basal 2–3 mm — 19

19 Stipules blunt, not inrolled, tubular in the basal 2–3 mm (***26***). Fruit not muricate on the dorsal side (***28***)
83 *P. pusillus*

19 Stipules tightly inrolled and open at the base (***25***). Fruit muricate on the dorsal side (***27***)
84 *P. trichoides*

20 Plants with only floating leaves — 21

20 Plants with at least some submerged leaves — 27

21 Leaves rather thin, translucent with distinct primary and secondary veins
68 *P. coloratus*

21 Leaves firm, coriaceous, not translucent — 22

22 Petiole with a 1–2 cm long, slightly swollen, flexible, discoloured joint between the petiole and the lamina (***29***)
66 *P. natans*

22 Petiole without a discoloured joint between the petiole and the lamina — 23

23 Stipule 6–14 cm long. Fruits never develop
86 *P. ×schreberi* *(P. natans × nodosus)*

23 Stipule 1–6 cm long. Fruits develop — 24

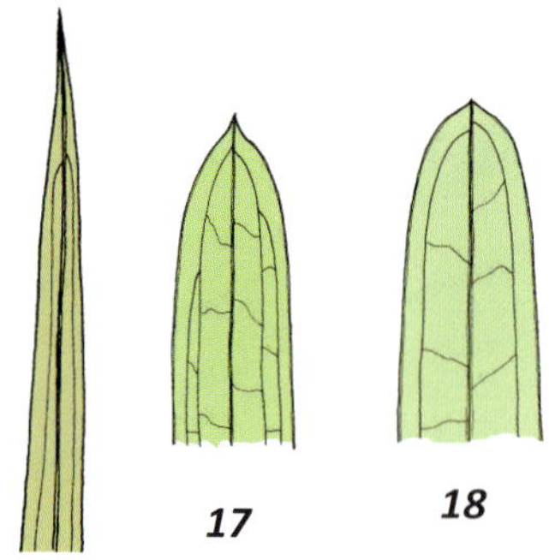
16 ***17*** ***18***

19

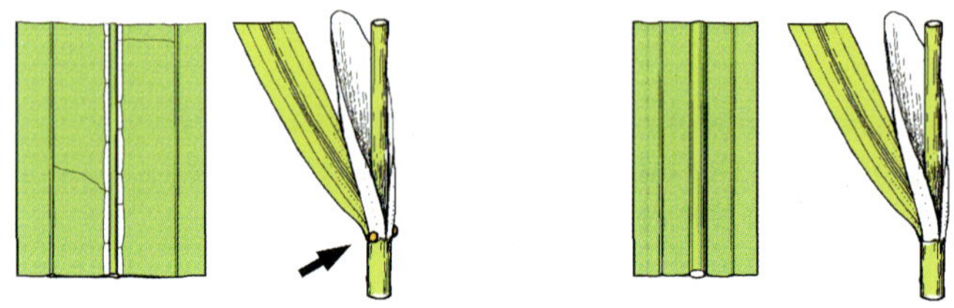
20 ***21*** ***22*** ***23***

24 ***25*** ***26***

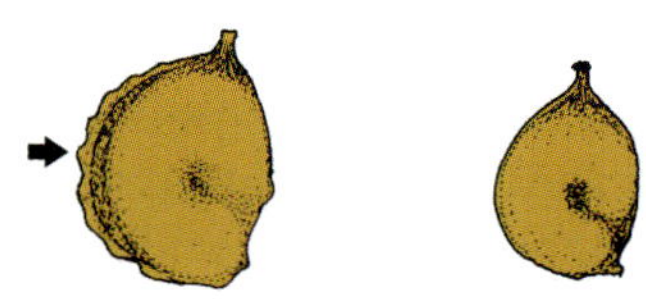
27 ***28***

29

24 Peduncle becoming broader and spongy towards the inflorescence spike (***30***). Fruit 2.4–3.1 mm long
70 *P. gramineus*
24 Peduncle of more or less uniform diameter and texture throughout 25

25 Fruits 1.9–2.6 mm long
67 *P. polygonifolius*
25 Fruits 2.6–4.1 mm long 26

26 Stipules 1.5–3.5(–4.5) cm long, the 2 most prominent veins not forming distinct ridges. Leaves with reddish tinge which intensifies when dried
71 *P. alpinus*
26 Stipules 3–8(–12) cm long, the 2 most prominent veins forming ridges (***31***). Leaves without a reddish tinge which intensifies when dried
69 *P. nodosus*

27 Leaves of two kinds present: petiolate floating leaves with oblong-elliptic to ovate lamina and lanceolate to linear submerged leaves 28
27 Only lanceolate to linear submerged leaves present 39

28 Petiole with a 1–2 cm long, slightly swollen, flexible, discoloured joint between the petiole and the lamina (***29***). Submerged leaves phyllodial, linear, semi-terete without distinct midrib and lateral veins (***7***)
66 *P. natans*
28 Petiole of floating leaves without a discoloured joint between the petiole and the lamina. Submerged leaves lanceolate to linear, flat (***8***), with distinct veins 29

29 Floating leaves translucent with prominent primary and secondary veins, not very different from the submerged leaves
68 *P. coloratus*
29 Floating leaves coriaceous, not translucent and very different in shape and structure to the submerged leaves 30

30 Submerged leaves linear with parallel margins - sometimes slightly wider towards the apex 31
30 At least the upper submerged leaves lanceolate 32

31 Submerged leaves sessile. Stem somewhat compressed or almost terete towards the apex of flowering stems
76 *P. epihydrus*
31 Submerged leaves petiolate. Stem terete
88 *P. ×sparganiifolius* *(P. gramineus × natans)*
or ***P. ×vepsicus*** *(P. natans × praelongus)*
(*P. ×sparganiifolius* and *P. ×vepsicus* can not be separated without use of DNA sequencing, but see p. 226.)

32 Margins of submerged leaves entire 33
32 Margins of submerged leaves denticulate (use a hand lens and look carefully, especially toward the apex) 37

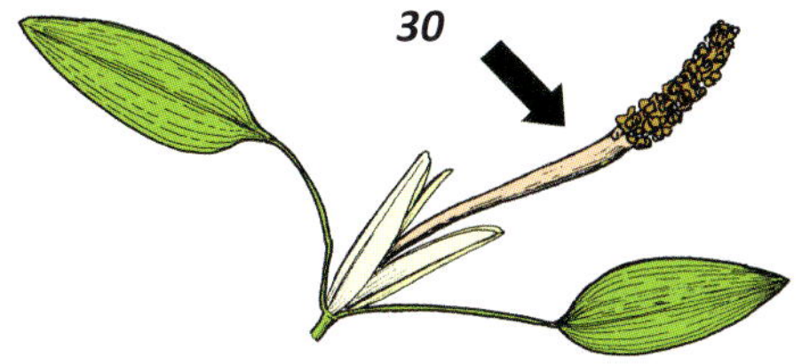
30

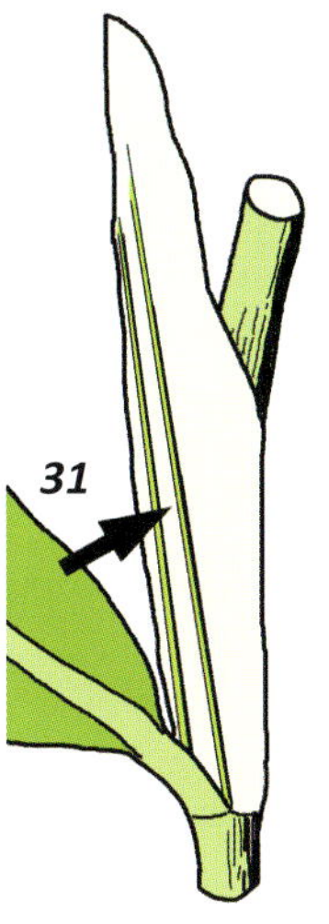
31

33 Submerged leaves all petiolate 34
33 Submerged leaves all sessile or the upper very rarely shortly petiolate
71 *P. alpinus*

34 Stem cross section with subepidermal and/or interlacunar bundles (***33***) (use a microscope) 35
34 Stem cross section without subepidermal and interlacunar bundles (***32***) (use a microscope)
69 *P. nodosus*

35 Margins of floating leaves somewhat inrolled (***34***). Upper submerged leaves with a mucro or an extended midrib
87 *P. ×fluitans* *(P. lucens × natans)*
35 Margins of floating leaves more or less flat. Apices of upper submerged leaves acute or obtuse, never with a mucro or extended midrib 36

36 Submerged leaves all linear elliptical or narrowly elliptical
67 *P. polygonifolius*
36 Basal submerged leaves phyllodial, linear, semi-terete, without distinct midrib and lateral veins (***35***), whereas the rest are linear elliptical with a lamina
86 *P. ×schreberi* *(P. natans × nodosus)*

37 Submerged leaves gradually tapering to the base 38
37 Submerged leaves rounded or more or less amplexicaul at base
90 *P. ×nitens* *(P. gramineus × perfoliatus)*

38 Submerged leaves usually less than 12 mm wide, the upper sessile
70 *P. gramineus*
38 Submerged leaves usually more than 12 mm wide, the upper petiolate
89 *P. ×angustifolius* *(P. gramineus × lucens)*

39 Submerged leaves distinctly petiolate 40
39 Submerged leaves sessile or some of them with a 1–5 mm long petiole 46

40 Submerged leaves with a mucro or an extended midrib. The 2 most prominent veins on the stipule appearing winged on the lower half of the abaxial side (***31***) 41
40 Submerged leaves acute or obtuse at apex, never with a mucro or an extended midrib. The 2 most prominent veins on the stipule not appearing winged on the lower half of the abaxial side 43

41 Petiole 2.5–10 cm long
87 *P. ×fluitans* *(P. lucens × natans)*
41 Petiole less than 2.5 cm long 42

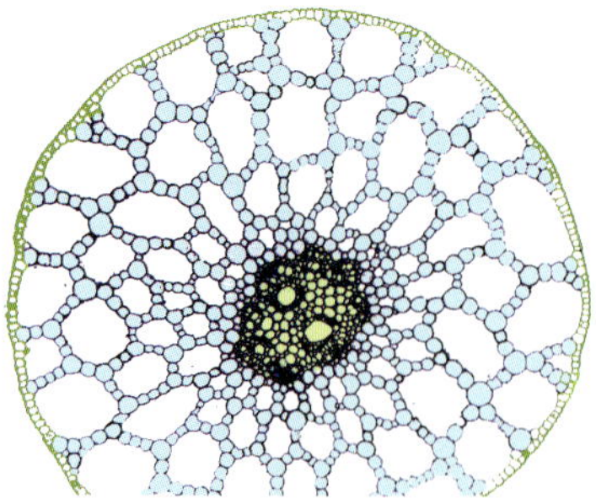

32

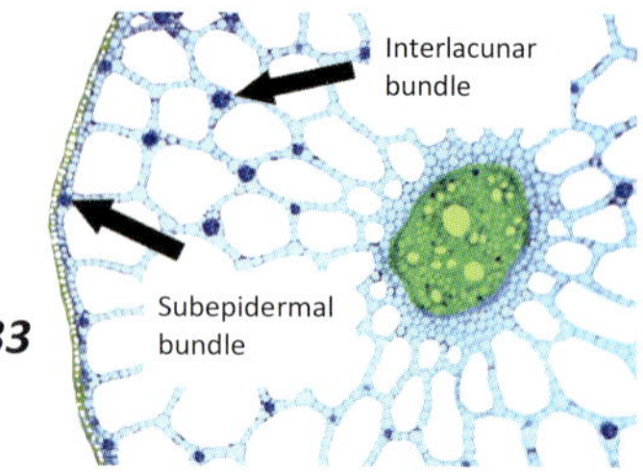

33

34 Lower side of leaf margins inrolled

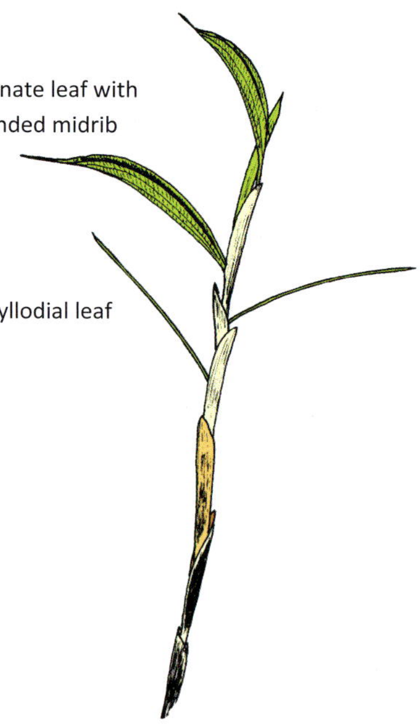

35 Phyllodial leaf

42 Leaves on main stem and branches of more or less equal size. All leaves with 0.2–0.7 cm long petioles. Stipules on main stem 3–8 cm long
73 *P. lucens*

42 Leaves on main stem larger than those on branches (***36***). Most leaves sessile, but in the upper part of the plant there are usually leaves with long petioles. Stipules on main stem 2–5 cm long
89 *P. ×angustifolius* *(P. gramineus × lucens)*

43 Stem cross section with 2–3 rows of interlacunar bundles (***33***) (use a microscope)
86 *P. ×schreberi* (*P. natans × nodosus*)

43 Stem cross section without interlacunar bundles (***32***) (use a microscope) 44

44 Stem cross section without subepidermal bundles (***32***) (use a microscope). Young leaves with minutely denticulate margin (use a hand lens)
69 *P. nodosus*

44 Stem cross section with subepidermal bundles (***33***) (use a microscope). Young leaves without minute teeth (use a hand lens) 45

45 Submerged leaves 2–5 cm wide, petiole 1.5–6.5 cm long. Fruits 1.3–1.9 mm long. Predominantly in alkaline water
68 *P. coloratus*

45 Submerged leaves 0.3–2.4 cm wide, petiole 1.5–8(–15) cm long. Fruits 1.9–2.6 mm long. Predominantly in non-alkaline water
67 *P. polygonifolius*

46 Lamina of submerged leaves gradually tapering to the base 47

46 Lamina of submerged leaves rounded or more or less amplexicaul at base 50

47 Submerged leaves with entire margin
71 *P. alpinus*

47 Submerged leaves with denticulate margin (use a hand lens and look carefully, especially toward the apex) 48

48 Most or all leaves on main stem less than 12 mm wide
70 *P. gramineus*

48 Most or all leaves on main stem greater than 12 mm wide 49

49 Submerged leaves with petioles 0.2–0.7 cm long, not recurved. Leaves on main stem not much larger than those on branches. Floating and transitional leaves absent
73 *P. lucens*

49 Submerged leaves sessile, often recurved, the upper often shortly petiolate. Leaves on main stem distinctly larger than those on branches. Floating and transitional leaves often present
89 *P. ×angustifolius* *(P. gramineus × lucens)*

50 Stem terete or nearly so (***10***) 51

50 Stem slightly compressed with a shallow groove running down on one or both of the broader sides (***3***) 56

36

51 Submerged leaves with denticulate margin (use a hand lens and look carefully, especially toward the apex) 52
51 Submerged leaves with entire margin 55

52 Stipules fugacious, present only on young leaves. Leaves hooded or slightly hooded at apex 53
52 Stipules persistent. Leaves flat at apex 54

53 Leaf margin densely denticulate. Stipules present only on the youngest leaves
74 *P. perfoliatus*
53 Leaf margin sparsely denticulate. Stipules present on most of the younger leaves
96 *P. ×cognatus* *(P. perfoliatus × praelongus)*

54 Submerged leaves widest at base. The 2 most prominent veins on the stipule not appearing winged on the lower half of the abaxial side
90 *P. ×nitens* *(P. gramineus × perfoliatus)*
54 Submerged leaves widest at the middle. The 2 most prominent veins on the stipule appearing winged on the lower half of the abaxial side
95 *P. ×salicifolius* *(P. lucens × perfoliatus)*

55 Stem with a characteristic zigzag shape (***37***). Stipules long persistent
72 *P. praelongus*
55 Stem more or less straight. Stipules degrading early and only present on the younger leaves
96 *P. ×cognatus* *(P. perfoliatus × praelongus)*

56 Leaves with 1–2 veins on each side of the midrib. Leaf margin denticulate to coarsely serrate (***4***) - visible without a hand lens and often undulate
75 *P. crispus*
56 Leaves with 2–6 veins on each side of the midrib. Leaf margin entire or sparsely denticulate, undulate or not 57

57 Leaf margin entire or very obscurely denticulate (use a hand lens). Leaves 5–15 cm long
93 *P. ×undulatus* *(P. crispus × praelongus)*
57 Leaf margin denticulate (use a hand lens and look carefully especially towards the apex). Leaves 2.5–6 cm long
92 *P. ×cooperi* *(P. crispus × perfoliatus)*

37

Potamogeton ×nitens. Flora Danica. Booklet 31, plate 1807. Published by J.W.Hornemann, Copenhagen 1825.

Tabular key to *Potamogeton* species and most common hybrids

● yes ○ no ●○ yes or no ( ) rare - irrelevant	*66 P. natans*	*67 P. polygonifolius*	*68 P. coloratus*	*69 P. nodosus*	*70 P. gramineus*	*71 P. alpinus*	*72 P. praelongus*	*73 P. lucens*	*74 P. perfoliatus*	*75 P. crispus*
Floating leaves present	●○	●(○)	●○	●○	●○	●○	○	○	○	○
Floating leaves thin, translucent	○	○	●	○	○	○	○	○	○	-
Petiole of floating leaves with a soft, reddish section at the junction with the lamina. Submerged leaves as phyllodes, linear, semi-terete	●	○	○	○	○	○	○	○	○	○
Plant with a creeping rhizome	●	●	●	●	●	●	●	●	●	●
Stem terete	●	●	●	●	●	●	●	●	●	○
Stem with 1-2 furrows	○	○	○	○	○	○	○	○	○	●
Stem flat to winged	○	○	○	○	○	○	○	○	○	●
Stem winged	○	○	○	○	○	○	○	○	○	○
Submerged leaves with convex margins	○	●	●	●	●	●	●	●	●	●○
Submerged leaves with parallel or almost parallel margins	●	○	○	○	○	○	○	○	○	○
Submerged leaves with small teeth (use hand lens)	○	○	○	○	●	○	○	●	●	●
Submerged leaves petiolate, at least some of them	●	●	●	●	○	●	○	●	○	○
Submerged leaves with a band of lacunae on each side of the midrib. Band narrow • wide ●	○	•	•		•	●	•	○	•	•
Submerged leaves with nodal glands at base. Distinct ● indistinct •	○	○	○	○	○	○	○	○	○	○
Submerged leaves with blunt or rounded apex	○	○	○	○	○	○	●	○	●	●
Submerged leaves < 12 mm wide on main stem	●	●○	(●)○		●(○)	●○	○	○	○	●
Submerged leaves widest below the middle	○	●○	●○		○	○	●	○	●	○
Stipules fused in the lower part (look carefully at young leaves)	○	○	○		○	○	○	○	○	○
Stipules persistent or splitting into fibres	●	●	●		●	●○	●	●	○	(●)○
Stipules splitting into fibres	○	○	○		○	○	●	○	○	●
Number of carpels per flower	4	4	4		(3–)4 (–5)	4	(2–)4	4(–6)	4	(2–)4
Length of ripe fruit in mm	3.5–5.0	1.9–2.6	1.3–1.9		2.4–3.1	2.6–3.7	4.5–5.5	3.2–4.5	2.5–3.5	4.0–6.0

96 P. ×cognatus	○	-	-	●	●	○	○	○	●	○	(●)○	○	•	○	●	○	●	○	(●)○	○	4	-
95 P. ×salicifolius	○	-	-	●	●	○	○	○	●	○	●	○	○(•)	○	●	○	●○	○	●○	○	(3–)4	-
93 P. ×undulatus	○	-	-	●	○	●	○	○	●(○)	(●)○	(●)○	○	•	○	●	(●)○	●○	○	●	●○	4	-
90 P. ×nitens	●○	○	○	●	●	○	○	○	●	○	●	○	•	○	●○	○	●○	○	●(○)	○	4–5	-
89 P. ×angustifolius	(●)○	(●)○	○	●	●	○	○	○	●	○	●	(●)○	•	○	○	●○	○	○	●	○	(3–)4	2.7–3.4
88 P. ×sparganifolius	●(○)	○	○	●	●	○	○	○	(●)○	●(○)	(●)○	●	•	○	○	●	○	○	●	○	4	-
87 P. ×fluitans	●	○	(●)○	●	●	○	○	○	●	○	(●)○	●	○	○	○	●○	○	○	●	○	4(–5)	-
84 P. rutilus	○	-	-	○	(●)○	○	●	○	○	●	○	○	○	○(•)	○	●	○	●○	●		(2–)4	2.0–2.1
83 P. trichoides	○	-	-	○	●	○	○	○	○	●	○	○	○	○(•)	○	●	○	○	●	○	1(–2)	2.5–3.2
82 P. pusillus	○	-	-	○	●	○	○	○	○	●	○	○	○	○	○	●	○	●	●	○	4(–5)	1.8–2.4
81 P. berchtoldii	○	-	-	○	●	○	○	○	○	●	○	○	●•	●	○	●	○	○	●○	○	4–5	1.8–2.8
80 P. friesii	○	-	-	○	○	○	●	○	○	●	○	○	•	●	○	●	○	●○	●	●	4	2.4–3.0
79 P. obtusifolius	○	-	-	○	○	○	●	○	○	●	○	○	●•	●	●	●	○	○	●	○	(3–)4(–5)	2.6–3.6
78 P. compressus	○	-	-	○	○	○	●	●	○	●	○	○	•	○	○	●	○	○	●	●	(1–)2(–3)	3.4–4.0
77 P. acutifolius	○	-	-	○	○	○	●	○	○	●	○	○	•	○	○	●	○	○	●	●	1	3.0–4.0

Potamogetonaceae

65 *Groenlandia densa* (L.) Fourr.

Syn.: *Potamogeton densus* L.
DK: Tæt Vandaks **N:** Kranstjønnaks **S:** Tätnate **FIN:** - **IS:** -
GB: Opposite-leaved Pondweed **NL:** Dichtbladig fonteinkruid **F:** Potamot dense
D: Fischkraut **CZ:** Rdest hustolistý **PL:** Grenlandia gęsta
EST: - **LV:** - **LT:** Tankialapė grenlandija **Ru:** Гренландия густая

The name *Groenlandia densa* refers to the dense pair of leaves. In addition, the species forms dense populations and covers large areas in colonised habitats. Subbæk Mill near Viborg, Denmark. Photo JCS.

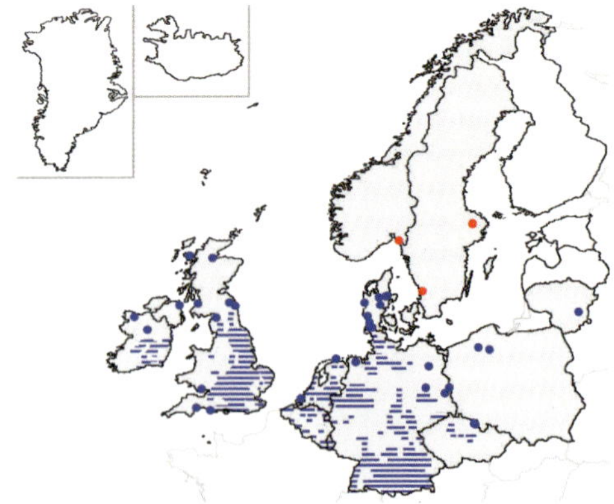

Red dots indicate extinct populations.

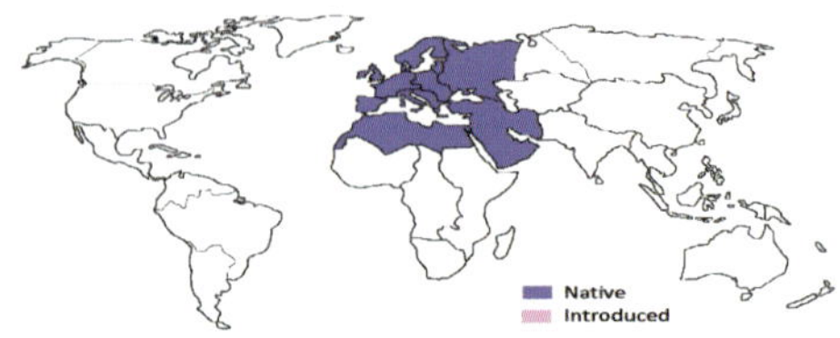

Perennial, submerged hydrophyte. Rhizome fairly robust with ascending shoots from every second internode and often horizontal on or immediately below the substrate.

Stem: Up to 65 cm long, terete, bright green to light yellowish brown, more or less branched, and unlike *Potamogeton* species *G. densa* often develops roots from the lower nodes.

Floating leaves: Absent.

Submerged leaves: Up to 4.2 cm long and 1.2 cm wide, shiny, bright to mid-green, eventually with a brownish tinge on the upper part of the shoots, translucent, more or less broadly lanceolate, with 5–7 veins, often somewhat undulating and recurved, with delicate denticulate margin, acute, sessile, amplexicaul. Leaves normally in two rows, appearing opposite, only separated by very short internodes. The distance between leaf pairs varies, but is normally short, making the vegetative shoots look dense and compact.

Stipules: Delicate and translucent, 2–4 mm long, and only present on leaves supporting branches or peduncles.

Inflorescence: With 2 flowers, each with 4 carpels. Peduncle developing from a leaf axil, erect in flower, becoming recurved in fruit.

Fruits: 3–4 mm long, brown, with distinct keel and a 0.5–0.9 mm long beak.

Flowering: July-September.

Biology: Floating or emergent flowers are wind pollinated, while submerged flowers are water pollinated. Pollen becomes attached to the stigmas, small air bubbles adhering to the pollen grains. Vegetative propagation occurs by division of the rhizome. *G. densa* does not form turions but overwinters as leafy shoots.

Habitats: Shallow, clear, base-rich water, often influenced by groundwater. Ditches, canals, streams, old marl pits and mill ponds. It is less frequent in lakes and reservoirs. In slow-flowing (or still) water, it may cover large areas

Groenlandia densa is a beautiful species, always neat and without epiphytes. Subbaek Mill, Denmark. Photo JCS.

Groenlandia densa. The peduncle develops from the axil of a leaf, and becomes strongly recurved in fruit. Subbaek Mill pond, Denmark. Photo JCS.

and continue to grow until late autumn. In streams, it can create dense populations of compact vegetative shoots. *Groenlandia densa* can endure some eutrophication and is quite tolerant of ochre pollution.

Distribution within the region: Native. Widespread in the southwest of the region, but declining at least in Britain. Extinct in Norway and Sweden. Introduced in Lithuania.

Characteristics and similar species: *Groenlandia densa* can be recognised by the lanceolate, apparently opposite leaves, with often somewhat undulating leaf margins, and amplexicaul base. It is never covered with epiphytic algae.

54 *Elodea canadensis* may resemble *G. densa* but differs in leaf shape and has leaves arranged in whorls.

Forms of **75 *Potamogeton crispus*** with only slightly undulate leaves can be mistaken for *G. densa*, but it has a compressed stem with a shallow groove running down one or both of the broader sides, and leaves that only occasionally are opposite, but always have distinct lateral transverse veins and fruits with a long beak.

Hybrids: *Groenlandia densa* is not known to form hybrids.

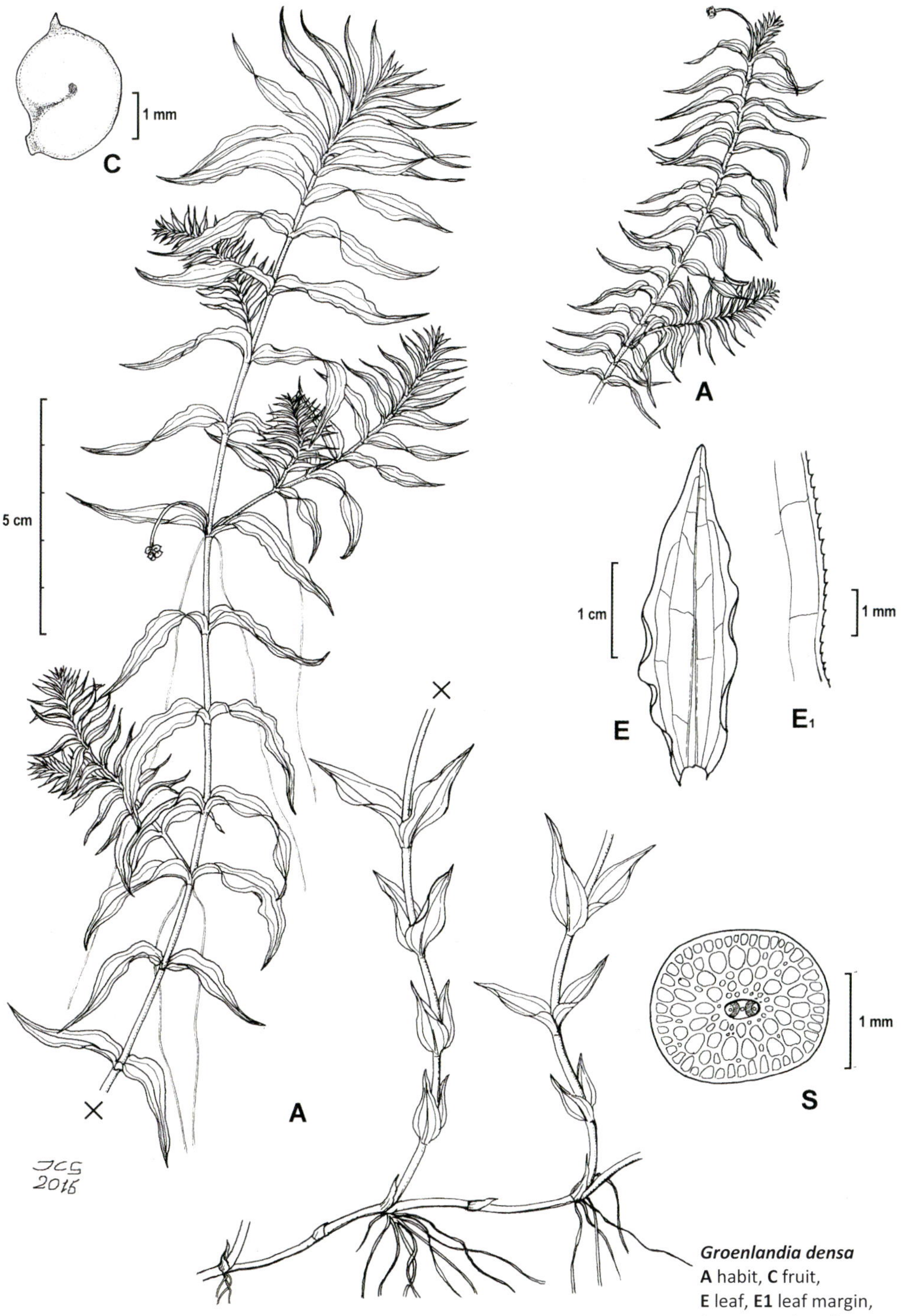

Groenlandia densa
A habit, **C** fruit,
E leaf, **E1** leaf margin,
S cross section of stem.

Potamogetonaceae

66 *Potamogeton natans* L.

DK: Svømmende Vandaks **N:** Vanleg tjørnaks **S:** Gäddnate **FIN:** Uistinvita **IS:** Blöðkunykra
GB: Broad-leaved Pondweed **NL:** Drijvend fonteinkruid **F:** Potamot nageant
D: Schwimmendes Laichkraut **CZ:** Rdest plovouci **PL:** Rdestnica pływająca
EST: Ujuv penikeel **LV:** Peldošā glīvene **LT:** Plūduriuojančioji plūdė **Ru:** Рдест плавающий

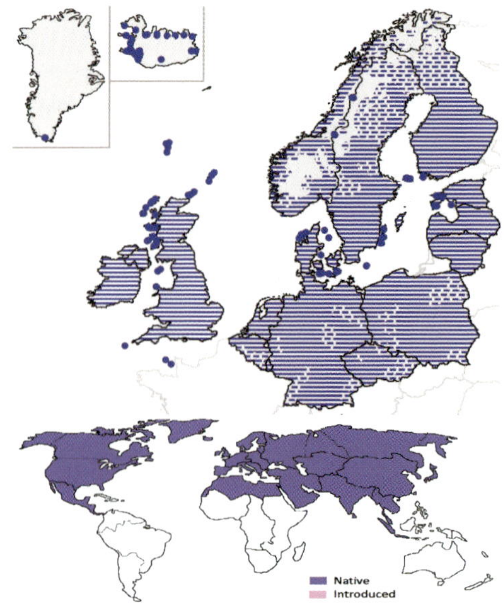

Like all other *Potamogeton* species, *P. natans* is pollinated above the water, but after fertilisation the peduncle bends down and the fruits mature under water. Pond at Bornholm, Denmark. Photo JCS.

Perennial hydrophyte. Rhizome horizontal, branched, with short internodes and shoots from every second node.

Stem: From just a few centimetres long in shallow water up to several metres long in deep or flowing water, terete, unbranched or sparingly branched.

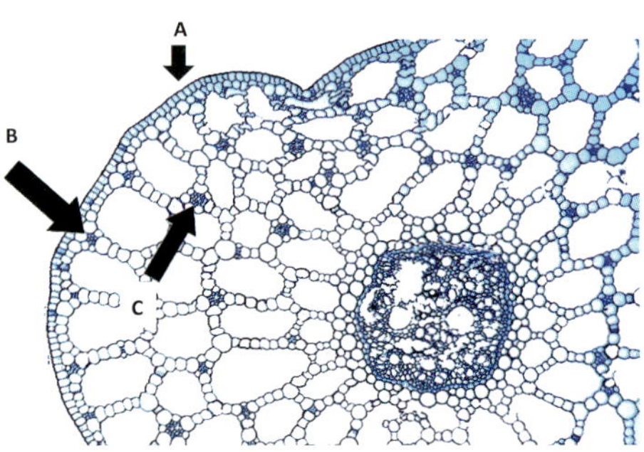

Stem cross section with 1–2 layers of pseudohypodermis (A), subepidermal bundles (B) and 3–4 rows of interlacunar bundles (C). Photo KvdW.

Petiole linear, slightly compressed, semi-terete - flattened on one side, convex on the other, 5-veined, firm, the uppermost 1–2 cm with a slightly swollen, flexible, pinkish section just below the lamina.

Submerged leaves: Linear, semi-terete, flat or shallowly concave on the adaxial side and convex on the abaxial side, 1–3 veins, firm, coriaceous, acuminate.

Stipules: Persistent, 4–17 cm long, rigid, acute to narrowly acuminate at the apex, colourless to whitish, opaque, veins close together, 2 more prominent than the others.

Inflorescence: 2–6 cm long, cylindrical. Peduncle 4–10 cm long. Usually with only one inflorescence on each shoot.

Flowers: Contiguous, with 4 carpels.

Fruits: 3.8–5.0 mm long. Beak 0.3–0.8 mm long.

Flowering: June-August.

Biology: Capable of rapid growth to cover a large area. In ponds, lakes and streams, the species produces shoots with floating leaves, but in terrestrial or semi-terrestrial habitats, forms with very short shoots develop. *P. natans* has high dispersal capacity, presumably through dispersal of seeds by birds. The species does not develop turions, but overwinters as rhizomes and short vegetative shoots with submerged leaves.

Habitats: In virtually all aquatic habitats and water types except brackish water.

Distribution within the region:
Native. Widespread and abundant more or less throughout the regions, but rare in Greenland.

Floating leaves: Lamina 4–10(–14) cm long, oblong ovate to broadly elliptical, acute to obtuse at the apex, with rounded or subcordate base, shortly and narrowly decurrent down the petiole, firm, more or less coriaceous, not translucent. Young leaves coppercoloured, later green to dark olive green and often with a brownish tinge. With 8–12(–15) translucent lateral veins on each side of the midrib, and numerous dark secondary veins.

In rivers *P. natans* appear as large waving populations near the bank. Skjernå, Denmark. Photo JCS.

The soft, discoloured section of the petiole in the junction with the lamina distinguishes *P. natans* from all other *Potamogeton* species, but sometimes this feature is absent. Photo JCS.

Section from the base of the stipule of *P. natans* (left) and *P. polygonifolius* (right) at the same scale. Notice that *P. natans* has numerous closely set veins, making the texture of the stipule more rigid and opaque than the translucent stipule of *P. polygonifolius*, in which the weaker, dispersed veins are less conspicuous - a good character for identification even on dried material. Photo JCS.

Characteristics and similar species: *Potamogeton natans* is typically recognised by the linear, semi-terete phyllodes, the long stipule and the 1–2 cm long, discoloured petiole section below the lamina.

It can be mistaken for **67 *P. polygonifolius,* 69 *P. nodosus*** *and* **68 *P. coloratus***, which all have submerged leaves with lamina, and the floating leaves lack the flexible, pinkish section between the petiole and the lamina.

With or without submerged leaves *P. natans* can be reliably separated by the stem cross section. *P. natans* has both subepidermal and interlacunar bundles, *P. polygonifolius* and *P. coloratus* only subepidermal bundles and *P. nodosus* neither subepidermal nor interlacunar bundles.

Potamogeton natans has larger fruits than *P. polygonifolius. P. natans* stipules have more closely set veins, which give them a more opaque appearance (see photo at top right of page), whereas *P. polygonifolius* has weaker and less densely set veins. By transmitted light, the lateral veins in floating leaves of *P. natans* appear translucent, in contrast to the darker veins of *P. polygonifolius.*

Potamogeton coloratus,* 70 *P. gramineus* and 71 *P. alpinus all have submerged leaves with broad lamina.

Hybrids: See page 221.

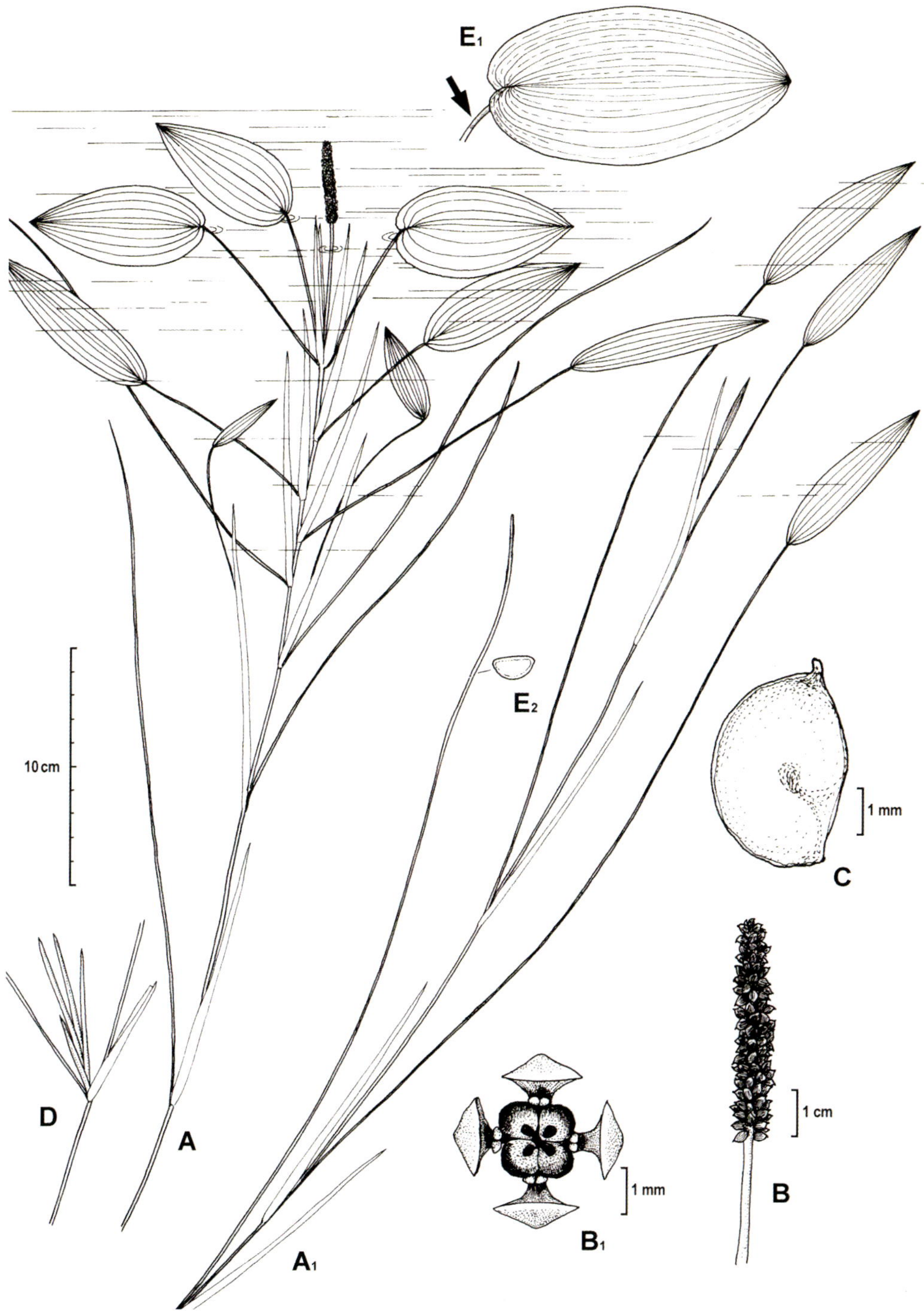

Potamogeton natans
A habit, **A1** habit flowing water, **B** infructescence, **B1** flower, **C** fruit, **D** short vegetative shoot in winter, **E1** floating leaf - arrow at the flexible part of the pedicel, **E2** cross section of submerged leaf.

Potamogetonaceae

67 *Potamogeton polygonifolius* Pourr.

Syn.: *P. oblongus* Viv.

DK: Aflangbladet Vandaks **N:** Kysttjørnaks **S:** Bäcknate **FIN:** Tatarvita **IS:** -
GB: Bog-Pondweed **NL:** Duizendknoopfonteinkruid **F:** Potamot à feuilles de renouée
D: Knöterich-Laichkraut **CZ:** Rdest rdesnolistý **PL:** Rdestnica podługowata
EST: - **LV:** - **LT:** Rūgčialapė plūdė **Ru:** Рдест гречихолистный

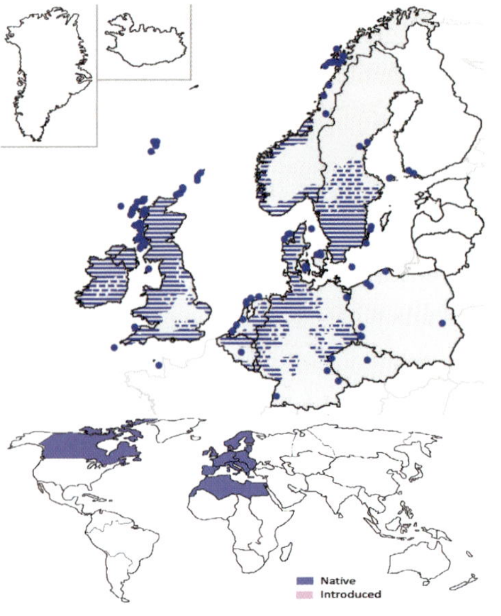

Potamogeton polygonifolius normally produces many floating leaves covering the water surface and preventing submerged leaves from developing. Shoots of *Utricularia intermedia* can be seen among the leaves. Hansted, Thy, Denmark. Photo JCS.

Perennial hydrophyte. Rhizome slender, terete, sometimes with terminal, scaly buds.

Stem: Up to 145 cm long, slender to robust, terete, unbranched or slightly branched.

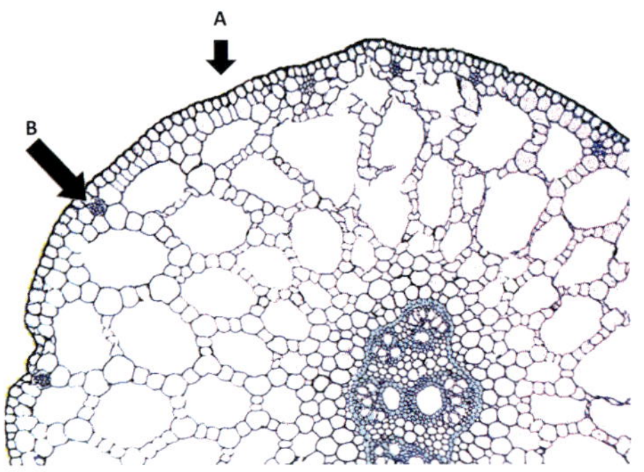

Stem cross section with 1–2 layers of pseudohypodermis (A), subepidermal bundles (B), but without interlacunar bundles. Photo KvdW.

Floating leaves: Lamina 2.5–7.5 cm long, 1.0–6.5 cm wide, oblong elliptical or ovate, acute to somewhat obtuse, with rounded or subcordate leaf base, coriaceous, not translucent, bright green to brownish green, young leaves with a reddish tinge. With 5–9 lateral veins on each side of the midrib and numerous more or less inconspicuous secondary veins. Petiole 3–15 cm long.

Submerged leaves: Often absent in shallow water or terrestrial habitats. Lamina 5–31 cm long, 0.5–2.5(–7.0) cm wide, narrowly lanceolate, entire margins, narrowly acute, gradually tapering to the base, delicate, translucent, light green to olive green often with a reddish tinge. With a narrow band of lacunae, 2–7 lateral veins and numerous secondary veins on each side of the midrib. Leaves transitional to floating leaves are sometimes produced. Petiole 1.5–8(–15) cm long.

Stipules: Persistent, 1–6.6 cm long, translucent, veins rather inconspicuous, 2 of them more prominent than the others.

Inflorescence: 1–4 cm long, cylindrical. Peduncle 2.5–10 cm long and of the same diameter as the stem.

Flowers: With 4 carpels.

Fruits: 1.9–2.6 mm long, reddish brown. Beak 0.1–0.2 mm long.

Flowering: March-November.

Biology: Forms with only floating leaves often occurring on moist ground, over sphagnum or on the drawdown zones of wetlands. It is rare in deep water, tolerant of ochre pollution and unable to take up bicarbonate ions from water.

Habitats: Neutral to strongly acidic, oligotrophic, clear water to dystrophic bogs, small lakes, peat cuttings, quaking bogs, ditches and streams in heathland and dune slacks. In Ireland and on Anglesey it is found growing together with *P. coloratus* in calciphilous blanket bogs with high CO_2 content.

Distribution within the region:
Native. Throughout most of the area, but rare or absent in the eastern and northern parts.

Potamogeton polygonifolius. Mature submerged leaves and a few floating leaves growing towards the water surface. Ditch west of Mosevrå Church, SW-Jutland. Denmark. Photo BM.

Potamogeton polygonifolius with floating leaves. Left: In clear water in a dune lake at Raabjerg Mile, North-Jutland, Denmark. Photo PH. Right: In an ochre-rich ditch, Borris Moor, Middle Jutland, Denmark. Photo JCS.

Characteristics and similar species:
Potamogeton polygonifolius normally has floating leaves. They are coriaceous, and not translucent. The submerged leaves are petiolate with narrowly lanceolate lamina and entire leaf margins.

When submerged leaves are lacking, it can be mistaken for **69 *P. nodosus*** *or* **66 *P. natans*** - see these.

Potamogeton nodosus and **68 *P. coloratus*** both grow in calcareous habitats - the latter with translucent floating leaves and broader submerged leaves.

In **71 *P. alpinus*** the submerged leaves are sessile. See also **70 *P. gramineus*.**

Hybrids: *P. ×gessnacensis* G.Fisch. *(P. natans × polygonifolius)* - a sterile hybrid known from Great Britain, Germany and Poland.

P. ×spathulatus Schrad. ex W.D.J.Koch & Ziz *(P. alpinus × polygonifolius)* - a sterile hybrid known from Germany and Sweden.

Potamogeton polygonifolius
A1-A2 habit - peat cutting, **A3** habit submerged plant with floating leaves reaching for the water surface, **A4** habit - dried-out ground, **C** fruit.

Potamogetonaceae

68 *Potamogeton coloratus* Hornem.

Syn.: *P. plantagineus* Du Croz
DK: Vejbred-Vandaks **N:** Kjeldetjørnaks **S:** Källnate **FIN:** - **IS:** -
GB: Fen Pondweed **NL:** Weegbree fonteinkruid **F:** Potamot coloré
D: Gefärbtes Laichkraut **CZ:** Rdest zbarvený **PL:** Rdestnica zabarwiona
EST: - **LV:** -**LT:** - **Ru:** -

Submerged *P. coloratus*. Pool at Schadbruch, Kempen/Lower Rhine, Germany. Photo KvdW.

Perennial hydrophyte. Rhizome rather slender, terete.

Stem: 10–70 cm long, unbranched or slightly branched, terete.

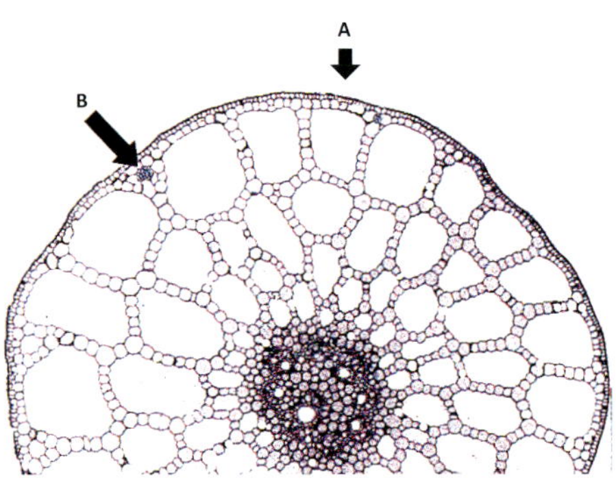

Stem cross section with 1–2 layers of pseudohypodermis (A), some subepidermal bundles (B), but without interlacunar bundles. Photo KvdW.

Floating leaves: Lamina 2.5–8.5 cm long, 1.5–5.5 cm wide, broadly elliptical to broadly ovate, more or less acute to obtuse, with rounded or cordate base, more or less delicate, translucent, light green or with a reddish or brownish tinge. With 6–10 lateral veins on each side of the midrib and numerous very conspicuous secondary veins. Petiole shorter than the lamina.

Submerged leaves: Lamina 7.0–17.5 cm long, 1.0–4.5 cm wide, narrowly elliptical to lanceolate, entire, more or less acute, with rounded or gradually tapering base, delicate, translucent, light green or with a reddish or brownish tinge. With a narrow band of lacunae, 4–8 lateral veins and numerous conspicuous secondary veins on each side of the midrib. Petiole 1.5–6.5 cm long.

Stipules: Persistent, 2.0–6.5 cm long, obtuse, translucent.

Inflorescence: 1.5–4.5 cm long, cylindrical. Peduncle 3–8 cm long and of the same diameter as the stem.

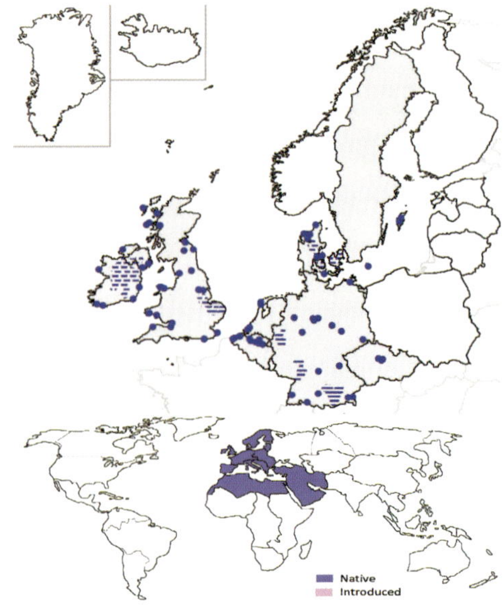

Flowers: With 4 carpels.

Fruits: 1.3–1.9 mm long. Beak 0.2–0.3 mm long.

Flowering: June-August.

Biology: The flowers are wind pollinated but also able to self-pollinate. Vegetative propagation is by division of the rhizome or by overwintering buds. Terrestrial forms are frequently found on moss or mud and mainly produce floating leaves.

Habitats: Shallow ponds, ditches, pools with nutrient-poor to nutrient-rich water; predominantly in alkaline water. *P. coloratus* is unable to take up bicarbonate ions from water and therefore restricted to water with suitable CO_2 content.

Distribution within the region:
Native. Scattered within the region

Potamogeton coloratus has delicate, translucent floating leaves which resemble the submerged leaves. Older leaves, especially in sunny habitats, tend to have a brownish colour. Denmark. Photo JCS.

P. coloratus is recognised by the delicate, translucent floating leaves, which differ only in shape from the submerged leaves. The conspicuous secondary veins and the lateral veins create a reticulate net. Denmark. Photos JCS.

from Britain and Ireland, East Scotland, to Gotland, Germany and the Czech Republic.

Characteristics and similar species: Superficially, *P. coloratus* may resemble several other broad-leaved *Potamogeton* species, but differs by the delicate, translucent floating leaves (often looking a little drowned) and by the fairly broad submerged leaves with numerous conspicuous secondary veins. The semi-terrestrial form produces only floating leaves and retains the delicate, translucent structure, which separates it from other species.

If floating leaves are absent, it can be mistaken for **71 *P. alpinus*,** which has narrow, sessile leaves, and **73 *P. lucens*,** which has denticulate leaf margins and a more robust habit.

69 *Potamogeton nodosus* and **67 *P. polygonifolius*** both have opaque, coriaceous floating leaves. The stem cross section of *P. nodosus* lacks subepidermal and interlacunar bundles.

Hybrids: Hybridises with **70 *P. gramineus*** and **81 *P. berchtoldii*.** Both hybrids are sterile.

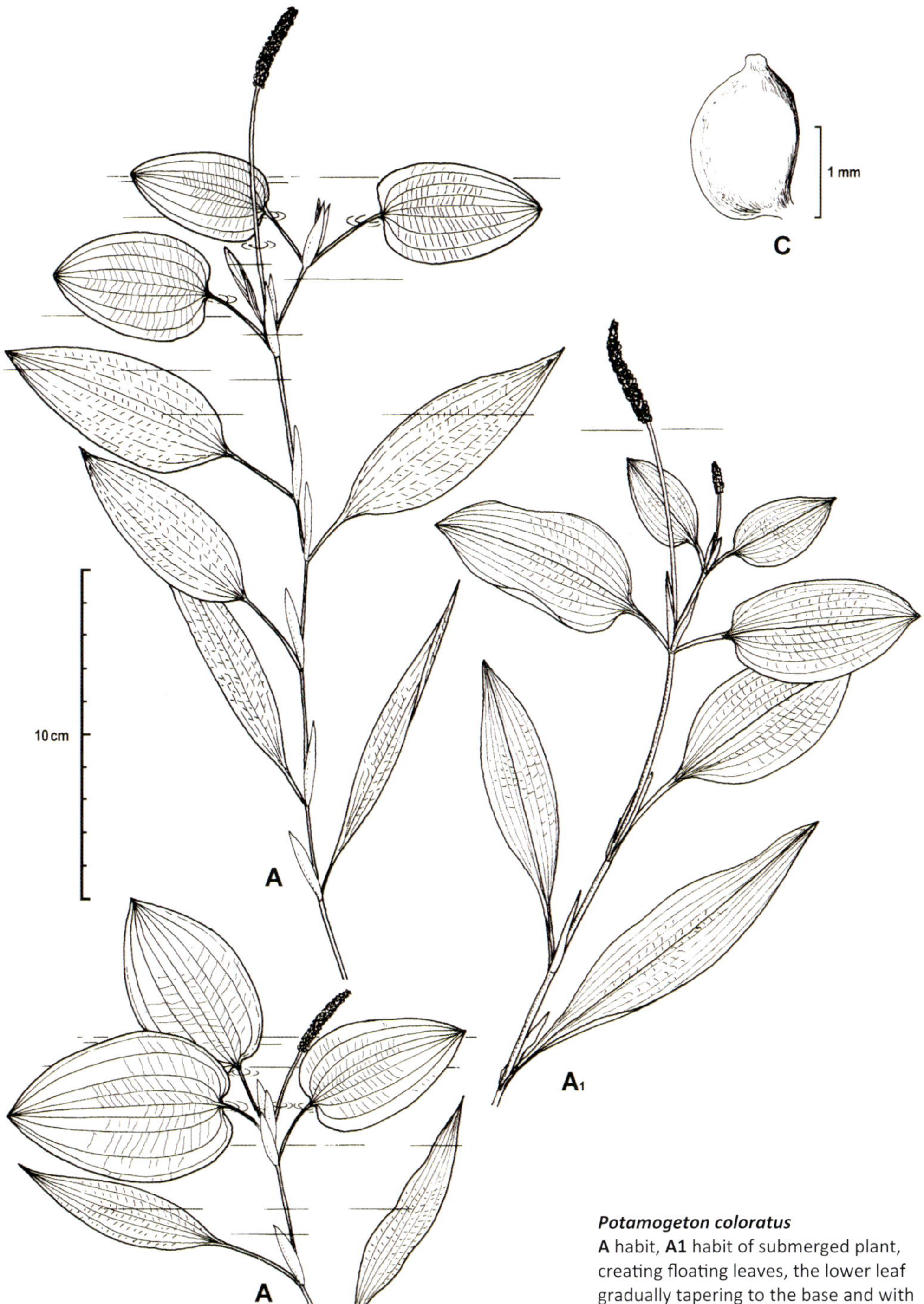

Potamogeton coloratus
A habit, **A1** habit of submerged plant, creating floating leaves, the lower leaf gradually tapering to the base and with a short petiole, **C** fruit.

Potamogetonaceae

69 *Potamogeton nodosus* Poir.

Syn.: *P. fluitans* auct.
DK: - **N:** - **S:** - **FIN:** - **IS:** -
GB: Loddon Pondweed **NL:** Rivierfonteinkruid **F:** Potamot des rivières
D: Flutendes Laichkraut **CZ:** Rdest uzlinatý **PL:** Rdestnica nawodna
EST: - **LV:** -**LT:** Nariuotoji plūdė **Ru:** Рдест узловатый

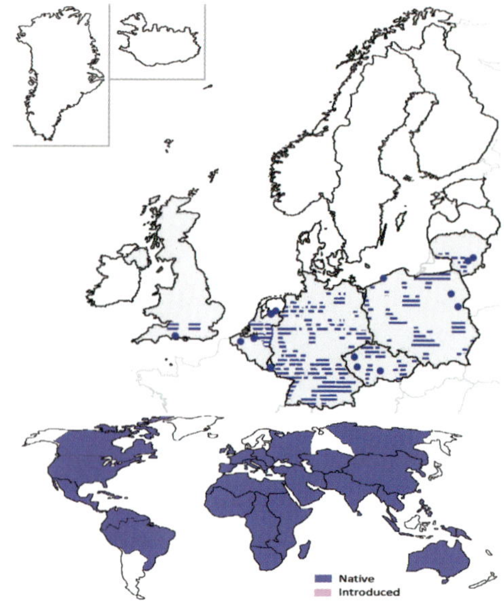

Potamogeton nodosus is mainly found in eutrophic rivers and streams. Form with floating leaves in River Lippe, Germany. Photo KvdW.

Perennial hydrophyte. Rhizome horizontal, branched.

Stem: Up to 2.5 metres long, terete, unbranched or little branched.

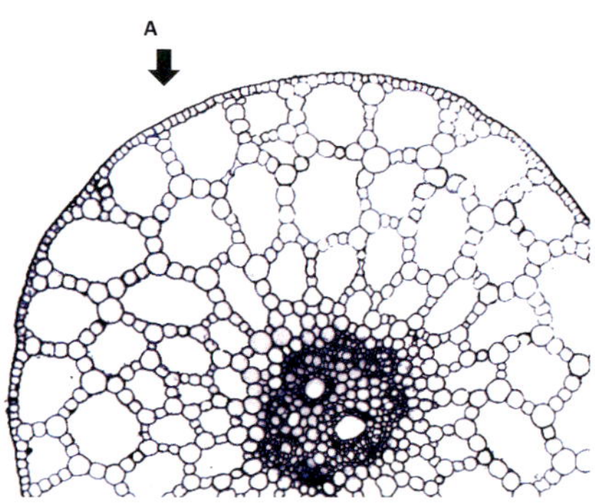

Stem cross section with 0–1 layers of pseudohypodermis (A), without subepidermal and interlacunar bundles. Photo KvdW.

Floating leaves: Lamina 5–20 cm long, 2–5 cm wide, ovate oblong to broadly elliptical, acute at the apex, rounded to cuneate at the base, firm, almost coriaceous, not translucent, green to reddish brown. With 7–11 lateral veins, and numerous, obscure secondary veins on each side of the midrib. Petiole 3–20 cm long, shorter or longer than the lamina.

Submerged leaves: Lamina 6–30 cm long, 2–5 cm wide, leaves lanceolate to elliptical, translucent, petiolate, acute at the apex, green to reddish brown. On each side of the midrib with 5–10 lateral veins, alternating strong and weaker veins. Petiole 7–21 cm long.

Stipules: Persistent, 3–8(–12) cm long, wide, open, flexible, translucent when fresh, green to brown when dry, veins inconspicuous, but 2 are more prominent than the others and form ridges.

Inflorescence: 2–7 cm long, cylindrical. Peduncle up to 13 cm long.

Flowers: Contiguous, with 4, rarely 2–5, carpels.

Fruits: 2.7–4.1 mm long, 2–3 mm wide, reddish to chestnut brown, with a keel. Beak 0.3–0.8 mm long.

Flowering: May-September.

Biology: Capable of rapid expansion to cover a large area. Normally, the species develops shoots with floating and/

Potamogeton nodosus can be found submerged in oligo-mesotrophic lakes. Lake Monbag, Germany. Photo KvdW.

Potamogeton nodosus growing submerged in River Lippe, Germany. Photo KvdW.

or submerged leaves. *P. nodosus* overwinters by turions and probably by rhizomes.

Habitats: In oligo-eutrophic calcareous rivers, streams, channels and lakes. It can tolerate slightly saline water.

Distribution within the region:
Native. Common in the middle European part of the area, rather common in southern Britain and Lithuania.

Characteristics and similar species:
Potamogeton nodosus resembles **67 *P. polygonifolius*, 68 *P. coloratus*** and **66 *P. natans***. *P. natans* has a soft discoloured section just below the lamina. *P. polygonifolius* primarily grows in soft water of low pH, *P. nodosus* grows in calcareous water. The most reliable character to separate *P. nodosus* from other species and hybrids is the cross section of the stem. *P. natans* has both subepidermal and interlacunar bundles. *P. polygonifolius* and *P. coloratus* have only subepidermal bundles.

Hybrids: 86 *P. ×schreberi* G.Fisch. (*P. natans × P. nodosus*).

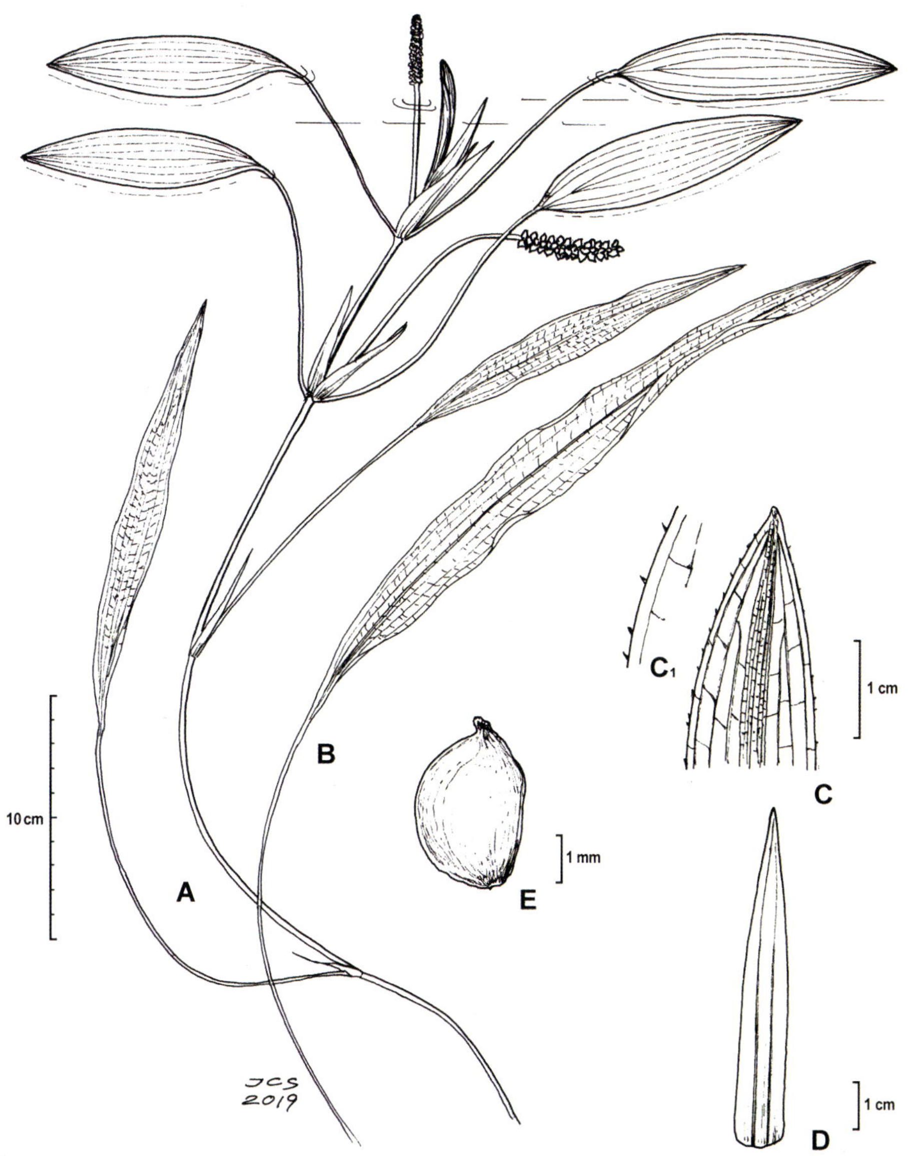

Potamogeton nodosus
A habit, **B** submerged leaf, **C** leaf tip, **C1** leaf margin, **D** stipule, **E** fruit.

Potamogetonaceae

70 *Potamogeton gramineus* L.

Syn.: *P. heterophyllus* Schreb.

DK: Græsbladet Vandaks **N:** Grastjørnaks **S:** Gräsnate **FIN:** Heinävita **IS:** Grasnykra
GB: Various-leaved Pondweed **NL:** Grasfonteinkruid **F:** Potamot graminée
D: Grasartiges Laichkraut **CZ:** Rdest trávolistý **PL:** Rdestnica trawiasta
EST: Hein-penikeel **LV:** Zālainā glīvene **LT:** Siauralapė plūdė **Ru:** Рдест злаковый

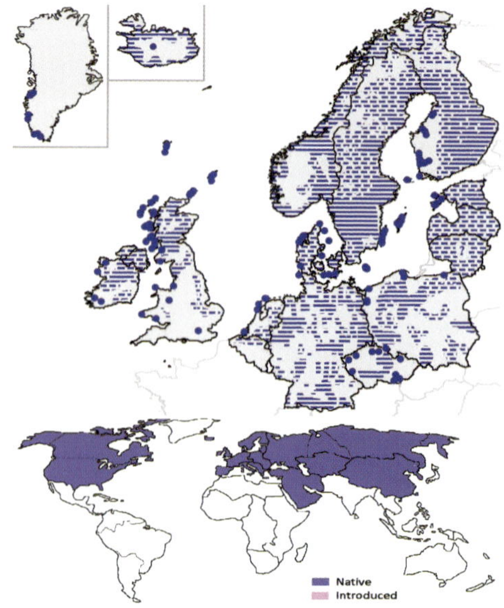

In shallow water *P. gramineus* may become richly branched with short, recurved, undulate submerged leaves and often forming floating leaves. Førby Sø, Denmark. Photo JCS.

Perennial hydrophyte. Rhizome slender, terete. 2–3 internodes near the rhizome apex become thicker and starch filled before the winter.

Stem: 10–80 cm long, in lakes with clear water up to several metres long, slender to fairly robust, terete, often richly branched.

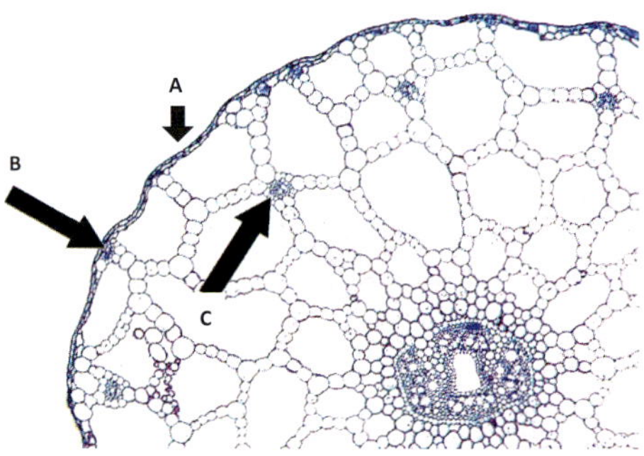

Stem cross section without or with 1 incomplete layer of pseudohypodermis (A), with or without subepidermal bundles (B), with 1 row of interlacunar bundles (C). Photo KvdW.

Floating leaves: Not always present. Lamina 1.5–7.0 cm long, 0.8–3.5 cm wide, broadly oblong to elliptical, acute to somewhat obtuse, with broad cuneate or rounded base, coriaceous, not translucent, yellowish green to dark green. With 5–10 lateral veins on each side of the midrib and numerous more or less conspicuous secondary veins. Petiole 1.8–6.0 cm long.

Submerged leaves: Lamina 3.5–9(–17) cm long, 0.5–1.2(–1.7) cm wide, narrowly oblong to narrowly lanceolate, often recurved, undulate, with a delicate denticulate margin, acute to somewhat obtuse often mucronate at the apex, gradually tapering to the base, delicate, translucent, light to dark green or a little brownish. With a narrow band of lacunae, (2–)3–4 lateral veins and inconspicuous secondary veins on each side of the midrib. The submerged leaves are sessile, but the upper leaves may occasionally be petiolate, representing transitional forms to floating leaves.

Stipules: Persistent, 1.0–2.5(–3.5) cm long, translucent, inrolled and typically protruding at a 45 degree angle to the stem.

Inflorescence: 1.5–4.0 cm long, cylindrical. Peduncle 4–10 cm long, often very robust, broader towards the spike. Normally with several inflorescences on each shoot.

Flowers: With 4, rarely 3–5, carpels.

Fruits: 2.4–3.1 mm long. Olive green, seldom with a pinkish tinge. Beak 0.2–0.3 mm long.

Flowering: July-August.

Biology: *P. gramineus* does not form turions. In the winter, the stems dieback and only the rhizome survives as short, starch-filled and much thickened internodes. On the drawdown zones of water bodies, it may only form leaves similar to floating leaves.

Habitats: Weakly-acidic to alkaline water. In nutrient-poor to moderately eutrophic lakes with clear water and sandy substrate, ponds in sand dunes, wet areas in bogs and meadows, in bays and lagoons with very low salinity.

Submerged *P. gramineus.* Lake Nors, Thy, Denmark. Photo JCS.

Potamogeton gramineus is a very variable species but normally quite easy to recognise. The submerged leaves are sessile, narrowly oblong to narrowly lanceolate with denticulate margin and often mucronate at the apex. The peduncle characteristically widens towards the spike. Left: Plant from Lake Filsø with a pinkish tinge. Right: Plant from shallow water with many small, floating leaves and fewer submerged leaves. Lake Navn. Denmark. Photo JCS.

It can grow in shallow as well as deep water (to 1 m). Rarely in streams.

Distribution within the region: Native. Widespread and abundant throughout much of the region, more scattered in the south.

Characteristics and similar species: *Potamogeton gramineus* is a very variable species, but differs from similar species such as ***66 P. natans*, 67 *P. polygonifolius*** and ***71 P. alpinus*** by the sessile, oblong elliptical to lanceolate leaves with denticulate margins, often mucronate at the apex and normally with recurved submerged leaves. Also by the short, quite rigid stipule typically projecting at an acute angle from the stem, and by the peduncle, which is much broader beneath the spike.

It can be confused with **90 *P. ×nitens***, which has semi-amplexicaul leaves. Submerged forms can also be confused with **75 *P. crispus***, which has an obtuse leaf apex, serrate leaf margins and compressed stem.

Hybrids: *Potamogeton gramineus* contributes to many hybrids. See page 221.

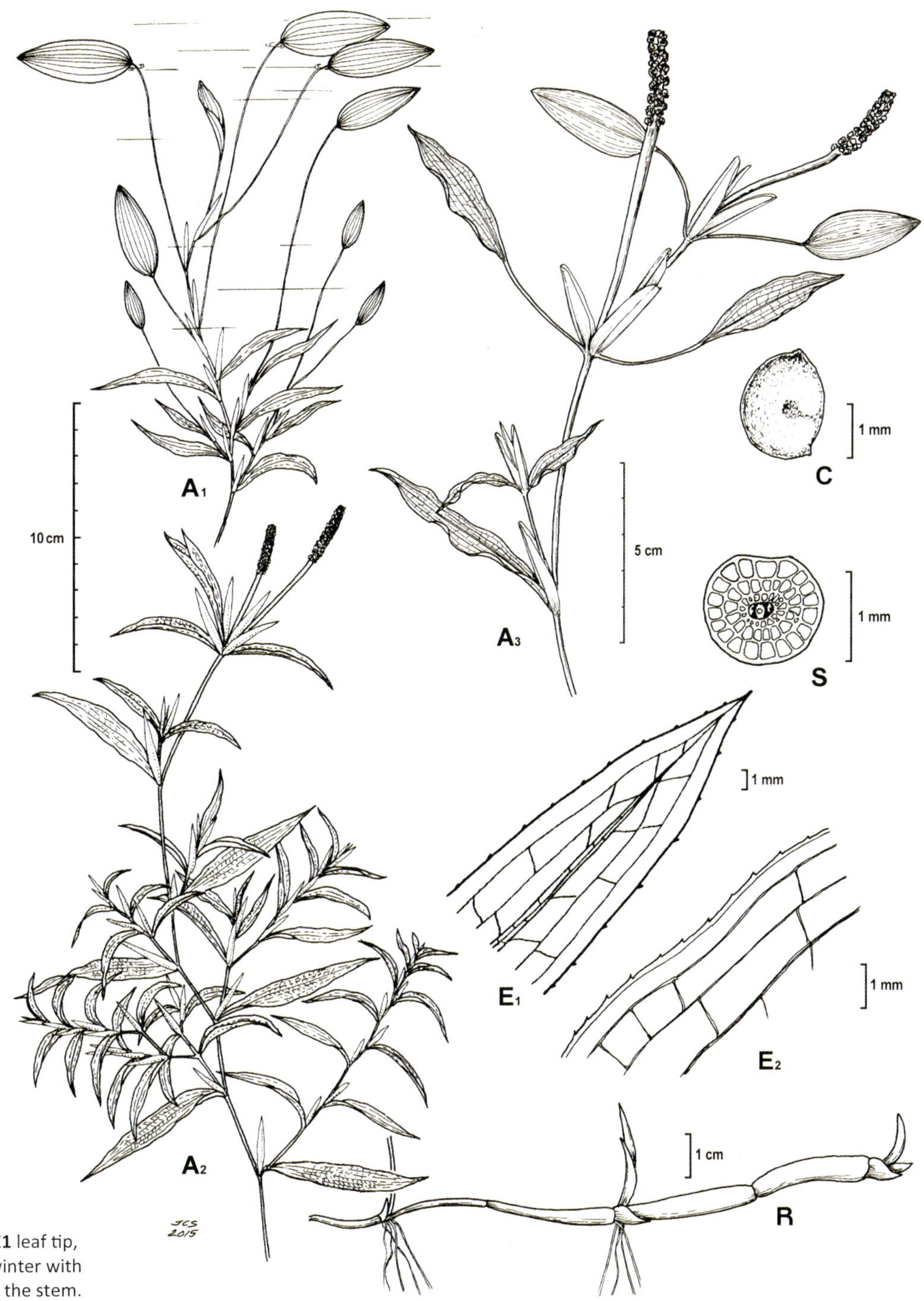

Potamogeton gramineus
A1, **A2** habit, **A3** upper part of a plant, **C** fruit, **E1** leaf tip, **E2** denticulated leaf margin, **R** rhizome in the winter with thick starch-filled internodes, **S** cross section of the stem.

Potamogetonaceae

71 *Potamogeton alpinus* Balb.

DK: Rust-Vandaks **N:** Rusttjørnaks **S:** Rostnate **FIN:** Purovita **IS:** Fjallnykra
GB: Red Pondweed **NL:** Rossig fonteinkruid **F:** Potamot alpin
D: Alpen-Laichkraut **CZ:** Rdest alpský **PL:** Rdestnica alpejska
EST: Ruske penikeel **LV:** Alpu glīvene **LT:** Alpinė plūdė **Ru:** Рдест альпийский

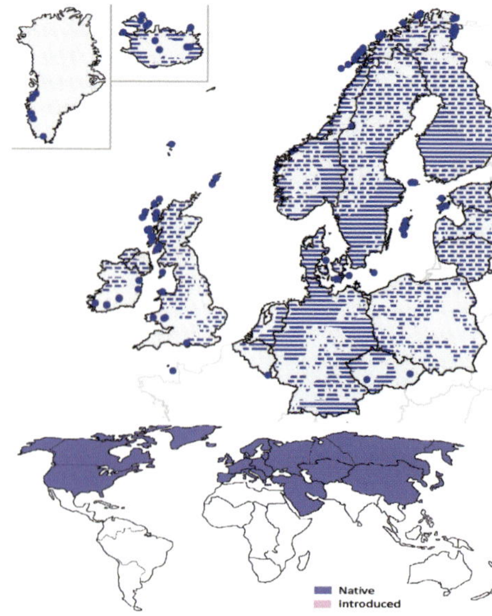

Potamogeton alpinus often develops coriaceous floating leaves in slow-flowing or stagnant water. River Skjernå, Denmark. Photo JCS.

Perennial hydrophyte. Rhizome slender, terete. Long, slender stolons develop from the lower leaf axils during the growing season. It develops stolons and rhizomes with an ovoid scaly bud at the apex.

Stem: Up to 2.5 metres long, slender to rather robust, terete, unbranched.

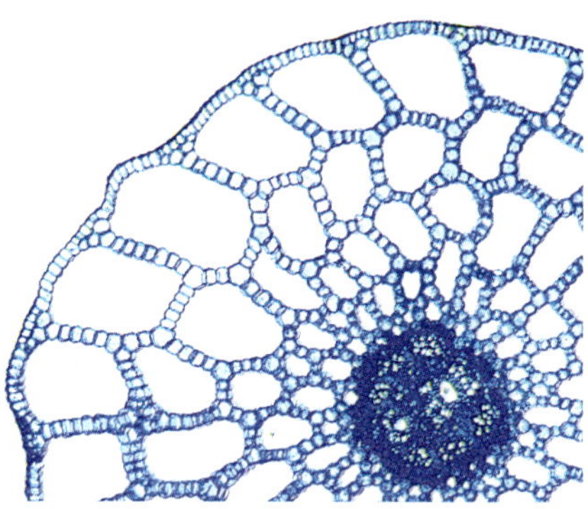

Stem cross section without pseudohypodermis, subepidermal or interlacunar bundles. Photo KvdW.

Floating leaves: Not always present. Lamina 4–9(–15) cm long, 1.0–2.5(–3.5) cm wide, oblong to elliptical or ovate, obtuse, with wedge-shaped base, more or less coriaceous, not translucent, yellowish green to green often with a rust-red tinge, which intensifies when dried. With 4–9 lateral veins and numerous more or less obscure secondary veins on each side of the midrib. Petioles 1.0–3.5 cm long.

Submerged leaves: Lamina 7–18(–30) cm long, 1.0–2.5 cm wide, oblong lanceolate, entire and shallowly undulate at the margin, narrowly obtuse and often with a hooded apex, gradually tapering to the base. Translucent, light green to green or reddish brown, often with a reddish tinge, which strongly intensifies when dried. With a narrow band of lacunae, 4–7 lateral veins and rather inconspicuous secondary veins on each side of the midrib. The submerged leaves are sessile or with a short petiole. However, occasionally leaves with a longer petiole are formed in the upper part of the plant as transitions to floating leaves. In semi-terrestrial habitats, submerged leaves may be absent.

Stipules: Not persistent, decaying, 1.5–3.5(–4.5) cm long, with inconspicuous veins - the 2 most prominent not forming distinct ridges, translucent, obtuse at the apex.

Inflorescence: 1.5–4.5 cm long, cylindrical. Peduncle 3–15 cm long and the same diameter as the stem. Normally with several inflorescences on each shoot.

Flowers: With 4 carpels.

Fruits: 2.6–3.7 mm long. With a distinct keel, pale brown and often with a pinkish tinge. Beak 0.5 mm long.

Flowering: July-August.

Biology: Often with floating leaves in slow-flowing water. The buds at the apex of rhizomes and stolons function as vegetative propagules during the growing season and resting buds during winter.

Without floating leaves *P. alpinus* can be mistaken for *P. praelongus*. Indfjorden at Nees, Denmark. Photo JCS.

In *P. alpinus* (left) the apex of the leaf is flat or slightly hooded. In *P. praelongus* (right) the apex is always distinctly hooded. Photos JCS.

Potamogeton alpinus. Section of leaf with lacunae on each side of the midrib. Photo BM.

◀ *Potamogeton alpinus* (left) and *P. praelongus* (right) both have long, sessile or rather short petiolate submerged leaves, but the leaf base of *P. alpinus* tapers into the base and is not rounded as at *P. praelongus*. Photos JCS.

Habitats: Weakly acidic to mildly alkaline, moderately eutrophic lakes, ponds, gravel pits, marl pits, ditches, canals and streams.

Distribution within the region:
Native. Widespread and abundant throughout much of the region, more scattered in the south.

Characteristics and similar species: *Potamogeton alpinus* can be recognised by the unbranched stem, the sessile or very shortly petiolate leaves with entire margins and obtuse apex, the delicate stipules with inconspicuous veins, and the reddish tinge which intensifies when dried, especially in apical leaves. It can be mistaken for **72 *P. praelongus*** (see photos above).

Hybrids: 91 *P. ×olivaceus* Baagøe ex G.Fisch. (*P. alpinus x P. crispus).*

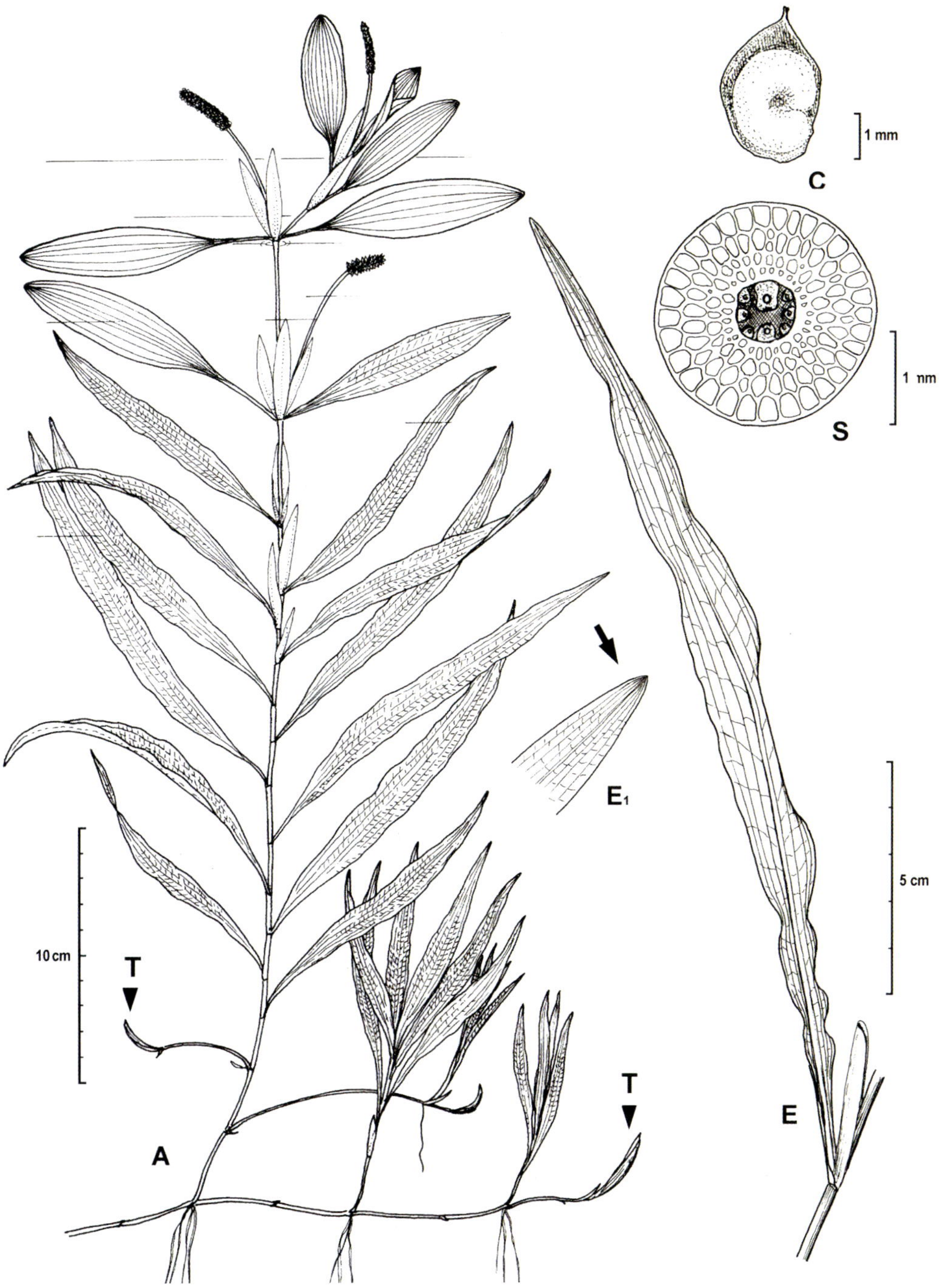

Potamogeton alpinus
A habit, **C** fruit, **E** submerged leaf, **E1** tip of submerged leaf, **S** cross section of the stem, **T** apical scaly bud.

Potamogetonaceae

72 *Potamogeton praelongus* Wulfen

DK: Langbladet Vandaks **N:** Nykketjørnaks **S:** Långnate **FIN:** Pitkälehtivita **IS:** Langnykra
GB: Long-stalked Pondweed **NL:** Langstengelig fonteinkruid **F:** Potamot allongé
D: Langblättriges Laichkraut **CZ:** Rdest dlouholistý **PL:** Rdestnica zabarwiona
EST: Pikk penikeel **LV:** Visgarā glīvene **LT:** Ilgoji plūdė **Ru:** Рдест длиннейший

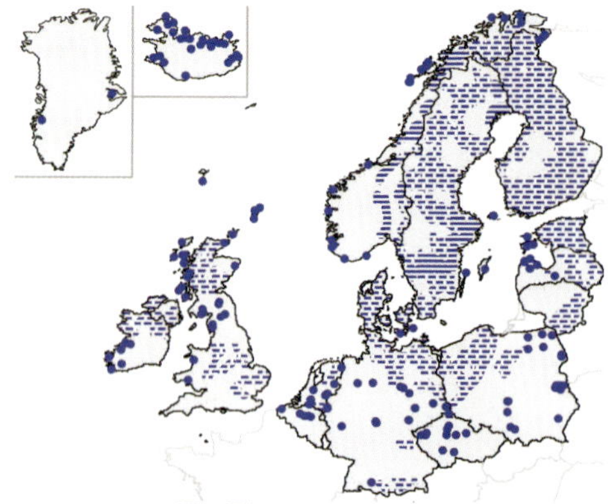

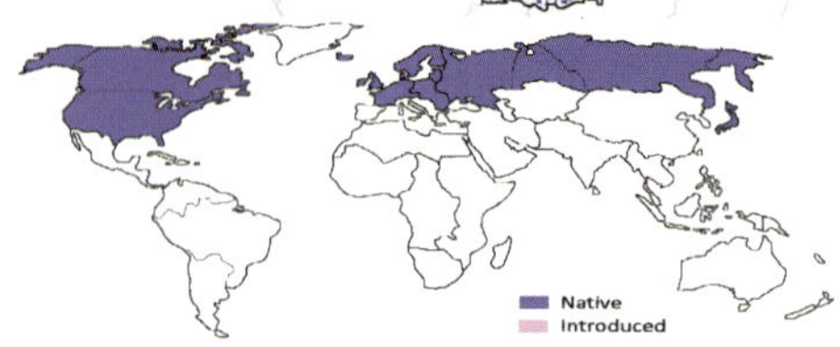

Potamogeton praelongus. Laacher See, Germany. Photo KvdW.

Perennial hydrophyte. Rhizomes robust, terete, with a large apical bud in the winter.

Stem: Terete, unbranched to richly branched, up to 3 metres long in deep or flowing water. In specimens with short internodes the stem has a zigzag form, while stems with longer internodes tend to be straight.

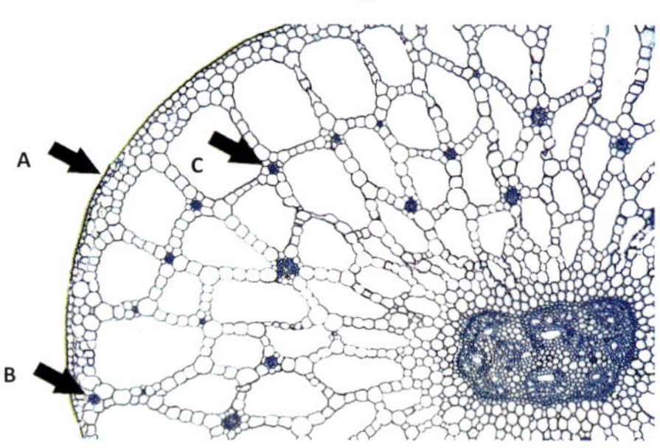

Stem cross section with 1–3 layers of pseudohypodermis (A), often only few subepidermal bundels (B) and 2–3 rows of interlacunar bundles (C). Photo KvdW.

Floating leaves: Absent.

Submerged leaves: Sessile, 6–18 cm long, 1.5–4 cm wide, lanceolate to oblong lanceolate, plane, often with undulate margins, entire, obtuse and broadly hooded at the apex, semi-amplexicaul at the base, translucent, bright to dark green, sometimes tinged pink along the midrib. Young leaves often with a brownish-red tinge. With a band of lacunae and 5–9 lateral veins of which 1–2 are more prominent, as well as numerous secondary veins on each side of the midrib.

Stipules: Persistent, 1–8 cm long, obtuse, whitish to slightly reddish or brownish, translucent when fresh, but opaque when dry.

Inflorescence: 2–8 cm long, cylindrical. Peduncle 8–25(–35) cm long and of the same diameter as the stem or slightly broader. Normally with several inflorescences on each shoot.

Flowers: With (2–)4 carpels.

Fruits: 4.5–5.5 mm long with distinct dorsal keel and a 0.6–0.8 mm beak.

Flowering: July-August.

Biology: *Potamogeton praelongus* normally overwinters as a rhizome and a large apical winter bud, occasionally also with leafy shoots. The plant continues to grow late in the autumn, forming new leaves.

Habitats: Larger water bodies. In mesotrophic lakes, rivers, canals and fenland drains.

Distribution within the region:
Native. Scattered throughout the region but nowhere abundant, rare and declining.

The long peduncles of *P. praelongus* are prominent when present. River Nørreå near Randers, Denmark. Photo BM.

Potamogeton praelongus has long, green, undulating leaves that are obtuse and broadly hooded at the apex and semi-amplexicaul at the base. The fresh stipules are translucent with numerous inconspicuous veins of which two are more prominent than the others, but not forming distinct ribs. Notice the zigzag form of the stem. Grund Fjord at Randers, Denmark. Photo JCS.

Leaf bases of *Potamogeton alpinus* (left) and *Potamogeton praelongus* (right). Photos JCS.

Characteristics and similar species: *Potamogeton praelongus* does not vary much, except for the length and size of the internodes. It can be recognised by the long, semi-amplexicaul, bright to dark green leaves with an obtuse, broadly hooded apex.

It may be mistaken for **71 *P. alpinus*** (see this and the photo above right). Sometimes it has also been confused with **93 *P. ×undulatus*** and **96 *P. ×cognatus*** (see these).

Hybrids: *Potamogeton longifolius* is involved in several hybrid combinations:

93 *P. ×undulatus*

96 *P. ×cognatus*

P. ×vepsicus A.A.Bobrov & Chemeris (*P. natans* × *praelongus*) - see p. 226.

P. ×jutlandicus Zalewska-Gałosz (*P. lucens* × *praelongus*) - a sterile hybrid known only from Denmark and Estonia.

P. ×griffithii A.Benn. (*P. alpinus* × *praelongus*) - a sterile hybrid known from Ireland, Britain.

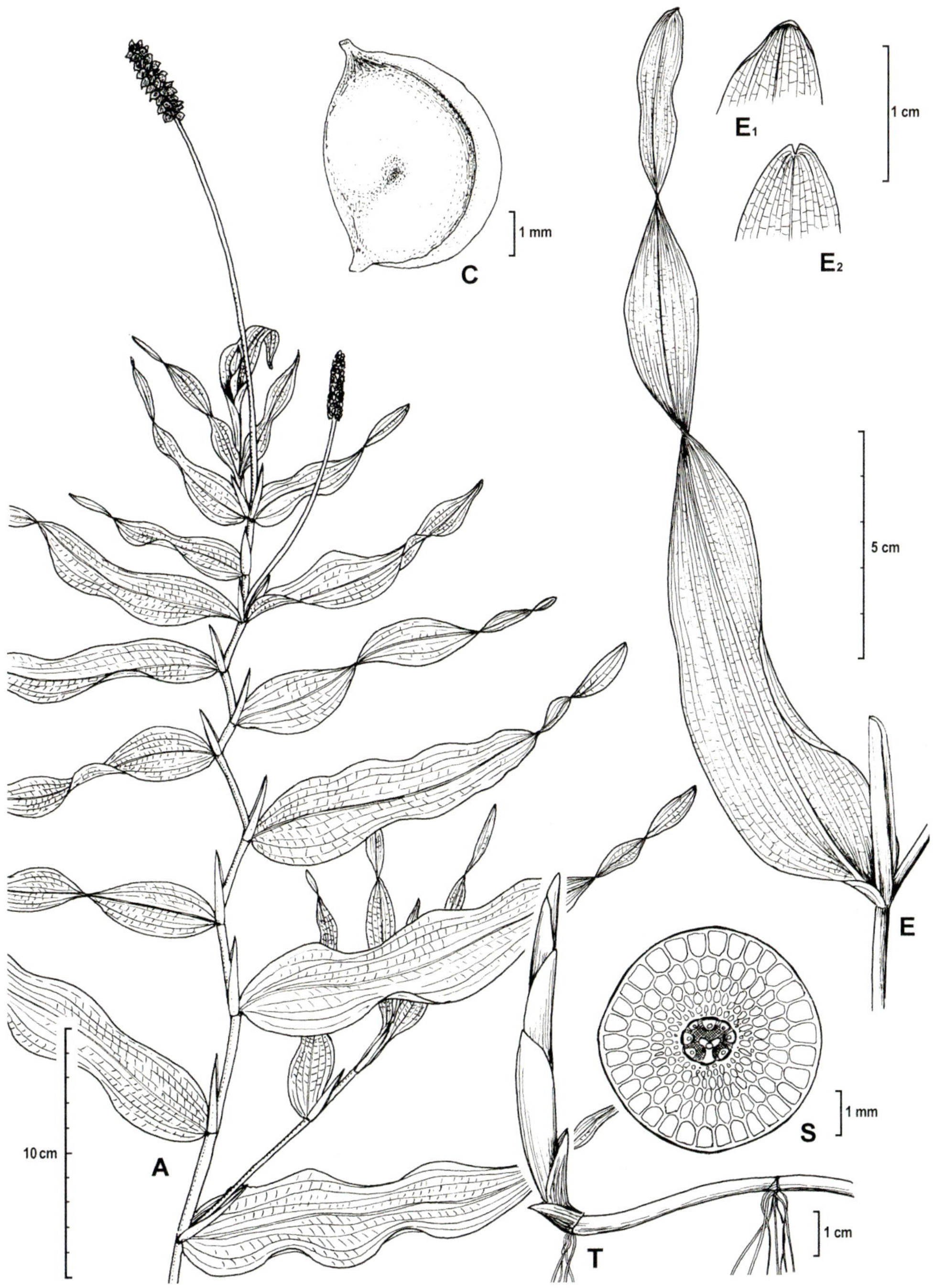

Potamogeton praelongus
A habit, **C** fruit, **E** leaf with stipules,
E1 hooded leaf tip, **E2** flattened leaf tip splitting up as a V, **S** stem cross section,
T rhizome with apical bud.

Potamogetonaceae

73 *Potamogeton lucens* L.

DK: Glinsende Vandaks **N:** Blanktjørnaks **S:** Grovnate **FIN:** Välkevita **IS:** -
GB: Shining Pondweed **NL:** Glanzig fonteinkruid **F:** Potamot luisant
D: Glänzendes Laichkraut **CZ:** Rdest světlý **PL:** Rdestnica połyskująca
EST: Läik-penikeel **LV:** Spožā glīvene **LT:** Blizgančioji plūdė **Ru:** Рдест блестящий

When the shoots of *P. lucens* reach the water surface they start branching, and soon the large, broad leaves will more or less cover the surface. In the photo *P. lucens* is mixed with *P. perfoliatus*. Gudenå near Bjerringbro, Denmark. Photo JCS.

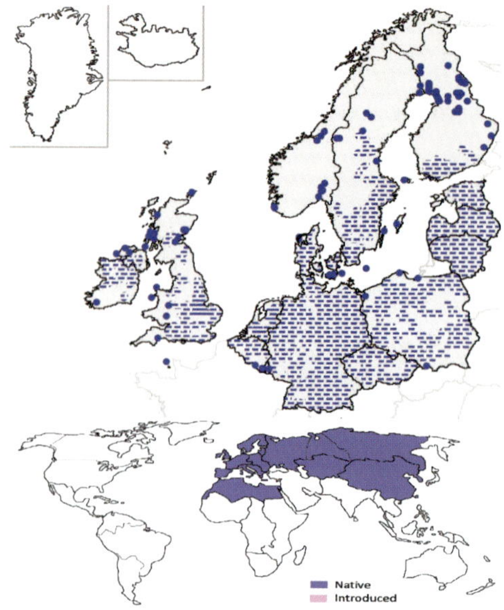

Perennial hydrophyte. Rhizome robust, terete, with short internodes. With swollen internodes constricted at the joints in winter.

Stem: Up to several metres long in deep or flowing water, terete, and more or less branched, especially in the upper part.

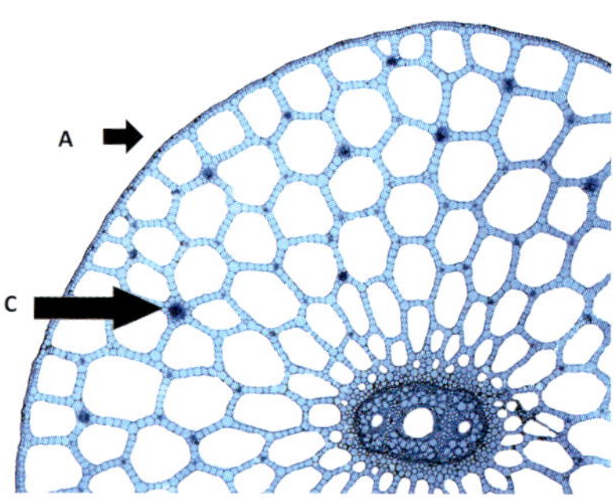

Stem cross section with 0–1 layer of pseudohypodermis (A), 1–3 rows of interlacunar bundles (C), with or without subepidermal bundles. Photo KvdW.

Floating leaves: Absent.

Submerged leaves: 7–20(–30) cm long, 2.5–6 cm wide, lanceolate to broadly elliptical, translucent, glossy, yellowish green to bright green, occasionally with a reddish tinge, the margins minutely undulate and denticulate. Midrib prominent, without lacunae, and often excurrent in the rounded to acuminate apex, base cuneate. The leaves at the top of vegetative shoots may be reduced, appearing as a prominent midrib with a very narrow lamina. With 4–5(–6) lateral veins, and numerous, well-developed, somewhat inclined secondary veins on each side of the midrib, which form a slanting network of rectangular cells. The leaves on the lower part of the stem and near the flowers are sometimes reduced to just the midrib. The petiole is short and seldom more than 0.7 cm.

Stipules: Persistent, (2–)3–8 cm long, rigid, obtuse to rounded at the apex, greenish, translucent, with many veins of which two are more prominent than the others and forming conspicuously winged ribs on at least the basal part of the abaxial side of the stipule.

Inflorescence: 2.5–7.0 cm long, cylindrical. Peduncle 5–20(–35) cm long, broader than the stem and expanded towards the inflorescence. Normally with several inflorescences on each shoot.

Flowers: Contiguous, with 4(–6) carpels.

Fruits: 3.2–4.5 mm long with a distinct rib and a 0.5–0.8 mm long beak.

Flowering: July-August.

Biology: In the winter, the stems wither away and only the rhizome survives, as short, starch-filled and much thickened internodes. Turions absent.

Habitats: In neutral to weakly alkaline water in moderately eutrophic lakes, rivers and canals.

Potamogeton lucens. The large submerged young leaves are bright green. The secondary veins form an oblong reticulation slanting towards the midrib. The rigid stipule has 2 prominent, abaxially winged veins, which are more than half the length of the stipule. Photos JCS and BM.

Potamogeton lucens. The very robust peduncle is broader towards the apex. Notice the reduced leaves with the prominent long excurrent midrib. Gudenå, Randers, Denmark. Photo JCS.

Potamogeton lucens. Completely submerged plants in slowly flowing water in River Storå close to Vemb, Denmark. Photo JCS.

Distribution within the region: Native. Widespread and locally abundant in the south, rare in northern Scandinavia and in Scotland. Absent from Iceland and Greenland.

Characteristics and similar species: Leaf shape and size vary widely in *Potamogeton lucens* shows, but it can be distinguished from other *Potamogeton* species by the translucent, glossy green leaves with distinct secondary veins forming a reticulate pattern, and in particular by the rigid stipules with 2 prominent winged veins on the abaxial side, which are more than half the length of the stipule.

Hybrids: *Potamogeton lucens* is involved in several hybrid combinations and closely resembles **89 *P. ×angustifolius*** and **95 *P. ×salicifolius***. In contrast, **87 *P. ×fluitans*** typically has floating leaves.

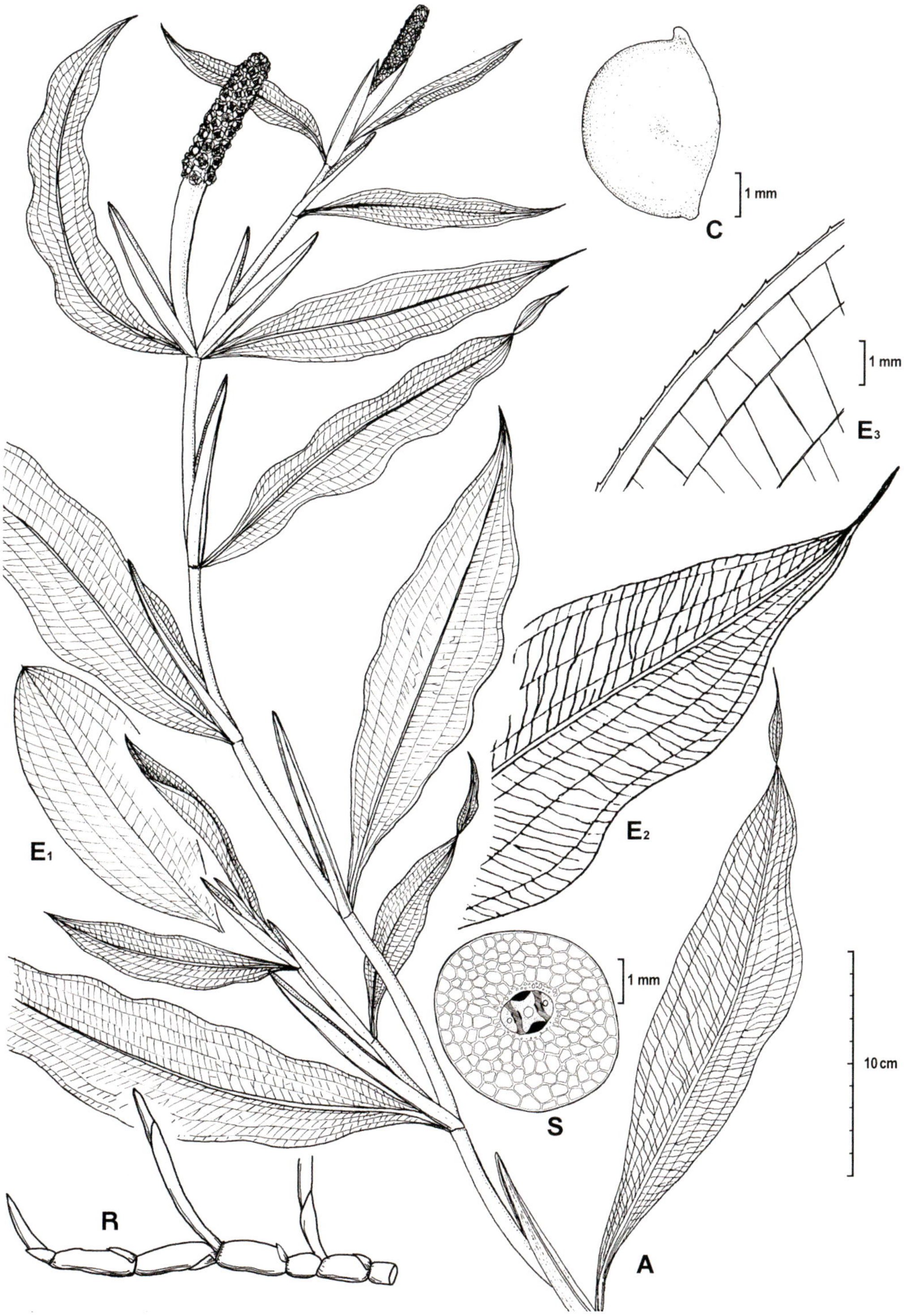

Potamogeton lucens
A habit, **C** fruit, **E1** rounded leaf with mucronate apex, **E2** leaf tip with excurrent midrib, **E3** denticulate leaf margin, **R** rhizome in the winter (Raunkiær 1890), **S** stem cross section.

Potamogetonaceae

74 *Potamogeton perfoliatus* L.

DK: Hjertebladet Vandaks **N:** Hjartetjørnaks **S:** Ålnate **FIN:** Ahvenvita **IS:** Hjartanykra
GB: Perfoliate Pondweed **NL:** Doorgroeid fonteinkruid **F:** Potamot perfolié
D: Durchwachsenes Laichkraut **CZ:** Rdest prorostlý **PL:** Rdestnica przeszyta
EST: Kaelus-penikeel **LV:** Skaujošā glīvene **LT:** Permautalape plūdė **Ru:** Рдест пронзённолистный

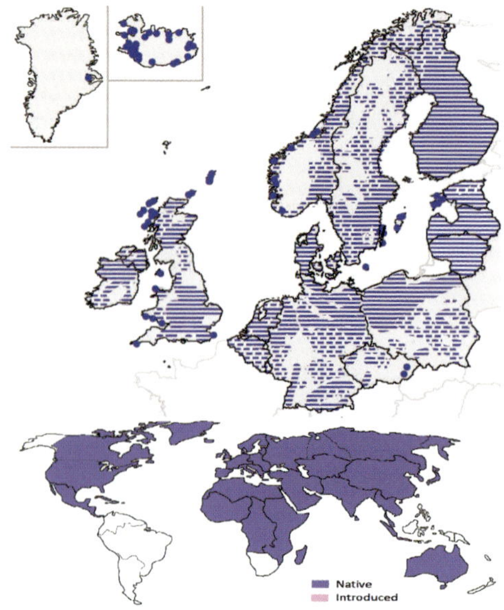

Potamogeton perfoliatus from River Gudenå at Bjerringbro, Denmark. The several metres long vegetative shoots cover large areas close to the bank. Photo JCS.

Perennial hydrophyte. Rhizome robust, terete.

Stem: Up to several metres long, unbranched to richly branched, terete.

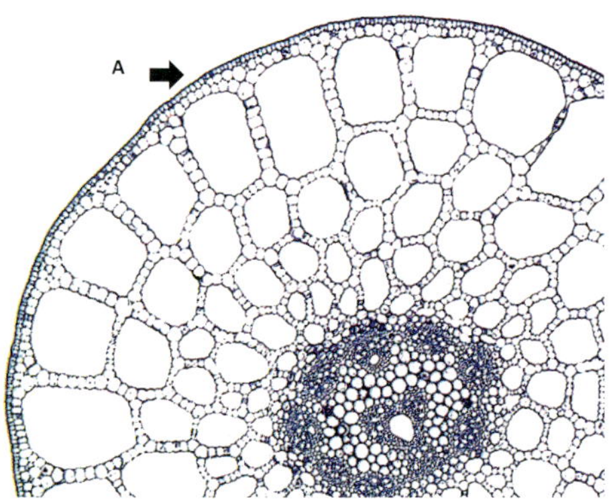

Stem cross section with 1 layer of pseudohypodermis (A), lacking subepidermal and interlacunar bundles. Photo KvdW.

Floating leaves: Absent.

Submerged leaves: 2–10(–11.5) cm long, (0.7–)1–4 cm wide, narrowly elliptical to widely ovate, the margins minutely undulate and denticulate, often slightly hooded at the apex, sessile, amplexicaul, translucent, bright green, olive or brownish green - sometimes tinged pink. With a narrow band of lacunae, 5–12 lateral veins of which 1–3 are more prominent and numerous transverse secondary veins on each side of the midrib.

Stipules: Fugacious, and usually only present on young leaves, 0.3–2 cm long, rounded at the apex, translucent.

Inflorescence: 1.2–2.5 cm long, cylindrical. Peduncle 2–20 cm long and of the same diameter as the stem.

Flowers: Contiguous, with 4 carpels.

Fruits: 2.5–3.5 mm long. Beak 0.4–0.6 mm, apical and straight.

Flowering: July-August.

Habitats: Lakes, ponds, rivers, canals, with weakly acidic to alkaline, moderately eutrophic water. The species can tolerate weakly brackish water.

Distribution within the region:
Native. Widespread and locally frequent throughout much of the region, but very rare in Greenland. Rare and declining in the Czech Republic.

Characteristics and similar species: Leaf shape varies widely in *Potamogeton perfoliatus*, but it is easily recognisable by the amplexicaul leaves with undulate and denticulate margins, slightly hooded leaf apex and the fugacious stipules.

Plants with long leaves might be mistaken for **72 *Potamogeton praelongus***, but this species is distinguished by its larger and more rigid, persistent stipules with two prominent veins, and entire leaf margins.

Hybrids: Hybrids involving *Potamogeton perfoliatus* are often misidentified for this species, however the hybrids do not usually set fruit and

Potamogeton perfoliatus. In lakes with transparent water, the vegetative shoots can reach the surface from depths of 3–4 meters. AQUA, Silkeborg, Denmark. Photo JCS.

Potamogeton perfoliatus. Inflorescence. Photo JCS.

Potamogeton perfoliatus shows great variability in leaf form, size and colour. Thy, Denmark. Photos JCS.

typically have more persistent stipules.

90 ***P. ×nitens*** (see this), **96** ***P. ×cognatus*** (see this), **92** ***P. ×cooperi*** has a somewhat compressed stem and only 3–5(–6) veins on each side of the midrib, **95** ***P. ×salicifolius*** resembles **73** ***P. lucens.***

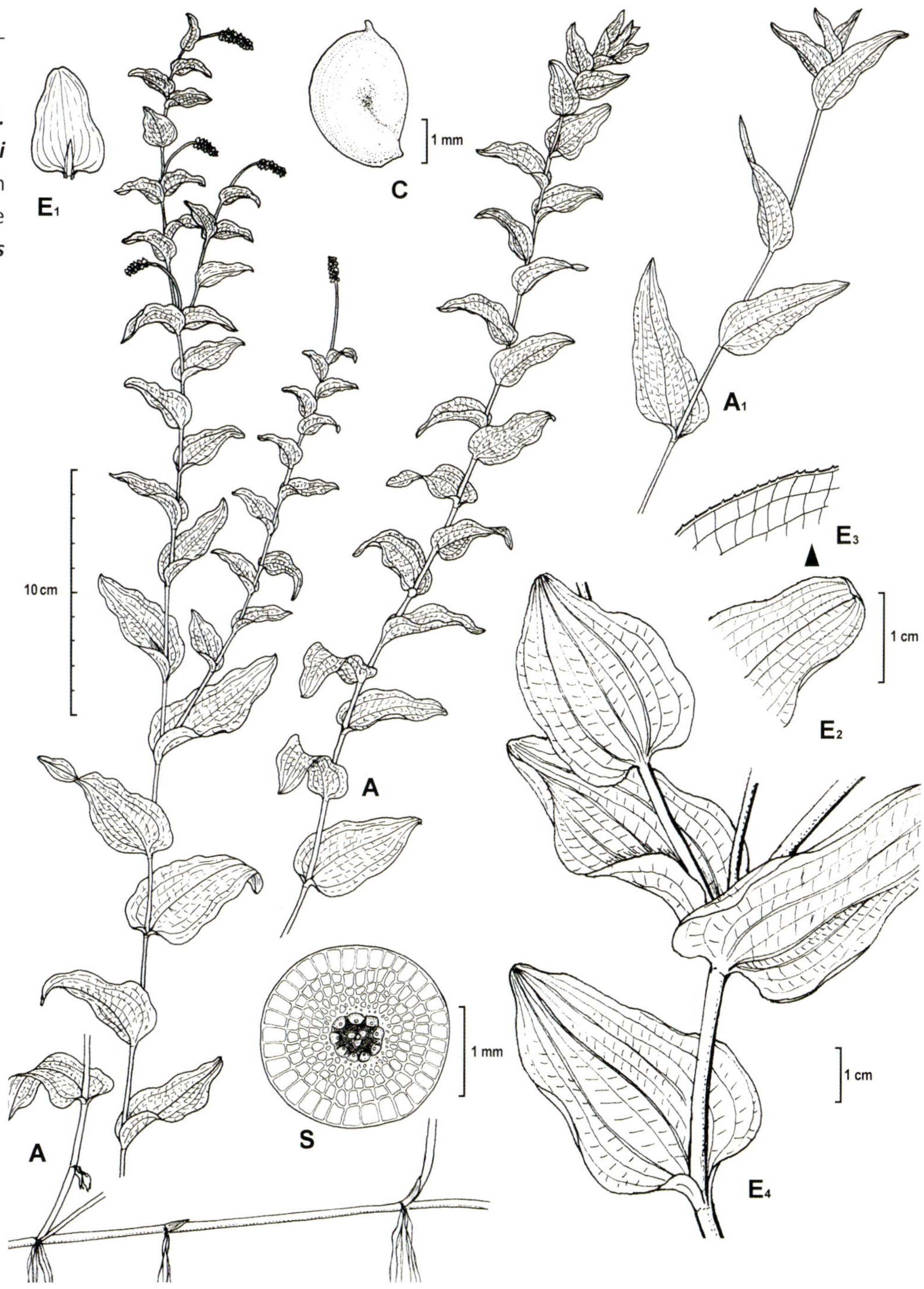

Potamogeton perfoliatus
A, **A1** habit, **C** fruit, **E1** leaf with stipules, **E2** hooded leaf tip, **E3** leaf margin, **E4** part of shoot, **S** stem cross section.

Potamogetonaceae

75 *Potamogeton crispus* L.

DK: Kruset Vandaks **N:** Krustjørnaks **S:** Krusnate **FIN:** Poimu vita **IS:** -
GB: Curled Pondweed **NL:** Gekroesd fonteinkruid **F:** Potamot crépu
D: Krauses Laichkraut **CZ:** Rdest kadeřavý **PL:** Rdestnica kędzierzawa
EST: Kähar penikeel **LV:** Krokainā glīvene **LT:** Garbiniuotoji plūdė **Ru:** Рдест курчавый

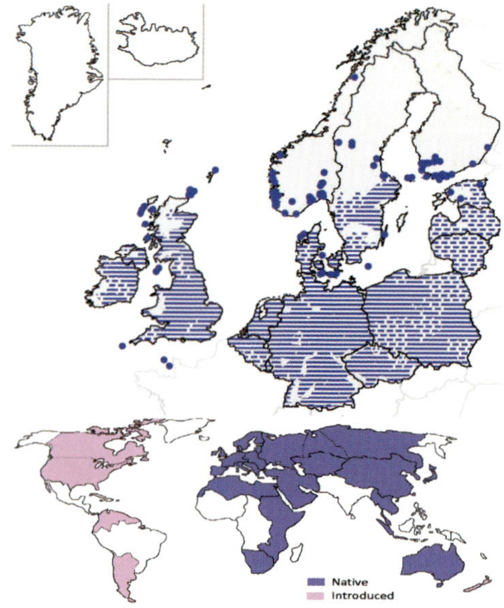

Potamogeton crispus. Sand pit, Krefeld, Germany. Photo KvdW.

Perennial hydrophyte. Rhizome very thin or slender, compressed.

Stem: Up to several metres long, compressed, with a shallow groove running down on one or both of the broader sides, slender to robust, often unbranched to sparsely branched.

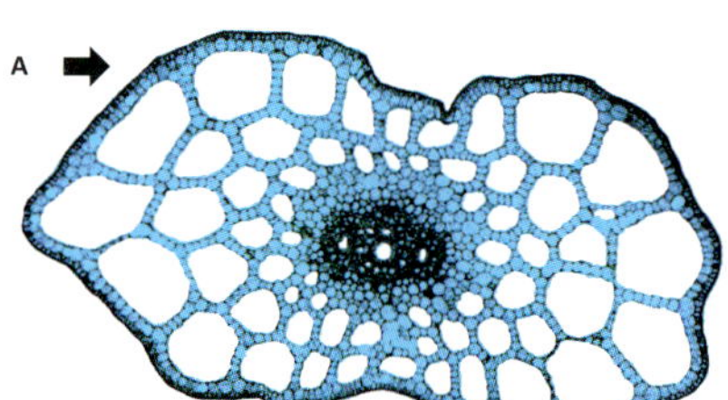

Stem cross section with 0–1 layer of pseudohypodermis (A), without subepidermal and interlacunar bundles. Photo KvdW.

Floating leaves: Absent.

Submerged leaves: 2.5–9.5 cm long and 0.5–1.2 cm wide, obtuse to semi-acute, broadly linear with parallel sides, sessile, rounded to cuneate at base, firm, not translucent, bright to dark green or brownish green, often with a reddish tinge at the midrib. With a narrow band of lacunae, 1–2(–3) lateral veins and a small number of secondary veins on each side of the midrib. The upper leaves typically undulate, whereas the lower and younger leaves may be more or less flat but always with a distinctly serrate margin.

Stipules: Fugacious, 0.4–1.5 cm long, apex truncate or emarginate, faintly reddish brown, translucent, veins inconspicuous.

Inflorescence: 0.5–1.6 cm long, cylindrical. Peduncle 1.5–7.0 cm long, somewhat compressed and of the same diameter as the stem. Normally with several inflorescences on each shoot.

Flowers: Contiguous, with (2–)4 carpels.

Fruits: 4–6 mm long, reddish brown. Beak 1.5–2.3 mm long - more than half the length of the rest of the fruit.

Flowering: June-August.

Biology: *P. crispus* is partially wintergreen, overwintering as buds or leafy shoots from the rhizome and as 1–5 cm axillary or apical turions. Turions are very variable and develop over a long period from early autumn into the winter. They are characterised by the hardened leaf base and the short, thickened, often congested, rigid leaves.

Habitats: *Potamogeton crispus* grows in a wide range of habitats, such as weakly acidic to alkaline, moderately eutrophic to very eutrophic lakes, gravel pits, ponds, streams, rivers, canals and ditches. It can also grow in weakly brackish water and ditches with strong ochre deposits.

Potamogeton crispus. In a small lake with clear water the vegetative shoots reach for the surface. Silkeborg, Denmark. Photo JCS.

Despite the high variability of leaf shape *P. crispus* is relatively easy to recognise by the minutely denticulate leaf margin and the compressed, shallowly grooved stems. Photo BM.

Potamogeton crispus. The long-beaked fruits are characteristic for the species. Photo JCS.

Potamogeton crispus. The distinct denticulate leaves with only 1–2 lateral veins on each side of the midrib. Photo JCS.

Distribution within the region: Native. Widespread and abundant in the south, scarce or absent from the north, including Norway, the northern part of Sweden and Finland. Absent from Iceland and Greenland.

Characteristics and similar species: *Potamogeton crispus* does not always have undulate leaves, but can always be recognised by the serrate leaf margins and compressed stem, which always has a shallow groove running down at least one of the broader sides.

Hybrids: *Potamogeton crispus* is involved in several hybrid combinations: **93 *P. ×undulatus***, **91 *P. ×olivaceus*** and **92 *P. ×cooperi***.

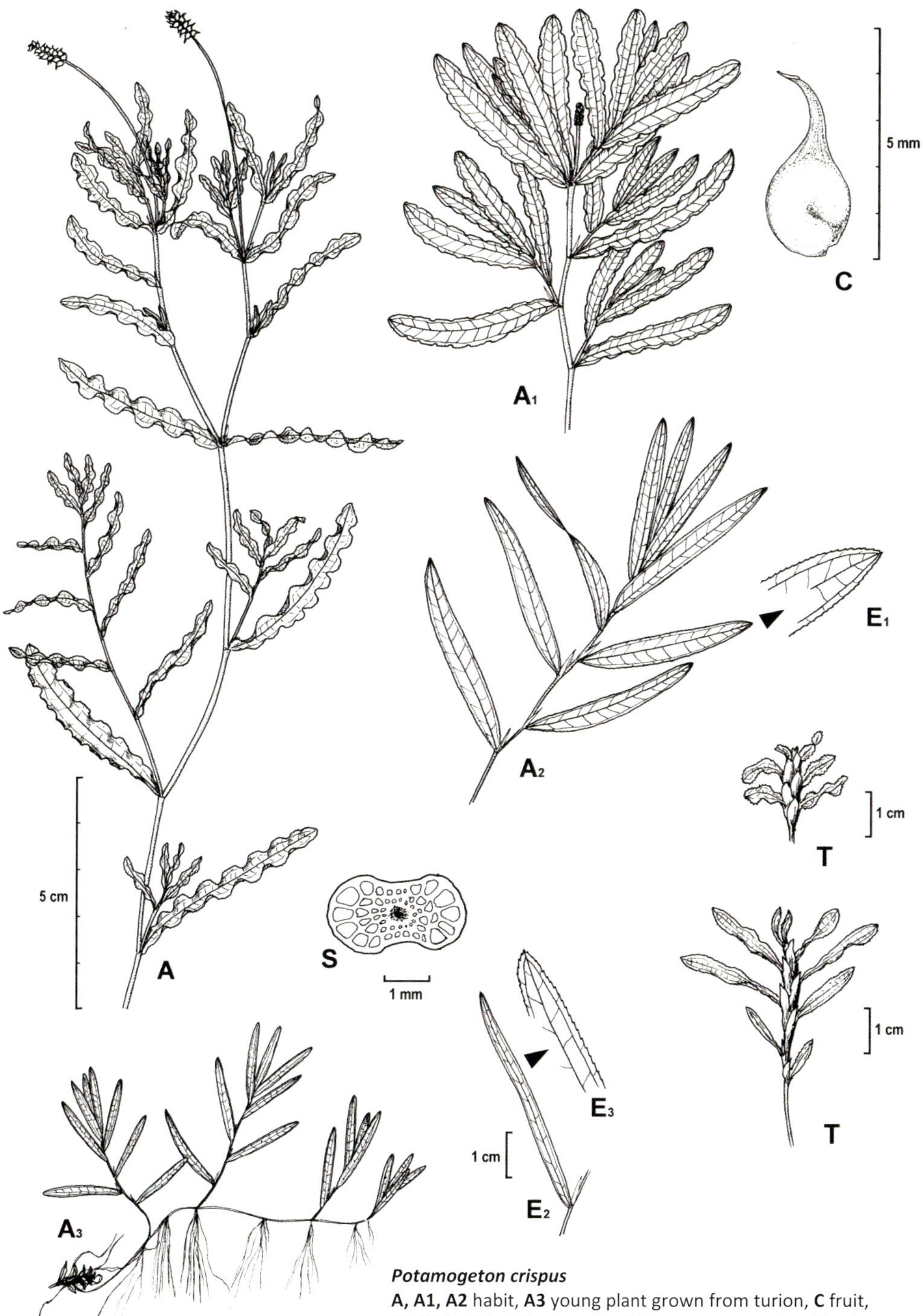

Potamogeton crispus
A, A1, A2 habit, **A3** young plant grown from turion, **C** fruit, **E1, E3** leaf tip, **E2** narrow leaf, **S** stem cross section, **T** turions.

Potamogetonaceae

76 *Potamogeton epihydrus* Raf.

Syn.: *P. nuttallii* Cham. & Schltdl.; *P. pensylvanicus* Willd. ex Cham. & Schltdl.
DK: - **N:** - **S:** - **FIN:** - **IS:** -
GB: American Pondweed **NL:** - **F:** -
D: - **CZ:** - **PL:** -
EST: - **LV:** - **LT:** - **Ru:** -

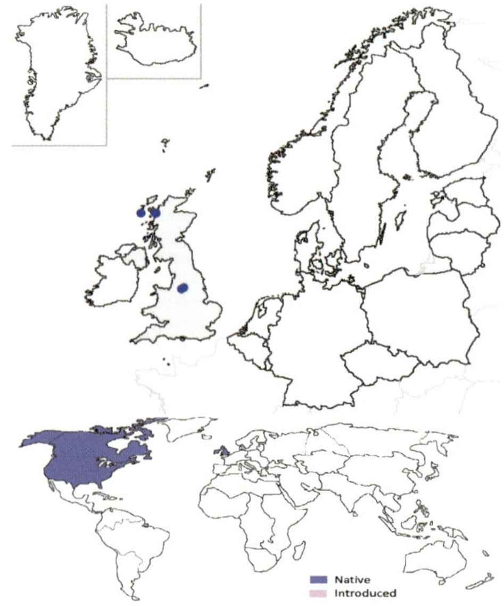

Potamogeton epihydrus at 1–2 metres depth in a very clean, oligotrophic lake. Loch Crocabhat, South Uist, Outer Hebrides, Scotland. Photo Claudia Ferguson-Smyth.

Perennial hydrophyte. Rhizome slender to robust, more or less compressed, sometimes with terminal, scaly buds.

Stem: Up to 185 cm long, slender to robust, unbranched or slightly branched, somewhat compressed or almost terete towards the apex. Stem cross section shows no subepidermal or interlacunar bundles.

Floating leaves: Lamina 3.5–7.5 cm long, 0.7–2.2 cm wide, ovate to oblong elliptical, apex obtuse, base cuneate, coriaceous, not translucent, brownish-green, young leaves with a pinkish tinge. With 4–10 lateral veins and numerous more or less inconspicuous secondary veins on each side of the midrib. Petiole 2–6(–9) cm long.

Submerged leaves: Lamina 6.5–24 cm long, 2.5–11.0 mm wide, 18–30(–60) times as long as wide, sessile, linear, pale to olive green often with a pinkish tinge, delicate, translucent, entire, plane or with margins a little undulated towards the narrow obtuse to acute leaf tip, base cuneate. With a broad band of lacunae and 2–4(–6) lateral veins on each side of the midrib, secondary veins irregular and more or less transverse. The submerged leaves are often distichous on vegetative shoots.

Stipules: Open, soon decaying, 1–4.5 cm long, convolute at the base, translucent.

Inflorescence: 1–3 cm long, cylindrical. Peduncle 2.5–9 cm long and of the same diameter as the stem.

Flowers: With 4 carpels.

Fruits: 2.5–3.1(–4.0) mm long, lacking a distinct dorsal keel, olive green or brownish. Beak 0.1–0.4 mm long.

Flowering: June-August.

Biology: Reproduces by fruits and vegetatively by stolons and buds.

Habitats: In still, shallow, oligotrophic water in lochans. Introduced in mesotrophic canals.

Distribution within the region:
In Europe, *Potamogeton epihydrus* is native only on South Uist in the Outer Hebrides. A record from Skye needs further verification. *P. epihydrus* has been recorded from canals in Yorkshire and Lancashire, England, since 1907, but this is an introduction of uncertain origins. An additional record from Ireland needs confirmation. Native and widespread in boreal and temperate USA and Canada.

Characteristics and similar species: *Potamogeton epihydrus* differs from other European *Potamogeton* species with both floating and submerged leaves, by the compressed stem and the delicate, linear to ribbon-like submerged leaves with up to 4 veins and a broad band of lacunae on each side of the midrib.

Hybrids: *Potamogeton epihydrus* is not involved in any hybrid combinations in Europe but is known to hybridise with

Potamogeton epihydrus. The submerged vegetative shoots of *P. epihydrus* have distichously arranged peach-tinged pink leaves. Photo Claudia Ferguson-Smyth.

In Loch Crocabhat, South Uist, the floating leaves of *P. epihydrus* seldom reach the surface. Photo Claudia Ferguson-Smyth.

P. gramineus and *P. nodosus* in North America.

Potamogeton epihydrus. Floating leaf (top left) and submerged leaf (right), the enlargement (bottom left) shows the lacunae more clearly.
Photo Claudia Ferguson-Smyth.

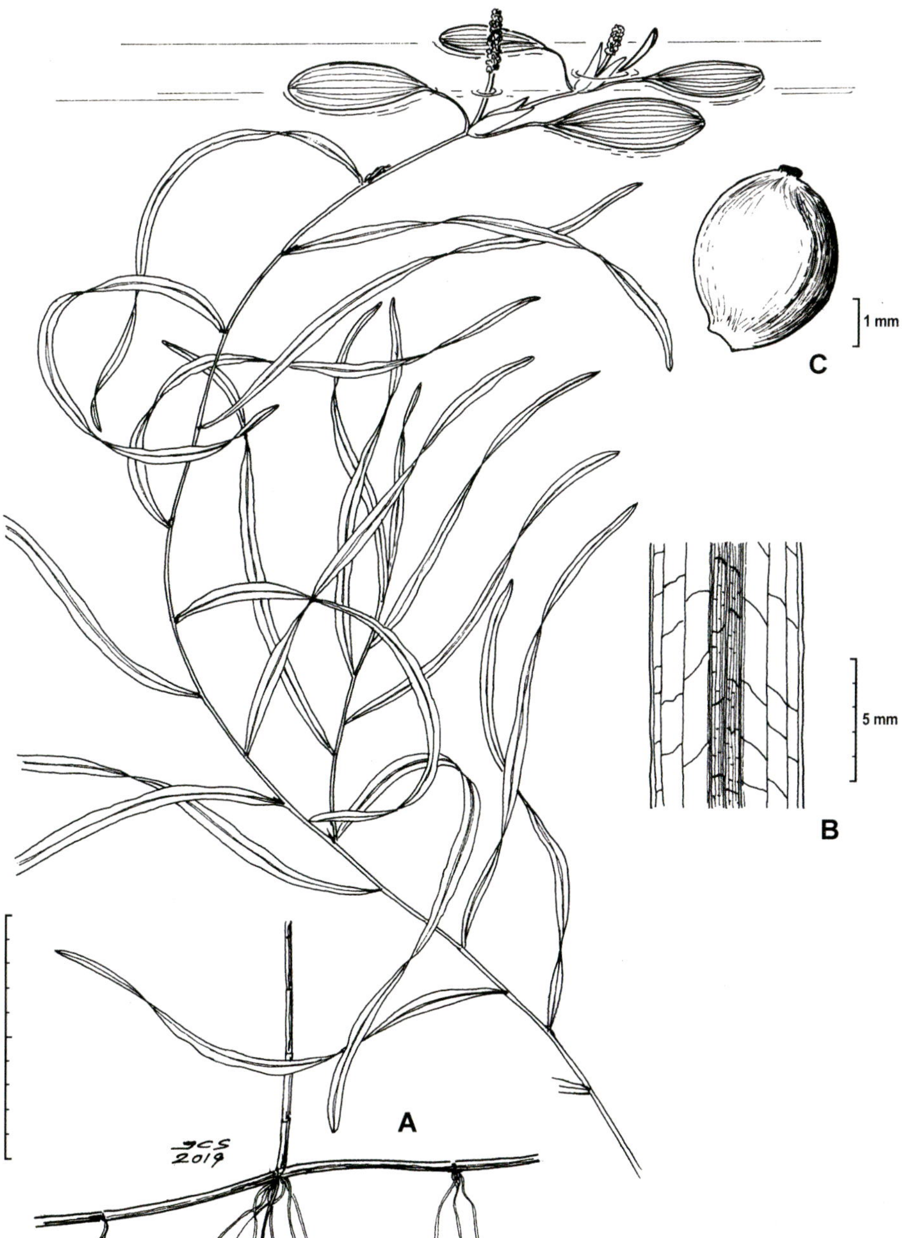

Potamogeton epihydrus
A habit, **B** middle part of submerged leaf, **C** fruit.

Potamogetonaceae

77 *Potamogeton acutifolius* Link.

DK: Spidsbladet Vandaks **N:** Spiss-tjørnaks **S:** Spetsnate **FIN:** - **IS:** -
GB: Sharp-Leaved Pondweed **NL:** Spits fonteinkruid **F:** Potamot à feuilles aiguës
D: Spitzblättriges Laichkraut **CZ:** Rdest ostrolistý **PL:** Rdestnica ostrolistna
EST: Teravalehine penikeel **LV:** Smaillapu glīvene **LT:** Smailialapė plūdė **Ru:** Рдест остролистный

Potamogeton acutifolius often grows in smaller lakes and ponds. Tønder, Denmark. Photo JCS.

Perennial hydrophyte. Rhizome short or absent, very delicate and not overwintering.

Stem: Up to 1 metre long, 0.4–3.0 mm wide, compressed to flat, not winged, slender to fairly robust, more or less branched. Nodal glands lacking.

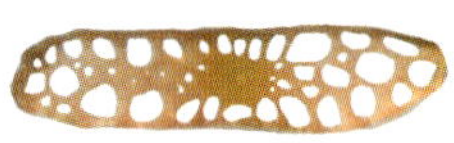

Stem cross section. Photo JCS.

Floating leaves: Absent.

Submerged leaves: Lamina 3.5–8(–12) cm long, 1.5–4.0 mm wide, sessile, linear, entire, leaf margins bordered by a strong marginal vein, tapering abruptly to an acuminate apex, base cuneate, dark green more rarely with a reddish or brownish tinge. With a narrow band of lacunae on each side of the midrib, 1 lateral vein and several more or less inconspicuous sclerenchymatous strands.

Stipules: Persistent but soon decaying to filamentous strands, 1.0–2.1 cm long, obtuse at the apex, translucent, whitish or pale brown, open.

Inflorescence: 0.4–0.8 cm long, somewhat compressed, with 4–7 flowers. Peduncle 0.5–2.0(–2.5) cm long, compressed, about 1–2 times the length of the inflorescence and of the same diameter as the stem. Normally with several inflorescences on each shoot.

Flowers: With 1 carpel.

Fruits: 3–4 mm long, reddish brown to brown, with a distinct keel on the dorsal side, and with a single blunt tooth on the ventral side. The beak is 0.4–1.0 mm long.

Flowering: July-August.

Biology: *P. acutifolius* overwinters as turions 2.0–3.5 cm long, which develop in August at the apices of vegetative shoots and lateral branches, or in leaf axils. Turions consist of short, compressed, mucronate leaves, above 2–3 short, erecto patent leaves.

Habitats: Neutral to alkaline, moderately nutrient-rich to nutrient-rich, shallow ponds, small lakes and grazing marsh ditch complexes. Often in newly established or restored ponds and lakes.

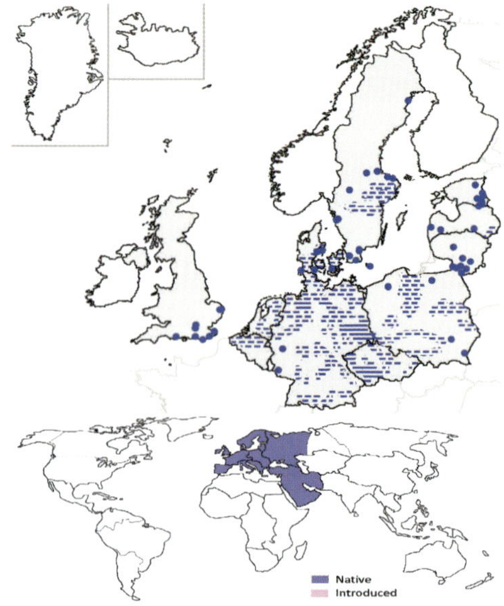

Distribution within the region:
Native. Locally frequent in the Netherlands and Germany but otherwise scarce and local throughout, north to Östergötland in southern Sweden.

Characteristics and similar species:
Potamogeton acutifolius can be recognised by the 3-veined, quite short, acuminate leaves, the compressed but not winged stem, and by the short inflorescence with a short peduncle, often of the same length as the inflorescence.

Potamogeton acutifolius has a short inflorescence and stipules, which soon split up into filamentous strands, or divide in two along the two prominent veins (in the middle of the photo). Photo JCS.

Potamogeton acutifolius. Flowering shoot. Photo KvdW.

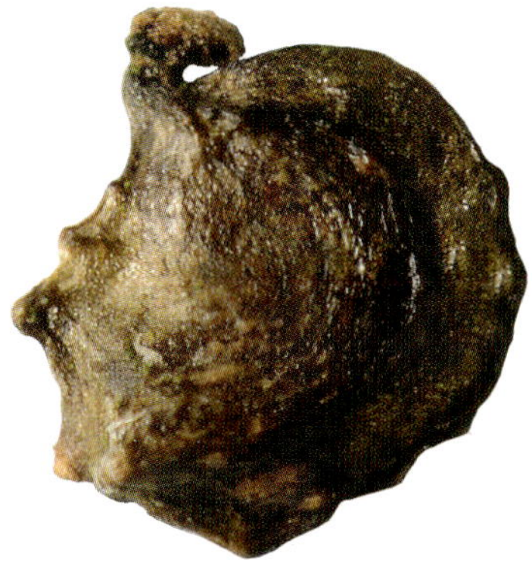

Potamogeton acutifolius. Fruit in lateral view has a distinct keel on the dorsal side (right), and at the ventral side a single blunt tooth (left). Photo JCS.

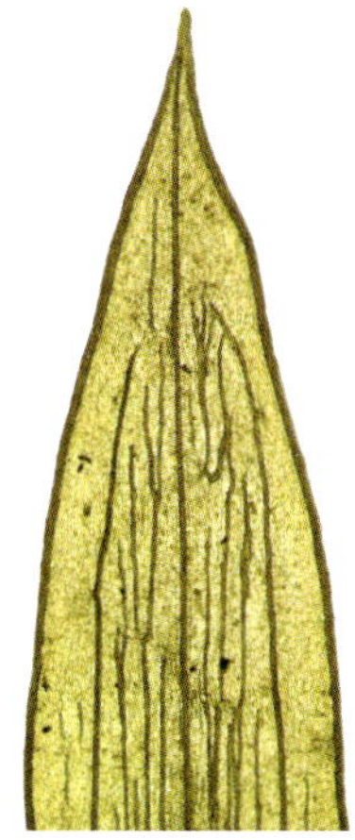

Potamogeton acutifolius. The leaf apex tapers abruptly to an acuminate apex, almost like a nib. The leaf has 3 veins and several more or less inconspicuous sclerenchymatous strands. Such strands are also present in *P. compressus*, but not in any other narrow-leaved *Potamogeton* species. Photo KvdW.

It could be mistaken for **78 *P. compressus***, which has 5-veined leaves and a conspicuously winged stem, with a longer peduncle and inflorescence. The fruit of *P. compressus* also lacks the tooth on the ventral side. *Potamogeton acutifolius* and *P. compressus* differ from all other narrow-leaved *Potamogeton* species by the presence of sclerenchymatous strands in the lamina (best seen with backlighting) and flat or almost winged stem cross section.

97 *P. ×bambergensis* G.Fisch. (*P. acutifolius × compressus*) is very similar - see this.

The hybrid *P. compressus × berchtoldii* is also very similar, but is only recognisable by DNA analysis.

Hybrids: See p. 221.

Note: The literature states that the peduncles of *P. acutifolius* have longer epidermal cells than those of *P. compressus*, however, measurement of material from many Danish populations has shown that the variation is too great for it to serve as an identification character, at least for Danish plants.

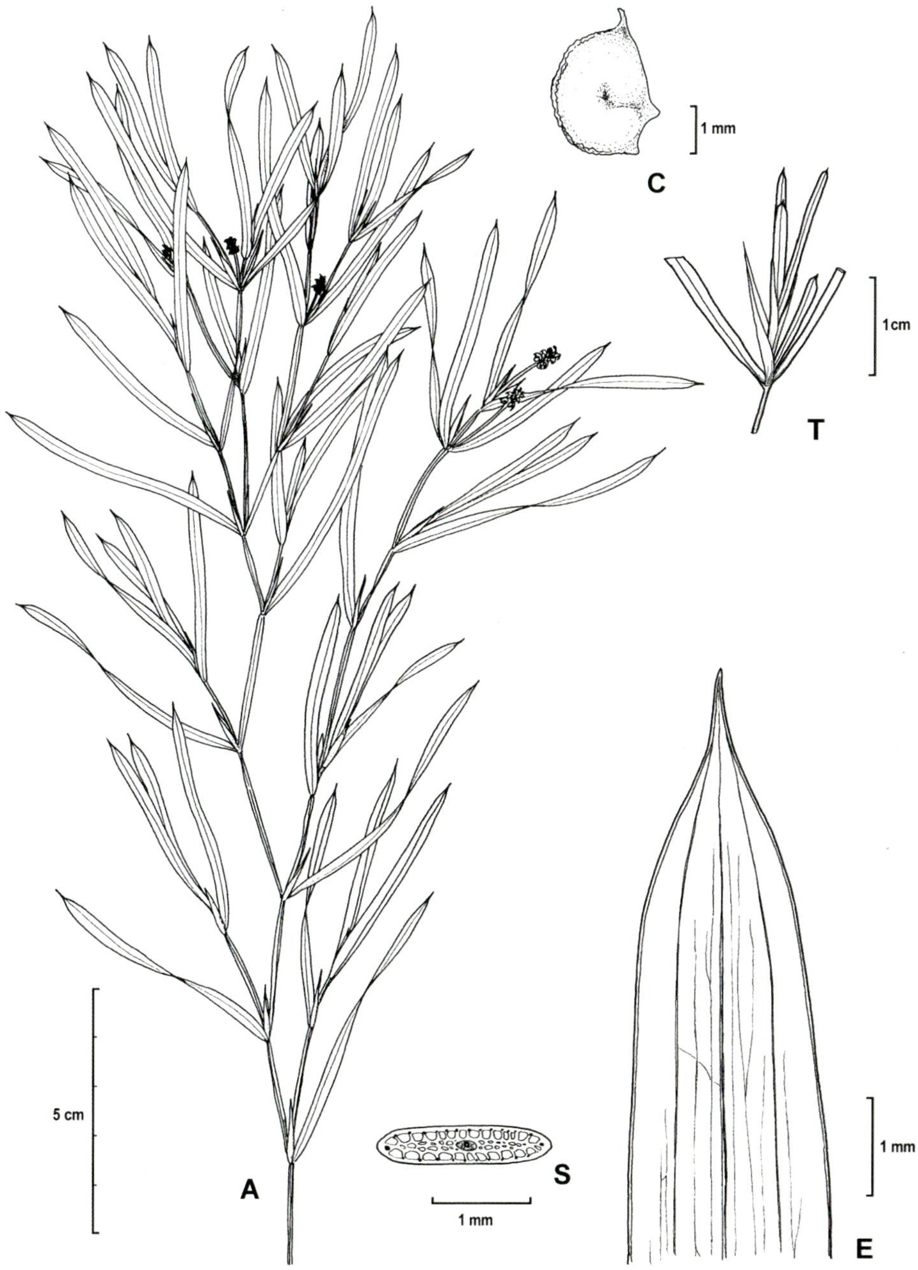

Potamogeton acutifolius
A habit, **C** fruit, **E** leaf tip with 3 veins and several sclerenchymatous strands, **S** stem cross section, **T** turion.

Potamogetonaceae

78 *Potamogeton compressus* L.

Syn.: *P. zosterifolius* Schum.

DK: Bændel-Vandaks **N:** Bendeltjørnaks **S:** Bandnate **FIN:** Litteävita **IS:** Sverðnykra
GB: Grass-wrack Pondweed **NL:** Plat fonteinkruid **F:** Potamot comprimé
D: Flachstengliges Laichkraut **CZ:** Rdest smáčknutý **PL:** Rdestnica ściśniona
EST: Lapik penikeel **LV:** Plakanā glīvene **LT:** Plokščioji plūdė **Ru:** Рдест сплюснутый

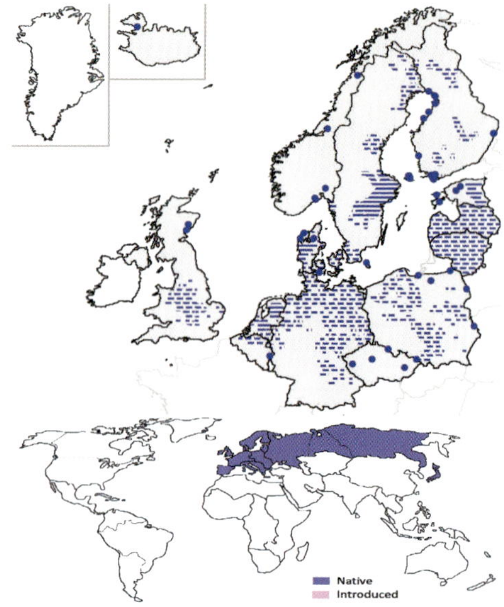

Potamogeton compressus has long ribbon-like leaves. Grund Fjord, Randers, Denmark. Photo JCS.

Perennial hydrophyte. Rhizome short or absent, delicate and not overwintering.

Stem: Up to 1 metre long, 0.5–3.5 mm wide, compressed, winged, slender to robust, usually richly branched. Nodal glands lacking.

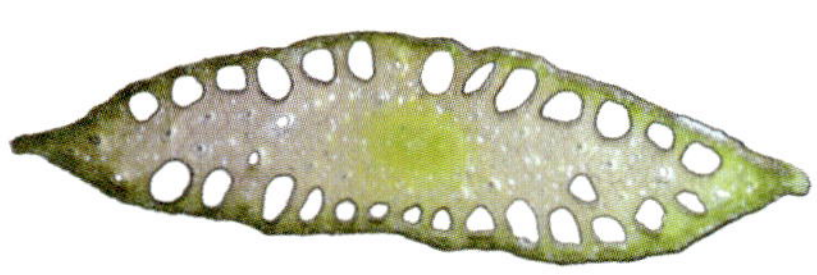

Stem cross section.

Floating leaves: Absent.

Submerged leaves: Lamina 8.5–24 cm long, 3–6 mm wide, sessile, linear, entire leaf margins bordered by a strong marginal vein, apex more or less shortly acute and mucronate, base cuneate, olive green to dark green and often with a reddish tinge along the midrib. With a narrow band of lacunae, 2 lateral veins and several inconspicuous sclerenchymatous strands on each side of the midrib.

Stipules: Persistent, but soon decaying to filamentous strands, 2.0–4.5 cm long, obtuse, translucent, whitish or with a faint brown tinge, open and convolute at the base. The 2 most prominent veins are green and remain, when the rest of the stipule has decayed.

Inflorescence: 1.5–3.5 cm long, cylindrical, with 10–20 flowers. Peduncle 2.8–7.0 cm long, about 2–5 times the length of the inflorescence, of the same diameter as the stem and slightly compressed. Normally with several inflorescences on each shoot.

Flowers: With (1–)2(–3) carpels.

Fruits: 3.4–4.0 mm long, reddish brown to brown, with a distinct keel on the dorsal side and without a tooth on the ventral side. The beak is 0.4–0.7 mm long.

Flowering: June-August.

Biology: *Potamogeton compressus* overwinters as turions, 2.5–4.5 cm long, which develop at the apex of the vegetative shoots, on lateral and axillary branches. The turions develop in August and consist of short, appressed, mucronate leaves with 2–4 short, erecto-patent leaves near the base.

Habitats: Neutral to alkaline, moderately nutrient-rich to nutrient-rich, larger lakes, rivers, streams, canals and ditches.

Distribution within the region:
Native. Scattered to locally rather common throughout most of the region.

Potamogeton compressus is often richly branched. Canal at Grund Fjord, Randers, Denmark. Photo JCS.

Potamogeton compressus. The obtuse stipules have 2 prominent green veins, which remain when the rest of the stipule has decayed. Photo JCS.

Potamogeton compressus. Fruit with distinct keel on the dorsal side (right), and no tooth at the ventral side (left). Photo JCS.

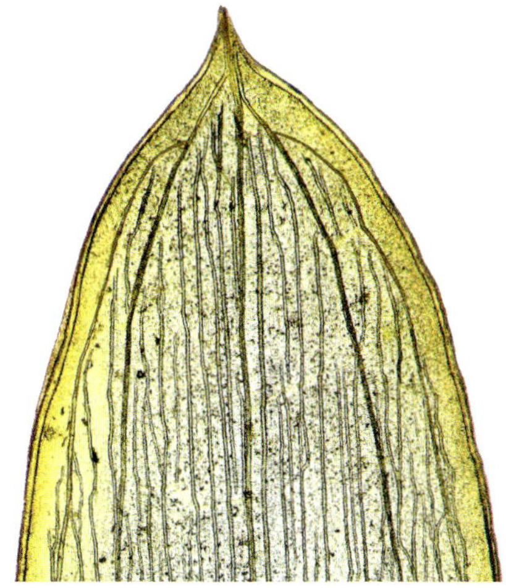

Potamogeton compressus. Leaf tip with 5 primary veins and many sclerenchymatous strands. Photo KvdW.

Absent from Ireland and Greenland.

Characteristics and similar species: *Potamogeton compressus* can be recognised by the long, linear, 5-veined leaves which are more or less shortly acute and mucronate at the apex, the compressed, winged stem and the peduncle, which is 2–5 times the length of the inflorescence.

77 *Potamogeton acutifolius* has 3-veined leaves, the stem is compressed but not winged and the peduncle shorter.

Potamogeton compressus and *P. acutifolius* differ from all other narrow-leaved *Potamogetons* by the presence of sclerenchymatous strands in the lamina (best seen with backlighting) and flat or more or less winged stem cross section.

Hybrids: *Potamogeton compressus* is involved in several hybrid combinations:

97 *P. ×bambergensis*.

P. ×ripensis Baagöe (*P. compressus x trichoides*) has long leaves, only 1.5–1.8 mm wide and compressed stem.

P. compressus x berchtoldii.

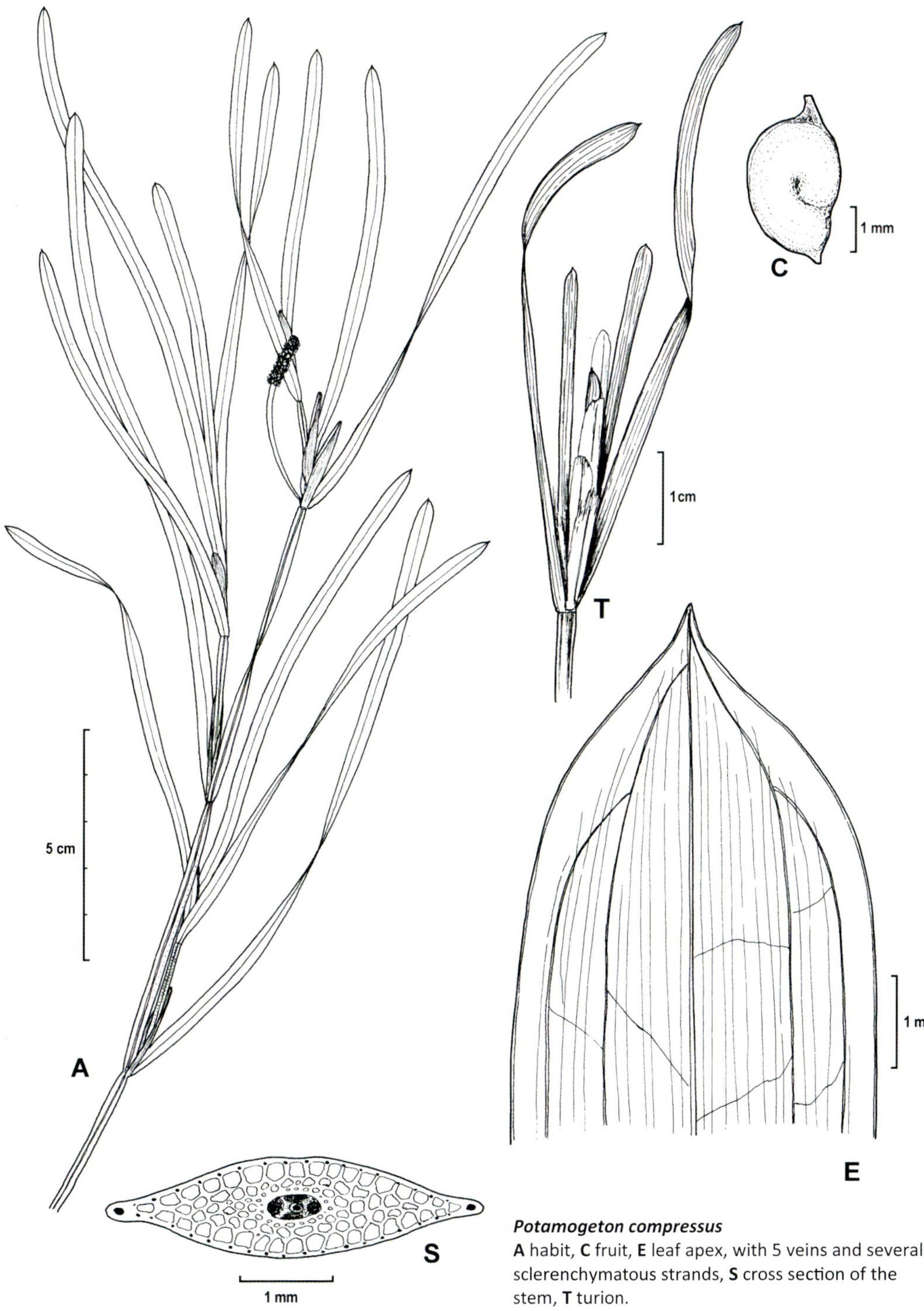

Potamogeton compressus
A habit, **C** fruit, **E** leaf apex, with 5 veins and several sclerenchymatous strands, **S** cross section of the stem, **T** turion.

Potamogetonaceae

79 *Potamogeton obtusifolius* Mert. & W.D.J.Koch

DK: Butbladet Vandaks **N:** Buttjørnaks **S:** Trubnate **FIN:** Tylppälehtivita **IS:** Seilunykra
GB: Blunt-leaved Pondweed **NL:** Stomp fonteinkruid **F:** Potamot à feuilles obtuses
D: Stumpfblättriges Laichkraut **CZ:** Rdest tupolistý **PL:** Rdestnica stępiona
EST: Tömbilehine penikeel **LV:** Struplapu glīvene **LT:** Bukalapė plūdė **Ru:** Рдест туполистный

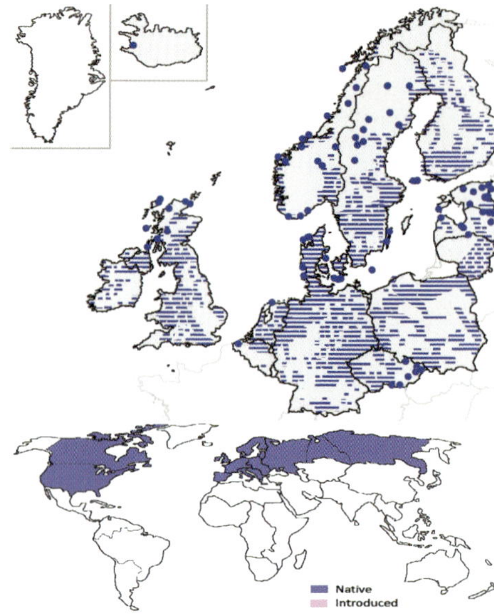

Potamogeton obtusifolius is often richly branched in the upper part, developing many leaves just below the water surface. Pond close to Videbaek, Jutland, Denmark. Photo BM.

Perennial hydrophyte. With or without a short, delicate, rhizome which does not overwinter.

Stem: More or less compressed, with rounded sides, slender, normally richly branched, particularly in the upper part of shoots. Nodal glands conspicuous and of irregular shape.

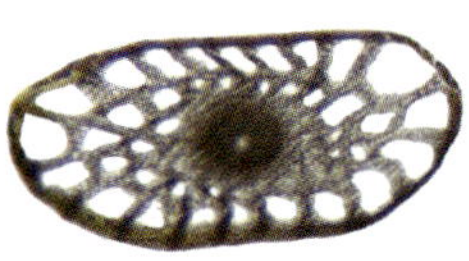

Stem cross section.

Floating leaves: Absent.

Submerged leaves: Lamina 5.0–8.5 cm long, 2.5–3.5 mm wide, sessile, linear, entire, leaf margins without a strong marginal vein, apex rounded or obtuse, mucronate, base cuneate, light to dark green often with a reddish tinge, delicate, almost translucent. With a narrow band of lacunae, and 1(–2) obscure lateral veins on each side of the midrib; the second vein often between the primary vein and the midrib.

Stipules: Persistent, 1–3 cm long, open and convolute, rounded or obtuse at the apex, translucent when fresh, light beige coloured when dry. With (8–)10–17 fairly obscure veins between the 2 prominent veins.

Inflorescence: 0.7–1.3 cm long, short cylindrical, rather dense and with 6–8 flowers. Peduncle 0.5–2.0(–4.0) cm long, 1–4 times as long as the inflorescence, somewhat compressed and of the same diameter as the stem. Usually with several inflorescences on each shoot.

Flowers: With (3–)4(–5) carpels.

Fruits: 2.6–3.6 mm long, greenish brown to brown, with a more or less distinct keel on the dorsal side. The beak is 0.2–0.5 mm long.

Flowering: July-August.

Biology: *Potamogeton obtusifolius* overwinters as turions 2.5–4.0 cm long, which develop at the apices of the vegetative shoots and lateral branches. The turions develop in early August, and consist of short, compressed, obtuse, mucronate leaves with 3–5 short erecto-patent leaves near the base.

Habitats: Weakly acidic to weakly alkaline, moderately nutrient-rich lakes, ponds and small water bodies. In northern and western parts of Britain, also in species-poor oligotrophic acidic lakes. Rare in the downstream parts of streams, canals and ditches.

Potamogeton obtusifolius. The leaves are delicate, almost translucent and often with a reddish tinge. Filsoe, Jutland, Denmark. Photo JCS.

Potamogeton obtusifolius
Fruit. Turion, Photos JCS.

Potamogeton obtusifolius. The stipules are open and convoluted, rounded or more or less obtuse and do not divide to a V when they decay. Photo KvdW.

Potamogeton obtusifolius. The obtuse, mucronate leaves have rather obscure lateral veins – usually there is only one on each side of the midrib, but sometimes two; in that case the distal vein is obscure. Photo JCS.

Distribution within the region:
Native. Widespread and locally abundant more or less throughout mainland areas of the region.

Characteristics and similar species:
Potamogeton obtusifolius can be recognised by the flattened, rounded stem, the 2.5–3.5 mm wide, normally 3-veined (rarely 5-veined) obtuse leaves. The stipules are convolute, with a rounded or obtuse apex, and the peduncle is short. *P. obtusifolius* is often densely branched in the apical parts of the shoots.

It can be mistaken for **80 *P. friesii*,** which has lateral branches occurring throughout the stem rather than particularly toward the apex, longer peduncle, stipules that are connate at the base when young and with two very prominent veins, soon splitting up like an acute V.

Broad-leaved specimens of **81 *P. berchtoldii*** differ in having a more open habit, 2–5 flowers, smaller fruits and a more slender peduncle, shorter leaves, rarely more than 2 mm wide, which are typically green, without a reddish tinge, and stipules with only 4–8 obscure veins between the two main veins.

Hybrids:
Potamogeton ×semifructus A.Benn. (*P. friesii x obtusifolius*).

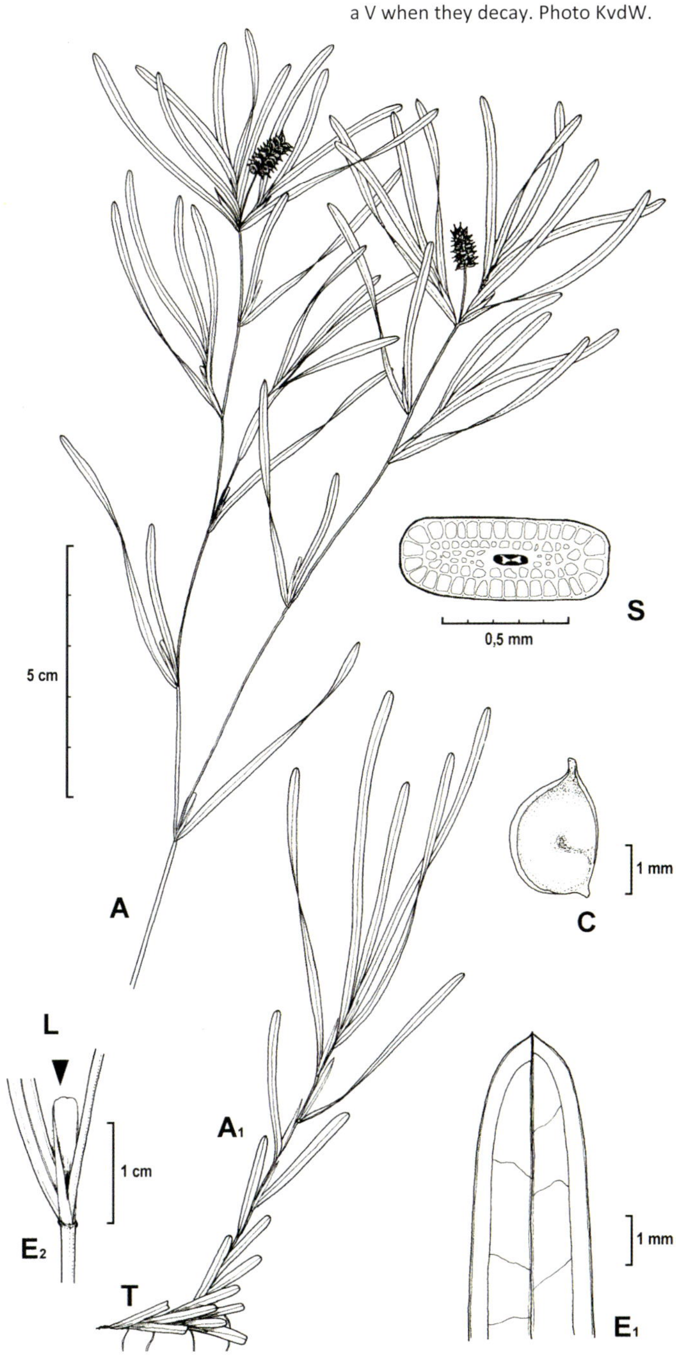

Potamogeton obtusifolius
A habit, **A1** young plant developed from a turion, **C** fruit, **E1** leaf tip, **E2** base of leaf, **L** stipule, **S** cross section of the stem, **T** turion.

Potamogetonaceae

80 *Potamogeton friesii* Rupr.

Syn.: *P. mucronatus* Schrad. ex Roem. & Schult.
DK: Brodbladet Vandaks **N:** Broddtjørnaks **S:** Uddnate **FIN:** Otalehtivita **IS:** -
GB: Flat-stalked Pondweed **NL:** Puntig fonteinkruid **F:** Potamot mucroné
D: Stachelspitziges Laichkraut **CZ:** Rdest hrotitý **PL:** Rdestnica szczeciolistna
EST: Ogaterav penikeel **LV:** Frīza glīvene **LT:** Dygliaviršūnė plūdė **Ru:** Рдест Фриса

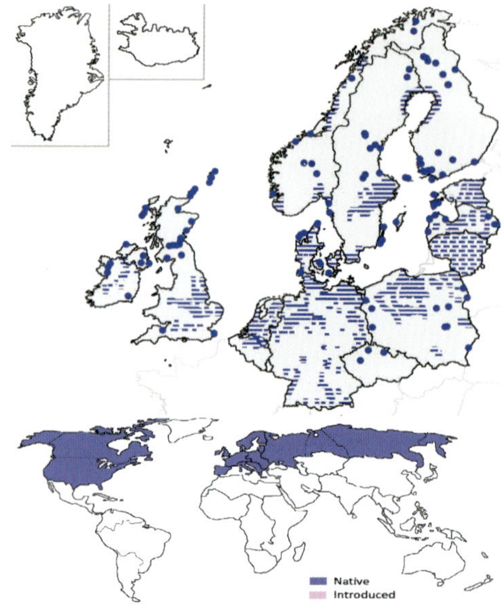

Potamogeton friesii. Typical plants have short lateral branches and an "open" habit. Indfjorden, Nees Jutland, Denmark. Photo JCS.

Perennial hydrophyte. With or without a short, delicate, rhizome which does not overwinter.

Stem: Somewhat compressed, ovate in cross section, slender, with lateral branches occurring throughout the stem. Nodal glands strongly developed and regularly semi-globose.

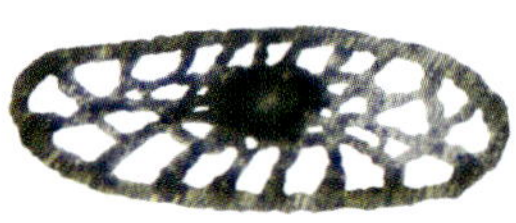

Stem cross section.

Floating leaves: Absent.

Submerged leaves: Lamina 4.0–8.5 cm long, 1.5–3.5 mm wide, sessile, linear, entire leaf margins bordered by a delicate marginal vein, apex obtusely mucronate, base cuneate, light green to dark green sometimes with a reddish tinge. With a narrow band of lacunae on each side of the midrib and (1–)2(–3) lateral veins.

Stipules: Persistent, 1.0–2.5 cm long, obtuse at the apex, whitish to light brownish, shortly tubular and connate at the base. With (8–)10–17 somewhat indistinct strands between two prominent, green veins. The stipule soon splits at the apex, and divides gradually to a narrow V shape, with the edges marked by two prominent veins.

Inflorescence: 0.7–1.4 cm long, shortly cylindrical, rather loose. Peduncle 1.5–4.0 cm long, 1–5 times as long as the inflorescence, the same diameter as the stem and slightly compressed. With 4–8 flowers. Usually with several inflorescences on each shoot.

Flowers: With 4 carpels.

Fruits: 2.4–3.0 mm long, greenish brown to brown, without a distinct keel. The beak is 0.3–0.5 mm long.

Flowering: July-August.

Biology: *Potamogeton friesii* overwinters as 1.2–2.3 cm long turions, which develop at the apices of lateral branches and in leaf axils. These emerge in early August, and are composed of short, compressed, dark green mucronate leaves, with bright green stipules arranged like a fan and perpendicular to the surrounding leaves (when viewed from above).

Habitats: Moderately nutrient-rich, neutral to alkaline, lakes, ponds, canals and ditches in marshland - rarely in streams.

Distribution within the region:
Native. Scattered to rather rare more or less throughout mainland areas of the region.

Characteristics and similar species:
See the text and photos at the top of the opposite page.
Potamogeton friesii is most often confused with **79 *P. obtusifolius*** (see this) but also with large individuals of **81 *P.***

Potamogeton friesii has relatively long peduncles. Ny Frederikskog at Højer, Jutland, Denmark. Photo JCS.

The combination of mucronate leaf apex and the structure of the stipule separates *P. friesii* from other *Potamogeton* species with narrow leaves. Note the semi-globose nodal glands. Photos KvdW and JCS.

Potamogeton friesii. The stipule has two prominent veins that soon split up at the apex and divide gradually to a narrow V shape, in which the two prominent veins mark the edges. Photos JCS and KvdW.

berchtoldii; slender individuals of **77 *P. acutifolius*** and **78 *P. compressus*** may also cause identification problems.

Hybrids: *Potamogeton friesii* is involved in several hybrid combinations:

94 *P. ×lintonii*.

P. ×confinis Hagstr. (*P. friesii* × *pusillus*).

P. ×pusilliformis Hagstr. (*P. friesii* × *berchtoldii*).

P. ×semifructus A.Benn. (*P. friesii* × *obtusifolius*).

Potamogeton friesii. The stipules of the turions are characteristically arranged like a fan and perpendicular to the surrounding leaves. Ny Frederikskog, Højer, Jutland, Denmark. Photo JCS.

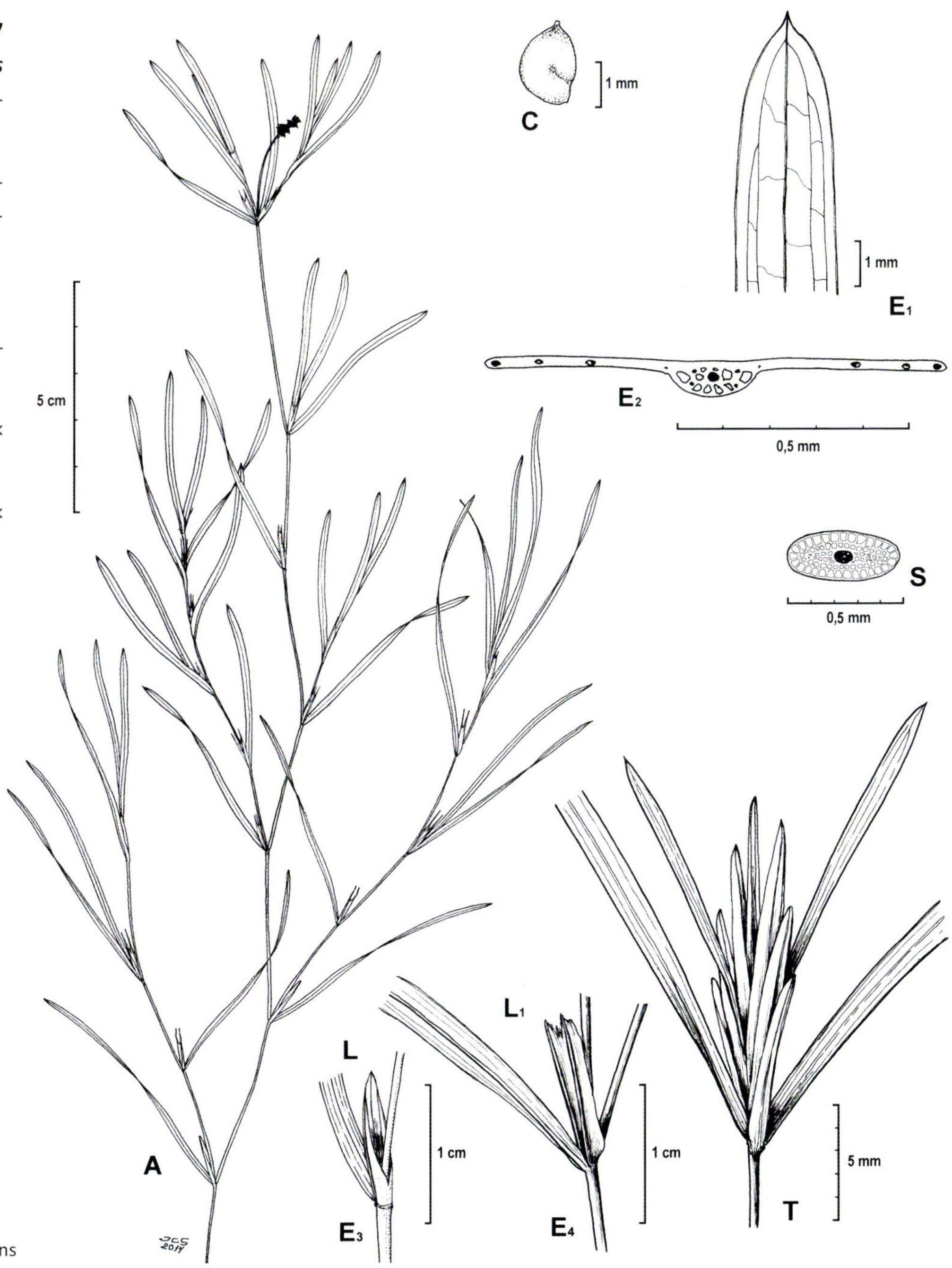

Potamogeton friesii
A habit, **C** fruit, **E1** leaf tip, **E2** cross section of leaf, **E3, E4** leaf-base, **L** stipule ventral view, **L1** stipule dorsal view, **S** cross section stem, **T** turion.

Potamogetonaceae

81 *Potamogeton berchtoldii* Fieber

Syn.: *P. pusillus* auct. Mult., non L. sec. Dandy & Taylor
DK: Liden Vandaks **N:** Småtjørnaks **S:** Gropnate **FIN:** Tylppälehtivita **IS:** Smånykra
GB: Small Pondweed **NL:** Klein fonteinkruid **F:** Potamot de Berchtold
D: Berchtolds Zwerg-Laichkraut **CZ:** Rdest Berchtoldův **PL:** Rdestnica Berchtolda
EST: Muda-penikeel **LV:** Berhtolda glīvene **LT:** Berchtoldo plūdė **Ru:** Рдест Берхтольда

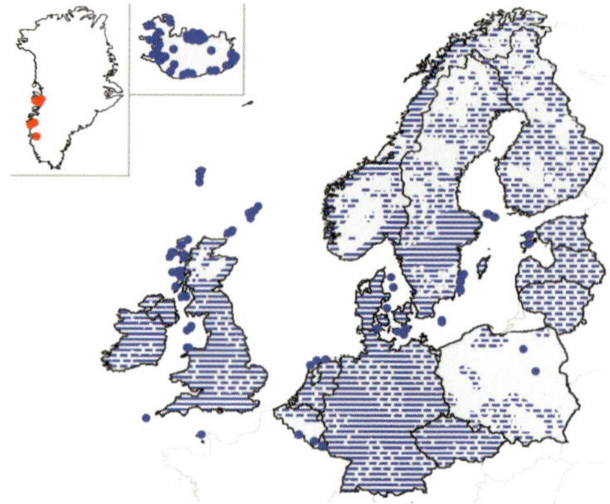

Red dots = *P. groenlandicus* Hagstr.

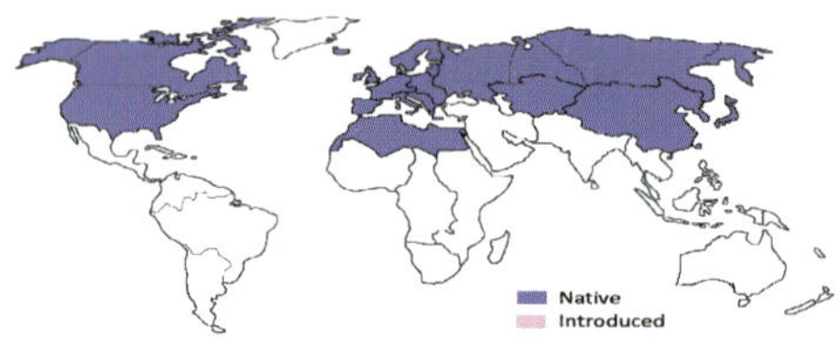

Flowering *P. berchtoldii* with fresh green leaves and inflorescence. Gudenåparken, Randers, Denmark. Photo JCS.

Perennial hydrophyte. Rhizome absent or short, terete, delicate, developing late in the autumn and not overwintering.

Stem: Up to 1 metre long, quite delicate, terete to slightly compressed, bright green, more or less branched. With 2 well-developed glands at each node.

Floating leaves: Regular floating leaves are not developed, but leaves surrounding the inflorescence are often somewhat spathulate and float.

Submerged leaves: Lamina 2.5–5.0 (–7.0) cm long, 0.8–1.8(–2.3) mm wide, linear, rather flaccid, entire, leaf margins not bordered by a marginal vein, acute or somewhat obtuse, tapering into the base, green - often brownish tinged, especially at the leaf apex. With one lateral vein and sometimes some additional very fine veins and a distinct band of pale tissue (lacunae) on each side of the midrib. The leaves are sessile.

Leaf cross section. Photo KvdW.

Stipules: Persistent or completely degraded on older leaves, 0.5–1.5 cm long, open, rounded to obtuse, membranous, translucent, more or less green or slightly reddish. Veins inconspicuous, the two primary veins only slightly stronger than the other veins.

Inflorescence: Short, cylindrical, 0.4–0.9 cm long, rather loose, with 2–4 flowers. Peduncle 1–3 cm long, of the same diameter as the stem and slightly compressed. Usually with several inflorescences on each shoot.

Flowers: With 4–5 carpels.

Fruits: 1.8–2.8 mm long, greenish brown, without a dorsal keel. The beak is 0.2–0.6 mm long.

Flowering: June-September.

Biology: Overwintering as dark green to brownish-green turions developed from early September at the apex of the main shoots and axillary branches. The turions are 8–24 mm long and 0.8–1.6 mm wide, fusiform, and with 2–4 acute erect to erecto-patent leaves in the lower part.
Potamogeton berchtoldii often forms large, dense populations, which in smaller lakes and canals may cover the entire surface in extensive patches.

Habitats: In more or less nutrient-rich lakes, ponds, ditches, streams and canals. Rare in large rivers.

Distribution within the region:
Native. Widespread and abundant throughout much of the region, absent from Greenland.

Flowering *P. berchtoldii* with reddish tones. Lake Hestholm, Skjern, Denmark. Photo JCS.

Characteristics and similar species: *Potamogeton berchtoldii* varies in shape and size depending on the habitat. In shallow water the plants can be very small and slender, whereas in flowing or deep water they may have long vegetative shoots and broader leaves. It differs from other narrow-leaved species by the combination of open stipules, well-developed nodal glands and the band of pale tissue (lacunae) along the weakly prominent midrib.

The similar **83 *P. pusillus*** differs by tubular stipules connate for 2–3 mm at the base, but this character is not always easy to use in practice, as the stipule is delicate and soon splits. *P. pusillus* also lacks nodal glands and lacunae and the leaves are more rigid with a marginal vein and a prominent midrib, clearly arching on the abaxial side of the leaf. Compare also **84 *P. trichoides*** and **80 *P. friesii.***

Hybrids: ***P. ×dualis*** Hagstr. (*P. berchtoldii x pusillus*), ***P. berchtoldii x compressus***, ***P. ×pusilliformis*** Hagstr. (*P. berchtoldii x friesii*), ***P. ×franconicus*** Fisch. (*P. berchtoldii x trichoides*).

82 *P. groenlandicus* Hagstr.

Syn.: *P. pusillus* L. subsp. *groenlandicus* (Hagstr.) Böch.

In terms of habit resembles *P. berchtoldii*, but the leaves with 7–11 veins and fruits on average larger, 2.6–2.7 mm long. Stipules open, convolute.

Habitats: Ponds and lakes.

Distribution: West-Greenland (red dots on the map).

Note: The taxonomic rank and origin of this taxon are still uncertain and await further investigation. It may be a geographically isolated morphotype of *P. sibiricus* A.Benn.

Leaf size varies widely in *Potamogeton berchtoldii*, but it usually is easy to distinguish from other narrow-leaved species. It is quite small, has a more or less terete stem with 2 nodal glands (black arrows) and leaves with 3 main veins, sometimes some additional very fine veins. The band of pale tissue (lacunae) (white arrows) is best seen in backlight. Valbjerg Sande in Thy, Denmark. Photos JCS.

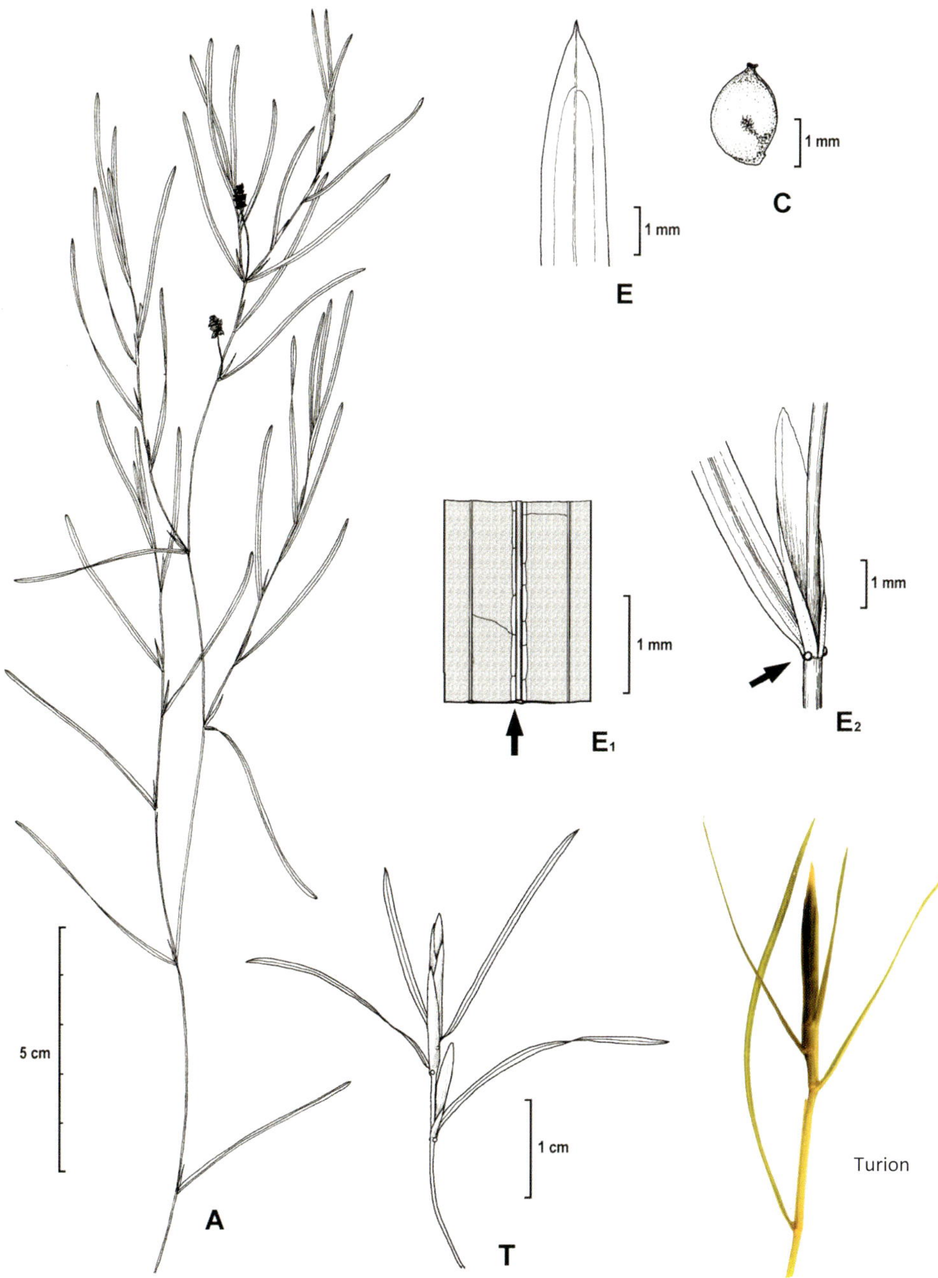

Potamogeton berchtoldii
A habit, **C** fruit, **E** leaf tip, **E1** part of leaf with lacunae bordering the midrib, **E2** open stipule and nodulae at base of leaf, **T** turion.

Potamogetonaceae

83 *Potamogeton pusillus* L.

Syn.: *P. panormitanus* Biv.

DK: Spinkel Vandaks **N:** Granntjørnaks **S:** Spädnate **FIN:** Hentovita **IS:** -
GB: Lesser Pondweed **NL:** Tenger fonteinkruid **F:** Potamot de Palerme
D: Kleines Laichkraut **CZ:** Rdest maličký **PL:** Rdestnica drobna
EST: Väike penikeel **LV:** Sīkā glīvene **LT:** Smulkioji plūdė **Ru:** Рдест маленький

Potamogeton pusillus together with *P. perfoliatus* in a ditch in Tøndermarsken, Denmark. Photo JCS.

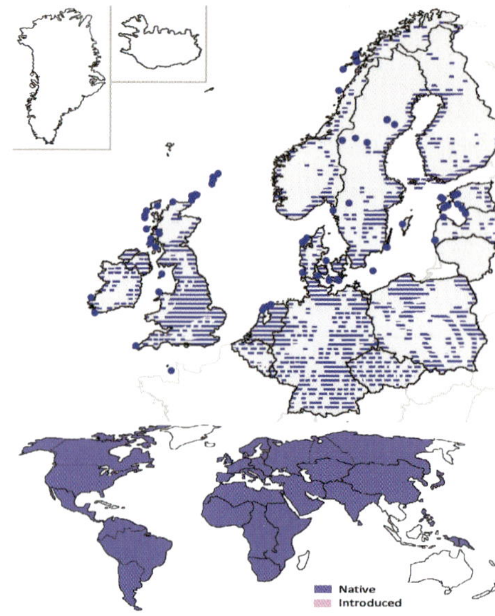

Perennial hydrophyte. Rhizome absent or short, delicate, developing late in the autumn and not overwintering.

Stem: Up to 1 metre long, quite delicate, terete to slightly compressed, bright green, more or less branched. Without or rarely with weakly developed nodal glands.

Floating leaves: Absent.

Submerged leaves: Lamina 2.0–5.0(–8.0) cm long, 0.8–1.4(–1.9) mm wide, linear, often somewhat rigid, entire, leaf margins bordered by sclerenchymatous strands, acute, with narrow wedge-shaped base, green to olive green to brownish. Midrib prominent and abaxially elevated, with one lateral vein on each side and without, or rarely, with a narrow band of pale tissue (lacunae) near the leaf apex. The leaves are sessile.

Leaf cross section. Photo KVDW.

Stipules: Persistent, 0.5–1.7 cm long, rounded to obtuse, membranous, translucent, pale brown to slightly green, tubular and connate for at least 2–3 mm at the base when young, but soon splitting. Veins inconspicuous and the two main veins are only slightly more distinct than the others.

Inflorescence: Shortly cylindrical, 0.6–1.3 cm long, rather loose, with 3–6 flowers. Peduncle 1–4 cm long, of the same diameter as the stem, terete or slightly compressed. Usually with several inflorescences on each shoot.

Flowers: With 4(–5) carpels.

Fruits: 1.8–2.4 mm long, bright greenish brown, without a dorsal keel. The beak is 0.2–0.4 mm long.

Flowering: June-September.

Biology: Overwintering as dark green to brown-green turions developed in the leaf axils from late July - later also from the apex of the main shoots and axillary branches. The turions are 10–16(–22) mm long and 0.5–1.0 mm wide, fusiform, and with 2–3 free, erecto-patent leaves in the lower part.

Habitats: In more or less nutrient-rich, ponds, lakes, ditches and canals, rarely in brackish water.

Distribution within the region:
Native. Widespread and locally abundant throughout most of the region. Less frequent in the north and absent from Iceland and Greenland.

Characteristics and similar species:
Potamogeton pusillus can be recognised by the narrow leaves with an acute apex and a distinct, strong, elevated midrib on the abaxial leaf surface, which is not normally bordered by lacunae.

In general *P. pusillus* has more rigid leaves than *P. berchtoldii*. Indfjorden at Nees, Jutland, Denmark. Photo JCS.

Flowering *P. pusillus* in very shallow water in Lake Filsø, Jutland, Denmark. Photo JCS.

Potamogeton pusillus. The stipule is tubular and connate at the base (arrow), but soon splits. The stems lack nodal glands. The midrib is clearly elevated on the abaxial leaf surface and without or only rarely bordered by lacunae. Photo JCS.

Normally without nodal glands. The stipule is tubular and connate for at least 2–3 mm at the base.

The very similar **81 *P. berchtoldii*** has well-developed nodal glands, more flaccid leaves not bordered by sclerenchymatous strands and a less distinct midrib, which is not much elevated on the abaxial leaf surface, but bordered by lacunae. In addition, the stipules are open.

Potamogeton pusillus can be very delicate and small, especially in shallow water, where the vegetative shoots are short and the leaves very narrow. Such specimens can easily be confused with **84 *P. trichoides*** - see this.

85 *Potamogeton rutilus* can be quite similar, but differs in leaves more gradually tapering to a sharply pointed apex and stipules with several prominent veins.

Hybrids: *P.* ×*dualis* Hagstr. (*P. berchtoldii x pusillus*), ***P.* ×*confinis*** Hagstr. (*P. friesii x pusillus*), ***P.* ×*grovesii*** Dandy & G.Taylor (*P. pusillus x trichoides*).

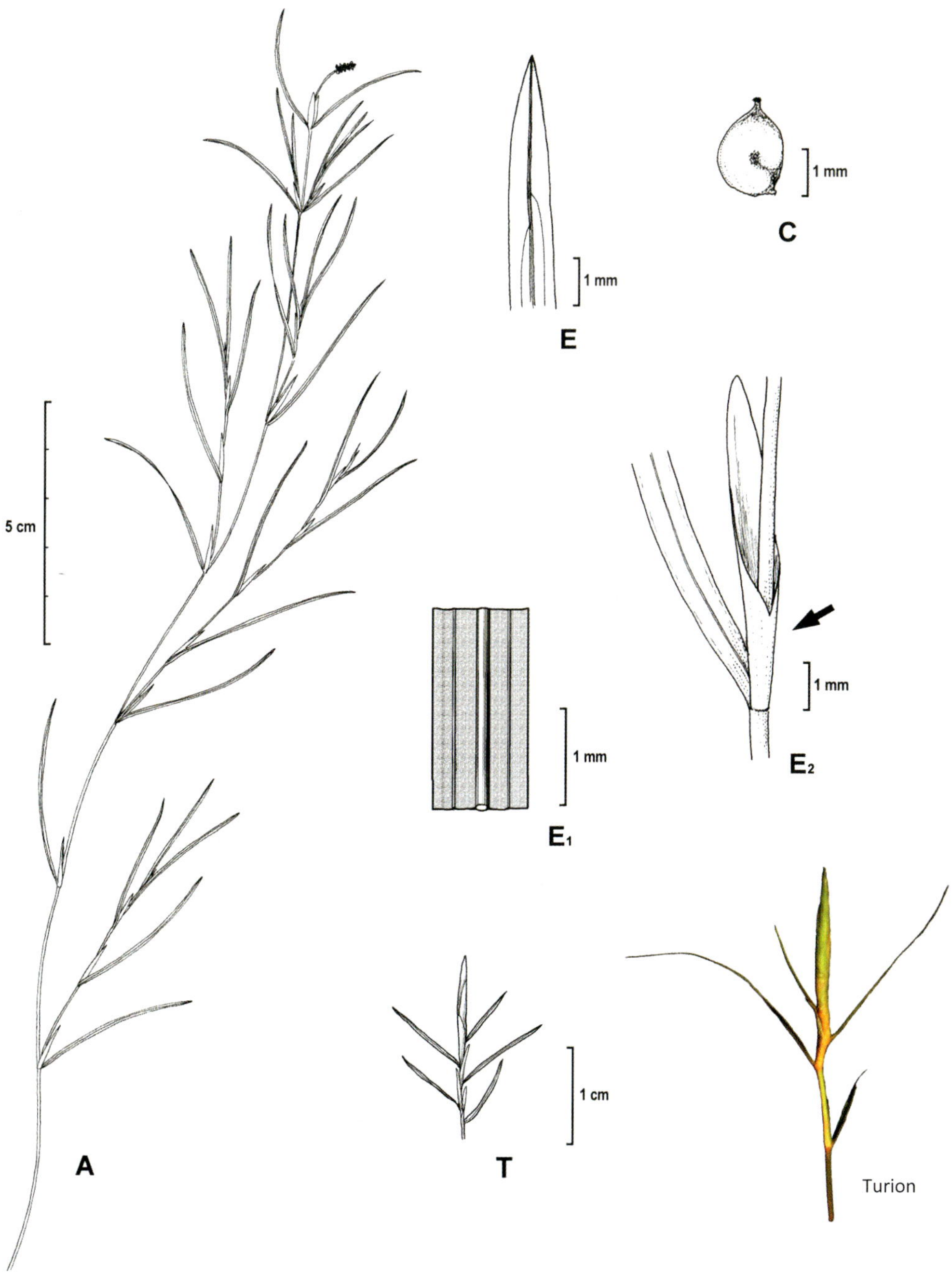

Potamogeton pusillus
A habit, **C** fruit, **E** leaf tip, **E1** section of leaf with prominent midrib, **E2** the stipule is tubular and connate at the base, **T** turion.

Potamogetonaceae

84 ***Potamogeton trichoides*** Cham. & Schlecht.

DK: Hårfin Vandaks **N:** - **S:** Knölnate **FIN:** - **IS:** -
GB: Hairlike Pondweed **NL:** Haarfijn fonteinkruid **F:** Potamot capillaire
D: Haarförmiges Laichkraut **CZ:** Rdest vláskovitý **PL:** Rdestnica włosowata
EST: Juus-penikeel **LV:** Matveida glīvene **LT:** Siūlinė plūdė **Ru:** Рдест волосовидный

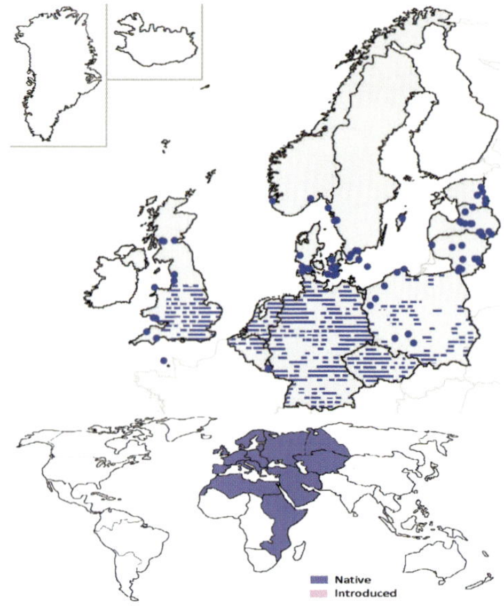

Potamogeton trichoides with turions and epiphytic microalgae. Stubberup, Faxe, Zealand, Denmark. Photo JCS.

Perennial hydrophyte. Rhizome absent or short, filamentous, developing late in the year and not overwintering.

Stem: Up to 1(–1.5) metres long, quite delicate, terete or slightly compressed, bright green, often very branched toward the apex. Without or rarely with pairs of weakly developed nodal glands.

Floating leaves: Absent.

Submerged leaves: Lamina 1.5–8.0 (–13.0) cm long, 0.3–1.0(–1.8) mm wide, linear, rigid, leaf margins not bordered by sclerenchymatous strands, acute, with narrow wedge-shaped base, fresh green, sometimes with a little brownish colour. With one lateral vein and normally without a band of pale tissue (lacunae) on each side of the strongly elevated midrib. The leaves are sessile.

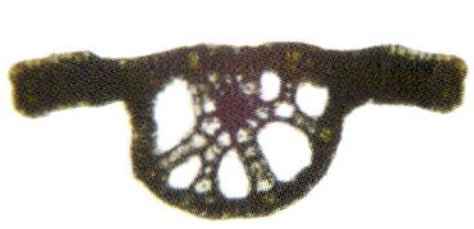

Leaf cross section. Photo KVDW.

Stipules: Persistent, 0.5–3.0 cm long, obtuse at the apex - but often appearing to be acute, because of the tightly inrolled edges, translucent, more or less green, open at the base. With 4–7 indistinct veins between the two primary veins.

Inflorescence: Short cylindrical, 0.5–1.0 cm long, rather loose, with a 1–7 cm long peduncle of the same diameter as the stem, terete or slightly compressed, with 3–5 flowers. Normally with several inflorescences on each shoot.

Flowers: With 1(–2) carpels.

Fruits: 2.5–3.2 mm long, greenish brown, with more or less distinct bumps on the dorsal side and a single tooth on the ventral side. The beak is 0.3–0.5 mm long.

Flowering: June-September.

Biology: Overwinters as dark green to brownish-green turions formed in the summer from vegetative shoots. The turions are 9–20 mm long and 1.5–1.8 mm thick, fusiform, and in the lower part with 2–3 free, erect to erecto-patent leaves.

Habitats: In more or less nutrient-rich ponds, lakes, gravel pits and ditches in marshes.

Distribution within the region: Native. Scarce in Scotland, Sweden, Denmark and the Baltic states, widespread and locally abundant from England east to Poland.

Characteristics and similar species: *Potamogeton trichoides* can be identified by the narrow leaves with acute apex and distinct, strong, elevated midrib, which is prominent on the abaxial side of the leaf. Tightly inrolled stipules, open at the base. The fruits are 2.5–3.2 mm long and typically have small bumps on the dorsal side.

P. trichoides is easily overlooked and

The very delicate habit of *P. trichoides* is not easy to show in photos. Gammel Frederikskog, Hoejer, Jutland, Denmark. Photo JCS.

Potamogeton trichoides.
The stipule is open and obtuse, but appears to be acute, because of the tightly inrolled edges.

The leaf tip is acute or acuminating and the midrib is very prominent, especially at the base.

The fruits have more or less distinct bumps on the dorsal side (right), and a single tooth on the ventral side. Photos JCS.

Turion at the apex of a lateral branch. Photo JCS.

can be mistaken for very delicate forms of **83 *P. pusillus***, which has tubular stipules connate at the base (look carefully) and fruits without bumps on the dorsal side.

81 *Potamogeton berchtoldii* has leaves with lacunae on each side of the midrib, which is not prominent, and well-developed nodal glands.

Hybrids: ***P. ×franconicus*** Fisch. (*P. berchtoldii x trichoides*), ***P. ×grovesii*** Dandy & G.Taylor (*P. pusillus x trichoides*), ***P. ×ripensis*** Baagöe (*P. compressus x trichoides*).

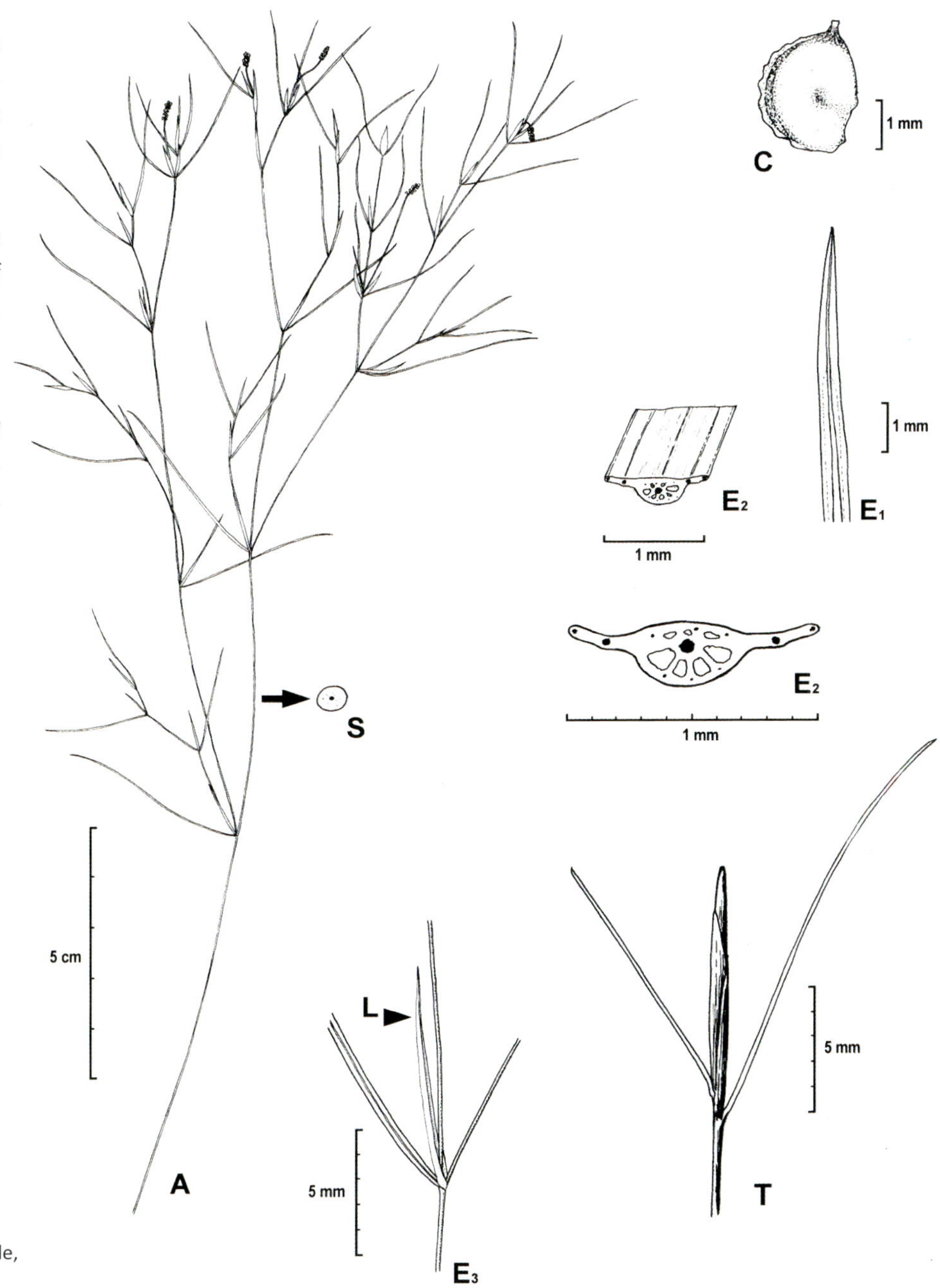

Potamogeton trichoides
A habit, **C** fruit, **E1** leaf tip, **E2** cross section from middle part of a leaf, **E3** base of leaf, with **L** stipule, **S** stem cross section, **T** turion.

Potamogetonaceae

85 *Potamogeton rutilus* Wolfg.

DK: Rødlig Vandaks **N:** Stivtjørnaks **S:** Styvnate **FIN:** - **IS:** -
GB: Shetland Pondweed **NL:** - **F:** -
D: Rötliches Laichkraut **CZ:** - **PL:** Rdestnica błyszcząca
EST: Punakas penikeel **LV:** Iesārtā glīvene **LT:** Rausvoji plūdė **Ru:** Рдест красноватый

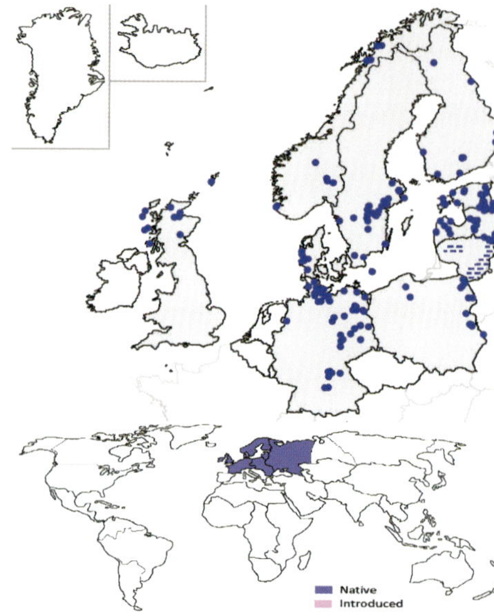

Potamogeton rutilus is a rare species, recognised by the rigid leaves gradually tapering to a very sharply pointed apex. Byn at Nees, Jutland, Denmark. Photo BM.

Perennial hydrophyte. Rhizome absent or short, filamentous, developing late in the year and not overwintering.

Stem: Up to 60 cm long, a little compressed, quite delicate, with few or no short branches. With or without weakly developed nodal glands.

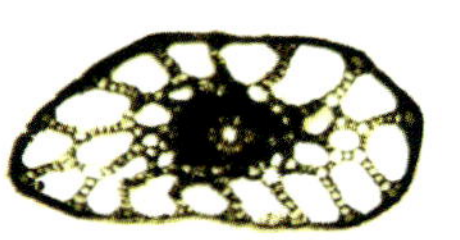

Stem cross section. Photo KvdW.

Floating leaves: Absent.

Submerged leaves: Lamina 3.0–7.5 cm long, 0.5–1.1 mm wide, linear, rather rigid, leaf margins bordered by sclerenchymatous strands, gradually tapering to a very sharply pointed apex, with narrow wedge-shaped base, bright green or more often tinged reddish brown. With one distinct lateral vein, but without a band of pale tissue (lacunae) on each side of the prominent midrib. The leaves are sessile.

Stipules: Persistent, translucent, 1.5–2.0 cm long, obtuse, with many veins, tubular in the basal 2–4 mm, but soon splitting. Whitish with prominent veins when dry.

Inflorescence: Cylindrical, 0.4–1.0 cm long with 5–6 flowers. Peduncle 1–3 cm long, of the same diameter as the stem and somewhat compressed. Normally with several inflorescences on each shoot.

Flowers: With (2–)4 carpels.

Fruits: 2.0–2.1 mm long, olive green to somewhat reddish brown, without a keel. The beak is 0.3–0.4 mm long.

Flowering: July-August.

Biology: Overwinters as turions developing in August-September from the leaf axils or at the apex of old vegetative shoots. The turions are very long, 3.0–7.5 cm, with somewhat recurved leaves at the base, the upper more erect.

Habitats: In shallow, clear water in nutrient-poor to weakly nutrient-rich lakes.

Distribution within the region:
Native. Scattered from Scotland east through Germany and southern Scandinavia to the Baltic states and Poland. Absent from the rest of the region. Rare and declining in many countries.

Characteristics and similar species:
Potamogeton rutilus has long, rigid leaves gradually tapering to a very sharply pointed apex. The stipule is tubular at the base, but soon splits, with many veins (especially prominent when dry and whitish).

It can be mistaken for **84 *P. trichoides***, which has stipules that are open at the base, with indistinct veins, flowers with only 1–2 carpels, and fruits often with bumps on the dorsal side.

Potamogeton rutilus often has a reddish-violet tinge, which also can be seen in other *Potamogeton* species. Byn at Nees, Denmark. Photo JCS.

Potamogeton rutilus has remarkably long turions. Photo JCS.

Potamogeton rutilus.
The fruit is c. 2 mm long and without a distinct keel. Photo JCS.

Potamogeton rutilus. The stipule turns whitish when dry and the numerous veins become very distinct. Photo JCS.

It differs from **83 *P. pusillus*** by the very sharply pointed leaf apex and the prominent veins on the stipule.

Hybrids: *P. ×maëmetsiae* Zalewska-Gałosz & M.Ronikier (*P. friesii × rutilus*).

Potamogeton rutilus. The leaves are very sharply pointed. Photo JCS.

Potamogeton rutilus
A habit, **A1** habit with turions in late autumn, **C** fruit, **E1** leaf tip, **E2** leaf cross section, **E3** leaf base, with **L** stipule, **S** stem cross section, **T** turion.

Potamogetonaceae

Potamogeton hybrids

Potamogeton hybrids have been recognised since the beginning of the twentieth century based on their sterility and unusual morphology. However, both these characters may be strongly influenced by environmental factors. Sterility may be caused by unfavourable conditions, while morphological characters may be considerably altered by phenotypic plasticity. Another challenge was the identification of the parent species. *Potamogeton* hybrids are not necessarily intermediate between their parent species. They may be similar to one of the parent species, or they may show morphological overlap with both parent species, in particular when the parent species share the predominant growth form, or they form specific morphotypes which cannot be derived from any assumed parent species. Four methods have been used to distinguish hybrid taxa:

1. Morphology. In broad-leaved species morphological characters alone are often decisive. Characters can be compared in tabular form, leading to a decision about hybridity in many cases. However, some species are too similar, e.g., *P. perfoliatus* vs. *P. praelongus*, to unambiguously define a hybrid in morphological terms alone. In other cases, particularly in rivers, atypical morphotypes may occur which resemble neither parent species, e.g., *P. natans* × *P. gramineus*. In addition, some hybrids are as likely to be confused with other species as with their parents, e.g., *P. natans* × *lucens*, which resembles *P. nodosus*.

2. Stem anatomy. Anatomical characters of the stem are helpful when the assumed parent species belong to different stem anatomy types. In broad-leaved species there is a great diversity in stellar types, in the shape of the endodermis as well as in size and arrangement of interlacunar and subepidermal bundles. The approach does not work for narrow-leaved species, all of which belong to the same strongly reduced anatomical type.

3. Cytology. Chromosome numbers are regularly distributed among species, with 2n = 26, 28 and 52 being normal. Numbers such as 27 (from 26 x 28) and 39 (from 26 x 52) indicate origin from reduced gametes of the parent species, while numbers such as 78 (from 52 x 26) or 65 (52 x 13) indicate the origin from unreduced gametes of one parent species.

4. Genetics and genomics. Several genetic markers have been used in combination with multivariate statistics to identify hybrids. Use of single markers will usually fail to identify the hybrid nature of a population in *Potamogeton* for a variety of reasons. A combination of at least three markers with different genetic functions and properties (e.g., chloroplast, ribosomal, nuclear) should be used. Maybe in the future next-generation sequencing will become the method of choice.

On the following pages we present a selection of hybrids from the area, which it should be possible to identify by using a stereo microscope to study the stem anatomy and other morphological characters.
The selected hybrids have, with one exception, broad-leaved species as at least one of the parents. Hybrids known only from few localities are omitted.

The narrow-leaved *Potamogeton* species offers only few morphological and anatomical characters useful for identification, and these are to some extent obscured by a wide range of phenotypic plasticity within the species. Recognition of hybrids between the narrow-leaved species is in general very difficult and identification problematic even with the use of DNA technology.

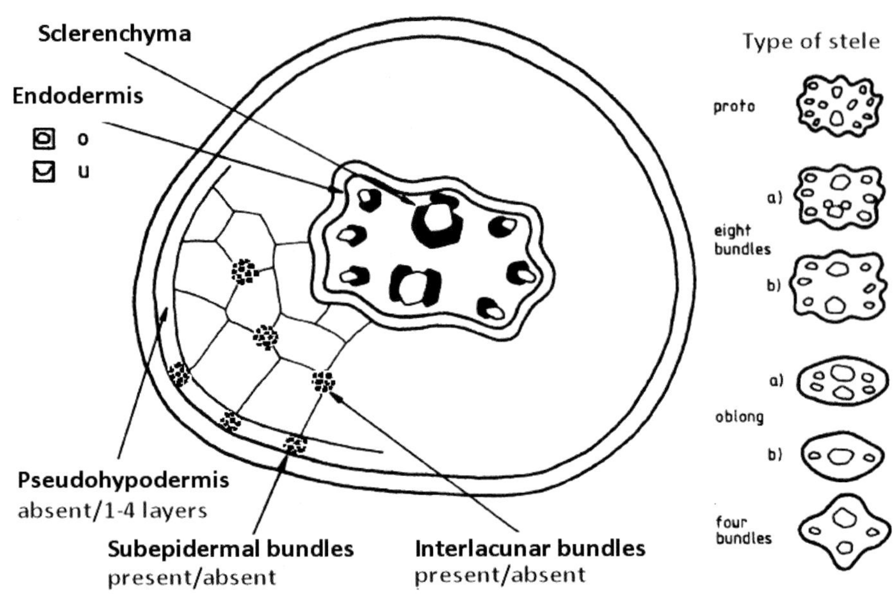

***Potamogeton* hybrids found in the study area:**
Hybrids marked with a number have full descriptions in the text, including their distribution within the region.

P. ×sudermannicus Hagstr. (*P. acutifolius × berchtoldii*): Britain, Netherlands, Germany, Sweden.

97 *P. ×bambergensis* G.Fisch.: (*P. acutifolius × compressus*)

P. ×pseudofriesii Dandy & G.Taylor (*P. acutifolius × friesii*): Britain, Germany.

P. acutifolius × trichoides: Czech Republic.

91 *P. ×olivaceus* Baagøe ex G.Fisch.(*P. alpinus × crispus*):

P. ×nericius Hagstr. (*P. alpinus × gramineus)*: Iceland, Germany, Poland, Sweden, Norway.

P. ×exilis Z.Kaplan & Uotila (*P. alpinus × natans*): A very rare hybrid hitherto only known from Finland.

P. ×prussicus Hagstr. (*P. alpinus × perfoliatus*): Ireland, Britain, Czech Republic, Lithuania, Sweden, Norway.

P. ×spathulatus Schrad. ex W.D.J.Koch & Ziz (*P. alpinus × polygonifolius*): Germany, Sweden.

P. ×griffithii A.Benn (*P. alpinus × praelongus*): Ireland, Britain.

P. ×lanceolatus Sm. (*P. berchtoldii × coloratus*): Ireland, Britain.

P. berchtoldii × compressus: Denmark, Sweden.

P. ×pusilliformis Hagstr. (*P. berchtoldii × friesii*): Denmark, Sweden.

P. ×variifolius Thore (*P. berchtoldii × natans*): Ireland, Germany.

P. x saxonicus Hagstr. (*P. berchtoldii × obtusifolius*): Germany, Sweden.

P. ×dualis Hagstr. (*P. berchtoldii × pusillus*): Denmark ?, Sweden, Estonia.

P. ×franconicus G.Fisch. (*P. berchtoldii × trichoides*): Germany, Denmark, Sweden.

P. ×billupsii Fryer (*P. coloratus × gramineus*): Britain, Sweden.

P. ×bifrons Hagstr. ap Holmberg (*P. compressus × obtusifolius*): Denmark.

P. ×ripensis Baagøe ex G.Fisch. (*P. compressus × trichoides*): Denmark.

94 *P. ×lintonii* Fryer (*P. crispus × friesii*):

P. ×cadburyae Dandy & G.Taylor (*P. crispus × lucens*): Britain.

92 *P. ×cooperi* Fryer (*P. crispus × perfoliatus*):

93 *P. ×undulatus* Wollfg. (*P. crispus × praelongus*):

P. ×bennetii Fryer (*P. crispus × trichoides*): Britain.

P. ×confinis Hagstr. (*P. friesii × pusillus*): Belgium, Denmark, Sweden.

P. ×maëmetsiae Zalewska-Gałosz & M.Ronikier (*P. friesii × rutilus*): Estonia, Lithuania.

89 *P. ×angustifolius* J.Presl (*P. gramineus × lucens*):

P. ×torsanderi (Tis.) G.Fisch. (*P. gramineus × lucens x perfoliatus*): Sweden, Estonia.

88 *P. ×sparganiifolius* Laest. ex Fr. (*P. gramineus × natans*):

P. ×lanceolatifolius (Tis.) C.D.Preston (*P. gramineus × nodosus*): Sweden.

90 *P. ×nitens* Weber (*P. gramineus × perfoliatus*):

87 *P. ×fluitans* Roth (*P. lucens × natans*):

P. ×subrufus Hagstr. (*P. lucens × nodosus*): Poland, Denmark.

95 *P. ×salicifolius* Wolfg. (*P. lucens × perfoliatus*):

P. ×jutlandicus Zalewska-Gałosz (*P. lucens × praelongus*): Denmark, Estonia.

86 *P. ×schreberi* G.Fisch. (*P. natans × nodosus*):

P. ×gessnacensis G.Fisch. (*P. natans × polygonifolius*): Britain, Germany, Poland.

P. ×vepsicus A.A.Bobrov & Chemeris (*P. natans × praelongus*): Ireland, Poland, Denmark.

P. ×assidens Z.Kaplan, Zalewska-Gałosz & M.Ronikier (*P. nodosus × perfoliatus*): Poland, Lithuania.

96 *P. ×cognatus* Asch. & Graebn. (*P. perfoliatus × praelongus*)

P. ×grovesii Dandy & G.Taylor (*P. pusillus × trichoides*): Britain, Denmark.

Potamogetonaceae

86 *Potamogeton ×schreberi* G.Fisch
(P. natans × nodosus)

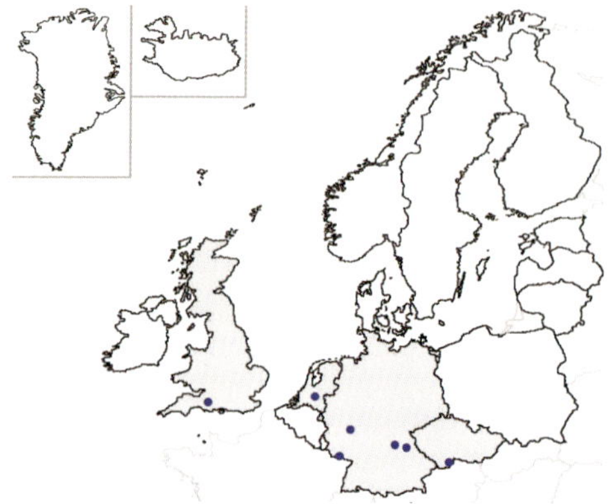

Potamogeton ×schreberi in a canal at Neunergraben, Austria. Photo KvdW.

Perennial hydrophyte. Rhizome robust to very robust.

Stem: Up to 2.0 metres long, slender to robust, terete to slightly compressed. Unbranched.

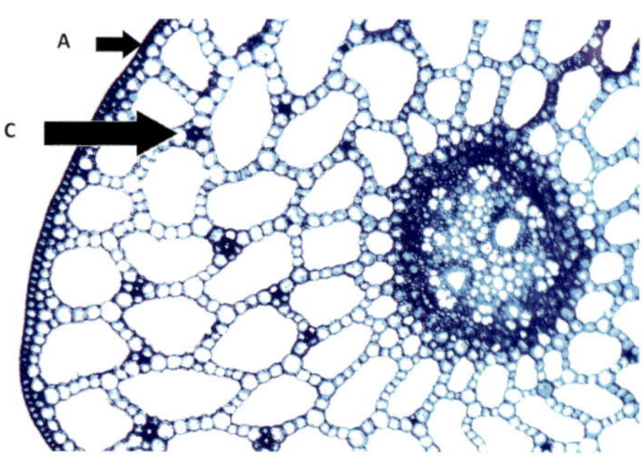

Stem cross section with 1 layer of pseudohypodermis (A), with or without subepidermal bundles, with 1–2(–3) rows of interlacunar bundles (C). Photo KvdW.

Floating leaves: Lamina 7–14 cm long, 1.5–4.5 cm wide, elliptical to oblong elliptical, acute or somewhat obtuse, gradually tapering to the base, opaque, coriaceous, brownish green when young, mature leaves dark olive green. With 5–10 pale lateral veins and inconspicuous secondary veins ascending near the midrib on each side of the midrib. Petiole 9–18(–33) cm, usually without a flexible, discoloured section between the petiole and the lamina - if present faint and inconspicuous. Often with transitional leaves.

Submerged leaves: Lamina 3–18(–30) cm long, 0.2–1.5 cm wide, linear to elliptical, gradually tapering into the petiole, translucent, pale brown when young, green when mature. Margin entire. With a band of pale tissue (lacunae) on each side the midrib, 1–4 (–5) veins, secondary veins more or less transverse or ascending. Petioles (6–) 10–42 cm long. The lowest leaves often represented by linear, semi-terete, phyllodes, flat or shallowly concave on the upper side and convex on the lower side.

Stipules: Persistent, 6–14 cm long, stiff, obtuse to rounded at the apex, but often rolled, green, translucent, with many veins of which two are more prominent than the others.

Inflorescence: Cylindrical, 2.2–2.8 cm long, with peduncles 4.5–8.0 cm long, robust, slightly compressed and not tapered.

Flowers: Contiguous, with 4(–5) carpels.

Fruits: Not developed.

Flowering: June-August.

Biology: Sterile hybrid. Propagation by rhizome fragments.

Potamogeton ×schreberi. Den Bosch, Netherlands. Herbarium John Bruinsma. Photo KvdW.

The submerged leaves of *P. ×schreberi* have linear to narrowly lanceolate lamina and long petioles. Canal at Neunergraben, Austria. Photo KvdW.

Potamogeton ×schreberi. Submerged leaves with 1–4 veins and a band of lacunae on each side of the midrib. Den Bosch, Netherlands. Photo KvdW.

Habitats: In calcareous and moderately eutrophic rivers, slow-flowing streams, ponds and ditches with still water.

Distribution within the region:
Native. From the Mediterranean region extending to Northern Europe and Estonia. Rare in the region and known only from England, the Netherlands, Germany and the Czech Republic.

Characteristics and similar species:
The basal leaves of *P. ×schreberi* which are reduced to phyllodes, combined with fully developed laminar submerged leaves, distinguish it from both parents. In ***66 P. natans*** all submerged leaves are reduced to phyllodes while **69 *P. nodosus*** never has phyllodes and the submerged leaves are slightly denticulate.

The floating leaves of *P. ×schreberi* lack the flexible, discoloured section between the petiole and the lamina normally found in *P. natans.*

P. ×schreberi differs from **87 *P. ×fluitans*** by having stipules with ridges which are not winged on the abaxial side and from **88 *P. ×sparganiifolius*** by the longer petiole of the submerged leaves.

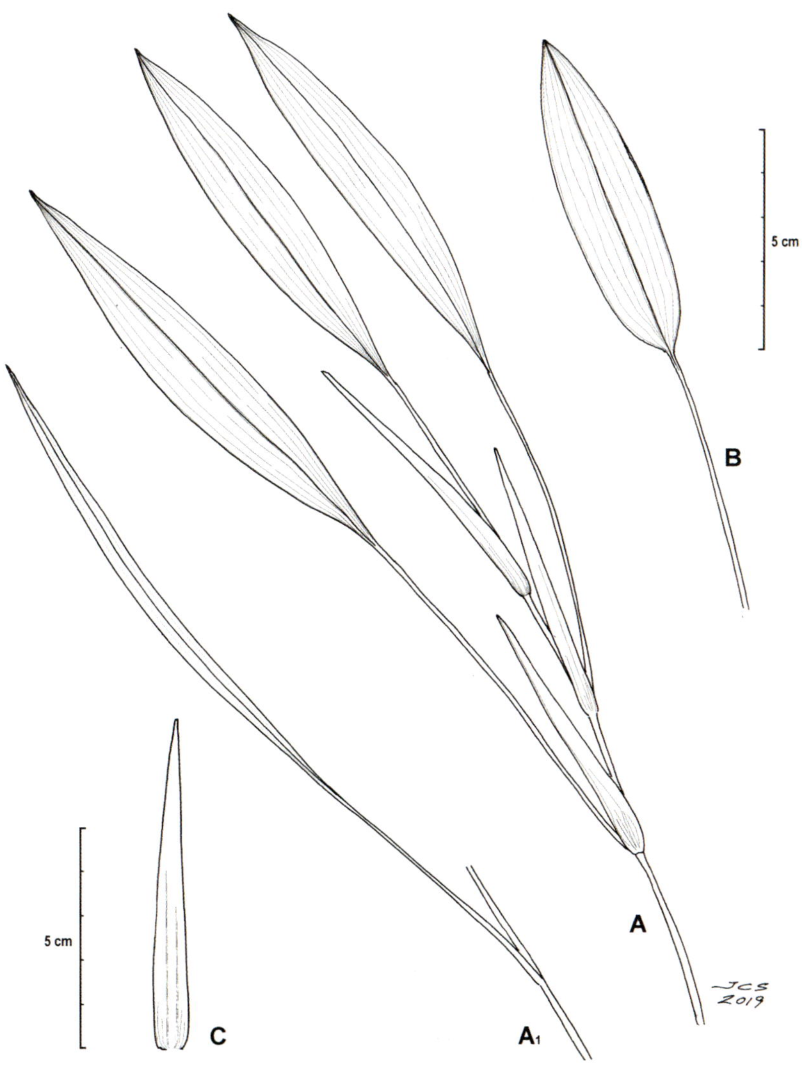

Potamogeton ×schreberi
A Upper part of stem with floating leaves, **A1** submerged leaf, **B** floating leaf, **C** stipule.

Potamogetonaceae

87 *Potamogeton ×fluitans* Roth

(P. lucens × natans)

Syn.: *P. ×sterilis* Hagstr.

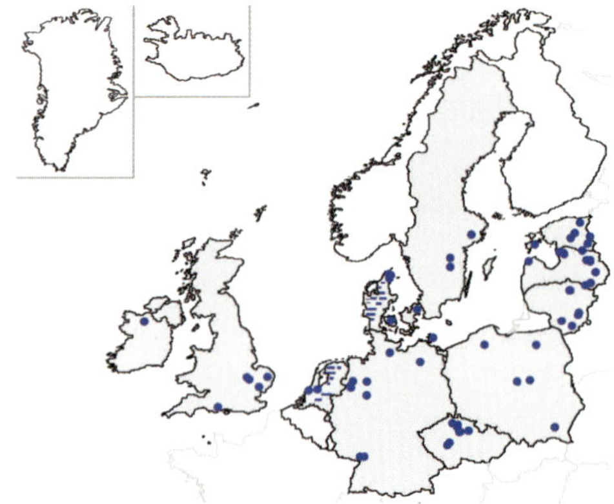

Perennial hydrophyte. Rhizome robust to very robust, terete.

Stem: Up to 2.5 metres long, robust to very robust. Unbranched or sparingly branched.

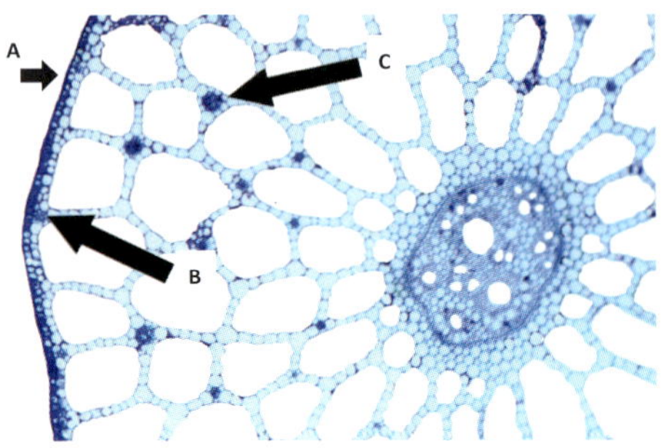

Stem cross section with 1 layer of pseudohypodermis (A), subepidermal bundles (B) and 2–3 rows of interlacunar bundles (C). Photo KvdW.

Floating leaves: Usually present. Lamina 7–15 cm long, 1.5–5.5 cm wide, oblong to narrowly elliptical or ovate, acute to obtuse and often mucronate, with rounded to narrowly wedge-shaped leaf base, more or less coriaceous, not or only slightly translucent, yellowish- to dark green, often with a brownish tinge when young, leaf margins often slightly inrolled. With 5–11 lateral veins and numerous more or less distinct secondary veins on each side of the midrib. Petiole 2.5–10 cm long, without the discoloured joint at the junction with the petiole.

Submerged leaves: Lamina 6–22 cm long, 0.5–3.2 cm wide, oblong ovoid to lanceolate, often with slightly wavy margins, entire or with very delicate serrated margins, apex acute or the margins concave for 3–4 mm below the apex, sometimes with a short excurrent midrib, the peduncle gradually tapering to the base, translucent, green to olive green. Petiole 2.5–10 cm long. With 3–7 lateral veins and numerous slanting secondary veins, but without a band of pale tissue (lacunae) on each side of the midrib. The upper leaves are typically represented by transitional forms between submerged and floating leaves.

Stipules: Persistent, 3–10 cm long, stiff, obtuse to rounded at the apex, dark to brownish green, translucent, with many veins of which two are more prominent than the others and form more or less conspicuously winged ribs on at least the lower part of the abaxial side of the stipule.

Inflorescence: Cylindrical, 2–6 cm long, with a robust to very robust peduncle 4–10 cm long, which is broader than the stem.

Flowers: With 4(–5) carpels.

Fruits: Normally not developed.

Flowering: July-August.

Biology: Normally a sterile hybrid but plants with few fruits have been observed in the Moors River, a shallow calcareous stream in Dorset and Hampshire, southern England (Preston 1995). Propagation is by rhizome fragments and slender branches with narrow leaves developed in the leaf axils and producing roots.

Habitats: In neutral to weakly alkaline, nutrient-poor streams and rivers.

Potamogeton ×fluitans often forms large and uniform populations. River Varde Å, SW-Jutland, Denmark. Photo BM.

Distribution within the region:
Native. Scattered to rather rare throughout most of the region. Absent from Belgium, Luxembourg, Norway, Finland, Iceland and Greenland.

Characteristics and similar species:
The hybrid is rather variable with forms that may be close to *P. natans* and *P. lucens*.

66 *P. natans* has the submerged leaves represented by phyllodes which are semi-terete and lack a lamina, and floating leaves often with a 1–2 cm long flexible discoloured section between the petiole and the lamina.

73 *P. lucens* never develops floating leaves, and the petioles on the submerged leaves are short.

88 *P. ×sparganiifolius* and ***P. ×vepsicus*** A.A.Bobrov & Chemeris (*P. natans* × *praelongus*) have more slender submerged leaves and stipules without winged ribs.

71 *P. alpinus* has sessile or very shortly pedicellate submerged leaves.

67 *P. polygonifolius* differs in habit, with shorter stipules and floating leaves usually without inrolled margins.

Potamogeton ×fluitans. Especially the floating leaves have rolled margins. River Tungelroy, Netherlands. Photo KvdW.

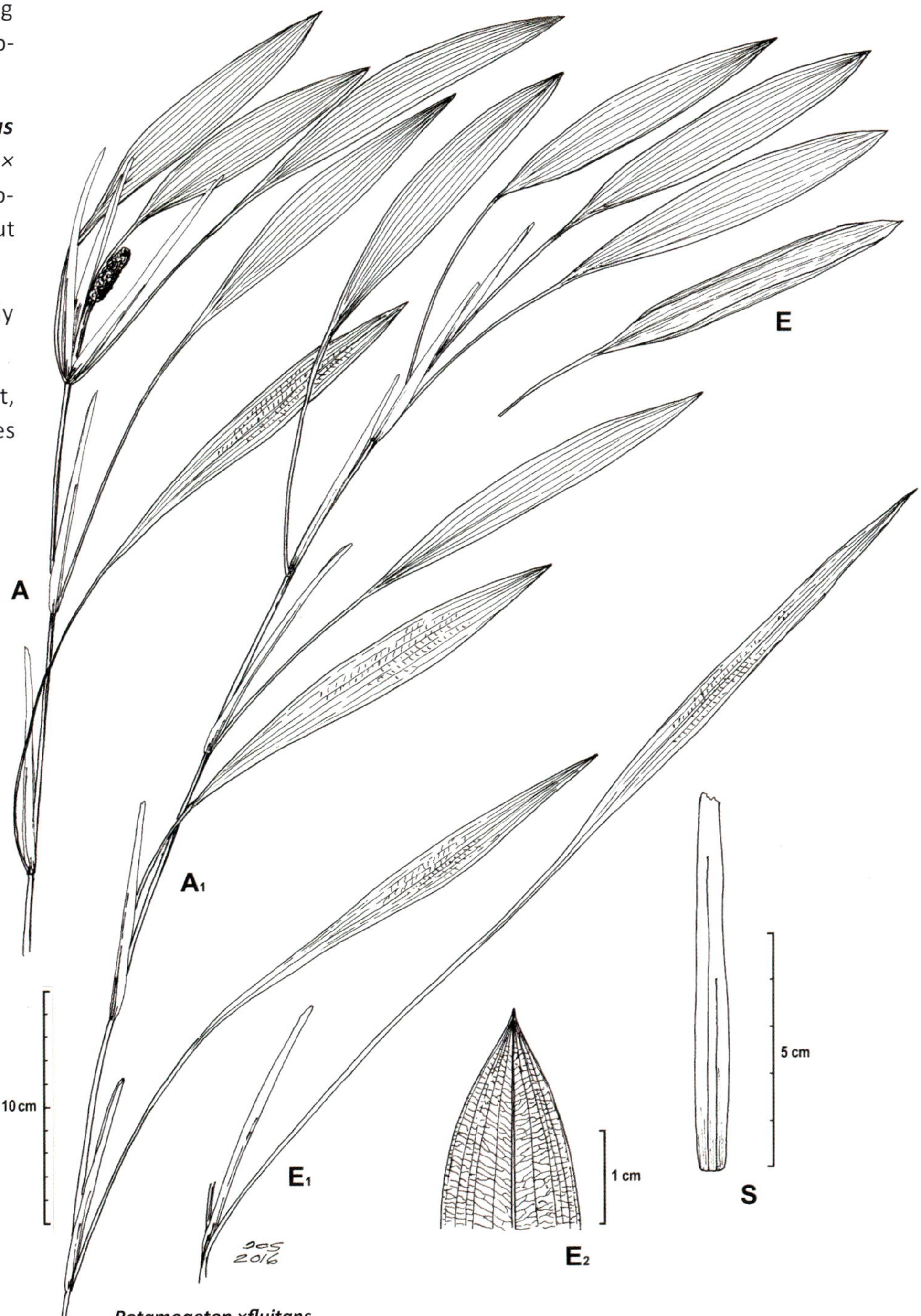

Potamogeton ×fluitans
A, **A1** part of branch, **E** floating leaf with rolled margins seen from below, **E1** submerged leaf from middle part of the stem, **E2** apex submerged leaf, **S** stipule - note the two prominent ribs.

Potamogeton ×fluitans Stipule with winged ribs. Photo KvdW.

Potamogetonaceae

88 *Potamogeton ×sparganiifolius* Laest. ex Fr.

(P. gramineus × natans)

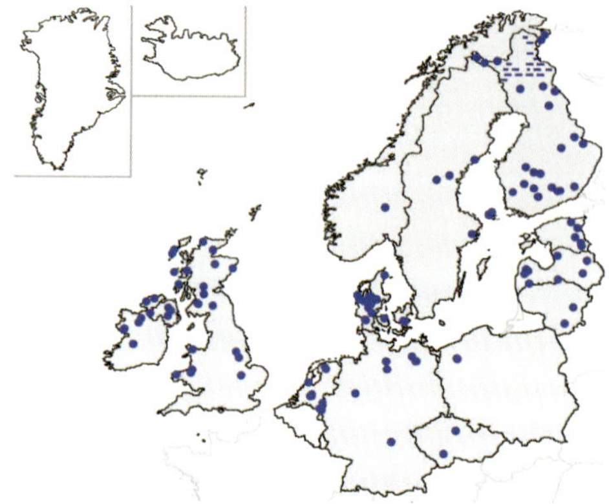

Potamogeton ×sparganiifolius. Vlijmen, Netherlands. Photo KvdW.

Perennial hydrophyte. Rhizome slender to robust, terete.

Stem: Up to 1.5 metres long, slender to robust. More or less branched.

Floating leaves: Lamina 4.0–12 cm long, 1.0–3.5 cm wide, narrowly lanceolate to oblong elliptical, apex obtuse to acute or mucronate, base rounded to wedge shaped, tapering - rarely subcordate, coriaceous, yellowish- to dark green, often with a brownish tinge. With (4–)6–12 lateral veins and numerous less distinct secondary veins on each side of the midrib. Petiole 3–25 cm long, sometimes with a short, faint, reddish to pinkish section at the connection to the lamina.

Submerged leaves: Lamina 10–50 cm long, 0.2–0.9(–1.2) cm wide, flat, thin, semi-translucent or opaque, pale green to dark green, ribbon-like or very narrowly oblanceolate, widest towards the apex. Margin entire or with sparse, scattered teeth. With a band of pale tissue (lacunae), 1–3(–6) veins - some more prominent than others and weaker transverse veins on each side of the midrib. Apex bluntly acuminate or acute. Gradually tapering to sessile base or into a petiole up to 5.5 cm long.

Lowest leaves often represented by linear, semi-terete, phyllodes, flat or shallowly concave on the upper side and convex on the lower side.

Stipules: Persistent, 1.8–9.5 cm long, translucent and hyaline when fresh, more or less opaque when dry, obtuse, 2 veins more prominent than the others and sometimes forming ridges along the back of the stipule.

Inflorescence: Cylindrical, 2.5–4.5 cm long, with a 3–12 cm long peduncle, terete, sometimes widening towards the spike.

Flowers: Numerous, with 4 carpels. Pollen sterile.

Fruits: Not developed.

Flowering: July-August.

Biology: Turions absent. Summer green, overwintering with stolons.

Habitats: In neutral to weakly alkaline, moderately nutrient-rich streams, canals, rivers and lakes.

Distribution within the region:
Native. Scattered to rare throughout most of the region. Absent from Belgium, Luxembourg, Iceland and Greenland.

Characteristics and similar species:
The hybrid is highly variable. In still water it can develop many floating leaves, while the submerged leaves are short and narrowly oblanceolate. In flowing water there may be many floating leaves, but the submerged leaves are usually long and ribbon-like.

It can be mistaken for **66 *P. natans***, which has submerged leaves consisting of linear, semi-terete phyllodes without lamina.

87 *P. ×fluitans* has floating leaves with a shorter petiole, wider submerged leaves and 2 prominent veins forming ridges on the stipules.

67 *P. polygonifolius* and **71 *P. alpinus*** both have submerged leaves with a broader, lanceolate lamina.

P. ×vepsicus A.A.Bobrov & Chemeris (*natans × praelongus*) cannot be separated reliably from *P. ×sparganiifolius* using morphological characters, but cultivation experiments have revealed a hooded apex to the floating leaves, inherited from *P. praelongus*. It has so far been recorded from Ireland, Denmark, Poland, Austria and North European Russia.

Potamogeton ×vepsicus. The floating leaves are slightly hooded at the apex, but this is unfortunately not very clear in the photo. Photo KvdW.

Potamogeton ×sparganiifolius. River Laxford in northwest Scotland. Photo RVL.

Potamogeton ×sparganiifolius. Ribbon-like submerged leaf - slightly wider in the upper part of the leaf. Canal at Lake Ferring, W-Jutland. Denmark. Photo JCS.

Potamogeton ×sparganiifolius
A habit, **A1** rhizome with young shoots.

Potamogetonaceae

89 *Potamogeton ×angustifolius* J.Presl

(P. gramineus × lucens)

Syn.: *P. ×zizii* Mert. & Koch.

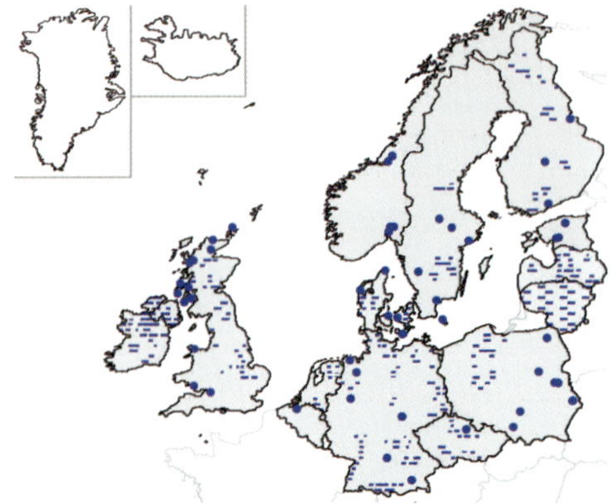

Potamogeton ×angustifolius. In 4 m deep water of a main drain in the fens near Spalding, Lincolnshire, England. Photo RVL.

Perennial hydrophyte. Rhizome robust to very robust, terete.

Stem: Up to 2.0 metres long, terete, slender to quite robust, more or less branched.

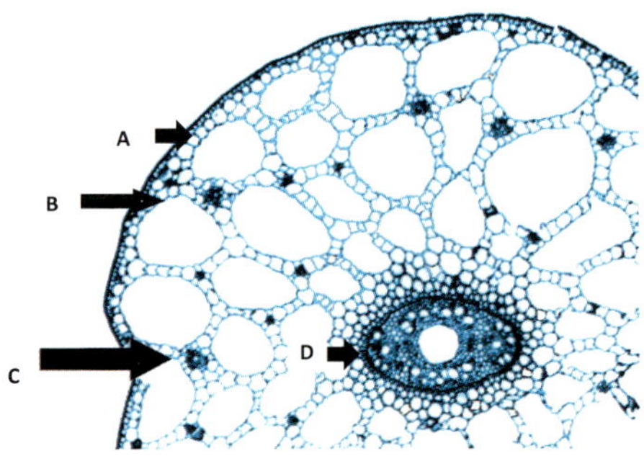

Stem cross section with 1 layer of pseudohypodermis (A), subepidermal bundles (B), 1–3 rows of interlacunar bundles (C) and oblong stele (D). Photo KvdW.

Floating leaves: Rarely present. Lamina 5.5–10 cm long, 2.3–4.0 cm wide, elliptical to broad elliptical, slightly obtuse to mucronate, tapering to a more or less rounded or slightly cordate base, coriaceous, not translucent, light green. With 6–9 lateral veins and numerous conspicuous secondary veins on each side of the midrib. Petiole 2.0–6.5 cm long.

Submerged leaves: Stem leaves and main branches with lamina 4.5–15 cm long, 0.6–2.9 cm wide, markedly larger than the leaves on the short branches, lanceolate to oblong elliptical, often recurved, slightly undulating, with delicate serrated margin, acute or a little obtuse and often mucronate, gradually tapering to the base, translucent, shining, green or brownish-green, sometimes with a reddish tinge.

With a narrow band of pale tissue (lacunae), 4–6(–7) lateral veins and numerous quite prominent secondary veins on each side of the midrib. The lowest leaves are often reduced to phyllodes, consisting in extreme cases only of the petiole. The majority of the submerged leaves are sessile, but in the upper part of the plant there are usually long-petiolate leaves.

Stipules: Persistent, 2–5 cm long on the stem - shorter on the branches, translucent, greenish to reddish-brown, obtuse, 2 veins very distinct and more prominent than the others, sometimes appearing winged.

Inflorescence: Cylindrical, 2–5 cm long, with a 4–15 cm long, terete peduncle, which becomes wider towards the apex, and often of the same diameter or even wider than the stem.

Flowers: With (3–)4 carpels.

Fruits: 2.7–3.4 mm long, with a 0.2–0.3 mm long beak.

Flowering: July-August.

Biology: Unlike other *Potamogeton* hybrids *P. ×angustifolius* is able to produce fertile fruits. It does not develop turions, but overwinters as short, much widened internodes on the stolon.

Habitats: In neutral to weakly alkaline, nutrient-poor to nutrient-rich lakes and streams.

Distribution within the region:
Native. Widespread and locally frequent throughout most of the region. Absent from Iceland and Greenland.

Flowering *P. ×angustifolius*. Note the partly sessile and partly petiolated submerged leaves, and stipules with 2 prominent keeled/winged veins. Søvigsund, Denmark. Photo JCS.

Characteristics and similar species:
This hybrid is rather variable and can be intermediary between the two parent species or occur in forms similar to one or the other of the parents. The quite robust stem leaves resemble a slender **73 *P. lucens***, but the branch leaves are smaller and more like the leaves of **70 *P. gramineus***.

It differs from *P. gramineus* by larger submerged leaves, the upper often with long petioles. Stipules larger and with 2 major veins of a winged appearance.

All leaves of *P. lucens* have 0.2–0.7 cm long petioles and floating leaves never develop.

Potamogeton ×angustifolius. Submerged leaves with rather long petioles. Oxbow lake Lingenfeld, Germany. Foto KvdW.

Potamogeton ×angustifolius
A, A2 habit - plant with submerged leaves only, **A1** upper part of plant with floating leaves, **B** rhizome with young shoot in spring.

Potamogetonaceae

90 *Potamogeton* ×*nitens* Weber

(P. gramineus × perfoliatus)

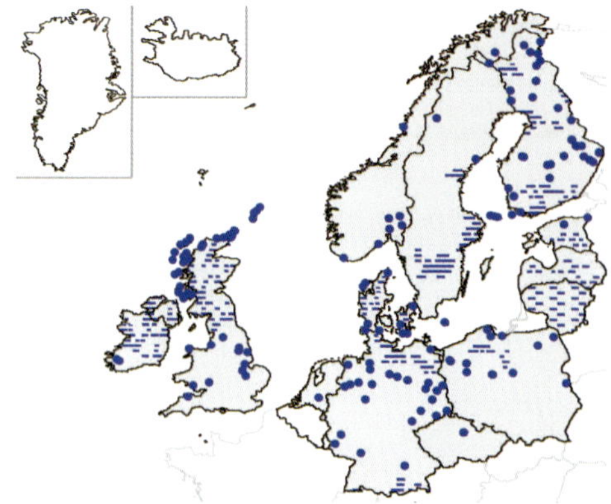

Potamogeton ×nitens. Lough Bane, Ireland. Photo KvdW.

Perennial hydrophyte. Rhizome slender to robust, terete.

Stem: Up to 2.5 metres long, slender to somewhat robust, terete, more or less branched.

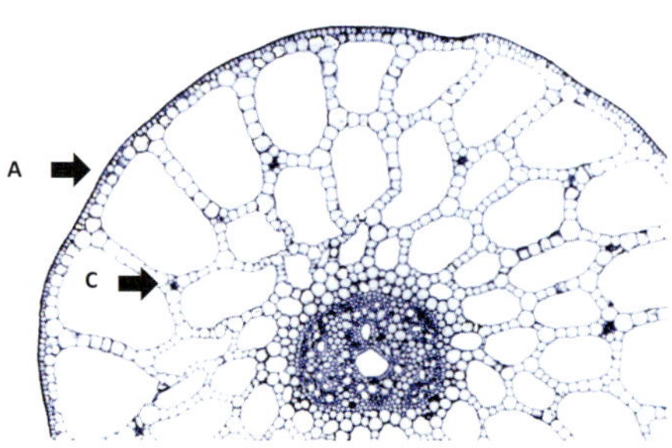

Stem cross section with 1 layer of pseudohypodermis (A), 1 row of interlacunar bundles (C) but without subepidermal bundles. Photo KvdW.

Floating leaves: The hybrid is very variable, and plants close to *P. gramineus* often develop floating leaves. These with lamina 3.5–6.5 cm long and 1.0–2.5 cm wide, elliptical to broadly elliptical, acute, tapering toward base, not or only slightly translucent, coriaceous, green to brownish green. With 5–7 lateral veins and numerous secondary veins on each side of the midrib. Petioles 1.2–4.0 cm long.

Submerged leaves: Lamina 4–12 cm long, 0.7–2.3 cm wide, lanceolate to broadly elliptical, often recurved, slightly undulate, with delicately serrate margin, acute or rarely obtuse or slightly hooded, broad and amplexicaul at the base, thin, translucent, yellowish green to brown, sometimes with a reddish tinge. With a narrow band of pale tissue (lacunae), 3–8 lateral veins of which 1 or 2 are prominent and with obscure secondary veins on each side of the midrib.

Stipules: Persistent for some time, but often degraded on older leaves, 1–3 cm long, translucent, obtuse, 2 veins often more prominent than the others, but not appearing winged.

Inflorescence: Cylindrical, 0.5–2.5 cm long. Peduncle 1.5–8 cm long, widening upwards and often of the same diameter or wider than the stem, terete.

Flowers: With 4–5 carpels. Pollen sterile.

Fruits: Normally not developed. Rarely with immature and infertile fruits.

Flowering: July-August.

Biology: The species does not form turions, but overwinters as much thickened buds at the apex of the rhizome.

Habitats: In neutral to weakly alkaline, moderately nutrient-rich streams, canals and lakes.

Distribution within the region: Native. Widespread and locally frequent throughout most of the region.

Potamogeton ×nitens as normally seen without floating leaves. Lake Nors Sø, Denmark. Photo JCS.

Potamogeton ×nitens. Viskan, Västra Götaland, Sweden. Photo BM.

Potamogeton ×nitens. Leaves with more veins than *P. gramineus*. Denmark. Photo BM.

Absent from Belgium, Luxembourg, Iceland and Greenland.

Characteristics and similar species: The hybrid is rather variable, intermediate between the parent species or similar to one or the other.

It differs from **74 *P. perfoliatus*** by the acute submerged leaves, the fairly persistent stipules and the capacity to develop floating leaves.

It can be distinguished from **70 *P. gramineus*** by the number of veins in the submerged leaves, these with a rounded and more or less amplexicaul base. The floating leaves are less coriaceous and wedge shaped tapering into the petiole.

72 *P. praelongus* is more robust and with leaves broadly hooded at the apex.

92 *P. ×cooperi* differs in its compressed stem, acute submerged leaves and fugacious stipules.

Potamogeton ×nitens. Leaves rounded and more or less amplexicaul at base. Denmark. Photo JCS.

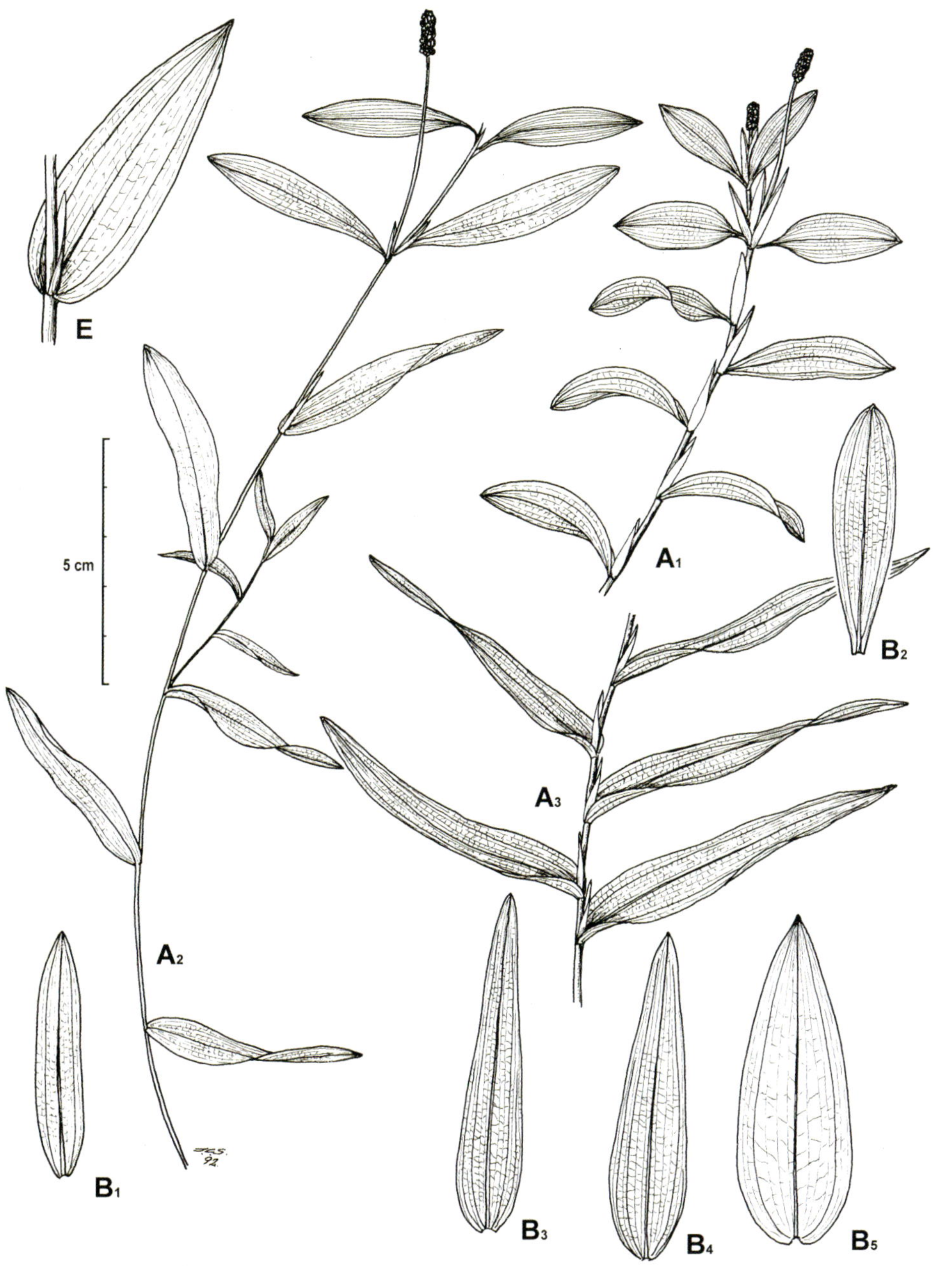

Potamogeton ×nitens
A1-A3 habit, **B1-B5** submerged leaves, **E** stem with leaf and stipules.

Potamogetonaceae

91 ***Potamogeton ×olivaceus*** Baagøe ex G.Fisch.

(P. alpinus × crispus)

Syn.: *P. ×venustus* Baagøe ex A.Benn.

Potamogeton ×olivaceus in River Tweed, Britain. Photo KvdW.

Perennial hydrophyte. Rhizome slender.

Stem: Up to 1 metre long, somewhat compressed, shallowly furrowed on one or both of the broader sides, slender to robust, unbranched to sparsely branched.

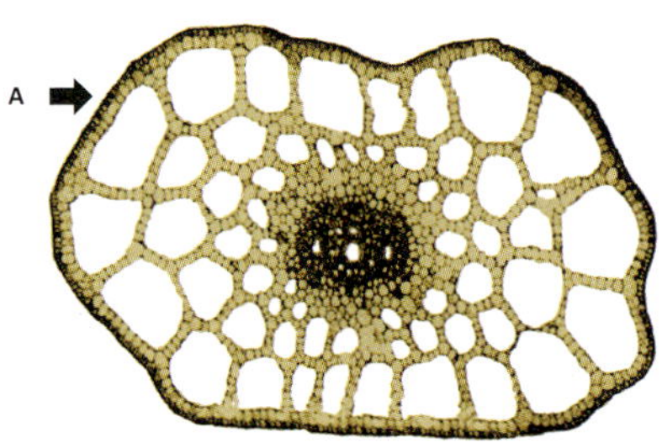

Stem cross section with 0–1 layer of pseudohypodermis without subepidermal or interlacunar bundles. Photo JCS.

Floating leaves: Absent or very rarely present, oblong to elliptical, tapering to a short petiole.

Submerged leaves: Lamina 4.5–14.0 cm long, 0.6–1.5 mm wide, linear to oblong lanceolate, apex subacute to obtuse - not hooded, sessile, gradually tapering to a slightly auriculate base. Leaf margins undulate, more or less entire or with very obscure teeth towards the apex. Translucent, light green to bright green or olive green, often with a reddish tinge, which intensifies when dried. The midrib is reddish tinged and bordered on each side by a band of pale tissue (lacunae) and 2–3 lateral veins of which the outermost is weak and close to the margin, secondary veins transverse or ascending.

Stipules: Persistent though sometimes torn on lower leaves, 1.0–2.5 mm long, translucent, hyaline with a brownish tinge, obtuse to truncate or a little emarginated at the apex, veins inconspicuous.

Inflorescence: 0.5–1 cm long, cylindrical. Peduncle 1.0–6.0(–10.5) cm long, terete to slightly compressed.

Flowers: Contiguous, 10–12, with 3(–4) carpels.

Fruits: Not developed.

Flowering: June-August.

Biology: Probably dispersed within water bodies by rhizome fragments. Turions have never been observed.

Habitats: Moderately eutrophic rivers, canals and streams.

Distribution within the region:
Native. Very rare in Britain, Germany, Poland. Probably extinct in Denmark and the Czech Republic.

Potamogeton ×olivaceus. The midrib and veins of the leaves are rather prominent. River Tweed, Britain. Photo KvdW.

Potamogeton ×olivaceus
Leaf base and stipule.

Leaf tip.

Midrib bordered by lacunae. Photos KvdW.

Characteristics and similar species:
P. ×olivaceus is more or less intermediate between the parents and varies only a little in the serration of the leaf margins and in leaf shape from oblong lanceolate to linear. Floating leaves are usually absent.

P. ×olivaceus can be separated from **71** ***P. alpinus*** by the compressed, furrowed stem and narrower leaves with only 2–3 veins on each side of the midrib. In *P. alpinus* the stem is terete and the leaves have 4–7 lateral veins on each side of the midrib.

It differs from **75** ***P. crispus*** in the more or less entire leaves, at most with very obscure teeth and often with 3 pairs of lateral veins on each side of the midrib.

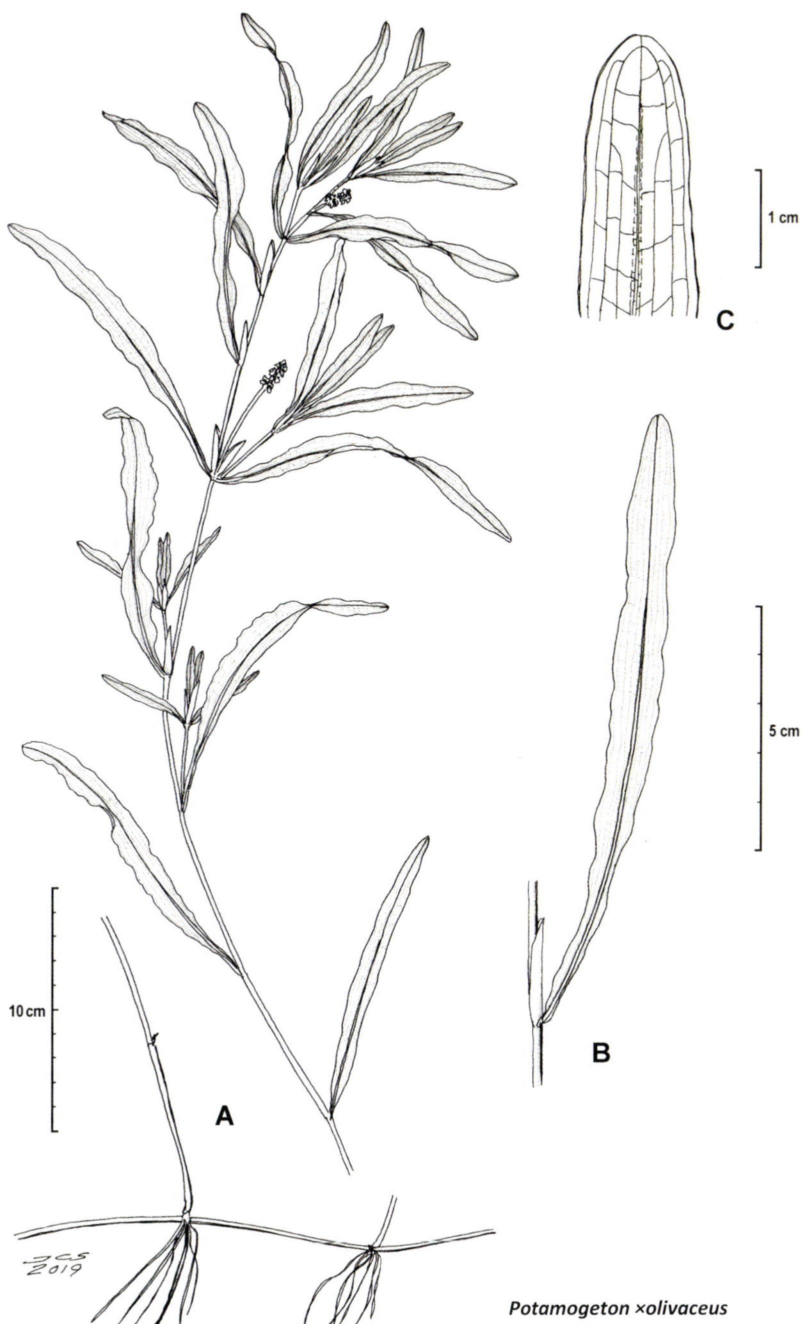

Potamogeton ×olivaceus
A habit, **B** leaf, **C** leaf tip.

Potamogetonaceae

92 ***Potamogeton ×cooperi*** (Fryer) Fryer

(P. crispus × perfoliatus)

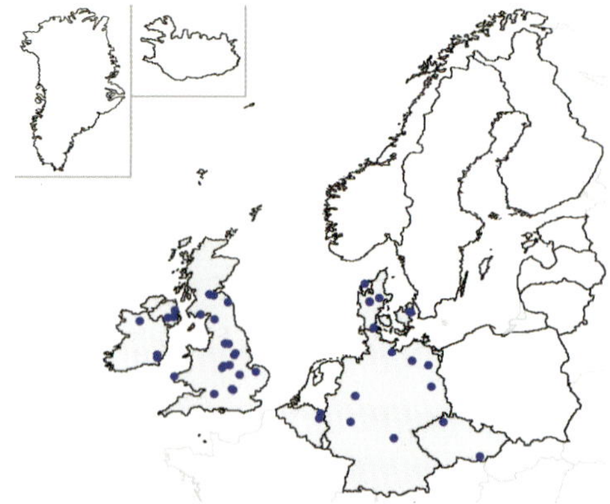

Potamogeton ×cooperi. Solva, Wales. Photo RVL.

Perennial hydrophyte. Rhizome slender to robust.

Stem: Up to 1.5(–4) metres long, rather robust, slightly compressed, shallowly furrowed on both of the broader sides, usually unbranched or with short branches from the leaf axils.

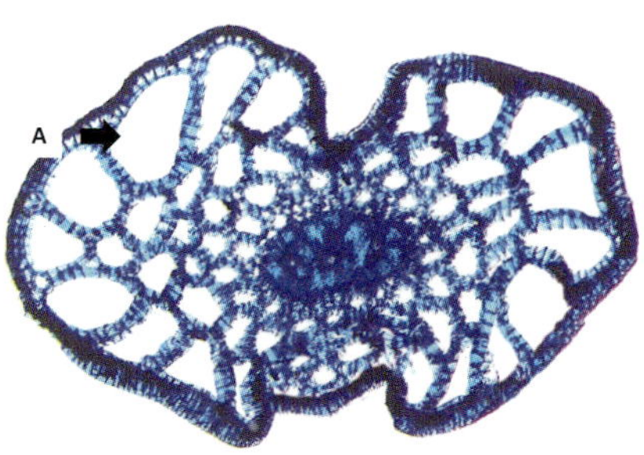

Cross section of stem from dried material. The furrows on the broad sides appear much more prominent than seen on living plants. 1 layer of pseudohypodermis, without subepidermal or interlacunar bundles. Photo KvdW.

Floating leaves: Absent.

Submerged leaves: Lamina 2.0–8.5 cm long, 0.8–2.5 cm wide, length/width ratio (1.9–)2.3–6.6, linear lanceolate to ovate, sessile, semi-amplexicaul at the base, subacute to obtuse, sometimes slightly hooded at the apex, translucent, light green to brownish green or dark green, eventually with a reddish tinge at the midrib. Particularly towards the apex, leaf margins minutely denticulate or irregularly serrulate and sometimes slightly undulate. With a band of pale tissue (lacunae), 3–5(–6) lateral veins and transverse secondary veins on each side of the midrib.

Stipules: Fugacious, 0.4–2.0 cm long, open, clasping the stem, translucent, delicate, truncate or emarginate at the apex, veins inconspicuous.

Inflorescence: Cylindrical, 0.4–1.3 cm long. Peduncle 3–10 cm long, terete, often of the same diameter as the stem, slightly compressed towards the apex.

Flowers: Contiguous, 4–15, with 3(–4) carpels. Remain closed.

Fruits: Not developed.

Flowering: June-August.

Biology: Vegetative propagation is by detached rhizome and stem fragments. Overwinters as leafy shoots and 0.7–5.5 cm long, slender turions produced in leaf axils. The turions resemble those produced by *P. crispus.*

Habitats: Moderately eutrophic rivers, canals, ponds and streams.

Distribution within the region:
Native. Scattered from Ireland east to the Czech Republic.

Characteristics and similar species:
Potamogeton ×cooperi is a variable hybrid, at times resembling one or the other of the parents. Shoots produced early in the year often have linear-

Flowering *P. ×cooperi.* Solva, Wales. Photo RVL.

oblong leaves with subacute apices like those of **75 *P. crispus***, whereas flowering plants more closely resemble **74 *P. perfoliatus*** with broader and more obtuse leaves.

P. ×cooperi differs from *P. perfoliatus* by having a compressed, shallowly furrowed stem, semi-amplexicaul leaf base and leaves with only 3–5(–6) veins on each side of the midrib.

P. crispus has (4–)6–12 mm wide leaves, with broadly cuneate to auriculate leaf base and only 1–2(–3) veins on each side of the midrib. The teeth on the leaf margin of *P. ×cooperi* are almost invisible to the naked eye.

P. ×cooperi is normally easily separated from *P. crispus* by the broader leaves with more lateral veins, and leaf margins with inconspicuous teeth almost invisible to the naked eye, whereas *P. crispus* always has clearly visible, distinct denticulate leaf margins.

Forms of **90 *P. ×nitens*** without floating leaves differ in having a terete stem, persistent stipules and more acute submerged leaves.

The leaf base of *P. ×cooperi* is semi-amplexicaul. Solva, Wales. Photo RVL.

Potamogeton ×cooperi. The leaf tip is often hooded and splitting a little when pressed. Photo KvdW.

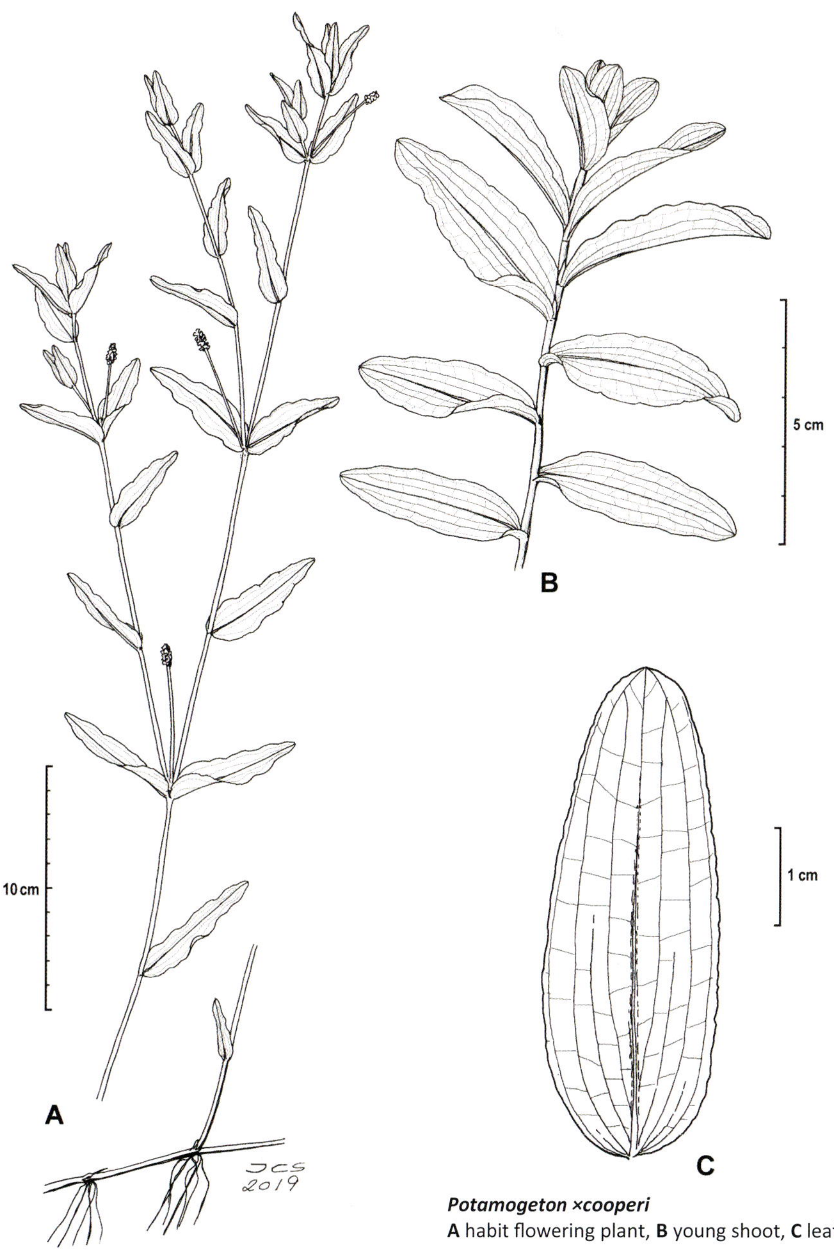

Potamogeton ×cooperi
A habit flowering plant, **B** young shoot, **C** leaf.

Potamogetonaceae

93 *Potamogeton ×undulatus* Wollfg.

(P. crispus × praelongus)

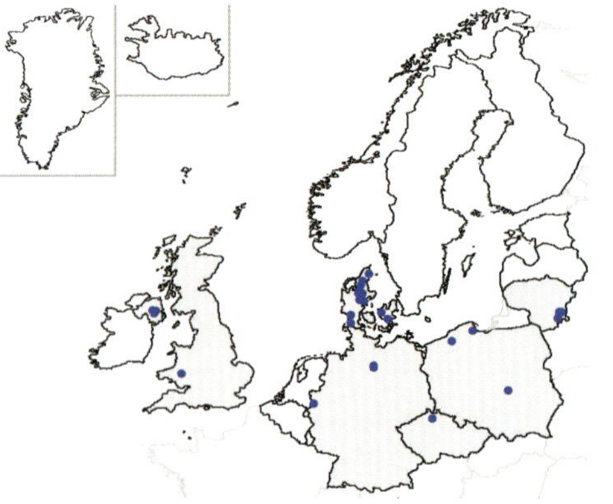

Potamogeton ×undulatus. River Gudenå, Denmark. Photo KvdW.

Perennial hydrophyte. Rhizome robust.

Stem: Up to 1.5 metres long, 1–3 mm diameter, slightly compressed and shallowly furrowed on both sides, sparsely branched at the top. The stem may have a zigzag form.

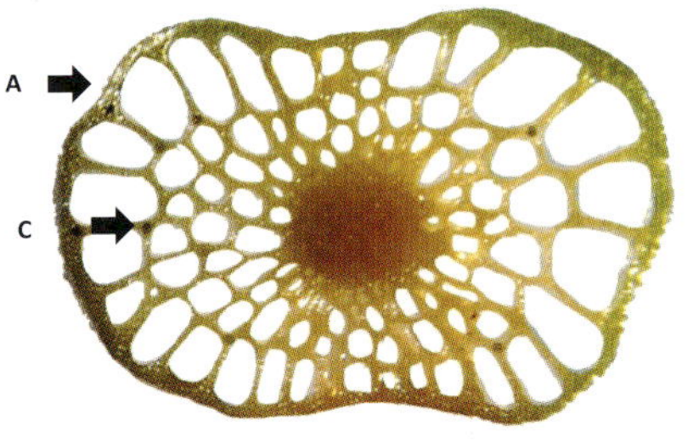

Stem cross section with 1–2 layers of pseudohypodermis (A), with or without subepidermal bundles, with 1–2 rows of interlacunar bundles (C). Note the shallow groove on both sides. Photo JCS.

Floating leaves: Absent.

Submerged leaves: Lamina 5–15 cm long, 0.8–2.3 cm wide, oblong ovoid to oblong lanceolate, with undulate margins, entire or with obscurely denticulate margins, especially towards the apex, obtuse, sometimes slightly hooded, but not splitting when pressed, with rounded to semi-amplexicaul base, translucent, bright green to dark green, often with a reddish tinge at the midrib. With a band of pale tissue (lacunae),
2–4 lateral veins of which one usually is more strongly developed than the others and inconspicuous transverse veins on each side of the midrib.

Stipules: Persistent, but gradually decaying to fibrous remnants, 0.6–2.5 cm long, obtuse to truncate, whitish to coriaceous, translucent when fresh, opaque when dry.

Inflorescence: Rarely developed, cylindrical, 1–2 cm long, with a 2–12 cm long peduncle, which is more slender than the stem.

Flowers: With 4 carpels.

Fruits: Not developed.

Flowering: July-August.

Biology: A sterile hybrid often forming large populations. Overwinters as rhizomes and axillary turions. The turions resemble those produced by *P. crispus*.

Habitats: In moderately nutrient-rich streams. Previously found in lakes.

Distribution within the region:
Native. Scattered but locally common in Denmark, Germany, Poland, Czech Republic and Lithuania, very rare in Britain.

Characteristics and similar species:
This hybrid can be recognised by the slightly compressed stem with shallow

Potamogeton ×undulatus. Occasionally, it is possible to find plants with inflorescence, but the hybrid is like most other hybrids sterile and does not develop fruits. Sønderup Å, Jutland, Denmark. Photo JCS.

Potamogeton ×undulatus. Plant with a reddish tinge on stem and veins. Sønderup Å, Jutland, Denmark. Photo JCS.

Potamogeton ×undulatus. Typical plants with leaves close together in the upper part, undulating and somewhat recurved. River Gudenå, Ulstrup, Denmark. Photo JCS.

furrows on both sides, and the long undulate leaves, which are normally entire or with scattered, short teeth, especially towards the obtuse or sometimes slightly hooded apex.

71 *P. alpinus* and **72 *P. praelongus*** have a terete stem without furrows and entire margins with 4–9 lateral veins on each side of the midrib.

91 *P. ×olivaceus* is difficult to separate from *P. ×undulatus*, but has leaves more gradually tapering to a less amplexicaul base.

92 *P. ×cooperi* has 2.5–6 cm long, not or only slightly undulate leaves with serrate margins, especially towards apex.

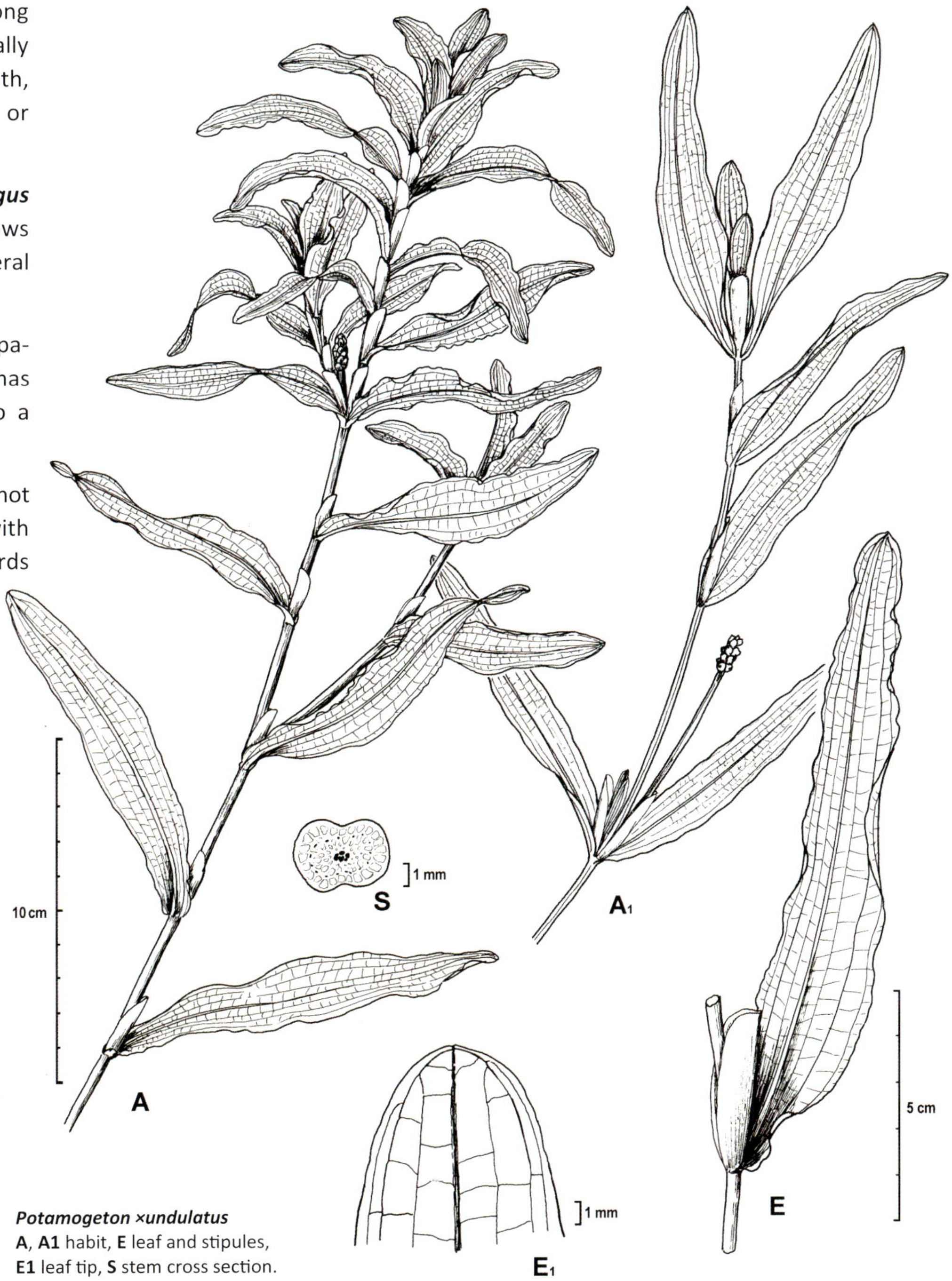

Potamogeton ×undulatus
A, **A1** habit, **E** leaf and stipules, **E1** leaf tip, **S** stem cross section.

Potamogetonaceae

94 *Potamogeton ×lintonii* Fryer

(P. crispus × friesii)

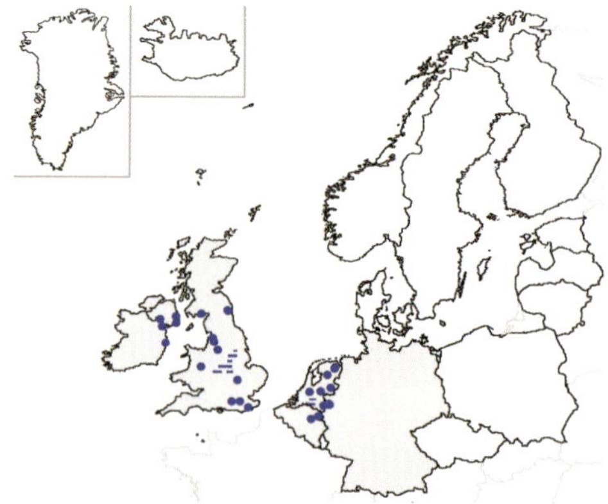

Potamogeton ×lintonii. Pool close to River Niers, Wetten, Germany. Photo KvdW.

Perennial hydrophyte. Rhizome very slender, up to 1.5 mm thick.

Stem: 0.2–1.1 metres long, up to 1.7 mm diameter, compressed, on robust stems shallowly furrowed on one or both of the broader sides. Young stems mostly unbranched, the older stems with few to many branches. Minute nodal glands only present on some nodes.

Floating leaves: Absent.

Submerged leaves: Lamina 2.5–6.0 cm long, 0.2–0.5 cm wide, sessile, linear to narrowly oblong, rounded, obtuse or acute at the apex, tapering at base, plane or slightly undulate and twisted on older stems. Margins usually entire close to the base and denticulate towards the apex, rarely serrate or more or less entire throughout. The leaves are dark green to brownish green with a pinkish midrib. The midrib is bordered on each side by a band of pale tissue (lacunae), which is broad at the base but narrows towards the apex, lateral veins 1–2 on each side, secondary veins few.

Stipules: Persistent, but the truncate apex soon decays to leave fibrous strands, 0.6–1.2 cm long, hyaline, translucent, pale brown, with numerous, distinct veins, 2 of them green and more prominent than the others, forming weak ribs. Young stipules tubular at base.

Inflorescence: Cylindrical, 0.3–1.1 cm long, with a 0.8–4.3 cm long peduncle, slender, not tapered, slightly compressed.

Flowers: With 3(–4) carpels.

Fruits: Not developed.

Flowering: June-August.

Biology: Sterile hybrid. Vegetative propagation by turions produced in the leaf axils of older stems. Turions very variable in size, 1.2–7.5 cm long, slender, with 3–12 short, well-spaced, narrow, erect or slightly recurved leaves.

Habitats: In shallow water in lakes, canals, rivers, streams and flooded marl pits.

Distribution within the region:
Native. Scattered to locally common in Ireland, Britain, the Netherlands, Belgium and Germany.

Characteristics and similar species:
Potamogeton ×lintonii does not vary

Potamogeton ×lintonii. The twisted, dark green to brown leaves with pinkish-tinged midrib is characteristic for mature shoots. Ditch, Kerken, Germany. Photo KvdW.

Leaf tips. From left *P. crispus, P. ×lintonii* and *P. friesii.* The teeth on the margins are only just visible (detail right). Photos KvdW.

much. Young shoots generally have narrower and less undulate leaves than older shoots. Leaf width and serration can vary depending on which of the parents the hybrid resembles the most. *P. ×lintonii* resembles a narrow-leaved **75 *P. crispus***, but with more obscure teeth than is normal for that species. The stipules are closed at the base (look carefully on young specimens) like **80 *P. friesii***, from which it has also inherited the 2 weak, green ribs.

P. ×lintonii plants with inconspicuous teeth on the leaf margins can be mistaken for *P. friesii* or **79 *P. obtusifolius***, especially in the field, but both of these have well-developed nodal glands and flat stems without shallow furrows.

Potamogeton ×lintonii. Pool close to River Niers, Wetten, Germany. Photo KvdW.

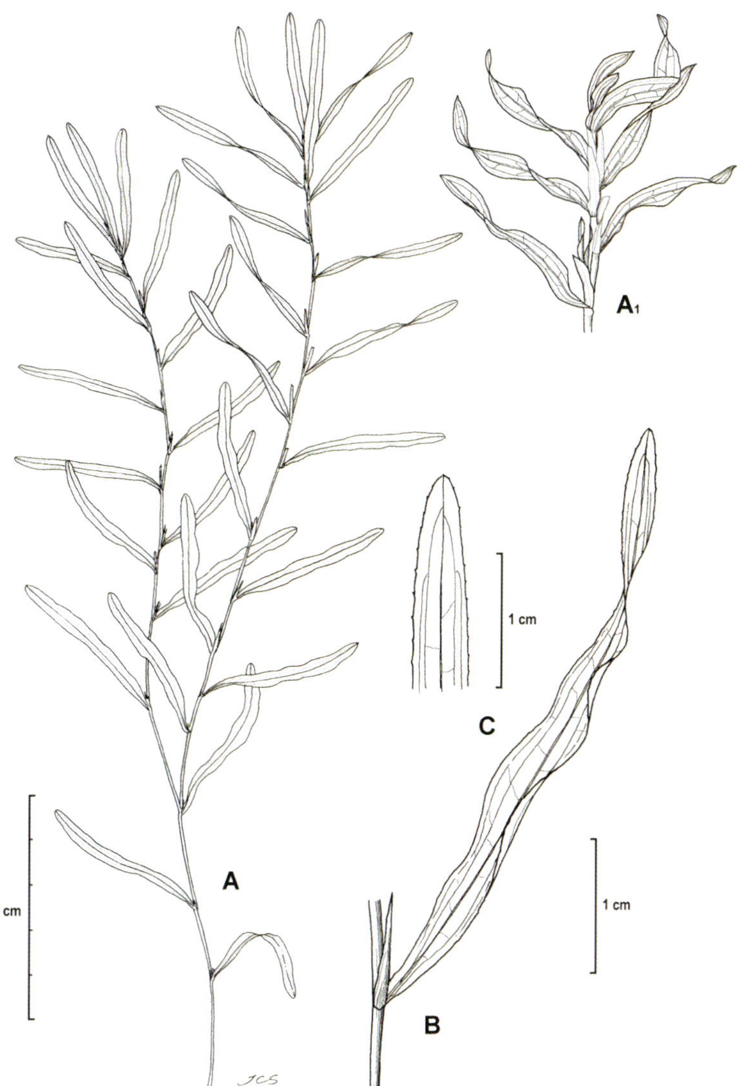

Potamogeton ×lintonii
A habit young plant, **A1** part of mature shoot, **B** leaf with stipule, **C** leaf tip.

Potamogetonaceae

95 ***Potamogeton ×salicifolius*** Wolfg.

(P. lucens × perfoliatus)

Syn.: *P. ×decipiens* Nolte ex Koch

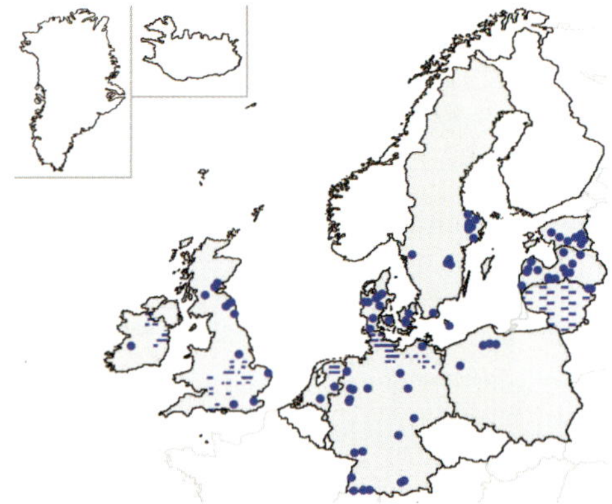

Potamogeton ×salicifolius. Kupa River, Croatia and Slovenia. Photo KvdW.

Perennial hydrophyte. Rhizome robust, terete.

Stem: Up to 3 metres long, robust, terete, often zigzag curved, more or less branched.

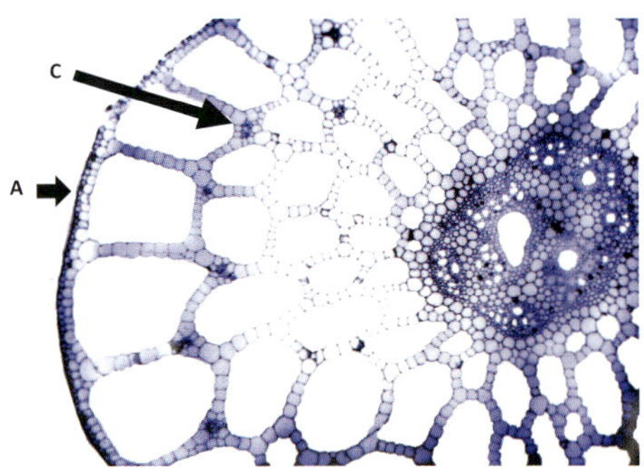

Stem cross section with 1 layer of pseudohypodermis (A), with or without subepidermal bundles, with 1–2 rows of interlacunar bundles (C). Photo KvdW.

Floating leaves: Absent.

Submerged leaves: Lamina 6–12 cm long, 1.5–4.0 cm wide, lanceolate to elliptical, undulate and often a little recurved, leaf margins very delicately denticulate, apex obtuse or rounded and apiculate, leaf base rounded or slightly semi-amplexicaul, translucent, yellowish green to dark green, occasionally with a pinkish tinge at the midrib or especially in young leaves on the whole lamina. With 4–8 lateral veins of which 1–2 are more prominent than the others and numerous, quite robust transverse veins on each side of the midrib. Sometimes with a band of pale tissue (lacunae) close to the midrib. The leaves are sessile.

Stipules: Normally persistent, 2–5 cm long, obtuse, hyaline, sometimes with a pinkish tinge, translucent when fresh. With many veins of which 2 are more prominent than the others and often more or less narrowly winged at least on the lower half on the abaxial side.

Inflorescence: Cylindrical, 1.5–4.0 cm long, with a 3–10 cm long peduncle, which often is broader than the stem and a little compressed.

Flowers: With (3–)4 carpels.

Fruits: Normally not developed.

Flowering: July-August.

Biology: Sterile hybrid, normally overwintering as thickened, starch-filled internodes.

Habitats: In moderately nutrient-poor water. Streams, canals and lakes.

Distribution within the region:
Native. Scattered, but locally frequent in Britain, Ireland, the Netherlands, Germany, Denmark, Sweden, Poland and the Baltic states.

Leaves of *P. lucens* (left) and *P. ×salicifolius* (right). The white arrows point at the reticulated veins forming a web-like structure along the midrib. In *P. lucens* these are oblong rhombic or oblong rectangular, but in *P. ×salicifolius* they are almost equilateral rhombic to square shaped. Photo JCS.

In *P. ×salicifolius* the stipules are persistent for some time, hyaline, not as rigid as in *P. lucens*, and the 2 most prominent veins do not form conspicuously winged keels as in *P. lucens*.

The leaves have very delicate denticulated leaf margins, but sometimes the leaf margins are entire, which is strange because both parent species have denticulated leaf margins.

River Gudenå, Denmark. Photos JCS.

Characteristics and similar species:
This hybrid typically strongly resembles small plants of **73 *P. lucens***, but can be recognised by the sessile leaves, which are rounded at apex and base, and by the equilaterally rhomboid to square-shaped reticulated veins along the midrib. The midrib never extends beyond the tip of the lamina.

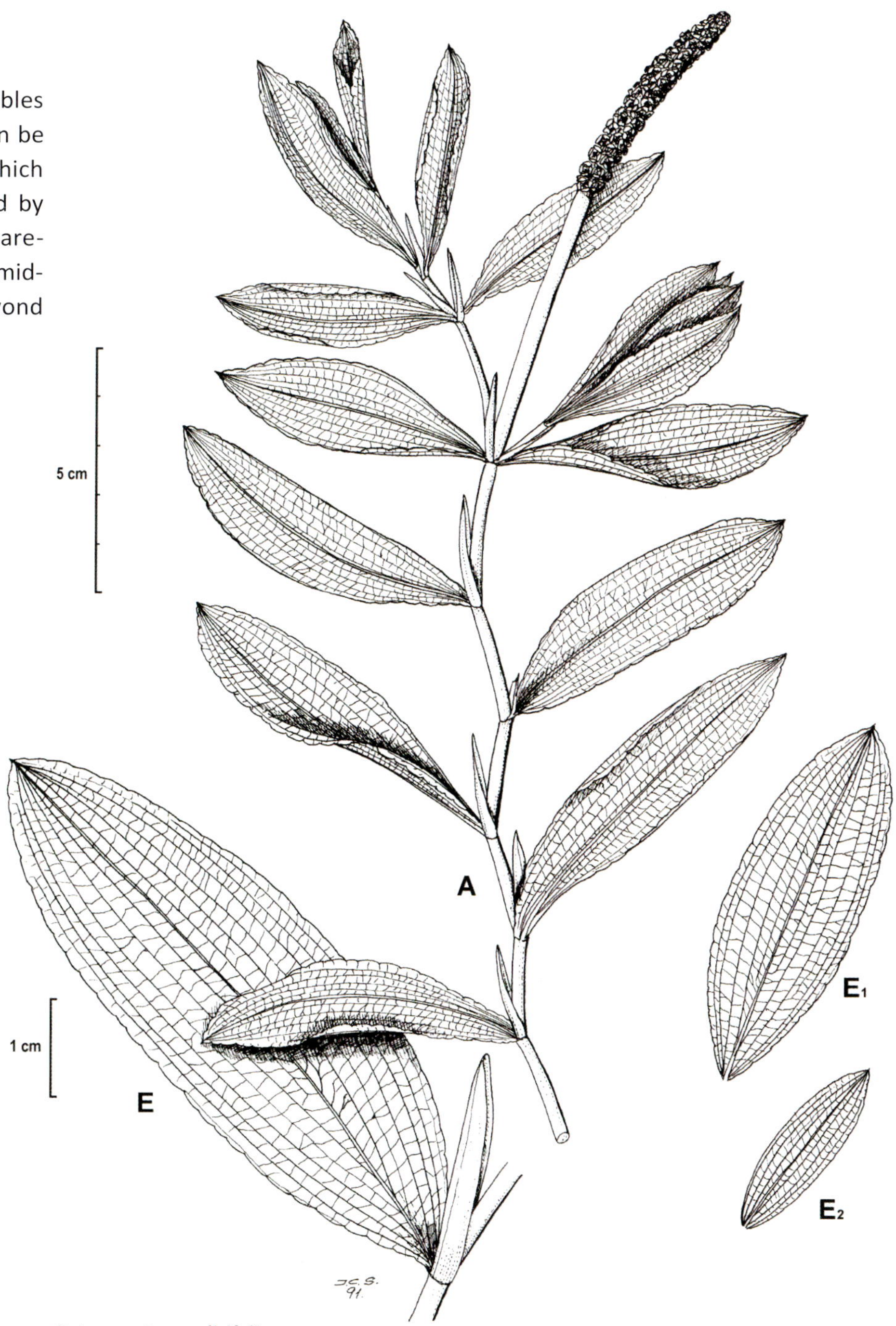

Potamogeton ×salicifolius
A habit, **E** leaf with stipules, **E1-E2** leaves.

Potamogetonaceae

96 ***Potamogeton ×cognatus*** Asch. & Graebn.

(P. perfoliatus × praelongus)

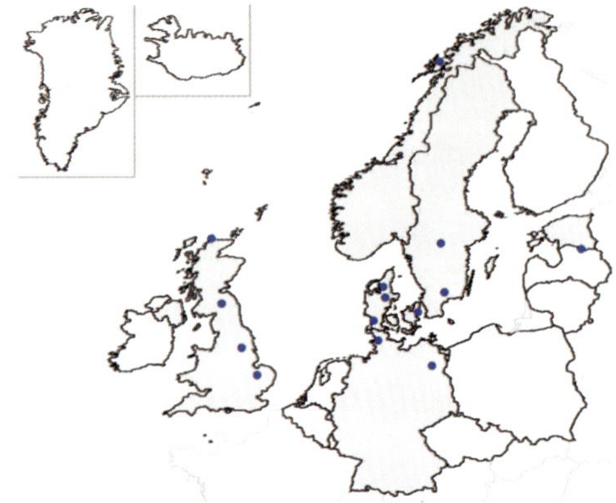

Potamogeton ×cognatus - form with long leaves resembling *P. praelongus*. Ivösjö, Skåne, Sweden. Photo BM.

Perennial hydrophyte. Rhizome robust, terete.

Stem: Up to 1.3 metres long, robust, terete, unbranched or sparingly branched.

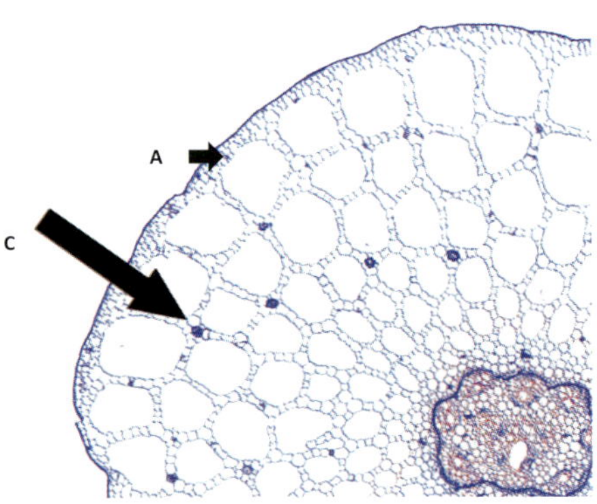

Stem cross section with 1–2 layers of pseudohypodermis (A), 1–2 rows of interlacunar bundles (C), with or without subepidermal bundles. Photo KvdW.

Floating leaves: Absent.

Submerged leaves: Lamina 5–11 cm long, 1.5–4.0 cm wide, translucent, bright green to dark green, ovate lanceolate to oblong lanceolate, margins slightly undulate, entire or sparsely denticulate, obtuse with a hooded apex which splits into a V shape when pressed, leaf base rounded to amplexicaul.

Stipules: Persisting for some time on young leaves but soon decaying without leaving fibrous remains, 1.5–6.0 cm long, rounded, whitish, sometimes with a delicate reddish tinge, with very inconspicuous veins, translucent when fresh, but opaque when dry.

Inflorescence: Cylindrical, 0.8–2.5 cm long. Peduncle 3–20 cm long, terete, normally more slender than the stem.

Flowers: With 4 carpels.

Fruits: Not developed.

Flowering: July-August.

Biology: Sterile hybrid. Overwintering as a rhizome, without shoots or leaves.

Habitats: In clean to moderately nutrient-rich water. Lakes and streams.

Distribution within the region:
Native. Scarce and local in Britain, Denmark, Norway, Sweden, Germany and Estonia.

Characteristics and similar species:
The hybrid is intermediate between the parents, but leaf length and shape can approach either parent species.

It can be distinguished from **72 *P. praelongus*** by the stipules having even more inconspicuous veins and mostly being present only on young leaves, by the stem tending to form less of a zigzag and, at least on some plants, by the leaf margins being delicately denticulate.

Potamogeton ×cognatus from the River Gudenå, Langå, Denmark. Note the dark green leaves with very bright and very conspicuous veins. Photo JCS.

Potamogeton ×cognatus. Plant with broad leaves. The hooded apex splits up when pressed - marked with a white arrow. River Gudenå, Denmark. Photos JCS.

74 ***P. perfoliatus*** is a highly variable species, including forms reminiscent of the hybrid. It has denticulate leaf margins and fugacious stipules only present on the youngest leaves.

90 ***P. ×nitens*** differs by having acute or mucronate submerged leaves with 3–8 veins on each side of the midrib, persistent stipules and the ability to form floating leaves.

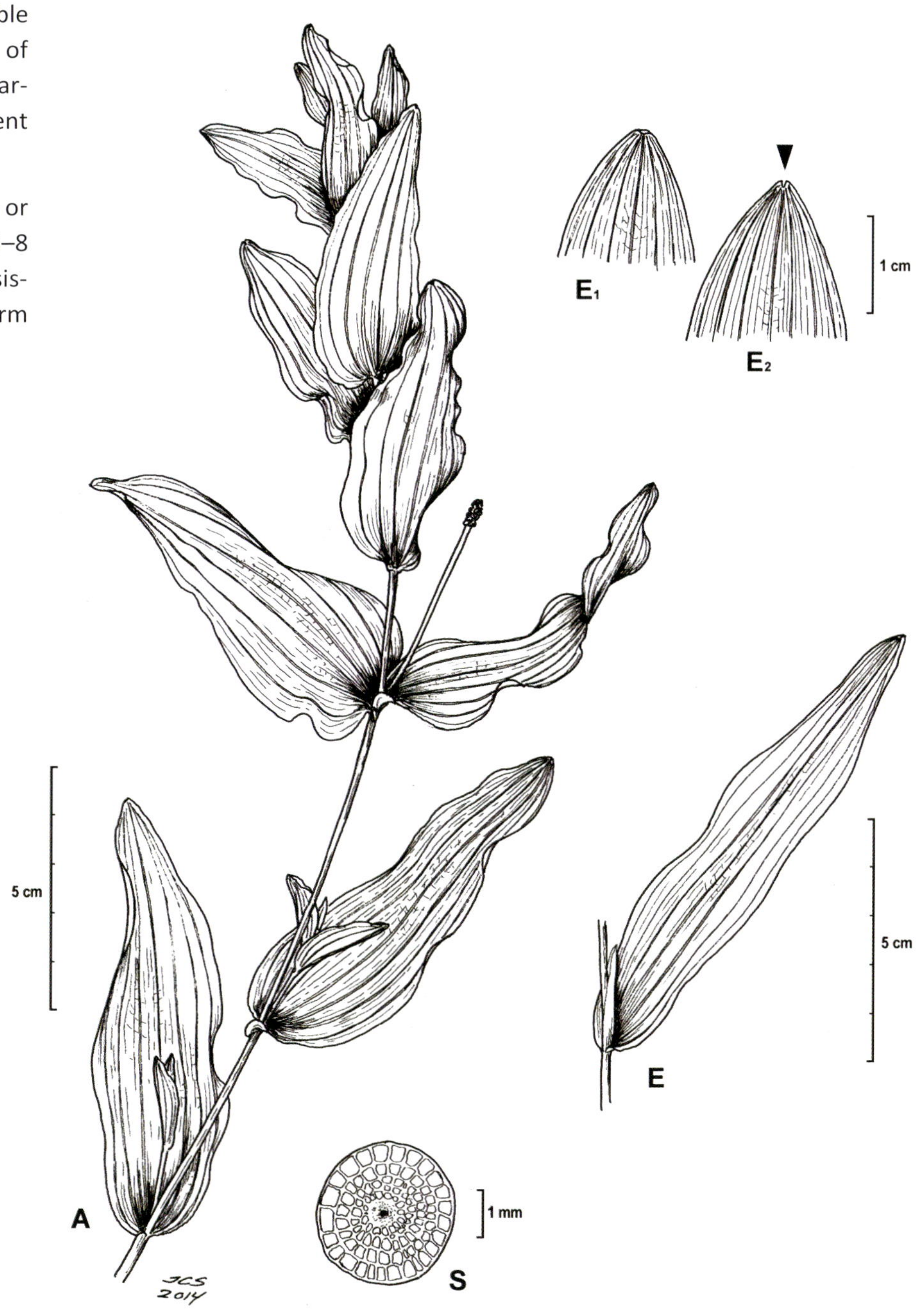

Potamogeton ×cognatus
A habit, **E** leaf with stipules, **E1** leaf apex somewhat hooded, **E2** leaf tip splitting after pressure, **S** cross section of stem.

Potamogetonaceae

97 *Potamogeton ×bambergensis* G.Fisch.

(P. acutifolius × compressus)

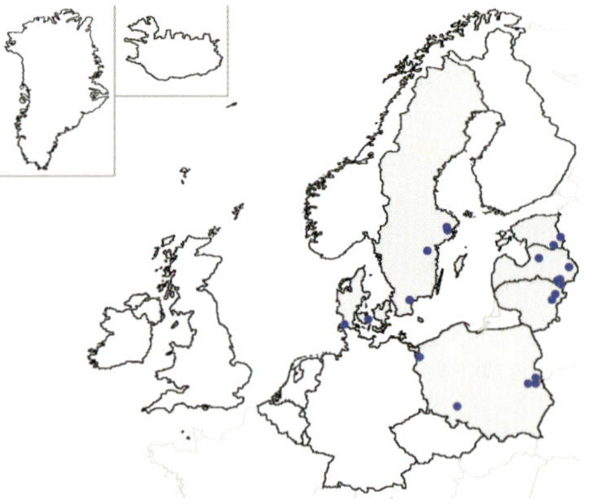

Perennial hydrophyte. The rhizome is absent or short, very delicate and does not overwinter.

Stem: Up to 1 metre long, 0.5–3.0 mm wide, compressed to flat, sometimes slightly winged, slender to fairly robust, more or less branched. Nodal glands absent.

Stem cross section. Ny Frederikskog, SW-Jutland, Denmark. Photo JCS.

Floating leaves: Absent.

Submerged leaves: Lamina (5.4–)6.5–10.0(–14.0) cm long, 1.5–3.5(–4.0) mm wide, linear, entire, leaf margins bordered by a strong marginal vein, apex acuminate to mucronate, olive green to dark green sometimes with a brownish tinge. With a narrow band of pale tissue (lacunae), 1 or 2 lateral veins and several more or less inconspicuous sclerenchymatous strands on each side of the midrib. The leaves are sessile.

Stipules: Persistent but soon decaying to leave filamentous strands, 1.5–3.0 cm long, obtuse at the apex, translucent, with a whitish or weakly brown tinge, open.

Inflorescence: 0.5–2 cm long, cylindrical to slightly compressed, with 7–10 (–15) flowers. Peduncle (1.7–)3.0–4.0 (–7.0) cm long, compressed. Normally with several inflorescences on each shoot.

Flowers: With 1(–2) carpels.

Fruits: Usually not developed. If present then resembling *P. acutifolius* in shape.

Flowering: July-August.

Biology: The hybrid overwinters as turions which develop at the apex of vegetative shoots and lateral branches, or in leaf axils. The turions develop in August and consist of short, compressed, mucronate leaves, below which are 2–4 short erecto-patent leaves.

Habitats: In neutral to alkaline, moderately nutrient-rich to nutrient-rich water. Shallow ponds, gravel pits, lakes and ditches.

Distribution within the region:
Native. Hitherto known from Denmark, Sweden, Poland, Lithuania, Latvia and Estonia, but probably overlooked in other countries.

Characteristics and similar species:
Like its parents the hybrid has leaves with numerous sclerenchymatous strands, flattened or sometimes winged stem and 1(–2) carpels per flower. It is normally sterile or produces only one (seldom a few) ripe fruit.

The hybrid is rather variable and can be intermediate between the parents or more or less similar to one or the other.

Potamogeton ×bambergensis from Højer, Denmark. Confirmed by sequencing of ITS of DNA. Photo JCS.

Potamogeton ×bambergensis is in terms of habit like *P. acutifolius*, but sterile and with up to 4 cm long peduncles. Ny Frederikskog, SW-Jutland, Denmark. Photo JCS.

When trying to identify the hybrid, the following characters might be useful:

P. acutifolius: Stem flat. Spike 0.4–0.8 cm long, somewhat compressed, with 4–7 flowers. Peduncle 0.5–2.0(–2.5) long. Leaves with 3 veins.

P. compressus:
Stem winged. Spike 1.5–3.5 cm long, cylindrical, with 10–20 flowers. Peduncle 2.8–7.0 cm long. Leaves with 5 veins.

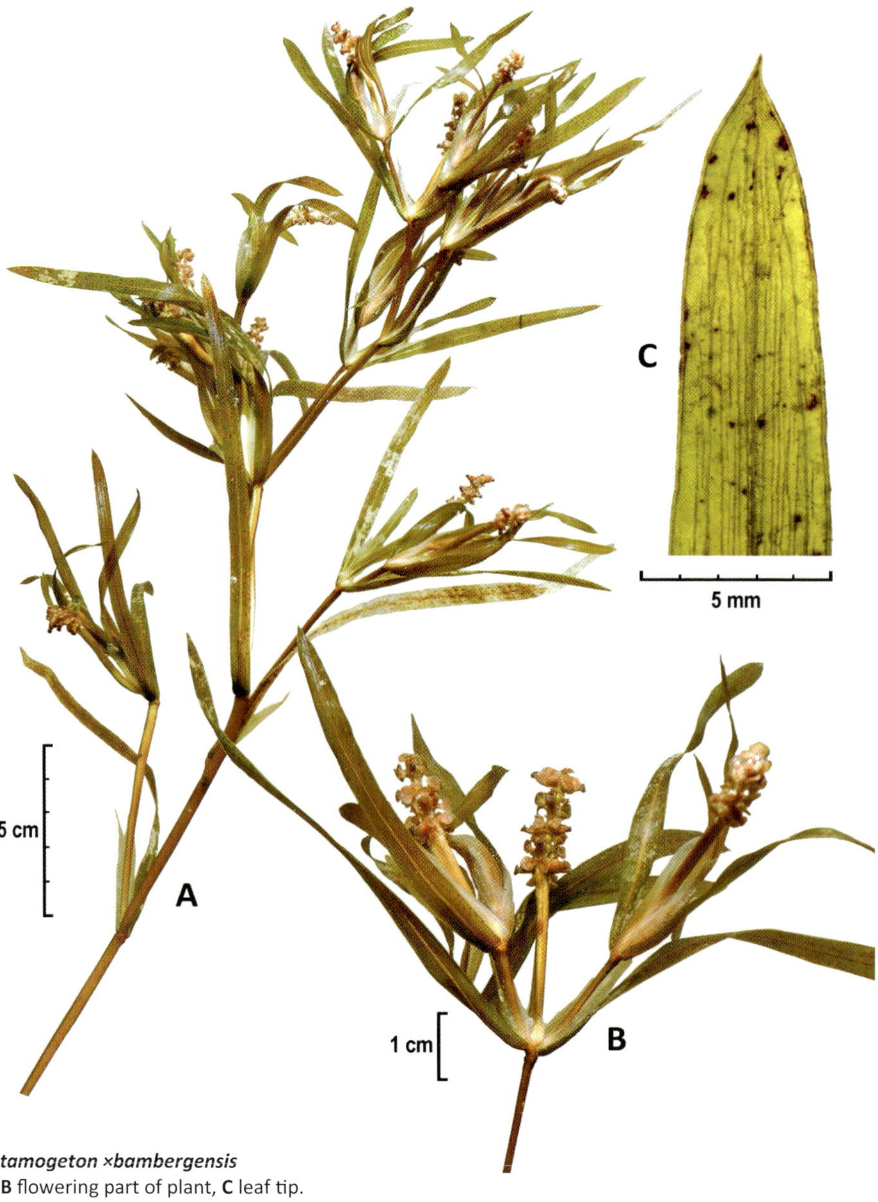

Potamogeton ×bambergensis
A, B flowering part of plant, **C** leaf tip.

Potamogetonaceae

Key to *Stuckenia*

In *Stuckenia* the leaf sheath is adnate to the leaf base in the lower part and free in the upper, forming a ligule. The rhizome develop shoots from every second node.

Because of great variation in the parental species and a wide range of phenotypic plasticity, recognition and proper determination of hybrids is often very difficult. The key below is not an optimal solution, but the best we can do.

1 All leaf sheaths open with overlapping edges (convolute) (***1***) 2

1 Leaf sheaths tubular at base (connate) at least when young for 2–3 mm or more (***2***) 4

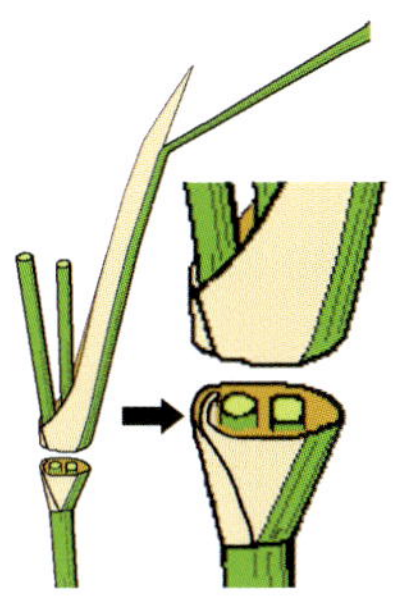

1

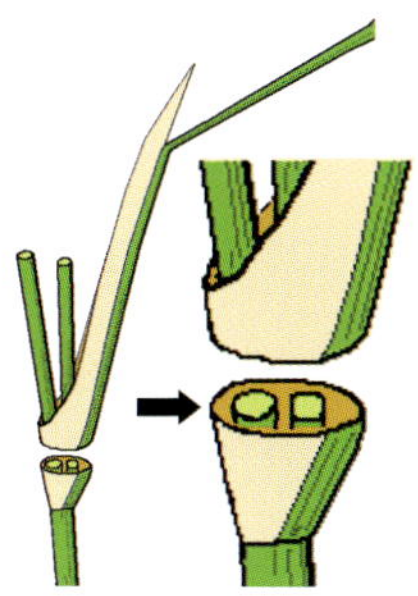

2

2 Leaf apex acute to acuminate (***3***). Lower sheaths on main stem less than 1.5 times as wide as the stem. Ligules 1–9(–14) mm long. Fruits 3.2–4.5 mm long. Stigma with a distinct style 0.2 mm long

99 *S. pectinata*

2 Leaf apex truncate to obtuse, rarely acute (***4***). Lower sheaths on main stem usually more than 1.5 times as wide as the stem. Fruits 2.6–3.4 mm long or not developed. Stigma sessile, without distinct style 3

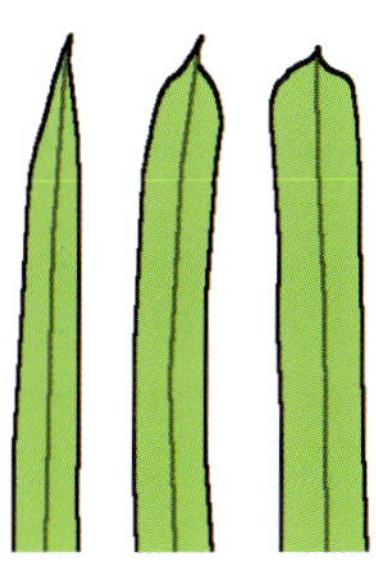

3

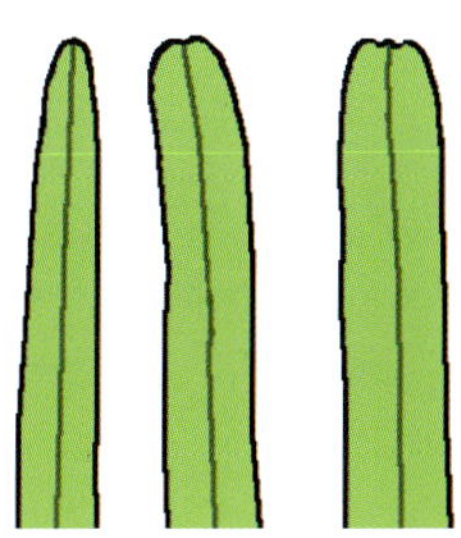

4

3 Leaves on main stem 1–6 cm long, often shorter than sheath. Ligules (or free part of sheaths) 1–1.5(–2.9) mm long on main stems, often caducous. Fruits 2.6–3.4 mm long

100 *S. vaginata*

3 Leaves on main stem longer than 6 cm, longer than sheath. Ligules (or free part of sheaths) 6–17 mm long on main stems. Fruits not developed

101 *S. ×bottnica* (*S. pectinata × vaginata*)

4 Stem richly branched at base, mostly unbranched above (***5***). Leaf apex rounded. No leaves with short lamina on the lower part of the stem. Fruits 2.2–2.8(–3.2) mm long

98 *S. filiformis*

4 Stem branched above the base (***6***). Leaf apex shape variable. Some leaves on the lower part of the stem with short lamina. Fruits not developed 5

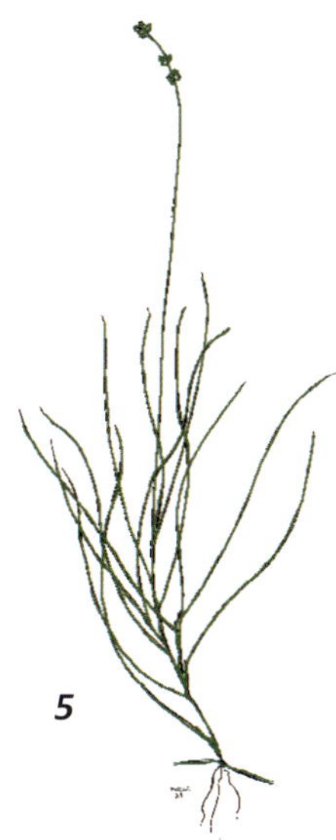

5

6

5 Fully developed leaves up to 16 cm long. Leaf apex obtuse to acuminate. Lower sheaths on main stem less than twice as wide as the stem

103 *S. ×suecica* (*S. filiformis × pectinata*)

5 Fully developed leaves more than 20 cm long. Leaf apex truncate, rounded or subretuse. Lower sheaths on main stem more than twice as wide as the stem

102 *S. ×fennica* (*S. filiformis × vaginata*)

Tabular key to *Stuckenia* taxa

	S. filiformis	***S. ×suecica*** *filiformis x pectinata*	***S. pectinata***	***S. ×bottnica*** *pectinata x vaginata*	***S. ×fennica*** *filiformis x vaginata*	***S. vaginata***
Branching	usually richly branched from base, unbranched or sparingly branched above	branched near the base, sparsely to richly branched above	usually richly branched, especially in the upper stem	sparsely branched in the upper part, without or only with a few branches below	sparsely branched	richly branched, especially above
Number of branches from sheaths on main stem	1	2	1(–2)	2–3	2–3	1–5
Sheaths with reduced lamina in the basal part of the stem	absent	present	present	present	present	present
Leaf sheaths	tubular at least on young sheaths	some tubular, some open	open	open	some tubular, some open	open
Length of leaf sheath	0.8–3.0 cm	1–5(–6) cm	1–7 cm on main stems	2–8 cm on main stems	2–10 cm	2–8 cm on main stems
Width of leaf sheath	0.5–1.8 mm	0.8–3.5 mm	0.5–4.0 mm on main stems	2.1–5.5 mm	2–6 mm	1.4–7.1(–10) mm on main stems
Length of leaves	3–16 cm	(3–)5–16(–22) cm	(2.5–)4.0–12.5(–20) cm on main stem, 3–9 cm on branches	6–20(–25) cm	(5–)7–15 cm on main stem, 15–25(–30) on branches	1–6 cm on main stem, 3.7–12.5 cm on branches
Width of leaves	0.2–1.2 mm	0.3–2.3 mm	0.3–4.0 mm on main stem, 0.2–2.8 mm on branches	2.0–3.5 mm on main stem, 0.8–3.2 mm on branches, 3–7 mm on vegetative shoots	(1–)1.5–2.5 mm in upper leaves, 3.0–4.5 mm in lower leaves	0.7–3.0 mm on main stem, 0.2–1.0 mm on branches
Leaf apex	obtuse to acute	obtuse to acuminate	rounded to obtuse and mucronate on broad leaves acute to acuminate on narrow leaves	truncate or obtuse to mucronate, often asymmetric, rarely acute	truncate, rounded or subretuse and often asymmetrical	obtuse to truncate or retuse on broad leaves obtuse to subacute on narrow leaves
Length of ligules	5–15 cm	7–24 mm	5–10(–15) mm	6–17 mm	3–15 mm	0–1.5(–4) mm
Whorls of flowers	(2–)3–5 usually distant groups	3–7 distant groups	(2–)4–5, especially downwards, distant groups	3–6, especially downwards, somewhat separated groups	6–9 more or less evenly spaced	(5–)7–9(–11) more or less evenly spaced
Stigma	sessile		with a 0.2 mm long, distinct style	sessile or with a short style	sessile	sessile
Fruits	2.2–2.8(–3.2) mm long	not developed	3.2–4.5 mm long	not developed	not developed	2.6–3.4 mm long

Potamogetonaceae

98 *Stuckenia filiformis* (Pers.) Börner

Syn.: *Potamogeton filiformis* Pers.

DK: Tråd-Vandaks **N:** Trådtjørnaks **S:** Trådnate **FIN:** Merivita **IS:** Þráðnykra
GB: Slender-leaved Pondweed **NL:** Draadfonteinkruid **F:** Potamot filiforme
D: Faden-Laichkraut **CZ:** - **PL:** Rdestnica nitkowata
EST: Niitjas penikeel **LV:** Pavedienu glīvene **LT:** Laiboji plūdė **Ru:** Рдест нитевидный

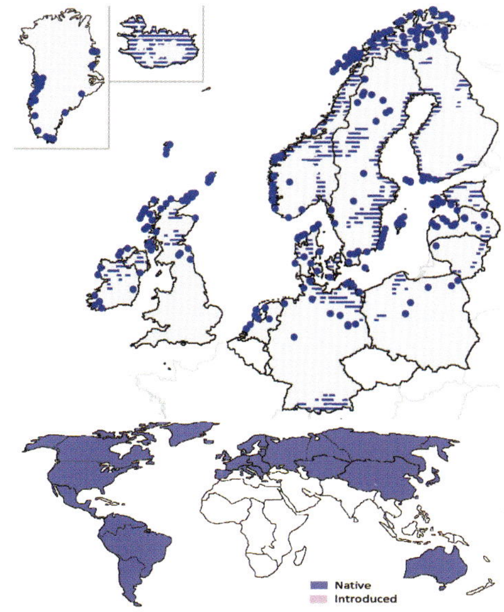

Perennial hydrophyte. Rhizome slender, terete.

Stem: 20–50 cm long, slender, terete, pale brown, eventually with a faint reddish tinge, usually richly branched from base, unbranched or sparingly branched above.

Floating leaves: Not developed.

Submerged leaves: 3.0–16 cm long, 0.2 –1.2 mm wide, filiform, semi-terete, slightly stiff, green, with entire margins and obtuse to acute apex, sessile. With one large and some smaller air channels on each side of the midrib, the lateral veins, one on each side close to the margin, inconspicuous. Leaf sheaths 0.8–3.0 cm long, 0.5–1.8 mm wide, tubular at base when young, but often split on older leaves, green to reddish brown - hyaline on the side opposite the leaf junction.

Ligules: Persistent, 5–15 mm long, obtuse to truncate, hyaline, often with a reddish-brown tinge.

Inflorescence: 1.5–7.5(–9) cm long, in (2–)3–5 usually quite distant groups, each with 1–2(–3) flowers. The two groups at the top often contiguous. Peduncle 5–15 cm long, terete, slender and flexuous - the inflorescence normally raised above the leaves.

Flowers: With 4–6 carpels. Stigma sessile without style.

Fruits: 2.2–2.8(–3.2) mm long, greenish brown to brown, rounded, without keel. Beak 0.1–0.2 mm long.

Flowering: June-August.

Biology: *Stuckenia filiformis* often produces mature fruits as early as June. It overwinters as 3–5 mm long swollen tubers developed from the distal part of the rhizome and with buds from the lower shoots. The tubers serve for regrowth and vegetative propagation.

Habitats: In shallow water in unpolluted, weakly acidic to alkaline, nutrient-poor to moderately nutrient-rich lakes, reservoirs, flooded gravel pits, streams and ditches.

Distribution within the region:
Native. In our area it has a temperate, boreal and arctic distribution with a disjoint occurrence in the Alps. Absent from Belgium, Luxembourg and Czech Republic.

Characteristics and similar species:
Stuckenia filiformis is a very variable species, but can be recognised by the stem being richly branched at the base and mostly unbranched above, the leaf apex obtuse and sheaths tubular for 2–3 mm or more at base.

Stuckenia filiformis in dense formations with *Ranunculus baudotii*. The inflorescences are raised high above the leaves. Thy, Denmark. Photo JCS.

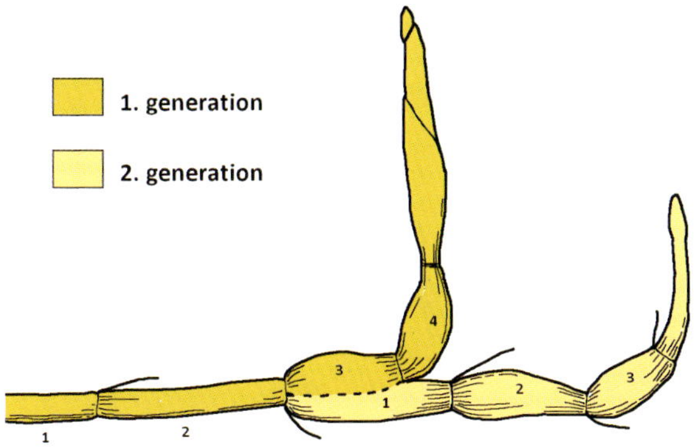

Stuckenia filiformis. Overwintering tubers consist of two generations of shoots in bud state. The 3. and 4. nodes of the 1. generation are swollen, and so are all four of the 2. generation. The 3. node of the 1. generation and the 1. node of the 2. generation are partly fused. Total length of both generations is 15 mm. Drawing based on Hagström 1916.

Fruiting *Stuckenia filiformis* in shallow, calcareous water. Norrsund, Fårø, Gotland, Sweden. Photo JCS.

99 *Stuckenia pectinata* differs in acute to acuminate leaf apex, stem branched at the top and normally much longer shoots. **108-109 *Ruppia* spp.** have leaf sheaths but lack ligules and shoots from every node at the rhizome.

Hybrids: 103 *Stuckenia ×suecica* is as variable as *S. pectinata* and easily mistaken for this. The hybrid differs from *S. pectinata* by having at least some sheaths tubular with the basal part closed (dissection with a stereo microscope is necessary) and by never producing fruits. From *S. filiformis* it is separated by being more branched in the upper part and having flowers with distinct styles.

Flowering *S. filiformis.* Selenter See, Germany. Photo KvdW.

Stuckenia filiformis. Fruiting head. Note the missing style and sessile stigmas. Thy, Denmark. Photo JCS.

Stuckenia filiformis. Leaf tips. Photo KvdW.

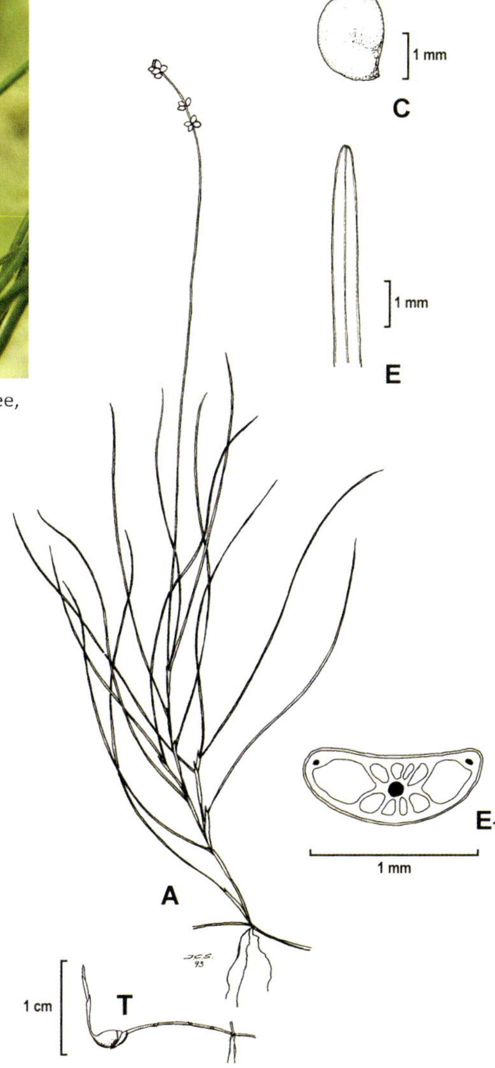

Stuckenia filiformis
A habit, **C** fruit, **E** leaf tip, **E1** cross section of leaf, **T** tuber.

Potamogetonaceae

99 *Stuckenia pectinata* (L.) Börner

Syn.: *Potamogeton pectinatus* L.

DK: Børstebladet Vandaks **N:** Bust-tjørnaks **S:** Borstnate **FIN:** Hapsivita **IS:** -
GB: Fennel Pondweed **NL:** Kam-fonteinkruid **F:** Potamot pectiné
D: Kamm-Laichkraut **CZ:** Rdest hřebenitý **PL:** Rdestnica grzebieniasta
EST: Kamm-penikeel **LV:** Ķemmveida glīvene **LT:** Šukinė plūdė **Ru:** Рдест гребенчатый

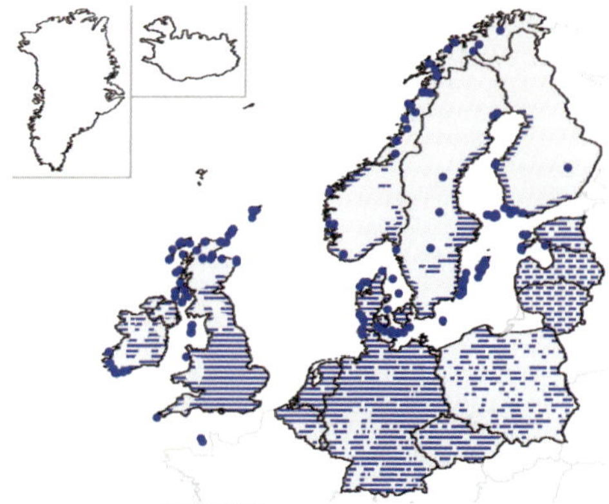

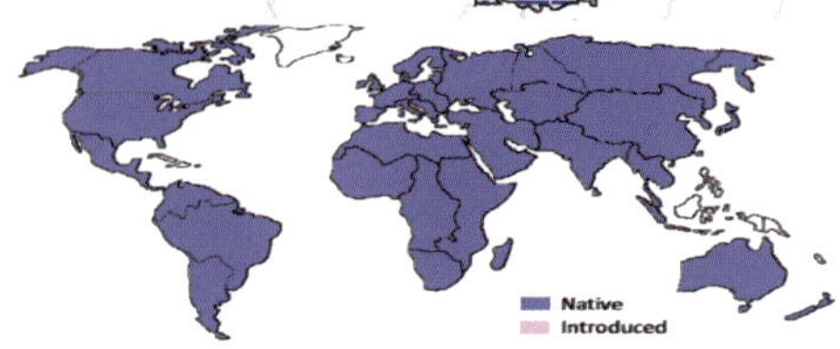

Stuckenia pectinata in still water with spread-out branches and rather stiff leaves. Tjele Langsø, Jutland, Denmark. Photo JCS.

Perennial hydrophyte. Rhizome slender to robust, terete.

Stem: In still water 0.2–1.0 metre long, but in flowing water significantly longer, slender to quite robust, terete, often zigzag shaped, mainly richly branched above with 1(–2) branches emerging from the sheaths of the main stem.

Floating leaves: Not developed.

Submerged leaves: Lamina (2.5–)4.0–12.5(–20) cm long, 0.2–4.0 mm wide, filiform to linear, semi-circular to canaliculate, quite rigid, entire leaf margins, broad leaves with rounded to obtuse and mucronate leaf apex, narrow leaves with acute to finely acuminate leaf apex, green, often with a brown tinge. With one inconspicuous lateral vein and one large and several smaller air channels on each side of the midrib. The leaves are sessile, with a 1–7 cm long, 0.5–4.0 mm wide, open sheath on main stems, which is green on the outer side and hyaline on the inner side.

Ligules: 5–10(–15) mm long on main stems, obtuse to oblique, hyaline, often with a pale green or brown tone. Decaying after some time.

Inflorescence: 1.5–4.5(–6.0) cm long, with (2–)4–5 groups of 1–2 flowers, the lower quite distant. Peduncle 2–10 cm long, terete, slender and flexuous.

Flowers: With 4 carpels. Stigma with a distinct style 0.2 mm long.

Fruits: 3.2–4.5 mm long, pale brown – becoming brown, rounded, without keel. Beak 0.2–0.4 mm long.

Flowering: June-September.

Biology: *Stuckenia pectinata* is sometimes wintergreen or overwinters with 3–5 mm long, much-thickened tubers developed on the distal part of the rhizome, and from buds on the lower shoots. The tubers also serve as vegetative propagules.

Habitats: In neutral to alkaline, moderately nutrient-rich to nutrient-rich lakes, flooded gravel pits, ditches, canals and streams. The species is tolerant of salt and is therefore found in coastal lagoons and brackish water.

Distribution within the region:
Native. Common throughout most of the region. Absent from Iceland and Greenland.

Characteristics and similar species:
Stuckenia pectinata is very variable - ranging from very slender plants with filiform leaves to robust plants with branches several metres long and 1–2 (–4) mm wide leaves in flowing water. Many forms and varieties have been described, such as var. *scoparius* Wallr., var. *helveticus* (G.Fisch.) Glück and var. *interruptus* (Kit.) Asch., but these are of no taxonomical value.

The species is recognisable by the acute leaves with sheaths and ligules.

Stuckenia pectinata with flaccid stem and leaves in River Gudenå, Jutland, Denmark. Photo JCS.

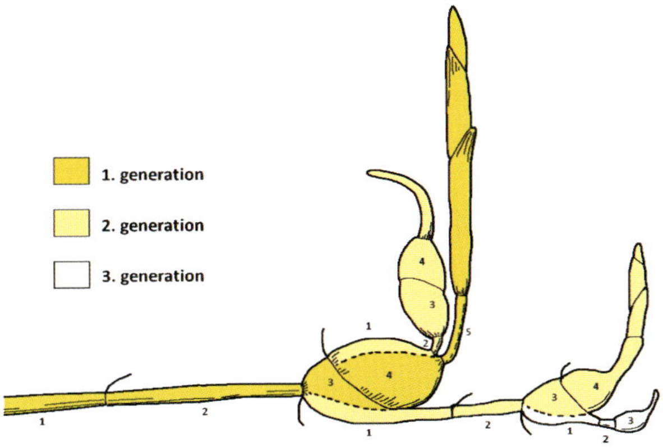

Stuckenia pectinate. Overwintering tubers consist of three or more generations of shoots in bud state. The 3. and 4. nodes of all generations are swollen but not the 1. and 2. nodes. As indicated by dotted lines some nodes are fused. Total length of all generations is 6–7 cm. Drawing based on Hagström 1916.

Stuckenia pectinate. Fruiting specimens in shallow water. Note the short, but distinct style on the fruits to the right. Lake Filsø, SW-Jutland, Denmark. Photo JCS.

Slender plants can be mistaken for **98 *Stuckenia filiformis***, which has obtuse leaves and is branched at the base. **108-109 *Ruppia*** spp. have leaf sheaths, but lack ligules; their rhizomes have ascending shoots from every node. Robust forms can be mistaken for **101 *S. ×bottnica*** – see this. Submerged plants of **163 *Isolepis fluitans*** lack ligules.

Hybrids: 103 *Stuckenia ×suecica*, 101 *S. ×bottnica*.

Stuckenia pectinata. Wummsee, Germany. Photo KvdW.

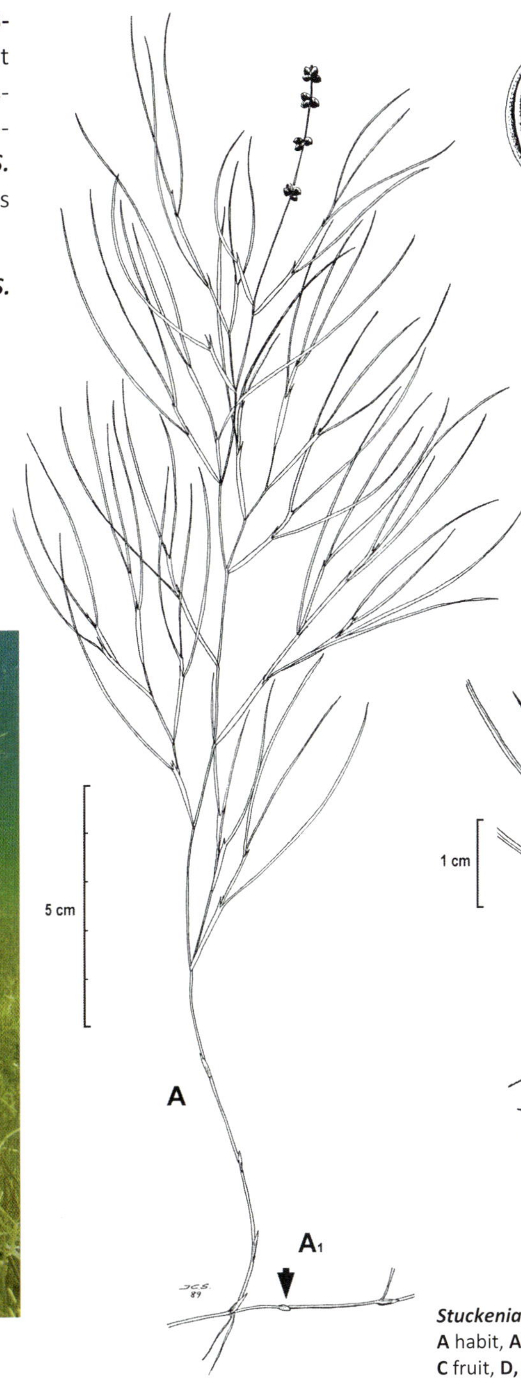

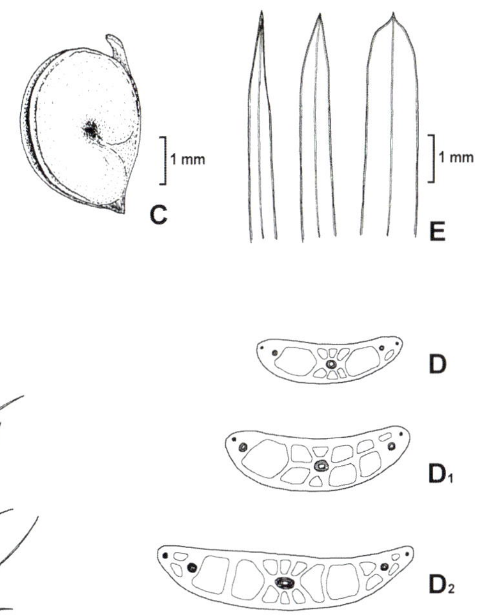

Tuber. Photo JCS.

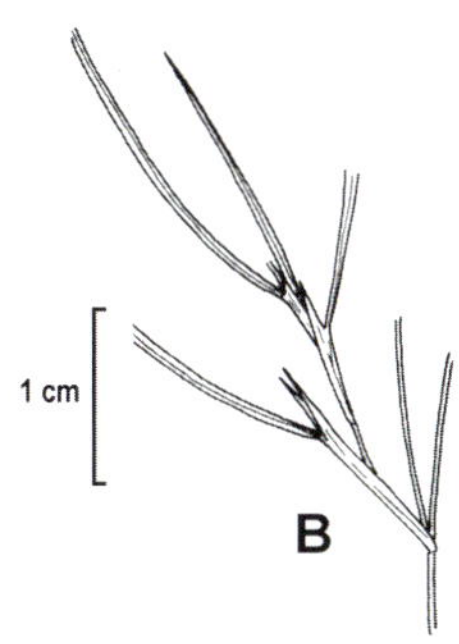

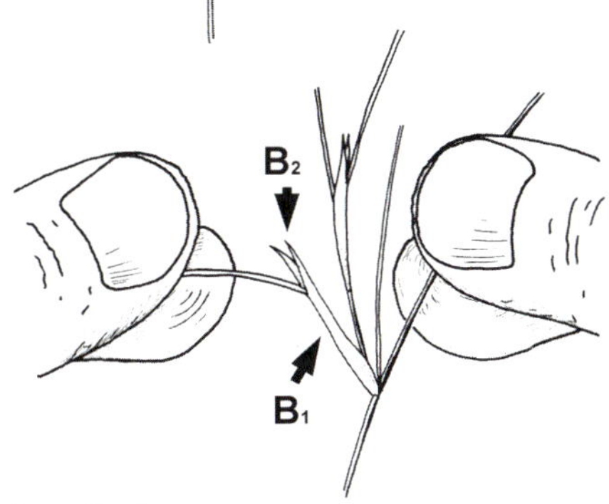

Stuckenia pectinata
A habit, **A1** rhizome, **B** branch, **B1** leaf sheath, **B2** ligule, **C** fruit, **D, D1, D2** cross sections leaves, **E** leaf tips.

Potamogetonaceae

100 ***Stuckenia vaginata*** (Turcz.) Holub

Syn.: *Potamogeton vaginatus* Turcz.

DK: - **N:** Sliretjørnaks **S:** Slidnate **FIN:** Tuppivita **IS:** -
GB: - **NL:** - **F:** -
D: - **CZ:** - **PL:** -
EST: - **LV:** - **LT:** - **Ru:** -

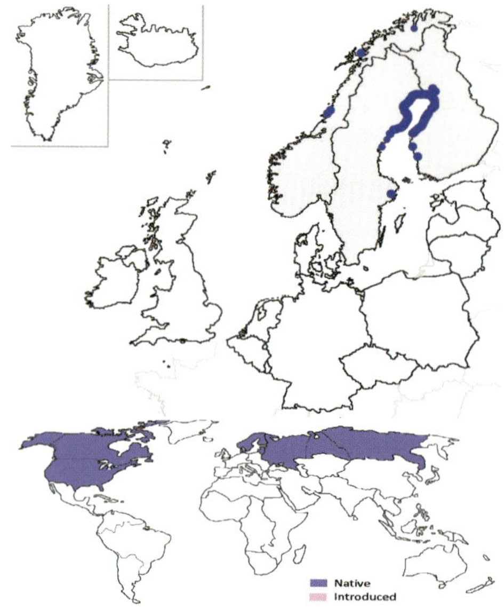

Stuckenia vaginata. Torget, Brønnøy, Norway. Photo Klaus Høiland.

Perennial hydrophyte. Rhizome 2–4 mm thick, terete.

Stem: Up to 4 metres long depending on the depth of the water, slender to quite robust, terete, richly branched, especially above, with 1–5 branches emerging from each sheath of the main stem.

Floating leaves: Not developed.

Submerged leaves: Lamina on main stems of more or less equal length, 1–6 cm long, 0.7–3 mm wide, often shorter than the sheath, linear, apex obtuse to truncate or retuse. Lamina on branches 3.7–12.5 cm long, 0.2–1.0 mm wide, linear, apex obtuse to subacute. Broad and medium-width leaves with several small air channels and 2 lateral veins on each side of the midrib. The vein closest to the margin inconspicuous and only visible by cross section of the leaf and use of a stereo microscope. The leaves are sessile, olive green to greyish brown, with a very robust, stiff, 2–8 cm long, 1.4–7.1(–10) mm wide, open sheath on main stems, much wider than those on the branches. Leaf sheaths olive green to greyish brown, truncate and hyaline at apex.

Ligules: 0–1.5(–2.9) mm long on main stems, indistinct, hyaline, yellow brownish.

Inflorescence: 1.5–4.5(–7.0) cm long, with (5–)7–9(–11) more or less evenly spaced whorls of (1–)2 flowers. Peduncle 2–20 cm long, terete, slender and flexuous.

Flowers: With 4 carpels. Style inconspicuous, stigma horizontal.

Fruits: 2.6–3.4(–3.8) mm long, brown, rounded, without keel. Beak inconspicuous.

Flowering: July-August.

Biology: *Stuckenia vaginata* overwinters as swollen tubers developed on the distal part of the rhizome, and from buds on the lower shoots. The tubers also serve as vegetative propagules.

Habitats: In the Gulf of Bothnia in brackish water with soft sediment to a depth of 3 metres. In freshwater in neutral to alkaline lakes, ponds or slow-flowing rivers.

Stuckenia vaginata is richly branched with 1–5 branches emerging from each sheath of the main stem. Hailuoto, Finland. Photo KvdW.

Stuckenia vaginata. Overwintering tubers consisting of two or three generations of shoots in bud state. The 3. and 4. nodes of the 1. generation are swollen. and so are all 4 of the 2. and 3. generation. As indicated by dotted lines some nodes of two generations are fused. Total length of all generations is 4–5 cm. Drawing based on Hagström 1916.

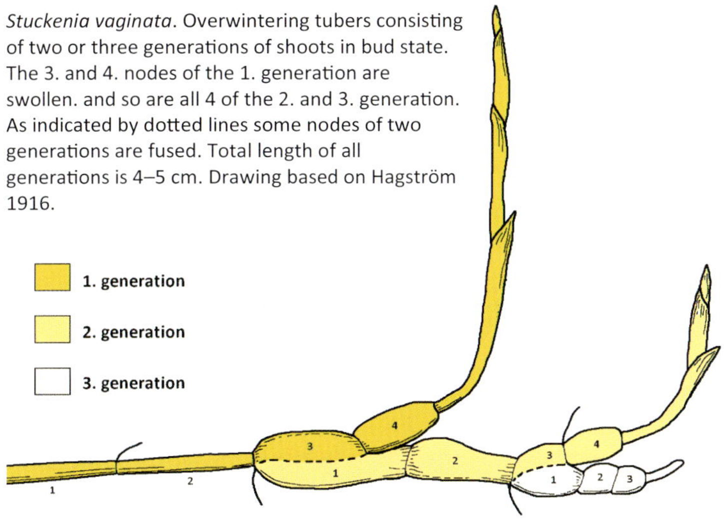

Stuckenia vaginata. The whorls of flowers are more or less evenly distributed. Photo JCS.

Distribution within the region:
Native. Along the coast of northern Norway, and in the Gulf of Bothnia in Sweden and Finland.

Characteristics and similar species:
Stuckenia vaginata can be recognised by the (5–)7–9(–11) more or less evenly spaced whorls of (1–)2 flowers. Leaves on the main stem short, 1–6 cm long, their sheaths much wider than those on the branches and with (1–)2–4(–5) branches emerging from each sheath.

Robust forms of **99 *Stuckenia pectinata*** differ by (2–)4–5 unevenly spaced whorls of flowers, fruits larger, 3.2–4.5 mm long and 1–(2) branches emerging from each leaf sheath on the main stem.

Hybrids: 102 *S. ×fennica*, 101 *S. ×bottnica*.

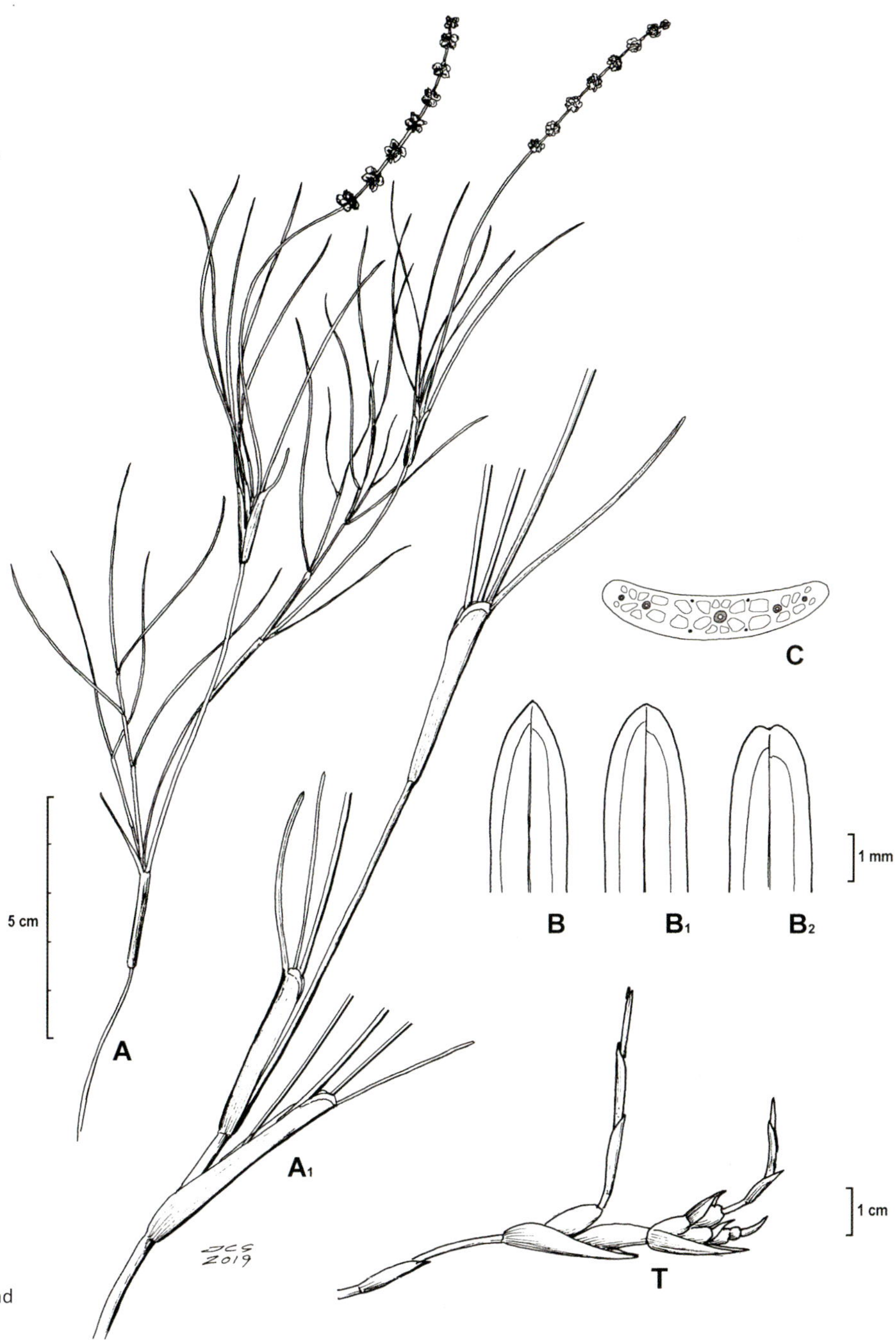

Stuckenia vaginata
A flowering branches, **A1** main stem with leaf sheaths and branches, **B, B1, B2** leaf tips, **C** cross section of leaf, **T** overwintering rhizome with swollen nodes and tubers.

Potamogetonaceae

101 *Stuckenia ×bottnica* (Hagstr.) Holub

(S. pectinata × vaginata)

Syn.: *Potamogeton ×bottnicus* Hagstr.; *P. pectinatus* L. x *P. vaginatus* Turcz.

Perennial hydrophyte. Rhizome slender to very robust, terete.

Stem: 20–150 cm long, slender to robust, pale brown, eventually with an obscure reddish tinge, sparsely branched in the upper part, without or only with a few branches below.

Floating leaves: Not developed, but broad-leaved submerged leaves are often at the surface.

Submerged leaves: Basal leaves reduced to scales. Lower leaves on flowering stems 2.5–20 cm long, 2.0–3.5 mm wide, upper leaves 10–20(–25) cm long, 0.8–3.2 mm wide, canaliculate, oblique, apex truncate to rounded or obtuse to mucronate, rarely acute, often asymmetrical. Vegetative shoots with leaves up to 5–7 mm wide. The leaves are linear, entire, quite rigid, canaliculate, green to dark green, and close to the water surface often with a brownish tinge. With 1–2 obscure lateral veins and several air channels on each side of the midrib. Sheaths 2–8 cm long, 2.1–5.5 mm wide, often more than twice as wide as stem, open, green with hyaline edges.

Ligules: Persistent, 0.6–1.7 cm long, obtuse to oblique, hyaline.

Inflorescence: 1–5 cm long, with 3–6 groups of 1–2 flowers, particularly the lower ones somewhat distant. Peduncle 3.5–10 cm long, terete, slender and quite flexible.

Flowers: With 4 carpels. Pollen sterile. Stigma sessile.

Fruits: Never developed.

Flowering: June-August.

Biology: Sterile hybrid. Vegetative propagation by short to long rhizomes. It overwinters as 5–6 mm long swollen tubers, which develop at the distal part of the rhizome. The tubers also serve as vegetative propagules.

Habitats: In neutral to alkaline and nutrient-rich streams.

Distribution within the region: Native. Known from Denmark, Finland, Sweden and Britain.

Characteristics and similar species: *Stuckenia ×bottnica* can be confused with broad-leaved forms of the very variable **99** ***Stuckenia pectinata***. *S. pectinata* has acute to finely acuminate leaf apices on the vegetative shoots, while the leaves at *S. ×bottnica* are truncate to rounded, mucronate and often asymmetrical. The hybrid does

Stuckenia ×bottnica in a stream. Aars, Jutland, Denmark. Photo BM.

Stuckenia ×bottnica. Lower stem leaves much broader than the upper. Aars, Jutland, Denmark. Photo JCS.

Stuckenia ×bottnica. Leaf tip blunt, short mucronate and usually asymmetric. Photo BM.

not produce fertile pollen and never develops fruits. The stigma is sessile.

Stuckenia ×bottnica.
Tubers on distal part of rhizome.
Aars, Jutland, Denmark.
Photo JCS.

Stuckenia ×bottnica
A flowering shoot, **A1** vegetative shoot,
B branch from lower part of the stem,
E leaf tip.

Potamogetonaceae

102 *Stuckenia ×fennica* (Hagstr.) Holub

(S. filiformis × vaginata)

Syn.: *Potamogeton ×fennicus* Hagstr.; *P. filiformis* Pers. x *P. vaginatus* Turcz.

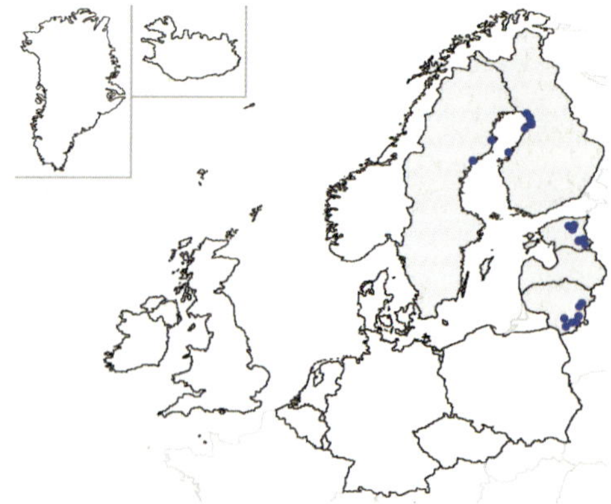

Stuckenia ×fennica. Upper part with inflorescence. River Emtsa, Arkhangelsk region, north of European Russia. Photo Alexander A. Bobrov.

Perennial hydrophyte. Rhizome slender to very robust, terete.

Stem: Up to 2.5 m long, slender to robust, not or only sparingly branched below.

Floating leaves: Not developed.

Submerged leaves: Basal leaves reduced to scales. Lower leaves on overwintering or flowering stems (5–)7–15 cm long, 3.0–4.5 mm wide, upper leaves 15–25(–30) cm long, (1–)1.5–2.5 mm wide. Leaves green to dark green, when close to the water surface, often with a brownish tinge, truncate, rounded or subretuse and often asymmetrical at apex. On each side of the midrib with 1–2 obscure lateral veins and several air channels. Sheaths 2–10 cm long, 2–6 mm wide, dark green with hyaline brownish edges, basal part tubular for 2–10 mm, but soon splitting (look at young sheaths from the upper part of the plant).

Ligules: Persistent, 0.3–1.5 cm long, hyaline, truncate or rounded at apex.

Inflorescence: 4–8 cm long, with 6–9 more or less evenly spaced whorls of (1–)2 flowers. Peduncle 5–15 cm long, terete, slender and quite flexible.

Flowers: With 4 carpels. Pollen sterile. Stigma sessile.

Fruits: Never developed.

Flowering: June-August.

Biology: Sterile hybrid. Vegetative propagation by rhizomes and stem fragments and by small plants developed on short axillary stolons. Tubers not seen.

Habitats: In neutral to alkaline, fresh or slightly brackish coastal waters. Also in alkaline waters in fast-flowing rivers in sandur landscapes with exposed carbonate rocks, mostly in postglacial areas.

Distribution within the region:
Native. Known from Sweden, Finland, Estonia and Lithuania.

Characteristics and similar species:
A variable hybrid. *Stuckenia ×fennica* can be identified by the young leaf sheaths being tubular at the base for 2–3 mm or more, by mature leaves more than 20 cm long, with truncate, rounded or subretuse apex. Lower sheaths on main stem more than twice as wide as the stem. The hybrid does not produce fertile pollen and never develops fruits. The stigma is sessile.

Stuckenia ×fennica. Luukkaanperä, Tornio, Outer Ostrobothnia, Finland. Photo RVL.

Dense stand of *S. ×fennica* in river. Luukkaanperä, Tornio, Outer Ostrobothnia, Finland. Photo RVL.

Stuckenia ×fennica. Luukkaanperä, Tornio, Outer Ostrobothnia, Finland. Photo RVL.

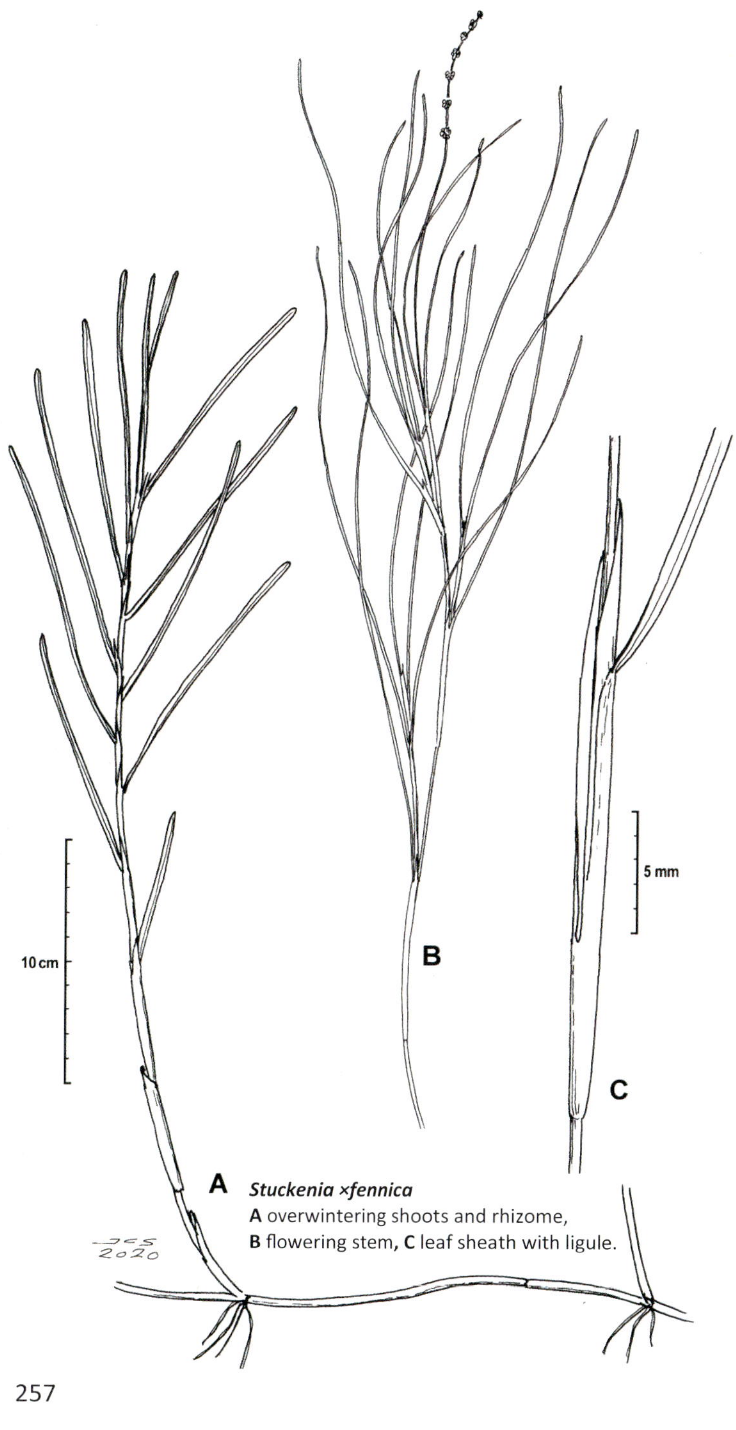

Stuckenia ×fennica
A overwintering shoots and rhizome, **B** flowering stem, **C** leaf sheath with ligule.

Potamogetonaceae

103 *Stuckenia ×suecica* (Hagstr.) Holub

(S. filiformis × pectinata)

Syn.: *Potamogeton ×suecicus* Hagstr.; *P. filiformis* Pers. x *P. pectinatus* L.

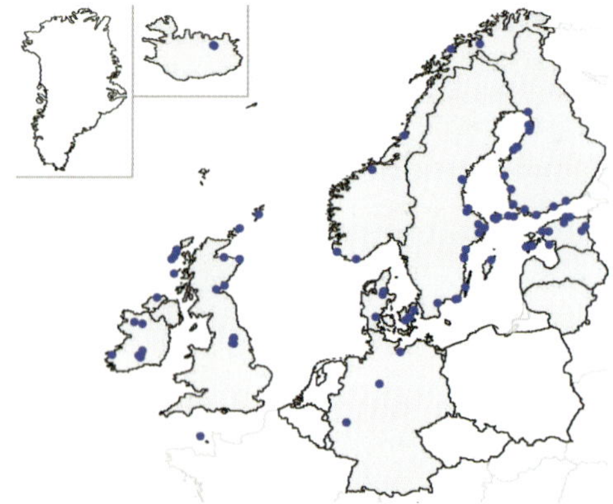

Stuckenia ×suecica. Laacher See, Rhineland-Palatinate, Germany. Photo KvdW.

Flowers: With 4 carpels. Pollen sterile. tigma sessile or with short style.

Fruits: Never developed.

Flowering: June-August.

Biology: Sterile hybrid. Vegetative propagation by rhizomes and stem fragments. Tubers unknown.

Habitats: In unpolluted, weakly acidic to alkaline, nutrient-poor to moderately nutrient-rich lakes, pool, streams and rivers.

Perennial hydrophyte. Rhizome slender to very robust, terete.

Stem: 20–130 cm long, slender to robust, branched at base and more or less above.

Floating leaves: Not developed.

Submerged leaves: Basal leaves short. Leaves (3–)5–16(–22) cm long, 0.3–2.3 mm wide, green to dark green, obtuse to acuminate at apex. With 1 lateral vein, 0–4 large and several small air channels on each side of the midrib. Sheaths 1–5(–6) cm long, 0.8–3.5 mm wide, green with hyaline edges, some of them with tubular basal part for 2–5(–12) mm, but soon splitting (look at young sheaths from the upper part of the plant).

Ligules: Persistent, 0.7–2.4 cm long, 0.8–3.5 mm wide, obtuse to oblique, hyaline.

Inflorescence: 1.5–2.5 cm long, with 3–7 distant groups of 1–3 flowers. Peduncle 3.5–22 cm long, terete, slender and flexible.

Stuckenia ×suecica. Santower See, Mecklenburg-Vorpommern, Germany. Photo KvdW.

Leaf tips. *S. ×suecica*. Santower See, Mecklenburg-Vorpommern, Germany. Photo KvdW.

Distribution within the region:
Native. Scattered and local from Ireland to southern Finland and Estonia.

Characteristics and similar species:
A very variable hybrid not easily recognised and probably overlooked. It can be identified by young leaf sheaths being tubular at base for 2–3 mm or more, mature leaves less than 16 cm long, with obtuse to acuminate apex and sheaths less than twice as wide as stem.

Stuckenia ×suecica is more branched above than **98** ***S. filiformis*** and has leaf sheaths with a short lamina at base. **99** ***S. pectinata*** has acute to finely acuminate leaves and all sheaths open, convolute. The hybrid does not produce fertile pollen and never develops fruits. The stigma is sessile.

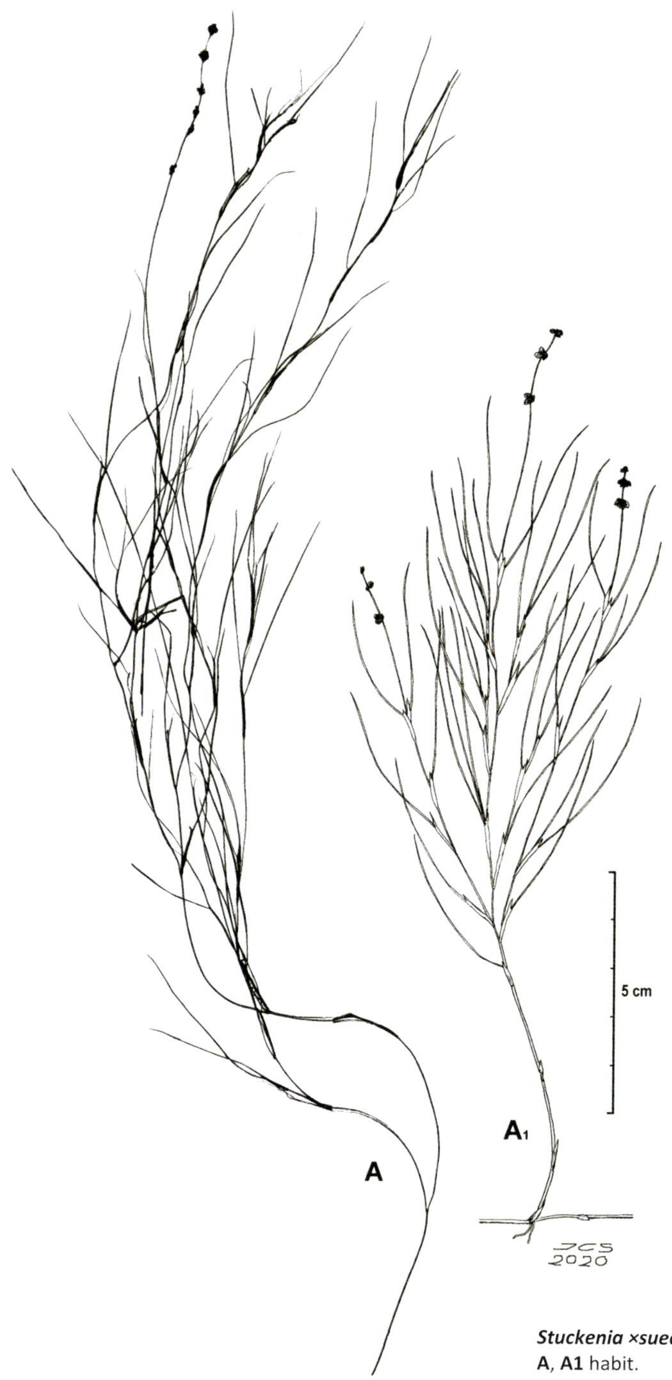

Stuckenia ×suecica
A, **A1** habit.

Potamogetonaceae

Key to *Zannichellia*

1 Male and female flowers at the same node (with rare exceptions) (***1***). Filaments short (< 12 mm), not elongating as they mature, anthers with 2 (rarely 3–4) pollen sacs. Leaves biconvex with air channels (= *Z. palustris* L. (agg.)) 2

1 Male and female flowers at different nodes (***2***). Filaments long (up to 70 mm), elongating as they mature, anthers with 4 pollen sacs. Leaves biconvex or flat, with or without air channels 4

2 Achenes (2.5–)3.0–4.5 mm long, (3–)4–6(–8), pedicels < 0.8 mm long (***3***). Leaves (0.8–)1.0–2.0 mm wide. Plant perennial, salt tolerant 5‰ to 20‰, in coastal brackish waters

107 *Z. palustris* subsp. ***major***

2 Achenes 1.5–2.5(–3.0) mm long, (1–)2–6, pedicels up to 2.6 mm long. Leaves < 1 mm wide. Plant mostly annual, in brackish or inland waters 3

3 Pedicels < 0.5 mm long, fruits (2–)4–6, style length 25%–50% of the length of the achene (***4***). Salt tolerant to 7‰–8‰, mostly in inland waters

106 *Z. palustris* var. ***palustris***

3 Pedicels 1.5–2.5 mm long, fruits 1–4, style length 60%–80% of the length of the achene. Salt tolerant to 20 ‰, in inland and coastal habitats

106 *Z. palustris* var. ***pedicellata***

4 Leaves obtuse (***5***), translucent, flat, without conspicuous air channels. Anthers conspicuous, 1.5–2.5 mm

104 *Z. obtusifolia*

4 Leaves acute (***6***), opaque, biconvex in cross section, with several air channels. Anthers small, 1.4–1.9 mm

105 *Z. peltata*

1

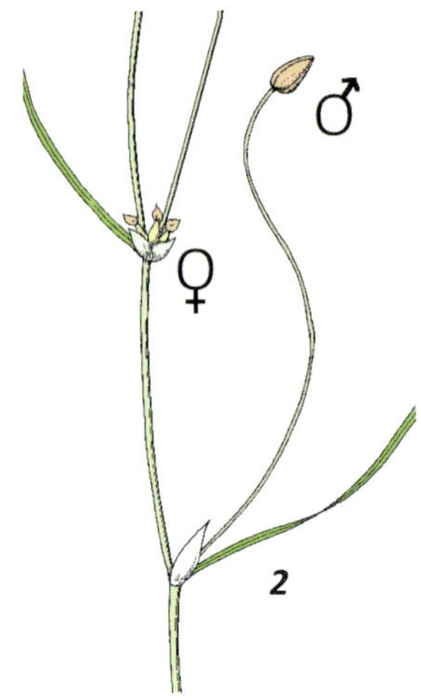

2

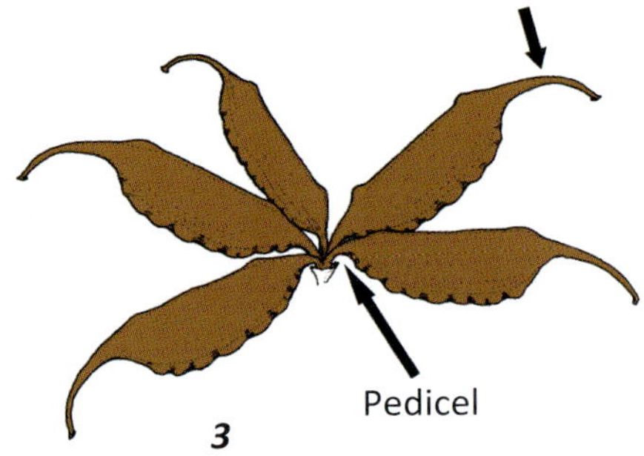

3

4

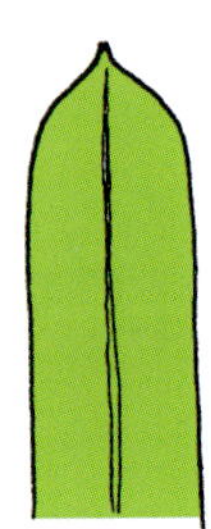

5

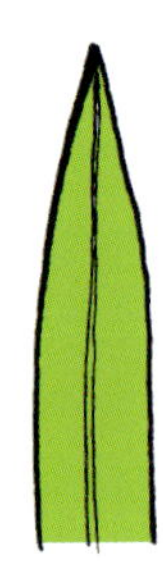

6

Potamogetonaceae

104 *Zannichellia obtusifolia* Talavera, Garcia-Muríllo & Smit

Perennial hydrophyte. **Stem** up to 50 cm long. **Leaves** linear, up to 1.5 mm wide and 45 mm long, flat, without air channels, apex obtuse. **Flowers** axillary, male and female at separate nodes. Male flower naked with 1.7–2.5 mm long anther and 25–33(–70) mm long filament, elongating as it matures. Female flower with short pedicel.

Fruit: 2(–3–5) achenes on 0.3–1.5 mm long pedicels. Achene (1.5–)1.7–2.2 (–2.5) mm long, dorsal side curved, with a row of connected teeth forming a low ridge, style 1–1.5(–2) mm.

Distribution within the region:
Native to France, Spain, Portugal, Sardinia, Sicily, Morocco. Recently recorded and possibly native near Colchester, Britain.

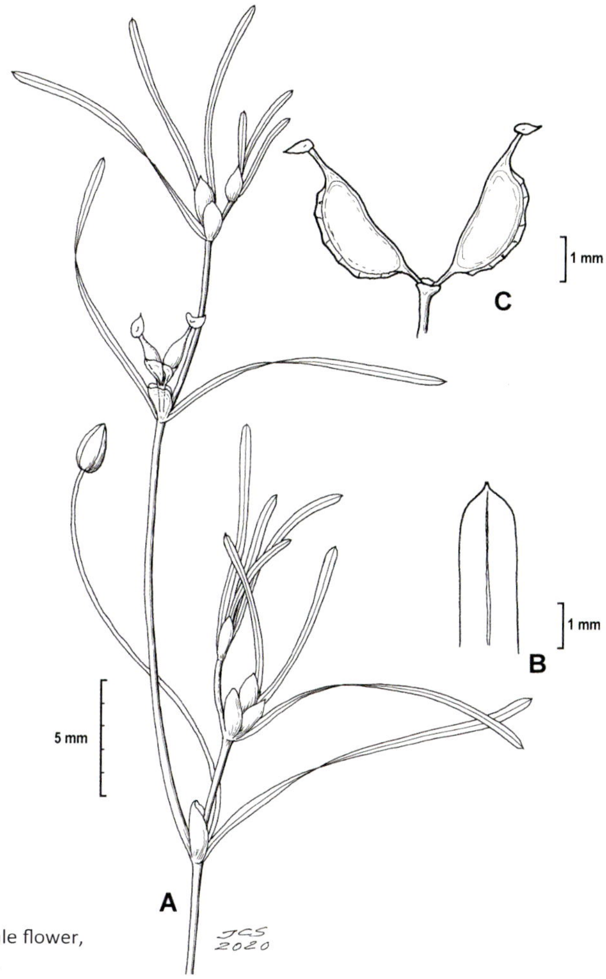

Zannichellia obtusifolia
A part of stem with male and female flower, **B** leaf apex, **C** fruit with 2 achenes.

Potamogetonaceae

105 *Zannichellia peltata* Bertol.

Syn.: *Z. macrostemon* J.Gay ex Willk.

Perennial hydrophyte. **Stem** up to 50 cm long. **Leaves** linear, up to 0.5 mm wide and 75 mm long, biconvex, with 2 air channels, apex acute. **Flowers** axillary, male and female at separate nodes. Male flower naked with anther 1–1.7 mm long and filament 10–30(–40) mm long, elongating as they mature. Female flower with short pedicel.

Fruit 2(–3–5) achenes on 0.2–0.5 mm long pedicels. Achene (2–)2.5–3 mm long, dorsal side curved, with a row of connected teeth forming a low ridge, style 1–2 mm.

Distribution within the region:
Introduced and later disappeared from Bad Salzungen, Germany. Native to France, Spain, Portugal, Italy, Algeria.

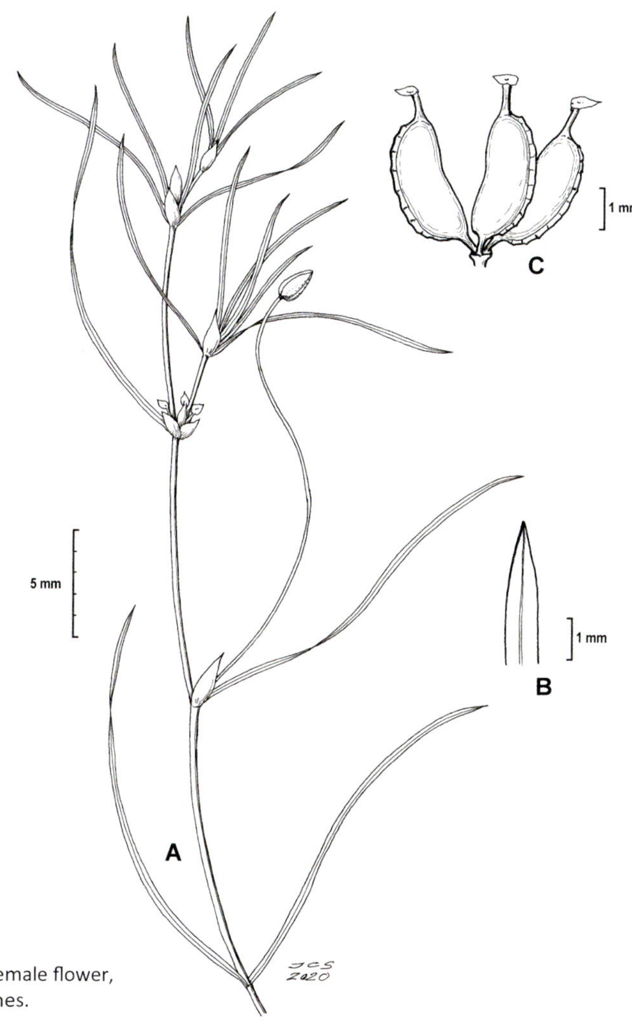

Zannichellia peltata
A part of stem with male and female flower, **B** leaf apex, **C** fruit with 3 achenes.

Potamogetonaceae

106 *Zannichellia palustris* L. subsp. *palustris*

DK: Vandkrans **N:** Vasskrans **S:** Hårsärv **FIN:** Pikkuhaura **IS:** Hnotsörvi
GB: Horned Pondweed **NL:** - **F:** Zannichellie des marais
D: Sumpf-Teichfaden **CZ:** Šejdračka bahenní **PL:** Zamętnica błotna typowa
EST: Harilik hanehein **LV:** Purva diedzene **LT:** Pelkinė vandensargė **Ru:** Занникеллия болотная

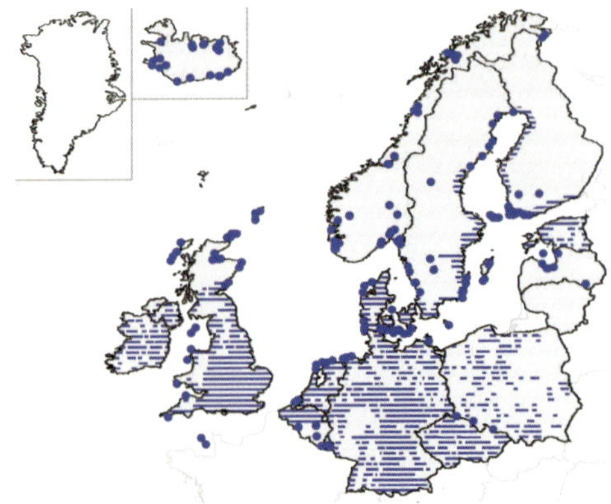

Zannichellia palustris L. subsp. *palustris*

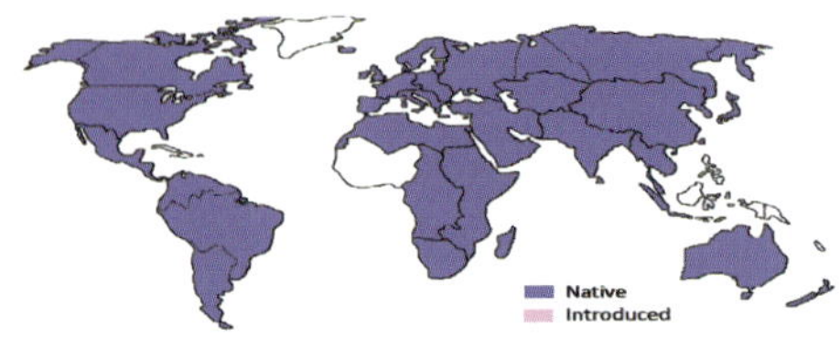

World-map showing *Zannichellia palustris* s.lat.

Zannichellia palustris in flowing water. Støvring, Denmark. Photo BM.

Perennial hydrophyte. Rhizome slender, terete, not always distinctly differentiated from the stem.

Stem: 5–50 cm long - up to 150 cm in flowing water, slender, terete, whitish to light greenish, rooting from the lower nodes, more or less branched.

Leaves: 2–10 cm long, 0.3–1.2 mm wide, linear, flat, entire, light green to green, tapering to a bluntly acuminate to acute apex, midrib more or less inconspicuous. Leaves sessile, opposite or apparently in whorls of 3–4 on fertile shoots, opposite or rarely alternate on sterile shoots. Leaf base with a free, stipule-like ligule, 1.5–3.5 mm long, obtuse.

Inflorescence: Flowers solitary in leaf axils.

Flowers: Male flower with 1(–2) stamens, naked, positioned close to and at the same node as the female flower. Female flower with 2–6 carpels with 1 ovule and surrounded by a cup-shaped, hyaline perianth. Stigma peltate. Style 0.3–1.5 mm, straight or curved.

Fruits: In groups of 2–6, sessile or pedicellate achenes. Achenes 2.0–2.6 mm long, light brown to greenish or dark brown, terete or laterally a little compressed. Ventral side of achene more or less straight or slightly curved, sometimes with some short teeth. The dorsal side is more or less curved, with or without teeth, the teeth sometimes connected, forming a low ridge.

Flowering: July-September.

Biology: *Zannichellia* is pollinated under water.

Habitats: In eutrophic, neutral to alkaline, fresh or brackish water, with a salinity of up to 20 ‰. Ponds, water-filled gravel pits or limestone quarries, lakes, streams, small water bodies in saline marshes, in shallow water along inland seashores.

Distribution within the region:
Native. Common throughout most of the region. Absent from Greenland.

Characteristics and similar species:
Zannichellia palustris L. subsp. ***palustris*** differs from **107 *Z. palustris*** subsp. ***major*** by the leaves being 1–2 mm wide and with a prominent midrib. The fruits are larger - achenes 3–4.5 mm long, on the dorsal side with a row of connected teeth forming a low ridge.

108-109 *Ruppia* spp. have alternate leaves with a sheathing base but without stipules and finely denticulate leaf apex (use a hand lens).

Stuckenia and narrow-leaved ***Potamogeton*** species have apical/terminal inflorescences and alternate leaves on the lower part of the stem.

Zannichellia palustris var. *palustris*. Blekinge, Sweden. Photo JCS.

Zannichellia palustris L. subsp. ***palustris*** is very variable and often split into two varieties as shown below. Such a division is not always possible because of great overlap in the characters.

var. ***palustris***

Syn.: *Z. repens* Boenn.; *Z. palustris* var. *repens* (Boenn.) W. D. J. Koch;
DK: Krybende Vandkrans **N:** Liten vasskrans **S:** Småsärv **FIN:** Merihaura **IS:** -
GB: - **NL:** Zittende zannichellia **F:** Zannichellie des marais
D: Sumpf-Teichfaden **CZ:** - **PL:** Zamętnica błotna typowa
EST: - **LV:** - **LT:** - **Ru:**

Leaves very narrow, 0.2-0.3 mm wide in dry material, up to 0.5 mm in fresh material, light green.
Fruit usually with 4-6 achenes. Achenes about 2 mm long, more or less sessile, stalk 0.2-0.5 mm long, style 0.3-0.7 mm long, 25%-50% of the length of the achene.
Salt tolerant to 7‰-8‰, mostly in inland waters.

The achenes are slightly curved to almost straight, terete, usually without or sometimes with low, short teeth on the dorsal margin - more or less resembling small sausages.

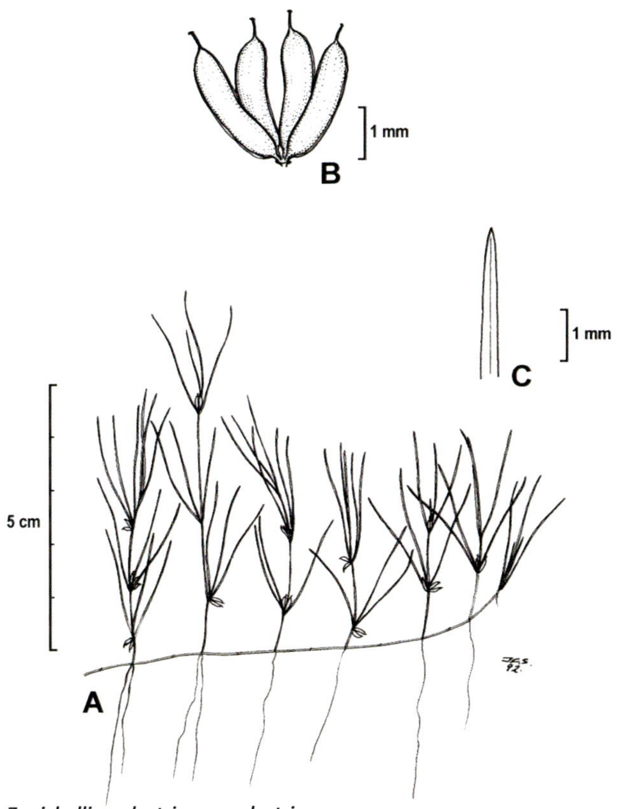

Zanichellia palustris var. ***palustris***
A habit, **B** achenes, **C** leaf apex.

var. ***pedicellata*** Wahlenb. & Rosén

Syn.: *Z. pedicellata* (Wahlenb. & Rosén) Fr.; *Z. pedunculata* Rchb.
DK: Stilk-Vandkrans **N:** Stilkvasskrans **S:** Skaftsärv **FIN:** Otohaura **IS:** -
GB: - **NL:** Gesteelde zannichellia **F:** Zannichellie pédicellée
D: Gestielter Sumpf-Teichfaden **CZ:** - **PL:** Zamętnica błotna trzoneczkowata
EST: - **LV:** - **LT:** - **Ru:** Занникеллия стебельчатая

Leaves narrow, more than 0.3 mm wide in dry material, up to 1.2 mm in fresh material, green.
Fruit usually with 2-4(-5) achenes. Achenes about 2-2.6 mm long, stalk 1-2 mm long, style 1.0-1.5 mm long, (40%-)60%-80% of the length of the achene.
Salt tolerant to 20‰, in inland and coastal habitats.

The achenes are more or less curved, somewhat laterally compressed, often with a row of distinct teeth on the dorsal edge, which can be connected to form a low ridge, but sometimes totally untoothed.

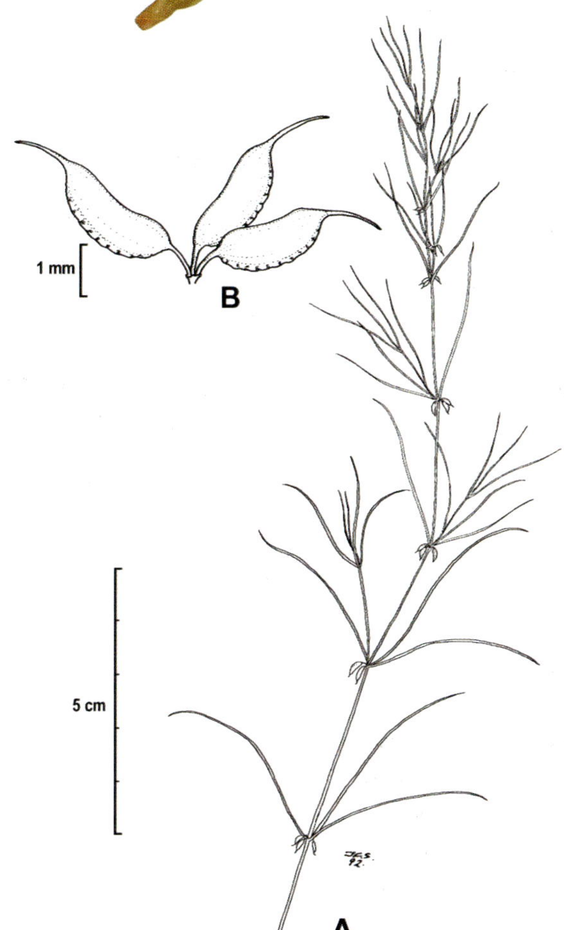

Zanichellia palustris var. ***pedicellata***
A habit, **B** achenes.

Potamogetonaceae

107 *Zannichellia palustris* L. subsp. *major* (Hartm.) Oostr. & Reichg.

Syn.: *Zannichellia palustris* var. *major* Hartm.; *Z. palustris* subsp. *polycarpa* (Rchb.) K.Richt.; *Z. marina* Niels.
DK: Stor Vandkrans **N:** Stor vasskrans **S:** Storsärv **FIN:** Isohaura **IS:** - **GB:** - **NL:** Brede zannichellia **F:** -
D: Vielfrüchtiger Teichfaden **CZ:** - **PL:** -
EST: Harilik hanehein **LV:** Purva diedzene **LT:** Pelkinė vandensargė **Ru:** Заннинкеллия большая

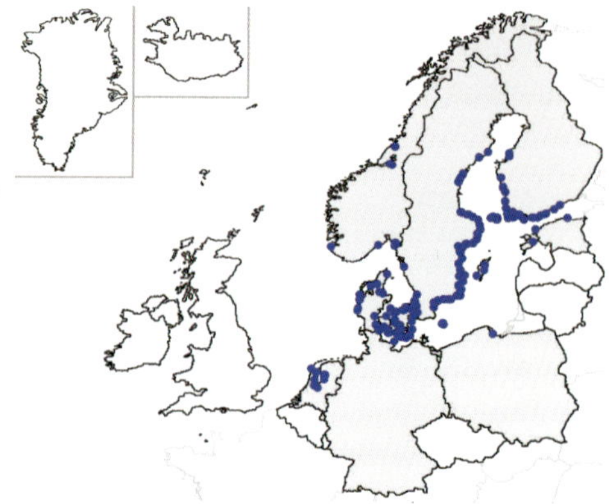

Zannichellia palustris subsp. *major*. Island of Hiddensee, Baltic Sea. Photo Sven Dahlke.

Perennial hydrophyte. Rhizome slender, terete, not always distinctly differentiated from the stem.

Stem: 20–45 cm long, slender, terete, pale brown to greenish, rooting from the lower nodes, not or sparsely branched below.

Leaves: 3–9,5 cm long, (0,5–)1,0–1,9 mm wide, linear, flat, rather thick, entire, dark green, with prominent midrib, gradually tapering to a bluntly acute or obtuse apex. With an air channel on each side of the midrib.

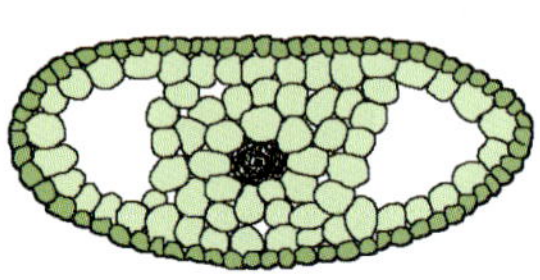

Stem cross section. Drawing JCS.

Leaves sessile, opposite or sometimes alternate on sterile shoots, on fertile shoots apparently in whorls of 3–4. Leaf base with a free, stipule-like ligule, 2.5–3.5 mm long, obtuse.

Inflorescence: Flowers solitary in leaf axils.

Flowers: Male flower with 1(–2) stamens, naked, positioned close to and at the same node as the female flower. Female flower with (3–)5–8 carpels with 1 ovule and surrounded by a cup-shaped, hyaline perianth. Stigma peltate or lingulate. Style 1–2 mm, curved.

Fruits: In groups of (3–)5–8, short pedicellate achenes. Achenes 3–4 mm long, reddish brown, laterally a little compressed. Ventral side of achene more or less straight or slightly curved, sometimes with two small teeth. The dorsal side is curved, with a row of connected teeth forming a low ridge which often extends to the basal part of the 1–2 mm long, curved style.

Flowering: July-September.

Biology: *Zannichellia* is pollinated under water.

Habitats: In brackish or salt water with a salinity of 5‰ to 20‰. It is usually found at depths of 0.3–1.5 metres, but depending on the amount of light sometimes to 5 metres depth.

Distribution within the region:
Native. Coastal areas formerly from the Netherlands through Denmark and Germany to the Gulf of Bothnia and possibly to Estonia and Lithuania? Extinct in the Netherlands.

Characteristics and similar species:
Fresh material of *Z. palustris* subsp. *major* can be distinguished from **106** ***Z. palustris*** subsp. ***palustris*** by the leaves of the latter less than 1 mm wide, with an inconspicuous midrib and the fruits smaller, achenes 1.5–2.5(–3.0) mm long.

108-109 ***Ruppia*** spp. have alternate leaves with a sheathing base but without stipules and finely denticulate leaf apices (use a hand lens).

99 ***Stuckenia pectinata*** has alternate leaves with sheath and stipules.

Zannichellia palustris subsp. *major* in shallow water. East-Jutland, Denmark. Photo JCS.

In *Zannichellia* the inflorescence is situated in the leaf axils. The single naked male and the single female flower are surrounded by a hyaline, cup-shaped perianth formed by remnants of the ligules from the 3 leaves in the whorl. The anther is raised above the 4 peltate stigmas of the flower. *Z. palustris* subsp. *major*. Mariager Fjord, Denmark. Photo JCS.

Zannichellia palustris subsp. *major*. Inflorescence with 5 achenes. Lemvig, Jutland, Denmark. Photo JCS.

Ripe achene. Dorsal side (facing down) with a row of connected teeth. Ventral side almost straight with 2-3 teeth. Photo JCS.

Zannichellia palustris subsp. *major*. Leaf tips. Photo JCS.

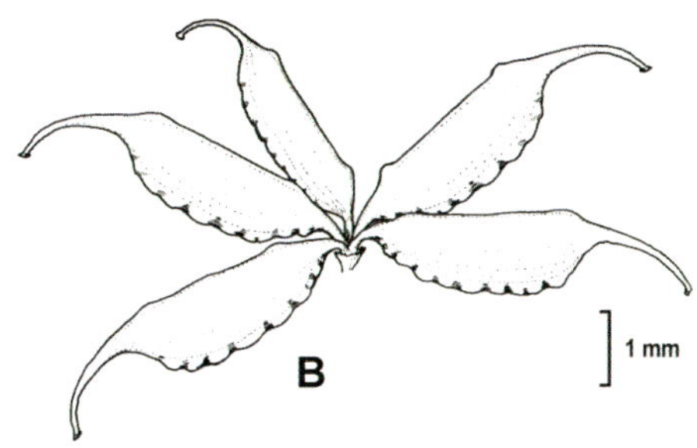

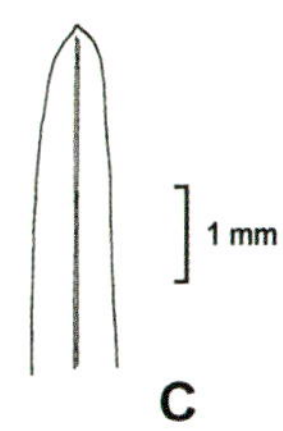

***Zannichellia palustris* subsp. *major*.**
A habit, **B** group of 5 achenes, **C** leaf apex.

Ruppiaceae

108 *Ruppia cirrhosa* (Petagna) Grande

Syn.: *R. spiralis* L. ex Dumort.

DK: Langstilket Havgræs **N:** Skruehavgras **S:** Skruvnating **FIN:** Kiertohapsikka **IS:** -
GB: Spiral Tasselweed **NL:** Spiraalruppia **F:** Ruppia spiralée
D: Schraubige Salde **CZ:** - **PL:** -
EST: Keerd-heinmuda **LV:** - **LT:** Ilgakotė rupija **Ru:** Руппия спиральная

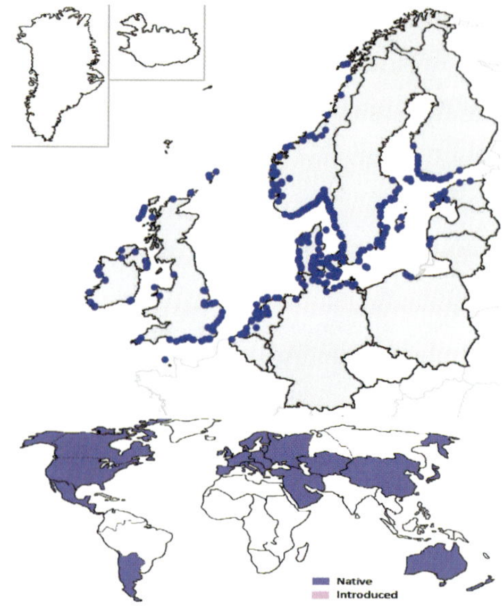

Fruiting *Ruppia cirrhosa* is easily recognised by the long, spirally coiled peduncles. Mariager Fjord, Denmark. Photo JCS.

Perennial hydrophyte. Rhizome monopodial, about 1 mm in diameter, terete, whitish to almost transparent, with 1–2 roots and ascending shoots from every node.

Stem: Up to 40(–60) cm long, slender, terete, branched.

Leaves: Dark green, often with a reddish tinge, filiform to linear, about 1 mm wide when fresh, narrower when dry, elliptic to subterete in cross section near the base, flatter towards the apex, entire except for the obtuse to rounded, symmetrical, finely denticulate apex. With one large and some small air channels on each side of the midrib. Leaves sessile, alternate, but may appear opposite when subtending an inflorescence or side shoot. Leaf sheaths 1–2.5 cm long, open, convolute, partly hyaline, without a ligule. Sheaths of leaves subtending inflorescences hyaline and dilated.

Inflorescence: A 2-flowered spike. Peduncle slender, very long - 5 to 30 cm or more, spirally coiled after fertilisation.

Flowers: Naked, each with 4 carpels and 2 anthers with 2 lobes (thus appearing as 4).

Fruits: In 2 groups of 4 drupelets. These appear almost sessile at anthesis, but as they mature the carpel stalks elongate up to 3.2 cm, forming an umbel-like group at the end of the peduncle. Drupelets 2.4–3.0 mm long, more or less asymmetrical, pyriform with a short beak.

Flowering: July-September.

Biology: The peduncle lifts the inflorescence to the water surface, where the finely mucronate, sausage-like pollen grains are released and later caught on the ring-shaped stigmas. Self-pollination is probably possible. After fertilisation the peduncle coils up and the fruits ripen under water. Vegetative propagation is by fragmentation of the rhizome.

Habitats: In brackish and saline waters, salt tolerant to 5‰–18‰. On sandy or clay mixed with sandy substrate in sheltered locations along coasts, in fjords, inlets and coastal lagoons, in channels and pools in salt marshes. In shallow water to a depth of about 2 metres or more.

Distribution within the region:
Native. Widespread and fairly frequent in coastal areas throughout most of the region except parts of Britain, northern Norway, the Gulf of Bothnia, Iceland and Greenland.

Characteristics and similar species:
Ruppia cirrhosa can be recognised by the obtuse, more or less symmetrical, finely denticulate leaf apices, by the

Ruppia cirrhosa reproduces vegetatively with creeping rhizomes. Fehmarn, Germany. Photo KvdW.

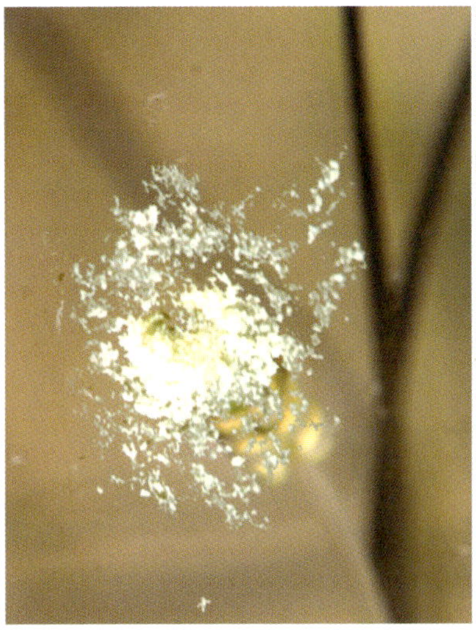

Ruppia cirrhosa.
Inflorescence with unopened anthers. Pollen realeased on the water surface. Immature fruits ripening below water. Photos JCS.

dilated sheaths of leaves subtending the inflorescence, and by the long, soon spirally coiled peduncle with long-stalked drupelets.

109 *Ruppia maritima* differs in much shorter peduncles - up to 3.5 cm, narrower leaves and acute, somewhat asymmetrical leaf tips.

99 *Stuckenia pectinata* has leaf sheaths with ligules, acuminate to acute entire leaf apices and the rhizome with shoots arising from every second node. **110-112 *Zostera*** spp. have wider leaves with rounded or emarginate, entire leaf apices and yellowish to brownish rhizomes.

106-107 *Zannichellia palustris* has opposite or whorled leaves, and very different fruits.

Hybrids: *R. cirrhosa × maritima.*

Ruppia cirrhosa. Leaf tip. Photo KvdW.

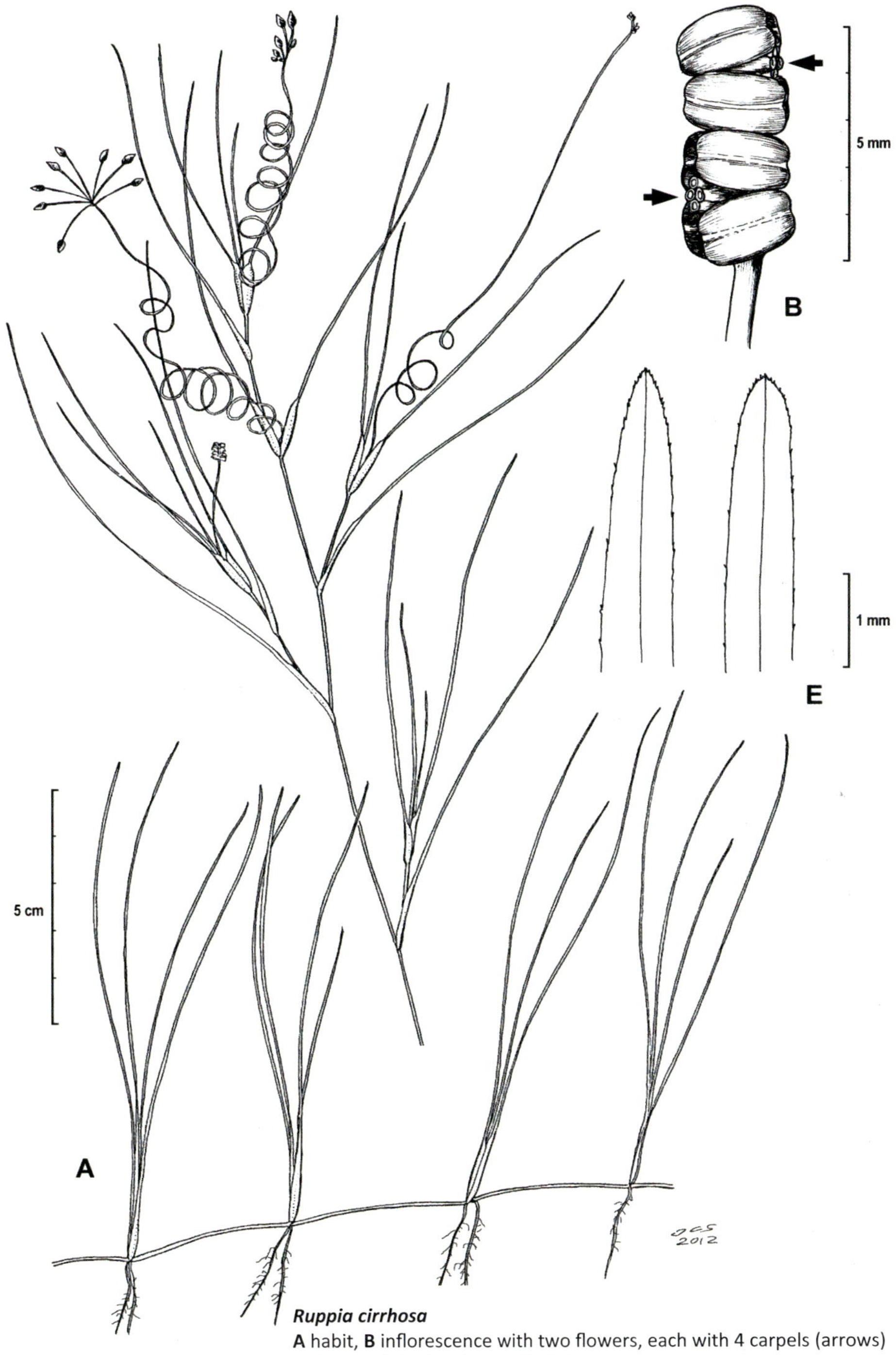

Ruppia cirrhosa
A habit, **B** inflorescence with two flowers, each with 4 carpels (arrows) and two 2-lobed anthers, **E** leaf tips.

Ruppiaceae

109 *Ruppia maritima* L.

Syn.: *R. rostellata* W.D.J.Koch; *R. brachypus* J.Gay
DK: Almindelig Havgræs **N:** Småhavgras **S:** Hårnating **FIN:** Merihapsikka **IS:** Lónajurt
GB: Beaked Tasselweed **NL:** Zee-ruppia **F:** Ruppia maritime
D: Meeres-Salde **CZ:** - **PL:** Rupia morska
EST: Harilik heinmuda **LV:** Jūras rupija **LT:** Jūrinė rupija **Ru:** Руппия морская

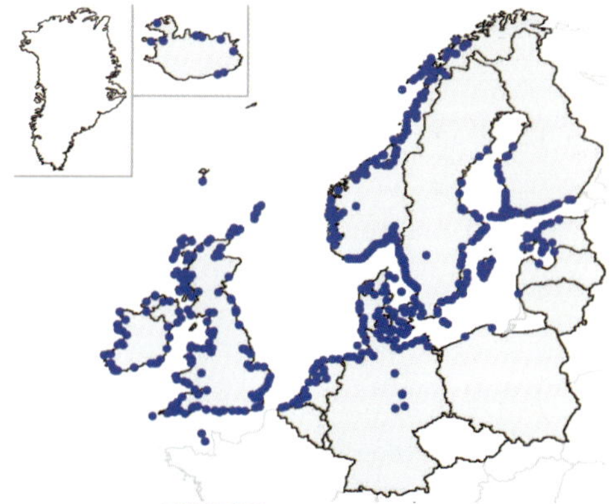

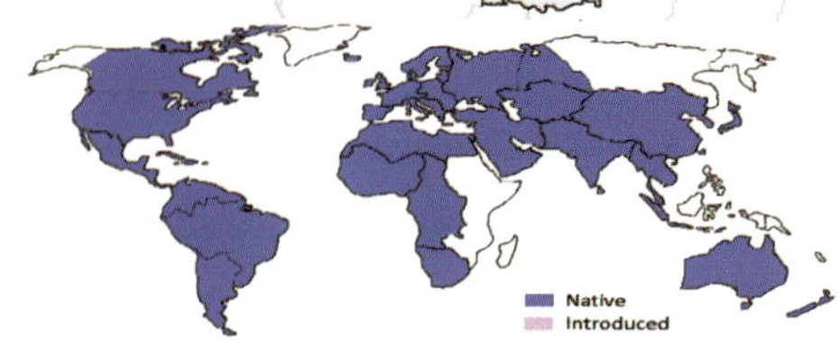

Fruiting *R. maritima.* Note the short peduncle and the long carpel stalks. Risgårde, Himmerland, Denmark. Photo Birgit Knudsen.

Perennial hydrophyte. Rhizome monopodial, about 1 mm in diameter, terete, whitish to almost translucent, with 1–2 roots and ascending shoots from every node.

Stem: Up to 40 cm long, slender, terete, more or less branched.

Leaves: Green, eventually with a reddish tinge, filiform to linear, about 0.5 mm wide when fresh, narrower when dry, base elliptic to subterete in cross section near the base, flatter and thin towards the apex, entire except for the acute, asymmetrical, irregularly denticulate apex. With an air channel on each side of the midrib. Leaves sessile, alternate, but may appear opposite when subtending an inflorescence or side shoot. Leaf sheaths 1–2.2 cm long, open, convolute, partly hyaline, without a ligule. Sheaths of leaves subtending inflorescences hyaline and slightly dilated.

Inflorescence: A 2-flowered spike. Peduncle slender, short, 1.0–2.6 cm.

Flowers: Naked, each with 4 carpels and 2 anthers with 2 lobes (thus appearing as 4).

Fruits: In 2 groups of 4 drupelets. These appear more or less sessile at anthesis, but as they mature the carpel stalks elongate up to 3.5 cm, forming an umbel-like group at the end of the peduncle. Drupelets 1.5–2.7 mm long, asymmetrical, pyriform with a short beak.

Flowering: July-September.

Biology: Pollination as in *Ruppia cirrhosa*. Self pollination is possible with pollen being released before the flowers emerge from the leaf sheath. Vegetative propagation is by fragmentation of the rhizome.

Habitats: In brackish and saline waters, salt tolerant to 3‰–18‰. On sandy or clay mixed with sandy substrate in sheltered locations along the coasts, in fjords, inlets and coastal lagoons, as well as channels and pools in salt marshes. In shallow water to a depth of 1 metre, sometimes to 4.5 metres.

Distribution within the region:
Native. Scattered to locally frequent in coastal areas throughout most of the region, except Greenland. Rarely inland.

Characteristics and similar species:
Ruppia maritima can be recognised by the narrow leaves with asymmetrical,

Ruppia maritima. In var. *brevirostris* C.Agardh the carpel stalks are about as long as the drupelets. Als Odde, Jutland, Denmark. Photo JCS.

Ruppia maritima. Fruit with long-stalked drupelets. Risgårde, Himmerland, Denmark. Photo Birgit Knudsen.

Ruppia maritima. Vegetative plants on sandy substrate. Insel Fehmarn, Germany. Photo KvdW.

irregularly denticulate apices, peduncles 1.0–2.6 cm with up to 8 drupelets on 3–35 mm long stalks.

When not fruiting, it can be mistaken for **108 *R. cirrhosa***, but differs in the narrower leaves with acute, asymmetrical, irregularly denticulate apices.

99 *Stuckenia pectinata* has leaf sheaths with ligules, acuminate to acute, entire leaf apices and rhizome with shoots arising from every second node.

110-112 *Zostera* spp. have wider leaves with rounded or emarginate, entire leaf apices and yellowish to brownish rhizomes.

106-107 *Zannichellia palustris* have opposite or whorled leaves, and very different fruits.

Hybrids: *R. cirrhosa × maritima.*

Ruppia maritima. Leaf tip. Photo KvdW.

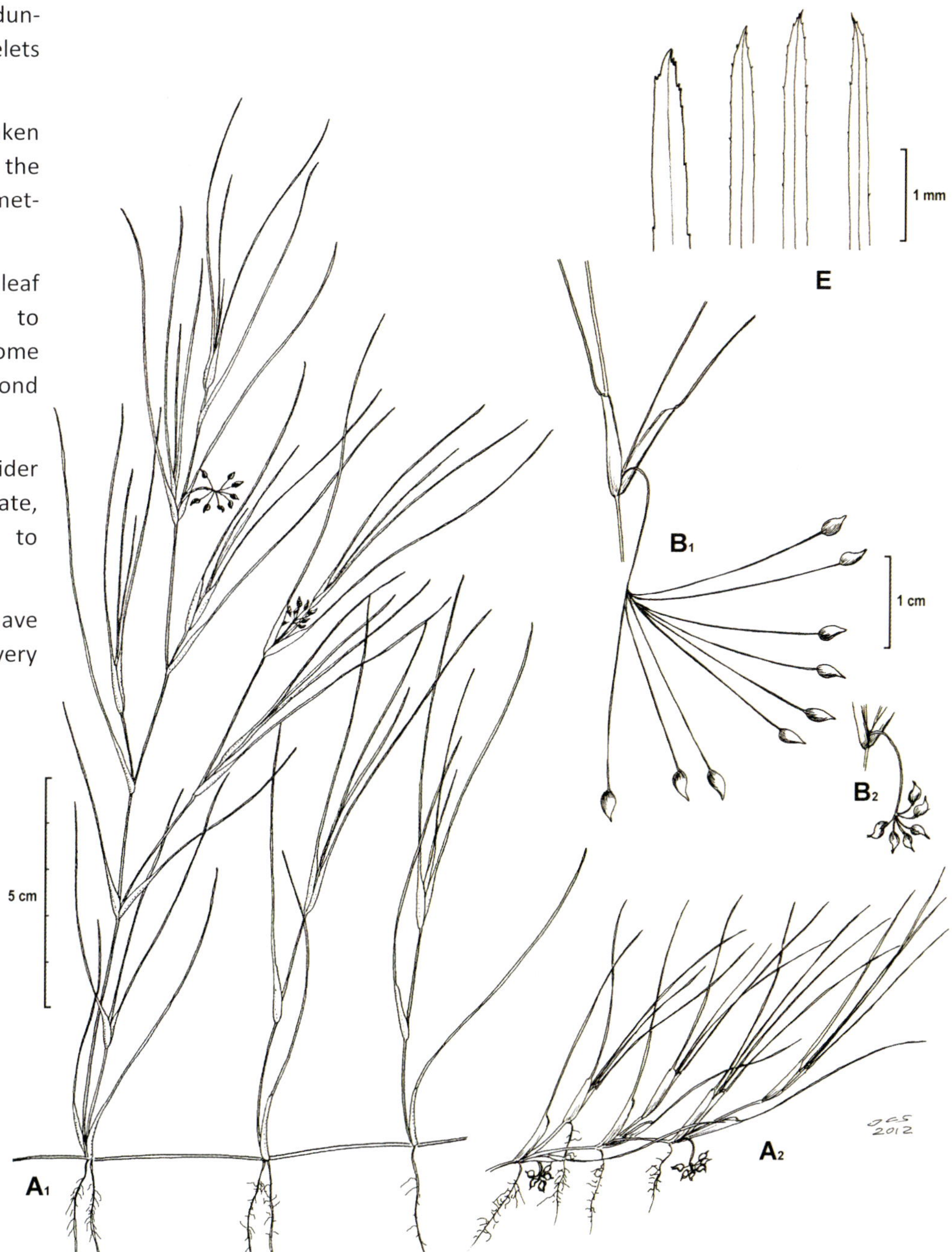

Ruppia maritima
A1 habit var. *maritima*, **A2** hab var. *brevirostris*, **B1** fruiting head var. *maritima*, **B2** fruiting head var. *brevirostris*, **E** leaf tips.

Zosteraceae

110 *Zostera marina* L.

DK: Almindelig Bændeltang **N:** Ålegras **S:** Bandtång **FIN:** Meriajokas **IS:** -
GB: Common Eelgrass **NL:** Gewoon zeegras **F:** Zostère marine
D: Echtes Seegras **CZ:** - **PL:** Zostera morska
EST: Pikk merihein **LV:** Parastā jūraszāle **LT:** Jūrinis andras **Ru:** Взморник морской

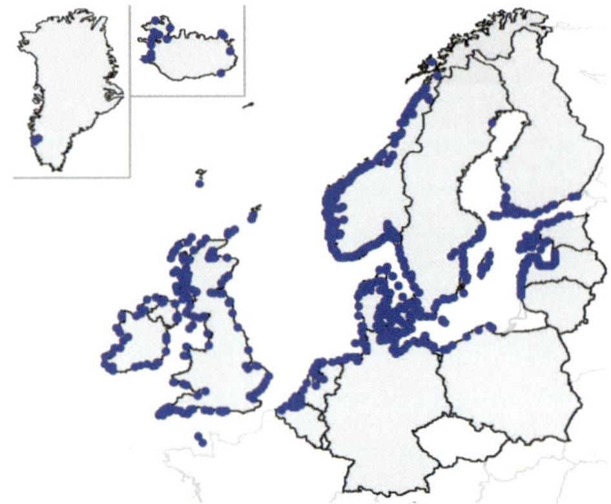

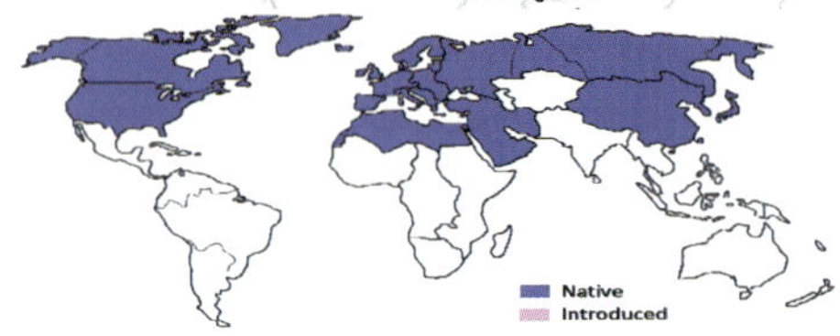

Zostera marina in dense formation. Insel Fehmarn, Germany. Photo KvdW.

Perennial hydrophyte. Rhizome monopodial, creeping, 2–5 mm in diameter, somewhat compressed, greenish to brownish. All nodes with a leaf, a short axillary shoot and numerous roots.

Stem: Vegetative and flowering stems lateral or terminal, short with several distichous leaves.

Leaves: Dark green, becoming black when dried, firm, linear, with 5–9(–11) veins (including marginal veins). Leaf apex rounded when young, sometimes with a short mucro, older leaves somewhat emarginate. Leaves of vegetative shoots typically 5–10(–12) mm wide and 20–50 cm long, but under favourable conditions up to 1 m or longer. Leaves of fertile shoots are narrower and shorter.

Inflorescence: Flowering shoots up to 60(–150) cm long, branched, with numerous spathes. The distichous inflorescence (spadix), with alternating male and female flowers opposite each other, is enclosed in a 4–8 cm long, 3–4 mm wide sheath. This has auricles and a short ligule at the top, green on the dorsal side, hyaline on the ventral side, gradually splitting at anthesis. The lamina of the spathe is constricted at the base, up to 20 cm long and narrower than the other leaves. Peduncle 2–3 (–4) cm long.

Flowers: Naked. Male flower with 1 anther, soon splitting into two halves. Female flower with 2 stigmas almost twice as long as the style.

Fruits: Sessile drupes, 3–3.5 mm long (excluding style) at maturity, with a scarious pericarp, which when removed or eroded reveals the light brown seed, with prominent longitudinal ribs.

Flowering: July-August.

Biology: Fruiting rather sparse and propagation mainly vegetative by fragmentation of the rhizome. For details on pollination see **111 *Zostera angustifolia***. The limited sexual reproduction may influence resistance to *Labyrinthula zosterae*, a microscopic slime mold, which wiped out 90% of eelgrass populations in Europe and along the North American East Coast in the 1930s–1940s. Only populations in brackish water survived.

Habitats: In saline or brackish water at salt concentrations ranging from 5‰–40‰. On muddy, sandy or stony substrate on sea coasts, from low water mark to a depth of 4(–10) metres.

Distribution within the region:
Native. Widespread and locally abundant in coastal areas throughout most of the region, except Greenland where it is rare; absent from the Gulf of Bothnia and Iceland.

Characteristics and similar species:
Zostera marina can be recognised by the 5–9(–11) veins (including marginal veins), dark green, firm, linear leaves, with blunt or very shortly mucronate apices, by the stigmas being about twice as long as the style, and by the 3–3.5 mm long, longitudinally ribbed fruits.

Var. ***stenophylla*** Asch. & Graebn. with 2–4 mm wide, 3-veined leaves (plus

Zostera marina. Rhizome. Als, Denmark. Photo JCS.

Zostera marina. Inflorescence with distichous arranged flowers enclosed in the spathal sheath. Inserted: Immature fruit. Helgenæs, Denmark. Photo JCS.

marginal veins). The lateral veins almost as close to the margin as to the midrib. It can easily be mistaken for **111 *Z. angustifolia***, which differs in the shorter fruits and stigmas and by the stigma and style of equal length.

112 *Zostera noltii* has 0.5–1.5 mm wide, 1(–3) veined leaves, and dark brown fruits without ribs.

Note: In Germany, the Netherlands and Belgium the *Zostera marina* aggregate is treated as one species only, while in Britain, Norway, Sweden and Denmark it is split into two species *marina* and *angustifolia* based on several constant vegetative and reproductive features, and by being ecologically distinct. Material from Greenland, Alaska and Canada shows the same differences. Transitional forms between the two species are not observed, where they are mixed.

Because of the different treatment, distribution data for Germany, the Netherlands, Belgium and Greenland is incomplete. In this work we accept both taxa as species.

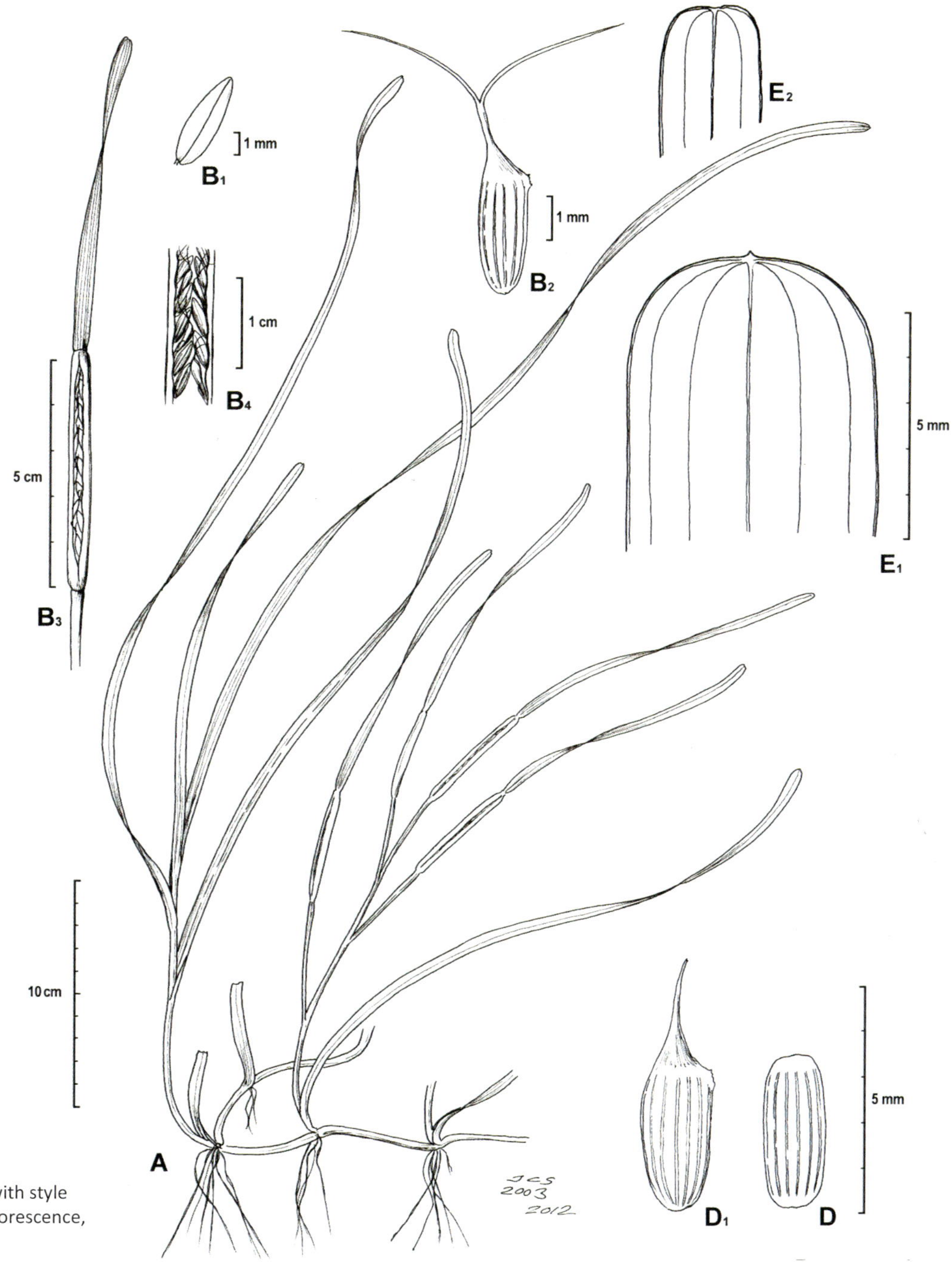

Zostera marina
A habit, **B1** male flower, **B2** female flower with style and stigma, **B3** inflorescence, **B4** part of inflorescence, **D** drupe, **D1** fruit with style, **E1, E2** leaf tips.

Zosteraceae

111 *Zostera angustifolia* (Hornem.) Rchb.

Syn.: *Z. marina* L. var. *angustifolia* Hornem.; *Z. hornemanniana* Tutin; *Z. marina* subsp. *hornemanniana* (Tutin) Lemke
DK: Smalbladet Bændeltang **N:** Smalt ålegras **S:** Smal bandtång **FIN:** - **IS:** Marhálmur
GB: Narrow-leaved Eelgrass **NL:** - **F:** Zostère à feuilles étroites
D: Schmalblättriges Seegras **CZ:** - **PL:** -
EST: -**LV:** - **LT:** - **Ru:** -

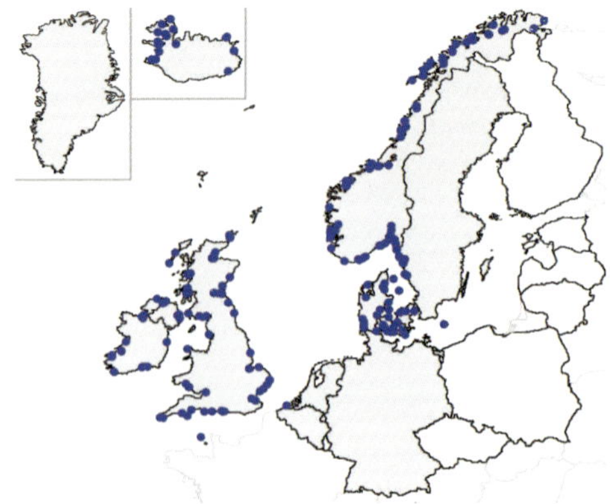

Map incomplete for Greenland, Germany, Netherlands and Belgium.

Zostera angustifolia in shallow water on sandy, muddy substrate. The Wadden Sea close to the Rømø Dam, Denmark. Photo JCS.

Perennial hydrophyte. Rhizome monopodial, creeping, 1–2 mm in diameter, somewhat compressed, greenish to brownish. All nodes with a leaf, a short axillary shoot and numerous roots.

Leaves: Light green to green, becoming black when dry, flaccid, soft, linear, with 5 veins (including marginal veins). The lateral veins closer to the margin than to the midrib. Leaf apex rounded when young, somewhat emarginate when older. Leaves of vegetative shoots typically 15–30 cm long and 2 mm wide, in winter a little more than 1 mm wide. Leaves of fertile shoots 4–15 cm long, and up to 3 mm wide.

Inflorescence: Flowering shoots 10–30 cm long, branched. The distichous inflorescence (spadix), with alternating male and female flowers opposite each other, is enclosed in a 4–5(–6) cm long, 2–3 mm wide sheath. This with auricles and a short ligule at the top, green on the dorsal side, hyaline on the ventral side, gradually splitting at anthesis. The lamina of the spathe constricted at the base, 4–8(–15) cm long and up to 3 mm wide. Peduncle 2–3(–4) cm long.

Flowers: Naked. Male flower with 1 anther, soon splitting into two halves. Female flower with 2 stigmas more or less the same length as the style.

Fruits: Sessile drupes, 2.2–2.7(–3) mm long (excluding style) at maturity, with a scarious pericarp, which when removed or eroded reveals the light brown seed, with prominent longitudinal ribs.

Flowering: June-November.

Biology: Vegetative propagation is by fragmentation of the rhizome. Typically abundantly fruiting. Flowering under water, starting with a female stage, where the stigmas extend from the sheath to catch drifting pollen. When the stigmas wither the male stage takes over with the release of all the thread-like pollen grains at one time.

Habitats: In saline or brackish water at salt concentrations of 9‰ or above. On soft muddy or sandy substrate on the coast and in estuaries, typically in low-lying areas that retain standing water at low tide. From the half-tide mark to the low-tide mark, sometimes to a depth of 1.5 m, rarely more.

Distribution within the region:
Native. Along the coasts of southwest Greenland, Iceland, Norway, Sweden, Denmark, Britain, Ireland, the Netherlands and Germany.

Zostera angustifolia fruits. The left one is free of the scarious pericarp. The Wadden Sea, Denmark. Photo JCS.

Leaf tips of all 3 *Zostera* species photographed with strong backlighting.
Left: *Z. noltii* with midrib and marginal veins only.
Centre: *Z. angustifolia* with 3 main veins and marginal veins.
Right: Rather narrow-leaved form of *Z. marina* with 5 main veins and marginal veins.
Photos JCS.

Characteristics and similar species: *Zostera angustifolia* can be recognised by the 1–2 mm wide, 5-veined (including marginal veins) leaves with blunt to emarginate apex, by the stigmas being about as long as the style, and by the 2.2–2.7(–3) mm long, longitudinally ribbed fruits.

110 ***Zostera marina*** normally has wider leaves with 5–9(–11) veins (including marginal veins), 3–3.5 mm long fruits and stigmas which are about twice as long as the style. ***Z. marina*** var. ***stenophylla*** Asch. & Graebn. has narrow leaves with 5 veins (including marginal veins), but in addition to the fruit and stigma/style characters, it differs vegetatively by the lateral veins of leaves almost as close to the margin as to the midrib (this character should not be used in isolation).

112 ***Zostera noltii*** has 0.5–1.5 mm wide, 1-veined leaves, and dark brown fruits without ribs.

108-109 ***Ruppia*** spp. have finely toothed leaf apices (use a hand lens).

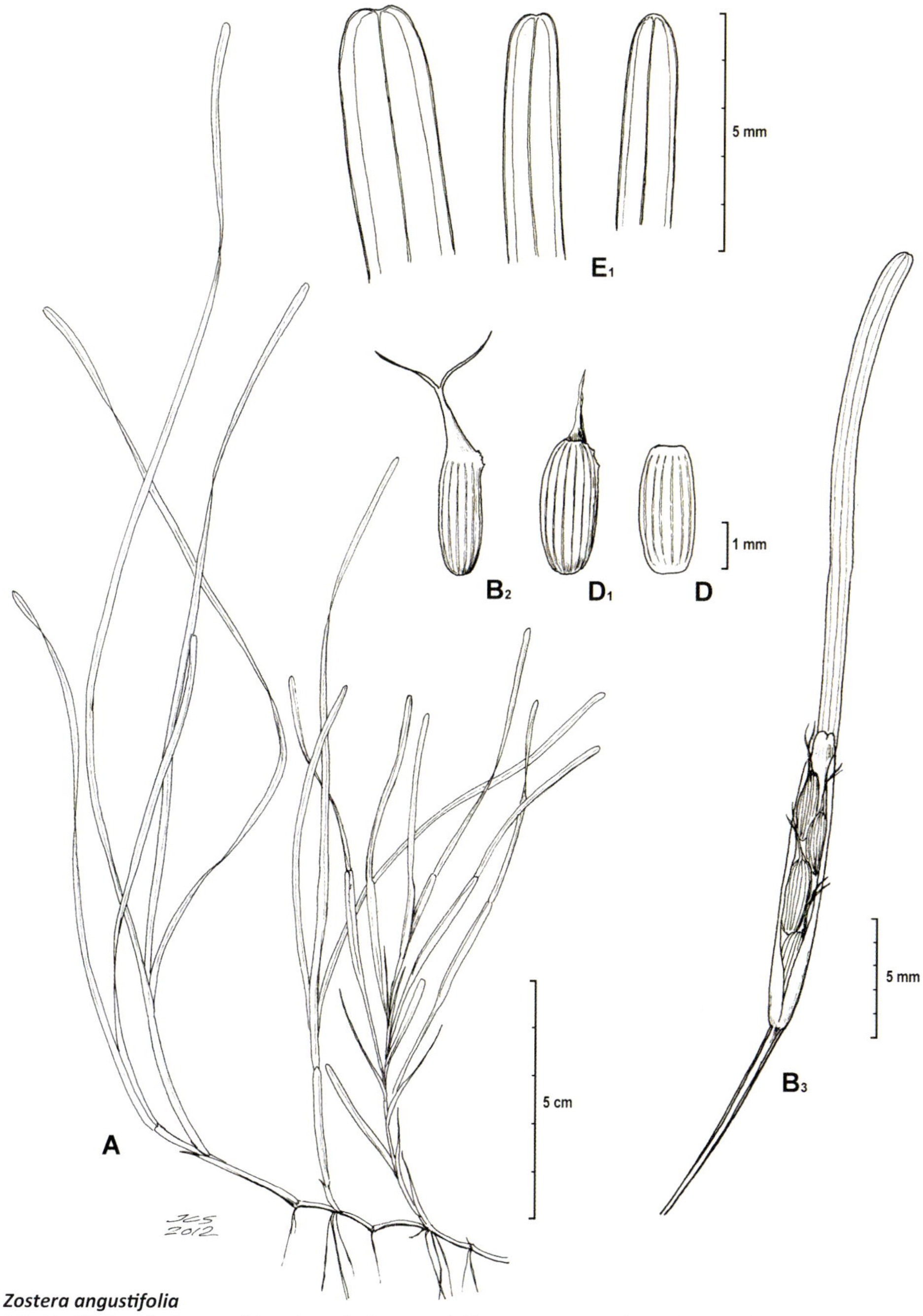

Zostera angustifolia
A habit, **B2** female flower with style and stigma, **B3** inflorescence, **D** seed, **D1** drupe with style, **E1** leaf tips.

Zosteraceae

112 *Zostera noltii* Hornem.

Syn.: *Z. minor* Nolte ex Rchb.; *Z. nana* Roth. p.p.
DK: Dværg-Bændeltang **N:** Dvergålegras **S:** Dvärgbandtång **FIN:** Pikkuajokas **IS:** -
GB: Dwarf Eelgrass **NL:** Klein zeegras **F:** Zostère naine
D: Zwerg-Seegras **CZ:** - **PL:** Zostera drobna
EST: -**LV:** - **LT:** Mažasis andras **Ru:** Взморник малый

In Hornemann's original description in *Flora Danica* 1832 tab. MMXLI the species was named *Zostera noltii,* but in floras later often spelled as *Zostera noltei*.

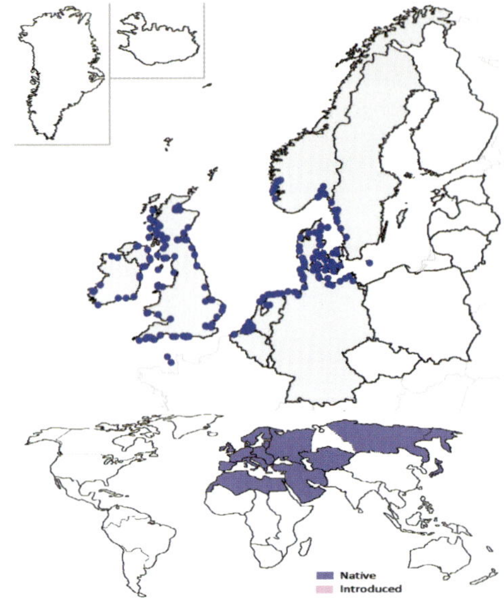

Zostera noltii on muddy sand at low tide. Vlissingen, Netherlands. Photo KvdW.

Perennial hydrophyte. Rhizome monopodial, creeping, 0.5–1 mm thick, terete, yellow reddish. All nodes with a leaf, a short axillary shoot and a number of roots.

Leaves: Yellowish green to green, becoming black when dry, linear, up to 20 cm long, 0.5–1.5 mm wide, flaccid, soft, with 3 veins (including marginal veins). Leaf apex rounded when young, but becoming emarginate with age.

Inflorescence: Flowering shoots unbranched, with 1–4(–6) spathes. The distichous inflorescence (spadix) is enclosed in a 1–2 cm long, 1.5–2 mm wide sheath. Spadix with membranous projections (retinacula) and 4–5 male and 4–5 female flowers. Male and female flowers alternating, opposite each other. The sheaths of the spathe with auricles and a very short ligule at the top, green on the dorsal side, hyaline on the ventral side, gradually splitting at anthesis. Lamina of spathes 3–8 cm long and up to 1.5 mm wide. Peduncles 0.5–1.5 cm long.

Flowers: Naked. Male flower with 1 anther, soon splitting into two halves. Female flower with 2 stigmas about as long as the style.

Fruits: Sessile drupes. Dark brown, dull, without ribs, 1.5–2 mm long (excluding style) with a scarious pericarp, which when removed or eroded reveals the reddish brown seed without ribs.

Flowering: June-August.

Biology: As **111** ***Z. angustifolia***.

Habitats: In saline or brackish water at salt concentrations of 9‰ or above. On soft, muddy or sandy substrate on the coast and in estuaries, from half-tide mark to low-tide mark, sometimes to a depth of 1 metre.

Distribution within the region:
Native. Widespread and locally abundant on the coasts of Ireland, Britain, the Netherlands, Belgium, Germany, Denmark, southern Norway and Sweden.

Zostera noltii. Open spathes each with 4 fruits. Those in front without epicarp, showing the shiny, reddish-brown seeds. Wadden Sea, Denmark. Photo JCS.

Zostera noltii uprooted to show rhizome. Dokkedal, Jutland, Denmark. Photo JCS.

Zostera noltii. Spathes. Spadix with distichously arranged flowers and retinacula is visible. Photo JCS.

Characteristics and similar species:
Zostera noltii can be recognised by the 0.5–1.5 mm wide leaves with emarginate apex, with only a midrib and marginal veins. The dark brown fruits without ribs are also characteristic.

It may be confused with young or narrow-leaved forms of **110** ***Zostera marina*** and especially **111** ***Z. angustifolia*** with which it often occurs. Both of these, however, have leaves with 3 or more veins (plus marginal veins) and ribbed fruits.

108-109 ***Ruppia*** spp. have finely toothed leaf apices (use a hand lens).

106-107 ***Zannichellia palustris*** has opposite or whorled leaves, very different fruits and rhizome rooting from every second node.

99 ***Stuckenia pectinata*** has the leaf-sheath with a ligule, acuminate to acute, entire leaf apex and rhizome with shoots from every second node.

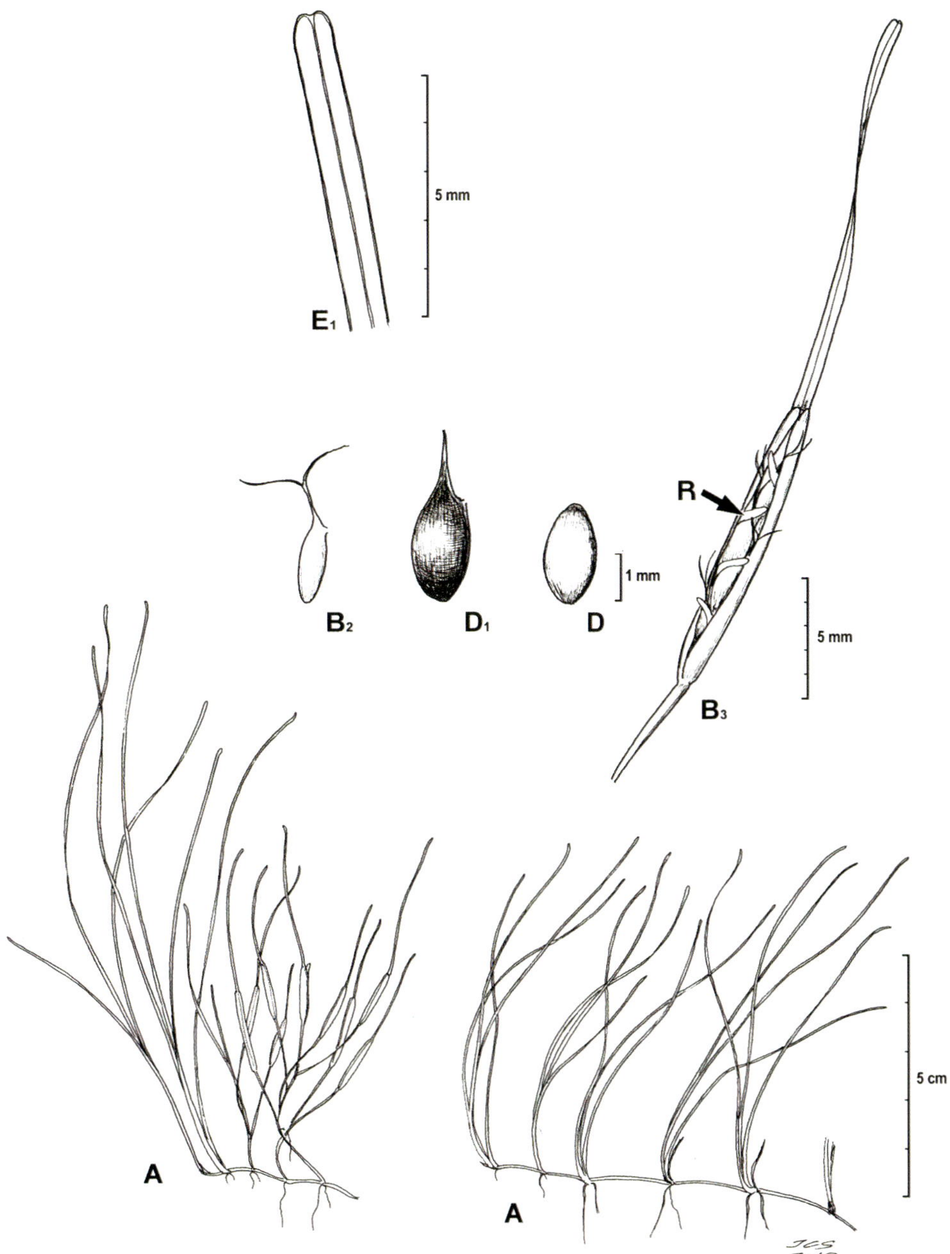

Zostera noltii
A habit, **B2** female flower with style and stigmas, **B3** inflorescence, **D1** drupe with style, **D** seed, **E1** leaf tip, **R** retinacula.

Iridaceae

113 *Iris pseudacorus* L.

DK: Gul Iris **N:** Sverdlilje **S:** Svärdslilja **FIN:** Keltakurjenmiekka **IS:** -
GB: Yellow Iris **NL:** Gele lis **F:** Iris jaune
D: Sumpf-Schwertlilie **CZ:** Kosatec žlutý **PL:** Kosaciec żólty
EST: Kollane võhumõõk **LV:** Purva skalbe **LT:** Geltonasis vilkdalgis **Ru:** Ирис ложноаировый

Iris pseudacorus. Lake Kielstrup, Mariager Fjord, Denmark. Photo JCS.

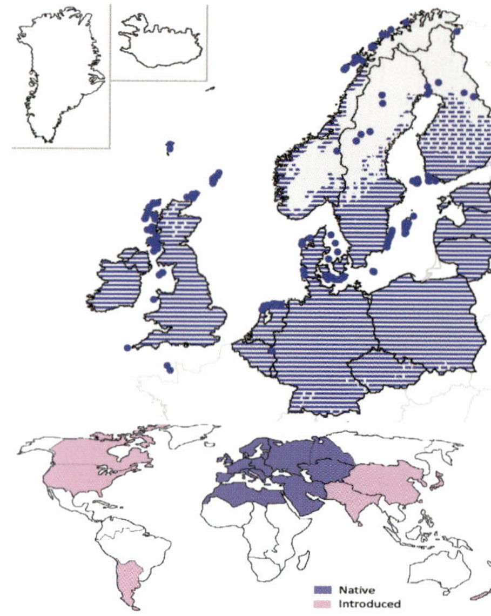

Perennial helophyte. Rhizome robust, richly branched, up to 3 cm in diameter and typically covered in layers of the fibrous remains of dead leaves. Roots whitish to light brown.

Stem: 60–120 cm long, slightly compressed.

Leaves: Up to 90 cm long, 1.5–3 cm wide, sessile and laterally compressed (isobilateral) with identical faces, greyish green or green, ensiform with a robust and prominent midrib. Leaves distichous, emerging like a fan from the terminal end of the rhizome. The lower part of each leaf forms a narrow sheath that encloses the base of the neighbouring younger leaf. Stem leaves alternate with same shape as the basal leaves.

Inflorescence: With few, robust and upright flowering branches, the lower ones typically long. Flowers 2–3 together in tassels emerging from 2 short, green bracts.

Flowers: Yellow, 7–10 cm in diameter. Perianth tube 1–1.5 cm long. Outer perianth segments 6–7 cm long, narrow at base, but widening to a 2–3 cm wide, downward-reflexed, lip-shaped outer part. Inner perianth segments 2–3 cm long, 4–8 mm wide, more or less upright. Style divided into 3 tepal-like branches which are fused with the inner perianth segments, forming three tubes that guide insects towards the perianth tube. Anthers fused with the abaxial side of the style branches.

Fruits: Capsules 3.8–5.6 cm long, oblong, with 6 longitudinal furrows and a short beak, obtuse triangular in cross section. Seeds approximately 7 mm long, flat, D shaped to circular, brown with smooth surface at maturity, densely stacked in the capsule.

Flowering: June-August.

Biology: Pollination by insects. Seeds float, enabling dispersal by water. Vigorous vegetative propagation by branching rhizomes in the period after flowering. Often covering large areas in wetlands. All parts of the plant are poisonous, especially the rhizome. Not typically wintergreen, but short leaves may remain green when covered by water or in mild winters.

Habitats: Nutrient-poor to nutrient-rich mineral as well as organic soils, in the littoral zone of lakes, ponds, streams and rivers, on canal and ditch margins, meadow and marshes, including in woodland.

Distribution within the region: Native. Widespread and abundant throughout most of the region, less frequent in the north. Absent from Iceland and Greenland.

Characteristics and similar species: Flowering plants are very distinctive and easily distinguished from all other species. Sterile plants are primarily characterised by the glaucous or greyish green, ensiform leaves arranged in a fan. It is not possible to separate ster-

Iris pseudacorus. The oblong and obtusely triangular capsules overhanging the water. Photo JCS.

Iris pseudacorus. The glaucous, ensiform leaves are characteristic. Randers, Denmark. Photo JCS.

Iris pseudacorus bordering a small stream. Elbæk northwest of Engesvang, Denmark. Photo BM.

Iris pseudacorus. Richly branched rhizome with many scars – the residues from dead leaves and roots. Photo BM.

ile specimens of *Iris pseudacorus* from other *Iris* spp.

9 ***Acorus calamus*** has yellowish-green leaves and when crushed the whole plant has a characteristic aromatic smell.

Iris pseudacorus. The shoots are flat and the leaves are Y or V shaped in cross section at the base (with younger leaves arising within the fork of the Y. Photo JCS.

Iris pseu.dacorus The large yellow outer tepals serve as landing strips for bumblebees and other insects with a long proboscis. The brownish marks guide the insects towards the tube that is formed by the fused style branch and the sepal, leading to the nectar. Photo JCS.

Iris pseudacorus
A rhizome and lower part of stem, **A1** inflorescence, **C-C1** open and closed capsule**, E** part of leaf and leaf sheath, **F** seed.

Typhaceae

Key to *Sparganium* and *Typha*

Several *Sparganium* hybrids are recognised within the study area and presented on pages 300-301. They cannot be identified by use of the key below.

1 Flowers and fruits in globose unisexual heads in spikes or panicles. Leaves flat or keeled (***1***)
  Sparganium 2

1 Flowers in dense racemes, forming a cylindrical spadix. Leaves in cross section flat convex-concave (***2***)
  Typha 15

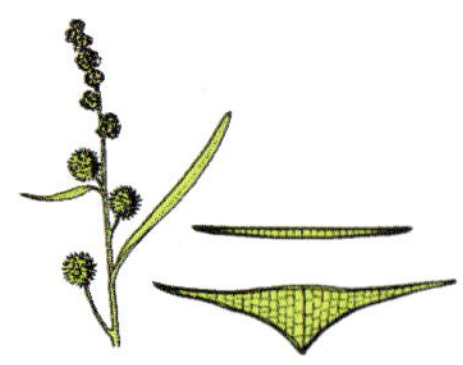
1

2

Sparganium

2 Inflorescence branched with several heads on main stem and branches (***3***, ***3a***) 3

2 Inflorescence not branched (***4***) 9

3 Leaves 7–20 mm wide, erect or ascending, sometimes floating. Main axis of inflorescence almost straight (***3***)
  120-122 *S. erectum* aggregate
  (more or less ripe fruits needed for determination) 4

3 Leaves 2–3 mm wide, floating. Main axis of inflorescence S-shaped (***3a***)
  117 *S. gramineum*

3 3a

4

4 Fewer than half of stigmas bifid, unbranched (***5***). Fruit medium sized to large (*S. erectum* group) 5

4 More than half of stigmas bifid (***6***). Fruits large, regularly up to 10 mm long
  124 *S. eurycarpum* group
  (not previously recognised within the region) 8

5

6

5 Fruit obpyramidal, upper part flat, pyramid shaped, contracting into a short beak, with distinct lateral ridges, angled (***7***). Perianth segments dark at the top, not visible in mature fruit heads
  120 *S. erectum*

5 Fruit with domed or rounded upper part, lateral ridges fairly inconspicuous, indistinctly angled (***8***, ***9***). Perianth segments not conspicuously dark at the top, visible or not in mature fruit heads 6

6 Fruit fusiform with inconspicuous shoulder and inconspicuous lateral ridges, upper and lower part alike in form and texture (***8***)
  122 *S. neglectum*

6 Fruit fusiform to ellipsoidal, slightly constricted below conspicuous shoulder, with domed upper part tapering into the beak 7

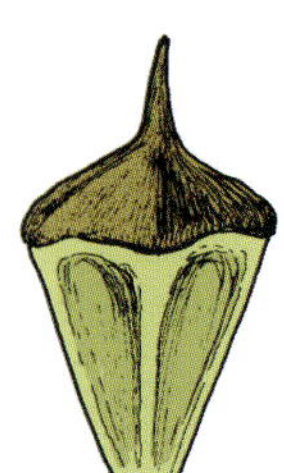
7

8

9

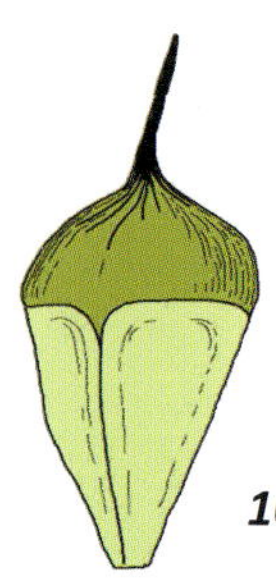
10

7 Upper part of mature fruit dark brown or black, beak up to 2 mm long (***9***). Pedicels up to 1.5 mm long
121 *S. microcarpum*

7 Upper part of mature fruit light coloured or brown, beak up to 4 mm long (***10***). Pedicels missing
123 *S. stoloniferum*
(not yet recognised within the region)

8 Fruit obpyramidal, upper part flattened when mature (***11***)
124 *S. eurycarpum*
(not yet recognised within the region)

8 Fruit rhomboidal to obovoid, upper part conical or pyramidal when mature (***12***)
125 *S. coreanum*
(not yet recognised within the region)

9 Stem leaves keeled (***13***) 10

9 Stem leaves not keeled, sometimes inflated (***14***) 11

10 Male spikes 3–10. Female spikes not contiguous (***15***)
118 *S. emersum*

10 Male spikes 1(–2). Upper female spikes contiguous (***16***)
119 *S. glomeratum*

11 Fruit with beak (***18, 19***) 12

11 Fruit without beak (***17***)
114 *S. hyperboreum*

12 Lowest leaf-like bract shorter than 10 cm and at most slightly longer than the inflorescence (***20***). Usually with 1 male spike
115 *S. natans*

12 Lowest leaf-like bract more than 10 cm long and at least twice as long as the inflorescence (***21***). 2–8 male spikes (sometimes very close together) 13

13 Ripe fruits with a straight beak (***18***). Inflorescence straight or a little curved (***21***)
116 *S. angustifolium*

13 Ripe fruits with a curved beak (***19***). Inflorescence S shaped (***3a***)
117 *S. gramineum*

11

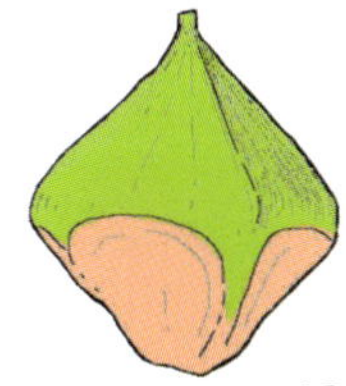
12

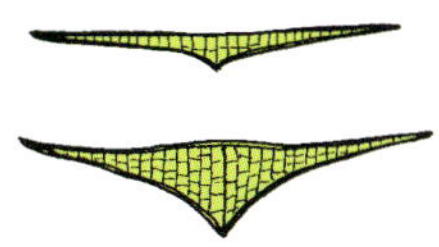
13

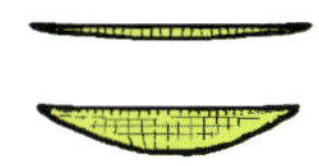
14

15

16

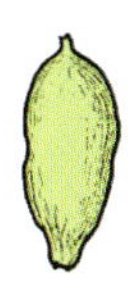
17

18

19

20

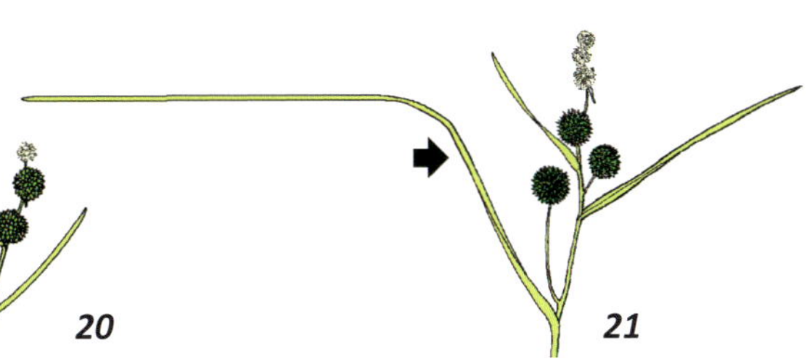
21

Typha

1 Flowering stem without leaves, but with several usually bladeless sheaths at base (***1***). Leaves 1–2(–3) mm wide. Hairs of female flowers with swollen tips (***2***)
 126 *T. minima*

1 Flowering stem with leaves and some bladeless sheaths at base. Leaves (2–)3–20 mm wide. Hairs of female flowers without swollen tips 2

2 Female flowers with bracteoles (***3***). Male and female part of spike remote, rarely contiguous 3

2 Female flowers without bracteoles. Male and female part of spike contiguous or remote 5

3 Basal 1–2 cm of older leaf blades without orange-brown glands on the adaxial side, but with brown glands higher up (***4***). Female flowers with brown bracteoles 4

3 Basal 1–2 cm of older leaf blades with orange-brown glands on the adaxial side (***5***). Female flowers with straw-coloured to light brown bracteoles
 130 *T. domingensis* (only planted)

4 Leaves yellowish green to green. Pedicels 0.5(–0.8) mm long. Pollen solitary
 127 *T. angustifolia*

4 Leaves bluish green. Pedicels 0.9–1.5(–1.7) mm long. Pollen in groups of 1, 2, 3 or 4
 128 *T. ×glauca*

5 Male and female part of spike more or less contiguous, < 10 mm apart (***6***) 6

5 Male and female part of spike remote, > 10 mm apart (***7***) 7

6 Female part of spike dark brown when mature, about as long as the male part
 129 *T. latifolia*

6 Female part of spike silvery grey when mature, distinctly longer than the male part (***8***)
 131 *T. shuttleworthii*

7 Leaves 6–13(–19) mm wide, bluish green
 128 *T. ×glauca*

7 Leaves 2–4(–7) mm wide, yellow green to green
 132 *T. laxmannii*

1

2

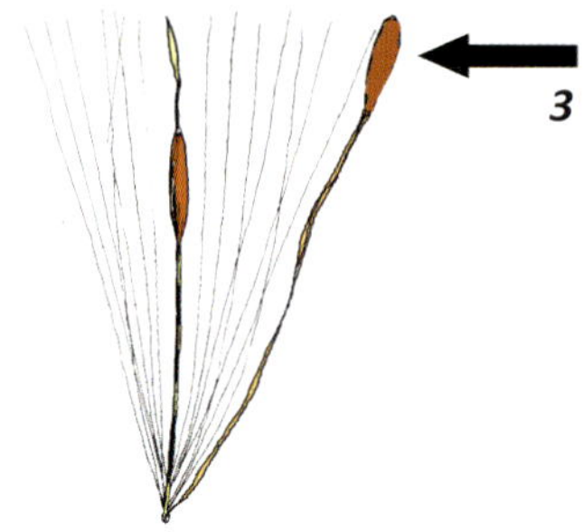
3

4

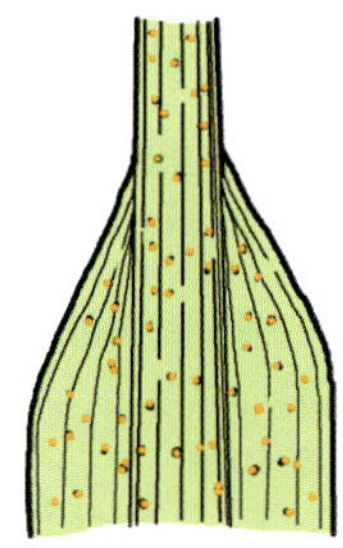
5

6

7

8

S. ×oligocarpon Ångstr. (*S. angustifolium × natans*). Flora Danica. Supplementary volume 3, plate 172. 1874.

Typhaceae

114 ***Sparganium hyperboreum*** Laest. ex Beurl.

Syn.: *S. submuticum* (Hartm.) Neuman

DK: Fjeld-Pindsvineknop **N:** Fjellpiggknopp **S:** Fjälligelknopp **FIN:** Pohjanpalpakko **IS:** Mógrafabrúsi
GB: - **NL:** - **F:** Rubanier hyperboréal
D: Nordischer Igelkolben **CZ:** - **PL:** -
EST: - **LV:** - **LT:** - **Ru:** Ежеголовник северный

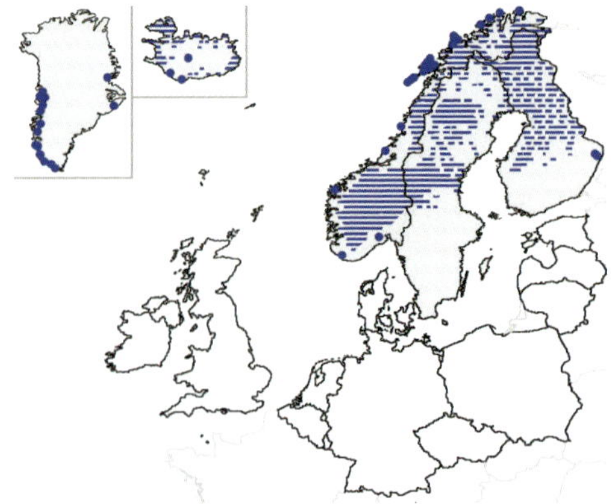

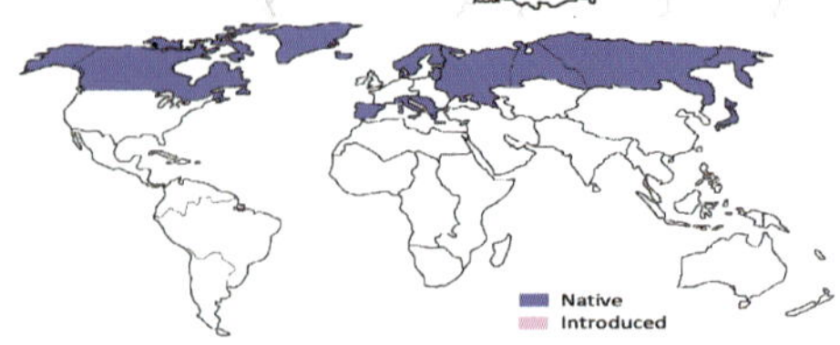

Sparganium hyperboreum. Raubergstulen, Jotumheimen, Norway. Photo Jan-Thomas Johansson.

Perennial hydrophyte or helophyte. Rhizome short, up to 10 cm long, 1–1.2 mm thick, creeping, with scale leaves up to 1 cm long.

Stem: 10–40 cm long, slender, 1–2.5 mm diameter, ascending or procumbent, flexuous, often floating and up to 80 cm long.

Leaves: Yellowish green to green, (1–)2–3(–5) mm wide, often longer than stem, thick, opaque, yellowish, linear, tapering to a blunt apex, without distinct midrib, upper part often floating, flat on the adaxial side, rounded and slightly inflated in the basal part of the abaxial side.

Inflorescence: 2.5–5(–8) cm long, unbranched, erect or floating. Bracts ascending, somewhat inflated and with wide, whitish, hyaline margins at base. Lowest bract exceeding the inflorescence. Male head solitary, contiguous with or 2–5 mm from uppermost female head. Female heads (1–)2–4, in fruit 0.5–0.9(–1.0) cm across, light yellowish, the lowest often supra-axillary, with the peduncle fused with the stem in the basal part. The upper female heads sessile or supra-axillary, contiguous or more or less remote.

Flowers: Male flowers with filaments 3–4 mm long and anthers about 0.5 mm long. Female flowers with thin, pale, translucent, spathulate, erose perianth segments. Stigmas 0.5–1 mm long, rounded to oblong. Style absent or almost so.

Fruits: (2–)3–4 mm long (including beak), smooth, rather dull, dark yellowish to brownish yellow at maturity, obovate, slightly constricted around the middle, rounded at the top and at most with a 0.3 mm long beak. 1-seeded.

Flowering: June-August.

Biology: The flowers are windpollinated and the fruits are dispersed by water. Vegetative propagation may occur by fragmentation of the short rhizome.

Habitats: In moderately acidic to alkaline, oligo- to mesotrophic waters, such as pools, puddles or small streams in mires.

Distribution within the region: Native. Widespread and fairly abundant in boreal, subarctic and alpine areas of Norway, Sweden, Finland, Iceland and Greenland. Circumpolar.

Characteristics and similar species: Similar to **115 *S. natans***, but differs by fruits almost without a beak, lowest bract longer than the inflorescence and the lowest female head normally being supra-axillary, with the basal part of the peduncle fused with stem. See also **116 *S. angustifolium*.**

Hybrids: See page 301.

Sparganium hyperboreum. Snowshoe Lake, west of Glenallen, Alaska. Photo RVL.

Sparganium hyperboreum resembles *S. natans*, but the lowest bract is much longer than the inflorescence and the fruits are almost without a beak. Utajarvi, Finland. Photo RVL.

Sparganium hyperboreum. Fruiting head - note the lowest supra-axillary head, with the peduncle partly fused with the stem. Alaska. Photo RVL.

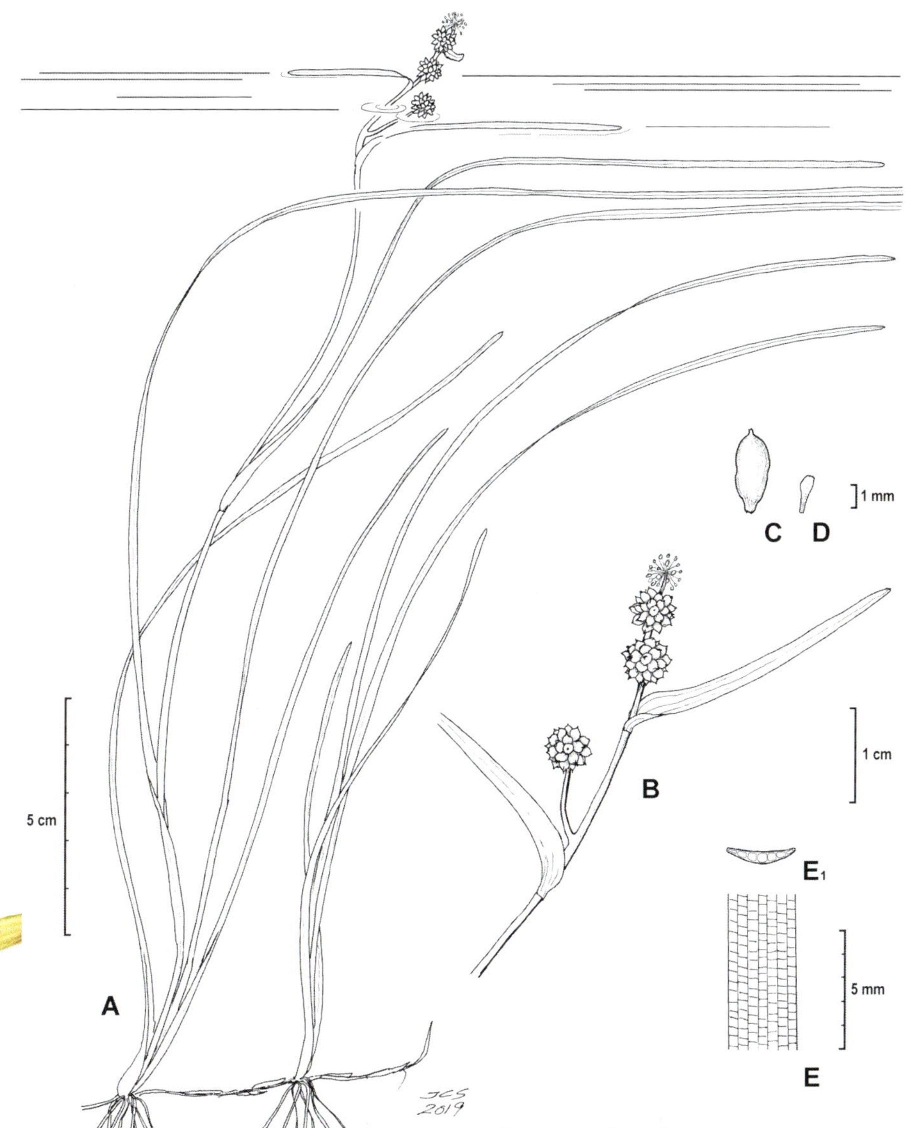

Sparganium hyperboreum
A habit, **B** fruiting head, **C** fruit, **D** perianth segment, **E** leaf section, **E1** cross section of leaf.

Typhaceae

115 *Sparganium natans* L.

Syn.: *S. minimum* Wallr.

DK: Spæd Pindsvineknop **N:** Småpiggknopp **S:** Dvärgigelknopp **FIN:** Pikkupalpakko **IS:** Tjarnabrúsi
GB: Least Bur-reed **NL:** Kleinste egelskop **F:** Petit rubanier
D: Zwerg-Igelkolben **CZ:** Zevar nejmenší **PL:** Jeżogłówka namniejsza
EST: Väike jõgitakjas **LV:** Mazā ežgalvīte **LT:** Mažasis šiurpis **Ru:** Ежеголовник малый

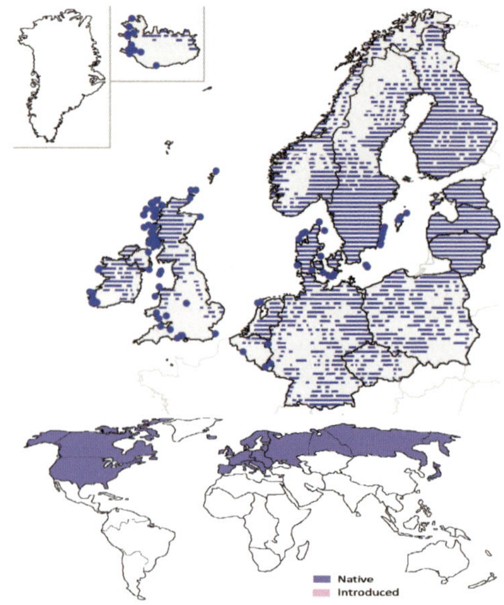

Flowering *S. natans*. Insh Marshes, Scotland. Photo RVL.

Perennial hydrophyte or helophyte. Rhizome short, creeping, 1–2 mm diameter, with scale leaves up to 1.2 cm long.

Stem: 10–40 cm long, slender, 1.5–3 (–4) mm diameter, erect, ascending or procumbent, sometimes floating and up to 100 cm long.

Leaves: Yellowish green to green, (1.5–)3–5(–10) mm wide, often longer than stem, linear, tapering to a blunt apex, flat to concave, thin, in cross section with 1–2 layers of air chambers, without distinct midrib or keel even on the sheath. Leaf sheaths not inflated. 10–30 cm long, erecto-patent with drooping tips, or flaccid and flat to the substrate in terrestrial forms. Flaccid, ribbon-like, permanently submerged or partly floating, 30–100 (–250) cm long in aquatic forms.

Inflorescence: 1.5–8 cm long, unbranched, erect or floating. Bracts spreading to ascending, not or only slightly inflated with hyaline margins at base. Lowest bract less than 10 cm long and at most slightly longer than the inflorescence. Male head usually solitary, sometimes 2 close together, separated by 0.5–1 cm from the uppermost female head. Female heads 1–3(–4), in fruit 0.7–1.1(–1.4) cm across, light yellowish, axillary, sessile or the lowermost shortly pedunculate.

Flowers: Male flowers with filaments 3–5 mm long and anthers 0.5–0.7 mm long. Female flowers with thin, pale, translucent, spathulate, erose perianth segments. Stigma 0.5–0.8 mm long, oblong. Style up to 1 mm long.

Fruits: 3.5–6.5 mm long (including beak), smooth, rather dull, green to greenish brown, spindle shaped, not constricted around the middle at maturity, tapering to a sessile or very shortly pedicellate base below, tapering to a 0.5–1 mm long, straight beak at apex. 1-seeded.

Flowering: June-August.

Biology: The flowers are wind pollinated and the fruits dispersed by water. Vegetative propagation is by iterative branching from the lower leaf axils. Overwintering as leafless rhizomes in the substrate.

Habitats: In moderately acidic to alkaline, oligo- to mesotrophic waters, such as pools, small lakes and sheltered bays of larger lakes, ditches and small streams. Terrestrial forms grow on moist ground near pools and streams.

Distribution within the region:
Native. Widespread and at least locally abundant throughout most of the region. Absent from Greenland.

Submerged *S. natans* in still water in an oligotrophic lake. Råbjerg Mile, Jutland, Denmark. Photo JCS.

Submerged *S. natans* in flowing water. Kidmose Bæk, Fasterholt, Jutland, Denmark. Photo BM.

Fruiting *S. natans*. Katbjerg, Jutland, Denmark. Photo JCS.

Characteristics and similar species: Similar to **114 *S. hyperboreum***, but differs by the fruits tapering to a short, straight beak, lowest bract usually shorter than inflorescence and female heads axillary.

Distinguished from other *Sparganium* species by the simple inflorescence with 1–3 female and 1(–2) male heads, by the flaccid, flat leaves, without a midrib even at the base. Plants from flowing water with long and wide leaves can easily be mistaken for **116 *S. angustifolium*** and **118 *S. emersum***, but both of these have a distinct and often prominent midrib at least close to base.

Hybrids: See page 301.

Sparganium natans. Inflorescence with 1 male and 1 female head. Photo JCS.

Sparganium natans. Fruiting head. Photo KvdW.

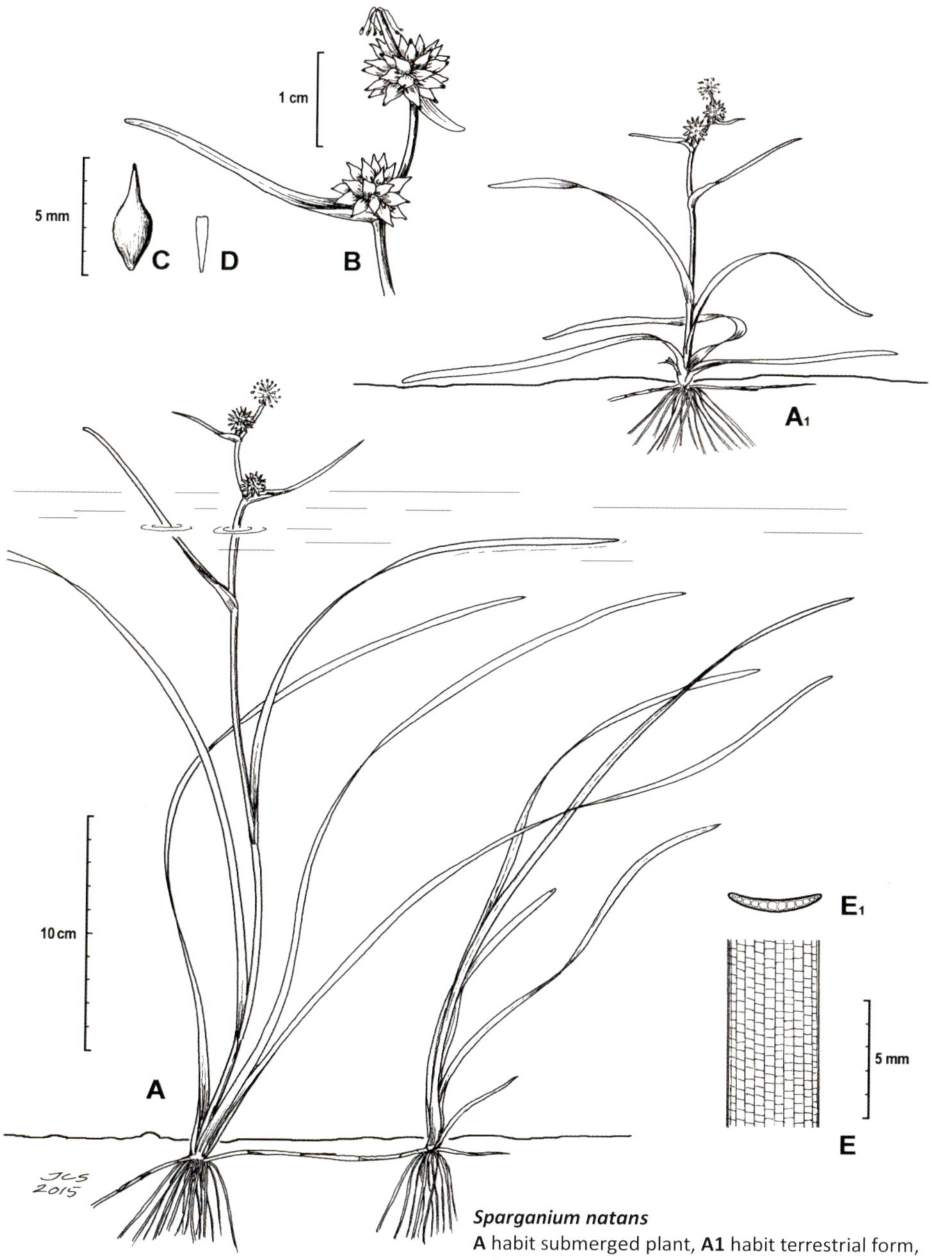

Sparganium natans
A habit submerged plant, **A1** habit terrestrial form, **B** fruiting head, **C** fruit, **D** perianth segment, **E** leaf section, **E1** cross section of leaf.

Typhaceae

120 *Sparganium erectum* L. s. str.

Syn.: *S. erectum* L. subsp. *erectum*; *S. ramosum* subsp. *polyedrum* (Graebn.) Schinz & Thell.
DK: Sodfrugtet Pindsvineknop **N:** Nystepiggknopp **S:** Sotigelknopp **FIN:** Isopalpakko **IS:** -
GB: - **NL:** Grote egelskop **F:** Ruban-d'eau
D: Ästiger Igelkolben **CZ:** Zevar vzpřímený pravý **PL:** Jeżogłówka gałęzista
EST: Haruline jõgitakjas **LV:** Stāvā ežgalvīte **LT:** Šakotasis šiurpis **Ru:** Ежеголовник прямой

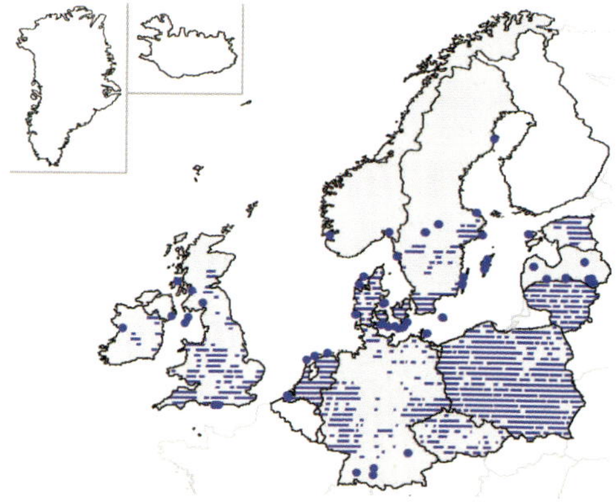

Sparganium erectum s. str. with fully developed green fruit (left) and ripe fruits (right). Left: Stroudwater Canal, Stroud, Gloucestershire, Britain. Right: Cantley Marshes, Norfolk, Britain. Photos RVL.

Flowers: Perianth segments spathulate, more or less entire, blackish at apex.

Fruits: 6–10(–12) mm long including the beak, which is up to 2.0 mm long, smooth, somewhat shiny - becoming dull over time, with a distinct shoulder separating the upper and lower part. The upper part at first pyramidal, green to greyish green, but as the fruits ripen becoming flatter and dark to blackish brown. The lower part obconical, pale brown, with 3–6 distinct angles in transverse section.

Distribution within the region:
Native. Widespread and abundant throughout much of the south, less frequent in the north; absent from Belgium, Luxembourg, Finland, Iceland and Greenland.

Sparganium erectum s. str. The fruits are dark to blackish brown when ripe. Semi-ripe fruits have a greyish-green to yellow-green colour. Filsø, W-Jutland, Denmark. Photos JCS.

Submerged *S. natans* in flowing water. Kidmose Bæk, Fasterholt, Jutland, Denmark. Photo BM.

Fruiting *S. natans*. Katbjerg, Jutland, Denmark. Photo JCS.

Characteristics and similar species: Similar to **114** ***S. hyperboreum***, but differs by the fruits tapering to a short, straight beak, lowest bract usually shorter than inflorescence and female heads axillary.

Distinguished from other *Sparganium* species by the simple inflorescence with 1–3 female and 1(–2) male heads, by the flaccid, flat leaves, without a midrib even at the base. Plants from flowing water with long and wide leaves can easily be mistaken for **116** ***S. angustifolium*** and **118** ***S. emersum***, but both of these have a distinct and often prominent midrib at least close to base.

Hybrids: See page 301.

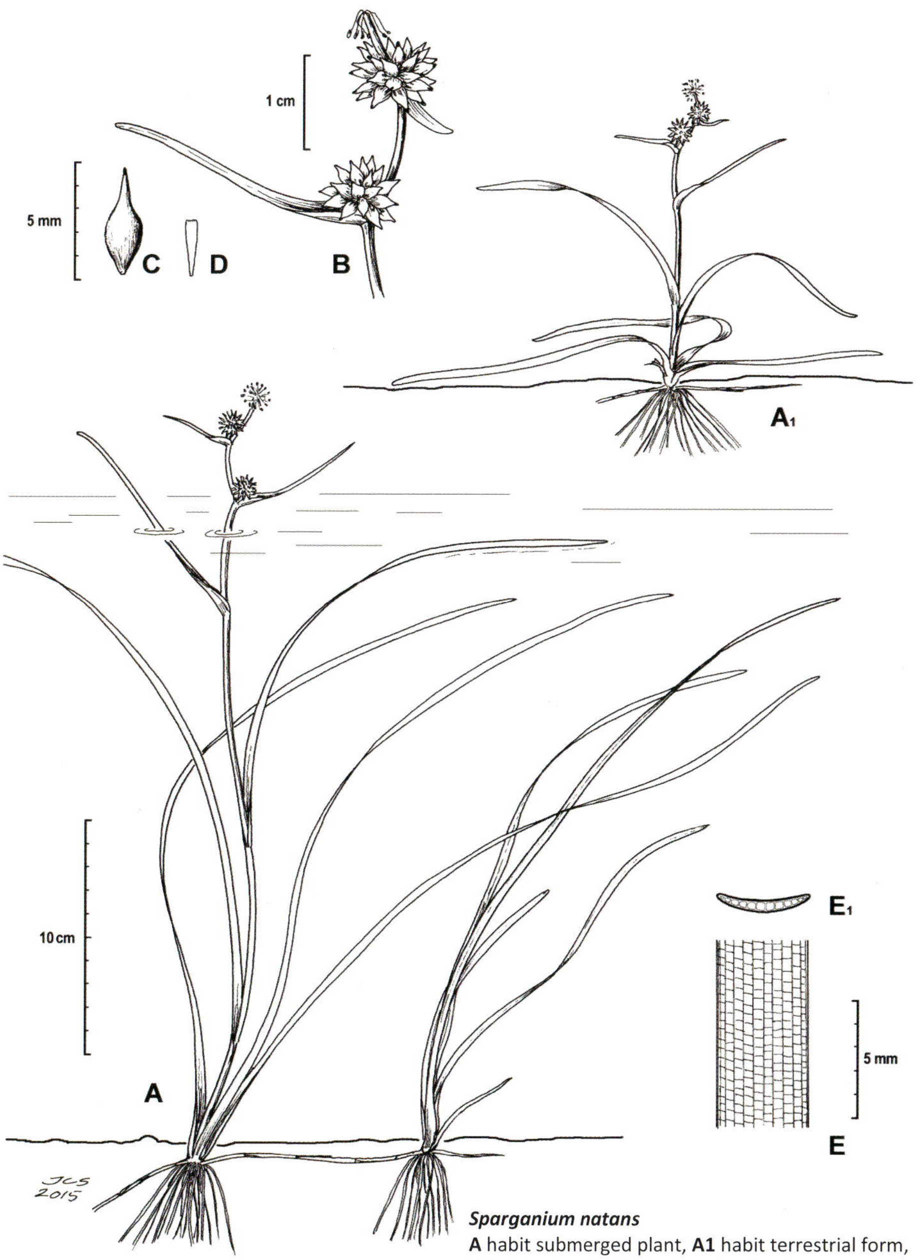

Sparganium natans
A habit submerged plant, **A1** habit terrestrial form, **B** fruiting head, **C** fruit, **D** perianth segment, **E** leaf section, **E1** cross section of leaf.

Sparganium natans. Inflorescence with 1 male and 1 female head. Photo JCS.

Sparganium natans. Fruiting head. Photo KvdW.

Typhaceae

116 *Sparganium angustifolium* Michx.

DK: Smalbladet Pindsvineknop **N:** Flotgras **S:** Plattbladig igelknopp **FIN:** Kaitapalpakko **IS:** Trjónubrúsi
GB: Floating Bur-reed **NL:** Drijvende egelskop **F:** Rubanier à feuilles étroites
D: Schmalblättriger Igelkolben **CZ:** Zevar úzkolistý **PL:** Jeżogłówka pokrewna
EST: Lamedalehine jõgitakjas **LV:** Šaurlapu ežgalvīte **LT:** Siauralapis šiurpis **Ru:** Ежеголовник скученный

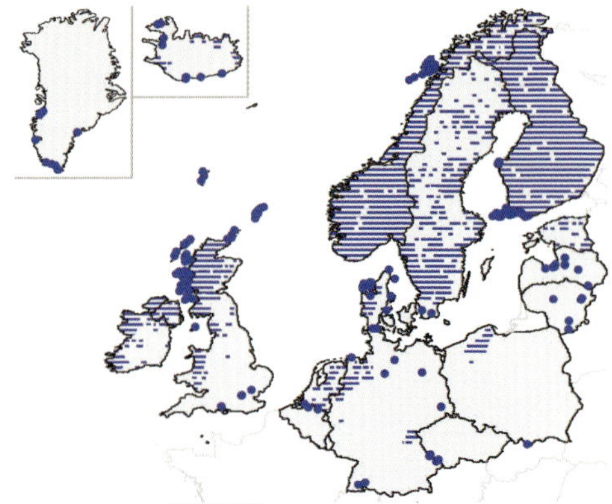

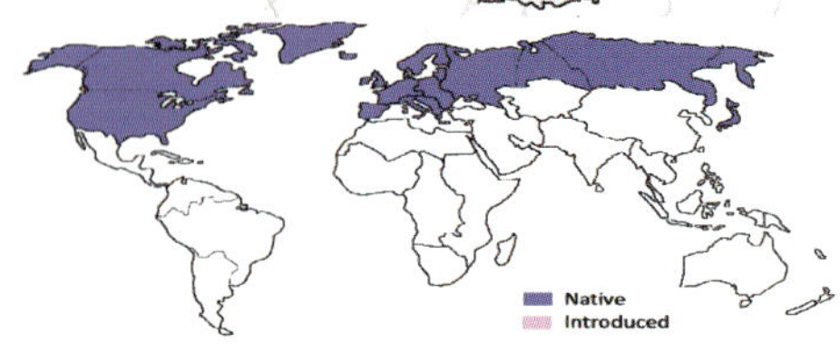

Sparganium angustifolium with narrow, floating leaves. Grube Vande, Thy, Denmark. Photo JCS.

Perennial hydrophyte or helophyte. Rhizome short, creeping, 2.5–3 mm diameter, with scale leaves up to 1.5 cm long.

Stem: 2–5 mm diameter, ascending and normally floating, up to 170 cm long, flexuous, erect or ascending in terrestrial forms.

Leaves: Yellowish green to green, floating, 30–150(–250) cm long, 1.5–4(–10) mm wide, linear, tapering to a blunt apex. Floating leaves flat or slightly rounded on the adaxial (upper) side. The abaxial (lower) side slightly rounded in the proximal part of the leaves, more swollen, inflated, rounded to obtusely keeled towards the base, midrib rather prominent. With 3–4 layers of air chambers in cross section.

Inflorescence: 2–9 cm long, unbranched, erect or floating. Bracts spreading to ascending, inflated and with hyaline margins at the base. Lowest bract more than 10 cm long and more than twice as long as the inflorescence. Male heads (1–)2–3(–5), contiguous, forming an elongated head, typically separated by 1–2 cm from the uppermost female head. Female heads (1–)2–3(–4), (0.7–)1.4–2.3 cm diameter in fruit, pale greenish, remote, the upper usually sessile, the lower axillary or supra axillary, more or less pedunculate.

Flowers: Male flowers with filaments 4–6 mm long and anthers 0.9–1.0(–1.2) mm long. Female flowers with thin, pale brown, translucent, spathulate, erose perianth segments. Stigma about 1 mm long, oblong. Style up to 1.5–2 mm long.

Fruits: 5–7 mm long (including beak), greyish green to brownish, sometimes with a reddish-brown tinge at maturity, somewhat shiny, ellipsoid to fusiform, usually somewhat constricted around the middle, obconical towards the base, tapering to a 1–2 mm long pedicel, gradually narrowed upwards into a 1.5–2 mm long, straight beak.

Flowering: July-August.

Biology: The flowers are windpollinated and the fruits dispersed by water. Vegetative propagation is by fragmentation of the rhizome. Overwintering as leafless rhizomes in the substrate.

Habitats: In acidic to alkaline, oligo- to moderately eutrophic waters. Pools, ditches and small lakes in dunes and moorland, lakes, occasionally in flowing water.

Distribution within the region:
Native. Widespread and abundant in the north, more scattered and locally absent in the south.

Characteristics and similar species:
Sparganium angustifolium can be recognised by its long, narrow, floating leaves with 3–4 layers of air chambers in cross section, the unbranched inflorescence with 2–3 more or less remote

Dense formation of *S. angustifolium*. Øje Sø, Vesthimmerland, Denmark. Photo JCS.

female heads and 2–3 male heads very close together often united into one oblong head. The lowest bract more than twice as long as the inflorescence.

117 *S. gramineum* is very similar in habit - see this.

118 *S. emersum* differs in having 4–8 male heads, fruits not constricted around the middle and usually wider leaves.

It is not possible to distinguish *S. angustifolium* from *S. gramineum and S. emersum* using vegetative characters alone.

114 *S. hyperboreum* and **115 *S. natans*** usually have only 1 male head, floating leaves with an indistinct midrib and smaller female heads 1 cm or less in diameter.

Hybrids: See page 301.

Sparganium angustifolium. As with all other *Sparganium* species, the female flowers bloom before the male flowers at the top of the inflorescence to ensure cross-pollination. Grube Vande, Thy, Denmark. Photo JCS.

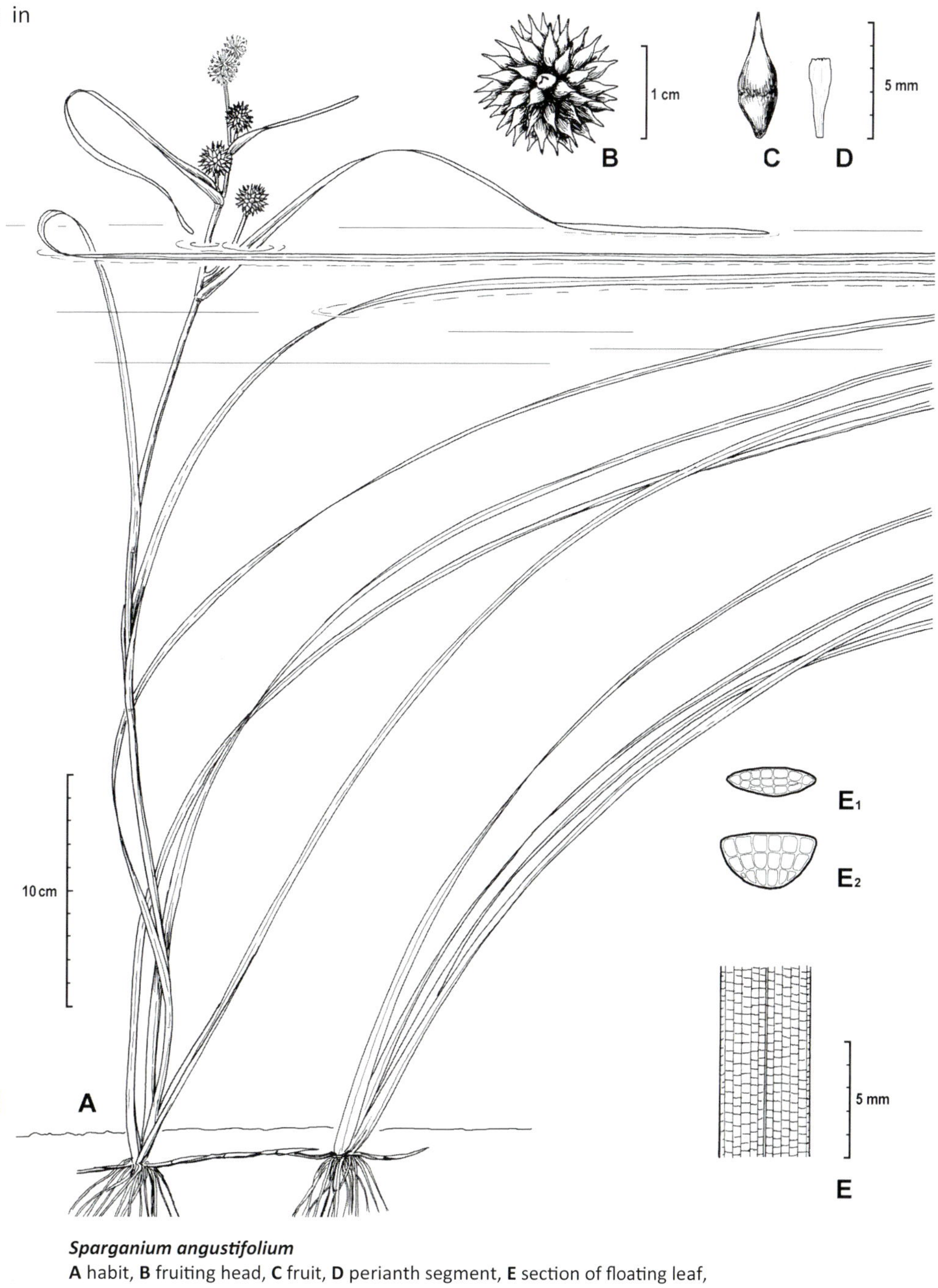

Sparganium angustifolium
A habit, **B** fruiting head, **C** fruit, **D** perianth segment, **E** section of floating leaf, **E1** cross section of leaf close to apex, **E2** cross section of leaf close to base.

Sparganium angustifolium. Female heads with almost ripe fruits. Spinsterberg, Netherlands. Photo KvdW.

Typhaceae

117 ***Sparganium gramineum*** Georgi

Syn.: *S. friesii* Laest. ex Beurl.

DK: Sø-Pindsvineknop **N:** Sjöpiggknopp **S:** Flotagräs **FIN:** - **IS:** -
GB: - **NL:** - **F:** -
D: Grasblättriger Igelkolben **CZ:** - **PL:** -
EST: Ujuv jõgitakjas **LV:** Zālainā ežgalvīte **LT:** Siūlinis šiurpis **Ru:** Ежеголовник злаколистный

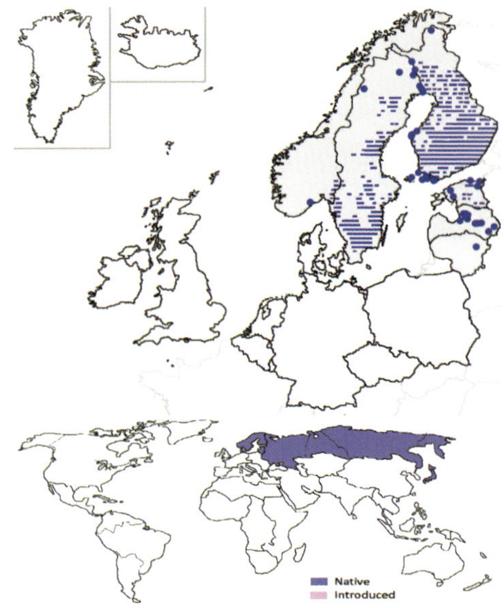

Perennial hydrophyte or helophyte. Rhizome short, creeping, 2.5–3 mm diameter, with scale leaves about 1 cm long.

Stem: 2–5 mm diameter, ascending and floating, 1–2 m long, flexuous.

Leaves: Yellowish green to green, floating, 1–3 m long, (1–)2–3 mm wide, linear, tapering to a blunt apex. Floating leaves flat on the adaxial (upper) side. The abaxial (lower) side flat to slightly rounded in the proximal part of the leaves, gradually narrowing towards the base and semi-terete to obtusely keeled, midrib rather prominent. In cross section with 2–4 layers of air chambers.

Inflorescence: 10–30 cm long, characteristically S shaped, usually branched from the lowest 2 or 3 bracts. Bracts spreading to ascending, inflated and with hyaline margins at base. The lower 2 or 3 bracts with lamina wider than the floating leaves and at least the lowest distinctly longer than the inflorescence, the upper bracts more or less reduced. Branches with 1–4 female heads and 0–4 male heads above. Unbranched specimens may have long pedunculate lower female heads. The male portion of the main stem is elongate and raised above the surface, with 3–8 more or less remote male heads, the upper often contiguous. Female heads 3–5 on main stem, (0.7–)1.0–1.6 cm diameter in fruit, remote, the lower often long pedunculate.

Flowers: Male flowers with filaments 3–4 mm long and anthers about 1 mm long. Female flowers with thin, pale brown, translucent, spathulate, erose perianth segments. Stigma about 0.5–0.7 mm long, oblong. Style up to 1.5–2 mm long.

Fruits: About 5 mm long (including beak), dull, reddish brown at maturity, ellipsoid, not constricted around the middle, obconical towards the base, tapering to a short pedicel, gradually narrowed upwards into a 1.5–2 mm long, curved beak.

Flowering: July-August.

Biology: The flowers are windpollinated and the fruits dispersed by water. Vegetative propagation is by fragmentation of the rhizome. Overwintering as leafless rhizomes in the substrate.

Habitats: In acidic, oligotrophic to meso-oligotrophic water typically at depths of 1–2 metres. Mostly found in larger lakes, rarely in pools and ditches.

Distribution within the region:
Native. Scattered to locally abundant in Sweden, Finland, Estonia and Latvia, rare in Norway and Lithuania.

Characteristics and similar species:
In habit very similar to **116 *S. angustifolium***, but differs in the branched, sinuous, S-shaped inflorescence, the more remote female heads and at maturity by reddish-brown fruits with curved beak.

Sparganium gramineum in an early stage of flowering. Note the sinuous, S-shaped inflorescence. Vederslövssjön, Vaxjö, Sweden. Photo JCS.

Sparganium gramineum. Vederslövssjön, Vaxjö, Sweden. Photo JCS.

Sparganium gramineum. Male heads raised above the surface for wind pollination. Vederslövssjön, Vaxjö, Sweden. Photo BM.

118 *S. emersum* differs in having 4–8 male heads and usually wider leaves.

It is not possible to distinguish *S. gramineum* from *S. angustifolium and S. emersum* using vegetative characters alone.

114 *S. hyperboreum* and **115 *S. natans*** usually have only 1 male head, floating leaves with 1–2 layers of air chambers and indistinct midrib. Female heads smaller, 1 cm or less in diameter.

Hybrids: See page 301.

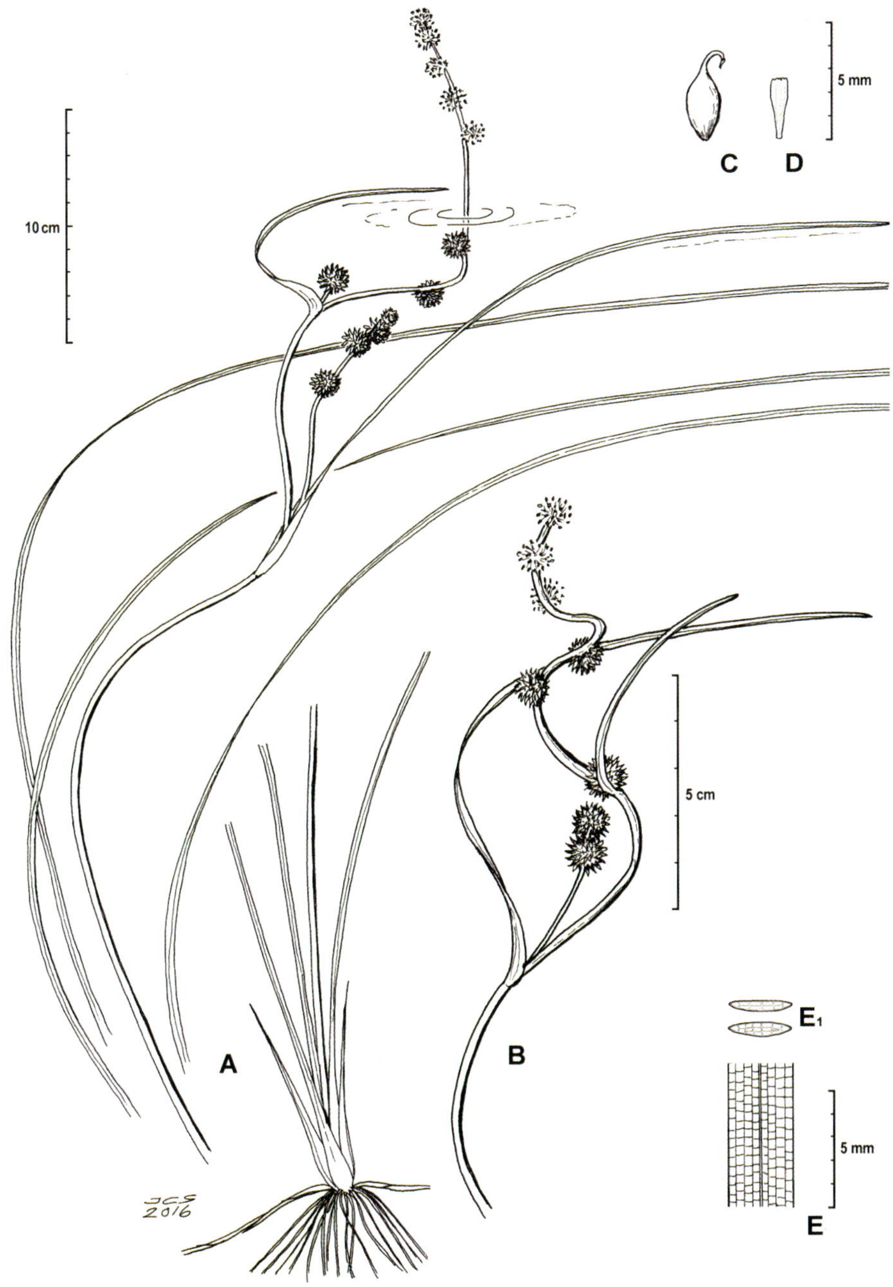

Sparganium gramineum
A habit, **B** inflorescence, **C** fruit, **D** perianth segment, **E** section of floating leaf, **E1** cross sections of leaf.

Typhaceae

118 *Sparganium emersum* Rehmann

Syn: *S. simplex* Huds.

DK: Enkelt Pindsvineknop **N:** Stautpiggknopp **S:** Gles igelknopp **FIN:** Rantapalpakko **IS:** -
GB: Unbranched Bur-reed **NL:** Kleine egelskop **F:** Rubanier simple
D: Einfacher Igelkolben **CZ:** Zevar jednoduchý **PL:** Jeżogłówka pojedyncza
EST: Liht-jõgitakjas **LV:** Vienkāršā ežgālvīte **LT:** Paprastasis šiurpis **Ru:** Ежеголовник всплывающий

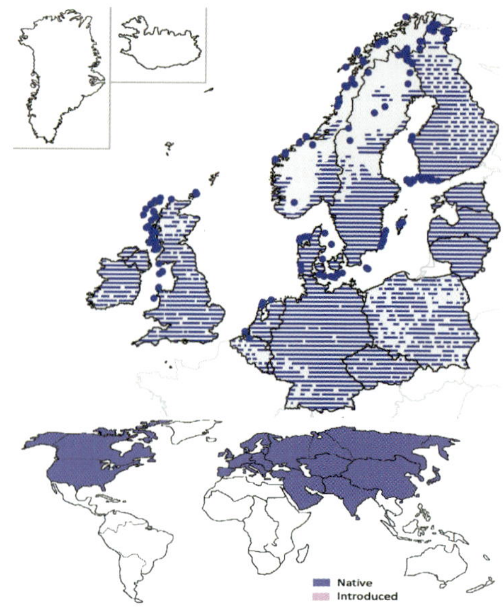

Perennial hydrophyte or helophyte. Rhizome short, creeping, up to 10 mm diameter, with scale leaves up to 1.5 cm long.

Stem: 5–10 mm diameter, erect, 25–50 cm long in terrestrial forms, ascending and floating, up to 2 m long in aquatic forms.

Leaves: Aerial leaves 20–60 cm long, 4–10(–12) mm wide, erect or spreading, keeled throughout their length, somewhat flaccid and often drooping distally, thick but not inflated at base, apex blunt. Submerged or floating leaves 30–150 (–250) cm long, 4–8(–16) mm wide, linear, flaccid, flat, on the abaxial (lower) side sometimes with a slightly prominent midrib, somewhat keeled at base, narrowing to a blunt apex, yellowish-green to green, often with a brownish tinge.

Inflorescence: (10–)15–30 cm long, unbranched, erect. Bracts erect, inflated and with hyaline margins at base. Lowest bract at most slightly longer than the inflorescence. Male heads (3–) 4–8(–10), those at top close together, the lower ones more remote and separated by 0.5–2 cm from the uppermost female head. Female heads (1–)3–5(–6), in fruit 1.6–2.5(–3.2) cm across, the upper usually sessile, the lower with up to 4(–8) cm long peduncles, axillary or supra-axillary.

Flowers: Male flowers with filaments 5–7 mm long and anthers 1.0–1.5(–2.0) mm long. Female flowers with thin, pale brown, translucent, spathulate, erose perianth segments. Stigma about 1.5–2.0 mm long, oblong ovate. Style 2–3 mm long.

Fruits: 7–10 mm long (including beak), somewhat shiny, yellowish green to greyish green, brownish at maturity, ellipsoid, sometimes a little constricted around the middle, obconical towards the base, tapering to a pedicel up to 4 mm long, gradually narrowed upwards into a 2.0–4.5 mm long, straight beak.

Flowering: July-August.

Biology: The flowers are wind-pollinated and the fruits dispersed by water. Vegetative propagation is by division of the rhizome. Overwinters as leafless rhizomes in the substrate.

Habitats: In neutral to alkaline, meso- to eutrophic water in rivers, streams, canals and ditches, less often in lakes and ponds.

Distribution within the region:
Native. Common throughout the region except Iceland and Greenland.

Characteristics and similar species:
Fertile specimens of *S. emersum* can be recognised by the unbranched inflorescence typically with 3–5 female and 4–8 male heads, the bracts only slightly longer than the inflorescence.
S. emersum is very common in flowing water and here usually present with long, 4–8 mm wide, submerged or partly floating leaves only. Such plants may be mistaken for other plants with long, ribbon-like leaves, but differ like other *Sparganium* species in the almost square reticulation of the leaves (fig. **E** opposite) - compare with leaves of **45 *Sagittaria sagittifolia*** and **145-146 *Schoenoplectus*** spp.

Sparganium emersum. Leaves partly floating. Inserted: Section of submerged leaf. River Skjern Å, Tarm, Denmark. Photo JCS.

Dense formations of submerged *S. emersum*. River Glen, Lincolnshire, Britain. Photo RVL.

Sparganium emersum. Flowering plant with both aerial and floating leaves. Left: Plant in female stage. Right: Plant in male stage. River Skals Å, Jutland, Denmark. Photo JCS.

Sparganium emersum. Fruiting plants in shallow water - only aerial leaves developed. Lake Filsø, SW-Jutland, Denmark. Photo JCS.

120-122 ***Sparganium erectum*** agg. differs by the inflorescence with male and/or female heads on both main stem and branches.

116 ***Sparganium angustifolium*** is rare in flowing water and usually has narrower leaves - see this.

119 ***Sparganium glomeratum*** has 1–2 male heads, and 3–5(–6) female heads - the upper of these contiguous.

It is not possible with certainty to distinguish *S. emersum* from most other *Sparganium* species using vegetative characters alone. However, see *S. erectum* aggr.

Hybrids: See page 301.

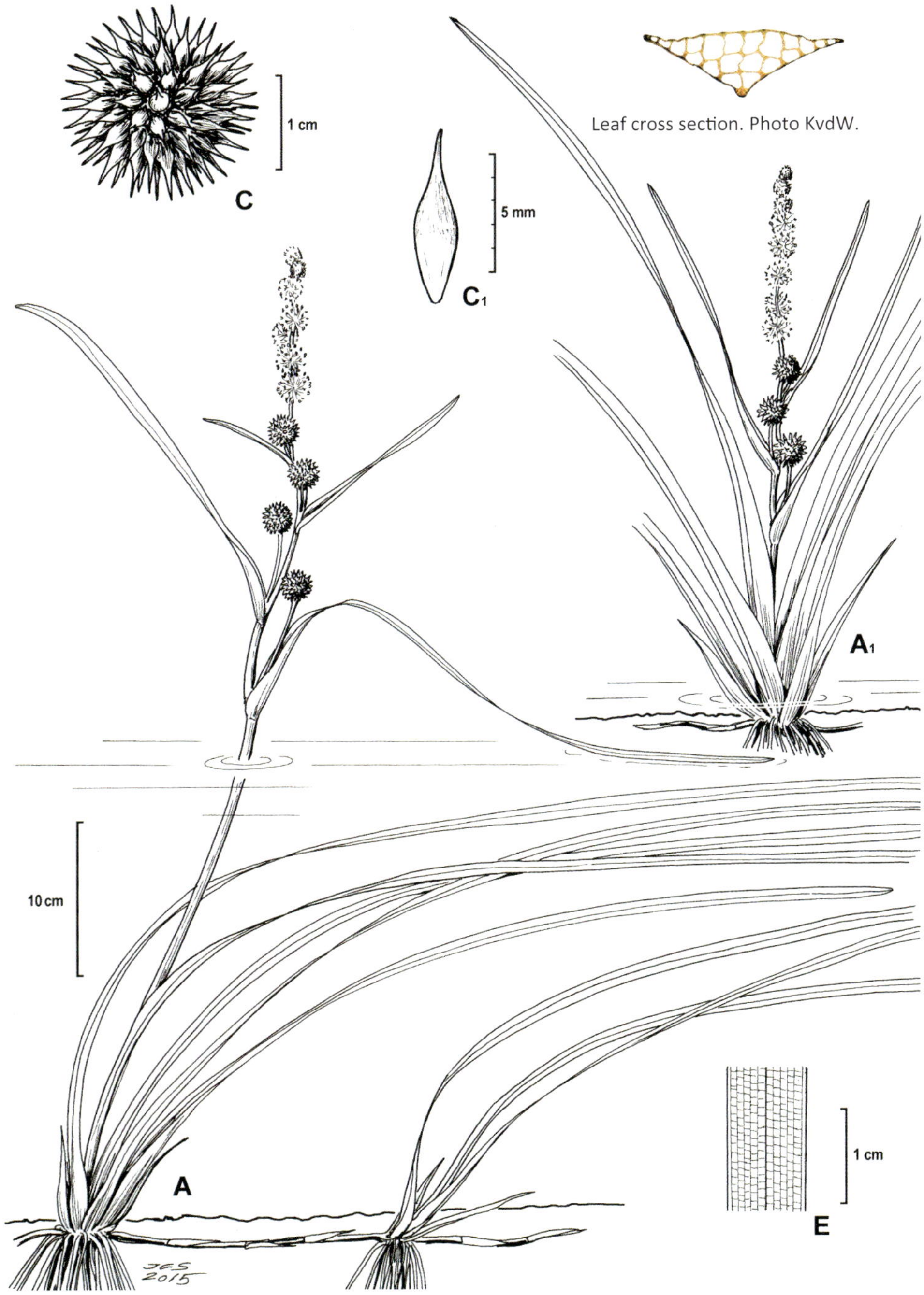

Sparganium emersum
A habit partly submerged plant, **A1** habit terrestrial plant, **C** fruit head, **C1** fruit, **E** submerged leaf section.

Sparganium emersum. Female heads with almost ripe fruits. Photo KvdW.

Typhaceae

119 *Sparganium glomeratum* (Laest.) Neuman

Syn.: *S. fluitans* Fr.

DK: Tæt Pindsvineknop **N:** Nystepiggknopp **S:** Gyttrad igelknopp **FIN:** - **IS:** -
GB: - **NL:** - **F:** -
D: - **CZ:** - **PL:** -
EST: Kera-jõgitakjas **LV:** Kamolainā ežgalvīte **LT:** Tankiažiedis šiurpis **Ru:** Ежеголовник скученный

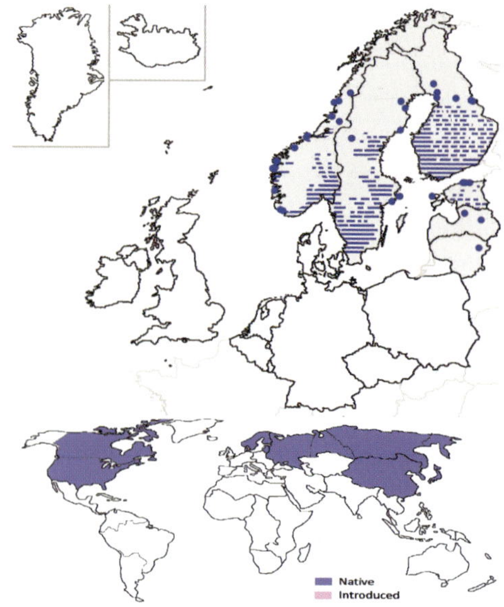

Sparganium glomeratum. Robertsfors, Västergötland, Sweden. Photo KvdW.

Perennial helophyte. Rhizome short, creeping, 3–4 mm diameter.

Stem: 4–8 mm diameter, robust, 10–40 (–60) cm long, erect.

Leaves: Aerial leaves 20–60 cm long, markedly longer than stem, 4–10 mm wide, erect, thick, triangular in cross section and sharply keeled at base, gradually narrower and thinner towards the tip, flat on the adaxial (upper) side, abaxial (lower) side with a distinct keel. Without submerged leaves, but sometimes with flaccid, floating leaves.

Inflorescence: 6–14 cm long, unbranched. Bracts erect, inflated and with hyaline margins at the base, much longer than the inflorescence. Male head normally solitary, rarely 2 close together. Female heads 3–5(–6), 1.4–1.8 cm diameter in fruit, the upper usually sessile and close together, the lower more distant with up to 3–6 cm long peduncles, axillary or often supra-axillary.

Flowers: Male flowers with filaments approximately 3 mm long and anthers 1 mm long. Female flowers with thin, pale brown, translucent, spathulate to lanceolate, erose perianth segments. Stigma 0.5–1.0 mm long, oblong ovate. Style 1–1.5 mm long.

Fruits: 7–10 mm long (including beak), somewhat shiny, dark greyish green to greyish yellow, greenish brown at maturity, ellipsoid, somewhat constricted below the middle, obconical towards the base, tapering to a pedicel 1–2 mm long, gradually narrowed upwards into a 1.0–1.5 mm long, straight beak.

Flowering: July-August.

Biology: The flowers are wind-pollinated and the fruits dispersed by water. Vegetative propagation is by division of the rhizome. Overwinters as leafless rhizomes in the substrate.

Habitats: In acidic to neutral, mesotrophic, shallow water or on muddy substrate. Ditches, canals, small streams and pools.

Distribution within the region:
Native. Scattered to locally frequent in parts of southern Scandinavia and Estonia. Rare in Latvia and Lithuania.

Sparganium glomeratum. Female heads close together; the male head on top has withered. Hautaniemi, Finland. Photo RVL.

Floating leaves of *S. glomeratum*. Robertsfors, Västergötland, Sweden. Photo KvdW.

Sparganium glomeratum. Male head flowering. Robertsfors, Västergötland, Sweden. Photo RVL.

Characteristics and similar species:
In habit, *Sparganium glomeratum* is somewhat similar to **118 *S. emersum***, which however has 4–8 male heads and remote female heads.

Hybrids: See page 301.

Sparganium glomeratum. Fruiting head. Photo JCS.

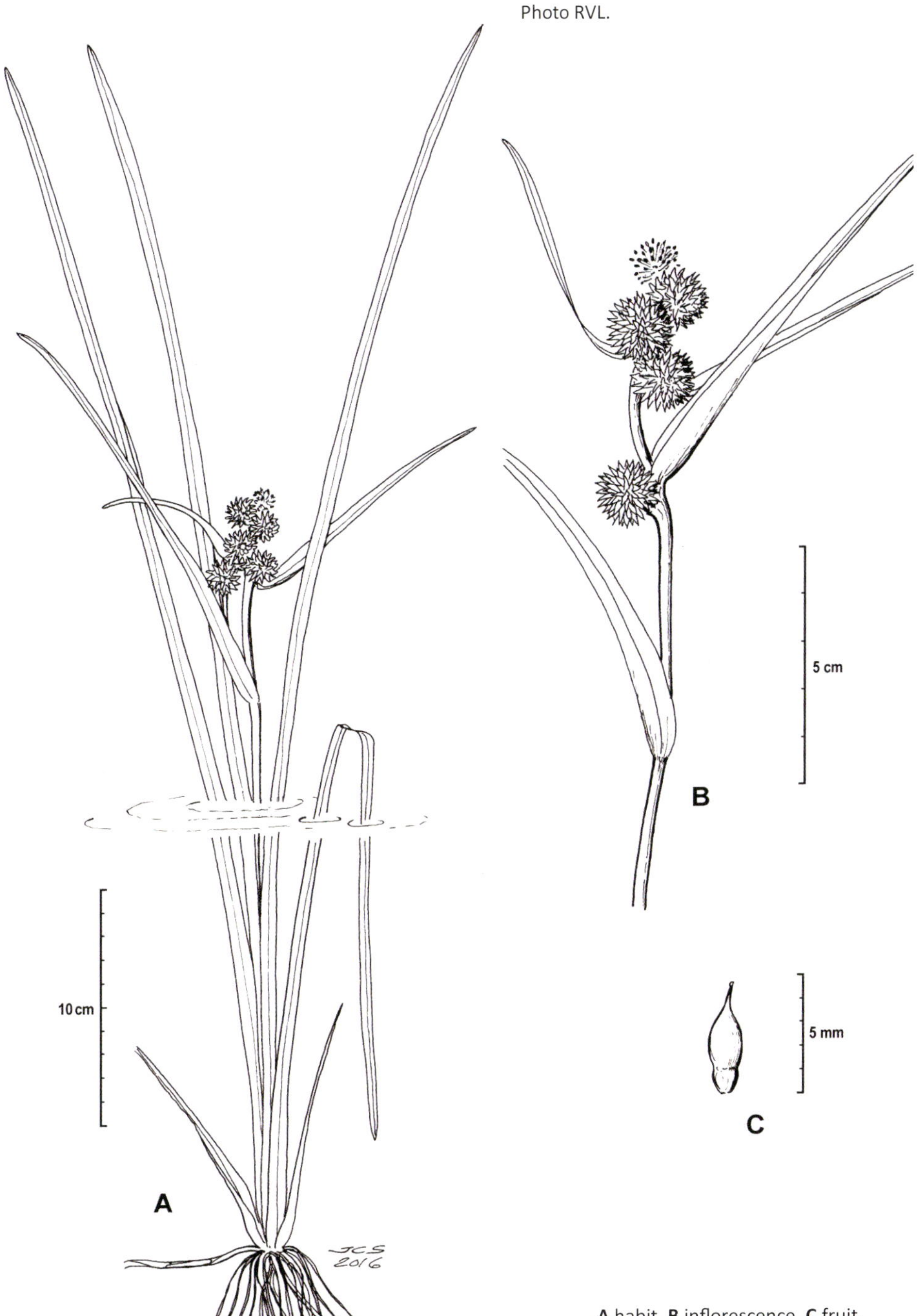

A habit, **B** inflorescence, **C** fruit.

Typhaceae

Sparganium erectum L. aggregate

Syn.: *S. ramosum* Huds.

DK: Grenet Pindsvineknop **N:** Kjempepiggknopp **S:** Storigelknopp **FIN:** Haarapalpakko **IS:** -
GB: Branched Bur-reed **NL:** Grote egelskop **F:** Ruban-d'eau
D: Ästiger Igelkolben **CZ:** Zevar vzpřímený **PL:** Jeżogłówka gałęzista
EST: Haruline jõgitakjas **LV:** Stāvā ežgalvīte **LT:** Šakotasis šiurpis **Ru:** Ежеголовник прямой

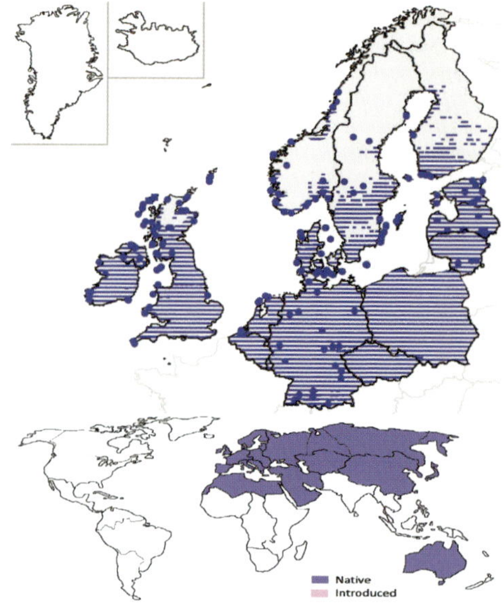

Sparganium erectum with female heads blooming. Note the long, white stigmas. Nørrestrand, Horsens, Jutland, Denmark. Photo JCS.

Perennial helophyte or more rarely hydrophyte. Rhizomes up to 60 cm long, up to 10 mm diameter, creeping, underground, with scale leaves up to 1.5 cm long.

Stem: 8–20 mm diameter, stiff, erect or ascending 30–100(–150) cm long.

Leaves: Aerial leaves (30–)50–150 cm long, 10–20(–30) mm wide, erect or spreading, stiff, keeled throughout their length, flat and tapering outwards, thick and almost triangular in cross section at the base, apex blunt. Submerged leaves, up to 2 m long, 7–12(–18) mm wide, linear, a little stiff, flat, on the abaxial (lower) side with a somewhat prominent midrib, somewhat keeled at the base, narrowing outwards to a blunt apex, yellowish green to green, and like other *Sparganium* species with more or less square reticulation.

Inflorescence: (10–)15–30 cm long, branched, erect. Bracts erect, inflated and with hyaline margins at the base. Lowest bract usually longer than the inflorescence. Branches 3–6(–8), axillary, with 1–4 sessile, female heads at the proximal end and 5–20 male heads at the distal end. The terminal part of the inflorescence above the uppermost branch with 5–15 male heads.

Flowers: Male flowers with filaments about 5 mm long and anthers 1.2–1.5 mm long. Stigma 2–4 mm long, oblong, bifid or entire.

Fruits: The *Sparganium erectum* complex exhibits strong fruit polymorphism, and a total of 10 taxa have been described within the aggregate. Three of these have been recognised within the study area, and characters for distinguishing them are described on the following pages.
Three non-native species which could occur in the region are briefly described on page 301.

Flowering: July-August.

Biology: The flowers are windpollinated and the fruits dispersed by water. Vegetative propagation is by rhizome division. Overwinters as leafless rhizomes in the substrate.

Habitats: In neutral to moderately alkaline, meso- to eutrophic water. Ponds, ditches, small streams, swamps, marshy meadows, along the shores of canals, rivers and lakes.

Distribution within the region:
Native. Widespread and abundant throughout the south, less frequent or absent in the north. Absent from Iceland and Greenland.

Characteristics and similar species:
Species of the *S. erectum* aggregate are recognised by the branched inflorescence and 7–20 mm wide leaves. In streams and rivers they are often fertile along the margins, but sterile with only submerged or floating leaves in deeper water.

Sparganium erectum. Pond near Østerlars Church, Bornholm, Denmark. Photo JCS.

The submerged leaves of full grown *S. erectum* (greenish) usually appear stiffer than those of *S. emersum* (brownish). Photo BM.

Vegetative specimens of full grown plants can in many cases be distinguished from **118. *S. emersum*** by the stiffer, often twisted leaves.

Sparganium species have more or less square reticulation, which separates them from other genera with submerged, ribbon-like leaves. Compare with leaves of **45 *Sagittaria sagittifolia,* 145-146 *Schoenoplectus*** spp. and **50 *Butomus umbellatus***.

Sparganium erectum **agg.**
A habit emergent form, **E1-E3** cross sections of aerial leaf, **E4** submerged leaf.

Typhaceae

120 *Sparganium erectum* L. s. str.

Syn.: *S. erectum* L. subsp. *erectum*; *S. ramosum* subsp. *polyedrum* (Graebn.) Schinz & Thell.
DK: Sodfrugtet Pindsvineknop **N:** Nystepiggknopp **S:** Sotigelknopp **FIN:** Isopalpakko **IS:** -
GB: - **NL:** Grote egelskop **F:** Ruban-d'eau
D: Ästiger Igelkolben **CZ:** Zevar vzpřímený pravý **PL:** Jeżogłówka gałęzista
EST: Haruline jõgitakjas **LV:** Stāvā ežgalvīte **LT:** Šakotasis šiurpis **Ru:** Ежеголовник прямой

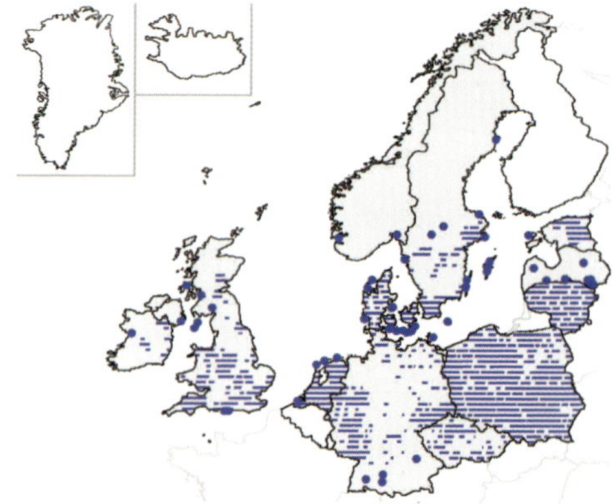

Sparganium erectum s. str. with fully developed green fruit (left) and ripe fruits (right). Left: Stroudwater Canal, Stroud, Gloucestershire, Britain. Right: Cantley Marshes, Norfolk, Britain. Photos RVL.

Flowers: Perianth segments spathulate, more or less entire, blackish at apex.

Fruits: 6–10(–12) mm long including the beak, which is up to 2.0 mm long, smooth, somewhat shiny - becoming dull over time, with a distinct shoulder separating the upper and lower part. The upper part at first pyramidal, green to greyish green, but as the fruits ripen becoming flatter and dark to blackish brown. The lower part obconical, pale brown, with 3–6 distinct angles in transverse section.

Distribution within the region:
Native. Widespread and abundant throughout much of the south, less frequent in the north; absent from Belgium, Luxembourg, Finland, Iceland and Greenland.

Sparganium erectum s. str. The fruits are dark to blackish brown when ripe. Semi-ripe fruits have a greyish-green to yellow-green colour. Filsø, W-Jutland, Denmark. Photos JCS.

Typhaceae

121 *Sparganium microcarpum* (Neuman) Čelak.

Syn.: *S. erectum* L. subsp. *microcarpum* (Neuman) Domin

DK: Småfrugtet Pindsvineknop **N:** Blankpiggknopp **S:** Vanlig storigelknopp **FIN:** Ojapalpakko **IS:** - **GB:** - **NL:** - **F:** -
D: Kleinfrüchtiger Igelkolben **CZ:** Zevar vzpřímený drobnoplodý **PL:** -
EST: Väikeseviljane jõgitakjas **LV:** Sīkaugļu ežgalvīte **LT:** Šakotojo šiurpio smulkiavaisis porūšis
Ru: Ежеголовник мелкоплодный

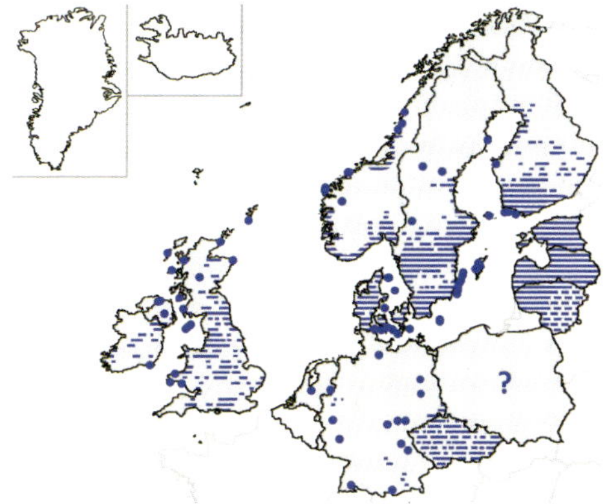

Sparganium microcarpum with semi-ripe and ripe fruits. Left: Skåne, Sweden. Photo Jan-Thomas Johansson. Right: Stroudwater Canal, Stroud, Gloucestershire, Britain. Photo RVL.

Flowers: Perianth segments spathulate, more or less entire, brownish at apex.

Fruits: 6–7(–9) mm long including the beak, which is up to 2.0 mm long, somewhat shiny, with a distinct shoulder separating the upper and lower parts. The upper part domed, initially yellowish green, but as the fruits ripen becoming pale brown to brown or blackish brown. The lower part distinctly narrower than the upper part, pale brown, narrowly obconical with 3–6 distinct angles in transverse section.

Distribution within the region:
Native. Scattered to common throughout most of the region; absent from Belgium, Luxembourg, Iceland and Greenland.

Characteristics and similar species:
S. microcarpum is always pure white at the base of the stem and leaves, while in **120 *S. erectum*** and **122 *S. neglectum***, including the 3 hybrids within the *S. erectum* aggregate p. 300, these parts have pinkish or reddish tones.

Sparganium microcarpum Semi-ripe fruits and fruiting head. River Skjern Å, Arnborg, Jutland. Photo JCS.

Typhaceae

122 *Sparganium neglectum* Beeby

Syn.: *S. erectum* L. subsp. *neglectum* (Beeby) K.Richt.
DK: Ten Pindsvineknop **N:** Nebbpiggknopp **S:** Glansigelknopp **FIN:** Jokipalpakko **IS:** -
GB: - **NL:** Blonde egelskop **F:** Rubanier négligé
D: Unbeachteter Igelkolben **CZ:** Zevar vzpřímený přehlížený **PL:** Jeżogłówka zapoznana
EST: Põlu-jõgitakjas **LV:** Necilā ežgalvīte **LT:** Šakotojo šiurpio gelsvasis porūšis
Ru: Ежеголодник незамеченный

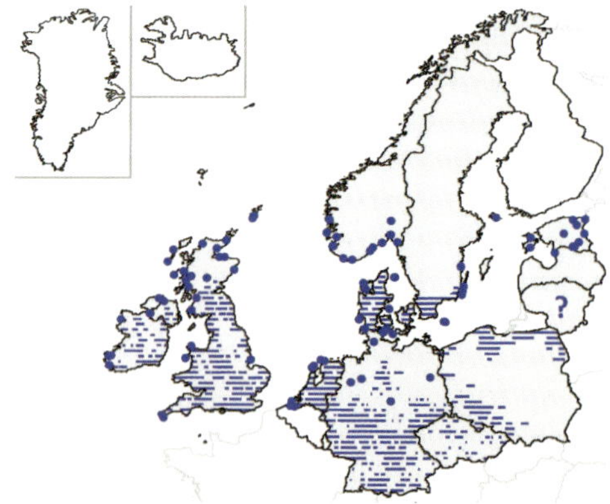

Sparganium neglectum with semi-ripe and ripe fruits. Left: Hobro, Jutland, Denmark. Right: Filsø, SW-Jutland, Denmark. Photos JCS.

Distribution within the region:
Native. Scattered to rather common throughout much of the region, absent from Belgium, Luxembourg, Finland, Iceland and Greenland.

Flowers: Perianth segments spathulate, more or less entire, brownish to dark brown at apex.

Fruits: 8–11(–13) mm long including the beak, which is 2.5–3.5 mm long, somewhat shiny, fusiform, without a distinct shoulder separating the upper and lower parts. The upper part conical, yellowish green to greyish green, but as the fruits ripen, pale brown to straw coloured. The lower part obconical, indistinctly angled, smooth, pale brown when ripe - not much different from the upper part.

Sparganium neglectum. Lower part of stem and leaves with a pinkish tone. Bråstorp, Sweden. Photo Åke Widgren.

Sparganium neglectum. Fruits and fruiting head. Filsø, SW-Jutland, Denmark. Photos JCS.

Sparganium species that are potentially introduced

123 *Sparganium stoloniferum*

Hamilton ex Graebner

Syn.: *S. erectum* L. subsp. *stoloniferum* (Buch.-Ham. ex Graebn.) H.Hara

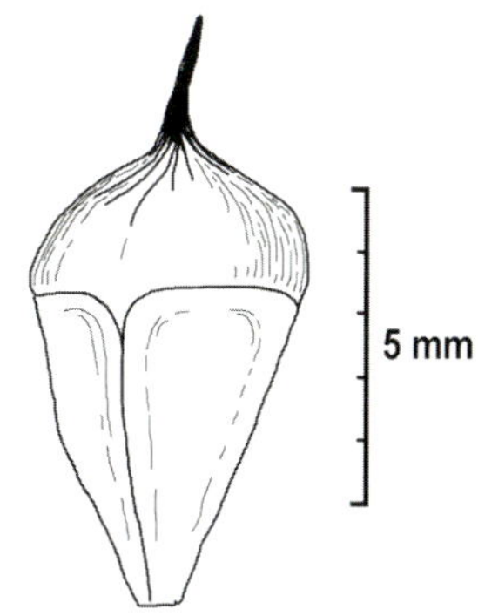

Lower part of fruit obpyramidal, distinctly 3–6 angled, upper part domed to conical, 6–9(–10) mm long, light brown, shiny, beak 2–3(–4) mm long.

Disjunct distribution from the Caucasus to Iran and Central Asia, as well as in northern China, Korea and Japan. It also occurs in Australia and the Pacific coast of North America. In the latter two regions it may be introduced.

124 *Sparganium eurycarpum*

Engelm.

Very similar to **120 *S. erectum*** in fruit shape and colour of the perianth segments, but differs in stigmas bifid, slightly larger fruits and narrower leaves.

Distributed in North America from Canada to central Mexico.

Sparganium eurycarpum.
Toronto Botanical Garden, Canada. Photo RVL.

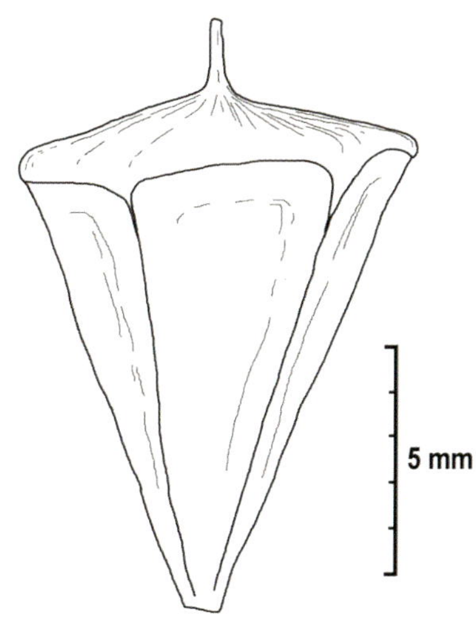

125 *Sparganium coreanum* Léveillé

Syn.: *S. eurycarpum* Engelm. subsp. *coreanum* (Léveillé) C.D.K.Cook & M.S.Nicholls

Most stigmas bifid. Fruits very large, trapezoidal to obovoid. The shoulders less conspicuous than in **124 *S. eurycarpum***. The fruit shape is similar to that of *S. ×oocarpum,* which differs from *S. eurycarpum* not only in the shape of the fruit but also in the branching pattern of the inflorescence. The two lowermost branches of the inflorescence of *S. coreanum* bear only a solitary female head.

Distributed in East Siberia, Russian Far East, Northern China, Korea and Japan.

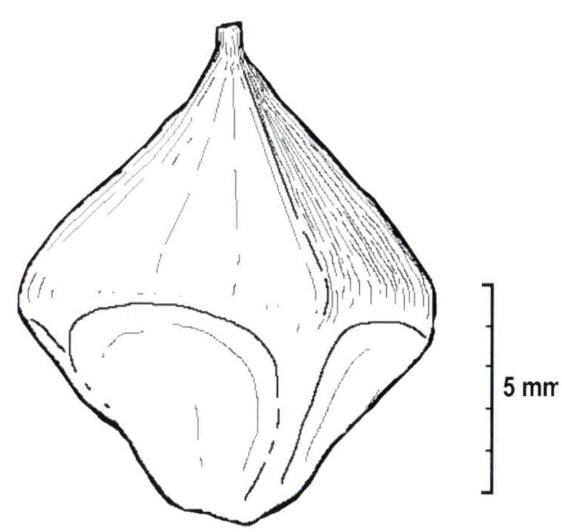

Typhaceae

Sparganium hybrids recognised within the area

Suspicion of hybridisation among *Sparganium* species is often based on full or partial sterility and diverging morphology, but both of these may have other causes. Hybridisation seems to be rather common when more species grow together, but reliable determination of such is often problematic, even with good material at hand. In the field it is important to note possible parents in the surroundings and collect proper material with flowers and fruits (if developed). Most of the hybrids listed here need confirmation by use of DNA technology.

Subgenus *Sparganium*

Hybrids within the subgenus are usually found as single specimens in mixed populations of the parental species.

S. erectum × microcarpum

Heads highly sterile. Fruits intermediate between the parental species, less angled than those of *erectum*, but wider than those of *microcarpum*. Reported from Czech Republic, Denmark and Sweden.

Siesjö, Blekinge, Sweden. Photo Åke Widgren.

S. ×oocarpum (Čelak.) Fritsch

(*S. erectum × neglectum* and back crosses with *neglectum*)

Heads highly sterile. Fruits ovoid to almost circular, beak short. Reported from Baltic States, Czech Republic, Denmark, Germany, Britain, Austria, Bulgaria, European Russia, France, Iraq, Portugal, Romania, Slovakia, Spain, Switzerland, Transcaucasia, Turkey, Ukraine, Yugoslavia.

Skivarp, Skåne, Sweden. Photo Jan-Thomas Johansson.

S. microcarpum × neglectum

Heads highly sterile. Fruits intermediate between the parental species, widest above the middle. Reported from Denmark, Sweden.

Vedeby bäck, Karlskrona, Sweden. Photo Åke Widgren.

Subgenus *Sparganium* × *Xanthosparganium*

S. emersum* × *erectum agg.
With uneven pollen and failed fruiting, but the hybrid has not been confirmed. Reported from Denmark, Finland, Norway? and Sweden.

Subgenus *Xanthosparganium*

S. ×diversifolium Graebn.
(*S. angustifolium* × *emersum*)
Intermediate, with uneven pollen and uneven fruiting. Probably backcrosses. Reported from Denmark, Finland, Germany, Norway, Sweden and Russia.

S. angustifolium* × *glomeratum
With poor pollen and almost no fruit developed. Reported from Norway and Sweden.

S. ×speirocephalum Neum.
(*S. angustifolium* × *gramineum*)
Intermediate, with uneven pollen but fairly good fruiting. Probably backcrosses. Fruit somewhat shiny, distinctly constricted below the middle, beak straight or slightly curved. Reported from Denmark?, Finland, Norway, Sweden and Russia.

S. angustifolium* × *hyperboreum
Reported from Finland, Norway and Sweden.

S. ×oligocarpon Ångstr.
(*S. angustifolium* × *natans*)
Reported from Finland, Norway and Sweden. - See illustration p. 281.

S. emersum* × *glomeratum
With poor pollen and very sparse fruiting. Reported from Finland, Norway, Sweden and Russia.

S. ×longifolium Turcz ex Ledeb.
(*S. emersum* × *gramineum*)
Similar to *S. ×speirocephalum*, but with wider leaves, longer and narrower stigmas, usually several male heads and longer anthers. Reported from Finland, Norway and Sweden.

S. emersum* × *hyperboreum
Rare and reported from scattered locations in Norway and Sweden.

S. emersum* × *natans
Reported from Finland, Norway, Sweden and Russia.

S. glomeratum* × *natans
Reported from Norway.

S. hyperboreum* × *natans
Like *hyperboreum* with supra-axillary female heads, but leaves wider, 2–5 mm, and fruits light green to brown with 0.5–1 mm long beaks. Appears to be fertile. Rare and reported from scattered locations in Finland, Norway, Sweden and Canada.

Sparganium ×diversifolium (S. angustifolium × emersum). Heerenveen, Netherlands. Photo KvdW.

Sparganium ×speirocephalum (S. angustifolium × gramineum). Högfjärden, Västergötland, Sweden. Photo KvdW.

Typhaceae

126 *Typha minima* Funck

Syn.: *Rohrbachia minima* (Funck ex Hoppe) Mavrodiev; *Typha martinii* Jord.; *T. minima* subsp. *gracilis* (Jord. ex Ducommun) K.Richt.
DK: Dværg-Dunhammer **N:** - **S:** - **FIN:** Pikkuosmankäämi **IS:** -
GB: Dwarf Bulrush **NL:** Dwerglisdodde **F:** Petite Massette
D: Kleiner Rohrkolben **CZ:** orobinec nejmenší **PL:** Pałka drobna
EST: - **LV:** - **LT:** -

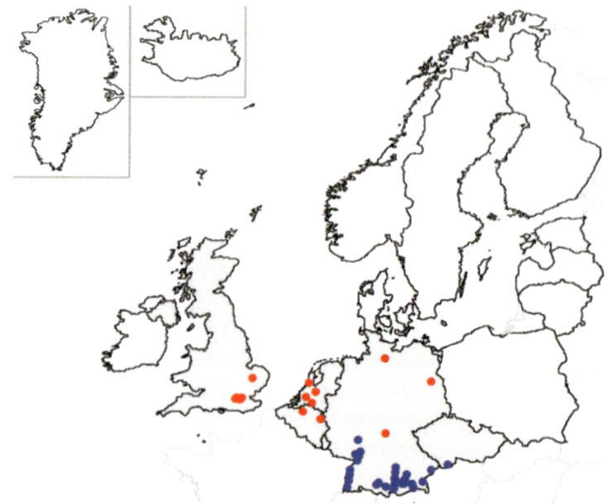

Red dots indicate non-native occurrences.

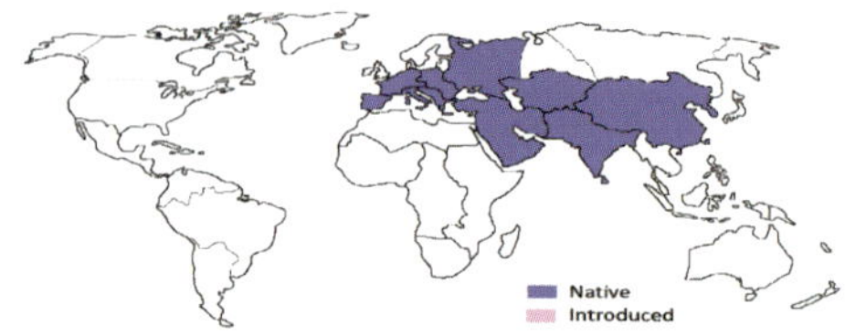

Typha minima. Plants from Lechaue, Bayern, Germany - cultivated in Botanical Garden Boku. Photo Karl Georg Bernhardt.

Perennial helophyte. Rhizome long, creeping, underground.

Stem: 30–80 cm long, slender, erect, terete, bluish green, smooth, with several usually bladeless sheaths at base, but otherwise leafless.

Leaves: 15–40 cm long on sterile shoots, distinctly shorter than the stem, 1–2(–3) mm wide, linear, straight to slightly arched, semi-terete, in cross section with 4 air chambers, bluish green.

Inflorescence: A spike with male flowers at the top and female flowers below. Male part of spike 3–8 cm long, with a deciduous bract at the base. Female part of spike separated by 0.5–3 cm from the male part, 1.5–4.5 cm long, almost spherical to shortly cylindrical, dark reddish brown, with a deciduous bract at the base.

Flowers: Unisexual, hypogynous. Male flowers without hairs, stamens 1(–2–3), anthers about 1.5 mm long, pollen in groups of four. Female flowers with brown, spathulate bracteoles of the same length as the hairs. Hairs with swollen, almost spherical tips, not exceeding the style. Ovary fusiform with stalk, 4–6 mm long. Styles about 0.5 mm long, stigma linear, 0.5 mm long.

Fruits: Achenes ellipsoid, faintly longitudinally striate, brown, 0.7–0.9 mm long and about 0.3 mm wide.

Flowering: May-June.

Biology: *Typha minima* is a poor competitor, a pioneer plant of riparian flood-plains and probably dependent on some disturbance, such as fluctuating water levels. The flowers are wind-pollinated and the fruits are dispersed by wind. Fruit production is generally very high, but depends on successful cross-pollination, as fruit set is reduced to about 25% by self-pollination. Germination of seeds decreases over time; after 8 months it is reduced to less than 50% and after 16 months to zero. Vegetative propagation is by division of the rhizome. Overwinters as leafless rhizomes in the substrate.

Habitats: On sandbanks or in shallow water of braided channels of slow-flowing, oligotrophic, calcareous rivers, oxbows, ditches and ponds.

Distribution within the region:
Native. Native only in southern Germany and the Czech Republic. Declining and threatened in Europe. It is cultivated as an ornamental plant for garden ponds and increasingly in landscape planting, naturalised in Britain, the Netherlands and Belgium.

Characteristics and similar species:
Typha minima can be recognised by its leafless stem with only bladeless sheaths at base. Stem longer than leaves of sterile shoots. Hairs of female flowers with swollen tips, male flowers without hairs.

Typha minima is often confused with dwarf forms of **132 *T. laxmannii***, but differs by having leafless stems, leaves of sterile shoots shorter than the stem, less than 3 mm wide and by the hairs of female flowers swollen at apex.

Typha minima. Trowbridge, Wiltshire, Britain. Photo RVL.

Typha minima. Cultivation, garden centre, Gloucestershire, Britain. Photo RVL.

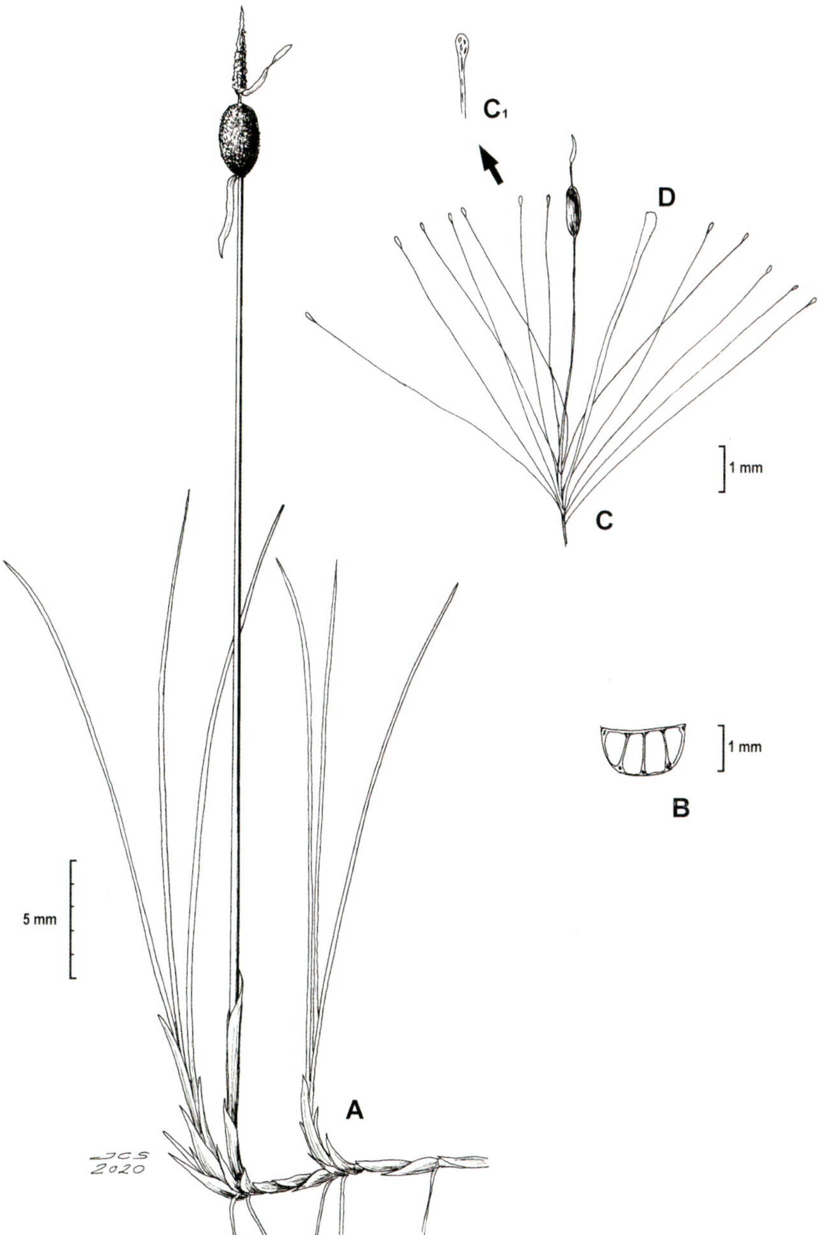

Typha minima
A habit, **B** cross section of leaf, **C** female flower, **C1** tip of hairs, **D** bracteole.

Typhaceae

127 *Typha angustifolia* L.

DK: Smalbladet Dunhammer **N:** Smalt dunkjevle **S:** Smalkaveldun **FIN:** kapeaosmankäämi **IS:** -
GB: Lesser Bulrush **NL:** Kleine lisdodde **F:** Massette à feuilles étroites
D: Schmalblättriger Rohrkolben **CZ:** Orobinec úzkolistý **PL:** Pałka wąskolistna
EST: Ahtalehine hundinui **LV:** Šaurlapu vilkvālīte **LT:** Siauralapis švendras

Typha angustifolia. Gross Glienicker See, Berlin, Germany. Photo KvdW.

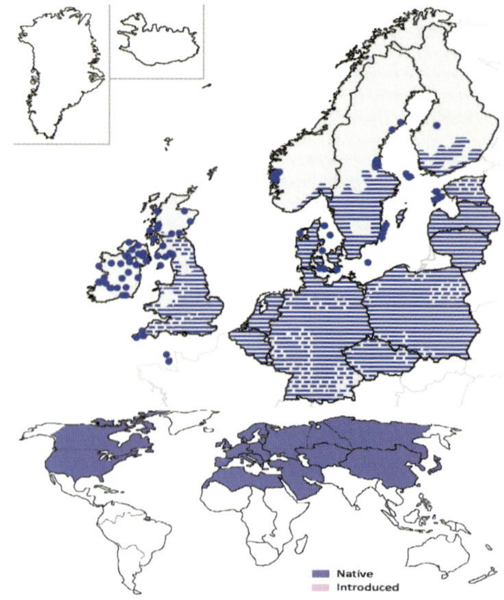

Perennial helophyte. Rhizome creeping with long, underground stolons.

Stem: (1–)2–3 m long, erect, terete, rather robust, 2–3(–4) mm in diameter beneath the spike, 5–12 mm in the middle, bluish green, smooth.

Leaves: Less than 10. Up to 3.5 m long, often exceeding the inflorescence, 3–8(–11) mm wide, stiff, yellow green to fresh green, twisted 1 or 1.5 times, the adaxial side flat or somewhat concave, the abaxial side convex.
Leaf sheaths long, thick, clasping flowering stems and usually closed at the throat, with auricles. Mucilage glands on the inner side of the sheath brown, not extended to the basal part of the leaf. Leaves with cavities throughout. With up to 15 air chambers in cross section.
In slow-flowing rivers it may occasionally appear with floating or in deep water with only submerged, ribbon-like leaves.

Inflorescence: A spike, up to 50 cm long with a male part at the top and a female part below, each with a deciduous bract at base. Male part of spike 16–30 (–35) cm long. Female part separated by 2–9(–12) cm from the male part, 8–30(–35) cm long, 13–20 mm wide, greenish brown to dark cinnamon brown when mature. Female flowers densely packed, arranged up to 9 together on 0.5(–0.8) mm long, staircase-like, compound pedicels.

Flowers: Unisexual, hypogynous. Male flowers with 1–3 stamens, 2.2–3 mm long anthers and thin, simple or forked hairs. Pollen solitary. Female flowers with 25–60, 5–9 mm long, white, brownish-tipped hairs and dark brown, spathulate bracteoles. Each group of female flowers with a mixture of fertile flowers with cylindrical ovaries and brown to dark brown, linear, persistent stigmas, and sterile flowers with an ellipsoid aborted ovary.

Fruits: Achenes 1.0–1.3 mm long, cylindrical, dehiscent.

Flowering: July.

Biology: The flowers are wind-pollinated and the fruits dispersed by wind. Vegetative propagation is by division of the rhizome. Overwinters as leafless rhizomes in the substrate.

Habitats: In meso- to eutrophic water. Reedbeds and on the margins of lakes, ponds, rivers, streams and ditches, growing to a depth of 2 metres.

Typha angustifolia with floating leaves. Iteltelch, Walkenried, Germany. Photo KvdW.
Inserted: Leaf section of floating leaf. Gudenå, Denmark. Photo JCS.

Typha angustifolia. Female part of spike with flowers removed, revealing the short, compound pedicels. Lille Vildmose, Denmark. Photo JCS.

Typha angustifolia. Uprooted stolons. Chertsey, West London, England. Photo RVL.

Distribution within the region:
Native. Widespread and abundant in the south, less common and locally absent in the north.

Characteristics and similar species:
Typha angustifolia can be recognised by its long, slender spike with distinctly separated male and female parts, 3–6 mm wide leaves and flowers on short compound pedicels.

129 *T. latifolia* differs in its wider leaves, broader, dark brown female spike, which is contiguous with the male part.

Easily confused with **128 *T. ×glauca***, which is more or less intermediate between the parental species. See next page.

132 *T. laxmannii* is also very similar but lacks bracteoles on the female flowers. See this.

The long, narrow leaves of submerged or young specimens differ from the leaves of **118 *Sparganium emersum*** and **145-146 *Schoenoplectus*** spp. in being thicker and more firm, with up to 15 longitudinal air chambers forming a rectangular reticulation when lit from behind.

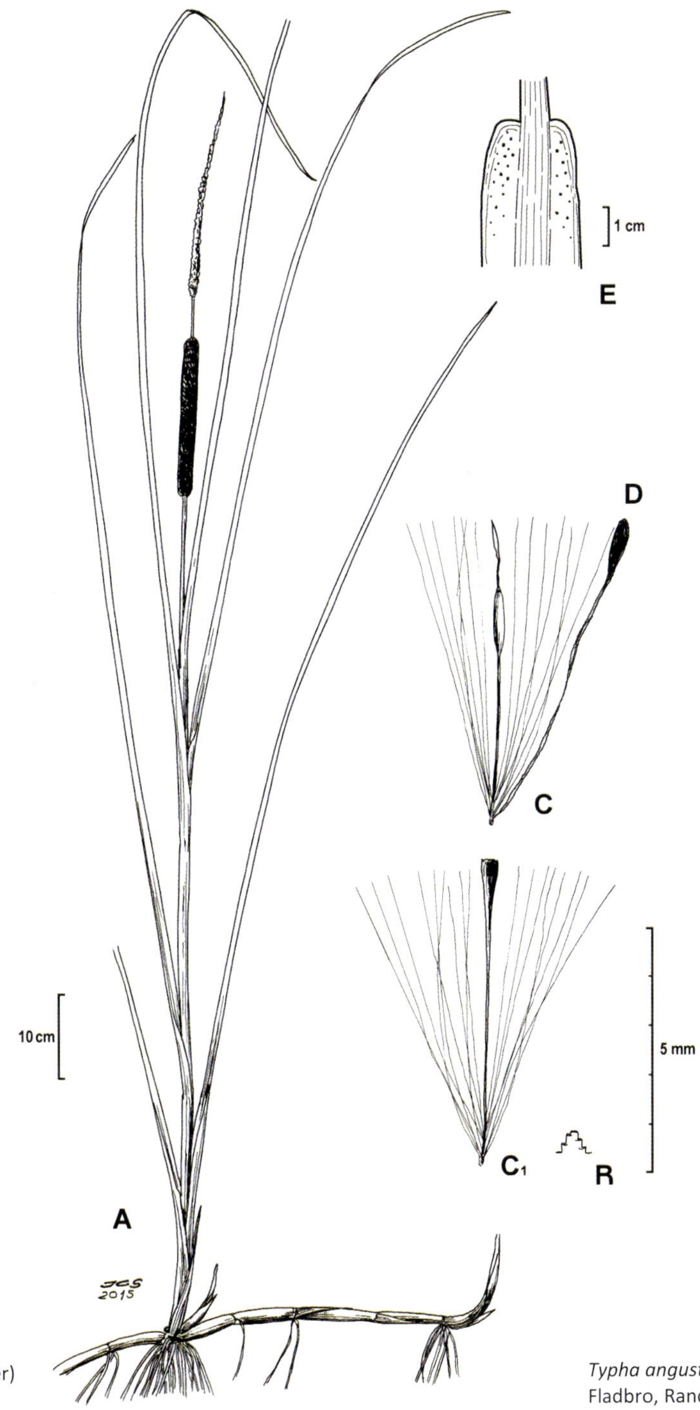

Typha angustifolia
A habit, **C** female flower,
C1 sterile female flower, **D** bracteola, **E** adaxial (inner) side of leaf base/sheath, **R** compound pedicel.

Typha angustifolia.
Fladbro, Randers, Denmark. Photo JCS.

Typhaceae

128 ***Typha ×glauca*** Godr. (*T. angustifolia × latifolia*)

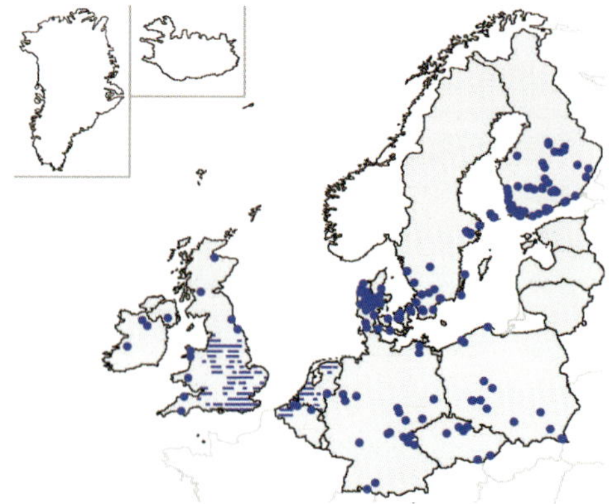

Vigorous growth of *T. ×glauca*. River Gudenå, Silkeborg, Denmark. Photo BM.

Typha ×glauca is generally closer in habit to *T. angustifolia* than to *T. latifolia*, but otherwise morphologically more or less intermediate between the parental species.

In the field the hybrid can be recognised by a shorter space between the male and female parts of the spike and by intermediate leaf width. The most important character for separation from the parental species is the length of the compound pedicels of the female spike, these 0.9–1.5(–1.7) mm long. Female flowers with or without dark brown bracteoles. Pollen solitary, in pairs or in groups of four.

Biology: New research has shown that the hybrid is not always fully sterile, and to some extent backcrosses. It is also capable of forming large, monospecific stands through vegetative propagation.

Distribution within the region:
Native. Scattered to locally frequent throughout much of the region, except Norway, Iceland and Greenland.

In North America the hybrid has arisen from introduced *T. angustifolia* and native *T. latifolia*, and is regarded as invasive.

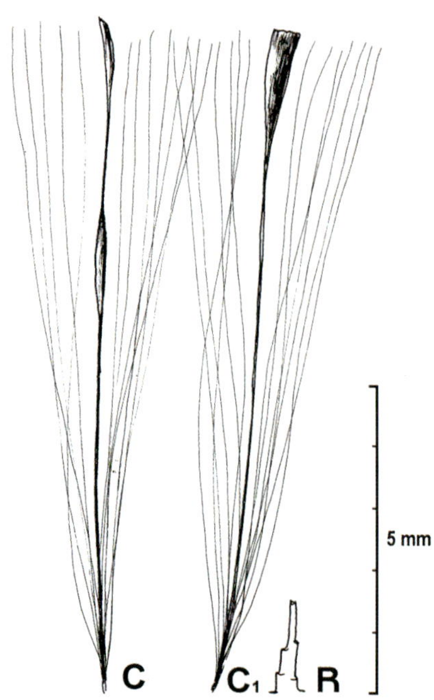

Typha ×glauca
C female flower, **C1** sterile female flower, **R** compound pedicel.

Typha ×glauca. The male and female part of the spike are only shortly separated. Photos RVL.

Diagnosis of native *T. ×glauca* Godron

Observations of Grace & Harrison (1986), van de Weyer (1996), Nowińska et al. (2014) and Bernhardt & Gregor (2019), combined with our own observations in East Germany and western Poland.

Character	*T. angustifolia*	*T. ×glauca*	*T. latifolia*
Stem length	(1–)2–3 m	1–2(–2.3) m	1–2.5 m
Leaf length	Up to 3.5 m	Up to 2.2(–2.5) m	Up to 2.5 m
Leaf exceeding inflorescence	Regularly, up to 30 cm	Regularly, up to 20 cm	Rarely, if so, less than 10 cm
Leaf width	3–8(–11) mm	6–13(–19) mm	10–20(–23) mm
Leaf sheath	With auricles	Without auricles	With auricles
Leaf colour	Yellow to fresh green	Bluish green (glaucous)	Greyish green
Total length of inflorescence	Up to 50 cm	Up to 30 cm	Up to 40 cm
Length of male part	16–30(–35) cm	10–15 cm	10–15(–20) cm
Length of female part	8–30(–35) cm, as long as female part	10–20 cm, longer than male part	10–20 cm, as long as the male part
Interval within female spike	Absent	Sometimes present	Absent
Length of interval between male and female part of spike	20–90(–120) mm	(4–)5–16 mm	0–5 mm
Compound pedicel length	0.5(–0.8) mm	0.9–1.5(–1.7) mm	1.5–3(–4) mm
Bracteoles of female flowers	Present, spathulate	Present, hair-like, or mostly absent	Absent
Pollen	Solitary	Solitary, in pairs or in groups of three or four	In groups of four; aborted solitary grains may occur
Fertility	Fully fertile	Reduced	Fully fertile

Important field characters

Decisive floral characters

In contrast to the North American populations *T. ×glauca* does not show heterosis in Europe; on the contrary, plants are typically smaller than the parent species. Plants can resemble either parent species but are mostly recognisable by several intermediate or unique characters.

Typhaceae

129 *Typha latifolia* L.

DK: Bredbladet Dunhammer **N:** Breitt dunkjevle **S:** Bredkaveldun **FIN:** Leveäosmankäämi **IS:** -
GB: Bulrush **NL:** Grote lisdodde **F:** Massette a larges feuilles
D: Breitblättriger Rohrkolben **CZ:** Orobinec širokolistý **PL:** Pałka szerokolistna
EST: Laialehine hundinu **LV:** Platlapu vilkvālīte **LT:** Plačialapis švendras

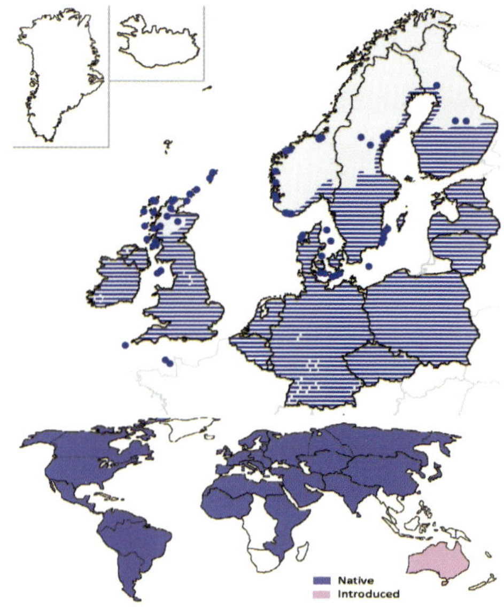

Extensive stands of *T. latifolia* growing in an area periodically inundated by freshwater on the coast of Vendsyssel, Denmark. Photo JCS.

Perennial helophyte. Rhizome creeping with long, underground stolons.

Stem: 1–2.5 m long, erect, terete, robust, 3–7 mm in diameter beneath the spike, 10–20 mm in the middle, bluish green, smooth.

Leaves: Up to 2.5 m long, equal to or slightly exceeding the inflorescence, 10–20(–25) mm wide, greyish green, twisted 1 or 1.5 times, adaxial side flat or somewhat concave, abaxial side convex. Leaf sheaths long, thick, clasping flowering stems, open at the throat, with auricles. Inner side of the sheath-blade transition with colourless, obscure mucilage glands, these absent from sheath centre and blade. Leaves with cavities throughout. With up to 20 air chambers in cross section.

Inflorescence: A spike up to 40 cm long with a male part at the top and a female part below. Male part of spike 5–15(–20) cm long. Female part contiguous, rarely separated by up to 1(–1.5) cm from the male part, (8–)10–20 cm long, 18–30 mm wide, greenish brown, dark brown when mature, with a deciduous bract at the base. Female flowers densely packed, arranged up to 9 together on 1.5–3(–4) mm long, narrow, staircase-like, compound pedicels.

Flowers: Unisexual, hypogynous. Male flowers with 1–3 stamens, anthers 2.5–3.5 mm long, and thin, simple or forked hairs. Pollen in groups of four. Female flowers with 25–50 white hairs, 7–11 mm long, but without bracteoles. Each group of female flowers with a mix of fertile flowers with cylindrical ovary and brown to dark brown, lanceolate ovate, fleshy, persistent stigma, and sterile flowers with an ellipsoid aborted ovary.

Fruits: Achenes 1.2–1.5 mm long, cylindrical, dehiscent.

Flowering: June-July.

Biology: Like other *Typha* species the female flowers have their stigma exposed several days before the male flowers start to shed pollen. The stigmas remain receptive for some weeks and self-pollination is frequent. A single spike may produce up to 100,000 female flowers. The flowers are wind-pollinated and the fruits are dispersed

Typha latifolia. Cross section of shoot. Chertsey, West London, England. Photo RVL.

Typha latifolia. Seedlings. Alney Isand, Gloucestershire, England. Photo RVL.

Female part of spike with flowers removed, revealing the short, compound pedicels. Hobro, Denmark. Photo JCS.

Fruiting plants. Alney Island, Gloucestershire, Britain. Photo RVL.

Typha latifolia. Male and female part of spike contiguous. Photos JCS and RVL.

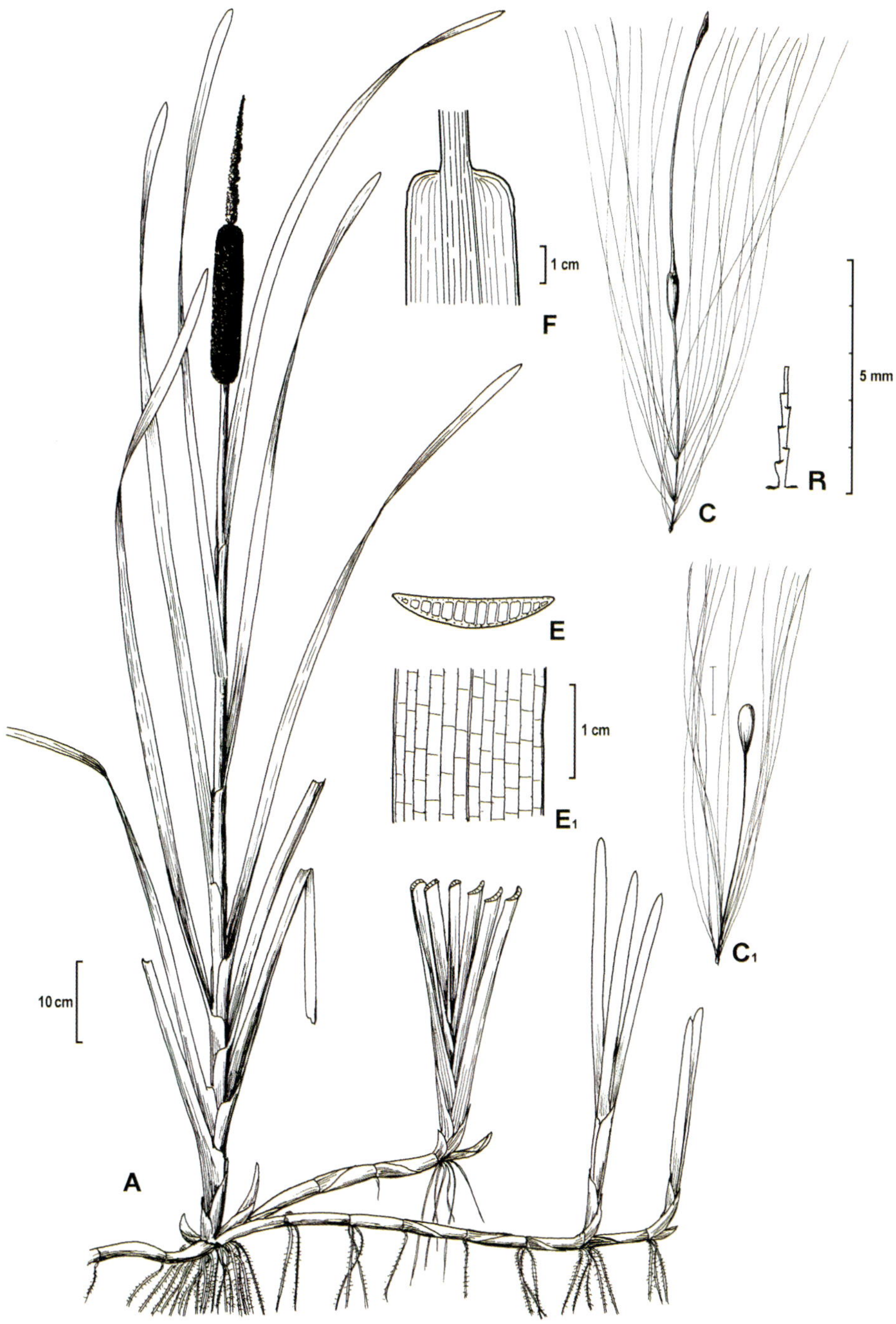

Typha latifolia
A habit, **C** female flower, **C1** sterile female flower, **R** compound pedicel, **E** leaf cross section, **E1** leaf section, **F** adaxial (inner) side of leaf base/sheath, mucilage glands colourless, obscure.

Typha latifolia with aerial and floating leaves. Rainwater reservoir. Stavtrup, Aarhus, Denmark. Photo BM.

Typha latifolia. The long, narrow leaves of floating, submerged or young specimens differ from the leaves of *Sparganium* spp. and *Schoenoplectus* spp. in being thicker and firmer, with up to 20 longitudinal air chambers forming a rectangular reticulation when seen in backlight.
Photo Mogens Holmen.

by wind. The long hairs on the flower stalk provide buoyancy when the fruits are spread with the wind and keep the achene floating on contact with water. Vegetative propagation is by division of the rhizome. Overwinters as leafless rhizomes in the substrate.

Habitats: In meso- to eutrophic waters. Reedbeds and along the shores of lakes, ponds, rivers, streams and ditches, growing to a depth of 2 metres.

Distribution within the region:
Native. Widespread and abundant throughout except in the north.

Characteristics and similar species:
Typha latifolia can be recognised by the broad, dark brown female part of the spike, with flowers on 2–4 mm long, narrow, staircase-like, compound pedicels.

131 *Typha shuttleworthii* is similar in habit, but differs in the female part of the inflorescence - silvery grey when mature, and distinctly longer than the male part.

For distinguishing from **127 *T. angustifolia*** and **128 *T. ×glauca*** - see these.

Typha capensis (Rohrbach) N.E.Brown is very similar to *T. latifolia* in leaf colour, width and length. The leaves usually remain shorter than the stem. The male part of the inflorescence is often considerably shorter than the female part (as in *T. shuttleworthii*) and sometimes interrupted by large, membranous bracts, which are also found in the female part. Male and female parts are contiguous or separated by a very short distance. The female bracteoles are mostly absent, and the pollen is solitary. *T. capensis* might be a hybrid of *T. latifolia* with an unknown species. Rarely introduced and difficult to recognise in the field.

Typhaceae

130 *Typha domingensis* Pers.

Syn.: *T. angustifolia* subsp. *domingensis* (Pers.) Rohrb.; *T. australis* Schumach.; *T. angustata* Bory & Chaub.
DK: - **N:** - **S:** - **FIN:** - **IS:** -
GB: - **NL:** - **F:** Massette de Dominique
D: Südlicher Rohrkolben **CZ:** - **PL:** -
EST: -**LV:** - **LT:** - **Ru:** Рогоз доминиканский

Typha domingensis. Male part of spike usually exceeding the leaves. Tuz Golu, southern Turkey. Photo RVL.

Perennial helophyte. Rhizome creeping with long, underground stolons.

Stem: 1.5–4 m long, erect, terete, rather robust, 3–4 mm in diameter beneath the spike, 10–20 mm in the middle.

Leaves: More than 10. Equal to or shorter than the inflorescence, 6–18 mm wide, pale green, flat. Initially colourless, but soon orange-brown mucilage glands occur on the inner side of the sheath extended to the basal 1–10 cm of the blade. Leaves with cavities throughout. Upper leaf sheath open at the throat, without auricles.

Inflorescence: With a male part at the top and a female part below. Both with a deciduous bract at base. Male part of spike 12–38 cm long. Female part separated by 1.5–8 cm from the male part, 6–35 cm long, 5–8 mm wide when flowering, 15–25 mm wide in fruit, cinnamon brown with whitish stigmas when fresh, brown when dry or mature. Female flowers densely packed, arranged on 0.5–0.8 mm long, blunt, staircase-like, compound pedicels.

Flowers: Unisexual, hypogynous. Male flowers with anthers 2–2.5 mm long. Pollen solitary. Female flowers with 25–55, 5–7 mm long, orange-brown-tipped hairs and straw-coloured to light brown, lanceolate bracteoles.

Fruits: Achenes about 1 mm long, cylindrical, dehiscent.

Flowering: June-July.

Biology: The flowers are wind-pollinated and the fruits are dispersed by wind. Vegetative propagation is by dividing of the rhizome. Overwinters as leafless rhizomes in the substrate.

Habitats: In meso- to eutrophic, fresh or brackish water. Reedbeds and along the shores of lakes, ponds, rivers, streams, ditches and channels.

Distribution within the region:
Alien. Introduced but not naturalised in Germany and Luxembourg.

Characteristics and similar species:
Similar in habit to **127 *Typha angustifolia***, but differs by the more than 10 pale green leaves, usually shorter than the inflorescence, by orange glands on the inner side of older leaf sheaths extending to the basal 1–10 cm of the blade and by the female flowers with straw-coloured to light brown bracteoles.

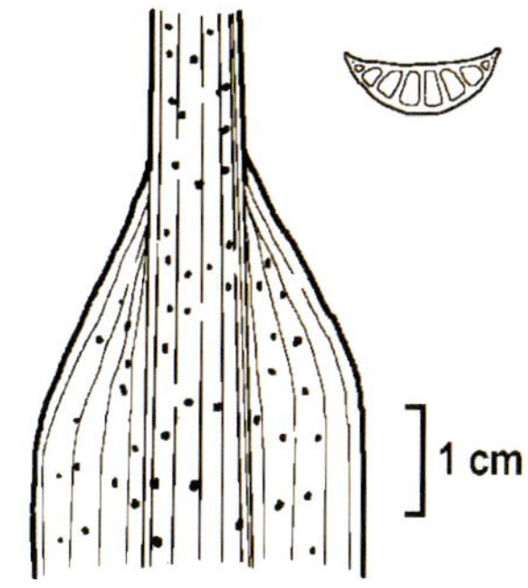

Typha domingensis. Adaxial (inner) side of leaf base/sheath with orange-brown mucilage glands extended to the blade. Cross section of leaf.

Typha domingensis
The male and female part of the spike are distinctly separated. Photos RVL.

Typhaceae

131 *Typha shuttleworthii* W.D.J.Koch & Sond.

Syn.: *T. transsilvanica* Schur
DK: - **N:** - **S:** - **FIN:** - **IS:** -
GB: - **NL:** - **F:** Massette de Shuttleworth
D: Shuttleworth-Rohrkolben **CZ:** Orobinec stříbrošedý **PL:** -
EST: - **LV:** - **LT:** - **Ru:** Рогоз Шаттлворта

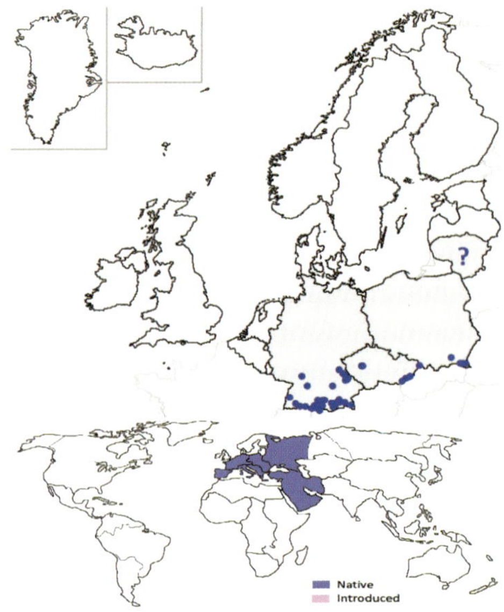

Perennial helophyte. Rhizome creeping with long, underground stolons.

Stem: 0.8–1.5(–1.9) m long, erect, terete, robust, yellowish green to pale green, smooth.

Leaves: Equal to or slightly exceeding the inflorescence, (5–)7–10(–15) mm wide, yellowish green to pale green. Leaf sheaths open at the throat. Leaves with cavities throughout. With 10–12 air chambers in cross section.

Inflorescence: Spike with a male part at the top and a female part below, both with a deciduous bract at base. Male part of spike 4–6(–10) cm long, distinctly shorter than the female part. Female part contiguous with male part, (6–)9–11(–17) cm long, brown when young, silvery grey with dark spots when mature, because the white perianth hairs exceed the dark stigma. Female flowers densely packed on 1–1.5 mm long, staircase-like, compound pedicels.

Flowers: Unisexual, hypogynous. Male flowers with 1–2(–5) stamens, anthers 1.5–2 mm long and narrow, lanceolate hairs. Pollen in groups of four. Female flowers with 5–25, white hairs, but without bracteoles. Stigmas dark brown, lanceolate to spathulate, fleshy, persistent.

Fruits: Achenes 0.7–0.9 mm long, cylindrical, dehiscent.

Flowering: June-August.

Biology: The flowers are windpollinated and the fruits dispersed by wind. Vegetative propagation is by division of the rhizome. Overwinters as leafless rhizomes in the substrate.

Habitats: In still or slow-flowing water along the shores of streams and rivers and in ditches.

Typha shuttleworthii. Czech Republic. Photo Martin Dančák.

Typha shuttleworthii. Lidečko, Czech Republic. Photo David Hlisnikovský.

Distribution within the region:
Native. Scattered to locally frequent in southern Germany, Poland and the Czech Republic. Occurrence in Lithuania is unconfirmed.

Characteristics and similar species:
Typha shuttleworthii is similar to **129** ***T. latifolia*** in habit, but differs by the male part of the inflorescence distinctly shorter than the female part, by narrower leaves and shorter anthers.

Typha shuttleworthii. Třeboň, Jindřichův Hradec-district, Czech Republic. Photo Jana Navrátilová.

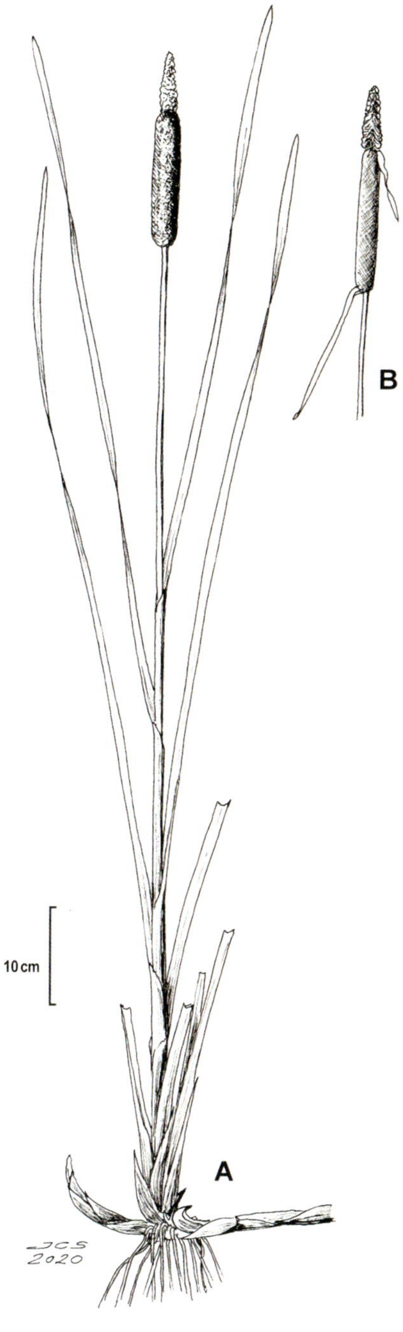

Typha shuttleworthii
A habit, **B** inflorescence with bracts.

Typhaceae

132 *Typha laxmannii* Lepech.

Syn.: *T. minima* subsp. *laxmannii* (Lepech.) Nyman, *T. lugdunensis* auct. non P.Chabert
DK: - **N:** - **S:** - **FIN:** - **IS:** -
GB: Laxman's Bulrush **NL:** - **F:** Massette de Laxmann
D: Laxmanns Rohrkolben **CZ:** Orobinec sítinovitý **PL:** Pałka wysmukła
EST: - **LV:** - **LT:** - **Ru:** Рогоз Лаксмана

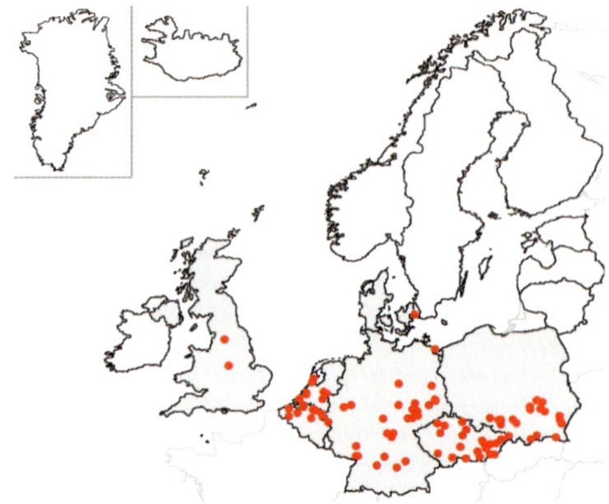

Red dots indicate non-native occurrences.

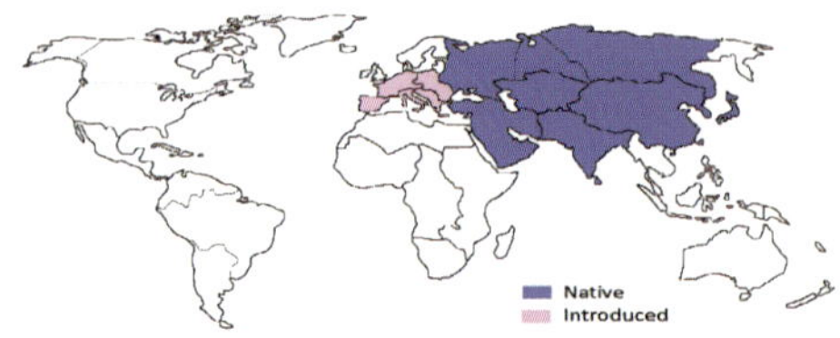

Typha laxmannii. Tullner Feld, Pixendorf, Österreich. Photo Karl Georg Bernhardt.

Perennial helophyte. Rhizome creeping with long, underground stolons.

Stem: 0.6–1.3(–2) m long, erect, terete, rather robust, yellowish-green, smooth.

Leaves: Exceeding the inflorescence, 2–4(–7) mm wide, yellowish-green to green, semi-terete. Upper leaf sheaths usually open at the throat, with auricles. Leaves with cavities throughout, and with about 8 air chambers in cross section.

Inflorescence: Spike with a male part at the top and a female part below, both with a deciduous bract at the base. Male part of spike 8–10(–18) cm long, about 6 mm wide. Female part separated by up to (0.5–)2–4(–6) cm from the male part, (2–)3–5(–13) cm long, up to 25 mm wide, light brown to brown when mature. Female flowers densely packed, arranged on compound pedicels less than 1 mm long.

Flowers: Unisexual, hypogynous. Male flowers with 1–2(–5) stamens, anthers 1–1.8 mm long, fine hairs and bracteoles. Pollen solitary. Female flowers without bracteoles, with white hairs up to 10 mm long. Each group of female flowers with a mix of fertile flowers with cylindrical ovaries, brownish, spathulate, persistent stigmas longer than the hairs, and sterile flowers each with an ellipsoid aborted ovary.

Fruits: Achenes 1–1.2(–1.4) mm long, cylindrical, dehiscent.

Flowering: July-August.

Biology: The flowers are wind-pollinated and self-pollination is very frequent. The fruits are dispersed by wind. Vegetative propagation is by division of the rhizome. Overwinters as leafless rhizomes in the substrate.

Habitats: In shallow, meso- to eutrophic water. In water bodies in abandoned sand and gravel pits, artificial ponds, canals, channels in reedbeds and at the outlet of rivers.

Distribution within the region:
Alien. Introduced, naturalised and widely established in the Netherlands, Germany and southern Poland, with an increasing number of records in Britain.

Typha laxmannii. Tullner Feld, Pixendorf, Österreich. Photo Karl Georg Bernhardt.

Characteristics and similar species: *Typha laxmannii* can be recognised by its leaves, which are 2–7 mm wide and longer than the inflorescence, by the male part of the spike 2–4(–6) cm separated from and much longer than the female part - this short, cylindrical, 3–5 (–13) mm long and light brown to brown.

It may be confused with **126 *Typha minima*,** but differs in the flowering stems with leaves, and hairs of the female flowers not swollen at the apex.

127 *T. angustifolia* differs by female flowers with bracteoles, the inner side of the sheath with mucilage glands and by the longer female part of the spike.

Typha laxmannii. Spikes in different stages. Photo RVL.

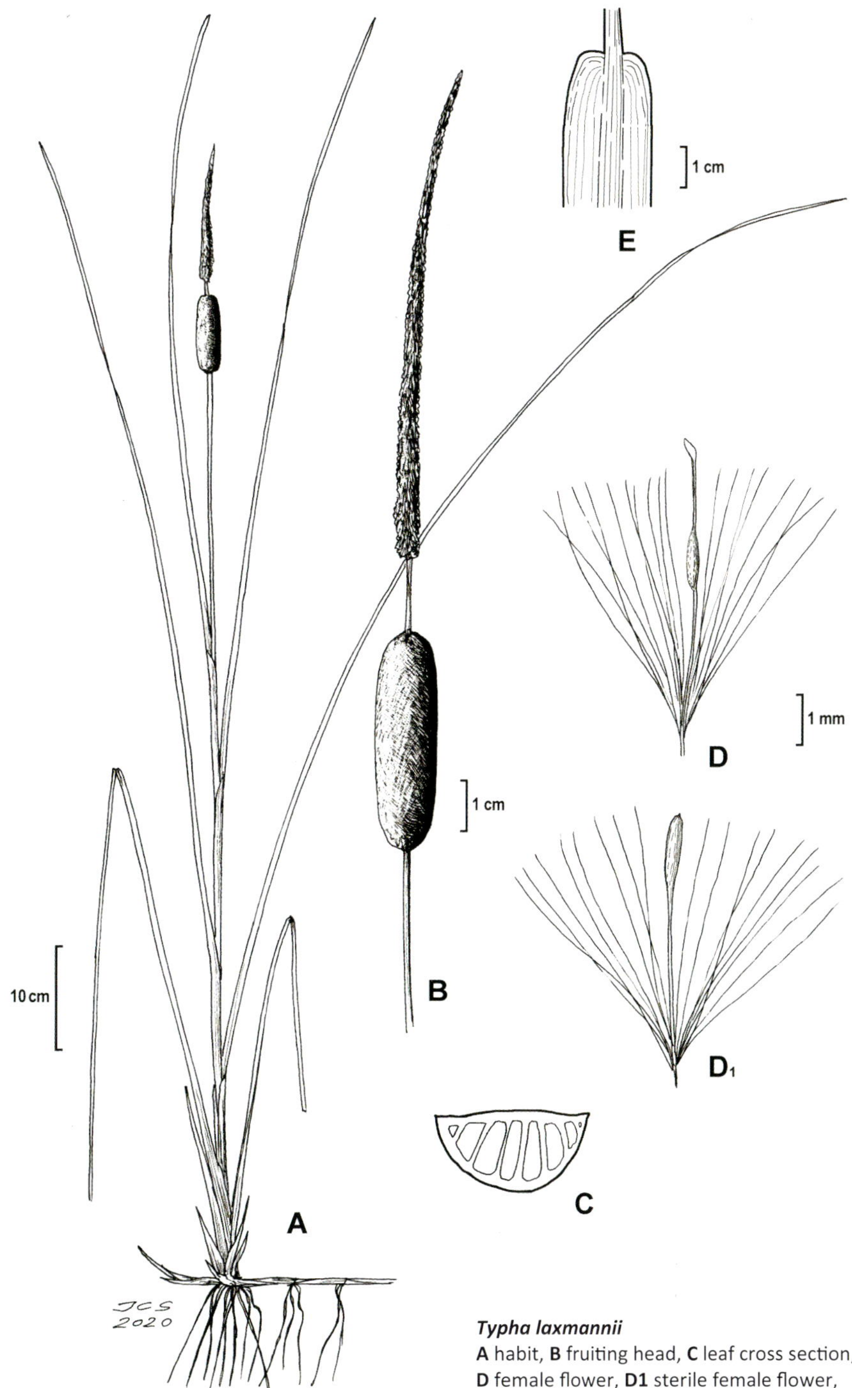

Typha laxmannii
A habit, **B** fruiting head, **C** leaf cross section, **D** female flower, **D1** sterile female flower, **E** leaf sheath.

Eriocaulaceae

133 *Eriocaulon aquaticum* (Hill) Druce

Syn.: *E. septangulare* With.; *Cespa aquatica* Hill; *Eriocaulon articulatum* (Hudson) Morong
DK: - **N:** - **S:** - **FIN:** - **IS:** -
GB: Pipewort **NL:** - **F:** Ériocaulon à sept angles
D: - **CZ:** - **PL:** -
EST: - **LV:** - **LT:** -

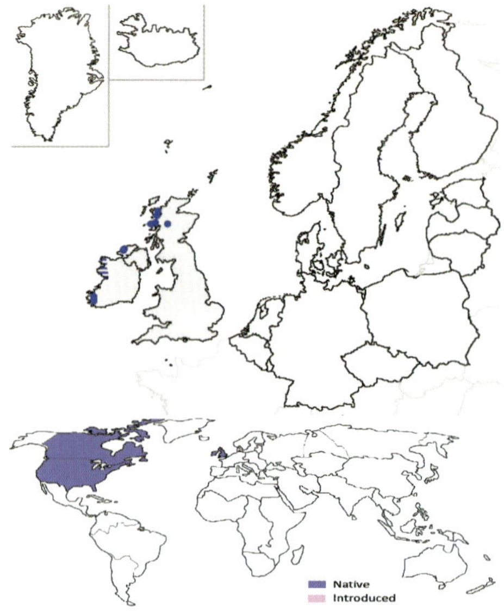

Perennial hydrophyte, forming small clumps or mats. Rhizome short, creeping. Roots white, appearing articulated due to brownish striations.

Stem: In shallow water it is normally 5–30 cm tall, but in deep water it may reach 150 cm to lift the inflorescence above the water. About 1 mm in diameter, hollow, yellowish green, slender, erect, smooth, with 5–8 prominent ridges, somewhat twisted, leafless, borne from an inflated sheath in the rosette centre. Typically with only one stem per rosette.

Leaves: Sessile, forming a basal rosette. Up to 10 cm long or even longer when submerged, 1–2.5 mm wide, entire, linear, broadest at the base and gradually tapering to a fine, subulate tip, very thin and translucent, with 3–9 conspicuous veins and cross-veins.

Inflorescence: Emergent. A terminal button-shaped, white to pale grey head, 4–10(–15) mm across and supported by obovate, blunt, greyish bracts. Bracteoles blackish. The inflorescence carries both male and female flowers.

Flowers: With 4 perianth segments and white club-shaped hairs at the margins and on the abaxial apical surface. Male flowers with 2 outer segments, which are fused at the base, 1.5 mm long, oblong linear, greyish. The 2 inner segments are 0.5 mm long and with a nectar gland close to the tip. 4 stamens with black anthers. Female flowers resemble the male flowers, but the 2 inner segments are longer, 1.5 mm. Stigmas 2, filiform.

Fruits: Seeds light brown, ovoid, 0.5 mm long, finely reticulate.

Flowering: July-September.

Biology: The nectar glands on the inner perianth segments are believed to attract insects for pollination, but autogamy is also possible. Vegetative reproduction is possible by dichotomous branching of the shoot tip. Photosynthesis is supported by root uptake from the sediment, while oxygen produced is released to the sediment.

The flower heads of *E. aquaticum* raise above the water surface in a small pool. Moore, Ireland. Photo KvdW.

Eriocaulon aquaticum. The roots are white and soft and look almost articulated, worm-like. Moore, Ireland. Photo KvdW.

The thin leaves in rosettes, and gradually tapering to a subulate tip, are very distinct. Because of vegetative reproduction, *E. aquaticum* often forms small clumps or mats. Moore, Ireland. Photo KvdW.

Habitats: In acidic or more rarely slightly alkaline water on acidic substrates; such as peat, gravel and sand. From periodically flooded ground above the water line it descends to a depth of 1 metre or more. In pools, smaller lakes and ditches. Often associated with **334 *Lobelia dortmanna***, **137 *Juncus bulbosus*** and **296 *Littorella uniflora***.

Distribution within the region:
Native. Western parts of Scotland and Ireland. Elsewhere in Canada and USA.

Characteristics and similar species:
The rosette of *E. aquaticum* is clearly distinct from those of other rosette species in the area. The thin, translucent leaves gradually tapering to a subulate tip and with a prominent reticulate venation eliminate confusion with **359-360 *Isoëtes*** spp., **254 *Subularia aquatic*** and **37 *Baldellia ranunculoides***.

Note: The European and North American plants may prove to be two separate species. According to Löve & Löve (1958), the European plants have a chromosome number of 2n = 64, whereas the American plants have 2n = 32.

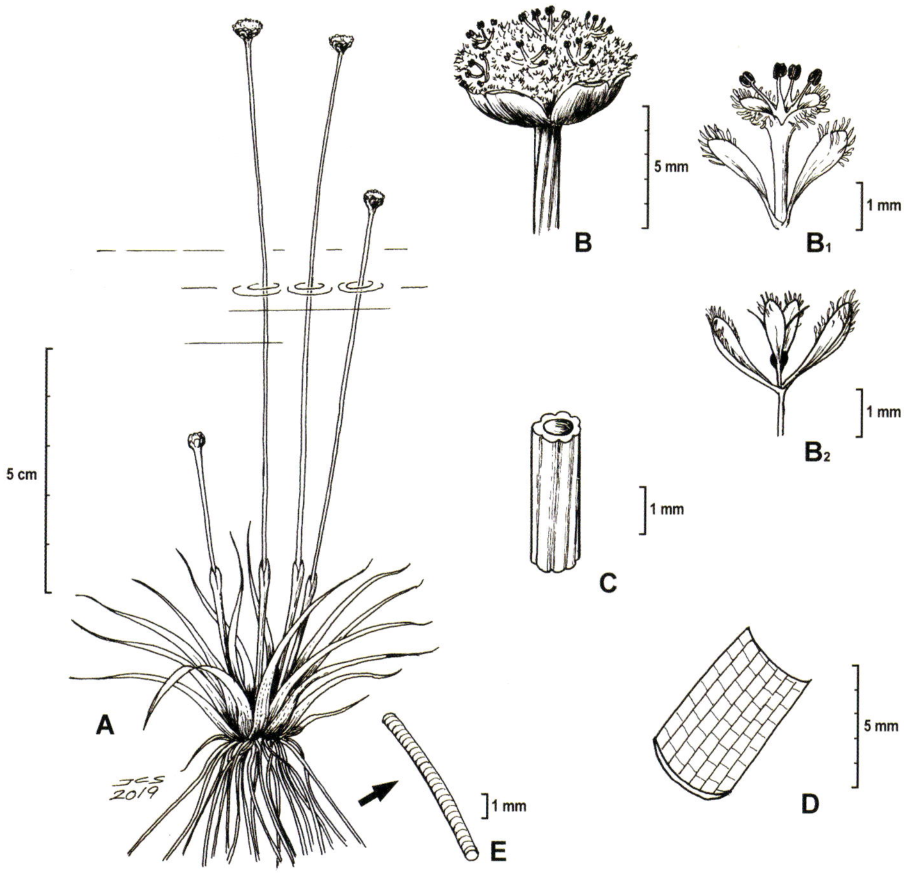

Eriocaulon aquaticum
A habit, **B** inflorescence, **B1** male flower, **B2** female flower, **C** stem section, **D** leaf section, **E** root.

Key to grass-like plants in vegetative state

Determination of grass-like plants without floral characters is a difficult task and requires solid investigation of the plant. The key below may help in deciding which family a given plant belongs to.

1 Leaves similar to stem or absent, sometimes with scales or bladeless sheaths at base ... 2
1 Leaves not similar to stem ... 4

2 Stem < 5 mm in diameter ... 3
2 Stem > 5 mm in diameter
Schoenoplectus p. 337

3 Basal sheaths open (***1***)
Juncus p. 319
3 Basal sheaths closed (***2***)
Eleocharis p. 329

4 Ligules or auricles present (***3-7***) ... 5
4 Ligules or auricles absent ... 8

5 Ligule present - eventually as a line of hairs (***5, 6, 7***) ... 6
5 Ligule absent, auricles present (***3, 4***)
Juncus p. 319

6 Ligule partly attached to blade (***5***)
Cyperaceae p. 328
6 Ligule free of blade - eventually in form of a line of hairs (***6, 7***) ... 7

7 Ligule rather thick, not hyaline
Triglochin or ***Scheuchzeria***
(not treated in this book)
7 Ligule thin, hyaline or as a line of hairs (***6, 7***)
Poaceae p. 412

8 Leaf margins with long, light hairs (***8***)
Luzula (not treated in this book)
8 Leaf margins without long, light hairs ... 9

9 Margins of leaf blade rough
Cyperaceae p. 328
9 Margins of leaf blade smooth ... 10

10 Basal leaves present ... 11
10 Only stem leaves present
Cyperaceae p. 328

11 Leaf sheaths open (***3***)
Juncus p. 319
11 Leaf sheaths closed (***5***)
Cyperaceae p. 328

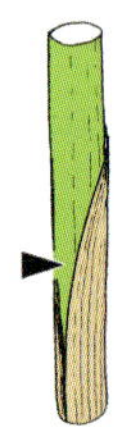

1

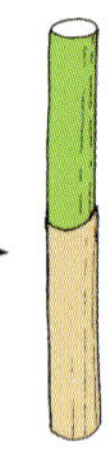

2

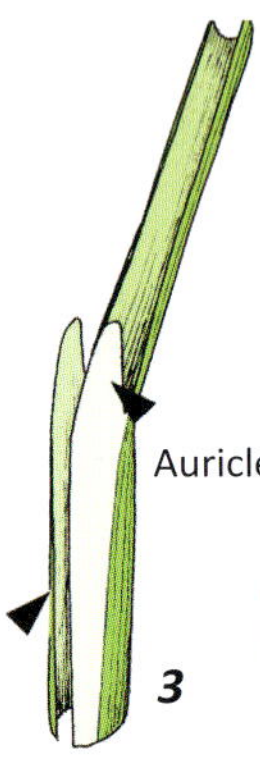

3

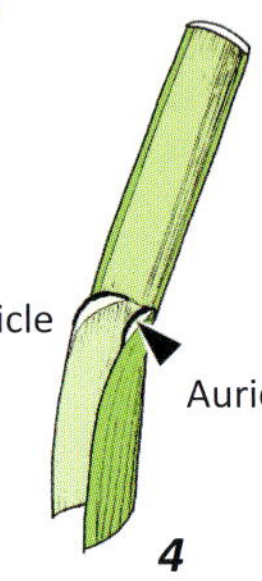

4

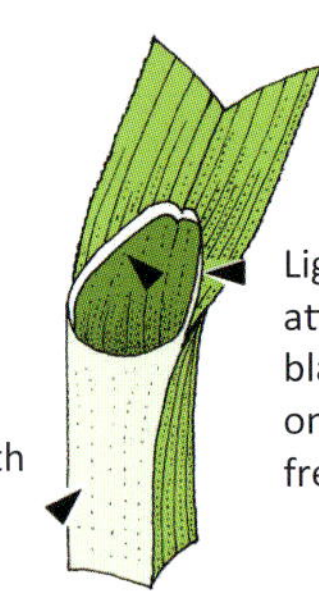

5

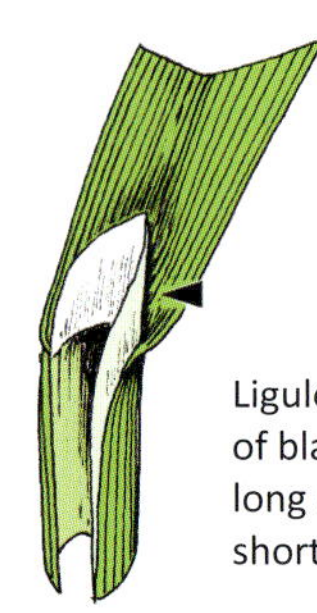

6

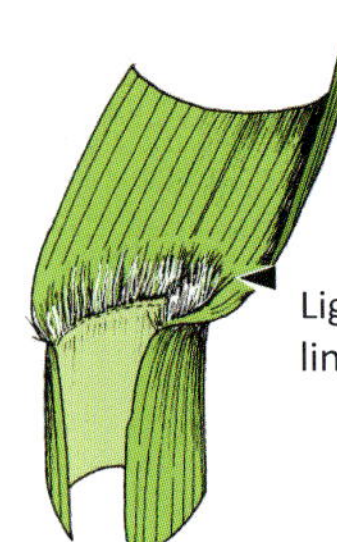

7

8

Key to vegetative *Juncus* growing in water

1 Leaves with distinct septa (***1***) - feels like small bumps when squeezing the leaf between two fingers and pulling them towards the tip.
Stem without swollen base ... 2

1 Leaves with indistinct septa (***2***) - no bumps when squeezing the leaf between two fingers and pulling them towards the tip.
Stem with or without swollen base
137 *Juncus bulbosus*

2 Leaves with a ring of about 15 air tubes, surrounding a large central air tube and a central vascular bundle in cross section (***3***)
134 *Juncus subnodulosus*

2 Leaves with one large airtube in cross section (***4***) ... 3

3 Leaves with 5–10 septa per 5 cm (***1***).
Rhizome 2–3 mm thick. Internodes normally 0.1–0.5 cm long
135 *Juncus articulatus*

3 Leaves with 1(–2) septa per 5 cm (***4***).
Rhizome 5–8 mm thick. Internodes normally 0.5–2 cm cm long
136 *Juncus acutiflorus*

1

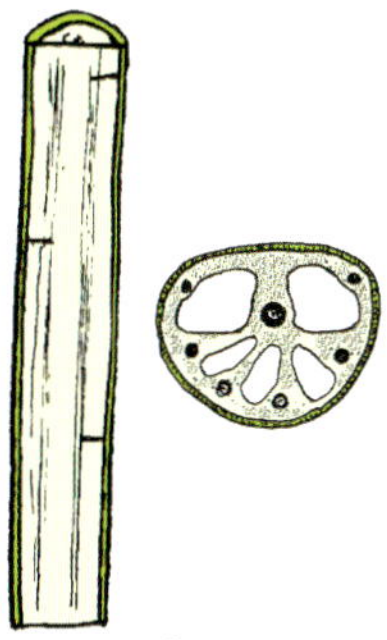

2

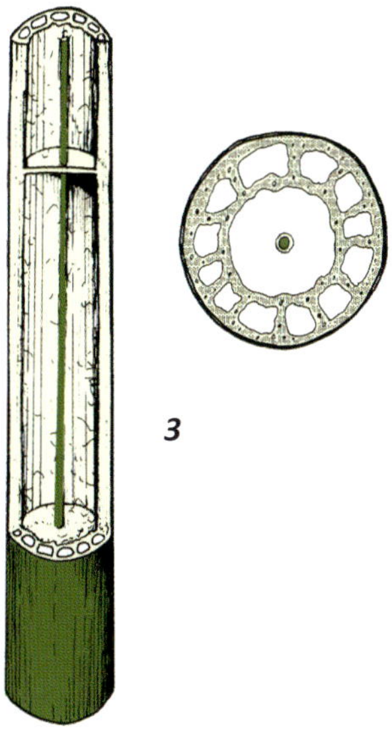

3

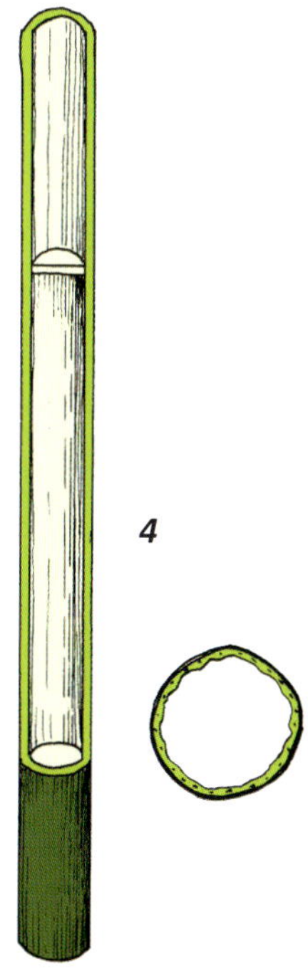

4

Several other *Juncus* species, both annuals and perennials, live in the wet environment of lakes and streams without forming special aquatic forms. They bloom and bear fruit and can be named using common floras.

Juncus effusus growing in a pond. Hobro, Denmark. Photo JCS.

Juncus biglumis in a brook at an altitude of 1200m. Besstrond, Norway. Photo JCS.

Juncaceae

134 *Juncus subnodulosus* Schrank

Syn.: *J. obtusifolius* Ehrh. ex Hoffm.
DK: Butblomstret Siv **N:** - **S:** Trubbtåg **FIN:** - **IS:** -
GB: Blunt-flowered Rush **NL:** Padderus **F:** Jonc à fleurs obtuses
D: Stumpfblütige Binse **CZ:** Sítina slatinná **PL:** Sit tępokwiatowy
EST: Tömbiõiene luga **LV:** Strupais donis **LT:** - **Ru:** Ситник узловатый

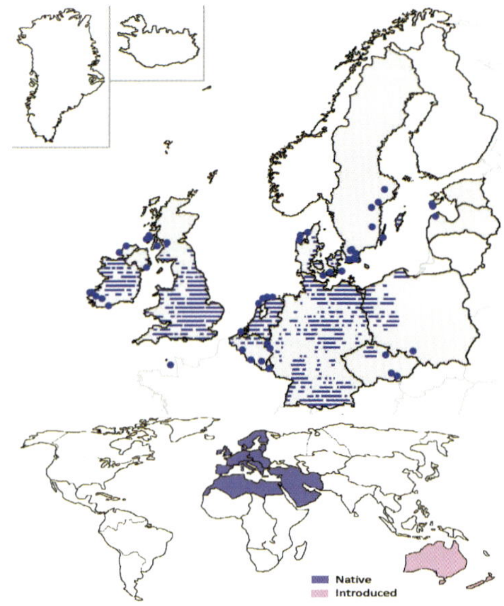

Juncus subnodulosus. Inflorescences. Mariager, Denmark. Photo JCS.

Perennial helophyte. Rhizome long, creeping, underground, sparsely branched, robust, 7–10 mm in diameter, with internodes 1–2 cm long.

Stem: 30–100 cm long, 2–4 mm in diameter, erect, somewhat shiny, terete or slightly compressed, hollow, greyish green, often ochre at base.

Leaves: With 3–5 bladeless leaf sheaths at the base of the stem, the uppermost up to 10 cm long. Cauline leaves 1–2, with blades 2–4 mm in diameter and up to 50 cm long. On vegetative shoots the blades can reach a length of 1 m. Blades bright green to grey green, terete, thick walled, with very distinct transverse septa. With a ring of about 15 air tubes, surrounding the large central air tube and the central vascular bundle in cross section. Leaf sheaths with short, blunt auricles.

Inflorescence: 4–10 cm long, richly branched, with 10–50(–100) hemispherical to subglobose pale heads, each with 3–12 flowers. The branches spreading at a right angle or an obtuse angle. Bracts scarious.

Flowers: Bisexual. Bracteoles absent. Perianth segments 6, equal, 2–2.5 mm long, greenish to light brown, sometimes with reddish-brown tones, with wide, whitish margins, obtuse, the outer vaulted - boat shaped. Stamens 6, anthers yellow, 0.8–1 mm long with shorter filaments. Stigmas 3.

Fruits: Capsules 2.5–3 mm long, shiny, light brown, sharply triangular in cross section, ovate, tapering to a short beak. Seeds about 0.5 mm long, ovoid, with a distinct reticulation of about 20 longitudinal and transverse ribs.

Flowering: July-September.

Biology: The flowers are windpollinated, the fruits are dispersed by wind, water and animals. Vegetative propagation is by division of the rhizome. Often forming dense stands. In the Nordic countries mature capsules are mostly seen in warm summers and develop late in the year.

Habitats: On moist, peaty substrates in calcareous fens, marshes and dune slacks. In the pre-alpine region it often forms fully submerged, dense stands in oligotrophic groundwater-fed streams, ditches and gravel pits, associated with *Potamogeton coloratus*, *Sparganium natans* and Characeae.

Distribution within the region:
Native. Widespread and locally abundant from Ireland east through Denmark, Germany and the Czech Republic to western Poland; rare or local in southern Sweden, Estonia and Latvia.

Submerged *J. subnodulosus*. Pizikotbach, Oberau, Germany. Photo KvdW.

Characteristics and similar species:
Juncus subnodulosus can be recognised by its slender, erect growth, inflorescence pale with widely spreading, often reflexed branches, perianth segments with obtuse apex and wide, whitish margins. Leaves with a ring of about 15 air tubes, surrounding a large central air tube and a central vascular bundle in cross section. Basal sheaths 3–5, long and yellowish.

136 *J. acutiflorus* differs by acute perianth segments and thin-walled leaves with only one tube.

Juncus subnodulosus. Fruiting head. Note the blunt perianth segments. Photo JCS.

Juncus subnodulosus.
Longitudinal sections of stem. Photo RVL.

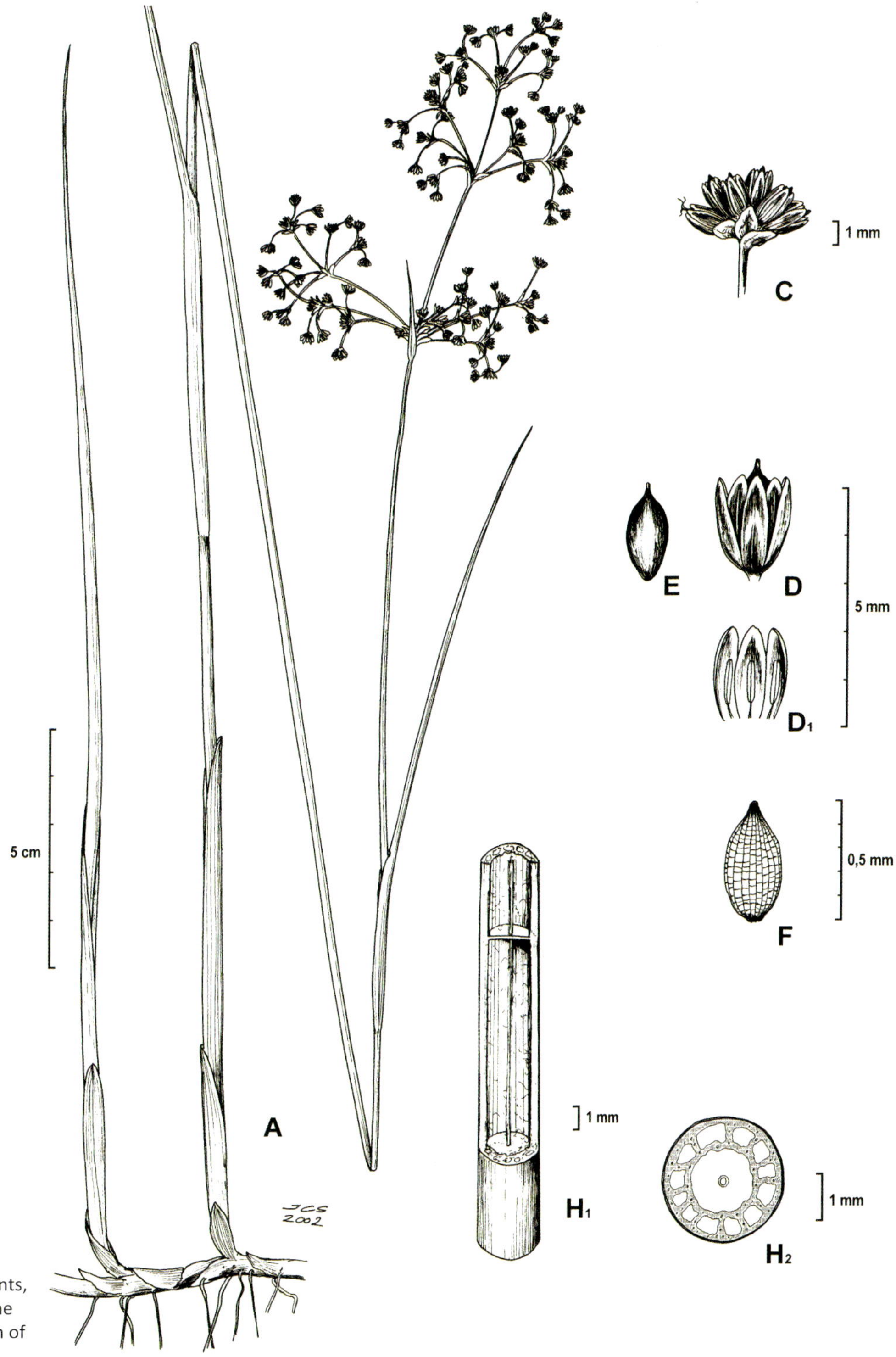

Juncus subnodulosus
A habit, **C** head, **D** capsule with perianth segments, **D1** perianth segments and anthers seen from the inside, **E** capsule, **F** seed, **H1** longitudinal section of leaf, **H2** cross section of leaf.

Juncaceae

135 *Juncus articulatus* L.

Syn.: *J. lamprocarpus* Ehrh. ex Hoffm.
DK: Glanskapslet Siv **N:** Ryllsiv **S:** Ryltåg **FIN:** Solmuvihvilä **IS:** Laugasef
GB: Jointed Rush **NL:** waterrus **F:** Jonc articulé
D: Glanzfrüchtige Binse **CZ:** Sítina článkovaná **PL:** Sit członowaty
EST: Läikviljane luga **LV:** Posmainais donis **LT:** Nariuotalapis vikšris **Ru:** Ситник членистый

Juncus articulatus in a dune slack. North Jutland, Denmark. Photo JCS.

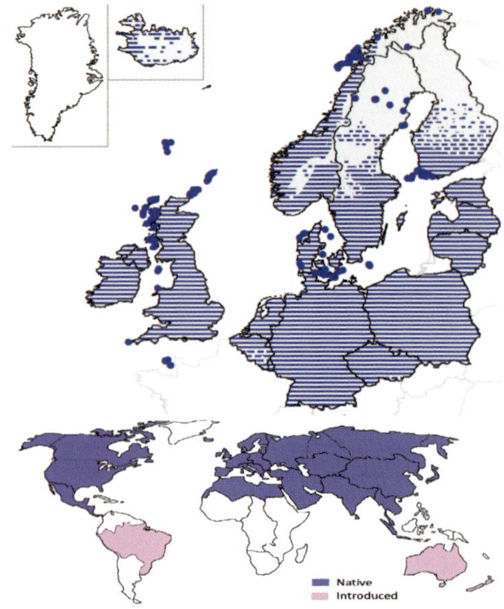

Perennial helophyte. Rhizome 2–3 mm thick, sparsely branched, in caespitose plants short, with 0.1–0.5 mm long internodes, or especially on moist substrates with longer internodes and more scattered shoots.

Stem: 5–40(–70) cm long, 2–3 mm thick, dark green, somewhat striate, erect, ascending or procumbent and rooting at the nodes, sometimes floating in water.

Leaves: With 1–2 bladeless leaf sheaths at base of stem. Cauline leaves 2–7, with blades about 2 mm diameter, dark green, terete or somewhat compressed, acute, with distinct septa and one central air tube. Leaf sheath with short, blunt auricles.

Inflorescence: 2–15 cm long, richly branched, with 5–80(–150) hemispherical to subglobose heads, each with 3–20 flowers. Richly branched, with erect to erecto-patent branches in several levels. Bracts scarious, the lowermost leafy.

Flowers: Bisexual. Bracteoles absent. Perianth segments equal or the outer slightly longer than the inner, equal to or a little shorter than the mature capsule, 2.5–3.5(–4) mm long, greenish brown to dark brown. The outer boat-shaped, distinctly acute or mucronate, with narrow scabrous margins. The inner wider, often with broad scarious margins, acute or obtuse, sometimes mucronate. Stamens 6, anthers yellowish, 0.5–1 mm long, about as long as filaments. Stigmas 3.

Fruits: Capsules 3–4 mm long, shiny, light brown to dark chestnut brown or almost black, sharply triangular in cross section, ovoid, smoothly or abruptly tapering to a 0.1–0.3 mm long beak. Seeds 0.5–0.6 mm long, ovoid, with a reticulation of about 25 longitudinal ribs and weaker transverse ribs.

Flowering: July-October.

Biology: The flowers are windpollinated and the fruits dispersed by wind, water and animals. Vegetative propagation is by division of the rhizome. Often among the first species to establish on new freshwater habitats.

Habitats: In wet meadows, marshes, swamps, dune slacks and ditches, the margins of ponds, lakes and streams, forming fully submerged stands in streams in soft water and acid conditions.

Distribution within the region:
Native. Common throughout the region, except the north; absent from Greenland.

Characteristics and similar species:
Very variable in size and habit. Despite great variation is it possible to recognise *J. articulatus* by the acute outer perianth segments, a little longer than or equal to the acute or obtuse inner,

Juncus articulatus. Flowering and fruiting heads. Photo JCS.

Juncus articulatus. Form with prostrate stems. Filsø, Jutland, Denmark. Photo JCS.

Juncus articulatus. Submerged plants. Germany. Photo KvdW.

these with wide, scarious margins. The leaves with distinct septa.

Very robust plants may be mistaken for **136 *J. acutiflorus***, but differ in the wide, scarious margins of the inner perianth segments, and vegetatively by a thinner rhizome.

137 *J. bulbosus* has leaves with several air tubes and indistinct septa.

Hybrids: See *J. acutiflorus*.

Juncus articulatus with proliferating inflorescences. N-Jutland, Denmark. Photo JCS.

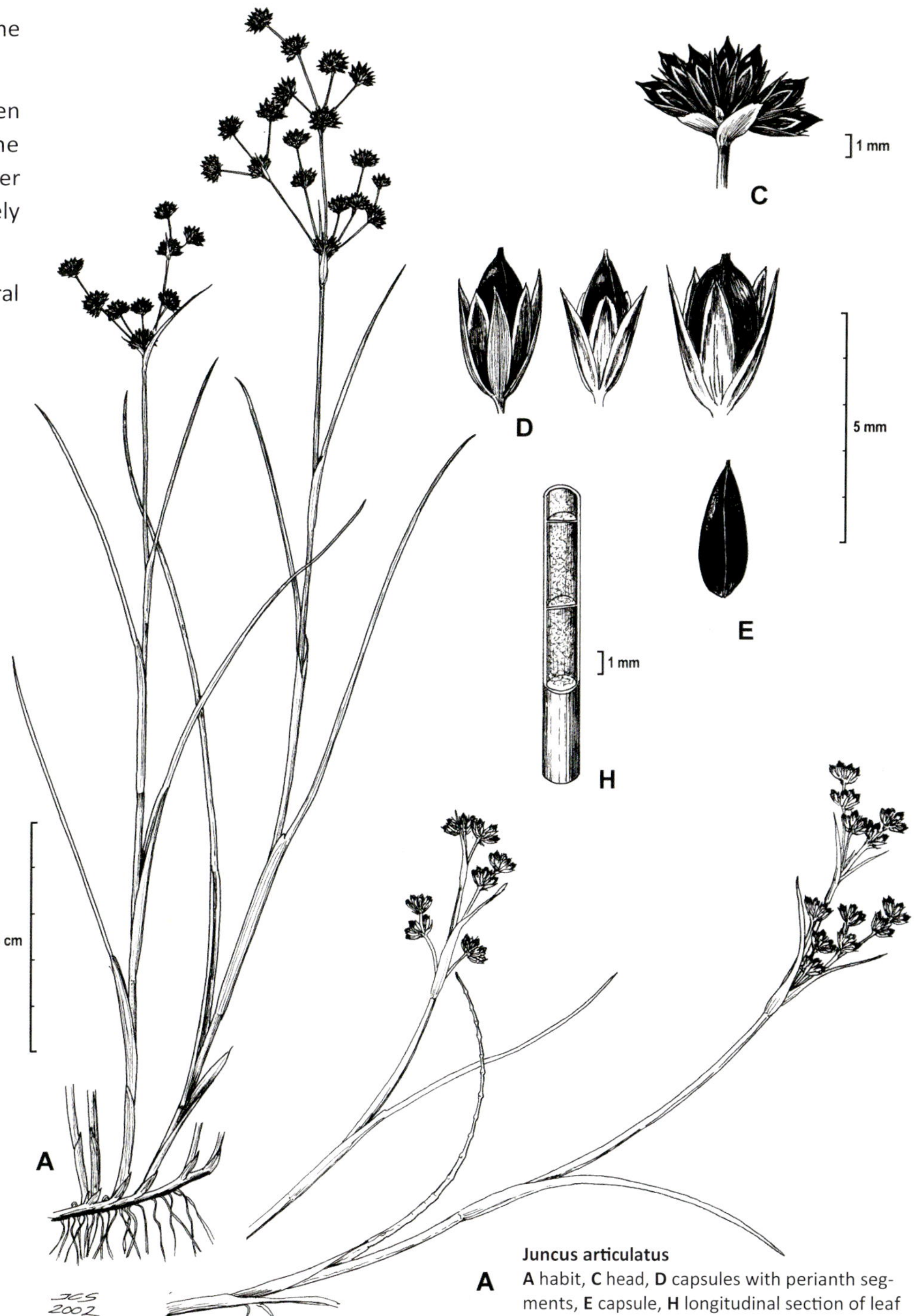

Juncus articulatus
A habit, **C** head, **D** capsules with perianth segments, **E** capsule, **H** longitudinal section of leaf (note the transverse septa).

Juncaceae

136 *Juncus acutiflorus* Ehrh. ex Hoffm.

Syn.: *J. sylvaticus* Reich.
DK: Spidsblomstret Siv **N:** Spiss-siv **S:** Spetståg **FIN:** - **IS:** -
GB: Sharp-flowered Rush **NL:** Bosrus **F:** Jonc acutiflore
D: Spitzblütige Binse **CZ:** Sítina ostrokvětá **PL:** Sit ostrokwiatowy
EST: - **LV:** - **LT:** Smailiažiedis vikšris **Ru:** Ситник остроцветковый

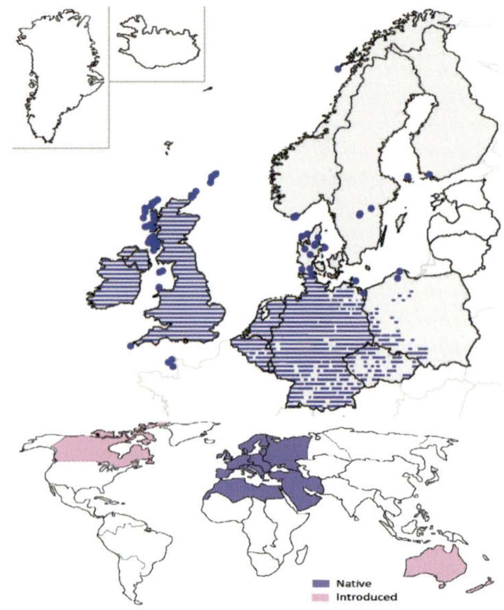

Juncus acutiflorus is usually a high and robust plant. Hobro, Jutland, Denmark. Photo JCS.

Perennial helophyte. Rhizome long, creeping, sparsely branched, robust, 5–8 mm diameter, with internodes 0.5–2 cm long. Often forming dense stands.

Stem: 30–110 cm long, 2–3 mm diameter, erect, somewhat shiny, terete and bright green above, somewhat compressed and often ochrous yellow towards the base.

Leaves: With 2–3 bladeless leaf sheaths at the base of the stem. Cauline leaves 2–4, with blades about 2 mm diameter and up to 50 cm long. Blades bright green, terete, thin walled, with a large central air tube and distinct transverse septa. Leaf sheaths with short, blunt auricles.

Inflorescence: 5–15 cm long, richly branched, with (10–)50–80(–150) hemispherical to sub globose heads, each with 3–10 flowers. With erect to erecto-patent branches at several levels. Bracts scarious, the lowermost leafy.

Flowers: Bisexual. Bracteoles absent. Perianth segments 6, 2–3 mm long, chestnut brown with greenish centre and narrow, light brown margins, subulate, somewhat divergent, the inner slightly longer than the outer. Stamens 6, anthers yellow, 0.7–1 mm long with shorter filaments. Stigmas 3.

Fruits: Capsules 2.5–3.5 mm long, light brown, sharply triangular in cross section, tapering from the rounded base to a short, 0.5–1 mm long beak. Seeds about 0.4 mm long, ovoid, with a distinct reticulation of about 25 longitudinal ribs and weaker transverse ribs.

Flowering: July-September.

Biology: The flowers are windpollinated and the fruits dispersed by wind, water and animals. Vegetative propagation is by division of the rhizome. In the Nordic countries mature capsules are mostly seen in warm summers and develop late in the year. Elsewhere it is abundantly fertile.

Habitats: On wet, oligo- to mesotrophic substrates in meadows, fens, marshes, bogs, swampy woods, and on the margins of ponds, lakes and rivers - always in places with emerging groundwater.

Distribution within the region:
Native. Widespread and abundant in Ireland, Britain, the Netherlands, Belgium, Luxembourg and Germany. Locally abundant to scattered in Denmark,

Juncus acutiflorus. Head and flower. Photo JCS.

Juncus acutiflorus. Inflorescences. Thy, Denmark. Photo JCS.

Juncus acutiflorus. Submerged plant. New Forest, Britain. Photo KvdW.

the Czech Republic and western Poland. Rare in Norway, Sweden and Finland.

Characteristics and similar species: *Juncus acutiflorus* can be recognised by the long, slender, erect stems, subulate perianth segments with narrow, brownish margins, the inner longer than the outer. The leaves septate, thin walled with a large central air tube only.

134 *J. subnodulosus* is similar in habit, but the perianth segments are obtuse with wide, scarious margins, the leaves with thick walls and several small air tubes around the large central air tube.

Large specimens of **135 *Juncus articulatus*** differ in inner perianth segments with wide, scarious margins and a thinner rhizome.

Hybrids:

Juncus ×montserratensis Marcet

Syn.: *J. ×surrejanus* Druce ex Stace & Lambion.
(*J. acutiflorus × articulatus*)

Variable in habit and intermediate in the floral characters. It may be sterile and may continue developing additional flowers in the heads, which therefore become rather large. Capsules are formed, but in most cases without seeds.

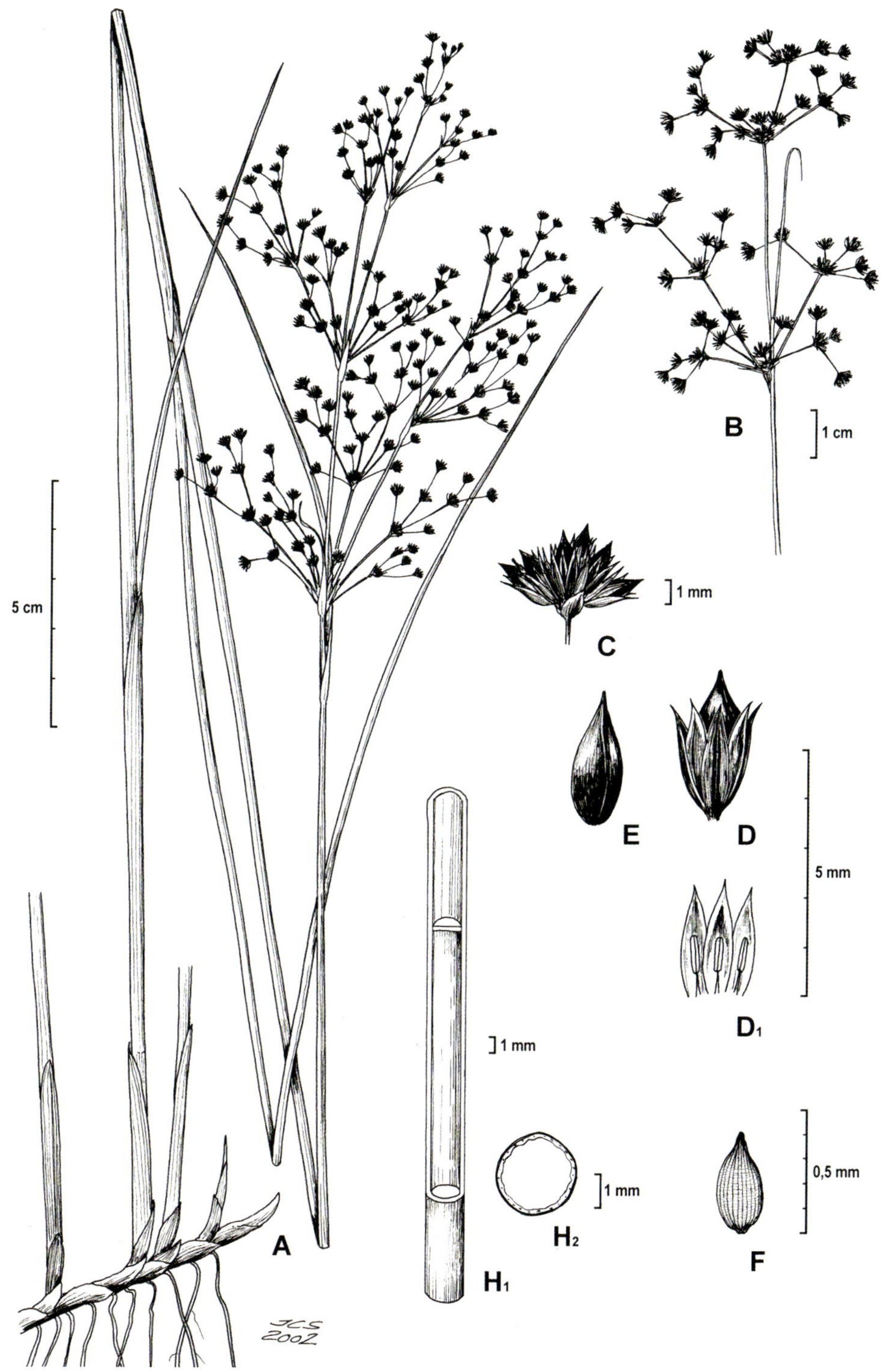

Juncus acutiflorus
A habit, **B** inflorescence, **C** head,
D capsule with perianth segments, **D1** perianth segments and anthers seen from the inside,
F seed, **H1** longitudinal section of leaf,
H2 cross section of leaf.

Juncaceae

137 *Juncus bulbosus* L.

Syn.: *J. supinus* Moench
DK: Liden Siv **N:** Krypsiv **S:** Löktåg **FIN:** Rentovihvilä **IS:** Hnúðsef
GB: Bulbous Rush **NL:** Knolrus **F:** Jonc bulbeux
D: Zwiebel-Binse **CZ:** Sítina cibulkatá **PL:** Sit drobny
EST: Madal luga **LV:** Sīpoliņu donis **LT:** Liūninis vikšris **Ru:** Ситник луковичный

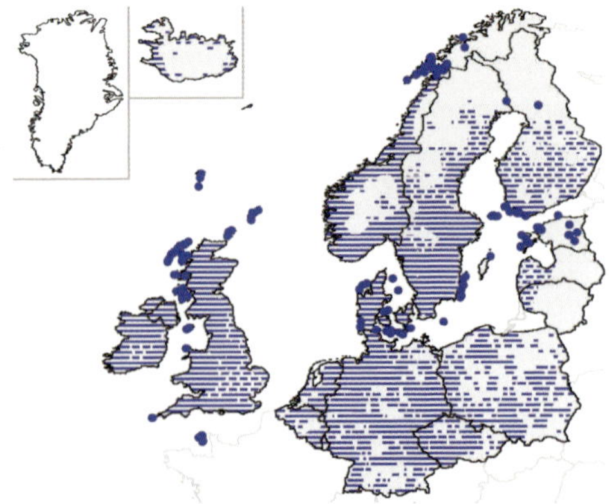

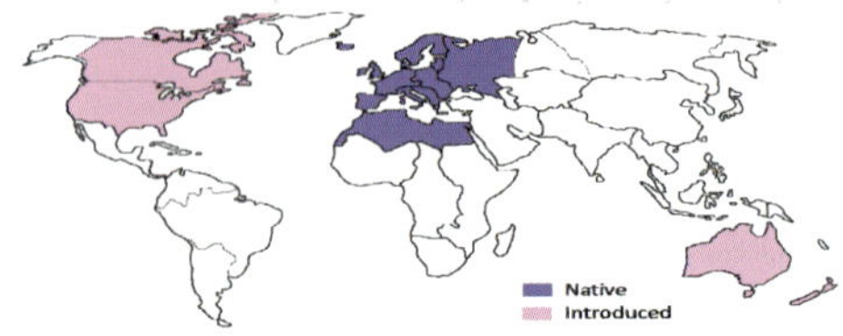

Submerged *J. bulbosus* floating and flowering at the water surface. The red colours are caused by anthocyanidins, which intensify in sunlight. Arnborg, Jutland, Denmark. Photo JCS.

Perennial helophyte, often flowering first year. Without rhizome.

Stem: Usually swollen at base, about 1 mm thick, terete, green or reddish to brownish. Terrestrial and partly submerged plants with 1–20 cm long, erect or procumbent, often rooting stems, or longer stems floating in the water. Submerged forms with prolonged, 0.5–1 (–3) metre long stems, often rooting from the nodes.

Leaves: In terrestrial forms most of the leaves are basal, while in partly or fully submerged forms they are produced in numbers at the upper nodes. Submerged plants at the bottom of lakes usually with thread-like leaves in rosettes only. Leaves 1–2 mm thick, terete or slightly furrowed on the adaxial side, green or reddish to brownish, with several longitudinal and very indistinct transverse septa. Leaf sheaths with 1–2 mm long, obtuse auricles.

Inflorescence: With sparse, suberect to patent branches and flowers in 1–15 (–30) roundish heads, each with 2–10 flowers. In submerged forms some or all flowers are often replaced by adventitious shoots. Bracts membranous.

Flowers: Bisexual. Bracteoles absent. Perianth segments 6, 2–3 mm long, green or brownish to reddish, with broad, scarious margins, obtuse or the outer somewhat acute, equal of length or the inner slightly longer than the outer. Stamens 3–6, anthers yellow, 0.3–1.0 mm long, filaments about the same length or longer. Stigmas 3.

Fruits: Capsule 2.2–3.5 mm long, brown to reddish, terete or bluntly triangular in cross section, obtuse to truncate at apex, beak very short. Seeds 0.5–0.6 mm long, oblong ovate, with about 25 longitudinal ribs and weaker transverse ribs.

Flowering: June-September.

Biology: The flowers are windpollinated and the fruits are dispersed by wind, water and animals. Self-pollination occurs frequently. Vegetative propagation by rooting shoots from inflorescence or stem nodes. The plant mostly reproduces by pseudovivipary. In acid sites this is the only form of reproduction. Instead of flowers small bulbs are formed, which grow to new plants remaining attached to the old one and forming adventitious roots.

Habitats: On moist or wet oligotrophic to moderately eutrophic substrate in dune slacks, heathland and peat bogs, on the banks of ponds, ditches, streams

Submerged *J. bulbosus* in a mining lake in Saxony, Germany. Photo KvdW.

Juncus bulbosus. The swollen, almost onion-like base of the stem is normally present at flowering plants and a very useful character. Thy, Denmark. Photo JCS.

Juncus bulbosus on moist sand. Stenbjerg, Thy, Denmark. Photo JCS.

and lakes, and fully submerged in acid streams and mining lakes (down to a pH of 2.8).

Distribution within the region:
Native. Common throughout most of the region. Scattered to rather rare in the Baltic states. Absent in Greenland.

Characteristics and similar species:
The terrestrial and partly submerged form is recognised by the swollen base of the stem, by the terete leaves with weak transverse septation (hard to feel with the fingers, but visible with a magnifying glass, when the leaves are dissected), and by the blunt, inner perianth segments and the obtuse to truncate capsule.

The aquatic form may be confused with **163 *Isolepis fluitans***, which has green, flat leaves, without septa and never with chains of leaf rosettes.

See also **160 *Eleocharis acicularis***.

Plants with 6 stamens and anthers 0.3–0.5 mm long, about half as long as the filaments, have been described as subsp. ***kochii*** (F.W.Schultz) Reichg.

Juncus bulbosus.
Fruit head. Photo JCS.

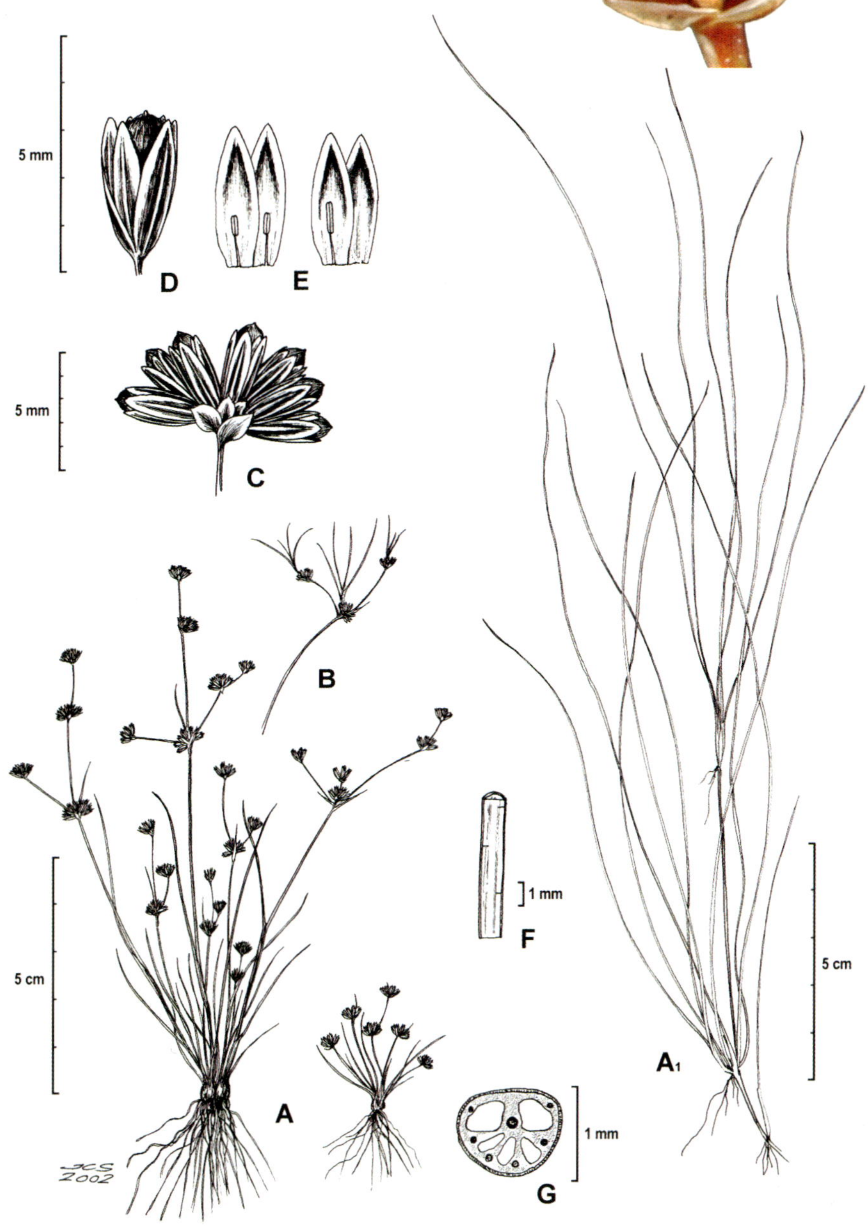

Juncus bulbosus
A habit, **A1** habit submerged plant,
B inflorescence with pseudovivipary shoots,
C head, **D** capsule with perianth segments,
E perianth segments and anthers seen from the inside, **F** longitudinal section of leaf (note the weak transverse septa). **G** cross section of leaf.

Key to vegetative Cyperaceae growing in water

1. Leaves reduced, sheath-like, tubular, sometimes with a very short and narrow blade (***1***)
   Subkey A
1. Some or all leaves with normally developed blades and sheaths (***2***) ... 2

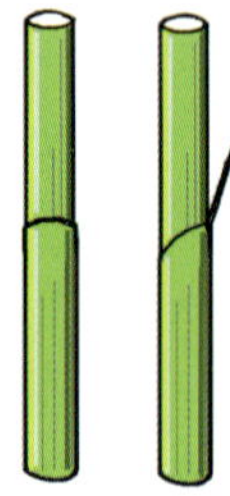

1

2

2. Annual plants forming small tufts without remnants of old, withering leaves and sheaths. Rhizome absent. Roots thin
   Subkey B
2. Perennial plants with remnants of old, withering leaves and sheaths. Rhizome short, erect or creeping ... 3

3. Leaf sheaths without ligule (***3***), but sometimes thickened around the opening
   Subkey C
3. Leaf sheaths with ligule (***4***) ... 4

3

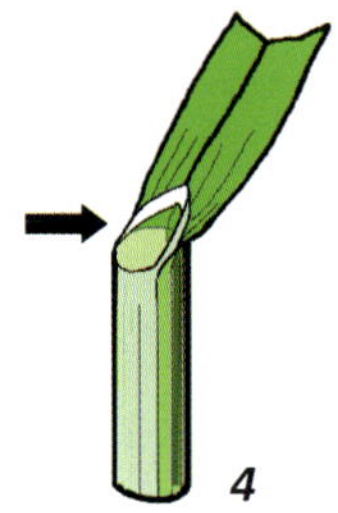

4

4. Leaves < 1.5 mm wide
   Subkey D
4. Leaves > 1.5 mm wide ... 5

5. Stolons absent. Shoots more or less densely tufted (***5***) ... 6
5. Stolons present. Shoots solitary or few together from the rhizome (***6***) ... 7

6. Plants forming dense tussocks with a peaty base (***7***)
   Subkey E
6. Plants forming more or less dense tufts
   Subkey F

7. Leaf sheaths splitting into fibres (***8***)
   Subkey G
7. Leaf sheaths not splitting into fibres (***9***)
   Subkey H

7

5

6

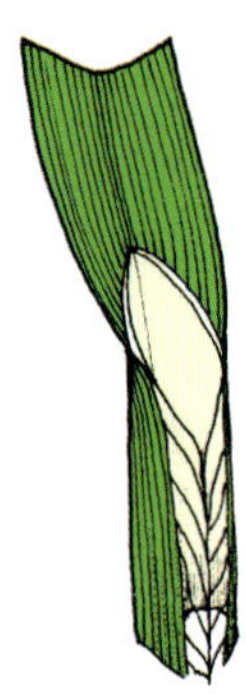

8

9

Subkey A

Leaves reduced, sheath-like - tubular, without blade.

1 Annual plants forming small tufts, without rhizome or remnants of old stems
152 *Eleocharis ovata*, 153 *E. obtusa* and 154 *E. engelmannii*

1 Perennial plants with long or short underground rhizome 2

2 Stems sharply trigonous 3

2 Stems terete or bluntly trigonous above 4

3 Plant caespitose
151 *Schoenoplectiella mucronata*

3 Plant with scattered shoots from a creeping rhizome
149 *Schoenoplectus triqueter*

4 Stems > 5 mm in diameter 5

4 Stems < 5 mm in diameter 6

5 Stems dull, grey green, up to 1.5 m long
146 *Schoenoplectus tabernaemontani*

5 Stems dull to somewhat shiny, green to dark green, up to 3.5 m long
145 *Schoenoplectus lacustris*

6 Underground stolons absent 7

6 Underground stolons present (***10***) 8

7 Stems 1–1.5 mm in diameter, numerous together in dense tufts. Roots whitish to yellowish, 0.5–1 mm in diameter
158 *Eleocharis multicaulis*

7 Stems 0.4–1 mm in diameter, at most 10 together in small tufts. Roots brownish, 0.1–0.5 mm in diameter
162 *Eleocharis quinqueflora*

8 Stems > 0.5 mm in diameter 9

8 Stems < 0.5 mm in diameter 11

9 Stems weak, easily broken
156 *Eleocharis mamillata*

9 Stems firm, not easily broken 10

10 Stolons > 1 mm in diameter.
155 *Eleocharis palustris*
157 *Eleocharis uniglumis*
(These two species cannot be differentiated reliably by vegetative characters)

10 Stolons < 1 mm in diameter
162 *Eleocharis quinqueflora*

11 Stems 4-angular (***11***). Stolons without terminal tubers
160 *Eleocharis acicularis*

11 Stems terete (***12***). Stolons with terminal tubers (***13***) 12

10

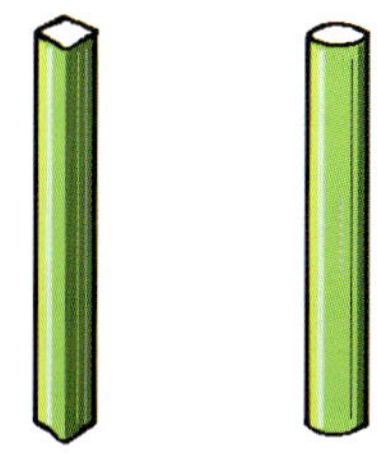

11 ***12***

12 Stems septate (***14***). Tubers 2–4 mm long
161 *Eleocharis parvula*

12 Stems without septa (***15***). Tubers c. 1 cm long
162 *Eleocharis quinqueflora*

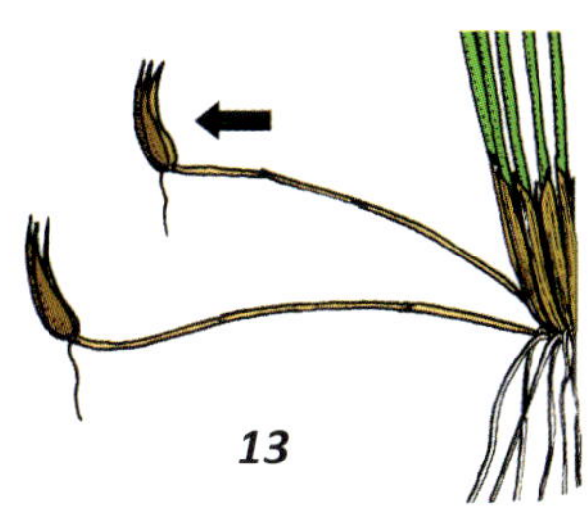

13

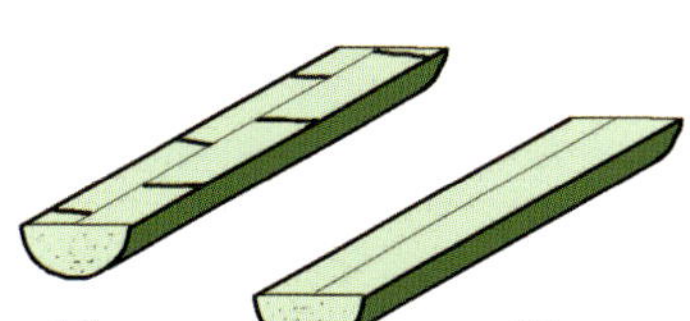

14 ***15***

Subkey B

Annual plants forming small tufts without remnants of old, withering leaves and sheaths. Leaves with blade and leafsheaths. Rhizome absent. Roots thin.

1 Ligule absent or at most 1 mm long. Leaves arising from the basal part of the stem 2

1 Ligule 2–4 mm long. Leaves arising from the upper 2/3 of the stem
170 *Carex bohemica*

2 Leaves 0.2–0.7 mm wide, almost to terete
Isolepis setacea (L.) R.Br.
(not treated in this book; photo to the right)

2 Leaves 1–4(–5) mm wide, flat to channelled
164 *Cyperus fuscus*

Isolepis setacea (L.) R.Br. Voldsted, Jylland, Denmark. Photo JCS.

Subkey C

Perennial plants without ligule.

1 Leaf margins with sharp teeth visible with the naked eye (***16***). Leaves grey green, very long
167 *Cladium mariscus*

1 Leaf margins without teeth visible with the naked eye but sometimes somewhat rough. Leaves yellowish green to green or dark green 2

2 Base of aerial shoots swollen (***17***)
5 species which can not be separated reliably using vegetative characters.
Bolboschoenus sp.
(5 species which cannot be differentiated reliably using vegetative characters. If fruits are present use the key p. 336)

2 Base of aerial shoots not swollen 3

3 Leaves > 4 mm wide. Large plants 4

3 Leaves < 3 mm wide. Plants creeping on mud or floating in water (***18***)
163 *Isolepis fluitans*

4 Underground stolons present 5

4 Underground stolons absent. Leaves dark green. Vegetative growth from arching and rooting stem or inflorescence nodes
139 *Scirpus radicans*

5 Leaves light green to green, at least some more than 10 mm wide
138 *Scirpus sylvaticus*

5 Leaves green to grey green, not more than 10 mm wide
165 *Cyperus longus*

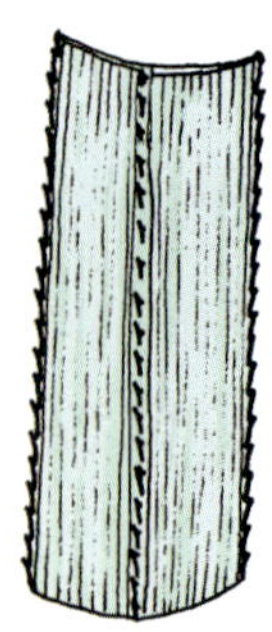

16

17

18

Subkey D

Leaves < 1.5 mm wide, channelled to inrolled or flat.

1 Leaf sheaths splitting into fibres (***8***). Scales red brown to purplish brown
171 *Carex lasiocarpa*

1 Leaf sheaths not splitting into fibres (***9***). Scales without red-brown to purplish-brown tones 2

2 Rhizome short, with short, ascending stolons usually 1–2 cm long. Plants loosely tufted. Leaves 1–2 mm wide, with stomata on the lower side
Carex diandra Schrank
(not treated in this book; photo below)

2 Rhizome creeping. Stolons rather long. Plants with scattered shoots or tufted with long stolons. Leaves 1.5–3 mm wide, with stomata on the upper side (***19***)
Carex nigra (L.) Reichard.
(not treated in this book; photo below)

19. Stomata are only visible using a 10-20x hand lens. They will then appear as small pale or greyish dots between the leaf ribs.

Carex nigra is one of many smaller sedges growing in wet conditions close to open water. It differs from similar species by having the most stomata on the upper side of the leaves. Var. ***recta*** forms rather dense tufts. Hobro, Denmark. Photo JCS.

Carex diandra may be mistaken for young plants of *C. appropinquata* (see next page) but differ by being only loosely tufted with basal scales not splitting into fibres and by roots less than 1 mm in diameter. *C. paniculata* differs in (3–)4–7 mm wide leaves and in forming large tussocks. Himmerland, Denmark. Photo JCS.

Subkey E

Leaves > 1.5 mm wide. Plants forming dense tussocks. Stolons absent.

1 Lower leaf sheaths splitting into fibres (***8***). At base with straw-coloured, ± shiny, keeled scales (***20***)
  179 *Carex elata*

1 Lower leaf sheaths not splitting into fibres (***9***) 2

2 Leaves with stomata on the upper side (***19***)
  Carex nigra (L.) Reichard. var. ***recta*** (Fleisch.) Hyl.
  (not treated in this book; photo page 331)

2 Leaves with stomata on the lower side (***19***) 3

3 Basal leaf sheaths and scales with light brown ribs. Leaves flaccid
  169 *Carex remota*

3 Basal leaf sheaths and scales with dark brown to blackish brown ribs. Leaves rather rigid 4

4 Leaves 3–6 mm wide, with 9–12 ribs on both side of the midrib. Basal scales not splitting into horsehair-like fibres (***21***)
  168 *Carex paniculata*

4 Leaves 1.5–3 mm wide, with c. 6 ribs on both side of the midrib. Basal scales splitting into horsehair-like fibres (***22***)
  Carex appropinquata Schumach.
  (not treated in this book; photo below)

20

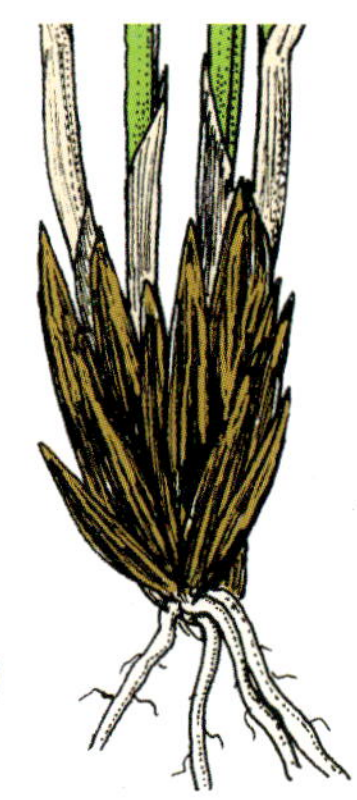

21

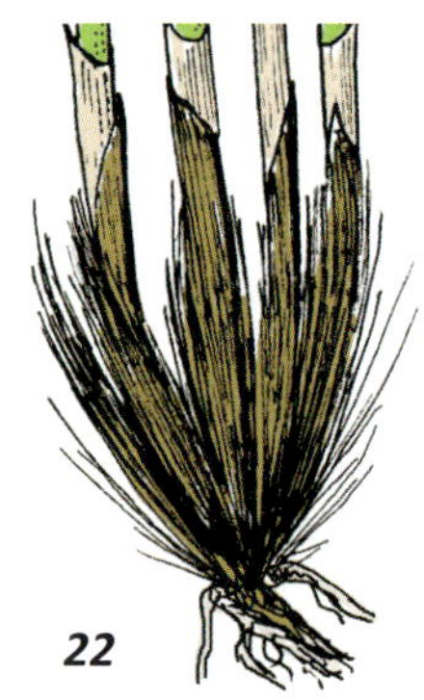

22

Carex appropinquata forms large tussocks, which can be similar in size to those of **168 *C. paniculata*** with which it often hybridises. The leaves of *C. appropinquata* are (1–)1.5–3 mm wide, and the basal scales split characteristically into horsehair-like fibres. Mariager, Denmark. Photos JCS.

Subkey F

Leaves > 1.5 mm wide. Plants in more or less dense tufts or tussocks. Stolons absent.

1 Leaves with stomata on the upper side (***19***)
Carex canescens L.
(not treated in this book; photo below)
1 Leaves without stomata on the upper side 2

2 Leaves ≤ 4 mm wide 3
2 Leaves > 4 mm wide 6

3 Leaves with distinct hollows in cross section (use a hand lens) (***23***) 4
3 Leaves without hollows in cross section
169 *Carex remota*

4 Basal scales splitting into horsehair-like fibres (***22***)
Carex appropinquata Schumach.
(not treated in this book; photo p. 332)
4 Basal scales not splitting into horsehair-like fibres (***21***) 5

5 Leaves < 3 mm wide.keeled
Carex diandra Schrank
(not treated in this book; photo p. 331)
5 Leaves ≥ 3 mm wide, channelled
168 *Carex paniculata*

6 Lower side of leaves glaucous
179 *Carex elata*
6 Lower side of leaves not glaucous 7

7 Leaves channelled to flat (***23, 25***)
168 *Carex paniculata*
7 Leaves keeled to plicate (***24***)
174 *Carex pseudocyperus*

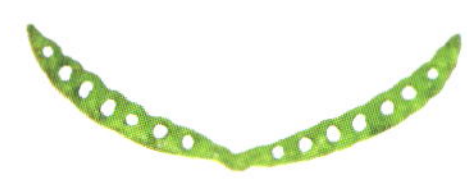

23

Carex canescens is recognised by the greyish-green leaves with stomata on the upper side and sharply trigonous stems. When fertile the 4-8 pale green, ovoid spikes are characteristic. Rold, Jutland, Denmark. Photos JCS.

Subkey G

Leaves > 1.5 mm wide. Stolons present. Shoots solitary or few together from the rhizome. Leaf-sheaths fibrously splitting.

1 Leaves with stomata on the upper side 2
1 Leaves without stomata on the upper side 3

2 Leaves with a trigonous point 2–6 cm long. Sterile shoots often forming false stems (a stem-like structure formed by leaf sheaths)
175 *Carex rostrata*
2 Leaves with a flat point. Sterile shoots not forming false stems.
178 *Carex aquatilis*

3 Leaves more or less of the same colour on both sides 4
3 Leaves dark green on the upper side and glaucous on the lower side 6

4 Leaves plicate (***24***) 5
4 Leaves channelled to inrolled (***23, 25***), 1–2 mm wide
171 *Carex lasiocarpa*

5 Leaves light green to green. Stolons short
176 *Carex vesicaria*
5 Leaves bluish green. Stolons long
173 *Carex riparia*

24

25

6 Leaves 3–10(–12) mm wide. With 10–16 ribs on each side of the midrib
172 *Carex acutiformis*
6 Leaves 8–15(–20) mm wide. With 18–22 ribs on each side of the midrib
173 *Carex riparia*

Eriophorum gracile is recognised by the 1–2 mm wide leaves, trigonous for almost its entire length. Peduncles triquetrous, with a clothing of fine, forward-directed hairs. SW-Jutland, Denmark. Photo JCS.

Subkey H

Leaves > 1.5 mm wide. Stolons present. Shoots solitary or few together from the rhizome. Leaf sheaths not splitting into fibres.

1 Leaves usually less than 30 cm long — 2
1 Leaves usually more than 30 cm long — 6

2 Shoots with laminate leaves in the upper one-half of stem only. Shoots bamboo-like. Seen from above the leaves appear in three vertical rows along the stem (***26***)
166 *Dulichium arundinaceum*
2 Shoots with laminar leaves in the lower one-half of stem. Shoots not bamboo-like — 3

3 Stems less than 2 mm wide, firm — 4
3 Stems more than 2 mm wide, spongy — 5

4 Leaves flat to channelled, with stomata on the upper side.
Carex nigra (L.) Reichard.
(not treated in this book; photo p. 331)
4 Leaves trigonous almost throughout their length, with stomata on the lower side
Eriophorum gracile W.D.J.Koch ex Roth.
(not treated in this book; photo p. 334)

5 Uppermost leaf sheath without or with a short lamina. Stem 3–8 mm in diameter
149 *Schoenoplectus triqueter*
5 Uppermost leaf sheath with a long lamina, up to 30 cm long. Stem 2–5 mm in diameter
150 *Schoenoplectus pungens*

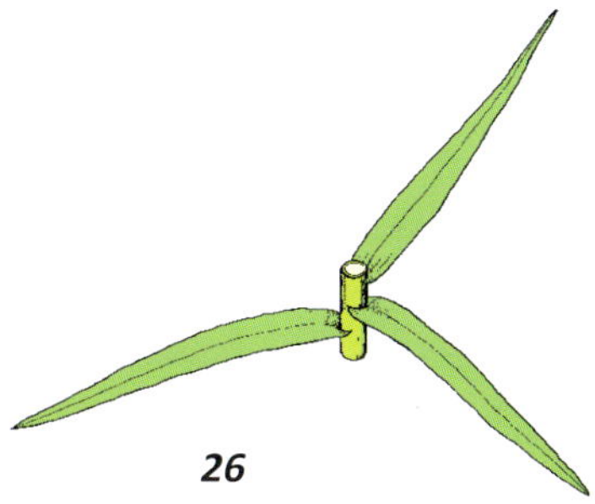

26

6 Leaves trigonous in the distal one-third, channelled in the basal part
Eriophorum angustifolium Honck.
(not treated in this book; photo below)
6 Leaves at most trigonous at apex — 7

7 Leaves 8–15(–20) mm wide
173 *Carex riparia*
7 Leaves 2.5–10(–12) mm wide
177 *Carex acuta*

Eriophorum angustifolium is recognised by its long, channelled leaves being trigonous in the distal one-third. Gjeller Odde, Lemvig, Jutland, Denmark. Photo JCS.

Key to *Bolboschoenus*

For proper identification of *Bolboschoenus* species, cross sections of fully developed fruits are needed to observe the thickness and structure of the 3 layers of the pericarp.

1 Inflorescence more or less capitate, formed by a group of sessile spikelets and sometimes 1 or 2 rays each bearing 1–2(–4) spikelets (***1***). Rays less than twice as long as spikelets. Nuts light brown to reddish brown, plano-convex to lenticular or faintly triangular in cross section. Exocarp as thick as or thicker than mesocarp — 2

1 Inflorescence branched with a central group of sessile spikelets and 2–7 rays each bearing 1–4 spikelets (***2***). Rays more than twice as long as spikelets. Nuts dark brown to almost black, trigonous in cross section. Exocarp much thinner than mesocarp — 3

2 Nuts convex on the abaxial side, lenticular to faintly trigonous in cross section (***3***). Exocarp about twice as thick as mesocarp. Most styles trifid
140 *B. maritimus*

2 Nuts concave on the abaxial side, plano-convex in cross section (***4***). Exocarp about as thick as mesocarp. Most styles bifid
141 *B. planiculmis*

3 Nut 3.1–4.0 mm long — 4

3 Nut 2.0–2.5 mm long
144 *B. glaucus*

4 Nut 1.6–1.8 mm wide, equilaterally triangular in cross section (***5***). Exocarp very thin; less than 1/10 as thick as the mesocarp
143 *B. yagara*

4 Nut 2.0–2.4 mm wide, obtusely triangular in cross section, the abaxial side longer than the other sides (***6***). Exocarp about 1/3 as thick as the mesocarp
142 *B. laticarpus*

1

2

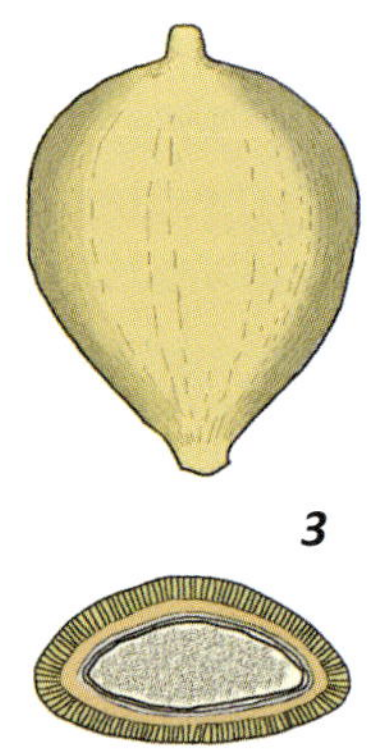
3

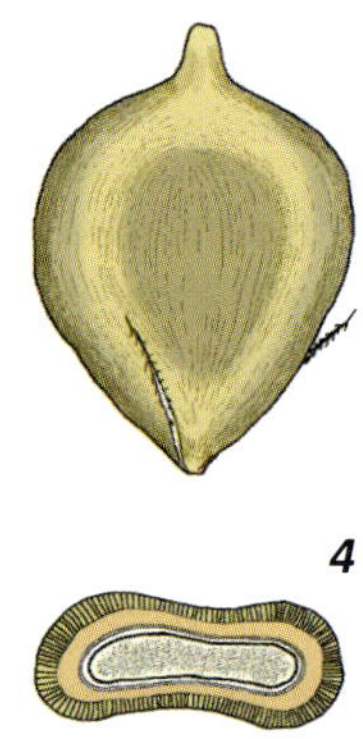
4

5

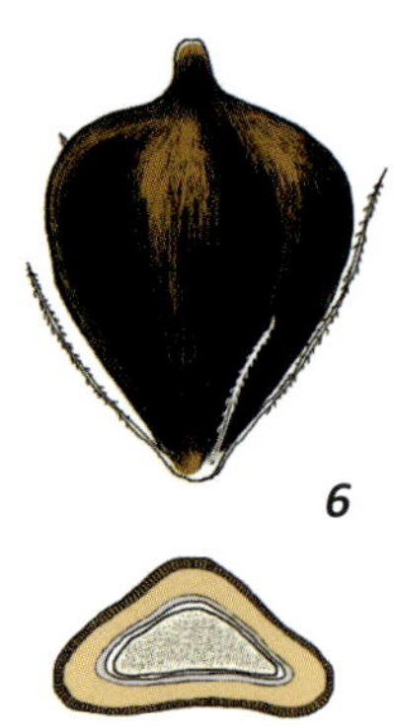
6

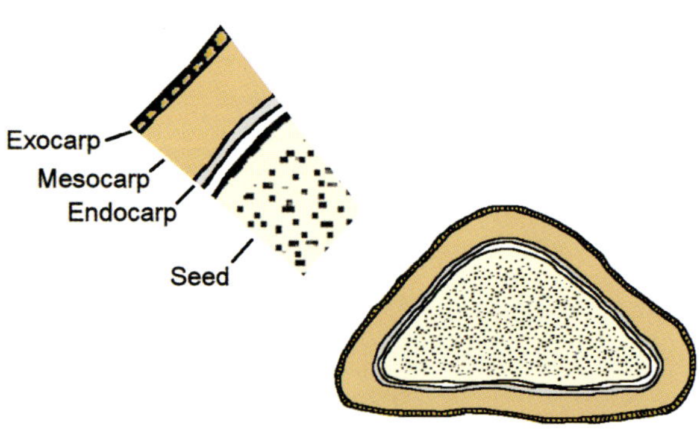

Key to *Schoenoplectus*

1 Stem bluntly trigonous to sharply triquetrous, at least in the upper part 2

1 Stem terete, including in the upper part 4

2 Plant with scattered shoots from a creeping rhizome. Styles bifid 3

2 Plant caespitose. Styles trifid
151 *S. mucronata*

3 Inflorescence with sessile spikelets only (***1***). Stem with 2 or 3 laminate leaves. Glumes with acute lateral lobes
150 *S. pungens*

3 Inflorescence with sessile spikelets and some pedunculated clusters of spikelets (***2***). Stem with lamina on uppermost sheath only. Glumes with obtuse lateral lobes
149 *S. triqueter*

4 Glumes smooth - without papillae (***3***). Styles trifid (***4***). Anthers with broad and rounded apex with a fringe of hair (***5***)
145 *S. lacustris*

4 Glumes with reddish papillae (***6***). Styles bifid (***7***). Anthers with narrow, tapering apex (***8***)
146 *S. tabernaemontani*

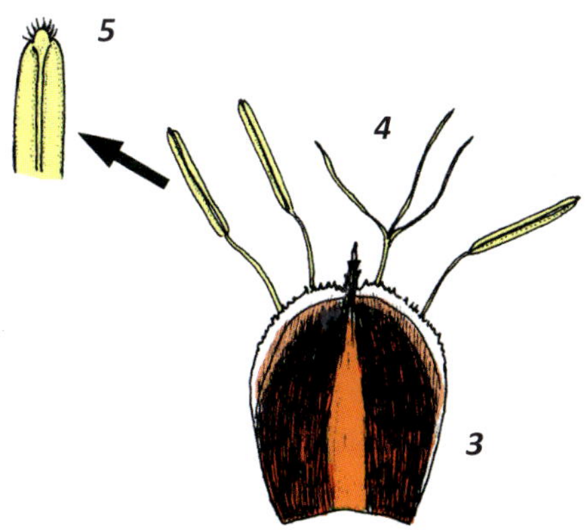

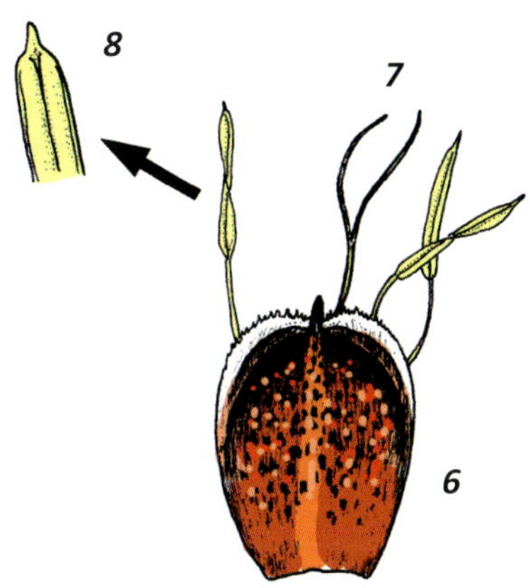

Cyperaceae

138 *Scirpus sylvaticus* L.

DK: Skov-Kogleaks **N:** Skogsivaks **S:** Skogssäv **FIN:** Korpikaisla **IS:** -
GB: Wood Club-rush **NL:** Bosbies **F:** Scirpe des bois
D: Wald-Simse **CZ:** Skřípina lesní **PL:** Sitowie leśne
EST: Metskõrkjas **LV:** Meža meldrs **LT:** Liekninis viksvameldis **Ru:** Камыш лесной

Scirpus sylvaticus abundant in a forest pool. Hadsund, Denmark. Photo JCS.

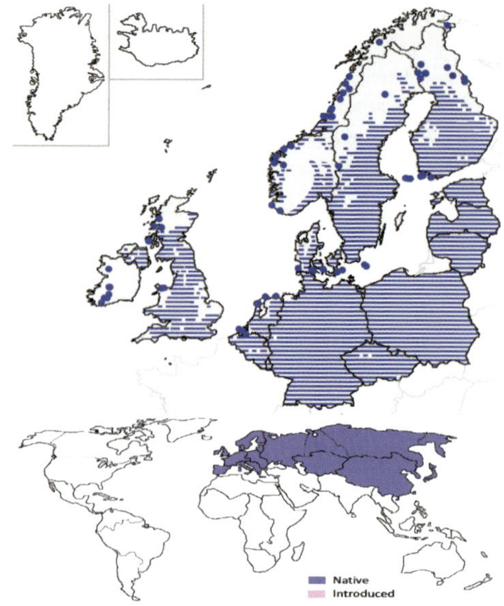

Perennial helophyte, forming large stands. Shoots single from the rhizome. Rhizome creeping, 3–5 mm in diameter, covered with pale brown to reddish-brown scales. Roots greyish to brownish, up to 2 mm in diameter.

Stem: 50–120 cm long, rather robust, 4–8 mm diameter at the base, green, trigonous and somewhat rough above, terete below.

Leaves: With fibrous remnants of old leaves and several leaves up to 120 cm long at the base. Cauline leaves 2–7, 8–20 mm wide, up to 45 cm long. All leaves bright green to fresh green, keeled to plicate, rough on keel and margins. Hypostomous. Leaf sheaths yellowish green to brownish, forming pseudostems on vegetative shoots. Inner face hyaline, splitting like parchment. Ligule absent.

Inflorescence: Terminal, 2–3 times branched, almost hemispherical in outline. Primary branches 3–8 of unequal length, up to 10 cm long, trigonous and rough; secondary and tertiary branches 0.5–5 cm long. Involucral bracts of secondary and tertiary branches more or less membranous, greenish to brownish, those of the primary branches leaf-like, the lowest up to 50 cm long. Spikelets 3–5 mm long, ovoid, green to brownish, in clusters of 1–5. Glumes 1.5–2.5 mm long, green to blackish green with a green midrib often extending into a short mucro.

Flowers: Bisexual, with trifid style and 3 stamens. Anthers 1 mm long. Perianth bristles 6, retrorsely scabrid.

Fruits: Nuts 0.7–1.4 mm long, obovoid, bluntly triangular in cross section, dull, whitish yellow to pale green. Beak short.

Flowering: June-July.

Biology: Propagation by seed and vegetatively by division of the rhizome.

Habitats: In mesotrophic to eutrophic, neutral, non-calcareous water. On wet or moist, peaty or muddy substrates by pools, lakes, rivers and streams, in wet pastures and forest swamps.

Distribution within the region:
Native. Common throughout most of the area, rather rare in Ireland. Absent from Greenland and Iceland.

Characteristics and similar species:
Very distinctive when flowering and, within the study area, only likely to be confused with **139 *S. radicans*** - see this. In the vegetative state it could be confused with **173 *C. riparia*** - see this.

Scirpus sylvaticus.
Leaf sheath. Photo JCS.

Vegetative *S. sylvaticus*. Photo KvdW.

Cyperaceae

139 *Scirpus radicans* Schkuhr

DK: Bølget Kogleaks **N:** Bogsivaks **S:** Bogsäv **FIN:** Juurtokaisla **IS:** -
GB: Creeping Club-rush **NL:** - **F:** Scirpe radicant
D: Wurzelnde Simse **CZ:** Skřípina kořenující **PL:** Sitowie korzenioczepne
EST: Juurduv kõrkjas **LV:** Sakņojošais meldrs **LT:** Pelkinis viksvameldis **Ru:** Камыш укореняющийся

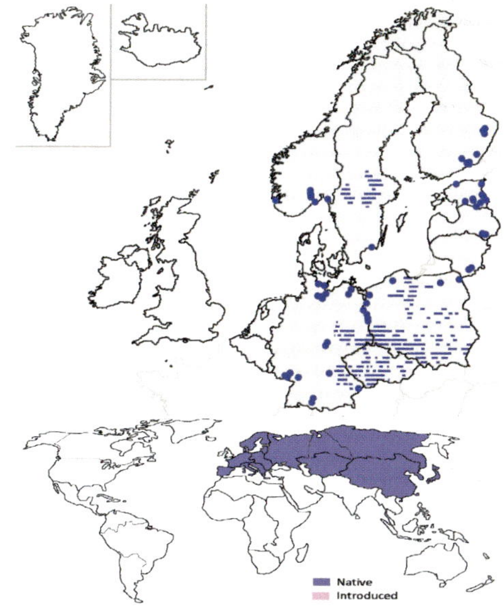

Somewhat similar in habit to **138 *S. sylvaticus*** but without underground stolons. Leaves 6–10 mm wide, dark green. Spikelets solitary, up to 6–7 mm long. Perianth bristles 2–3 times longer than nut. Vegetative growth by rooting from the upper stem nodes and inflorescence.

A pioneer plant of eutrophic, humus-rich, muddy substrates in or close to water. On disturbed, often periodically flooded habitats, such as ditches and the margins of lakes, streams and reservoirs.

Native. Scattered to locally common in Germany, Poland, the Czech Republic, the Baltic states, Norway, Sweden and Finland.

Scirpus cyperinus (L.) Kunth and ***S. georgianus*** R.M.Harper are two introduced species which are superficially similar to *S. radicans.* Both are very rare within the study area and not covered here.

Scirpus sylvaticus
A habit, **B** spikelet, **C** nut with bristles, **E** flower and glume.

Scirpus radicans
A1 habit, **B1** spikelet, **C1** nut with bristles.

Cyperaceae

140 *Bolboschoenus maritimus* (L.) Palla

Syn.: *Schoenoplectus maritimus* (L.) Lye ; *Scirpus maritimus* L.
DK: Strand-Kogleaks **N:** Havsivaks **S:** Havssäv **FIN:** Merikaisla **IS:** -
GB: Sea Club-rush **NL:** Heen **F:** Souchet maritime
D: Gewöhnliche Strandsimse **CZ:** Kamyšník přímořský **PL:** Sitowiec nadmorski
EST: Meri-mugulkõrkjas **LV:** Jūrmalas gumumeldrs **LT:** Pajūrinis liūnmeldis **Ru:** Клубнекамыш морской

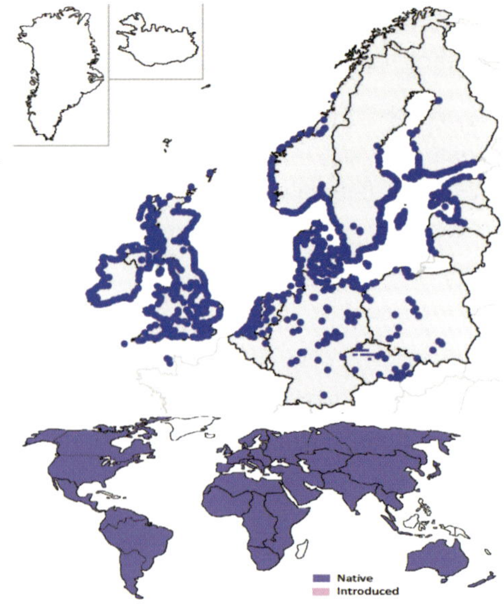

Bolboschoenus maritimus growing in slightly brackish water. Kielstrup Sø, Mariager, Denmark. Photo JCS.

Perennial helophyte. Rhizome underground, extensive, branching, 2–5 mm in diameter, covered with brown to blackish, sheath-like scales. Base of aerial shoots swollen, forming ellipsoid tubers 2–4 cm in diameter. Roots fleshy, grey brown, about 1 mm in diameter.

Stem: 30–110 cm long, 3–8 mm diameter, solid, trigonous, becoming sharply angled to almost winged in cross section towards the top. Upper 1/3–1/2 leafless.

Leaves: 3–7 mm wide, up to 40 cm long, light green, with scaberulous margins and keel, tapering gradually to a triangular point. Lower stem leaves with relatively short blades, some reduced to sheaths. Sheath up to 20 cm long, tight, ventral side membranous, red dotted. Ligule absent. Seedlings sometimes submerged with 10–30 cm long, 2–3 mm wide, thin, yellowish-green, ribbon-like leaves, with a pattern of rectangular air chambers.

Inflorescence: Terminal, either capitate and composed of 1–10, sessile spikelets, or with a central head of 3–7 sessile spikelets and 1–2(–3) rays up to 5 cm long, each with 1–3 spikelets. Spikelets 1–2(–4) cm long, 5–8 mm wide, ovoid to elliptic, dull, brownish. Involucral bracts leaf-like, 2–3, the lowest up to 25 cm long. Glumes 5.5–7.0 mm long, somewhat glossy, reddish brown to chestnut, glabrous to pubescent. Midrib prominent, light brown elongated into a scabrid awn.

Flowers: Bisexual, with trifid (rarely bifid) style and 3 stamens. Anthers 4–6 mm long.

Fruits: Nuts 3.0–3.2 mm long, 2.0–2.3 mm wide, broadly obovate in outline, abruptly tapering to a beak up to 0.5 mm long, light brown to rusty brown, lenticular to plano-convex or rounded trigonous in cross section. Surface of nut appears cellular at 20x magnification. Exocarp about twice as thick as the mesocarp. Perianth bristles 0–2(–4), caducous, retrorsely scabrid.

Flowering: June-August.

Biology: Nuts dispersed by water and birds. Vegetative propagation by division of the rhizome. The shoots decay in autumn and new shoots arise from the swollen parts of the rhizome in spring.

Habitats: Wet, silty substrates. Saline or brackish marshes by the sea, also inland in saline, brackish or alkaline channels, ditches, rivers, ponds and lakes.

Distribution within the region:
Native. Common in coastal areas throughout most of the region,

Bolboschoenus maritimus. Flowering plants. Bornholm, Denmark. Photo JCS.

scattered to locally frequent inland. Absent from Iceland and Greenland.

Characteristics and similar species: *Bolboschoenus* species differ from ***Schoenoplectus*** and ***Carex*** species by the swollen, ellipsoid tubers at base of the shoots.

Young submerged plants with thin, ribbon-like leaves resemble those of ***Schoenoplectus*** and should be determined by looking for older specimens close by.

To distinguish the five ***Bolboschoenus*** species, ripe nuts are needed - please use the key on p. 336.

Bolboschoenus maritimus. Inflorescence. Løgstør, Denmark. Photo JCS.

Bolboschoenus maritimus. Nut in side view and cross section. Photo JCS.

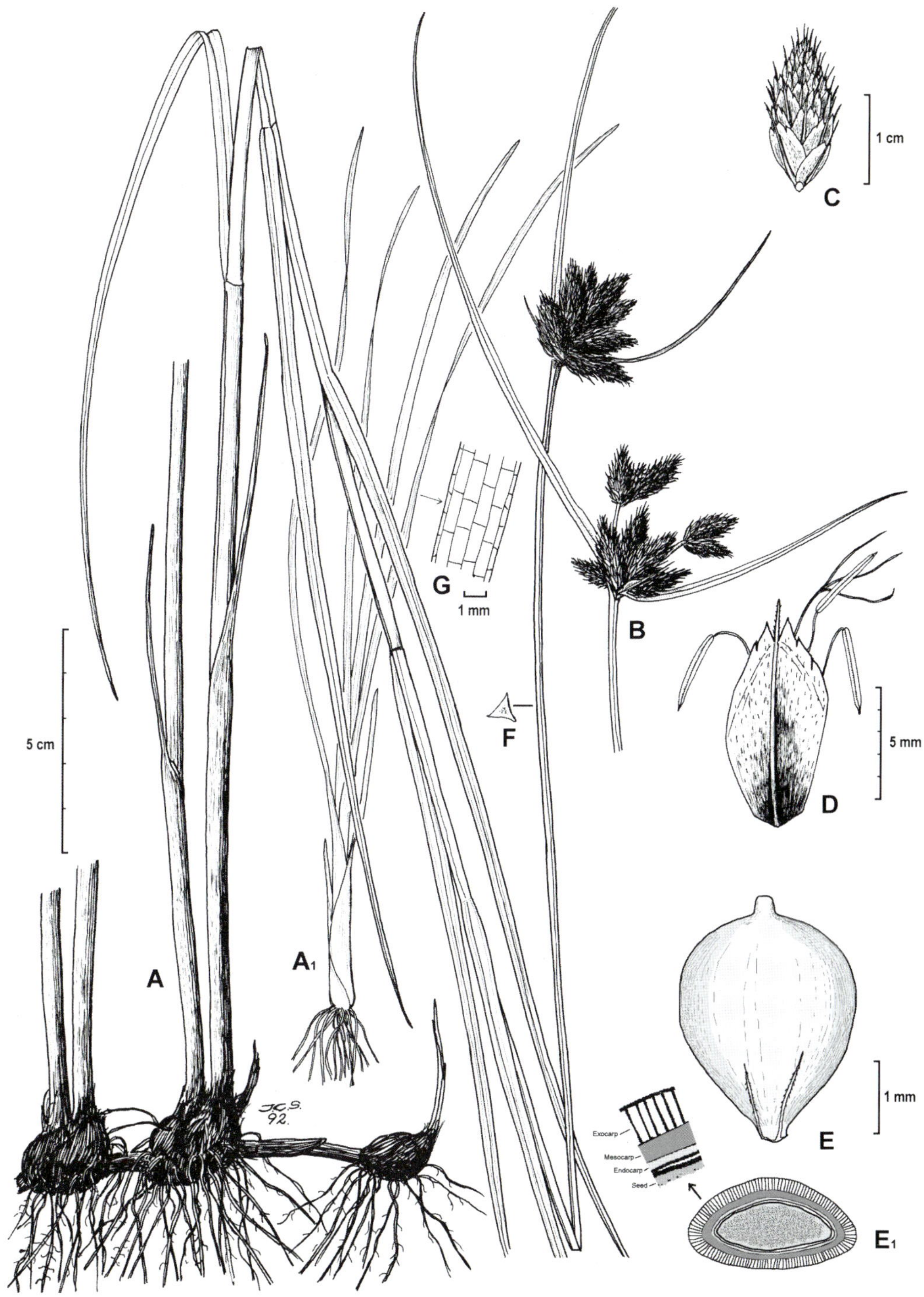

Bolboschoenus maritimus
A habit, **A1** habit submerged young plant, **B** inflorescence, **C** spikelet, **D** flower and glume, **E** nut with bristles, **E1** cross section of nut with enlarged detail, **F** stem cross section, **G** part of submerged leaf.

Cyperaceae

141 *Bolboschoenus planiculmis* (F.Schmidt) T.V.Egorova

Syn.: *Bolboschoenus maritimus* var. *planiculmis* (F.Schmidt) Jauzein
DK: - **N:** - **S:** - **FIN:** - **IS:** -
GB: - **NL:** Oostelijke bies **F:** -
D: Flachfrüchtige Strandsimse **CZ:** Kamyšník polní **PL:** -
EST: -**LV:** - **LT:** - **Ru:** Камыш плоскостебельный

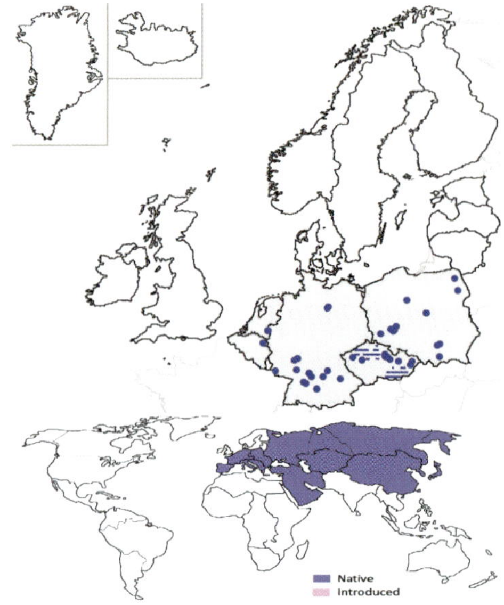

The inflorescences of *B. planiculmis* are quite similar to those of *B. maritimus*. Edenryd, Netherlands. Photo KvdW.

Perennial helophyte. Rhizome underground, extensive, branching, 2–5 mm in diameter, covered with brown to blackish, sheath-like scales. Base of aerial shoots swollen, forming ellipsoid tubers up to 1.5 cm in diameter. Roots fleshy, grey brown, 0.1 mm in diameter.

Stem: 50–90(–110) cm long, 3–8 mm in diameter, solid, trigonous, becoming sharply angled to almost winged in cross section towards the top. Upper 1/3–1/2 leafless.

Leaves: 3–8 mm wide, up to 40 cm long, light green, with scaberulous margins and keel, gradually tapering to a triangular point. Lower stem leaves with relatively short blades, some reduced to sheaths. Sheath up to 20 cm long, tight, ventral side membranous, often red dotted. Ligule absent.

Inflorescence: Terminal. With a central group of 3–6(–10) sessile spikelets and 0–3 rays, each with a cluster of 1–2(–4) spikelets. Spikelets 0.8–1.8 (–2.1) cm long, 5–8 mm wide, ovoid to elliptic, dull, brownish. Involucral bracts leaf-like, 2–3, the lowest up to 25 cm long. Glumes 4.5–7.0 mm long, somewhat glossy, light brown to reddish brown, glabrous to pubescent. Midrib prominent, greenish to light brown, elongated into a scabrid awn.

Flowers: Bisexual, with bifid style and 3 stamens. Anthers 4–6 mm long.

Fruits: Nuts 3.1–3.8 mm long, 2.2–2.5 mm wide, obovate to broadly obovate in outline, abruptly tapering to a short beak, light brown to rusty brown when ripe, flattened oval with more or less concave faces in cross section. The surface of the nut appears cellular at 20x magnification. Exocarp about as thick as the mesocarp. Perianth bristles 0–2(–4), caducous, retrorsely scabrid.

Flowering: June-August.

Biology: Nuts dispersed by water and birds. Vegetative propagation by division of the rhizome. The shoots die-back in autumn, and new shoots arise from the swollen parts of the rhizome in spring.

Habitats: In fresh, acidic to neutral water. Mostly in wet, human-managed habitats, preferably with fluctuating water level; fishponds, pools, ditches, lakes and riverbanks.

Bolboschoenus planiculmis. Parc Naturel Régional de la Brenne, France. Photo RVL.

Distribution within the region:
Native. Scattered to locally frequent in the Netherlands, Belgium, Germany, the Czech Republic and Poland.

Characteristics and similar species:
Bolboschoenus species differ from ***Schoenoplectus*** and ***Carex*** species by the swollen, ellipsoid tubers at base of the shoots.

Young submerged plants with thin, ribbon like leaves resemble those of ***Schoenoplectus*** and should be determined by looking for older specimens close by.

To distinguish the five ***Bolboschoenus*** species, ripe nuts are needed - please use the key on p. 336.

Bolboschoenus planiculmis. Nut in side view and cross section. Photo JCS.

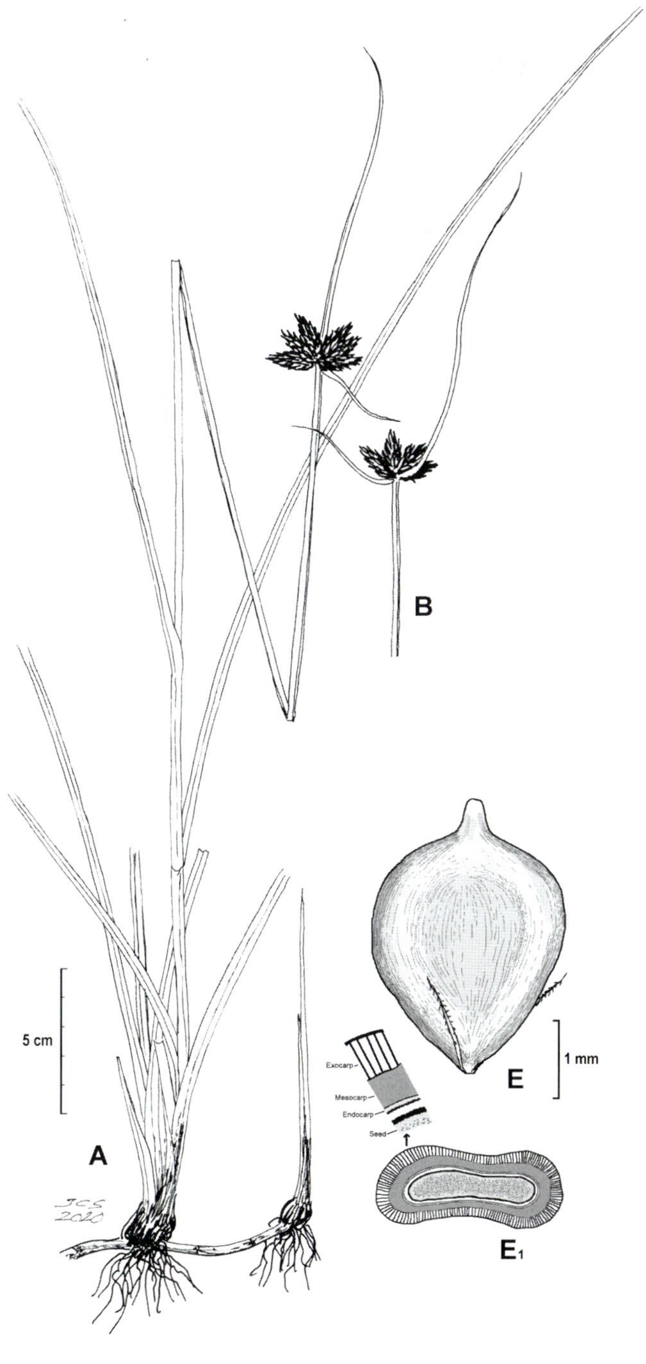

Bolboschoenus planiculmis
A habit, **B** inflorescence, **E** nut with bristles, **E1** cross section of nut with enlarged detail.

Cyperaceae

142 *Bolboschoenus laticarpus* Marhold, Hroudová, Duchácek & Zákr.

Syn.: *Bolboschoenus maritimus* var. *cymosus* (Rchb.) Kit Tan & Oteng-Yeb.
DK: Mørkfrugtet Kogleaks **N:** - **S:** - **FIN:** - **IS:** -
GB: Pedunculate Club-rush **NL:** Oeverbies **F:** -
D: Breitfrüchtige Strandsimse **CZ:** Kamyšník širokoplodý **PL:** -
EST: -**LV:** - **LT:** - **Ru:** -

Bolboschoenus laticarpus. Abbey Lake, Thorpe Park, Staines, England. Photo RVL.

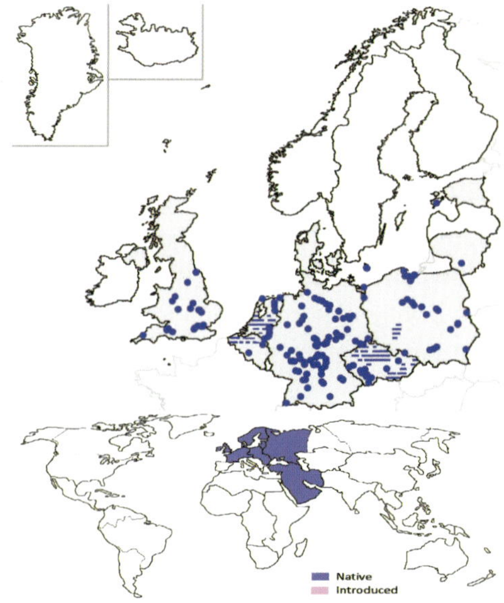

Perennial helophyte. Rhizome underground, extensive, branching, 2–5 mm in diameter, covered with brown to blackish, sheath-like scales. Base of aerial shoots swollen, forming 2–3 cm thick, ellipsoid tubers. Roots fleshy, grey brown, about 1 mm in diameter.

Stem: 30–110(–150) cm long, 4–10 mm in diameter, solid, trigonous, becoming sharply triangular in cross section towards the top. Upper 1/3 leafless.

Leaves: 3–10 mm wide, up to 40 cm long, light green to green, with scaberulous margins and keel, gradually tapering into a triangular point. Lower stem leaves with relatively short blades, some reduced to sheaths. Sheath up to 20 cm long, tight, ventral side membranous. Ligule absent.

Inflorescence: Terminal. With a central group of 2–7(–13) sessile spikelets and 1–5(–7) rays 1–6 cm long, each with a cluster of (1–)2–4(–8) spikelets. Spikelets (0.6–)1.0–1.8(–2.1) cm long, 5–8 mm wide, ovoid to elliptic, dull, brownish. Involucral bracts leaf-like, 2–3, the lowest up to 25 cm long. Glumes 5–8 mm long, somewhat glossy, light brown to reddish brown, glabrous to pubescent. Midrib prominent, light brown elongated into a scabrid awn.

Flowers: Bisexual, with trifid or sometimes bifid style and 3 stamens. Anthers 4–6 mm long.

Fruits: Nuts 3.1–3.7 mm long, 2–2.4 mm wide, obovate in outline, abruptly narrowed to a short beak, dark brown to almost black when ripe, obtusely trigonous in cross section, the abaxial side longer than the other sides, rarely flattened (when style bifid). Surface of nut looks smooth at 20x magnification. Exocarp about 1/3 as thick as the mesocarp. Beak up to 0.5 mm long. Perianth bristles (0–)3–6, persistent to partly caducous, retrorsely scabrid.

Flowering: June-September.

Biology: Nuts dispersed by water and birds. Vegetative propagation by division of the rhizome. The shoots die back in autumn and new shoots arise from the swollen parts of the rhizome in spring.

Habitats: In fresh, weakly acidic to weakly alkaline water in the littoral zone of large rivers and standing waters, such as oxbows, fishponds, ditches and reservoirs. It is somewhat tolerant of salt, and can be found in inland saline waters and in brackish water at river outlets.

Distribution within the region:
Native. Scattered to locally frequent in Britain, the Netherlands, Belgium, Germany, the Czech Republic and Poland. Rare in Lithuania and Estonia.

Characteristics and similar species:
Bolboschoenus species differ from ***Schoenoplectus*** and ***Carex*** species by the swollen, ellipsoid tubers at base of the shoots.

Bolboschoenus laticarpus. Inflorescence with long rays. Abbey Lake, Thorpe Park, Staines, England. Photo RVL.

Young submerged plants with thin, ribbon like leaves resemble those of ***Schoenoplectus*** and should be determined by looking for older specimens close by.

To distinguish the five ***Bolboschoenus*** species, ripe nuts are needed - please use the key on p. 336.

Flowering *B. laticarpus*. Geiseltalsee, Sachsen-Anhalt, Germany. Photo KvdW.

Bolboschoenus laticarpus. Inflorescence. Photo JCS.

Bolboschoenus laticarpus. Nut in side view and cross section. Photo JCS.

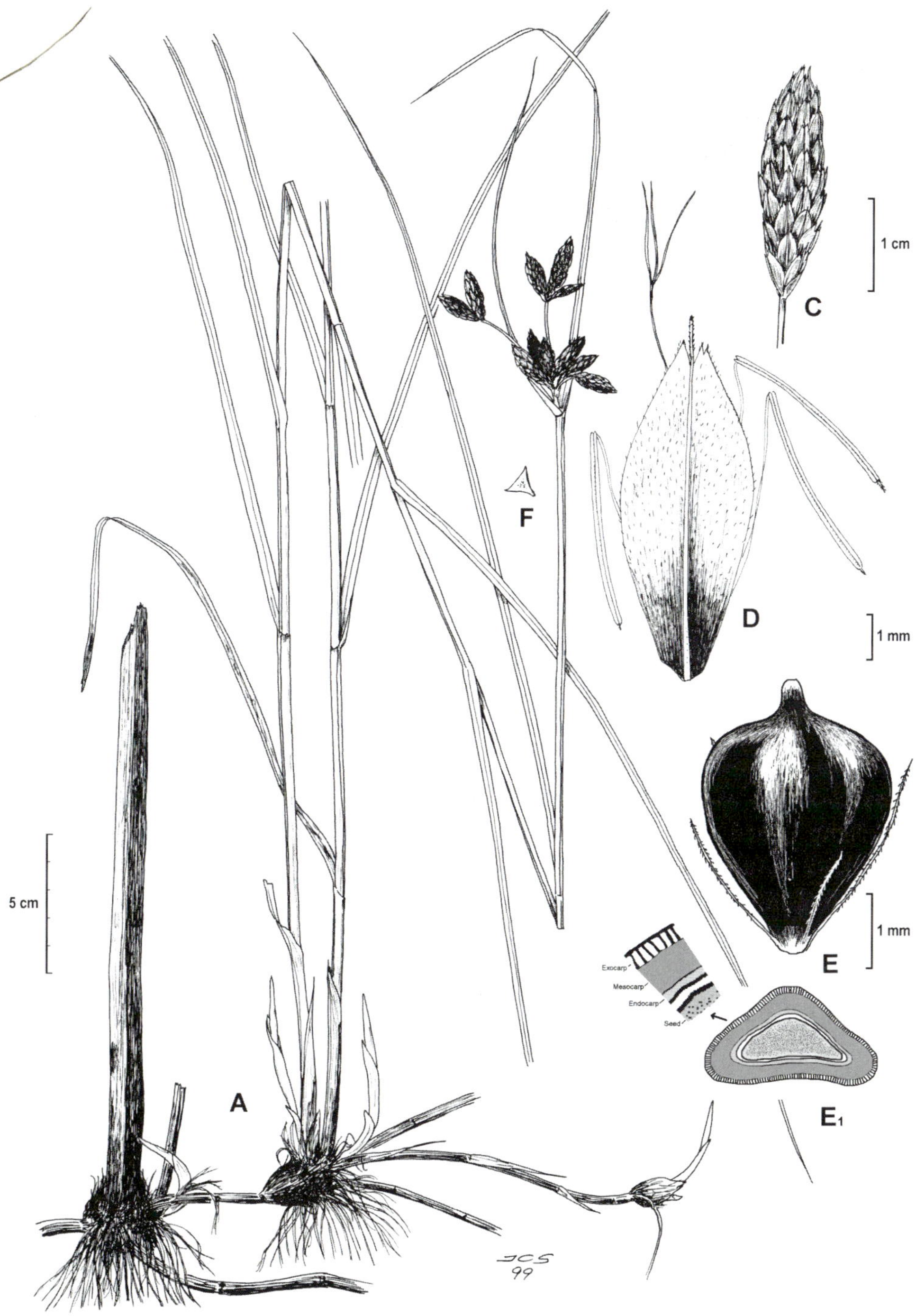

Bolboschoenus laticarpus
A habit, **C** spikelet,
D flower and glume, **E** nut with bristles,
E1 cross section of nut (with enlarged detail), **F** cross section of stem.

Cyperaceae

143 *Bolboschoenus yagara* (Ohwi) Y.C.Yang & M.Zhan

Syn.: *Bolboschoenus maritimus* subsp. *yagara* (Ohwi) Jauzein
DK: - **N:** - **S:** - **FIN:** - **IS:** -
GB: - **NL:** - **F:** -
D: Yagara Strandsimse **CZ:** Kamyšník vrcholičnatý **PL:** -
EST: -**LV:** - **LT:** - **Ru:** Клубнекамыш ягара

Bolboschoenus yagara. Sechtem, Köln, Germany. Photo Michael Hassler.

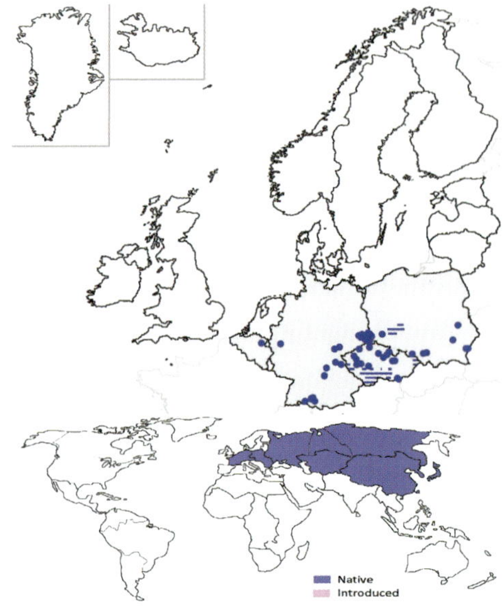

Perennial helophyte. Rhizome underground, far creeping, branching, 2–5 mm in diameter, covered with brown to blackish, sheath-like scales. Base of aerial shoots swollen, forming ellipsoid tubers 3–4 cm in diameter. Roots fleshy, grey brown, about 1 mm in diameter.

Stem: 80–130(–160) cm long, 4–7 mm in diameter, solid, trigonous, becoming sharply triangular in cross section towards the top. Upper 1/5–1/4 leafless.

Leaves: 3–8 mm wide, up to 50 cm long, light green to green, with scaberulous margins and keel, gradually tapering into a triangular point. Lower stem leaves with relatively short blades, some reduced to sheaths. Sheath up to 20 cm long, tight, ventral side membranous. Ligule absent.

Inflorescence: Terminal. With a central group of 2–4(–8) sessile spikelets and (2–)5–7(–9), 3–6 cm long rays, each with a cluster of (1–)2–4 spikelets. Spikelets 1.0–1.8(–2.1) cm long, 5–8 mm wide, ovoid to elliptic, dull, brownish. Involucral bracts leaf-like, 2–3, the lowest up to 25 cm long. Glumes 4–6 mm long, somewhat glossy, light brown to reddish brown, glabrous to somewhat pubescent. Midrib prominent, greenish to light brown, elongated into a scabrid awn.

Flowers: Bisexual, with trifid style and 3 stamens. Anthers 4–6 mm long.

Fruits: Nuts 3.2–4.0 mm long, 1.6–1.8 mm wide, narrowly obovate in outline, gradually tapering to a beak, dark brown to almost black when ripe, equilaterally triangular in cross section. Surface of nut appears smooth and glossy at 20x magnification. Exocarp very thin, less than 1/10 as thick as the mesocarp. Perianth bristles (2–)4–6, persistent, retrorsely scabrid.

Flowering: July-October.

Biology: Nuts dispersed by water and birds. Vegetative propagation by division of the rhizome. The shoots die-back in autumn and new shoots arise from the swollen parts of the rhizome in spring.

Habitats: In fresh, acid to neutral water. In the littoral zone of fishponds, pools, oxbows and shallow bays of rivers.

Distribution within the region:
Native. Scattered to locally frequent in Belgium, Germany, the Czech Republic and Poland.

Characteristics and similar species:
Bolboschoenus species differ from ***Schoenoplectus*** and ***Carex*** species by the swollen, ellipsoid tubers at the base of the shoots.

Young submerged plants with thin, ribbon like leaves resemble those of ***Schoenoplectus*** and should be determined by looking for older specimens close by.

Bolboschoenus yagara. Sechtem, Köln, Germany. Photo Michael Hassler.

To distinguish the five ***Bolboschoenus*** species, ripe nuts are needed - please use the key on p. 336.

Bolboschoenus yagara. Nut in side view and cross section. Photo JCS.

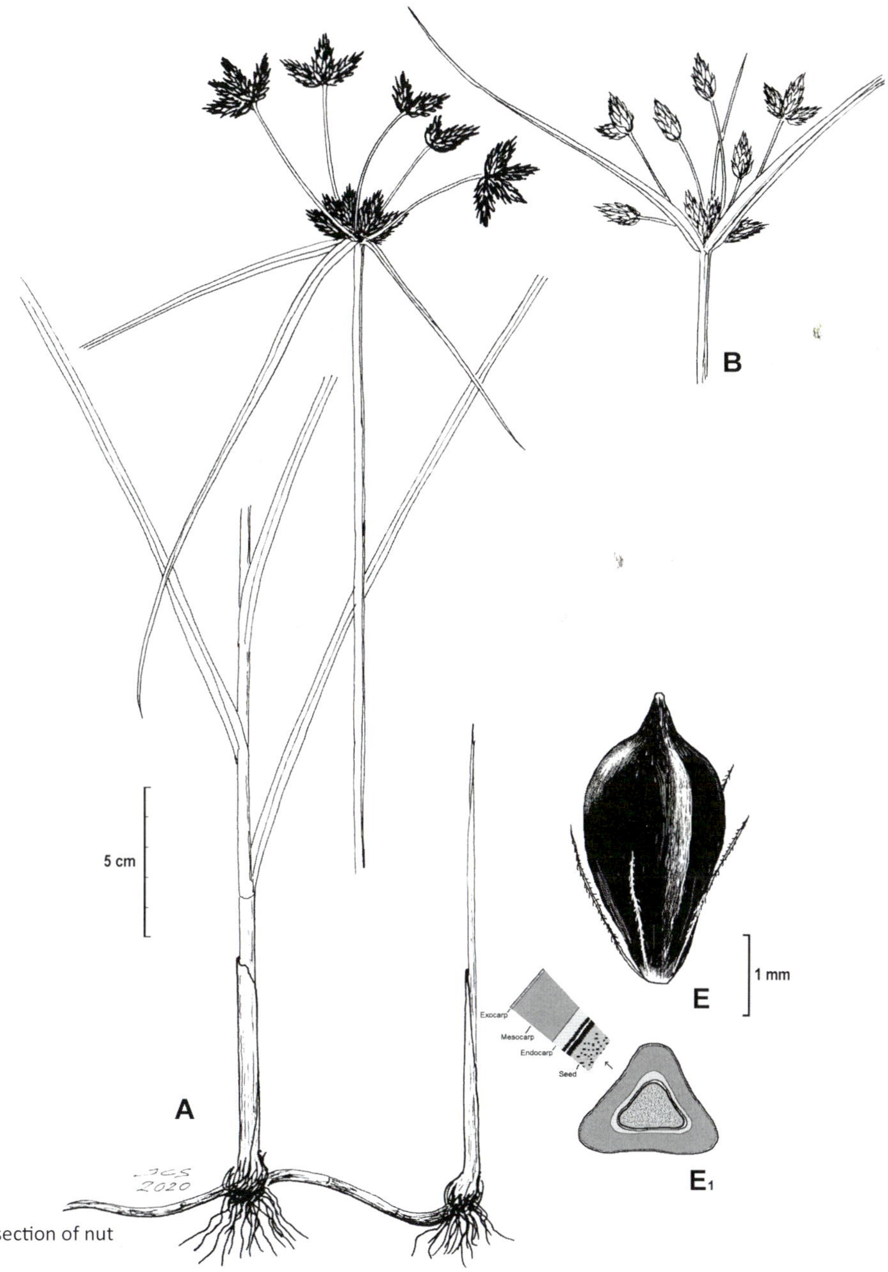

Bolboschoenus yagara
A habit, **B** inflorescence, **E** nut with bristles, **E1** cross section of nut (with enlarged detail).

Cyperaceae

144 *Bolboschoenus glaucus* (Lam.) S.G.Sm.

Syn.: *Bolboschoenus maritimus* subsp. *macrostachys* (Willd.) Soják

DK: - N: - S: - FIN: - IS: - GB: - NL: - F: - D: - CZ: - PL: - EST: -LV: - LT: - Ru:

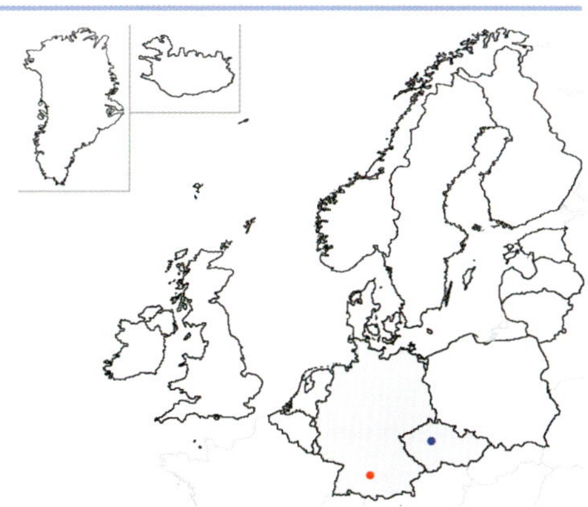

Differs from similar species by the smaller, 2.2–2.5 mm long, 1.3–1.7 mm wide, dark brown to blackish brown, trigonous nuts rounded at the angles. The surface of the nut appears cellular at 20x magnification. Exocarp 1/3 as thick as the mesocarp.

Distribution within the region:
Introduced and naturalised in Prague, Czech Republic. Casually introduced in Ulm, Germany.

Bolboschoenus glaucus.
Oued Tizguite, Morocco. Photo RVL.

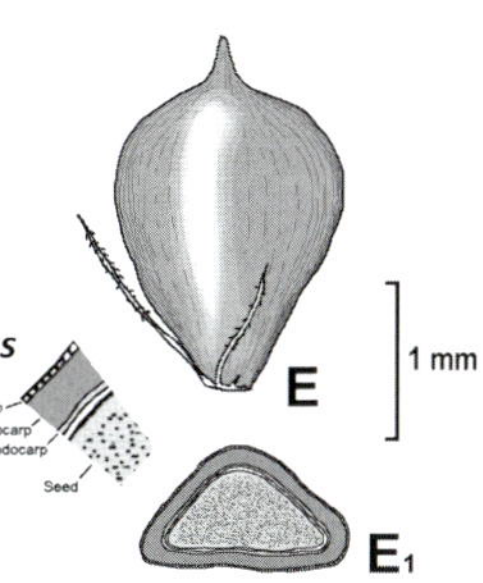

Bolboschoenus glaucus
E nut with bristles,
E1 cross section of nut with (enlarged detail).

Cyperaceae

145 *Schoenoplectus lacustris* (L.) Palla

Syn.: *Scirpus lacustris* L.

DK: Sø-Kogleaks **N:** Sjösivaks **S:** Säv **FIN:** Järvikaisla **IS:** -
GB: Common Club-rush **NL:** Mattenbies **F:** Scirpe lacustre
D: Gewöhnliche Teichsimse **CZ:** Skřípinec jezerní **PL:** Oczeret jeziorny
EST: Järvkaisel **LV:** Ezera lielmeldrs **LT:** Ežerinis meldas **Ru:** Камыш озёрный

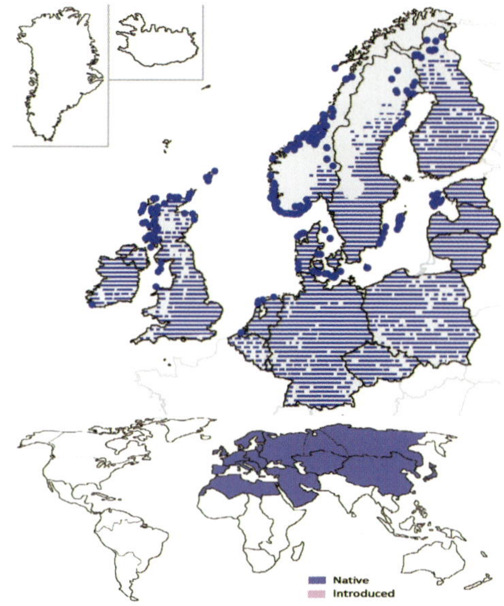

Schoenoplectus lacustris in the River Gudenå close to Randers, Denmark. Photo JCS.

Perennial helophyte or hydrophyte. Rhizome underground, creeping, branched, 1–2(–3) cm in diameter, covered with reddish-brown, sheath-like scales.

Stem: 1–2(–3.5) m long, erect, terete, 7–15(–25) mm in diameter, green to dark green, smooth, somewhat glossy, solid, spongy. Emergent stems terminating in an inflorescence or when sterile with one or few bracts. Fully submerged plants without stems.

Leaves: Emergent stems with 1–3 bladeless sheaths at base and above these a sheathing leaf with reduced, linear lamina up to 15 cm long, blunt at the apex. Plants with both emergent stems and ribbon-like submerged leaves often occur, likewise submerged plants occur with light green to dark green, ribbon-like leaves up to 1 m long, 2–4 mm wide, with a pattern of oblong rectangular air chambers. Submerged leaves more or less sheathing at base and finely serrated at apex.

Inflorescence: Composed of numerous spikelets forming a congested head or a lax, one or two times branched panicle with 3–10, up to 10 cm long, scabrous rays, each with (1–)2–6 spikelets. Spikelets 6–15 mm long, ovoid to cylindrical, brownish. Involucral bracts 2, the lower up to 10 cm long (occasionally longer), stem-like, making the inflorescence appear lateral, the upper brownish, more or less glume-like. Glumes 2.5–4 mm long, glabrous, smooth, somewhat glossy, reddish-brown, hyaline to membranous at the finely fringed margin. Midrib often excurrent in a scabrid mucro.

Flowers: Bisexual, with trifid (rarely bifid in some flowers) style and 3 stamens. Anthers 1.8–2.5 mm long, apex broad and rounded with a fringe of teeth.

Fruits: Nuts 2–3 mm long, 1.5–2.5 mm wide, obovate in outline, abruptly tapering to a short beak, yellowish to greyish brown when ripe, trigonous in cross section. Perianth bristles 4–6, reddish brown, retrorsely scabrid, shorter or slightly longer than nut, persistent.

Flowering: June-July.

Biology: Nuts dispersed by water and birds. Vegetative propagation by division of the rhizome. Emergent stems develop on moist substrates and in relatively shallow water, where ribbon-like leaves may also appear. In deep and fast-flowing water plants often only have shoots with ribbon-like leaves.

Habitats: In fresh, more or less eutrophic water. In lakes and rivers from the shore to a depth of several metres, also in canals and ditches.

Distribution within the region:
Native. Common throughout most of the region, less frequent in the north. Absent from Iceland and Greenland.

Characteristics and similar species:
Schoenoplectus lacustris differs from **146** ***S. tabernaemontani*** in smooth glumes without papillae, trifid styles and anthers with a broad, rounded apex with a fringe of teeth.

Submerged plants with only ribbon-like leaves may be mistaken for

Schoenoplectus lacustris. Asserbo, Denmark. Photo JCS.

Plant with emergent stems and ribbon-like leaves. River Gudenå, Denmark. Photo JCS.

Laacher See, Germany. Tip of submerged leaves serrated. Photos KvdW.

115-118 ***Sparganium*** spp.; these however, have square air chambers.

42-46 ***Sagittaria*** spp. have 10–20 mm wide submerged leaves with different vein pattern and exude latex when damaged.

Submerged **146** ***S. tabernaemontani*** and **140-144** ***Bolboschoenus*** spp. rarely produce ribbon-like leaves. Determination of these must be based on a search for emergent plants nearby.

Hybrids: The hybrid ***Schoenoplectus ×flevensis*** (D.Bakker) Lansdown & Rumsey (*S. lacustris × tabernaemontani*) is not easily recognised. It is more or less intermediate between the parents, has very tall, glaucous stems, bifid to trifid styles and 1.2–1.3 mm long anthers with variable apex.

147 ***Schoenoplectus ×carinatus***.

Submerged, ribbon-like leaves with a pattern of oblong rectangular air chambers. Photo JCS.

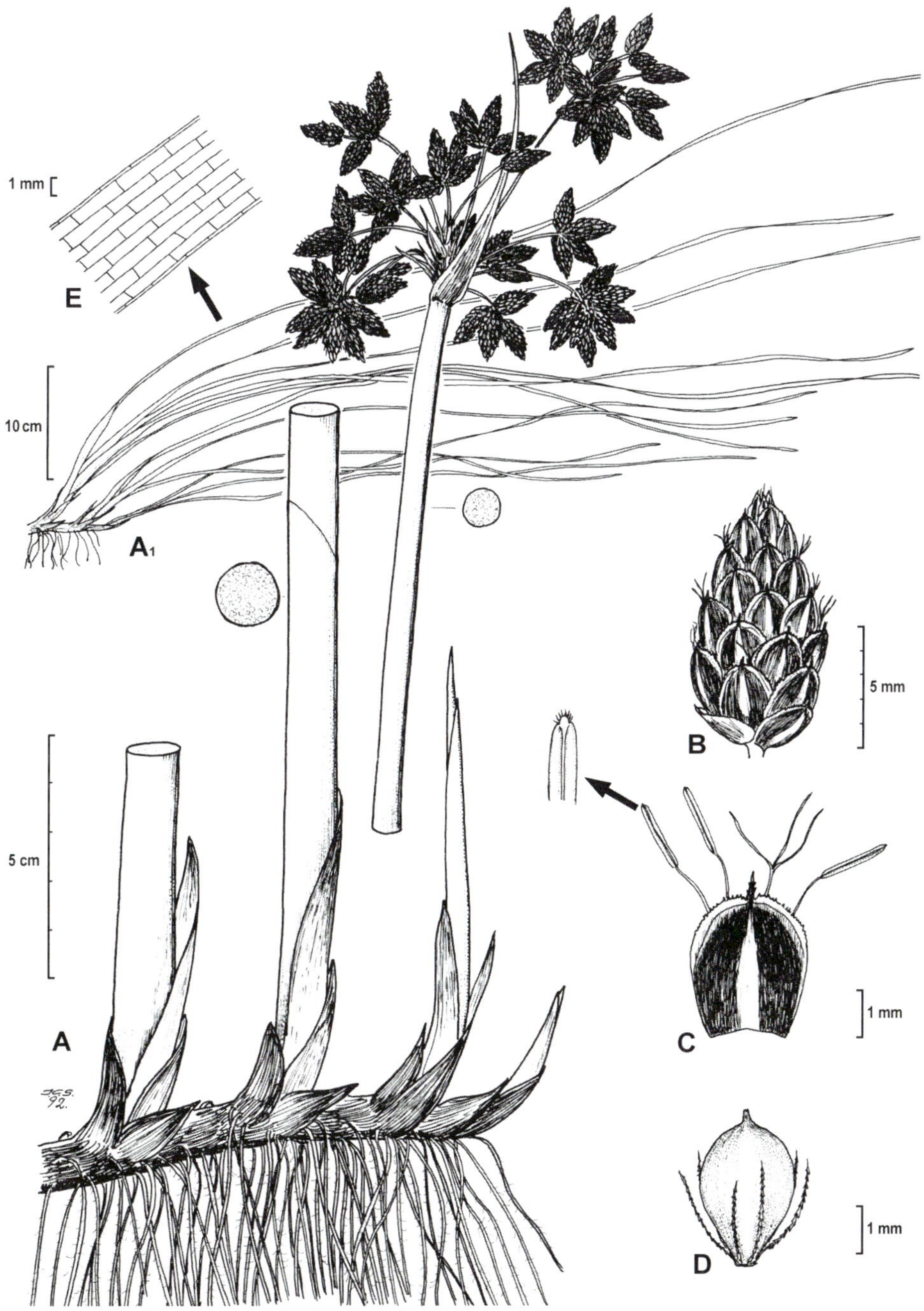

Schoenoplectus lacustris
A rhizome and flowering stem, **A1** submerged plant, **B** spikelet, **C** flower and glume - apex of anther magnified, **D** nut with bristles, **E** part of submerged leaf.

Cyperaceae

146 *Schoenoplectus tabernaemontani* (C.C.Gmel.) Palla

Syn.: *Scirpus tabernaemontani* C.C.Gmel.
DK: Blågrøn Kogleaks **N:** Pollsivaks **S:** Blåsäv **FIN:** Sinikaisla **IS:** -
GB: Grey Club-rush **NL:** Ruwe bies **F:** -Schoenoplectus de Tabernaemontanus
D: Salz-Teichsimse **CZ:** Skřípinec Tabernaemontanův **PL:** Oczeret Tabernemontana
EST: Kare kaisel **LV:** Zilganais lielmeldrs **LT:** Melsvasis meldas **Ru:** Камыш Табернемонтана

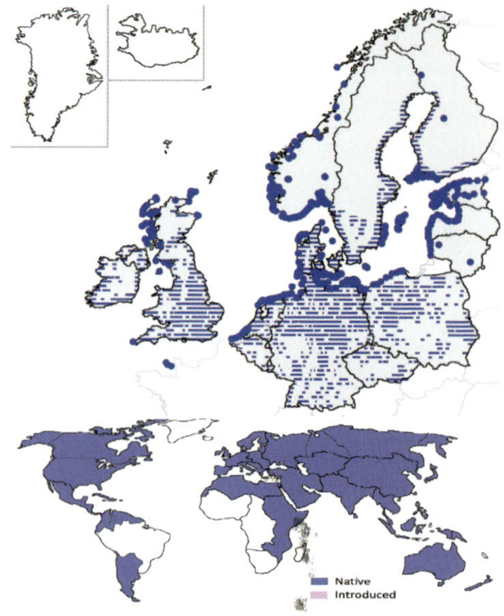

Schoenoplectus tabernaemontani colonizing a reestablished lake. Filsø, SW-Jutland. Photo JCS.

Perennial helophyte or hydrophyte. Rhizome underground, creeping, branched, 1–2 cm in diameter, covered with reddish-brown, sheath-like scales.

Stem: 0.5–1.5 m long, erect, terete sometimes bluntly trigonous at the top, 4–12(–20) mm in diameter, bluish grey to glaucous, smooth, somewhat dull, solid, spongy. Emergent stems terminating in an inflorescence or when sterile with one or few bracts. Fully submerged plants without stems.

Leaves: Emergent stems with 1–3 bladeless sheaths at base and above these a sheathing leaf with reduced, linear lamina, blunt at the apex. Plants with both emergent stems and ribbon-like submerged leaves are rather rare, likewise submerged plants with light green to green, ribbon-like leaves up to 75 cm long, 2–4 mm wide, with a pattern of oblong rectangular air chambers.

Inflorescence: Composed of numerous spikelets, forming a congested head or a lax, one or two times branched panicle with 3–10, up to 5 cm long, scabrous rays, each with 1–6 spikelets. Spikelets 5–11 mm long, ovoid to cylindrical, brownish. Involucral bracts 2, the lower up to 6 cm long, rarely longer, stem-like, making the inflorescence appear lateral, the upper brownish, more or less glume like. Glumes 2.5–4 mm long, dull, dark reddish-brown, with numerous red-brown more or less hooked papillae, hyaline to membranous at the finely fringed margin. Midrib often excurrent in a scabrid mucro.

Flowers: Bisexual, with bifid (rarely trifid in some flowers) style and 3 stamens. Anthers 1.5–3 mm long, apex narrow and tapered with scattered teeth all around.

Fruits: Nuts 2–3 mm long, 1.5–2.5 mm wide, obovate in outline, abruptly tapering to a short beak, brownish grey, more or less smooth when ripe, compressed or compressed trigonous in cross section. Perianth bristles usually 6, reddish brown, retrorsely scabrid, shorter or slightly longer than nut, persistent.

Flowering: June-July.

Biology: Nuts dispersed by water and birds. Vegetative propagation by division of the rhizome. Emergent stems develop on moist substrates and in relatively shallow water, where ribbon-like leaves may also appear. Submerged plants with only ribbon-like leaves are rare.

Habitats: In brackish to freshwater.

Distinguishing *S. tabernaemontani* (photo) from *S. lacustris* based on stem colour alone is uncertain. Hobro, Denmark. Photo JCS.

Flowering *S. tabermaemontani.* Styles bifid with white stigmas. Botanical Garden, Aarhus, Denmark. Photo JCS.

Flowering *S. tabermaemontani.* Northwest of Vitskøl Monestary, Jutland, Denmark. Photo JCS.

Mostly coastal in salt marshes and tidal flats. Inland habitats include eutrophic, base-rich, wet meadows, ditches, village ponds and calcareous lakes.

Distribution within the region:
Native. Widespread and abundant throughout much of the centre of the region, less frequent in the south and absent from many areas in the north, including Iceland and Greenland.

Characteristics and similar species:
Schoenoplectus tabernaemontani differs from **145** ***S. lacustris*** in the glumes, with minute, reddish papillae, bifid styles and anthers with narrow, tapering apex. Submerged plants with ribbon-like leaves are identical to those of **145** ***S. lacustris*** and may be mistaken for **115-118** ***Sparganium*** spp., these however have square air chambers. **42-46** ***Sagittaria*** spp. have 10–20 mm wide submerged leaves with different vein-pattern and exude latex when damaged.

Hybrids: ***Schoenoplectus ×flevensis*** (D.Bakker) Lansdown & Rumsey (*S. lacustris × tabernaemontani*) - please see **145** ***S. lacustris.***

148 ***Schoenoplectus ×kuekenthalianus.***

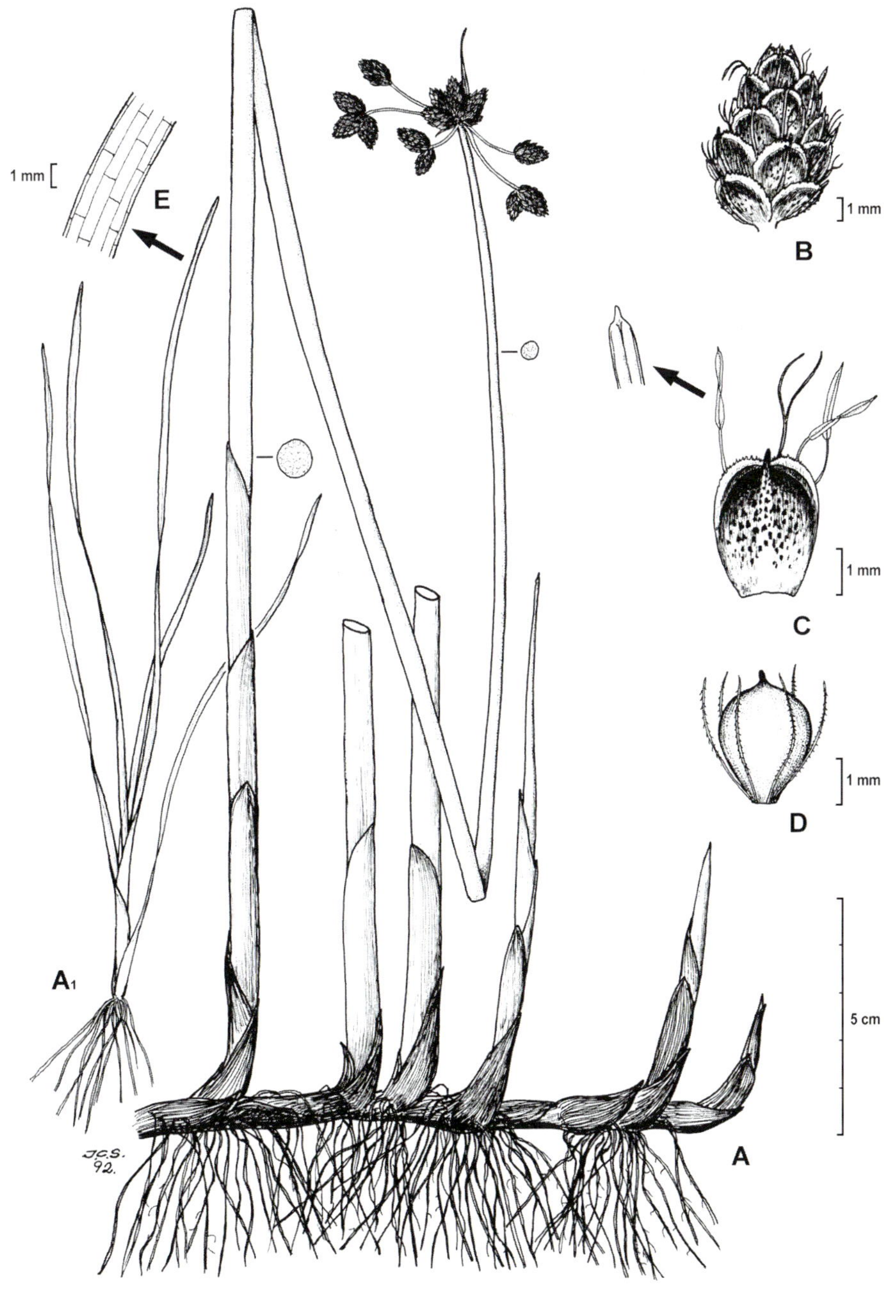

Schoenoplectus tabernaemontani
A rhizome and flowering stem, **A1** submerged plant, **B** spikelet, **C** flower and glume - apex of anther magnified, **D** nut with bristles, **E** part of submerged leaf.

Cyperaceae

147 *Schoenoplectus ×carinatus* (Sm.) Palla

S. lacustris × triqueter

DK: - **N:** - **S:** - **FIN:** - **IS:** -
GB: - **NL:** Bastaardbies **F:** Scirpe caréné
D: Gekielte Teichsimse **CZ:** - **PL:** -
EST: -**LV:** - **LT:** - **Ru:** -

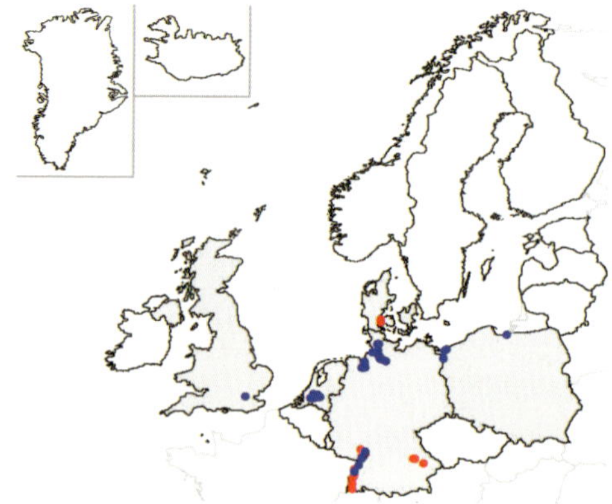

Schoenoplectus ×carinatus. Inflorescence. Lek Krimpen, Netherlands. Photo KvdW.

Sterile hybrid more or less intermediate between the parental species.

The stem is up to 200 cm tall, green to dark green, terete in the lower part, but gradually becoming more trigonous towards the top. The lower bract is rather stout, often stem-like. Glumes smooth without or with only a few papillae. Styles bifid.

Differs from **145.** ***lacustris*** by the stem being trigonous at top, stouter lower bract and bifid styles. Differs from **148** ***Schoenoplectus ×kuekenthalianus*** by the green to dark green stem colour and smooth glumes.

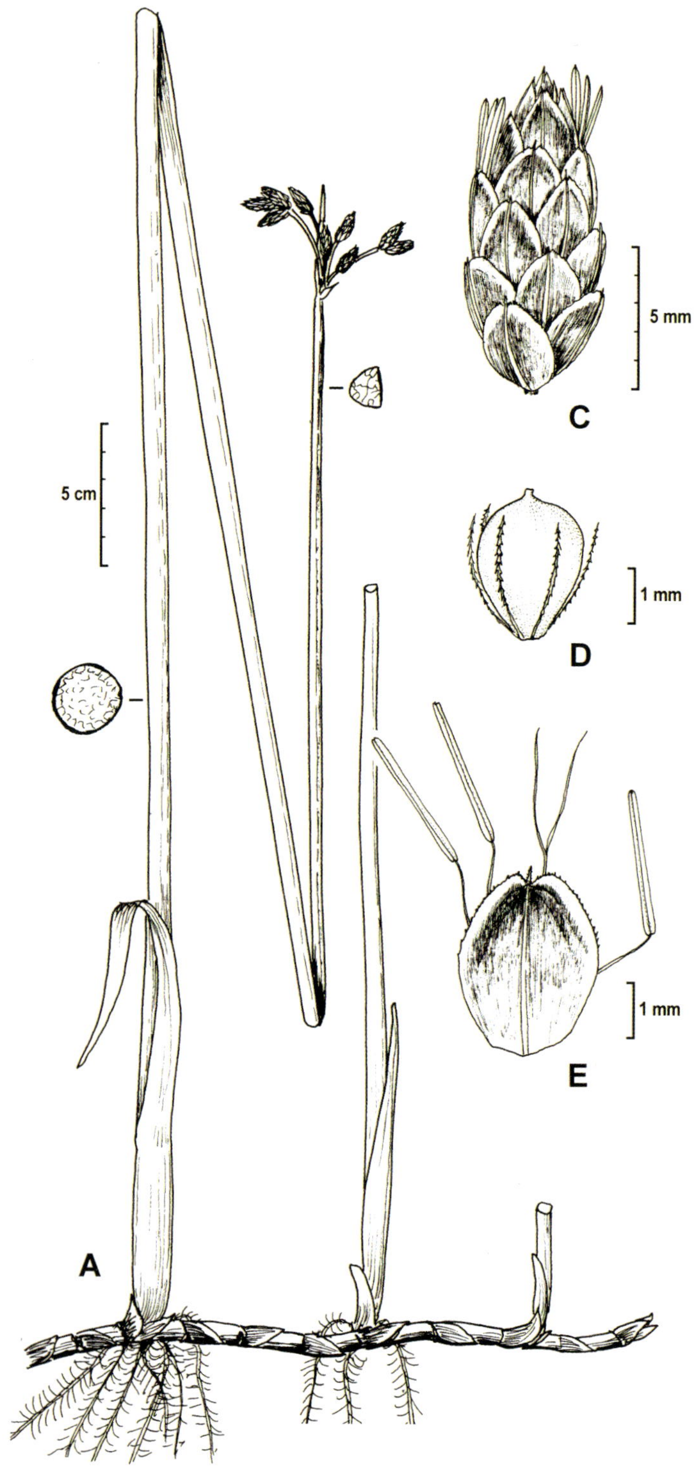

Schoenoplectus ×carinatus
A rhizome and flowering stem, **C** spikelet, **D** nut with bristles, **E** flower and glume.

Cyperaceae

148 *Schoenoplectus ×kuekenthalianus* (Junge) D.H.Kent

S. tabernaemontani × triqueter

DK: - **N:** - **S:** - **FIN:** - **IS:** -
GB: - **NL:** - **F:** -
D: Kükenthal Teichsimse **CZ:** - **PL:** -
EST: -**LV:** - **LT:** - **Ru:** -

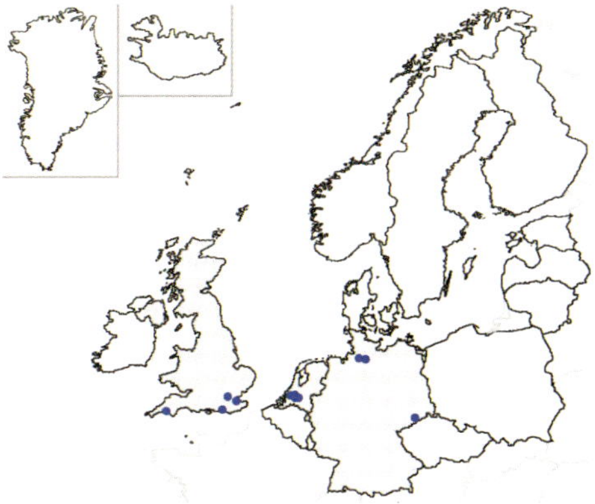

Schoenoplectus ×kuekenthalianus. Inflorescence. Arun, England. Photo RVL.

Sterile hybrid more or less intermediate between the parental species.

The stem is up to 200 cm tall, glaucous, terete in the lower part, but gradually becoming more trigonous towards the top. The lower bract is rather stout often stem-like. Glumes with reddish papillae. Styles bifid.

Differs from **146 *S. tabernaemontani*** by the stem being trigonous at top and the lower bract exceeding the inflorescence. Differs from **147 *Schoenoplectus ×carinatus*** by the glaucous stem colour and papillate glumes.

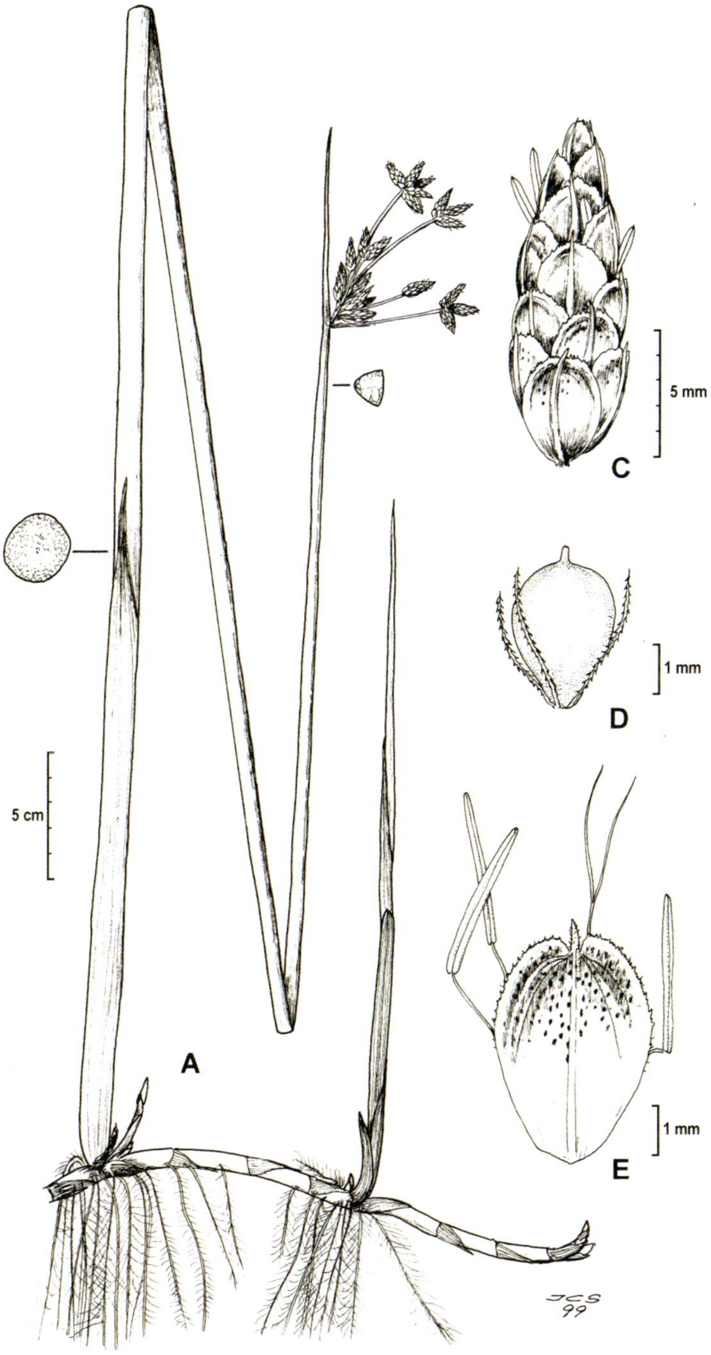

Schoenoplectus ×kuekenthalianus
A rhizome and flowering stem, **C** spikelet, **D** nut with bristles, **E** flower and glume.

Cyperaceae

149 *Schoenoplectus triqueter* (L.) Palla

Syn.: *Scirpus triqueter* L.
DK: - **N:** - **S:** - **FIN:** - **IS:** -
GB: Triangular Club-rush **NL:** Driekantige bies **F:** Scirpe triquètre
D: Dreikant-Teichsimse **CZ:** Skřípinec trojhranný **PL:** Sitowie trójgraniaste
EST: -**LV:** - **LT:** - **Ru:** Камыш трёхгранный

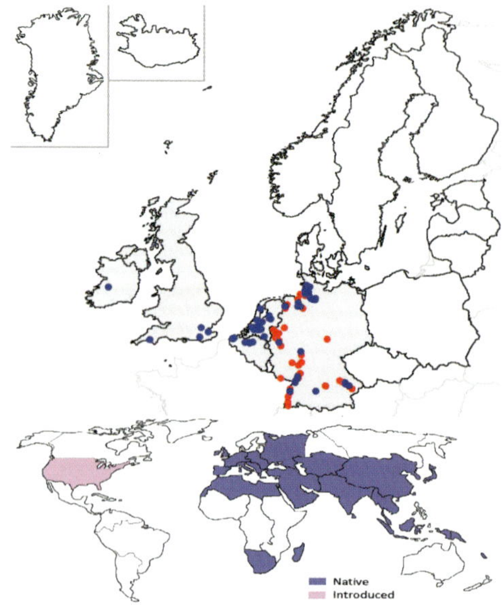

Perennial helophyte. Rhizome underground, creeping, branched, 0.5–1 cm in diameter, covered with dark or reddish-brown sheath-like scales.

Stem: 0.5–1.5 m long, erect, sharply trigonous, 3–8 mm in diameter, green to glaucous, smooth, solid, spongy. Emergent stems terminating in an inflorescence or when sterile with one or few bracts.

Leaves: Emergent stems with 1–3 bladeless sheaths at the base and above these 1 or 2 sheathing leaves with a short, 1–20 cm long lamina. Ligule present.

Inflorescence: A more or less congested, head-like cluster composed of few to numerous sessile spikelets and often with 2–5 spikelets on up to 2 cm long peduncles. Spikelets 5–10 mm long, ovoid to cylindrical, brownish. Involucral bracts 2, the lower erect, almost stem-like, up to 10 cm long and always longer than the inflorescence, the upper much shorter, greenish or glume-like. Glumes 3–4.5 mm long, dull, reddish brown, hyaline to membranous at the finely fringed margin, emarginated at apex and with obtuse lateral lobes. Midrib greenish, often excurrent in a short mucro.

Flowers: Bisexual, with bifid style and 3 stamens. Anthers 1.2–1.5 mm long, connective tip smooth.

Fruits: Nuts 2.5–3 mm long, 1.5–2.5 mm wide, obovate in outline, abruptly tapering to a short beak, olive brown to reddish brown, smooth, shiny, in cross section compressed. Perianth bristles 4–6, brownish, retrorsely scabrid, about as long as the nut, persistent.

Flowering: June-July.

Biology: Nuts dispersed by water and birds. Vegetative propagation by division of the rhizome. Submerged or flooded plants are not known to produce ribbon-like leaves.

Habitats: In fresh or brackish, eutrophic water. In ditches, channels, muddy pools, by lakes and big rivers, in estuaries and in the tidal zone of rivers.

Distribution within the region:
Native. Rare in Ireland, the Netherlands and Belgium, rare and declining in Germany. Extinct as a native in England

Schoenoplectus triqueter. Aalst, Netherlands. Photo KvdW.

The stout, erect lower bracts make the inflorescences of *S. triqueter* appear lateral. Aalst, Netherlands. Photo KvdW.

where the only surviving populations are from a re-introductions project.

Characteristics and similar species: *Schoenoplectus triqueter* can be recognised by the combination of creeping rhizome, sharply trigonous stem, upper sheaths with a lamina, inflorescence usually with some pedunculate spikelets, clusters and glumes with obtuse lateral lobes.

150 *Schoenoplectus pungens* is very similar, but differs in all spikelets sessile, glumes with acute lateral lobes, nut without or with very short bristles.

151 *Schoenoplectiella mucronata* is densely caespitose.

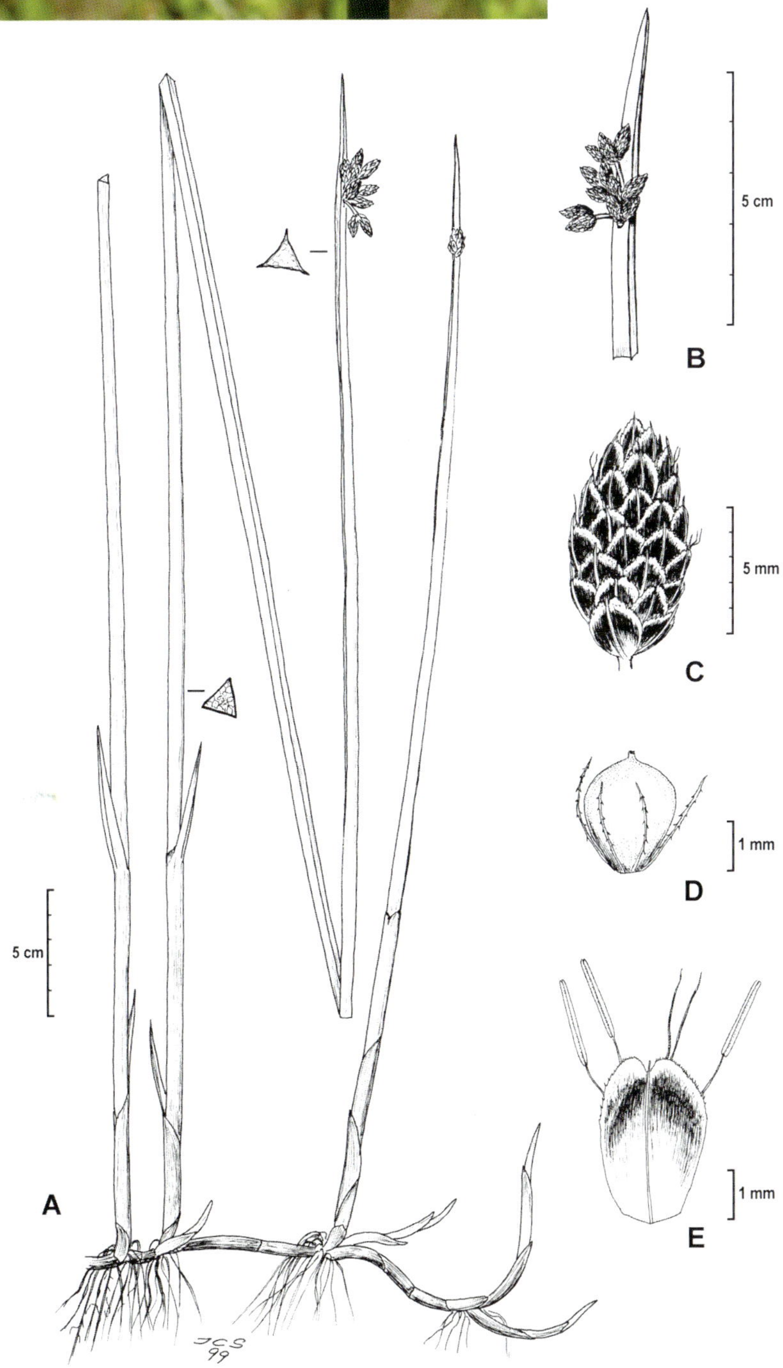

Schoenoplectus triqueter
A rhizome and flowering stem, **B** inflorescence, **C** spikelet, **D** nut with bristles, **E** flower and glume.

Cyperaceae

150 *Schoenoplectus pungens* (Vahl) Palla

Syn.: *Scirpus pungens* Vahl: *S. americanus* var. *pungens* (Vahl) Barros & Osten
DK: - **N:** - **S:** - **FIN:** - **IS:** -
GB: Sharp Club-rush **NL:** Stekende bies **F:** Scirpe piquant
D: Amerikanische Teichsimse **CZ:** Skřípinec pichlavý **PL:** Oczeret amerykański
EST: -**LV:** - **LT:** - **Ru:** -

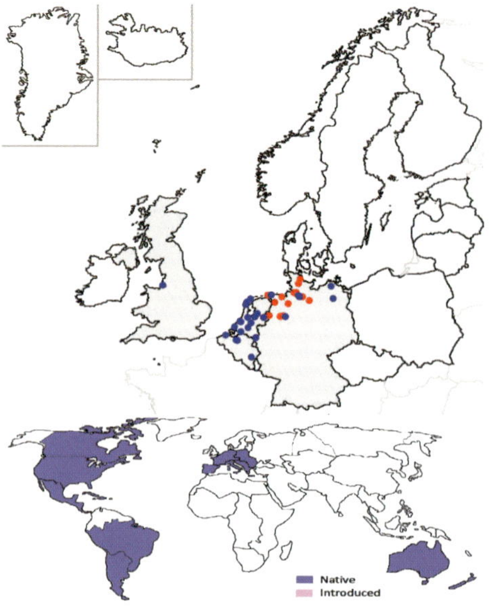

Perennial helophyte with solitary stems. Rhizome underground, creeping, branched, about 5 mm in diameter, covered with dark or brownish sheath-like scales.

Stem: 30–60 cm long, erect, sharply trigonous, 2–5 mm wide, green, smooth, solid, spongy. Emergent stems terminating in an inflorescence or when sterile with one or few bracts.

Leaves: Emergent stems with few bladeless sheaths at the base and above these 2 or 3 sheathing leaves each with a lamina 3–30 cm long.
Ligule present.

Inflorescence: A dense cluster of 1–6 sessile spikelets. Spikelets 6–11 mm long, ovoid, subacute at the apex, brownish to dark reddish brown. Involucral bracts 2, the lower erect, stem-like, 3–12 cm long and always longer than the inflorescence, the upper much shorter, brown, glume-like. Glumes 3.5–4.0 mm long, smooth, somewhat glossy, reddish brown, membranous, margins hyaline sometimes finely fringed, emarginate at apex and with acute lateral lobes. Midrib pale brown, excurrent in a short mucro.

Flowers: Bisexual, with bifid style and 3 stamens. Anthers 2.5–3.5 mm long.

Fruits: Nuts 2–2.5 mm long, 1–1.5 mm wide, obovate in outline, abruptly tapering to a short beak, greyish brown, smooth, shiny, compressed in cross section. Perianth bristles absent or up to 6, rudimentary, less than half as long as the nut.

Schoenoplectus pungens at the muddy shore of Lek Everdingen, Netherlands. Photo KvdW.

Flowering: July-August.

Biology: Nuts dispersed by water and birds. Vegetative propagation by division of the rhizome. Submerged plants are not known to produce ribbon-like leaves.

Habitats: In fresh or brackish, mesotrophic to eutrophic water. In areas influenced by tidal water, such as salt marshes, coastal reed swamps, wet dune slacks, silty shores of ponds and lakes.

Distribution within the region:

Native. Rare in Great Britain, Netherlands and Belgium, rare and declining in Germany.

Characteristics and similar species:

Schoenoplectus pungens can be recognised by the combination of creeping rhizome, sharply trigonous stem, upper sheaths with lamina, inflorescence a dense cluster of sessile spikelets, and glumes with acute lateral lobes.

149 *Schoenoplectus triqueter* is very similar, but differs in a more open inflorescence with some pedunculate spikelet clusters, glumes with obtuse lateral lobes and nut with bristles about as long as the nut.

151 *Schoenoplectiella mucronata* is densely caespitose.

Schoenoplectus pungens. Lek Everdingen, Netherlands. Photo KvdW.

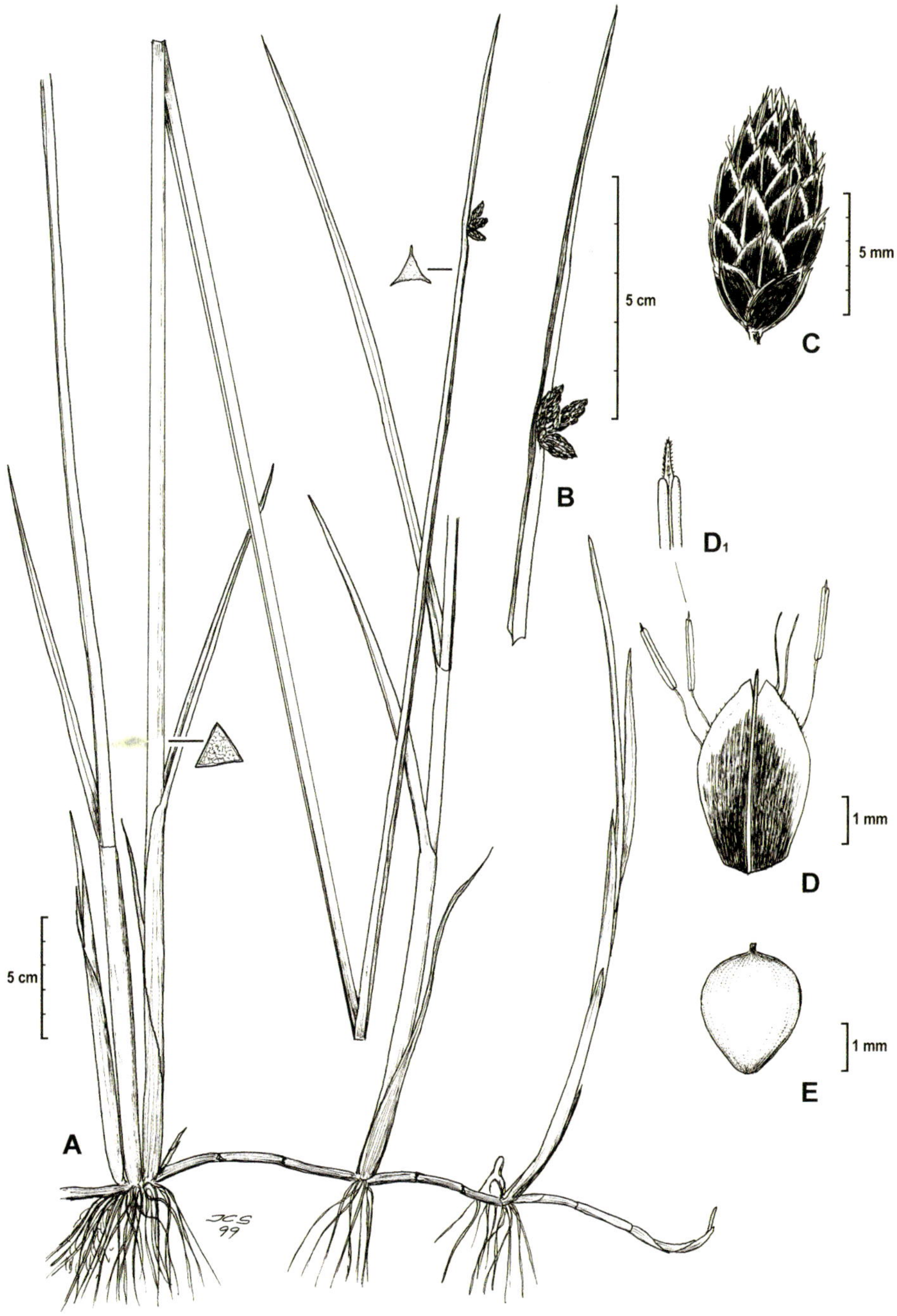

Schoenoplectus pungens
A rhizome and flowering stem, **B** inflorescence, **C** spikelet, **D** flower and glume, **D1** tip of anther, **E** nut without bristles

Cyperaceae

151 *Schoenoplectiella mucronata* (L.) J.Jung & H.K.Choi

Syn.: *Schoenoplectus mucronatus* (L.) Palla

DK: - **N:** - **S:** - **FIN:** - **IS:** -

GB: Rough-seed Club-rush **NL:** Ribbelbies **F:** Scirpe mucroné

D: Stachelspitzige Teichsimse **CZ:** Skřípinec osinkatý **PL:** Oczeret sztyletowaty

EST: -**LV:** - **LT:** - **Ru:** Камыш остроконечный

Schoenoplectiella mucronata Cherine, Parc Naturel Régional de la Brenne, Indre, France. Photo RVL.

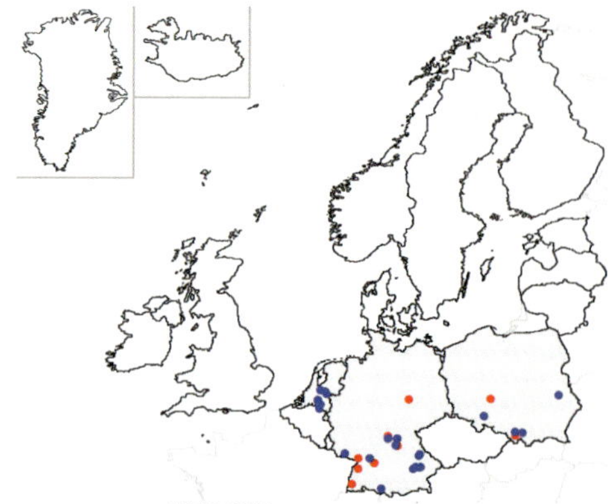

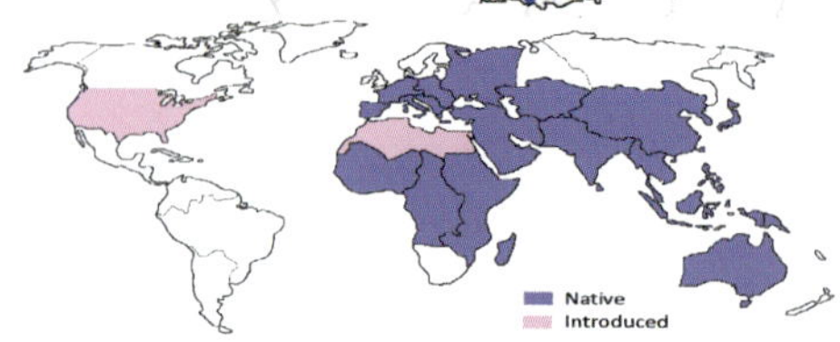

Perennial helophyte, forming dense tussocks. Rhizome very short, more or less vertical, stolons absent.

Stem: 40–100 cm long, 2–8 mm wide, sharply trigonous to winged, green, glabrous, somewhat dull, solid, spongy. Emergent stems terminating in an inflorescence or when sterile with one or few bracts.

Leaves: Emergent stems with 1–3 bladeless, brownish to greenish sheaths at base. Sheaths with lamina absent.

Inflorescence: A dense, head-like cluster of 2–20 sessile spikelets. Spikelets 4–12 mm long when flowering, up to 30 mm long when fruiting, ovoid to oblong, brownish. The lowest bract up to 15 cm long, green, sharply trigonous, stem-like, rather abruptly tapering into a short point, at first erect, making the inflorescence appear lateral, later increasingly reflexed, becoming almost perpendicular to the stem. The other bracts brownish, more or less glume-like. Glumes 3–4.5 mm long, obtuse, obovate, glabrous, pale green to pale brown, with greenish midrib excurrent in a very short mucro.

Flowers: Bisexual, with trifid, rarely bifid style and 3 stamens. Anthers about 1 mm long.

Fruits: Nuts 1.5–2.5 mm long, 1–1.5 mm wide, obovoid, abruptly tapering to a short beak, blackish brown, transversely wrinkled, bluntly trigonous in cross section. Perianth bristles (4–)6, brownish, retrorsely scabrid, as long as or slightly longer than nut, persistent.

Flowering: August-October.

Biology: Nuts dispersed by water and birds. Submerged or flooded plants are not known to produce ribbon-like leaves. Wintergreen.

Habitats: In fresh, mildly acid to neutral, soft, moderately eutrophic to eutrophic water. Sandy or muddy banks of fish-ponds and pools, along riverbanks, in wet meadows, marshes and swamps.

Distribution within the region:

Native. Local, rare and declining in the Netherlands, Germany and Poland.

Characteristics and similar species:

Schoenoplectiella mucronata can be recognised by its densely caespitose habit, with sharply trigonous stem, all sheaths lacking a lamina, the trifid styles and by the transversely wrinkled nut.

149 ***S. triqueter*** and **150** ***S. pungens*** also have sharply trigonous stems but differ in having creeping rhizomes, at least one sheath with a lamina, bifid styles and nuts without wrinkles.

Flowering *S. mucronatus*. Tilburg, Netherlands. Photos KvdW.

Schoenoplectiella mucronata. Cherine, Parc Naturel Régional de la Brenne, Indre, France. Photo RVL.

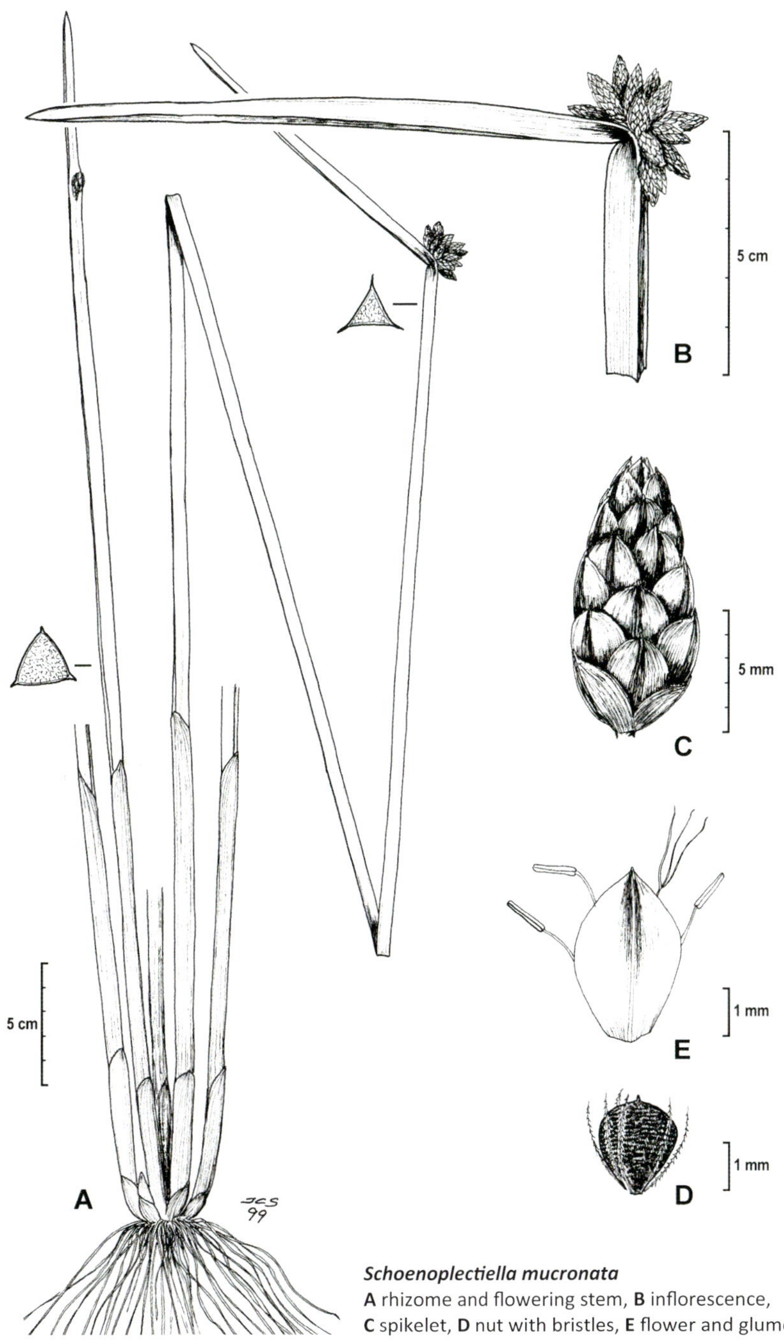

Schoenoplectiella mucronata
A rhizome and flowering stem, **B** inflorescence, **C** spikelet, **D** nut with bristles, **E** flower and glume.

Cyperaceae

152 *Eleocharis ovata* (Roth) Roem. & Schult.

Syn.: *Scirpus ovatus* Roth

DK: - **N:** - **S:** - **FIN:** Soikkoluikka **IS:** -
GB: Ovate Spikerush **NL:** Eivormige waterbies **F:** Héléocharis ovale
D: Eiförmige Sumpfbinse **CZ:** Bahnička vejčitá **PL:** Ponikło jajowate
EST: Munajas alss **LV:** Olveida pameldrs **LT:** - **Ru:** Ситняк яйцевидный

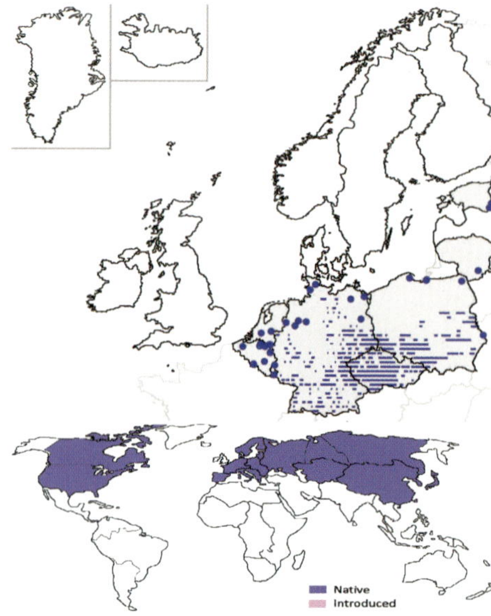

Eleocharis ovata. Fishpond near Zonhoven, Belgium. Photo KvdW.

Annual helophyte with numerous stems forming dense tufts.

Stem: 2–35(–50) cm long, of varying length, 0.3–1.0(–2.0) mm in diameter, erect to ascending, yellow green to green, smooth, almost translucent, quite robust.

Leaves: Leaves reduced to two basal, green to straw-coloured or reddish, bladeless sheaths at the base of each stem, the upper with a tooth up to 0.2 mm long.

Inflorescence: A single terminal spikelet 3–8(–13) mm long, spherical to oblong ovate or cylindrical, rounded to blunt at the top, brownish, with 15–100 flowers. Glumes 1.5–2.5 mm long, acute to obtuse, brown to purplish brown with green midrib and scarious margins. The lowest empty glume encircling about 2/3 of the stem.

Flowers: Bisexual. Stigmas 2(–3). Stamens 2(–3), anthers c. 0.3 mm long, brown.

Fruits: Nuts 1.0–1.3 mm long (including style base), obovate, elliptic in cross section, straw coloured to dark reddish brown, smooth, shiny. Perianth bristles 6–7, longer than the fruit. Style base conical triangular, 0.3–0.5 mm wide, 1/2–2/3 as wide as the nut, as long as wide, persistent, not constricted at base.

Flowering: June-August.

Biology: Propagation by seeds.

Habitats: On wet, muddy, meso- to eutrophic substrate. The drawdown zones of lakes and streams, disturbed ground, aquaculture basins when these are drained.

Distribution within the region:
Native. In the south of the region, rare in the Netherlands, Belgium, Luxembourg, Estonia and Lithuania. Scattered to locally frequent in Germany, Poland and the Czech Republic.

Characteristics and similar species:
Eleocharis ovata is characterised by the shiny, smooth, urn-shaped nuts, perianth bristles longer than the nut, as well as the style base 1/2–2/3 as wide as the nut and normally as long as wide. **153 *E. obtusa*** and **154 *E. engelmannii*** differ in being more robust and with style bases almost as wide as the nut. Both **158 *E. multicaulis*** and **159 *E. carniolica*** are perennials.

Eleocharis ovata. Parc Naturel Régional de la Brenne, Indre, France. Photo RVL.

Eleocharis obtusa. Maui Nui Botanical garden, Hawaii. Photo Jan-Thomas Johansson.

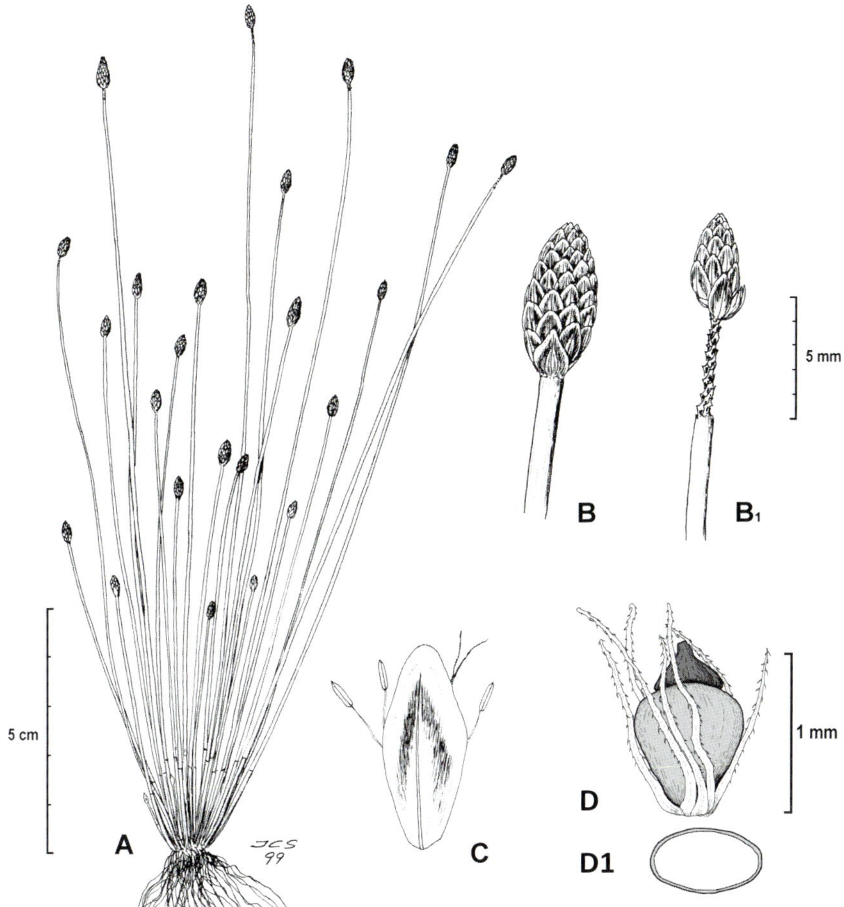

Eleocharis ovata
A habit, **B**, **B1** inflorescence, **C** glume,
D nut with style base (dark) at top and perianth bristles,
D1 cross section of nut.

Cyperaceae

153 *Eleocharis obtusa* (Willd.) Schult.

Caespitose annual (rarely perennial). Stems erect, robust, stiff. Style base 0.35–0.5 mm high, (0.4–)0.5–0.8 mm wide, 2/3–9/10 as wide as nut. Perianth bristles (5–)6–7, much longer than the nut.

Distribution within the region:
Non-native. Scattered in the Netherlands, Belgium and Germany. Native to North America and Argentina.

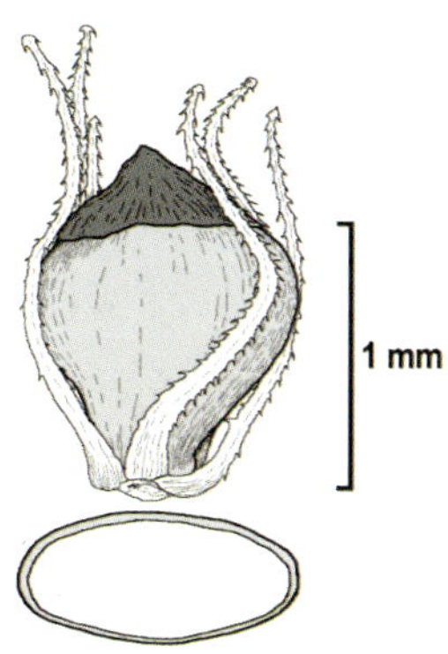

Nut with style base (dark) at top and perianth bristles

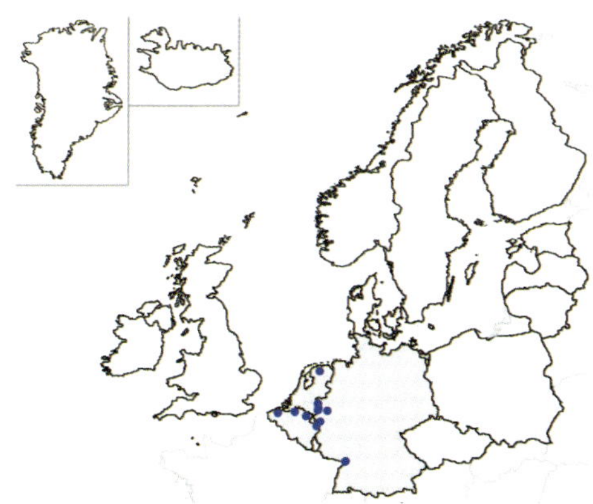

Cyperaceae

154 *Eleocharis engelmannii* Steud.

Caespitose annual (rarely perennial). Stems erect, robust, stiff. Style base 0.1–0.3(–0.4) mm high, 0.6–0.9 (–1) mm wide, 9/10 as wide as nut. Perianth bristles 5–8, shorter or slightly longer than the nut.

Distribution within the region:
Non-native. Scattered in the Netherlands, Belgium, and Germany. Native to southern Canada and USA.

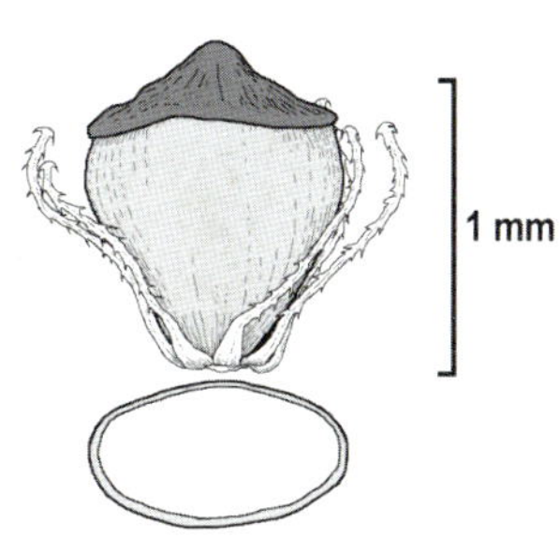

Nut with style base (dark) at top and perianth bristles

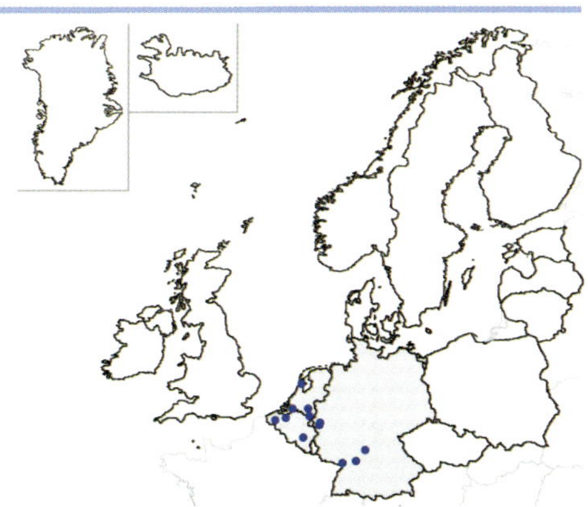

Cyperaceae

155 *Eleocharis palustris* (L.) Roemer & Schultes

Syn.: *Scirpus palustris* L.

DK: Almindelig Sumpstrå **N:** Sumpsivaks **S:** Knappsäv **FIN:** Rantaluikka **IS:** Vatnsnál
GB: Common Spike-rush **NL:** Gewone waterbies **F:** Souchet des marais
D: Gewöhnliche Sumpfsimse **CZ:** Bahnička bahenní **PL:** Ponikło błotne
EST: Sooalss **LV:** Purva pameldrs **LT:** Pelkinis duonis **Ru:** Болотница болотная

Eleocharis palustris. Whitlaw Moss, southern Scotland. Photo RVL.

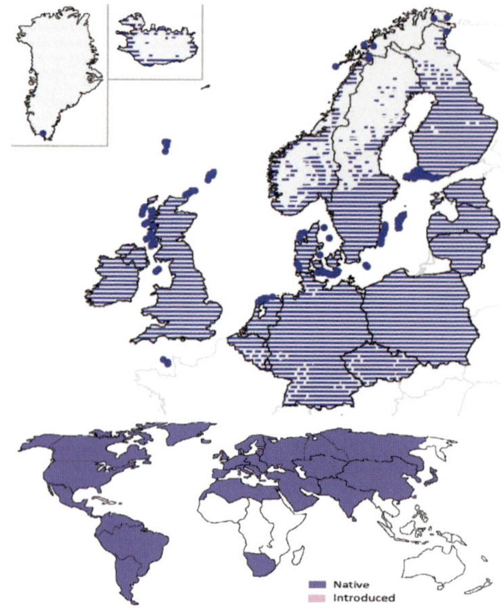

Perennial helophyte. Rhizome long, creeping, underground, 2–6 mm in diameter, light brown to blackish brown often with a purple tinge, with a short sheathing scale at each node. Roots pale brown to greyish brown, up to 1.5 mm wide.

Stem: 1 to several from each rhizome node, 5–50 cm long, submerged stems up to 1 m, 1–3.5 mm in diameter, firm, terete or slightly compressed, somewhat shiny, green to dark green or bluish green, with more than 20 vascular bundles not forming distinct ridges when the stem is dried. Stomata guard cells shorter than the subsidiary cells. Emergent stems fertile or sterile. Submerged stems mostly sterile and terminating in a few small glumes.

Leaves: With 2 wine-red to blackish-red bladeless sheaths at base of stem. The sheaths are more or less transversely truncate.

Inflorescence: A single terminal spikelet 5–20 mm long, oblong ovate to cylindrical, brownish to blackish brown. The 2 lowest glumes are normally empty and about half encircle the stem.

Flowers: Bisexual. Stigmas 2. Stamens 3, anthers 1.6–3.0 mm long.

Fruits: Nuts yellow brown to brown, obovate, somewhat compressed, slightly shiny, with a pattern of very fine, oblong hollows.

Flowering: June-August.

Biology: Reproduces by seed and vegetative reproduction is by division of the rhizome.

Habitats: In mesotrophic to eutrophic water in or by ponds, lakes, ditches and rivers; in dune slacks and oligotrophic, acidic lakes on heaths, in wet meadows and mires, as well as at low salinity even in marshy coastal habitats.

Distribution within the region:
Native. Common throughout most of the region, less common in the north and very rare in Greenland.

Characteristics and similar species:
Eleocharis palustris can be recognised by the long, creeping rhizome, the relatively long spikelets with the 2 sterile lower glumes only half encircling the stem.

156 *E. mamillata* differs in the bright green stems circular in crosssection, fragile and easily compressed, the 5–8 perianth bristles usually much longer than the nut and in the stomata guard cells longer than the subsidiary cells (see pp. 363 and 366 for comparison).

Eleocharis palustris subsp. *waltersii* and *Hydrocotyle vulgaris* at a depth of 30-40 cm in an oligotrophic lake. Madum Sø, Himmerland, Denmark. Photo JCS.

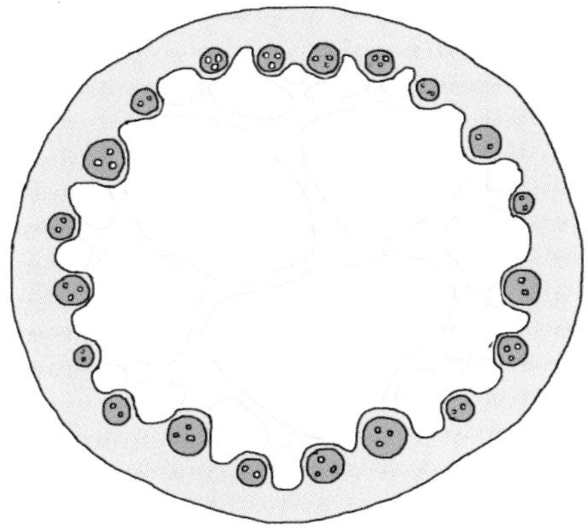

Stem cross section with 20 or more vascular bundles . The bundles do not form very obvious ridges when the stem is dried. (Source: H. Lindberg 1902.)

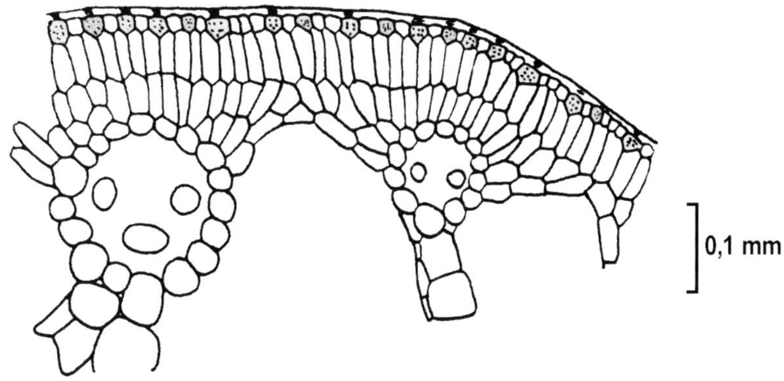

Stem cross section in detail. The epidermis with fibre cells (grey and dotted) more or less alternating with clear cells. There are two layers of palisade cells below the epidermis. (Source: H. Lindberg 1902.)

157 ***E. uniglumis*** differs in only the lowest glume empty and completely encircling the stem. Two subspecies are recognised, which are presented on the next pages with characters differentiating them. The variation within both subspecies is rather broad.

Hybrids: Partly fertile specimens more or less intermediate with *E. uniglumis* or *E. mamillata* are regarded as hybrids.

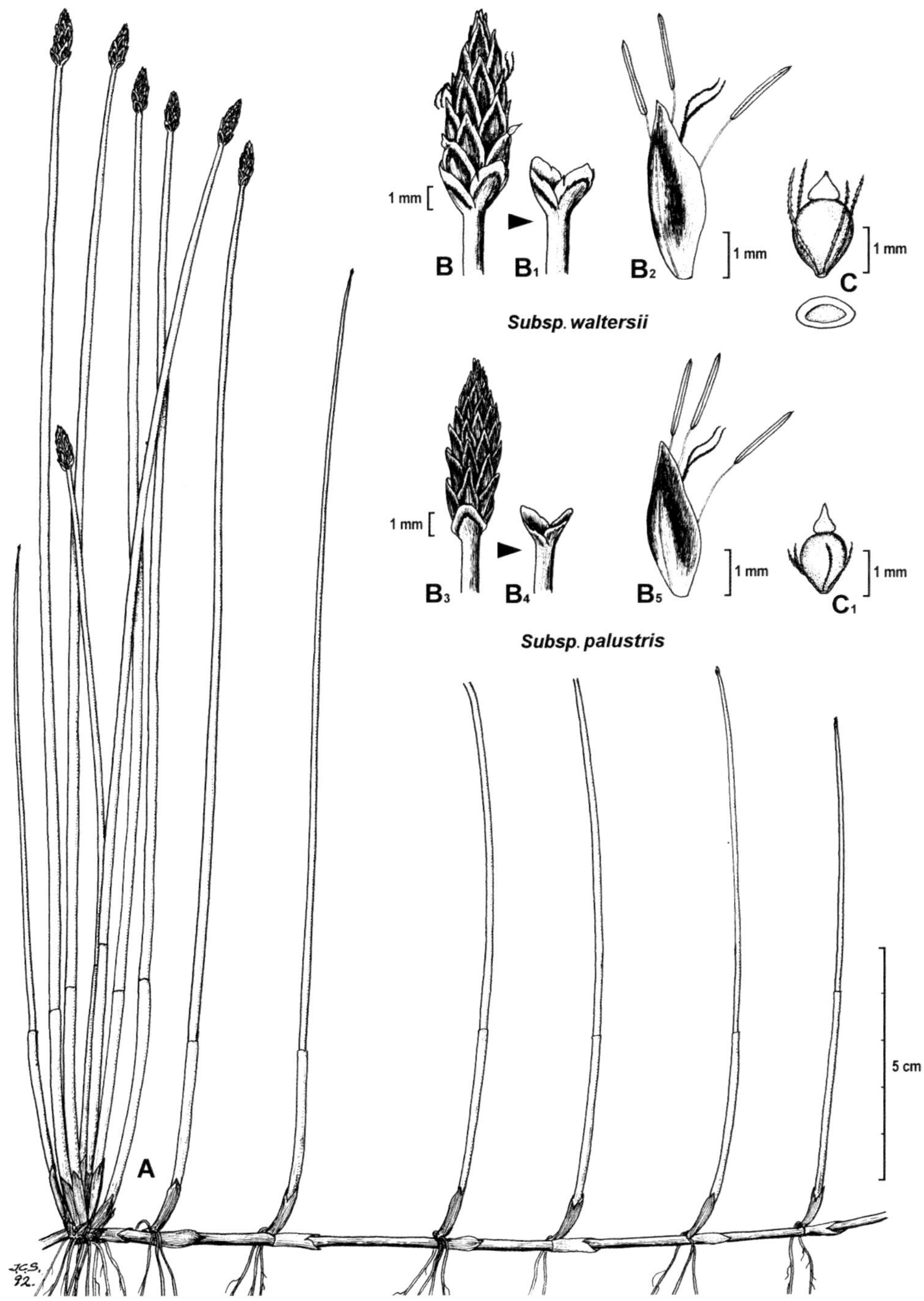

Eleocharis palustris
A habit. Subsp. ***waltersii:*** **B** spikelet, **B1** lower glumes, **B2** glume with flower, **C** nut with perianth bristles and style base. Subsp. ***palustris:*** **B3** spikelet, **B4** lower glumes, **B5** glume with flower, **C1** nut with perianth bristles and style base.

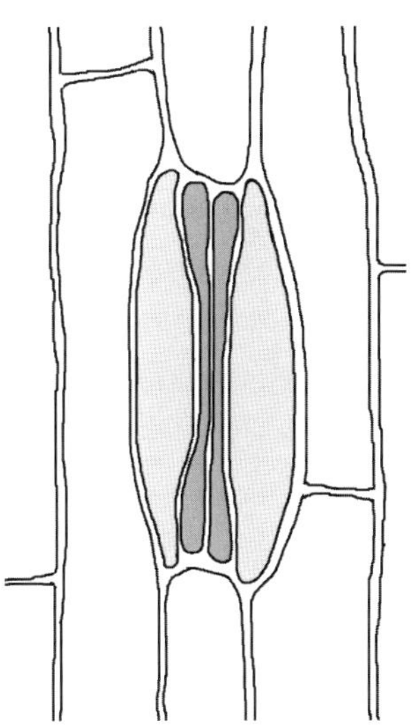

In both subspecies the stomata guard cells (dark grey) are shorter than the subsidiary cells (light grey).

Cyperaceae

155a *Eleocharis palustris* subsp. *palustris*

Syn.: *E. palustris* subsp. *microcarpa* Walters
DK: Sydlig Sumpstrå **N:** - **S:** Nordknappsäv **FIN:** - **IS:** -
GB: - **NL:** - **F:** -
D: Echte Sumpfsimse **CZ:** - **PL:** -
EST: -**LV:** - **LT:** - **Ru:** -

Stem: Shallowly furrowed, dull, greyish green. Mean stomata length 35–56 µm.

Inflorescence: Spikelet with 40–70 flowers. Glumes from the middle of the spikelet less than 3.5 mm long and normally with a narrow hyaline margin.

Fruits: Nuts (1.1–)1.2–1.5(–1.6) mm long (excluding style base). Style base (0.3–)0.4–0.8(–1.0) mm long and often longer than wide. Perianth bristles 0–4, shorter or slightly longer than the nut.

Distribution within the region:
The distribution of the subspecies is poorly documented in some countries.

Eleocharis palustris subsp. *palustris*.
Nut with style base and perianth bristles.
Photo JCS.

Eleocharis palustris subsp. *palustris*.
The stem is dull greyish green.
Photo Jan-Thomas Johansson.

Cyperaceae

155b *Eleocharis palustris* subsp. *waltersii* Bures & Danihelka

Syn.: *E. palustris* subsp. *vulgaris* Walters
DK: Glat Sumpstrå **N:** - **S:** Sydknappsäv **FIN:** - **IS:** -
GB: - **NL:** - **F:** -
D: Gewöhnliche Sumpfsimse **CZ:** - **PL:** -
EST: -**LV:** - **LT:** - **Ru:** -

Stem: Smooth, somewhat shiny, green. Mean stomata length 50–77 µm.

Inflorescence: Spikelet with 20–40 flowers. Glumes from the middle of the spikelet more than 3.5 mm long and normally with a rather wide hyaline margin.

Fruits: Nuts (1.3–)1.4–1.8(–1.9) mm long (excluding style base). Style base (0.4–)0.5–0.7(–0.9) mm long and often wider than long. Perianth bristles 4, slightly to much longer than the nut.

Distribution within the region:
The distribution of the subspecies is poorly documented in some countries.

Eleocharis palustris subsp. *waltersii*.
Nut with style base and perianth bristles.
Photo JCS.

Eleocharis palustris subsp. *waltersii*.
The stem is green and somewhat shiny.
Photo JCS.

Cyperaceae

156 *Eleocharis mamillata* H.Lindb.

Syn.: *Scirpus mamillatus* (H.Lindb.) H.Lindb.; *S. palustris* subsp. *mamillatus* (H.Lindb.) Cajander
DK: Vorte-Sumpstrå **N:** Mjuksivaks **S:** Veksäv **FIN:** Mutaluikka **IS:** -
GB: - **NL:** - **F:** Éléocharis à têtons
D: Zitzen-Sumpfsimse **CZ:** Bahnička bradavkatá **PL:** Ponikło sutkowate
EST: Muda-alss **LV:** Iežmaugtais pameldrs **LT:** Karpotasis duonis **Ru:** Ситняк сосочковый

Eleocharis mamillata subsp. *mamillata.* The entire plant may have a tousled look because of the fragile stems. Nordsjælland, Denmark. Photo JCS.

Eleocharis mamillata subsp. *austriaca*. Ribblesdale, Peak Distict, England. Photo RVL.

Perennial helophyte. Rhizome weak, slender, creeping, underground, 1.5–3 mm in diameter, light brown, bare, with only a short sheathing scale at each node. Roots pale- to greyish brown, up to 1.5 mm diameter.

Stem: 1 to several from each rhizome node, 10–50 cm long, 1.3–2.5 mm in diameter, weak, easily cracked, terete or slightly compressed towards the spikelet, light green, somewhat translucent, with 8–16 vascular bundles forming distinct ridges when the stem is dried. Stomata guard cells longer than the subsidiary cells. Emergent stems fertile or sterile. Submerged stems mostly sterile and terminating in a few small glumes.

Leaves: With 2 yellowish-brown to reddish bladeless sheaths at the base of the stem. The sheaths are more or less transversely truncate.

Inflorescence: A single terminal spikelet 5–20 mm long, oblong ovate to cylindrical, brownish. Glumes 3–3.5 mm long, dark greyish brown with a distinct midrib and more or less wide hyaline margins. The 2 lowest glumes are normally empty and encircle about half of the stem.

Flowers: Bisexual. Stigmas 2. Stamens 3, anthers 1.2–1.8 mm long.

Flowering: June-August.

Biology: *Eleocharis mamillata* is an early colonist often occurring patchily but not for long in any one place. It reproduces mainly by seeds.

Distribution within the region:
Native. Locally frequent throughout much of the region but known from only a handful of sites in Britain.

Characteristics and similar species:
Similar in habit to **155** ***E. palustris*** and **157** ***E. uniglumis***, but it differs from both by the bright green, weak, somewhat translucent, easily cracked stems and perianth bristles with relatively longer teeth. The stomata guard cells are longer than the subsidiary cells, and the stem has only one layer of palisade cells and 8–16 vascular bundles in cross section.

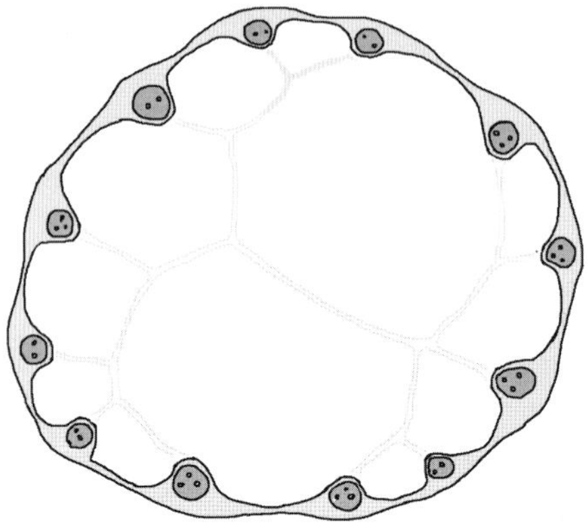

Eleocharis mamillata. Stem cross section with 12 vascular bundles. The bundles become obvious as ridges when the stem is dried. (Source: H. Lindberg 1902.)

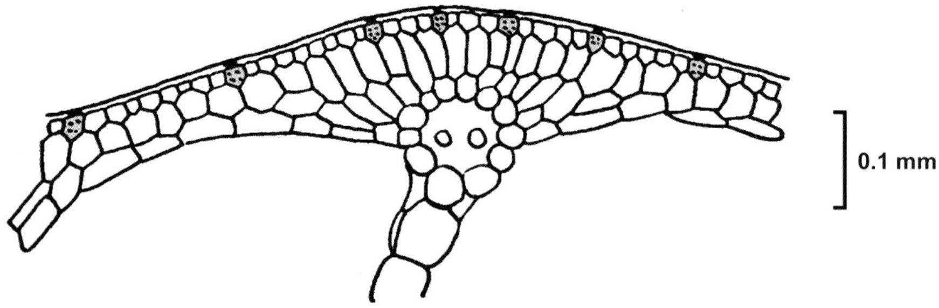

Eleocharis mamillata. Stem cross section in detail. The epidermis with fibre cells (grey and dotted) separated by 3 or more clear cells. Below the epidermis one layer of palisade cells is present. (Source: H. Lindberg 1902.)

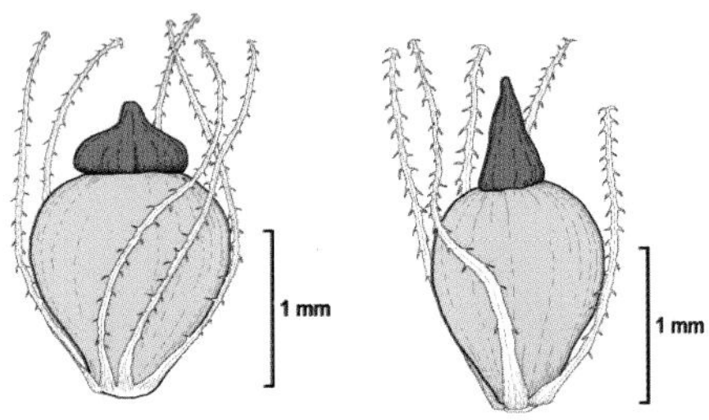

Subsp. *mamillata* Subsp. *austriaca*

The shape of the style base offers the best character for distinguishing between the two subspecies, but the number of perianth bristles can also be useful.

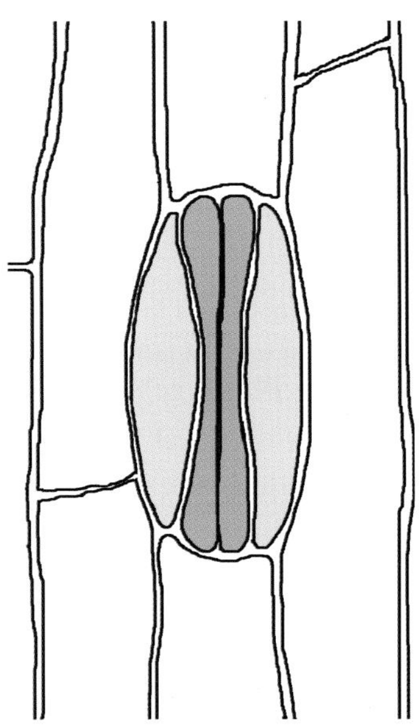

Eleocharis mamillata. In both subspecies the stomata guard cells (dark grey) are longer than the subsidiary cells (light grey).

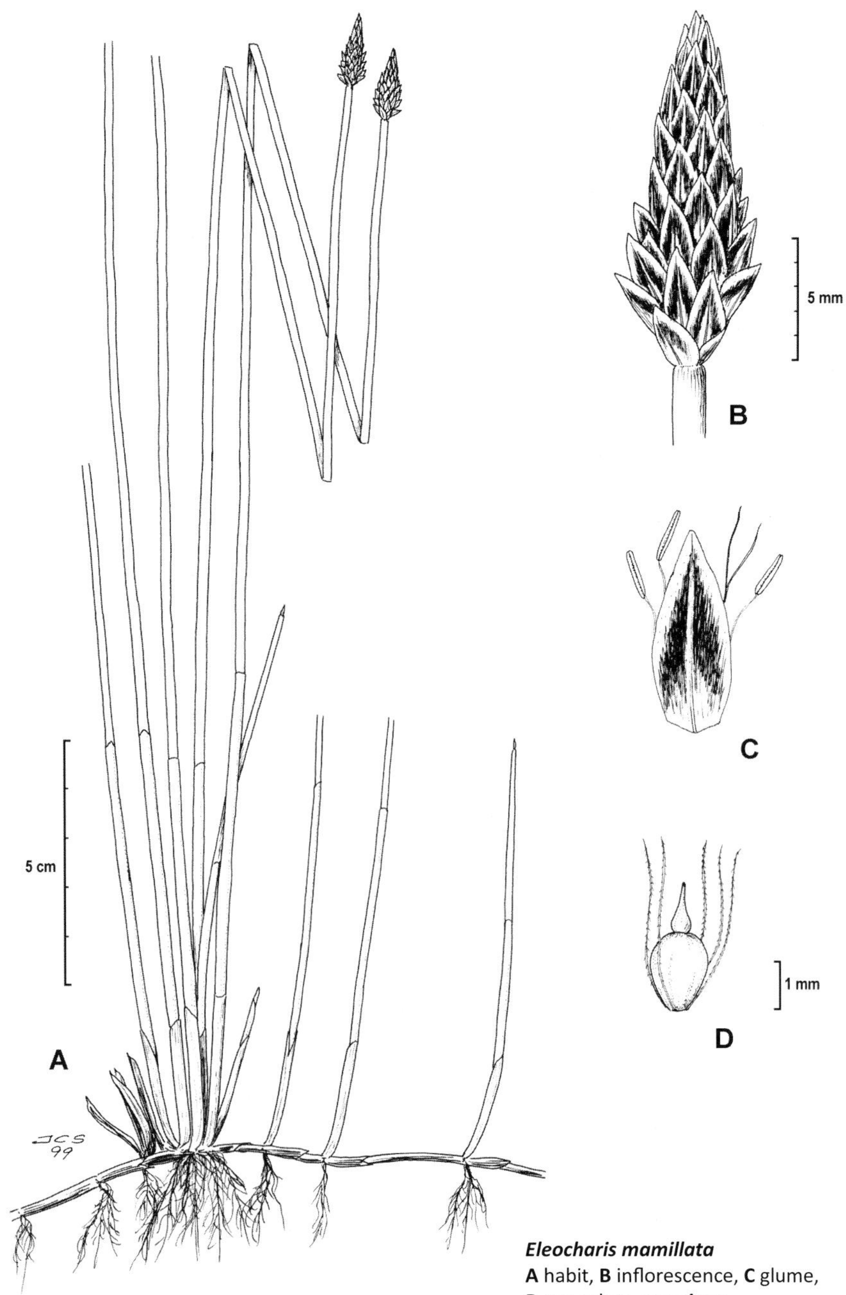

Eleocharis mamillata
A habit, **B** inflorescence, **C** glume, **D** nut subsp. ***austriaca***.

Cyperaceae

156a *Eleocharis mamillata* subsp. *mamillata*

Syn.:
DK: - **N:** - **S:** - **FIN:** - **IS:** -
GB: - **NL:** - **F:** -
D: - **CZ:** - **PL:** -
EST: -**LV:** - **LT:** - **Ru:** -

Eleocharis mamillata subsp. *mamillata*. Nut with stylebase and bristles. Spikelet. Photos JCS.

Fruits: Nuts 1.2–1.4(–1.5) mm long (excluding style base), yellowish brown to brown, obovoid to almost spherical, slightly compressed, somewhat shiny, with a pattern of very fine hollows. Style base mammiform, wider than long, 0.3–0.5 mm long, (0.4–)0.5–0.7 (–0.8) mm wide, distinctly constricted at junction with the nut. Perianth segments 5–6(–8), with long, backward-curved teeth, extending above the style base and often even longer.

Habitats: A weak competitor colonising newly established ponds, ditches and lakes in non-calcareous, more or less oligotrophic areas.

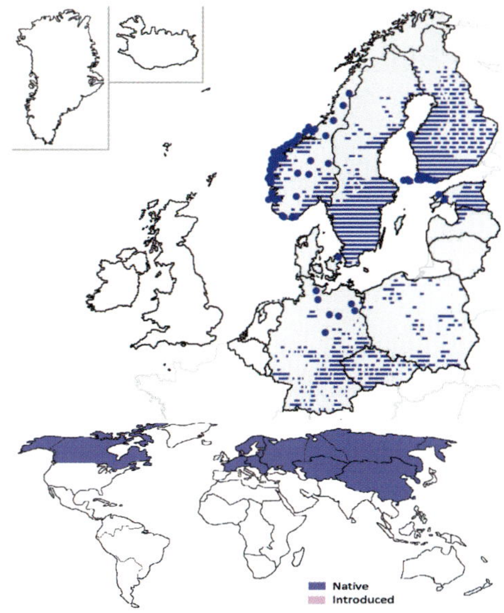

Distribution within the region: Widespread and abundant in parts of Norway, Sweden, Finland, Germany, the Czech Republic, Poland and Estonia. Rare in Denmark and Latvia.

Cyperaceae

156b *Eleocharis mamillata* subsp. *austriaca* (Hayek) Strandh.

Syn.: *E. austriaca* Hayek; *E. benedicta* Beauverd; *E. palustris* subsp. *austriaca* (Hayek) Podp.
DK: - **N:** - **S:** - **FIN:** - **IS:** -
GB: Northern Spike-Rush **NL:** - **F:** -
D: - **CZ:** - **PL:** -
EST: -**LV:** - **LT:** - **Ru:** -

Eleocharis mamillata subsp. *austriaca*. Nut with style base and bristles. Spikelet. Photo JCS and RVL.

Fruits: Nuts 1.0–1.5 mm long (excluding style base), yellowish brown to dark brown or reddish brown, obovoid to broadly elliptical, slightly compressed, somewhat shiny, with a pattern of very fine hollows. Style base narrowly conical, longer than wide, 0.5–0.8 mm long, 0.3–0.5 mm wide, with a narrow constriction at the junction with the nut. Perianth segments (4–)5(–6), with long, backward-curved teeth, extending above the style base and often even longer.

Habitats: A weak competitor colonising open habitats mainly in mountainous, calcareous areas. In shallow backwaters, oxbow pools, riverbanks, marshes and lakeshores.

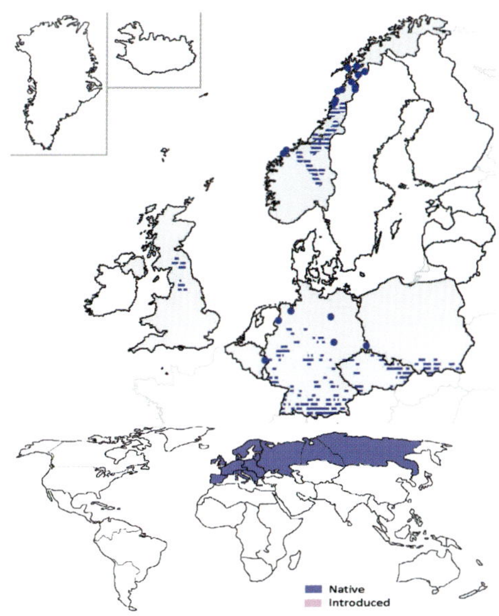

Distribution within the region: Scattered to rather common in Norway, Germany, the Czech Republic and Poland, rare in Belgium, very local in Britain.

Cyperaceae

157 *Eleocharis uniglumis* (Link) Schultes

Syn.: *E. palustris* subsp. *uniglumis* (Link) Hartm.; *Scirpus uniglumis* Link; *S. palustris* subsp. *uniglumis* (Link) Syme
DK: Enskællet Sumpstrå **N:** Fjøresivaks **S:** Vanlig agnsäv **FIN:** Meriluikka **IS:** Vætuskúfur
GB: Slender Spike-rush **NL:** Slanke vaterbies **F:** Souchet à une glume
D: Einspelzige Sumpfsimse **CZ:** Bahnička jednoplevá **PL:** Ponikło jednoprzysadkowe
EST: Soomusalss **LV:** Vienplēksnes pameldrs **LT:** Pievinis duonis **Ru:** Ситняк одночешуйный

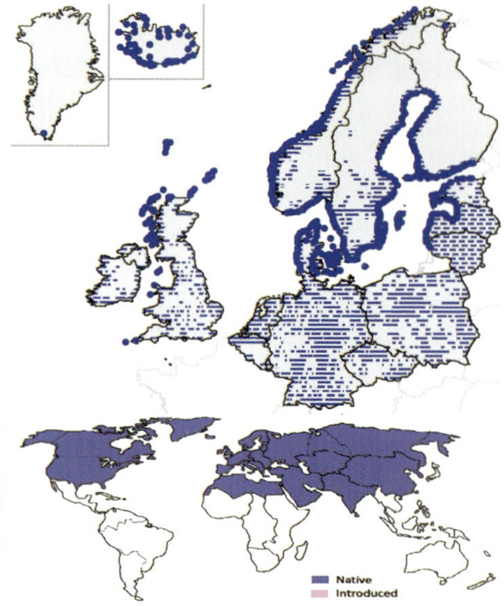

Eleocharis uniglumis growing on moist to wet substrate in a marsh at Mariager Fjord, Denmark. Photo JCS.

Perennial helophyte. Rhizome long, creeping, 2–6 mm diameter, light- to blackish brown, often with purplish tones, bare, with only a short sheathing scale at each node. Roots pale brown to dark greyish brown, up to 1.5 mm wide.

Stem: One to several from each rhizome node, 5–40(–70) cm long, 0.5–2 mm in diameter in the middle, firm, terete, somewhat shiny, dark green to greyish green, vascular bundles not forming distinct ridges when the stem is dried. Stomata cells about 57 µm, the guard cells shorter than subsidiary cells (see figure for **155 *E. palustris*** p. 363 bottom left).

Leaves: With 2 brown to purplish-red bladeless sheaths at the base of the stem. The sheaths are more or less transversely truncate.

Inflorescence: A single terminal spikelet 5–20 mm long, oblong ovoid to cylindrical, brownish to blackish brown. Glumes somewhat shiny, purplish brown to dark reddish brown, with a paler brown midrib and narrow hyaline margins. The lowest glumes normally empty and encircling the stem, with a wide, whitish hyaline margin.

Flowers: Bisexual. Stigmas 2. Stamens 3, anthers 2.0–2.7 mm long. Pollen grains about 46 x 35 µm.

Fruits: Nuts 1.4–1.8 mm long (excluding style base), obovoid, laterally compressed, somewhat shiny, yellowish brown to brownish, with a pattern of relatively large, roundish hollows. Perianth bristles absent or up to 4, rudimentary, slender, sometimes extending to the style base. Style base about 0.6 mm long, conical.

Flowering: June-July.

Biology: Reproduces by seed, as well as vegetative reproduction by division of the rhizome.

Habitats: Mostly coastal in dune slacks, brackish pools, salt marshes and tidal flats. Inland habitats include eutrophic, base-rich, wet meadows, ditches, village ponds, and calcareous marshes.

Distribution within the region:
Native. Widespread and abundant throughout much of the region, mostly coastal in the Nordic part. Rare in Greenland.

Characteristics and similar species:
Similar in many ways to **155 *E. palustris*** but recognised by the lowest glume entirely encircling the stem and by the often absent or rudimentary perianth bristles - at most reaching the style base.

158 *E. multicaulis* lacks the long, creeping rhizome, has 3 stigmas, trigonous nut and lowest glume usually with a distinct notch.

Hybrids: Supposed hybrids with *E. palustris* are intermediate and sometimes present with the parents.

157a *Eleocharis uniglumis* subsp. ***sterneri*** Strandhede

Differ in longer stomata cells - mean length 70 µm, and pollen grains about 55 x 38 µm.

Only known from Öland and Gotland, Sweden.

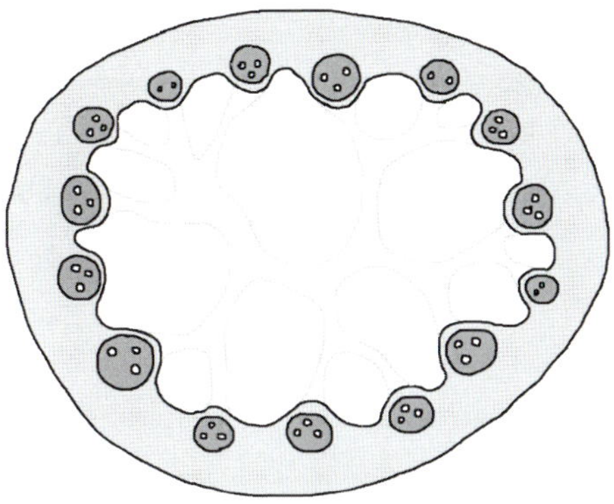

Eleocharis uniglumis. Stem cross section with vascular bundles. The bundles do not become very obvious as ridges when the stem is dried. (Source: H. Lindberg 1902).

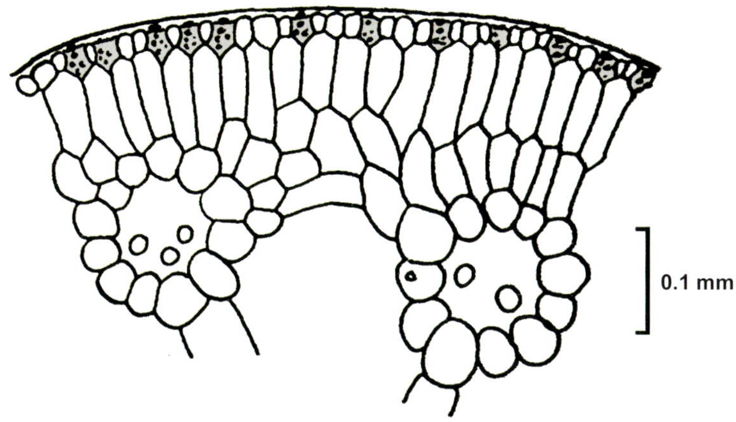

Eleocharis uniglumis. Stem cross section in detail. The epidermis with fibre cells (grey and dotted) more or less alternating with clear cells. Below the epidermis there are two layers of palisade cells. (Source: H. Lindberg 1902).

Eleocharis uniglumis. Spikelet. Photo JCS.

Eleocharis uniglumis. Nuts with style base and bristles/rudiments. Photo JCS.

Eleocharis uniglumis
A, **A1** habit, **B** spikelet, **B1** lowest glume, **B2** glume with flower, **C** nut with rudimentary perianth bristles and cross section of nut below.

Cyperaceae

158 *Eleocharis multicaulis* (Sm.) Sm.

Syn.: *Scirpus multicaulis* Sm.
DK: Mangestænglet Sumpstrå **N:** Buntsivaks **S:** Dysäv **FIN:** - **IS:** -
GB: Many-stalked Spike-rush **NL:** Veelstengelige waterbies **F:** Héléocharis à tiges nombreuses
D: Vielstängelige Sumpfsimse **CZ:** - **PL:** Ponikło wielołodygowe
EST: - **LV:** Daudzstublāju pameldrs **LT:** - **Ru:** Ситняк многостебельный

Red dots - *Eleocharis carniolica*

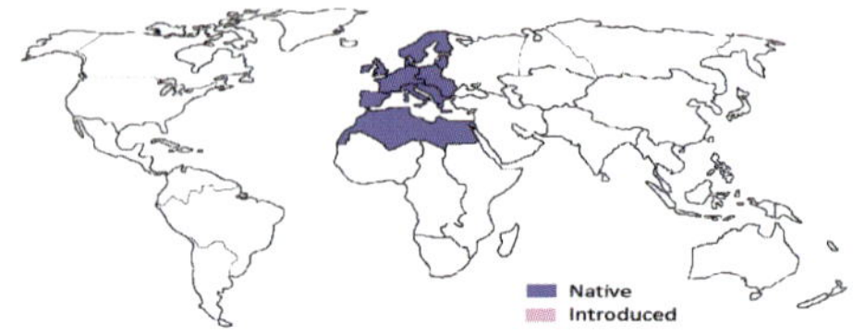

Eleocharis multicaulis in shallow water by an oligotrophic lake. Vangså, Thy, Denmark. Photo JCS.

Perennial helophyte. Rhizome short, creeping, robust, densely branched. Roots whitish to pale brown, up to 1 mm thick.

Stem: Usually several from each rhizome node, 10–30(–50) cm long, slender, rigid, terete, about 1 mm diameter, erect to descending or recurved, yellowish green to green, striate. Some stems sterile and terminating in a few small glumes.

Leaves: With pale brown to yellowish-green or whitish, sometimes purplish, bladeless sheaths at the base of the stem. The sheaths are obliquely truncate, usually with a short point.

Inflorescence: A single terminal spikelet, 4–15 mm long, oblong ovate to cylindrical, brownish, with 10–30 flowers. Glumes acute to obtuse at apex, reddish brown to dark brown with greenish to yellowish midrib and brownish scarious margins. The lowest glume empty, about 1/4 of the length of the spikelet, usually with a prominent notch at the apex and not completely encircling the stem at the base.

Flowers: Bisexual. Stigmas 3. Stamens 3, anthers 1.8–2.5(–2.8) mm long.

Fruits: Nuts 1.7–2.0 mm long (including style base), obovate, triangular in cross section, dull to slightly shiny, greenish to dark brown. Perianth bristles 6, longer than the nut. Style base triangular, trigonous, about 0.5 mm long, persistent, not confluent with the nut.

Flowering: June-August.

Eleocharis multicaulis with *Littorella uniflora* on inundated soil. New plants are developed from the floating inflorescences. Hvidbjerg Klitplantage, Thy, Denmark. Photo JCS.

Biology: Wind pollinated. Plants growing in water or on very moist soil often develop pseudoviviparous shoots from the lower glumes in the inflorescence of collapsed or floating stems. Wintergreen.

Habitats: In soft, acidic, oligotrophic water and periodically inundated sand and peat. Coastal dune slacks, lake margins, wet heaths, floodplains and upland mires.

Distribution within the region:
Native. Scattered to locally frequent in southern and western parts of the region, from Ireland north and east to southern Norway and Sweden and Poland where it is rare.

Characteristics and similar species:
Eleocharis multicaulis can be recognised by the dense tufts without underground stolons, slender, often recurved stems, 3 stigmas, triangular nuts, lowest glume often with a distinct, apical notch and upper sheath with a short point.

157 ***E. uniglumis*** has underground stolons, 2 stigmas, flat nuts, and lower glumes without an apical notch.

162 ***E. quinqueflora*** differs in the 2 lowest glumes being more than half as long as the spikelet, larger nuts, fewer stems and upper sheath without a short point.

152-154 ***E. ovata, obtusa*** and ***engelmannii*** are annuals with biconvex, somewhat compressed nuts (not triangular in cross section).

Trichophorum species are similar in habit but have a short blade on the uppermost basal sheath.

159 *Eleocharis carniolica*
W.D.J.Koch

Similar in habit, but differs in having compressed, biconvex nuts with acute margins and usually 2 (rarely 3) stigmas. It grows in periodically flooded places by rivers, pools and reservoirs, as well as in wet and marshy meadows.

Distribution within the region:
Only known from a few locations in southeast Poland.

Eleocharis multicaulis. Pseudoviviparous shoots from the lower glumes in the inflorescence.

Eleocharis multicaulis. The lowest glume is empty and often with a prominent apical notch. Photos JCS.

Eleocharis multicaulis
A habit, **B** inflorescence, **B1** lowest glume with notch at apex, **B2** flower and glume, **C** nut with perianth bristles, **D** rooting shoots from the lower glumes.

Cyperaceae

160 *Eleocharis acicularis* (L.) Roemer & Schultes

Syn.: *Scirpus acicularis* L.

DK: Nåle-Sumpstrå **N:** Nålesivaks **S:** Nålsäv **FIN:** Hapsiluikka **IS:** Efjuskúfur
GB: Needle Spike-rush **NL:** Naaldwaterbies **F:** Héléocharis Epingle
D: Nadelsimse **CZ:** Bahnička jehlovitá **PL:** Ponikło austriackie
EST: Nõelalss **LV:** Adatu pameldrs **LT:** Adatinis duonis **Ru:** Болотница игольчатая

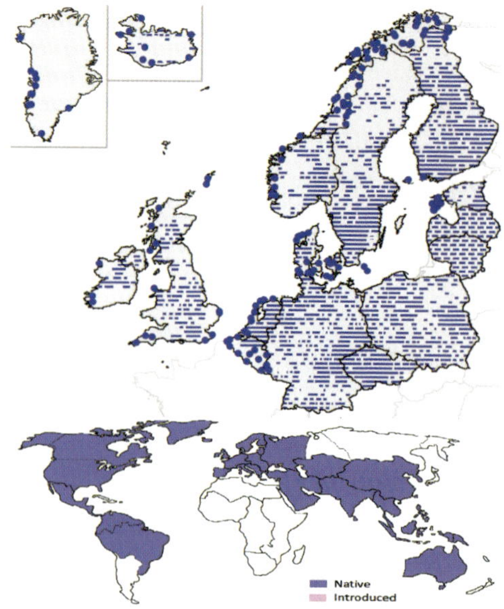

Uprooted *E. acicularis* floating in the water. Lake Hampen, Jutland, Denmark. Photo JCS.

Perennial hydrophyte or helophyte. Rhizome creeping, somewhat branched, very fine, 0.2–0.5 mm diameter, white to pale brown, naked, with a single, very thin, shortly sheathing scale at each node. Roots very thin, white to pale brown.

Stem: One to several from each rhizome node, 2–15(–50) cm long, very fine, 0.1–0.5 mm thick, erect, yellowish green to green, septate, sulcate, 4-angular. Emergent stems fertile or sterile. Submerged stems usually sterile, 2–20(–30) cm long, terete, yellowish green, somewhat translucent. Epidermal cells long and narrow, with straight walls.

Leaves: With 2 thin, bladeless sheaths at the base of the stem. Often streaked with purple in terrestrial plants, greenish in submerged plants. Cauline leaves absent.

Inflorescence: A single terminal spikelet 2–4(–7) mm long, oblong ovate, greenish to brownish, with 4–10(–12) flowers. Glumes acute to obtuse, reddish brown to dark brown with green midrib and scarious margins. The lowest glume encircling the stem, up to half as long as the spikelet and normally empty.

Flowers: Bisexual. Stigmas 3–4. Stamens 3, anthers 0.7–1.3 mm long.

Fruits: Nuts 0.8–1.0 mm long (including style base), fusiform to oblong obovate, yellowish white to pale brown, with numerous, fine longitudinal ribs and oblong epidermal cells arranged in dense latitudinal lines. Perianth bristles 0–4, caducous, shorter than the nut,

Eleocharis acicularis growing submerged in a ditch in Wachtendonk, Germany. Photo KvdW.

Formation of *E. acicularis* at 20-30 cm depth. Førby Sø, Thy, Denmark. Photo JCS.

Flowering *E. acicularis*. Førby Sø, Thy, Denmark. Photos JCS.

often very short to rudimentary. Style base narrow, flat conical triangular, 0.1–0.2 mm long, persistent, not confluent with the nut.

Flowering: July-October.

Biology: Emergent plants produce abundant flowers, whereas fully submerged plants are always sterile. Vegetative propagation by division of the rhizome. Overwinters as rhizomes in the substrate, sometimes wintergreen.

Habitats: In oligo-, meso- or eutrophic, mildly acidic to alkaline water. On moist sandy, gravelly or clay substrate from the shoreline to a depth of 1(3) metres in pools, ditches, streams and lakes. Locally even in weakly brackish water in pools in salt marshes or coastal inlets, sometimes associated with **161 *E. parvula***.

Distribution within the region: Native. Widespread and locally frequent throughout the region.

Characteristics and similar species: Vegetative plants may be mistaken for **161 *E. parvula***, which differs in the stems with shorter and wider epidermal cells with wavy walls, and by the capillary stolons with whitish to pale brown tubers 2–3 mm long, comma-like and easily detached.

Vegetative plants of **162 *E. quinqueflorus*** have stems without septae, a short, brownish rhizome and buds at the ends of stolons about 1 cm long.

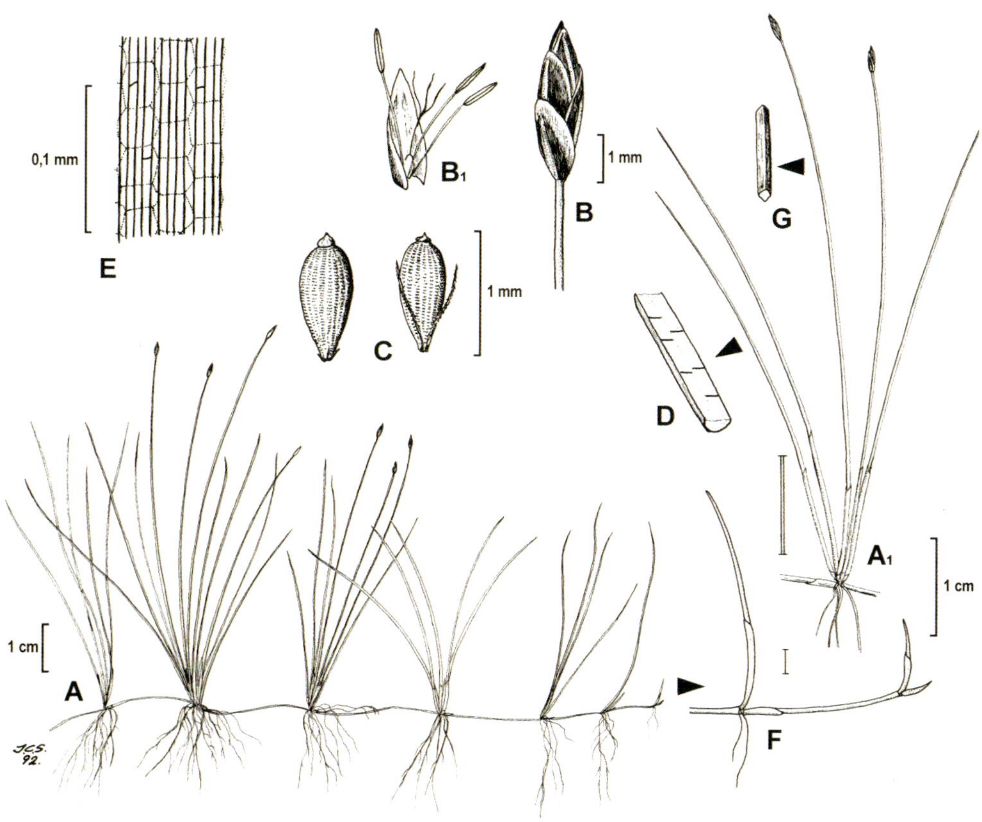

Eleocharis acicularis
A, **A1** habit, **B** inflorescence, **B1** flower and glume, **C** nut with or without perianth bristles, **D** longitudinal section of stem, **E** long, rectangular epidermal cells, **F** apical part of rhizome, **G** section of stem.

Cyperaceae

161 *Eleocharis parvula* (Roemer & Schultes) Bluff, Ness & Schauer

Syn.: *Scirpus parvulus* Roemer & Schultes
DK: Lav Sumpstrå **N:** Dvergsivaks **S:** Dvärgsäv **FIN:** Pikkuluikka **IS:** -
GB: Dwarf Spike-rush **NL:** - **F:** Petit Souchet
D: Zwerg-Sumpfsimse **CZ:** - **PL:** Ponikło maleńkie
EST: Väike alss **LV:** Sīkais pameldrs **LT:** Smulkusis duonis **Ru:** Ситняк маленький

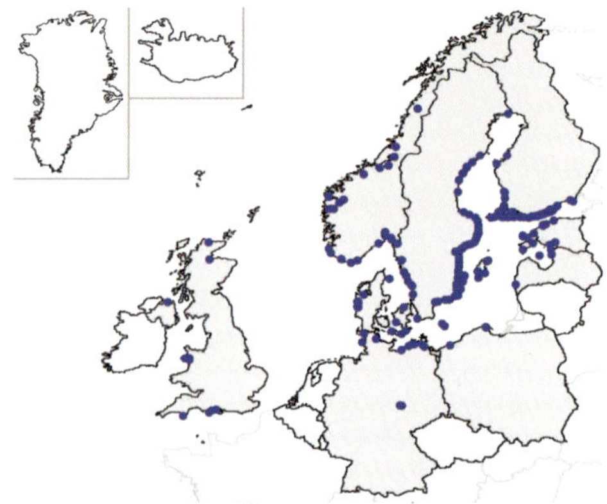

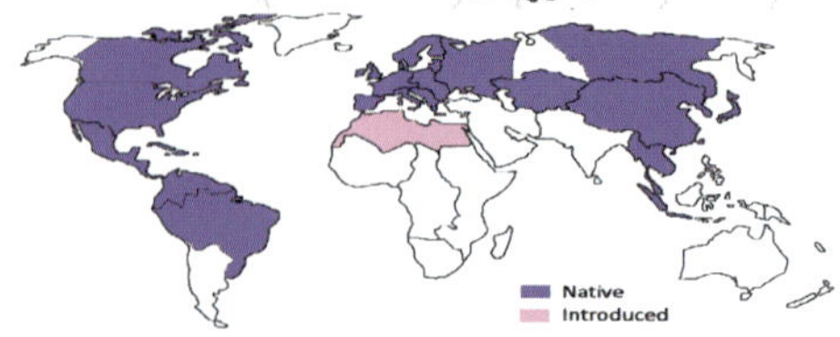

Flowering *E. parvula*. Nissum Fjord, W-Jutland, Denmark. Photo JCS.

Perennial hydrophyte or helophyte. Rhizome creeping, sparsely branched, capillary, internodes 0.5–5 cm long, white to pale brown, naked, with a single, very thin, shortly sheathing scale at each node, terminating in a 2–3 mm long, crooked, whitish to brownish tuber from late summer.

Stem: One to several from each node, 2–6 cm long, 0.3–0.5 mm in diameter, green, paler towards the base, somewhat translucent, septate, terete. Emergent stems fertile or sterile. Submerged stems sterile, evenly tapering towards the apex and usually somewhat curved. Epidermal cells long and narrow, with wavy walls.

Leaves: With 2 thin bladeless sheaths at the base of the stem. Often streaked with reddish in terrestrial plants, pale in submerged plants. Cauline leaves absent.

Inflorescence: A single terminal spikelet 2.0–3.5 mm long, oblong ovate, greenish to light brownish, with 2–6(–10) flowers. Glumes acute to obtuse, with green midrib, yellowish-green to brownish sides and scarious margins. The lowest glume encircling the stem, about half as long as the spikelet and normally empty.

Flowers: Bisexual. Stigmas 3. Stamens 3, anthers 0.9–1.2 mm long.

Fruits: Nuts about 1.1–1.4 mm long (including style base), obovate to broadly fusiform, triangular in cross section, smooth, shining, yellowish brown. Perianth bristles (2–5–)6, equal to or longer than the nut. Style base triangular, 0.2–0.3 mm long, persistent, confluent with the nut.

Flowering: July-August.

Biology: Emergent plants on wet substrate sometimes flower, whereas fully submerged plants are always sterile. *E. parvula* forms small summer-green tufts which last only a single season; it overwinters as tubers developed at the apex of stolons from late summer. Vegetative reproduction is by tubers and rhizome fragmentation.

Habitats: On muddy and sandy substrates in brackish water from the shoreline to a depth of 1 metre. Brackish lakes, sea inlets, bays, lagoons, estuaries.

Distribution within the region:
Native. Scattered to rather rare along the coasts of Northern Ireland, Britain, Norway, Sweden, Finland, Denmark, Poland, Estonia, Latvia and Germany where an inland population has also been recorded.

Eleocharis parvula. Submerged plants are always sterile and characterised by the rows of small tufts with outward-curved stems. Nees, W-Jutland, Denmark. Photo JCS.

Eleocharis parvula. Spikelets. Skanörs Ljung, Skåne, Sweden. Photo Jan-Thomas Johansson.

Eleocharis parvula. The 2-3 mm long tubers at the end of the stolons are an important character for distinguishing from *E. acicularis*. Baltic Sea, Blekinge, Sweden. Photo JCS.

Characteristics and similar species: *Eleocharis parvula* is often mistaken for **160 *E. acicularis***, which differs in the sulcate, 4-angled stem, epidermal cells with straight walls, smaller nuts and lack of tubers.

162 *E. quinqueflora* is only rarely submerged, it is taller, has stems without septa, and has tubers about 1 cm long from late in the summer.

Eleocharis parvula. The stems are septate - visible when backlit. Photo JCS.

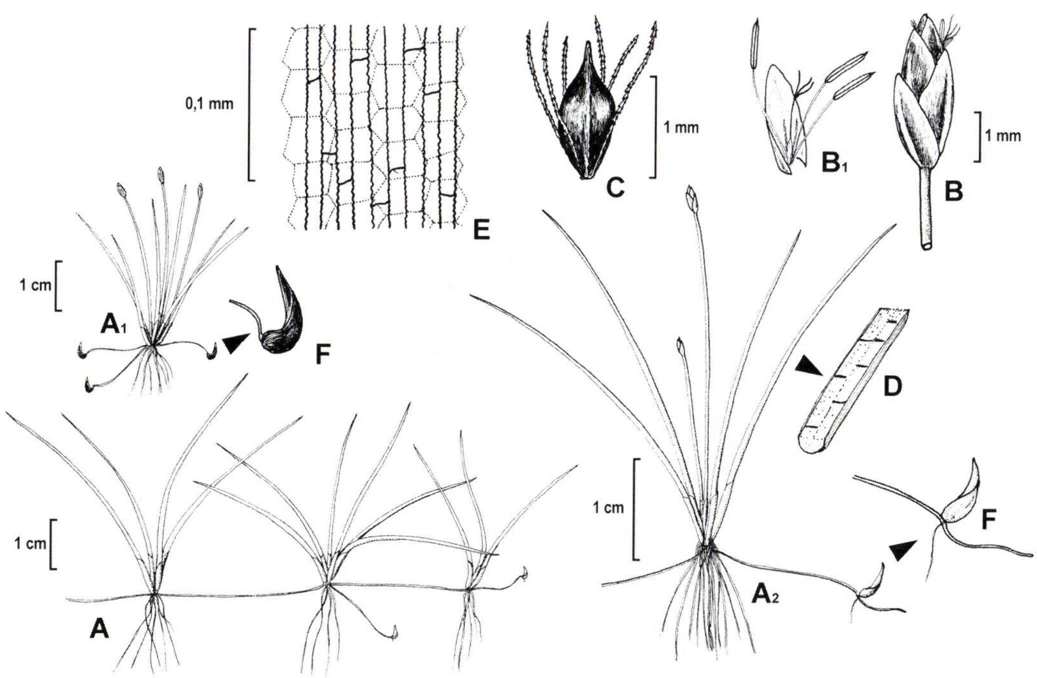

Eleocharis parvula
A habit sterile plants, **A1-A2** habit flowering plant, **B** inflorescence, **B1** flower and glume, **C** nut with perianth bristles, **D** longitudinal section of stem, **E** epidermal cells, **F** tuber.

Cyperaceae

162 *Eleocharis quinqueflora* (Hartmann) O.Schwarz

Syn.: *Scirpus quinqueflorus* Hartmann; *S. pauciflorus* Lightf.
DK: Fåblomstret Sumpstrå **N:** Småsivaks **S:** Tagelsäv **FIN:** Jouhiluikka **IS:** Fitjaskúfur
GB: Few-flowered Spike-rush **NL:** Armbloemige waterbies **F:** Souchet pauciflore
D: Wenigblütige Sumpfsimse **CZ:** Bahnička chudokvětá **PL:** Ponikło skąpokwiatowe
EST: Õievähene alss **LV:** Mazziedu pameldrs **LT:** Ganyklinis duonis **Ru:** Ситняк малоцветковый

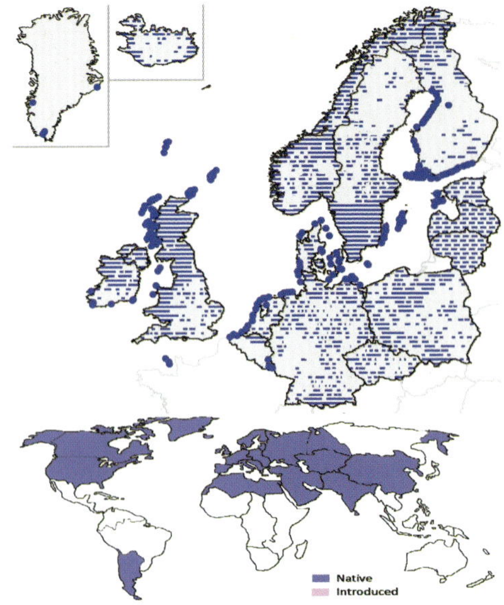

Eleocharis quinqueflora in shallow, calcareous water. Small brook at Stora Alvaret, Öland, Sweden. Photo JCS.

Perennial helophyte. Rhizome short, brownish. In late summer, underground stolons are developed 2–10 cm long, about 1 mm in diameter with an apical bud of 4–6 young shoots, about 1 cm long, somewhat acute and brownish. Roots reddish brown to dark brown, 0.1–0.5 mm in diameter.

Stem: (3–)7–20(–40) cm long, erect, (0.3–)0.5–1.0 mm in diameter, more or less terete, striate, green to greyish green, not translucent, without septa, up to 10 together in small tufts. Some stems sterile and terminating in a few small glumes.

Leaves: With 2 greenish to pale brown or reddish-brown bladeless sheaths at the base of the stem. The lower sheath obliquely truncate, while the upper one is almost transversely truncate, often with a dark border.

Inflorescence: Ovate, greenish-brown to brown spikelet 4–8(–10) mm long, with 5–8 flowers. Glumes reddish brown to dark brown with green to brownish midrib and broad, scarious margin. The lowest glume encircling the stem, more than half as long as the spikelet and normally enclosing a flower.

Flowers: Bisexual. Stigmas 3. Stamens 3, anthers 1.8–2.5 mm long.

Fruits: Nuts 2.0–2.9 mm long (including style base), oblong obovate, triangular in cross section, shiny, black when fresh, becoming light grey when dry. Epidermal cells forming a light reticulate pattern on top of the darker cells. Perianth bristles 4–6, as long as or longer than the nut, sometimes rudimentary. Style base triangular about 0.5 mm long, black at the top, persistent, confluent with the nut.

Flowering: June-August.

Biology: Vegetative propagation and overwintering as buds of 4–6 young shoots at the tips of stolons.

Eleocharis quinqueflora. The lowest glume is more than half as long as the spikelet, and very dark brown compared to similar species. Skarpa Alby Alvar, Öland, Sweden. Photo JCS.

Eleocharis quinqueflora. Flowering plants at the shore of a small lake. Jutland, Denmark. Photo JCS.

Habitats: In flushes, springs and shallow standing water over calcareous substrates, fens, dune depressions with seashell fragments, sandy shores of lakes - also in acidic mires in heathland.

Distribution within the region:
Native. Widespread and locally abundant throughout the region, but rare in Greenland.

Characteristics and similar species:
Eleocharis quinqueflora may be mistaken for **160 *E. acicularis,*** which has 4-angular, septate stems and creeping rhizome. **158 *E. multicaulis*** is a more robust plant with numerous stems from the rhizome and differs further in the lower glume being emarginate and less than half the length of the spikelet.

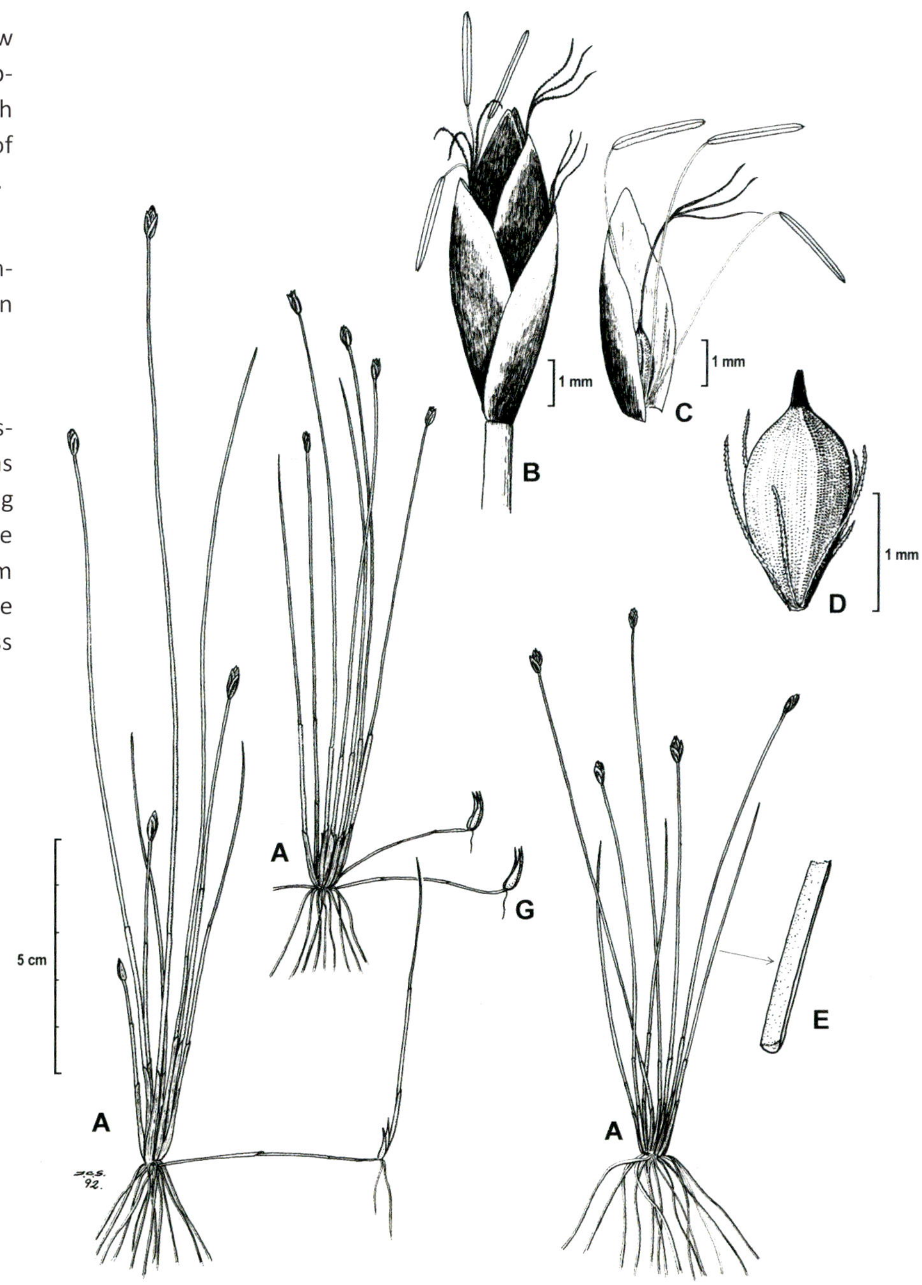

Eleocharis quinqueflora
A habit, **B** inflorescence, **C** flower and glume, **D** nut with perianth bristles, **E** longitudinal section of stem, **G** bud of young shoots.

Cyperaceae

163 *Isolepis fluitans* (L.) R.Br.

Syn.: *Scirpus fluitans* L.; *Eleogiton fluitans* (L.)Link
DK: Flydende Kogleaks **N:** Flytesivaks **S:** Flytsäv **FIN:** - **IS:** -
GB: Floating Club-rush **NL:** - **F:** -
D: Flutende Schuppensimse **CZ:** - **PL:** -
EST: -**LV:** - **LT:** - **Ru:** -

Flowering *I. fluitans*. Grubevande, Torup Klitplantage. Thy, Denmark. Photo JCS.

Perennial hydrophyte or helophyte with long shoots rooting at the nodes. Stem and rhizome not clearly differentiated, the latter though with shorter internodes.

Stem: 5–40(–100) cm long, elongate, usually repeatedly branched, solid, somewhat compressed, yellowish green, striate. Plants creeping on moist substrate have short internodes and short branches. Floating plants form shoots up to 40 cm long with elongated branches and spreading leaves. Permanently submerged plants usually have long internodes.

Leaves: 2–10 cm long, short and often recurved in terrestrial forms, otherwise more or less straight, up to 1 mm wide, narrowly linear, flat or slightly crescent shaped in section near the base, bright green, smooth. Sheaths 3–25 mm long, yellowish green to brownish and sometimes purplish red at the base. Ligule absent.

Inflorescence: Spikelets terminal on peduncles up 10 cm long arising from the stem or branches, 2–5 mm long, ovoid, greenish, with 3–8 flowers. Bracts glume-like.
Glumes 1.5–3.0 mm long, with several ribs, ovate, obtuse to acute, bright green, green in the middle and with hyaline, sometimes reddish or purplish-tinged margins.

Flowers: Bisexual, with 2(–3) stigmas and 2–3 stamens. Anthers 0.5–1.6 mm long.

Fruits: Nuts 0.6–1.0 mm long, obovoid, greyish brown. Beak short. Perianth bristles absent.

Flowering: July-August.

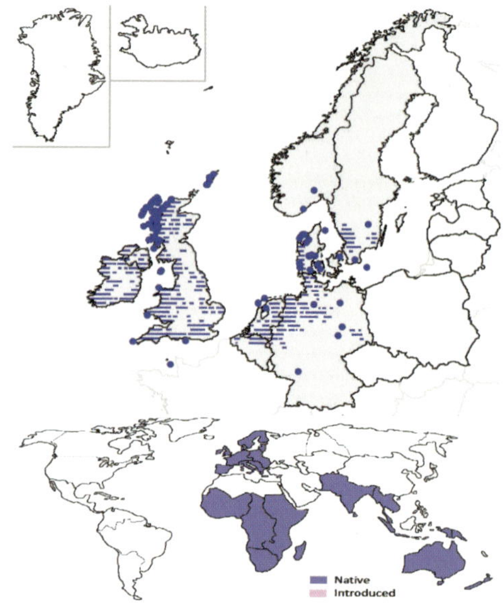

Biology: Nuts dispersed by water. Vegetative propagation by division of the rhizome. Not wintergreen.

Habitats: In acidic to mildly alkaline, soft, oligotrophic water. On moist, sandy or silty substrate from the drawdown zone of lakes and pools and floating or submerged further out to a depth of several metres. Also in bog pools, ditches and streams - sometimes even in fast-flowing water.

Distribution within the region:
Native. Widespread to locally frequent in the west, from Ireland east to southern Scandinavia and Germany.

Characteristics and similar species:
Flowering specimens are not likely to be confused with other species. Submerged plants can be mistaken for **99**

Isolepis fluitans. Submerged plant in still water. Hatchett Pond, the New Forst, England. Photo RVL.

Isolepis fluitans. The uncombed look is characteristic for submerged plants in still water and for terrestrial plants. Pool, Elmpt, Germany. Photo KvdW.

Isolepis fluitans. Byn, W-Jutland, Denmark. Photo BM.

Stuckenia pectinate, which differs in having ligules.

137 *Juncus bulbosus* has flowers in heads and semi-terete leaves with indistinct transverse septa in rosettes from the stem nodes.

104-107 *Zannichellia* spp. have opposite leaves in pairs or 3–4 together in whorls.

108-109 *Ruppia* spp. grow in brackish or saline water, have creeping rhizomes, fine-toothed leaf apices and very different-looking inflorescences.

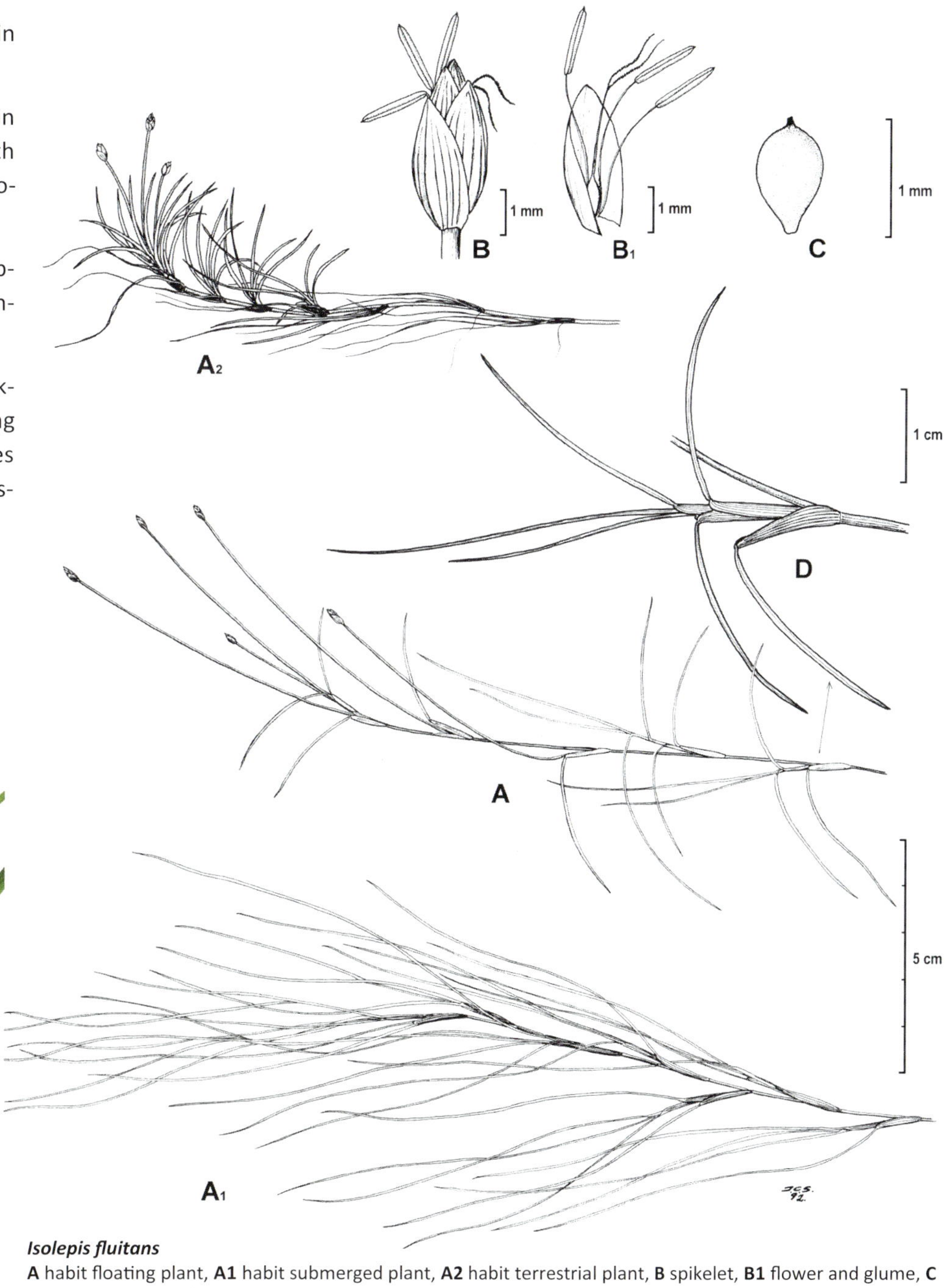

Isolepis fluitans
A habit floating plant, **A1** habit submerged plant, **A2** habit terrestrial plant, **B** spikelet, **B1** flower and glume, **C** nut, **D** leaves with sheaths.

Isolepis fluitans. Leaf sheath. Photo KvdW.

Cyperaceae

164 *Cyperus fuscus* L.

DK: Brun Fladaks **N:** Dvergak **S:** Dvärgag **FIN:** - **IS:** -
GB: Brown Galingale **NL:** Bruin cypergras **F:** Souchet brun
D: Braunes Zypergras **CZ:** šáchor hnědý **PL:** Cibora brunatna
EST: Pruun lõikhein **LV:** Brūnais dižmeldrs **LT:** Rudoji viksvuolė **Ru:** Сыть бурая

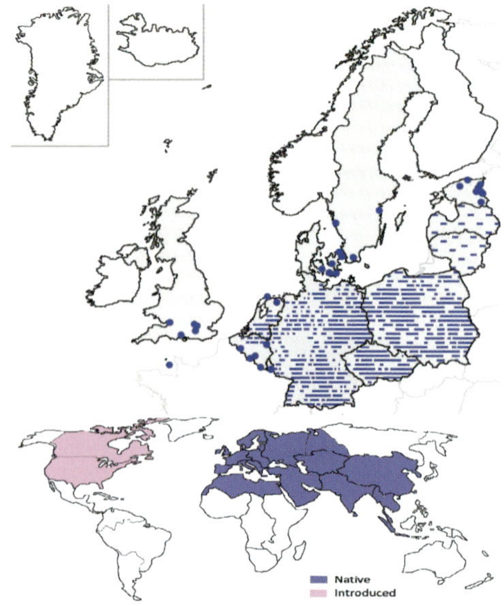

Cyperus fuscus. Fishpond, Zonhoven, Belgium. Photo KvdW.

Annual. Roots pale brownish to reddish, up to 1 mm in diameter.

Stem: 3–15(–20) cm long, spreading, ascending to erect, 1–3 mm diameter, triquetrous with concave sides, bright green, somewhat shiny, glabrous.

Leaves: Arising from the basal part of the stem. Shorter than or equal to stem, 3.5–8.3 cm long, 1–4(–5) mm wide, flat, acuminate, bright green, somewhat shiny, very rough - especially towards the trigonous tip. Hypostomous. Ligule about 1 mm long.

Inflorescence: Usually rather compact, composed of 1–4 umbel-like spikes on 0–1.5 cm long rays. Each spike with 3–12(–15) spikelets. Involucral bracts (2–)3(–4), leaf-like, of unequal length, the lower far exceeding the inflorescence. Spikelets 3–8 mm long, sessile, flat, oblong, green to dark brown, with 10–18 distichously arranged glumes. Glumes 1.0–1.5 mm long, 3-ribbed, ovate, obtuse at apex, glabrous, green with dark blackish brown sides and hyaline margin. Midrib excurrent in a minute mucro.

Flowers: Bisexual, with trifid style and 2 stamens. Anthers 0.3–0.4 mm long.

Fruits: Nuts 0.8–1.0 mm long, elliptical to obovoid, trigonous, pale brown to golden brown, abruptly tapering to a short beak. Perianth bristles absent.

Flowering: June-September.

Biology: Forming small tufts. The high number of seeds produced by a single plant maintains a very large soil seed bank, from which new plants may arise when conditions become favourable. The seeds remain viable for many years.

Habitats: On eutrophic, temporarily flooded, moist, muddy or sandy substrates in the littoral zone of ponds, fishponds, lakes and floodplains.

Distribution within the region:
Native. Rare in Britain, Denmark and Sweden, scattered to locally frequent in the Netherlands, Belgium, Luxembourg, Germany, the Czech republic, Poland and the Baltic states.

Cyperus fuscus. Inflorescence. Vombsjön, Sweden. Photo JCS.

Characteristics and similar species: *Cyperus fuscus* can be recognised by its small, tufted habit with usually 3 long, involucral bracts and numerous, flat, many-flowered spikelets gathered in 1–4 umbel-like heads.

At a distance *C. fuscus* may be mistaken for **170 *Carex bohemica***, which differs in the more dense, greenish inflorescence, with much longer glumes and fruit represented by utricles.

Cyperus fuscus. Ermelunden, Sjælland, Denmark. Photo JCS.

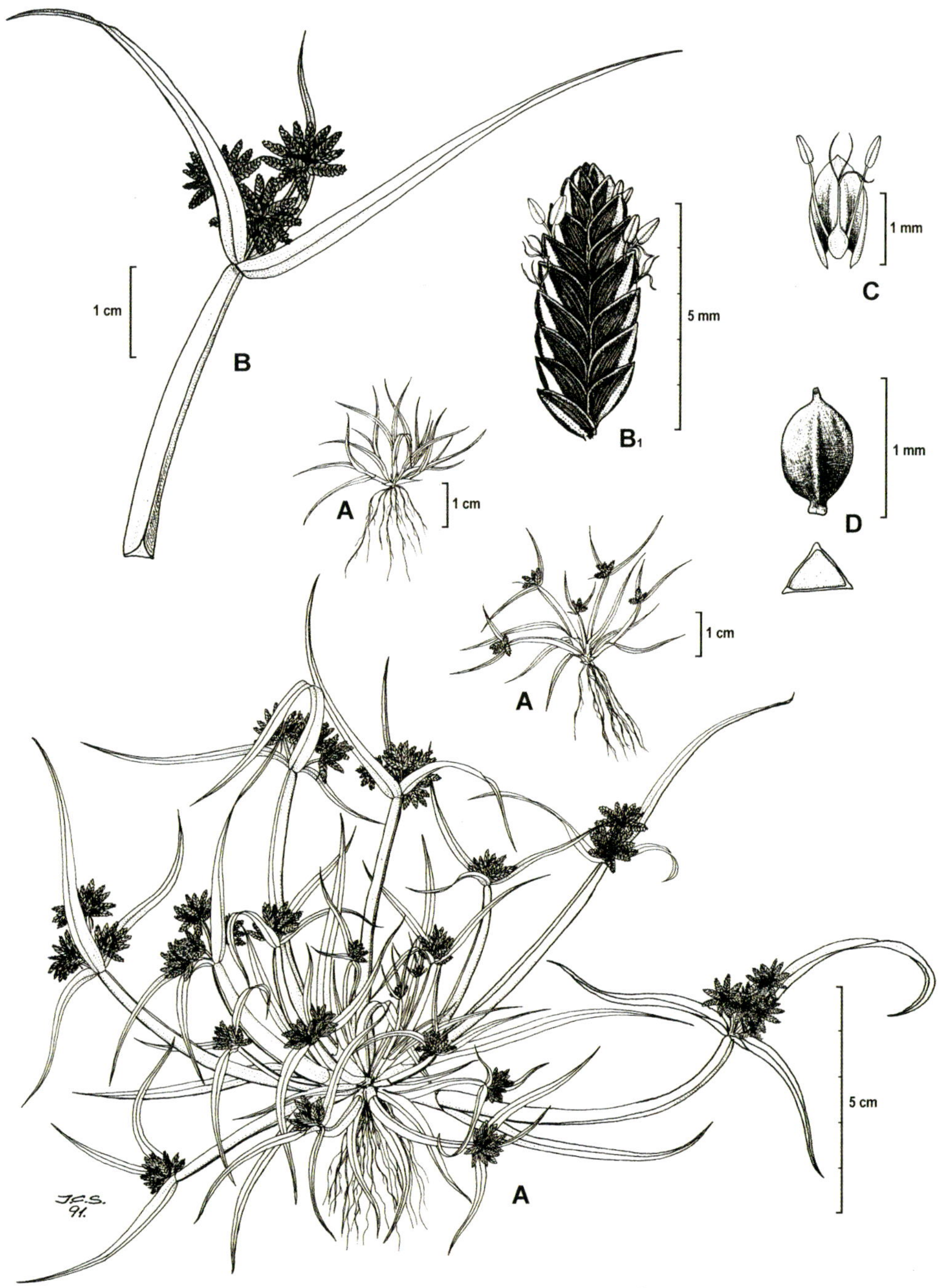

Cyperus fuscus. Fruits and glumes. Photo JCS.

Cyperus fuscus
A habit, **B** inflorescence with 3 umbel-like spikes, **B1** spikelet, **C** flower and glume, **D** nut and cross section of nut.

Cyperaceae

165 *Cyperus longus* L.

DK: Smalbladet Fladaks **N:** - **S:** Dammag **FIN:** - **IS:** -
GB: Galingale **NL:** Wilde galigaan **F:** Souchet long
D: Hohes Zypergras **CZ:** - **PL:** -
EST: - **LV:** -**LT:** - **Ru:** Сыть длинная

Cyperus longus. Bakel, Netherlands. Photo KvdW.

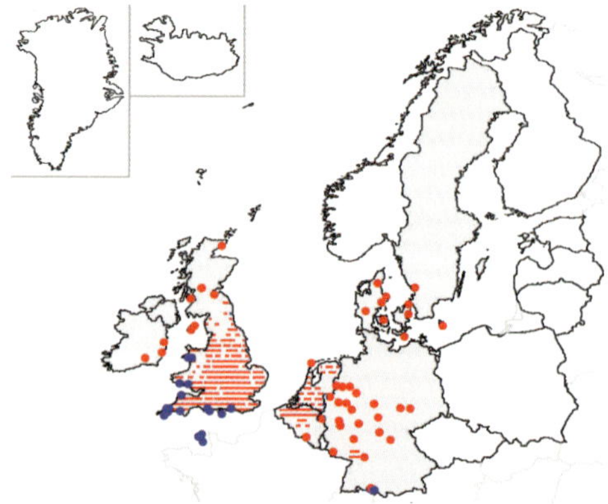

Red dots/shading indicates non-native occurrences.

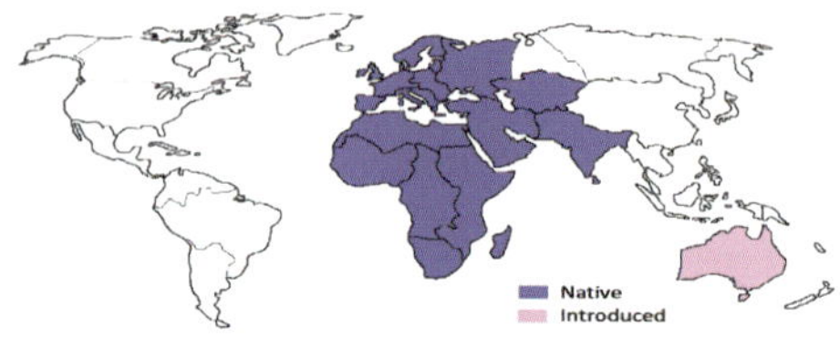

Distribution within the region:
Native in Ireland, southwestern Britain, the Netherlands, Belgium and Germany. Introduced in Denmark, Sweden and much of Britain.

Characteristics and similar species:
Cyperus longus can be recognised by the creeping rhizome, the large, compound umbel-like inflorescence subtended by up to 3–6 long, leaf-like, involucral bracts, and by the long, flat spikelets.

Vegetatively, it may be mistaken for several ***Carex*** species, but it can be rec-

Perennial helophyte. Rhizome underground, creeping, branched, c. 8 mm in diameter, covered with purplish brown sheath-like scales, splitting into fibres with age. Roots pale brownish to dark brown, up to 2 mm in diameter with numerous rootlets.

Stem: 70–130(–190) cm long, 3–5 mm diameter, stiff, erect, bluntly trigonous below, triquetrous above, green, glabrous.

Leaves: Usually shorter than stems, 4–10 mm wide, flattish keeled to M shaped in cross section, green to greyish green, somewhat shiny, scabrid - especially towards the long, acuminated tip. Hypostomous. Leaf sheaths reddish brown, splitting, forming pseudostems on vegetative shoots. Ligule absent.

Inflorescence: Diffuse, quite large, composed of 6–12 umbel-like spikes on rays up to 35 cm long. Each spike with 4–15 spikelets. Involucral bracts 3–6, leaf-like, of unequal length, the lower up to 70 cm long, far exceeding the inflorescence. Spikelets 10–25 mm long, sessile, flat, narrowly oblong, acute at apex, brownish, with 12–30 distichously arranged glumes. Glumes 2.5–3 mm long, 1.5–2 mm wide, indistinctly 5-ribbed, ovate, obtuse at apex, glabrous, midrib pale green, sides reddish brown, margins hyaline. Midrib excurrent in a minute mucro.

Flowers: Bisexual, with 3-fid style and 3 stamens. Anthers c. 1 mm long.

Fruits: Nuts 1.2–1.4 mm long, oblong ellipsoid to oblong obovoid, trigonous, brown, abruptly tapering to a short beak. Perianth bristles absent.

Flowering:
July-September.

Biology: Forming loose tufts. Propagation by seed and vegetatively by division of the rhizome.

Habitats: In acid to alkaline, mesotrophic to eutrophic water. On wet, muddy or sandy substrates along the shores of ponds, reservoirs, ditches and lakes.

Spike of *C. longus*. Aalborg, Denmark. Photo JCS.

Cyperus longus. Cultivated. Bakel, Netherlands. Photo KvdW.

ognised by the absence of ligules and presence of so-called pseudo-stems - stem-like shoots formed only of sheaths.

Several other *Cyperus* species are introduced and locally naturalised in wet areas. For identification of such, please consult special taxonomic keys.

Cyperus longus
A habit, **B** spikelet, **C** flower and glume, **D** nut, **E** cross section of leaf.

Cyperaceae

166 *Dulichium arundinaceum* (L.) Britton

DK: Vifteaks **N:** - **S:** Spärrag **FIN:** - **IS:** -
GB: - **NL:** - **F:** -
D: - **CZ:** - **PL:** -
EST: - **LV:** - **LT:** - **Ru:** -

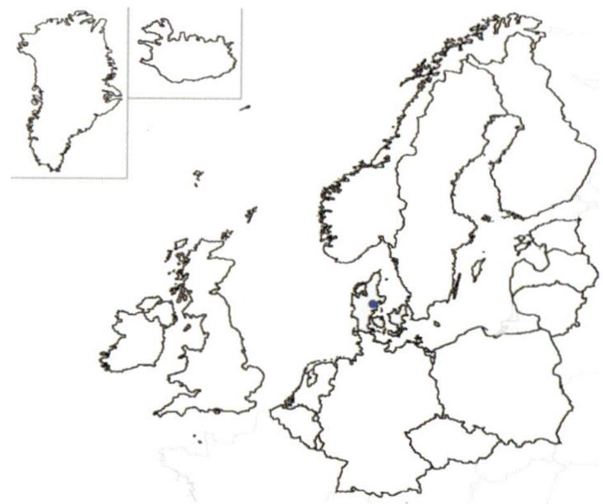

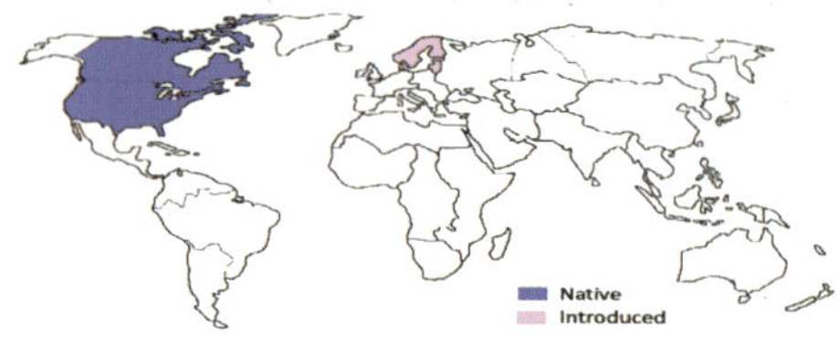

Stand of *Dulichium arundinaceum* competing with *Typha latifolia*. Stavtrup, Aarhus, Denmark. Photo JCS.

Perennial helophyte. Rhizome underground, creeping, branched, 3–6 mm in diameter, internodes 2–5 cm long, covered with dark brown sheath-like scales. Roots pale brown, up to 1 mm in diameter.

Stem: 20–120 cm long, 3–6 mm diameter, stiff, erect, terete to bluntly triangular, hollow, yellowish green to green, glabrous.

Leaves: Alternate, with 2–14 greyish to yellowish-green, bladeless sheaths in the basal half of the stem and laminar leaves gradually becoming longer upwards and soon functioning as bracts in the upper half. Lamina 3–10(–15) cm long, 3.5–8 mm wide, linear lanceolate, flat, firm, yellowish green to green, with acute apex and clasping base. The leaves appear from above to be in three vertical rows along the stem. Hypostomous. Leaf sheaths firm, inner face thinner, becoming brownish hyaline and convex at apex. Ligule 0.5–1 mm long, obtuse.

Inflorescence: Axillary spikes develop along the upper 6–30 cm of the stem. Spikes almost sessile to 5 cm long pedicellate, fan shaped, composed of 5–8 (–20) spikelets. Spikelets 1.5–3 cm long, 1.5–2.2 mm wide, sessile, somewhat compressed, narrowly oblong, acute at apex, yellowish green, reddish brown at maturity, with 12–30 distichously arranged glumes. Glumes 5–9 mm long, 1.4–2.5 mm wide, 5–9 ribbed, midrib green, sides yellow-green, margins yellow-orange hyaline.

Flowers: Bisexual. Style 3.5–6 mm long, bifid. Stamens 3. Anthers 3–4 mm long.

Fruits: Nut 2–4 mm long, excluding remnants of style, 0.8–1.1 mm wide, flat ovate in cross section, oblong. Perianth bristles 6–9, 4–7.5 mm long, pale, retrorsely scabrid.

Flowering: July-October.

Biology: Forming large stands. Propagation by seed and vegetatively by division of the rhizome. The extensive rhizomatic growth may contribute to the stabilisation of shorelines. The seeds are eaten by waterfowl.

The plant is sometimes used as an ornamental plant and sold under names such as dwarf water bamboo.

In earlier interglacial periods *Dulichium arundinaceum* had a wider distribution including Europe, but subsequently became extinct.

Dulichium arundinaceum. Flowering spike. Stavtrup, Aarhus, Denmark. Photo JCS.

Habitats: In weakly acidic to neutral, mesotrophic to eutrophic water, on wet, muddy or sandy substrates along the shoreline of ponds, lakes and streams, to a depth of less than 1 m.

Distribution within the region:
Introduced and naturalised in Denmark. Native to Canada and the United States.

Characteristics and similar species:
The alternate leaves arranged in 3 ranks along the terete or bluntly triangular stem and the axillary, fan-shaped spikes make *Dulichium* easy to recognise.

Dulichium arundinaceum. Fruit with bristles. Photo JCS.

Dulichium arundinaceum. Stavtrup, Aarhus, Denmark. Photo JCS.

Dulichium arundinaceum
A habit, **B** spike and bract, **C** flower with perianth bristles, **D** leaves positioned in 3 ranks along the stem.

Cyperaceae

167 *Cladium mariscus* L.

DK: Avneknippe **N:** Storak **S:** Ag **FIN:** Taarna **IS:** -
GB: Great Fen-sedge **NL:** Galigaan **F:** Cladium marisque
D: Schneide **CZ:** Mařice pilovitá **PL:** Kłoć wiechowata
EST: Lääne-mõõkrohi **LV:** Dižā aslape **LT:** Šakatoji ratainytė **Ru:** Меч-осока обыкновенная

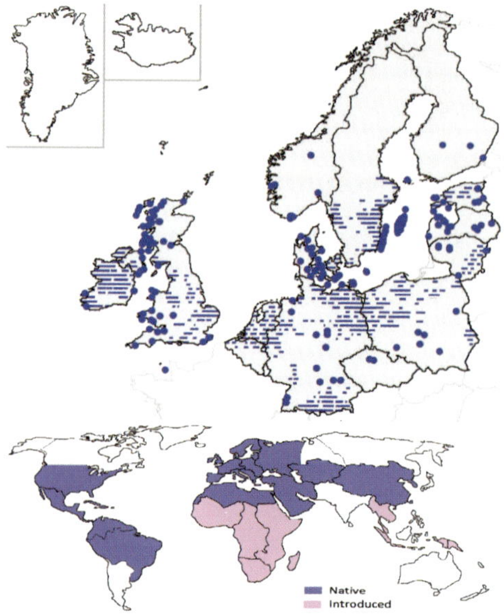

Cladium mariscus in a wet depression on calcareous ground. Öland, Sweden. Photo JCS.

Characteristics and similar species:
Even in a vegetative state *Cladium mariscus* is easily recognised by the long, firm, flat to channelled, grey-green leaves with very sharply toothed margins.

Perennial helophyte. Rhizome long, creeping, 4–8 mm in diameter, covered with dark brown to blackish-brown scales. Roots dark brown, 1–3 mm in diameter.

Stem: 60–200 cm long, 6–14 mm in diameter at the base, erect, green to greyish green, stiff, hollow, terete to bluntly trigonous in the upper part.

Leaves: Base of stem with dark brown to blackish remnants of old sheaths. Leaves up to 2 m long, 8–15 mm wide, grey green, channelled, stiff, gradually tapering to a trigonous apex, margins very rough with forward-pointed teeth. Amphistomous. Leaf sheaths reddish brown. Inner face hyaline, not splitting into fibres. Ligule absent.

Inflorescence: A leafy panicle up to 70 cm long, composed of corymbose partial inflorescences subtended by more or less distant bracts. Primary branches 4–12 cm long, secondary branches 0.5–6 cm long, terminating in subglobose clusters of many ovoid, 2–4 flowered spikelets 2–4 mm long. Involucral bracts leaf-like, the upper with reduced blade, the lower up to 90 cm long. Glumes 3.5–4 mm long, orange brown, with a prominent midrib and hyaline margins.

Flowers: Bisexual. Style 3.5–6 mm long, trifid or rarely bifid. Stamens 3. Anthers 3 mm long.

Fruits: Nuts 2.5–3.5 mm long (including beak), c. 1.5 mm wide, ovoid to ellipsoid, trigonous at base, shining, brown - darker towards the conical beak. Perianth bristles absent.

Flowering: July-August.

Biology: Shoots single or few together in tufts from the rhizome, forming large stands. Propagation by seed and vegetatively by division of the rhizome.

Habitats: In acidic or alkaline, oligotrophic to mesotrophic water. On wet, calcareous, peaty or silty substrates in reedbeds and fens, by ponds, pools, lakes, and streams.

Distribution within the region:
Native. Scattered to locally frequent in the south, from Ireland east to the Baltic states, rare or absent further north. Rare in Norway, Finland and Czech Republic, absent from Greenland and Iceland.

Fruiting *C. mariscus*. Gotland, Sweden. Photo JCS.

Flowering *C. mariscus*. Öland, Sweden. Photo JCS.

Submerged parts of *C. mariscus*. Lankauer See, Germany. Photo KvdW.

Cladium mariscus. Nut. Photo JCS.

Cladium mariscus. Leaf. Photo JCS.

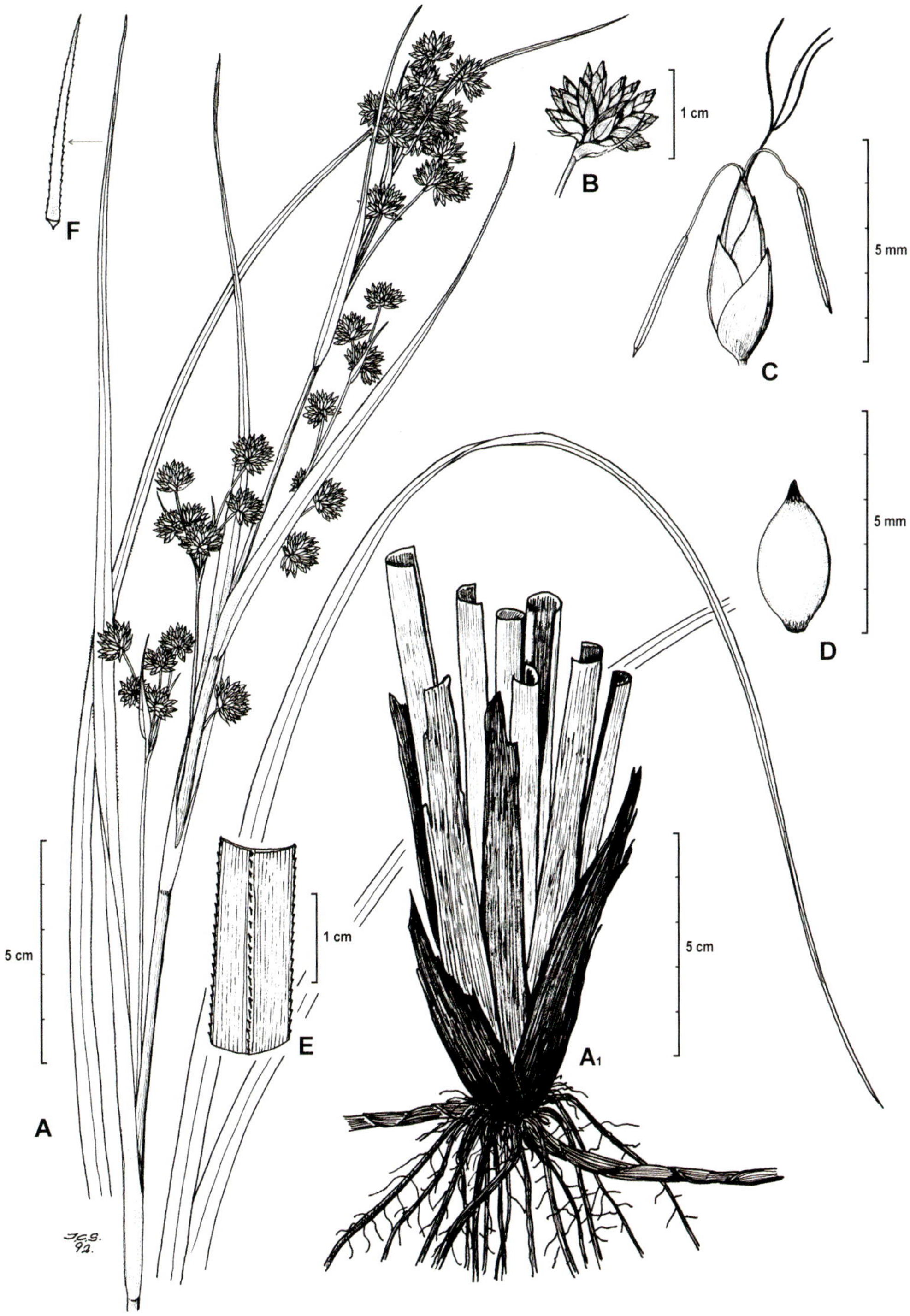

Cladium mariscus
A, A1 habit, **B** cluster of spikes, **C** flowering spike, **D** nut, **E** part of leaf, **F** leaf tip.

Cyperaceae

168 *Carex paniculata* L.

DK: Top-Star **N:** Toppstorr **S:** Vippstarr **FIN:** Lähdesara **IS:** -
GB: Greater Tussock Sedge **NL:** Pluimzegge **F:** Laiche paniculée
D: Rispen-Segge **CZ:** Ostřice latnatá **PL:** Turzyca prosowa
EST: Pööristarn **LV:** Skarainais grīslis **LT:** Šluotelinė viksva **Ru:** Осока метельчатая

Carex paniculata forms large tussocks. De Wittsee, Germany. Photo KvdW.

Perennial helophyte. Rhizome very short, more or less erect. Roots greyish brown to dark brown, 2–3 mm in diameter.

Stem: 50–100(–150) cm long, 6–10 mm in diameter at base, erect to spreading, green, stiff, trigonous, rough.

Leaves: Base of stem with dark brown scales not splitting into fibres. Leaves shorter than to more or less equalling stems, (3–)4–7 mm wide, dark green, flat to channelled, stiff, abruptly tapering to a trigonous apex, margins rough. Hypostomous. Leaf sheaths persistent, brownish. Inner face hyaline, not splitting into fibres. Ligule 2–5 mm long, rounded.

Inflorescence: A compact panicle 5–10 (–15) cm long, often with longer branches, each with several few-flowered spikes 5–8 mm long. Spikes with male flowers apical and female flowers below. Bracts greenish, setaceous or glume-like with hyaline margins and long excurrent midrib. Male glumes 3–4 mm long, orange brown, with pale midrib and hyaline margins. Female glumes 3–4 mm long, orange brown, with pale midrib and wide, hyaline margins.

Flowers: Female flowers with 2 stigmas. Male flowers with 3 2.5–3 mm long anthers.

Fruits: Utricles 2.5–3.5 mm long (including beak), 0.9–1.2 mm wide, green to dark brown when ripe, slightly shining, with 8–10 ribs on each side. Beak 1–1.2 mm long, bidentate, with broad, serrulate wings.

Flowering: May-June.

Biology: Forming large, dense tussocks up to 1.5 m high. Propagation by seed.

Habitats: In neutral to weakly alkaline, mesotrophic to eutrophic water. On wet, peaty or silty substrates in or by ponds, lakes, canals, ditches and streams, as well as in fens and marshes.

Distribution within the region:
Native. Widespread and abundant throughout most of the region except the northern part of Fennoscandia, Iceland and Greenland.

Characteristics and similar species:
Carex paniculata is one of the few sedges forming large tussocks. It is most likely to be confused with ***C. appropinquata***, with which it hybridises.

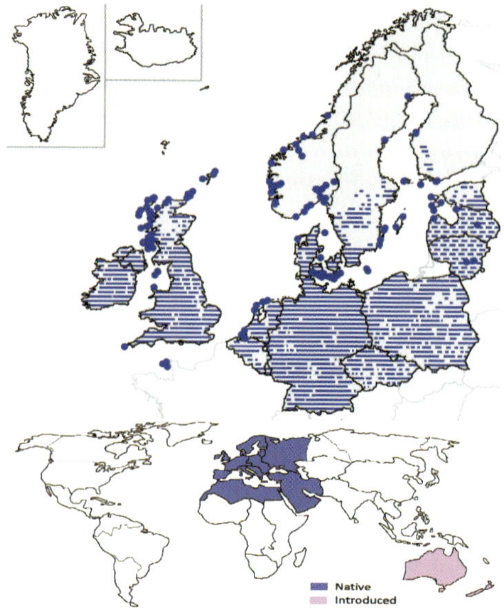

C. appropinquata differs by narrower, (1–)1.5–3 mm wide leaves, basal sheaths and scales splitting into fibres and by unwinged utricles.

Vegetative specimens of **179 *C. elata*** differ by glaucous leaves and by the basal sheaths and scales splitting into fibrous strands.

Hybrids: *Carex paniculata* forms hybrids with ***C. appropinquata*** and **169 *C. remota*** and other *Carex* species. Specialist literature is needed for identification of *Carex* hybrids.

Carex paniculata. Basal scales and sheaths. Photo JCS.

Carex paniculata. Utricles with or without glume. Photo JCS.

Inflorescences of *C. paniculata*. Ljyngsjön, Kristianstad, Sweden. Photo JCS.

Carex paniculata. Leaf sheaths. Photo JCS.

Carex paniculata
A habit, **B** inflorescence, **C** utricle, **D** female flower and glume, **E** male flower and glume, **F** upper part of sheath with ligule.

Cyperaceae

169 *Carex remota* L.

DK: Akselblomstret Star **N:** Slakkstorr **S:** Skärmstarr **FIN:** Hajasara **IS:** -
GB: Remote Sedge **NL:** Ijle zegge **F:** Laiche espacée
D: Winkel-Segge **CZ:** Ostřice řídkoklasá **PL:** Turzyca rzadkokłosa
EST: Varjutarn **LV:** Attālvārpu grīslis **LT:** Retavarpė viksva **Ru:** Осока раздвинутая

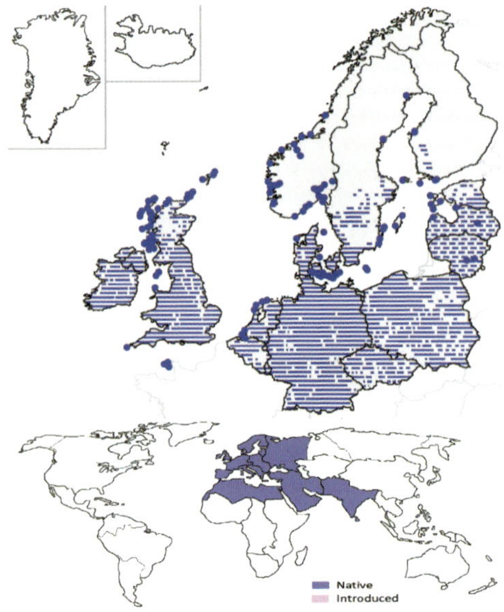

Carex remota is easily recognised by its very long, leaf-like bracts. Hadsund, Denmark. Photo JCS.

Perennial helophyte. Rhizome very short, erect. Roots greyish brown to pale purplish brown, 0.5–1 mm in diameter.

Stem: 30–60 cm long, 1.5–2.5 mm in diameter at base, arching or spreading, green, bluntly trigonous, slender, flaccid, somewhat rough above.

Leaves: Base of stem with pale, greyish-brown, more or less fibrous scales and some sheaths with short blades. Stem leaves equalling or shorter than stems, 1.5–2 mm wide, flat to channelled, flaccid, slowly tapering to a fine apex. Hypostomous. Leaf sheaths long, on sterile shoots forming pseudostems. Ligule 1–2 mm long, tubular, obtuse.

Inflorescence: About 10 cm long, composed of 5–7(–10) pale green, sessile, obovate spikes, contiguous at top, but gradually becoming more and more remote downwards. Upper spikes with female flowers above and male flowers below, the lower spikes purely female. The upper bracts are glume-like, the lower 10–20 cm long and leaf-like. Both male and female glumes about 2.5 mm long, hyaline, often with a brownish tinge and green midrib.

Flowers: Female flowers with 2 stigmas.

Fruits: Utricles 3–3.5 mm long (including beak), 1.2–1.4 mm wide, greenish to buff when ripe, slightly shining, with 5–7 ribs on the dorsal side, 0–3 ribs on the ventral side. Beak about 1 mm long, conical, bidentate, with a cleft on both sides - longest on the dorsal side.

Flowering: June-July.

Biology: Forming dense tufts or peaty tussocks up to 30 cm high. Propagation by seed.

Habitats: In neutral to weakly alkaline, mesotrophic to eutrophic water. On wet, peaty or clay substrates, usually in shaded places along forest roads or tracks, in or by forest streams and marshes, as well as in alder or willow carrs.

Distribution within the region:
Native. Widespread and abundant throughout most of the area except the northern part of Fennoscandia, Iceland and Greenland.

Characteristics and similar species:
Carex remota can be recognised by its

Carex remota growing in a rocky stream. Brede, Sjælland, Denmark. Photo JCS.

Carex remota. Utricles with or without glume. Photo JCS.

The inflorescence of *C. remota* is very distinctive. Photo JCS.

tufted or tussocky habit and remote lower spikes subtended by very long bracts.

Hybrids: *Carex remota* forms hybrids with **168 *C. paniculata*** and other *Carex* species. Specialist literature is needed for identification of *Carex* hybrids.

Carex remota. Leaf sheath. Photo JCS.

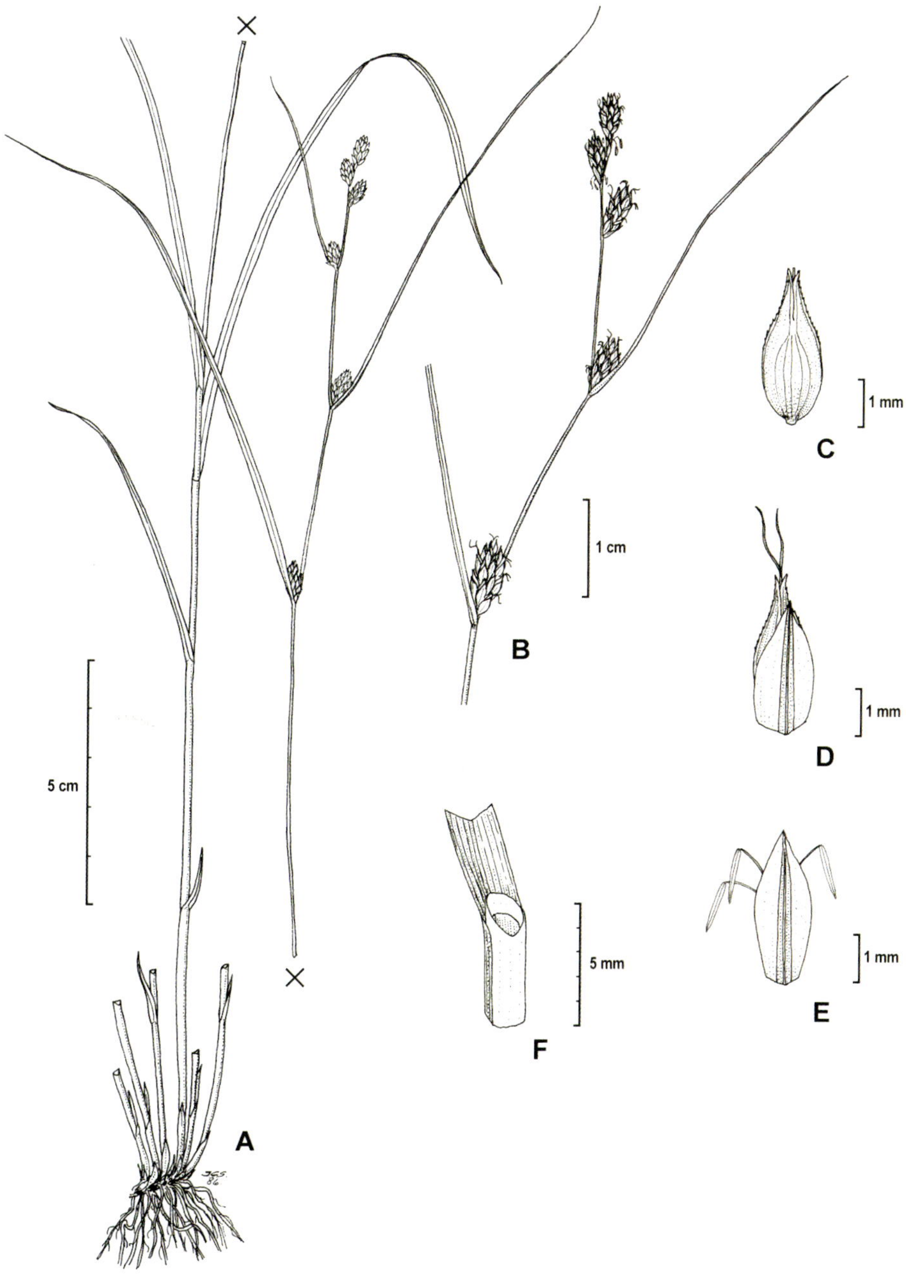

Carex remota
A habit, **B** inflorescence, **C** utricle, **D** female flower and glume, **E** male flower and glume, **F** upper part of sheath with ligule.

Cyperaceae

170 ***Carex bohemica*** Schreb.

Syn.: *C. cyperoides* L.
DK: Fladaks-Star **N:** Skjermstorr **S:** Svepestarr **FIN:** Mykerösara **IS:** -
GB: Bohemian Sedge **NL:** - **F:** Laiche de Bohême
D: Zypergras-Segge **CZ:** Ostřice šáchorovitá **PL:** Turzyca ciborowata
EST: - **LV:** - **LT:** - **Ru:** Осока богемская

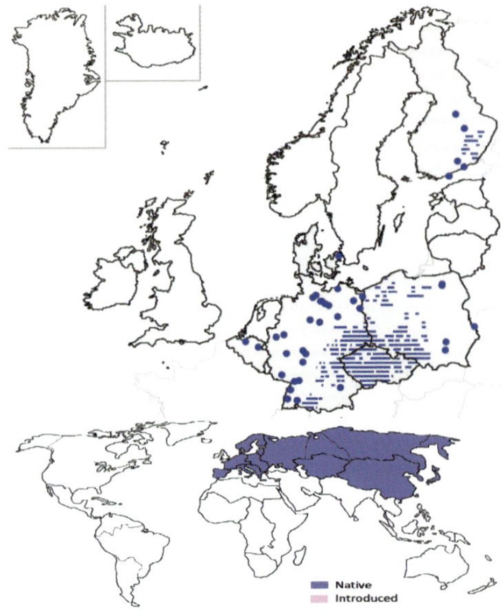

Carex bohemica forms small tufts. It is a so-called meteoric plant, arising from seed only when the conditions are right. Etang Mouton, Parc Naturel Régional de la Brenne, Indre, France. Photo RVL.

Annual or rarely biannual. Roots pale brown, c. 0.5 mm in diameter.

Stem: 10–40 cm long, erect to ascending, light green, bluntly trigonous, glabrous or slightly scabrid above.

Leaves: Basal sheaths few, thin, pale brown. Stem leaves usually shorter than the stem, 1.5–3 mm wide, light green, flat to keeled, gradually tapering to a very fine point, margins scabrid. Hypostomous. Ligule 2–4 mm long, obtuse.

Inflorescence: 1–2 cm long, ovate to semi-globose, composed of several densely packed spikes. Each spike with female flowers above and male flowers below. Involucral bracts 3–4, leaf-like, with broad hyaline margins clasping the stem at the base. Glumes acute, pale green, male c. 0.5 mm long, female c. 0.4 mm long.

Flowers: Female flowers with 2 stigmas.

Fruits: Utricles 7–10 mm long (including beak), broader than glumes, pale yellowish green, glabrous, glossy and silky, with more or less hyaline walls, dentate margin and inconspicuous ribs. Beak 5.5–6.5 mm long, deeply bifid, margins dentate.

Flowering: June-September.

Biology: Forming small tufts. Not consistently growing at any individual site, with new plants arising from the seed bank when conditions become favourable. The seeds remain viable for many years.

Habitats: On eutrophic, temporarily flooded, moist, muddy or sandy substrates in the littoral zone of ponds, aquaculture ponds, lakes and floodplains.

Distribution within the region:
Native. Scattered to locally frequent in Belgium, Germany, the Czech Republic, Poland and Finland.

Characteristics and similar species:
Carex bohemica does not resemble other species within the area except for maybe **164 *Cyperus fuscus.***

Carex bohemica. Etang Mouton, Parc Naturel Régional de la Brenne, Indre, France. Photo RVL.

Carex bohemica. Utricles. Photo JCS.

The inflorescence of *C. bohemica* is very distinctive. Etang Mouton, Parc Naturel Régional de la Brenne, Indre, France. Photo RVL.

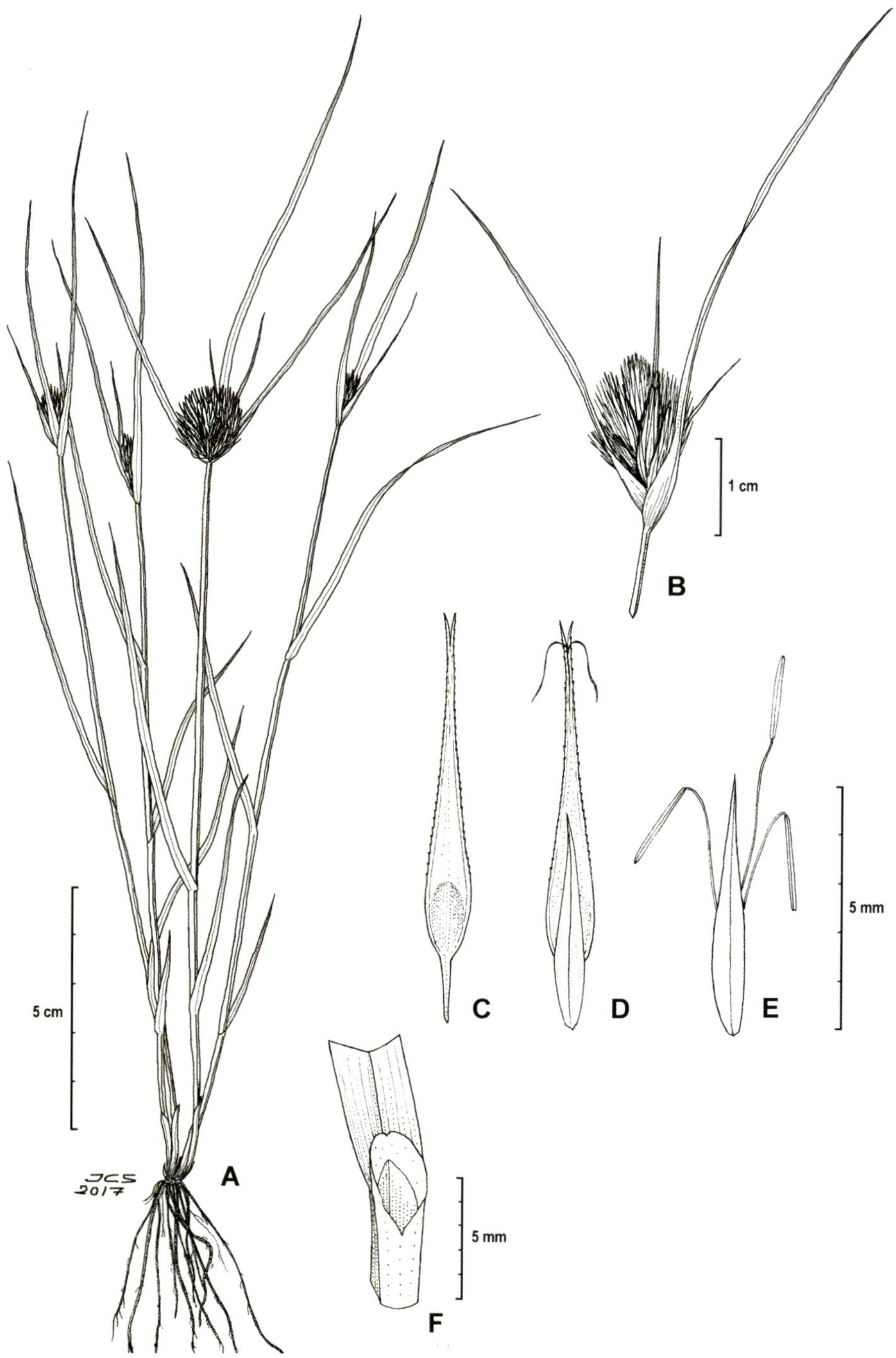

Carex bohemica
A habit, **B** inflorescence, **C** utricle, **D** female flower and glume, **E** male flower and glume, **F** upper part of sheath with ligule.

Cyperaceae

171 *Carex lasiocarpa* L.

Syn.: *C. filiformis* auct. non L.
DK: Tråd-Star **N:** Trådstorr **S:** Trådstarr **FIN:** Jouhisara **IS:** -
GB: Slender Sedge **NL:** Draadzegge **F:** Laîche filiforme
D: Faden Segge **CZ:** Ostřice plstnatoplodá **PL:** Turzyca nitkowata
EST: Niitjas tarn **LV:** Pūkaugļu grīslis **LT:** Laiboji viksva **Ru:** Осока волосистоплодная

Carex lasiocarpa in a peat bog. Mariager, Denmark. Photo JCS.

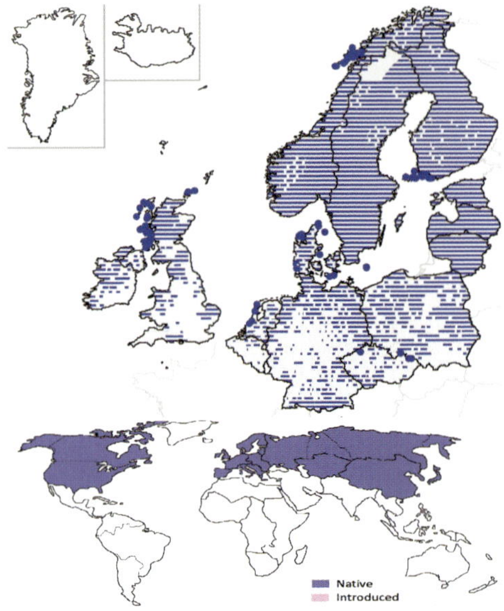

Perennial helophyte. Rhizome long, creeping, 2–4 mm in diameter, covered with pale to orange-brown scales. Roots pale brown up to c. 1 mm in diameter.

Stem: 40–100 cm long, rather slender, 2–4 mm in diameter at the base, stiff, erect, green, bluntly trigonous, striate, smooth or somewhat rough above.

Leaves: Base of stem with greyish-brown to reddish-brown or purplish scales. Stem leaves usually longer than stem, 30–100 cm long, 1–2 mm wide, grey green, stiff, upright, flat or inrolled, gradually tapering to a 10–30 cm long, fine, trigonous point. Hypostomous. Leaf sheaths terete, often with a purplish tinge, soon decaying, inner face hyaline, splitting into fibres. Ligule 2–3 mm long, obtuse, with a short free margin.

Inflorescence: The upper 1–3 spikes entirely male, the lower (1–)2–3 female. Male spikes 1.5–7 cm long, reddish brown. Female spikes 1–3 cm long, oblong to shortly cylindrical, more or less distant or contiguous, erect, the upper sessile and sometimes with some male flowers at the top, the lower more or less pedunculate.

Bracts leaf-like, the lower often exceeding the inflorescence. Male glumes 4–6 mm long, acute at the apex, dark reddish brown with a green or pale midrib. Female glumes 3.5–5 mm long, acute to acuminate at the apex, dark reddish brown, with a green midrib.

Flowers: Female flowers with 3 stigmas.

Fruits: Utricles 3.5–5.5 mm long (including beak), 1.5–2 mm wide, broader than glume, ovoid, bluntly triangular in cross section, grey green to greyish brown, densely greyish tomentose, ribs not visible. Beak 0.7–1.0 mm long, deeply bifid.

Flowering: June-July.

Biology: Forming large stands. Shoots few to several together from the rhizome. Propagation by seed and vegetatively by division of the rhizome.

Habitats: In mesotrophic to eutrophic, acidic or alkaline water. Peat bogs, swamps, fens, marshes and on the margins of rivers or lakes.

Distribution within the region:
Native. Common in Fennoscandia, the Baltic states and in Scotland. Scattered to locally frequent throughout the rest of the region. Absent in Iceland and Greenland.

Dense formation of *C. lasiocarpa*. Note the long, curved, whip-like leaf tips. Lüsekamp, Germany. Photo KvdW.

Inflorescences of *C. lasiocarpa*. Denmark. Photo JCS.

Carex lasiocarpa.
Leaf sheath and cross section. Female spike. Utricles with or without glumes. Photos JCS.

Characteristics and similar species: Recognised by tomentose utricles and rather slender habit. Easily overlooked when vegetative, but recognised by the narrow, grey-green, erect leaves with an almost whip-like, long, trigonous point which is often held horizontally and by the reddish-brown or purplish scales at base of stem.

In the vegetative state it may be mistaken for **175 *C. rostrata***, which differs in being epistomous and lacking long, trigonous leaf point.

Carex lasiocarpa
A habit, **B** female spike, **C** utricle, **D** female flower and glume, **E** male flower and glume, **F, F1** leaf-sheath and ligule.

Cyperaceae

172 *Carex acutiformis* L.

Syn.: *C. paludosa* Gooden.

DK: Kær-Star **N:** Stautstorr **S:** Brunstarr **FIN:** Hetesara **IS:** -
GB: Lesser Pond-sedge **NL:** Moeraszegge **F:** Laîche des marais
D: Sumpf-Segge **CZ:** Ostřice ostrá **PL:** Turzyca błotna
EST: Sootarn **LV:** Krastmalas grīslis **LT:** Pelkinė viksva **Ru:** Осока островатая

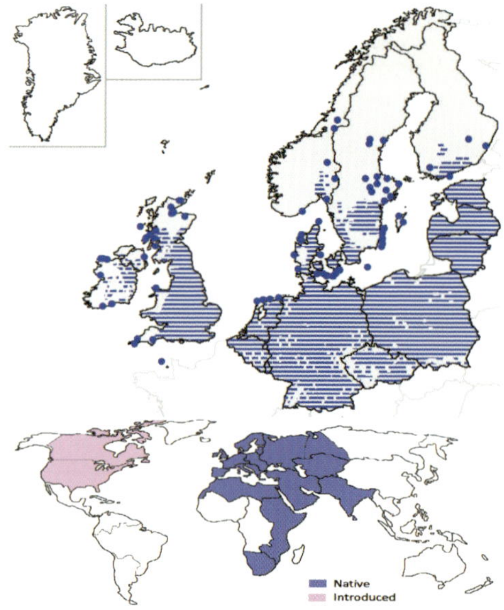

Carex acutiformis in an old moat. Fussingø, Jutland, Denmark. Photo JCS.

Perennial helophyte. Rhizome long, creeping, 4–6 mm in diameter, covered with greyish brown, fibrous scales. Roots pale brown to dark brown up to c. 2 mm in diameter.

Stem: 50–120 cm long, robust, 10–15 mm in diameter at the base, stiff, erect, bluish green, sharply trigonous, rough above, usually smooth below.

Leaves: Base of stem with dark brown to blackish-brown scales and remnants of last year's sheaths. Stem leaves as long as or longer than stem, up to 150 cm long, 4–10(–25) mm wide, dark green above, glaucous and papillose below, arching, keeled or plicate, gradually tapering to a fine point. Hypostomous. Leaf sheaths yellow brown to brown, sometimes red streaked, ribs and secondary ribs forming an irregular pattern of oblong rectangular cells. Inner face brownish hyaline, more or less distinctly splitting into fibres. Ligule 5–20 mm long, acute, with a free margin.

Inflorescence: The upper (1–)2–3(–4) spikes entirely male, the lower 2–3 female. Male spikes 1–5 cm long, 4–6 mm in diameter, cylindrical, clustered, reddish brown to brown. Female spikes 2–5(–7) cm long, 6–8 mm in diameter, cylindrical, dense, distant, erect, the upper sessile and often with some male flowers at the top, the lower shortly pedunculate. Bracts leaf-like, the lower exceeding the inflorescence.

Male glumes 5–6 mm long, blunt to subacute, dark brown to blackish brown, with a pale midrib. Female glumes 4–5 mm long, acute to acuminate at apex, red brown to dark brown, with a pale or green midrib.

Flowers: Female flowers with 3 stigmas, rarely 2.

Fruits: Utricles 3.5–4.5 mm long (including beak), 1.5–2.2 mm wide, broader than glume, ovoid, bluntly triangular in cross section, grey green to brown, dull, papillose, with 8–9 distinct ribs on each side. Beak 0.4–0.6 mm long, notched or shallowly bidentate.

Flowering: May-June.

Biology: Forming large stands. Shoots few or several together in tufts from the rhizome. Propagation by seed and vegetatively by division of the rhizome.

Habitats: In mesotrophic, weakly acid or alkaline water. On wet, muddy substrates on the landward side of reedbeds, in sedge swamps, forest swamps, by pools, lakes, streams and rivers.

Carex acutiformis. Germany. Photo KvdW.

Inflorescences of *C. acutiformis*.
Denmark. Photo JCS.

Carex acutiformis.
Leaf sheaths soon splitting into fibres.

Upper and lower side of leaf.

Utricles with or without glumes. Photos JCS.

Distribution within the region:
Native. Widespread and abundant throughout most of the area except the northern part of Fennoscandia, Iceland and Greenland.

Characteristics and similar species:
Differs from the rather similar **173 *C. riparia*** by fewer male spikes, shorter, dull, grey-green utricles with a narrower beak and the pattern of oblong rectangular cells formed by ribs and secondary ribs on the leaf sheath.

177 *C. acuta* differs in female flowers normally with 2 stigmas, flattish utricles and leaf sheaths not splitting into fibres.

179 *C. elata* differs in female flowers normally with 2 stigmas, the lack of rhizome and in forming dense tufts or tussocks.

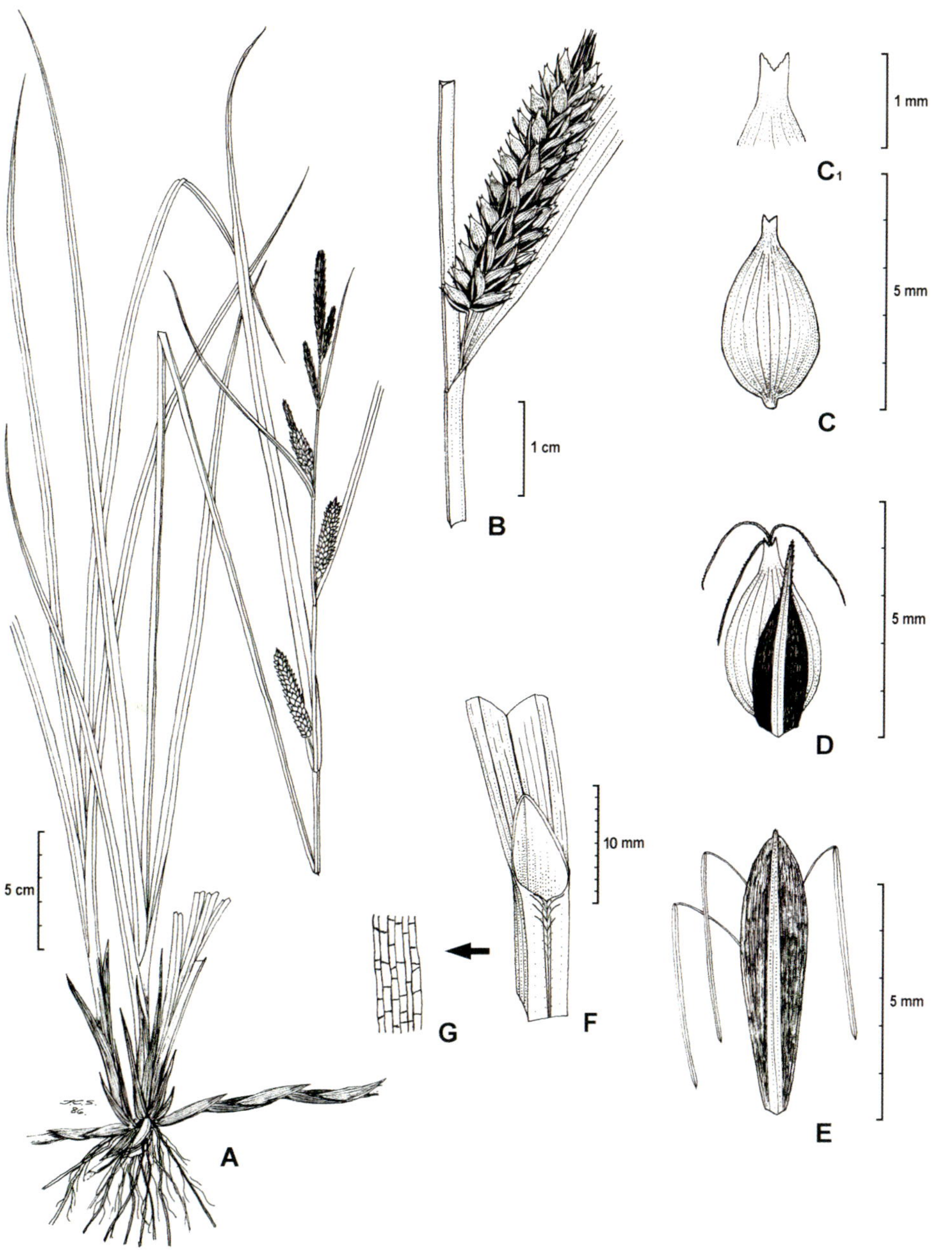

Carex acutiformis
A habit, **B** female spike, **C** utricle, **C1** beak, **D** female flower and glume, **E** male flower and glume, **F** leaf sheath and ligule, **G** pattern of ribs on sheath.

Cyperaceae

173 *Carex riparia* Curtis

DK: Tykakset Star **N:** Kjempestorr **S:** Jättestarr **FIN:** Jouhisara **IS:** -
GB: Greater Pond-sedge **NL:** Oeverzegge **F:** Laîche des rives
D: Ufer-Segge **CZ:** Ostřice pobřežní **PL:** Turzyca brzegowa
EST: Kallastarn **LV:** Krasta grīslis **LT:** Pakrantinė viksva **Ru:** Осока береговая

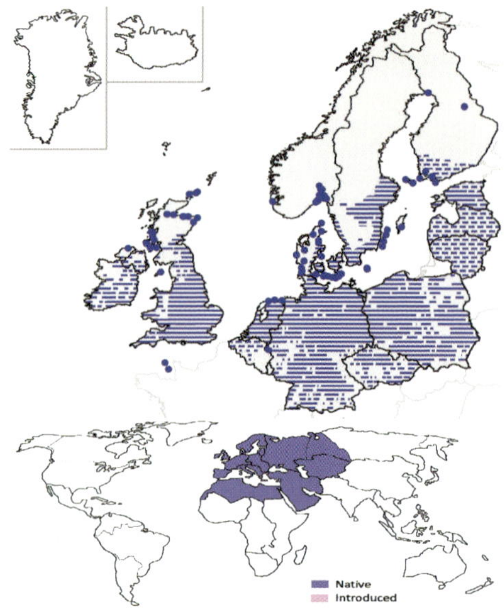

Carex riparia in a drainage ditch. Hadsund, Denmark. Photo JCS.

Perennial helophyte. Rhizome long, creeping, 4–6 mm in diameter, covered with brownish, fibrous scales. Roots pale brown, up to c. 2 mm in diameter.

Stem: 60–120 cm long, robust, 8–15 mm diameter at base, stiff, erect, green, sharply trigonous, rough above, usually smooth below.

Leaves: Base of stem with blackish brown, often red-tinged remnants of previous year's sheaths. Stem leaves as long as or longer than stem, up to 160 cm long, 7–20(–24) mm wide, dark green above, glaucous below, arching, keeled or plicate, shortly tapering to a fine trigonous point. Hypostomous. Leaf sheaths yellowish brown to reddish brown, often purplish tinged, ribs and secondary ribs forming an irregular pattern of oblong rectangular cells. Inner face brownish hyaline, more or less distinctly splitting into fibres. Ligule 5–10 mm long, blunt to rounded, with a short free margin.

Inflorescence: The upper 3–4(–6) spikes entirely male, the lower (2–)3–4(–5) female. Male spikes 2–6 cm long, 4–8 mm in diameter, cylindrical, densely clustered, brown. Female spikes 3–6(–10) cm long, 10–12 mm in diameter, cylindrical, dense, more or less contiguous, erect, the upper sessile or shortly pedunculate sometimes with some male flowers at the top, the lower pedunculate. Bracts leaf-like, the lower exceeding the inflorescence. Male glumes 7–9 mm long, acuminate at the apex, dark brown, with a pale midrib. Female glumes 6–10 mm long, acute to acuminate at the apex, dark reddish brown to dark brown, with a pale or greenish midrib.

Flowers: Female flowers with 3 stigmas.

Fruits: Utricles (4.5–)5–7(–8) mm long (including beak), 1.5–2.0 mm wide, broader than glume, ovoid, bluntly triangular in cross section, yellowish green to light brown to brown, slightly glossy, with 9–11 rather indistinct ribs on each side. Beak 0.5–0.7 mm long, more than 0.5 mm wide, bidentate, with 0.3–0.4 mm long, divergent, slightly outward-directed tips.

Flowering: June-July.

Biology: Forming large stands. Shoots single or a few together from the rhizome. Propagation by seed and vegetatively by division of the rhizome.

Habitats: In mesotrophic, neutral to alkaline water. On wet, muddy substrate on the landward side of reedbeds, in sedge swamps, forest swamps, by pools, lakes, streams and rivers.

Carex riparia. Massenveense Plassen, Netherlands. Photo KvdW.

Carex riparia. Inflorescences. Öland, Sweden. Photo JCS.

Carex riparia. Leaf sheath. Photo JCS.

Carex riparia. Utricles with or without glumes. Photo JCS.

Distribution within the region:
Native. Widespread and abundant throughout most of the area except the northern part of Fennoscandia, Iceland and Greenland.

Characteristics and similar species: Recognised by the 3–4 thick, tightly clustered, brownish male spikes and usually 3–4, 10–12 mm diameter cylindrical female spikes and by the beak more than 0.5 mm wide.

May be confused with **172 *C. acutiformis*** and **177 *C. acuta*** - see these.

138 *Scirpus sylvaticus* differs by an umbel-like inflorescence with 3–5 mm long spikelets and vegetatively by the leaves yellowish green to green on both sides, without a ligule and with leaf sheaths splitting like parchment.

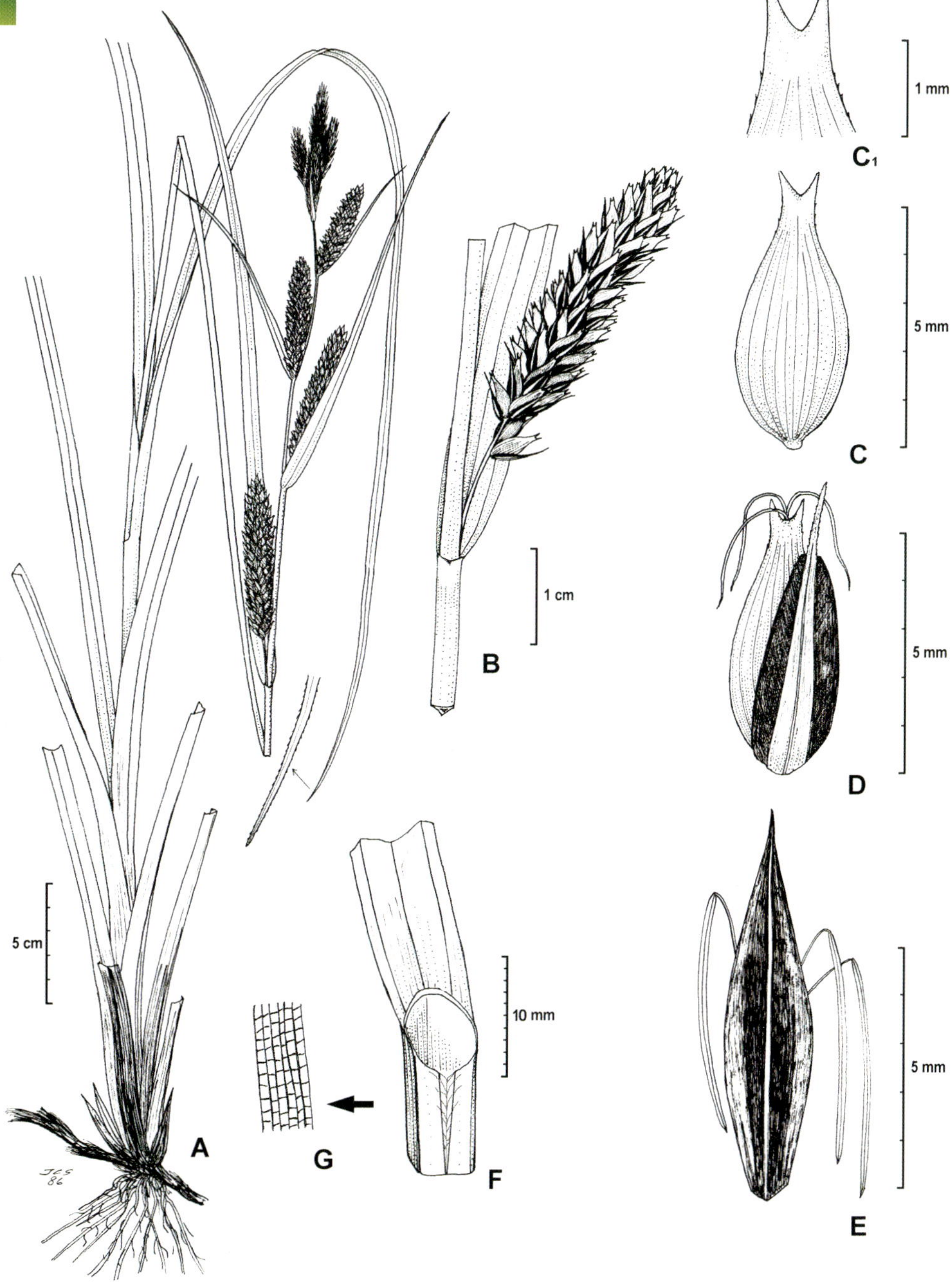

Carex riparia
A habit, **B** female spike, **C** utricle, **C1** beak, **D** female flower and glume, **E** male flower and glume, **F** leaf sheath and ligule, **G** pattern of ribs on sheath.

Cyperaceae

174 *Carex pseudocyperus* L.

DK: Knippe-Star **N:** Dronningstorr **S:** Slokstarr **FIN:** Varstasara **IS:** -
GB: Cyperus Sedge **NL:** Cyperzegge **F:** Laiche faux-souchet
D: Scheinzypergras-Segge **CZ:** Ostřice nedošáchor **PL:** Turzyca nibyciborowata
EST: Kraavtarn **LV:** Dižmeldru grīslis **LT:** Šiurkščioja viksva **Ru:** Осока ложносытевая

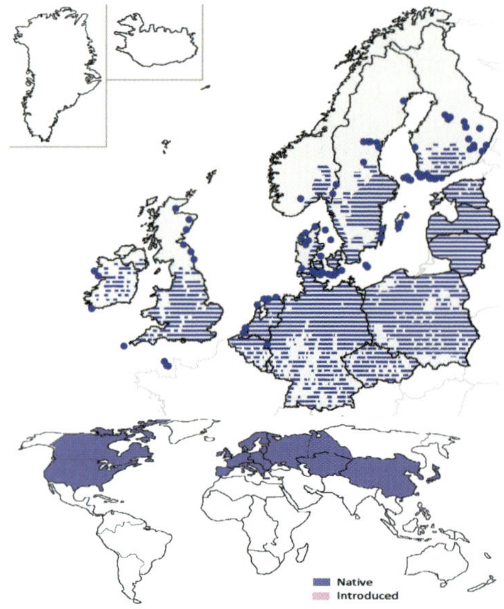

Carex pseudocyperus in a drainage ditch. Lille Vildmose, Jutland, Denmark. Photo JCS.

Perennial helophyte. Rhizome short, erect. Roots orange brown, up to 1.5 mm in diameter.

Stem: 50–80 cm long, solid, 5–10 mm thick at base, stiff, erect, green, sharply trigonous, rough on angles.

Leaves: Base of stem with light brown scales and remnants of previous year's sheaths. Stem leaves longer than stem, up to 120 cm long, 5–12 mm wide, yellowish green to green, rigid, rough on margins and keel, plicate, gradually tapering to a long and very fine point. Hypostomous. Leaf sheaths yellowish green becoming greyish brown or reddish. Inner face brownish hyaline, more or less distinctly splitting into fibres. Ligule 10–15 mm long, obtuse, with a short free margin.

Inflorescence: The upper spike entirely male, the lower 3–5 female. Male spike 2–6 cm long, 3–6 mm in diameter, cylindrical, greenish to pale brown, pedunculate. Female spikes 3–7 cm long, 8–12 mm in diameter, cylindrical, dense, clustered, pendent, peduncles rough and rather long. Bracts leaf-like, the lower long, exceeding the inflorescence. Male glumes 5–7 mm long, gradually acuminate at apex, pale brownish with a pale midrib. Female glumes 4–10 mm long, acute to acuminate at the apex, greenish to brownish hyaline, with a green midrib extended into a long, ciliate arista.

Flowers: Female flowers with 3 stigmas.

Fruits: Utricles 4.5–5.8 mm long (including beak), 1.2–1.5 mm wide, broader than glume, ovoid, inflated, bluntly triangular in cross section, light green to yellowish green, smooth, faintly shining, ribs distinct, 7–9 dorsally, 6–7 ventrally between the edges. Beak 1.5–2.0 mm long, bidentate with 0.8–1.0 mm long, erect or slightly outward-directed tips.

Flowering: June-July.

Biology: Forming dense tufts. Propagation by seed.

Habitats: In mesotrophic to eutrophic, acid to weakly alkaline water. On wet or moist, muddy substrate on the landward side of reedbeds, in forest swamps, on the sides of ponds, moats, ditches, lakes and streams.

Distribution within the region:
Native. Widespread and abundant throughout most of the area except northern Britain, Fennoscandia, Iceland and Greenland.

Carex pseudocyperus. Boschbeek, Netherlands. Photo KvdW.

Carex pseudocyperus. Inflorescence. Denmark. Photo JCS.

Carex pseudocyperus. Leaf tip and sheath. Photo JCS.

Carex pseudocyperus. Utricles with or without glumes. Photo JCS.

Characteristics and similar species:
Recognised by its yellow-green leaves and stem and pendent, bristly female spikes close to each other below the male spike. In vegetative state it differs from other *Carex* species of wetland habitats by the tufted habit without stolons, plicate leaves and yellowish-green colour.

Carex pseudocyperus. Base of stem. Photo JCS.

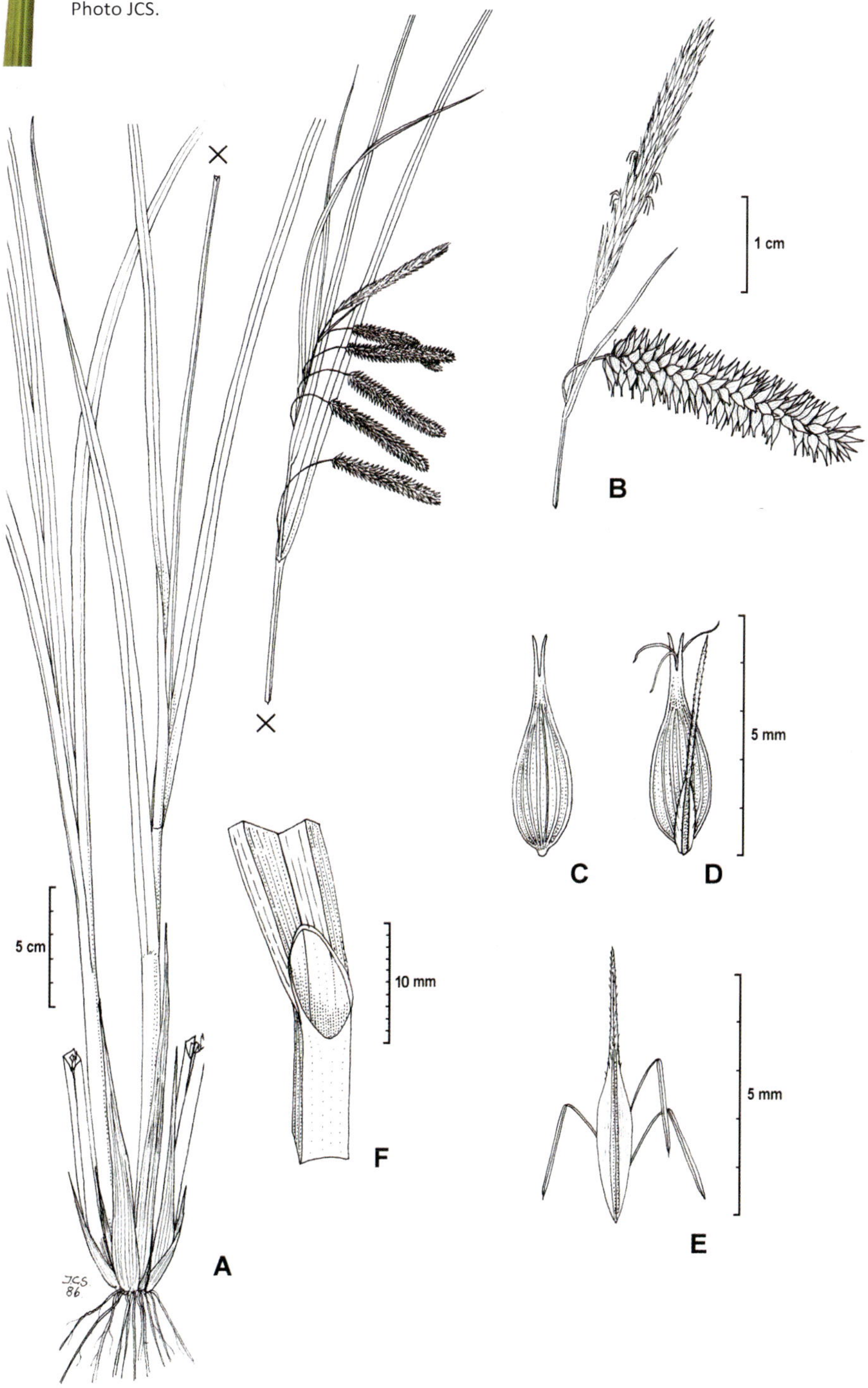

Carex pseudocyperus
A habit, **B** male and female spike, **C** utricle, **D** female flower and glume, **E** male flower and glume, **F** leaf sheath and ligule.

Cyperaceae

175 *Carex rostrata* Stokes

Syn.: *C. inflata* Huds.; *C. ampullacea* Gooden.

DK: Næb-Star **N:** Flaskestorr **S:** Flaskstarr **FIN:** Pullosara **IS:** Tjarnastör
GB: Beaked Sedge **NL:** Snavelzegge **F:** Laiche rostrée
D: Schnabel Segge **CZ:** Ostřice zobánkatá **PL:** Turzyca dzióbkowata
EST: Pudeltarn **LV:** Uzpūstais grīslis **LT:** Snapuotoji viksva **Ru:** Осока вздутая

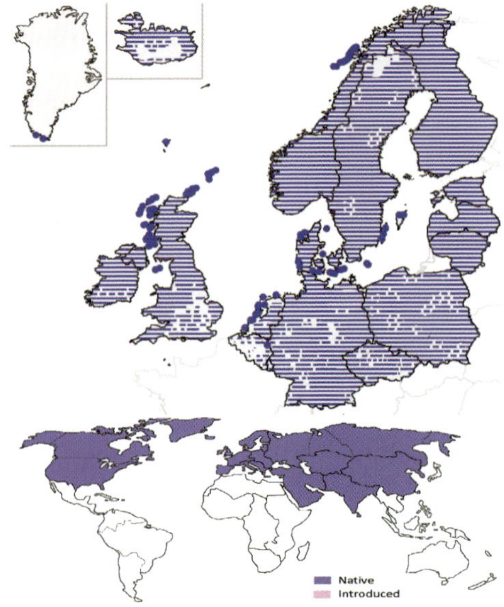

Carex rostrata expanding from the outside of a quaking bog into an acidic, oligotrophic lake with brownish water. Jutland, Denmark. Photo JCS.

Perennial helophyte. Shoots single or a few together in tufts from the rhizome. Rhizome long, creeping, 2–5 mm in diameter, covered with yellowish-grey to greyish-brown, fibrous scales. Roots orange brown to dark brown, up to 1.5 mm in diameter.

Stem: 20–100 cm long, slender, 5–10 mm thick at base, stiff, erect, bluish green, bluntly trigonous and rough above, subterete and smooth below.

Leaves: Base of stem with scales and brown to dark brown, often purplish-streaked remnants of previous year's sheaths. Stem leaves longer than stem, up to 120 cm long, 2–6(–10) mm wide, glaucous above, green and papillose below, rough on margins, keeled, plicate or inrolled, gradually tapering to a 2–6 cm long trigonous, acicular point. Epistomous, but in some populations with up to 25% of the stomata on the lower side of leaves. Leaf sheaths spongy and thickened when under water, brownish, often with a red or pink tinge. Inner face brownish-hyaline, indistinctly splitting into fibres. Ligule 2–3 mm long, rounded, with a short free margin.

Inflorescence: The upper 2–4 spikes entirely male, the lower 2–3(–5) female. Male spikes 1–6 cm long, 3–5 mm in diameter, cylindrical, the uppermost pedunculate, the others sessile, brownish. Female spikes 2–6(–10) cm long, 7–10 mm in diameter, cylindrical, dense, contiguous or the lowermost distant, suberect, subsessile or the lower shortly pedunculate.
Bracts of male spikes glume-like or setaceous, of female spikes leaf-like - the lowest as long as or exceeding the inflorescence. Male glumes 4–6 mm long, blunt to subacute at apex, brownish, with a pale midrib. Female glumes 3–5 mm long, acute at apex, brownish, with a pale or greenish midrib.

Flowers: Female flowers with 3 stigmas.

Fruits: Utricles (3–)4–5(–6) mm long (including beak), 1.6–2.5 mm wide, broader than glume, ovoid to globular, strongly inflated, elliptic or round in cross section, yellowish green becoming shining yellow to orange when mature, with 4–5 more or less distinct ribs on each side. Beak 0.8–1.5 mm long, more or less linear, bidentate, with
0.3–0.5 mm long, slightly outward-directed tips.

Flowering: May-July.

Biology: Forming large stands. Propagation by seed and vegetatively by division of the rhizome.

Habitats: In oligotrophic to mesotrophic, acid to neutral water. On wet, peaty or muddy substrates in transition mires, quaking mires, by pools, lakes and streams, in ditches and swampy fens.

Carex rostrata. The densely packed utricles are directed almost perpendicular to the axis. Denmark. Photo JCS.

Carex rostrata. Jutland, Denmark. Photo JCS.

Carex rostrata.
Leaf sheath. Photo JCS.

Carex rostrata. Utricles with or without glumes. Photo JCS.

Distribution within the region:
Native. Widespread and abundant throughout the region, but rare in Greenland.

Characteristics and similar species:
176 *C. vesicaria* differs in 6–8 mm long utricles, trigonous stem, it is hypostomous and has wider, yellowish-green leaves and 5–8 mm long, acute ligules.

In vegetative state *C. rostrata* may be mistaken for **171 *C. lasiocarpa*** - see this.

Hybrids: *Carex rostrata* forms hybrids with **174 *C. pseudocyperus*** and ***C. vesicaria.***

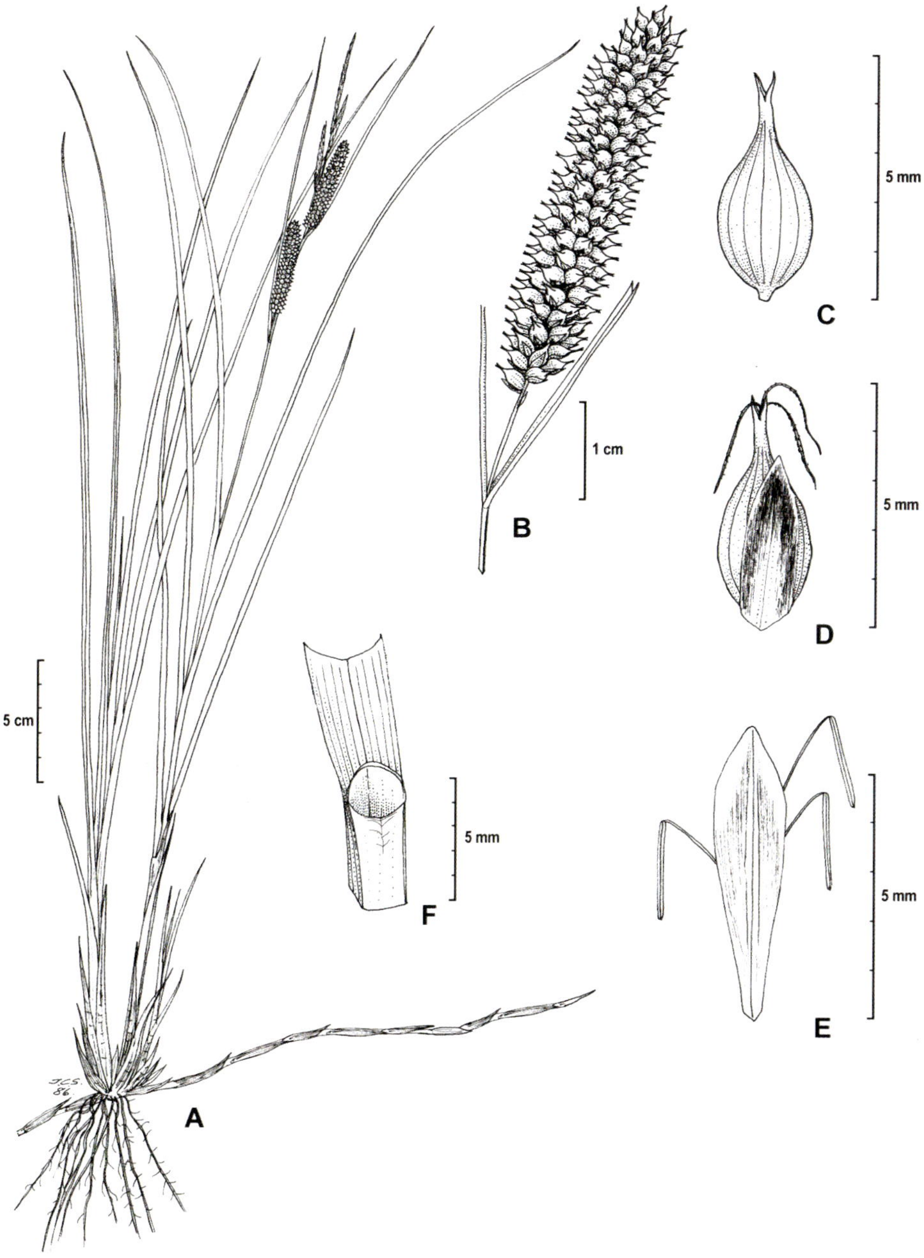

Carex rostrata
A habit, **B** female spike, **C** utricle, **D** female flower and glume, **E** male flower and glume, **F** leaf sheath and ligule.

Cyperaceae

176 *Carex vesicaria* L.

DK: Blære-Star **N:** Sennegras **S:** Blåsstarr **FIN:** Luhtasara **IS:** -
GB: Bladder-sedge **NL:** Blaaszegge **F:** Laîche vésiculeuse
D: Blasen-Segge **CZ:** Ostřice měchýřkatá **PL:** Turzyca pęcherzykowata
EST: Põistarn **LV:** Pūslīšu grīslis **LT:** Pūslėtoji viksva **Ru:** Осока пузырчатая

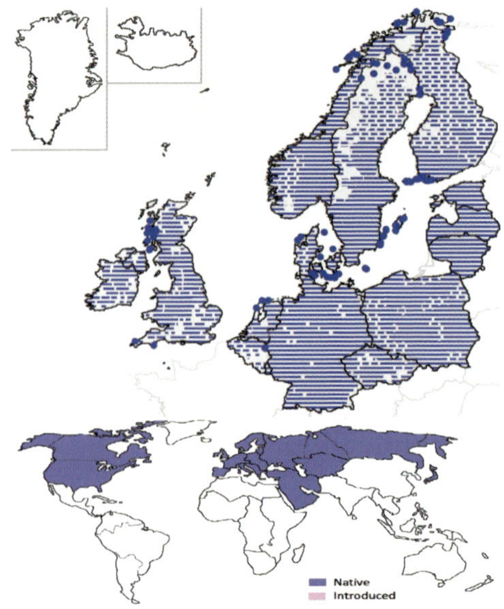

Carex vesicaria. Vlijmen, Netherlands. Photo KvdW.

Perennial helophyte. Shoots few together in tufts from the rhizome. Rhizome short, creeping, 2–5 mm in diameter, covered with brown to purplish scales. Roots pale yellowish brown up to 1.5 mm in diameter.

Stem: 30–70(–120) cm long, rather robust, 4–8 mm diameter at base, green, sharply trigonous, rough on the angles above, usually smooth below.

Leaves: Base of stem with purplish-red to purplish-black scales and remnants of previous year's sheaths. Stem leaves as long as or longer than stem, up to 150 cm long, 4–8 mm wide, yellowish green, rather rigid, plicate, rough on keel and margins, gradually tapering to a fine point. Hypostomous. Leaf sheaths thin, purplish tinged. Inner face hyaline, splitting into fibres. Ligule 5–12 mm long, acute, with a free margin.

Inflorescence: The upper 2–3(–4) spikes entirely male, the lower 1–3 female. Male spikes 1–5 cm long, 2–5 mm in diameter, cylindrical, contiguous, reddish brown to brown. Female spikes 3–7 cm long, 11–17 mm in diameter, cylindrical, dense, contiguous, erect to pendent. Bracts of male spikes glume-like or setaceous, of female spikes leaf-like, the lowest exceeding the inflorescence. Male glumes 4–7 mm long, more or less acute, reddish brown, with a pale midrib. Female glumes 4–6 mm long, acute to acuminate at apex, reddish brown to purplish brown, with a pale or green midrib.

Flowers: Female flowers with 3 stigmas.

Fruits: Utricles (5–)6–8 mm long (including beak), 2.5–3 mm wide, broader than glume, ovoid, strongly inflated, bluntly elliptic in cross section, dull olive green, becoming shining yellow when mature, with about 4 indistinct ribs on each side. Beak 1.5–2.3 mm long, more or less linear, tip deeply bidentate, with 0.5–1.0 mm long, slightly outward-directed tips.

Flowering: May-June.

Biology: Forming large stands. Propagation by seed and vegetatively by division of the rhizome.

Habitats: In mesotrophic to eutrophic, neutral to weakly alkaline water. On wet or moist, peaty or muddy substrates by pools, lakes and streams, in mires and forest swamps.

Distribution within the region:
Native. Widespread and abundant throughout most of the region. Absent from Iceland and Greenland.

Characteristics and similar species:
The slender, trigonous shoots with yellowish-green to green, hypostomous leaves, fibrillose, purplish-tinged leaf sheaths and acute ligule are characteristic of this species in a vegetative state.

Often confused with to **175** ***C. rostrata***, which differs in being epistomous with leaves glaucous on the upper side and dark green on the lower side, and smaller fruits rather abruptly tapering to an almost linear beak.

Carex vesicaria. The utricles are positioned at an oblique angle to the axis. Denmark. Photo JCS.

Flowering *C. vesicaria.* Hobro, Denmark. Photo JCS.

Carex vesicaria. Leaf sheath. Photo JCS.

Carex vesicaria. Utricles with or without glumes. Photo JCS.

Hybrids: *Carex vesicaria* forms hybrids with ***C. rostrata*** and **173 *C. riparia.*** Specialist literature is needed for identification of *Carex* hybrids.

Carex vesicaria
A habit, **B** part of female spike, **C** utricle, **D** female flower and glume, **E** male flower and glume, **F** leaf sheath and ligule.

Cyperaceae

177 *Carex acuta* L.

Syn.: *C. gracilis* Curt.
DK: Nikkende Star **N:** Kvass-storr **S:** Vassstarr **FIN:** Viiltosara **IS:** -
GB: Slender Tufted-sedge **NL:** Scherpe zegge **F:** Laiche grêle
D: Schlanke Segge **CZ:** Ostřice štíhlá **PL:** Turzyca zaostrzona
EST: Sale tarn **LV:** Slaidais grīslis **LT:** Lieknoji viksva **Ru:** Осока буровлагалищная

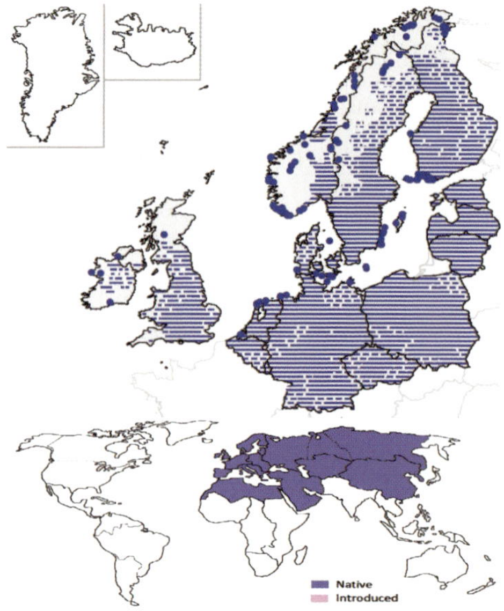

Carex acuta. River Gudenå, Jutland, Denmark. Photo JCS.

Perennial helophyte. Rhizome long, creeping, 2–3 mm in diameter, covered with soon splitting, greyish-brown to reddish-brown scales. Roots yellowish brown to reddish brown, up to 1.5 mm in diameter.

Stem: 50–120 cm long, 5–15 mm diameter at base, erect, often somewhat nodding, green, sharply trigonous and rough distally, glabrous and subterete at base.

Leaves: Base of stem with dark brown remnants of previous year's sheaths not splitting into fibres. Stem leaves usually longer than stem, 30–140 cm long, (3–)4–10 mm wide, dark green above, glaucous below, distinctly plicate, gradually tapering to a fine point, margins rough. Hypostomous. Leaf sheaths persistent, inner face hyaline, not splitting into fibres. Ligule 4–6 mm long, obtuse or truncate, with a short free margin.

Inflorescence: The upper 2–4 spikes entirely male, the lower 2–4 entirely female or with some male flowers distally. Male spikes 1.5–6 cm long, dark brown. Female spikes 2.5–10 cm long, cylindrical, narrowing towards apex, erect or nodding, the upper sessile, the lower shortly pedunculate and usually lax flowered at the base. Bracts leaf-like, the lower exceeding the inflorescence. Male glumes c. 5 mm long, obtuse to acute at apex, dark brown to purplish black, with a green midrib. Female glumes 2.5–4 mm long, usually longer than utricles, obtuse to acute at apex, dark brown to purplish black, with a green midrib.

Flowers: Female flowers with 2 stigmas.

Fruits: Utricles 2.5–3.5 mm long (including beak), 1–1.8 mm wide, broader than glume, ellipsoid to obovoid, biconvex to flat in cross section, glabrous, with 6–8 rather faint ribs on each side. Beak 0.1–0.2 mm long, truncate at apex.

Flowering: May-June.

Biology: Forming large, loose tufts. Propagation by seed and vegetatively by division of the rhizome.

Habitats: In weakly acid to alkaline, mesotrophic to eutrophic water. On sandy, peaty or clayey substrates by ponds, ditches and rivers or in fens.

Distribution within the region:
Native. Widespread and abundant throughout most of the region. Absent from Iceland and Greenland.

Characteristics and similar species:
A tall, slender rhizomatic plant forming loose tufts, the lower bracts exceeding the inflorescence, female flowers with 2 stigmas and the sheaths not splitting into fibres.

Carex acuta. Inflorescence. Randers, Denmark. Photo JCS.

Carex acuta. Utricles with or without glume. Photo JCS.

Carex acuta among tall grasses and herbs in a wet fen. Randers, Denmark. Photo JCS.

Similar in habit to **178 *C. aquatilis,*** which differs in being epistomous, the leaves light shiny green on the lower side, and in the female spikes being widest at the apex.

172 *C. acutiformis* and **173 *C. riparia*** differ by female flowers with 3 stigmas (rarely 2 stigmas), utricles trigonous (rarely flattish biconvex), distinctly ribbed, papillose, with a distinctly notched beak. Female glumes with mid-rib often long excurrent and leaf sheaths splitting into fibres.

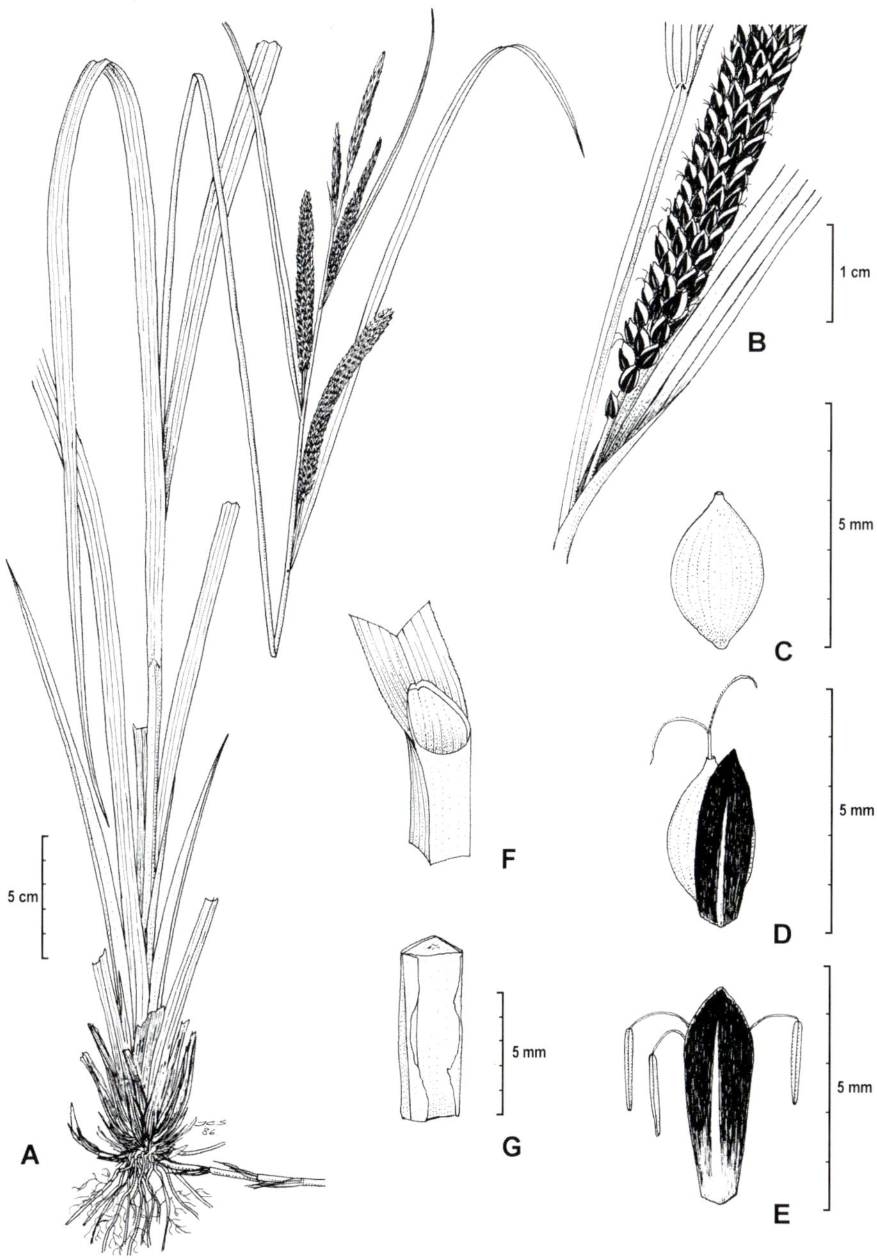

Carex acuta
A habit, **B** lower part of female spike, **C** utricle, **D** female flower and glume, **E** male flower and glume, **F** upper part of sheath with ligule, **G** sheath splitting up.

Cyperaceae

178 *Carex aquatilis* Wahlenb.

DK: Vand-Star **N:** Nordlandsstorr **S:** Norrlandsstarr **FIN:** Vesisara **IS:** -
GB: Water Sedge **NL:** Noordse zegge **F:** Laiche aquatique
D: Wasser-Segge **CZ:** - **PL:** -
EST: Vesitarn **LV:** Ūdeņu grīslis **LT:** Vandeninė viksva **Ru:** Осока водяная

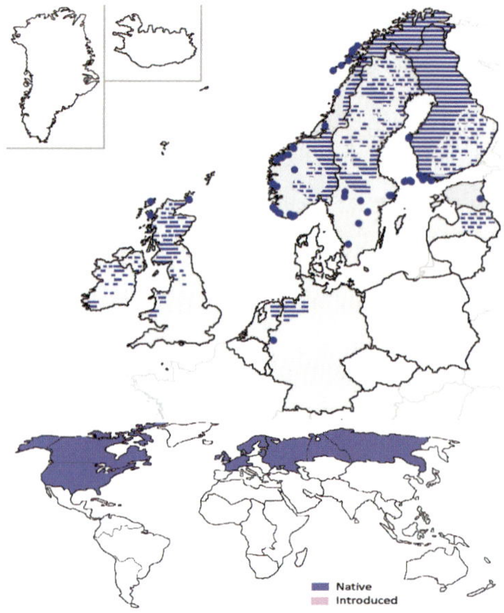

Carex aquatilis. Margin of River Spey, Boat of Garten, Speyside, Scotland. Photo RVL.

Perennial helophyte. Rhizome long, creeping, 3–4 mm in diameter, covered with persistent, brown to reddish-brown scales. Roots yellowish brown to reddish brown, up to 2 mm in diameter.

Stem: 30–90(–130) cm long, erect, often somewhat nodding, green, bluntly trigonous and smooth distally, subterete at base, brittle.

Leaves: Base of stem with brown to dark purplish-brown remnants of previous year's sheaths, not splitting into fibres. Stem leaves usually longer than stem, 20–130 cm long, 3–5(–8) mm wide, dull, papillose and more or less glaucous above, green and shiny below, flat or keeled, gradually tapering to a flat point. Epistomous. Leaf sheaths persistent, inner face hyaline, not splitting into fibres. Ligule 2–15 mm long, obtuse or acute, with a short free margin.

Inflorescence: The upper 2–4 spikes entirely male, the lower 2–5 entirely female or with some male flowers distally. Male spikes 1–5 cm long, dark brown. Female spikes 1.5–7 cm long, cylindrical, broadest towards apex, erect or nodding, the upper sessile, the lower shortly pedunculate and usually lax flowered at base. Bracts leaf-like - the lower up to 35 cm long and exceeding the inflorescence. Male glumes 3–4 mm long, obtuse at apex, brownish, with a pale midrib. Female glumes 2–3 mm long, shorter than or almost as long as utricles, obtuse to acute at apex, brown to dark brown, with a pale or greenish midrib.

Flowers: Female flowers with 2 stigmas.

Fruits: Utricles 2.5–3 mm long (including beak), 1.2–1.8 mm wide, broader than glume, ellipsoid to obovoid, biconvex to flat in cross section, green, often with small red dots, glabrous, without ribs. Beak 0.1–0.3 mm long, truncate at apex.

Flowering: June-July.

Biology: Forming large, loose tufts or extensive stands. Propagation by seed and vegetatively by division of the rhizome.

Habitats: In oligotrophic to mesotrophic, acid to neutral water. Wet or flooded meadows with other tall sedges, in the littoral zone of ponds, lakes and rivers, in the Nordic area also on seashores.

Distribution within the region:
Native. Scattered to locally frequent in northwestern Germany, the Netherlands, Ireland, northern and western Britain, Latvia and Estonia. Widespread and abundant in Norway, Finland, central and northern Sweden.

Characteristics and similar species:
Recognised by creeping rhizome, clumped heads with very long lowest

Carex aquatilis. Margin of River Spey, Boat of Garten, Speyside, Scotland. Photo RVL.

Carex aquatilis. Margin of Loch Morlich, Speyside, Scotland. Photo RVL.

bract, epistomous leaves which are glaucous above and shining green below, female spikes widest towards the apex and lax flowered at the base, and flat utricles.

Similar in habit to **177 *C. acuta***, which differs in being hypostomous, with leaves glaucous on the lower side, distally rough stem and female spikes with acute apex.

Carex aquatilis. Utricles with or without glumes. Photo JCS.

Carex aquatilis
A habit, **B** female spike, **C** utricle, **D** female flower and glume, **E** male flower and glume, **F** splitting sheath with ligule.

Cyperaceae

179 *Carex elata* L.

Syn.: *C.stricta* Gooden.; *C. hudsonii* A.Benn.
DK: Stiv Star **N:** Bunkestorr **S:** Bunkestarr **FIN:** Piukkasara **IS:** -
GB: Tufted Sedge **NL:** Stijve zegge **F:** Laiche élevée
D: Steif Segge **CZ:** Ostřice vyvýšená **PL:** Turzyca sztywna
EST: Luhttarn **LV:** Augstais grīslis **LT:** Aukštoji viksva **Ru:** Осока высокая

Carex elata forming a large tussock in a pond. Randers, Denmark. Photo JCS.

Perennial helophyte. Rhizome very short, more or less erect. Roots greyish-brown to purplish brown, up to 2 mm in diameter.

Stem: 30–120 cm long, 6–12 mm diameter at base, erect, often somewhat nodding, dark green, sharply trigonous, rough.

Leaves: Base of stem with straw-coloured, sharply keeled scales up to 10 cm long, surrounded by decaying remnants of previous year's sheaths. Stem leaves shorter than stem, 40–100 cm long, 2.5–6(–8) mm wide, glaucous on both sides, plicate, gradually tapering to a fine point, margins rough. Papillae on the lower side of leaves distinctly protruding from epidermis. Hypostomous. Leaf sheaths persistent, becoming straw-coloured, inner face hyaline, splitting into fibres. Ligule 4–12 mm long, acute, with a short free margin.

Inflorescence: The upper 1–2 spikes entirely male, the lower 2–4 entirely female or the uppermost with some male flowers distally. Male spikes 1.5–5.5 cm long, blackish brown. Female spikes 2–6 cm long, cylindrical, erect, sessile. Bracts leaf-like, the lowest 3–10 cm long, much shorter than the inflorescence. Male glumes c. 5 mm long, acute at apex, dark brown to blackish brown, with a yellow midrib. Female glumes c. 3 mm long, acute at apex, dark brown to blackish brown, with a yellowish midrib.

Flowers: Female flowers with 2 stigmas.

Fruits: Utricles 2.5–3(–3.5) mm long (including beak), 1.2–1.9 mm wide, broader than glume, ellipsoid to ovoid, biconvex to flat in cross section, papillose, with 3–4 distinct ribs on each side. Beak 0.2–0.3 mm long, truncate at apex.

Flowering: April-May.

Biology: Forming large, dense tufts or tussocks. Propagation by seed.

Habitats: In oligotrophic to eutrophic, acid to alkaline water. On wet, often seasonal flooded, silty or muddy substrate in ditches, rivers, ponds and lakes.

Distribution within the region:
Native. Rather common to common throughout most of the area except northern Britain, Fennoscandia, Iceland and Greenland.

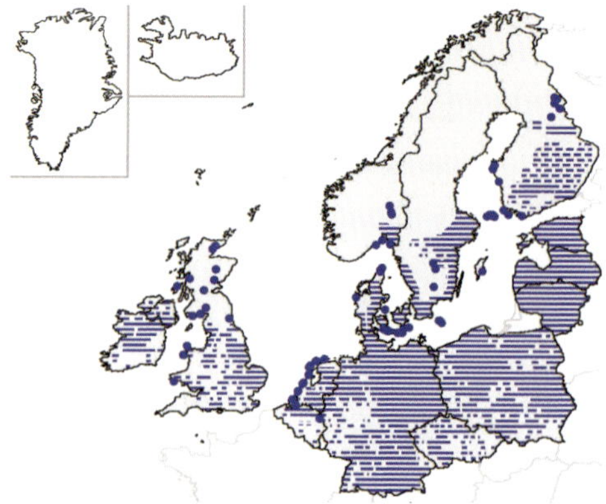

Distribution of *C. elata* including subsp. *omskiana*.

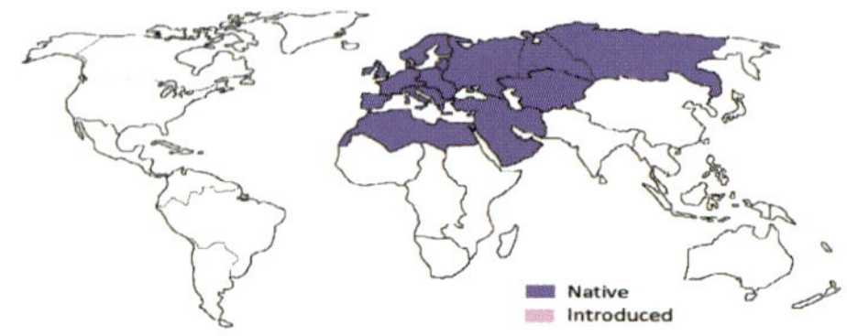

Characteristics and similar species:
A tall plant forming dense tufts or tussocks. The fibrillose sheaths and long straw-coloured scales at base of stem separate *Carex elata* from similar species in the study area,

Hybrids: Hybridises with **177 *C. acuta*** and ***C. nigra.***

Carex elata. Base of shoots with long, straw-coloured scales and old sheaths. Photo JCS.

Carex elata. Leaf sheaths splitting into fibres. Photo JCS.

Carex elata. Utricles with or without glumes. Photo JCS.

Carex elata. Inflorescence. Photo JCS.

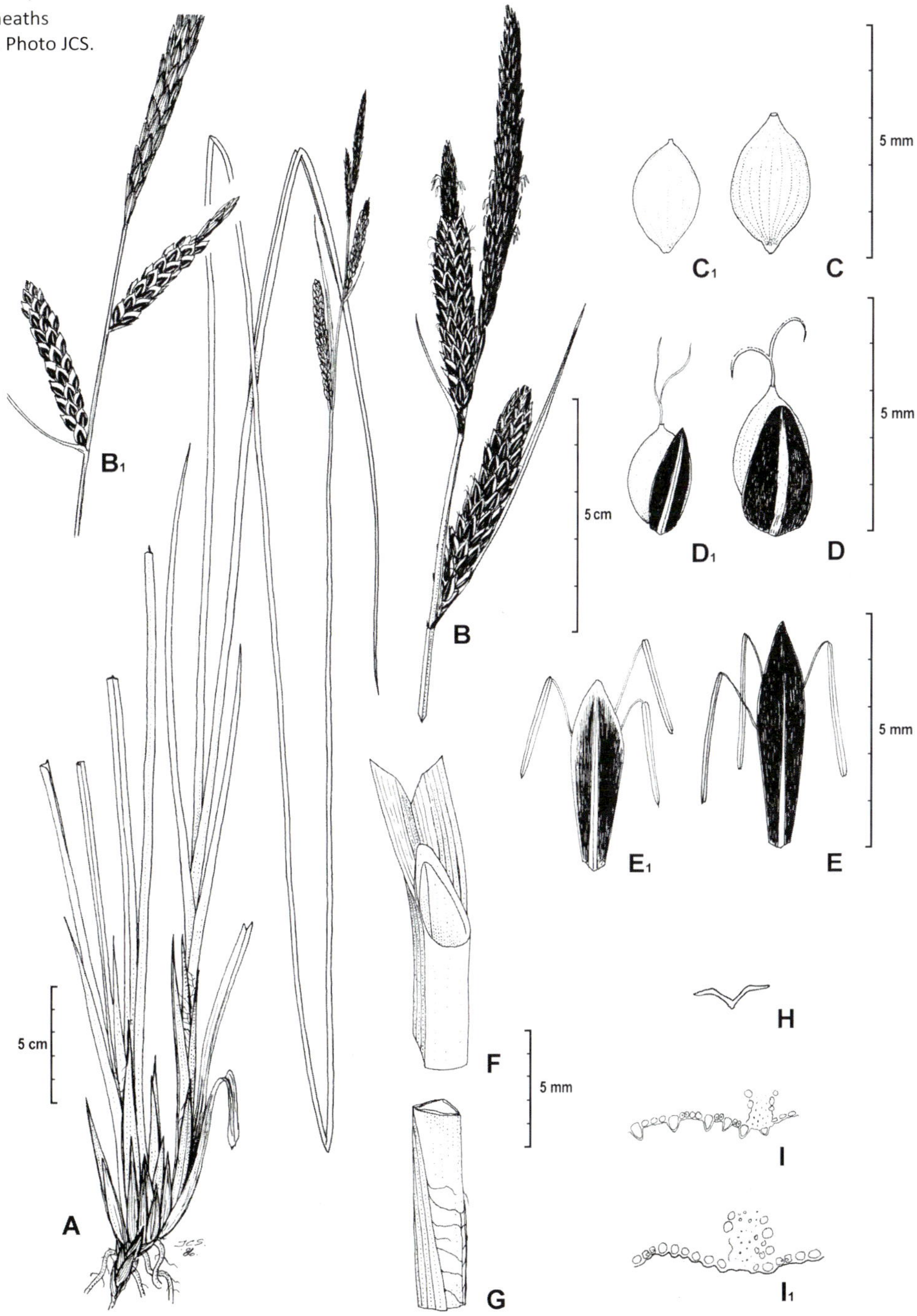

Subsp. ***omskiana*** (Meinsh.) Jalas

Differs in stem 30–80 cm long, leaves 2–3.5(–5) mm wide, papillae on the lower side only slightly protruding from epidermis. Utricles faintly ribbed.

Distribution within the region:
Finland, Baltic states, Poland, Germany.

Carex elata subsp. ***elata***
A habit, **B** inflorescence, **C** utricle, **D** female flower and glume, **E** male flower and glume, **F** upper part of sheath with ligule, **G** sheath splitting into fibres, **H** cross section of leaf, **I** magnified cross section of lower leaf side.
Carex elata subsp. ***omskiana***
B1 inflorescence, **C1** utricle, **D1** female flower and glume, **E1** male flower and glume, **F** upper part of sheath with ligule, **G** sheath splitting into fibres, **I1**magnified cross section of lower leaf side.

Key to aquatic *Poaceae* in vegetative state

1 All leaves with blades only 1–2 cm long. Upper leaf sheaths strongly inflated (***1***)
186 *Coleanthus subtilis*

1 At least some leaves more than 2 cm long. Upper leaf sheaths not or at most moderately inflated 2

2 Ligule present as a fringe of hairs (***2***) 18

2 Ligule membranous, thin 3

3 Leaves folded when young (***3***) 4

3 Leaves rolled when young (***4***) 13

4 Leaves 1–2 mm wide, bristle-like. Lower leaf sheaths open with overlapping margins
180 *Agrostis canina*

4 Leaves at least 3 mm wide. Lower leaf sheaths closed at least halfway up - if accidently split, then margins not overlapping 5

5 Leaves with cross-veins (***5***) 6

5 Leaves without cross-veins (***6***) 10

6 Leaf tip blunt, often asymmetric, slightly hooded (***7***). Leaves with small, irregular air cavities in cross section
185 *Catabrosa aquatica*

6 Leaf tip acute or rarely blunt, symmetric, hooded. Leaves with almost uniform, rectangular air cavities in cross section (***9***) 7

7 Ligule blunt with a fine, median tooth (***8***). Leaves 7–20 mm wide, upper side not ribbed. Shoots stout to robust
187 *Glyceria maxima*

7 Ligule blunt or acute, but without a median tooth. Leaves 2–14 mm wide, upper side ribbed. Shoots rather slender 8

8 Leaves glaucous to grey green or purplish tinged, 3–5(–9) mm wide. Cross-veins indistinct
189 *Glyceria declinata*

8 Leaves yellowish-green to dark green, 4–10(–15) mm wide. Cross-veins distinct 9

9 Youngest ligules much longer than the width of the associated leaf (investigate only very young, not yet unfolded leaves) (***10***). Leaves shallowly ribbed. Leaf sheaths usually reddish
188 *Glyceria fluitans*

9 Youngest ligules only slightly longer than the width of the associated leaf (***11***). Leaves deeply ribbed. Leaf sheaths not reddish
190 *Glyceria notata*

1

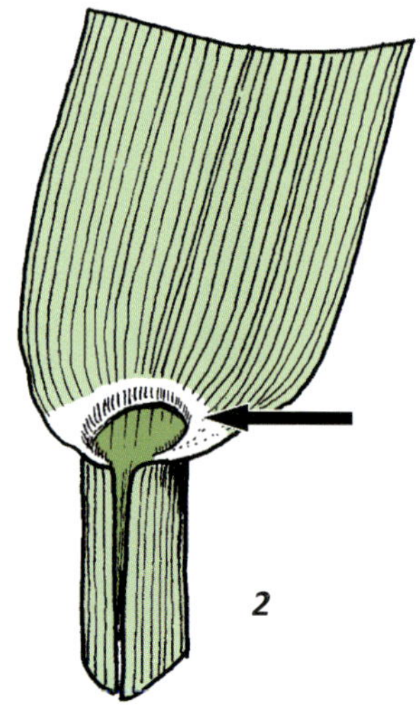
2

3

4

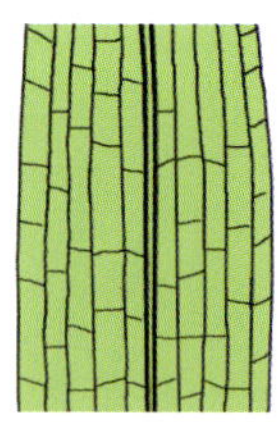
5

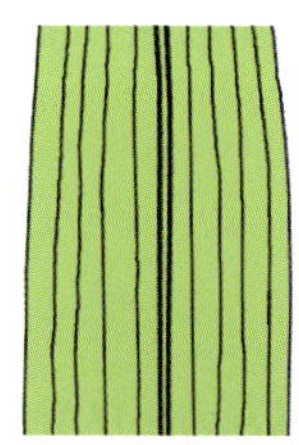
6

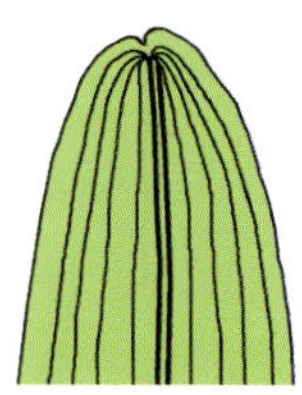
7

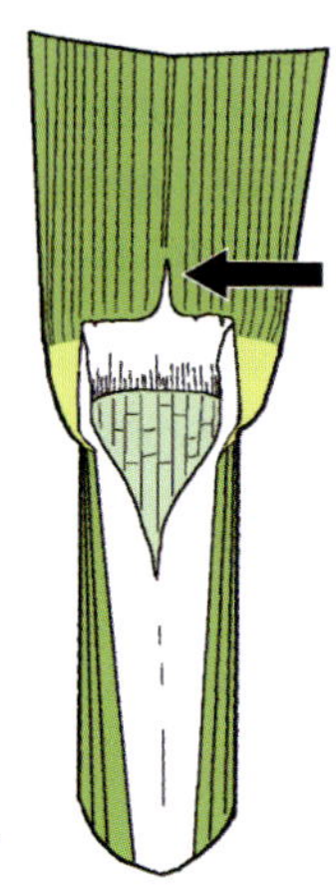
8

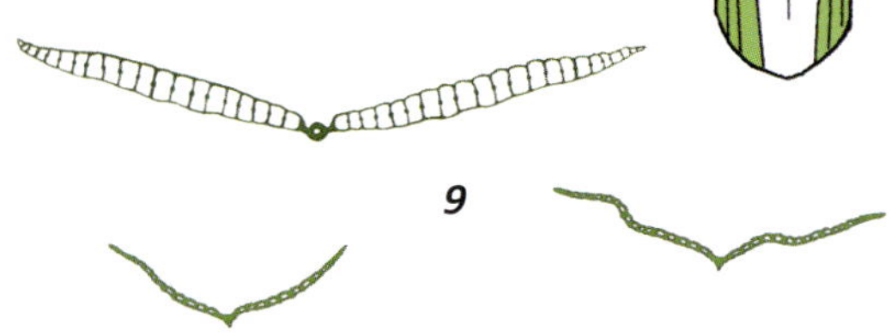
9

10 Leaves with more or less uniform, rectangular air cavities in cross section (***9***).
Leaves 3–5(–9) mm wide
189 *Glyceria declinata*

10 Leaves without or with small, irregular air cavities in cross section.
Leaves 2–4 mm wide 11

11 Plant with underground stolons. Ligules less than 1 mm long. Leaves abruptly pointed, hooded
Poa pratensis L.
(not treated in this book)

11 Plant without underground stolons. Ligules 2–10 mm long. 12

12 Leaf sheaths rough. Leaves abruptly pointed, hooded
195 *Poa trivialis*

12 Leaf sheaths smooth. Leaves gradually tapering
194 *Poa palustris*

13 Plants large. Stems 60–250 m long. Stolons 4–8 mm in diameter 14

13 Plants small or medium sized, slender. Stems 15–120 m long. Stolons 1–2 mm in diameter 15

14 Leaves stiff, greyish green
192 *Phalaris arundinacea*

14 Leaves rather flaccid, fresh green
196 *Scolochloa arundinacea*

15 Leaf sheaths, leaf margins and midribs rough with retrorse teeth on lower side of leaves (***12***)
191 *Leersia oryzoides*

15 Leaf sheaths, leaf margins and midribs not particularly rough on lower side of leaves 16

16 Plants with long underground stolons clothed with more than 3 scales. Leaves (2–)4–10 mm wide
181 *Agrostis gigantea*

16 Plants without or with very short underground stolons clothed with 1–3 scales. Above-ground, creeping stolons present or not. Leaves 2–5(–6) mm wide 17

17 Uppermost leaf sheath somewhat inflated. Blades not keeled beneath.
184 *Alopecurus geniculatus*
183 *Alopecurus aequalis*
(2 species which can not be separated reliably using vegetative characters)

17 Uppermost leaf sheath not inflated. Blades slightly keeled beneath
182 *Agrostis stolonifera*

18 Leaf margins papillate, somewhat rough, especially towards apex.
Plants of freshwater habitats 19

18 Leaf margins smooth, cartilaginous. Plants of salt marshes
Spartina species
(not treated in this book)

19 Leaves 20–60 cm long, more or less glaucous below
193 *Phragmites australis*

19 Leaves 60–100 cm long, shiny green below
Spartina pectinata Bosc ex Link
(not treated in this book)

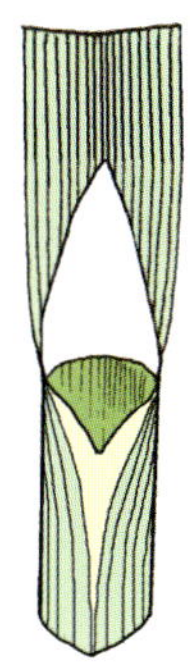

10

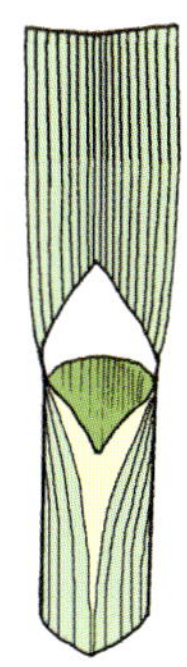

11

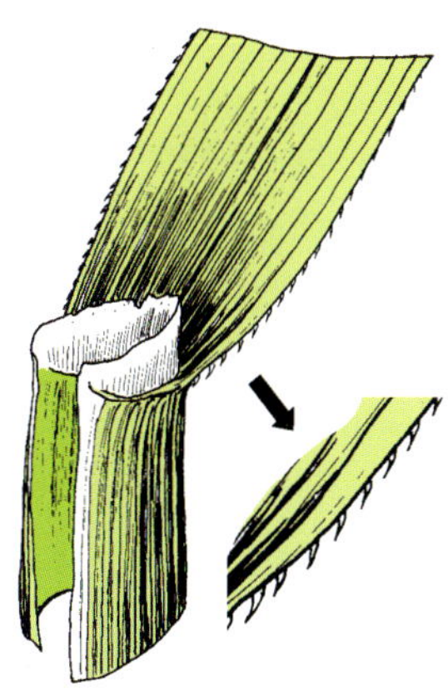

12

Poaceae

180 *Agrostis canina* L.

DK: Hunde-Hvene **N:** Hundekvein **S:** Brunven **FIN:** Luhtarölli **IS:** -
GB: Velvet Bent **NL:** Kruipend struisgras **F:** Agrostide canine
D: Hunds-Straußgras **CZ:** Psineček psí **PL:** Mietlica psia
EST: Soo-kastehein **LV:** Suņu smilga **LT:** Šuninė smilga **Ru:** Полевица собачья

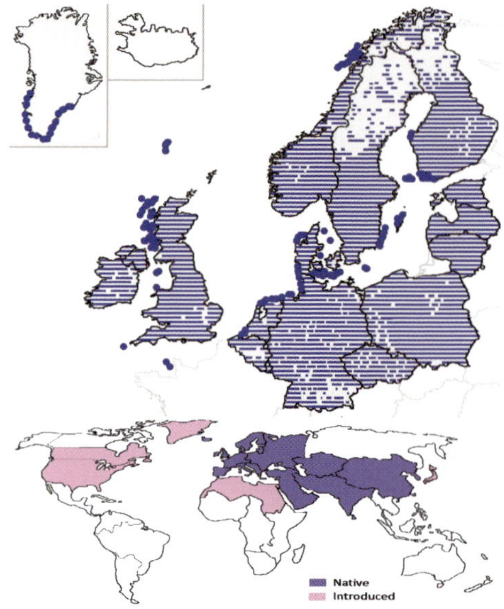

Agrostis canina on sandy, temporarily flooded substrate in a dune depression. Thy, Denmark. Photo JCS.

Perennial helophyte. Underground rhizomes absent.

Stem: 15–60 cm long, smooth, slender, rather stiff with 2–4 nodes, erect or ascending from a bent or prostrate base.

Leaves: 3–15 cm long, 1–2(–3) mm wide, gradually tapering to a fine point, green to greyish green, flat to rolled, rough, finely ribbed above. Rolled and bristle-like on vegetative shoots. Leaf sheaths open, smooth, but the upper sometimes rough. Ligule 2–4 mm long, membranous, blunt or subacute at apex.

Inflorescence: A panicle, 3–15 cm long, up to 7 cm wide, purplish brown to grey brown, ovate in outline during flowering, later more dense and lanceolate in outline. Branches rough or smooth, very fine, of different lengths, the lowest usually in a cluster of 3–12.

Spikelets: 1.7–3.0 mm long (excluding awn), 1-flowered, lanceolate, purplish to greyish brown. At maturity the seed falls enveloped in the thin lemma and palea, while the glumes remain.

Glumes: 1.7–3.0 mm long, the lower slightly longer than the upper, lanceolate, pointed, 1-veined, membranous, hairless, minutely rough on the keel.

Lemma: 1.5–1.8 mm long, 5-veined, glabrous but finely pubescent at base, broadly lanceolate, blunt, thin. Awn arising from the basal half of lemma, 2.0–4.5 mm long, bent, twisted in the lower part or absent.

Palea: 0.2–0.3 mm long, at most 1/5 the length of lemma.

Flowers: Stamens 3, anthers 1.0–1.5 mm long.

Fruits: Caryopsis c. 1.2 mm long, ellipsoid, flat.

Flowering: May-August.

Biology: The seeds are dispersed by wind. Vegetative propagation by long, branched, above-ground stolons rooting at the nodes and producing greyish-green tufts or turfs of new, leafy shoots.

Habitats: In acid, oligotrophic water. On moist or wet substrate in peaty meadows, mires, marshes, wet heathland, gravel pits, fens, ditches and lakeshores.

Distribution within the region:
Native. Widespread and abundant throughout the region, except Iceland.

Characteristics and similar species:
182 *A. stolonifera* differs by having 2–8 mm wide leaves, awnless spikelets and flowers with relatively much longer palea - 2/3–3/4 the length of lemma.

Agrostis canina. New Forest, Britain. Photo RVL.

Hybrids: The following combinations are reported from the study area:

A. canina × capillaris, A. canina × stolonifera.

Agrostis canina. Pool, Lüsekamp, Niederkrüchten, Germany. Photo KvdW.

Agrostis canina. Thy, Denmark. Photo JCS.

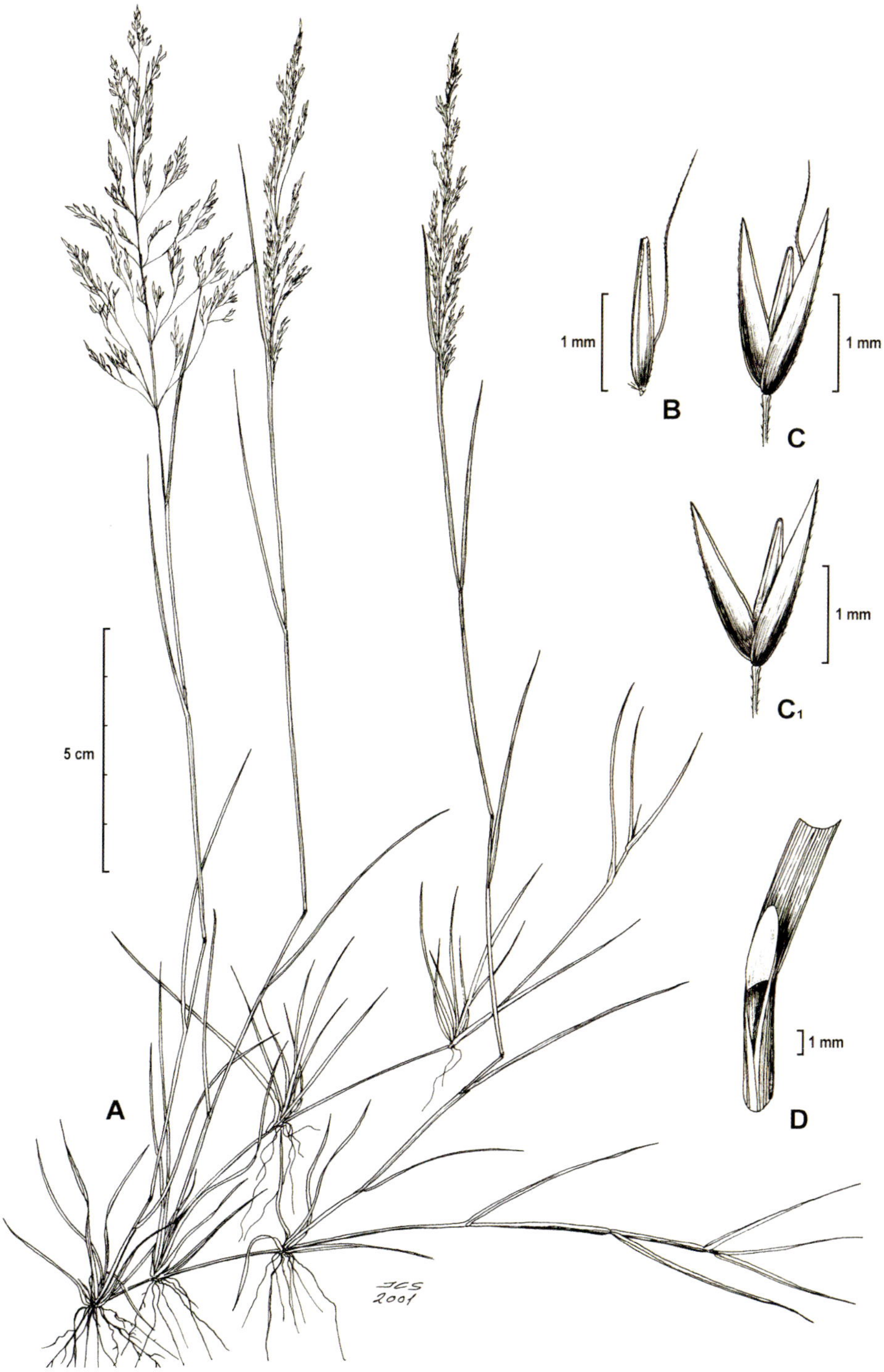

Agrostis canina
A habit, **B** lemma with awn,
C, C1 spikelet with or without awn, **D** leaf sheath.

Poaceae

181 *Agrostis gigantea* L.

DK: Stortoppet Hvene **N:** Storkvein **S:** Storven **FIN:** Isorölli **IS:** -
GB: Black Bent **NL:** Groot struisgras **F:** Agrostide géante
D: Riesen-Straußgras **CZ:** Psineček veliký **PL:** Mietlica olbrzymia
EST: Suur kastehein **LV:** Lielā smilga **LT:** Didžioji smilga **Ru:** Полевица гигантская

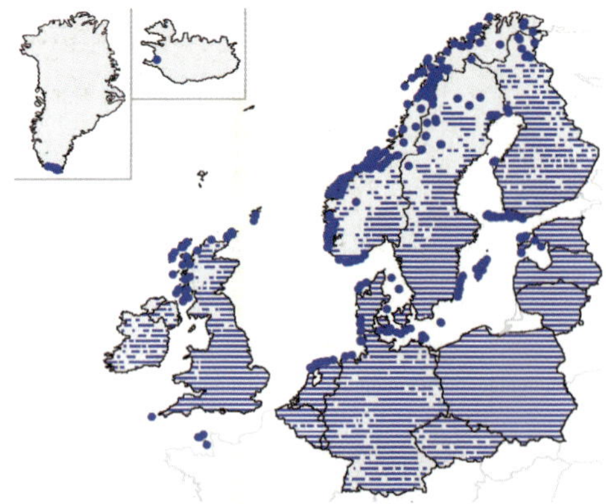

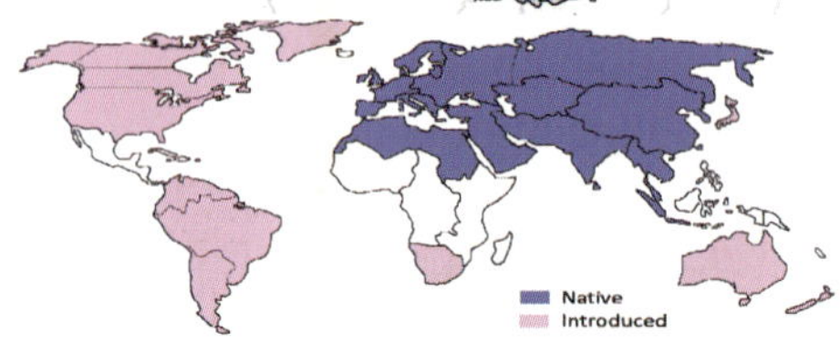

Agrostis gigantea on gravelly substrate at the shore of an artificial pond. Hobro, Denmark. Photo JCS.

Perennial helophyte. Underground rhizomes long, tough, 1–2 mm in diameter, branched, yellowish, with more than 3 scales.

Stem: 40–120 cm long, smooth or slightly rough, slender, rather stiff with 4–8 nodes, erect or ascending from a bent or prostrate, often rooting base.

Leaves: 5–20 cm long, (2–)4–10 mm wide, gradually tapering, to a fine point, dull green, flat or rolled, especially when young, rough, finely ribbed on the upper side. Leaf sheaths rounded on the back, open, smooth or rough. Ligule 2–6 mm long, membranous, toothed, blunt at the apex.

Inflorescence: A panicle, 8–25 cm long, 3–15 cm wide, green to purplish brown, open and ovate in outline during flowering, later more dense and lanceolate in outline. Branches rough, fine, of different lengths, the lowest usually in a cluster of 3–7.

Spikelets: 2.0–3.2 mm long (excluding awn), 1-flowered, lanceolate, purplish brown to greenish. At maturity the seed falls enveloped in lemma and palea, while the glumes remain.

Glumes 2.0–3.0 mm long, the lower slightly longer than the upper, lanceolate, pointed, 1-veined, membranous, hairless, minutely rough on the keel.

Lemma 1.6–2.2 mm long, 5-veined, glabrous, broadly lanceolate, blunt, thin. Awn absent or rarely present, very short and arising from the apical part of the lemma.

Palea 0.8–1.4 mm long, 1/2–2/3 the length of lemma.

Flowers: Stamens 3, anthers 1.0–1.5 mm long.

Fruits: Caryopsis c. 1 mm long, terete.

Flowering: June-August.

Biology: The seeds are dispersed by wind. Vegetative propagation by underground rhizomes forming loose tufts. Cultivated for hay and used in lawns.

Habitats: In neutral, mesotrophic to eutrophic water. On moist to dry substrate in meadows, grassland, roadsides and other disturbed places, along brooks and streams, lake margins.

Distribution within the region:
Native. Widespread and abundant throughout the region, but rare in Iceland and Greenland.

Characteristics and similar species:
Similar in habit to **182** ***A. stolonifera***, which differs by long, rooting stolons above ground, but usually without an underground rhizome - if this is present it is short and with 1–3 scales only. ***A. capillaris*** (not treated in this book) differs by short, 0.3–1.0 mm long ligules.

Agrostis gigantea. Inflorescence. Hobro, Denmark Photo JCS.

Agrostis gigantea. Sheath with ligule. Snæbum, Hobro, Denmark. Photo JCS.

Hybrids: The following combinations are reported from the region:

A. capillaris × gigantea, A. gigantea × stolonifera.

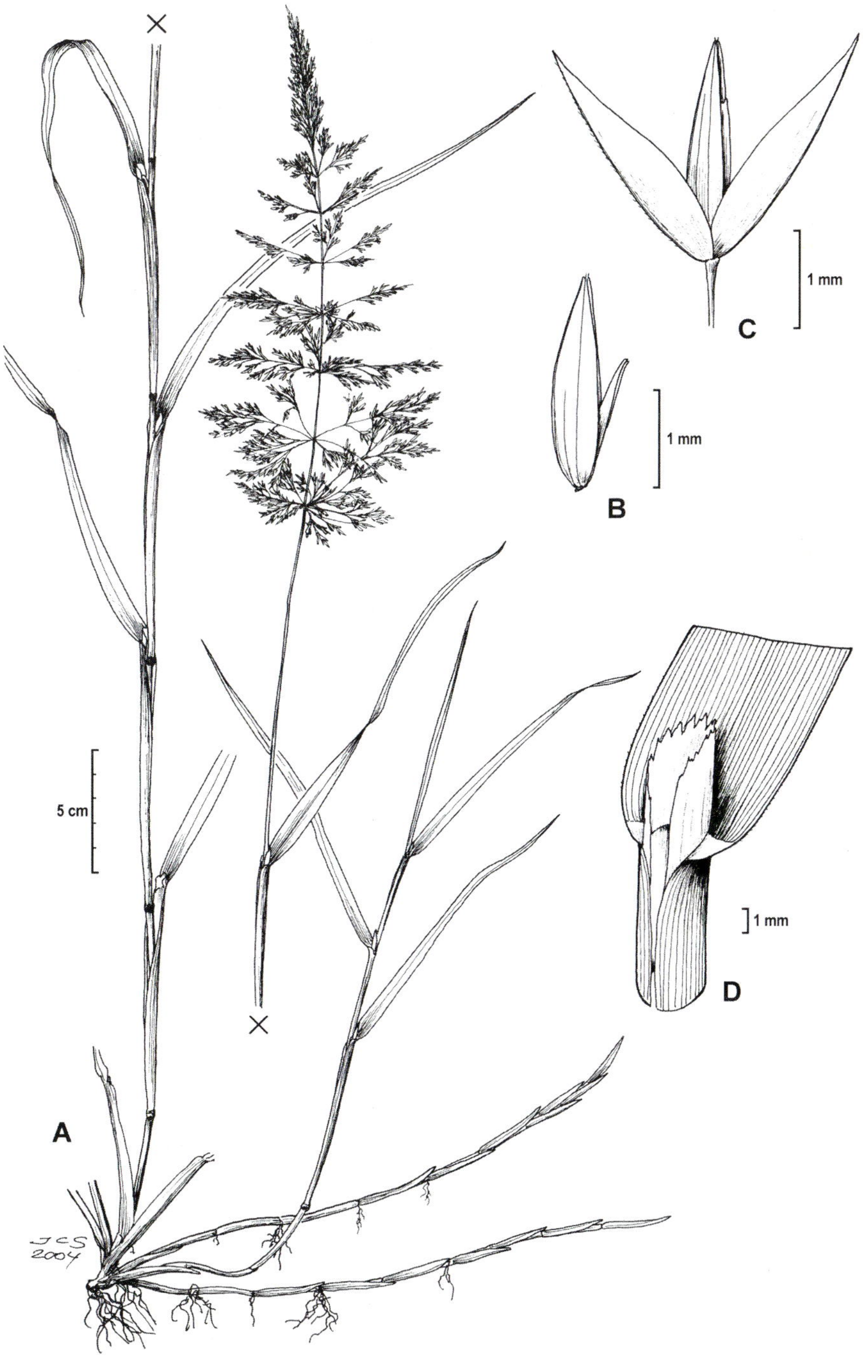

Agrostis gigantea
A habit, **B** lemma and palea, **C** spikelet, **D** leaf sheath.

Poaceae

182 *Agrostis stolonifera* L.

DK: Kryb-Hvene **N:** Krypkvein **S:** Krypven **FIN:** Rönsyrölli **IS:** Skriðlíngresi
GB: Creeping Bent **NL:** kruipgras **F:** Agrostide stolonifère
D: Weißes Straußgras **CZ:** Psineček výběžkatý **PL:** Mietlica rozłogowa
EST: Valge kastehein **LV:** Ložņu smilga **LT:** Baltoji smilga **Ru:** Полевица побегообразующая

Agrostis stolonifera stretching its shoots into the water. Ditch, Dilborner Benden, Brüggen, Germany. Photo KvdW.

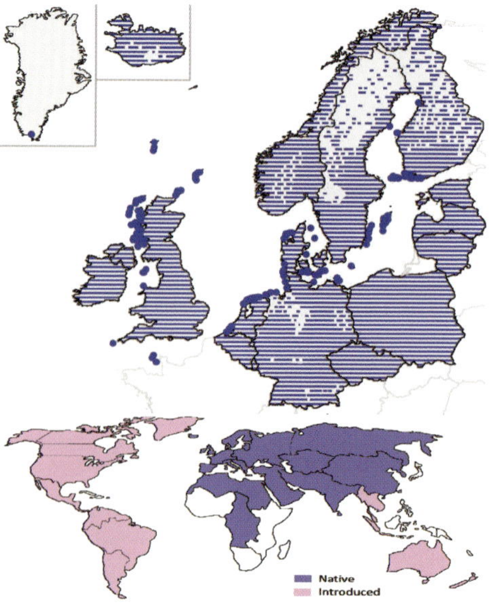

Perennial helophyte. Stolons long, branched, above ground, rooting at the nodes and producing new shoots. Underground rhizomes very short, with 1–3 scales.

Stems: 10–50 cm long, glabrous, rarely slightly rough, slender with 2–5 nodes, erect or ascending from a bent or prostrate, rooting base.

Leaves: 2–15 cm long, 2–8 mm wide, flat or rolled, especially when young, slightly keeled, finely pointed, dark green to glaucous, closely ribbed on the upper side, somewhat rough, hairless. Leaf sheaths rounded on the back, open, usually smooth. Ligule 2–6 mm long, membranous, entire, blunt, at apex but often torn.

Inflorescence: A panicle 2–15 cm long, up to 6 cm wide, greenish to dull purplish brown, oblong and open during flowering, before and after contracted and dense or open only towards the base. Branches rough, fine, of different lengths, the lowest usually in a cluster of 3–7.

Spikelets: 2.0–3.2 mm long (excluding awn), 1-flowered, lanceolate, purplish brown to greenish. At maturity the seed falls enveloped in lemma and palea, while the glumes remain.

Glumes: 1.7–3.2 mm long, the lower slightly longer than the upper, lanceolate, pointed, 1-veined, membranous, hairless, minutely rough on the keel.

Lemma: 1.6–2.2 mm long, 5-veined, glabrous, broadly lanceolate, blunt, thin. Awn rarely present, 0.5–3 mm long and arising from the apical part of lemma or absent.

Palea: 0.8–1.3 mm long, 2/3–3/4 the length of lemma.

Flowers: Stamens 3, anthers 1.0–2.5 mm long.

Fruits: Caryopsis c. 1.2 mm long, terete.

Flowering: July-August.

Biology: The seeds are dispersed by wind. Vegetative propagation is by above-ground, creeping stolons, often forming large mats. Cultivated as a lawn grass and to bind soil on roadsides and slopes.

Very variable. In dense vegetation the characteristic creeping stolons can be very short, while plants growing on open, wet substrates or on the margins of small streams or pools may produce stolons up to 2 metres long, growing out over the surface of the water or forming submerged stands.

Habitats: In mesotrophic to highly eutrophic water. On wet to dry substrates in a variety of habitats ranging from saltmarshes, heathland, wet meadows, cliffs, woodland and the margins of

The inflorescence of *A. stolonifera* is contracted before and after flowering. Hobro, Denmark. Photo JCS.

Agrostis stolonifera L. with up to 2 metre long stolons on wet substrate in a dune depression. Uggerby, North-Jutland, Denmark. Photo JCS.

lakes and streams to gravel pits, roadsides and waste places.

Distribution within the region:
Native. Widespread and abundant more or less throughout but rare in Greenland.

Characteristics and similar species:
Vegetative plants are easily mistaken for **184 *Alopecurus geniculatus*** and **183 *A. aequalis***, but both of these differ by leaves not keeled, upper leaf sheaths inflated and often with a whitish, waxy covering.

Vegetative **185 *Catabrosa aquatica*** differs by more or less blunt, hooded leaf apex and leaf blades with two grooves along the midrib.

181 *Agrostis gigantea* differs by long, yellowish underground rhizomes always with at least 3 scales.

180 *A. canina* has narrower leaves, folded when young.

Hybrids: *Agrostis* hybrids are not easily identified, and more knowledge of regional variation in the parent species is needed. They often have very thin anthers with failed pollen. The following combinations are reported from the region:

A. canina × stolonifera, A. capillaris × stolonifera, A. gigantea × stolonifera.

Agrostis canina
A, **A1** habit, **B** lemma and palea, **C** spikelet, **D** leaf sheath.

Poaceae

183 *Alopecurus aequalis* Sobol.

Syn.: *A. fulvus* Sm.
DK: Gul Rævehale **N:** Vassreverumpe **S:** Gulkavle **FIN:** Rantapuntarpää **IS:** Vatnsliðagras
GB: Orange Foxtail **NL:** Rosse vossenstaart **F:** Vulpin fauve
D: Rotgelber Fuchsschwanz **CZ:** Psárka plavá **PL:** Wyczyniec czerwonożółty
EST: Ruuge rebasesaba **LV:** Līdzīgā lapsaste **LT:** Glotnusis pašiaušėlis **Ru:** Лисохвост равный

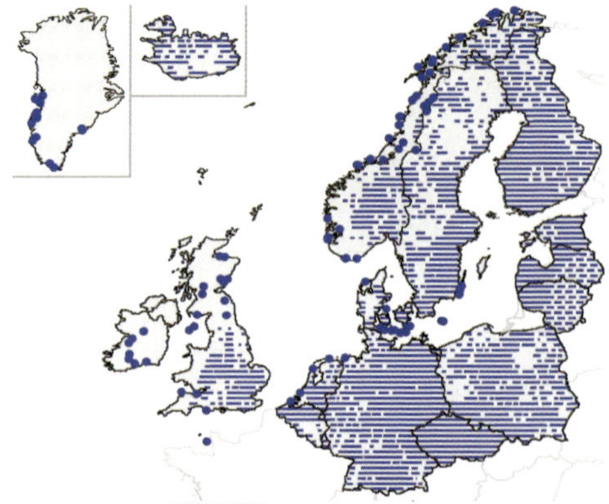

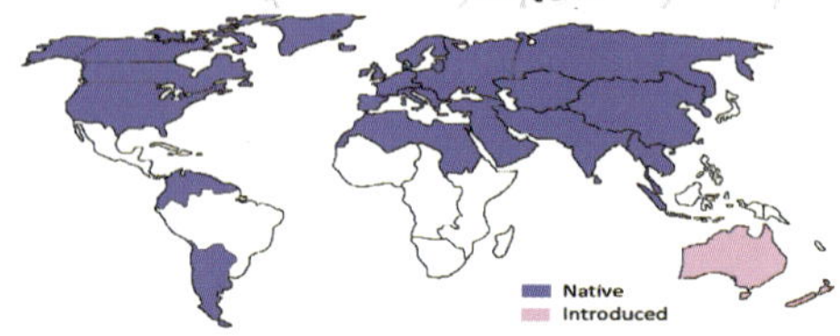

Alopecurus aequalis stretching out into the water. Vlijmen, Netherlands. Photo KvdW.

Annual or short-lived perennial helophyte. Underground rhizomes absent.

Stems: 10–35 cm long, glabrous, slender, whitish glaucous towards the inflorescence, ascending from a bent or prostrate base, with 5–8 nodes, the nodes usually yellow.

Leaves: 2–8(–12) cm long, 2–5 mm wide, flat, not keeled, gradually tapering to the apex, green to greyish, glabrous, rough on the upper side, smooth below, with 6–7 ribs on each side of the midrib. Leaf sheaths open, glabrous, terete. The upper somewhat inflated, greenish with a waxy, whitish covering, sometimes faintly streaked with purple. Ligule 2–5 mm long, membranous, entire, blunt.

Inflorescence: A dense, spike-like panicle, 2–7 cm long, 3–6 mm wide, narrowly cylindrical, pale green to greyish green, main axis hairy, pedicels very short.

Spikelets: 2.2–2.5 mm long (excluding awn), 1-flowered, oblong to lanceolate, flat. The entire spikelet falls at maturity.

Glumes: 2.2–2.5 mm long, equal in size and shape, fused at base, 3-veined, blunt or rounded at apex, keeled, with a fringe of hairs 0.5 mm long on the keel, with shorter hairs on the sides.

Lemma: 2.0–2.6 mm long, 4-veined, smooth, broadly elliptical, obtuse at apex, membranous, margins fused in the lower half. Awn 1–2 mm long, more or less straight, arising on the back of lemma just below the middle, and not or only shortly protruding from the glumes.

Palea: Absent.

Flowers: Stamens 3, anthers 1–1.2 mm long, at first yellowish white, later orange.

Fruits: Caryopsis 0.8–1.0 mm long.

Flowering: May-August.

Biology: Self-sterile. The seeds are dispersed by wind. The plant forms small tufts or loose mats. Perennial through vegetative propagation by aboveground, rooting stolons and short underground rhizomes. Sometimes with leaves floating on the water.

Tufts of *A. aeqaulis* on temporarily flooded, muddy substrate by a pool. Djurle Våtmark, Växsjö, Sweden. Photo JCS.

Alopecurus aequalis usually has yellow nodes, but this is not a reliable character for differentiation from *A. geniculatus*. Hobro, Denmark. Photo JCS.

Alopecurus aequalis. The yellowish-white anthers turning orange over time is a good character, but should be supported by examination of the awn. Vlijmen, Netherlands. Photo KvdW.

Distribution within the region: Native. Widespread and abundant throughout most of the region. Scattered in Ireland, northern Britain and Greenland.

Characteristics and similar species: The whitish-yellow anthers becoming orange over time when flowering is a good character for differentiation from the similar **184 *A. geniculatus***, which also differs in the bent awn protruding 2–3 mm from the glumes, anthers 1.2–1.8 mm long, purple or yellow, becoming brown with age. It is not possible to distinguish vegetative ***A. aequalis*** from ***A. geniculatus*** reliably.

Vegetative plants may be confused with **182 *Agrostis stolonifera***, but in this species the upper sheaths are not inflated and the leaves are slightly keeled.

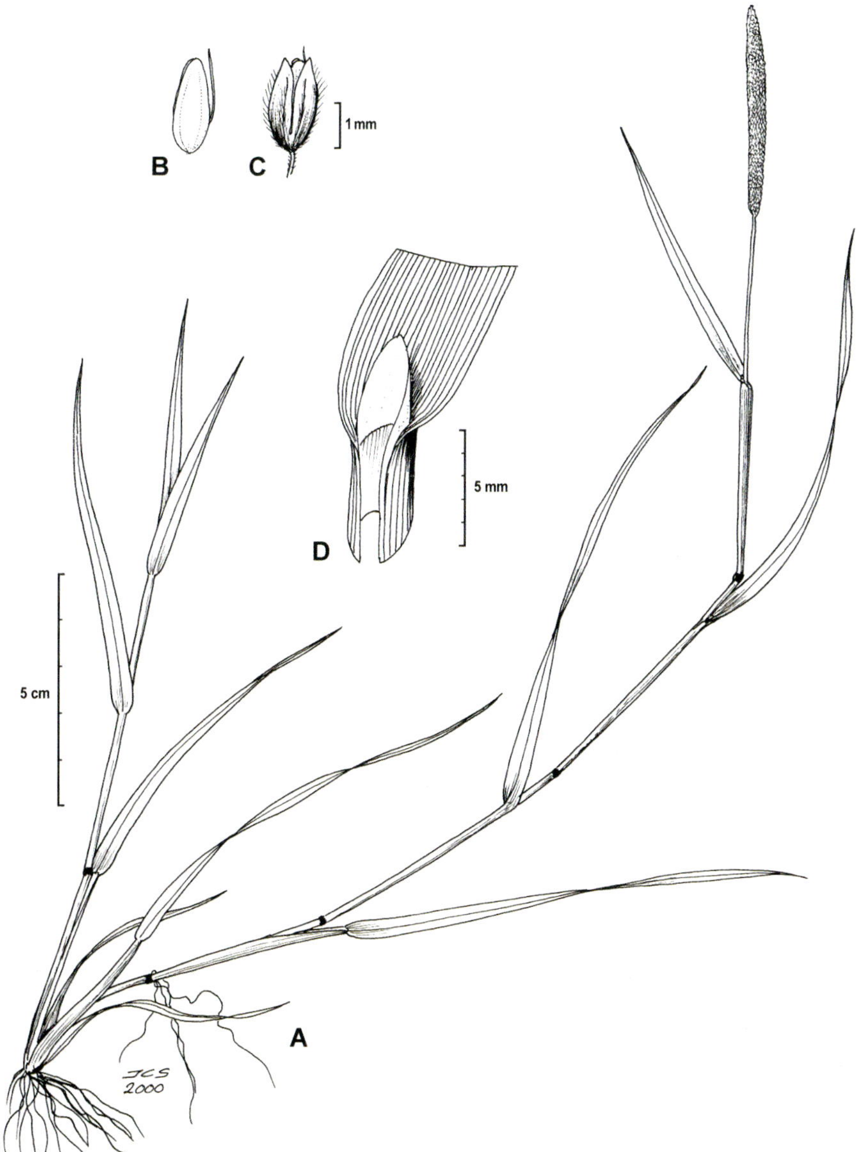

Alopecurus aequalis
A habit, **B** lemma and palea, **C** spikelet, **D** leaf sheath.

Poaceae

184 *Alopecurus geniculatus* L.

DK: Knøbøjet Rævehale **N:** Kneverumpe **S:** Kärrkavle **FIN:** Polvipuntarpää **IS:** Knjáliðagras
GB: Marsh Foxtail **NL:** Geknikte vossenstaart **F:** Vulpin genouillé
D: Knickfuchsschwanz **CZ:** Psárka kolénkatá **PL:** Wyczyniec kolankowy
EST: Põlvjas rebasesaba **LV:** Liektā lapsaste **LT:** Nariuotasis pašiaušėlis **Ru:** Лисохвост коленчатый

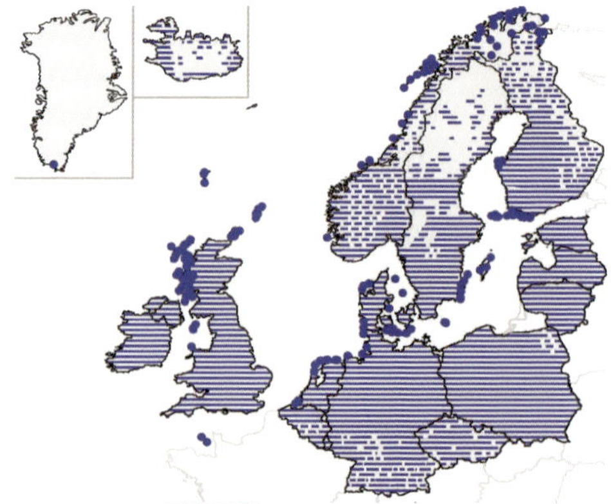

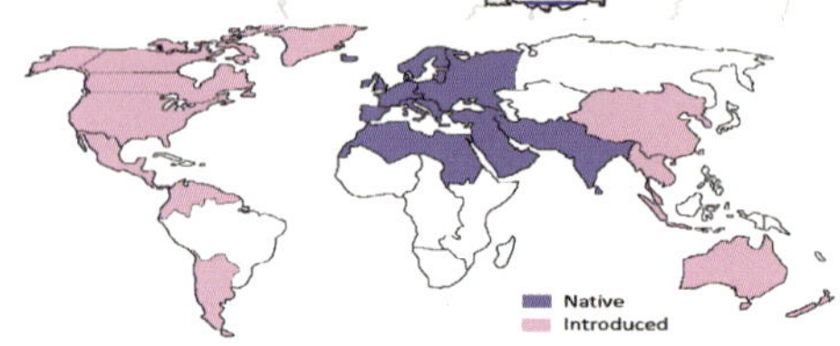

Alopecurus geniculatus. Constructed wetland near Åbybro, Denmark. Photo BM.

Perennial helophyte. Underground rhizomes absent.

Stems: 10–45 cm long, glabrous, slender, whitish glaucous towards the inflorescence, ascending from a bent or prostrate base, with 5–8 nodes, the nodes often purplish but sometimes yellow.

Leaves: 2–12(–20) cm long, 2–6 mm wide, flat, not keeled, gradually tapered to a long point, glabrous, green to greyish, rough on the upper side, smooth below, with 5–8 ribs on each side of the midrib. Leaf sheaths open, glabrous, terete. The upper somewhat inflated, greenish with a waxy, whitish covering, sometimes faintly streaked with purple. Ligule 2–4(–6) mm long, membranous, entire, blunt.

Inflorescence: A dense, spike-like panicle, 1–5(–7) cm long, 3–6 mm wide, narrowly cylindrical, pale green to purplish green, main axis hairy, pedicels very short.

Spikelets: 2.2–3.5 mm long (excluding awn), 1-flowered, oblong to lanceolate, flat. The entire spikelet falls at maturity.

Glumes: 2.2–3.5 mm long, equal in size and shape, fused at base, 3-veined, blunt or rounded at apex, keeled, with a fringe of 0.5–0.8 mm long hairs on the keel, with shorter hairs on the sides.

Lemma: 2.0–2.6 mm long, 4-veined, smooth, broadly elliptical, obtuse at apex, membranous, margins fused in the lower third. Awn 3–5 mm long, bent, arising on the back of lemma in the lower quarter, protruding 2–3 mm from the glumes.

Palea: Absent.

Flowers: Stamens 3, anthers 1.2–1.8 mm long, initially violet or yellow, becoming yellowish brown to brown.

Fruits: Caryopsis 1.2–1.5 mm long.

Flowering: May-August.

Biology: Self-sterile. The seeds are dispersed by wind. Vegetative propagation by above-ground, rooting stolons and short underground rhizomes forming tufts. Sometimes with leaves floating on water.

Distribution within the region:
Native. Widespread and abundant throughout the region, but rare in Greenland.

Characteristics and similar species:
Please see **183 *A. aequalis*** for characters distinguishing ***A. aequalis*** from ***A. geniculatus***.

Alopecurus geniculatus. Constructed wetland near Åbybro, Denmark. Photo BM.

Alopecurus geniculatus. Upper side of leaf with ribs - a cross section of these is magnified above. Photos KvdW.

Alopecurus geniculatus showing inflated upper leaf sheath, brownish anthers and purplish nodes. The colour of the nodes varies from purple to yellow and is of no use for differentiating *A. geniculatus* and *A. aeqaulis*. Hobro, Denmark. Photo JCS.

Vegetative plants may be confused with **182 *Agrostis stolonifera***, but in this species the upper sheaths are not inflated and the leaves are slightly keeled.

Alopecurus geniculatus. Inflorescence with purplish anthers and some older ones already turned brown. Photo JCS.

Alopecurus geniculatus
A habit, **B** lemma, **C** spikelet, **D** leaf sheath.

Poaceae

185 *Catabrosa aquatica* L.

DK: Tæppegræs **N:** Kjeldegras **S:** Källgräs **FIN:** Vesihilpi **IS:** Vatnsnarfagras
GB: Whorl-grass **NL:** Watergras **F:** Catabrosa aquatique
D: Quellgras **CZ:** Odemka vodní **PL:** Brodobrzanka wodna
EST: Vesi-tarnhein **LV:** Ūdeņu avotene **LT:** Šaltininė smilgaitė **Ru:** Поручейница водяная

Stand of *C. aquatica* in shallow water in a slow-flowing stream. Gettrup, Hobro, Denmark. Photo JCS.

Perennial helophyte. Stolons aboveground, branched, rooting at the nodes. Underground rhizomes absent.

Stem: 10–40(–75) cm long, erect or ascending from a bent base, rather flaccid, glabrous, often thickened and somewhat spongy at base.

Leaves: 5–25 cm long, 3–12 mm wide, folded in bud, flat, soft, dull, pale green to greyish green, often somewhat transversely wrinkled, glabrous, smooth or a little scabrid on margins, blunt and slightly hooded at apex. Upper side of leaves with two grooves along the midrib. Leaf sheaths closed in the lower half, more or less smooth, glabrous, somewhat compressed and keeled, the lower often purplish. Ligule rounded to blunt, 2–4 mm long, membranous, sometimes toothed.

Inflorescence: Panicle 5–30 cm long, 3–10 cm wide, erect, loose, ovate to pyramidal in outline, violet to greenish. Branches minutely rough, of unequal length, spreading, the lowest usually in a cluster of 3–5. Pedicels short.

Spikelets: 2.2–4.5 mm long, 1–2(–3) flowered, ovate to oblong, green to yellowish brown and often with a purplish tinge. Rachilla glabrous, at maturity breaking up below each lemma.

Glumes: Persistent, of unequal length, smooth, whitish to purplish tinged, membranous. Lower glume 0.8–1.5 mm long, narrowly ovate, without a vein. Upper glume 1.5–2.5 mm long, broadly ovate, with 1 vein.

Lemma: 2–3 mm long, 3-veined, broadly ovate, truncate at apex, rounded on the back, green to purplish tinged, with wide hyaline margins. Sometimes with short hairs on the prominent veins.

Palea: As long as lemma, ellipsoid, truncate at apex, smooth, with 2 veins, sometimes with short hairs on veins.

Flowers: Stamens 3, anthers 0.8–1.8 mm long.

Fruits: Caryopsis 1.8–2.2 mm long, smooth, elliptic, loosely enclosed between lemma and palea.

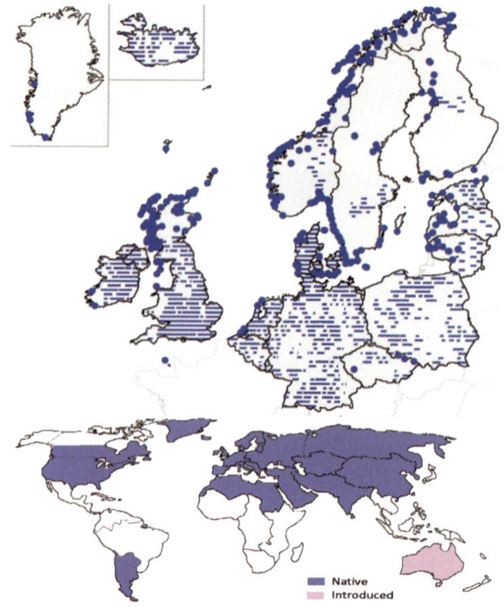

Flowering: June-August.

Biology: The seeds are dispersed by wind and float on the water surface. Vegetative propagation is by long, creeping or often floating stolons forming bright green mats.

Habitats: In neutral to alkaline, mesotrophic to eutrophic water. On wet, muddy substrate in wells, along slow-flowing streams, in ditches, pools, village ponds and occasionally in canals.

Catabrosa aquatica. The inflorescence resembles that of some *Poa* species, but the spikelets are usually 1-2 flowered with lemmas rounded on the back. Gettrup, Hobro, Denmark. Photo JCS.

Leaf blade of *C. aquatica* in backlight, showing transverse veins. Photo KvdW.

Anchored at the bank of a slow-flowing river, *C. aquatica* stretches its long stolons out into the water. Aarhus Å, Denmark. Photo BM.

Distribution within the region:

Native. Scattered to locally frequent throughout most of the region, but more sparse to the north.

Characteristics and similar species:

Recognised by the wide, dull, light green leaves with transverse veins and blunt, more or less boat-shaped apex, purplish leaf sheaths, rather flaccid stem and purplish-tinged spikelets.

Catabrosa aquatica may be mistaken for species of **194-195** ***Poa***, which differ in spikelet with scabrid or hairy keels and leaves without cross-veins.

The blunt, slightly hooded leaf tip of *C. aquatica*. Photo JCS.

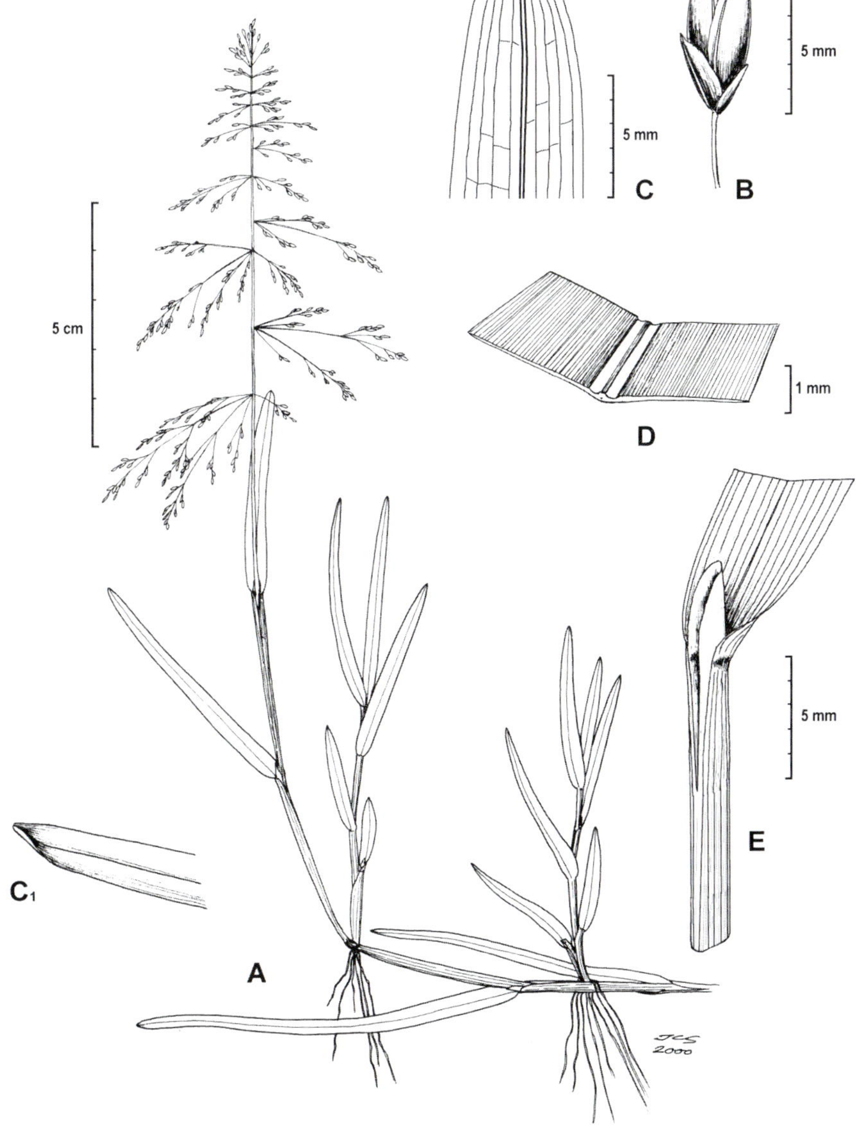

Catabrosa aquatica
A habit, **B** spikelet, **C, C1** leaf apex, **D** cross section of leaf, **E** leaf sheath.

Poaceae

186 *Coleanthus subtilis* (Tratt.) Seidel ex Roem. & Schult.

Syn.: *Schmidtia subtilis* Trattinnick
DK: - **N:** - **S:** - **FIN:** - **IS:** -
GB: - **NL:** - **F:** Coléanthe subtil
D: Scheidenblütgras **CZ:** Puchýřka útlá **PL:** Koleantus delikatny
EST: - **LV:** - **LT:** - **Ru:** Влагалищецветник маленький

Locations from which the species is known to have disappeared are marked in red.

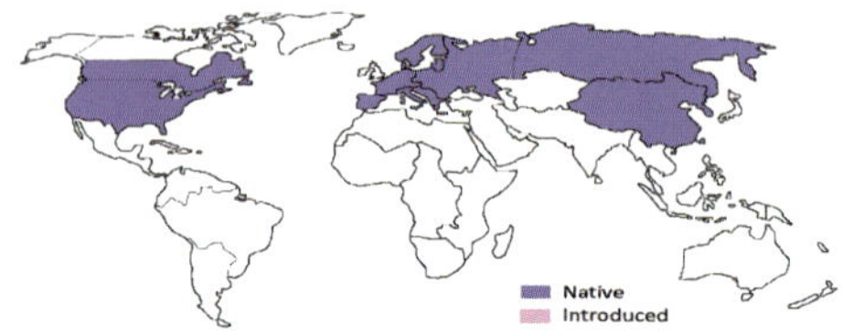

Coleanthus subtilis. Maxteich Lohsa, Germany. Photo KvdW.

Annual plant. Roots fibrous.

Stem: 2–8(–10) cm long, filiform, procumbent or ascending, with 2–3 nodes, branched at the lower nodes.

Leaves: 1–2 cm long, 0.5–2 mm wide, linear, flat, smooth, glabrous, usually recurved. Leaf sheaths closed in the lower half, strongly inflated - especially the upper ones, smooth, glabrous, green with a brownish or purplish tinge. Ligule 0.5–3 mm long, rounded, membranous.

Inflorescence: Panicle 1–5 cm long, branched, with whorls of 10–20(–40) spreading spikelets in distant umbel-like clusters. Pedicels 1–2 mm long, hairy. At maturity the seed falls, while lemma and palea remains.

Spikelets: c. 1 mm long, 1-flowered, lanceolate, flattened.

Glumes: Absent.

Lemma: 0.8–1.3 mm long, membranous, ovate, keeled, the scabrid vein extended into an awn-like apex.

Palea: Half as long as lemma, membranous, with 2 prolonged veins/keels.

Flowers: Stamens 2, anthers 0.3–0.4 mm long.

Fruits: Caryopsis 0.7–0.9 mm long, spindle shaped, dark brown.

Flowering: June-October.

Biology: The plants form small tufts or mats and the seeds are dispersed by wind. Populations are very unstable and may suddenly disappear after only a short number of years even when conditions appear favourable.

Habitats: In mesotrophic to eutrophic water. On exposed wet, muddy substrate in reservoirs, drained pools and ponds with a high variation in water levels, on the margins of lakes, streams and rivers.

Distribution within the region:
Very rare in Germany and Poland, rare but with many locations in the Czech Republic.

Coleanthus subtilis. Maxteich Lohsa, Germany. Photo KvdW.

Coleanthus subtilis. Note the strongly inflated upper leaf sheaths. Maxteich Lohsa, Germany. Photos KvdW.

Characteristics and similar species:
Its small size, strongly inflated upper leaf sheaths and open panicle of spikelets in distant umbel-like clusters make *Coleanthus subtilis* unlikely to be confused with other grasses.

Coleanthus subtilis. Maxteich Lohsa, Germany. Photos KvdW.

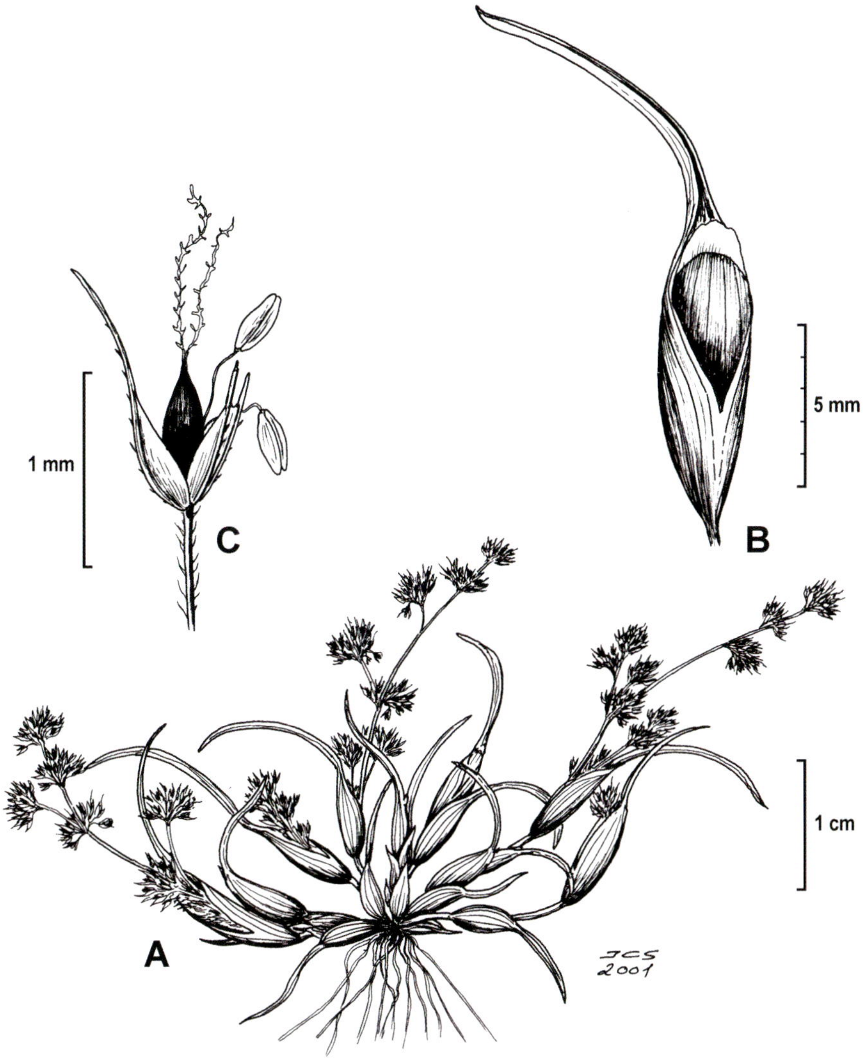

Coleanthus subtilis
A habit, **B** upper leaf sheath, **C** spikelet.

Poaceae

187 *Glyceria maxima* (Hartm.) Holmb.

Syn.: *G. aquatica* (L.) Wahlenb.

DK: Høj Sødgræs **N:** Kjempesøtgras **S:** Jättegröe **FIN:** Isosorsimo **IS:** -
GB: Reed Sweet-grass **NL:** Gewoon vlotgras **F:** Glycérie géante
D: Wasser-Schwaden **CZ:** Zblochan vodní **PL:** Manna wodna
EST: Suur parthein **LV:** Dižā ūdenszāle **LT:** Vandeninė monažolė **Ru:** Манник большой

Flowering *G. maxima* beside a river. Gudenå, Bjerringbro, Jutland, Denmark. Photo JCS.

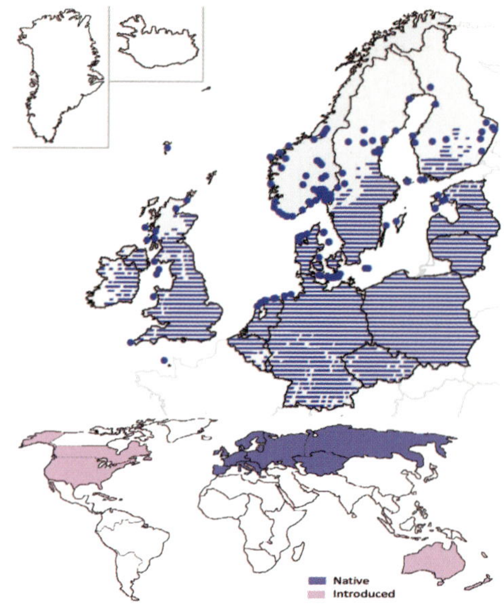

Perennial helophyte. Rhizomes long and stout. Shoots extravaginally branched from the lower sheaths.

Stems: 80–250 cm long, erect, stiff and stout to robust, up to 10 mm in diameter, smooth or slightly rough beneath the inflorescence, with 4–5 nodes.

Leaves: 30–60 cm long, 6–16 mm wide, linear, folded when young, becoming flat later, yellowish green to dark green, with distinct cross-veins, glabrous, upper side smooth, with two grooves along the midvein, lower side and margins rough, keel very prominent, apex abruptly pointed, boat shaped, aerial leaves rather stiff, submerged leaves flaccid. Leaf sheaths entire when young, but soon splitting, smooth or slightly rough, keeled. Ligule 2–4 mm long, entire, rounded with a central point, membranous.

Inflorescence: Panicle 15–40 cm long, open and pyramid shaped during flowering, slightly nodding above, richly branched, yellowish green to greenish brown or streaked with purple. Branches rough, in whorls of 5–11 from the main axis, the lower up to 20 cm long. Pedicels 1–10 mm.

Spikelets: 5–10 mm long, 2.0–3.5 mm in diameter, elongate cylindrical, gradually slightly compressed, green and often brown or purplish tinged, with 4–9 flowers. At maturity the seeds fall separately, enveloped between lemma and palea while the glumes remain.

Glumes: More or less equal, the lower 2.0–3.0 mm long, the upper 3.0–3.5 mm long, both 1-veined, broadly elliptical to oblong, subacute at apex, smooth, pale green or purplish tinged with wide, whitish margins.

Lemma: 3.0–4.0 mm long, with 7 prominent veins, minutely rough on veins, ovate, with entire, wide, whitish, membranous margins, tapering to a rather blunt apex.

Palea: As long as lemma, with 2 rough veins.

Flowers: Stamens 3. Anthers 1.5–2.0 mm long, usually violet.

Fruits: Caryopsis 1.5–2.0 mm long, dark brown, obovoid to ellipsoid.

Flowering: June-August.

Biology: Often forming extensive stands with many flowering stems and numerous erect, leafy shoots. *Glycerla maxlma* may grow to a depth of about 75 cm. In fast flowing water it can form large stands, consisting only of floating leaves and submerged leaves.

The seeds are dispersed by wind and also float. Vigorous vegetative propagation occurs by the long branching rhizome. Overwinters as green, leafy shoots.

Habitats: In mesotrophic to eutrophic, moderately acidic to alkaline water. In still or slow-flowing water on the margins of ditches, ponds, pools, lakes, streams and rivers, in forest swamps and wet meadows.

Glyceria maxima. Leaf apex boat shaped. Stroudwater Canal, Chalford, Gloucesterhire, England. Photo RVL.

The long rhizomes of *G. maxima* extend into the water. Stroudwater Canal, Stroud, Gloucestershire, England. Photo RVL.

Glyceria maxima. Floating leaves. Schaagbach, Effeld, Germany. Photo KvdW.

Distribution within the region:
Native. Widespread and abundant throughout most of the area except the northern part of Fennoscandia, Iceland and Greenland.

Characteristics and similar species:
This species typically forms large stands with numerous very tall, flowering shoots and can be recognised by the richly branched inflorescence with elongate cylindrical, awnless spikelets. Vegetative plants can be recognised by the wide leaf with boat-shaped apex.

193 ***Phragmites australis*** has grey-green leaves and ligule as a fringe of hairs.

Glyceria maxima spikelets.
Asserbo, Sjælland, Denmark. Photo JCS.

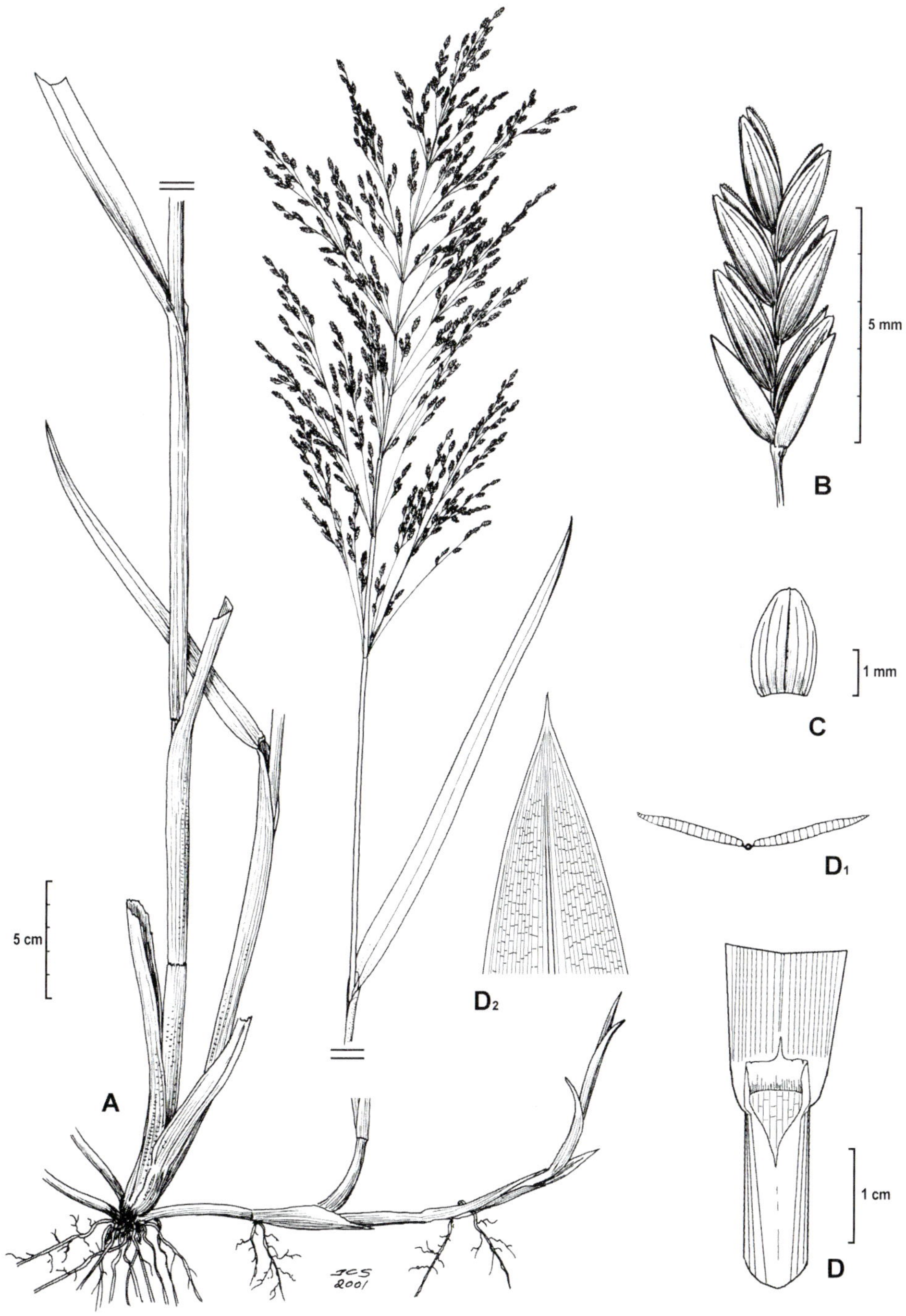

Glyceria maxima
A habit, **B** spikelet, **C** lemma, **D** leaf sheath, **D1** cross section of leaf, **D2** leaf apex.

Poaceae

188 *Glyceria fluitans* (L.) R.Br.

DK: Manna-Sødgræs **N:** Mannasøtgras **S:** Mannagräs **FIN:** Ojasorsimo **IS:** Flóðapuntur
GB: Floating Sweet-grass **NL:** Mannagras **F:** Glycérie flottante
D: Flutender Schwaden **CZ:** Zblochan vzplývavý **PL:** Manna jadalna
EST: Harilik parthein **LV:** Peldošā ūdenszāle **LT:** Paprastoji monažolė **Ru:** Манник наплывающий

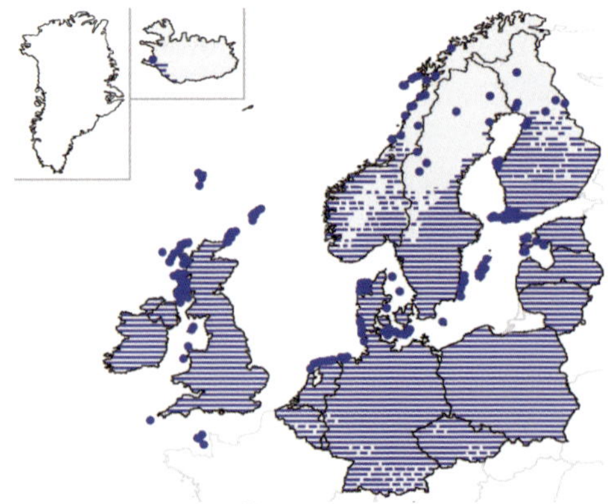

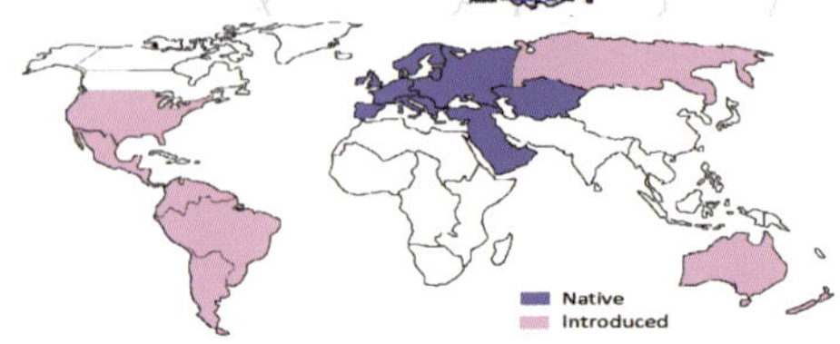

Flowering *G. fluitans* in shallow water in a stream. Norra Kvinnaby, Öland, Sweden. Photo JCS.

Perennial helophyte. Underground rhizomes absent. Shoots extravaginally branched from the lower sheaths.

Stems: 30–100 cm long, erect to ascending from a sometimes prostrate or floating base, terete, rather slender, smooth, with 2–3 nodes.

Leaves: 8–24(–50) cm long, 3–10 mm wide, linear, folded when young, becoming flat later, light green to greyish green, sometimes purplish tinged, with indistinct cross-veins, glabrous, rough on margins, apex abruptly pointed, boat shaped. Upper side of leaves with two grooves along the midvein. Leaf sheaths entire, smooth, keeled. Ligule 5–15 mm long, pointed, membranous, often torn.

Inflorescence: Panicle 10–50 cm long, erect or curved, one-sided, open during flowering, otherwise contracted and narrow. Branches somewhat rough. The upper solitary or in pairs, the lower in whorls of 1–2(–4) from the main axis.

Spikelets: 15–32 mm long, 2.0–3.5 mm thick, elongate cylindrical, gradually slightly compressed, light green and often violet tinged, with 7–16 flowers, usually spaced at a distance of 3–4 mm. At maturity the fruits fall separately, enveloped between lemma and palea, while the glumes remain.

Glumes: Unequal, the lower 2–3 mm long, the upper 3–5 mm long, both usually 1–3-veined, elliptic to broadly lanceolate, blunt, smooth, membranous, pale green with wide, whitish margins.

Lemma: 6.0–7.5 mm long, with 7 prominent veins, minutely rough, oblong elliptic, with entire, wide, whitish, membranous margins, tapering to a rather blunt apex.

Palea: As long as or shorter than lemma, with 2 narrowly winged veins extended into 2 short teeth.

Flowers: Stamens 3. Anthers 1.5–3.0 mm long, violet or yellow.

Fruits: Caryopsis 2–3 mm long, reddish brown.

Flowering: June-August.

Biology: The seeds are dispersed by wind or water. Vegetative propagation is by long, creeping stems rooting from the lower nodes. The plant forms loose tufts or large patches in shallow water. Permanently submerged stands occur in flowing or more rarely in standing water. Overwinters as floating or submerged green shoots. The grains, called "manna", have a sweet, almond-like taste and were formerly collected for food.

Glyceria fluitans. Floating leaves boat shaped at apex. Jutland, Denmark. Photo JCS.

Habitats: In oligotrophic to eutrophic, moderately acid to moderately alkaline water. On wet, more or less muddy substrates or in shallow water beside ponds, pools, lakes, ditches, canals and streams, occasionally in inundated areas as well as in marshes, bogs and forest swamps.

Distribution within the region:
Native. Widespread and abundant throughout most of the region except the northern part of Fennoscandia and Iceland. Absent from Greenland.

Characteristics and similar species:
Recognised by lemmas 6.0–7.5 mm long, 3–4 mm apart, entire, tapering to a rather blunt apex. Anthers 1.5–3.0 mm long, violet or yellow.

190 *G. notata* differs by 3.5–5.0 mm long, very blunt to truncate lemmas 1.0–1.2 mm apart, and a more richly branched panicle.

189 *G. declinata* differs by 3.5–5.0 mm long, distinctly toothed lemmas and 0.8–1.0 mm long anthers.

Vegetative plants may be mistaken for **185 *Catabrosa aquatica***, which differs in the blunt, often asymmetrical, slightly hooded leaf apex, and leaves with small, irregular air cavities in cross section.

Hybrids: The hybrids ***G. fluitans × declinata*** and ***G. fluitans × notata*** (**G. × *pedicellata*** F.Towns.) are both sterile with small, 1–2 mm long anthers remaining closed and usually hidden beneath the 4–6 mm long, somewhat toothed lemmas.

Hybrids cannot be reliably distinguished using only morphological characters, and a search for assumed parental species in the surroundings is therefore recommended.

Submerged shoots of *G. fluitans* in a small stream. Randers, Denmark. Photo JCS.

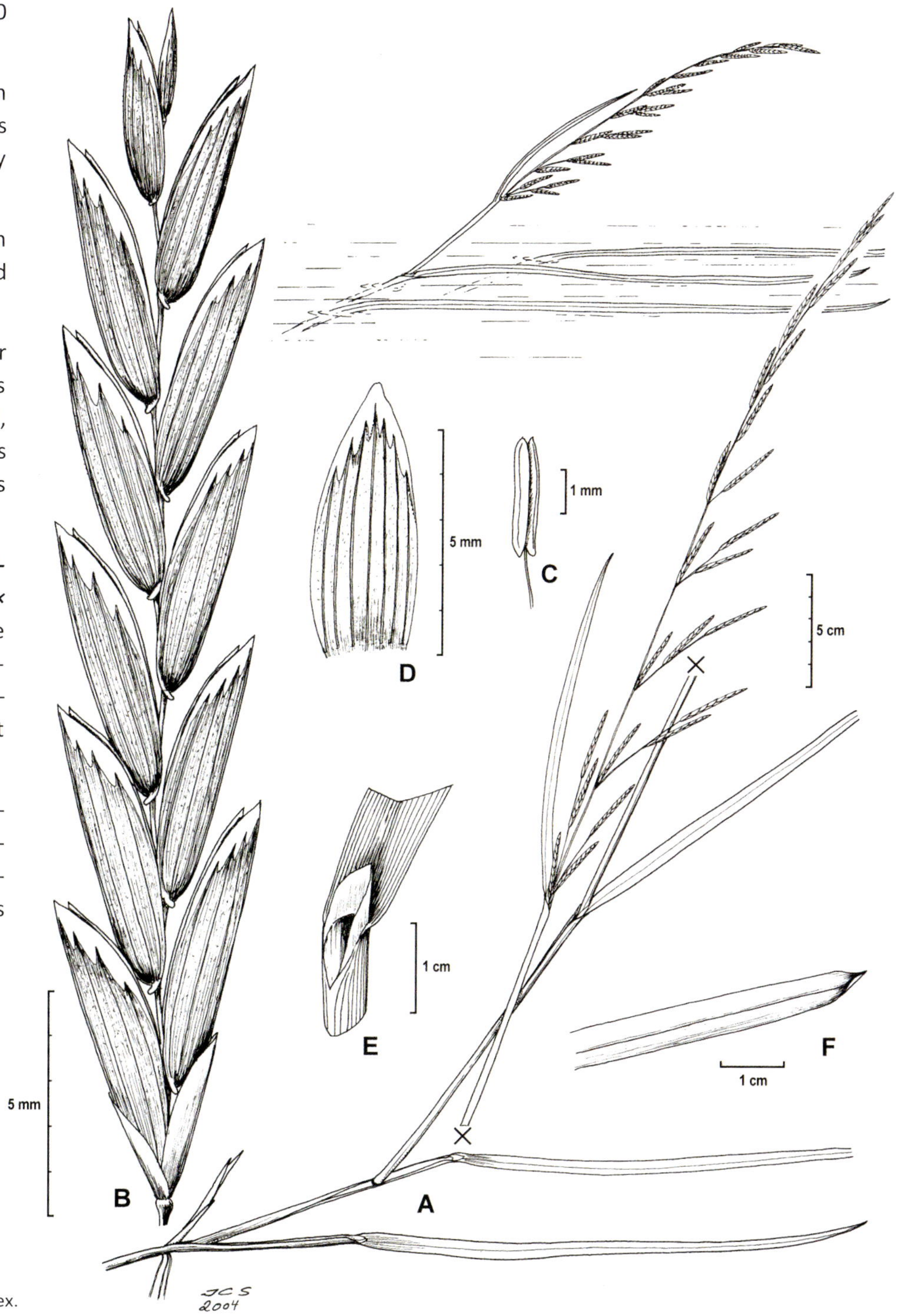

Glyceria fluitans
A habit, **B** spikelet, **C** anther, **D** lemma,
E leaf sheath, **F** floating leaf with boat-shaped apex.

Poaceae

189 *Glyceria declinata* Bréb.

DK: Tandet Sødgræs **N:** Bogesøtgras **S:** Blågrönt Mannagräs **FIN:** Pikkusorsimo **IS:** -
GB: Small Sweet-grass **NL:** Getand vlotgras **F:** Glycérie inclinée
D: Blaugrüner Schwaden **CZ:** Zblochan zoubkatý **PL:** Manna długoząbkowa
EST: Hallikas parthein **LV:** - **LT:** - **Ru:** Манник поникающий

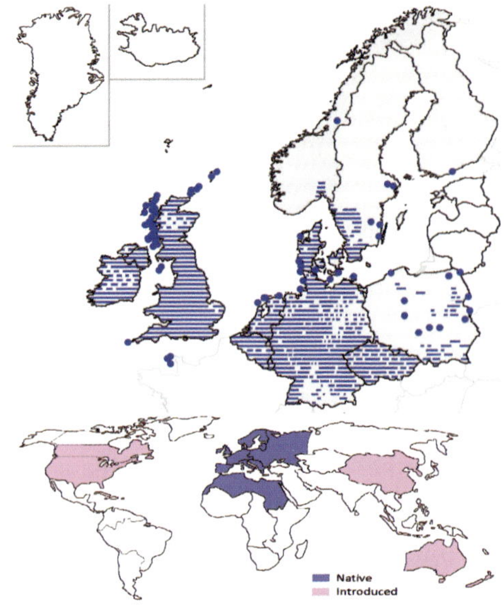

Glyceria declinata. Washpool, Cleeve Common, Gloucestershire, England. Photo RVL.

Perennial helophyte. Underground rhizomes absent. Shoots extravaginally branched from the lower sheaths.

Stems: 10–45(–60) cm long, erect or ascending from a curved or bent, sometimes prostrate base, terete, rather slender, smooth, with 1–3 nodes.

Leaves: 4–18 cm long, 3–5(–9) mm wide, linear, folded when young, becoming flat, glaucous to grey green or purplish tinged, with indistinct cross-veins, glabrous, rough on margins, apex abruptly pointed, boat shaped. Upper side of leaves with two grooves along the midvein. Leaf sheaths entire, smooth, keeled. Ligule 4–9 mm long, pointed, membranous, often torn.

Inflorescence: Panicle 5–30 cm long, open, erect to one-sided curved, greenish to purplish tinged. When flowering, the branches spread at an angle of more than 45° to the main axis. Branches smooth, solitary, each carrying 1–2 spikelets or the lowest in whorls of 1–3 and carrying 2–6 spikelets. Pedicels 1–4 mm.

Spikelets: 12–24 mm long, 1.5–2 mm thick, elongate cylindrical, gradually slightly compressed, green and often violet tinged, with 8–15 closely spaced flowers, usually separated by a distance of c. 1.5 mm. At maturity the fruits fall separately, enveloped between lemma and palea, while the glumes remain.

Glumes: Unequal, the lower 1.5–2.5 mm long, the upper 2.5–3.0 mm long, both usually 1-veined, broadly elliptic to ovate, blunt, smooth, membranous, pale green with wide, whitish margins.

Lemma; 3.5–5.0 mm long, with 7 prominent veins, minutely rough, broadly elliptic to ovate, distinctly 3-5-toothed, with wide, whitish, membranous margins, the middle rib protruding above the notches.

Palea: As long as or shorter than lemma, with 2 rough, narrowly winged veins, which in the upper, divided part are almost parallel.

Flowers: Stamens 3. Anthers 0.8–1.0 mm long, violet or yellow.

Fruits: Caryopsis 1.5–2.3 mm long, red-brown.

Flowering: June-August.

Biology: The seeds are dispersed by wind or water. Vegetative propagation is by stems rooting from the lower nodes, forming loose tufts.

Habitats: In mesotrophic to eutrophic shallow water or on wet, muddy often trampled substrate on the margins of ponds, pools, ditches, lakes, canals, streams and rivers.

Spikelet of *G. declinata* - note the distinctly toothed lemmas. Nibe, Denmark. Photo JCS.

Distribution within the region:
Native. Widespread and locally frequent from Ireland east to southern Sweden and Poland. Rare in Norway and Finland.

Characteristics and similar species:
G. declinata can be recognised by the glaucous to grey-green or purplish-tinged leaves with indistinct cross-veins, elongate cylindrical spikelets and strongly toothed lemmas.

Often mistaken for the usually larger, more richly branched **190 *Glyceria notata***, which differs in shallowly toothed lemmas and palea with the 2 veins converging in the upper divided part.

188 *G. fluitans* differs by 6.0–7.5 mm long, bluntly pointed, entire lemmas and 1.5–3.0 mm long anthers.

Hybrids: *G. fluitans × declinata*. Sterile hybrid with small, 1–2 mm long anthers remaining closed and usually hidden beneath the 5.0–5.5 mm long, somewhat toothed lemmas.

Glyceria declinata. Floating leaves with boat-shaped apex. Photo KvdW.

Glyceria declinata. Panicle. Nibe, Denmark. Photo JCS.

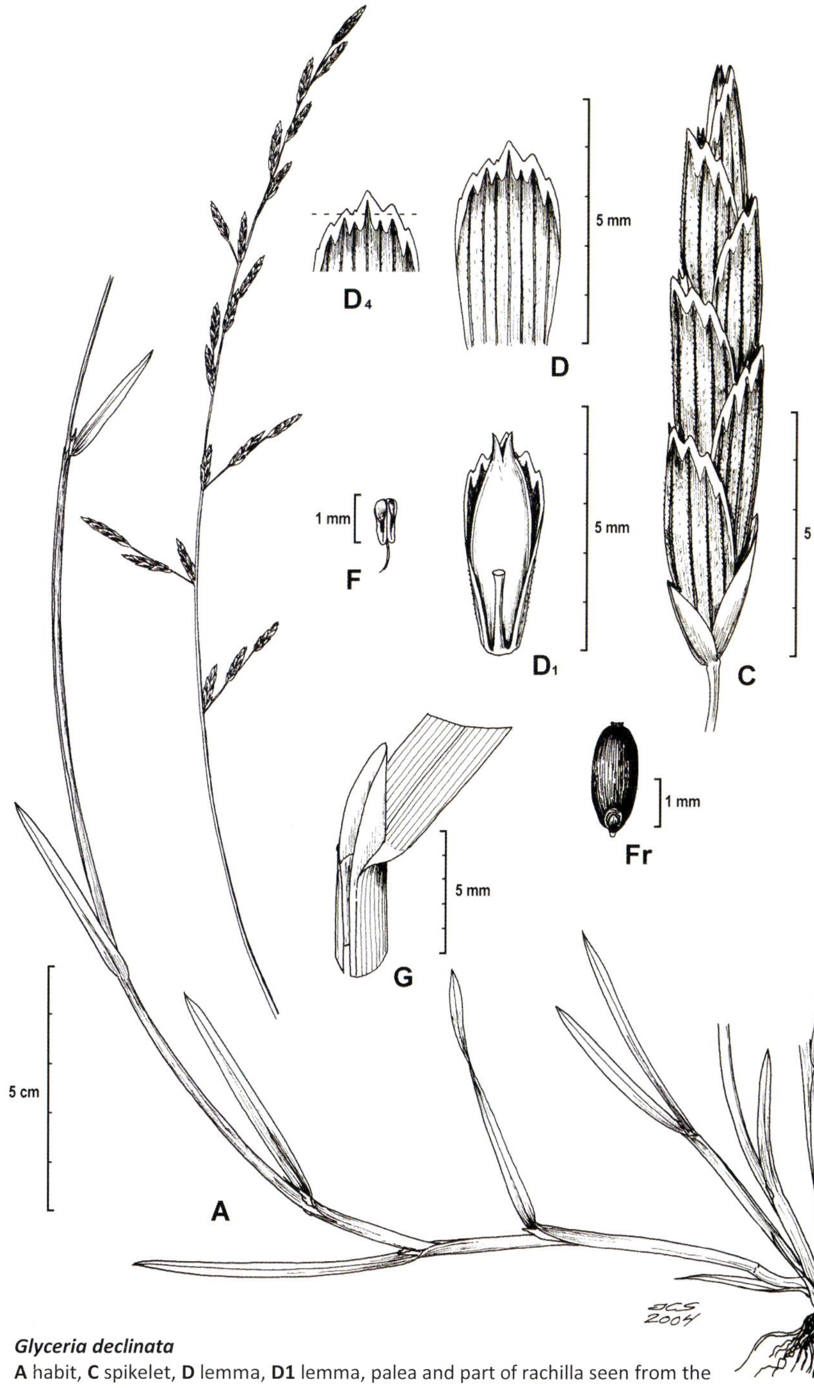

Glyceria declinata
A habit, **C** spikelet, **D** lemma, **D1** lemma, palea and part of rachilla seen from the inside, **D4** apical part of lemma, **F** anther, **Fr** caryopsis, **G** leaf sheath.

Poaceae

190 *Glyceria notata* L.

Syn.: *G. plicata* (Fr.) Fr.

DK: Butblomstret Sødgræs **N:** Sprikjesøtgras **S:** Skånskt mannagräs **FIN:** Savisorsimo **IS:** -
GB: Plicate Sweet-grass **NL:** Geplooid vlotgras **F:** Glycérie plissée
D: Gefalteter Schwaden **CZ:** Zblochan řasnatý **PL:** Manna fałdowana
EST: Voldine parthein **LV:** Krokainā ūdenszāle **LT:** Bukažvynė monažolė **Ru:** Манник складчатый

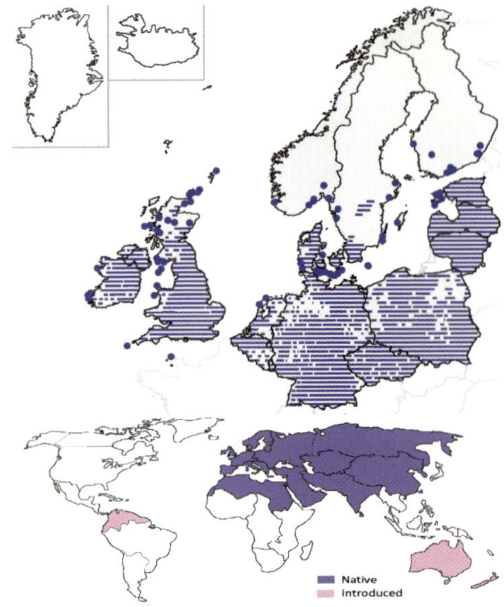

Glyceria notata in a storm water retention pond near Aarhus, Denmark. Photo BM.

Perennial helophyte. Underground rhizomes absent. Shoots extravaginally branched from the lower sheaths.

Stems: 30–80 cm long, erect or ascending from a curved or bent, sometimes prostrate base, terete, rather slender, smooth, with 3–4 nodes.

Leaves: 5–30 cm long, 3–12 mm wide, linear, folded when young, becoming flat later, green to grey green, with indistinct cross-veins, glabrous, rough on lower sides and margins, usually smooth and with two grooves along the midvein on the upper side, apex abruptly pointed, boat shaped. Leaf sheaths entire, smooth, keeled. Ligule 2–6 mm long, blunt, membranous.

Inflorescence: Panicle 10–45 cm long, open, often rather wide, erect or curved to oneside, flowering branches spreading at an angle of less than 45° to the main axis. Branches rough, in whorls of 1–5, with one of them longer than the other and up to 12 cm long. Pedicels 1–6 mm.

Spikelets: 12–25 mm long, 1.5–2 mm in diameter, elongate cylindrical, gradually slightly compressed, green and often violet tinged, with 5–16 closely spaced flowers, usually separated by a distance of 1.0–1.2 mm. At maturity the fruits fall separately, enveloped between lemma and palea, while the glumes remain.

Glumes: Unequal, the lower 1.5–2.5 mm long, the upper 2.5–4.0 mm long, both usually 1-veined, broadly elliptic to obovate, blunt to truncate, smooth, membranous, pale green with wide, whitish margins.

Lemma: 3.5–5.0 mm long, with 7 prominent veins, minutely rough, broadly elliptic to ovate, very blunt, shallowly 3-5-toothed, with wide, whitish, membranous margins, the middle vein not protruding above the notches.

Palea: As long as or shorter than lemma, with 2 rough, narrowly winged veins, which converge in the upper part.

Flowers: Stamens 3. Anthers 0.8–1.5 mm long, light yellow.

Fruits: Caryopsis c. 2 mm long, greyish brown.

Flowering: June-August.

Biology: The seeds are dispersed by wind or water. Vegetative propagation is by stems rooting from the lower nodes, forming loose tufts of prostrate shoots rooting at the nodes.

Habitats: In mesotrophic to eutrophic shallow water or on wet, swampy substrate on the margins of ponds, ditches, lakes, canals, streams and rivers.

Distribution within the region:
Native. Rare in Norway and Finland; locally common in the southern part of Sweden, widespread and abundant in the rest of the region, but absent from Iceland and Greenland.

Characteristics and similar species:
Recognised by large panicles with elongate cylindrical spikelets, blunt to truncate, shallowly toothed lemmas and yellow anthers. Leaves green to grey green, indistinctly cross-veined.

189 *Glyceria declinata* differs by distinctly 3–5-toothed lemmas with the middle vein protruding above the notches, and palea with the 2 veins parallel in the upper divided part.

188 *G. fluitans* differs by 6.0–7.5 mm long, bluntly pointed, entire lemmas and 1.5–3.0 mm long anthers.

Hybrids: G. × *pedicellata* F. Towns. (*G. fluitans × notata*). Sterile hybrid with small, 1–2 mm long anthers remaining closed and usually hidden beneath the 4.0–6.0 mm long, almost entire lemmas, very blunt at apex.

Glyceria notata. The inflorescence is open and usually with the flowering branches spreading less than 45° to the main axis. Ajstrup, Hadsund, Denmark. Photo JCS.

Glyceria notata. Typical spikelets with shallowly toothed lemmas. Hobro, Denmark. Photo JCS.

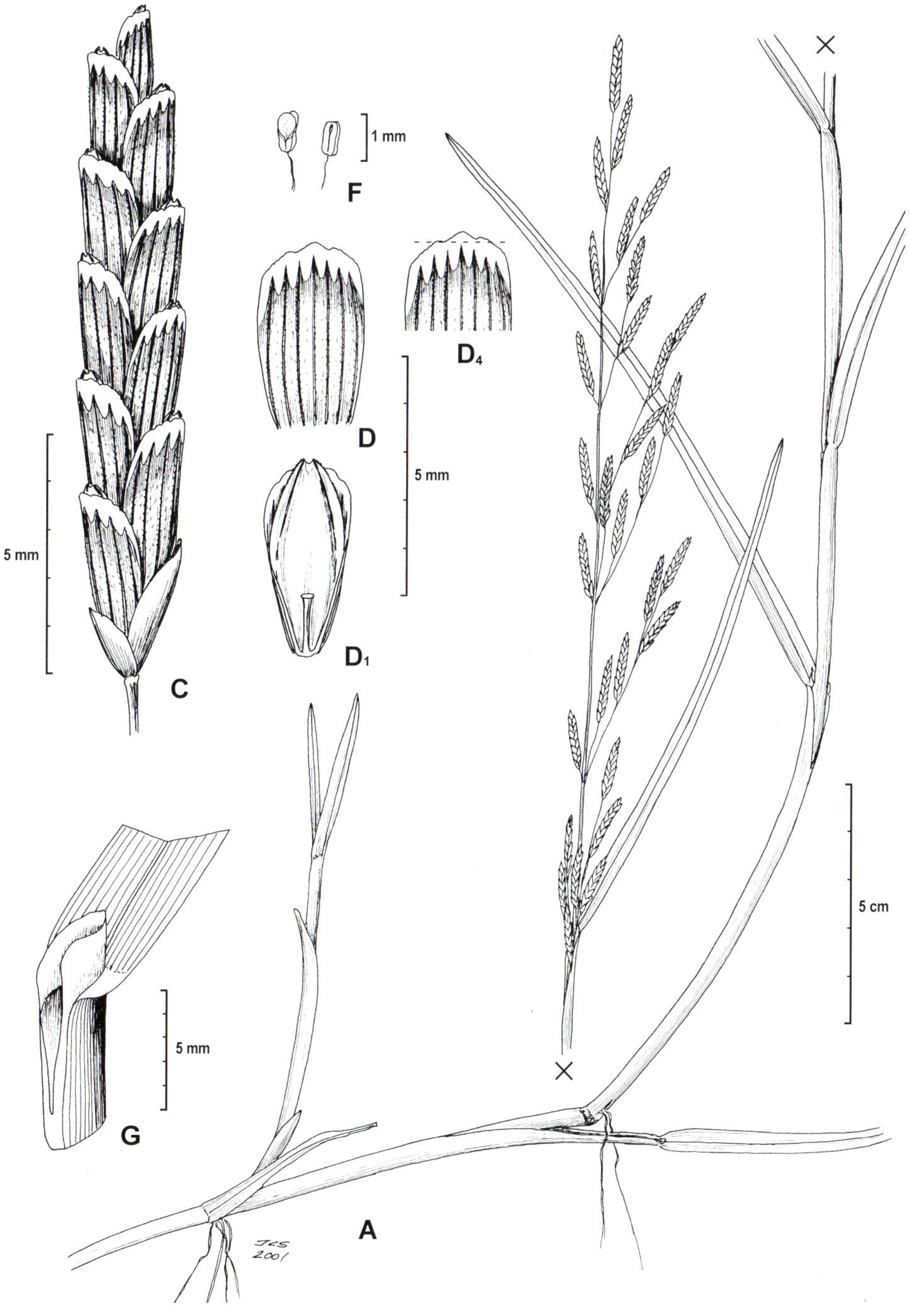

Glyceria notata
A habit, **C** spikelet, **D** lemma, **D1** lemma, palea and part of rachilla seen from inside, **D4** apical part of lemma, **F** anthers, **G** leaf sheath.

Poaceae

191 ***Leersia oryzoides*** (L.) Swartz

DK: Risgræs **N:** - **S:** Vildris **FIN:** Hukkariisi **IS:** -
GB: Cut grass **NL:** Rijstgras **F:** Leersie faux-riz
D: Reisquecke **CZ:** Tajnička rýžovitá **PL:** Zamokrzyca ryżowa
EST: Jõgi-metsriis **LV:** Parastais parīss **LT:** Ryžinė ravenė **Ru:** Леерсия рисовидная

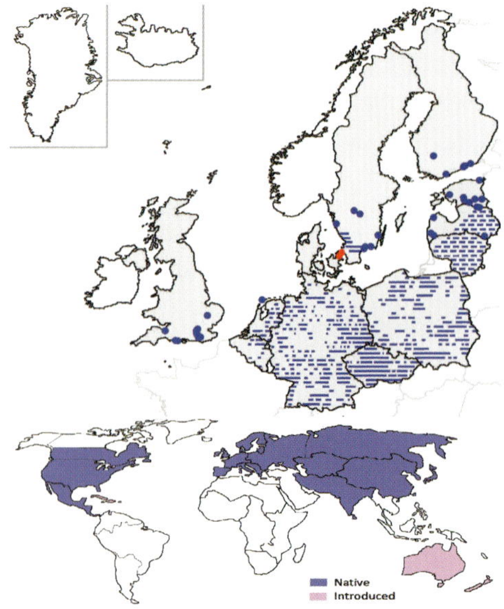

Leersia oryzoides with panicle fully emerged from the leaf sheath. Roadside ditch near Mezières-en-Brenne, Parc Naturel Régional de la Brenne, Indre, France. Photo RVL.

Perennial helophyte. Underground rhizomes long, slender, branched. Shoots extravaginally branched from the lower sheaths.

Stems: 30–120(–150) cm long, erect or ascending, slender, rigid, smooth, hairy at nodes, often branched and with more panicles.

Leaves: 8–30 cm long, 5–12 mm wide, flat, long acuminate into a fine point, yellowish green, with 7–11 distinct veins and some indistinct veins in between, rough with minute hairs on both sides, margins and lower side of the prominent middle vein with bristles, which are retrorse in the basal third of the leaf. Leaf sheaths open, terete, finely ribbed, with very short, retrorse stiff hairs, particularly on the upper ones. Ligule 1.0–1.5 mm long, entire or shallowly toothed.

Inflorescence: Panicles fully or partly enclosed within the leaf sheaths or more rarely completely emerged from them, 5–20 cm long, 6–14 cm wide, open, yellow green or brownish at maturity. Branches rough, flexuous. Pedicels 0.3–1 mm long.

Spikelets: 4–6 mm long, 1.5–2.0 mm wide, oblong to elliptic, somewhat compressed, 1-flowered. The entire spikelet falls at maturity.

Glumes: Rudimentary.

Lemma: 4–6 mm long, 1.2–1.4 mm wide, 5-veined, oblong, translucent yellowish green, greyish brown at maturity, sides with scattered, appressed hairs and stiff, prominent hairs on veins and keel.

Palea: As long as lemma, but narrower, 3-veined, with scattered appressed hairs and stiff, prominent hairs on the middle vein.

Flowers: Stamens 3. Anthers 1.5–1.8 mm long on chasmogamous flowers, 0.5–0.8 mm long on cleistogamous flowers, white to yellowish.

Fruits: Caryopsis about 3 mm long, 1 mm wide.

Flowering: August-September.

Biology: In colder climates, such as in the Nordic countries and Britain, the panicles rarely emerge from the leaf sheaths. Flowers which remain enclosed in the leaf sheath are normally cleistogamous, while those of fully emerged panicles are chasmogamous. The seeds

Leersia oryzoides. Leaf sheath. Note the retrorse bristles at the leaf margins. Roadside ditch near Mezières-en-Brenne, Indre, France. Photo RVL.

are dispersed by animals, water and wind. Vegetative propagation is by long, underground rhizomes, resulting in formation of loose tufts or patches.

Habitats: In mesotrophic to eutrophic waters. On wet, muddy substrates in aquaculture ponds, oxbow lakes, pools, ditches, reservoirs, as well as on the margins of ditches, streams and rivers.

Distribution within the region:
Native. Scattered in southern parts of the region from England east to southern Sweden, Finland and the Baltic states. Nowhere common. Extinct in Denmark.

Characteristics and similar species:
Leersia oryzoides can be recognised by its yellowish-green, rough leaves with bristly margins, panicles often only partly emerged from the leaf sheaths and 1-flowered spikelets with stiff, prominent hairs on lemma and palea. It is very easy to overlook, particularly among other tall marginal grasses, but is easily distinguished from all other wetland species.

Vegetative tufts of *L. oryzoides*. Duisburg, Germany. Photo KvdW.

Leersia oryzoides. Spikelets. Sweden. Photo JCS.

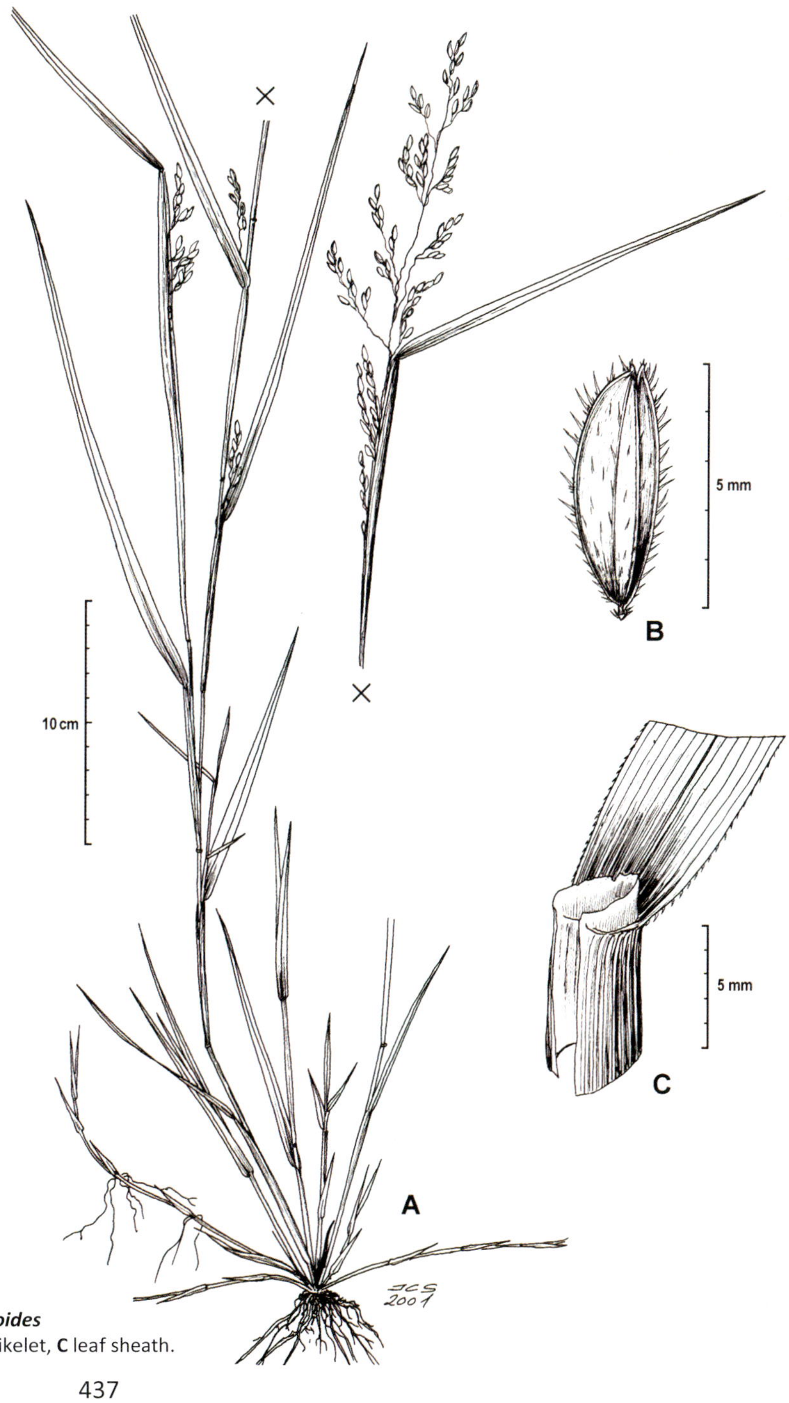

Leersia oryzoides
A habit, **B** spikelet, **C** leaf sheath.

Poaceae

192 *Phalaris arundinacea* L.

Syn.: *Baldingera arundinacea* (L.) Dumort.
DK: Rørgræs **N:** Strandrøyr **S:** Rörflem **FIN:** Ruokohelpi **IS:** -
GB: Reed Canary-grass **NL:** Rietgras **F:** Baldingère faux-roseau
D: Rohr-Glanzgras **CZ:** Chrastice rákosovitá **PL:** Mozga trzcinowata
EST: Päideroog **LV:** Parastais miežubrālis **LT:** Nendrinis dryžutis **Ru:** Канареечник тростниковидный

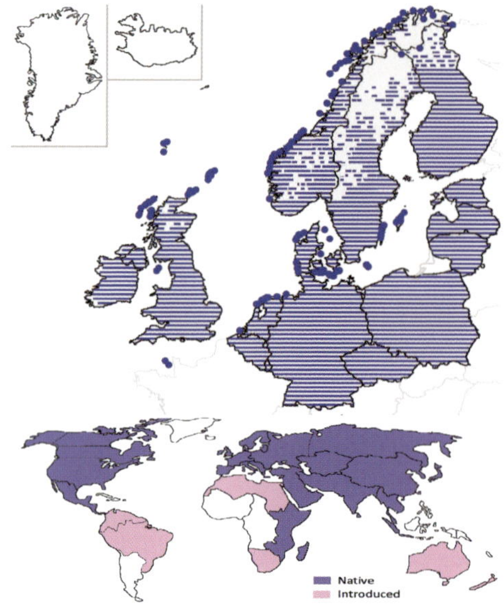

Phalaris arundinacea. River Swalm, the Netherlands. Photo KvdW.

Perennial helophyte. Underground rhizomes up to 2 m long. Shoots extravaginally branched from the lower sheaths.

Stems: 60–200 cm long, erect, stiff, glabrous, smooth or slightly rough below the panicle, with 4–6 nodes.

Leaves: 10–30 cm long, 6–20 mm wide, flat, long acuminate, grey green, glabrous, rough on the upper side and at the margins. Leaf sheaths terete, tightly enveloping the stem, open, glabrous, smooth or slightly rough, with distinct cross-veins, green to grey green. Ligule 3–10(–15) mm long, blunt to truncate, membranous, entire but often becoming torn.

Inflorescence: Panicle 7–30 cm long, 1–4 cm wide, pale green or often purplish tinged. Spreading during flowering, otherwise contracted and more or less cylindrical, sometimes lobed in the lower part. Branches rough, densely divided, with terminal clusters of spikelets. Pedicels short.

Spikelets: 5.0–6.5 mm long, with 1 fertile flower and 2 rudimentary, sterile, scale-like flowers at the base. At maturity the fruits fall separately, enveloped between lemma and palea, while the glumes remain.

Glumes: More or less equal, 5–7 mm long, laterally compressed, 3-veined, lanceolate, pointed, keel rough, distinct, sometimes appearing winged.

Lemma: Of the fertile flower 3–4 mm long, 5-veined, lanceolate, hairy in the upper part, becoming shiny and hard at maturity, densely enveloping the fruit. Lemmas of the two basal, sterile flowers 1.0–1.5 mm long, shortly hairy and without veins.

Palea: As long as lemma.

Flowers: 3 stamens, anthers 2.5–3.0 mm long, yellow or reddish.

Fruits: Caryopsis 2.0–2.5 mm long, laterally compressed.

Flowering: June-August.

Biology: The seeds are dispersed by wind and also float. Vegetative propagation is by long, underground rhizomes, often forming extensive stands. This species usually grows above the water line, but tolerates periodic inundation. It can even persist for years in a submerged form, completely detached from any stoloniferous structure connecting to the banks. It dies back in the autumn and overwinters as rhizomes.

Habitats: In moderately oligotrophic to eutrophic water. On dry to moist or wet substrates on the margins of streams, rivers, lakes, ponds, pools, ditches, forest swamps and marshes. It can grow in

Flowering *P. arundinacea*. Randers, Denmark. Photo JCS.

heavily polluted flowing waters and is often the last remaining species before complete extinction of aquatic plants. *P. arundinacea* will grow not only along the margins of rivers and streams, but in the river at considerable depths.

Distribution within the region:
Native. Widespread and abundant almost throughout but absent from Greenland.

Characteristics and similar species:
A tall grass with a panicle that is dense and cylindrical or more open when flowering, bearing 1-flowered spikelets densely clustered on the much divided branches.

193 ***Phragmites australis*** differs by much longer floral branches, spikelets with long hairs and leaves with a ligule represented by a fringe of hairs.

Note: A cultivated form called *picta* is locally naturalised. It differs by the leaves, which are striped lengthwise with light green and white.

Partly submerged *P. arundinacea*. River Swalm, the Netherlands. Photo KvdW.

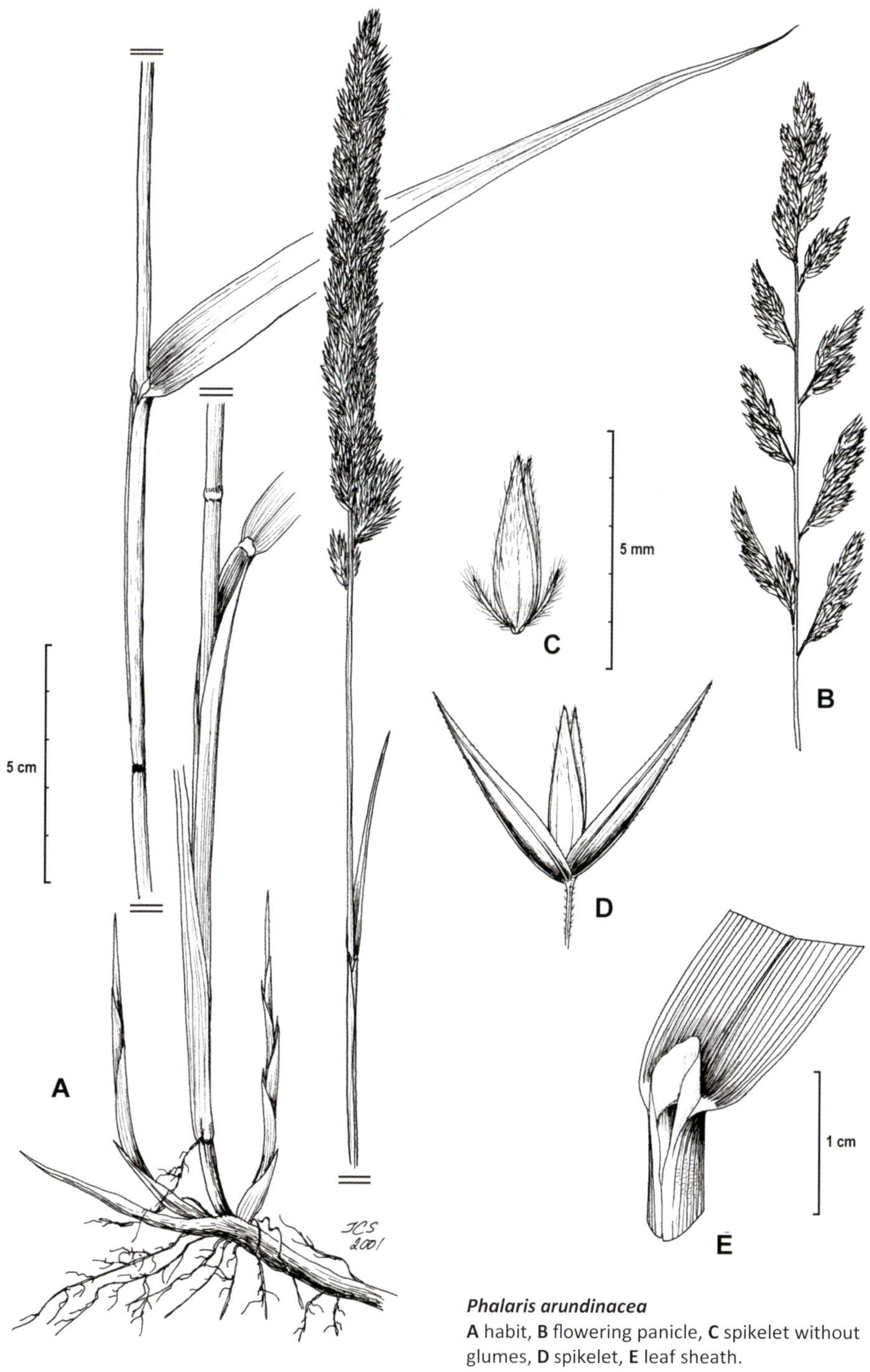

Phalaris arundinacea
A habit, **B** flowering panicle, **C** spikelet without glumes, **D** spikelet, **E** leaf sheath.

Phalaris arundinacea. Leaf base with ligule. Photo KvdW.

Poaceae

193 *Phragmites australis* (Cav.) Trin. ex Steud.

Syn.: *P. communis* Trin.
DK: Tagrør **N:** Takrøyr **S:** Vass **FIN:** Järviruoko **IS:** -
GB: Common Reed **NL:** Riet **F:** Roseau
D: Schilf **CZ:** Rákos obecný **PL:** Trzcina pospolita
EST: Harilik pilliroog **LV:** Parastā niedre **LT:** Paprastoji nendrė **Ru:** Тростник обыкновенный

Phragmites australis multiplies efficiently by branching of the rhizome and quickly forms extensive stands. Denmark. Photo JCS.

Perennial helophyte. Rhizome richly branched, 0.5–3 cm in diameter, covered with sheath-like scales. Stolons underground or above ground, far creeping or sometimes floating, up to or sometimes more than 10 m long, rooting at the nodes. Roots 2–4 mm in diameter.

Stems: 0.3–4.0(–10.0) m long, erect, stiff, stout, up to 10 mm in diameter, hollow, smooth, glabrous, with many nodes and leaves throughout, with yellowish-grey scales at base.

Leaves: 30–60 cm long, 4–30 mm wide, stiff, firm, flat, often with 1–3 constrictions across the middle of the blade, contracted or rounded at the base, long tapering to a very fine and often flexuous tip, grey green, glabrous, with many veins and a prominent central vein, margins more or less rough and when young with long, silky hairs at the wing-like auricles. The leaves fall in autumn leaving the leaf sheaths. Leaf sheaths open, with overlapping margins, smooth, glabrous, terete and narrow. Ligule represented by a fringe of hairs 0.8–1.2 mm long.

Inflorescence: Panicle 10–40 cm long, pyramid shaped, erect or gradually nodding to one side, open to dense, richly branched, violet to brownish or rarely yellowish green, silky. Branches smooth or almost smooth, slightly hairy at ramifications. Pedicels short.

Spikelets: 10–16 mm long, lanceolate, brownish, often violet tinged, with (2–)3–5 (–8) flowers, the lowest male and sterile, the rest bisexual and fertile. Rachilla with silky hairs 6–9 mm long. At maturity the fruits fall separately, enveloped between lemma and palea, while the glumes and the lowest male flower remain.

Glumes: Unequal, the lower 3.0–4.5 mm long, the upper 5–7 mm long, both 3-5-veined, smooth, lanceolate, pointed, brownish and often violet tinged.

Lemma 9–13 mm long, that of the male flower narrowly lanceolate, pointed, smooth, 3-veined, those of the bisexual flowers 1–3-veined, more or less linear, rolled like a tuber at apex.

Palea: 3–4 mm long.

Flowers: Male flower with 1–2 stamens, bisexual flowers with 3 stamens. Anthers 1.3–2 mm long, yellowish.

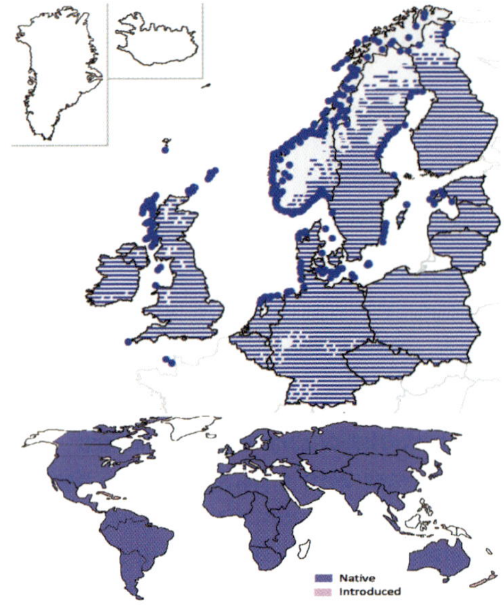

Fruits: Caryopsis 1.8–2 mm long, oblong. Ripe in December–January.

Flowering: July-October.

Biology: An extremely polymorphic species. The numerous seeds are dispersed by wind or water. Vegetative propagation is by division of rhizome and stolons. Forms extensive stands from shoreline to a depth of 2.5 metres and can form floating populations.

Habitats: In oligotrophic to eutrophic, mildly acid to mildly alkaline, fresh to brackish water. On moist to wet or dry substrate in reedbeds and ditches, on the margins of lakes, streams and inlets, in fens, salt marshes, dunes, dune slacks and on the seashore.

Phragmites australis. The leaves are grey green and the ligule represented by a fringe of very short hairs (just visible in the photo). With long, silky hairs at the yellowish auricles when young. Photo JCS.

Distribution within the region:
Native. Widespread and abundant throughout most of the region. Absent from Iceland and Greenland.

Characteristics and similar species:
This species can be recognised by the large grey-green stands with numerous erect shoots with stiffly protruding, long tapering leaves and violet or brownish inflorescences, together with the ligule represented by a fringe of hairs.

187 ***Glyceria maxima*** differs by glossy, green leaves with a short, boat-shaped tip and a membranous ligule 2–4 mm long.

192 ***Phalaris arundinacea*** has a short-branched inflorescence with clustered spikelets and a membranous ligule 3–10 mm long.

Phragmites australis. Stolons very long and sometimes floating. Filsø, SW-Jutland, Denmark. Photo JCS.

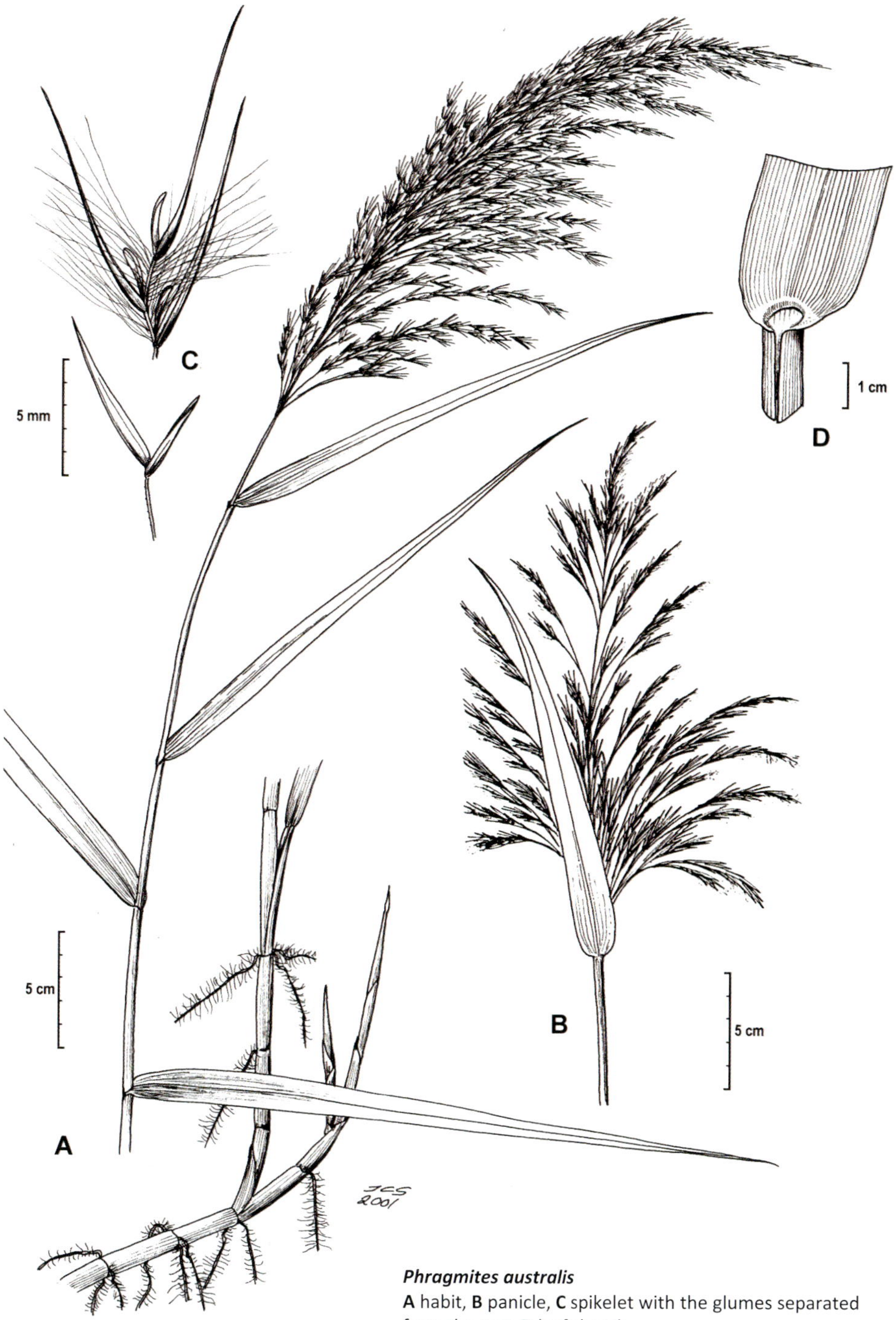

Phragmites australis
A habit, **B** panicle, **C** spikelet with the glumes separated from the rest, **D** leaf sheath.

Phragmites australis. Plant with yellow-green panicle. North-Jutland, Denmark. Photo JCS.

Poaceae

194 *Poa palustris* L.

DK: Stortoppet Rapgræs **N:** Myrrapp **S:** Sengröe **FIN:** Rantanurmikka **IS:** -
GB: Swamp Meadow Grass **NL:** Moerasbeemdgras **F:** Pâturin des marais
D: Sumpf-Rispengras **CZ:** Lipnice bahenní **PL:** Wiechlina błotna
EST: Soonurmikas **LV:** Purva skarene **LT:** Pelkinė miglė **Ru:** Мятлик болотный

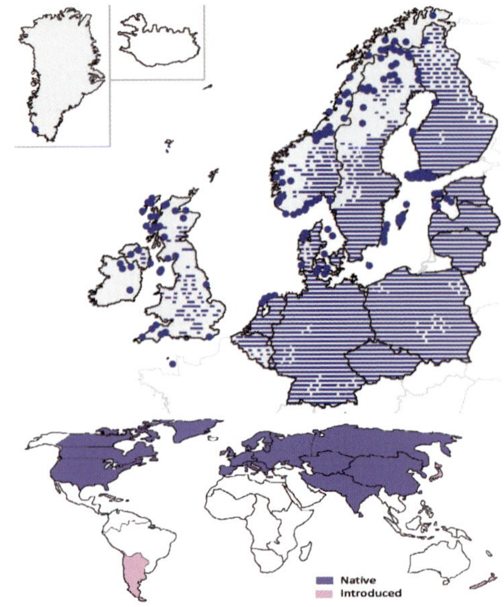

Poa palustris. Grund Fjord, Randers, Denmark. Photo JCS.

Short-lived, perennial hemicryptophyte. Underground rhizome short. Shoots extravaginally branched from the lower sheaths.

Stems: 20–150 cm long, smooth, erect, the basal part often prostrate, rooting and branched at the nodes.

Leaves: 5–20 cm long, 2–4 mm wide, linear, folded when young, later becoming flat, green, glabrous, slightly rough on the margins, gradually tapering to an acute point. Upper side of leaves with two grooves along the midvein. Leaf sheaths open, smooth, terete or the lower slightly keeled. Ligule 2–5 mm long, blunt, membranous.

Inflorescence: Panicle 10–20 cm long, erect or somewhat nodding, loose, spreading, yellowish green to brownish violet. Branches in whorls of 3–6 from the main axis, of different length and undivided in the lower part. Pedicels 1–5 mm long. Inflorescences born from the basal side branches are smaller and later flowering.

Spikelets: 3–5 mm long, oblong to ovate, yellowish brown or brownish violet tinged, 2–5-flowered. Rachilla glabrous. At maturity the fruits fall separately, enveloped between lemma and palea, while the glumes remain.

Glumes: Slightly unequal, the lower 2.0–3.0 mm long, 1–3-veined, the upper 2.3–3.4 mm long, 3-veined, both lanceolate, sharply pointed, apically rough on keels, green to violet tinged, margins membranous.

Lemma: 2.5–3.0 mm long, indistinctly 5-veined, oblong ovate, green to violet tinged, blunt and often golden brownish at apex, keeled, with short hairs on keel and marginal veins and long, soft and curly hairs at the base.

Palea: Slightly shorter than lemma, with 2 hairy keels.

Flowers: 3 stamens. Anthers 1.0–1.5 mm long, white, often aborted.

Fruits: Caryopsis 1.5 mm long.

Flowering: June-July.

Biology: Facultative apomictic. The seeds are dispersed by wind or water. Vegetative propagation is by stems rooting from the lower nodes, forming loose tufts.

Habitats: In mesotrophic to eutrophic, mildly acid to alkaline water. On moist to wet substrate on the margins of lakes, slow-flowing streams and rivers, in ditches, marshes, fens, and reedbeds, in damp forests, gravel pits, railway embankments, roadsides, waste ground, etc.

Distribution within the region:
Native. Common throughout most of the continental part of the region, though less common in the north. Very scarce and scattered in Britain and Ire-

Poa palustris. Spikelet. Note the aborted, empty anthers. Hobro, Denmark. Photo JCS.

Poa palustris. The ligule is 2-5 mm long. Hobro, Denmark. Photo JCS.

land. Rare in Greenland, absent from Iceland.

Characteristics and similar species:
Poa palustris is very similar to ***P. nemoralis*** but differs by the longer ligules.

195 ***P. trivialis*** has somewhat rough leaf sheaths and acute ligules.

Poa palustris
A habit, **B** panicle, **C** spikelet, **C4** spikelet without glumes, **G** leaf sheath.

Poaceae

195 *Poa trivialis* L.

DK: Almindelig Rapgræs **N:** Markrapp **S:** Kärrgröe **FIN:** Karheanurmikka **IS:** Hásveifgras
GB: Rough Meadow Grass **NL:** Ruw beemdgras **F:** Pâturin commun
D: Gewöhnliches Rispengras **CZ:** Lipnice obecná **PL:** Wiechlina zwyczajna
EST: Harilik nurmikas **LV:** Parastā skarene **LT:** Paprastoji miglė **Ru:** Мятлик обыкновенный

Poa trivialis. Hobro, Denmark. Photo JCS.

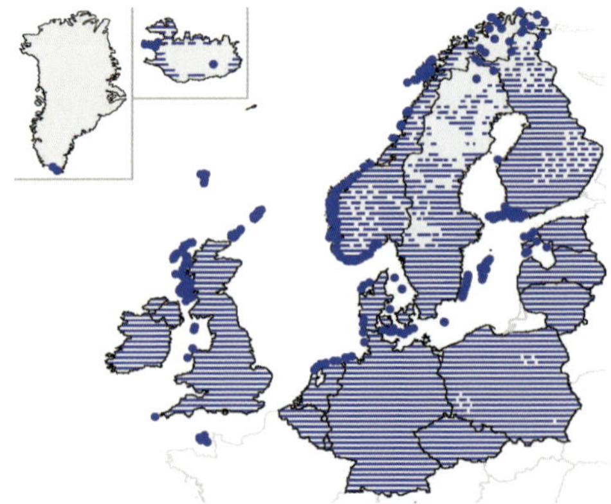

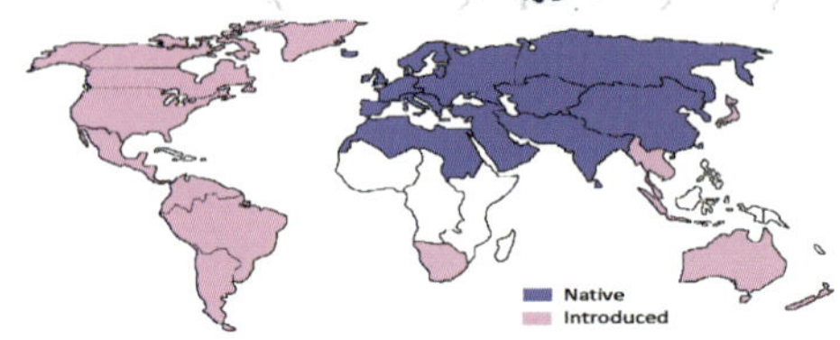

Perennial hemicryptophyte. Rhizome absent. New shoots extravaginally branched from the lower sheaths and rooting at the nodes.

Stems: 20–100 cm long, erect or ascending from a procumbent base, smooth but often slightly rough below the inflorescence, with 3–5 nodes.

Leaves: 1–20 cm long, 2–6 mm wide, linear, folded when young, later becoming flat, shortly tapering to a boat-shaped apex, green or with a violet tinge, glabrous and slightly glossy on the lower side, dull, somewhat rough and with two grooves along the midvein on the upper side. Leaf sheaths open, rough, terete or more often compressed and keeled. Ligule 4–10 mm long, acute, membranous.

Inflorescence: Panicle 10–25 cm long, erect or somewhat nodding, loose, spreading, green or with a violet tinge. Branches in whorls of 3–7 from the main axis, of different length and undivided in the lower part. Pedicels 0.5–3 mm long.

Spikelets: 3–4 mm long, oblong to ovate, green or violet tinged, 2–4-flowered. Rachilla glabrous. At maturity the fruits fall separately, enveloped between lemma and palea, while the glumes remain.

Glumes: Slightly unequal, the lower 2.0–3.0 mm long, 1-veined, the upper 2.5–3.5 mm long, 3-veined, both lanceolate, sharply pointed, apically rough on keels, green to violet tinged, margins membranous.

Lemma: 2.5–3.5 mm long, 5-veined, oblong ovate, green to violet tinged, blunt or somewhat pointed, with short hairs on keel and marginal veins, as well as long, soft and curly hairs at base.

Palea: A little shorter than lemma, with 2 hairy keels.

Flowers: 3 stamens. Anthers 1.5–2.0 mm long, yellowish to purplish.

Fruits: Caryopsis 1.3–1.6 mm long.

Flowering: June-July.

Biology: The seeds are dispersed by wind or water. Vegetative propagation is by stems rooting from the lower nodes, forming loose tufts. The species is included in seed mixtures for pastures and lawns.

Habitats: In oligotrophic to eutrophic, mildly acid to alkaline water. On moist to wet substrate on the margins of ponds, lakes and streams, in meadows, pastures, marshes and mires. On drier substrate in gravel pits, railway em-

Poa trivialis. Spikelets. Photo JCS.

Poa trivialis. The long, pointed ligule. Photo JCS.

bankments, roadsides, waste grounds, cultivated areas, etc.

Distribution within the region:
Native. Widespread and abundant through the region, but rare in Greenland.

Characteristics and similar species:
Poa trivialis differs from other similar *Poa* species by the rough leaf sheath and long, pointed ligules.

Plants without flowers may be mistaken for **191 *Leersia oryzoides***, **188-190 *Glyceria*** spp., **180-182 *Agrostis*** spp., **183-184 *Alopecurus*** spp. Please use the key for identification of vegetative plants.

Poa trivialis. Boat-shaped leaf apex. Mailscot Wood, Gloucestershire, England. Photo RVL.

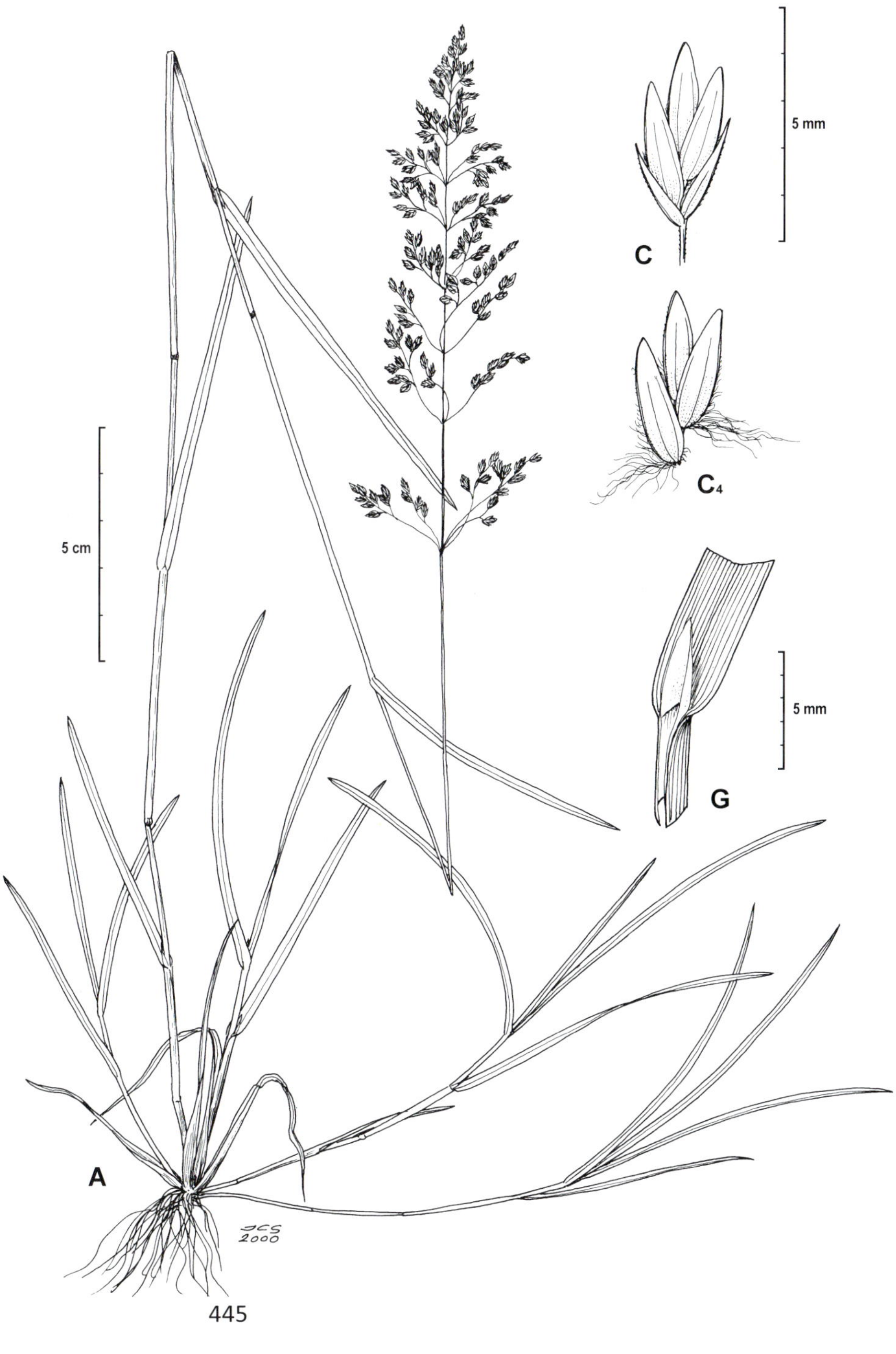

Poa trivialis
A habit, **C** spikelet, **C4** spikelet without glumes, **G** leaf sheath.

Poaceae

196 *Scolochloa festucacea* (Willd.) Link

Syn.: *Graphephorum arundinaceun* (Fr.) Ascherson
DK: Russersvingel **N:** Vass-svingel **S:** Kasgräs **FIN:** Piuru **IS:** -
GB: - **NL:** - **F:** -
D: Schwingelschilf **CZ:** - **PL:** Skokchloa trzcinowata
EST: Rooghein **LV:** Ūdeņu ērkšķuzāle **LT:** Eraičininė nendrūnė **Ru:** Тростянка овсяницевидная

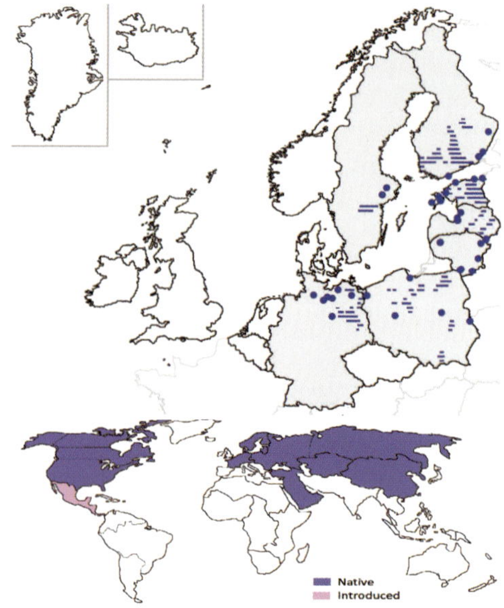

Scolochloa festucacea. Pool, Mecklenburg-West Pomerania, Germany. Photo KvdW.

Perennial helophyte. Rhizome long, branched, thick, underground.

Stem: 0.7–2 m long, 6–8 mm in diameter at base, terete, smooth, branched and rooting from the lower nodes.

Leaves: 30–55 cm long, 6–12 mm wide, flat, gradually tapering to a narrow apex, flaccid, arching with drooping tips, bright green, hairless, densely ribbed, rough on the upper side and margins, smooth below. Leaf sheaths open, terete, hairless, the lower usually spongy, smooth and shiny. Ligule 2–6(–10) mm long, truncate, often split.

Inflorescence: Panicle, 15–30 cm long, loose, light green to pale yellowish or brownish. Main axis rough. Branches initially erect, later spreading, rather flaccid, fine, terete, somewhat rough, the lower in whorls of 3–5. Pedicels 2–6 mm long, rough.

Spikelets: 8–11 mm long, laterally compressed, 3–4-flowered. At maturity the fruits falls separately, enveloped between lemma and palea, while the glumes remain.

Glumes: A little unequal, the lower 5–8 mm long, 3-veined, the upper 6–10 mm long, 5-veined, both membranous, rounded on the back, subacute at apex.

Lemma: 6–7.5 mm long, with 7 veins, rounded on the back, basally herbaceous, firm, smooth and shiny, gradually becoming membranous and rough towards the 3-toothed apex. Callus with 1–1.5 mm long hairs.

Palea: As long as lemma, with 2 minutely hairy keels and 2 teeth at apex.

Flowers: Stamens 3. Anthers 2.5–3.5 mm long, violet.

Fruits: Caryopsis about 2.5 mm long, hairy in the upper end.

Flowering: June-July.

Biology: The seeds are dispersed by wind or water. Vegetative propagation by branching rhizomes, consequently forming large stands of mostly vegetative shoots and sparse flowering shoots.

Habitats: In eutrophic still or slow-flowing water. On sandy, muddy or peaty substrates on the margins of lakes, streams and rivers, in reedbeds and wet fens.

Distribution within the region:
Native. Scattered to rather rare or locally common in Germany, Poland, the Baltic states, Finland and Sweden.

Scolochloa festucacea inflorescence. Lake Alksnas, Dukstas, Lithuania. Photo Jan-Thomas Johansson.

Scolochloa festucacea. Top of inflorescence. Lake Alksnas, Dukstas, Lithuania. Photo Jan-Thomas Johansson.

Scolochloa festucacea.
Leaf margin. Photo KvdW.

Characteristics and similar species:
With its wide, bright green, flaccid leaves *S. festucacea* may be mistaken for **187** ***Glyceria maxima***, which differs in the entire, keeled leaf sheaths with longer ligule.

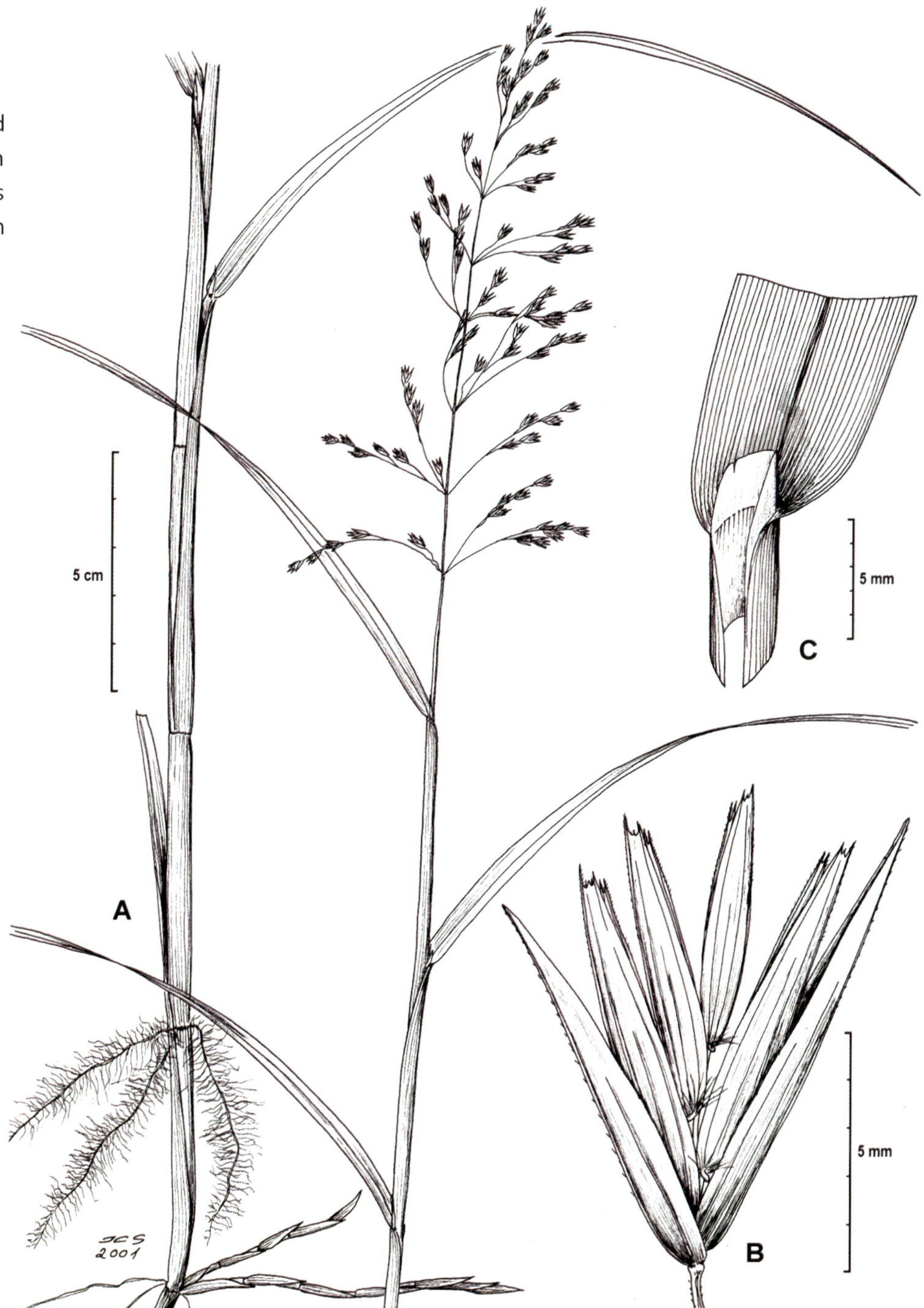

Scolochloa festucacea
A habit, **B** spikelet, **C** leaf sheath.

Ceratophyllaceae

197 *Ceratophyllum demersum* L.

DK: Tornfrøet Hornblad **N:** Hornblad **S:** Hornsärv **FIN:** Tankeakarvalehti **IS:** -
GB: Rigid Hornwort **NL:** Gedoornd hoornblad **F:** Cératophylle immergé
D: Raues Hornblatt **CZ:** Růžkatec ostnitý **PL:** Rogatek sztywny
EST: Räni-kardhein **LV:** Iegrimusī raglape **LT:** Paprastoji nertis **Ru:** Роголистник погружённый

Ceratophyllum demersum. Oxbow lake, Neuhofen, Germany. Photo KvdW.

Perennial hydrophyte without roots or rhizome, growing only fully submerged or forming dense mats at the surface. Anchored to sediment by a horizontal stem with short, stiff, claw-like leaves, or free floating.

Stem: Up to 3 metres long, prostrate or upright, thin to robust, stiff to flexible, brittle or not, unevenly branched, internodes' length decreasing towards the shoot apex.

Leaves: 6–10 together in whorls, each leaf 1–2(–3) times forked, rather stiff and robust, less often soft and delicate, commonly curved, dark to greyish green, often with small, red dots, the segment margins with many prominent teeth 0.1–0.4 mm long.

Inflorescence: Monoecious. Flowers solitary in leaf axils, mainly on the upper parts of the stems and branches. Male flowers 1–3 in each leaf whorl and commonly present at several whorls in a row. Female flowers sparse and never more than 1 in each whorl.

Flowers: Male flowers up to 3.5 mm wide, cup shaped, surrounded by 8–12 narrow bracts and up to 20 stamens with sessile anthers 1.0–1.5 mm long. Female flowers surrounded by 8–12 bracts, each containing 1 oval ovule, up to 1 mm long with a long, filamentous style.

Fruits: Nuts oval, 4.0–5.5 mm long, with a straight or curved, spine-like remnant of the style, up to 7.5 mm long at the tip and 2 downward-pointing spines at the base up to 6 mm long, or without spines. The surface of the fruit may be smooth or minutely warty.

Flowering: July-September.

Biology: Water pollinated - for details see **198 *Ceratophyllum submersum.*** Fruit development is generally sparse and most abundant in warm conditions, especially in shallow waters. Fruit dispersal by birds. Vegetative dispersal by detached stem fragments within and between water bodies. Overwinters as conical turions consisting of dense whorls of short, stiff leaves.

Habitats: Mesotrophic to eutrophic, neutral to alkaline, mainly standing waters - lakes, ponds, gravel pits and storm water retention ponds, less commonly in flowing waters – canals, ditches, slow-flowing streams and rivers.

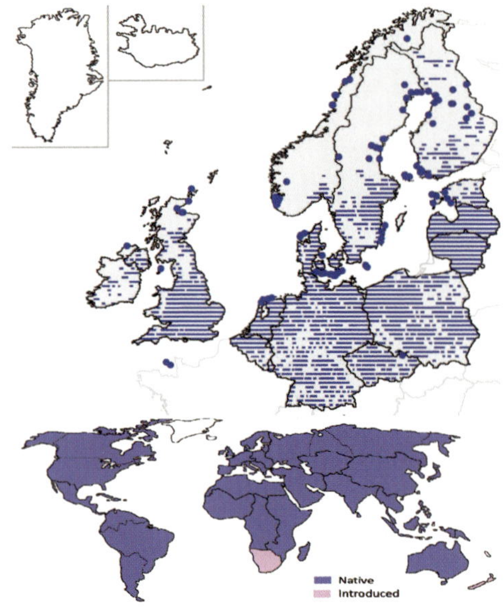

Distribution within the region:
Native. Sub-cosmopolitan, widespread and abundant throughout the region.

Characteristics and similar species:
C. demersum rarely fruits in the region, but although it is extremely variable, it can always be recognised by the whorls of 1–2 (occasionally 3) times forked leaves with dentate segments. The leaves are usually stiff, but in some conditions they may be soft and delicate.

199 *Ceratophyllum submersum* has leaves which are 3–4 times dichotomously forked with fewer, less conspicuous teeth on the segment margins.

226-232 *Myriophyllum* species differ from *Ceratophyllum species* by the whorls of pinnate leaves without dentate segments. See also **216 *Ranunculus circinatus.***

Ceratophyllum demersum overwinters by turions consisting of dense whorls of short and stiff leaves. Sand pit, Rheinberg-Budberg, Germany. Photo KvdW.

Ceratophyllum demersum. Left: Typical fruit with 2 long spines at the base. Højer, Jutland, Denmark. Middle: Fruit with 2 short basal spines Faxe, Denmark. Right: Fruit without basal spines. Gatten, Jutland, Denmark. Photos JCS.

Note: Different floras recognise 2 or 3 varieties within *C. demersum*; however, treatment is inconsistent between floras and very few records are identified to variety level, mainly because fruiting material is rarely encountered. Plants with spiny fruits are named var. ***demersum***. Stace 2019) furthermore recognise var. ***apiculatum*** (Cham.) Asch. with tubercles and var. ***inerme*** J.Gay ex Radcl.-Sm. without spines or tubercles.

Ceratophyllum demersum.
Leaves 2–3 times divided. Photo KvdW.

198 *Ceratophyllum platyacanthum* Cham.

Resembles *C. demersum*, differing in the shape of the fruit with spines on the face and broad spiny wings, while the basal spines are flattened and point outwards.

C. platyacanthum is considered to be extinct in Germany. It is known from scattered locations in Central Europe east through Asia to Japan and Korea.

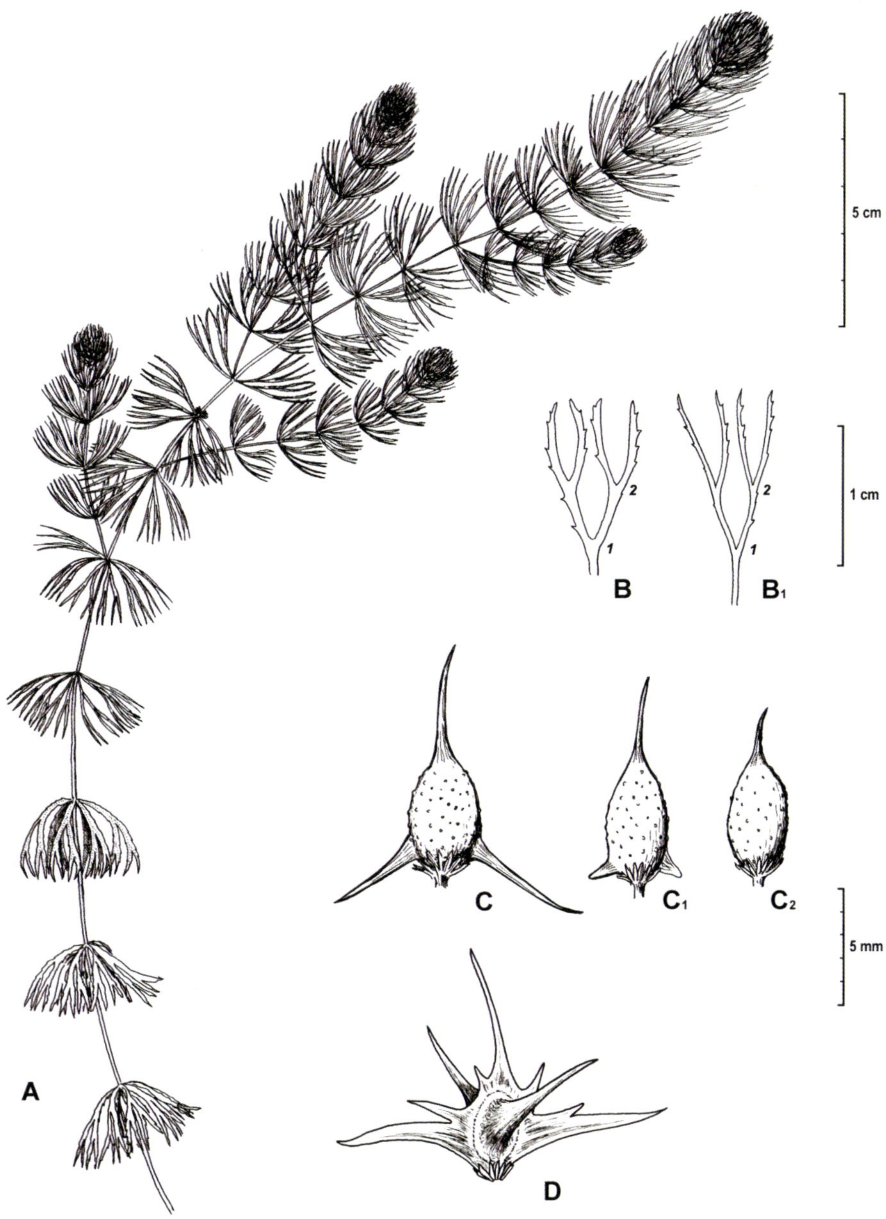

Ceratophyllum demersum
A habit, **B, B1** leaves, **C, C1** fruits, **C2** fruit var. *inerme*.
Ceratophyllum platyacanthum
D fruit.

Ceratophyllaceae

199 *Ceratophyllum submersum* L.

DK: Tornløs Hornblad **N:** Vorteblad **S:** Vårtsärv **FIN:** Hentokarvalehti **IS:** -
GB: Soft Hornwort **NL:** Fijn hoornblad **F:** Cératophylle submergé
D: Untergetauchtes Hornblatt **CZ:** Růžkatec bradavčitý **PL:** Rogatek krótkoszyjkowy
EST: Sile kardhein **LV:** Pusgrimusī raglape **LT:** Gležnalapė nertis **Ru:** Роголистник полупогружённый

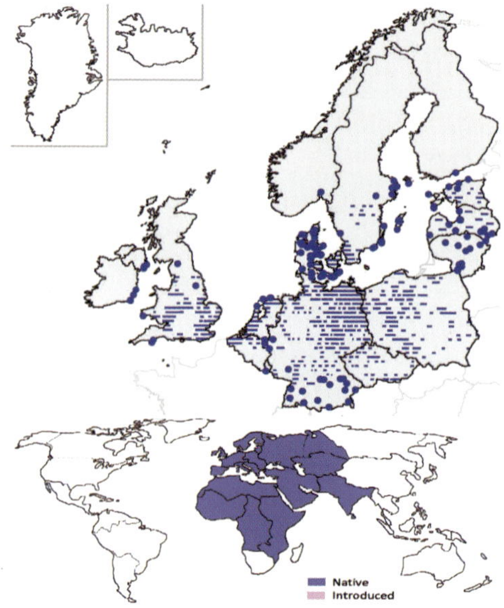

Ceratophyllum submersum in nutrient-rich water in a small lake near Hobro, Denmark. Photo JCS.

Perennial hydrophyte without roots and rhizome, growing only fully submerged. Anchored to sediment by a horizontal stem with short, stiff, claw-like leaves, or free-floating.

Stem: Up to 2 metres long, prostrate or upright, thin, flexible, brittle, unevenly branched, internodes' length decreasing towards the shoot apex.

Leaves: 6–10 leaves together in whorls, each leaf 3–4(–5) times forked, delicate and flexible, light to dark green, often with small, red dots, the segment margins with few rather inconspicuous teeth 0.1–0.2 mm long.

Inflorescence: Monoecious flowers solitary in leaf axils, mainly on the upper parts of the stems and branches. Male flowers 1–3 in each whorl and commonly present in several whorls in a row. Female flowers scarce and never more than 1 in each leaf whorl.

Flowers: Male flowers up to 3 mm wide, cup shaped, surrounded by 8–12 narrow bracts and with up to 12 stamens with sessile anthers 0.6–1.0 mm long. Female flowers surrounded by 8–12 bracts, each flower containing 1 oval ovule, up to 1 mm long with a long, filamentous style.

Fruits: Nuts ovoid, 4–5 mm long and commonly with a marginal wing, with a curved, spine-like remnant of the style 1.0–1.5 mm long at the tip and never with spines at the base.

Two varieties is recognised based on the surface of the fruit: Var. ***submersum*** - fruits smooth or warty. Var. ***haynaldianum*** (Borbas) Wilmot-Dear. - fruits with few to numerous minute and emarginate spines.

Flowering: June-September.

Biology: Water pollinated. At maturity the anthers become filled with air, detach from the flowers and rise to the surface where the pollen is released; this germinates and elongates before sinking and is thereby more easily caught on the style of female flowers. Fruiting is mostly abundant, mainly under warm conditions. Fruit dispersal by birds. Vegetative dispersal by detached stem fragments is important within water bodies. Overwinters as conical turions consisting of dense whorls of short, stiff leaves.

Habitats: Mostly eutrophic, neutral to alkaline, mainly standing waters - lakes, forest lakes, ponds, water-filled gravel pits and storm water retention ponds, less commonly in flowing waters –

The leaves of *C. submersum* are 3–4 times bifurcate and with only few, inconspicuous teeth on the leaf segments. Hobro, Denmark. Photo JCS.

Ceratophyllum submersum winterform with dense whorls of short and stiff leaves. Rheinauensee, Bonn, Germany. Photo KvdW.

Ceratophyllum submersum male flower with sessile anthers. Hobro, Denmark. Photo JCS.

Ceratophyllum submersum fruit with smooth surface and no spines. Hobro, Denmark. Photo JCS.

canals, ditches and slow-flowing streams and rivers.

Distribution within the region:
Native. Widespread and rather common in most of the region except Ireland, Scotland and northern Scandinavia. Absent from Iceland and Greenland.

Characteristics and similar species:
Ceratophyllum submersum is recognised by the whorls of flexible leaves, mostly 3–4 times forked leaves with inconspicuously dentate segments.

Can be confused with **197 *Ceratophyllum demersum***, which has leaves forked 1–2 times and these are typically stiffer with distinctly dentate filaments.

226-232 *Myriophyllum* species differ from *Ceratophyllum* species by the whorls of pinnate leaves without dentate segments. See also **216 *Ranunculus circinatus***.

Ceratophyllum submersum
Leaf 3–4 times divided.
Photo KvdW.

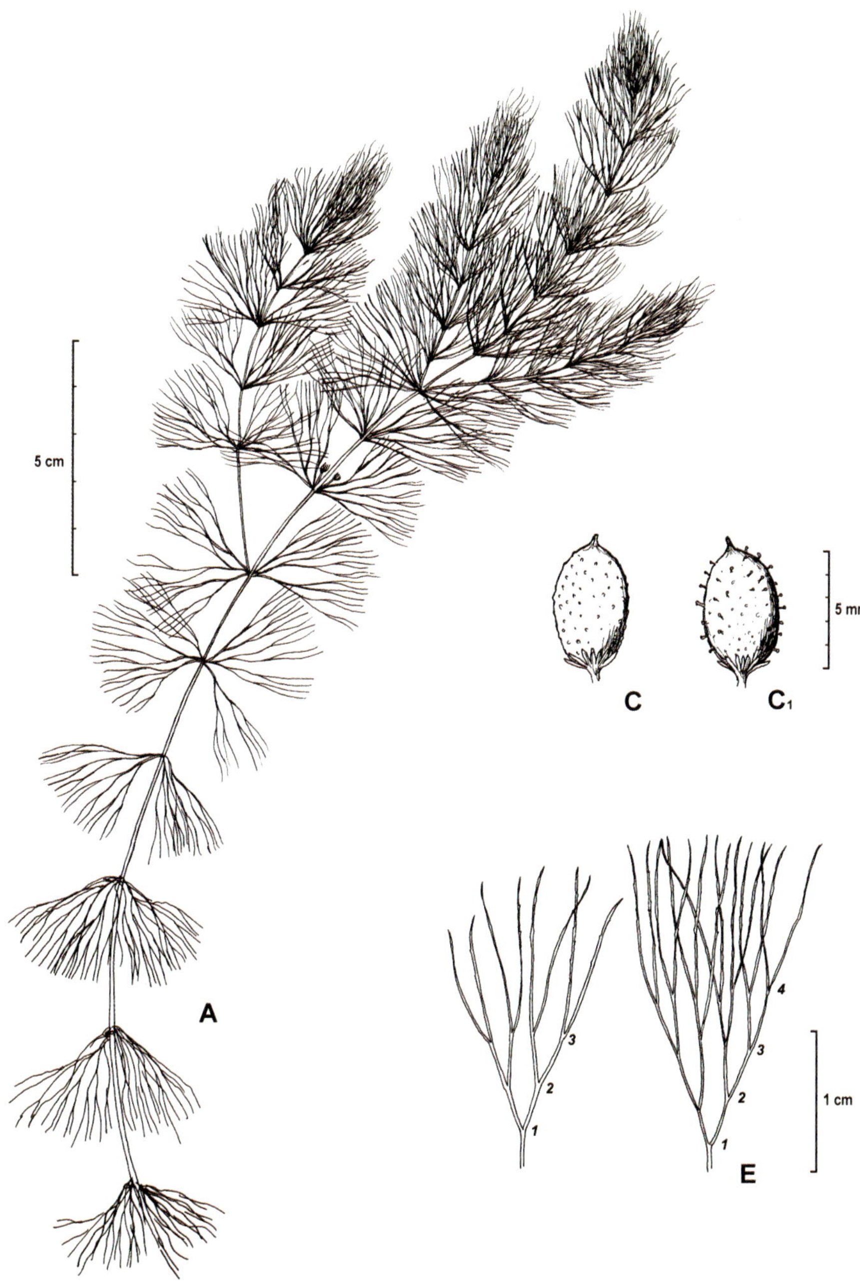

Ceratophyllum submersum
A habit, **C** fruit var. *submersum*, **C1** fruit var. *haynaldianum*, **E** leaves.

Key to Ranunculaceae.

The genus *Ranunculus* includes the white-flowered species in sect. *Batrachium* and the yellow-flowered in sect. *Ranunculus*. Identification of *Batrachium* taxa depends on examination of entire individuals. Most of the characters used can only be observed in generative (fertile) shoots. The middle and upper parts of several flowering shoots from a population should be studied. Vegetative shoots of perennial species are not representative and usually have longer capillary leaves, while important floral characters will be missing. Measurements of capillary leaves apply to the middle part of a generative shoot. Information on the total length of shoots, shoot types, branching pattern and structure of capillary leaves should be noted as these may not be preserved in herbarium specimens. Measurements of floral parts (petal length, fruit length) given here refer to dried herbarium material. Fragmentary specimens, terrestrial forms and hybrids cannot be identified with this key. The section "Useful characters for identification" (p. 464) offers additional information on how to work with *Ranunculus* section *Batrachium*. Besides the dichotomous key below, a synoptical key for *Ranunculus* section *Batrachium* is given on pp. 466-467.

1 Petals white with yellow claw at base
Ranunculus section _Batrachium_ 2

1 Petals yellow 23

2 Stipules less than 1/2 adnate to the petiole (***1***), free part conspicuous, often whitish, membranous. Capillary leaves present or absent; if present filiform. Laminar leaves present, alternate or opposite. Sepals reflexed. Petals small to medium-sized. Nectar pits lunate (***3***), 1 per petal 3

2 Stipules more than 1/2 adnate to the petiole (***2***), free part mostly small, greenish. Capillary leaves present, mostly persistent, rigid or flaccid. Laminar leaves present or absent, alternate. Sepals spreading or reflexed. Petals medium-sized to large. Nectar pits lunate, horseshoe-like, circular, triangular, or pyriform (***3-7***), 1(–4) per petal 5

3 Capillary and intermediate leaves absent. Laminar leaves 3–7-lobed, dissected up to 1/2 the width of the lamina, margin of primary lobes entire or crenulate (***8***). Sepals greenish. Petals 4–7 mm long, not contiguous during anthesis. Receptacle glabrous (***12***)
202 _R. omiophyllus_

3 Capillary leaves mostly present. Intermediate leaves rarely present. Laminar leaves 3(–5)-lobed, dissected to more than 2/3 the width of the lamina (***9***), margin of primary lobes crenate. Sepals bluish or greenish with a blue tip. Petals 1–15 mm long, contiguous or not at anthesis. Receptacle puberulent or pubescent (***13***) 4

4 Petals white with a small yellow base (***10***), 5–15 mm long, 2x sepal length, often contiguous during anthesis. Sepals greenish with a blue tip. Receptacle puberulent to pubescent. Intermediate leaves absent
204 _R. ololeucos_

4 Petals white with a yellow base (***11***), 3–5.5 mm long, as long as or slightly longer than sepals, not contiguous during anthesis. Sepals bluish. Receptacle pubescent. Intermediate leaves divided in threes, with 3–5 acute tips occasionally present
203 _R. tripartitus_

1

2

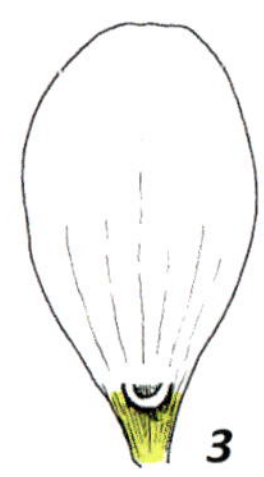
3

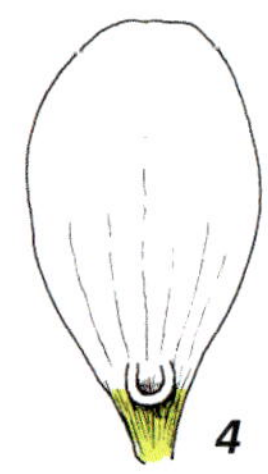
4

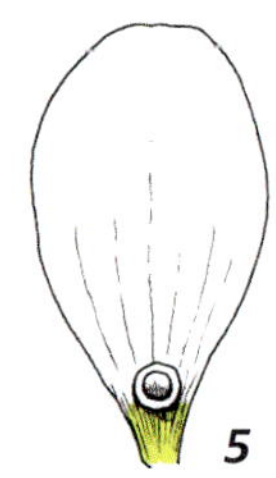
5

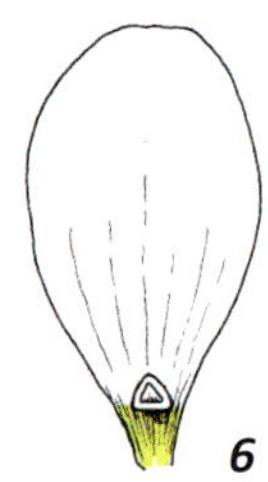
6

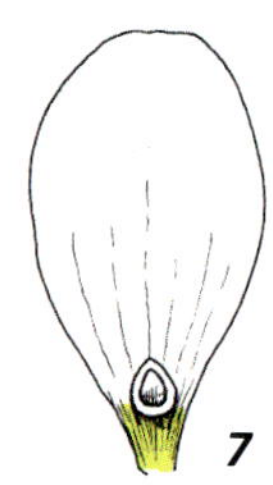
7

8

9

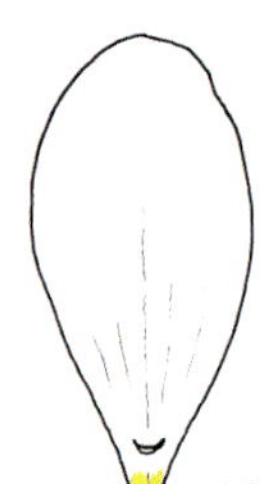
10

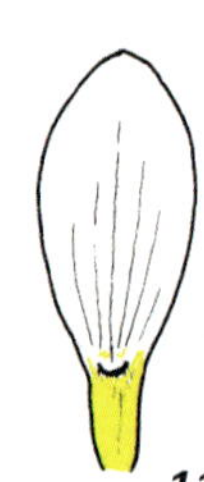
11

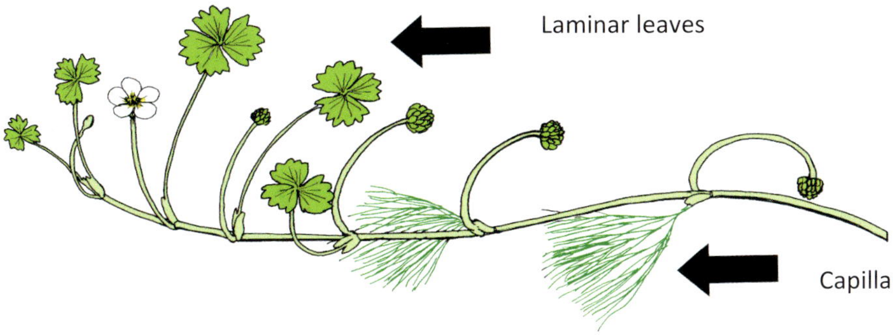

5 Capillary leaves absent. Laminar leaves reniform to cordate, shallowly 3(–5)-lobed, lobes broadest at base, margins rounded (***14***). Petals 1.2–5 mm long, not contiguous during anthesis
205 *R. hederaceus*

5 Capillary leaves present or absent. Laminar leaves reniform or truncate, 3–7-lobed, primary lobes broadest at the top, margins often crenulate or dentate. Petals (3–)5–23 mm long, contiguous during anthesis or not 6

6 Laminar leaves present. With or without capillary leaves (Batrachids) 7

6 Laminar leaves absent, occasionally a few intermediate leaves present. Capillary leaves present (Myriophyllids) 12

7 Nectar pits lunate (***3***), horseshoe-like (***4***) or triangular (***6***), mostly open at the top. Sepals bluetipped, usually reflexed. Pedicels elongated towards maturity, strongly curved or coiled 8

7 Nectar pits circular (***5***) or pyriform (***7***), mostly closed at the top. Sepals rarely bluetipped, spreading. Pedicels elongating under water during flowering or not elongated, straight or slightly curved 10

8 Fruits 1.0–1.5 mm long, ventrally and dorsally winged, glabrous at maturity (***16***). Receptacle hemispherical, elongating at maturity, pubescent, hairs longer than 2 mm. Stems fleshy, whitish. Laminar leaves 3(–5)-lobed, deeply dissected to 2/3 or 3/4 of the lamina (***15***). Intermediate leaves tripartite, mostly symmetrical with parallel-margined terminal segments. Capillary leaves rigid, often fleshy. Flowers medium -sized. Nectar pits lunate (***3***), rarely horseshoe-like (***7***)
209 *R. baudotii*

8 Fruits (1.2–)1.8–2.3 mm long, rarely ventrally winged, pubescent. Receptacle hemispherical, not or only slightly elongating at maturity, glabrous to puberulent, hairs shorter than 2 mm. Stems not fleshy, greenish. Laminar leaves 5–7-lobed, often truncate at base. Intermediate leaves with cuneate lobes and capillary segments. Capillary leaves flaccid, sometimes lacking at maturity. Flowers medium-sized to large. Nectar pits lunate (***3***), rarely triangular (***6***)
R. saniculifolius Viv.
(not in this book)

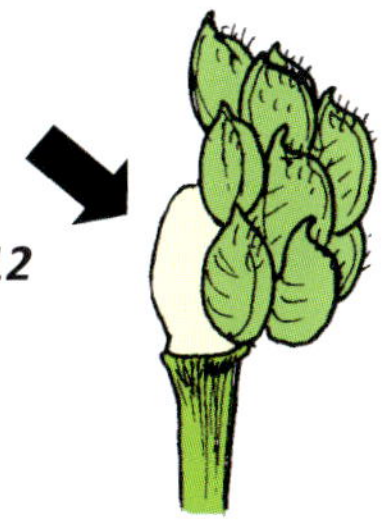
12

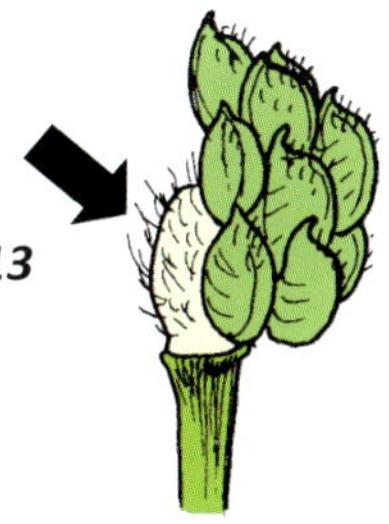
13

14

16

15

9 Nectar pits circular (***5***). Pedicels not elongating, shorter than 5 cm, shorter than or equal to opposite leaf (***17***). Laminar leaves with 3–7 primary lobes, up to 30 mm wide, reniform to circular, margins crenate or dentate (***18***). Intermediate leaves with basal capillary segments, often asymmetrical (***19***). Flowers medium-sized. Mature fruits with a few brush-like hairs
213 *R. aquatilis*

9 Nectar pits pyriform (***7***). Pedicels elongating under water, 5–20 cm long, mostly longer than opposite leaf. Laminar leaves with (3–)5(–7) primary lobes, up to 40 mm wide, margins entire or obtusely crenate. Intermediate leaves with apical capillary segments. Flowers large. Mature fruits glabrous or puberulent 10

10 Receptacle glabrous (***12***). Petals 5(–10), nectar pits often 2–4 per petal (***20***). Laminar leaves 5-lobed, often divided to more than 1/2 the width of the lamina, margins of secondary lobes acutely crenate or dentate. Lower capillary leaves long with exceedingly long petiole and long first segments, becoming gradually shorter towards the stem tip, upper capillary leaves condensed, with short terminal segments (***21***)
211 *R. schmalhausenii*

10 Receptacle pubescent (**13**). Petals 5, rarely 6–7, nectar pits 1 per petal. Laminar leaves 5–7-lobed, mostly shallowly divided, margins of secondary lobes rounded or crenate. Length of capillary leaves more regular along the stem, upper ones not condensed 11

11 Fertile shoots up to 2 m long. Capillary leaves 3–10 cm long, flaccid or rigid, divergent or penicillate, mostly shorter than adjacent internode. Pedicels not tapering towards the top. Stamens 15–30
210 *R. peltatus*

11 Fertile shoots up to 6 m long. Capillary leaves of fertile shoots 5–20(–30) cm long, mostly longer than adjacent internode, segments flaccid, subparallel to clasping. Pedicels tapering towards the top. Stamens (8–)20–40
208 *R. penicillatus*

12 Fertile shoots (1–)2–6 m long. Pedicels as wide as the stem, tapering toward the top. Capillary leaves 5–25(–50) cm long. Nectar pits pyriform (***7***). Usually in running waters and deep lakes 13

12 Fertile shoots 0.5–2 m long. Pedicels less wide than stem, not tapering toward the top. Capillary leaves 2–5(–7) mm long. Nectar pits lunate (***3***), circular (***5***) or pyriform (***7***). Usually in small streams, ponds and shallow lakes 16

17

18

19

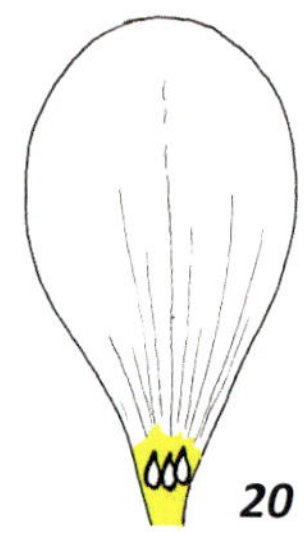

20

21

13 Capillary leaves up to 55 cm long, fleshy, segments subparallel to parallel, number of terminal segments low (up to 40) (***22***). Receptacle glabrous (***12***). Number of petals (5–)6–10(–12). Nectar pits pyriform (***7***)

206 *R. fluitans*

13 Capillary leaves up to 20(–28) cm long, fleshy or flaccid, segments divergent or subparallel, number of terminal segments 50–400(–900) (***23***). Receptacle pubescent (***13***). Number of petals mostly 5 (rarely 6–7). Nectar pits mostly pyriform(***7***), occasionally horseshoe-like (***4***) or lunate (***3***) 14

22

14 Capillary leaves not fleshy, mostly flaccid to subrigid, penicillate, in the lower and middle part of the stem longer than adjacent internode on fertile shoots. Flowers large. Nectar pits pyriform (***7***) 15

14 Capillary leaves fleshy, mostly subrigid to rigid, penicillate or divergent, mostly shorter than adjacent stem internode. Flowers medium-sized to large. Nectar pits lunate (***3***), horseshoe-like (***4***) or pyriform (***7***) 16

15 Branching of main shoot not dichotomous (***23***). Capillary leaves gradually becoming shorter from the base to the apex of the stem, uppermost leaves not condensed. Number of terminal segments 60–150 (rarely up to 900)

208 *R. penicillatus* (see couplet 11)

15 Branching of main shoot dichotomous (***21***). Lower capillary leaves long with exceedingly long petiole and long first segments, becoming shorter towards the apex of the stem, uppermost capillary leaves condensed, with short terminal segments. Number of terminal segments up to 300 in uppermost leaves

211 *R. schmalhausenii* (see couplet 10)

23

16 Flowers large, nectar pits pyriform (***7***). Fruits unwinged. Stem greenish or brownish. Capillary leaves regularly branched, number of terminal segments up to 350. Short side shoots in the axils of capillary leaves regularly present with short divergent leaves

207 *R. pseudofluitans*

16 Flowers medium-sized, nectar pits lunate (***3***) or horseshoe-like (***4***). Fruits ventrally and dorsally winged (***16***). Stem whitish. Capillary leaves unevenly branched, number of terminal segment 50–150(–200). Short shoots in leaf axils with similar leaves as on main shoot

209 *R. baudotii* (see couplet 8)

17 Capillary leaves orbicular, flat, lying in one plane, mostly considerably shorter than adjacent internode, segments rigid (***24***). Flowers medium-sized. Nectar pits lunate (***3***)
216 *R. circinatus*
17 Capillary leaves spherical, not lying in one plane, shorter or longer than adjacent internode, segments rigid or flaccid. Flowers small to large. Nectar pits lunate (***3***), circular (***5***) or pyriform (***7***) 18

18 Petals mostly shorter than 5.5(–7) mm, spathulate or lanceolate, not contiguous during anthesis. Nectar pits lunate (***3***) 19
18 Petals longer than 5.5(–7) mm, obovate, contiguous during anthesis. Nectar pits lunate (***3***), circular (***5***) or pyriform (***7***) 22

24

19 Plant annual, primary roots usually present. Petioles short (10 mm). Capillary leaves of equal size along the stem, broader than long. Pedicels less than 1 mm wide. Flowers small. Receptacle elongating in fruit. Carpels (20–)60(–90), 0.8–1.3 mm long
217 *R. rionii*
19 Plant perennial, with adventitious roots only. Petioles mostly longer than 1 cm. Capillary leaves longer at the base of the stem, longer than broad. Pedicels more than 1 mm wide. Receptacle not elongating in fruit. Carpels 5–45, 1.5–2.5 mm long 20

20 Plant small (< 40 cm long), prostrate, rooting at all nodes. Capillary leaves 10–35 mm long, very fine, terminal segments elongated. Carpels 5–15
214 *R. confervoides*
20 Plant small to medium-sized (up to 200 cm long), mostly erect, rooting only at lower nodes. Capillary leaves 20–80 mm long, mostly flaccid, rarely subrigid. Carpels > 15 21

21 Plant up to 1 m long. Capillary leaves rigid or flaccid, green or brown when dry. Pedicels less than 50 mm long. Petals 3–5(–6) mm long
212 *R. trichophyllus*
21 Plant up to 2 m long. Capillary leaves flaccid, yellowish green when dry. Pedicels longer than 50 mm. Petals 5–7(–8) mm long
215 *R. kauffmannii*

22 Nectar pits circular (***5***) or horseshoe-like (***4***). Pedicels less than 5 cm long, not elongating, mostly straight. Flowers medium-sized. Mature carpels with brush-like hairs
213 *R. aquatilis* (see couplet 9)
22 Nectar pits pyriform (***7***). Pedicels longer than 5 cm, elongating under water, often curved. Flowers large. Mature carpels sparsely hairy
210 *R. peltatus* (see couplet 11)

23 With only petal-like sepals, petals absent (***25***). Fruit a follicle
200-201 *Caltha palustris*
23 Both sepals and petals present. Fruit an achene 24

25

24 Leaves unlobed, entire or margin toothed 25
Ranunculus section _Flammula_

24 At least some leaves with lobes (***31, 32***)
Ranunculus section _Hecatonia_ 28

25 Flower 2.5–4.5 cm in diameter. Stem erect, 50–150 cm tall. Emerged leaves 15–25 cm long, lanceolate (***26***), submerged leaves 5–20 cm long, with ovate to oblong lamina and long petiole (***27***)
219 _Ranunculus lingua_

25 Flower 0.5–2.5 cm in diameter. Stem ascending procumbent or erect 26

26 Stem erect. Fruits tuberculate (***28***)
223 _R. ophioglossifolius_

26 Stem ascending or procumbent. Fruits smooth (***29, 30***) 27

27 Fruits with a straight beak, comprising 1/6–1/8 of the total length of the achene (***29***). Stem rooting at lower nodes
220 _Ranunculus flammula_

27 Fruits with a curved beak, comprising 1/4 of the total length of the achene (***30***). Stem rooting at almost all nodes
221 _Ranunculus reptans_

28 Annual. Stem erect, 5–50 cm tall. Flowers numerous, c. 10 mm in diameter
218 _Ranunculus sceleratus_

28 Perennial. Stem creeping, 2–5 cm tall plant (***32***). Flowers few, c. 5 mm in diameter
222 _Ranunculus hyperboreus_

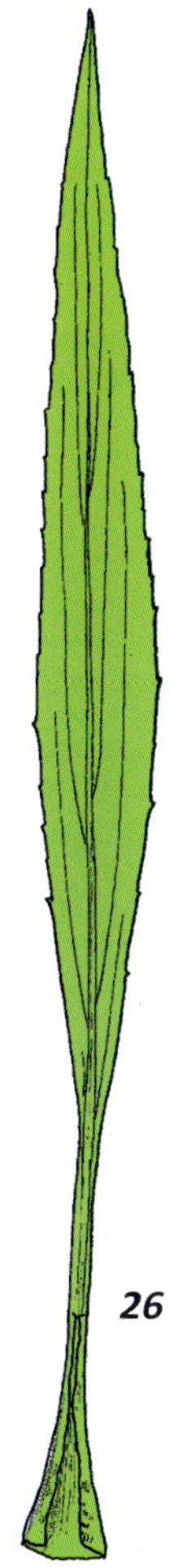
26

27

28

29

30

31

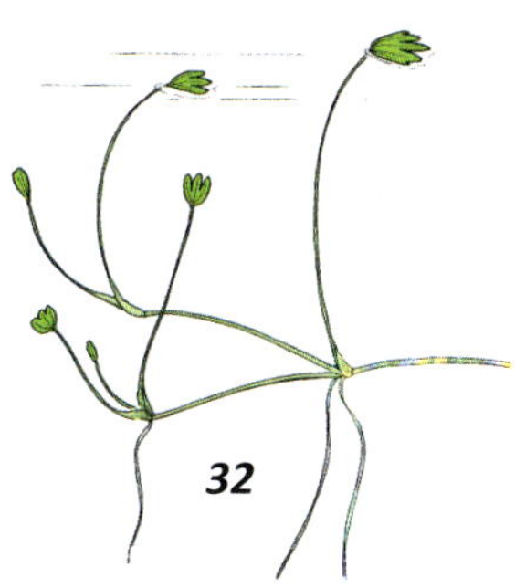
32

Ranunculaceae

200 *Caltha palustris* L. subsp. *palustris*

DK: Eng-Kabbeleje **N:** Stor Bekkeblom **S:** Vanlig kabbleka **FIN:** Rantarentukka **IS:** Hófsóley
GB: Marsh Marigold **NL:** Gewone dotterbloem **F:** Caltha des marais
D: Sumpfdotterblume **CZ:** Blatouch bahenní **PL:** Knieć błotna typowa
EST: Harilik varsakabi **LV:** Purva purene **LT:** Pelkinė puriena **Ru:** Калужница болотная

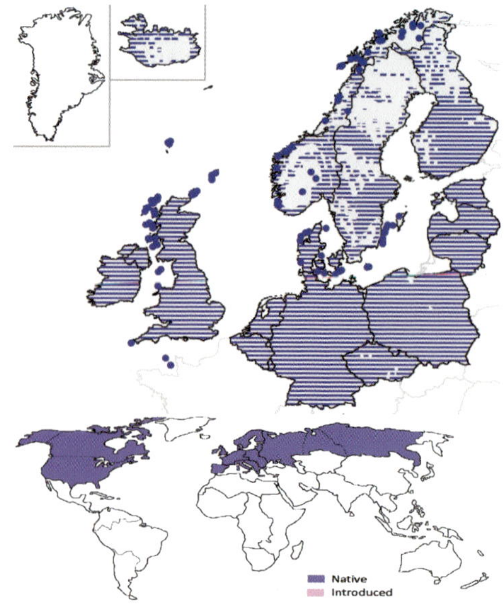

Flowering stems of *C. palustris* subsp. *palustris* extend from the shore into a pond. Hobro, Denmark. Photo JCS.

Perennial helophyte. Rhizome short, vertical and branched, without stolons.

Stem: 10–50(–60) cm long, up to 15 mm in diameter, prostrate to mainly upright, terete, glabrous, hollow, greyish green to reddish or purple, with few to many branches arising from the upper part of each stem. One or two stems arise from each rhizome. Prostrate stems never root at the nodes.

Leaves: Leaves dark green. Basal leaves long petiolate with petioles 1–5(–7) times as long as the lamina, with a membranous sheath with free stipules up to 3 cm long at the base. Lamina 2–10(–40) cm wide, more or less circular to reniform with a cordate base, enlarging to almost twice this size when plants are in fruit. The margins crenate to dentate, rarely entire. Stem leaves short petiolate or almost sessile with a short, membranous ochrea, the lamina reniform or almost triangular.

Inflorescence: With up to several short branches at the tip of each stem carrying up to 15 flowers.

Flowers: 1–5 cm in diameter, pedicels several cm long, terete and glabrous. Sepals (4–)5(–9), obovate, up to 25 mm long and 18 mm wide, with a bright to warm yellow upper side and green lower side, thus appearing like petals, which are absent! Sepals without nectar pits. Stamens numerous with yellow anthers, 0.9–2.4 mm long.

Fruits: Carpels 1–15, free. Follicles 3–19 mm long, oblong, slightly compressed, glabrous, with a straight or slightly curved beak 0.6–2.2 mm long and nectaries on the sides. Each follicle with up to 20 seeds in two rows. Seeds dark brown to blackish, 1.6–2.7 mm long, with a brownish, spongy flotation appendage.

Flowering: (March-) April-June.

Biology: The flowers are pollinated by a variety of insects, and are presumed to be self-incompatible. Seed dispersal is by water. Plants often develop from flooded ground in meadows or during high water at the margins of streams, ponds and lakes. *C. palustris* is highly variable, morphologically as well as in appearance. Permanently submerged plants are very rare, and no distinct

Caltha palustris subsp. *palustris* on muddy substrate in the margin of a stream. Hobro, Denmark. Photo JCS.

Caltha palustris subsp. *Palustris.* Both flowers and fruits are characteristic. Jutland, Denmark. Photos JCS.

submerged form exists, but plants subject to prolonged inundation in shallow water may develop erect stems with aerial leaves and flowers. The species often forms large stands of many individuals. Plants with only basal leaves are often seen in early spring and after flowering.

Habitats: Open, moist to wet ground with oligotrophic to nutrient-rich, alkaline to acidic soils and waters. In wet woodland, swamps and meadows as well as on the margins of lakes, ditches, canals and streams.

Distribution within the region: Native. Widespread and abundant throughout most of the region.

Characteristics and similar species: The large basal leaves and the large yellow flowers without petals separate *Caltha palustris* from all other water plants, helophytes as well as hydrophytes.

Caltha palustris is separated into two subspecies of which subsp. *palustris* is described above and subsp. *radicans* on the next pages.

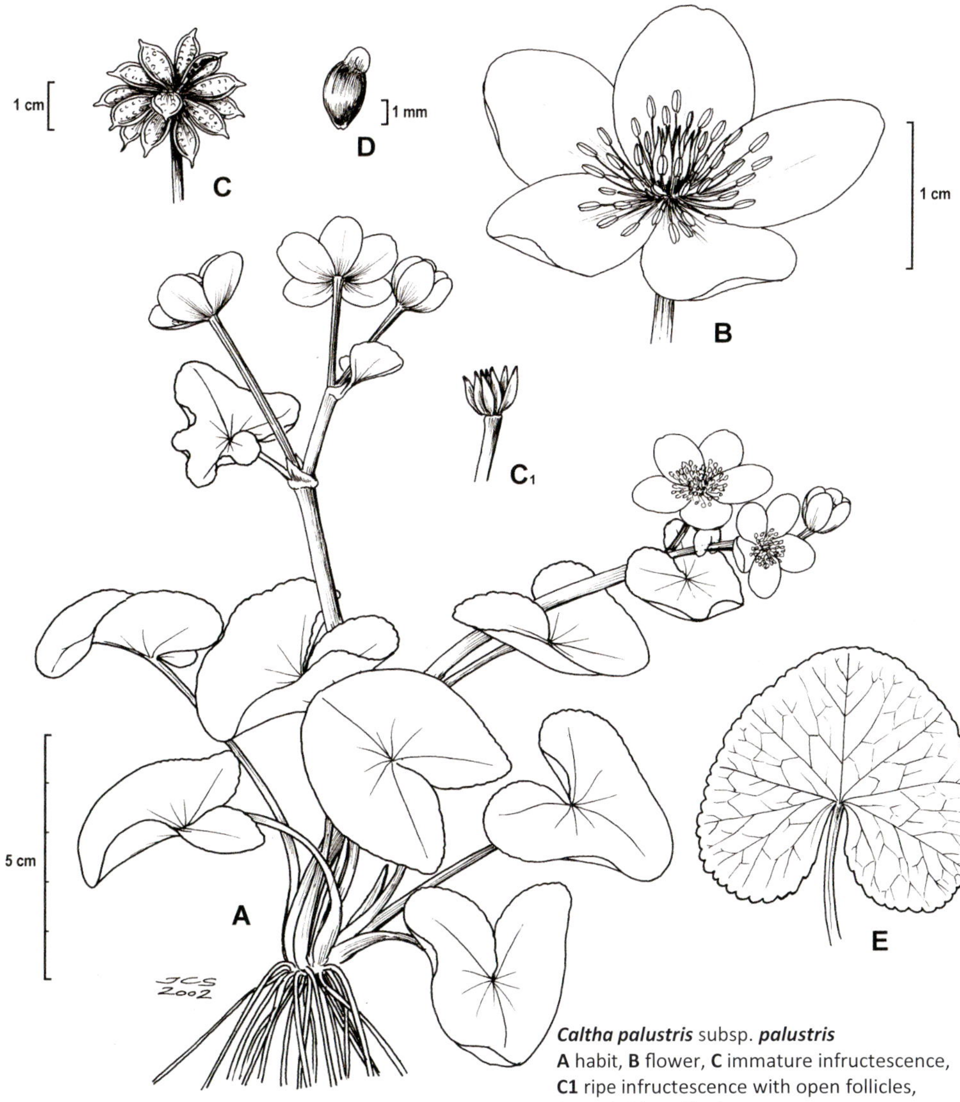

Caltha palustris subsp. ***palustris***
A habit, **B** flower, **C** immature infructescence, **C1** ripe infructescence with open follicles, **D** seed, **E** basal leaf.

Ranunculaceae

201 *Caltha palustris* L. subsp. *radicans* (T.F.Forst.) Syme

Syn.: *C. minor* auct.; *C. palustris* L. var. *radicans* (T.F.Forst.) Fr.
DK: Krybende Kabbeleje **N:** Fjellbekkeblom **S:** Revkabbleka **FIN:** Purorentukka **IS:** -
GB: - **NL:** - **F:** -
D: - **CZ:** - **PL:** Knieć błotna płożąca
EST: - **LV:** - **LT:** - **Ru:** -

Caltha palustris subsp. *radicans* in a small stream close to Lake Hald, Jutland, Denmark. Photo JCS.

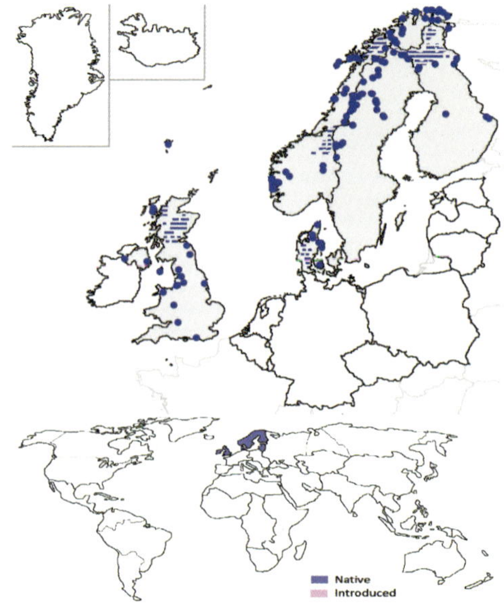

Perennial helophyte or hydrophyte with short, vertical, branched rhizome. Without stolons.

Stem: 10–50(–60) cm long, up to 8 mm in diameter, prostrate, rooting at the nodes, terete, glabrous, hollow, greyish green to reddish or purple.

Leaves: Basal leaves dark green. Petioles 4–5(–7) times as long as the lamina of fully developed leaves, with a membranous sheath with free stipules up to 3 cm long at the base. Lamina 7–12 cm wide, almost circular to reniform with a cordate base, the margin crenate to dentate, rarely entire. Stem leaves dark to yellowish green, petioles shorter than those of basal leaves.

Inflorescence: With 1–3(–5) flowers at the tip of each of the stems as well as from the nodes and daughter plants.

Flowers: 1–4 cm in diameter, pedicels 2–5 (or more) cm long, terete, glabrous. Sepals (4–)5(–9), obovate, up to 20 mm long and 14 mm wide, pale yellow above and green below, thus appearing like petals, which are absent! Sepals without nectar pits. Stamens numerous with yellow anthers, 0.9–2.4 mm long.

Fruits: Carpels 1–15, free. Follicles 3–13 mm long, oblong, slightly compressed, glabrous, with nectaries on the sides. Each follicle with up to 20 seeds. Seeds normally without a spongy appendage.

Flowering: (March-) April-July.

Biology: Insect pollinated; the pollination mechanism of submerged flowering plants unknown. Seeds do not float, but are dispersed by water movement and probably by animals. Vegetative propagation is by detached stem fragments and daughter plants. Basal leaves often develop from flooded ground in meadows or during high water at the margins of streams, ponds and lakes. Permanently submerged plants are common in some parts of the region, primarily in flowing waters and sometimes even forming large stands in deep flowing water. This subspecies often occurs in large stands, both submerged and emergent, due to the prostrate, rooting stems and the production of daughter plants. Plants often only have basal leaves in early spring and after flowering.

Habitats: Open and wooded swamps as well as springs, flushes and spring-fed meadows. Also on open, moist to wet ground with oligotrophic to mesotrophic soil and water on the margins and in the water of ponds, lakes, ditches, canals, smaller streams and montane flushes. It occurs in both alkaline and acidic waters, as well as on both humic and mineral soils.

Caltha palustris subsp. *Radicans.* Small specimen growing in a stony flush (700 m) on steep slopes above the Spittal of Glenshee in the Cairngorm Mountains of Scotland. Photo RVL.

Typically developed *C. palustris* subsp. *radicans* in a clayey ditch. Gram, Jutland, Denmark. Photo Inge Nagstrup.

Distribution within the region:
Native. Britain, Denmark, Norway, Sweden and Finland. May be overlooked or unrecorded elsewhere in the region.

Characteristics and similar species:
The basal leaves and the large, yellow flowers with large, pale yellow sepals and without petals separate *Caltha palustris* from all other water plants, helophytes as well as hydrophytes.

Caltha palustris subsp. *radicans* can only be confused with *Caltha palustris* subsp. *palustris*, from which it differs by a number of characters, especially the smaller size, the prostrate stems rooting and forming daughter plants from the nodes, the smaller and paler yellow flowers and the ability to form permanently submerged stands. Terrestrial plants may be very similar to *Caltha palustris* subsp. *palustris*, which however does not root from the stem nodes and typically grows in stands with more upright stems.

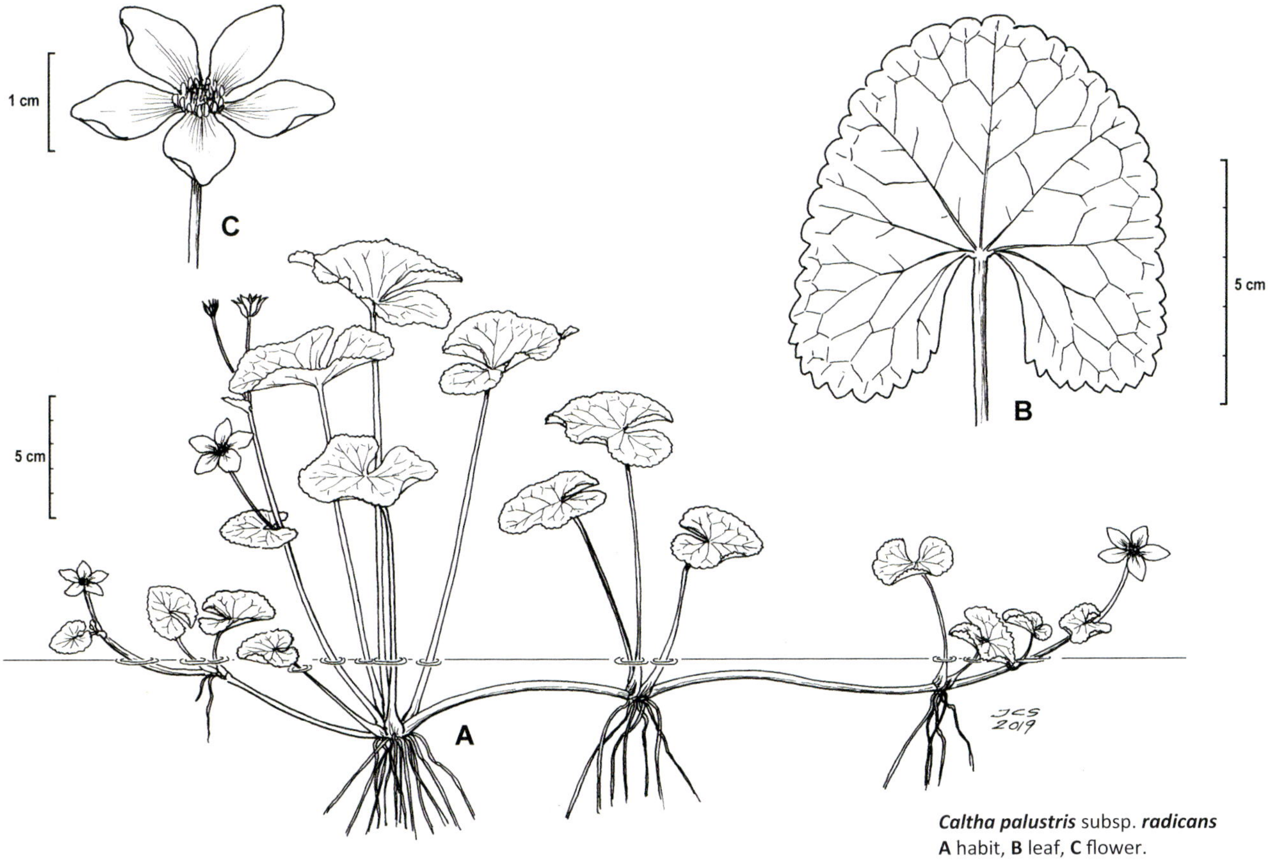

Caltha palustris **subsp.** ***radicans***
A habit, **B** leaf, **C** flower.

Introduction to *Ranunculus* sect. *Batrachium*

Flowering *R. baudotii* in running water. River Skals Å near Hobro, Denmark. Photo JCS.

The water-crowfoots - white-flowered *Ranunculus* of the section *Batrachium* - comprise the most difficult water plants in the region to identify. The reasons for this difficulty are outlined below.

The importance of plasticity

Batrachium species are characterised by a strong phenotypic variation, which can be observed both between and within sites, as well as between and within stands and even individual plants. During their evolution *Batrachium* have acquired multi-purpose genotypes, which are able to produce different phenotypes in relation to current velocity, temperature, and water level. River forms, deep water forms in lakes, shallow water forms in ponds, or terrestrial forms in damp places may look considerably different from one another. The development of an individual plant is seasonal and heteroblastic. During the growth period different types of leaves are formed in a deterministic sequence. This process is regulated by antagonistic plant hormones, such as abscisic acid (ABA) and ethylene. It may however be interrupted by drought, cold or mechanical destruction. On the other hand, warm winters may lead to survival of above-ground parts of the plants, eventually leading to a change in function, e.g., an erect shoot may become a prostrate "rhizome" in the subsequent year. Depending on the circumstances of regrowth different phenotypes will be produced by the same genotype. Plants grown from seeds, broken-off fragments, temporary terrestrial forms, or buried horizontal shoots will look significantly different. Most species, especially the large river species, do not grow well in cultivation. Phenological observations combined with long-term monitoring of sites under a variety of climatic conditions may allow assessment of the degree of plasticity of a single population. Phenotypic plasticity is the main reason for practical difficulties in specimen identification.

Breeding system

Fifteen of our sixteen species are involved in a variety of hybridisation processes. Hybridisation is unevenly distributed among species. Most hybridisation events occur between *R. fluitans*, *R. peltatus*, *R. trichophyllus* and *R. circinatus*. An-

Flowering terrestrial form of *R. baudotii* on dried-up bottom. Samsoe, Denmark. Photo JCS.

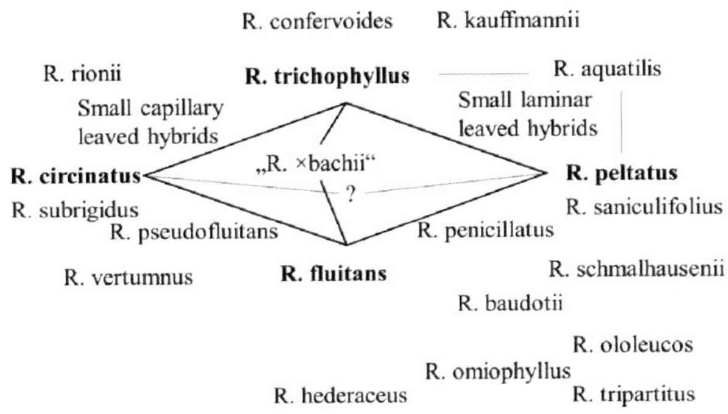

Morphological similarity of *Batrachium* species and preferential hybridisation pathways.

other group showing active hybridisation involves *R. peltatus*, *R. aquatilis* and *R. trichophyllus*. The extent to which *R. baudotii* is involved needs further study. Repeated crossing between two species may produce a partly fertile F1 generation. Backcrossing with both parent species will produce a continuum of forms (a hybrid swarm). In *Batrachium* backcrossing mostly occurs in the direction of the most frequent parent species in the region or catchment, leading to introgression swarms. Crossings among F1 individuals will produce a diverse F2 generation. Some forms may be stabilised as sterile strains (vegetative apomixis), others by formation of allopolyploid cytotypes. Somatic polyploidy as well as production and fusion of unreduced gametes may lead to formation of unbalanced polyploids. These processes can also be found in other aquatic plant groups. In *Batrachium* outcrossing of fertile F1 or F2 plants with a third species also occurs, which may be the starting point for new hybridisation processes. In addition, genetically polymorphic species may form intraspecific "hybrids" showing a different morphology.

A study showed that approximately 15% of all populations sampled were of hybrid origin. As a result, regionally frequent fertile species are surrounded by introgression swarms, which show uncorrelated variation of characters and often cannot be distinguished by morphological analysis alone. Genetic methods applied so far have failed to recognise true hybridisation history, especially in tetra- and hexaploid forms. Careful karyological studies can be helpful. In the region most species have one preferential, mostly tetraploid or hexaploid ploidy level. Autopolyploid triploids are found in *R. fluitans*. Autopolyploidisation does not necessarily produce new phenotypes but may give rise to new opportunities for hybridisation. Diploid species are more frequently found in areas which were not affected by the last glaciation, e.g., in the Mediterranean, in Central Asia and in North America. Epigenetic effects may be involved in the creation and preservation of morphological variation in unstable habitats. The temporary persistence of epimutations may be caused by changing selection pressure due to changing human impacts on water bodies, such as eutrophication, acidification, glyphosate concentration, morphological alteration, and restoration. By this mechanism populations may undergo rapid phenotypic change. Somatic mutations in clonal lineages and agamospermy are mentioned in the literature as well but not empirically confirmed.

Taxonomic concepts

Batrachium were first introduced to science in 1753 by Carl von Linné. He distinguished two species, namely *R. hederaceus* and *R. aquatilis* within the genus *Ranunculus*. Soon afterwards, the most frequent species still recognised today were separated from *R. aquatilis* (*R. fluitans*, *R. peltatus*, *R. circinatus* and *R. trichophyllus*). Many famous botanists such as J. B. de Lamarck and A. P. de Candolle, were involved in early *Batrachium* studies, and in 1827 Dumortier promoted *Batrachium* to the rank of a genus of its own right. The first monographic treatment was published by Hiern (1871). By 1880 all north-west-central European species had been described. Simultaneously, it became popular to describe numerous variants and forms, which are treated in detail by Glück (1936) in his *Freshwater Flora of Central Europe*. Since the beginning of the twentieth century, taxonomists have increasingly recognised hybrids. Christopher D. K. Cook (1966) eliminated most of the subspecific taxa with exceptions involving *R. penicillatus* and *R. trichophyllus*. Cook's circumscription of species differed considerably from the approaches of his predecessors. His definitions and delimitations of species such as *R. aquatilis* and *R. peltatus* are still valid. In the 1970s and 1980s N. T. H. Holmes and S. D. Webster continued research into British river *Batrachium* without succeeding in finding an acceptable solution to the *R. penicillatus* problem. A contribution of major importance was given in the work of José Pizarro (1995). The precision of his morphological observations and detailed drawings of specimens is still unrivalled. During the 1990s G. Dahlgren and her co-workers published a series of *Batrachium* studies, which contained a lot of useful information.

Accumulated information was summarised in the worldwide account of the section *Batrachium* published by Wiegleb et al. (2017). Cook (1966) had treated *Batrachium* at the level of a subgenus. Today *Batrachium* is considered to be closely related to yellow-flowered amphibious species, such as *R. sceleratus*. Therefore, the taxon was downgraded to the level of a section. The progress of science did not always have immediate consequences for regional and national treatments in floras, databases and distribution maps. In several countries, species are still treated under the genus name *Batrachium*. In many countries species lists include artificial taxa such as *R. aquatilis* sensu latu or *R. aquatilis* agg., "lumping" of similar species (e.g., *R. aquatilis* + *peltatus*, *R. aquatilis* + *baudotii*) is also widespread. *R. aquatilis* and *R. trichophyllus* are also often treated as varieties of *R. aquatilis*, or *R. baudotii* is treated as a subspecies of *R. peltatus,* even though it is genetically more similar to *R. fluitans*. A variety of species and hybrids have been included under the name *R. penicillatus*. A revision of *Batrachium* collections in Germany showed that specimens are very often stored under obsolete names, misplaced or misidentified.

For critical groups such as *Batrachium*, taxonomy and nomenclature strongly depend on the species concept. Evolutionary processes such as described above are not specific for *Batrachium* but can be found in many other plant groups such as *Mentha*, *Pilosella* or *Crataegus*. In these groups widely recognised species concepts have been adopted, even though the classical Biological Species Concept cannot be applied. For *Batrachium* no such general agreement exists. In our approach we are using a Morphological Species Concept, which guarantees both recognisability and identifiability of specimens and populations. The distinguished species morphologically are as coherent as possible. Delimitation is supported by other criteria, e.g., widespread inbreeding, a preferential ploidy level, a common descent, or a continuous distributional range. However, some variation in morphology, karyology and genetics must be accepted. Taxa which are solely based on karyological or genetic characters ("microspecies", "cryptospecies") are not recognised. "Primary hybrids" are included in "hybrid species" of the same hybrid combination.

Arrangement of species

The arrangement of species follows assumed phylogenetic relationships. Species can be grouped as follows:

R. tripartitus group: The group comprises amphibious species with large free stipules. *R. omiophyllus*, *R. tripartitus* and *R. ololeucos* are forming a natural group of common descent.

R. hederaceus group: The group is made up of the amphibious taxon *R. hederaceus*, which is characterised by complete absence of capillary leaves.

R. fluitans group: The group comprises large river species. The core species is *R. fluitans*. Despite the different growth it may be a close relative of *R. hederaceus*. Both share the fleshy structure of the stem. *R. fluitans* gave rise to two hybrid species (*R. penicillatus* and *R. pseudofluitans*). The two latter species are considered relatively young and show low morphological coherence. Morphologically, these species are partly overlapping with other *R. fluitans* hybrids.

R. peltatus group: This group comprises species with abundant laminar leaves. *R. peltatus* is a polymorphic species with numerous morphotypes, which do not correspond to published karyological and genetic information. The species is tetraploid, while the similar Mediterranean species *R. saniculifolius* is diploid. *R. schmalhausenii* is a Northern European species which has developed from *R. peltatus*-like ancestors. *R. baudotii* is represented in the study area by its tetraploid cytotype. Inland forms may differ considerably from the typical coastal morphotype.

R. trichophyllus group: The group comprises species with and without laminar leaves. *R. trichophyllus* is a complex species, of which numerous cytotypes and ITS-haplotypes have been described. Morphology does not coincide with the other sources of information. The morphotype with short rigid leaves shows some similarities to the *R. circinatus* group, while the morphotype with longer flaccid leaves seems to be related to the *R. peltatus* group. Whether this distinction is basal or has recently evolved by introgression is unknown. *R. aquatilis*, *R. confervoides* and *R. kauffmannii* are sibling species of *R. trichophyllus*. *R. aquatilis* arose from crossings of two rigid-leaved *R. trichophyllus* haplotypes, while *R. confervoides* and *R. kauffmannii* relate to the flaccid-leaved subgroup.

R. circinatus group: The group is characterised by short rigid leaves and a regular pattern of leaf length and shape. It comprises *R. circinatus* and *R. rionii*.

Useful characters for identification

Water-crowfoots have a variety of characters that can be used for determination. Due to the variation between plants and over time, not all characters can be considered constant. No species can be distinguished by means of a single differentiating character. Two different keys are presented, one dichotomous key and one tabular or multi-access key. The multi-access key comprises 12 vegetative and 17 floral characters or character states, which agree with the major alternatives in the dichotomous key. Both keys should be used in combination.

Only a few species are obligate annuals and most species are perennial. Whether a perennial plant is summer– or wintergreen depends on climatic conditions. Annual shoots can often be recognised by the presence of primary roots, while perennial shoots have only adventitious roots. Size and robustness of the generative shoots are important characters for rough classification. Plants with exceedingly long shoots (> 3 m) are usually restricted to fast-flowing streams and rivers (e.g., *R. fluitans*, *R. penicillatus* and *R. pseudofluitans*). Many species are medium-sized (1–2 m long), while others usually remain small (less than 1 m). However, individuals of small species can produce elongated shoots in deep lakes as well as in ponds and streams with strong water level fluctuations.

The leaves are alternate, except for the upper laminar leaves of amphibious species. The presence of laminar leaves is an artificial character which must be used with care. Laminar leaves developed independently in several groups. Within "laminar-leaved species" the laminar leaves may be absent due to an interruption of the growth cycle or unfavourable habitat conditions. Laminar leaves mainly differ by the depth of division of the lamina (more or less than 1/2), the number of primary lobes (3 to 7) and the basal sinus of the lamina. Four types of intermediate leaves are distinguished: the tripartitus type in *R. tripartitus* (with 5–7 acute lobes), the baudotii type in *R. baudotii* (tripartite symmetric structures with mostly rounded lobes), the *peltatus*-type in *R. peltatus* and *R. penicillatus* (with basal laminar and apical capillary part), and the *aquatilis* type in *R. aquatilis* (with basal capillary and apical laminar parts, often asymmetric). The "intermediate leaves" of some *R. fluitans* hybrids are just broadened segments without joints.

Capillary leaves are found in almost all species. Both length and structure are relevant. The relationship between leaf length and length of the adjacent internode is often characteristic, however both characters are under partial environmental control and can vary within one individual. The segments of the capillary leaves can be fleshy in some species (*R. fluitans*, *R. baudotii*); otherwise they are either rigid or flaccid. Several species comprise both rigid-leaved and flaccid-leaved strains. Rigid capillary leaf segments oriented in one plane are only found in *R. circinatus*. Several species can develop rigid capillary aerial leaves and are often confused with *R. circinatus*, however terrestrial forms without flowers and fruits cannot be identified to the species level. *Ranunculus* species do not have true stipules. In *Batrachium* basal appendages of the petioles are conventionally called "stipules". These can vary in size and the degree of adnation to the petiole (see above, *R. tripartitus* group).

All *Batrachium* species have solitary, leaf-opposite flowers. The shape and length of the pedicel is strongly dependent on environmental conditions when the plant is mature. In some species (e.g., *R. aquatilis* and *R. trichophyllus*) pedicels do not elongate. In other species (e.g., *R. peltatus* and *R. penicillatus*) pedicels elongate under water, allowing the flowers to reach the surface. In a third group (e.g., *R. baudotii* and *R. saniculifolius*) pedicels elongate and bend back towards maturity, depositing the ripe fruits in the sediment. Pedicels may also bend back without elongation (e.g., in *R. rionii*). Whether or not the pedicels are elongated and recurved is an important character.

The blue colour of the sepals is characteristic for *R. tripartitus*, while partly bluish sepals are found in more species than reported in much of the literature. Flower size is another important character for rough classification, even though petal size can vary within individuals and populations. Large-flowered species may also have individuals with medium-sized or small flowers, depending on environmental conditions. However, medium-sized and small-flowered species will never develop large flowers. If they appear to do so, this may indicate hybridisation. The petals are white with a more or less pronounced yellow base. The shape of the nectar pits is variable. In populations of species with pyriform nectar pits one may find single petals with circular or lunate nectar pits. Circular nectar pits are sometimes slightly elongated or open at the top. Nectar pits of species with lunate nectar pits will rarely develop beyond this stage. Most species have one nectar pit per petal, but up to four nectar pits are often found in *R. schmalhausenii*.

The number of carpels and average size of ripe achenes are negatively correlated. The highest number of small achenes is found in *R. rionii*. Achenes may also develop a conspicuous dorsal and ventral wing in *R. baudotii*. In most species the style is caducous, but in *R. rionii* it may be persistent. Other characters of the achenes (shape, position of style) are variable between populations, while hairiness changes as carpels mature. In some species (e.g., *R. baudotii* and *R. rionii*) the receptacle elongates in fruit, but this character must be treated with care, as some species can already have an elongate receptacle when immature (e.g., *R. fluitans*). The hairiness of the receptacle is distributed erratically among *Batrachium* species. It can, however, distinguish closely related similar pairs of species. While *R. omiophyllus*, *R. fluitans* and *R. saniculifolius* have a glabrous receptacle the similar species *R. tripartitus*, *R. pseudofluitans* and *R. peltatus*, respectively, have a pubescent receptacle. In contrast to the hairs on vegetative parts and fruits, the hairs on the receptacle are persistent and are still present after ripe achenes have been shed.

The identification of hybrids in field and herbarium studies is difficult due to the morphological overlaps between many species. Primary hybrids normally show lowered vitality and fertility. Sterility is never complete as otherwise the existence of hybrid or introgression swarms as well as outcrosses could not be explained. Sterility may also be caused by unfavourable conditions, such as high current velocity or disturbance events. Hybrids usually have an "untidy" growth pattern, forming atypical side branches with atypical leaf types. Pedicels often remain undeveloped or are aborted at early growth stages. Unopened flowers with deformed petals are also frequent. Nectar pits can have an intermediate shape or show deformations. Hybrids between heterophyllous and homophyllous species are often characterised by the abundant formation of intermediate leaves. Hybridisation can be assumed in case of an intermediate population growing in the same water body as the parent species. In many cases karyological or genetic studies are necessary for confirmation of the hybrid status.

Tabular key to *Ranunculus* sect. *Batrachium*

•• key character

• regular character

(•) rare

○ no or missing

- not relevant

	R. omiophyllus	*R. tripartitus*	*R. ololeucos*	*R. hederaceus*	*R. fluitans*	*R. pseudofluitans*	*R. penicillatus*	*R. baudotii*	*R. peltatus*	*R. saniculifolius*	*R. schmalhausenii*	*R. trichophyllus*	*R. aquatilis*	*R. confervoides*	*R. kauffmannii*	*R. circinatus*	*R. rionii*
Plant annual	(•)	(•)	(•)	(•)	○	○	○	(•)	(•)	••	○	(•)	(•)	•	○	○	••
Large river plants (mostly 2–6 m long)	○	○	○	○	••	••	••	(•)	(•)	○	••	○	○	○	•	○	○
Stem (and capillary leaves) fleshy	○	○	○	○	••	•	○	••	○	○	○	○	○	○	○	○	○
Adventitious roots at all nodes	○	○	○	•	○	○	○	○	○	(•)	○	○	○	••	○	○	○
Laminar leaves present	••	•	•	••	○	○	•	•	••	••	••	○	••	○	○	○	○
Laminar leaves deeply divided	○	••	••	○	-	-	○	••	○	○	•	-	(•)	-	-	-	-
Intermediate leaves present	○	(•)	○	○	(•)	○	••	••	••	•	(•)	(•)	••	○	(•)	○	○
Capillary leaves present	○	(•)	•	○	••	••	••	••	••	•	••	••	••	••	••	••	••
Capillary leaves fine, hair-like	-	•	•	-	○	○	○	○	○	○	○	○	○	••	○	○	○
Capillary leaves longer than adjacent internode	-	○	○	-	••	(•)	••	(•)	○	○	(•)	(•)	○	○	•	○	○
Segments of capillary leaves in one plane	○	○	○	○	○	○	○	○	○	○	○	○	○	○	○	••	○
Stipules adnate for < 1/2 of length	••	••	••	○	○	○	○	(•)	○	○	○	○	○	○	○	○	○
Sepals blue, at least at tip	○	••	•	○	○	○	○	•	(•)	•	○	(•)	(•)	○	○	○	○
Pedicels tapering to the top	○	○	○	○	••	•	(•)	(•)	(•)	○	(•)	○	○	○	○	○	○
Pedicels at maturity shorter than 5 cm	○	•	••	(•)	○	○	○	○	○	○	○	••	••	•	•	••	••
Pedicels at maturity recurved	○	•	•	•	○	○	○	••	•	••	○	○	○	••	○	○	••
Petals shorter than 5 mm, lanceolate or spathulate	•	••	○	••	○	○	○	○	○	○	○	••	○	••	•	○	••
Petals contiguous in anthesis	○	○	•	○	••	•	••	•	••	•	•	○	•	○	○	•	○

●● key character ● regular character (●) rare ○ no or missing - not relevant	*R. omiophyllus*	*R. tripartitus*	*R. ololeucos*	*R. hederaceus*	*R. fluitans*	*R. pseudofluitans*	*R. penicillatus*	*R. baudotii*	*R. peltatus*	*R. saniculifolius*	*R. schmalhausenii*	*R. trichophyllus*	*R. aquatilis*	*R. confervoides*	*R. kauffmannii*	*R. circinatus*	*R. rionii*
Petals more than 5	○	○	○	○	●●	●	●	○	●	●	●	○	○	○	○	○	○
Petals almost completely white	○	○	●●	○	○	○	○	○	○	(●)	○	○	○	○	○	○	○
Nectar pits lunate	●●	●●	●●	●●	○	(●)	○	●●	○	●●	○	●●	(●)	●●	●●	●●	●●
Nectar pits circular	○	○	○	○	○	(●)	○	(●)	○	(●)	○	○	●●	○	(●)	○	○
Nectar pits pyriform	○	○	○	○	●●	●	●●	○	●●	○	●●	○	○	○	○	○	○
Nectar pits more than 1 per petal	○	○	○	○	○	○	○	○	○	●	●●	○	○	○	○	○	○
Ripe fruits winged at both sides	○	○	○	○	○	○	○	●●	○	●	○	○	○	○	○	○	○
Ripe fruit small (< 1.2 mm)	○	○	○	○	○	○	○	○	○	●	○	○	○	○	○	○	●●
Style persistent	○	○	○	○	○	○	○	○	○	○	○	○	○	○	○	○	(●)
Receptacle glabrous	●●	○	○	●●	●●	○	○	○	○	●●	○	○	○	○	(●)	○	○
Receptacle elongating in fruit	○	○	○	○	○	○	○	●●	○	(●)	○	○	○	○	○	○	●●

Ranunculaceae

202 *Ranunculus omiophyllus* Ten.

Syn.: *R. lenormandii* F.W.Schultz, *Batrachium omiophyllum* (Ten.) C.D.K.Cook
DK: - **N:** - **S:** - **FIN:** - **IS:** -
GB: Round-leaved Water-crowfoot **NL:** Drijvende waterranonkel **F:** Renoncule de Lenormand
D: Lenormand-Wasserhahnenfuss **CZ:** - **PL:** -
EST: - **LV:** - **LT:** - **Ru:** -

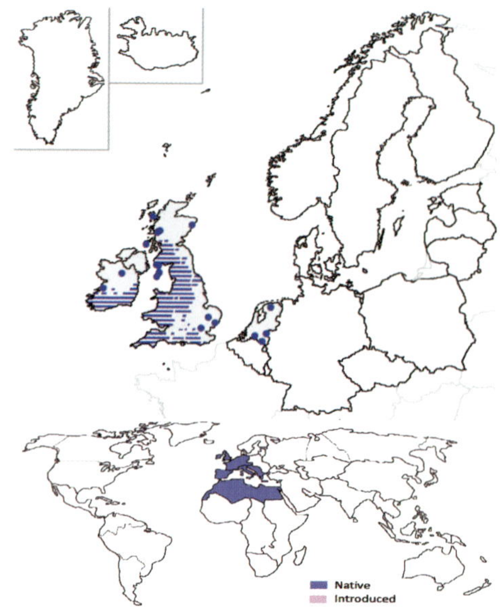

Ranunculus omiophyllus. New Forest, Britain. Photo KvdW.

Perennial or rarely annual amphiphyte. Aquatic and terrestrial plants prostrate, creeping or floating, with only laminar leaves.

Stem: Up to 0.5 m long, 2–3 mm in diameter, green, branched. Rooting at most nodes.

Laminar leaves: Always present, opposite or alternate, reniform or circular, basal sinus 0°–120°, with 3–7 primary lobes, up to 26 mm long and 32 mm wide, dissected to 1/2 of the lamina, primary lobes narrowest at base, margin entire or crenulate. Petioles 1.2–7.8 cm long, 3–6 times longer than the lamina. Intermediate leaves absent.

Capillary leaves: Absent.

Stipules: Less than 1/2 adnate to the petiole, apex of free part rounded, glabrous.

Pedicels: Opposite to laminar leaf, 1.5–6 cm long, equal to or shorter than petiole of opposite leaf, straight.

Flowers: Small to medium-sized. Sepals 5, 2–4 mm long, greenish, reflexed, caducous. Petals 5, 4–7(–9) mm long, longer than sepals, not contiguous at anthesis, white with yellow base. Nectar pits 1 per petal, lunate. Receptacle sub-globose, glabrous.

Fruits: Number of carpels 15–67. Ripe achenes 1.2–2.2 mm long, glabrous. Style short, caducous.

Flowering: March-November.

Biology: *R. omiophyllus* is mostly wintergreen, rarely summer green. Flowers are pollinated by insects. Vegetative reproduction by detached fragments is frequent.

Habitats: Springs, small streams, ponds, temporary pools, and ditches on muddy substrate or on peat, in and around oligotrophic, nutrient-poor water bodies. Shade tolerant.

Distribution within the region:
Native. Widespread in Ireland and Britain, extinct in the Netherlands.

The species occurs in Western Europe and Northern Africa.

Characteristics and similar species:
R. omiophyllus can be confused with **203 *R. tripartitus***, which differs in the smaller flowers, the more deeply divided laminar leaves and the presence of capillary leaves in aquatic forms.

Ranunculus omiophyllus. New Forest, Britain. Photo RVL.

Ranunculus omiophyllus. Corno di Bico, Portugal. Photo KvdW.

205 ***R. hederaceus*** has smaller flowers and lobes of laminar leaves broadest at base.

Hybrids: A rare hybrid is formed with *R. hederaceus*. A hybrid with *R. tripartitus* (= *R. lutarius* (Bouvet) Revel; *R. x novae-forestae* S.D.Webster) has been frequently reported from southern England. A hybrid with *R. ololeucos* has been reported from France.

The hybrid with *R. peltatus* (= *R. ×hiltonii* H. & J.Groves) is characterised by the formation of more dissected laminar leaves, a character that can be found in other *R. omiophyllus* hybrids. It can be confused with annual forms of *R. aquatilis* and *R. baudotii.*

Note: Most populations of *R. omiophyllus* are tetraploid (2n = 32). A diploid cytotype (2n = 16) is reported from Algeria, Sicily, and southern Italy.

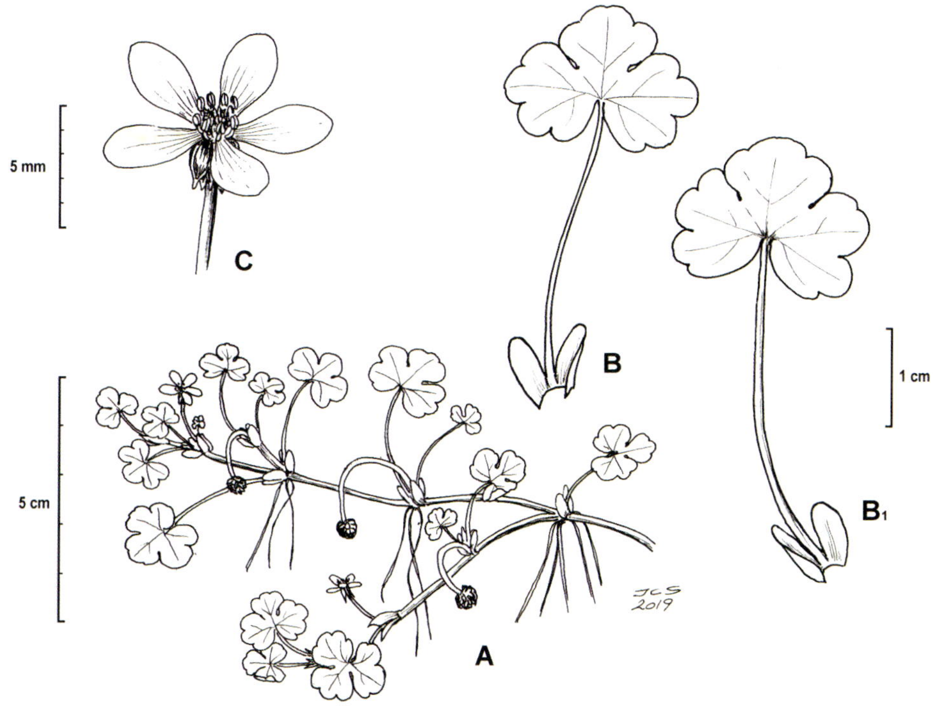

Ranunculus omiophyllus
A habit, **B, B1** leaf with stipules, **C** flower.

Ranunculaceae

203 *Ranunculus tripartitus* DC.

Syn.: *Batrachium tripartitum* (DC.) S.F.Gray; *R. baudotii* Nolte non Godr.

DK: - **N:** - **S:** - **FIN:** - **IS:** -

GB: Three-lobed Crowfoot **NL:** Driedelige waterranonkel **F:** Renoncule tripartite

D: Dreiteiliger Wasserhahnenfuss **CZ:** - **PL:** -

EST: - **LV:** - **LT:** - **Ru:** -

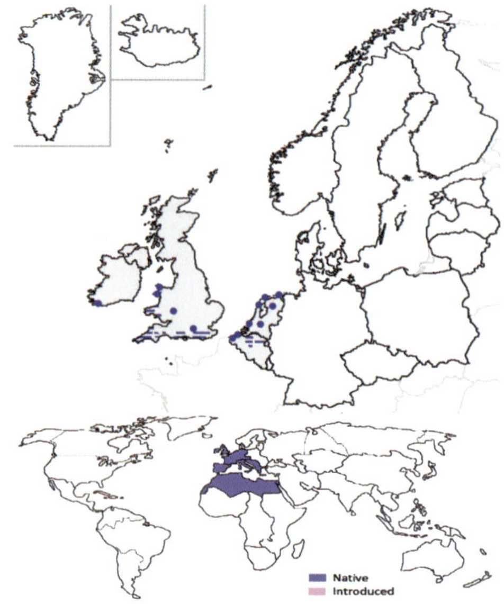

Ranunculus tripartitus. Ruan Pool, the Lizard, Cornwall, England. Photo RVL.

Perennial or annual amphiphyte. Aquatic plants erect, spreading, mostly with laminar and capillary leaves, rarely with laminar leaves or capillary leaves only.

Stem: Up to 0.5 m long, 2–3 mm in diameter, glabrous, green, branched. Rooting at lower nodes.

Laminar leaves: Opposite or alternate, up to 20 mm long and 40 mm wide, reniform or subcordate, basal sinus 90°–180°, deeply dissected, with 3–5 cuneate primary lobes, margins crenate. Petioles 1–7 cm long. Intermediate leaves sometimes present, of *tripartitus* type (radially symmetrical with 3–5 pointed lobes).

Capillary leaves: Present or often absent, confined to the lower nodes, alternate or opposite, 1–4 cm long, filiform, number of lamina divisions 3–5, number of terminal segments up to 90. Petioles 0.1–1 cm long.

Stipules: Adnate to petiole for less than 1/2 their length, apex of free part obtuse, glabrous.

Pedicels: Opposite to laminar leaf, rarely to capillary leaf, elongating in fruit, 2–5(–7) cm long, recurved, glabrous.

Flowers: Small. Sepals 5, 1–3 mm long, bluish, reflexed. Petals 4–5, 1–5 mm long, white with yellow base, obovate to elliptical, not contiguous at anthesis. Nectar pits 1 per petal, lunate. Receptacle globose, pubescent.

Fruits: Number of carpels 4–15(–27). Ripe achenes 1.5–2.0 mm long, glabrous or pubescent. Style short, caducous.

Flowering: December-April (-May).

Biology: *R. tripartitus* overwinters as prostrate plants with laminar leaves; flowering is frequent in the winter. It is mostly self-pollinated. Vegetative reproduction by detached fragments is also common.

Habitats: The species occurs in shallow seasonal pools and flushes, on the margins of ponds, ditches and small streams. It typically occurs in acid heathlands over basic clays, often in disturbed places where poaching by cattle goes through the acid upper horizons into basic clays below.

Distribution within the region:
Native. Locally abundant in southern Britain, rare in Belgium and Ireland, extinct in the Netherlands.

Ranunculus tripartitus with floating and capillary leaves. Ruan Pool, The Lizard, Cornwall, England. Photo RVL.

This species has an oceanic distribution in Western Europe and Morocco.

Characteristics and similar species: *R. tripartitus* is characterised by small flowers, with widely spreading petals and characteristic 3-lobed laminar leaves.

It may be confused with **202 *R. omiophyllus***, which differs in the much larger petals, less dissected laminar leaves, and the greenish sepals.

204 *R. ololeucos* has larger flowers. In its northern range it has almost white petals and laminar leaves with a reddish tinge.

In terrestrial state, it can be confused with dwarf forms of **210 *R. peltatus***, **213 *R. aquatilis*** and **209 *R. baudotii***, all of which have much larger flowers with petals contiguous at anthesis.

Hybrids: *R. tripartitus* hybridises with the closely related *R. omiophyllus* (= *R. lutarius* (Revel) Bouvet; *R. ×novae-forestae* S.D.Webster), forming morphologically intermediate populations which appear to backcross with *R. tripartitus*. In France it is regarded as a distinct hybrid species. Confirmation requires chromosome counts. A hybrid with *R. ololeucos* has been reported from France. Reported hybrids with *R. peltatus* and *R. aquatilis* are unconfirmed.

Note: *R. tripartitus* is hexaploid (2n = 48), deviating counts may suggest hybridisation.

Ranunculus tripartitus. Ruan Pool, The Lizard, Cornwall, England. Photo RVL.

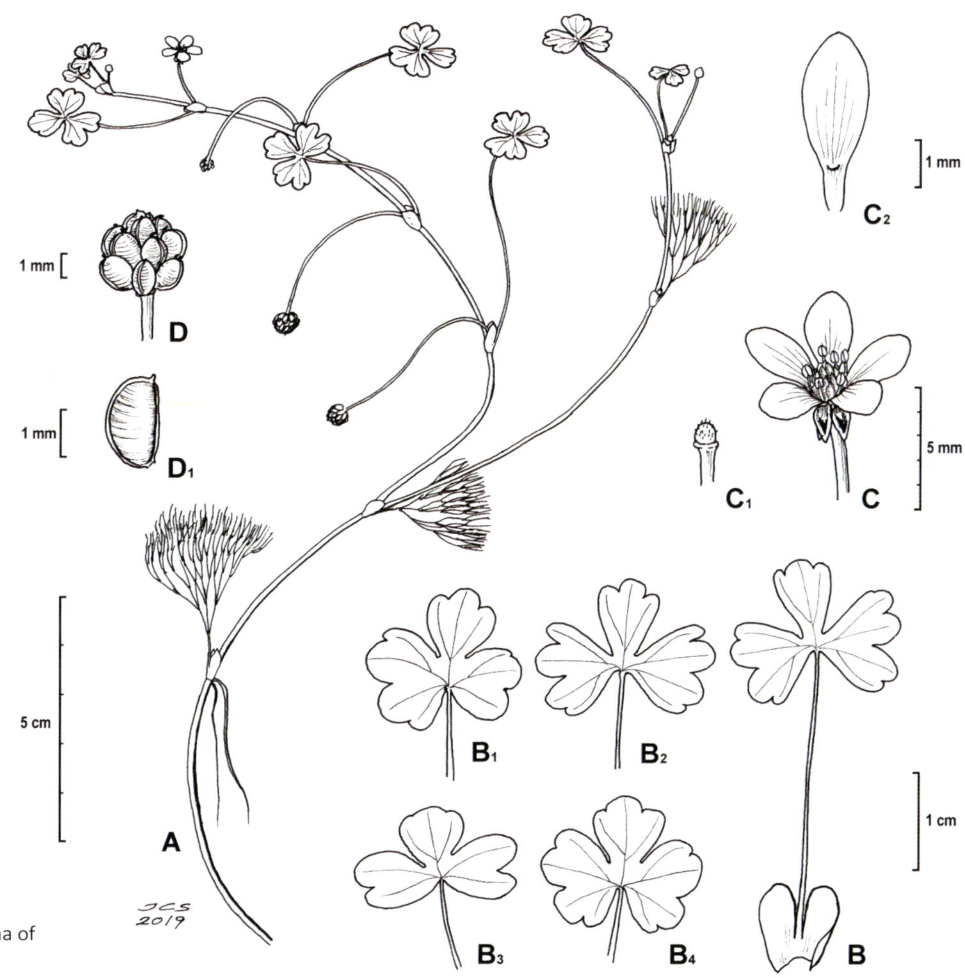

Ranunculus tripartitus
A habit, **B** floating leaf with stipules, **B1-B4** lamina of floating leaves, **C** flower, **C1** receptacle, **C2** petal, **D** infructescence, **D1** achene.

Ranunculaceae

204 *Ranunculus ololeucos* J.Lloyd

Syn.: *Batrachium ololeucum* (J.Lloyd) Bosch
DK: -**N:** - **S:** - **FIN:** - **IS:** -
GB: Ivy-leaved Water-crowfoot **NL:** Witte waterranonkel **F:** Renoncule toute blanche
D: Reinweißer Wasserhahnenfuss **CZ:** - **PL:** -
EST: - **LV:** - **LT:** - **Ru:** -

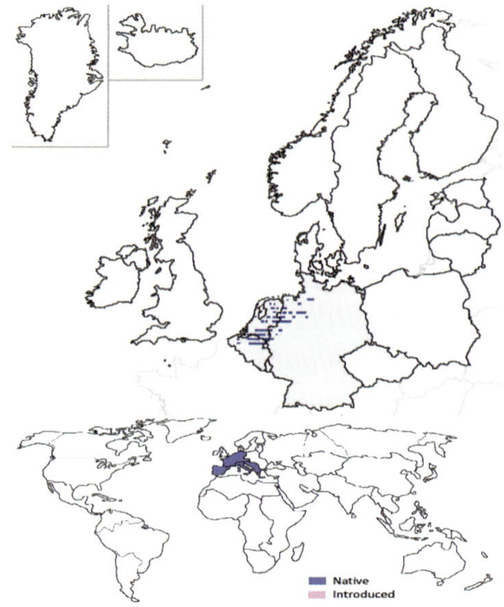

Perennial hydrophyte or annual amphiphyte. Aquatic plants erect, with laminar leaves or laminar and capillary leaves.

Stem: Up to 0.8 m long, 2–3 mm in diameter, green, branched, glabrous. Rooting at lower nodes.

Laminar leaves: Always present, alternate or opposite, up to 20 mm long and 30 mm wide, green, often with a reddish tinge, with 3–5 primary lobes, deeply dissected, secondary lobes profound, margin crenate, basal sinus of the lamina 60°–150°. Petioles 2–10 cm long. Intermediate leaves absent.

Capillary leaves: Alternate, up to 8 cm long, green, glabrous, segments filiform, clasping, middle part often missing (first branching dichotomous), laminar divisions 3–8, terminal segments up to 300. Petioles up to 2 cm long.

Stipules: Less than 1/2 adnate to the petiole, whitish to hyaline, apex of the free portion acute, glabrous.

Pedicels: Opposite to laminar leaf, 2.5–5.5 cm long, slender, often recurved.

Flowers: Medium-sized. Sepals 5, 2.5–6 mm long, blue tipped, reflexed. Petals 5, 5–15 mm long, obovate, at least twice as long as sepals, white with a very small yellow base, sometimes contiguous during anthesis. Nectar pits 1 per petal, lunate. Receptacle globose to elliptical, strongly pubescent.

Fruits: Number of carpels 15–30. Ripe achenes 1.1–2.0 mm long. Style short, caducous.

Flowering: March-May (-August).

Biology: The species is wintergreen. It generally flowers abundantly but is predominantly self-pollinating. Vegetative propagation occurs through detached stem fragments. Prostrate terrestrial forms are frequently formed, creeping on mud with both laminar and capillary leaves.

Habitats: This species occurs in lowland areas in shallow oligo- to dystrophic stagnant water bodies, which have become rare due to eutrophication and habitat destruction.

Distribution within the region:
Native. *R. ololeucos* has an oceanic distribution, occurring from Belgium where it is rare, to the Netherlands and West Germany; it is absent from the British Isles.

Outside the region it extends southwards to Portugal and Sardinia.

Characteristics and similar species:
R. ololeucos can be confused with **203** ***R. tripartitus***, which has smaller flowers and petals with a more pronounced yellow base.

Ranunculus ololeucos. Beira Litoral, Portugal. Photo KvdW.

Laminar leaves. Photo KvdW.

Ranunculus ololeucos. Peneda-Gerês, Portugal. Photo KvdW.

Petal. Note the very small yellow claw. Photo KvdW.

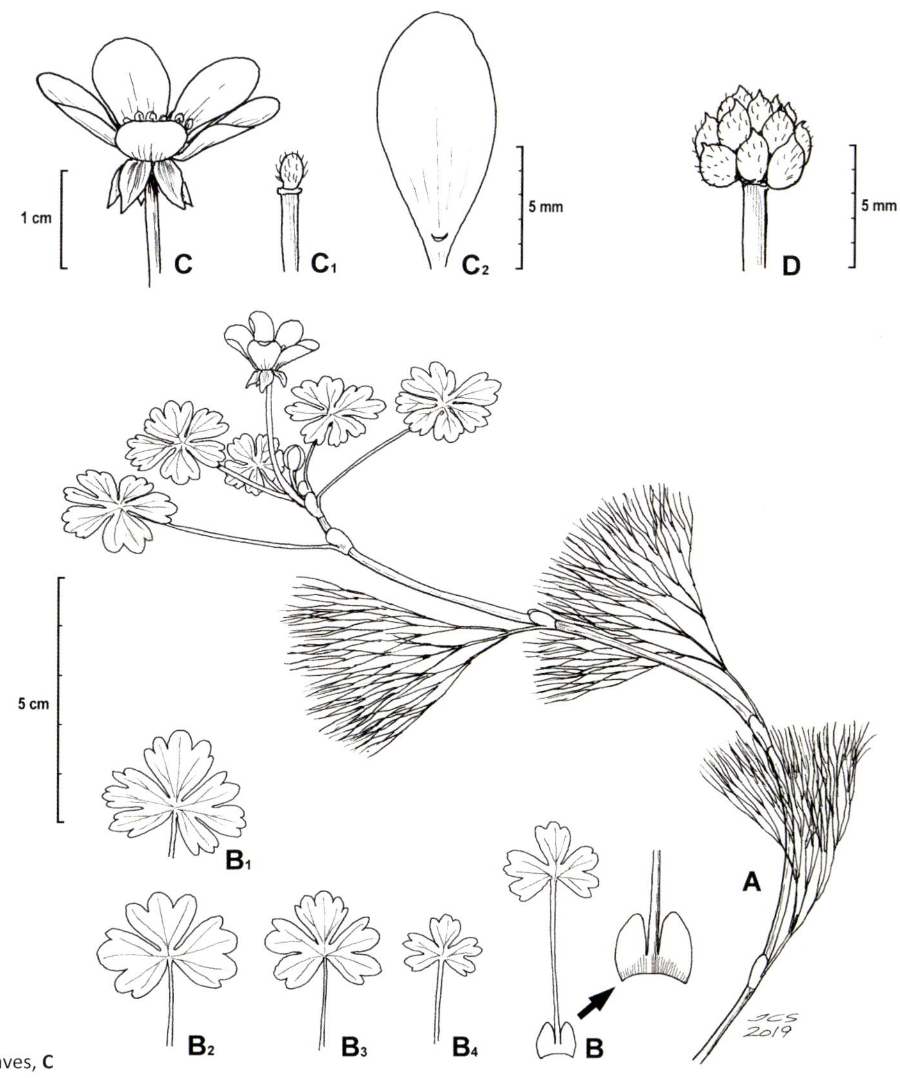

Individuals with a larger yellow base (*R. lusitanicus*) may be confused with ***R. saniculifolius*** Viv. (not treated in this book). This morphotype has not yet been found in the region.

Hybrids: No hybrids are reported from the region. Hybrids with *R. omiophyllus* and *R. tripartitus* are reported from France.

Note: Northern populations of *R. ololeucos* are diploid (2n = 16), while tetraploid plants (2n = 32) are reported from Spain.

Ranunculus ololeucos
A habit, **B** leaf with stipules, **B1-B4** lamina of floating leaves, **C** flower,
C1 receptacle, **C2** petal, **D** infructescence.

Ranunculaceae

205 *Ranunculus hederaceus* L.

Syn.: *Batrachium hederaceum* (L.) S.F.Gray
DK: Vedbend-Vandranunkel **N:** Leirsoleie **S:** Murgrönsmöja **FIN:** - **IS:** -
GB: Ivy-leaved Crowfoot **NL:** Klimopwaterranonkel **F:** Renoncule lierre
D: Efeu-Wasserhahnenfuss **CZ:** - **PL:** Jaskier bluszczolistny
EST: - **LV:** Efeju ūdensgundega **LT:** - **Ru:** Шелковник плющевидный

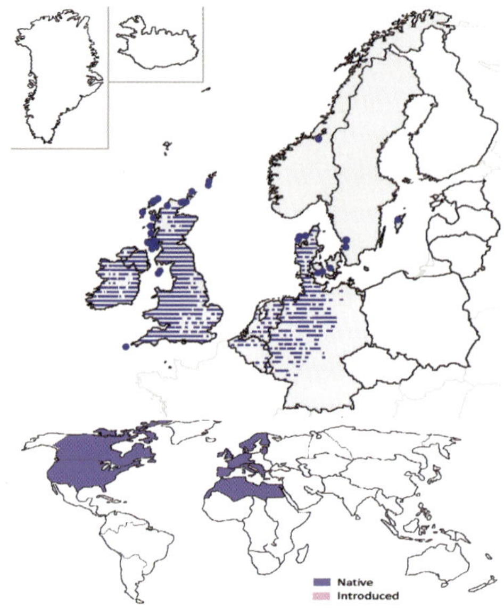

Ranunculus hederaceus creeping on muddy lake shore. Førby Sø, Thy, Denmark. Photo JCS.

Biennial or perennial, rarely annual amphiphyte. Aquatic and terrestrial forms are prostrate with laminar leaves only.

Stem: Up to 0.5 m long, 2–5 mm in diameter, green, branched, glabrous. Rooting at all nodes.

Laminar leaves: Always developed, opposite, up to 25 mm long and 35 mm wide, reniform to cordate, often floating on the surface; with 3–5 primary lobes, dissected to up to 1/3 of the lamina length, broadest at base, margin entire or rounded, often with blackish suffusion along main veins. Petioles 1.5–3(–7) cm long, 2–4 times longer than the lamina. Intermediate leaves absent.

Capillary leaves: Absent.

Stipules: 1/2 to 3/4 adnate to the petiole, apex of free part acute, glabrous.

Pedicels: Opposite to laminar leaf, 1–2 cm long at anthesis, equal in length to the petiole of the opposite leaf, elongating up to 7 cm in fruit, recurving.

Flowers: Small. Sepals 5, 1–3 mm long, greenish, spreading. Petals 5, 1.2–5 mm long, not much longer than the sepals, white with yellow base, spathulate to elliptical, not contiguous at anthesis. Nectar pits 1 per petal, lunate. Receptacle spherical or subspherical, glabrous, rarely puberulent.

Fruits: Number of carpels 9–43. Ripe achenes 1.2–2.0 mm long, glabrous, sometimes with a narrow wing. Style short, caducous.

Flowering: (March-) June-November.

Biology: In oceanic areas the species is green throughout the year but may dieback in seasonal wetlands. Under subcontinental influence it is either wintergreen or short-lived summer green. Flowers are pollinated by insects, but self-pollination also occurs. Mature pedicels curve, sinking the ripe achenes in the mud. Vegetative reproduction by clonal growth and fragmentation is frequent.

Habitats: In spring-fed wetlands, temporary pools, ponds, and along small streams, rarely submerged, mostly creeping on mud or floating, often in places which are poached by cattle; tolerates eutrophication.

Distribution within the region:
Native. It has a suboceanic-amphi-atlantic distribution, occurring in West-

Ranunculus hederaceus has small flowers with petals not contiguous during anthesis. Førby Sø, Thy, Denmark. Photo JCS.

Ranunculus hederaceus in a small ditch at Hærup west of Hobro, Denmark. Photo JCS.

ern Europe, Northern Africa and the eastern part of North America. Within the region it is extinct in Norway and is scattered from southern Sweden southwards to Germany and the British Isles.

Characteristics and similar species:
R. hederaceus is characterised by the small flowers and characteristic reniform leaves with lobes broadest at the base.

It may be confused with **202 *R. omiophyllus***, which has larger, free stipules and the lobes of laminar leaves narrowest at the base.

Hybrids: A hybrid with *R. omiophyllus* has been reported from Britain. The leaves closely resemble those of *R. omiophyllus,* but it has strongly curved pedicels.

Note: *R. hederaceus* appears to be diploid (2n = 16) throughout its range.

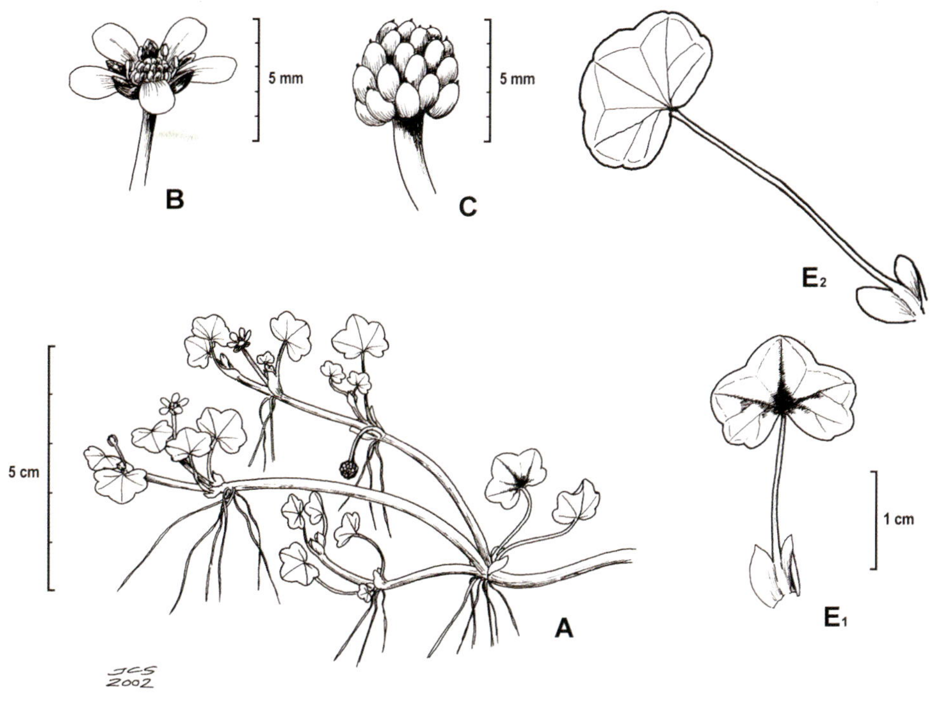

Ranunculus hederaceus
A habit, **B** flower, **C** infructescence, **E1, E2** leaves with stipules.

Ranunculaceae

206 *Ranunculus fluitans* Lam.

Syn.: *R. fluviatilis* Wigg.; *Batrachium fluitans* (Lam.) Wimm.
DK: Flod-Vandranunkel **N:** Kjempevassoleie **S:** Jättemöja **FIN:** - **IS:** -
GB: River Water-crowfoot **NL:** Vlottende waterranonkel **F:** Renoncule des rivières
D: Flutender Wasserhahnenfuss **CZ:** Lakušník vzplývavý **PL:** Włosienicznik rzeczny
EST: - **LV:** - **LT:** Plūdinė kurklė **Ru:** Шелковник плавающий

Ranunculus fluitans. River Swalm, the Netherlands. Photo KvdW.

Perennial hydrophyte. Aquatic plants with fleshy capillary leaves.

Stem: Up to 6 m long, 2–5(–7) mm in diameter, fleshy, branched, green, glabrous. Rooting at lower nodes.

Laminar leaves: Absent. Intermediate leaves are rarely formed and resemble flattened capillary segments; they lack a joint between the petiole and the laminar part.

Capillary leaves: Always present, alternate, up to 60 cm long, bright green, glabrous, segments rigid or flaccid, fleshy, divergent to subparallel, with 2–4 lamina divisions and fewer than 40 terminal segments. Petioles 5–22 cm long.

Stipules: More than 3/4 adnate to the petiole, ovate or oblong, apex of free part obtuse or truncate, glabrous.

Pedicels: Opposite a capillary leaf, 4–10 cm long, mostly straight, as thick as the stem at the base, tapering towards the flower, sometimes recurved at the base.

Flowers: Large. Sepals 5–6, 6–6.5 mm long, green, sometimes dark at apex, spreading. Petals 5–12, 7–18 mm long, white with yellow base, broadly obovate, contiguous during anthesis. Nectar pits 1 per petal, pyriform. Receptacle elliptical, glabrous or scarcely puberulent.

Fruits: Carpels 34–63. Ripe achenes 1.5–2.0 mm long, frequently glabrescent. Style short, caducous.

Flowering: June-August (-September).

Biology: *R. fluitans* is wintergreen. Shoots are dimorphic; short vegetative shoots bearing shorter, more rigid, dark green leaves occur in addition to the generative shoots, throughout the year. Vegetative shoots become dominant in autumn throughout the winter. Caespitose terrestrial forms with subrigid, spathulate aerial leaves form when plants are exposed by falling water levels. Aquatic plants which develop from terrestrial forms often continue to form flattened leaf segments. The species is mostly self-incompatible and remains sterile. Vegetative reproduction by detached fragments is frequent within water bodies.

Habitats: *R. fluitans* occurs in large lowland rivers and smaller upland streams with regular discharge and can tolerate high current velocities. It will also occur in large canals, oxbow lakes and deep pre-alpine lakes. It prefers sediments without an organic

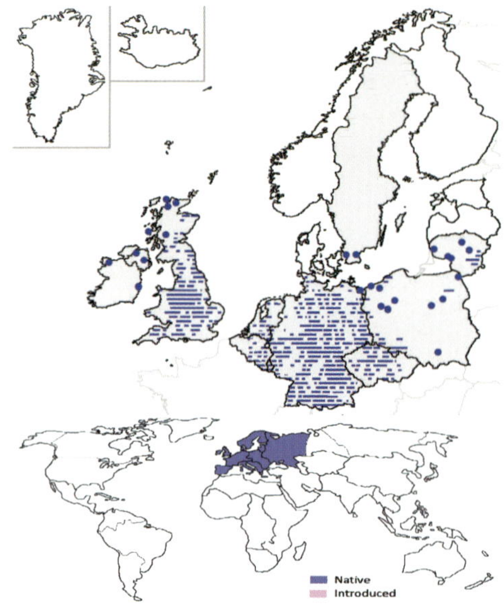

layer and will occur in both neutral and alkaline waters and can tolerate a high degree of eutrophication.

Distribution within the region:
Native. Scattered to rather common in the temperate part of the region. It is absent from Denmark, Norway, Finland, Estonia and Latvia.

Outside the region it occurs southwards to the Pyrenees, the Alps and the Hungarian plains, with isolated occurrences in Spain and Italy, but it is absent from the Balkans.

Characteristics and similar species:
R. fluitans is characterised by the long, fleshy capillary leaves, the higher number of petals, the tapering peduncle and the glabrous receptacle.

It can be confused with **207 *R. pseudofluitans***, which has shorter capillary leaves and a pubescent receptacle and typically shows more variation in

As elsewhere, Swedish plants usually have more than 5 petals and do not produce fruits. Sterility and clonal growth is typical; see description. Vramså, Sweden. Photo JCS.

Ranunculus fluitans. The leaves are much longer than the internodes and only 2-4 times divided. Vramså, Sweden. Photo BM.

the shape of both petals and nectar pits. *R. fluitans* populations with higher numbers of terminal segments, puberulent receptacle or occasional laminar leaves without a joint between the petiole and lamina may be the result of historical introgression.

Hybrids: Due to the large morphological overlap in *R. fluitans* hybrids, exact identification is difficult. Often the hybrid nature can be recognised but the second parent species cannot be identified morphologically. The most frequent *R. fluitans* hybrid is *R. fluitans* × *R. peltatus*, which gave rise to the hybrid species *R. penicillatus.* The hybrid species *R. pseudofluitans* originated from hybridisation of *R. fluitans* with *R. circinatus*.

The hybrid *R. fluitans* × *R. trichophyllus* has been reported from several countries (e.g., Britain, Germany, the Czech Republic and Austria). In Britain both *R. calcareous* and *R.* ×*bachii* were assumed to represent this hybrid (see under *R. pseudofluitans* and *R. baudotii*). *R. fluitans* hybrids with *R. baudotii* may have been overlooked for a long time. The names *R. bachii* var. *mosellana* Wirtgen, *R. fluitans* var. *heterophyllus* Glück and *R.* ×*kelchoensis* S.D.Webster may represent this taxon. It is characterised by the presence of tripartite laminar leaves with rounded lobes. In Britain the taxon *R.* ×*kelchoensis* ($2n$ = 40) has been regarded a primary hybrid of *R. fluitans* × *R. peltatus*, which is not supported by morphological and karyological data. A hybrid of *R. fluitans* with *R. aquatilis* has been reported from Britain, but never confirmed. It was associated with the names *R.* ×*bachii* and *R. pseudofluitans*.

Note: Continental populations are diploid ($2n$ = 16) or triploid ($2n$ = 24), while tetraploid plants ($2n$ = 32) have been reported from Britain.

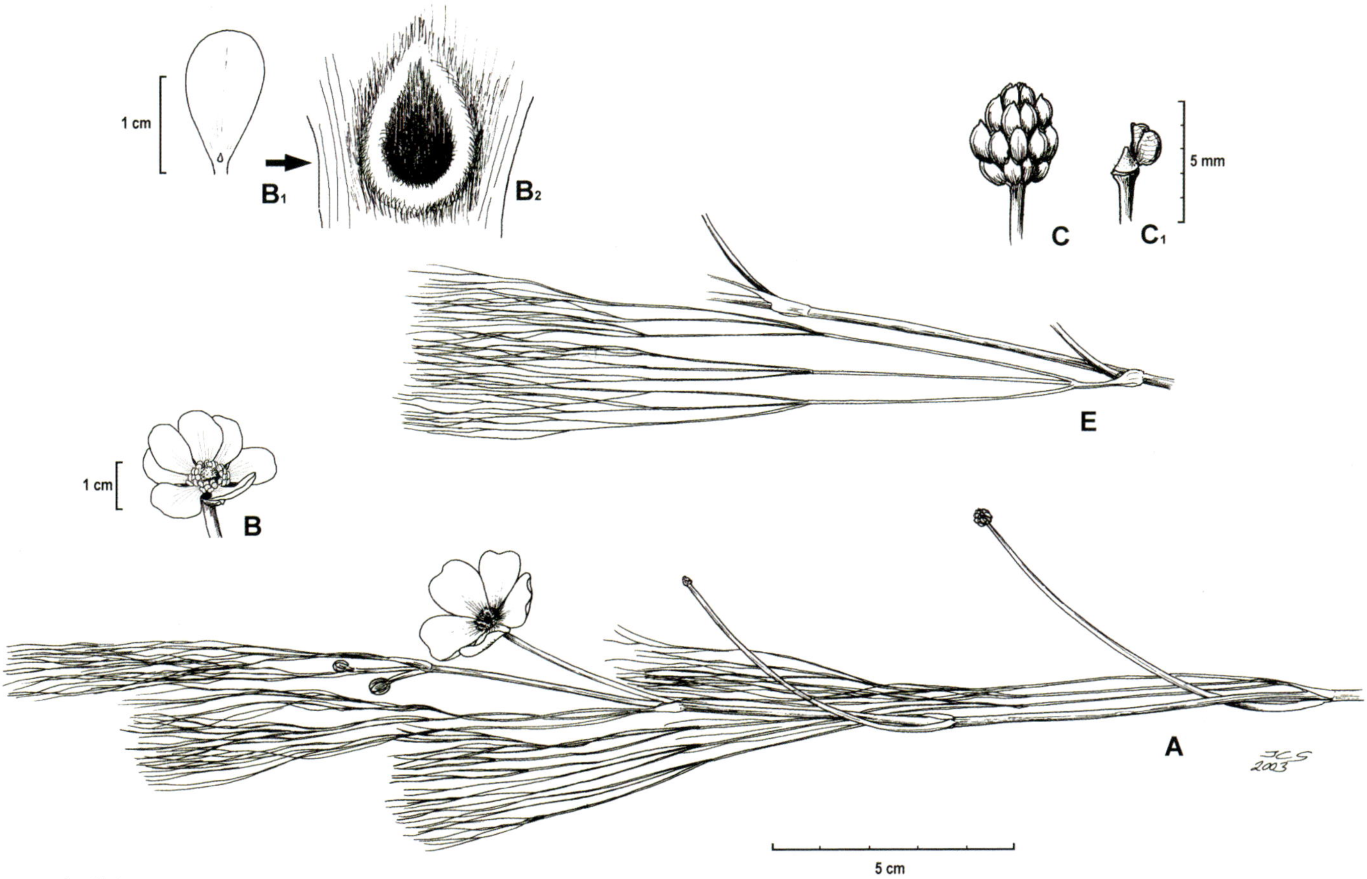

Ranunculus fluitans
A habit, **B** flower, **B1** petal, **B2** nectar pit, **C** fruiting head, **C1** receptacle with 2 achenes, **E** leaf and stem fragment.

Ranunculaceae

207 *Ranunculus pseudofluitans*

(Syme) Newbould ex Baker & Foggit

Syn.: *R. aquatilis* var. *pseudofluitans* Syme; *R. penicillatus* subsp. *pseudofluitans* (Syme) S.D.Webster p.p.
DK: Symes Vandranunkel **N:** - **S:** - **FIN:** - **IS:** -
GB: Stream Water-crowfoot **NL:** - **F:** Renoncule à pinceau
D: Flutender Pinselblättriger Wasserhahnenfuss **CZ:** - **PL:** -
EST: - **LV:** - **LT:** - **Ru:** Шелковник ложноплавучий

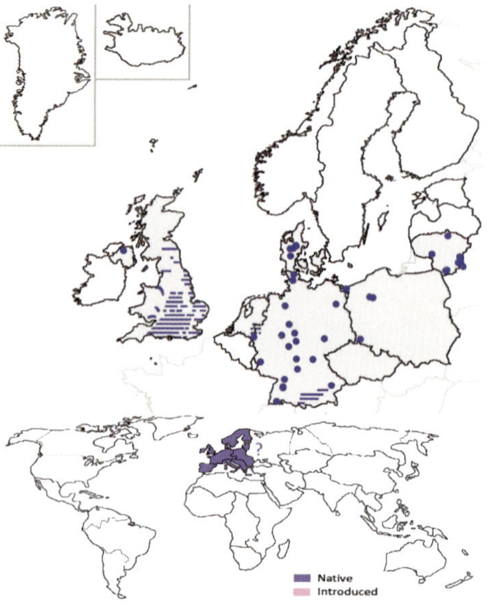

Perennial hydrophyte. Aquatic plants erect, supported by the water column, with capillary leaves.

Stem: Up to 2.5 m long, 2–3 mm in diameter, greenish, erect, branched. Rooting at almost all nodes.

Laminar leaves: Absent. Intermediate leaves rare.

Capillary leaves: Alternate, 2–8 cm long, shorter than or as long as the adjacent internode, subrigid to fleshy, green, glabrous, with 3–8 lamina divisions and 70–350 terminal segments. Petioles up to 15 cm long.

Stipules: More than 1/2 adnate to petiole, apex of free portion obtuse, glabrous.

Pedicels: Opposite a capillary leaf, 2.5–10 cm long, straight or slightly curved, glabrous.

Flowers: Medium-sized to large. Sepals 5, greenish or brownish, often reflexed. Petals 5–6, 5–15 mm long, white with yellow base, ovate to broadly obovate, contiguous during anthesis. Nectar pits 1 per petal, mostly pyriform. Receptacle subglobose, pubescent.

Fruits: 10–58 carpels. Ripe achenes 1.5–2.5 mm long, glabrous or hairy. Style short, caducous.

Flowering: May-August.

Biology: *R. pseudofluitans* is wintergreen, overwintering as short vegetative shoots. Shoots are dimorphic. Flower development is often low, while vegetative reproduction by means of detached shoots is frequent. No terrestrial form is known.

Habitats: Calcareous rivers and streams, rarely in lakes or large canals.

Distribution within the region:
Native. Scattered in Britain, Denmark, the Netherlands, Germany, Poland, Lithuania and the Czech Republic.

Outside the region in France, northern Italy, Switzerland and Austria.

Reports from Spain, Corsica, Sardinia and northwest European Russia are unconfirmed.

Characteristics and similar species:
Morphological variation in *R. pseudofluitans* is obscured by frequent confusion with other taxa. Some forms have long subparallel leaves, while others have shorter divergent leaves. Short axillary shoots with circinatus-like leaves are also often produced, which may be adnate to the petiole of the adjacent capillary leaf.

206 *R. fluitans* has longer, fleshier capillary leaves and a glabrous receptacle. Forms of **210 *R. peltatus*** and **208 *R. penicillatus*** lacking laminar leaves have larger flowers and less rigid capillary leaves than *R. pseudofluitans*.

River forms of **209 *R. baudotii*** have a whitish stem but cannot be distin-

Ranunculus pseudofluitans. River Gudenå, Denmark. Photo JCS.

Ranunculus pseudofluitans. River Tweed, Scotland. Photo RVL.

Flower and fruiting head. Photos JCS.

guished from *R. pseudofluitans* using vegetative morphological characters alone.

212 *R. trichophyllus* and **213 *R. aquatilis*** have smaller flowers, lunate or circular nectar pits, and a pedicel which does not taper toward the apex.

The distribution map is likely to include misidentified populations of species, such as *R. fluitans*, *R. baudotii*, *R. penicillatus*, *R. peltatus*, *R. aquatilis* and *R. trichophyllus*.

It also includes occurrences of an unresolved taxon, originally described from Britain as ***Ranunculus calcareus*** Butcher (*R. penicillatus* var. *calcareus* C.D.K. Cook, *R. pseudofluitans* var. *minor* Pearsall, *R. fluviatilis* Willd.). This taxon was originally described as a separate species, but later included within *R. pseudofluitans* by Webster (1988). Sell & Murrell (2018) treat it as a taxon of separate hybrid origin. It differs from *R. pseudofluitans* by the smaller overall size of plants (typically around 1 m long), pedicels which do not taper and flaccid capillary leaves. It resembles both *R. trichophyllus* and non-laminar-leaved forms of *R. aquatilis.* It is hexaploid (2*n* = 48).

Hybrids: *R. pseudofluitans* is a hybrid species derived from crossings between *R. fluitans* and *R. circinatus*. It may sometimes backcross with its parent species, while outcrossing with *R. baudotii* was reported from Denmark.

Note: Chromosome numbers of 2*n* = 24, 32 have been reported from the continent, while a count of 2*n* = 48 has also been reported from Britain (see *R. calcareus* below).

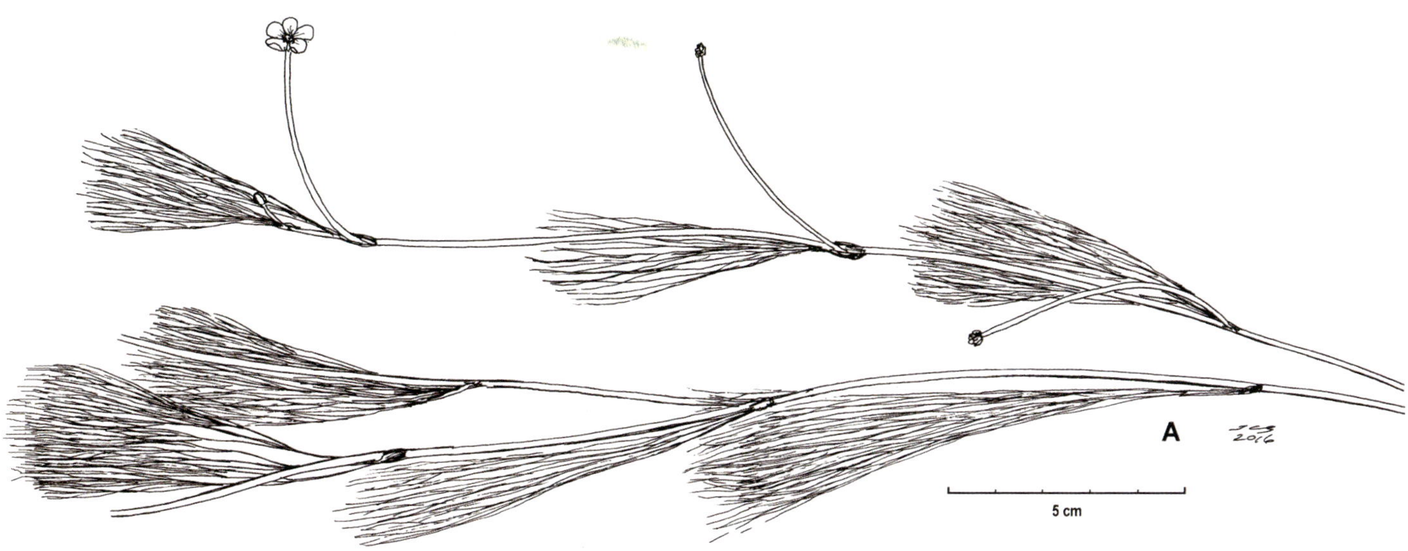

Ranunculus pseudofluitans
A habit.

Ranunculaceae

208 *Ranunculus penicillatus* (Dumort.) Bab.

Syn.: *R. penicillatus* var. *penicillatus* C.D.K.Cook; *R. penicillatus* (Dumort.) Bab. subsp. *penicillatus* S.D.Webster; *Batrachium penicillatum* Dumort.; *R. ×kelchoensis* S.D.Webster.
DK: Pensel-Vandranunkel **N:** Penselvassoleie **S:** Penselmöja **FIN:** - **IS:** -
GB: Stream Water-crowfoot **NL:** Penseelbladige waterranonkel **F:** Renoncule en pinceau
D: Pinselblättriger Wasserhahnenfuss **CZ:** Lakušník štětičkový **PL:** Włosienicznik pędzelkowaty
EST: - **LV:** - **LT:** - **Ru:** Шелковник кистелистный

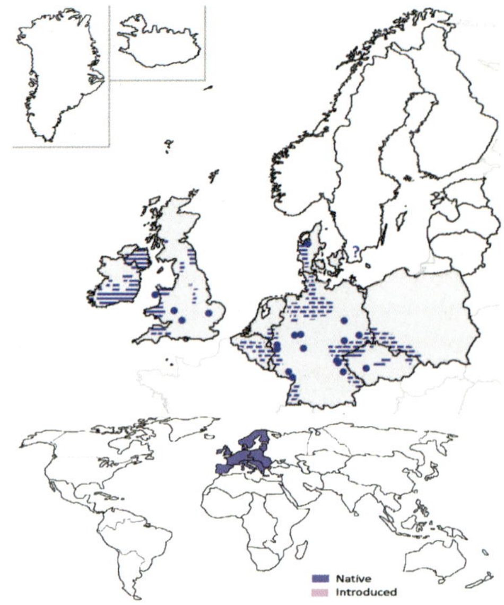

Ranunculus penicillatus. Rio Mira, Portugal. Photo KvdW.

Perennial hydrophyte. Aquatic plants supported by the water column with both laminar and capillary leaves or only capillary leaves.

Stem: Up to 5 m long, 2–5 mm in diameter, greenish, branched, glabrous. Rooting at lower nodes.

Laminar leaves: Alternate, occasionally absent, up to 25 mm long and 46 mm wide, reniform to semi-circular, with 3–5 primary lobes, margins crenate to dentate, basal sinus 90°–180°. Petioles 5–8 cm long. Intermediate leaves of *peltatus* type frequent.

Capillary leaves: Always present, greenish or yellowish, up to 30 cm long, longer than the adjacent internode in the middle part of the stem, with 3–6 (–7)lamina divisions and up to 150 terminal segments; segments flaccid, clasping. Petioles 5–8(–12) cm long.

Stipules: More than 1/2 adnate to the stipule, apex of free portion obtuse or rounded, glabrous.

Pedicels: Opposite to a laminar or capillary leaf, 5–20 cm long in fruit, straight, glabrous.

Flowers: Large. Sepals 5, 3–7 mm long, greenish, spreading. Petals 5–7, 10–25 mm long, white with yellow base, broadly obovate, contiguous at anthesis. Nectar pits 1 per petal, pyriform. Receptacle globose, densely pubescent.

Fruits: Carpels 15–80. Ripe achenes 1.2–2.0 mm long. Style short, caducous.

Flowering: May-August.

Biology: *R. penicillatus* is wintergreen. Shoots are dimorphic with short vegetative and longer generative shoots. Despite the large insect-pollinated flowers the species mainly reproduces by dispersal of detached fragments. A prostrate terrestrial form resembling the terrestrial form of *R. fluitans* develops when plants are exposed by falling water levels.

Habitats: Typically found in fast-flowing streams and rivers and only occurring in standing water bodies when these are or have been connected to a stream or river. It is not shade tolerant.

Distribution within the region:
Native. *R. penicillatus* is widespread and quite abundant in the British Isles, Denmark and Central Europe from Belgium to Poland.

Ranunculus penicillatus. Idom Å, West-Jutland, Denmark. Photo JCS.

Ranunculus penicillatus. Idom Å, West-Jutland, Denmark. Photo JCS.

It also occurs southwards to the Canary Islands, Iberian Peninsula and Sardinia, but is apparently absent from south-eastern Europe.

Characteristics and similar species: *R. penicillatus* is a large species with long capillary leaves, abundant intermediate leaves, large laminar leaves and large flowers.

It can be confused with river forms of **210 *R. peltatus***, which usually remain shorter, develop shorter capillary leaves and shorter, thinner pedicels.

There is some morphological overlap with **207 *R. pseudofluitans***, especially when laminar leaves are absent, but *R. pseudofluitans* usually has shorter, more rigid capillary leaves and smaller flowers with smaller petals.

211 *R. schmalhausenii* differs by the regular sequence of capillary leaves with penicillatus-like leaves which are only produced from the bottom to the middle of the stem, the more deeply-divided laminar leaves and dichotomously branched main shoots.

Ranunculus vertumnus (C.D.K.Cook) Luferov (*R. penicillatus* subsp. *pseudofluitans* var. *vertumnus* S.D.Webster) differs from *R. penicillatus* by the shorter capillary leaves, the higher number of terminal capillary segments (150–900) and smaller size (1–2.5 m). It can occasionally produce long capillary leaves as well as laminar leaves. It is hexaploid ($2n = 48$).

Hybrids: *R. penicillatus* is a hybrid species, derived from crossings between *R. fluitans* and *R. peltatus*. It frequently backcrosses with *R. peltatus,* forming introgression swarms. These morphotypes were reported in the literature under several informal names, such as *R. peltatus/penicillatus*, *river peltatus* or the like. Backcrossing with *R. fluitans* seems to be rare. Outcrossing with *R. aquatilis* was assumed from German and Danish material. Outcrossing with *R. circinatus* has been reported from Lithuania.

Note: A range of chromosome numbers ($2n = 32, 40, 48$) have been reported from the region.

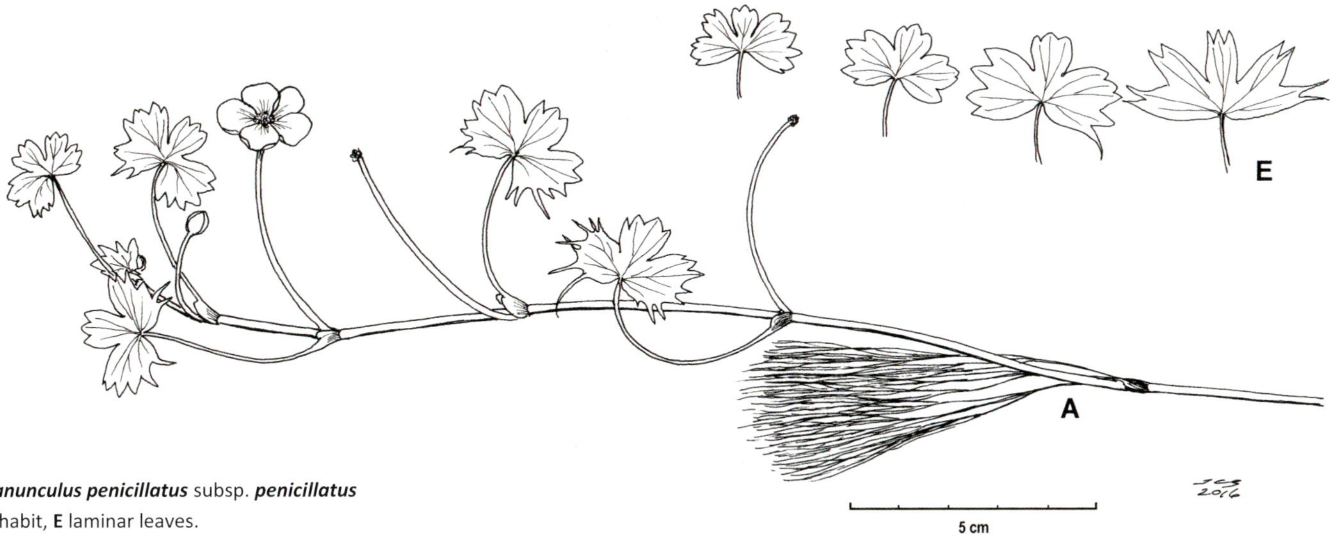

Ranunculus penicillatus subsp. ***penicillatus***
A habit, **E** laminar leaves.

Ranunculaceae

209 *Ranunculus baudotii* Godr.

Syn.: *R. peltatus* subsp. *baudotii* (Godr.) C.D.K.Cook; *R. confusus* Godr.; *Batrachium baudotii* (Godr.) Bosch; *B. marinum* Fr.
DK: Strand-Vandranunkel **N:** - **S:** Vitstjälksmöja **FIN:** Merisätkin **IS:** -
GB: Brackish Water-crowfoot **NL:** Zilte waterranonkel **F:** Renoncule de Baudot
D: Baudot-Wasserhahnenfuss **CZ:** Lakušník Baudotův **PL:** Włosienicznik Baudota
EST: Meri-särjesilm **LV:** Jūras ūdensgundega **LT:** - **Ru:** Шелковник Бодо

Ranunculus baudotii. Vandet Sø, Thy, Denmark. Photo JCS.

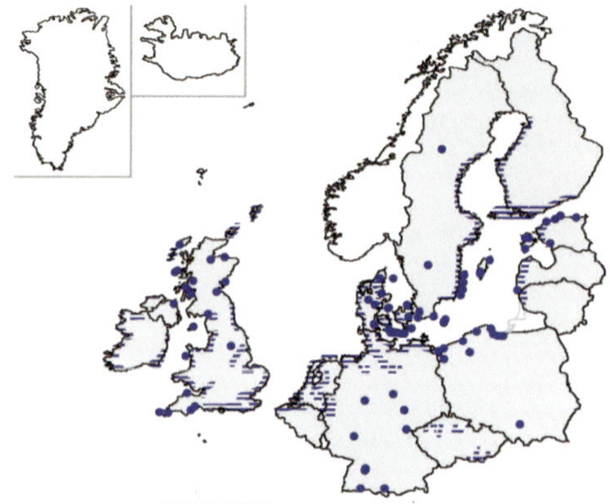

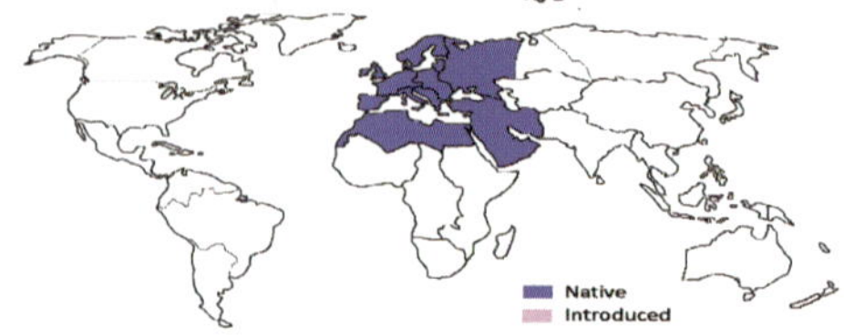

Perennial or rarely annual hydrophyte. Aquatic plants erect or spreading with both laminar and capillary leaves, with only capillary leaves in deep water.

Stem: Up to 1.5(–3) m long, 2–5(–6) mm in diameter, whitish to pale green, branched, fleshy, glabrous, internodes of the middle part as long as adjacent leaves. Rooting at lower nodes.

Laminar leaves: Present or absent, alternate or rarely opposite, up to 10 mm long and 25 mm wide, reniform to subcircular, with 3–5 primary lobes, deeply dissected to more than 1/2 of the lamina length, margin acute or crenate, basal sinus of the lamina 90°–180°. Petioles up to 8 cm long. Intermediate leaves often present, with subparallel segments and obtuse apex (*baudotii* type), symmetrical.

Capillary leaves: Present in aquatic plants, alternate, 2–10(–20) cm long, the basal leaf often significantly longer than subsequent leaves, green, glabrous, mostly rigid, fleshy, divergent, with 3–7 laminar divisions and 50–200 terminal segments. Petioles 5–20 cm long.

Stipules: Approximately 1/2 of the length adnate to petiole, free part obtuse, glabrous.

Pedicels: Opposite to a laminar or capillary leaf, elongating in fruit, 3.5–10 cm long, up to 10 times longer than petiole of adjacent leaf, recurved.

Flowers: Medium-sized. Sepals 5, 2–6 mm long, greenish with a blue tip, spreading. Petals 5, 5–13 mm long, broadly obovate, white with yellow base, mostly contiguous during anthesis. Nectar pits 1 per petal, lunate, rarely circular or horseshoe shaped. Receptacle subglobose, pubescent, elongating in fruit.

Fruits: Carpels 16–65(–100). Ripe achenes 1.2–1.8 mm long, with ventral and dorsal wings at maturity, glabrous. Style short, caducous.

Flowering: May-July (-September).

Biology: In coastal waters and rivers the plant is wintergreen, while in inland ponds it may be summer green. It flowers abundantly. Self-pollinated (in deep water) and pollinated by insects. Vegetative reproduction by detached shoots is frequent. A prostrate terrestrial form with laminar leaves and many small achenes develops on moist soil.

Habitats: Occurs mainly in brackish water, but also in highly alkaline freshwater (eventually with high sulphate content), along seashores, in coastal lagoons and marsh ditches, also in streams and rivers, scattered in inland lakes and ponds.

Distribution within the region:
Native. Widespread along the coasts of both the North Sea and Baltic; this species occurs throughout Europe mainly along the coasts. Inland occurrence is restricted to highly alkaline, eutrophic water, often along bird migration routes.

Characteristics and similar species:
R. baudotii is characterised by the fleshy stem and capillary leaves, the glabrous, winged achenes and the receptacle which elongates in fruit.

It is often confused with **210 *R. peltatus***, which has larger flowers, pyriform nectar pits and less deeply divided laminar leaves.

In coastal rivers it may be confused with **207 *R. pseudofluitans***, while in inland ponds it develops a morphotype resembling **213 *R. aquatilis***. Inland

Ranunculus baudotii. Fruiting head elongated - ovate. Achenes with a more or less prominent wing. Photo JCS.

Ranunculus baudotii with deeply divided laminar leaves. Left Filsø, SV-Jutland, Denmark. Right Langør, Stavns Fjord, Samsø, Denmark. Photos JCS.

forms of *R. baudotii* often lack typical characters, such as the winged achenes or the elongating receptacle.

Hybrids: In regions where its natural habitat overlaps with that of freshwater species, including the southern part of the Baltic Sea, hybrids with *R. peltatus*, *R. aquatilis* (*R.* ×*lamberti* A.Félix) and *R. trichophyllus* (*R.* ×*segreti* A.Félix) occur.

Other hybrids with species such as *R. pseudofluitans, R. schmalhausenii* and *R. confervoides* are rarely reported. A hybrid that may have been overlooked is *R. baudotii* × *R. fluitans*. This hybrid may frequently occur among stands of *R. fluitans* and *R. pseudofluitans* in inland waters of Central Europe and the British Isles. It may be confused with *R. pseudofluitans* if laminar leaves are lacking. It is easily confused with *R. penicillatus* when laminar leaves are present. Both intermediate and laminar leaves are, however, 3-lobed. As a rule, hybrids of *R. baudotii* are difficult to recognise as they resemble the corresponding *R. peltatus* hybrids.

Note: *R. baudotii* is tetraploid (2*n* = 32) in the northern part of its range, but Southern European populations are diploid (2*n* = 16) and show a greater similarity with *R. peltatus*.

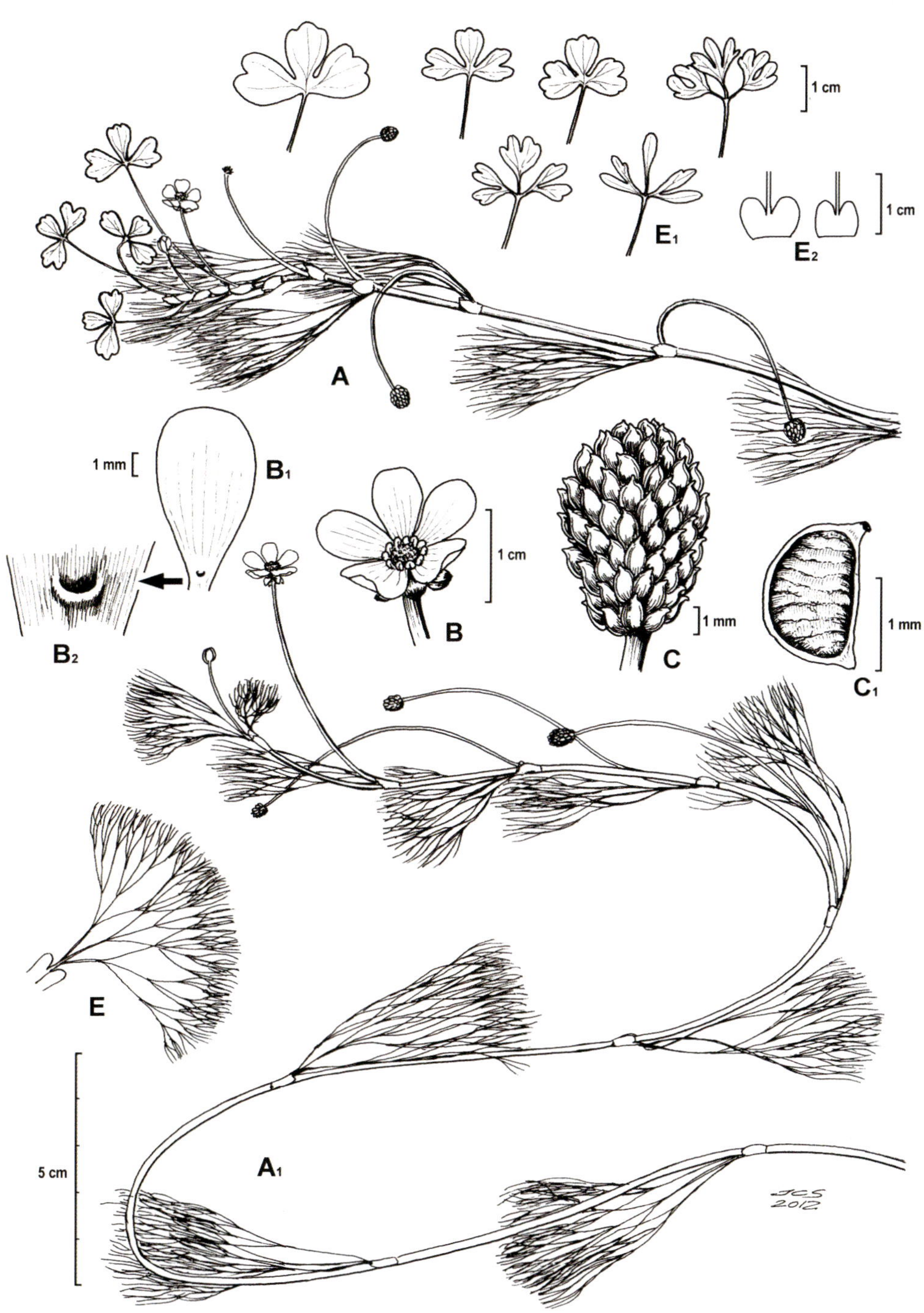

Ranunculus baudotii
A, A1 habit, **B** flower, **B1** petal, **B2** nectar pit, **C** fruiting head, **C1** achene, **E** capillary leaf, **E1** laminar leaves, **E2** stipules.

Ranunculaceae

210 ***Ranunculus peltatus*** Schrank

Syn.: *R. peltatus* Schrank subsp. *peltatus; Batrachium peltatum* (Schrank) V.V.Petrovsky; *R. floribundus* Bab.

DK: Storblomstret Vandranunkel **N:** Storvassoleie **S:** Sköldmöja **FIN:** - **IS:** -
GB: Pond Water-crowfoot **NL:** Grote waterranonkel **F:** Renoncule peltée
D: Schild-Wasserhahnenfuss **CZ:** Lakušník štítnatý **PL:** Włosienicznik tarczowaty
EST: Harkjas särjesilm **LV:** Trejlapu ūdensgundega **LT:** Skydalapė kurklė **Ru:** Шелковник щитовидный

Ranunculus peltatus. River Nuthe, Germany. Photo KvdW.

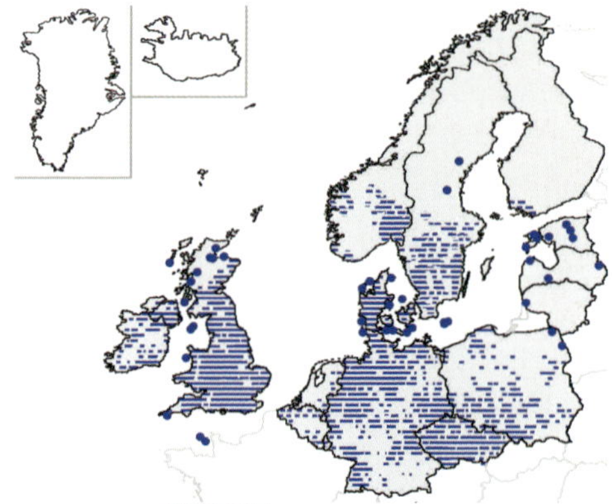

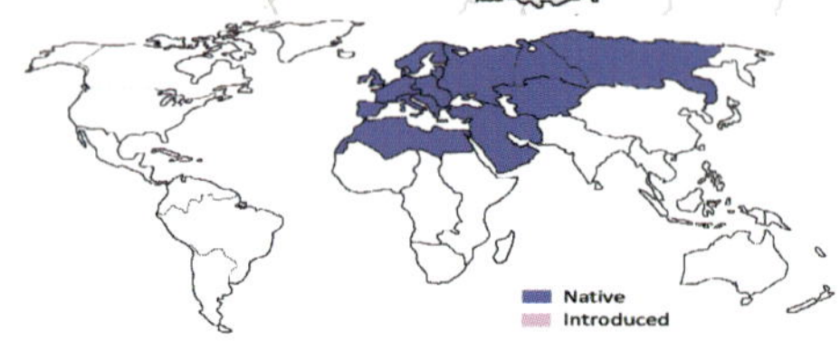

Perennial or rarely annual hydrophyte. Aquatic plants with both laminar and capillary leaves, rarely only with capillary leaves.

Stem: Up to 2(–2.5) m long, 2–6 mm in diameter, greenish or brownish, glabrous, internodes of the middle part longer than adjacent capillary leaves, rooting at the lower nodes.

Laminar leaves: Alternate, sometimes absent in streams, up to 25 mm long and 40 mm wide, reniform to circular, with 3–5 primary lobes, margins obtusely crenate, basal sinus of the lamina 30°–150°. Petioles 5–8 cm long. Intermediate leaves frequently present, of *peltatus* type with apical capillary segments, but truncate or triangular lobes also occur.

Capillary leaves: Alternate, always present, 2–8 cm long, in the middle part of the shoots, leaves are shorter than the adjacent internodes, 3–7 lamina divisions and up to 200 terminal segments. Petioles 5–25 mm long.

Stipules: 2/3 to 3/4 adnate to the petiole, free part oblong or triangular with obtuse apex, glabrous.

Pedicels: Opposite to laminar or capillary leaves, 5–20 cm long, elongate, straight or recurved.

Flowers: Large. Sepals 5, 3–6 mm long, greenish, sometimes blue tipped, spreading. Petals 5(–7), 9–23 mm long, white with yellow base, broadly obovate, sometimes crenulate at apex, contiguous during anthesis. Nectar pits 1 per petal, elongate to pyriform. Receptacle subglobose, pubescent.

Fruits: Carpels 25–60. Ripe achenes 1.5–2.0 mm long. Style short, caducous.

Flowering: May-July (-September).

Biology: Insect pollinated. *R. peltatus* shoots are dimorphic with different vegetative and generative shoots. Vegetative shoots are wintergreen and form the basis for regrowth in the spring. Vegetative reproduction is mainly by detached shoots. A prostrate terrestrial form with rigid capillary leaves develops when plants are exposed by falling water levels.

Habitats: It occurs in ponds, ditches, streams, the margins of smaller rivers and on lakeshores. It is most abundant in areas with acid rock or low-alkaline, sandy deposits and is tolerant of eutrophication, but sensitive to shading and turbidity.

Distribution within the region:
Native. Widespread and abundant throughout Europe but rare north of 60° northern latitude. It is the most frequent *Batrachium* species in most of the area except southern Sweden.

Characteristics and similar species: *R. peltatus* is a variable species, forming many different morphotypes. A small form without secondary lobes (*quinquelobus*) occurs in ponds.

It may be confused with **213 *R. aquatilis***, which has smaller flowers, circular nectar pits, more dentate laminar leaves and different intermediate leaves.

209 *R. baudotii* has more fleshy capillary leaves, shorter petals with lunate nectar pits, winged glabrous achenes and an elongating receptacle.

211 *R. schmalhausenii* has more deeply divided laminar leaves and a more regular sequence of capillary leaf formation with long flaccid leaves with subparallel segments followed by short subrigid, subglobose leaves.

208 *R. penicillatus* is generally larger and has capillary leaves which are longer than the adjacent internode.

The mainly Mediterranean species ***R. saniculifolius*** Viv. shows a large morphological overlap with *R. peltatus*. It differs by its annual life cycle, glabrous receptacle, cuneate lobes of intermediate leaves, and the occasional presence

Ranunculus peltatus. Vorgod Å, Jutland, Denmark. Photo JCS.

Flower and nectar pit. Foulum, Tjele, Jutland, Denmark. Photo JCS.

of more than 1 nectar pit per petal. It is diploid (2*n* = 16). It has been found in a few places in Germany.

Hybrids: Many hybrids have been reported. Hybridisation with *R. fluitans* gave rise to the hybrid species *R. penicillatus*. Introgression swarms between *R. peltatus* and *R. penicillatus* occur in several regions, sometimes referred to as *R. peltatus/ penicillatus*. The hybrid with *R. omiophyllus* (*R.* ×*hiltonii* H. & J.Groves) may produce deeply dissected intermediate leaves. A hybrid with *R. trichophyllus* (*R.* ×*grovesianus* Druce) is characterised by abundant formation of odd-shaped intermediate leaves. A hybrid with *R. aquatilis* (*R. x virzionensis* A.Félix) is often indistinguishable from *R. peltatus*. In still waters it can form almost circular leaves (basal sinus 0°) with crenate margins and mosaic-type intermediate leaves (with capillary segments both in basal and terminal position). Hybrids with *R. tripartitus, R. schmalhausenii* and *R. confervoides* require confirmation.

Note: Northern populations are usually tetraploid (2*n* = 32), with rare reports of aneuploid individuals. The identity of hexaploid forms (2*n* = 48) is unclear.

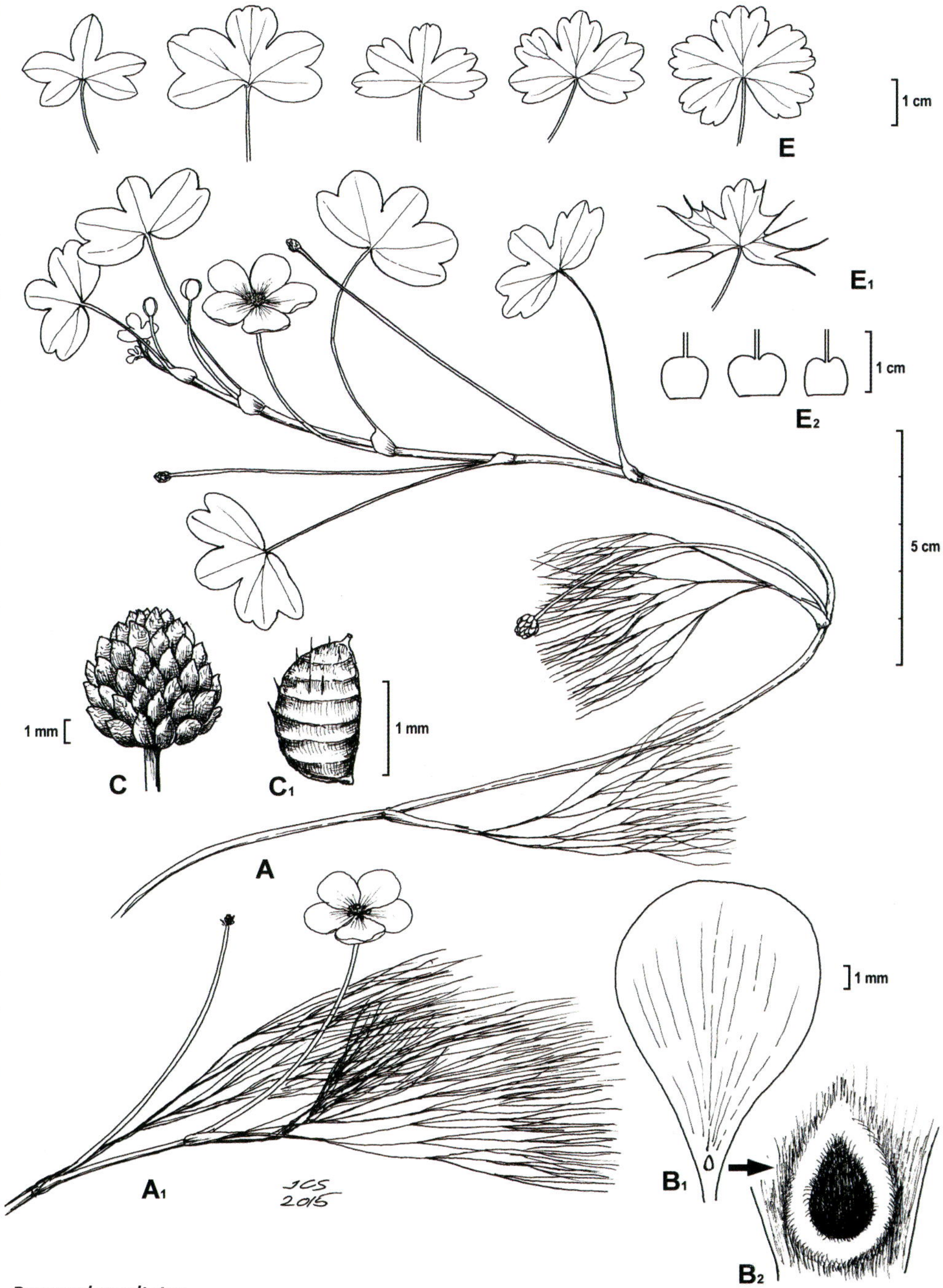

Ranunculus peltatus
A, A1 habit, **B1** petal, **B2** nectar pit, **C** fruiting head, **C1** achene, **E** laminar leaves, **E1** intermediate leaf, **E2** stipules.

Ranunculaceae

211 *Ranunculus schmalhausenii* Luferov

Syn.: *R. peltatus* var. *septentrionalis* H.Lindb.; *Batrachium nevense* Tzvelev; *B. dichotomum* (Schmalh.) Trautv.
DK: - **N:** - **S:** - **FIN:** Järvisätkin **IS:** -
GB: - **NL:** - **F:** -
D: Schmalhausen-Wasserhahnenfuß **CZ:** - **PL:** -
EST: - **LV:** - **LT:** - **Ru:** Шелковник Шмальгаузена

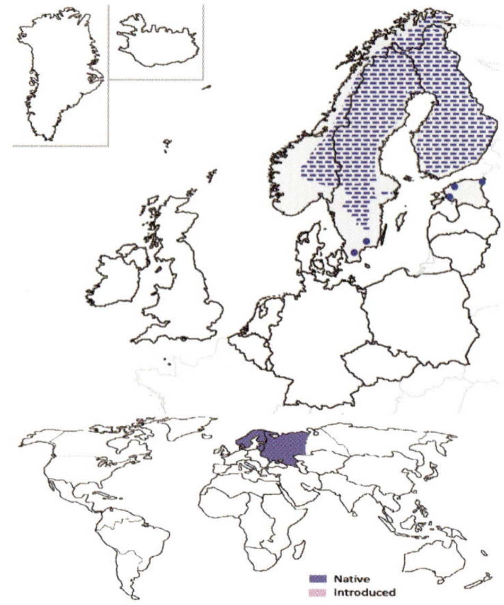

Perennial or rarely annual hydrophyte. Aquatic plants with both laminar and capillary leaves or only capillary leaves.

Stem: Up to 3 m long, 2–5 mm in diameter, light green to whitish, glabrous, dichotomously branched. Rooting at lower nodes.

Laminar leaves: Usually present, alternate, rarely the upper ones opposite, up to 30 mm long and 40 mm wide, with five primary lobes, dissected to 1/2 or 2/3 of the lamina, margin acutely crenate or dentate, basal sinus 30°–180°, glabrous. Petioles 2.5–15 cm long. Intermediate leaves with some more deeply divided lobes of *peltatus* type.

Capillary leaves: Alternate, the upper ones often opposite, 15–300 mm long, upper leaves mostly short and subglobose, lower leaves usually long with parallel segments, 3–10 lamina divisions and up to 300 terminal segments. Petioles 5–17 cm long.

Stipules: Adnate up to 3/4 to the petiole, apex of free portion obtuse, glabrous.

Pedicels: Opposite to laminar leaves, longer than the opposite leaf, 3–21 cm long, straight, glabrous.

Flowers: Large. Sepals 5, c. 5 mm long, greenish, blue tipped, spreading. Petals 5–12, 7–17 mm long, white with yellow base, broadly obovate, margins sometimes lobed, contiguous during anthesis, nectar pits up to 4 per petal, pyriform. Receptacle globose, glabrous or pubescent.

Fruits: Carpels 20–65. Ripe achenes 1.2–2.2 mm long, puberulent or glabrous. Style short, caducous.

Flowering: June-September.

Biology: *R. schmalhausenii* is wintergreen with dimorphic shoots. Short prostrate shoots persist during the winter. Despite abundant flowering the species mainly reproduces vegetatively by detached shoots. The terrestrial state is unknown.

Habitats: Deeper lakes and large rivers, mostly alkaline and nutrient poor.

Distribution within the region:
Native. Mainly in northern Scandinavia, north of 60° north latitude. In Sweden it

Ranunculus schmalhausenii in shallow water of Lake Bolshoe Severnoe, vicinity of village Sosnovyi, Russia. Photo Alexander A. Bobrov.

Ranunculus schmalhausenii in deep water of Lake Srednee Kuito, vicinity of town Kalevala, Russia. Photo Alexander A. Bobrov.

Ranunculus schmalhausenii in shallow water of Lake Bolshoe Severnoe, vicinity of village Sosnovyi, Russia. Photo Alexander A. Bobrov.

extends further southwards and shows a considerable overlap with *R. peltatus*. It also occurs in Estonia.

Otherwise the species is found in the adjacent northern part of European Russia. It has a similar range as *R. confervoides*.

Characteristics and similar species: *R. schmalhausenii* is characterised by the dichotomous branching pattern of the stem, the sequence of capillary leaves (from long to short), the deeply divided laminar leaves, often more than 5 petals and the multiple nectar pits per petal.

It may be confused with **210 *R. peltatus***, which has more rounded laminar leaves, more equal-sized capillary leaves and only 1 nectar pit per petal.

209 *R. baudotii* has a similar stem colour, but occurs more often without laminar leaves where their ranges overlap. The fruit is more prominently winged and always glabrous.

208 *R. penicillatus* has uniformly long capillary leaves and so does not show the regular sequence of the capillary leaves.

Hybrids: Hybridisation with *R. trichophyllus, R. aquatilis, R. confervoides* and *R. kauffmannii* has rarely been reported.

Note: *R. schmalhausenii* is tetraploid ($2n = 32$).

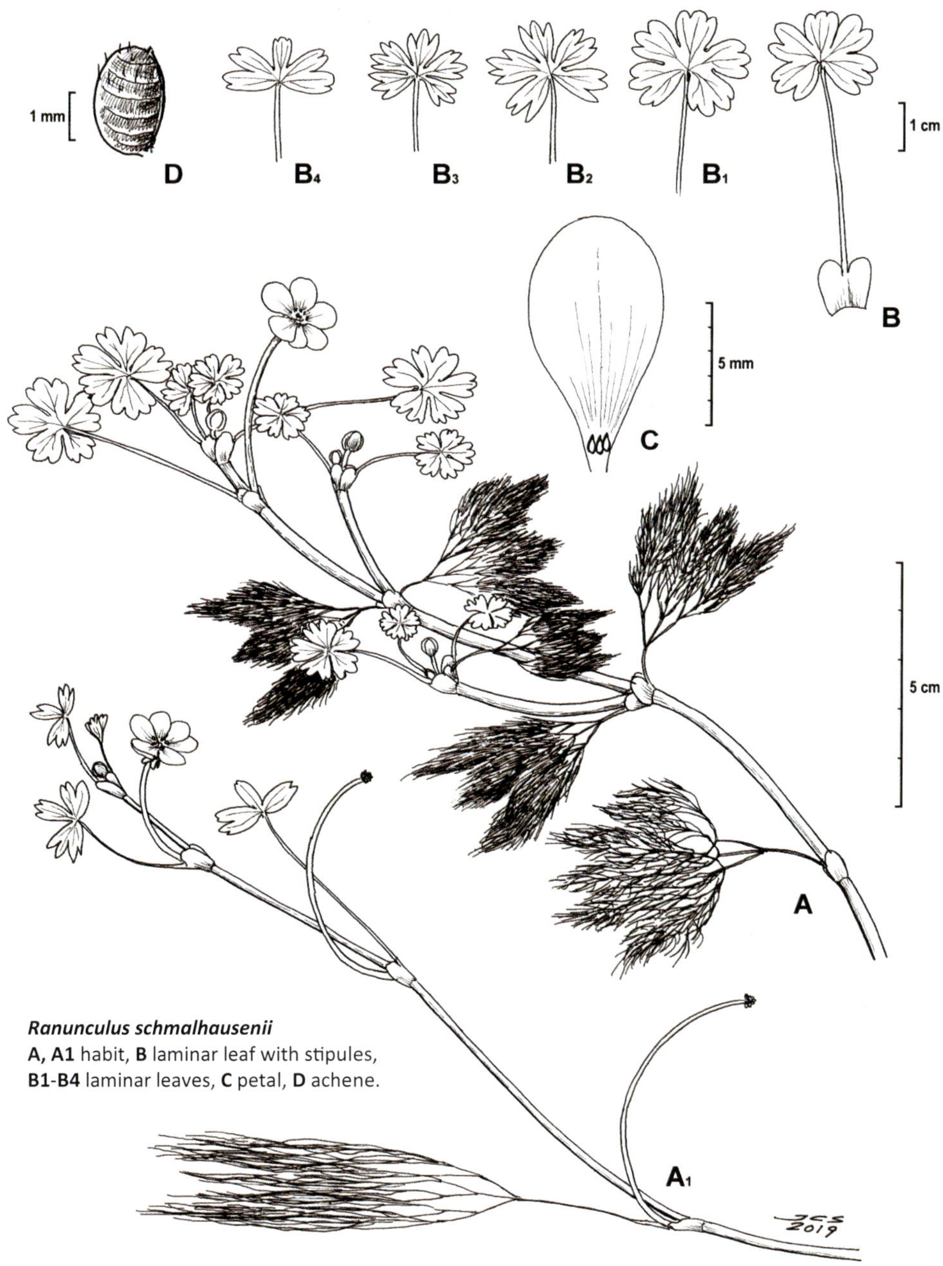

Ranunculus schmalhausenii
A, A1 habit, **B** laminar leaf with stipules, **B1-B4** laminar leaves, **C** petal, **D** achene.

Ranunculaceae

212 *Ranunculus trichophyllus* Chaix

Syn.: *R. aquatilis* var. *diffusus* With.; *R. flaccidus* Persoon; *R. drouetii* F.W.Schultz; *Batrachium trichophyllum* (Chaix) Bosch; *B. paucistamineum* (Tausch) F.W.Schultz
DK: Hårfliget Vandranunkel **N:** Småvassoleie **S:** Grodmöja **FIN:** Purosätkin **IS:** Lónasóley
GB: Thread-leaved Water-crowfoot **NL:** Haarblad-waterranonkel **F:** Renoncule à feuilles capillaires
D: Haarblättriger Wasserhahnenfuss **CZ:** Lakušník niťolistý **PL:** Włosienicznik skąpopręcikowy
EST: Jõgi-särjesilm **LV:** Spilvlapu ūdensgundega **LT:** Siūlalapė kurklė **Ru:** Шелковник волосолистный

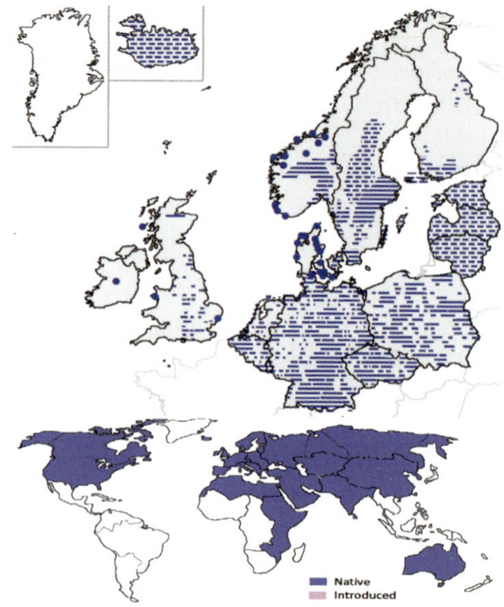

Ranunculus trichophyllus. Kongerslev, Jutland, Denmark. Photo JCS.

Perennial or rarely annual hydrophyte. Aquatic plants with capillary leaves.

Stem: Up to 1.5(–2.5) m long, 2–5 mm in diameter, green to greenish brown, erect to spreading; rooting at the lower nodes.

Laminar leaves: Absent. Small, palmate intermediate leaves are occasionally formed.

Capillary leaves: Always present, alternate, 2.5–8 cm long, flaccid or rigid, shorter than or as long as the adjacent internode, dark green to brownish, glabrous or puberulent, with 3–7 lamina divisions and up to 150 terminal segments. Petioles 5–20 mm long, occasionally longer in leaves at the base of the stem.

Stipules: Adnate to more than 2/3 of the petiole, apex of free part rounded, glabrous or hairy.

Pedicels: Opposite to a capillary leaf, 10–50 mm long, not elongating, slightly recurved.

Flowers: Small. Sepals 4–5, 2.5–3.5 mm long, greenish, rarely persisting after flowering. Petals 4–5(–6), 3–5(–6) mm long, lanceolate, elliptical or spathulate, white with yellow base, not contiguous during anthesis. Nectar pits 1 per petal, lunate. Receptacle pubescent, sometimes glabrous.

Fruits: Carpels 16–35. Ripe achenes 1.25–2.0 mm long, with some persistent brush-like hair near the apex. Style short, caducous.

Flowering: April-July (-September).

Biology: *R. trichophyllus* is wintergreen. Submerged flowers are mostly self-pollinated. Vegetative reproduction by detached shoots is frequent. A prostrate terrestrial form with short rigid capillary leaves develops in response to falling water levels.

Habitats: Mostly in nutrient-rich, alkaline waters, such as ponds, ditches, streams and margins of rivers. It also occurs in deep lakes, particularly in montane areas.

Distribution within the region:
Native. Scattered to rather common throughout the region except Greenland.

R. trichophyllus has a sub-cosmopolitan distribution occurring from Europe east to Central Asia and south to North Africa, as well as in North America, with disjunct reports of populations from South America, Yemen, Ethiopia, South Africa and Australia (including Tasmania).

Characteristics and similar species: *R. trichophyllus* does not have many characteristic features, and plants with only capillary leaves can easily be confused with other species. Plants can be delicate or robust, they can have flaccid or rigid leaves and they can be glabrous or pubescent.

217 *R. rionii* is mainly found in ponds. It is distinguished by its obligate annual life cycle, the more regular shape of the capillary leaves, smaller, more numerous achenes and the receptacle elongating in fruit.

215 *R. kauffmannii* is larger in all parts and restricted to fast-flowing watercourses.

Submerged forms of **210 *R. peltatus*** and **213 *R. aquatilis*** which lack laminar leaves have larger flowers with contiguous petals and pyriform or circular nectar pits, respectively.

In the mountains of Central Europe *R. trichophyllus* has been confused with the Scandinavian species **214 *R. confervoides***, which is usually smaller in all parts, has an obligate prostrate growth and roots at all nodes.

Hybrids: Hybridisation with *R. fluitans* has been reported and these hybrids have been erroneously referred to as *R. ×bachii* Wirtgen, often together with the hypothetical hybrid *R. fluitans* × *R. aquatilis*, while *R. calcareus* was assumed to be a hybrid species of similar

Ranunculus trichophyllus. Teichwiesen, Germany. Photo KvdW.

Ranunculus trichophyllus. Flower. Photo KvdW.

origin. The latter assumption may be correct. The hybrid is triploid or hexaploid (2*n* = 24, 48). The hybrid with *R. circinatus* (*R.* ×*gluckii* A.Félix) is largely sterile and occurs scattered among the parent species. The hybrid with *R. peltatus* (*R.* ×*grovesianus* Druce) occurs regularly in ecotones where both parent species coexist. It is characterised by abundant formation of intermediate leaves of variable shape. Repeated backcrossing with *R. peltatus* may occur. Hybridisation with *R. aquatilis* (*R.* ×*lutzii* A.Félix) has been assumed based on frequent intermediates between the two species.

Hybridisation with *R. baudotii*, *R. schmalhausenii*, and *R. rionii* may occur in some regions. The presence of intermediate leaves may generally indicate hybridisation or introgression.

Note: *R. trichophyllus* is tetraploid (2*n* = 32), with other chromosome numbers indicating hybridisation. Incongruous counts from southern Sweden (2*n* = 48) involve confusing forms of *R. aquatilis* that lack laminar leaves. Diploid counts (2*n* = 16) have been reported from the Iberian Peninsula and Russian Far East.

Nectar pit lunulate. Photo KvdW.

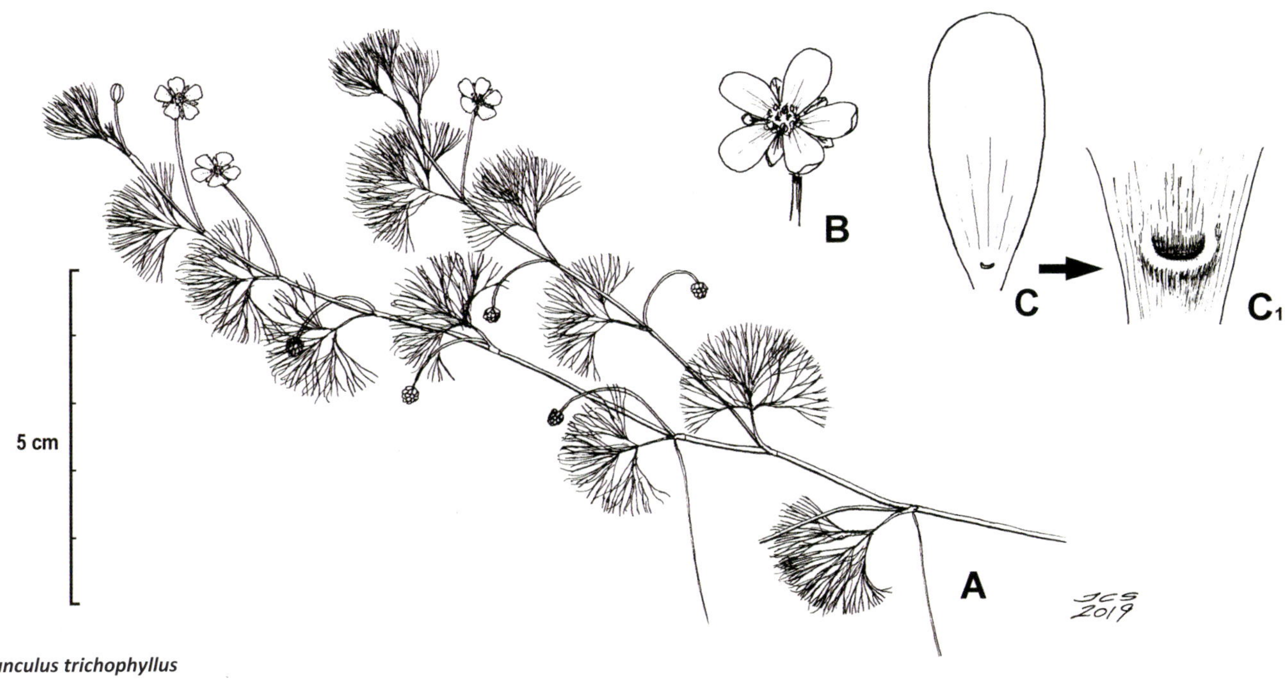

Ranunculus trichophyllus
A habit, **B** flower, **C** petal, **C1** nectar pit.

Ranunculaceae

213 *Ranunculus aquatilis* L.

Syn.: *R. aquatilis* L. var. *aquatilis*; *R. heterophyllus* Weber; *R. diversifolius* Schrank; *Batrachium aquatile* (L.) Dumort.; *B. radians* (Revel) Dumort.

DK: Almindelig Vandranunkel **N:** Småvassoleie **S:** Vattenmöja **FIN:** Vesisätkin **IS:** - **GB:** Common Water-crowfoot **NL:** Gewone waterranonkel **F:** Renoncule aquatique **D:** Gewöhnlicher Wasserhahnfuss **CZ:** Lakušník vodní **PL:** Włosienicznik wodny **EST:** Tume särjesilm **LV:** Parastā ūdensgundega **LT:** Paprastoji kurklė **Ru:** Шелковник водяной

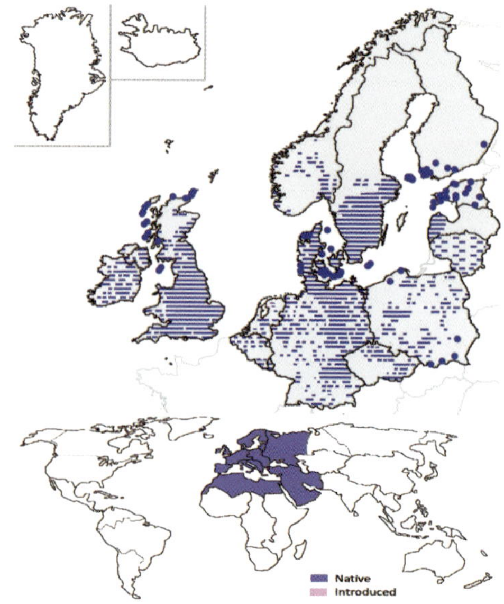

Ranunculus aquatilis. Pond at Katbjerg Odde, Jutland, Denmark. Photo JCS.

Perennial, rarely annual hydrophyte. Aquatic plants mainly with both laminar and capillary leaves, rarely with only capillary leaves.

Stem: Up to 1.5 m long, 2–4 mm in diameter, mostly erect, branched, greenish, glabrous. Rooting at lower nodes.

Laminar leaves: Frequent, alternate, reniform to semi-circular, with 3–7 primary lobes, often dissected to 1/2 the lamina length, secondary lobes 7–19, margin crenate or dentate, basal sinus of the lamina 0°–90°(–120°). Petioles up to 9 cm long. Intermediate leaves often present, of *aquatilis* type (with capillary segments at the base), asymmetrical.

Capillary leaves: Present, alternate, 2–6(–8) cm long, subrigid or flaccid, green, glabrous, shorter than or equal to the length of the adjacent internode, divided 5–6 times, the middle part shorter than or equal to the lateral parts, up to 100 terminal segments. Petioles up to 2.5 cm long.

Stipules: More than 2/3 adnate to petiole, free part triangular, acute, glabrous.

Pedicels: Opposite to laminar leaves, rarely opposite to capillary leaves; up to 5(–7) cm long, straight, not elongating in fruit, shorter than petiole of the opposite laminar leaf.

Flowers: Medium-sized. Sepals 5, 3–5 mm long, greenish, often with blue tip and membranous margin, spreading. Petals 5, 5–12 mm long, obovate, white with yellow base, usually contiguous during anthesis. Nectar pits 1 per petal, cup shaped, circular, rarely lunate or elongated. Receptacle subspherical, densely pubescent.

Fruits: Carpels 20–68. Mature achenes obovate, 1.2–1.8 mm long, with some brush-like persistent hairs at the apex. Style short, caducous.

Flowering: May-June (-August).

Biology: *R. aquatilis* is summer green, overwintering as submerged rooted stems, while capillary leaves only persist in warm winters. Shoots are dimorphic; vegetative shoots remain short; generative shoots flower abundantly; both self-pollination and cross-pollination by insects have been observed. Vegetative propagation by means of detached fragments is frequent. A prostrate terrestrial form with stomata and thick cuticle on rigid capillary leaves develops in response to falling water levels.

Habitats: Nutrient-rich, mostly alkaline waters, on lakeshores, in ponds, ditches, canals and small streams, mostly in shallow water; shade intolerant.

Ranunculus aquatilis - note some intermediate leaves. Amossen, Öland, Sweden. Photo JCS.

Ranunculus aquatilis. Jutland, Denmark. Photo BM.

Ranunculus aquatilis. Petal and flower. Kalø, Djursland, Denmark. Photo JCS.

Distribution within the region:
Native. *R. aquatilis* occurs in temperate part of the region eastwards to the Baltic countries and Poland. It is most frequent in southern Sweden.

In Karelia and the Russian Northwest it extends further eastwards than *R. peltatus*. Isolated populations occur in central Spain, Sicily, Greece (Thrakia) and Palestine. It is rare in areas of predominantly acid rocks and low alkaline glacial deposits.

Characteristics and similar species:
R. aquatilis is distinguished from other laminar-leaved species by the crenate or dentate laminar leaves, the characteristic intermediate leaves, short pedicels and circular nectar pits.

It is frequently confused with the more frequent **210 *R. peltatus***, which has larger flowers, pyriform nectar pits, laminar leaves with rounded secondary lobes, the intermediate leaves of *peltatus* type and the elongating pedicels, which are usually longer than the opposite leaf.

209 *R. baudotii* has achenes which are completely hairless and winged at maturity, more deeply dissected laminar leaves and pedicels which are mostly elongated and recurved.

212 *R. trichophyllus* differs from *R. aquatilis* by the absence of fully developed laminar leaves (although palmate intermediate leaves are occasionally present), shorter petals which are mainly lanceolate or spathulate and not contiguous during anthesis and lunate nectar pits. Intermediate forms between these two species are frequent in the region.

Hybrids: The species frequently hybridises with *R. peltatus* (*R. ×virzionensis* A.Félix) and *R. trichophyllus* (*R. ×lutzii* A.Félix), both forming swarms of intergrading morphotypes in some regions. Hybridisation has occasionally been reported with *R. tripartitus*, *R. fluitans*, *R. baudotii* (*R. ×lamberti* A.Félix), *R. schmalhausenii* and *R. kauffmannii*.

Note: *R. aquatilis* is hexaploid (2n = 48) throughout its range and has arisen from crossings of two rigid-leaved morphotypes of R. trichophyllus.

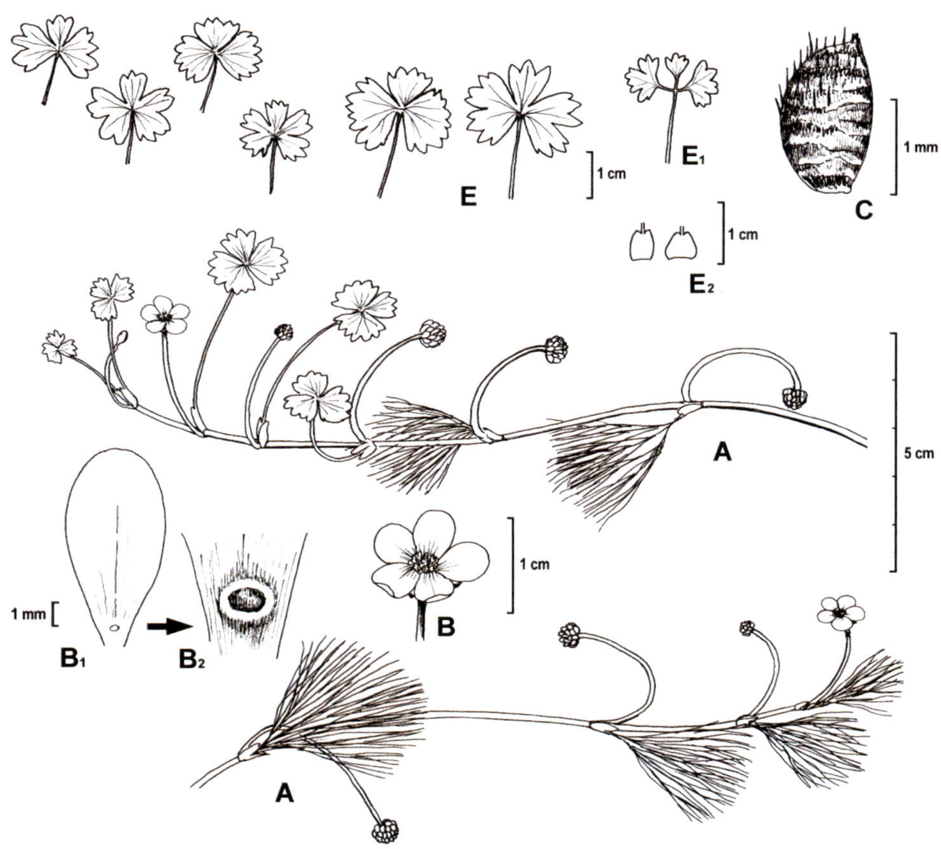

Ranunculus aquatilis
A habit, **B** flower, **B1** petal, **B2** nectar pit, **C** achene, **E** laminar leaves, **E1** intermediate leaf, **E2** stipules.

Ranunculaceae

214 *Ranunculus confervoides* (Fr.) Fr.

Syn.: *R. eradicatus* (Laestad.) F.Johansen, *R. trichophyllus* subsp. *eradicatus* (Laest.) C.D.K.Cook, *Batrachium confervoides* Fr.

DK: Dværg-Vandranunkel **N:** Dvergvassoleie **S:** Hårmöja **FIN:** Hentosätkin **IS:** Lónasóley
GB: - **NL:** - **F:** -
D: Wurzelnder Wasserhahnenfuss **CZ:** - **PL:** -
EST: Oja-särjesilm **LV:** - **LT:** - **Ru:** Шелковник неукореняющийся

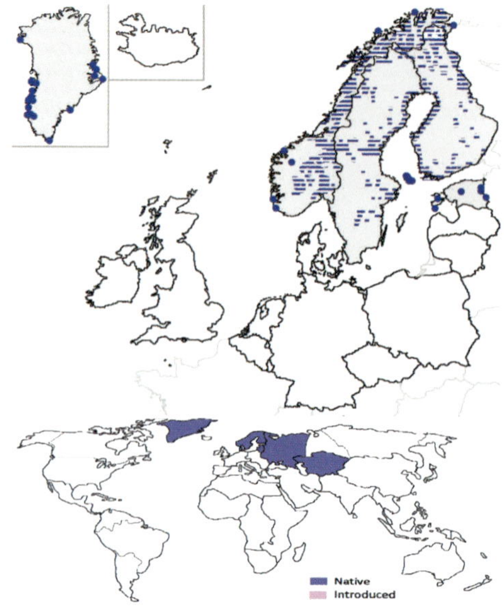

Ranunculus confervoides. Kilnäset, Sweden. Photo KvdW.

Perennial, rarely annual hydrophyte. Aquatic plants prostrate, spreading with only capillary leaves.

Stem: Up to 0.4 m (rarely to 1.5 m) long, 0.5–1.5 mm in diameter, glabrous, pale green to whitish, branched. Rooting at all nodes.

Laminar leaves: Absent. Intermediate leaves absent.

Capillary leaves: Present, alternate, 10–35 mm long, green, terminal segments extremely fine, flaccid, the ultimate ones elongated, narrowing outwards, with 2–4 lamina divisions and up to 80 terminal segments. Petioles 0.3–1.8 mm long.

Stipules: More than 2/3 adnate to the petiole, apex of free part rounded, glabrous.

Pedicels: Opposite to capillary leaves, 1–5 cm long, recurved.

Flowers: Small. Sepals 5, 2–3 mm long. Petals 5, 3–4.5 mm long, ovate to slightly obovate, white with yellow base, not contiguous during anthesis. Nectar pits 1 per petal, lunate. Receptacle sub-globose, pubescent.

Fruits: Carpels 5–15(–25). Ripe achenes 1.0–1.5 mm long, hairy. Style short, caducous.

Flowering: May-June(-August).

Biology: The species is wintergreen, growing with green shoots at the bottom of lakes and ponds. Flowers remain unopened in deep water. Vegetative reproduction commonly occurs by detached shoots. Flowering shoots sometimes float at the surface. A prostrate terrestrial form is frequent along lakeshores.

Habitats: Oligotrophic, slightly acidic to alkaline water bodies, also in slightly brackish habitats on the northern edge of the Baltic Sea.

Distribution within the region:
Native. *R. confervoides* is restricted to northern Scandinavia and adjacent parts of European Russia, sharing its range with *R. schmalhausenii*.

Records from Central Asian mountains and boreal parts of North America are based on confusion with other species (e.g., *R. trichophyllus*, *R. codyanus* B.Boivin and *R. subrigidus* W.B.Drew). Populations recorded from the Pyre-

Submerged *R. confervoides.* Varjakansaari, Finland. Photo RVL.

nees and Alps resemble small forms of *R. trichophyllus*. All specimens collected in the German Alps are *R. trichophyllus*.

Characteristics and similar species: *R. confervoides* is recognised by its small size, small flowers, the often elongated and curved pedicels, fine capillary leaves and adventitious roots at all nodes.

It is often confused with arctic and alpine morphotypes of **212 *R. trichophyllus*,** which share the small size, short petals, prostrate growth and cleistogamous pollination with *R. confervoides*. These characters are convergent adaptations to adverse habitat conditions in arctic and alpine regions.

Hybrids: Hybrids with *R. baudotii*, *R. peltatus* and *R. schmalhausenii* have been reported.

Note: *R. confervoides* is tetraploid ($2n$ = 32).

Partly submerged *R. confervoides.* Varjakansaari, Finland. Photo RVL.

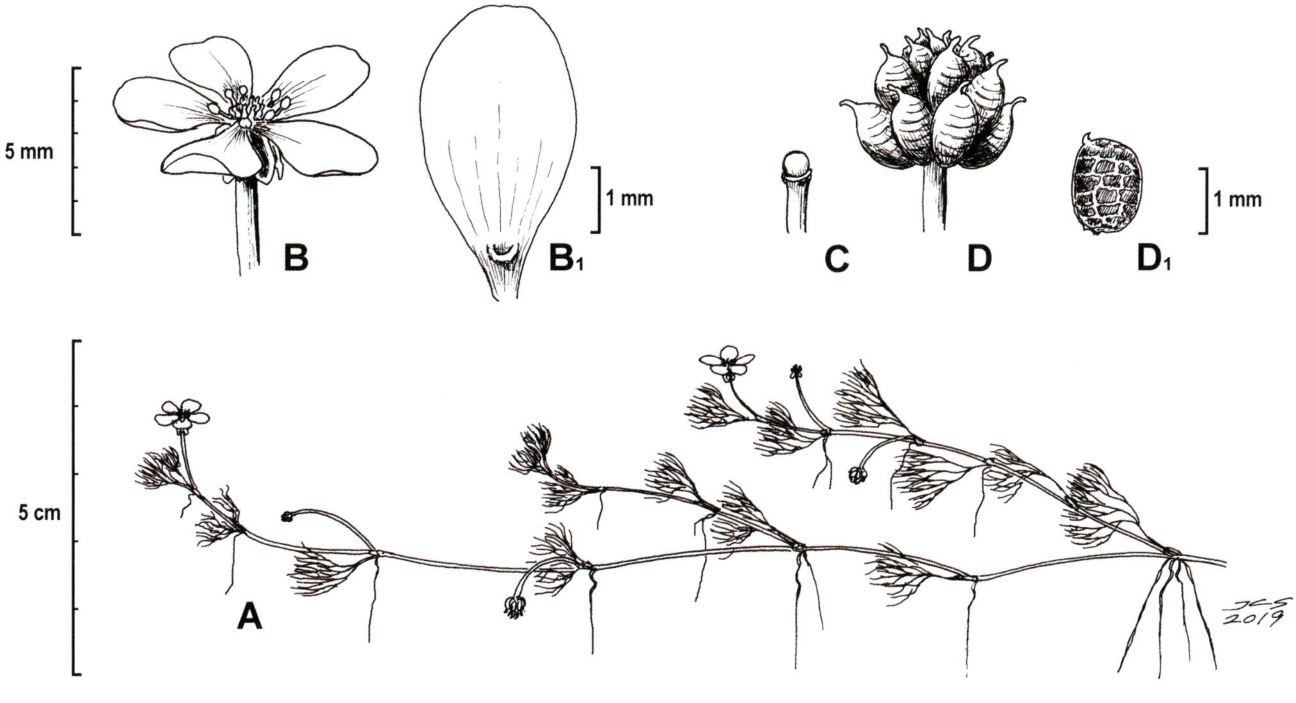

Ranunculus confervoides
A habit, **B** flower, **B1** petal, **C** receptacle, **D** fruiting head, **D1** achene.

Ranunculaceae

215 *Ranunculus kauffmannii* Clerc

Syn.: *Batrachium kauffmannii* (Clerc) V.Krecz.
DK: - **N:** - **S:** - **FIN:** Purosätkin **IS:** -
GB: - **NL:** - **F:** -
D: Kauffmann-Wasserhahnenfuss **CZ:** - **PL:** -
EST: Kaufmanni särjesilm **LV:** - **LT:** - **Ru:** Шелковник Кауфмана

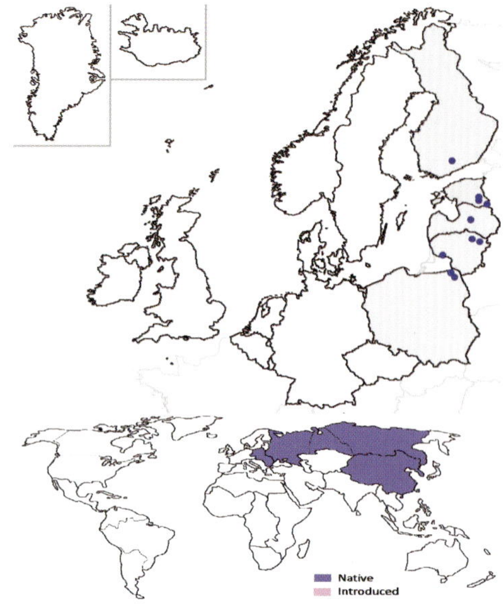

Perennial hydrophyte; aquatic form usually with only capillary leaves.

Stem: Up to 1.5(–2) m long, 2–3 mm in diameter, yellowish green, puberulent in upper part, branched, rooting at all nodes.

Laminar leaves: Absent. Small intermediate leaves rarely present, crescent shaped or truncate.

Capillary leaves: Always present. Alternate, 6–18 cm long, yellowish green (remaining yellowish green when dried), flaccid, subparallel, glabrous or hairy, with 3–5(–7) lamina divisions and up to 100(–200) terminal segments. Petioles 1–2 cm long.

Stipules: More than 2/3 adnate to the petiole, apex of free part rounded, pubescent in the upper part of the stem.

Pedicels: Opposite a capillary leaf, 2–6 cm long, straight, glabrous or hairy.

Flowers: Small to medium-sized. Sepals 5, 2–3 mm long, greenish, spreading or reflexed. Petals 5–8 mm long, white with a yellow base, narrowly obovate, not contiguous during anthesis. Nectar pits 1 per petal, lunate, rarely circular. Receptacle subglobose, pubescent, rarely glabrous.

Fruits: Carpels 15–40. Ripe achenes 1.7–2.2 mm long, puberulent. Style short, caducous.

Flowering: June-August.

Biology: The species is wintergreen. Shoots are dimorphic with short vegetative and longer generative ones. Flowers are mostly produced underwater and are self-pollinated. Vegetative reproduction by means of detaches shoots is frequent. A prostrate terres-

Ranunculus kauffmannii without flowers. Eastern Siberia. Photo Victor Chepinoga.

trial form develops in response to falling water levels.

Habitats: Restricted to fast-flowing streams and rivers, which are nutrient poor and low alkaline.

Distribution within the region:
Native. Restricted to the eastern part of the region, namely Finland, Latvia, Lithuania and Poland.

R. kauffmannii occurs mostly in the boreal zone of Eurasia eastwards to Lake Baikal. Records from northeastern China, the Russian Far East and Japan require confirmation.

Characteristics and similar species:
R. kauffmannii strongly resembles *R. trichophyllus*, but is larger in most parts and its capillary leaves are always flaccid.

It has been confused with **212 *R. trichophyllus***, which differs by the shorter, often subrigid capillary leaves and smaller flowers.

207 *R. pseudofluitans* differs by the thickened pedicels, usually rigid capillary leaves, and larger flowers with usually pyriform nectar pits.

Hybrids: A hybrid with *R. circinatus* has been reported. Hybrids with the Asian species *R. mongolicus* may occur in Lithuania, even though this species is unknown in this country. The presence of intermediate leaves in *R. kauffmannii* may indicate hybridisation with a laminar-leaved species, e.g., *R. aquatilis*, *R. peltatus* or *R. schmalhausenii*.

Note: *R. kauffmannii* is tetraploid ($2n = 32$) in its European range. It may have originated from crossings of two different capillary-leaved morphotypes of *R. trichophyllus*.

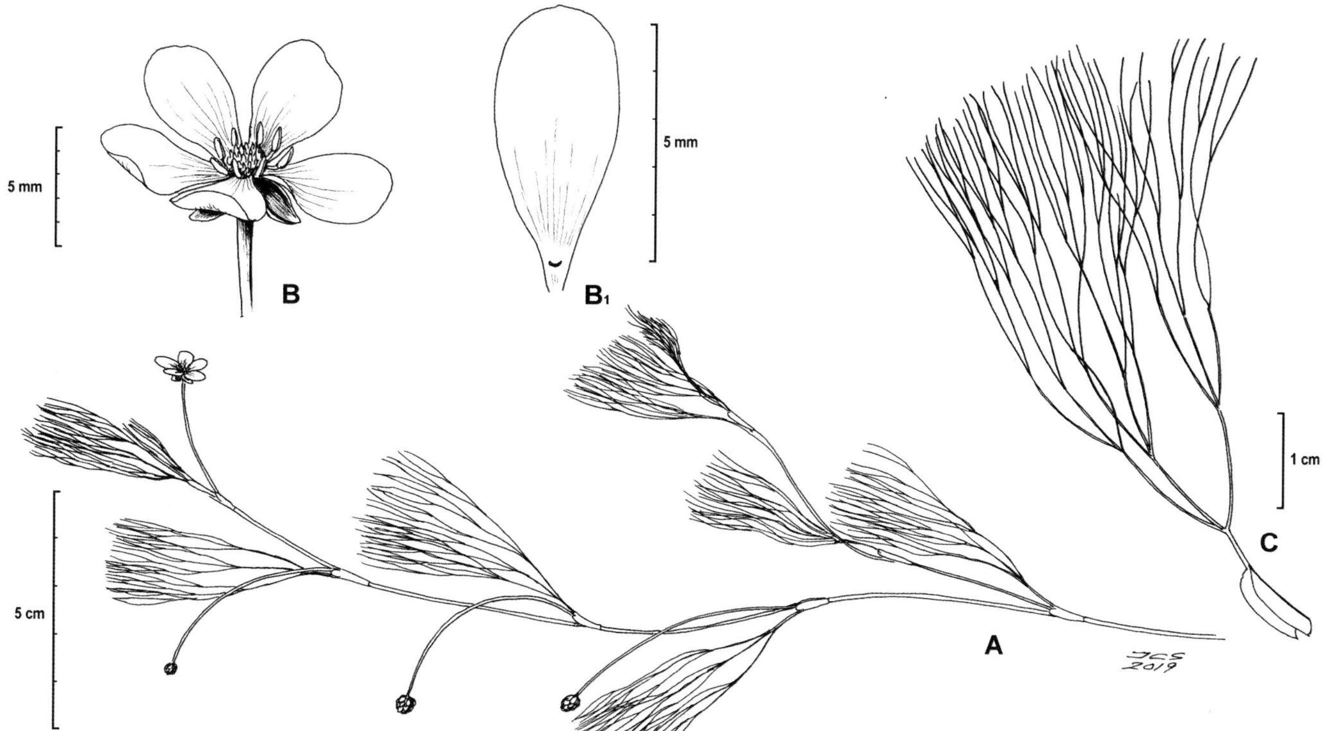

Ranunculus kauffmannii
A habit, **B** flower, **B1** petal, **C** leaf.

Ranunculaceae

216 *Ranunculus circinatus* Sibth.

Syn.: *R. foeniculaceus* auct., *R. divaricatus* auct.
DK: Kreds-Vandranunkel **N:** Hjulvassoleie **S:** Hjulmöja **FIN:** Pyörösätkin **IS:** -
GB: Fan-leaved Crowfoot **NL:** Stijve waterranonkel **F:** Renoncule divariquée
D: Spreizender Wasserhahnenfuss **CZ:** Lakušník okrouhlý **PL:** Włosienicznik krążkolistny
EST: Sõõr-särjesilm **LV:** Apaļlapu ūdensgundega **LT:** Standžialapė kurklė **Ru:** Шелковник жестколистный

Ranunculus circinatus never develops laminar leaves. Ny Frederiks Kog, SW-Jutland, Denmark. Photo BM.

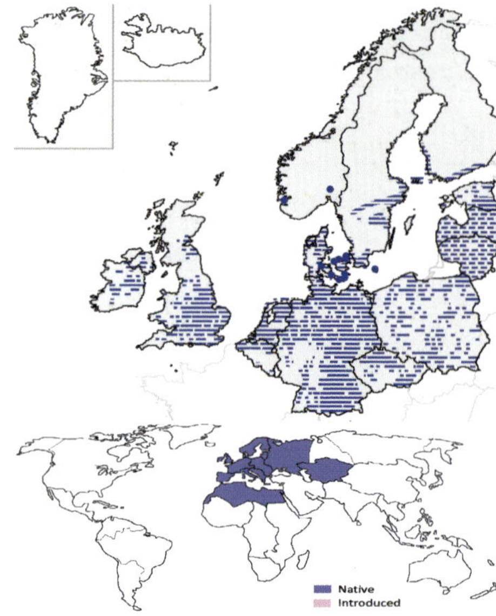

Perennial hydrophyte. Aquatic plants with only capillary leaves.

Stem: Up to 1 m, in deep water 3 m long, 1–3 mm in diameter, greenish, erect, sparsely branched, glabrous. Rooting at all nodes.

Laminar leaves: Always absent. Intermediate leaves always absent.

Capillary leaves: Always present. Alternate, 10–30 mm long, circular to semicircular, segments lying in one plane, with 3–6 lamina divisions and 80–100 terminal segments. Petioles 0–0.5 cm, rarely up to 2 cm long.

Stipules: Adnate to the petiole for more than 3/4, apex of free part obtuse, glabrous.

Pedicels: Opposite to capillary leaves, 2–10 cm long, elongating in fruit, straight.

Flowers: Medium-sized. Sepals 5, 4–6 mm long, green, spreading. Petals 5, 6–10 mm long, white with a yellow base, obovate, contiguous during anthesis. Nectar pits 1 per petal, lunate. Receptacle subglobose, pubescent.

Fruits: Carpels 30–56. Ripe achenes 1.0–1.6 mm long, mostly hairy. Style short, caducous or persistent.

Flowering: May-August.

Biology: The species develops dimorphic shoots. Vegetative shoots remain short and persist through the winter at the bottom of lakes, ponds and streams. Generative shoots flower abundantly. Flowers are insect pollinated. A prostrate terrestrial form with rigid capillary leaves develops in response to falling water levels.

Habitats: Mainly in alkaline water, frequent in deep lakes, but also found in ponds, ditches, canals and small streams.

Distribution within the region: Native. Frequent throughout most countries, except for Norway, northern Sweden and Finland.

R. circinatus has a predominantly European distribution, occurring eastwards to central Kazakhstan, with scattered populations around the Mediterranean.

Characteristics and similar species: *R. circinatus* is characterised by the short rigid capillary leaves which lie in a single plane. The segments of capillary leaves of rigid-leaved morphotypes of **212 *R. trichophyllus***, **213 *R. aquatilis*** and **209 *R. baudotii*** never lie in one plane. Rigid-leaved terrestrial forms of **213 *R. aquatilis***, **210 *R. peltatus*** and **212 *R. trichophyllus*** may superficially resemble *R. circinatus*, but the leaf segments are shorter, fleshy and not in one plane.

Ranunculus circinatus submerged. Germany. Photo KvdW.

Flowers. Ny Frederiks Kog, SW-Jutland, Denmark. Photo BM.

Terrestrial plants. White Mere, Shropshire, Britain. Photo RVL.

Segments of capillary leaves all lying in one plane. Grund Fjord, Randers, Denmark. Photo JCS.

Hybrids: Hybridisation with *R. trichophyllus* (*R.* ×*gluckii* A.Félix) has been reported. The hybrid has longer petioles and longer capillary leaves, the segments of which are not lying in one plane; it is sterile, usually with many undeveloped pedicels and flowers.

Rare hybridisation was recorded with **217 *R. rionii*** and **215 *R. kauffmannii***.

207 *R. pseudofluitans* is a hybrid species derived from crossings between **206 *R. fluitans*** and *R. circinatus*.

Note: *R. circinatus* is diploid (2*n* = 16) throughout its range.

Fruiting head. Ny Frederiks Kog, SW-Jutland, Denmark. Photo BM.

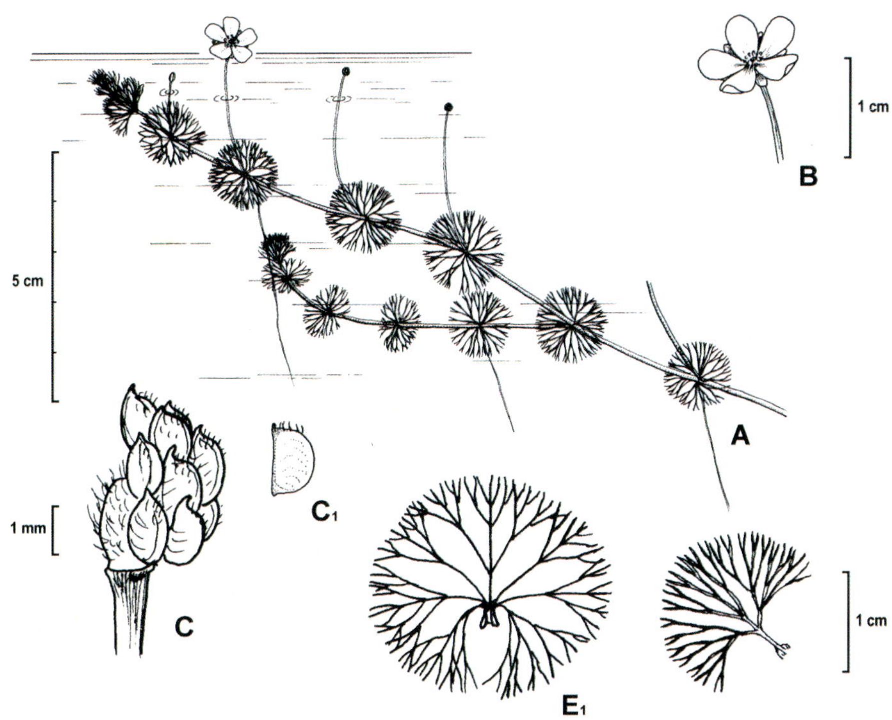

Ranunculus circinatus
A habit, **B** flower, **C** fruiting head - some achenes removed to show the hairy receptacle, **C1** achene, **E1** leaves.

Ranunculaceae

217 *Ranunculus rionii* Lagger

Syn.: *R. trichophyllus* subsp. *rionii* (Lagger) Soó, *Batrachium rionii* (Lagger) Nyman
DK: - **N:** - **S:** - **FIN:** - **IS:** -
GB: - **NL:** Rions waterranonkel **F:** Renoncule de Rion
D: Rion-Wasserhahnenfuss **CZ:** Lakušník Rionův **PL:** -
EST: - **LV:** - **LT:** - **Ru:** Шелковник Риона

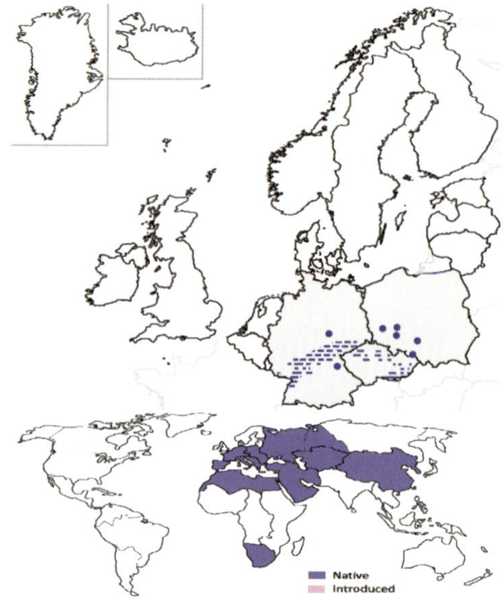

Annual hydrophyte, rarely amphiphyte. Aquatic plants with only capillary leaves.

Stem: Up to 0.5(–2) m long, 2–3 mm in diameter, erect or prostate, branched, green, glabrous, rooted at lower nodes, primary roots often still present.

Laminar leaves: Always absent. Intermediate leaves absent.

Capillary leaves: Always present. Alternate, 20–30 mm long, obconical (similar in shape and size from the bottom to the top of the stem), glabrous, flaccid to subrigid, with 3–6 lamina divisions and up to 150 terminal segments. Petioles up to 2 cm long.

Stipules: Adnate to c. 2/3 of the petiole, apex of free portion rounded, glabrous.

Pedicels: Opposite to capillary leaves, 3–5 cm long, 1 mm in diameter, curved, glabrous.

Flowers: Small. Sepals 5, 2–3 mm long, greenish, spreading. Petals 5, 3.5–5 mm long, white with yellow base, ovate to lanceolate. Nectar pits 1 per petal, lunate. Receptacle pubescent, elongating in fruit.

Fruits: Carpels (20–)60–90. Ripe achenes 0.8–1.3 mm long. Style short, often persistent.

Flowering: June-August(-September).

Biology: *R. rionii* is mostly annual in the region. Vegetative reproduction by detached shoots occurs within a catchment. Long-range dispersal may be facilitated by anthropogenic actions. A prostrate terrestrial form develops, producing many small achenes in response to falling water levels.

Habitats: *R. rionii* typically occurs in ponds, but also on the margins of lakes, reservoirs and larger rivers, in conditions where there is strong water level fluctuation and anthropogenic disturbance.

Distribution within the region:
Native. Within the region it occurs in southern Germany, the Czech Republic, and western Poland.

Ranunculus rionii. Kelbra, Saxony-Anhalt, Germany. Photo KvdW.

R. rionii occurs from southeastern Europe to Western and Central Asia with disjunct records from eastern China and South Africa.

Characteristics and similar species: *R. rionii* is characterised by its annual life cycle, the large number of small achenes, thin pedicels and elongating receptacle.

It may be confused with **212 *R. trichophyllus***, which differs in its perennial life cycle, larger achenes, thicker pedicels, and the receptacle remaining subglobose in fruit.

Hybrids: Hybridisation with *R. circinatus* and *R. trichophyllus* has occasionally been reported.

Note: *R. rionii* is diploid (2*n* = 16) throughout its range.

Ranunculus rionii. Kelbra, Saxony-Anhalt, Germany. Photo KvdW.

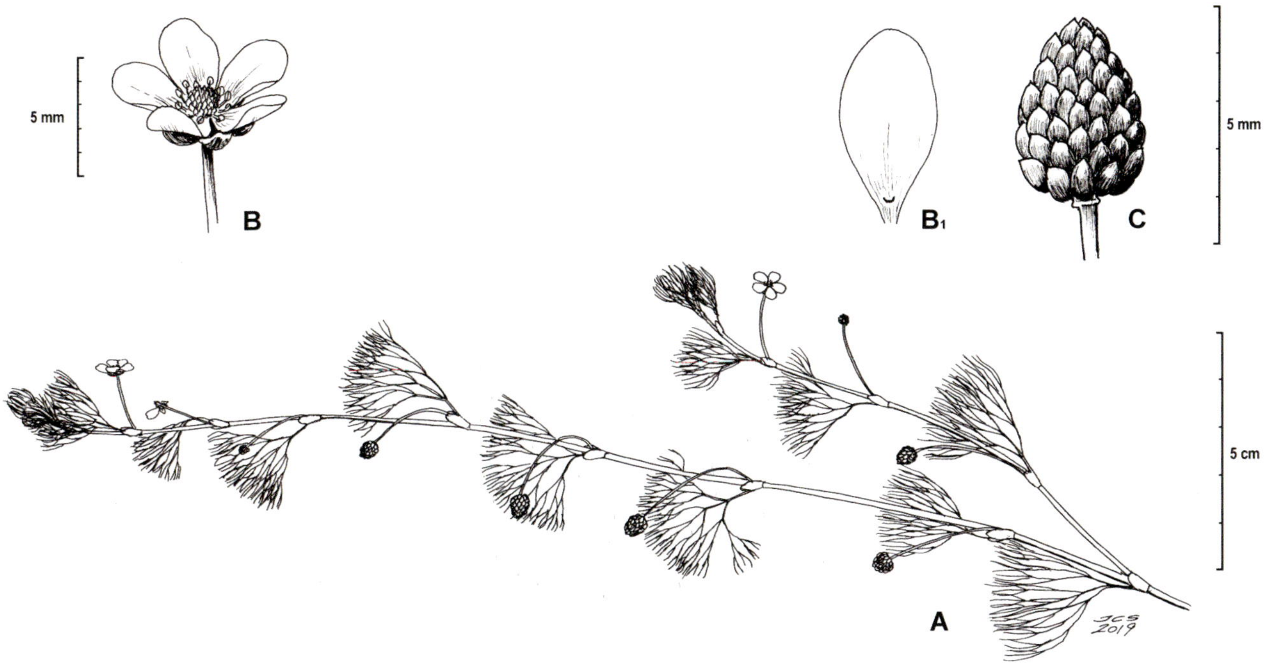

Ranunculus rionii
A habit, **B** flower, **B1** petal, **C** fruiting head.

Ranunculaceae

218 *Ranunculus sceleratus* L.

DK: Tigger-Ranunkel **N:** Tiggersoleie **S:** Tiggarranunkel **FIN:** Konnanleinikki **IS:** -
GB: Celery-leaved Buttercup **NL:** Blaartrekkende boterbloem **F:** Renoncule scélérate
D: Gift-Hahnenfuß **CZ:** Pryskyřník lítý **PL:** Jaskier jadowity
EST: Mürktulikas **LV:** Ļaunā gundega **LT:** Nuodingasis vėdrynas **Ru:** Лютик ядовитый

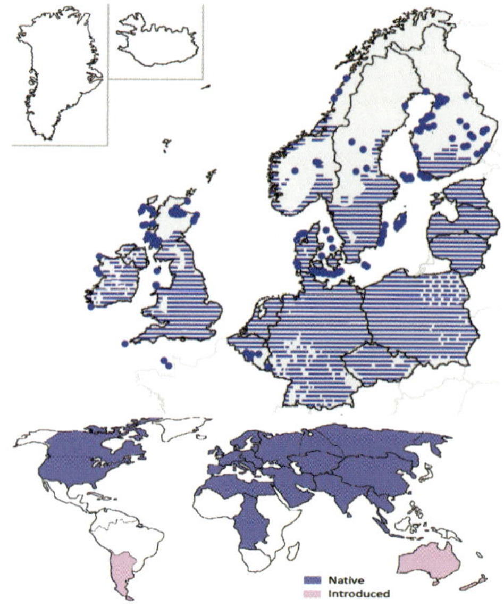

Ranunculus sceleratus in shallow water with *Bolboschoenus maritimus*. Vesthimmerland, Denmark. Photo JCS.

Short-lived summer or winter annual, without rhizomes or underground stolons.

Stem: 10–50 cm long, up to 12(–20) mm thick, upright, terete, furrowed, hollow, yellow to yellowish green, the upper part richly branched, glabrous or with scattered hairs.

Leaves: Green to yellowish green. Basal leaves numerous, in a rosette and often floating in shallow water. Petioles 2–25 cm long, amplexicaul at base. Lamina 2–6(–10) cm wide, with an almost pentagonal appearance in outline, deeply dissected into 3 primary lobes, both primary and secondary lobes obtuse. Upper leaves alternate with gradually shorter petioles, deeper dissection and even narrower, lanceolate lobes towards the stem apex. The uppermost leaves are usually sessile with entire, lanceolate lamina.

Inflorescence: Richly branched, with up to 50 flowers, these often opposite to a leaf.

Flowers: 5–10 mm in diameter, pedicels 2–6 cm long, furrowed, glabrous or hairy. Sepals 5, 2–4 mm long, ovate, yellow green, hairy on the back, reflexed and caducous. Petals 5, 3–4 mm long, only slightly longer than the sepals, ovate to spathulate, yellow, at the base pale yellow with a circular nectar pit. Stamens 12 to numerous with yellow anthers. Receptacle ovate, glabrous or with scattered hairs.

Fruits: Carpels 70–100(–200), forming an oblong, 5–6 mm wide fruiting head. Ripe achenes (nuts) 0.8–1.2 mm long, ovate, slightly compressed, smooth with weak, transverse wrinkles, the dorsal side with an inconspicuous wing. Beak 0.1–0.2 mm long, straight.

Flowering: June-September.

Biology: Poisonous. Pollinated by small insects attracted by the nectaries. Self-pollination may occur as a result of very small insects moving around between individual flowers. The fruits are dispersed by birds and animals or by water, facilitated by aerenchymatous cells at the edge of the nuts.

Habitats: Mainly in alkaline, but also slightly acidic water. On open, wet to moist or temporarily dry ground in re-

The rosette leaves of *R. sceleratus* are very different from the stem leaves and often float in shallow water. Plants germinating in spring flower and die the same year (summer annual), while those germinating in autumn overwinter as leaf rosettes and flower the following year (winter annual). Inglestone Common, Gloucestershire, Britain. Photo RVL.

Ranunculus sceleratus. Vigorous plant growing on the moist bed of a ditch. Lille Vildmose, Denmark. Photo JCS.

Fruiting *R. sceleratus.* Langør, Samsø, Denmark. Photo JCS.

ceding waters, moderately nutrient-poor to nutrient-rich soil in meadows and on the margins of ponds and lakes, ditches, canals and streams as well as coastal areas where freshwater discharges into the sea.

Distribution within the region:
Native. Common throughout the region except for northern Scandinavia, Iceland and Greenland.

Characteristics and similar species:
When flowering *R. sceleratus* is unlikely to be confused with other species.

Sterile aquatic plants with floating leaves or plants on wet mud may be mistaken for species of ***Ranunculus*** sect. ***Batrachium*** or **345 *Hydrocotyle ranunculoides***, which however have creeping stems, rooting at the nodes.

The stem of *R. sceleratus* is erect and the leaves arise as a basal rosette.

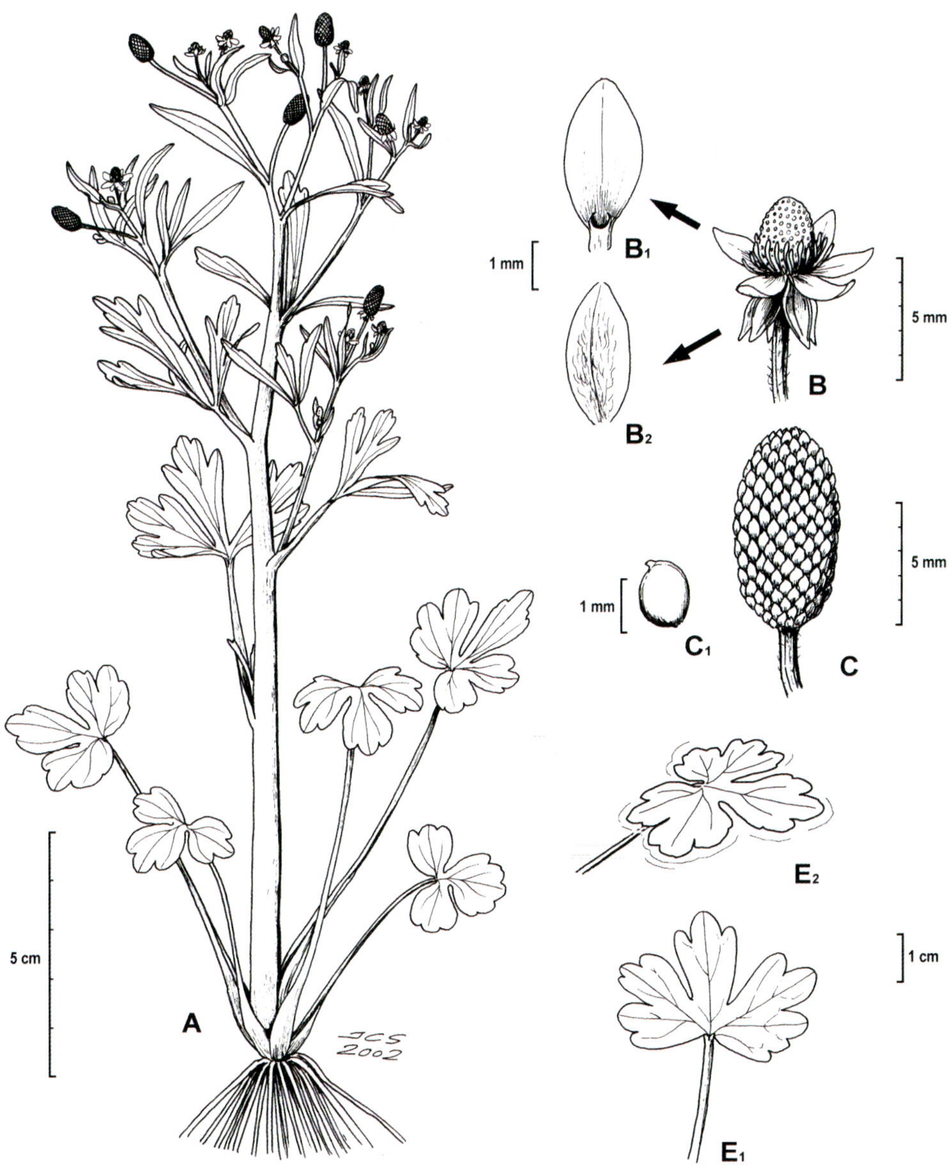

Ranunculus sceleratus
A habit, **B** flower, **B1** petal with nectar pit, **B2** sepal abaxial side, **C** fruiting head, **C1** achene, **E1, E2** basal leaf.

Ranunculaceae

219 *Ranunculus lingua* L.

DK: Langbladet Ranunkel **N:** Kjempesoleie **S:** Sjöranunkel **FIN:** Jokileinikki **IS:** -
GB: Greater Spearwort **NL:** Grote boterbloem **F:** Renoncule langue
D: Zungen-Hahnenfuß **CZ:** Pryskyřník **PL:** Jaskier wielki
EST: Suur tulikas **LV:** Garlapu gundega **LT:** Pelkinis vėdrynas **Ru:** Лютик длиннолистный

Ranunculus lingua. Stavtrup, Aarhus, Denmark. Photo BM.

Perennial helophyte with prostrate rhizome and hollow underground stolons up to 1 metre long.

Stem: 50–150 cm long, up to 18 mm in diameter, upright, terete, hollow, grey green, sometimes with a brownish or violet tinge, unbranched or with a few branches in the upper part, aerial parts hairy, submerged parts usually glabrous.

Leaves: With two very different types of leaves. Basal leaves normally develop under water at the end of one growing season or the beginning of the next. They are long petiolate with laminas ranging from broadly obovate with a cordate base to ovate oblong with a rounded base; in deep water they may be almost linear, up to 20 cm long and 8 cm wide, entire, yellowish green with a brownish or violet tinge. Aerial leaves 15–25 cm long, petioles of lower leaves short with semi-amplexicaul base, the upper leaves sessile, all narrowly lanceolate, gradually tapering towards the petioles, entire or with a minutely denticulate margin, lamina with short appressed hairs on both sides, greyish- to dark green.

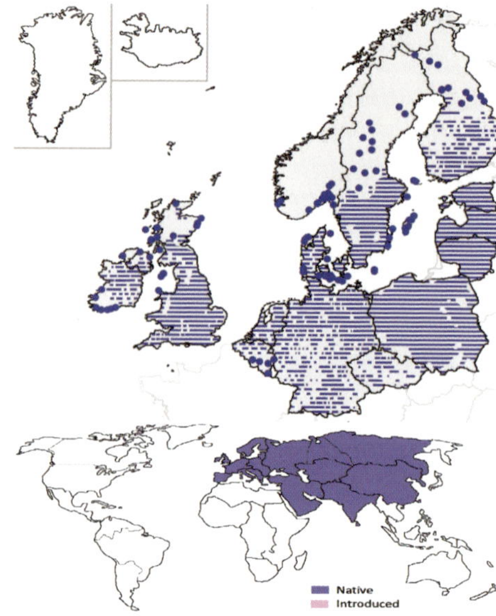

Inflorescence: With a few, short branches at the stem apex, bearing up to 8 flowers.

Flowers: 2.5–4.5 cm in diameter, pedicels 4–15 cm long, terete and hairy. Sepals 5, 5–10 mm long, ovate, yellow green, hairy, with a yellow, membranous margin. Petals 5, 10–20 mm long, oval to obovate, shiny yellow, with a nectar pit at the base. Stamens numerous with yellow anthers. Receptacle ovate, glabrous.

Fruits: Carpels numerous, forming a spherical fruiting head, 1 cm in diameter. Ripe achenes (nuts) 2.2–2.8 mm long, obovate, slightly compressed, winged, glabrous or sparsely hairy. Beak 0.5–0.8 mm long and slightly curved.

Flowering: June-August.

Biology: Vegetative propagation is prolific, partly by division of the rhizomes and underground runners, partly by short propagules developed from the

Ranunculus lingua. The large, yellow flowers attract insects. Aarhus, Denmark. Photo JCS.

Ranunculus lingua. Submerged leaves in spring. Pond near Aarhus, Denmark. Photo BM.

Ranunculus lingua with purplish stolons in the water. River Nørreå, Randers, Denmark. Photo BM.

lower part of the stem in the autumn, which detach and are dispersed by water. Self-pollinated or pollinated by a wide variety of insects which are attracted to the large yellow flowers with nectar pits. The fruits are dispersed by water despite poor buoyance. Submerged leaves develop in autumn, making the plants wintergreen in many waters.

Habitats: Open or densely vegetated, wet ground or standing water on nutrient-rich soil, in swamps and meadows and more commonly in ponds and on the margins of lakes, ditches, canals and streams. Mainly in alkaline habitats, but also in slightly acidic waters.

Distribution within the region: Native. Rather common throughout most of the region except for northern Scandinavia, Iceland and Greenland.

Characteristics and similar species: When flowering *R. lingua* may be mistaken for large-flowered forms of *R. flammula*, from which it can be distinguished by the smooth, terete pedicels. Non-flowering plants with submerged leaves with heart-shaped or cordate base are similarly unique and separate *R. lingua* from all other water plants.

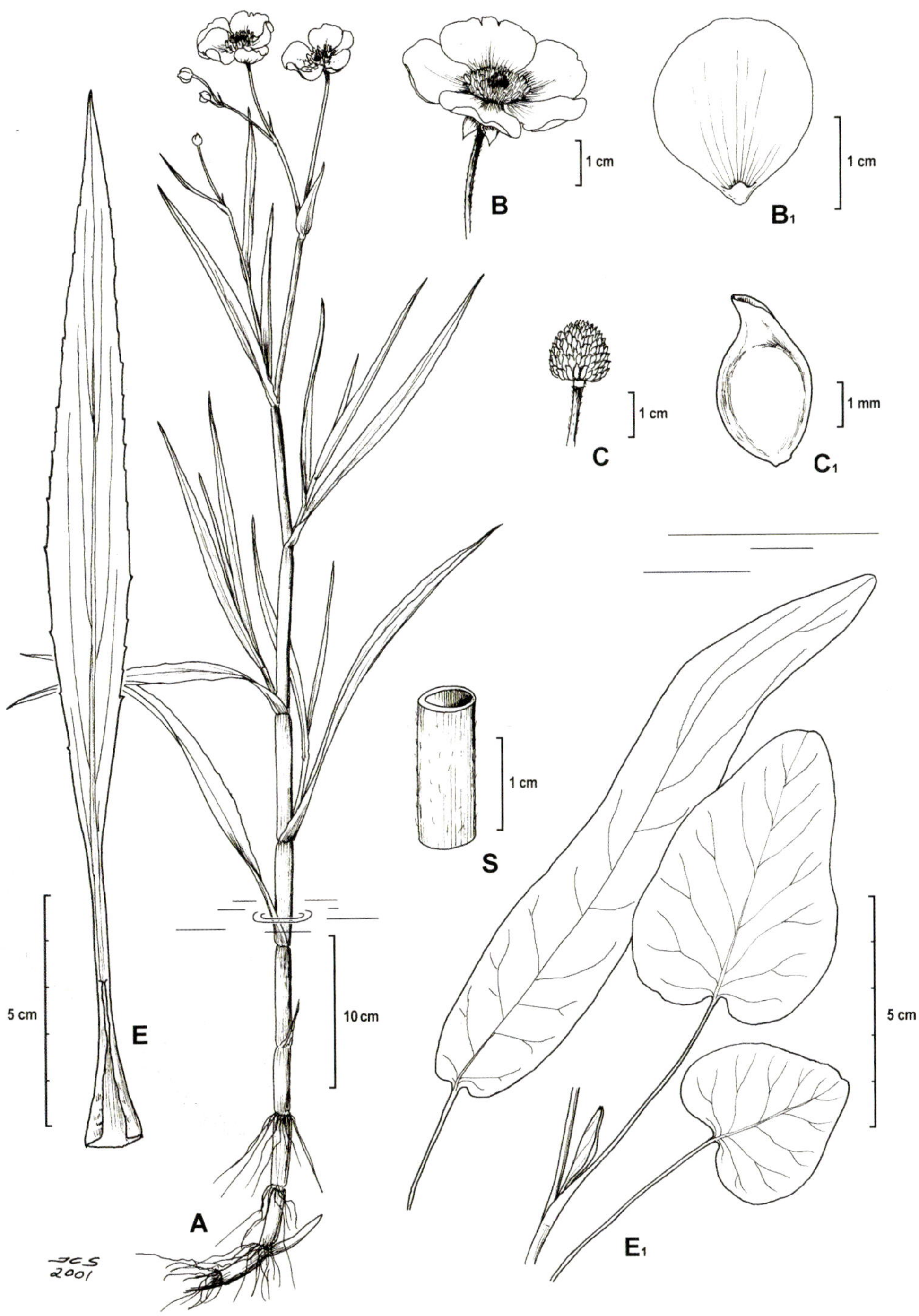

Ranunculus lingua
A habit, **B** flower, **B1** petal, **C** fruiting head, **C1** achene, **E** stem leaf, **E1** submerged leaves, **S** part of stem.

Ranunculaceae

220 *Ranunculus flammula* L.

DK: Kær-Ranunkel **N:** Grøftesoleie **S:** Ältranunkel **FIN:** Ojaleinikki **IS:** -
GB: Lesser Spearwort **NL:** Egelboterbloem **F:** Petite Douve
D: Brennender Hahnenfuß **CZ:** Pryskyřník plamének **PL:** Jaskier płomiennik
EST: Sootulikas **LV:** Rāvas gundega **LT:** Dedervinis vėdrynas **Ru:** Лютик жгучий

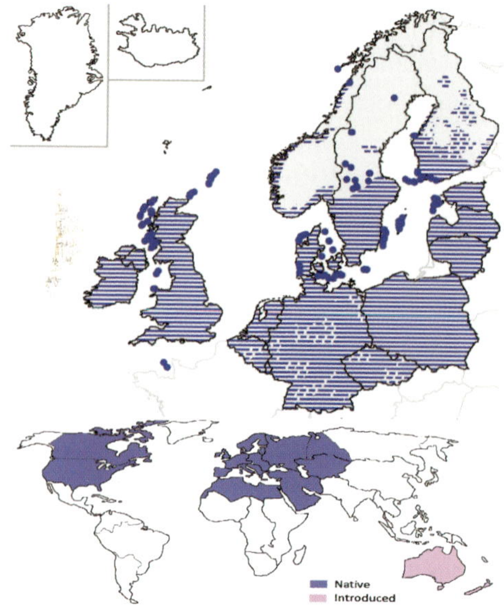

Ranunculus flammula. Pool. Lüsekamp, Germany. Photo KvdW.

Perennial helophyte or less commonly hydrophyte. Without rhizomes or underground stolons.

Stem: 10–40 cm long, 0.8–3.0 mm in diameter, terete, smooth or minutely furrowed, prostrate, ascending or upright, more or less branched, with short appressed hairs, green to grey green. Internodes straight or slightly curved. Commonly rooting from the lower stem nodes (except subsp. *scotticus*).

Leaves: Basal leaves petiolate with lamina up to 6 cm long, narrowly lanceolate to broadly ovate, bluntly glandular dentate on the margins, abruptly cordate to cuneate at base. Upper leaves 3–7 cm long, narrowly lanceolate, shortly petiolate to sessile, or occasionally amplexicaul. Lamina with bluntly, glandular-dentate margins, green to dark brown or blackish with pinnate to longitudinal venation and scattered appressed, short hairs.

Inflorescence: 1–25 flowers in loose clusters at the tip of stems and branches.

Flowers: 5–25 mm in diameter, pedicels 2–6 cm long, furrowed, with short appressed hairs. The 5 sepals 2–5 mm long, ovate, yellow green, with few appressed, short hairs, falling early. Petals 5–10, 4–10 mm long, approximately twice as long as the sepals, obovate, bright yellow and shiny. Nectar pits tiny, 1 per petal. Receptacle ovate and smooth.

Fruits: Carpels up to 20. Ripe achenes 1.6–2.3 mm long, broadly oval to obovate, slightly compressed, smooth, with or without a 0.1 mm wide wing. Beak 0.2–0.3 mm long, 1/6 to 1/8 of the total fruit length.

Flowering: June-September.

Biology: Pollination primarily by small insects, but self-pollination may also occur. Typically abundantly fertile with fruits dispersed by birds, mammals and possibly water. *R. flammula* often occurs as a terrestrial, although it often germinates and initially grows submerged with floating leaves on long, fine petioles. It gradually develops into a terrestrial form with flowers and aerial leaves as water levels drop. Less often, it will occur as permanently submerged, non-flowering plants with long-petiolate submerged and floating leaves.

Habitats: Open, wet to moist, temporarily dry ground, nutrient-poor to nutrient-rich soil in meadows, marshes, bogs, ponds, lakes, ditches and smaller streams as well as the inner and least saline parts of salt marshes.

Distribution within the region:
Native. Common throughout most of the region except for northern Scandinavia, Iceland and Greenland.

Characteristics and similar species:
Ranunculus flammula is highly variable in response to the hydrology and nutrient status of its environment. It can usually be recognised by the cordate to

Floating leaves reaching the water surface from a submerged stand of *R. flammula* growing at a depth of 10-30 cm in Lake Navn Sø, Denmark. Photo JCS.

Ranunculus flammula subsp. *scoticus*, Rannoch Station, Scotland. Photo RVL.

Ranunculus flammula var. *major*. Photo RVL.

lanceolate, bluntly glandular-dentate leaves with clusters of bright yellow flowers at the tips of stems and branches.

221 *Ranunculus reptans* differs in having all leaves extremely narrow with little distinction between the petiole and lamina and in the characteristic arching stem. The beak of the fruit of *R. reptans* is longer, equivalent to 1/4 of the total length of the fruit.

The very similar **223 *R. ophioglossifolius*** can only be reliably distinguished by the papillose fruits.

Large forms of *R. flammula*, especially var. *major,* may be confused with **219 *R. lingua***, which differs in the smooth pedicels and the characteristic sagittate submerged leaves.

Members of the Alismataceae (e.g., **33-35 *Alisma***, **37-38 *Baldellia*** and **41 *Luronium***) may have similar leaves, but differ in their parallel leaf venation.

Three subspecies are recognised: (1) Subsp. ***flammula*** is divided into two varieties, var. ***flammula*** and var. ***major*** Schulth.; the latter is characterised by extremely large flowers (> 18 mm in diameter) and is often as tall as *R. lingua*. The infraspecific taxa within *R. flammula* are not very well circumscribed, and there is a need for more work to clarify their taxonomic status and distribution. (2) Subsp. ***minimus*** (A.Benn.) Padmore has somewhat fleshy leaves on short, rigid petioles and broadly rounded or almost circular laminas. (3) Subsp. ***scotticus*** (E.S.Marshall) A.R.Clapham has very narrow basal leaves, showing little distinction between petiole and lamina. The stem leaves are always more or less typical for the species.

Hybrids: *R. ×levenensis* Druce ex Gornall *(flammula × reptans)*. Leaf width intermediate between those of the parent species. The most distinct feature of the hybrid is the length of the beak, which is 1/5 of the total length of the fruit. It is more creeping than typical *R. flammula* with more narrowly rhomboid leaves. The hybrid is fertile and because of back crossing with the parent species is often difficult to determine with certainty.

Within the region known from Ireland, Britain, Denmark, Norway, Sweden, Finland and Germany.

Ranunculus flammula
A habit, **A1** slender, prostrate form, **A2** plant with floating leaves, **B** flower, **C** fruiting head, **C1** achene, **E1-E2** leaves.

Ranunculaceae

221 *Ranunculus reptans* L.

DK: Krybende Ranunkel **N:** Evjesoleie **S:** Strandranunkel **FIN:** Rantaleinikki **IS:** Flagasóley
GB: Creeping Spearwort **NL:** - **F:** Renoncule radicante
D: Ufer-Hahnenfuß **CZ:** - **PL:** Jaskier leżący
EST: Kaartulikas **LV:** Plašā gundega **LT:** Rėpliojantis vėdrynas **Ru:** Лютик простёртый

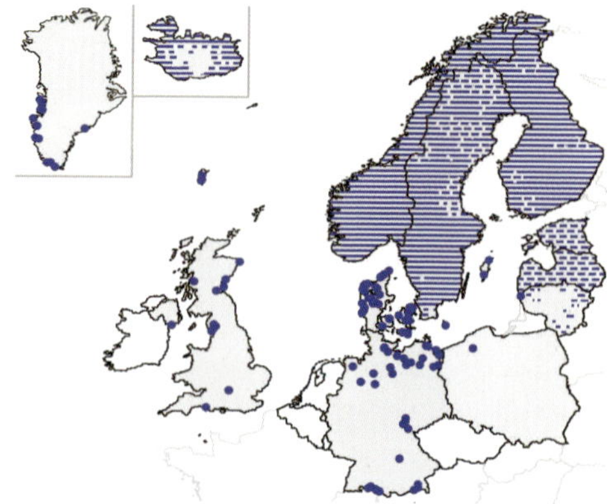

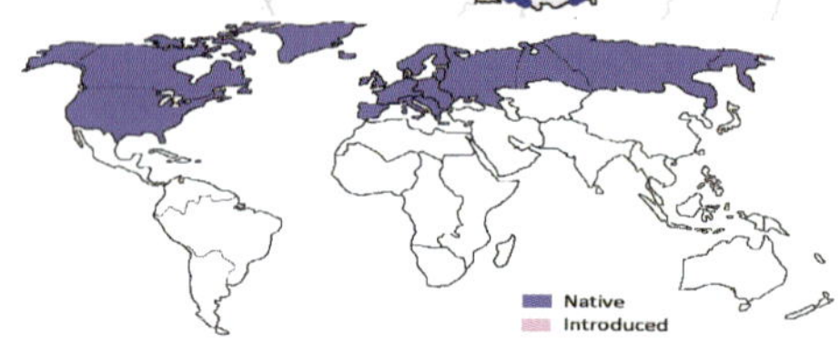

Ranunculus reptans forming a dense network of creeping and rooting stems on the dry shore of the recently restored Lake Filsø in SW-Jutland, Denmark. Photo JCS.

Perennial helophyte or hydrophyte. Without rhizomes or underground stolons.

Stem: 10–40 cm long, 0.5–1(–1.5) mm in diameter, terete, prostrate to slightly ascending, more or less branched, glabrous or with short, appressed hairs, green to pale green. Internodes distinctly arching. Usually rooting at almost all stem nodes.

Leaves: Terrestrial plants with 3–5(–10) leaves at the base of the stem and 1–3(–5) leaves from each of the nodes. Leaves 1–5 cm long, terete to subulate, 0.3–1.0 mm in diameter or narrowly spathulate, 1–4 mm wide, entire, often with a concave lamina. Submerged plants wintergreen with subulate, dark green leaves from each node. Leaves 1.2–5.0 cm long and 0.3–1.0 mm in diameter.

Inflorescence: Flowers solitary at the tips of stems and branches.

Flowers: 5–10 mm in diameter, pedicels 1–5 cm long with short appressed hairs. The 5 sepals 1.5–3.0 mm long, ovate, yellow green, glabrous, caducous. Petals 5–8, 2.5–5.0 mm long, approximately twice as long as the sepals, narrowly obovate, bright yellow and shiny, with a tiny nectar pit at the base. Receptacle more or less spherical and smooth.

Fruits: Carpels 10–20. Ripe achenes 1.3–1.8 mm long, broadly oval to ovate, slightly compressed, with or without an indistinct wing, smooth and with a 0.3–0.4 mm long, curved beak at the tip, equivalent to 1/4 of the total length of the fruit.

Flowering: June-September.

Biology: Pollination is primarily by small insects, but self-pollination is possible by raindrops hitting the flowers. Fruiting relatively abundant, the fruits being dispersed by birds and other animals, to some extent also by water. Permanently submerged plants are common and always sterile. Vegetative dispersal also occurs by detached stem fragments.

Habitats: In moderately acidic to moderately alkaline water. On open, wet to moist, temporarily dry ground, nutrient-poor to moderately nutrient-rich mineral soils (sand and fine gravel) in or by lakes. Favoured by fluctuating

Ranunculus reptans. The narrow petals have nectaries and attract small insects for pollination, but the species seems to some extent to be self-pollinating. Lake Navn Sø, Denmark. Photo JCS.

Ranunculus reptans. The aquatic form is wintergreen and sterile, normally growing to a depth of up to 100 cm with arching internodes, rooting nodes and subulate leaves. Lilla Håsjön near Hyssna, Sweden. Photo BM.

Ranunculus reptans. Fruiting head. Achene with curved beak. Umeälven, Degerfors, Sweden. Photo JCS.

water levels. Submerged plants can grow to approximately 100 cm depth.

Distribution within the region:
Native. Common in Fennoscandia, Iceland and southwestern Greenland. Scattered or rare in Britain, Denmark, Germany, Poland and the Baltic states.

Characteristics and similar species:
Terrestrial plants of *R. reptans* are characterised by the prostrate stem with arching internodes, rooting at all nodes, as well as the solitary flowers, achenes with a curved beak equivalent to 1/4 of the total length of the fruit and the very narrow aerial leaves.

220 *Ranunculus flammula* differs in the stem usually rooting only at the lower nodes and never arching between rooted nodes, achenes 1.6–2.3 mm long with the beak 1/6-1/8 of the total length of the achene.

Submerged plants with subulate leaves may be mistaken for **161 *Eleocharis parvula***, which however has leaves in rosettes connected by thin underground stolons or with terminal, onion-shaped winter buds.

Hybrids: *Ranunculus reptans* forms a fertile hybrid with *R. flammula* - see that species.

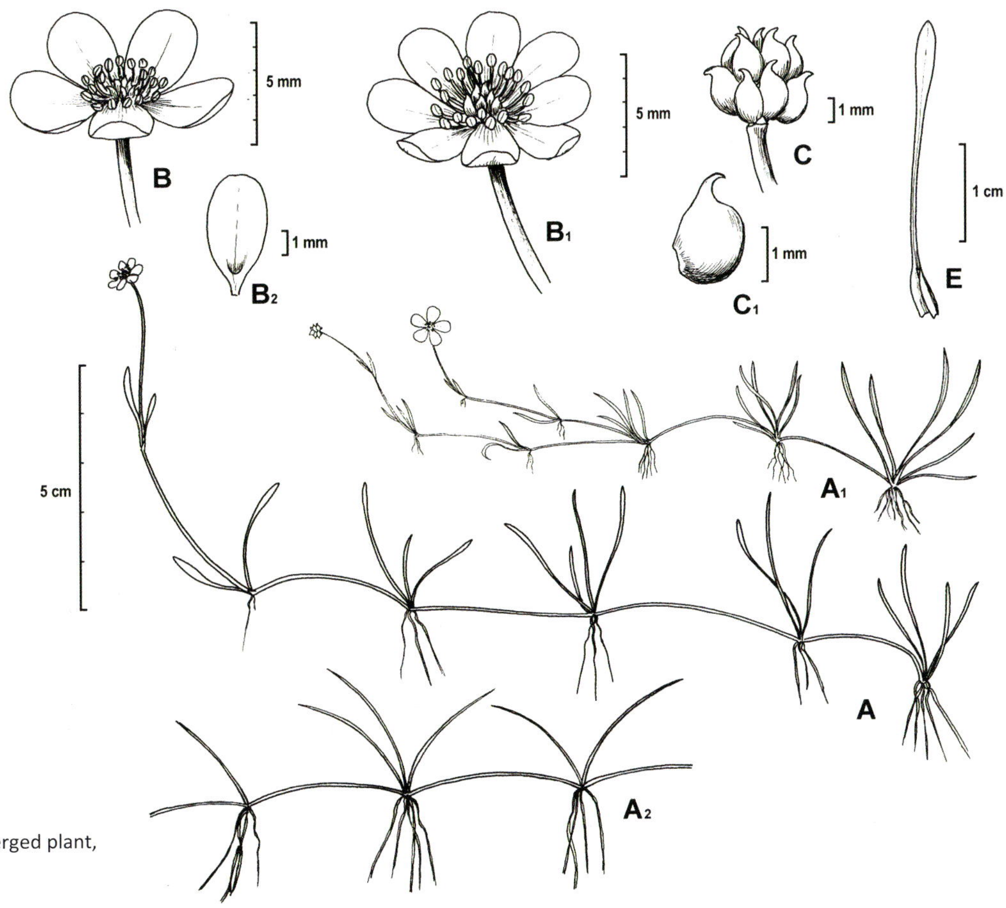

Ranunculus reptans
A-A1 habit terrestrial plant, **A2** habit submerged plant,
B-B1 flower, **B2** petal, **C** fruiting head,
C1 achene, **E** spathulate leaf.

Ranunculaceae

222 *Ranunculus hyperboreus* Rottb. subsp. *hyperboreus*

DK: Trefliget Ranunkel **N:** Setersoleie **S:** Jordranunkel **FIN:** Pohjanleinikki **IS:** Trefjasóley
GB: - **NL:** - **F:** -
D: - **CZ:** - **PL:** -
EST: - **LV:** - **LT:** - **Ru:** Лютик гиперборейский

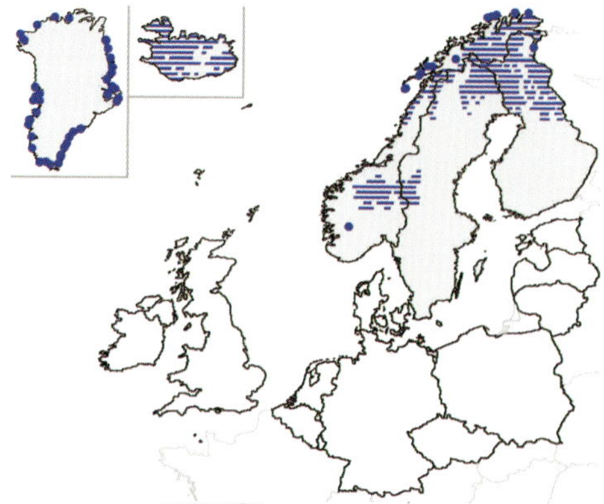

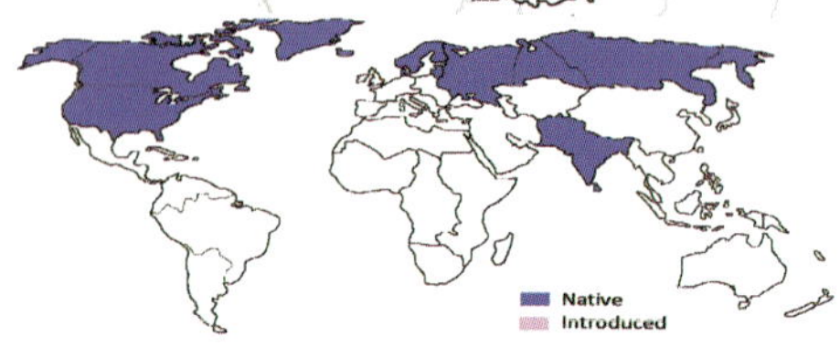

Ranunculus hyperboreus subsp. *hyperboreus* in shallow water. Abisko Östra, Torne Lappmark, Sweden. Photo Jan-Thomas Johansson.

Perennial, terrestrial or aquatic plant with a short vertical rhizome, without stolons.

Stem: 2–15 cm long in terrestrial plants, up to 40 cm in aquatic plants, slender, prostrate with ascending shoot tip, sometimes floating, terete, glabrous or with scattered white hairs, grey green, branches few to many, rooting at the nodes.

Leaves: All leaves similar in both terrestrial and aquatic plants, petioles 0.5–6.5 cm, longest in aquatic plants, with a semi-amplexicaul, often auriculate sheath. Lamina 0.3–1.6 cm long and 0.8–2.1 cm wide, green, 3(–5)-lobed, divided to about the middle of the lamina, lobes obtuse with entire margin, the lateral lobes often pointing forward, sometimes divided in two.

Inflorescence: Terrestrial plants and plants growing in very shallow water with 1–2(–4) flowers on each stem. Aquatic plants in deeper water normally sterile.

Flower: 3–7 mm in diameter, flat, pedicels 0.5–4 cm long, up to 7 cm after anthesis, curved, terete, furrowed and glabrous. Sepals usually 3, elliptic, often of unequal size, commonly deflexed, 2.3–3.4 mm long and 1.1–1.9 mm wide, yellowish green on the outer face, often with a violet tinge, caducous. Petals 2–3(–5), obovate, 1.7–3.2 mm long and 1.1–1.9 mm wide, as long as or shorter than the sepals, glabrous, yellow, at the base with nectar pits. Receptacles 2–3 mm, glabrous. Stamens few.

Fruits: Achenes numerous, 1.1–1.4 mm long and 0.8–1.1 mm wide, smooth and glabrous, not or only slightly compressed, beak 0.1 mm long, hook shaped. Fruiting head 2–5 mm long, semi-globose.

Flowering: July-September.

Biology: Primarily self-pollinated. Seed production abundant, dispersal by animals. Vegetative propagation as detached stem fragments. Terrestrial plants are normally solitary or a few together with relatively short stems. Aquatic plants often occur in large stands with long stems.

Habitats: Predominantly on wet, moist or temporarily wet grounds with clayey or sandy soil in or at the margins of shallow ponds, springs, small streams and lakes, primarily in mountainous areas. Also on disturbed ground in farmland, trails and ditches, etc. Indifferent to alkalinity.

Ranunculus hyperboreus subsp. *hyperboreus*. Store Saltsö, Kangerlussuaq, Greenland. Photo Jan-Thomas Johansson.

Distribution within the region:
Native. Arctic areas of the Northern Hemisphere. Scandinavia, Svalbard, Iceland, Greenland.

Characteristics and similar species:
R. hyperboreus is unlikely to be confused with other water plants, although the leaves of the aquatic form may superficially resemble the lobed floating leaves of species of *Ranunculus* section *Batrachium* – however, leaves of those species are 5-lobed and often accompanied by capillary submerged leaves.

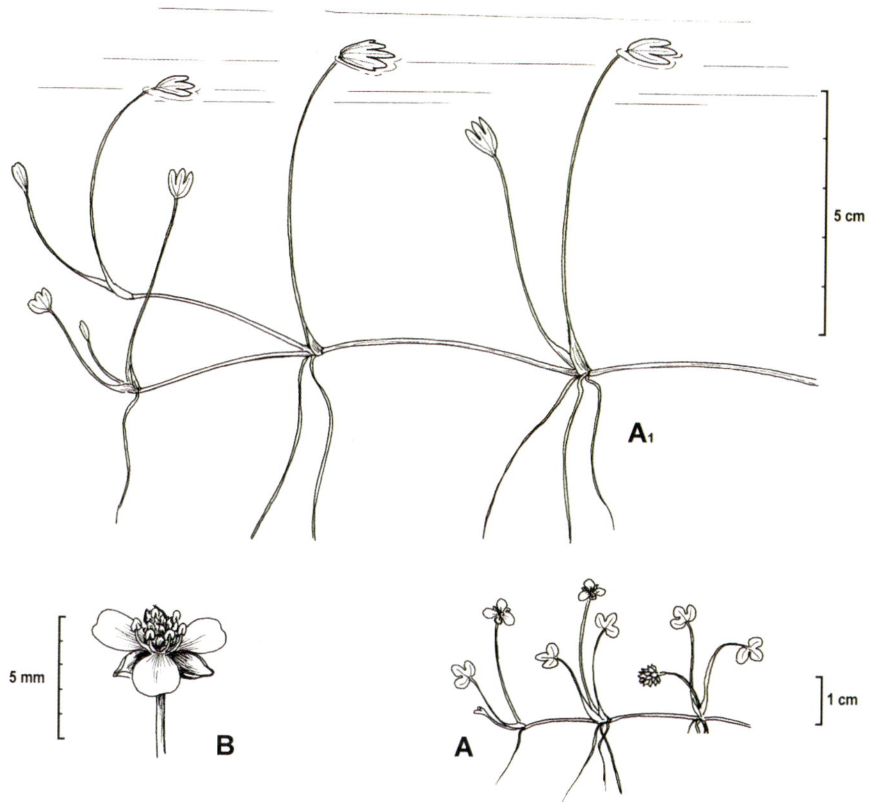

Ranunculus hyperboreus subsp. ***hyperboreus***
A habit terrestrial plant, **A1** habit submerged plant with floating leaves, **B** flower.

Ranunculaceae

222a *Ranunculus hyperboreus* Rottb. subsp. *arnellii* Scheutz

DK: - **N:** Tundrasoleie **S:** Mossranunkel **FIN:** - **IS:** -
GB: - **NL:** - **F:** -
D: - **CZ:** - **PL:** -
EST: - **LV:** - **LT:** - **Ru:** -

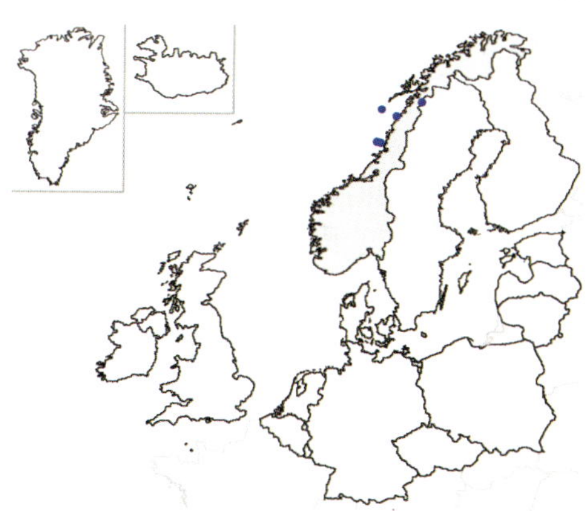

Ranunculus hyperboreus subsp. *arnellii*. Tverrdalen - Reindalen, Svalbard, Norway.
Photo Jan-Thomas Johansson.

Distribution within the region:
Native. Arctic areas of the Northern Hemisphere. Norway, Svalbard.

Differs from subsp. ***hyperboreus*** primarily by more slender and more richly branched stems, dark green leaves with deeply lobed laminas, the lobes often divided one more time, thus giving the lobes a more slender, narrow and pointed appearance. In addition the flowers are smaller with petals distinctly shorter than the sepals.

Ranunculaceae

223 *Ranunculus ophioglossifolius* Vill.

DK: Gotlands-Ranunkel **N:** Gotlandssoleie **S:** Gotlandsranunkel **FIN:** Gotlanninleinikki **IS:** -
GB: Adder's Tongue Spearwort **NL:** Addertongboterbloem **F:** Renoncule à feuilles d'ophioglosse
D: Natterzungblättriger Hahnenfuß **CZ:** - **PL:** -
EST: - **LV:** - **LT:** - **Ru:** Лютик ужовниколистный

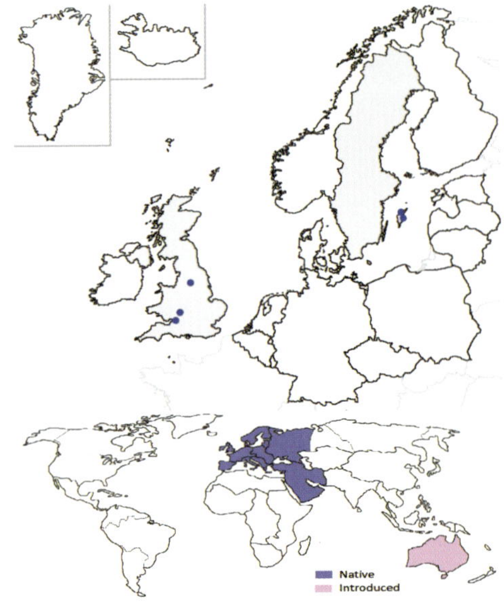

Ranunculus ophioglossifolius on dried out bottom by a small stream. Gotland, Sweden. Photo JCS.

Winter annual. Without rhizome or underground stolons.

Stem: 2–15(–40) cm long, up to 7 mm in diameter, hollow, terete to sulcate, erect to ascending, more or less branched, especially in the upper part, with short appressed hairs, green to grey green, sometimes with a brownish or purplish tinge. In water sometimes rooting from the lower stem nodes.

Leaves: Pale greenish yellow, entire or with a few shallow, rounded teeth and obtuse apex, with a sheath at the base - this c. 1 cm long, with hairy membranous margins. Lowest leaves with 4–8(–15) cm long petioles and lamina 7–20(–35) x 6–22 mm, broadly oval to suborbicular with more or less truncate to cordate base, floating to the surface when submerged. Stem leaves nearly sessile or with a flat petiole 1–2 cm long and lamina broadly to narrowly elliptic or lanceolate with a few shallow teeth.

Inflorescence: Flowers solitary or in few-flowered cymes from leaf axils.

Flowers: 4–10(–13) mm in diameter, pedicels 2–5 cm long, slightly sulcate, with short appressed hairs. Sepals 5, spreading, 2–3 mm long, ovate, greenish, with few appressed, short hairs. Petals 5, 2.5–5.5 x 2.2–3.0 mm, approximately twice as long as the sepals, obovate, pale to bright yellow and shiny. Nectar pits tiny, 1 per petal. Receptacle globose and glabrous.

Fruits: Carpels up to 20. Ripe achenes 0.9–1.3 x 0.7–1.2 mm, elliptical to obovoid, compressed, minutely tuberculate, brown when ripe. Wings green, broadest towards the apex. Beak 0.3–0.4 mm long.

Flowering: May-June (-September).

Biology: Pollination by a variety of insects. The fruits are dispersed by water, birds and mammals.

Plants from Gotland, Sweden are generally smaller than those from other parts of Europe, but plant size is strongly affected by growing conditions. The seeds can lie dormant for at least 30 years, but are sensitive to desiccation.

Habitats: *Ranunculus ophioglossifolius* is a poor competitor and requires open, winter-inundated, summer-dry, nutrient-poor, alkaline, muddy substrate on the margins of field ponds and marshy areas. Swedish plants in Gotland grow in winter-flooded depressions and brooklets.

Submerged *R. ophioglossifolius*. Seasonal pool, Lower Woods/Inglestone Common, Gloucestershire, Britain. Photo RVL.

Distribution within the region:
Native. Only present at three locations in England and two in Gotland, Sweden.

Characteristics and similar species:
Ranunculus ophioglossifolius can be confused with **220 *Ranunculus flammula***, from which it can be distinguished by the minutely tuberculate achenes, flowers which are usually smaller and a slightly paler yellow.

Ranunculus ophioglossifolius. Left: Flooded arable, Puebla de Don Rodrigo, Extremadura, Spain. Right: Seasonal pool, Lower Woods/Inglestone Common, Gloucestershire, Britain. Photos RVL.

Fruiting head with minutely tuberculate achenes and leaf from the lower part of the stem. Flooded arable, Puebla de Don Rodrigo, Extremadura, Spain. Photos RVL.

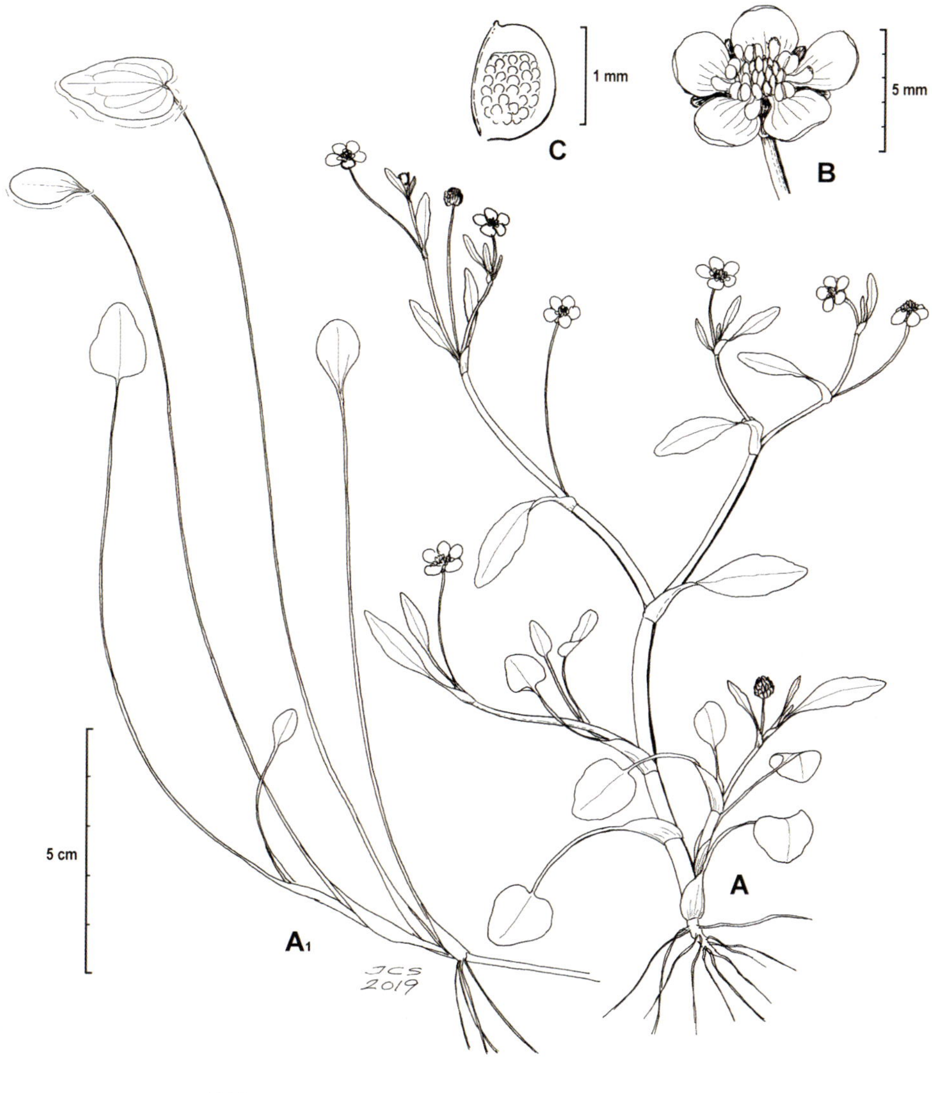

Ranunculus ophioglossifolius
A habit emergent plant, **A1** habit submerged plant, **B** flower, **C** achene.

Crassulaceae

224 *Crassula aquatica* (L.) Schönland

Syn.: *Tillaea aquatica* L.

DK: Korsarve **N:** Firling **S:** Fyrling **FIN:** Vesipaunikko **IS:** Vatnsögn
GB: Pigmyweed **NL:** - **F:** -
D: Wasser-Dickblatt **CZ:** Masnice vodní **PL:** Uwroć wodna
EST: Väike vesikas **LV:** Ūdeņu biezlapīte **LT:** Vandeninis storlapis **Ru:** Тиллея водная

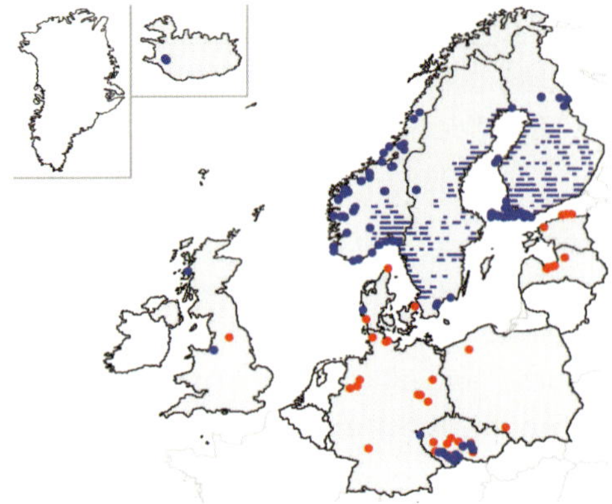

Locations from which the species is known to have disappeared are marked in red.

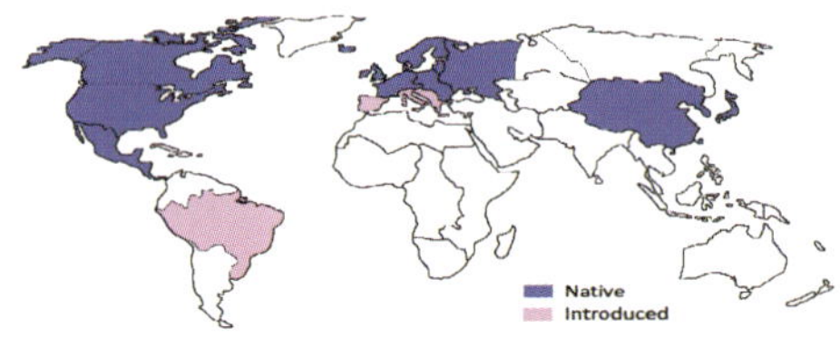

Seen from above the leaves of *C. aquatica* look like small crosses. Sandviken, Sweden. Photo KvdW.

Annual, amphiphyte with white roots from lower stem nodes, without rhizome.

Stem: Terete, glabrous, more or less branched, up to 10 cm long on submerged plants, erect, green, 0.8–3.0 cm long on emerged plants, prostrate to ascending, green, often reddish tinged.

Leaves: 3–6 mm long, 1.0–1.5 mm wide, 0.5–1.0 mm thick, succulent, linear lanceolate, acute, opposite, fused at the base. Each pair of leaves oriented at a right angle to the neighbouring pairs, giving the shoots a cross-like appearance when seen from above.

Inflorescence: Flowers solitary in leaf axils, only one flower from each pair of leaves. Submerged plants mainly flowering only in the upper part.

Flowers: 4-merous, rarely 3- or 5-merous, bisexual, sessile. Sepals c. 0.5 mm long, more or less fused at the base, obtuse or with rounded apex, green with a membranous edge. Petals 1.0–1.5 mm long, twice as long as sepals, upright, white, sometimes with a pinkish tinge, lanceolate, obtuse. Stamens 4.

Fruits: 4 follicles, 1.2–1.5 mm long, greenish, each containing 8–14 seeds. Seeds 0.3–0.6 mm long, oblong-oval-elliptic, dark brown, with fine longitudinal wrinkles.

Flowering: August-September.

Biology: Both emerged and submerged plants flower freely and are probably self-pollinated. Seed dispersal by water and wind as well as by animals. The seeds germinate on seasonally exposed sediments on the margins of water bodies or in seasonal pools, rarely in the water. Fully developed plants are mainly amphibious, but may become submerged by rising water levels. It normally occurs late in the growth season as littoral flats are exposed due to falling water levels.

Habitats: Oligotrophic to moderately eutrophic, slightly acidic to low-alkaline waters with sandy to silty substrate, from above the shoreline to a depth of 0.5 m. Lakes and ponds, riverbanks and river mouths, coastal ponds and the littoral zone of slightly brackish waters.

Distribution within the region: Native. From Britain north to Scandinavia and east through Siberia and Mongolia to the Russian Far East, Japan and China. It also occurs throughout North America and parts of southern South America. Within the region it is present in Britain, Norway, Sweden, Finland, Denmark and the Czech Republic. It is extinct in Estonia, Germany, Latvia and Poland.

Characteristics and similar species: *Crassula aquatica* is characterised by the small, opposite, succulent leaves and the very small and normally 4-merous, sessile flowers.

225 *Crassula helmsii* is larger and has distinctly pedicellate flowers.

235-238 *Elatine* species differ in non-succulent leaves of a different shape and by sub-globose capsules with an apical depression.

Submerged *C. aquatica* growing with *Eleocharis acicularis* at a depth of 30-40 cm. Søvigsund, SW-Jutland, Denmark. Photo JCS.

Crassula aquatica. The small whitish to faint pinkish flowers and fruits are usually 4-merous. Note the dark brown seed sticking to one of the follicles. Umeälv, Degerfors, Sweden. Photo JCS.

Typical amphibious specimen of *C. aquatica* on exposed wet soil. Søvigsund, SW-Jutland, Denmark. Photo BM.

Young specimens of **240** ***Lythrum portula*** can be very similar - see this. Beware also of possible confusion with species of *Sagina*, which may germinate and occur in similar habitats to *Crassula aquatica*.

Terrestrial forms of **281-292** ***Callitriche*** species, particularly *C. brutia*, can have similarly narrow leaves, but these are not succulent and the flowers are very different.

Crassula aquatica. Terrestrial plant. Umeälv, Degerfors, Sweden. Photo JCS.

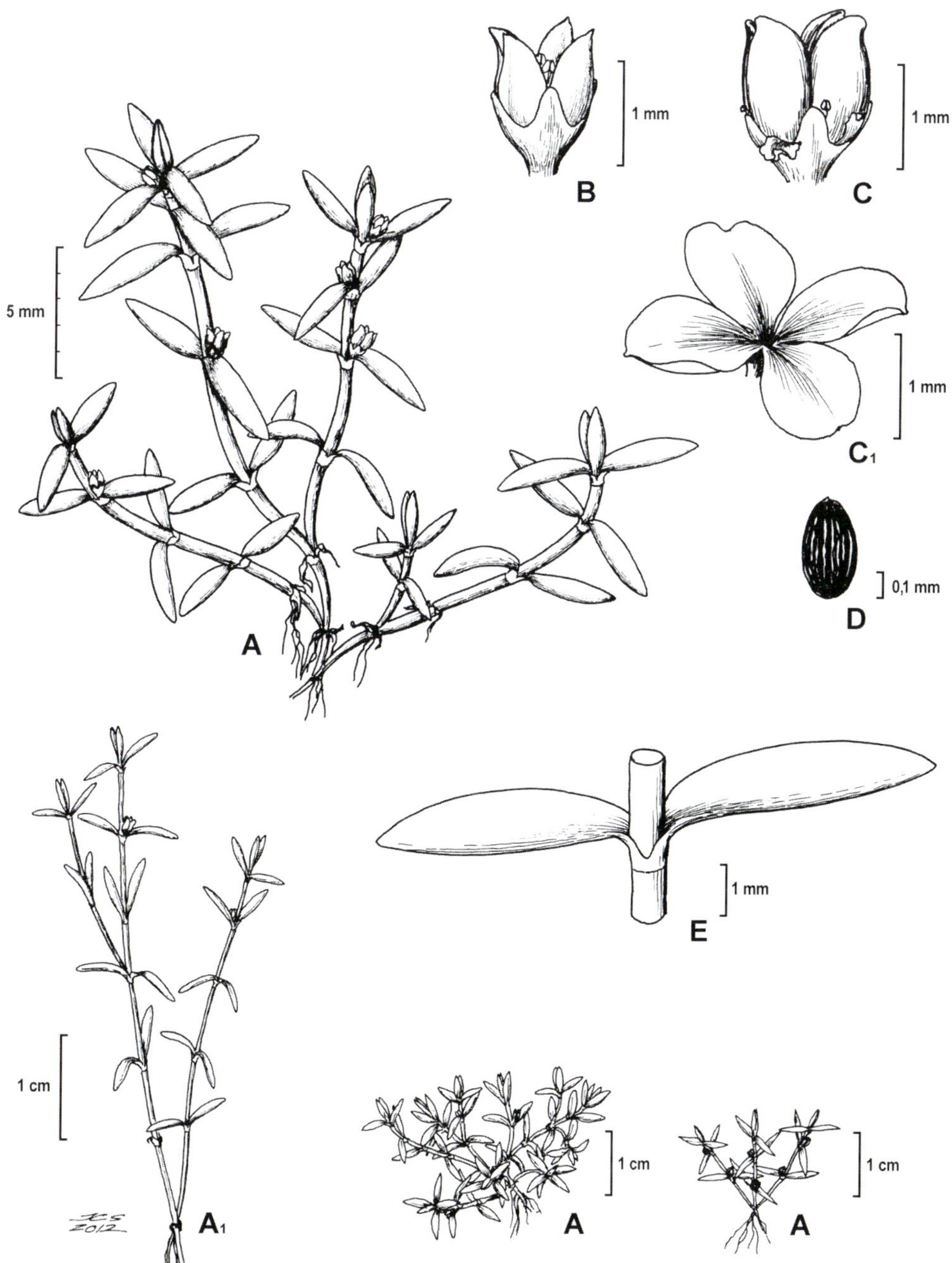

Crassula aquatica
A habit terrestrial plants, **A1** habit submerged plant, **B** flower, **C** fruit of 4 follicles, **C1** open follicles, **D** seed, **E** pair of leaves.

Crassulaceae

225 *Crassula helmsii* (Kirk) Cockayne

Syn.: *Tillaea recurva* (Hook. f.) Hook f.

DK: New Zealandsk Korsarve **N:** - **S:** Sydfyrling **FIN:** - **IS:** -
GB: New Zealand Pigmyweed **NL:** Watercrassula **F:** Crassule de Helms
D: Nadelkraut **CZ:** - **PL:** Grubosz helmsa
EST: - **LV:** - **LT:** - **Ru:** Толстянка Хелмса

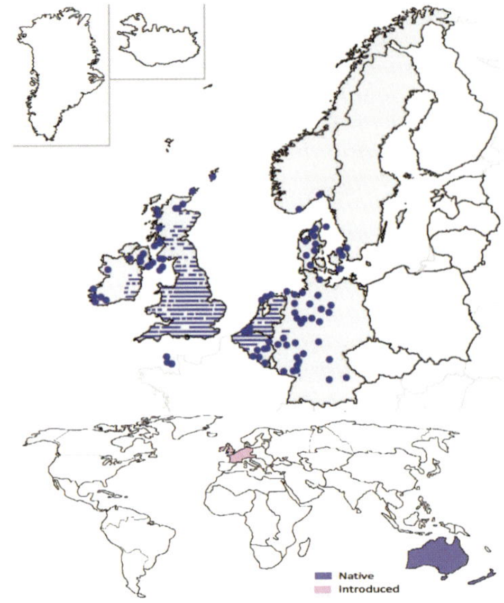

Large amphibious stand of *C. helmsii* in a small pond, clearly illustrating the competitive potential and invasive nature of the species. Chequers Pond, Surrey, Britain. Photo RVL.

Perennial amphiphyte or hydrophyte with white roots from lower stem nodes, without rhizome.

Stem: Terete, glabrous, more or less branched, 0.7–1.6 mm diameter. Submerged plants with ascending to erect, green stems up to more than 40 cm long. Emergent or floating stems 10–20 cm long, with long internodes, prostrate to ascending, green to pale green, often with a purplish tinge.

Leaves: 4–20 mm long, 1.0–1.5 mm wide and 0.5–1.0 mm in diameter, succulent, linear lanceolate, acute, opposite, fused at the base. Each pair of leaves oriented at a right angle to the neighbouring pairs, giving the shoots a cross-like appearance when seen from above. Submerged leaves with flatter lamina.

Inflorescence: Monoecious. Flowers solitary in leaf axils, only one flower from each pair of leaves. Permanently submerged plants lack flowers.

Flowers: 4-merous, bisexual, on pedicels > 2 mm long. Sepals c. 1 mm long, more or less fused at the base, obtuse to acute, green. Petals c. 1.5 mm long, slightly longer than the sepals, spreading, white, sometimes with a pinkish tinge, obovate, obtuse to acute. Stamens 4. Carpels gradually constricted into styles - 1/2 as long as carpels.

Fruits: 4 follicles, 1.0–1.5 mm long, green, each containing 2–5 seeds. Seeds 0.5 mm long, oblong elliptic, brown, dull.

Flowering: July-October.

Biology: Fruit set is rare in European plants, reproduction and dispersal is typically vegetative, even by small stem fragments. *C. helmsii* commonly forms large monospecific mats. Once established the plants are able to survive wide temperature variation, including extended periods of frost. Robust and very competitive in relation to other aquatic plants. In many countries considered an invasive species.

Habitats: A wide range of aquatic habitats - oligotrophic as well as eutrophic, acid as well as alkaline, still as well as flowing waters.

Distribution within the region: Imported as an ornamental plant and widespread in Northern Europe - Britain, Ireland, southern Scandinavia, Germany, Belgium, the Netherlands, France. Native to Tasmania and New Zealand.

Characteristics and similar species: *Crassula helmsii* is characterised by the opposite, succulent leaves and the small and normally 4-merous, petiolate flowers.

Crassula helmsii flowers richly over a long period, which along with its appealing green colour is the reason for its popularity among garden owners. Denmark. Photo JCS.

224 ***C. aquatica*** is smaller and has sessile flowers. **281-292** ***Callitriche*** spp. have leaves which are flattened (not succulent) and have a notched apex. **270-273** ***Galium*** spp. have leaves in whorls on square stems. **261-263** ***Montia*** spp. differ in non-succulent leaves and flowers 1–5 together.

Crassula peduncularis (Smith) Meigen is similar but with pedicels longer than the neighbouring leaves. Carpels abruptly constricted into short styles, usually less than 1/4 as long as carpels. Recorded in Brandenburg, Germany.

Crassula helmsii. Submerged plants resemble *Callitriche* spp. Shortwood Pond, Middlesex, Britain. Photo RVL.

Flower. Photo KvdW.

Crassula helmsii is a robust plant with petiolate flowers. Denmark. Photo JCS.

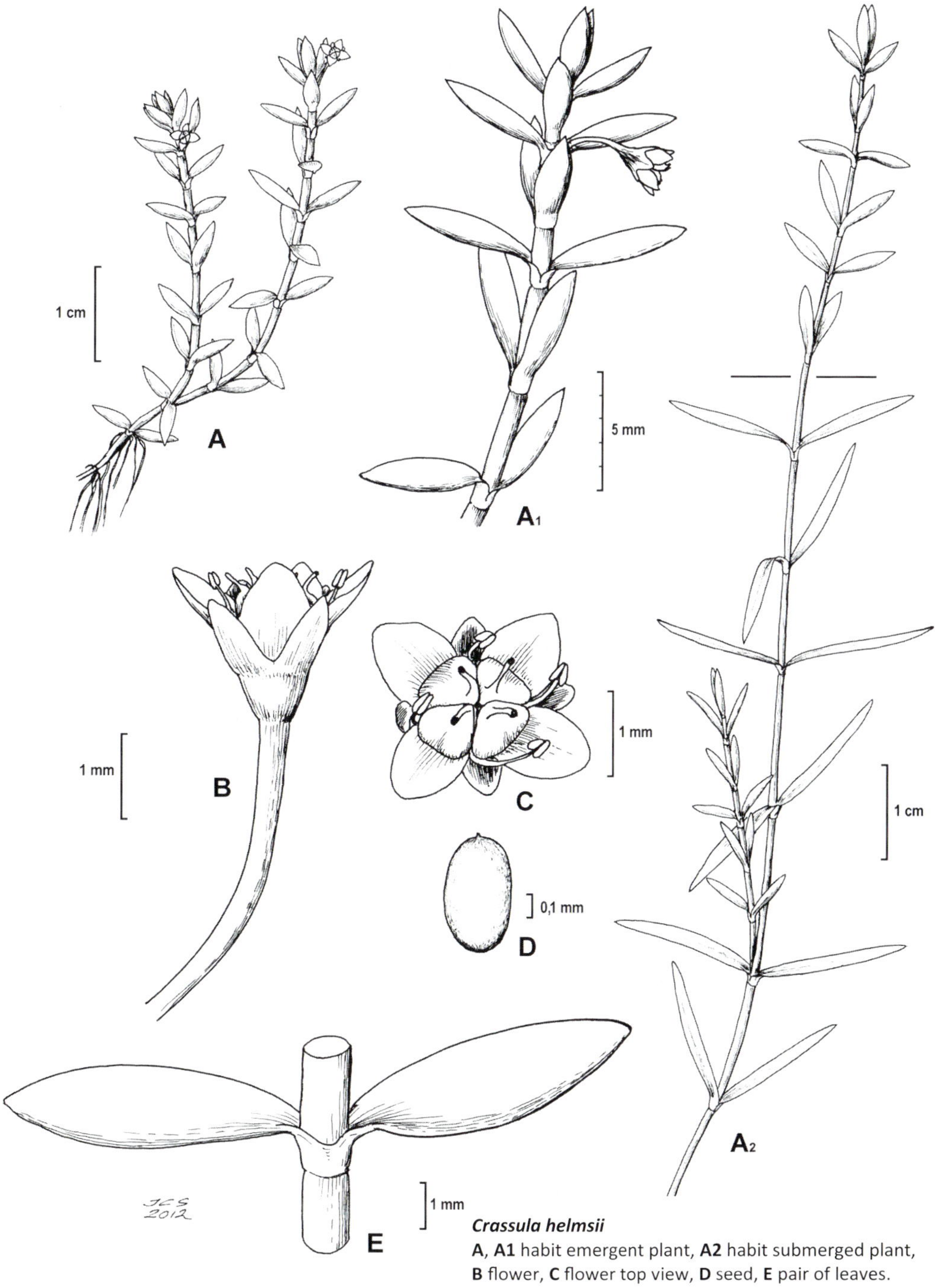

Crassula helmsii
A, **A1** habit emergent plant, **A2** habit submerged plant, **B** flower, **C** flower top view, **D** seed, **E** pair of leaves.

Haloragaceae

Key to *Myriophyllum*

Flowering specimens are necessary for reliable identification of *Myriophyllum* species. During field work, non-flowering plants are often encountered. Therefore, we present a key to non-flowering specimens, which allows identification of typical material of the species. The key cannot be perfect, as all species show considerably phenotypic variability in relation to climatic and hydrological conditions. Some species (e.g., *M. spicatum* and *M. sibiricum*) can form fertile hybrids. The morphological variation of non-native taxa is not fully known.

Flowering plants

1 Flowers solitary in the axils of upper pinnate leaves (***1***), resembling the lower ones in shape and outline. Emergent leaves blue green, rarely yellowish green, submerged leaves bright green or reddish, with whitish stipule-like outgrowth at the base (***2***), petioles 2–6(–8) mm long. Plants dioecious (only female plants are known in Europe). Turions absent

231 *M. aquaticum*

1 Flowers in emergent terminal spikes, mostly in the axils of bracts differing from submerged pinnate leaves, without whitish stipule-like outgrowth at the base. Submerged leaves olive green, bright or dark green, sometimes brown or reddish, not blue green when becoming emergent, petioles (0–)0.4–4 mm long. Plants monoecious (native species), rarely dioecious (introduced species). Turions present or absent 2

2 Bracts pectinate, pinnate or with serrate margins. Stamens 4 or 8 3

2 Bracts with entire margins. Stamens 8 6

3 Bracts pectinate or pinnate (***3***). Submerged leaves in whorls of 4–6, single alternate leaves absent. Stem moderately branched. Apical turions present, finally becoming club shaped (***9***). Turion leaves spathulate, with long segments with soft apices. Stamens 8

229 *M. verticillatum*

3 Bracts pinnately lobed or serrate. Leaves in whorls of 3–5, sometimes a few single alternate leaves present between whorls. Stem moderately to richly branched. Turions absent. Stamens 4 or 8 4

4 Bracts shorter than the flowers, pinnately lobed. Plants sometimes dioecious. Submerged leaves in whorls of 3–4, single alternate leaves absent. Stamens 8

M. verrucosum Lindl. *

4 Bracts longer than the flowers, serrate. Plants always monoecious. Submerged leaves in whorls of 4–5, single alternate leaves often present. Stamens 4 5

5 Bracts ovate lanceolate, tips acute, margins densely serrate (***11***)

230 *M. heterophyllum***

5 Bracts linear to linear lanceolate, tips rounded, margins remotely serrate (***12***)

230a *M. hippuroides* **

6 Male flowers all alternate (***5***), bracteolate. Female flowers alternate or in whorls, bracteolate or the lower ones in the axils of whorled submerged leaves. Spikes few flowered, initially curved. Submerged leaves 5–35 mm long, with 3–10 flaccid lateral segments, up to 0.1 mm wide, collapsing when removed from the water, often blackish when dried. Turions absent

226 *M. alterniflorum****

6 All flowers in bracteolate whorls (***6***). Spikes many flowered, upright, submerged leaves 10–50 mm long, with 3–24 more elastic lateral segments, up to 0.4 mm wide, slightly collapsing when removed from the water, remaining green when dried. Turions present or absent 7

7 Leaves mostly 3-dimensional, lateral segments 5–11(–13), proximal segment more than half as long as the leaf (***8***). Stem below inflorescence not swollen. Turions present, mostly apical, pointed, broadest at or below the middle (***10***). Turion leaves rhombic, thick with succulent apices

228 *M. sibiricum*

7 Leaves mostly flat, 2-dimensional, lateral segments (8–)12–24, proximal segment shorter than half the length of the leaf (***7***). Stem below inflorescence swollen to about double thickness. Turions absent

227 *M. spicatum*

* *M. verrucosum* is an Australian species traded as an aquarium plant. It was found in Britain in the 1940s, but did not persist.

** Different forms are traded under the name *M. hippuroides* as aquarium plants, including *M. heterophyllum* and *M. aquaticum*. The true *M. hippuroides* is closely related to *M. heterophyllum*, both originating from North America. Disagreement between cpDNA and ITS classification may indicate hybrid origin (Moody & Les 2010).

*** The taxon *Myriophyllum alterniflorum* var. *americanum* was cultivated by Aiken (1981). She found that this "variety" was a phenotypic modification without taxonomic value, developing under nutrient-poor conditions.

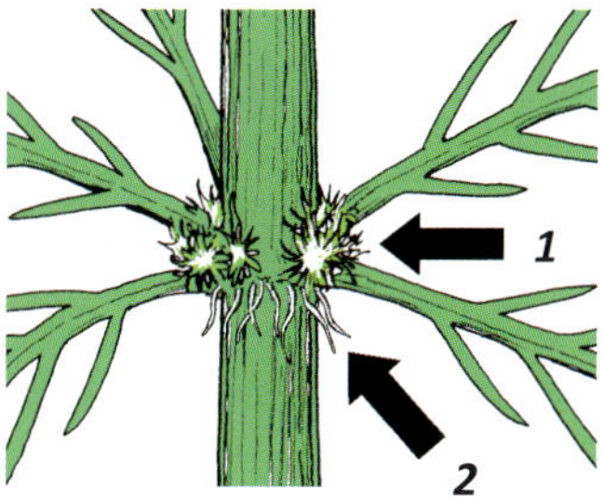

3

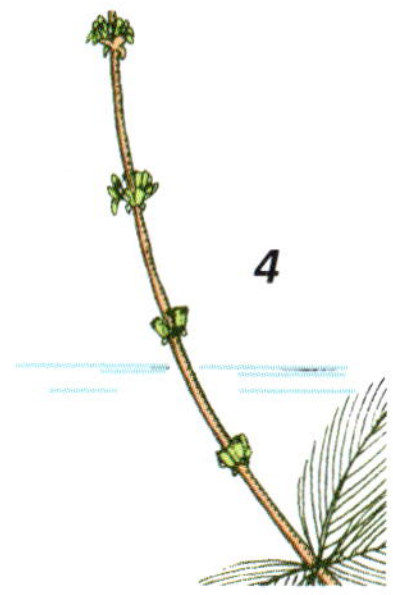

4

5

6

7

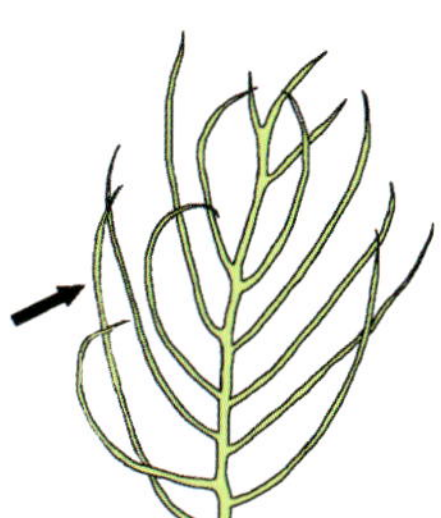

8

9

10

11

12

Non-flowering plants

1 Emergent leaves blue green, rarely yellowish green, submerged leaves bright green or reddish, with whitish stipule-like outgrowth at the base (***2***), petioles 2–6(–8) mm long. Turions absent
231 *M. aquaticum*

1 Emergent and submerged leaves olive green, bright or dark green, sometimes brownish or reddish, not blue green (except sometimes in *M. verrucosum**), whitish stipule-like outgrowth at base absent, petioles (0–)0.4–4 mm long. Turions present or absent 2

2 Solitary alternate leaves often present, submerged leaves in whorls of 4–5, petiole up to 4 mm long. Turions absent
230 *M. heterophyllum* or **230a *M. hippuroides*****

2 Single alternate leaves absent, submerged leaves in whorls of 3–6, petiole up to 2 mm long. Turions present or absent, if present with scale-like leaves 3

3 Apical, club-shaped turions present (***9***), or plant growing from last year's turion. Turion leaves spathulate, not scale-like. Leaves in whorls of 4–6, up to 60 mm long. Stem moderately branched, robust, up to 4 mm wide
229 *M. verticillatum*

3 Turion present or absent. Leaves in whorls of 3–5, up to 45 mm long. Stem richly branched, slender to robust, up to 3 mm wide 4

4 Submerged leaves in whorls of 3–4(–5), 5–35 mm long, with 3–10 pairs of lateral segments, alternate to subopposite, up to 0.1 mm wide, very flaccid, collapsing when removed from the water. Stem green to reddish. Stem and leaves often blackish when dried. Turions absent
226 *M. alterniflorum*

4 Submerged leaves in whorls of 4–5, 10–45 mm long, with 5–24 pairs of lateral segments, alternate or mostly subopposite, up to 0.2(–0.3) mm wide, flaccid, slightly collapsing when removed from the water. Stem light green, yellowish or reddish. Spindle-shaped turions present or absent 5

5 Submerged leaves mostly 3-dimensional, with 4–11(–13) pairs of lateral segments, segments up to 0.2 mm wide. Proximal segment almost as long as the leaf. Stem and leaves not olive green when dried. Turions present, turion leaves rhombic, scale-like
228 *M. sibiricum*

5 Submerged leaves mostly flat (pectinate), with (8–)12–24 pairs of lateral segments, segments up to 0.3 mm wide. Proximal segment much shorter than the leaf. Stem and leaves olive green when dried. Turions absent
227 *M. spicatum*

* *M. verrucosum* is omitted from the key as it cannot be identified when not flowering.

** The species cannot be distinguished in vegetative state.

Haloragaceae

226 *Myriophyllum alterniflorum* DC.

DK: Hår-Tusindblad **N:** Tusenblad **S:** Hårslinga **FIN:** Ruskoärviä **IS:** Síkjamari
GB: Alternate Water-milfoil **NL:** Teer vederkruid **F:** Myriophylle à fleurs alternes
D: Wechselblütiges Tausendblatt **CZ:** Stolístek střídavolistý **PL:** Wywłócznik skrętoległy
EST: Vahelduvaõiene vesikuusk **LV:** Pamīšziedu daudzlape **LT:** Pražangiažiedė plunksnalapė
Ru: Уруть очерёдноцветковая

Myriophyllum alterniflorum. Lake Nors Sø, Jutland, Denmark. Photo JCS.

Perennial hydrophyte. Rhizome subterranean, about 1 mm wide, brownish, branched and quite woody. Roots irregularly branched from the rhizome and lower stem nodes.

Stem: Up to more than 200 cm long and 0.5–1.8 mm in diameter, flexible, light brownish to reddish, ascending to upright, often richly branched, especially towards the tips and close to the water surface. Length of stem internodes decreasing towards the shoot tips.

Leaves: Submerged plants with pinnate, very finely dissected leaves in whorls of (3–)4(–5), each leaf 5–18 (–35) mm long with an irregular, and broadly ovate to elliptic outline with 3–8(–10) pairs of filamentous segments, 0.04–0.07(–0.1) mm wide and widely spaced, the terminal segment only slightly shorter than the others. Leaves sessile or very shortly petiolate, green to olive green or more often brownish to reddish, eventually very dark. Plants on dried-out sediments with stiffer, green leaves with fewer short, wide segments.

Inflorescence: Monoecious. Spikes aerial, up to 8 cm long, initially curved with nodding tips, later erect. Male flowers alternate, on the upper part of the spike. Female flowers in whorls on the lower part of the spike and sometimes in axils of the upper, submerged leaves. Each flower with one serrate, lanceolate bract, c. 1 mm long, and two ovate bracteoles, 0.3–0.5 mm long, with entire margins.

Flowers: Male flowers with 4 yellowish-white petals up to 2 mm long, with reddish stripes, soon decaying. Stamens 8 with 1.2–1.5 mm long, beige to yellowish anthers. Female flowers with an ovary up to 2 mm long and 4 red, cushion-shaped stigmas.

Fruits: Schizocarps, up to 2 mm, oblong ovate. At maturity the fruits split into 4 nutlets, each with 1 seed.

Flowering: June-August.

Biology: Wind pollinated, possibly also insect pollinated. Seeds dispersed by water and birds. Vegetative propagation by dispersal of detached stem fragments. Late in the year shoots typically die back and plants overwinter as buds at the root collar in or near the sediment surface. Turions are unknown. Sometimes overwintering as short green shoots, particularly in flowing water.

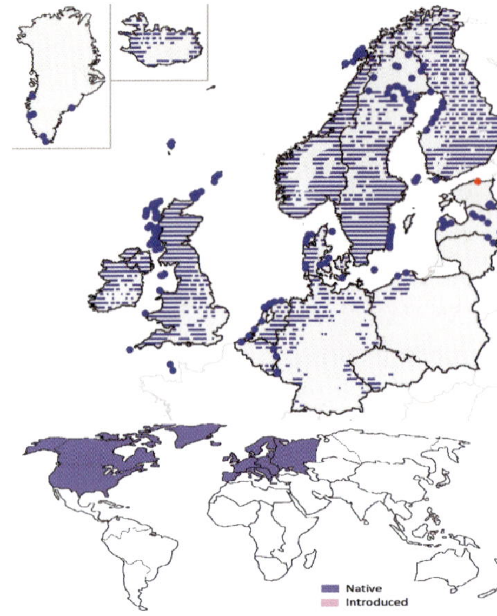

Habitats: Oligotrophic to mesotrophic, moderately acid to alkaline, clear but also humic waters. Streams and rivers as well as lakes and ponds, mainly with mineral sediments, less commonly on muddy or peaty soil. Rarely in slightly brackish water.

Distribution within the region: Native. Widespread and fairly abundant throughout the region, especially the northern and western parts, southwestern Greenland.

Characteristics and similar species: *Myriophyllum alterniflorum* varies morphologically due to different condi-

Myriophyllum alterniflorum.
Heidesee, Wankum, Germany. Photo KvdW.

Myriophyllum alterniflorum. Plants can survive short periods of air exposure during low water levels and then forming typical aerial shoots. Wolfssee, Duisburg, Germany. Photo KvdW.

Myriophyllum alterniflorum. Plants (brownish) in flowing water. River Teviot, Roxburghshire, Scotland. Photo RVL.

tions, but can be recognised by the short (5–18 mm long), very flaccid leaves, usually with fewer than 18 segments, the alternate male flowers and bracts shorter than the flowers. The whole plant turns black when dried.

229 ***Myriophyllum verticillatum*** differs in its larger and coarser leaves on unbranched or slightly branched stems, male flowers in whorls with pinnate bracts up to 2 cm long, and late in the season with distinct turions from the leaf axils.

227 ***Myriophyllum spicatum*** has stiffer, 2-dimensional leaves, usually with more than 18 segments, and male flowers in whorls.

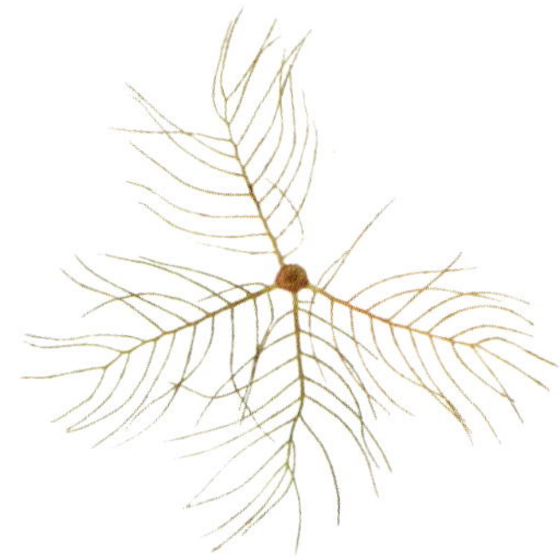

Myriophyllum alterniflorum usually has 4 leaves in a whorl, sometimes though 3 or even 5. Lake Hampen Sø, Denmark. Photo JCS.

Myriophyllum alterniflorum. Pollen released from the male flowers at the top of the spike is carried by wind to the reddish stigmas of the female flowers at the water surface. Lake Hampen Sø, Jutland, Denmark. Photo JCS.

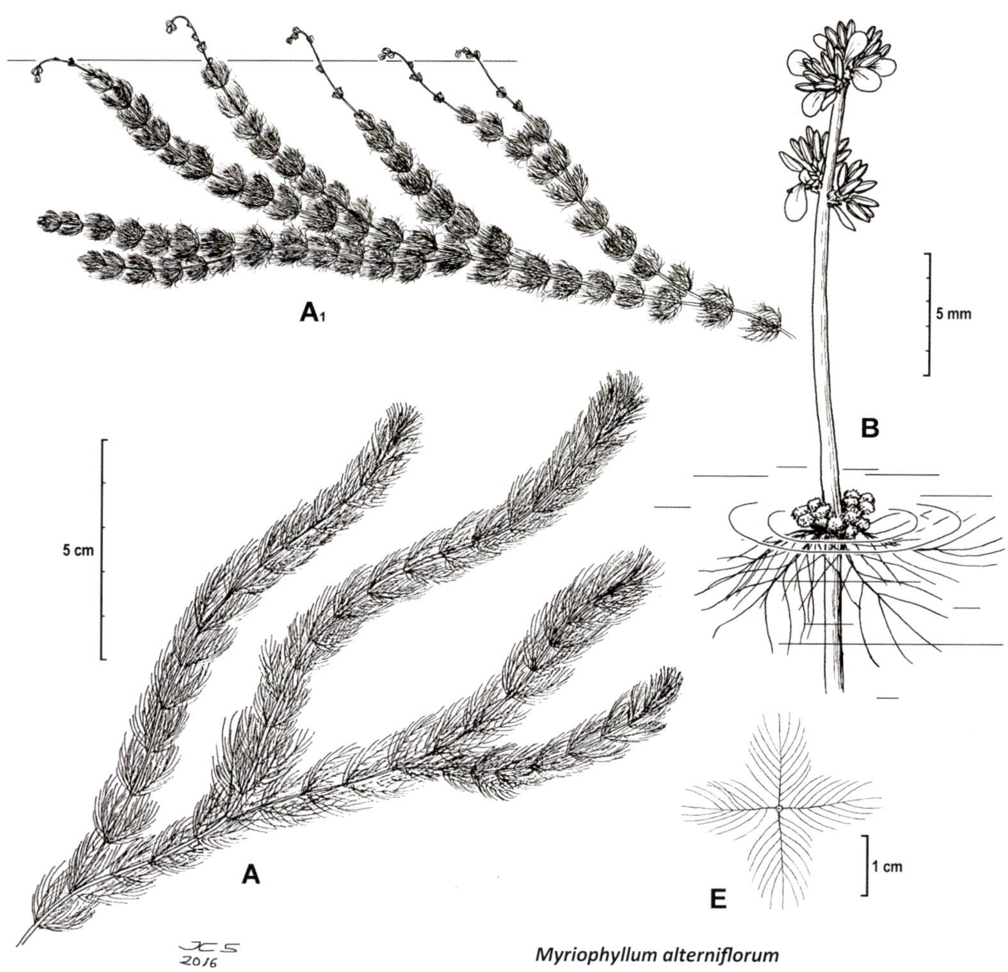

Myriophyllum alterniflorum
A, A1 habit, **B** inflorescence, **E** whorl of leaves.

Haloragaceae

227 *Myriophyllum spicatum* L.

DK: Aks-Tusindblad **N:** Akstusenblad **S:** Axslinga **FIN:** Tähkä-ärviä **IS:** -
GB: Spiked Water-milfoil **NL:** Aarvederkruid **F:** Myriophylle en épi
D: Ähren-Tausendblatt **CZ:** Stolístek klasnatý **PL:** Wywłócznik kłosowy
EST: Tähk-vesikuusk **LV:** Vārpainā daudzlape **LT:** Varpotoji plunksnalapė **Ru:** Уруть колосистая

Myriophyllum spicatum in flowing water. River Lippe, Germany. Photo KvdW.

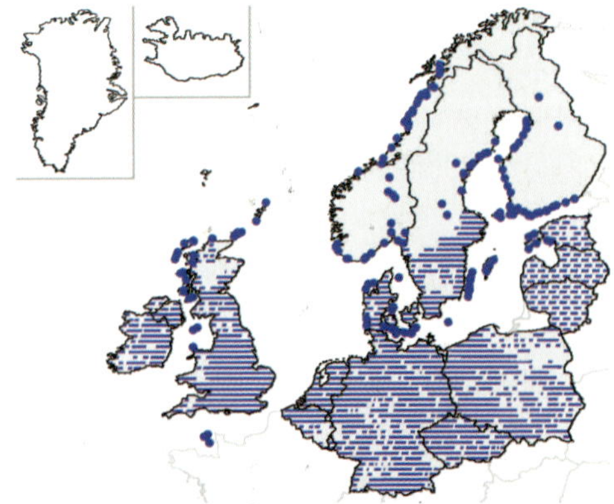

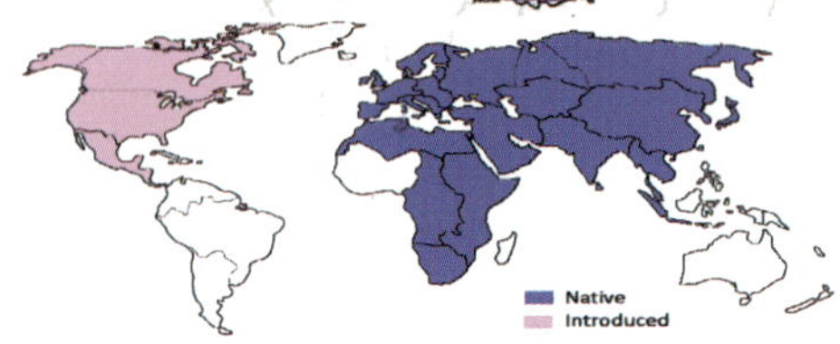

Perennial hydrophyte. Rhizome subterranean, about 2 mm wide, brownish, branched and quite woody. Roots irregularly branched primarily from the rhizome and from lower stem nodes.

Stem: Up to more than 300 cm long and 1.5–3.0 mm in diameter, flexible, reddish to greenish, ascending to upright, usually richly branched, especially towards the tips and near the water surface. Upper 5–20 stem nodes of flowering shoots commonly expanded to twice the diameter of the main stem. Length of stem internodes decreasing towards the tips of the shoots.

Leaves: Submerged plants with pinnate, finely dissected leaves in whorls of 4(–5) leaves from each stem node, each leaf 15–40(–45) mm long with a regular and oval to elliptic outline, (8–)12–24 pairs of filamentous, narrowly spaced segments, 0.1–0.3 mm wide, the terminal segment normally distinctly shorter than the other segments. Leaves sessile or very shortly petiolate, dark green, commonly with a reddish tinge towards the tips of the shoots. Plants growing on wet mud usually have stiffer, green or reddish leaves with fewer, shorter, wider lobes.

Inflorescence: Monoecious. Spikes aerial up to 15 cm long, upright, the lower part slightly curved. All flowers in whorls of 4, male flowers in the upper part of the spike, female flowers in the lower part. Flowers with one bract, 1.0–2.5 mm long, lanceolate to elongate with entire margins, and two bracteoles, 0.5–0.8 mm long, wider than long, with entire margins.

Flowers: Male flowers with 4 pinkish, soon decaying petals up to 2 mm long. 8 stamens with beige to yellowish anthers 1.2–1.8 mm long. Female flowers with an ovary up to 2 mm long and 4 white to pinkish recurved, brushy stigmas.

Fruits: Schizocarps, up to 2 mm, oval oblong. At maturity the fruits split into 4 nutlets, each with 1 seed.

Flowering: June-August.

Biology: Wind pollinated, possibly also insect pollinated. Seeds dispersed by water and birds. Vegetative propagation by dispersal of detached stem fragments or by rapidly rooting axillary buds. Late in the year shoots normally die off and plants overwinter as rhizomes in the sediment. Turions never develop. Plants may overwinter in deep water as short, green shoots. Plants exposed by falling water levels may develop a terrestrial form.

Habitats: Mesotrophic to eutrophic, neutral to alkaline waters. Streams, rivers, ditches and canals as well as lakes and ponds, natural as well as man-made sites, on mineral as well as soft and muddy sediment. Also in moderately brackish water.

Distribution within the region:
Native. Scattered to common throughout the region except Greenland and Iceland.

Characteristics and similar species:
Typical *Myriophyllum spicatum* is normally recognisable by the rather stiff and regularly feather-shaped leaves in

Myriophyllum spicatum in Monbagsee, Germany. Photo KvdW.

Myriophyllum spicatum has flat, propeller-like leaves in whorls, not clasping when removed from the water. Nors Sø, NW-Jutland, Denmark. Photo JCS.

Flowering *M. spicatum*. Bracts shorter than flower. Tversted, North-Jutland, Denmark. Photo JCS.

distant whorls, all flowers in whorls with bracts with entire margins and shorter than the flowers.

228 *Myriophyllum sibiricum* has leaves with a 3-dimensional appearance, leaf segments spaced (1–)1.5–4(–6) mm apart and develops turions late in the season.

229 *Myriophyllum verticillatum* has larger, flaccid leaves, unbranched or slightly branched stems, male flowers in whorls with pinnate bracts up to 2 cm long, and late in the season with turions.

226 *Myriophyllum alterniflorum* has very flaccid leaves with fewer than 18 leaf segments, and alternate male flowers.

Leaves 2-dimensional. Terminal segment usually much shorter than the others. Photo JCS.

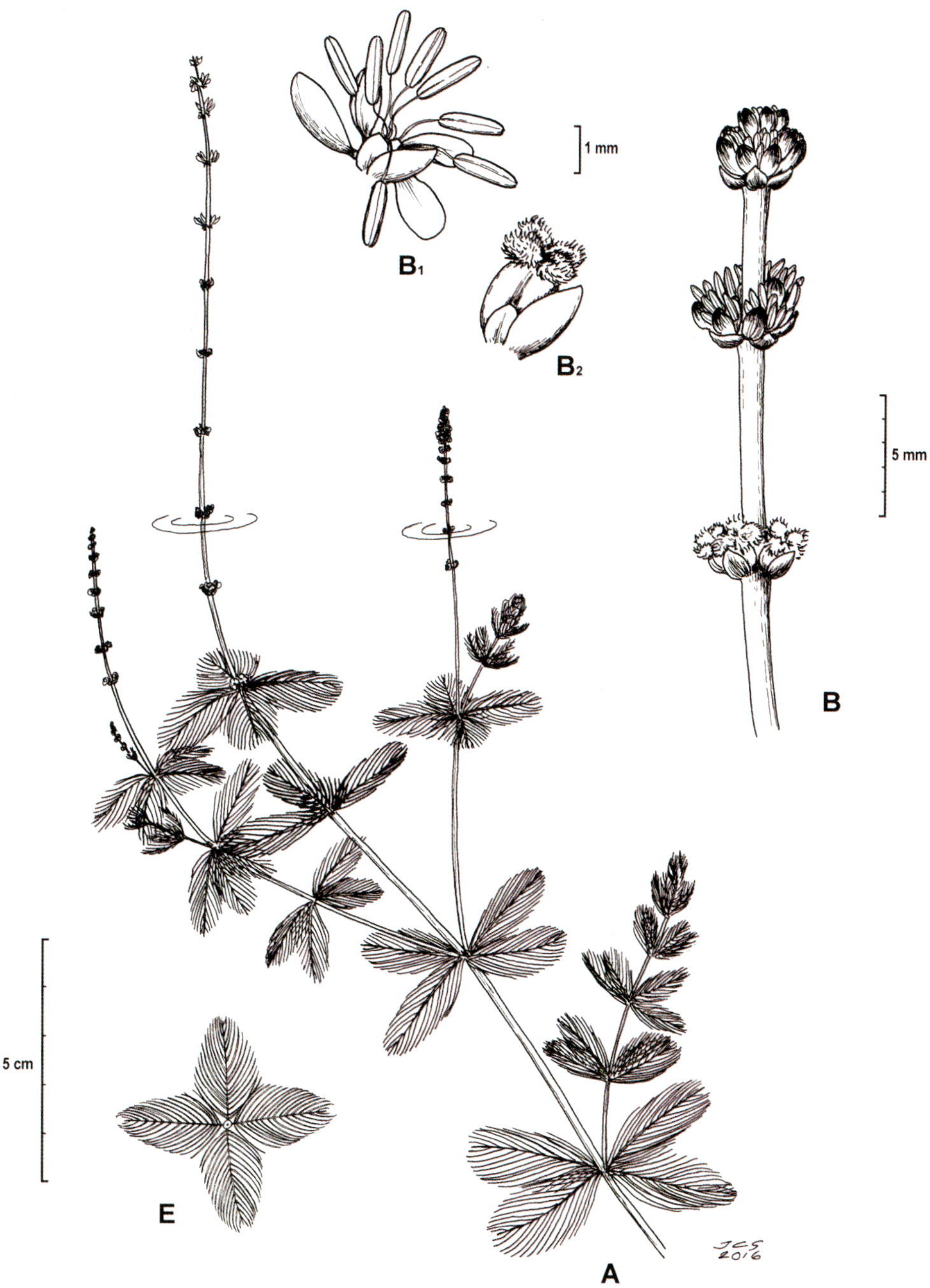

Myriophyllum spicatum
A habit, **B** inflorescence, **B1** male flower with bract and bracteole, **B2** female flower with bract and bracteole, **E** whorl of leaves.

Haloragaceae

228 *Myriophyllum sibiricum* Komarov

Syn.: *M. exalbescens* Fernald; *M. spicatum* var. *capillaceum* Lange;
DK: Grønlandsk tusindblad **N:** Kamtusenblad **S:** Knoppslinga **FIN:** Kalvasärviä **IS:** Vatnamari
GB: Northern Water-milfoil **NL:** - **F:** -
D: - **CZ:** - **PL:** -
EST: Siberi vesikuusk **LV:** - **LT:** - **Ru:** Урýть сибúрская

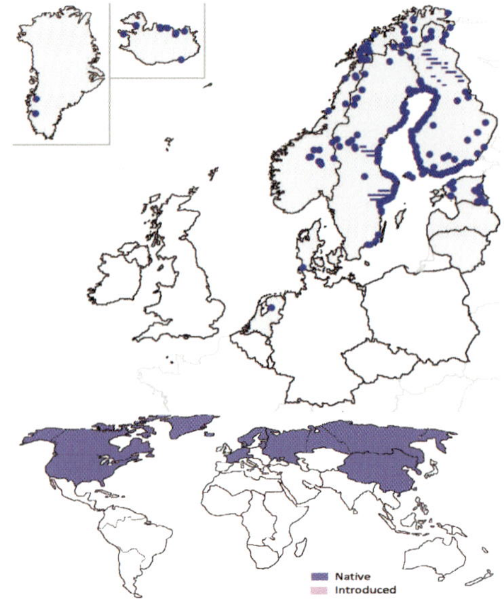

Myriophyllum sibiricum. Kalix älv near Kalix, Sweden. Photo RVL.

Perennial hydrophyte. Rhizome annual, subterranean, 2–3 mm diameter, brownish, branched and quite woody. Roots irregularly branched, in bundles from the rhizome and lower stem nodes.

Stem: (15–)40–80(–150) cm long, 1.5–3.0 mm in diameter, flexible, reddish to yellowish, often richly branched, especially from the lower stem. Upper stem nodes of flowering shoots commonly slightly wider, but not swollen. Stem nodes decreasing in length towards the tips of the shoots.

Leaves: Submerged plants with pinnate, finely dissected leaves in whorls of (3–) 4 from each node. Each leaf (10–)15–30 (–45) mm long, irregularly somewhat 3-dimensional in outline, with 3–13 pairs of linear, widely spaced segments (0.10–)0.15–0.30(–0.40) mm wide. The terminal segment normally distinctly shorter than the basal, which may be almost as long as the entire leaf. Leaves sessile or very shortly petiolate, initially reddish, becoming dark green and eventually very dark.

Inflorescence: Monoecious. Spikes, 2–11 cm long, upright, the lower part slightly curved. Flowers in whorls of 4, with male flowers above and female flowers below. Each flower with one bract, 1–2 mm long, lanceolate to elongate, the upper ones with entire margins, the lower ones with serrate margins. Bracteoles two, 0.7–1.0 mm long, ovate with entire margin.

Flowers: Male flowers with 4 pinkish, rapidly decaying petals up to 2 mm long. Stamens 8 with beige to yellowish anthers 1.0–1.4 mm long. Female flowers with an ovary up to 2 mm long and 4 pinkish, brush-shaped stigmas.

Fruits: Schizocarps, up to 2 mm long, ovate oblong. At maturity the fruit splits into 4 nutlets, each with 1 seed.

Flowering: June-August.

Biology: Wind pollinated, possibly also insect pollinated. Seed dispersal by water and birds. Vegetative propagation frequent, primarily by spindle-shaped turions, 15–80 mm long, that develop in late summer on shoot apices; also by dispersal of detached stem fragments. When the plants die back, the turions float and overwinter, enabling dispersal over considerable distances.

Habitats: Mesotrophic and neutral to alkaline waters, mainly lakes and ponds with soft and muddy substrate. Also frequently in running waters and moderately brackish water.

Distribution within the region:
Native. Circumpolar. In Europe it occurs in Iceland and Greenland, Netherlands,

Myriophyllum sibiricum. Jämtö-Avan, Baltic Sea. Photo KvdW.

The turions of *M. sibiricum* are more or less pointed; those of *M. verticillatum* are blunt. They are also widest close to the base or in the middle part. Photo KvdW.

Myriophyllum sibiricum. With turion. Kalix älv near Kalix, Sweden. Photo RVL.

Denmark, throughout Scandinavia and Estonia east into Russia.

Characteristics and similar species: Typical *M. sibiricum* has more or less pointed, spindle-shaped, apical turions which are widest at or below the middle, leaves with a 3-dimensional appearance and rather widely spaced leaf segments, the basal segments distinctly longer than the apical ones.

Typical **227** ***M. spicatum*** differs by the rather stiff, 2-dimensional leaves in distant whorls, undivided bracts, which are shorter than the flowers and with entire margins.

229 ***M. verticillatum*** develops blunt, club-shaped turions from the leaf axils late in the growth season, has large, coarser leaves on shorter and unbranched or slightly branched stems, male flowers in whorls with pinnate bracts up to 2 cm long.

226 ***M. alterniflorum*** has extremely flaccid and very thin leaf segments (usually 0.04–0.07 mm wide) and alternate male flowers.

Hybrids: *Myriophyllum sibiricum* hybridises with *M. spicatum*.

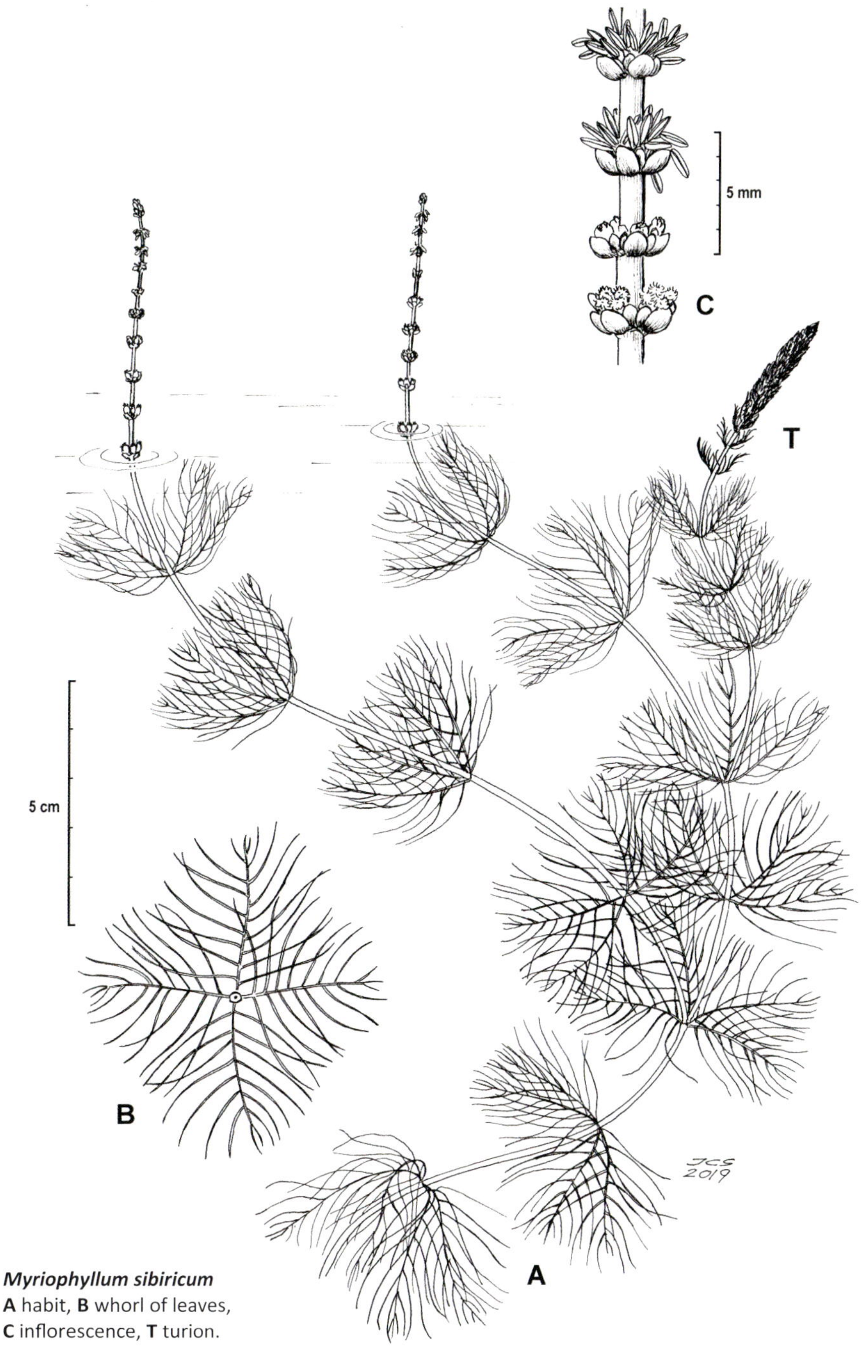

Myriophyllum sibiricum
A habit, **B** whorl of leaves,
C inflorescence, **T** turion.

Haloragaceae

229 *Myriophyllum verticillatum* L.

DK: Krans-Tusindblad **N:** Kranstusenblad **S:** Kransslinga **FIN:** Kiehkuraärviä **IS:** -
GB: Whorled Water-milfoil **NL:** Kransvederkruid **F:** Myriophylle verticillé
D: Quirl-Tausendblatt **CZ:** Stolístek přeslenatý **PL:** Wywłócznik okółkowy
EST: Männas-vesikuusk **LV:** Mieturu daudzlape **LT:** Menturinė plunksnalapė **Ru:** Уруть мутовчатая

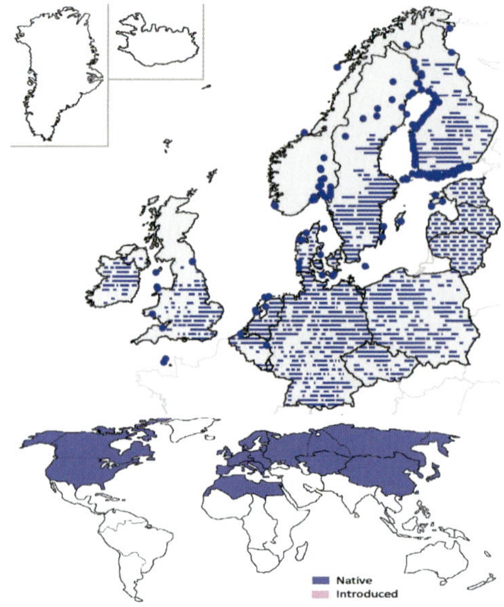

Dense stand of flowering *M. verticillatum* in shallow water. Kristianstad, Sweden. Photo BM.

Perennial hydrophyte with irregularly branched roots primarily from an annual rhizome, sometimes also from the lower stem nodes, the latter especially in dense stands and in very shallow water.

Stem: 0.5–1.5(–3) m long and 1.5–4.0 mm in diameter, flexible, bright green, unbranched or with few branches, in deeper water ascending to upright, in shallow water commonly prostrate to ascending. Rhizome 1.5–4 mm in diameter. Stem internodes decreasing in length towards the tips of the shoots.

Leaves: Leaves submerged, pinnate, in whorls of 4–5(–6) from each stem node, each leaf 20–45(–60) mm long with a somewhat irregular and broadly ovate outline with (7–)9–17 pairs of filamentous, normally densely spaced segments 0.1–0.2 mm wide, the terminal segment normally shorter than the other segments. Leaves sessile or very shortly petiolate, green, sometimes reddish. Plants exposed by falling water levels commonly have stiffer, green leaves with fewer and shorter but broader segments.

Inflorescence: Monoecious. Spikes aerial, up to 15 cm long, upright, the lower part often slightly curved, all flowers in whorls – 5 in each, male flowers in the upper part, female flowers in the lower part of the spike. Each flower with one bract, 0.2–2.5 cm long, pinnate with lanceolate outline, segments broader and stiffer than those of submerged or terrestrial plants, and two pinnate bracteoles, up to 0.2 mm long.

Flowers: Male flowers with 4 petals up to 2 mm long, yellowish green to reddish and rapidly decaying. 8 stamens with beige to yellowish anthers 1.2–1.7 mm long. Female flowers with an ovary up to 2.5 mm long and 4 white to pinkish, recurved, brushy stigmas.

Fruits: Schizocarps, up to 2 mm long, oval. At maturity the fruits split into 4 nutlets, each with 1 seed.

Flowering: June-August.

Biology: Wind pollinated, possibly also insect pollinated. Seeds dispersed by water and birds. Vegetative propagation is primarily by 15–35 mm long club-shaped turions that develop in late summer either on specialised short branches at the nodes, directly from lower stem nodes or from shoot apices. Plants die back in the autumn but turions overwinter, enabling dispersal over considerable distances. Plants exposed by falling water levels can survive as terrestrial forms with shorter, bright green leaves with shorter, wider leaf segments.

Habitats: Mesotrophic to eutrophic and slightly acid to alkaline water bodies on mineral as well as soft and muddy substrate, in standing and slow-flowing water, to a depth of 11 metres, in lakes, ponds, ditches and canals. Rarely in slightly brackish water in the Gulf of Bothnia.

Distribution within the region: Native. Scattered to common throughout the region except Greenland and Iceland.

Characteristics and similar species: *Myriophyllum verticillatum* is normally recognisable by the sparsely branched shoots, the large, flaccid, green leaves

The bracts of *M. verticillatum* are pinnate. Photo KvdW.

Myriophyllum verticillatum is sparsely branched with dense whorls of leaves more or less concealing the stem. Ans, Denmark. Photo JCS.

Myriophyllum verticillatum. Turions are developed late in the summer. Left: River Glen, near Spalding, Lincolnshire, Britain. Right: Filsø, SW-Jutland, Denmark. Photos RVL and JCS.

Myriophyllum verticillatum has leaves in whorls of 4-5. Photo JCS.

in dense whorls and the large club-shaped turions. Flowering plants can also be identified by all flowers in whorls with large pinnate, green bracts.

230 *Myriophyllum heterophyllum* have vegetative shoots rounded at the apex, lacks turions; flowering plants have serrate bracts.

226 *M. alterniflorum* has smaller leaves with fewer segments on richly branched stems, no turions but a woody rhizome, while flowering plants have alternate male flowers and very short, inconspicuous bracts.

Typical **227 *M. spicatum*** lacks turions, and flowering plants have very short bracts.

228 *M. sibiricum* rarely occurs with *M. verticillatum*, but when they do occur together, they are best differentiated by the shape of the turions.

197-199 *Ceratophyllum* species have whorls of forked leaves, while **293-295 *Hippuris*** have whorls of linear leaves with entire margin.

264 *Hottonia palustris* has alternate leaves; the upper leaves may appear to be in a rosette (pseudoverticillate), but they have wider pinnate segments.

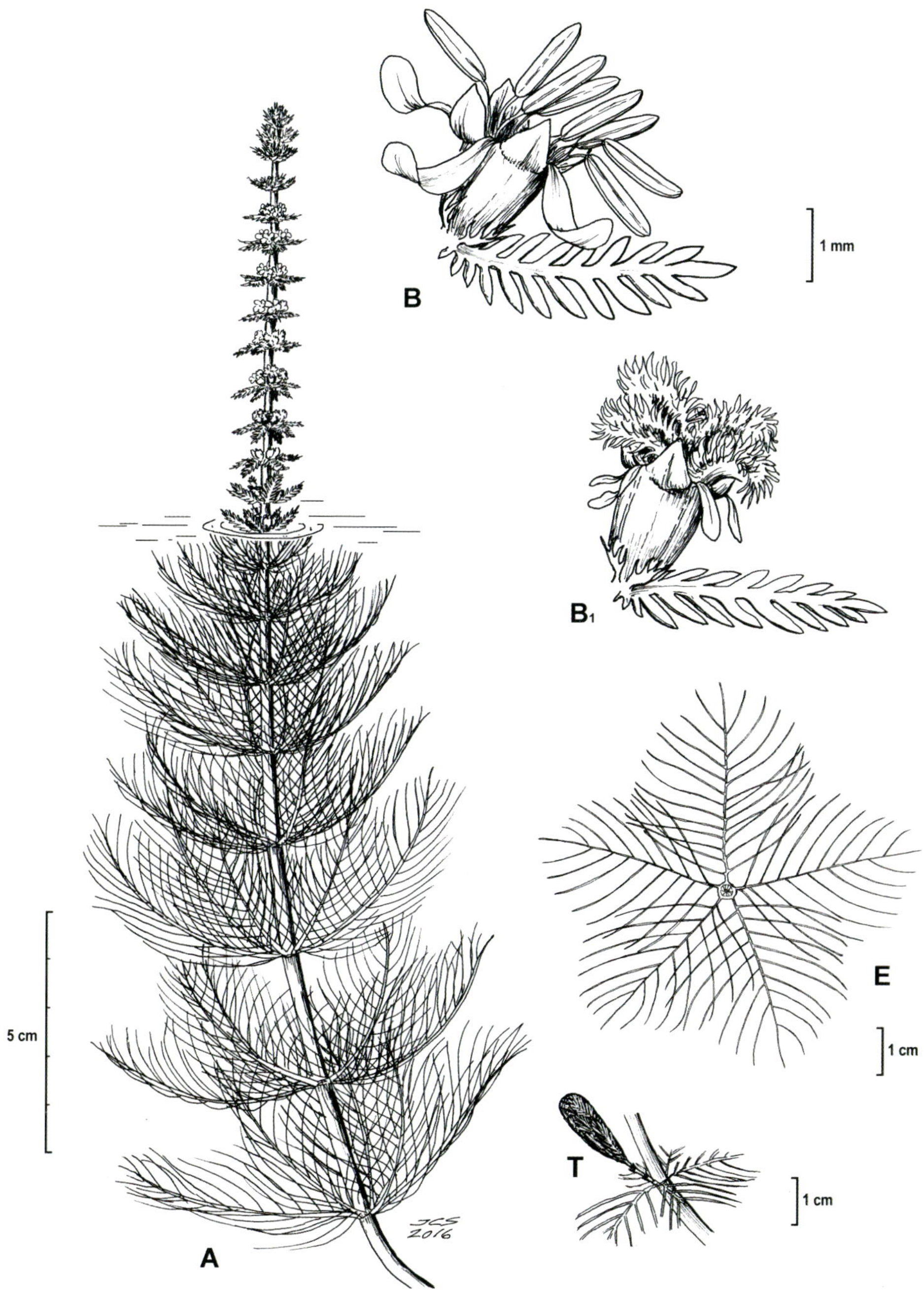

Myriophyllum verticillatum
A habit, **B** male flower with bract and bracteoles, **B1** female flower with bract and bracteoles, **E** whorl of leaves, **T** turion.

Haloragaceae

230 *Myriophyllum heterophyllum* Michx.

DK: - **N:** - **S:** - **FIN:** - **IS:** -
GB: Variable-leaf Water-milfoil **NL:** Ongelijkbladig vederkruid **F:** -
D: Verschiedenblättriges Tausendblatt **CZ:** - **PL:** -
EST: - **LV:** - **LT:** - **Ru:** Уруть разнолистная

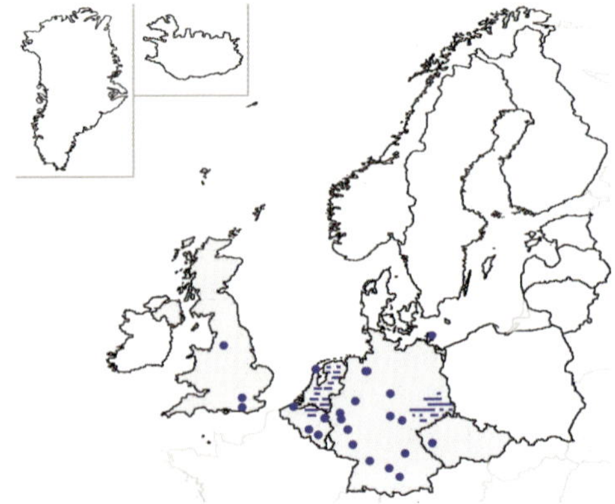

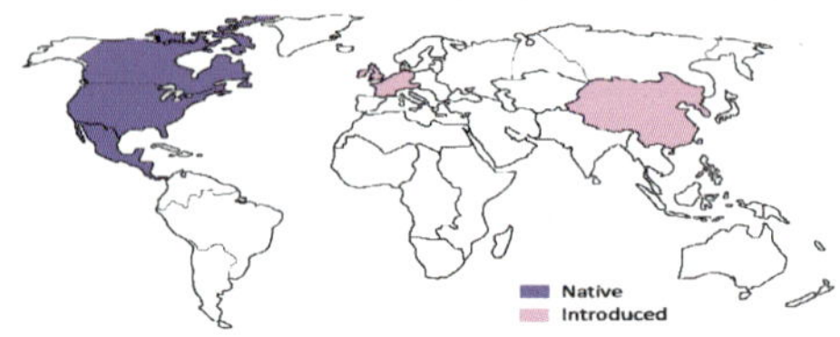

Flowering *M. heterophyllum*. Goitzsche-See, Germany. Photo KvdW.

Perennial hydrophyte with rhizome and adventitious roots arising from creeping stems.

Stem: Up to more than 1 m long and 3 mm in diameter, rather stiff, brownish to dark reddish, branched, upright in deeper water, creeping with ascending branches in shallow water and on wet substrates. Stem internodes short, decreasing in length towards the shoot tips.

Leaves: Submerged plants have finely dissected pinnate leaves in crowded whorls of 4–5 at each node in the upper part of the stem, while lower down the stem there may be additional alternate leaves scattered between the whorls, or all leaves alternate. Internodes less than 8(–10) mm. Leaves 20–50 mm long and 20–40 mm wide, elongate with a lanceolate or elliptic outline, 10–20 pairs of linear, widely spaced green segments, each about 20 mm long, the lower segments longer than the upper.

Inflorescence: Monoecious. Spikes 10–15(–35) cm long, all flowers with a single linear to lanceolate entire bract with finely serrate margins, up to 30 mm long and up to 5 mm wide. Upper flowers male, lower female. Each flower with 2 lobed bracteoles, up to 0.2 mm long.

Flowers: Small, c. 0.5 mm, male flowers with 4 very short whitish, membranous sepals, 4 petals and 4 short-petiolate stamens. Female flowers with 4 whitish, membranous petals 1.5–3.0 mm long and 4 filamentous stigmas.

Fruits: Schizocarps, up to 2 mm long, ovate oblong. At maturity the fruits split into 4 nutlets, each with 1 seed.

Flowering: June-August.

Biology: Pollination by small insects. Dispersal mainly by detached stem and rhizome fragments, less often by seeds. Heterophylly is temperature dependent; at low temperatures no aerial shoots and therefore no flowers are developed.

Habitats: Mesotrophic to eutrophic and slightly acid to alkaline water bodies, on mineral as well as soft, muddy sediment. Typically in standing water of lakes and ponds but also in ditches and

Submerged shoots of *M. heterophyllum* (left) and *M. verticillatum* with turion (right). Note the rounded shoot apex of *M. heterophyllum* versus the truncate apex of *M. verticillatum*. Heider Bergsee, Germany. Photo KvdW.

canals in still or slow-flowing water, rarely in faster-flowing water.

Distribution within the study area: Introduced and locally naturalised in the Netherlands, Belgium, Germany and the Czech Republic. Has occurred once in Britain but the population was eradicated.

Characteristics and similar species: Typical *M. heterophyllum* is characterised by aerial shoots with entire bracts with serrate margins and the leaves in whorls of 4–5 from each node; with some additional alternate leaves between whorls, or occasionally all leaves alternate.

Non-flowering plants may be confused with **227 *M. spicatum*** and **229 *M. verticillatum*,** but differs by having some alternate leaves between the whorls, and by shoots rounded at the apex (truncate in *M. spicatum and M. verticillatum*).

231 *Myriophyllum aquaticum* differs in being dioecious, with glaucous emergent leaves and submerged leaves with petioles 2–6(–8) mm.

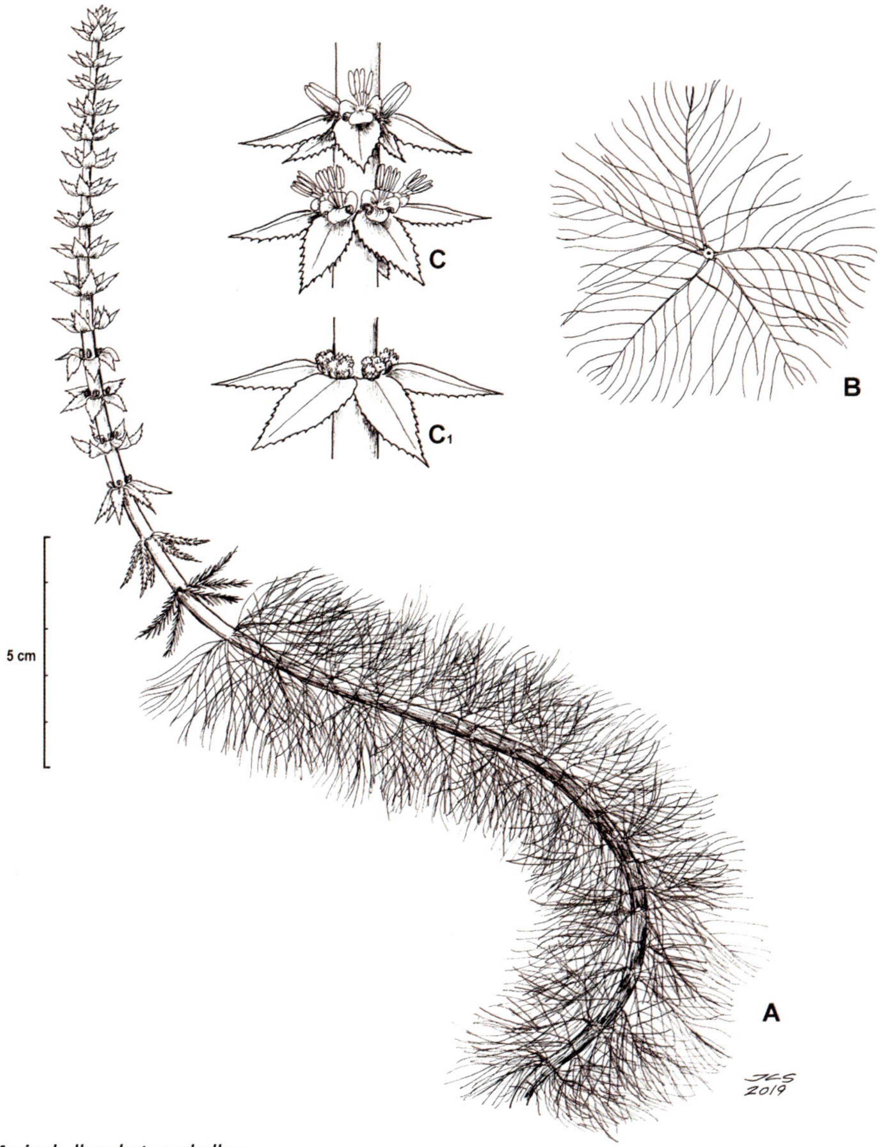

Myriophyllum heterophyllum
A habit, **B** whorl of leaves, **C, C1** part of inflorescence with male and female flowers.

Submerged shoots of *M. heterophyllum* appear very dense with short internodes. Photo KvdW.

Haloragaceae

230a *Myriophyllum hippuroides* Nutt. ex Torr. & A.Gray

S: Hästsvansslinga

Collected 1986 and 2008 from the river Alsterån, Kalmar län in southeastern Sweden, where it seemed to be fully naturalised. Often grown as an aquarium plant.

Differs from *M. heterophyllum* by linear to linear-lanceolate bracts with rounded tips and remotely serrate margin.

See also notes and figure on pages 516–517.

Myriophyllum heterophyllum. Leaves appear alternate, particularly in the lower part of the stem. Photo KvdW.

Haloragaceae

231 *Myriophyllum aquaticum* (Velloso) Verdc.

DK: Papegøjefjer **N:** - **S:** Storslinga **FIN:** - **IS:** -
GB: Parrot's-feather **NL:** Dicht federkruid **F:** Myriophylle aquatique
D: Brasilianisches Tausendblatt **CZ:** - **PL:** -
EST: - **LV:** - **LT:** - **Ru:** Уруть бразильская

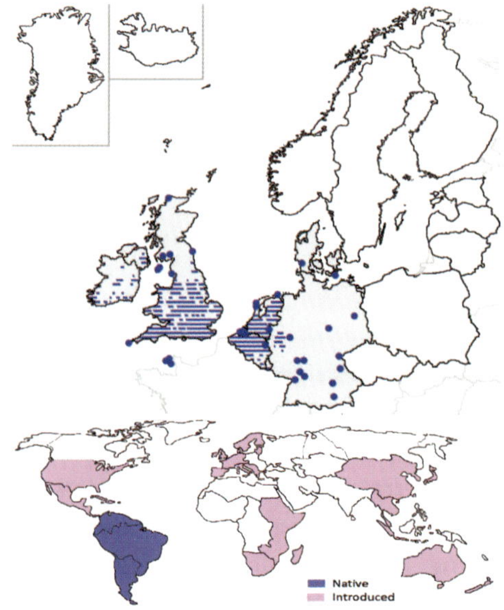

Submerged *M. aquaticum*. Signingstone, Glamorganshire, South Wales. Photo RVL.

Perennial hydrophyte with rhizome. Aquatic plants rooting from creeping stems, shoot apices typically float on the water surface with floating stems from which branches arise vertically to a height of up to 30 cm. Adventitious roots from creeping stems.

Stem: Up to 2 m long when submerged, 5 mm in diameter, rather stiff, bright to yellowish green or reddish, branched, upright through the water column in deeper water, creeping with ascending branches in shallow water and on wet mud. Stem internodes decreasing in length towards the shoot tips.

Leaves: Submerged and emergent shoots with finely dissected pinnate leaves in congested whorls of (4–)5–6 leaves from each stem node, 30–45 mm long and about 15 mm wide, elongate, lanceolate or elliptical in outline, with 10–18 pairs of linear segments, each 6–7 mm long, the lower segments shorter than the upper. Petioles of submerged leaves 2–6(–8) mm long. Aerial leaves glaucous, submerged leaves yellowish green or reddish. Submerged leaves commonly die off due to shading from the emergent shoots. 2–3 linear stipule-like growths at base of each leaf, which are white when young.

Inflorescence: Dioecious. Flowering spikes resemble aerial shoots, with flowers in the leaf axils of the upper whorls, each flower with 1 bract resembling the emergent leaves and 2 lobed bracteoles, up to 0.2 mm long.

Flowers: Male and female flowers small, c. 0.5 mm with whitish to membranous sepals, petals absent. Only female plants are known from the region.

Fruits: Unknown in the region as only female plants are present and so no fruits are produced.

Flowering: June-August.

Biology: The only means of propagation in the region is vegetative by stem and rhizome fragments. Stands may be permanently submerged in deeper water or dominated by dense mats of floating emergent aerial shoots due to shading of the submerged plant parts, often beached and rooting in the draw-down zone.

Habitats: Mesotrophic to eutrophic and slightly acid to alkaline water bodies, on mineral as well as soft and muddy substrates. Typically in standing water of ponds and lakes, as well as in rivers, ditches and canals tolerant of slow flow but less often in faster-flowing water.

Dense mat of emerged *M. aquaticum*. Signingstone, Glamorganshire, South Wales. Photo RVL.

Myriophyllum aquaticum. Top of emerged plant. Photo KvdW.

Myriophyllum aquaticum. Female flowers. Note the white stipule-like outgrowths. Photo KvdW.

Myriophyllum aquaticum. The white stipule-like outgrowths at base of leaves are diagnostic. Photo KvdW.

Distribution within the region:
Introduced. Widespread and locally invasive in Britain, the Netherlands and Belgium, scattered in Germany and local in Denmark.

Native to South America, but scattered throughout the world as a non-native.

Characteristics and similar species:
Within the region, typical *Myriophyllum aquaticum* is normally easily recognisable by the glaucous emergent stems and leaves. Submerged plants are recognised by the white stipules. Emergent and submerged leaves and bracts are similar.

Non-flowering plants may be confused with **227 *Myriophyllum spicatum***, in which the basal segments on each leaf are longer than the upper, while the opposite is the case for *Myriophyllum aquaticum*.

229 *Myriophyllum verticillatum* differs in being monoecious, having bracts very different to the submerged leaves, and forming turions late in the year.

264 *Hottonia palustris* has alternate leaves; the upper leaves may appear to be in a rosette (pseudoverticillate) but they have wider pinnate segments.

Myriophyllum rubricaule Valk. & Duist. is a newly described species known only from culture and as an escape in Netherlands and Belgium. It differs from *M. aquaticum* in its generally smaller size, purplish-red stems, emergent leaves green to bluish green but not glaucous (without waxy coating) and pinkish female flowers.

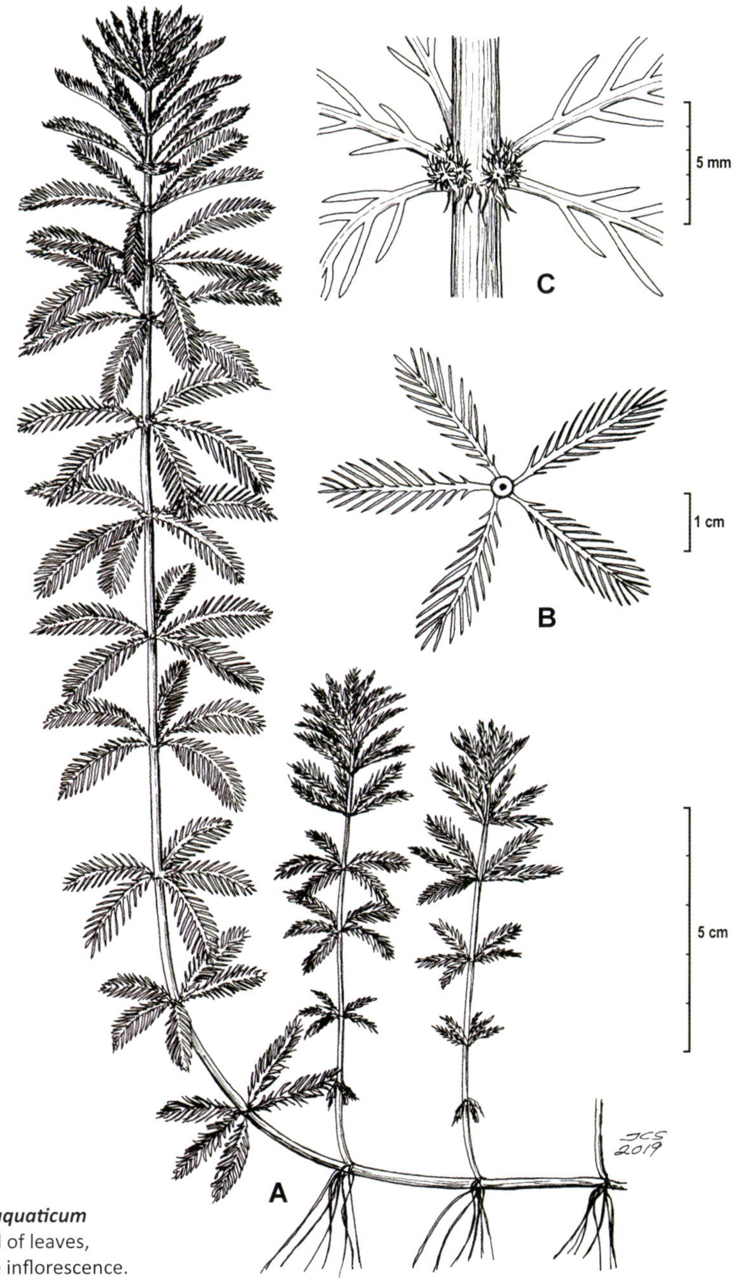

Myriophyllum aquaticum
A habit, **B** whorl of leaves,
C part of female inflorescence.

Rosaceae

232 *Comarum palustre* L.

Syn.: *Potentilla palustris* (L.) Scop.
DK: Kragefod **N:** Myrrhat **S:** Kråkklöver **FIN:** Kurjenjalka **IS:** Engjarós
GB: Marsh Cinquefoil **NL:** Wateraardbei **F:** Comaret des marais
D: Blutauge **CZ:** Zábělník bahenní **PL:** Siedmiopalecznik błotny
EST: Harilik soopihl **LV:** Purva vārnkāja **LT:** Pelkinis sūdras **Ru**: Сабельник болотный

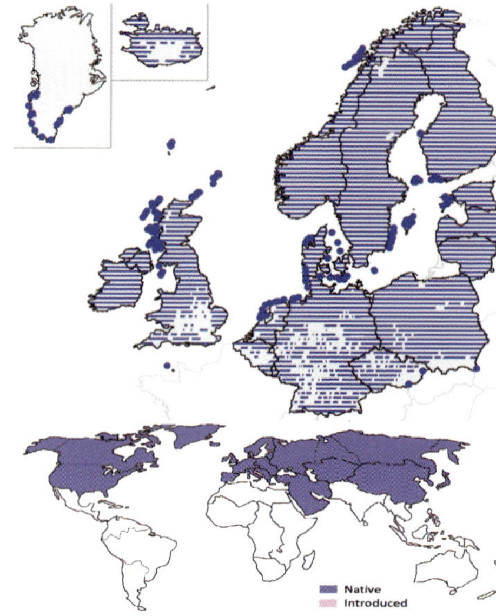

Perennial amphibious plant with a horizontal, woody rhizome up to more than 1 m long and 3–6 mm wide, with numerous roots. Creeping in and over wet soil or floating on the water surface.

Stem: 15–60 cm long, prostrate to ascending, green to purplish brown, terete with many simple and glandular hairs, the lower parts woody.

Leaves: With 2(–3) pairs of pinnate leaflets and one terminal leaflet, all with serrate margins, petioles short on upper leaves, longer on lower leaves, fused at the base with the two stipules, forming a semi-amplexicaul sheath. Leaflets with mid- to dark green, glabrous upper side and blue green lower side with fine, appressed hairs.

Inflorescence: A few-flowered apical cyme.

Flowers: 5-merous, bisexual, 18–22 mm in diameter. Sepals 1–1.5 cm long, acute, triangular, reddish brown with irregularly tangled simple and glandular hairs. Epicalyx segments, slender, triangular. Petals shorter than sepals, dark red, slender and acute. Stamens 15–25, filaments with hairy base. Anthers purple, 1.0–1.2 mm long. Stigmas 40–60, red.

Fruits: Achenes 1–1.5 mm long, ovate, semi-glossy, brown, attached to the swollen and spongy receptacle at maturity.

Flowering: June-August.

Biology: Pollination by insects attracted by the large red flowers and nectar pits at the base of the stamens; it is also capable of self-fertilisation. The relative importance of seed set and dispersal of floating fruits is unknown; in contrast, this species is capable of extensive spread by stolons over the soil surface or the surface of the water where dense floating mats can form, often with **335 *Menyanthes trifoliata,*** representing an important stage in the development of transition mires.

Rooted in the littoral zone of a shallow pond *C. palustre* sends its creeping stems out over the water surface, eventually contributing to the formation of transition mires. Hammer Bakker, Jutland, Denmark. Photo JCS.

Comarum palustre. Inflorescence with both open flowers with wide-spread sepals and fruits enclosed by the sepals. Hørby, Hobro, Denmark. Photo JCS.

Comarum palustre The small, narrow petals are situated in gaps between the much larger and much more conspicuous sepals. The nectar pits are discerned as green structures at the base of the red stamens. Lånum, Stoholm, Denmark. Photo JCS.

Habitats: On moist to wet, predominantly acid and humic soils in bogs and marshes as well as the margins of ponds and lakes.

Distribution within the region:
Common throughout the region.

Characteristics and similar species:
The creeping to prostrate habit, the commonly reddish stems and the characteristic flowers and leaves distinguish *C. palustre* from all other aquatic plants.

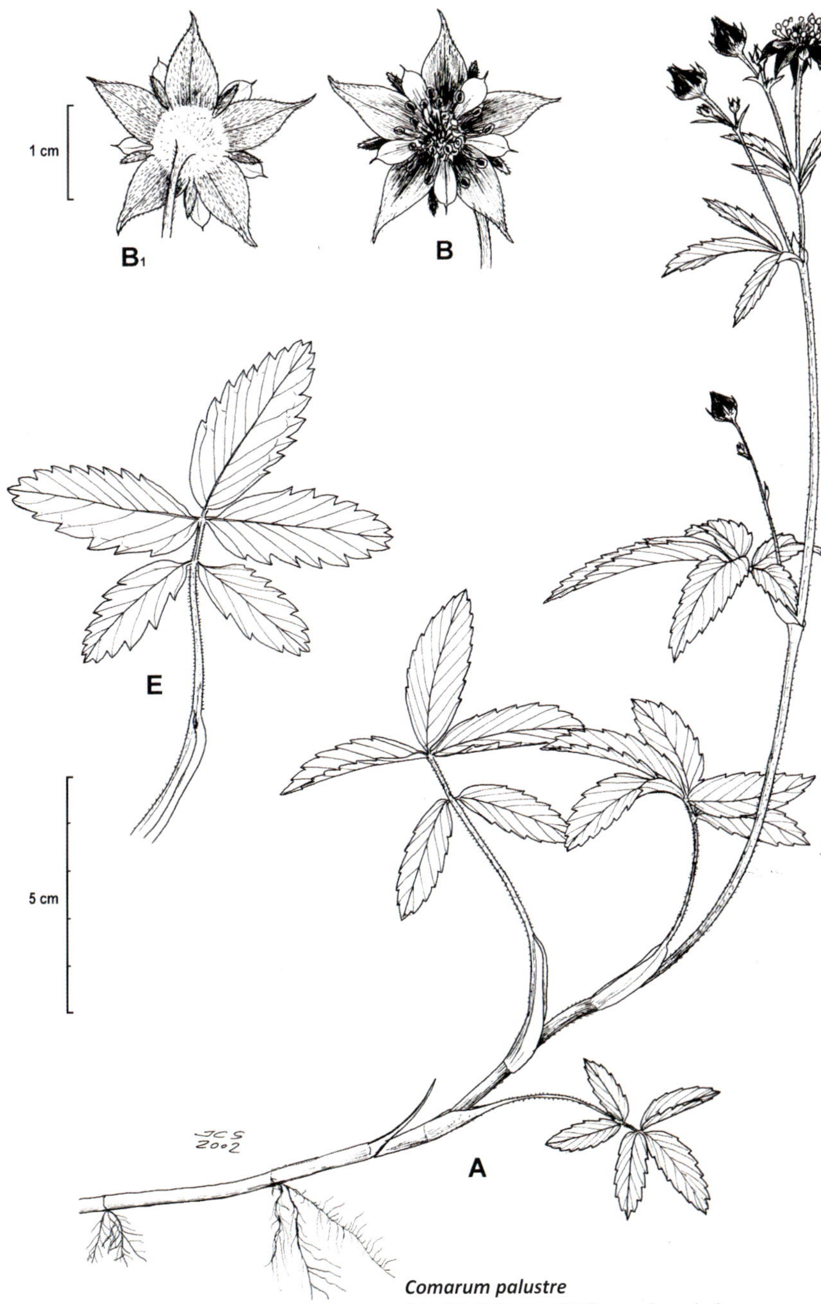

Comarum palustre
A habit, **B** flower, **B1** flower from below,
E leaf - stipules fused with the basal part of the petiole.

Elatinaceae

233 *Elatine alsinastrum* L.

DK: Krans-Bækarve **N:** Kransevjeblom **S:** Kransslamkrypa **FIN:** Isovesirikko **IS:** -
GB: - **NL:** Kranselatine **F:** Élatine fausse-alsine
D: Quirl-Tännel **CZ:** Úpor kuřičkovitý **PL:** Nadwodnik okółkowy
EST: -**LV:** - **LT:** - **Ru:** Повойничек мокричный

Elatine alsinastrum. In contrast to the stiff, green emergent leaves, the submerged leaves are flaccid and often brownish. Photo KvdW.

Annual or short-lived perennial (in southern parts of its range), hydrophyte or amphiphyte.

Stem: 5–30(–100) cm long, 2–7 mm in diameter, terete, slightly furrowed, erect or ascending, usually unbranched. Internodes on emergent stems 5–10 mm, on submerged stems up to 4 cm long.

Leaves: Emergent leaves in whorls of 3–5, 10–20 mm long, 4–9 mm wide, stiff, ovate to elliptic, with 5–12 more or less parallel veins. Small, brownish hydathodes along the leaf margins, the apical one distinct. Submerged leaves in whorls of 8–16, up to 50 mm long, 1.5 mm wide, linear, flaccid, with 1 vein and 1 apical hydathode. Often with intermediate leaves just below the water surface and often with only emergent leaves on wet substrates.

Elatine alsinastrum. The flowers are self-pollinated and 4-merous. Photo KvdW.

Stipules: Up to 2.5 mm long, linear or narrowly triangular, entire or digitate.

Inflorescence: Flowers solitary in leaf axils, up to 3 at each node.

Flowers: Sessile or with short pedicels, with 4 green sepals, 4 whitish-green petals 1–1.5 mm long, 8 stamens and 4 carpels.

Fruits: A subglobose capsule, with an apical depression. Seeds 0.7–0.8 mm long, slightly curved.

Flowering: June-September.

Biology: The flowers are self-pollinated and seeds are able to germinate immediately.

Habitats: On nutrient-rich, mostly calcium- and humus-poor, clayey substrate. In seasonal ponds and pools, ditches, fishponds, stream margins, gravel pits and temporarily inundated depressions in arable fields.

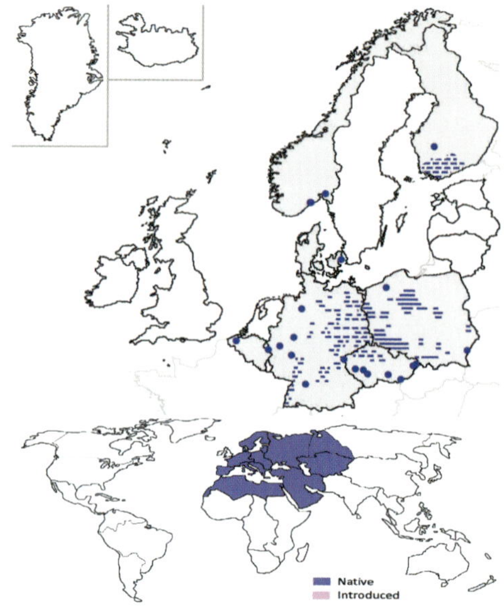

Distribution within the region:
Native and declining. Finnish populations are isolated from the main range. Extinct in Denmark and Belgium. Introduced to Norway in the 1880s but disappeared shortly thereafter.

Characteristics and similar species:
Although quite distinctive it may be confused with the emergent or flowering stems of **293 *Hippuris vulgaris***, from which it differs in the size and shape of the flowers and fruit. Submerged plants without flowers or fruits can be recognised by the lack of rhizomes and by the retuse leaf tips with

Elatine alsinastrum. Terrestrial forms seldom reach more than 10 cm in height. Photo KvdW.

Key to *Elatine*

1 Leaves in whorls
233 *E. alsinastrum*
1 Leaves opposite 2

2 Flowers with 6 stamens. Fruits with pedicels 3–10 mm long (***1***)
234 *E. hexandra*
2 Flowers with 3 or 8 stamens. Fruits sessile or with pedicels up to 4 mm long 3

3 Seeds hooked (***2***) or almost horseshoe shaped. Capsule with an apical depression. Flowers with 8 stamens (***5***)
236 *E. hydropiper*
3 Seeds only slightly curved (***3***). Capsule without an apical depression. Flowers with 3 or 8 stamens 4

4 Flower parts mostly in threes, with 3 stamens, 2–3 sepals, 3 petals (***4***) and 3 carpels
235 *E. triandra*
4 Flower parts in fours, with 8 stamens, 4 sepals, 4 petals (***5***) and 4 carpels
237 *E. orthosperma*

an apical hydathode. The submerged leaves of *Hippuris vulgaris* gradually taper into an acute tip. **226-231 *Myriophyllum*** species have pinnate submerged leaves.

Elatine alsinastrum. Fruit subglobose and depressed at apex. Photo KvdW.

Elatine alsinastrum only flowers on emergent stems. Photo KvdW.

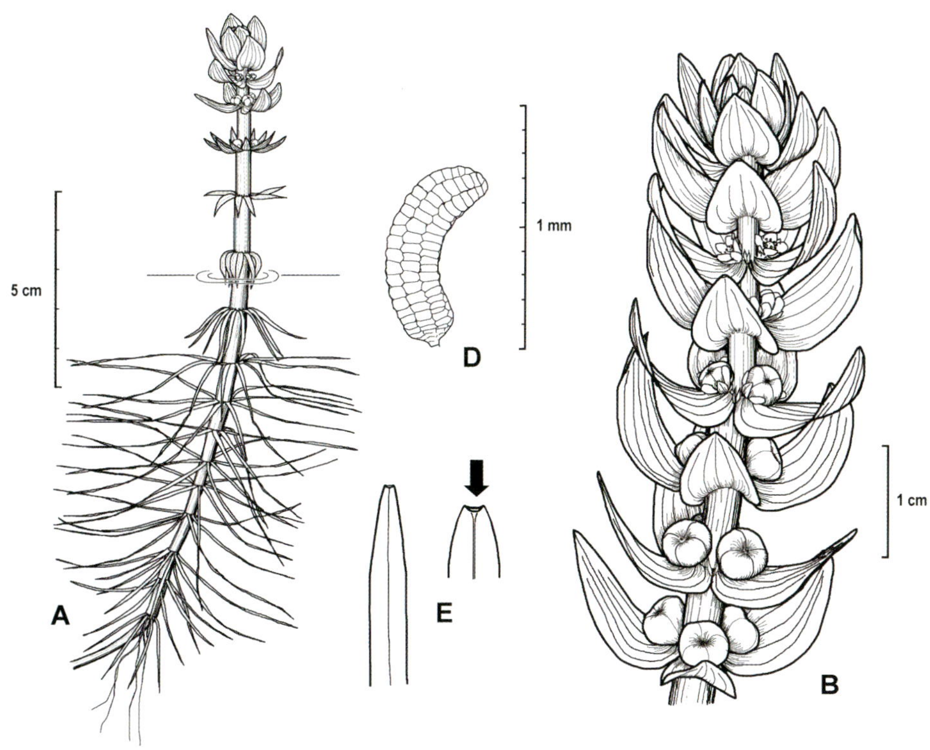

Elatine alsinastrum
A habit with emergent and submerged parts, **B** upper part of plant with flowers and fruits, **D** seed, **E** tip of a submerged leaf with apical hydathode (arrow).

Elatinaceae

234 *Elatine hexandra* (Lapierre) DC.

DK: Sekshannet Bækarve **N:** Skaftevjeblom **S:** Skaftslamkrypa **FIN:** Kuusihedevesirikko **IS:** -
GB: Six-stamened Waterwort **NL:** Gesteeld glaskroos **F:** Élatine à six étamines
D: Sechsmänniger Tännel **CZ:** Úpor šestimužný **PL:** Nadwodnik sześciopręcikoowy
EST: -**LV:** - **LT:** - **Ru:** Повойничек шестилепестный

Elatine hexandra. Flowers with 3 sepals, 3 petals and 6 stamens. Lake Hampen Sø, Jutland, Denmark. Photo JCS.

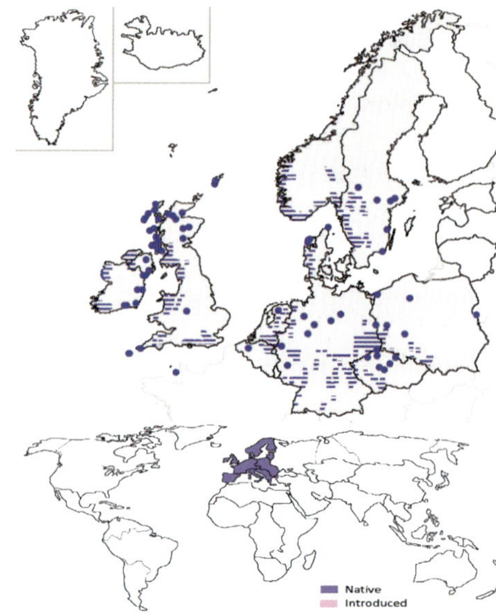

Annual amphiphyte or hydrophyte, forming small patches or sometimes extensive swards in deep water. Submerged plants may be short-lived perennials without a rhizome. Roots unbranched, whitish to brownish.

Stem: Up to 8 cm long, prostrate with erect or ascending branches, green, glabrous, terete, rooting at the nodes. Terrestrial plants 1–5 cm long, often red or purplish.

Leaves: Opposite, patent, 2–11 mm long, spathulate or lamina gradually tapering into the petiole, entire, pinnately veined, leaf tip obtuse to rounded, sometimes slightly notched. Hydathodes along the leaf margins small and indistinct. The apical hydathode blackish and usually distinct.

Stipules: Up to 1 mm long, membranous, linear or narrowly triangular, entire or with a few teeth.

Inflorescence: Flowers solitary in leaf axils.

Flowers: Flower parts in threes. Pedicels 3–10 mm long. Sepals 3, dimorphic, 1–2.5 mm long, connate at base, roundish, green or reddish in terrestrial forms. Petals 1.2–1.5 mm long, roundish, shorter than sepals, red or pinkish. With 6 stamens and 3 carpels.

Fruits: A capsule, subglobose to almost turban shaped with an apical depression, 1.2–1.5 mm wide. Seeds pale brown to blackish, 0.65–0.70 mm long, cylindrical, slightly curved, seed coat with (16–)18–21(–24) pits in the middle row.

Flowering: July-October.

Biology: The seeds remain viable for many years. They do not float, but are dispersed by water movement or stick to floating stem fragments or birds.

Submerged plants have cleistogamous self-pollinating flowers, while terrestrial plants have open chasmogamous flowers, probably pollinated by small insects.

Habitats: Acidic, oligotrophic to slightly eutrophic lakes and estuaries on clayey, sandy or gravelly substrate overlain by an organic layer. Often growing submerged from 10 to 100 cm depth, occasionally up to 3 m.

Distribution within the region:
Native. Scattered from Ireland east to western Poland and north into southern Norway and Sweden.

Elatine hexandra. Plants from dried-out pool. Note the reddish to purplish colouration and the apical depression on the capsules. Lake Skånsø, NE Vinderup, Jutland, Denmark. Photo JCS.

Characteristics and similar species:
E. hexandra is small and easily overlooked. It can superficially resemble new-sown cress, growing in small patches or mats.

Specimens of *E. hexandra* with short-petiolate flowers may be mistaken for **235 *E. triandra*,** which has very similar seeds, but only 3 stamens.

Separation from other species of *Elatine* depends on the characters given in the key (see p. 533).

Young specimens of **239 *Lythrum portula*** have blunt 4-angular to somewhat furrowed stems and leaves with a prominent and often basally red midrib.

261-263 *Montia fontana* does not root at the nodes.

281-292 *Callitriche* spp. leaves have different venation and lack stipules.

224 *Crassula aquatica* has narrower, acute, fleshy leaves.

225 *Crassula helmsii* has acute leaves and is much more robust.

Elatine hexandra. Submerged plants uprooted from the substrate. Lake Førby Sø, Thy, Denmark. Photo JCS.

3-merous flowers with long petioles and only slightly curved seeds is characteristic of *E. hexandra*. Lake Skånsø, Jutland, Denmark. Photos JCS.

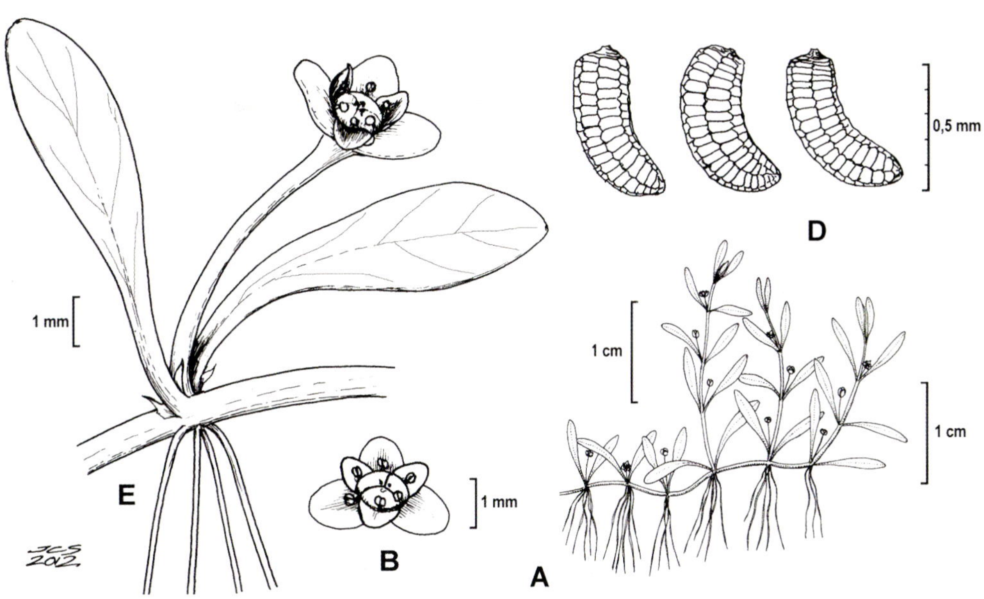

Elatine hexandra
A habit submerged plant, **B** flower, **D** seed, **E** rooting node with flower, pair of leaves and stipules.

Elatinaceae

235 *Elatine triandra* Schkuhr

Syn.: *Elatine callitrichoides* (Nyl.) Kauffm.

DK: Trehannet Bækarve **N:** Trefelt evjeblom **S:** Tretalig slamkrypa **FIN:** Kolmihedevisirikko **IS:** - **GB:** Three-stamen Waterwort **NL:** Drietallig glaskroos **F:** Élatine à trois étamines **D:** Dreimänniges Tännel **CZ:** Úpor trojmužný **PL:** Nadwodnik trójpręcikowy **EST:** - **LV:** - **LT:** - **Ru:** Повойничек трехтычинковый

Elatine triandra. Terrestrial plants are often obvious even at a distance because they spread horizontally over the substrate with branches and leaves more or less appressed to open ground. Germany. Photo KvdW.

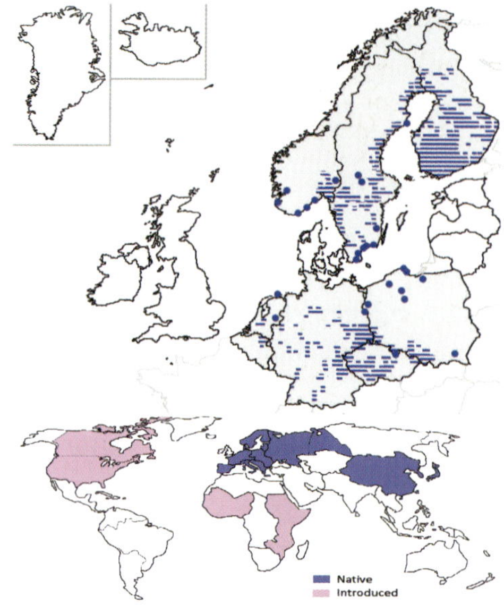

Annual amphiphyte or hydrophyte, forming small patches. Roots unbranched, whitish to brownish.

Stem: Green, glabrous, terete, rooting at the nodes. Submerged plants up to 18 cm long, more or less branched, flaccid. Terrestrial plants 1–10 cm long, prostrate, normally richly branched and often with red or purplish colours.

Leaves: Opposite, patent, 5–20 mm long, (1.5–)2.0–2.6(–3.3) mm wide, lanceolate, abruptly tapering at the base to a short, indistinct petiole, pinnately veined, entire or weakly dentate, especially when submerged, leaf tip notched although this may be difficult to see. Leaves usually green, but in submerged plants the lower side is greyish or silvery green; terrestrial plants may become red or purplish in colour. Hydathodes dark brown and usually distinct.

Stipules: About 1 mm long, membranous, triangular, dentate.

Inflorescence: Flowers solitary in leaf axils.

Flowers: 3-merous, sessile or with pedicels up to 3 mm long when fruiting. Sepals 2–3, dimorphic, with one smaller than the other two, up to 0.5 mm long, connate at base, elliptic, green or in terrestrial forms sometimes purplish. Petals up to 1 mm long, ovate, longer than sepals, white or reddish. With 3 stamens and 3 carpels.

Fruits: Capsule subglobose with an apical depression, 1.4–1.7 mm wide. Seeds light brownish, 0.5–0.6 mm long, cylindrical, slightly curved, seed coat with 20–24 pits in the middle row.

Flowering: (May-)July-October.

Biology: The seeds are dispersed by water or birds. Submerged plants have cleistogamous self-pollinating flowers and produce few flowers, while terrestrial plants have open chasmogamous flowers, probably pollinated by small insects.

Habitats: Mesotrophic to eutrophic water on carbonate-poor, more or less humus-rich substrate.

Submerged plants mainly occur in shallow water 10 to 30 cm deep, or occasionally up to 1.5 m deep, in lakes, ponds, and along slow-flowing rivers,

Elatine triandra. The small flowers have 3 sepals and 3 stamens. Note the membranous stipules on the upper leaves. Germany. Photo KvdW.

Elatine triandra. Submerged plants can easily be mistaken for *Callitriche*, but they differ in the leaf venation, the presence of petals and the somewhat globose fruit. Germany. Photo Thomas Franke.

Elatine triandra on moist substrate. Oulu, Finland. Photo RVL.

even in brackish coastal water in the Nordic region. Terrestrial plants occur in open vegetation on wet or temporarily dried-up substrate. It will also occur in man-made habitats, such as fishponds, temporarily inundated agricultural fields, small boat harbours, etc.

Distribution within the region:
Native. Declining in many regions due to eutrophication, drainage and loss of open habitats.

Scattered to locally frequent from Belgium and the Netherlands east to Poland, north into southern Norway, and throughout coastal areas of the Gulf of Bothnia and Finland.

Characteristics and similar species:
The size and shape of the leaves of *E. triandra* are very variable. It can be distinguished from other *Elatine* species by the 3-merous flowers with only 3 stamens, and by the shape of fruits and seeds. Even without flowers or fruits it may be identified by the linear to lanceolate leaves, which are distinctly notched or almost truncate at the apex with dark brown hydathodes.

Submerged plants with long internodes and long leaves with an apical notch may be mistaken for **281-292 *Callitriche*** species, but *E. triandra* differs in the pinnate venation of the leaves and the different flowers and fruits.

Elatine triandra. Seeds slightly curved. Photo JCS.

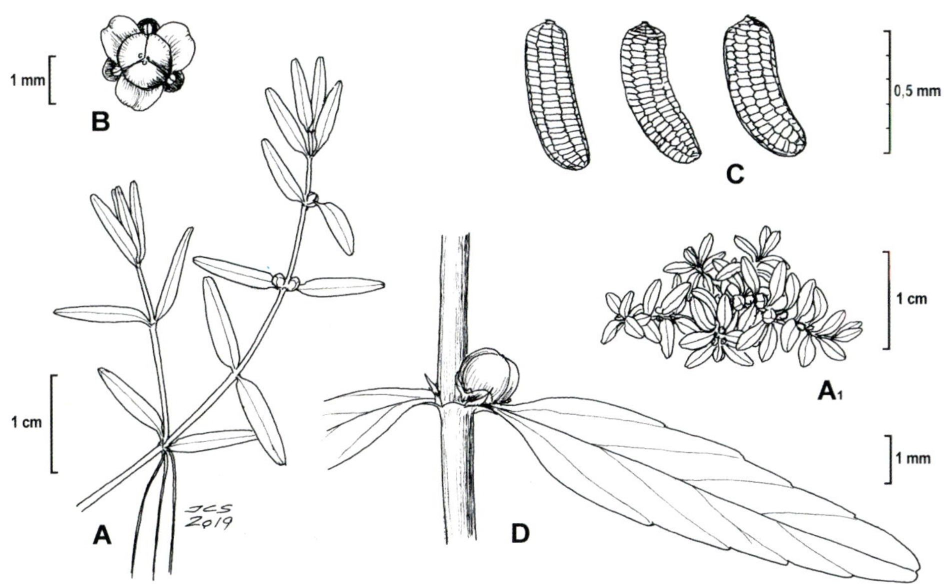

Elatine triandra
A habit submerged plant, **A1** habit terrestrial plant, **B** flower, **C** seeds, **D** leaf pair with fruit, submerged plant.

Elatinaceae

236 *Elatine hydropiper* L.

DK: Vandpeber-Bækarve **N:** Krossevjeblom **S:** Slamkrypa **FIN:** Katkeravesirikko **IS:** -
GB: Eight-stamened Waterwort **NL:** Klein glaskroos **F:** Élatine poivre d'eau
D: Wasserpfeffer-Tännel **CZ:** Úpor peprný **PL:** Nadwodnik naprzeciwlistny
EST: Mõru vesipipar **LV:** Ūdenspiparu sīkeglīte **LT:** Riestasėklė vandenė **Ru:** Повойничек согнутосемянный

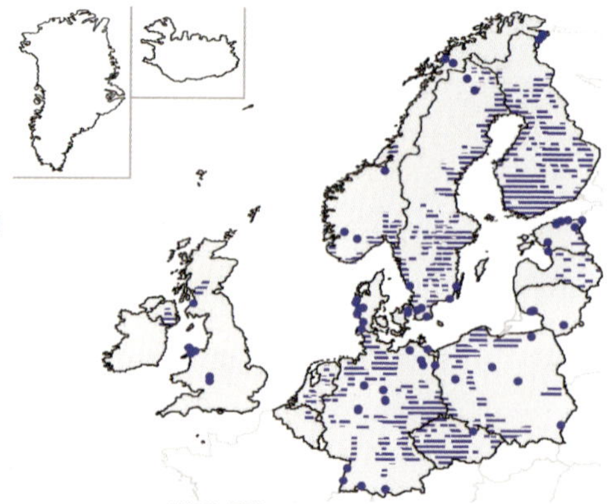

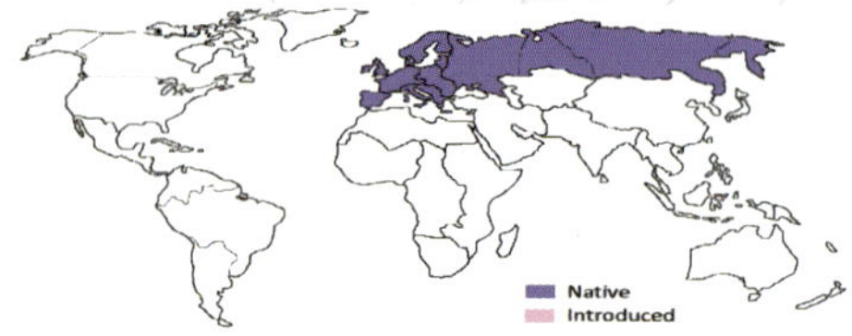

Elatine hydropiper has 4-merous flowers with 8 stamens. Indfjorden, Nees, Jutland, Denmark. Photo JCS.

Annual amphiphyte or hydrophyte, forming small patches. Roots unbranched, whitish to brownish.

Stem: Prostrate or erect, rather flaccid, terete, branched, rooting at the nodes. Terrestrial plants 1–9 cm long, prostrate, often reddish; submerged plants up to 16 cm long, creeping or partly erect, green.

Leaves: Opposite, patent to erecto-patent, 3–15(–20) mm long, petiolate. Petiole more or less abruptly tapering into the oval or ovate, often horizontally oriented lamina. Lamina entire, obtuse or slightly notched at the apex. Hydathodes dark, most distinct at the apex.

Stipules: More or less 1 mm long, membranous, narrowly triangular, acute, on one of the margins with a single tooth.

Inflorescence: Flowers solitary in leaf axils.

Flowers: 4-merous, sessile or sometimes shortly pedunculate, with a 0.3–1.0(–3.0) mm long pedicel when fruiting. Sepals uniform, 0.5–1.0 mm long, connate at base, elliptic, green or sometimes reddish in terrestrial plants. Petals 1.0–1.3 mm long, not much longer than the sepals, reddish, elliptic to obovate. With 8 stamens and 4 carpels.

Fruits: Capsule turbinate, with an apical depression, 1.0–1.6 mm wide. Seeds light brown to pale orange brown, 0.6–0.7 mm long, hooked or more or less horseshoe shaped. Seed coat with (22–)37–48(–62) inconspicuous pits in the middle row.

Flowering: June-October.

Biology: The seeds do not float, but are dispersed by water movement or by sticking to floating stem fragments or birds. They remain viable for many years.

Submerged plants have closed cleistogamous, self-pollinating flowers whereas terrestrial plants have open chasmogamous flowers, probably pollinated by small insects.

Habitats: Still or slowly flowing, mesotrophic to eutrophic or slightly brackish water (up to 2.5‰ salinity). Usually submerged to 50 cm depth, but occasionally up to 3 m. It prefers a clayey, silty or sandy, humus-rich substrate.

Elatine hydropiper in shallow water among *Eleocharis palustris*. Lake Filsoe, Denmark. Photo JCS.

Elatine hydropiper. Submerged plants with fruits and closed flowers in a drainage ditch. Ny Frederikskog, SW- Jutland, Denmark. Photo JCS.

Elatine hydropiper. Capsule usually sessile (rarely with pedicel up to 3 mm long), with an apical depression and 4 carpels, enclosing 12-20 hooked seeds. Indfjorden, Nees, W-Jutland, Denmark. Photos JCS.

Lakes, rivers, ditches, often on sites influenced by human activities. Terrestrial forms may occur on wet or temporarily dry, bare substrate close to the shoreline.

Distribution within the region:
Native. More or less throughout except Iceland and Greenland, scarce in Norway and Britain.

Characteristics and similar species:
Elatine hydropiper is small and easily overlooked and typically grows submerged. It strongly resembles **237 *E. orthosperma*** from which reliable identification must be based on the hooked seeds and capsules with an apical depression. For other similar species see **234 *E. hexandra***.

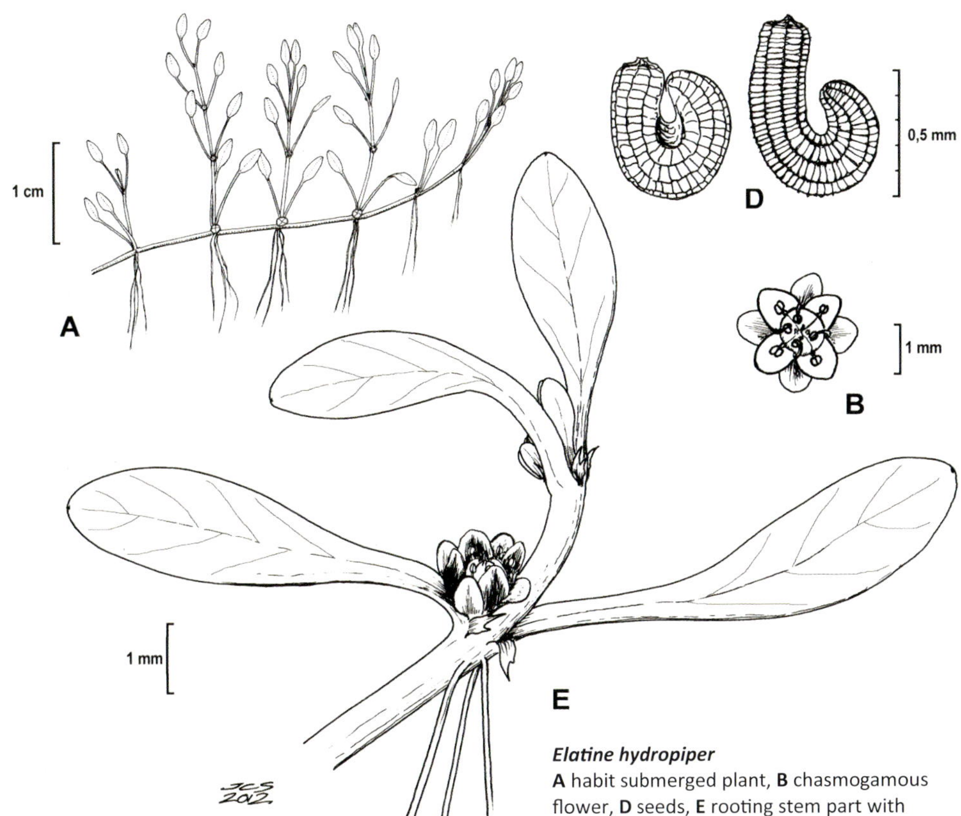

Elatine hydropiper
A habit submerged plant, **B** chasmogamous flower, **D** seeds, **E** rooting stem part with flower and a pair of leaves.

Elatinaceae

237 *Elatine orthosperma* Düben

Syn.: *E. hydropiper* subsp. *orthosperma* (Düben) F.Herm.

DK: Nordlig Bækarve **N:** Nordleg evjeblom **S:** Nordslamkrypa **FIN:** Oikovesirikko **IS:** - **GB:** - **NL:** - **F:** -
D: Geradsamiges Tännel **CZ:** Úpor rovnosemenný **PL:** -
EST: - **LV:** - **LT:** - **Ru:** Повойничек прямосемянный

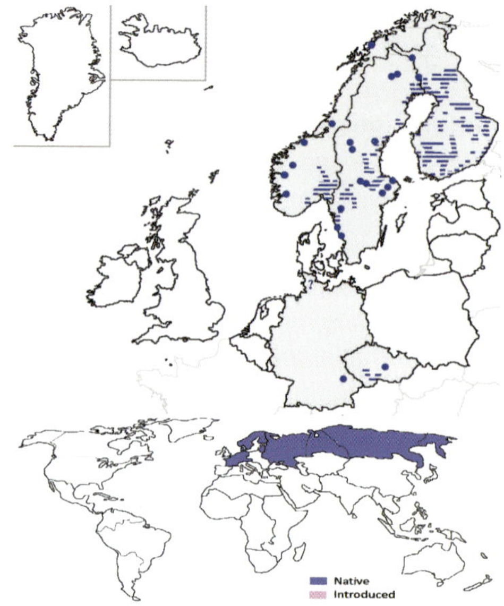

Elatine orthosperma. Tornio, Finland. Photo RVL.

Annual amphiphyte or hydrophyte, forming small patches. Roots unbranched, whitish to brownish.

Stem: Prostrate or erect, rather flaccid, terete, branched, rooting at the nodes. Terrestrial plants 1–5 cm long, prostrate, often reddish; submerged plants up to 10 cm long, creeping or partly erect, green.

Leaves: Opposite, patent to erecto-patent, 3–12 mm long, petiolate. Petiole more or less abruptly tapering into the oval or ovate, often horizontally oriented lamina. Lamina entire, obtuse or slightly notched at the apex. Hydathodes dark and most distinct at the apex.

Stipules: Up to 0.8 mm long, membranous, narrowly triangular, acute, with a single tooth on one of the margins.

Inflorescence: Flowers solitary in leaf axils.

Flowers: 4-merous, sessile or sometimes shortly pedunculate, when fruiting with a pedicel 0.1–1.5(–4.0) mm long. Sepals uniform, 0.4–0.8 mm long, connate at base, elliptic, green or sometimes reddish in terrestrial plants. Petals 0.6–1.0 mm long, not much longer than the sepals, reddish, elliptic to obovate. With 8 stamens and 4 carpels.

Fruits: Capsule cylindrical, 1.0–1.3 mm wide, with a slight apical depression. Seeds brown to pale orange brown, 0.77–0.85 mm long, cylindrical, nearly straight to slightly curved, resembling the letter J. Seed coat with (23–)32–38 (–47) inconspicuous pits in the middle row.

Flowering: June-October.

Biology: The seeds do not float, but are dispersed by water movement, by sticking to floating stem fragments or birds, remaining viable for many years. Submerged plants have closed cleistogamous, self-pollinating flowers whereas terrestrial plants have open chasmogamous flowers, probably pollinated by small insects.

Habitats: Still or slow-flowing, mesotrophic to eutrophic water. Usually submerged to 50 cm depth, but occasionally up to 1 m. It prefers a clayey, silty or sandy, humus-rich substrate. Lakes, rivers, river estuaries and seashores with very slightly brackish water, often on sites influenced by human activities. Terrestrial forms are rare, but may occur on wet or tempo-

Elatine orthosperma. Note capsule to the left. Bohuslän, Sweden. Photo JCS.

rarily dry, bare substrate close to the shoreline.

Distribution within the region:
Native. Throughout southern Norway, Sweden and Finland below the Arctic Circle and in the Alps.

Characteristics and similar species:
E. orthosperma is small, usually submerged and easily overlooked.

It is very similar to **236 *E. hydropiper*** but differs in the shape of the capsule and the size and form of the seeds.

234 *E. hexandra* has 3-merous flowers, 3 stamens and 3 carpels.

Taxonomy: *E. orthosperma* is often treated as a synonym or subspecies of *E. hydropiper*. In this book we regard it as a species in accordance with Pertti Uotila (2010) and Popiela et al. *(*2017).

Elatine orthosperma. Sangisfjärden, part of Sangis älv, Sweden. Photo RVL.

Elatine orthosperma. Seeds slightly curved. Photo JCS.

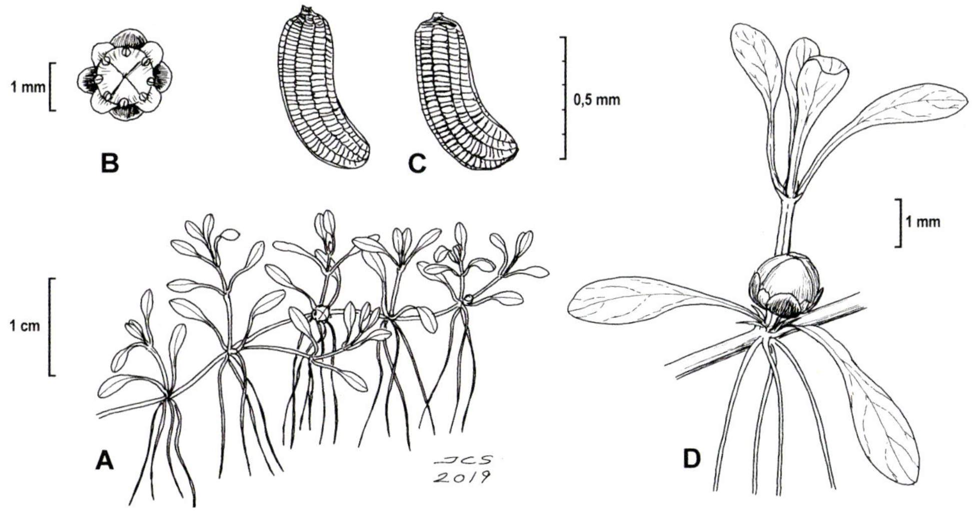

Elatine orthosperma
A habit submerged plant, **B** flower, **C** seeds, **D** part of stem with vertical shoot and fruit.

Hypericaceae

238 *Hypericum elodes* L.

DK: - **N:** - **S:** - **FIN:** - **IS:** -
GB: Marsh St. John's Wort **NL:** Moerashertshooi **F:** Millepertuis des marais
D: Sumpf-Johanniskraut **CZ:** - **PL:** -
EST: -**LV:** - **LT:** - **Ru:** -

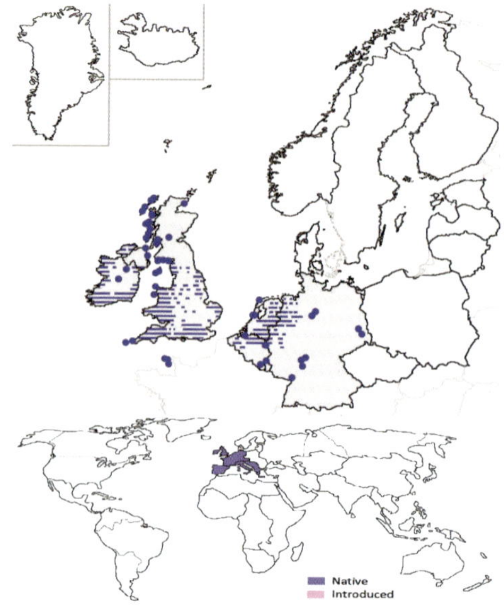

Hypericum elodes. Lüsekamp, Germany. Photo KvdW.

Perennial hydrophyte or amphiphyte. Rooting at the nodes.

Stem: 10–160 cm, 1–6 mm in diameter, branched or unbranched, terete, grooved, procumbent or ascending, glabrous in submerged plants, pubescent in floating and emergent plants.

Leaves: Opposite, circular to bluntly lanceolate when submerged, entire, 5.5–17(–23) mm long, 5–16(–19) mm wide, with 3–9 veins, leaf base cordate, semi–amplexicaul, axillary glands. Submerged leaves glabrous, green and shiny. Floating and emergent leaves pubescent and more or less glaucous, dotted.

Inflorescence: Flowers 1–13 in terminal panicles, sometimes also in leaf axils.

Flowers: 12–20 mm in diameter. Sepals ovate to elongate, acuminate, up to 3.5 mm long, 0.8–1.8 mm wide, with reddish glands. Petals yellow with green veins, persistent, triangular to obovate, 8–10 mm long, 2.7–3.0 mm wide, 3–4 times as long as the sepals. Stamens in 3 bundles of 5, connected in the basal half. Anthers yellow. Styles 3, white, 2–2.5 mm long slightly curved outwards. Stigmas capitate.

Fruits: Capsule ovoid to cylindric, 4–5 mm long, longer than the sepals, with 3 cylindrical styles. Seeds 0.8 mm long, ovate, dark yellow-brown, grooved, in a single row.

Flowering: June-September.

Biology: Submerged plants usually sterile, whereas floating and emergent plants flower. Self-pollinating or pollinated by insects. Reproduction is by seeds and rhizomes. Reproduction by fragments has not been reported but is very likely.

Habitats: Nutrient-poor, acidic, soft water, on sandy or peaty substrate. Bogs, pools, streams, ditches, lakes. As an amphiphyte *H. elodes* can grow in seasonally wet habitats.

Distribution within the region:
Native. Atlantic. From eastern Germany west to Britain and Ireland, where it is frequent in oceanic areas.

Characteristics and similar species:
Flowering emergent plants of *Hypericum elodes* are easily recognised by the densely pubescent stem and leaves.

Hypericum elodes. Floating plant. Lüsekamp, Germany. Photo KvdW.

Hypericum elodes. New Forest, Hampshire, Britain. Photo RVL.

Hypericum elodes. Submerged plant. Lüsekamp, Germany. Photo KvdW.

Non-flowering submerged plants can be confused with **225 *Crassula helmsii*** and **261-263 *Montia*** species, which differ in their more or less succulent leaves; leaves of *C. helmsii* are pointed while those of *M. fontana* are wider toward the leaf apex and arise from a square stem. The leaves of **281-292 *Callitriche*** species have veins which are more or less parallel; those of *Hypericum elodes* are divergent and branched from the base.

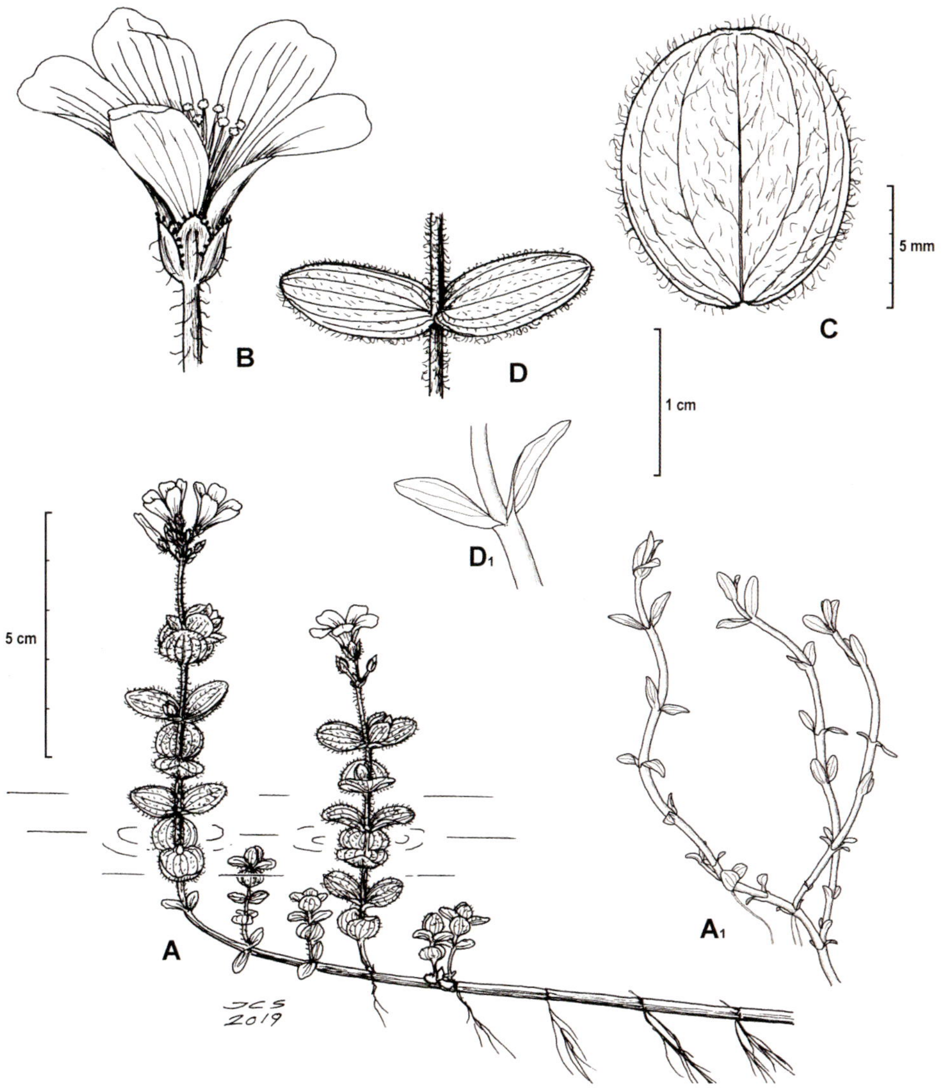

Hypericum elodes
A habit, **A1** habit submerged plant, **B** flower, **C** leaf, **D** stem and leaves, **D1** submerged stem and leaves.

Hypericum elodes. Flower. Lüsekamp, Germany. Photo KvdW.

Lythraceae

239 *Lythrum portula* (L.) D.A.Webb

Syn.: *Peplis portula* L.

DK: Vandportulak **N:** Vasskryp **S:** Rödlånke **FIN:** Ojakaali **IS:** -
GB: Water Purslane **NL:** Waterpostelein **F:** Pourpier d'eau
D: Gewöhnlicher Sumpfquendel **CZ:** Kalužník šruchový **PL:** Bebłek błotny
EST: Muda-ojakapsas **LV:** Portulaku pelcītis **LT:** Pelkinė šindra **Ru:** Бутерлак портулаковый

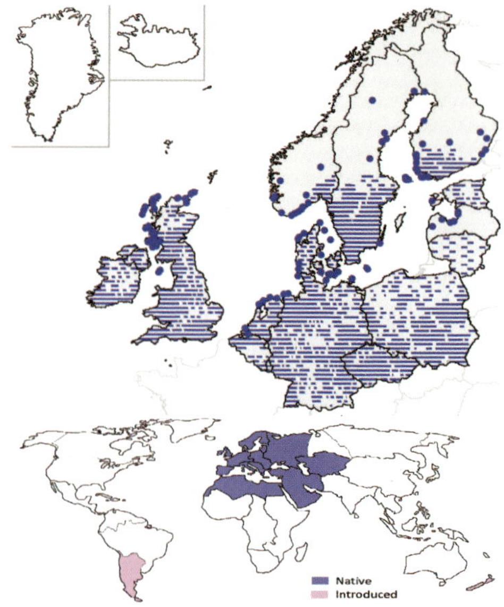

The amphibious form of *L. portula* often has reddish stems, flowers and leaf petioles. Førby Sø, Thy, Denmark. Photo JCS.

Annual amphibious or aquatic plant with whitish roots from the stem nodes. Rhizome absent.

Stem: 5–25 cm long, glabrous, square in cross section with rounded edges, more or less branched mainly near the base, green or more often distinctly reddish. Amphibious plants have prostrate stems and short internodes. Submerged plants have upright stems and long internodes, the length dependent on the depth.

Leaves: Opposite 5–15 mm long and up to 8 mm wide, petiolate, obovate with entire margin, glabrous and slightly glossy, green to brownish green, the petiole 2–3 mm long. Amphibious plants commonly have reddish petioles and reddish mid-veins, sometimes even completely reddish leaves; submerged plants often have pale greener leaves.

Inflorescence: Dioecious. Flowers solitary in the leaf axils, pedunculate with 2 bracteoles at the base of each peduncle. Amphibious plants have flowers more or less throughout the shoots; submerged plants lack flowers or have only few flowers in the upper parts.

Flowers: 6-merous, c. 3 mm in diameter with a 0.5–1.5 mm long pedicel. Hypanthium broadly bell shaped, calyx lobed with 6 acute, triangular lobes and 6 appendages, all pale green to reddish with 12 greenish to reddish veins. Petals up to 6, falling early and therefore often absent, c. 0.5 mm long, pinkish. 6 stamens with dark anthers; style short with semi-globular stigma.

Fruits: Globular capsules, 2.0–2.5 mm in diameter, commonly with a reddish tinge, slightly transparent; at maturity they split open irregularly. Seeds 0.5–0.7 mm long, light brown with one concave and one convex side.

Flowering: June-September.

Biology: Amphibious plants mainly germinate from seeds on dry or moist sediments as water levels fall. They are typically self-fertilised and flower abundantly, although insect pollination does occur. Seed dispersal by water and animals. Amphibious plants may be solitary, a few together or in large, dense stands. Germination of submerged plants from seed is uncommon as submerged plants flower much less abundantly; they more typically develop from terrestrial plants which are submerged by rising water levels. Plants die back late in the season and overwinter as seeds.

Habitats: Slightly acid to slightly alkaline and oligotrophic to mesotrophic water bodies. Mainly on dry or moist substrate in the littoral zone of permanent lakes and ponds, as well as in wet depressions and seasonal dune slacks. Plants show a positive response to physical disturbance of the substrate. Submerged plants have been recorded from depths of up to 2 metres, probably as a result of previous low water conditions, since germination in water seems to be the exception rather than the rule.

Plants with epicalyx segments up to 2 mm long have been named *L. portula* subsp. *longidentatum* (J.Gay) P.D.Sell, but are probably best treated as a variety. Dolifor, Elan Valley, Radnorshire, Britain. Photo RVL.

While typical *Lythrum portula* is normally distinctive, both submerged plants (left) and very young plants (right) are less so. Despite the longer internodes and paler leaves of submerged plants, the reddish tinge and the square cross section of the stem are diagnostic. The same applies to young and very small specimens. Råbjerg Mile, Jutland, Denmark. Photo JCS.

Distribution within the region:
Native. Throughout the region except for northern Scandinavia, Iceland and Greenland.

Characteristics and similar species:
The mainly prostrate habit of the amphibious form with ovate leaves, reddish stems with square cross section and reddish leaves are all characters that distinguish *L. portula* from most other similar aquatic plants with opposite leaves.

292 *Callitriche stagnalis* may have similar leaves, but has more or less parallel veins, the stem is terete and the flowers and fruits are very different.

Small *L. portula* specimens may be confused with species of **234-237 *Elatine***, which have terete stems, 3- or 4-merous flowers and different fruits, while **224-225 *Crassula*** species have terete stems and linear, succulent leaves.

Epilobium brunnescens (p. 549) may have a similar appearance.

The small flowers of *Lythrum portula* have 6 calyx lobes, between which the appendages radiate. The petals commonly fall off early and are thus rarely seen. Ringgive, Jutland, Denmark. Photo JCS.

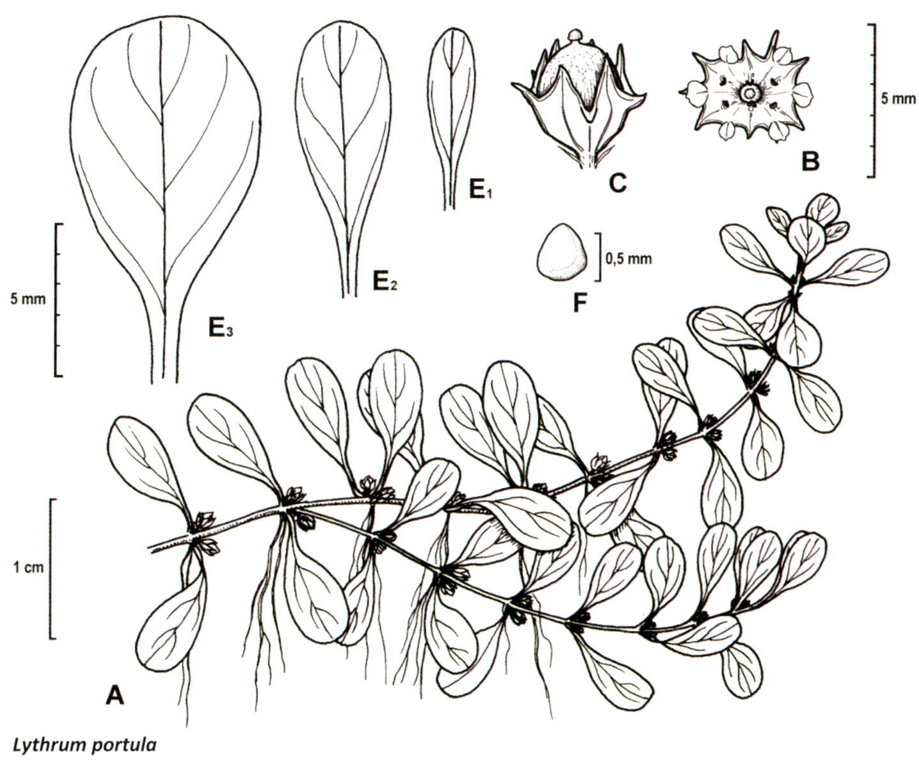

Lythrum portula
A habit, **B** flower top view, **C** capsule, **E1**, **E2**, **E3** leaves, **F** seed.

Lythraceae

240 *Trapa natans* L.

DK: Hornnød **N:** Vassnøtt **S:** Sjönöt **FIN:** Vesipähkinä **IS:** -
GB: Horn Nut **NL:** Waternoot **F:** Châtaigne d'eau
D: Wassernuß **CZ:** Kotvice plovoucí **PL:** Kotewka orzech wodny
EST: Vesipähkel **LV:** Peldošais ezerrieksts **LT:** Plūduriuojantis agaras **Ru:** Роrýльник плáвающи

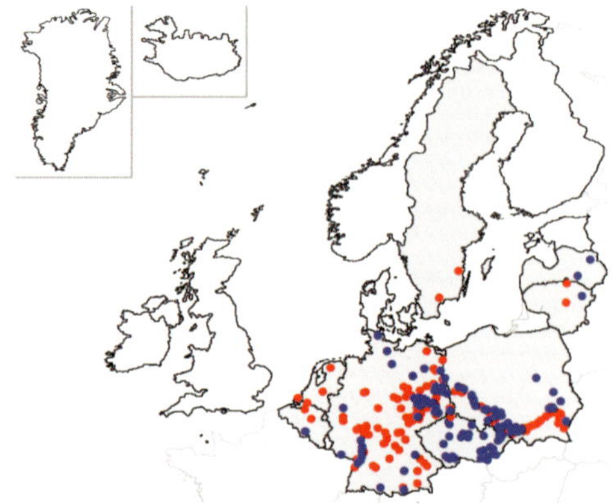

Red dots indicate historic records.

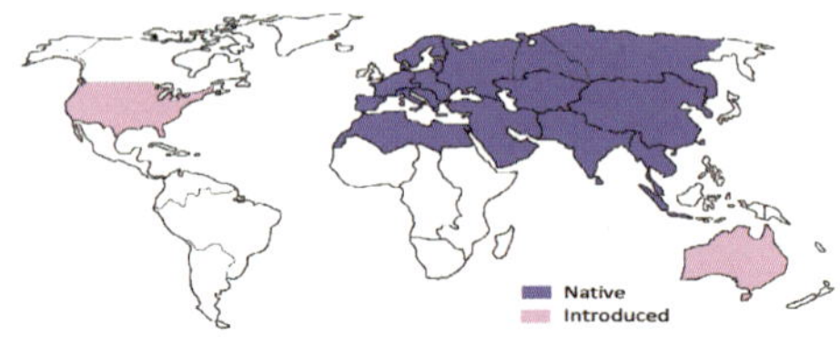

Flowering *Trapa natans*. Lake Klaucanu, Latvia. Photo Jan-Thomas Johansson.

Annual, rarely perennial hydrophyte with numerous long, fine, branched roots from the lower stem nodes.

Stem: Ascending to erect, 50–200(–500) cm long, up to 6 mm in diameter, branched or occasionally unbranched, flexible, pale- to olive green, often with a reddish tinge towards the end of the growing season.

Leaves: Submerged leaves opposite or alternate, sessile, up to 15 cm long, feathery and finely divided, with numerous segments, green to olive green, soon decaying as the floating leaves develop. Floating leaves in buoyant apical rosettes. Lamina up to 5 cm wide, broadly rhomboid, triangular, deltoid or broadly ovate with sharply serrate margin, green, leathery, glabrous and glossy above, olive green with prominent veins and short, stiff hairs below. Petioles of floating leaves 7–29 cm long, hairy, swollen, hollow and spongy in places, green to olive green. Leaves and petioles become increasingly pinkish to reddish toward the end of the growing season.

Inflorescence: Monoecious. Up to 20 solitary flowers in the axils of younger rosette leaves.

Flowers: 4-merous, shortly pedicellate. Sepals green, hairy, fused at base. Petals white to greenish, obovate, up to 8 mm long, soon decaying. Stamens 4. Style 1.

Fruits: 1-seeded, woody nuts, c. 2.5 cm wide, with (2–)4 distinct, sharp spines which develop from the sepals.

Flowering: June-August.

Biology: Wind or insect pollinated and self-pollinated. Seeds are dispersed by water. Ripe seeds sink to the bottom and usually germinate the following summer, when the water temperatures rise. Vegetative dispersal occurs widely by detached rosettes, which are able to produce seeds even when no longer rooted. Seeds can remain viable for up to 12 years in submerged sediments, but do not tolerate desiccation. Under favourable conditions dense stands are formed with rosettes covering the surface of the water and so excluding submerged aquatic plants.

Habitats: Mesotrophic to eutrophic, neutral to alkaline waters in open sunlit exposed ponds, lakes, canals, ditches and slow-flowing streams and rivers, as well as brackish estuaries, all mainly with soft and muddy substrate.

Dense floating mats of *Trapa natans*. Hungary. Photo KvdW.

Fruits. Lake Pokratas, Latvia. Photo Jan-Thomas Johansson.

Upturned specimen of *Trapa natans* showing lower side of floating leaves, fruits and submerged leaves. Hungary. Photo KvdW.

Distribution within the region:
The native distribution is obscured by historic planting and cultivation, with confirmed cultivation since at least the sixteenth century and speculation that it has been widely exploited since the Stone Age. It is local in Belgium, the Netherlands, Germany, the Czech Republic, Poland, Latvia, Lithuania and southern Finland, also formerly Sweden where it is now extinct.

Characteristics and similar species:
With its floating rosettes of rhomboid to triangular, sharply serrate leaves with swollen petioles and large, spiny fruits, fully developed *Trapa natans* can hardly be confused with any other aquatic plants.

When plants occur with only the finely divided submerged leaves it may be confused with, e.g., **226-231 *Myriophyllum*** spp., which have leaves in whorls, and with **264 *Hottonia palustris,*** which has alternate leaves usually with wider, flattened segments.

Taxonomy: The taxonomy of the genus *Trapa* appears to be complex, with one species recognised throughout much of Europe, but up to 15 species recognised further east.

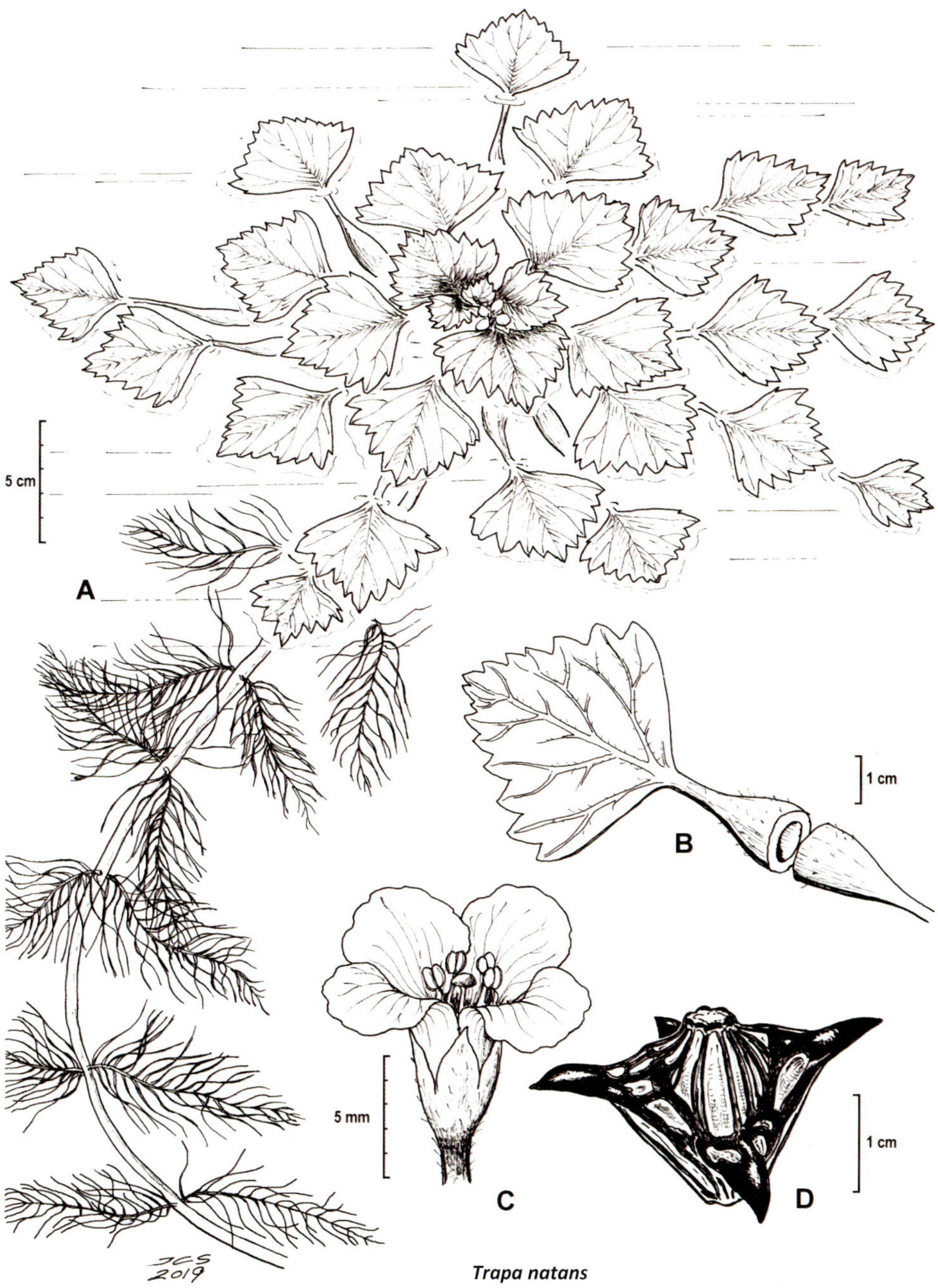

Trapa natans
A habit, **B** floating leaf, **C** flower, **D** fruit.

Onagraceae

241 *Epilobium hirsutum* L.

DK: Lådden Dueurt **N:** Stormjølke **S:** Rosendunört **FIN:** - **IS:** -
GB: Greater Willowherb **NL:** Haarig wilgeroosje **F:** Épilobe hirsute
D: Zottiges Weidenröschen **CZ:** Vrbovka chlupatá **PL:** Wierzbownica kosmata
EST: Karvane pajulill **LV:** Pūkainā kazroze **LT:** Plaukuotoji ožkarožė **Ru:** Кипрей волосистый

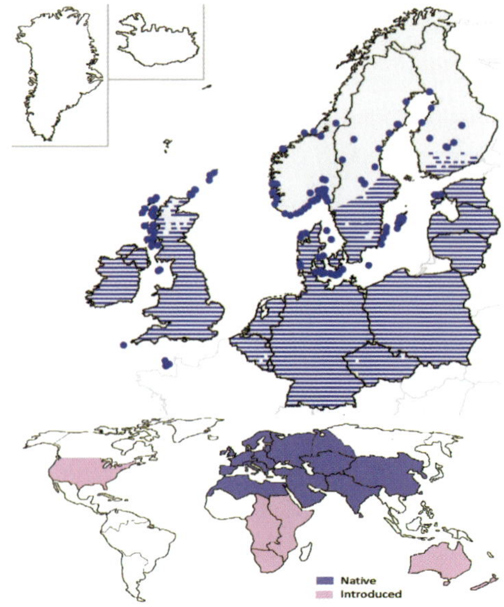

Epilobium hirsutum overwinters as thick stolons with apical rosettes of upright, glabrous emergent or submerged leaves. Hegedal, Hobro, Denmark. Photos JCS.

Perennial amphibious plant with creeping rhizome and numerous whitish roots.

Stem: 80–150(–200) cm long, erect, robust, the lower parts becoming woody with age, up to > 15 mm diameter, terete, densely hairy, especially in the upper parts, which are typically richly branched. Typically with numerous radiating, whitish subterranean stolons, as well as green to red stolons creeping over the substrate, both with robust bracts, each with a terminal rosette of short, upright, slightly fleshy and smooth, wintergreen leaves.

Leaves: Stem leaves 6–12(–14) cm long, up to 3 cm wide, opposite, the upper sometimes alternate, lanceolate, sessile, clasping, irregularly serrate and densely hairy, the hairs giving the green leaves a greyish look. Rosette leaves glabrous with crenate margins, bright green, commonly more or less reddish to purple on the lamina.

Inflorescence: Dioecious. Apical racemes with many pedicellate flowers on the main stem and branches.

Flowers: 4-merous, 15–28 mm diameter, on pedicels 10–15 mm long, arising from the axils of the bracts. Sepals 7–13 mm long, lanceolate, greenish to light brownish with a reddish tinge. Petals 10–20 mm long, bright pink to pinkish purple or white, deeply notched. 8 stamens, of unequal length. Style solitary with a deeply 4-lobed stigma.

Fruits: 4-locular capsules 6–9 cm long, the surface densely covered in long glandular hairs. Seeds papillose, 1.0–1.2 mm long, with silky hairs 8–12 mm long.

Flowering: June-September.

Biology: Pollinated by insects and by self-fertilisation. Seed set generally abundant, the seeds mainly wind-dispersed, but also dispersed by animals and water. Often forming dense stands through vigorous vegetative growth of numerous and often very long runners. Late in the season parent plants die back and plants overwinter as short-leaved rosettes on subterranean, submerged or creeping stolons which are commonly submerged during the winter. Seeds germinate on exposed substrate, and once established, the plants can grow with the roots and stem base submerged. Ditches and small streams often support dense, more or less monospecific stands.

Habitats: Predominantly eutrophic and neutral to alkaline soils within a wide range of wetlands, from relatively dry fields, meadows and riverbanks to ditches, canals, streams, rivers, ponds and lakes.

Epilobium hirsutum sending long runners out into the running water from the rooting point above the water line in the bank of a river. Gudenå, Denmark. Photo BM.

Distribution within the region:
Native. Throughout, except for Greenland and Iceland. In Norway, Sweden and Finland only widespread in the south, more scattered further north and absent from the extreme north.

Characteristics and similar species:
Flowering *Epilobium hirsutum* can be distinguished from other *Epilobium* species by the large flowers and the clasping, very pubescent leaves. Overwintering rosettes can be distinguished from those of other *Epilobium* species by the upright, 6–12 cm long, up to 3 cm wide and commonly purple-tinged leaves with irregularly serrate margins.

297 ***Veronica anagallis-aquatica*** and **300** ***Veronica catenata*** also have opposite leaves, but angular stems.

Flowering *E. hirsutum*. Fyrkat Enge, Hobro, Denmark. Photo JCS.

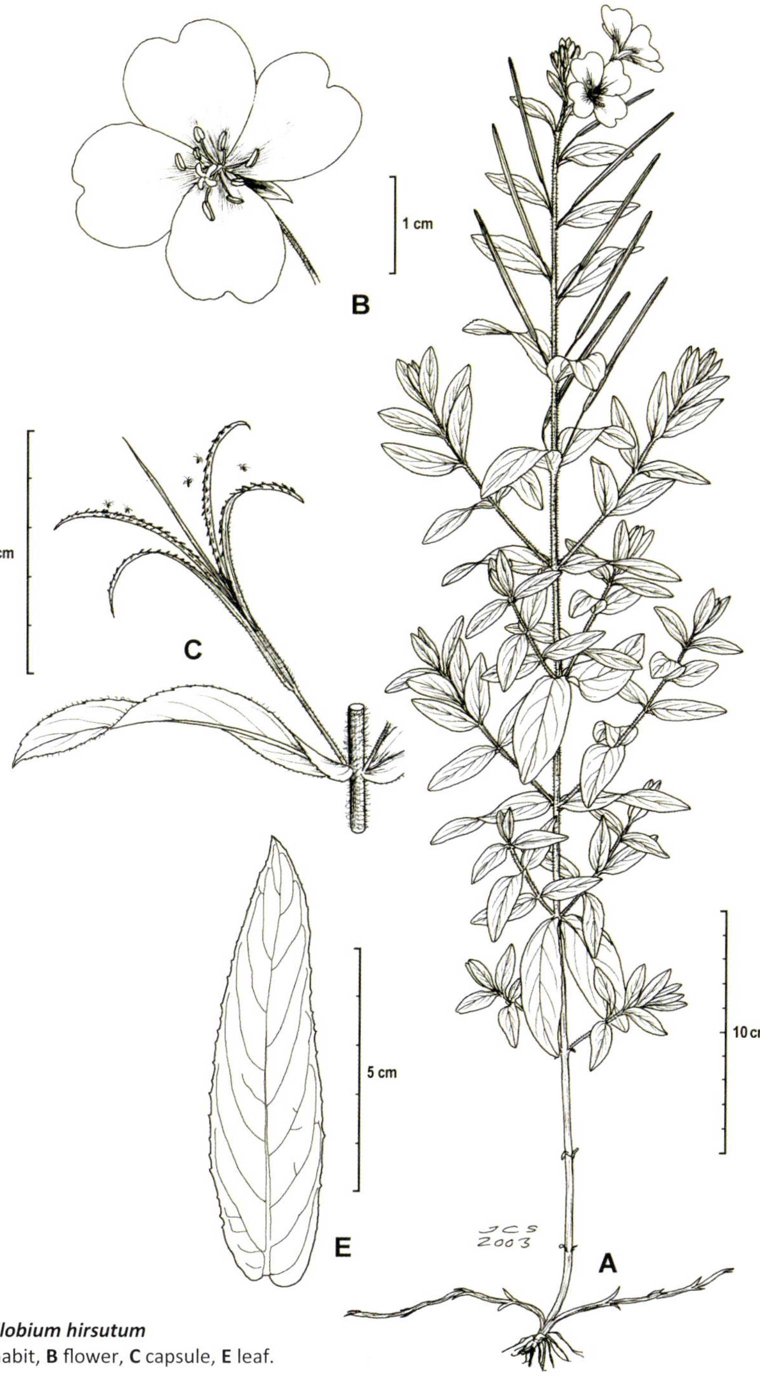

Epilobium hirsutum
A habit, **B** flower, **C** capsule, **E** leaf.

Several *Epilobium* species and hybrids can be found growing along the margins of streams and occasionally under water - with rosettes at least partly under water. Determination of these species requires flowering and fruiting material and is not treated in this book.

Epilobium brunnescens (Cockayne) Raven & Engelhorn. Cairnwell, Scotland. Photo RVL.

Epilobium hornemannii Rchb. Valdres Flya, Jotunheimen, Norway. Photo JCS.

Epilobium palustre L. Denmark. Photo JCS.

Onagraceae

242 *Ludwigia palustris* (L.) Elliott

Syn.: *Isnardia palustris* L.
DK: - **N:** - **S:** Ludwigia **FIN:** - **IS:** -
GB: Hampshire-purslane **NL:** Echt waterlepeltje **F:** Ludwigie de marais
D: Sumpf-Heusenkraut **CZ:** Zakucelka bahenní **PL:** Ludwigia błotna
EST: - **LV:** - **LT:** - **Ru:** Людвигия болотная

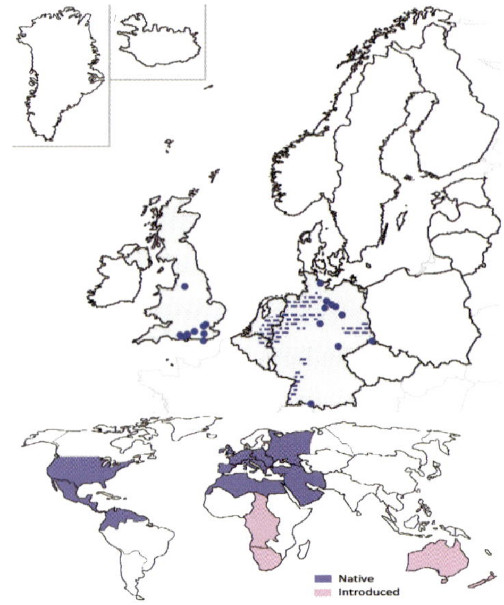

Ludwigia palustris growing on muddy substrate. Mol, Belgium. Photo KvdW.

Annual or short-lived perennial hydrophyte or amphiphyte. Creeping and rooting at the nodes.

Stem: 10–70 cm, branched, 4-angular, procumbent or ascending, often reddish.

Leaves: Emergent or floating, opposite, 10–30(–55) mm long, 6–15 mm wide, 1.5 times as long as wide, petiolate, ovate elliptic, entire, shiny to glossy, red veined. Petioles 5–20(–30) mm long. Stipules papillose.

Inflorescence: Flowers solitary in leaf axils.

Flowers: 2–5 mm diameter. Sepals 4, persistent, triangular, 1.4–2 mm long, 0.8–1.8 mm wide. Petals absent. Stamens 4, shorter than the sepals. Stigma capitate, 0.25–0.4 mm. Style pale green, 0.5–0.7 mm long.

Fruits: Capsule 3.5–5 mm long, 1.5–3 mm wide, pale yellow to pale brown with green ribs, slightly 4-angular. Seeds numerous, 0.5-0.7 mm long, ellipsoid, yellowish brown.

Flowering: June-September.

Biology: Aquatic plants tend to flower less frequently than terrestrial plants. Flowers are pollinated by water, beetles and flies, but are capable of self-pollination; it may also be viviparous and grow from detached shoots. Seeds require light for germination.

Habitats: Nutrient-rich, soft water, on sandy, gravelly or clayey substrate. Ponds, streams, ditches and shallow parts of lakes and gravel pits. As an amphiphyte *Ludwigia palustris* can grow on the drawdown zones of large water bodies and in seasonal wetlands.

Distribution within the region:
Native to southern England, Belgium, the Netherlands and Germany, occasionally occurring as an introduction elsewhere.

Characteristics and similar species:
Ludwigia species have entire leaves, whereas the leaves of similar *Veronica* species are toothed. **246 *Ludwigia peploides*** and **245 *L. grandiflora*** both have alternate leaves and large flowers, while *L. palustris,* **243 *L. ×kentiana*** and **244 *L. repens*** have opposite leaves and inconspicuous flowers.

L. palustris has no petals, *L. ×kentiana* has very short petals (0.5 mm long) and *L. repens* has longer petals (3–5 mm long).

Ludwigia palustris. Partly submerged plants. Beja Litoral, Portugal. Photo KvdW.

Ludwigia palustris has flowers without petals. Oberwater Floodplain, New Forest, Hampshire, Britain. Photos KvdW and RVL.

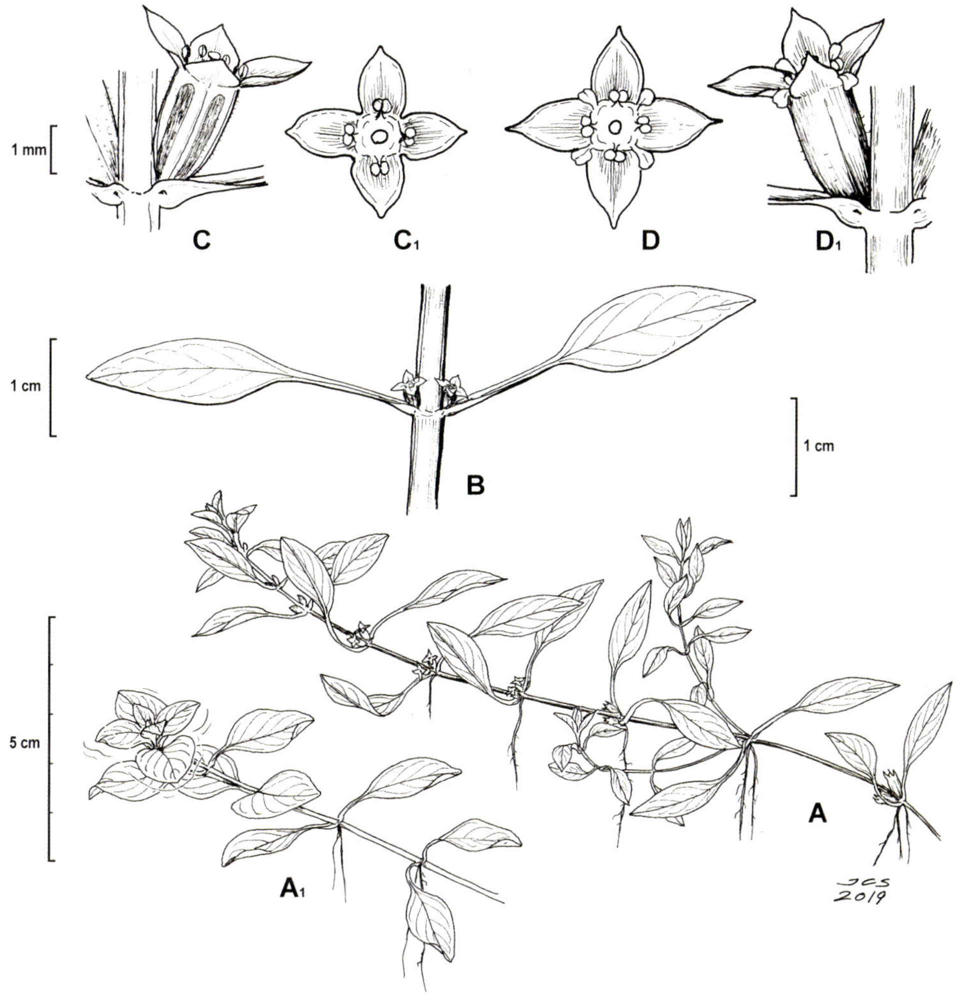

Ludwigia palustris
A, A1 habit, **B** opposite leaves, **C, C1** flower.
Ludwigia ×kentiana
D, D1 flower.

Onagraceae

243 *Ludwigia ×kentiana* E.J.Clement

L. palustris × repens

Leaves widest in the upper third. Bracteoles about 1 mm long. Sepals longer than wide. Pale yellow petals 0.5 mm long. More or less cylindrical deciduous, uniformly green capsule.

Introduced. Britain and Germany.

Ludwigia ×kentiana has small, yellow petals. Daxlanden, Upper Rhine, Germany. Photo KvdW.

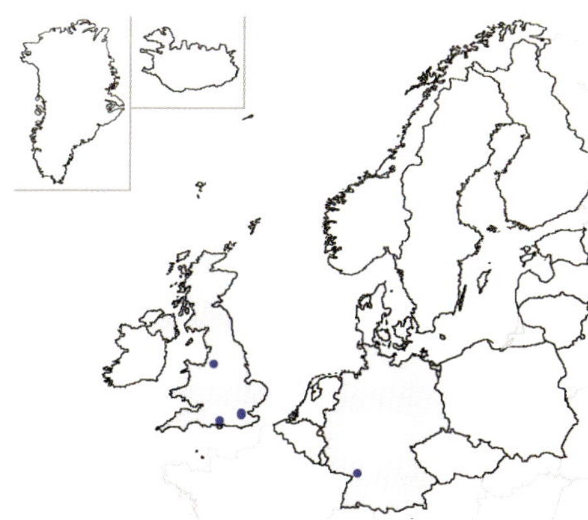

Onagraceae

244 *Ludwigia repens* J.R.Forst.

Syn.: *Ludwigia natans* Elliott

Leaves widest near the apex. Bracteoles 1-2 mm long, fused at the base. Sepals much longer than wide. Yellow petals 3-5 mm long. Cylindrical, persistent, uniformly yellow-green capsule.

Introduced. Hannover, Germany. Native in USA, Mexico to El Salvador and the Caribbean.

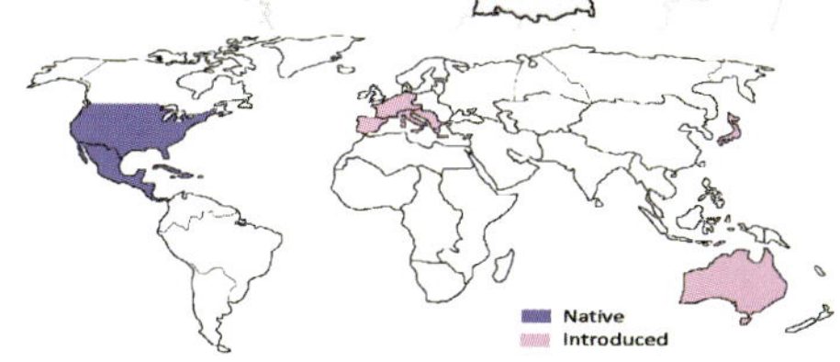

Onagraceae

245 *Ludwigia grandiflora* (Michx.) Greuter & Burdet

Syn.: *L. uruguayensis* Cambess. in A.St.-Hil.
DK: - **N:** - **S:** Sumpludwigia **FIN:** - **IS:** -
GB: Perennial Water Primrose **NL:** Waterteunisbloem **F:** Ludwigie à grandes fleurs
D: Grossblütiges Heusenkraut **CZ:** Úpor rovnosemenný **PL:** -
EST: - **LV:** - **LT:** - **Ru:**

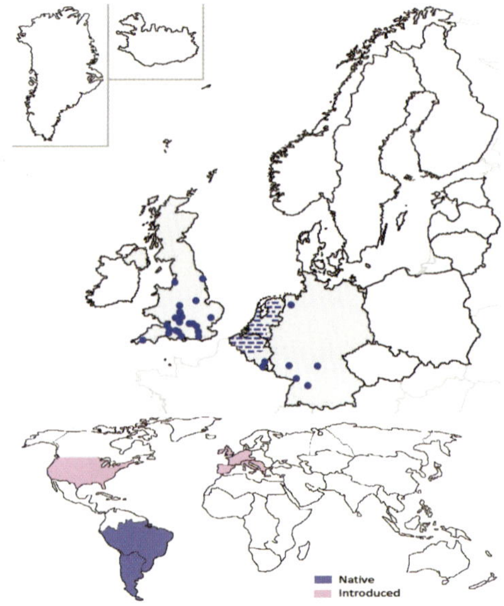

Extensive growth of *L. grandiflora* in a sand/gravel pit in the valley of River Maas. Stevensweert, Netherlands. Photo KvdW.

Perennial hydrophyte or amphiphyte. Rooting at the nodes.

Stem: 50–300 cm, branched, procumbent to ascending, glabrous or pubescent.

Leaves: Emergent or floating, alternate, 4–10(–13) cm long, 0.6–15(–27) mm wide, 5–6 times as long as wide, lanceolate to elliptic, acute, entire, dull and pubescent, with 8–13 veins on each side of the midrib. Floating leaves are very different to emergent leaves; they are glabrous and shiny, with an obtuse apex. Petiole 1.0–5.3 mm. Stipules triangular.

Inflorescence: Flowers solitary in leaf axils, peduncles approximately 2.5 cm long.

Flowers: 40–60 mm in diameter. Sepals 5 (rarely 6), pubescent or glabrous, 6–18 mm long, 1.5–3 mm wide. Petals 5, 18–25(–30) mm long, caducous, obovate, emarginate, bright yellow with an orange patch at the base. Stamens 10, in 2 whorls, the epipetalous ones shorter. Stigma capitate, 1.2–2 mm. Style glabrous or hairy.

Fruits: Capsule 1.3–2.5 x 3–4 mm, light-brown, pubescent, petiole 2.5–6 cm. Fruit containing 40–50 seeds. Seeds 1.5 mm long, embedded in a woody endocarp.

Flowering: August-October.

Biology: Most populations flower, but the primary means of reproduction is through clonal expansion by shoot fragments with sexual reproduction less important. Fragments are dispersed by animals, humans, and water. The flowers of *L. grandiflora* are self-incompatible, but the species has a very high seed output. Two different types of roots have been described: those which absorb nutrients and attach the plant to the soil and adventitious roots located along the stems, which ensure oxygen uptake and enable prostrate stems to root. *L. grandiflora* possesses allelopathic properties which increase its competitive ability by reducing competition from other plant species.

Habitats: Typically in eutrophic water bodies. On muddy, sandy or clay substrate in ponds, streams, canals, ditches and in shallow parts of lakes, reservoirs and gravel pits. *L. grandiflora* can grow to a depth of 3 metres, and up to 80 cm above the water level. As an amphiphyte *L. grandiflora* can grow in seasonal wetlands and the drawdown zones of permanent standing or flowing water bodies.

Ludwigia grandiflora. The large flowers are very attractive for the gardener. Stevensweert, Netherlands. Photo KvdW.

Ludwigia grandiflora. Plants with floating leaves. Stevensweert, Netherlands. Photo KvdW.

Ludwigia grandiflora.
Leaf base with stipule.
Photo KvdW.

Distribution within the region:
Introduced. Established in Belgium and the Netherlands, local in Germany. It has been recorded at many sites in Britain but most, if not all, populations have been eradicated. Introduced to Europe with the first observation in France in the 1830s where it is now very widespread and abundant.

Characteristics and similar species:
Ludwigia species can be differentiated from similar **297-302 *Veronica*** species by the entire leaves.

242 *L. palustris*, **244 *L. repens*** and **243 *L. ×kentiana*** have opposite leaves and inconspicuous flowers.

246 *L. peploides* and *L. grandiflora* have alternate leaves and large flowers. *L. grandiflora* has large flowers with sepals up to 18 mm long and petals 12–25 mm long and thereby different from *L. peploides,* which has sepals up to 10 mm long and petals 7–17 mm long. The stipules of *L. grandiflora* are triangular while *L. peploides* has obtuse to rounded stipules.

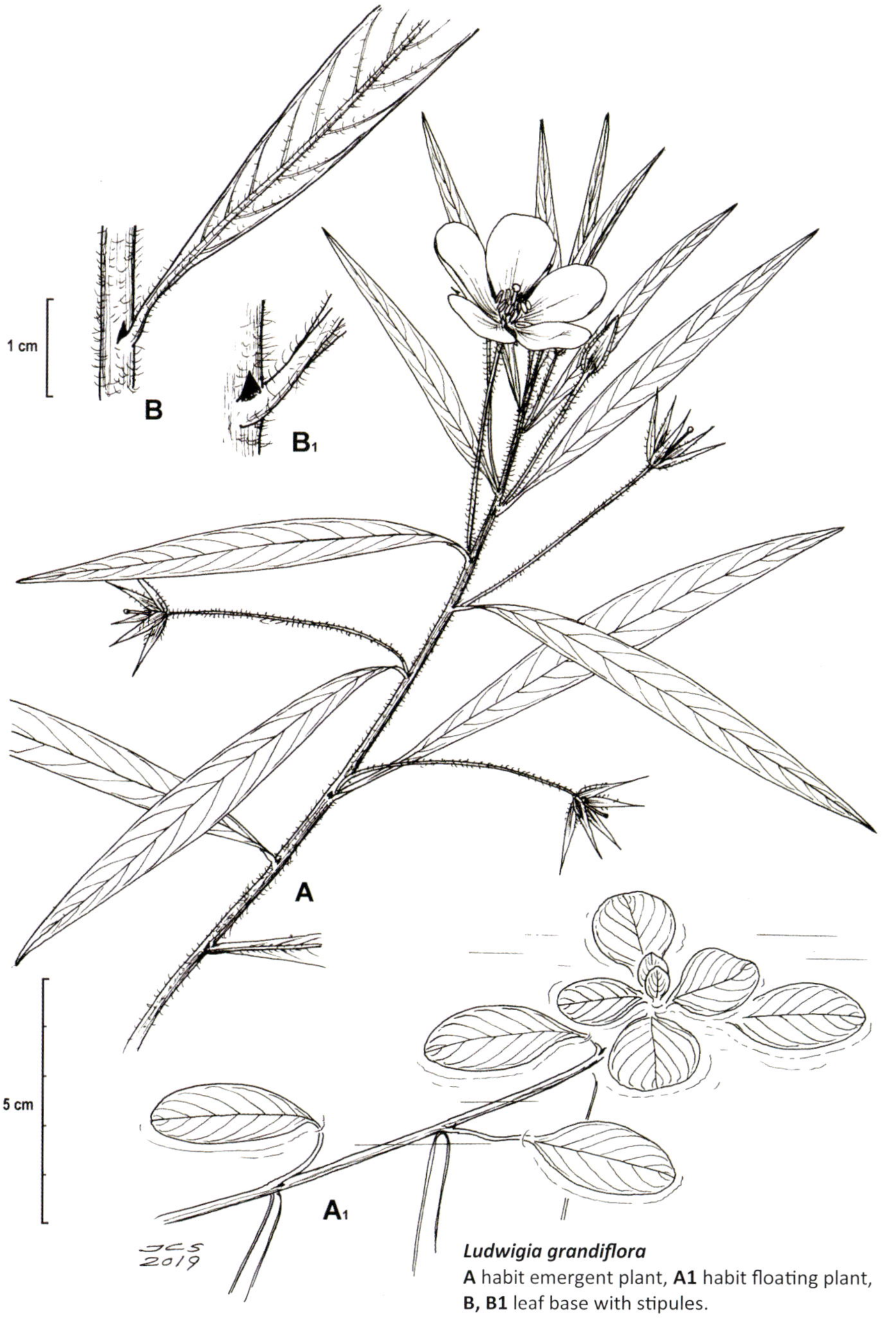

Ludwigia grandiflora
A habit emergent plant, **A1** habit floating plant, **B, B1** leaf base with stipules.

Onagraceae

246 *Ludwigia peploides* (Kunth) P.H.Raven

DK: - **N:** - **S:** - **FIN:** - **IS:** -
GB: Floating Water Primrose **NL:** Kleine Waterteunisbloem **F:** Ludwigie faux-pourpier
D: - **CZ:** - **PL:** -
EST: - **LV:** - **LT:** - **Ru:**

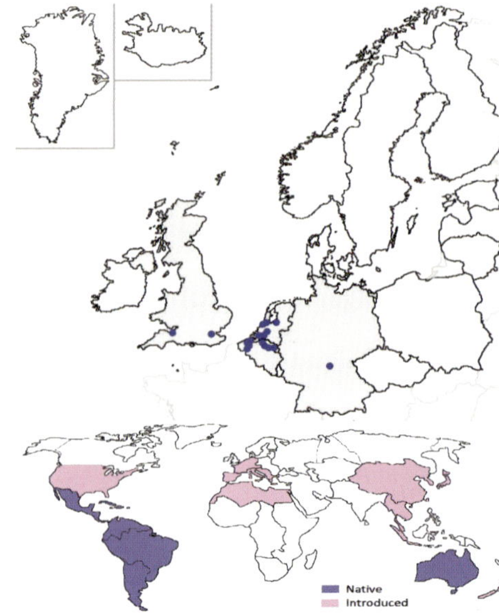

Ludwigia pepolides growing partly onshore and in the water, seasonally inundated hollows along the channel of the River Loire near Tours, France. Photo RVL.

Perennial hydrophyte or amphiphyte. Rooting at the nodes.

Stem: 30–200 cm, branched, procumbent, ascending or erect, glabrous or pubescent.

Leaves: Emergent or floating, alternate, lanceolate to elliptic and acute, entire, 3.5–6.0 cm long, 0.5–2.5 cm wide, 2–3 times as long as wide, emergent leaves dull and pubescent while floating leaves are shiny and glabrous, petiole 1.5–3.5 mm, stipules obtuse rounded.

Inflorescence: Flowers solitary in leaf axils, peduncles approximately 2.5 cm long.

Flowers: 20–30 mm in diameter. Sepals 5, pubescent or glabrous, (4–)7–8(–12) mm long, 1.5–2.5 mm wide. Petals 5, obovate, emarginate, bright yellow with an orange patch at the base, 12–16 mm long. Stamens 10, in 2 whorls, 4–5 mm long. Stigma capitate. Style thin, shorter than the stamen.

Fruits: Capsule 14–22 mm long, 5-angular, petiole 20–40 mm. Fruit containing 40–50 seeds 1.5 mm long, embedded in the inner fruit wall.

Flowering: June-October.

Biology: Most populations flower, but the primary means of reproduction is through clonal expansion by shoot fragments with sexual reproduction less important. Fragments are spread by animals, humans and water. Two different types of roots have been described: those which absorb nutrients and attach the plant to the soil, as well as adventitious roots located along the stems, which ensure oxygen uptake and enable prostrate stems to root. *L. peploides* possesses allelopathic properties which increase its competitive ability by reducing competition from other plant species.

Habitats: Typically in nutrient-rich water of ponds, streams, canals, ditches and rice fields. Also in shallow parts of lakes, reservoirs and gravel pits. On muddy, sandy or clay substrate.

L. peploides can grow up to a depth of 3 m, and up to 80 cm above water level, as well as on damp ground on riverbanks. As an amphiphyte *L. peploides* can grow in seasonal wetlands and the drawdown zones of permanent standing or flowing water bodies.

Flowering *L. peploides*. Nieuw-Loosdrecht, Netherlands. Photo KvdW.

Ludwigia peploides floating from plants rooted in the margin. Jardin Botanique, Conservtoire Botanique National de Brest, France. Photos RVL.

Ludwigia pepolides. Leaf base with stipule. Photo KvdW.

Distribution within the region:
Introduced. Scattered occurrences in Belgium and the Netherlands, very local in Germany; in Britain a few populations have been reported, but it is thought that all have been eradicated. Introduced to France in 1830 where it is widespread and abundant.

Characteristics and similar species:
Ludwigia species can be differentiated by the entire leaves from similar **297-302** ***Veronica*** species, which have toothed leaves.

242 ***L. palustris***, **244** ***L. repens*** and **243** ***L. ×kentiana*** have opposite leaves and inconspicuous flowers.

L. peploides and **245** ***L. grandiflora*** have alternate leaves and large flowers. *L. peploides* has flowers with sepals up to 10 mm long and petals 7–17 mm and thereby different from *L. grandiflora*, which has larger flowers with sepals up to 18 mm long and petals 12–25 mm long. The stipules of *L. peploides* are obtuse rounded while *L. grandiflora* has triangular stipules.

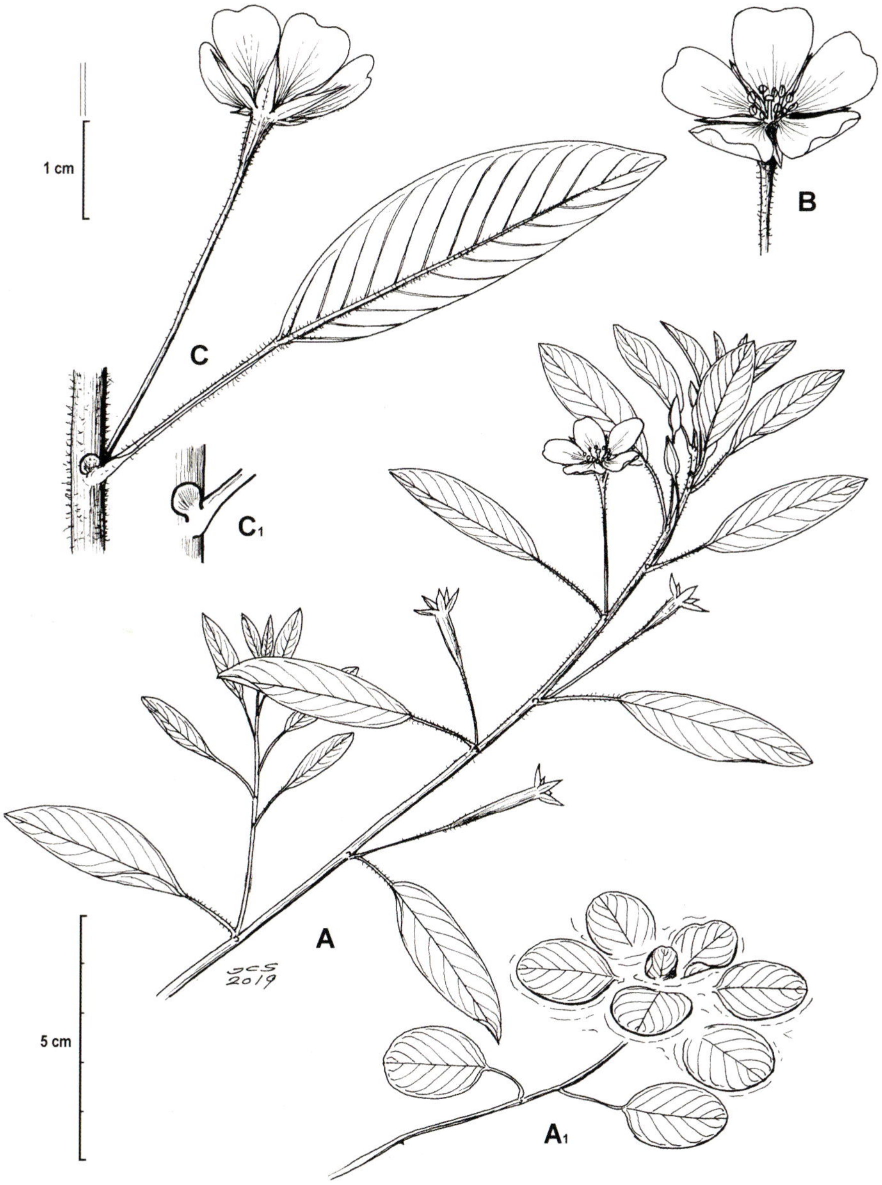

Ludwigia peploides
A habit emergent plant, **A1** habit floating plant, **B** flower, **C** flower and leaf with stipules, **C1** stipule.

Brassicaceae

247 *Cardamine amara* L.

DK: Vandkarse **N:** Bekkekarse **S:** Bäckbräsma **FIN:** Purolitukka **IS:** -
GB: Large Bitter-Cress **NL:** Bittere veldkers **F:** Cardamine amère
D: Bitteres Schaumkraut **CZ:** Řeřišnice hořká **PL:** Rzeżucha gorzka
EST: Mõru jürilill **LV:** Rūgtā ķērsa **LT:** Karčioji kartenė **Ru:** Сердечник горький

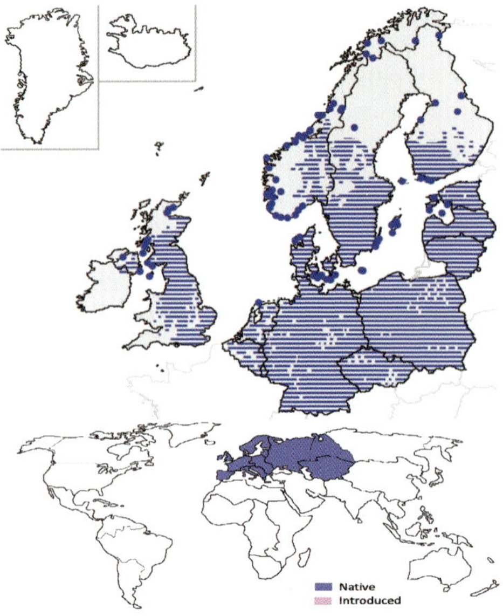

Cardamine amara. Rabis Bæk, Karup, Jutland, Denmark. Photo BM.

Perennial helophyte. Rhizome creeping, horizontal with numerous stolons bearing green leaves above ground, or below ground with short pale scales. Roots unbranched, white to brownish primarily from horizontal stem parts and lower stem nodes, the latter especially when plants are floating.

Stem: Aerial stems 15–50 cm long, ascending to upright from a prostrate rooting base, commonly unbranched, the upper parts glabrous, lower parts furrowed with short hairs.

Leaves: Stem leaves alternate, green, pinnate, with 2–4(–5) pairs of oval to ovate lobes and 1 terminal lobe, the lower leaves petiolate, the upper sessile, glabrous or hairy along the midrib and the margin of the lamina. Lower leaves never forming a rosette.

Inflorescence: Monoecious. Racemes with 9–24 flowers, emerging mainly from the upper leaf axils, eventually elongating during flowering.

Flowers: 4-merous. Bisexual. Flowers 12–16 mm in diameter with terete, green petioles. Sepals c. 4 mm long, glabrous, not joined, green, oblong ovate to obtuse with narrow membranous margin. Petals 6–9 mm long, not joined, white with greenish or more rarely violet veins, obovate. 4 long and 2 short stamens with green filaments and violet anthers. Style 2–3 mm long.

Fruits: Short-stalked siliquae, 20–40 mm long and 1–2 mm diameter, oblique upright. 1 row of seeds in each of the loculi. Pedicels 10–20 mm long.

Flowering: May-June.

Biology: Insect pollinated. Moist seeds are sticky; they are dispersed by birds and other animals. Vegetative propagation by long runners, thus often forming large stands. Wintergreen with

Cardamine amara. Submerged plants in the early spring. Aarhus, Denmark. Photo BM.

short shoots in or just above water. Permanently submerged plants are rare and mainly occur in springs.

Habitats: Mesotrophic and slightly acid to alkaline soil in the littoral zone of flowing as well as standing waters of springs, streams, rivers, canals and ditches, lakes and ponds. Also on wet soil in wet forest, meadows and spring fens.

Distribution within the region:
Native. Widespread and abundant throughout much of the region, but absent from Greenland, Iceland and northern Scandinavia and scarce in Ireland and southern Norway.

Characteristics and similar species:
The solid stems and large white flowers with violet anthers distinguish *Cardamine amara* from the two **248-250 *Nasturtium*** spp., which have hollow stems and smaller white flowers with yellow anthers.

C. pratensis L. differs from *C. amara* by bright violet to almost white petals, yellow anthers, hollow stems, basal leaves in rosettes and lack of stolons.

The violet anthers distinguish *C. amara* from *Nasturtium* spp. Photo KvdW.

Cardamine amara. The shape of the leaflets is characteristic. Stroudwater Canal, Gloucestershire, Britain. Photo RVL.

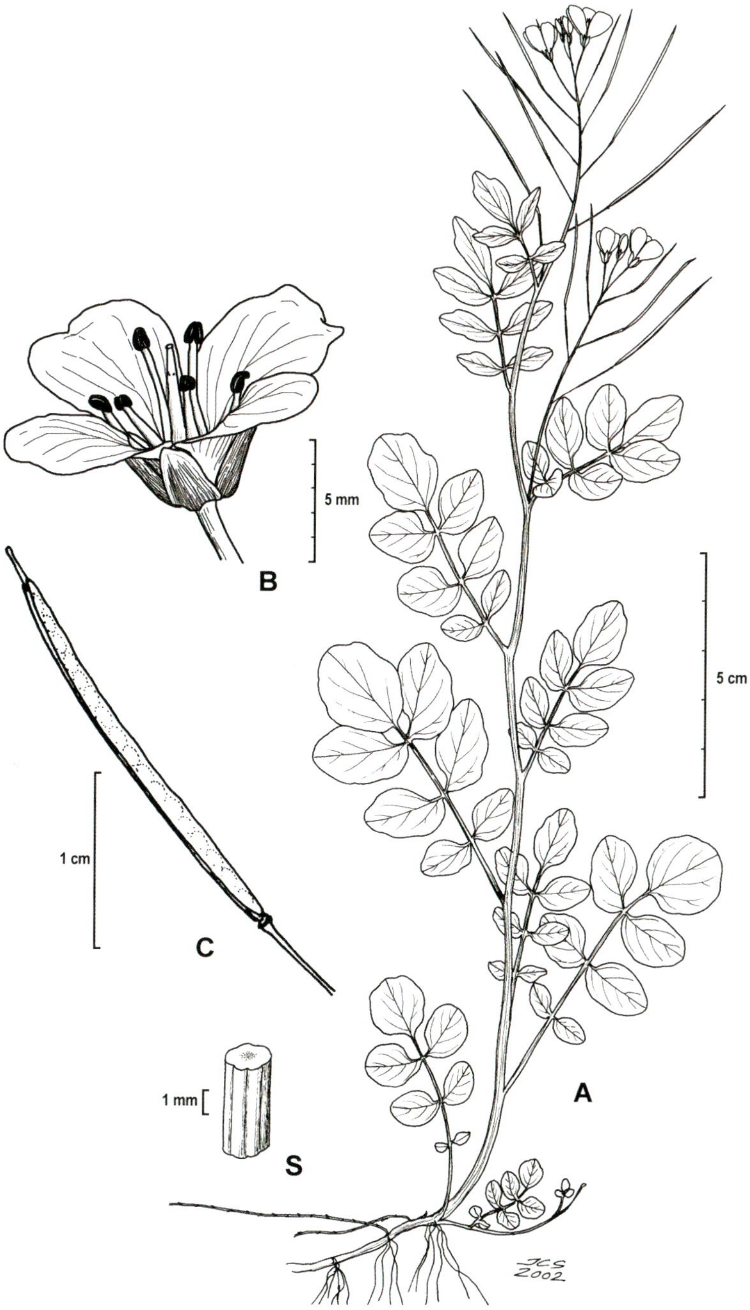

Cardamine amara
A habit, **B** flower, **C** siliqua, **S** stem section.

Brassicaceae

248 *Nasturtium microphyllum* (Boenn.) Rchb.

Syn.: *Rorippa microphylla* (Boenn.) Hyl.
DK: Tyndskulpet Brøndkarse **N:** Brun engelskkarse **S:** Bäckfräne **FIN:** - **IS:** -
GB: Narrow-fruited Water-cress **NL:** Slanke waterkers **F:** Cresson à petites feuilles
D: Kleinblättrige Brunnenkresse **CZ:** Potočnice drobnolistá **PL:** Rukiew drobnolistna
EST: -**LV:** - **LT:** - **Ru:** -

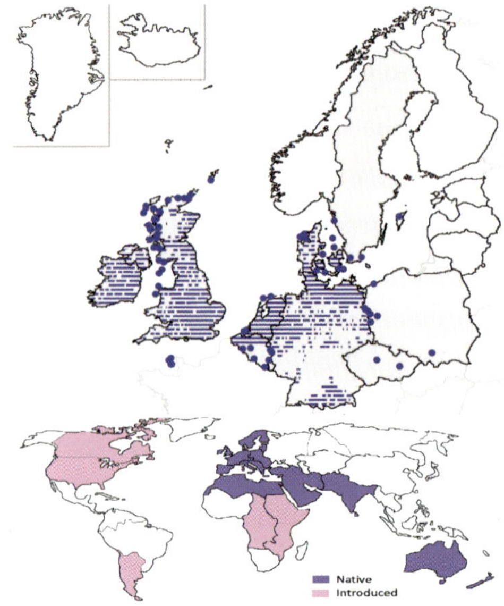

Nasturtium microphyllum. River Smedebæk, southern Jutland, Denmark. Photo BM.

Perennial helophyte with unbranched white to brownish roots, primarily from the creeping rhizome and lower nodes, sometimes also from the stem nodes.

Stem: 30–90 cm long, ascending from a prostrate base, 3–12(–15) mm in diameter, hollow, smooth and glabrous, angular, stiff to flexible, brittle, unevenly branched, the length of the internodes decreasing towards the tip of the shoots.

Leaves: Alternate, obliquely petiolate, pinnate, with 2–4 pairs of sessile oval to ovate lobes with entire to slightly serrate margin and 1 large terminal lobe, commonly with an elliptic base and entire to bluntly dentate margin. Leaves initially green, eventually with a purple or violet tinge, smooth, often with short hairs on the upper side of the petioles. Plants overwinter as short, compact submerged shoots with undivided leaves similar in shape to the terminal lobes of the stem leaves.

Inflorescence: Monoecious. Racemes with 30–60 flowers, emerging mainly from the upper leaf-bearing nodes of the stems and branches, elongating during flowering.

Flowers: 4-merous. Bisexual. 6–8 mm in diameter with c. 11–15 mm long terete, green pedicels, slightly hairy on the upper side. Sepals 2.5–3.5 mm long, glabrous, not joined, yellowish green, ovate oblong with narrow membranous margin. Petals 5–6 mm long, not joined, white, broadly ovate. Stamens, 4 long and 2 short, with bright violet filaments and yellow anthers. Style 1–3 mm long.

Fruits: Siliqa 16–24 mm long and 1.2–1.8 mm in diameter, shortly pedicellate, slender, curved with 1 row of seeds in both loculi. Seeds numerous, small and compressed; the broad side of each seed has a fine network of 100–150(–190) polygonal depressions; with 11–20 across the broadest width.

Flowering: June-September.

Biology: There is typically more than one flowering period per year alternating with non-flowering periods.

Mainly self-pollinated but also pollinated by insects. Newly released seeds float and are dispersed by water or birds and other animals. Dispersal by seed appears to be very effective, with rapid colonisation of newly created habitats, while stands are capable of rapid vegetative spread by runners. Seeds can remain dormant and viable in the seed bank for many years. Plant biomass culminates late in the year and growth terminates when all but the

Fertile shoot of *.microphyllum* with slender siliquae. Botanical Garden, Gotland, Sweden. Photo JCS.

submerged parts of plants are killed off by frosts. Only submerged parts survive through the winter, and these combine with germination from seed to form new stands in subsequent seasons. Stand size may vary widely between years, and it is capable of very rapid spread to the exclusion of other species. Plants may be contaminated by liver flukes.

Habitats: Mesotrophic and slightly acid to alkaline, mainly flowing water – springs, streams and rivers as well as canals and ditches, less commonly in standing water and then typically where there is inflow from groundwater.

Distribution within the region:
Native. Widespread and abundant in Britain and Ireland, east into Germany and Denmark, with a few scattered populations in southern Sweden, Poland and the Czech Republic.

Characteristics and similar species:
Nasturtium microphyllum is easily confused with **250 *Nasturtium officinale*** and **249 *Nasturtium ×sterile***, from which it can only be distinguished with certainty on the basis of the size and number of polygonal depressions on the broad sides of the seeds. One or two rows of seeds in the siliquae as well as the elliptic base of the apical leaf lobe should only be seen as guiding characters. Sterile plants cannot be identified with certainty.

247 *Cardamine amara* differs in having larger flowers with violet anthers and a different morphology.

Nasturtium microphyllum. Open siliqua showing 1 row of seeds. Photo KvdW.

249 ***Nasturtium ×sterile*** (Airy Shaw) Oefelein *(N. microphyllum × officinale)*

This hybrid is morphologically intermediate between the parents, with mainly sterile pollen, poor seed set and somewhat deformed siliquae, with 1–2 rows of seeds in each of the siliqua loculi. Not easily distinguished from the parent species, both of which may also display poor seed set or occur as sterile plants. Identification with certainty should be based only on fully developed seeds with a structure of 50–100 (–120) polygonal depressions showing in lateral view. The hybrid is cultivated as a vegetable and naturalised.

Nasturtium ×sterile. All seeds aborted. Photo RVL.

Seeds. From the left: *Nasturtium microphyllum, ×sterile, officinale*. Photo KvdW.

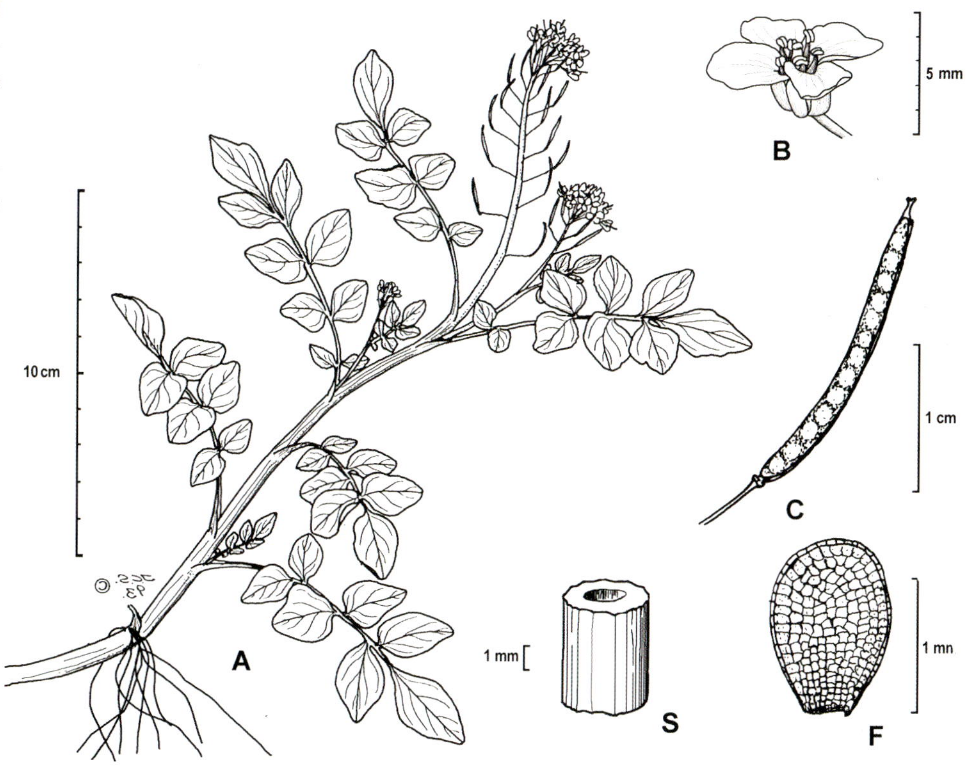

Nasturtium microphyllum
A habit, **B** flower, **C** siliqua, **F** seed, **S** stem section.

Brassicaceae

250 *Nasturtium officinale* W.T.Aiton

Syn.: *Rorippa nasturtium-aquaticum* (L.) Hayek

DK: Tykskulpet Brøndkarse **N:** Grøn engelskkarse **S:** Källfräne **FIN:** Isovesikrassi **IS:** -
GB: Water-cress **NL:** Witte waterkers **F:** Cresson officinal
D: Echte Brunnenkresse **CZ:** Potočnice lékařská **PL:** Rukiew wodna
EST: Ürt-allikkerss **LV:** - **LT:** Paprastasis rėžiukas **Ru:** Жеруха обыкновенная

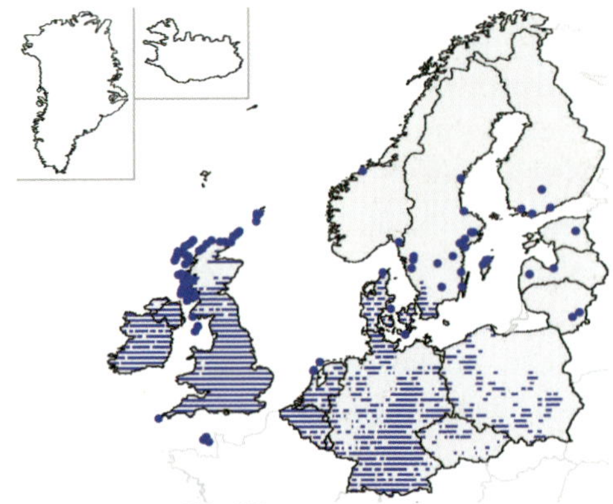

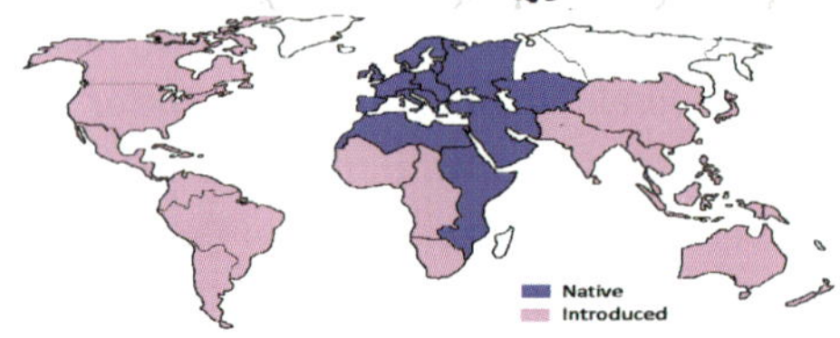

Nasturtium officinale in a fast-flowing stream. Rebild, Jutland, Denmark. Photo JCS.

Perennial helophyte with unbranched white to brownish roots, primarily from the creeping rhizome and lower nodes, sometimes also from the stem nodes.

Stem: 30–100 cm long, base prostrate, upper part ascending with roots, 3–12(–15) mm in diameter, hollow, smooth and glabrous, angular, brittle, unevenly branched, internode length decreasing towards the shoot tip.

Leaves: Alternate, obliquely petiolate, pinnate, 2–4 pairs of sessile oval to ovate lobes with entire to bluntly dentate margin and 1 large terminal lobe, often with a cordate base and entire to slightly serrate margin. Leaves green and smooth, the upper side of the petioles often with short hairs. Plants overwinter as short, compact submerged shoots with undivided leaves similar in shape to the terminal lobes of the stem leaves.

Inflorescence: Monoecious. Racemes with 30–60 flowers, emerging mainly from the upper leaf-bearing nodes of the stems and branches, elongating during flowering.

Flowers: 4-merous. Bisexual. 6–8 mm in diameter, with green, terete pedicels with sparse hairs on the upper side. Sepals 2–3 mm long, not joined, yellow green, ovate oblong with narrow membranous margin. Petals 4–6 mm long, not joined, white, broadly ovate. 4 long and 2 short stamens with violet filaments and yellow anthers. Style 1–3 mm long.

Fruits: Siliquae 12–20 mm long and 2.0–2.7 mm in diameter, shortly pedicellate, oblong elliptic, slightly curved. 2 rows of seeds in both loculi. Seeds numerous, small and compressed, the broad sides of each seed with a coarse structure of 25–50(–70) polygonal depressions, 6–12 across the broadest width.

Flowering: July-September (-October).

Biology: Mainly self-pollinated but also pollinated by insects. Newly released seeds float and are dispersed by water or birds and other animals. Dispersal by seed appears to be very effective, with rapid colonisation of newly created habitats, while stands are capable of rapid vegetative spread by runners. Seeds can remain dormant and viable in the seed bank for many years. Plant biomass culminates late in the year and growth terminates when all but the submerged parts of plants are killed off by frosts. Only submerged parts survive through the winter, and these combine with germination from seed to form new stands in subsequent seasons. Stand size may vary widely between years, and it is capable of very rapid spread to the exclusion of other species. Plants may be contaminated by liver flukes.

Habitats: Mesotrophic to eutrophic, neutral to alkaline, mainly flowing water – springs, streams and rivers as well as canals and ditches, less commonly standing waters and then typically where there is groundwater inflow.

Nasturtium officinale. Flowers with yellow anthers. Hobro, Denmark. Photo JCS.

Nasturtium officinale. The thick siliquae and the cordate base of the terminal leaf lobes are usable as guidelines, but should always be followed by investigation of the seeds. Skibsted, Denmark. Photo BM.

Submerged *Nasturtium* plants. Furlbach, Senne, Germany. Photo KvdW.

Distribution within the region:
Widespread and abundant from Ireland east to Denmark and southern Poland, scattered in southern Sweden, Finland, Latvia and Estonia. Increasing due to cultivation as a leaf vegetable.

Characteristics and similar species:
Nasturtium officinale is easily confused with **248 *Nasturtium microphyllum***, from which it can only distinguished with certainty by the size and number of polygonal depressions on the broad sides of the seeds. Two rows of seeds in the siliquae as well as the cordate base of the apical leaf lobe should only be seen as guiding characters. Sterile plants cannot be identified with certainty. Distinguishing from the hybrid **249 *Nasturtium ×sterile*** is only possible on basis of the structure of the seeds.

247 *Cardamine amara* differs by larger flowers with violet anthers and a distinctly different morphology.

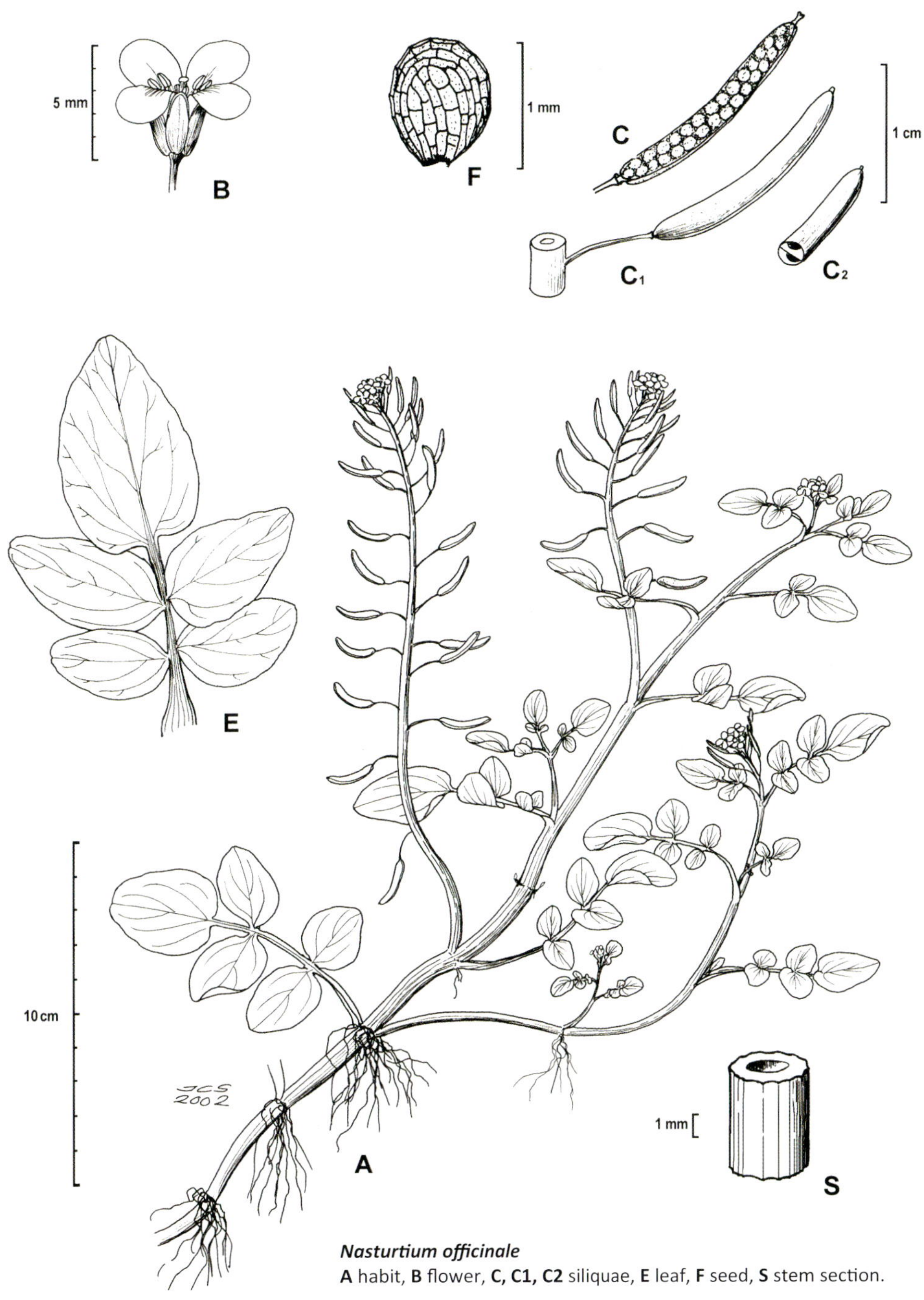

Nasturtium officinale
A habit, **B** flower, **C, C1, C2** siliquae, **E** leaf, **F** seed, **S** stem section.

Nasturtium officinale. Open siliqua showing 2 rows of seeds. Photo KvdW.

Brassicaceae

251 *Rorippa amphibia* (L.) Besser

Syn.: *Nasturtium amphibium* (L.) W.T.Aiton

DK: Vandpeberrod **N:** Vasskarse **S:** Vattenfräne **FIN:** Vesinenätti **IS:** -
GB: Great Yellow-cress **NL:** Gele waterkers **F:** Rorippe amphibie
D: Wasserkresse **CZ:** Rukev obojživelná **PL:** Rzepicha ziemnowodna
EST: Vesikerss **LV:** Abinieku paķērsa **LT:** Vandeninis čeriukas **Ru:** Жерушник земноводный

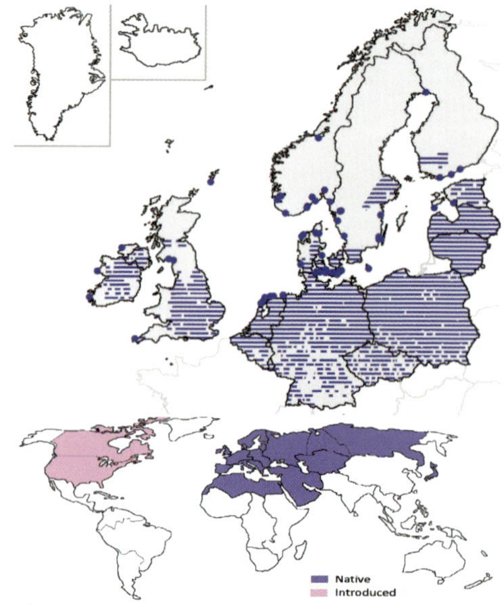

Rorippa amphibia. Vramsån, Skåne, Sweden. Photo Peter Holm.

Perennial helophyte with unbranched white to brownish roots, in bundles primarily from the horizontal stem parts and lower stem nodes, particularly on floating plants.

Stem: 40–150 cm long, ascending to upright from a basal horizontal and rooting part, often branched. Glabrous or more rarely with scattered hairs, furrowed, hollow and yellowish green to green. In deeper water plants may float on the surface with horizontal, inflated terete, hollow stems, up to 2 cm in diameter, from which ascending to upright stems arise, or rooted with vertical shoots rising to above the water surface.

Leaves: Morphologically very variable. Stem leaves alternate, the lower obliquely short petiolate, the upper sessile, glabrous, yellowish green to green, lamina narrowly to broadly elliptic, roughly and unevenly lobed or with entire to serrate margin. Submerged leaves entire or deeply and narrowly pinnatifid.

Inflorescence: Monoecious. Racemes with numerous flowers, emerging mainly from the upper leaf-bearing nodes of the stems and branches, elongating during flowering.

Flowers: 4-merous. Bisexual. Flowers 8–11 mm in diameter with c. 10 mm long terete, green pedicels. Sepals 2–3 mm long, not joined, yellow green, oblong ovate to obtuse with narrow membranous margin. Petals 4–5 mm long, not joined, golden yellow, obovate. 4 long and 2 short stamens with yellow filaments and yellow anthers. Style 1–2 mm long.

Fruits: Siliquae oblong elliptic, 3–7 mm long and 2–3 mm in diameter, shortly pedicellate. 2 rows of seeds in each of the loculi. Pedicels 6–18 mm long.

Flowering: June-July. Rarely more than one flowering period.

Biology: Self-incompatible with poor seed set. Germination takes place on moist soil, resulting in small leaf rosettes in the first year with flowering in the following year. Vegetative dispersal is by detached stem fragments, on which new shoots emerge from the leaf axils. Submerged leaves are mainly produced in spring when plants are

Rorippa amphibia. Various leaf shapes. Paderup, Jutland, Denmark. Photo JCS.

Rorippa amphibia with entire submerged leaves. Photo KvdW.

Rorippa amphibia. Stem. Germany. Photo KvdW.

Rorippa amphibia. Inflorescence. Silkeborg, Denmark. Photo JCS.

inundated by high water levels; in some regions they are rarely produced, while in others they are common. The stems often die back towards the end of the growing season, leaving plants with mainly entire or only partly lobed basal leaves.

Habitats: Mesotrophic to eutrophic and slightly acid to alkaline habitats, usually on silty soil in the littoral zone of flowing as well as standing waters – streams, rivers, canals, lakes and ponds.

Distribution within the region: Native. Widespread and abundant from Ireland east through southern Britain to Poland and the Baltic states, scattered in southern Scandinavia, absent in the north.

Characteristics and similar species: With its yellow flowers, *Rorippa amphibia* is unique among aquatic crucifers, although in marginal habitats it will occur with its close relatives **252** ***Rorippa palustris*** and **253** ***R. sylvestris*** from which it can be distinguished by the leaf shape and fruits.

It can be distinguished from **248-250** ***Nasturtium*** spp. by the leaf shape and the size of the siliquae.

Hybrids: ***Rorippa ×anceps*** - see next pages.

Rorippa amphibia
A habit submerged plant with deeply pinnatifid leaves, **A1** inflorescence, **B** flower, **C, C1** siliqua.

Brassicaceae

252 ***Rorippa palustris*** (L.) Besser

Syn.: *R. islandica* auct.; *Nasturtium palustre* (L.) Crantz
DK: Kær-Guldkarse **N:** Brønnkarse **S:** Sumpfräne **FIN:** Rantanenätti **IS:** -
GB: Marsh Yellow-cress **NL:** Moeraskers **F:** Rorippe faux-cresson
D: Wasserkresse **CZ:** Rukev bažinná **PL:** Rzepicha błotna
EST: Sookerss **LV:** Purva paķērsa **LT:** Pelkinis čeriukas **Ru:** Жерушник болотный

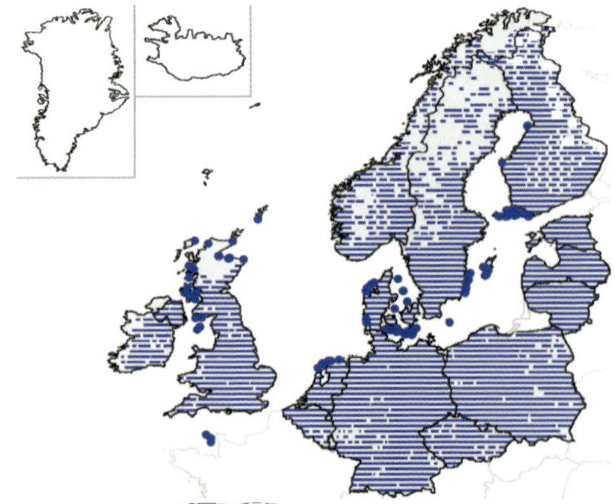

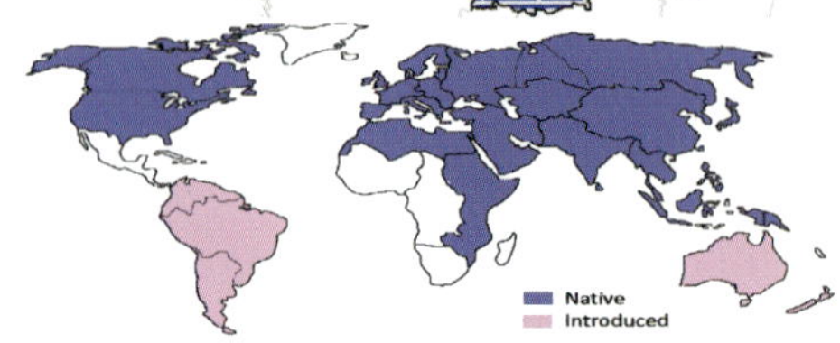

Rorippa palustris on the margin of River Gudenå, Denmark. Photo JCS.

Annual or short-lived perennial with a basal leaf rosette and taproot.

Stem: 5–60(–110) cm long, ascending to erect with spreading branches in the upper part, glabrous, stiff, hollow, angular.

Leaves: Rosette leaves pinnatifid, often dying back before flowering. Stem leaves alternate, the lower with a winged petiole, the upper more or less sessile, with distinct auricles. Leaves 1.5–12 cm long, light green to green, often with a reddish tinge, glabrous, pinnate, 1–6 pairs of lobes with irregularly serrate margins, and one terminal lobe, elliptic in outline and comprising 1/3–1/2 of the leaf length.

Inflorescence: Monoecious. Racemes terminal on stem and branches, 2–10 cm long, elongating during flowering. Flowering branches held with the fruits usually arranged evenly around the stem or branch, not swept down to one side.

Flowers: 4-merous. Bisexual. Flowers with pedicels 1.5–2.5 mm long, terete, green. Sepals 1.6–2.4 mm long, not joined, yellow green, ovate. Petals 1.7–2.8 mm long, not joined, yellow, spathulate to obovate. 4 long and 2 short stamens with yellow filaments and yellow anthers. Style 0.4–1.2 mm long.

Fruits: Siliquae 4–12 mm long, 1.7–3 mm in diameter, oblong to ellipsoid, somewhat curved. Valves firm, thick, not torulose. 2 rows of seeds in each loculus. Seeds irregularly elliptic, 0.6–0.9 mm long, brown to reddish brown, coarsely colliculate (with small rounded swellings), epidermal cells 35–55 µm wide. Pedicels 3.5–7 mm long, 1/3–1/2 as long as the fruit, horizontal to deflexed.

Flowering: June-September.

Biology: Mostly autogamous, spreading by seeds.

Habitats: Mesotrophic to eutrophic, acid to alkaline, wet to moist habitats with open, often disturbed, sandy or muddy substrate. In the littoral zone of lakes, pools, ditches and rivers, in wet meadows, wells, man-made habitats such as road banks, gravel pits and waste places.

Distribution within the region: Native. Throughout the region except Iceland and Greenland.

Characteristics and similar species: Recognised by the small, yellow flowers, short sausage-like siliquae and pinnate leaves.

The very similar ***R. islandica*** differs in seeds with epidermal cells 20–32 µm across (microscope needed). Ripe fruits 2–3 times as long as the pedicel, with thin valves, alternately swollen and pinched, formed by the contours of the seeds. Sepals less than 1.6 m long.

Other more or less similar ***Rorippa*** species differ in having stolons or rooting stem parts.

Subsp. ***hispida*** (Desv.) Jonsell

Differs by stem and leaves with dense, short, stiff hairs. Siliquae short, 3–5 mm long, ovoid or spherical.

Native to the eastern part of North America, and within the study area only known from an old record from Norway.

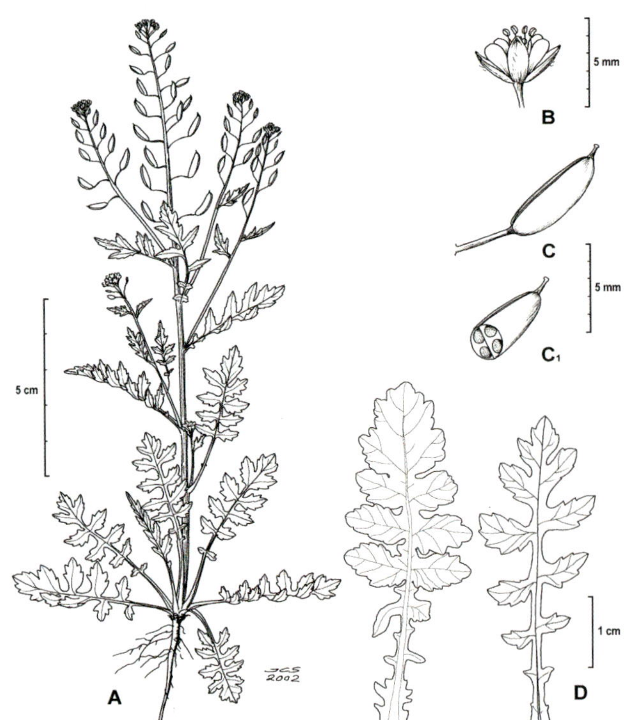

Rorippa palustris
A habit, **B** flower, **C, C1** siliqua, **D** leaves.

253 *Rorippa* hybrids

Rorippa hybrids are generally rare but occur locally in extensive stands spread by vegetative propagation. They often appear intermediate between the parent species, which may not necessarily be present.
Identification of *Rorippa* hybrids is difficult, requiring well developed, flowering specimens. Sterility can not be used as evidence for hybridity, because many species are highly self-sterile.

Rorippa ×anceps. Margin of the River Serven, Arley, Staffordshire, Britain. Photo RVL.

Rorippa ×erythrocaulis. Margin of the River Thames at Hammersmith, London, Britain. Photo RVL.

R. palustris × sylvestris. Unconfirmed; reported from scattered localities throughout the region.

R. ×anceps (Wahlenb.) Rchb.
(R. sylvestris × amphibia)
Silty river margins and bars. Most of Europe except the Baltic, absent from Iceland and Greenland.

R. ×armoracioides (Tausch) Fuss
(R. sylvestris × austriaca)
In ditches, banks of pools and streams, roadsides and waste ground. Most of Europe except Ireland, Netherlands, Belgium and Luxembourg. Absent from Iceland and Greenland.

R. ×erythrocaulis Borbas
(R. palustris × amphibia)
Silty river margins and bars. Confirmed from Britain and Sweden, scattered unconfirmed records elsewhere in the region.

R. ×hungarica Borbas
(R. amphibia × austriaca)
Known from Germany, Poland, Czech Republic and Britain.

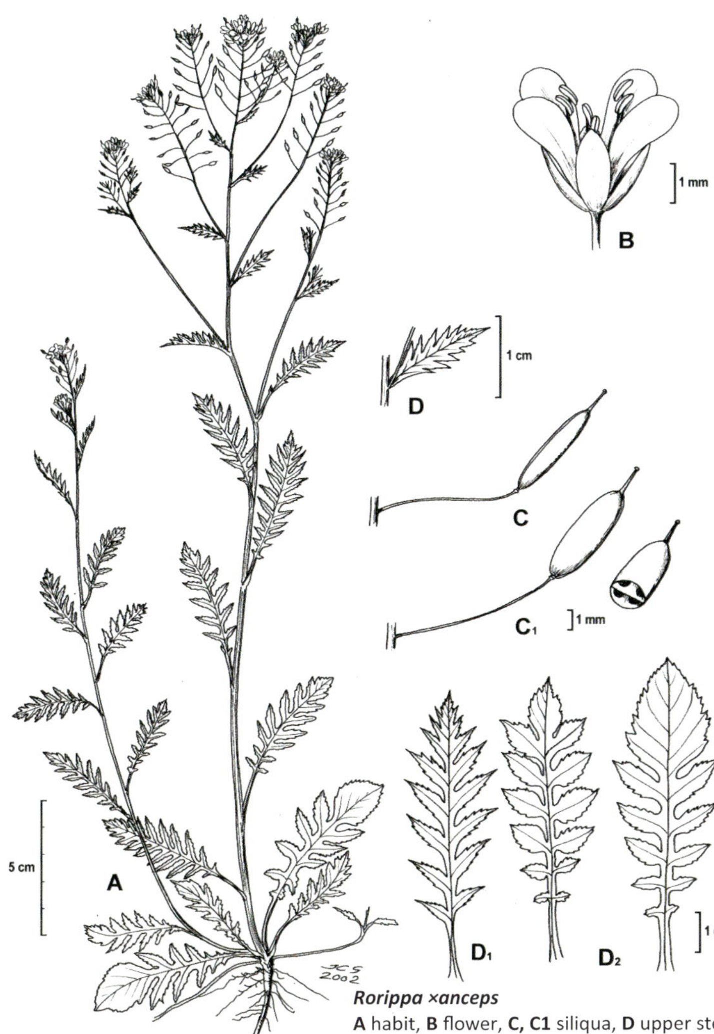

Rorippa ×anceps
A habit, **B** flower, **C, C1** siliqua, **D** upper stem leaf, **D1** lower stem leaf, **D2** basal leaves.

Brassicaceae

254 *Subularia aquatica* L.

DK: Sylblad **N:** Sylblad **S:** Sylört **FIN:** Äimäruoho **IS:** Alurt
GB: Awlwort **NL:** Priemkruid **F:** Subulaire aquatique
D: Pfriemenkresse **CZ:** - **PL:** -
EST: Vesi-naaskelleht **LV:** Ūdeņu subulārija **LT:** - **Ru:** Шильник водяной

Typical submerged *S. aquatica* with rosettes of acicular leaves with long-drawn tips. Södra Vixen, Eksjö, Sweden. Photo JCS.

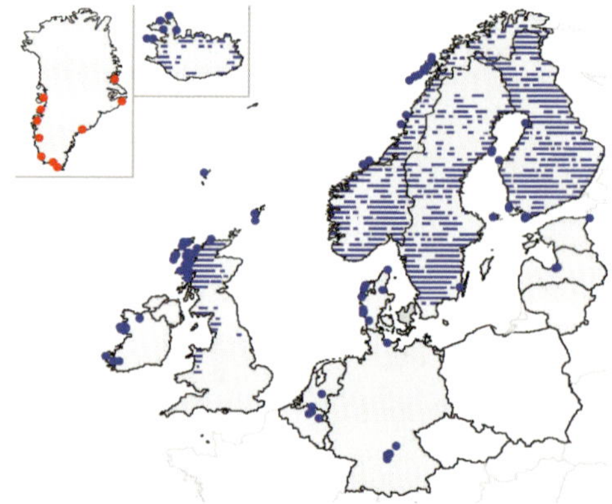

Red dots - subsp. ***americana***.

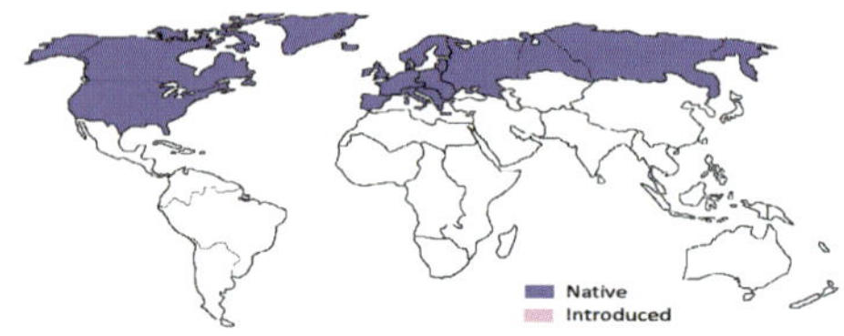

Annual hydrophyte. Roots white, unbranched.

Stem: 2–10 cm long, erect to prostrate, glabrous, terete, green, without leaves.

Leaves: In rosettes, up to 7 cm long and 2 mm in diameter, acicular with long tips, gradually tapering from the base, glabrous, green to dark green, compressed oval in cross section and solid, without lacunae.

Inflorescence: Monoecious. Racemes with 2–8 flowers, elongating during flowering.

Flowers: 4-merous. Bisexual. Flowers 1.5–2.5 mm in diameter with terete, green pedicels. Sepals 0.5–0.8 mm long, caducous, glabrous, not joined, green, ovate oblong to obtuse with narrow membranous margin. Petals 1.2–1.5 mm long, not joined, white, obovate. 4 long and 2 short stamens with pale green filaments and pale yellow anthers. Styles 0.1 mm long.

Fruits: Siliculae 3–5 mm long, 1.5–2.5 mm thick, ellipsoid in outline. Seeds 2–7 arranged in two rows in each loculus. Pedicels 0.5–1.5 times as long as the siliculae and ascending at angles of 50°–90° to the stem.

Flowering: July-October.

Biology: Flowering above and below water. The flowers of submerged plants are self-pollinated within the bud; they usually remain closed (cleistogamous) but occasionally open. Emergent plants produce open flowers which enable both cross- and self-pollination. Each plant produces up to 125 seeds, which when released from the siliculae sink to the sediment, where some may germinate immediately, while others remain dormant until the following summer. Plants germinating in autumn may overwinter with green leaves and flower the following summer.

Habitats: On sandy, gravelly or silty substrate in the littoral zone of oligotrophic and neutral to slightly acidic pools, lakes and rivers. It will grow as a terrestrial on damp substrate or to a maximum depth of 2–3 metres. In the Gulf of Bothnia and locally in tidal areas it will grow on muddy substrates in brackish water.

Distribution within the region:
Native. Widespread and abundant in parts of Western Ireland, northern Britain, Scandinavia and Iceland. Extinct in Netherlands, Belgium and Germany.

Characteristics and similar species:
This species is easily overlooked among other rosette-forming species and

Subularia aquatica is the only member of the Brassicaceae family flowering and fruiting while submerged and the only rosette-forming water plant with an annual life cycle. Førby Sø, Thy, Denmark. Photo JCS.

Emergent *S. aquatica* with open flowers on the shore of Lake Södra Vixen, Eksjö, Sweden. Photos JCS.

Subularia aquatica. Submerged plants. Lough Mhaimin, Ireland. Photo KvdW.

stands of submerged plants. When flowering or fruiting, *S. aquatica* can hardly be confused with any other water plant, and even the rosettes of solid, acicular leaves with long, fine tips are distinctive.

Non-flowering plants may be confused with a number of other rosette-forming plants from oligotrophic lakes:

296 *Littorella uniflora* differs in its stoloniferous growth and solid leaves with numerous air tubes.

Submerged forms of **221 *Ranunculus reptans*** do not form rosettes, and have stolons and thin needle-like leaves.

160 *Eleocharis acicularis* and **161 *Eleocharis parvula*** both differ in their filamentous stolons and narrower leaves.

137 *Juncus bulbosus* has rosettes of very fine, filamentous leaves with transverse septa.

334 *Lobelia dortmanna* has wider, blunt leaves with 2 large air tubes.

359-360 *Isoëtes* spp. differ in their leaves with 4 distinct air tubes.

Note: In Greenland and North America it is represented by subsp. ***americana*** G.A.Mulligan & Calder, which differs from subsp. *aquatica* in having persistent sepals, fruiting pedicels ascending at a narrower angle (30°–50°) to the stem and broadly obovoid siliculae.

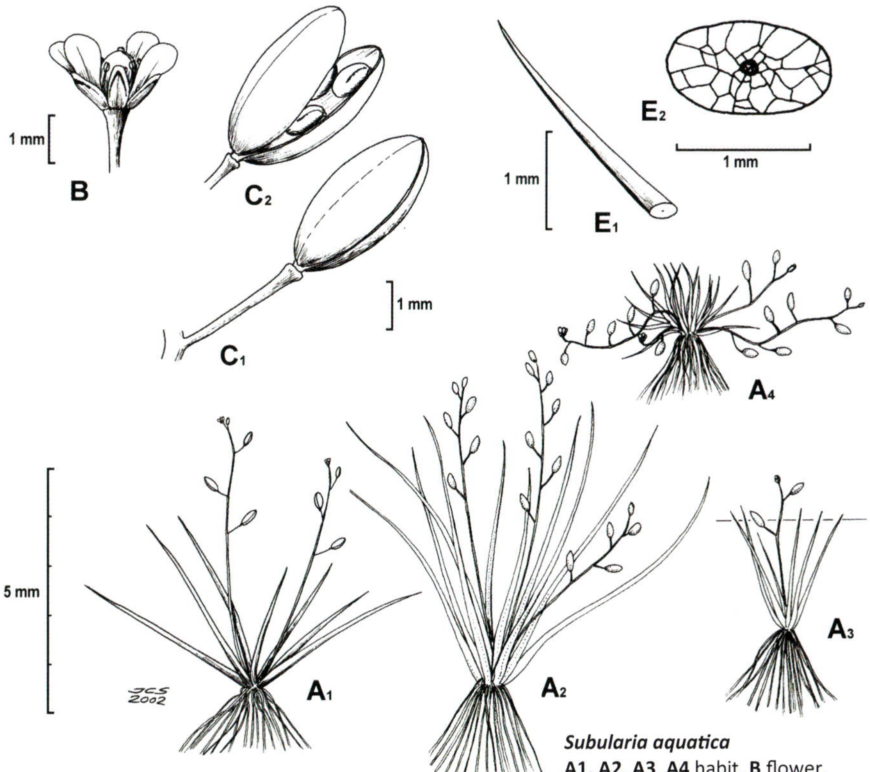

Subularia aquatica
A1, A2, A3, A4 habit, **B** flower, **C1, C2** silicula, **E1** cut leaf, **E2** cross section leaf.

Polygonaceae

255 *Persicaria amphibia* (L.) Gray

Syn.: *Polygonum amphibium* L.

DK: Vand-Pileurt **N:** Vass-slirekne **S:** Vattenpilört **FIN:** Vesitatar **IS:** Tjarnablaðka
GB: Amphibious Bistort **NL:** Watersmerte **F:** Persicaire amphibie
D: Wasser-Knöterich **CZ:** Rdesno obojživelné **PL:** Rdest ziemno-wodny
EST: Vesi-kirbutatar **LV:** Abinieku sūrene **LT:** Būdmainis rūgtis **Ru:** Горец земноводный

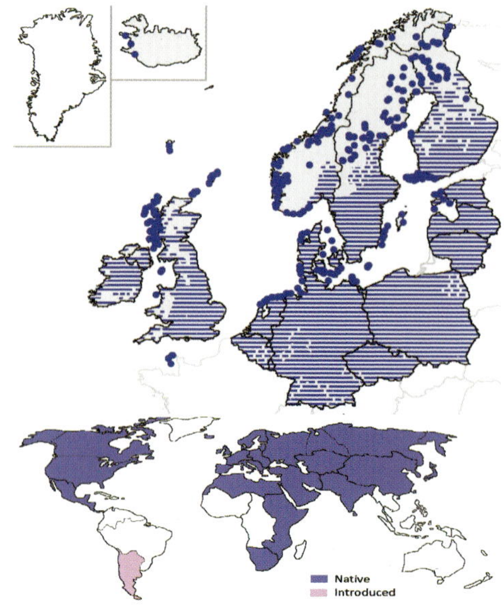

Dense stand of *P. amphibia* in a small shallow pond in the dunes at Valbjerg, Thy, Denmark. Photo JCS.

Perennial hydrophyte or helophyte/hemicryptophyte with horizontal, terete rhizome up to 10 m long and short underground runners. Roots long, branched and brown.

Stem: In aquatic plants up to c. 3 m long and 1.5–3 mm diameter, floating on the surface with the apex ascending to upright, terete, glabrous, unbranched or sparsely branched from the base. In terrestrial plants 0.2–1.0 m long, prostrate to ascending or erect, terete and hairy. Membranous ochrea enclosing the stem at nodes.

Leaves: In aquatic plants 5–20 cm long and 1.5–4.0 cm wide, lanceolate oblong ovate, obtuse with cordate base, glabrous and glossy, green, often with reddish tinge. Petiole long, arising from within the ochrea. Submerged leaves are often present but do not differ from floating leaves. Leaves of terrestrial plants are shortly petiolate, hairy, narrowly lanceolate, often with a dark patch on the upper side.

Inflorescence: Spikes 2–6 cm long, terminal, dense-flowered, oval to cylindrical. Peduncle 1–5 cm long, glabrous in aquatic plants or hairy in terrestrial plants.

Flowers: 5-merous, bisexual or functionally female, sessile. Tepals 4–6 mm long, fused at the base, lanceolate with prominent venation, acute or obtuse, pale to dark pink. 5 stamens and 2 styles. Some flowers with fertile stamens and short styles, others with sterile stamens and styles of varying length.

Fruits: Nuts, semi-globose 2.0–2.7 mm long and 1.7–2.2 mm wide, compressed, brownish black, glossy.

Flowering: June-September.

Biology: Aquatic plants often form large, dense, richly flowering stands; terrestrial plants typically produce fewer flowers. Flowers have nectaries and are insect pollinated; they are self-sterile, therefore stands consisting of only one clone do not set seeds. Seeds are dispersed by water and waterfowl and dormant during winter. Vegetative dispersal occurs by stem fragments.

Flowering aquatic *P. amphibia* with glabrous and glossy floating leaves and densely flowered spikes of pink flowers. Photo JCS.

Habitats: In oligotrophic to eutrophic, acidic to alkaline water. On organic as well as mineral substrates. Aquatic plants can grow to a depth of more than 2.5 metres, they occur in ponds, lakes, ditches and canals, less commonly in streams. Terrestrial plants occur in the margins and on banks of aquatic habitats but also in drier habitats, e.g., meadows and road margins.

Distribution within the region: Native. Common throughout most of the region, but scattered in northern parts of Scandinavia and Iceland, absent from Greenland.

Characteristics and similar species: Aquatic *Persicaria amphibia* without flowers may be confused with *Potamogeton* species with floating leaves, which however have pointed stipules and leaf veins parallel to the leaf margins as opposed to the ochreas and pinnate leaf veins of *Persicaria*.

Persicaria amphibia differs from other *Persicaria* species in being perennial with long rhizomes, with cordate leaf bases and often having the stamens exserted at anthesis.

Confusion with other aquatic species is unlikely.

Terrestrial *Persicaria amphibia* with more narrowly lanceolate and distinctly hairy leaves and stems, growing on moist-dry soil at Havnø near Hadsund, Denmark. Photo JCS.

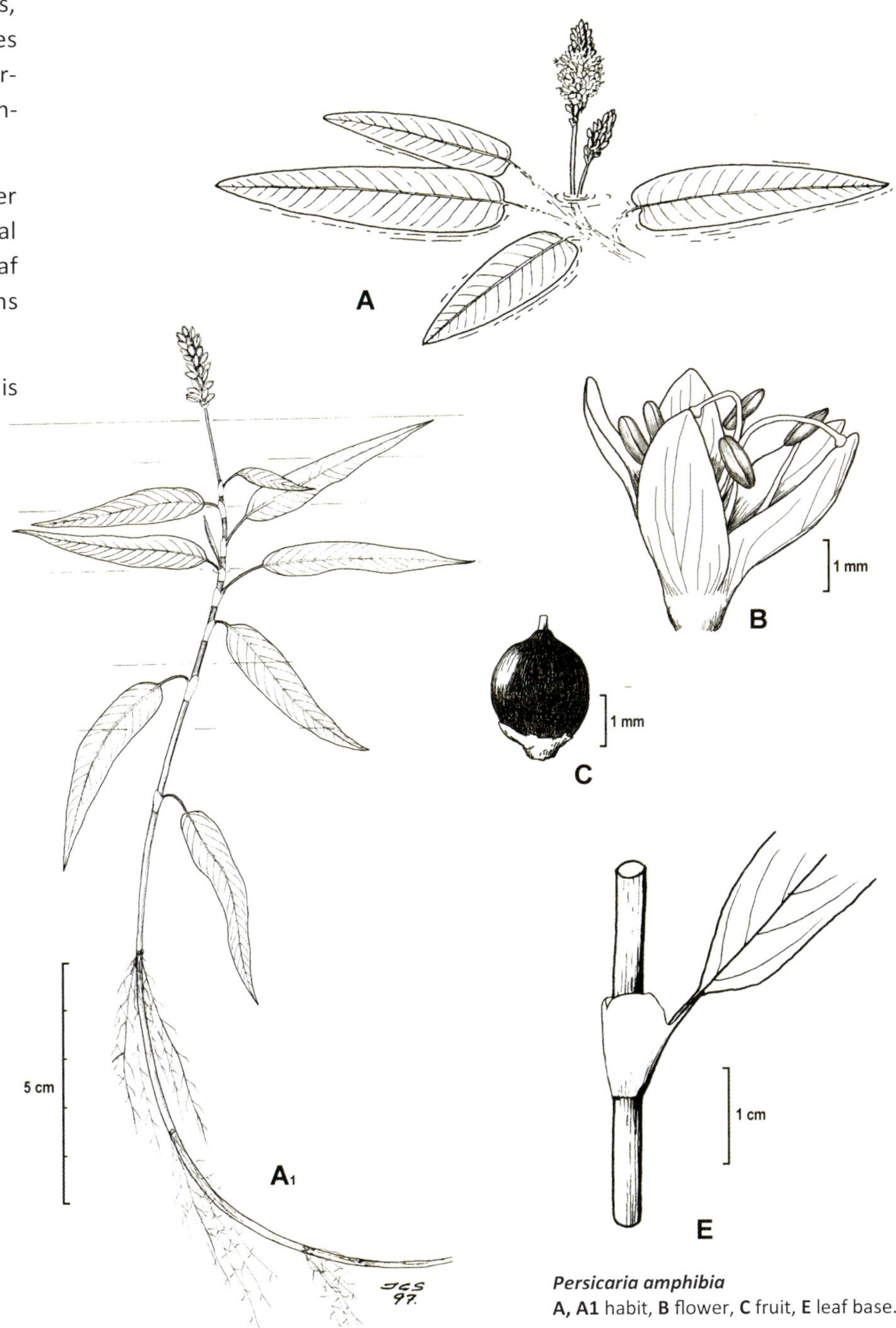

Persicaria amphibia
A, A1 habit, **B** flower, **C** fruit, **E** leaf base.

Persicaria amphibia. Floating leaf, backlit. Nors Sø, Thy, Denmark. Photo JCS.

Polygonaceae

256 ***Persicaria hydropiper*** (L.) Delarbre

Syn.: *Polygonum hydropiper* L.

DK: Bidende Pileurt **N:** Vasspepper **S:** Bitterpilört **FIN:** Katkeratatar **IS:** -
GB: Water Pepper **NL:** Waterpeper **F:** Poivre d'eau
D: Pfeffer-Knöterich **CZ:** Rdesno peprník **PL:** Rdest ostrogorzki
EST: Mõru kirbutatar **LV:** Ūdenspipars **LT:** Kartusis rūgtis **Ru:** Горéц пéречный

Persicaria hydropiper on the bank of River Gudenå, Denmark. Photo JCS.

Temporary, summer annual hydrophyte or helophyte with whitish fibrous, branched roots. Rhizome absent.

Stem: 10–60(–125) cm long and 2–4 mm thick, decumbent, ascending or erect, glabrous, normally purplish, less often pale green, the lower parts commonly richly branched.

Leaves: (1.5–)4.0–9.0(–12.5) cm long and (0.4–)0.8–2.0(–2.4) cm wide, alternate, lanceolate, lamina glabrous or with few short and stiff appressed hairs at the margins and the main veins at the lower side of the leaf, both sides normally with small, depressed, yellowish glands, green. Ochreas membranous, enclosing the stem, up to 1.5 cm long, glabrous or with few appressed hairs, normally with small, depressed, yellowish glands, purplish, less often green, the upper margins with whitish 0.5–2(–3) mm long, hair-like bristles. Petioles short, emerging from the base of the ochrea, green or purplish. No differentiated leaves, submerged leaves however commonly softer, more flaccid and brighter green.

Inflorescence: Spike-like, oblong and mostly nodding, lax and open with spaced flowers, rarely dense, sometimes with small leaves within the inflorescence. One or few flowers may also occur in the upper leaf axils. Peduncle and main axis with small, depressed, yellowish glands, green or purplish.

Flowers: (3–)4(–5)-merous, bisexual, sessile, occasional flowers also in upper leaf axils. Bracts fused into funnel-shaped, papery ochreolae below the perianth, the upper margins with whitish hair-like bristles. Tepals c. 3 mm long, fused at the base, lanceolate, outside green at the base, pink towards the apex; inside normally pinkish or whitish, with or without venation, the green part with small yellowish glands. 6(–8) stamens and 2(–3) styles, all shorter than the tepals.

Fruits: Nuts, one per flower, 2.5–3.5 mm long, lenticular or 3-angled, dull, black or dark brown.

Flowering: May-September.

Biology: Germinating from seeds on damp or wet substrate, as solitary plants or large, dense stands, sometimes permanently submerged, then however remaining sterile. Flowers with nectaries, pollinated by insects. Seeds dispersed by animals, dormant during winter. Favoured by physical disturbance of the soil, but also occurring in more stable conditions and plant associations. All parts of the plants with a burning peppery taste.

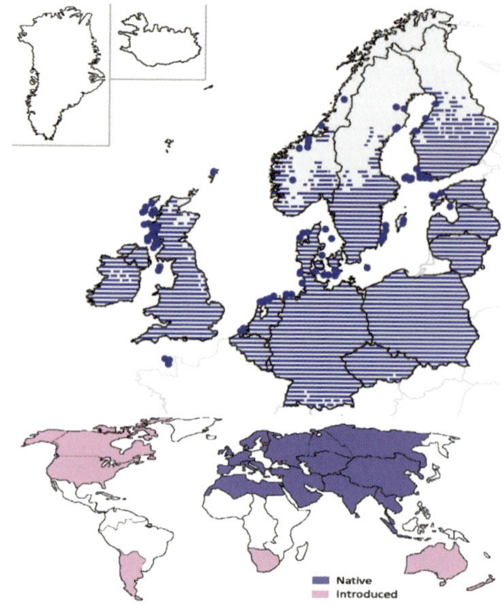

Habitats: In moderately acidic to slightly alkaline, mesotrophic to eutrophic water bodies on organic as well as mineral substrate. In the summer-dry littoral zone of ponds, lakes, ditches, canals and streams in open landscapes as well as open forests.

Distribution within the region:
Native. Common throughout except for the northern part of Scandinavia and Finland.

Characteristics and similar species: The open, slender and nodding inflorescences, glabrous or sparsely hairy cuneate aerial leaves without dark blotches and ochreae with whitish bristles on the margins distinguish *P. hydropiper* from **255 *P. amphibia***.

Persicaria hydropiper. Leaf base with ochrea. River Severn, Upper Lode, Gloucestershire, Britain. Photo RVL.

Persicaria hydropiper in flowing water. Photos KvdW.

Persicaria minor (Huds.) Opiz (see photo below) occurs in similar habitats, but is usually smaller, without the acrid taste and glands on the tepals.

The lack of dark blotches on the leaves and shape of the inflorescence distinguish *Persicaria hydropiper* from other morphologically similar but non-aquatic species, such as *P. maculosa* Gray and *P. lapathifolia* (L.) Gray, both of which may occur in the same biotopes as *P. hydropiper*.

Confusion with other aquatic plants is unlikely.

Hybridisation: *Persicaria* hybrids are rarely recorded, however it appears that they may be quite frequent in dense mixed stands.

Persicaria minor. Hobro, Denmark. Photo JCS.

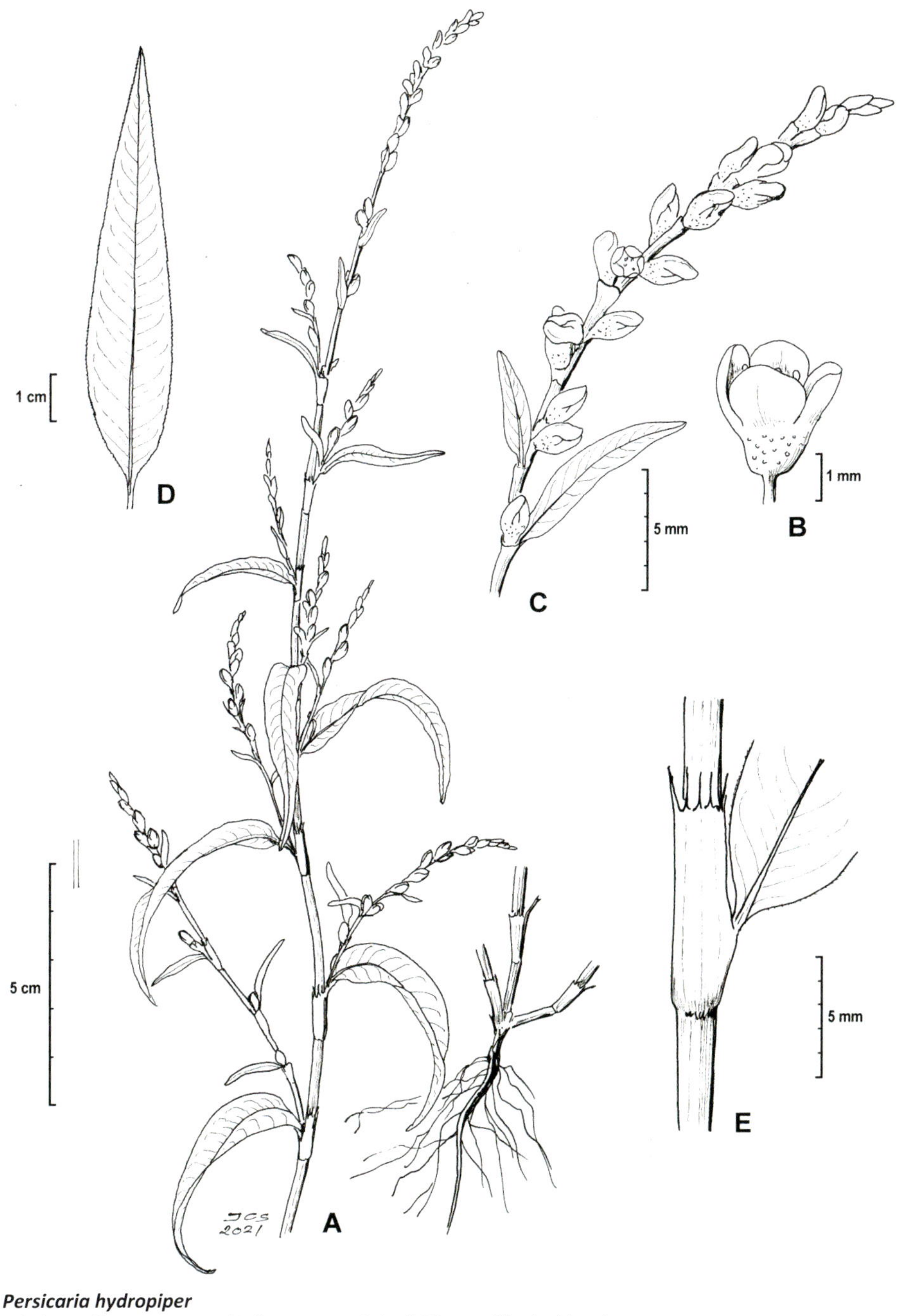

Persicaria hydropiper
A habit, **B** flower, **C** part of inflorescence, **D** leaf, **E** base of leaf with ochrea.

Polygonaceae

257 *Rumex hydrolapathum* Hudson

DK: Vand-Skræppe **N:** Kjempehøymole **S:** -Vattenskräppa **FIN:** Isohierakka **IS:** -
GB: Water Dock **NL:** Waterzuring **F:** Patience aquatique
D: Fluß-Ampfer **CZ:** Šťovík koňský **PL:** Szczaw lancetowaty
EST: Jõgioblikas **LV:** Krastmalas skābene **LT:** Rūgštynė gudažolė **Ru:** Щавель водနощавелевый

Rumex hydrolapathum at the bank of a pond. Paderup, Randers, Denmark. Photo JCS.

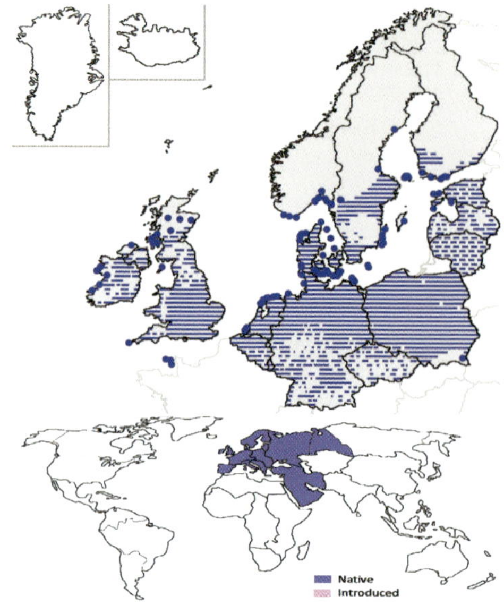

Perennial helophyte with a robust, semi-vertical, bulbous, branched, dark brown rhizome, covered by fibrous remains of old leaves. Roots branched, whitish.

Stem: Up to more than 2 m long and 5–12(–15) mm diameter, erect, later commonly collapsing onto adjacent vegetation or the water, glabrous, distinctly striate and furrowed, the wider parts hollow, otherwise solid, green, often with a brownish or purplish tinge with age.

Leaves: Alternate. Basal leaves petiolate, the petiole 5–30(–50) cm long. Lamina lanceolate or narrowly rhomboidal, widest near the middle, gradually tapering towards both ends, flat or with slightly undulate margin that is entire or irregularly dentate green, young leaves often suffused with red and older leaves becoming brown, up to 70(–100) cm long and 5–12(–15) cm wide, glabrous. Stem leaves sessile or shortly petiolate, narrowly and oblong lanceolate with cuneate to rounded base, becoming smaller and narrower towards the apex. No differentiated submerged leaves.

Inflorescence: Large and open compound racemes with 10–15(–30) obliquely ascending branches, one or more branches from each of the upper leaf axils. Branches without leaves or with only few leaves. Flowers in whorls.

Flowers: 3-merous, bisexual. 6 tepals, the 3 outer tepals narrow, acute and half as long as the inner tepals. The inner tepals are triangular, 5–7 mm long and 4.5–5.5 mm wide, firm, entire or slightly dentate at the base, initially green, eventually turning reddish or greyish brown, with distinct veins, each with an oblong and acute tubercle which is more than half as long as the tepal, initially pale greenish, later reddish. 6 stamens and 3 styles. Pedicels jointed, 2–10 mm long.

Fruits: Nuts, 3.5–4.5 mm long, pale brown and faintly glossy. Pedicels not more than twice as long as the nuts.

Flowering: June-September.

Biology: Wind pollinated and usually abundantly fertile. The tubercles make the nuts more buoyant, and they are easily dispersed by water (as well as wind). The seeds remain viable for several years. Plants do not flower until their second year. Although often rooted well below the water surface the plants do not develop underwater leaves but raise long petiolate basal leaves to the water surface. Long-distance dispersal is mainly by seeds. Plants solitary or forming extensive stands through spread of rhizomes. Vegetative dispersal may occur by uprooted rhizomes.

Habitats: Moderately acid to moderately alkaline, mesotrophic to eutrophic, organic as well as mineral substrate. In the wet margins of ponds, lakes, ditches, canals and streams in open landscapes. Less commonly in

Rumex hydrolapathum. River Gudenå, Ulstrup, Jutland, Denmark. Photo JCS.

Submerged *R. hydrolapathum* in a small pool early in the growth season. Stavtrup, Aarhus, Denmark. Photo BM.

Submerged *R. hydrolapathum*. Part of inflorescence. Djursland, Denmark. Photo JCS.

wet meadows. Mainly freshwater, less commonly in slightly saline waters.

Distribution within the region:
Native. Widespread and abundant from Ireland east to Denmark and the Baltic states, northward into southern Scandinavia.

Characteristics and similar species: The large basal leaves and long stem give this species a very robust appearance unlike all other aquatic plants except other large *Rumex* species. *R. aquaticus* (not in this book) is similar but differs in wider leaves with cordate base and fruits without tubercles.

Although the large oblong-lanceolate leaves are characteristic of *R. hydrolapathum*, identification must be based on the shape of mature fruits.

Hybrids: *Rumex hydrolapathum* is known to hybridise with several other Rumex species. The hybrids display characters which are intermediate between the parental species.

Rumex hydrolapathum. Flower. Photo JCS.

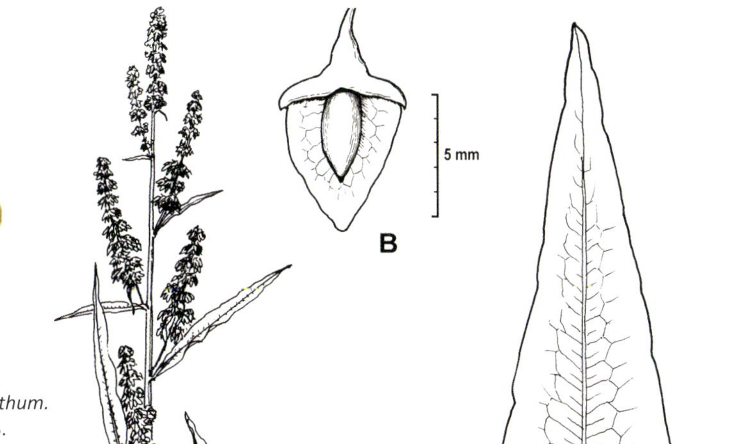
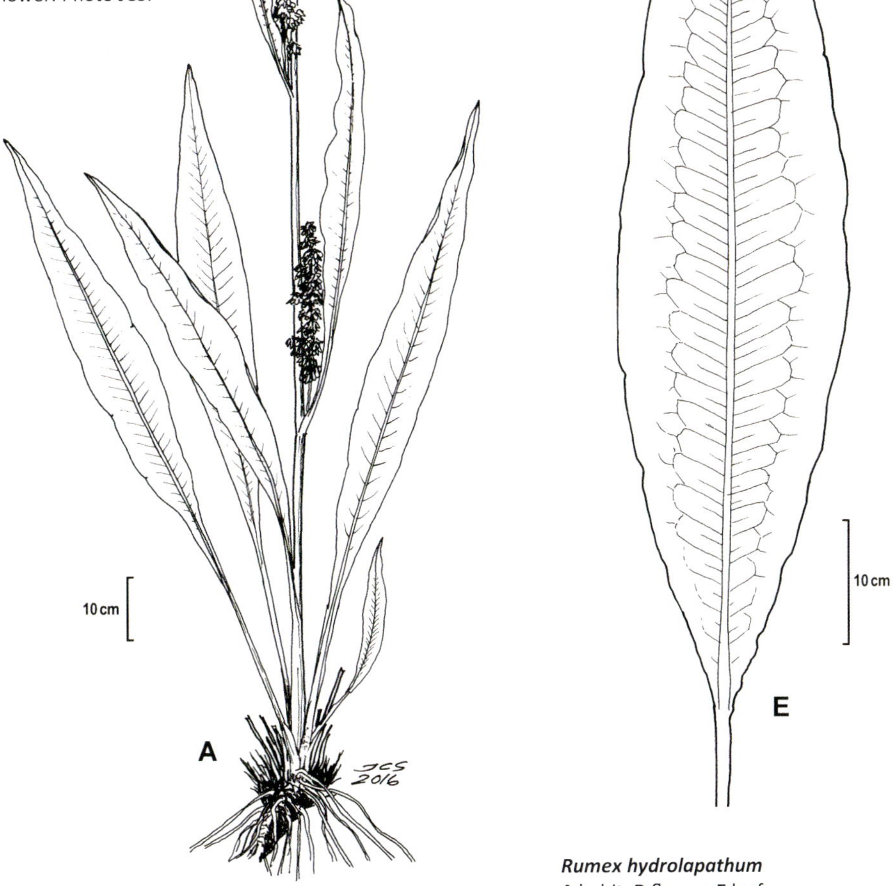

Rumex hydrolapathum
A habit, **B** flower, **E** leaf.

Polygonaceae

258 *Rumex conglomeratus* Murray

DK: Nøgle-Skræppe **N:** Hundehøymol **S:** Dikeskräppa **FIN:** Saksanhierakka **IS:** -
GB: Clustered Dock **NL:** Kluwenzuring **F:** Parelle
D: Knäuel-Ampfer **CZ:** Šťovík klubkatý **PL:** Szczaw skupiony
EST: -**LV:** - **LT:** Glaustažiedė rūgštynė **Ru:** Щавель клубковатый

Rumex conglomeratus in a wet meadow. Randers, Denmark. Photo JCS.

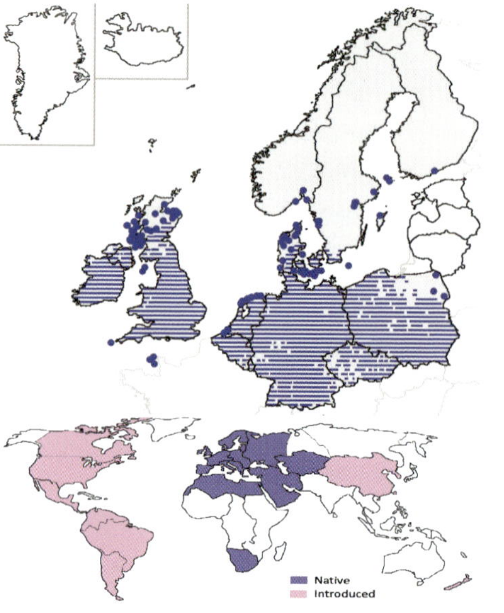

Perennial hemicryptophyte or helophyte with a slender to robust, ascending to vertical rhizome, dark brown. Roots branched, brownish.

Stem: (25–)40–120 cm long and 2.5–5.0 mm diameter, one or rarely a few from each rhizome, erect or commonly irregularly flexuous, slender, glabrous, almost smooth to striate, solid, grey green, often with a brownish or purplish tinge.

Leaves: Alternate. Lower stem leaves petiolate, the petiole 1–5(–20) cm long; lamina lanceolate to elliptic with slightly cordate base, glabrous and green, up to 2–15(–22) cm long and 1–5(–6) cm wide – upper basal leaves larger than the lower ones, flat or slightly undulate, with entire or irregularly dentate margin. Upper stem and branch leaves sessile or shortly petiolate, narrowly oblong lanceolate, numerous, becoming smaller and narrower towards the apex. No differentiated submerged leaves. Basal leaves decaying early.

Inflorescence: Delicate and narrow with 5–15(–25) irregularly, obliquely ascending, flexuous branches, emerging in narrow angles from leaf axils. Branches with numerous leaves. Flowers in remote whorls, each with 10–20 flowers, the apical whorls less distant.

Flowers: 3-merous, bisexual. 6 tepals, the 3 outer tepals c. half as long as the inner tepals that are lingulate to narrowly triangular, 2.5–3.0 mm long and 1.0–1.5 mm wide, subequal, entire or slightly dentate at the base, initially green, eventually becoming reddish, each with a large, oblong tubercle, all or at least the largest tubercle as wide as the tepal, initially pale greenish, later reddish white. 6 stamens and 3 styles. Pedicels filamentous and distinctly jointed, 1–4 mm long.

Fruits: Nuts, 1.5–1.8 mm long and 1.0–1.4 mm wide, dark reddish brown.

Flowering: June-September.

Biology: Wind pollinated, typically abundantly fertile. The tubercles provide buoyancy so that the nuts float and are easily dispersed by water. The seeds remain viable for several years. Plants do not flower until their second year. Long-distance dispersal is mainly by seeds. *R. conglomeratus* may occur as solitary plants or in stands.

Habitats: Moderately acid to moderately alkaline, organic as well as mineral soil in mesotrophic to eutrophic habitats, in moist or dry areas of the littoral zone of ponds, lakes, ditches, canals and streams in open landscapes as well as open forests. It also overwinters growing in water in seasonal pools in damp meadows or occasionally as a ruderal of drier, disturbed ground.

Distribution within the region: Native. Widespread and abundant from Ireland east through Denmark to Poland and Czech Republic, rare in Sweden. Introduced in Norway and Finland.

Rumex conglomeratus. Part of inflorescence. Randers, Jutland, Denmark. Photo JCS.

Characteristics and similar species: *Rumex conglomeratus* is characterised by the large and open inflorescence with remote whorls of flowers on distant branches along most of the stem length, together with large oblong tubercles on all 3 inner tepals.

R. sanguineus is similar in habit but differs by almost globose tubercles on one (rarely all) of the inner tepals.

Proper determination of *Rumex* species in general requires good material with well-developed fruits.

Confusion with other aquatic plants, including **257** ***Rumex hydrolapathum***, is unlikely.

Hybrids: *Rumex conglomeratus* is known to hybridise with several other *Rumex* species. The hybrids display characters which are intermediate between the parental species.

Rumex conglomeratus. Basal leaves. England. Photo RVL.

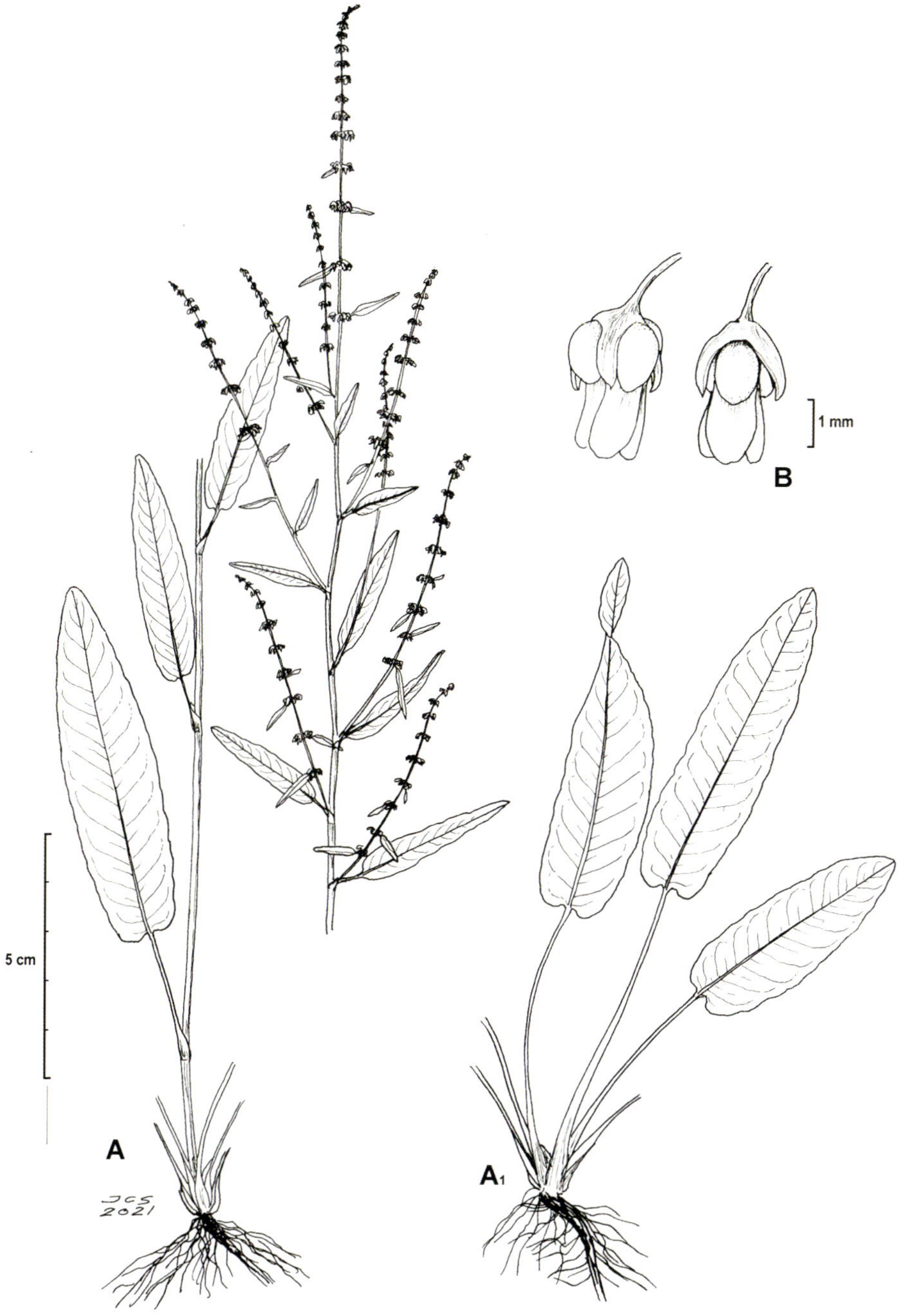

Rumex conglomeratus
A, A1 habit, **B** flowers.

Rumex conglomeratus. Inflorescences. Randers, Jutland, Denmark. Photo JCS.

Droseraceae

259 *Aldrovanda vesiculosa* L.

DK: - **N:** - **S:** - **FIN:** - **IS:** -
GB: Waterwheel **NL:** - **F:** Aldrovande vésiculeuse
D: Blasige Wasserfalle **CZ:** Aldrovandka měchýřkatá **PL:** Aldrowanda pęcherzykowata
EST: - **LV:** - **LT:** Pūslėtoji aldrūne **Ru:** Альдрованда пузерчатая

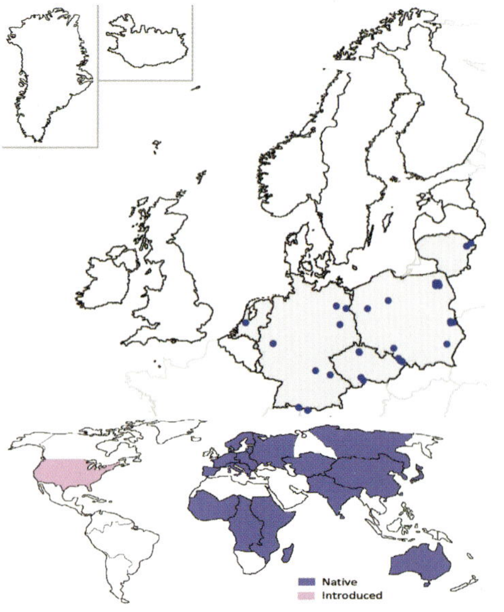

Aldrovanda vesiculosa floating among leaves of *Nuphar*. Lake Ruzas SE of Dukstas, Lithuania. Photo Jan-Thomas Johansson.

A perennial, submerged plant, free-floating just below the surface. Without roots.

Stem: Shoots sparsely branched, up to 6–20(–30) cm long, about 2 cm wide, whitish to greenish, glabrous, with 0.5–0.7 cm long internodes and numerous leaves in whorls of 6–10(-17 when supporting a flower).

Leaves: 7–15 mm long, light green to yellow green in Europe, but in Australia, East Timor, Africa and Hungary often reddish or purplish from anthocyanin under favourable growth conditions. Pedicel 4–8 mm long, flat, wedge shaped, expanded to about 2 mm. Lamina 3–5 mm long, flattened and forming a 2-lobed trap surrounded by 4–6, 6–8 mm long bristles. The lamina is twisted in a way that the trap opens outwards.

Inflorescence: Peduncle short, 5–15 mm long, thick, arising from a leaf axil and bearing a single flower.

Flowers: 5-merous, bisexual. Sepals ovate, 3–4 mm long, united at the base. Petals obovate, 4–5 mm long, greenish white or pinkish tinged. 5 stamens and 5 styles with fringed, terminal stigmas.

Fruits: Capsules ovate, 4–5.5 x 3–4 mm, with (1–)6–9 ripe seeds. Seeds black, glossy, ovoid, 1.2–1.4 mm long.

Flowering: July-September.

Biology: Carnivorous aquatic plants - see next page. In temperate parts of Europe plants very rarely flower and seed set is sparse. Each flower is only open for a few hours, and is probably self-pollinated. The seeds ripen under water.

In the region, propagation is almost exclusively through vegetative reproduction. In optimal conditions, stolons are produced at 3–4 cm intervals along the stem, forming in new plants as the tips continue to grow and the older parts die back.

Overwinters as dense, 4–6 mm long, broadly spindle-shaped turions developed from shoot apices in the autumn. In tropical areas growth continues through the year and turions are rarely produced.

Habitats: In still, shallow, dystrophic water among emergent vegetation, rarely in open water. Bogs, fens, peaty fishponds, backwaters and sheltered bays in lakes.

In optimal conditions *A. vesiculosa* has a stem growth rate of 4-9 mm per day. Lake Ruzas southeast of Dukstas, Lithuania. Photo Jan-Thomas Johansson.

The leaves of *Aldrovanda* are very specialised structures arranged in whorls on the stem. The wedge-shaped pedicel has air-filled chambers supporting buoyancy, and the long terminal bristles prevent floating plant parts from triggering the trap and may also be useful in forcing the prey towards the trap.

The almost circular lamina is transformed into a snap-trap with 2 semi-circular leaf lobes hinged by the mid-vein, and on the inside with approximately 20 trigger hairs on each lobe. The margins of the lobes are fringed by short, stiff, hair-like cilia, preventing prey (aquatic invertebrates) from escaping when caught. The imprisoned animals are digested by released enzymes. The closing of the trap takes just 10–20 milliseconds and is among the fastest plant movements known.

Lake Ruzas southeast of Dukstas, Lithuania. Photo Jan-Thomas Johansson.

Distribution within the region: Native. Globally threatened. Formerly more widespread, it now occurs at scattered sites in Germany, Poland, Lithuania and parts of European Russia. It has been reintroduced to the Czech Republic, the Netherlands and sites in Poland.

Characteristics and similar species: *Aldrovanda vesiculosa* is the only member of the genus, and it is unlike all other aquatic plants.

At a distance it may resemble **327-333 *Utricularia*** species, but the trap-bearing leaves in whorls and the structure of the traps are very different.

197-199 *Ceratophyllum* species have dichotomously divided leaves arranged in whorls and lack traps.

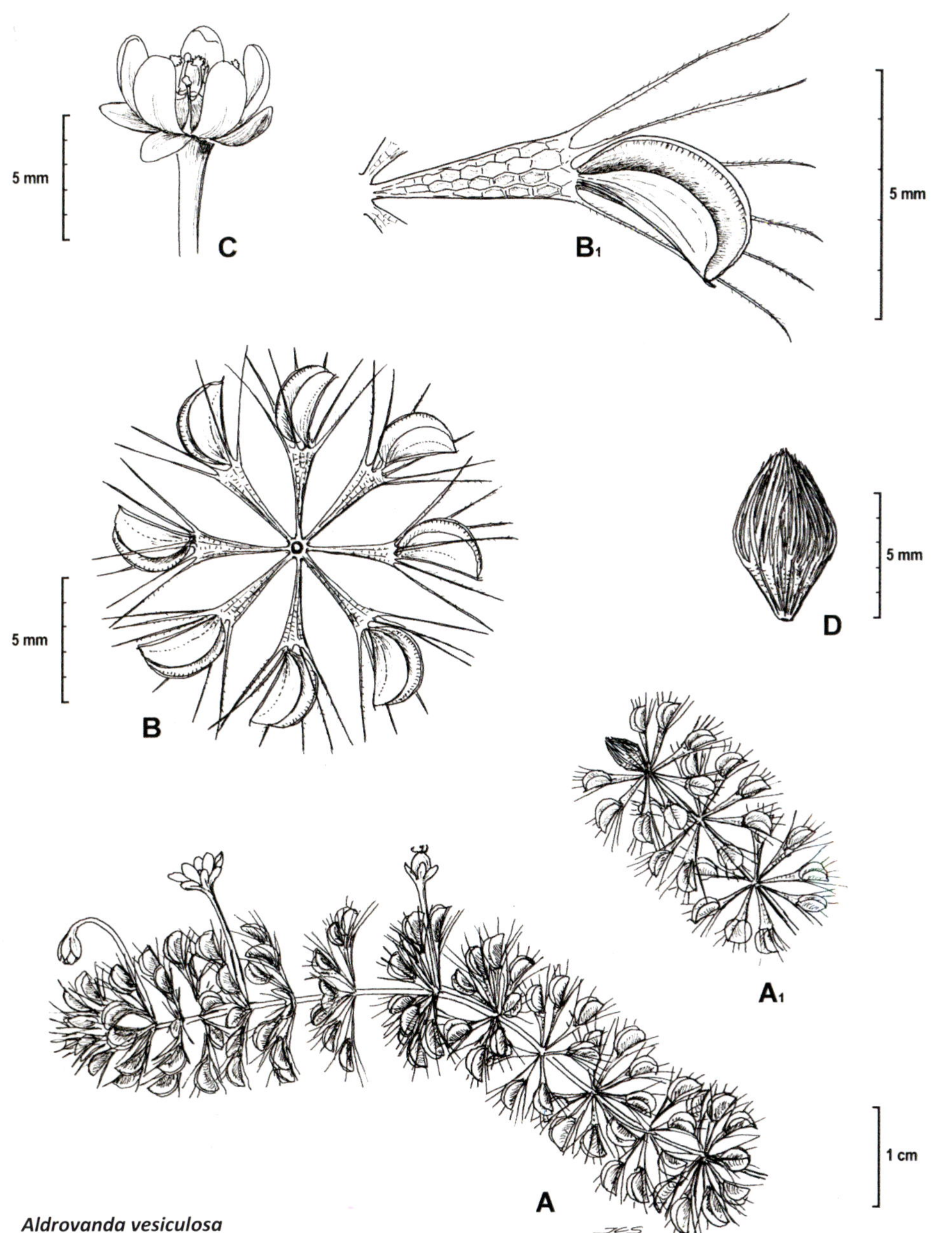

Aldrovanda vesiculosa
A habit, **A1** shoot with apical turion, **B** whorl of leaves, **B1** leaf with trap, **C** flower, **D** turion.

Caryophyllaceae

260 *Illecebrum verticillatum* L.

DK: Bruskbæger **N:** Bruskbeger **S:** Glimmerört **FIN:** Rustokki **IS:** -
GB: Coral necklace **NL:** Grondster **F:** Illecèbre verticillé
D: Quirlige Knorpelmiere **CZ:** Nehtovec přeslenitý **PL:** Goździenniec okółkowy
EST: - **LV:** - **LT:** - **Ru:** Иллецебрум мутовчатый

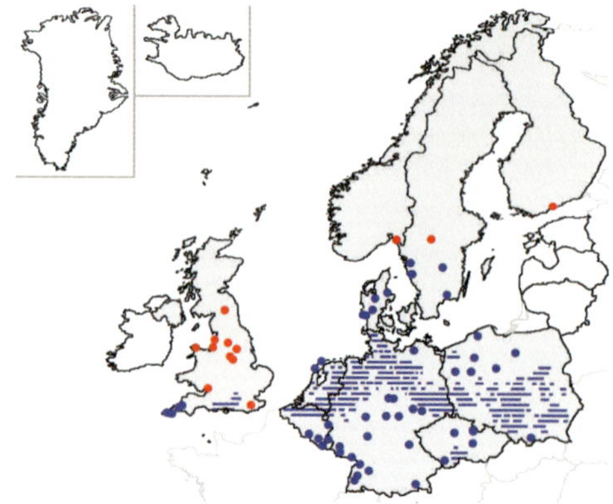

Red dots indicate historic records.

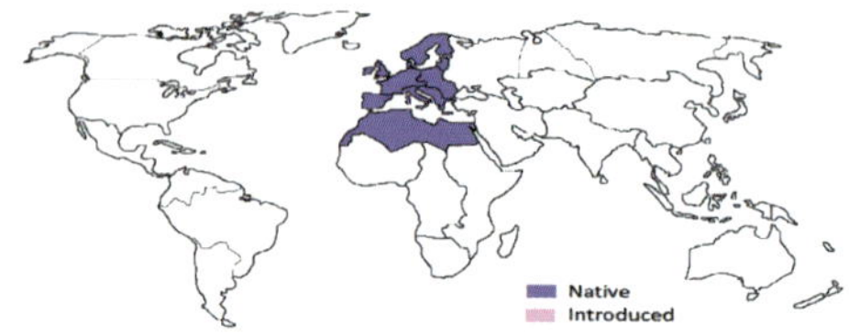

Illecebrum verticillatum growing fully submerged. Beira Litoral, Portugal. Photo KvdW.

Summer-annual herbaceous therophyte with white roots, adventitious roots from the lower stem nodes, no rhizome.

Stem: 5–30(–70) cm long, procumbent or decumbent, filamentous, four-angled in cross section, glabrous, pinkish red to red, richly branched from the base, forming a rosette of unevenly long branches, usually repeatedly branched. Submerged plants with long, mostly unbranched stems with long internodes, upright or ascending in the water column, sometimes floating on the surface.

Leaves: Opposite. Each pair of leaves oriented at a right angle to the neighbouring pairs, giving the shoots a cross-like appearance when seen from above. Lamina entire, 1.9–5.2(–6.3) mm long and 1.1–3.0(–3.4) mm wide, spathulate, ovate obovate or broadly lanceolate, obtuse to subacute, fleshy, glabrous, green red, sessile or with very short petiole. The leaves of submerged plants are more acute. Stipules 2, dry and membranous, 0.7–1.4 mm long, fused adaxially.

Inflorescence: 5–16 dense clusters of 3–6(–9) flowers, in leaf axils along the branches and thus appearing as whorls. Bracts silvery and triangular, 0.9–1.3 mm long.

Flowers: 5-merous, bisexual. Sessile, (1.8–)2.0(–2.5) mm long. Calyx shining white to pinkish, usually with small purple dots, 5 free oblong, hooded and connivent petal lobes, 1.8–2.8 mm long with a 0.4–0.8 mm long awn, becoming enlarged and spongy after flowering. Petals narrow and acute, 0.4–0.6 mm long, yellowish. Stamens 1–2, 0.2–0.4 mm long with very short filaments and globular yellowish anthers, c. 0.1 mm in diameter. Style reduced, 2 stigmas, c. 0.1 mm.

Fruits: Capsules yellowish, enclosed by the sepals, 1 seed per capsule, 0.8–0.9 mm, ovoid and slightly keeled, dark brown, smooth and glossy.

Flowering: July-September.

Biology: Self-pollinated and self-fertile. After flowering the dry capsules detach without splitting open. Seeds germinate when the soil in seasonally inundated habitats dries out during summer, often leading to wide population fluctuation between years. If plants become flooded during the growing season they may develop long ascending shoots with long internodes. After fruiting, plants die back, overwintering as seeds.

Illecebrum verticillatum in shallow water. Puebla de Don Rodrigo, Cuidad Real, Spain. Photo RVL.

Illecebrum verticillatum on moist substrate in a dune depression. SW-Jutland, Denmark. Photo JCS.

Illecebrum verticillatum. The peculiar-looking calyx with pinkish dots. SW- Jutland, Denmark. Photo JCS.

Dispersal mainly by seeds, although there may be limited vegetative dispersal by detached stem fragments. *I. verticillatum* is strongly associated with poaching by cattle or other actions, such as drying out of seasonal wetlands which suppress competition from other plants.

Habitats: Slightly acid to neutral, oligotrophic soils, mainly sand and fine gravel in the drawdown zone of ponds, lakes, ditches, canals and streams. Mainly in heathland and heathy grassland but also in seasonal pools and hollows in meadows and dune systems, in temporarily wet wheel ruts and in sand or gravel pits.

Distribution within the region:
Native. Locally abundant from southern England east to eastern Poland, scattered in Denmark, southern Sweden and southern Finland.

Characteristics and similar species: Typical plants with procumbent reddish stems and branches and whorls of white flowers distinguish this species from most other aquatic plants. Non-flowering submerged plants can usually be recognised by the long, filamentous stems with long internodes and the small, more acute leaves. However, they can be mistaken for **281-292 *Callitriche*** species and **239 *Lythrum portula***, which differ by terete stem and petiolate leaves. **238 *Hypericum elodes*** also has a terete stem, but leaves with translucent dots.

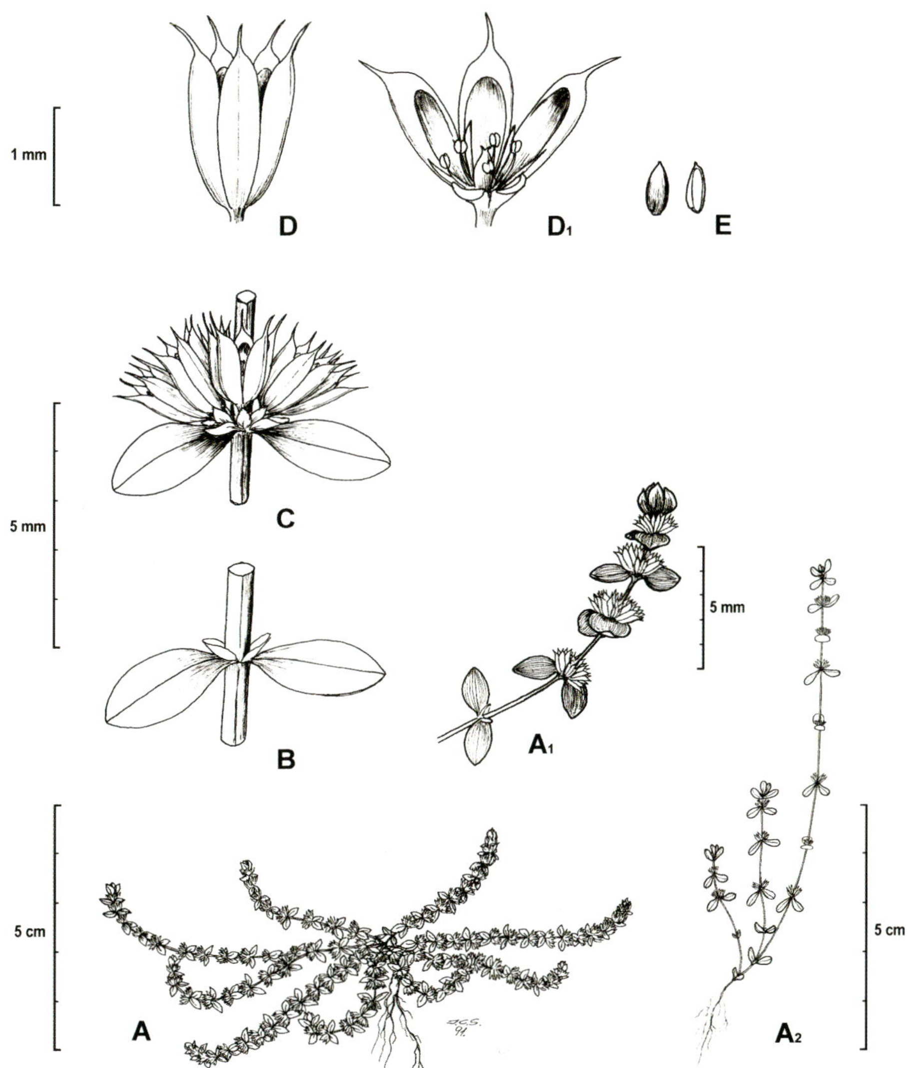

Illecebrum verticillatum

A habit emergent plant, **A1** part of inflorescence, **A2** habit submerged plant, **B** pair of leaves, **C** flowers in leaf axils, **D** flower, **D1** flower with part of calyx removed, **E** seeds.

Montiaceae

261 *Montia fontana* L.

DK: Stor vandarve **N:** Kjeldeurt **S:** Källört **FIN:** Hetekaali **IS:** Lækjagrýta
GB: Blinks **NL:** Groot bronkruid **F:** Montia des fontaines
D: Bach-Quellkraut **CZ:** Zdrojovka hladkosemenná **PL:** Zdrojek błyszczący
EST: Harilik allikrohi **LV:** Avotu montija **LT:** Šaltininė menuva **Ru:** Монтия родниковая

Montia fontana subsp. *amporitana*. Burley, New Forest, England. Photo RVL.

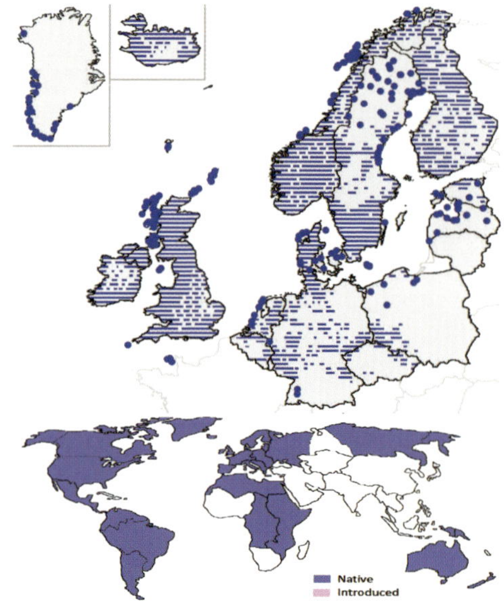

Annual or short-lived perennial helophyte or hydrophyte without rhizome. Roots whitish to pale brown.

Stem: 5–15(–25) cm long, creeping to ascending, erect in dry conditions, smooth, terete, more or less branched, rooting from the nodes.

Leaves: 5–20 mm long, 2–6 mm wide, opposite, spathulate to obovate, not or only partly fused at base, fleshy, green, shiny, blunt at apex, tapering to base.

Inflorescence: Flowers 1–5(–9) in apparently axillary cymes, which emerge from the tip of the shoots, but are overtopped by side branches.

Flowers: 5-merous, bisexual. With 2 oval, 2–2.5 mm long, broad, green sepals which enclose the flower. 5 petals unequal, 2 large and 3 small, up to 3 mm long, white, fused into a short tube which is split to the base at one point. 3 anthers and 3 stigmas.

Fruits: Capsule c. 2 mm, ovoid to globose, unilocular, 3-valved and with 3 seeds. Seeds c. 1 mm across, slightly flattened, almost circular, blackish brown, more or less shiny, smooth, with a delicate, more or less raised, reticulate pattern.

Flowering: June-September.

Biology: Insect pollinated or often cleistogamous and self-pollinated, especially when submerged.

Habitats: In acidic to alkaline, oligotrophic to moderately eutrophic water. In periodic or permanently wet habitats, in springs, flushes, ditches and the banks of streams and rivers, less commonly in meadows, etc. and usually only where there are seepages or springs.

Distribution within the region:
Native. Common in northern and western parts of the area, scattered in the southeast.

Montia fontana subsp. *fontana*. Germany. Photo KvdW.

Characteristics and similar species:
M. fontana is split into subspecies based on the seed sculpture (see page 336).

Stellaria alsine and ***Stellaria crassifolia*** (not in this book) differ from *Montia fontana* in 5-merous calyx and furrowed stem.

225 *Crassula helmsii* has solitary, 4-merous flowers in leaf axils, stems are slightly branched, with narrower, acute and fleshier leaves.

M. fontana subsp. *chondrosperma*. Parc Naturel Régional de la Brenne, Indre, France. Photo RVL.

Flower. *M. fontana* subsp. *fontana*. Glas Maol, Angus, Scotland. Photo RVL.

261a *Montia fontana* L.
subsp. ***fontana***

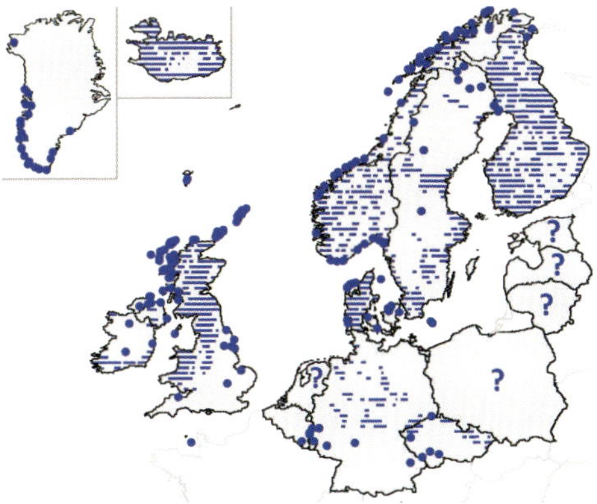

Seeds very shiny, smooth on faces and margin.

Distribution within the region:
Native. Distribution poorly known as it is rarely identified to subspecies, but widespread and abundant in the north, including Iceland and Greenland, less frequent and more scattered in the south.

262 *Montia fontana* L.
subsp. ***amporitana*** Sennen

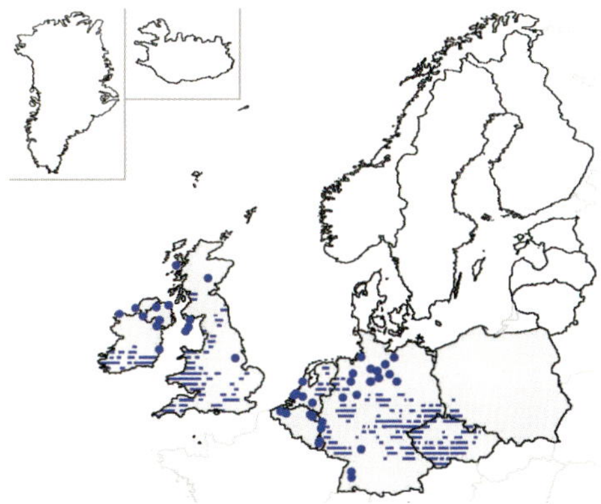

Seeds shiny, smooth on faces, but with 3–4(–5) rows of rather long, narrow pointed tubercles along the margin.

Distribution within the region:
Native. Frequent in the Czech Republic, Germany, the Netherlands, Britain and Ireland. Rare in Poland, Belgium and Luxembourg.

263 *Montia fontana*
subsp. ***chondrosperma*** (Fenzl.) Walters
Syn.: *M. minor* C.C.Gmel.; *M. arvensis* Wallr.

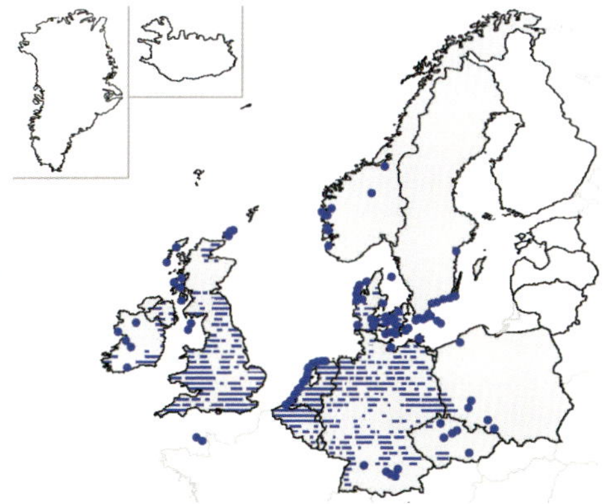

Seeds dull and coarsely tuberculate all over the faces and margin. Smaller and more erect in habit than subsp. *fontana*.

Distribution within the region:
Native. Widespread and abundant in southern Britain, the Netherlands, Belgium and Germany, occasionally in Ireland, the Czech Republic and Denmark. Rare in Poland, Sweden and Norway.

Hybrids:
Montia fontana subsp. *amporitana* × subsp. *fontana*
Syn.: *M. fontana* L. subsp. *variabilis* Walters

Seeds rather shiny, smooth on faces, but with 1-4 rows of rather short, less pointed tubercles along the margin.

Intermediate between the parent subspecies, occurring where the ranges of the parents overlap, as well as in Denmark.

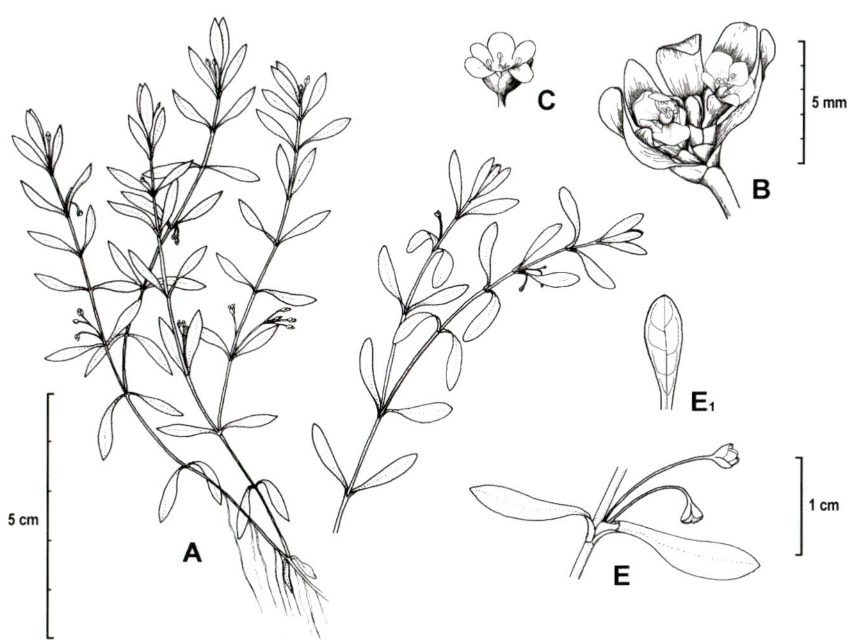

Montia fontana
A habit, **B** inflorescence, **C** flower, **E** leaf pair and flowers, **E1** leaf.

Primulaceae

264 *Hottonia palustris* L.

DK: Vandrøllike **N:** Vassblink **S:** Vattenblink **FIN:** Vesisulka **IS:** -
GB: Water-violet **NL:** Waterviolier **F:** Hottonie des marais
D: Wasserfeder **CZ:** Žebratka bahenní **PL:** Okrężnica bagienna
EST: Harilik vesisulg **LV:** Purva sermulīte **LT:** Plunksnalapė griovenė **Ru:** Турча болотная

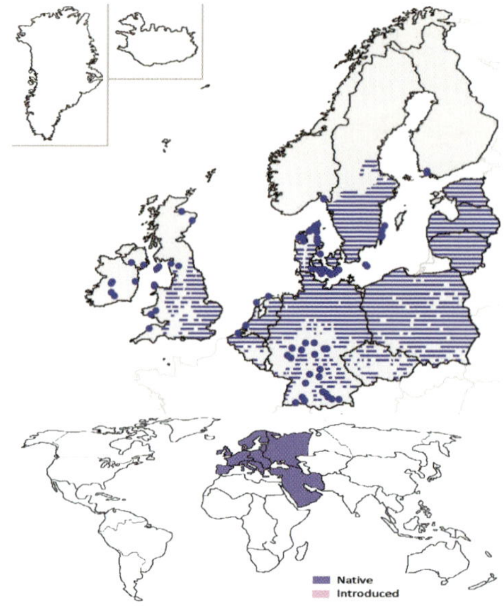

A dense mat of flowering *H. palustris* in a temporary pond. Stavtrup near Aarhus, Denmark. Photo JCS.

Perennial or annual hydrophyte or temporary helophyte without rhizome. Numerous white roots from lower stem nodes.

Stem: 30–90 cm long and 2.5–5.0 mm diameter when submerged, only a few cm long when emergent, terete and smooth, moderately branched, green.

Leaves: 2–8(–13) cm long, 2.5–5.0(–9.0) cm wide, alternate, oval to elliptic in outline, sessile, 1(–2)-pinnate with up to 40 linear to filiform segments, 10–40 (–50) mm long and 0.8–1.2(–1.5) mm wide. Leaves forming apical rosettes with very short internodes, light to dark green, rarely with a reddish tinge, smooth and glabrous on the upper side, with glandular hairs 0.1 mm long on lower side.

Inflorescence: 4–9 whorls of 3–7(–10) long-pedicellate flowers, carried high above the water surface by the 10–40 (–80) cm long leafless peduncle towards the apex, emerging from the tip of the shoot and commonly balanced by a number of radially oriented branches at or just below the surface.

Flowers: 5-merous, bisexual, 15–25 mm in diameter. Pedicels 5–20 mm long, terete, green, with numerous 0.1 mm long glandular hairs. Sepals 4–6 mm long, fused at the base, green, with 0.1 mm long glandular hairs. Petals fused into a 2.5–4.5 mm long corolla, with 5 flat protruding petal lobes. Petal lobes emarginate, 7–9 mm long, white to pink, rarely light purple, with a yellow base. 5 stamens with yellow anthers, 1 style. Two types of flowers (heterostyly), some with filaments shorter and style longer than the corolla, others with filaments longer and style shorter than the corolla; individual plants have only one type or the other.

Fruits: Obovate, 3–6 mm long capsules with five longitudinal seams, green with persistent style remains. Seeds numerous, brown, angular.

Flowering: May-June.

Biology: Typically insect pollinated; heterostyly serves to limit self-pollination, however self-fertilising cleistogamous flowers occur on submerged inflorescences. After flowering the peduncle bearing ripe fruits sinks below the water surface and the seeds are released as the capsule decays. Overwinters as submerged green shoots on the bed. In the early spring plants start to grow and simultaneously rise towards the surface, where they commonly branch to form dense mats or large individual plants, and eventu-

Submerged leaves of *H. palustris*. Photo RVL.

The flowers of *H. palustris* are most commonly pink with a yellow centre, but the colour may vary from white to light purple. The flowers shown here are the type with short styles and stamens with long filaments. Stavtrup near Aarhus, Denmark. Photo JCS.

Hottonia palustris. Part of fruiting inflorescence. Photo RVL.

ally flower. After flowering the plants typically sink towards the bottom.

The species is very variable depending on habitat, season and growing condition. When habitats dry out it can survive as a helophyte with short, very compact shoots with rosettes of shorter, fleshier leaves with fewer, broader segments.

Habitats: In slightly acid to moderately alkaline, mainly oligotrophic to mesotrophic water bodies. In ponds, shallow lakes, ditches and canals, seasonal pools or the drawdown zone of permanent water bodies.

Distribution within the region:
Widespread and abundant from England east to Poland and the Baltic states, including Denmark and southern Sweden. Scarce in Ireland and the southern part of the region.

Characteristics and similar species:
Flowering *Hottonia palustris* is unlikely to be confused with other water plants.

When flowers are lacking it may be confused with **226-231 *Myriophyllum*** spp., all of which have 1-pinnate leaves in whorls and often have reddish or brownish tinges. *H. palustris* rosettes may appear to be in whorls due to the very short internodes.

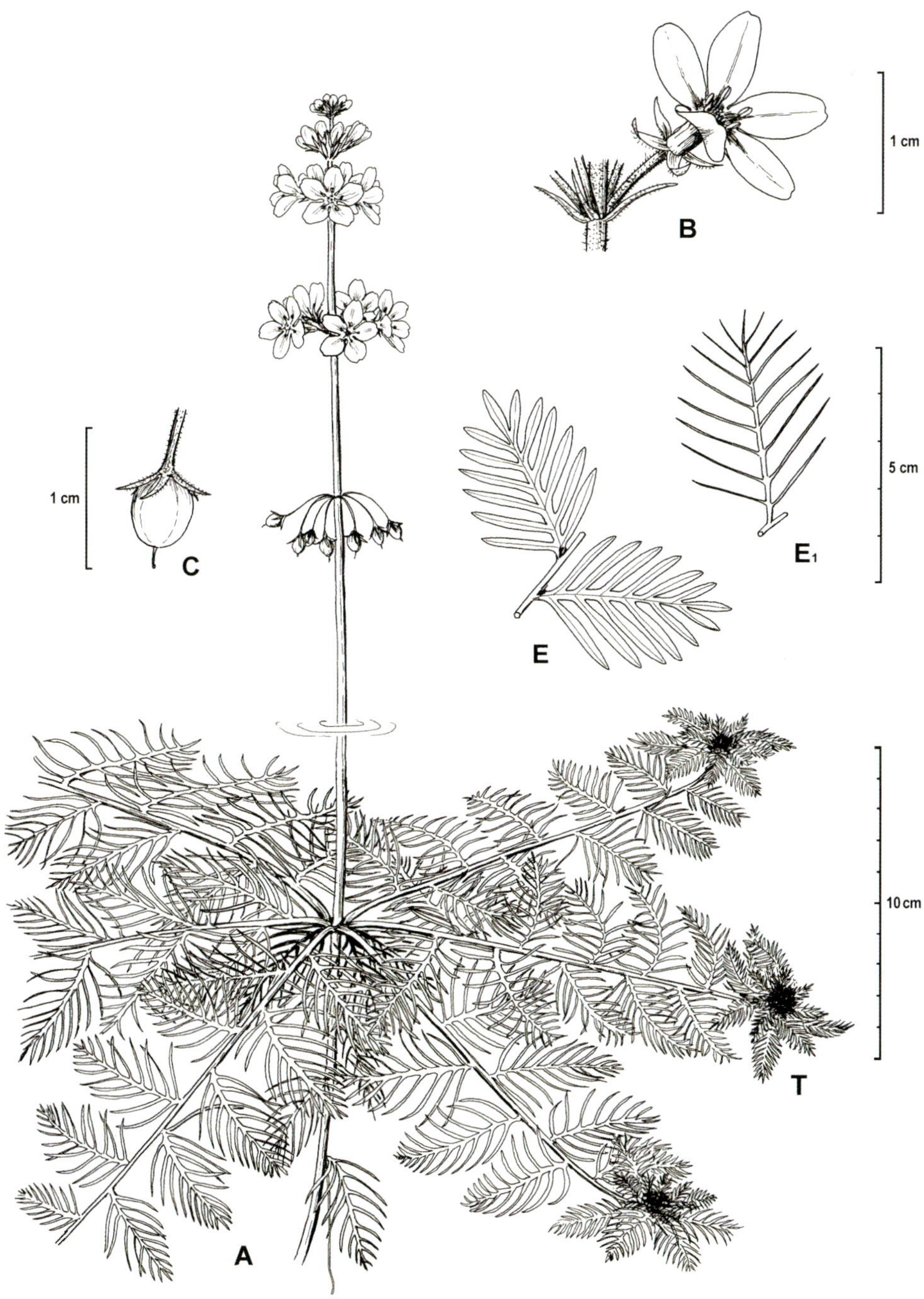

Hottonia palustris
A habit, **B** flower, **C** capsule, **E, E1** leaves, **T** turion.

Primulaceae

265 *Lysimachia tenella* L.

Syn.: *Anagallis tenella* (L.) L.
DK: - **N:** - **S:** Krypmire **FIN:** - **IS:** -
GB: Bog pimpernel **NL:** Teer guichelheil **F:** Mouron délicat
D: Zarter Gauchheil **CZ:** - **PL:** Kurzyślad wątły
EST: -**LV:** - **LT:** - **Ru:** Очный цвет нежны

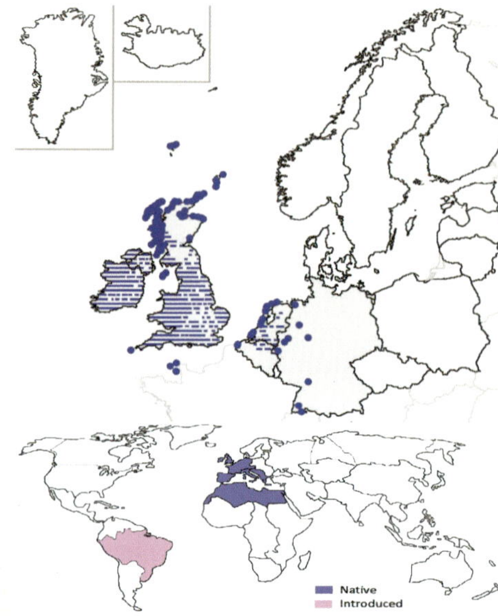

Lysimachia tenella. Cloon & Laghtanabba Bog E of Claddaghduff, Ireland. Photo Jan-Thomas Johansson.

Perennial helophyte with stem creeping above or below ground and rooting at the nodes. Roots thin, in the axils of the leaves.

Stem: 3–20 cm long, c. 0.5 mm diameter, slender, prostrate, sparsely branched, subterete to bluntly angular, glabrous or with scattered, glandular hairs on young shoots.

Leaves: Up to 2–6(–10) mm long, 1–8 mm wide, opposite, rarely some alternate, decussate, ovate to suborbicular, blunt to subacute, entire, glabrous, green, shortly petiolate. An interrupted black line around the margin of the undersides of the leaves.

Inflorescence: Flowers solitary on erect pedicels from leaf axils.

Flowers: 5-merous, bisexual, 12–16 mm in diameter. Pedicels 15–35 mm long, terete, with greenish, glandular hairs, 0.1 mm long. Sepals free or shortly fused at base, linear to lanceolate, acuminate, glabrous, green or slightly reddish tinged. Petals shortly fused at base, 6–10 mm long, 2–3 times as long as the sepals, pale pink with darker veins, rarely white. 5 stamens with tubular filaments fused at base and covered with multicellular hairs 1–2 mm long, anthers cream. 1 style with acute stigma.

Fruits: Capsule 2.5–3.0 mm long, globose, opening transversely. Seeds 6–8.

Flowering: June-September.

Biology: The flowers open in daylight and emit a strong honey-like smell which attracts insects for pollination. The seeds are wind dispersed. Vegetative propagation by fragmentation of the creeping stem. The plant remains green during winter. According to Glück (1936) aquatic forms are considerably reduced in size and lack flowers.

Habitats: Acidic to slightly alkaline peaty substrate, mostly in bogs, in pools, ditches and on the shoreline of lakes.

Distribution within the region:
Native. Widespread and abundant throughout much of Britain and Ireland, except for the Scottish Highlands. Common in Netherlands, scattered in Belgium, very rare and declining in Germany.

Lysimachia tenella. Clear Island, County Cork, Ireland. Photo RVL.

Lysimachia tenella flowers. Langstraat Waalwijk, Netherlands. Photo KvdW.

Lysimachia tenella flower. Note the numerous filamentous hairs. Cloon & Laghtanabba Bog E of Claddaghduff, Ireland. Photo Jan-Thomas Johansson.

Characteristics and similar species:
Even without the relatively large, pinkish flowers, *L. tenella* is unmistakable because of its small size and the creeping, more or less branched stem rooting at the nodes and with small, ovate, opposite leaves.

It could be mistaken for **268 *Lysimachia nummularia***, but that species never has the black margin to the underside of the leaves.

Lysimachia tenella
A habit, **B** flowers and part of stem, **C** flower, **D** stamen.

Primulaceae

266 *Lysimachia thyrsiflora* L.

DK: Dusk-Fredløs **N:** Gulldusk **S:** Topplösa **FIN:** Terttualpi **IS:** -
GB: Tufted loosestrife **NL:** Moeraswederik **F:** Lysimaque à fleurs en épi
D: Strauß-Gilbweiderich **CZ:** Vrbina kytkokvětá **PL:** Tojeść bukietowa
EST: Ussilill **LV:** Kekaru zeltene **LT:** Puokštinė poraistė **Ru:** Вербейник кистецветный

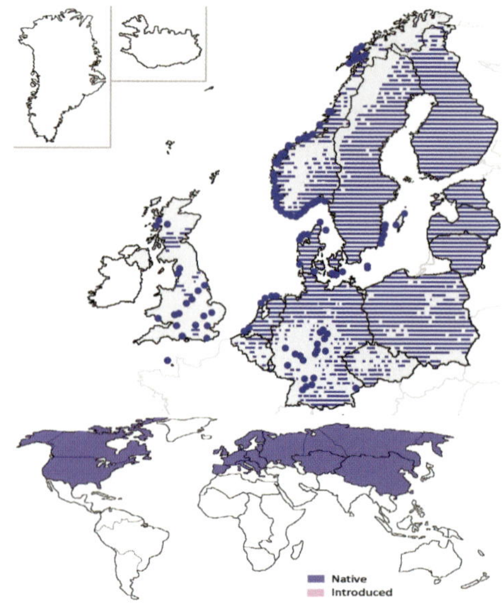

Dense stand of *Lysimachia thyrsiflora* on wet soil in the littoral zone of a small forest pond. Hørby Plantage near Hobro, Denmark. Photo JCS.

Perennial helophyte or temporary hydrophyte with horizontal rhizome and long underground stolons, roots fibrous and white.

Stem: 30–60(–80) cm long and 2–7 mm diameter, erect, straight, terete and hollow, light green to green with a reddish-brown tinge, commonly red spotted, base of stem glabrous, upper part with brownish hairs up to 2 mm long.

Leaves: Opposite, each pair oriented at a right angle to the neighbouring pairs, giving the shoots a cross-like appearance when seen from above, (2–)7–13 cm long and (0.5–)1.0–2.5 cm wide, linear elliptic to elliptic, acuminate, sessile and semi-amplexicaul. Upper side light to dark green, glabrous with numerous reddish glandular dots, lower side pale- to mid-green, glabrous or with sparse hairs on the midrib and margins. Leaf margins usually slightly recurved. Base of stem with small scale-like leaves. Submerged shoots with ovate-oblong leaves with an obtuse apex, 1.5–2.5 cm long, pale green, on the upper side with numerous depressed reddish glandular dots.

Inflorescence: Densely flowered semi-globular racemes, 1.5–2.5 cm long and wide. Peduncles up to 5 cm long, arising from the axils of the leaves on the middle part of the stem, green purple, usually hairy.

Flowers: Mostly 6-merous, sometimes 5- or 7-merous, bisexual. Pedicels c. 3 mm long, shorter than the linear-elliptic bracts, slightly hairy. Calyx with (5–)6(–7) free sepals, 2–4 mm long, linear elliptic, glabrous, green or yellowish. Corolla with (5–)6(–7) free petals, linear elliptic, golden yellow, with red dots at the apex. (5–)6(–7) stamens with filaments longer than the petals, anthers yellow. Rudimentary filamentous stamens between the petals. Style up to 8 mm long, filamentous.

Fruits: Capsules globular, 1.5–2.5(–3.0) mm long, shorter than the sepals, red dotted. Seeds relatively few, 1.0–1.5 mm, triangular, light brown.

Flowering: June-July.

Biology: Flowers self-sterile, pollinated by bees in the Melittidae family. Fruit and seed set commonly poor, seeds dispersed by wind and water. Vegetative propagation by creeping rhizomes and stolons, commonly leading to formation of large clonal stands. Plants typically rooted above or just below the water line and developing aerial shoots. Submerged plants have ascending shoots with long internodes and short narrow leaves with rounded apex.

Habitats: In acidic to weakly alkaline, oligotrophic to eutrophic water. In the margins of ponds, lakes, ditches, canals

Submerged shoots of *Lysimachia thyrsiflora* are slender with 1.5-2.5 cm long leaves with an obtuse apex. Inserted: The upper sides of the leaves have numerous reddish glandular dots, easily visible with a hand lens when backlit. Pond in coastal dune system at Råbjerg Mile, Jutland, Denmark. Photos JCS.

and streams in open landscapes and forests. Less often on moorland, meadows and freshwater marshes. Submerged plants usually occur in more or less acidic, oligotrophic ponds, lakes and streams.

Distribution within the region: Native. Widespread and abundant throughout most of the region, but very local in Britain, absent from Ireland, Iceland and Greenland.

Characteristics and similar species: Flowering *L. thyrsiflora* is easily distinguished from all other aquatic plants by the semi-globular inflorescences with golden-yellow flowers. Non-flowering plants can be distinguished by the red-dotted stems and the opposite leaves with reddish glandular dots on the upper side. Submerged plants can be recognised by the short, opposite leaves with reddish glandular dots on the upper side.

267 *Lysimachia vulgaris* only very rarely grows submerged, with shoots resembling those of *L. thyrsiflora*, which differs in the red glands on the upper leaf surface and the leaves slightly amplexicaul. *L. vulgaris* lacks glands, and the leaves have a short petiole (best seen by removing a leaf).

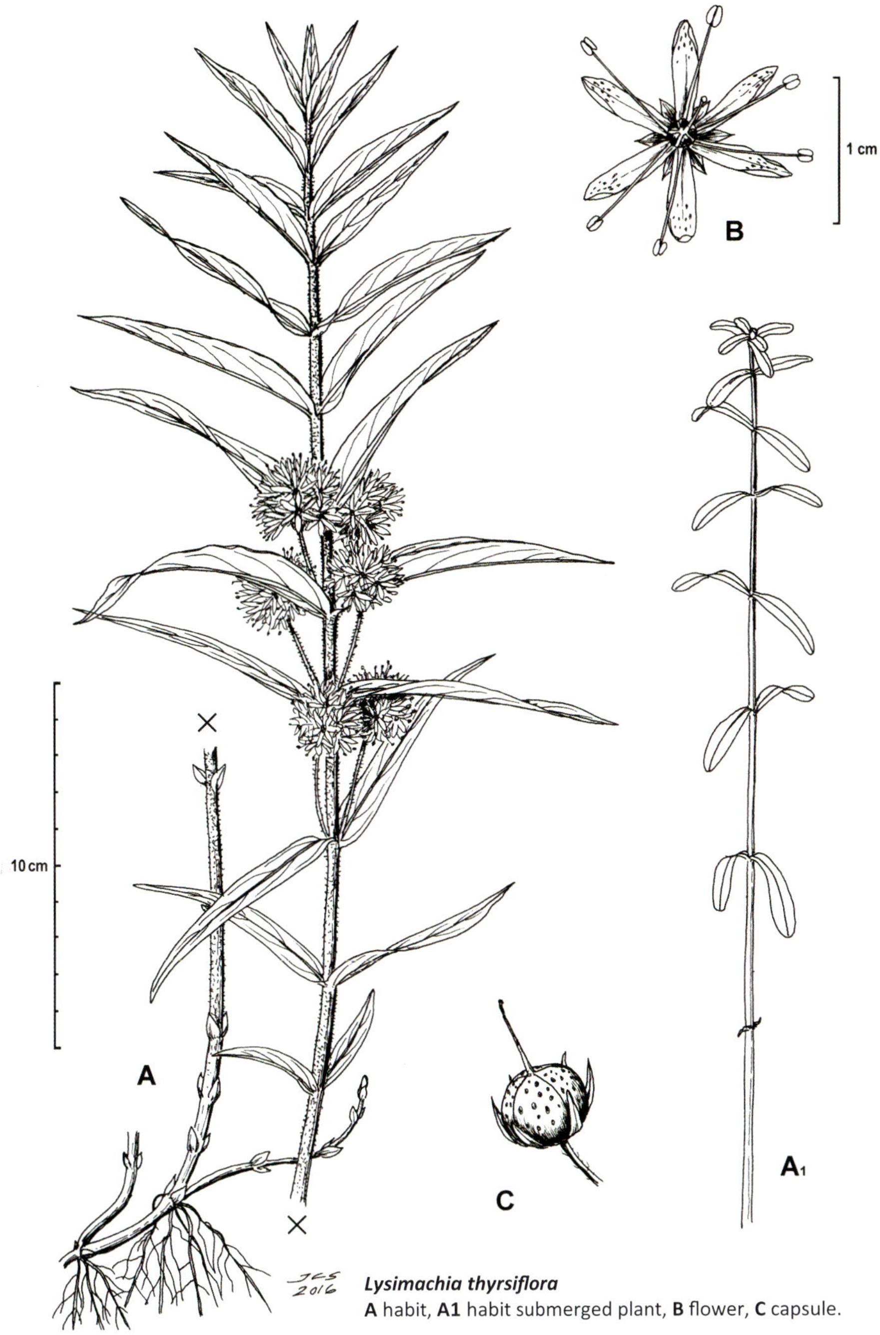

Lysimachia thyrsiflora
A habit, **A1** habit submerged plant, **B** flower, **C** capsule.

Primulaceae

267 *Lysimachia vulgaris* L.

DK: Almindelig Fredløs **N:** Fredløs **S:** Strandlysing **FIN:** Ranta-alpi **IS:** -
GB: Yellow Loosestrife **NL:** Grote wederik **F:** Lysimaque commune
D: Gewöhnlicher Gilbweiderich **CZ:** Vrbina obecná **PL:** Tojeść pospolita
EST: Harilik metsvits **LV:** Parastā zeltene **LT:** Paprastoji šilingė **Ru:** Вербейник обыкновенный

Perennial helophyte with underground rhizome and runners. Stem erect, 20–160 cm long, unbranched, upper part finely hairy and spotted. Leaves opposite or in whorls of up to 4, 7–12 cm long and 1.5–3.5 cm wide, subsessile, ovate lanceolate, acuminate, with dark spots, margin entire, underside finely hairy. Flowers in large terminal open clusters, corolla 8–16 mm in diameter with 5 yellow lanceolate petal lobes with reddish-brown margins, stamens 5, red or orange. Fruits globular, longer than the sepals. Found in marshes and meadows as well as the margins of ponds, lakes, streams and rivers; sometimes also in drier habitats. Very rarely submerged and then typically only during temporary flooding of all or part of stands.

Distribution within the region: Native. Widespread and abundant from Ireland east to the Baltic states and into southern Scandinavia, rare or absent further north.

Lysimachia vulgaris. Hadsund, Denmark. Photo JCS.

Primulaceae

268 *Lysimachia nummularia* L.

DK: Pengebladet Fredløs **N:** Krypfredløs **S:** Penningblad **FIN:** Suikeroalpi **IS:** -
GB: Creeping Jenny **NL:** Penningkruid **F:** Herbe aux écus
D: Pfennig-Gilbweiderich **CZ:** Vrbina penízková **PL:** Tojeść rozesłana
EST: Roomav metsvits **LV:** Pļavas zeltene **LT:** Šliaužiančioji šilingė **Ru:** Вербейник монетный

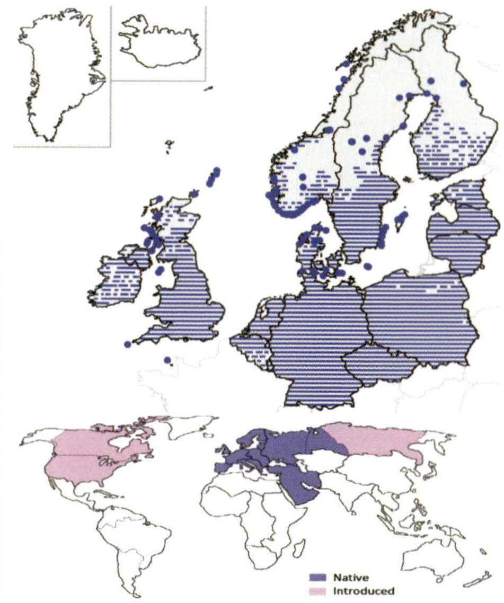

Lysimachia nummularia in grassy vegetation by an artificial pond. Thy, Denmark. Photo JCS.

Perennial helophyte or temporary hydrophyte with horizontal rhizome and fibrous, white roots at base and from the nodes of the stems.

Stem: 10–50(–100) cm long and 1.5–3 mm diameter, prostrate and creeping, angular, glabrous, richly branched, grey green or green.

Leaves: 12–20(–30) mm long and 11–20 (–28) mm wide, opposite, almost circular or ovate, obtuse with cordate base. Leaves on the basal part of stem smaller, 5–14 mm long and 3–15 mm wide, spathulate. Upper side of leaves yellowish green to green, glabrous with numerous reddish-black glandular dots. Submerged plants with leaves similar to those of aerial shoots. Petiole 2–5 mm long, robust.

Inflorescence: Flowers solitary or rarely in pairs, arising from leaf axils on the middle part of the stems, peduncles or pedicels relatively robust and barely as long as the adjacent leaf.

Flowers: 5-merous, bisexual. Calyx with 5 sepal lobes, (5–)7–9(–11) mm long, broadly elliptic with cordate base, glabrous, green with red dots, slightly overlapping. Corolla 15–30 mm in diameter, with 5 petal lobes, (8–)9–16(–18) mm long, obovate to broadly elliptic, obtuse, slightly overlapping, yellow with red to black dots, often with a brownish scale-like bract at the base. Stamens 5, with filaments shorter than the petal lobes, anthers yellow. Style up to 8 mm long, filamentous, as long as the stamens.

Fruits: Globular, red-dotted capsules, 3–5 mm long, shorter than the sepal lobes. Seeds few, 1.0–1.5 mm, irregularly triangular, dark brown.

Flowering: June-August.

Biology: Flowers self-sterile, pollinated by bees of the Melittideae family. Fruit and seed set usually poor, seeds dispersed by wind and water. Vegetative propagation by creeping rhizomes and rooting stems, commonly leading to formation of large clonal stands. Plants typically rooted above or just below the water line and developing aerial flowering shoots. Plants are usually only temporarily submerged during flooding of wetlands and the margins of ponds, lakes and rivers. Permanently submerged plants do occur; they are evergreen or semi-evergreen. Often used in aquariums and as an ornamental plant in gardens.

Lysimachia nummularia. Flower. Schadbruch, Kempen/Lower Rhine, Germany. Photo KvdW.

Lysimachia nummularia. Two submerged shoots (left) and a shoot penetrating the water surface (right). Pool, Schadbruch, Kempen/Lower Rhine, Germany. Photos KvdW.

Habitats: On wet moderately acid to moderately alkaline, oligotrophic to eutrophic, organic or mineral substrate. In the littoral zone of ponds, lakes, ditches, canals and streams in open landscapes as well as forests, on moorland, meadows and freshwater marshes. Submerged plants typically in in temporarily flooded areas. Also in drier habitats.

Distribution within the region: Native. Widespread and abundant throughout except the far north.

Characteristics and similar species: *Lysimachia nummularia* is easily distinguished from all other aquatic plants both with and without flowers due to the prostrate, mat-forming stems and the almost circular leaves.

It could be mistaken for **265 *Lysimachia tenella*** which is usually much smaller, occurs in more open habitats and has an interrupted black margin to the underside of the leaves.

Lysimachia nummularia. Part of leaf - the small, reddish, glandular dots are hardly visible to the naked eye. Photo KvdW.

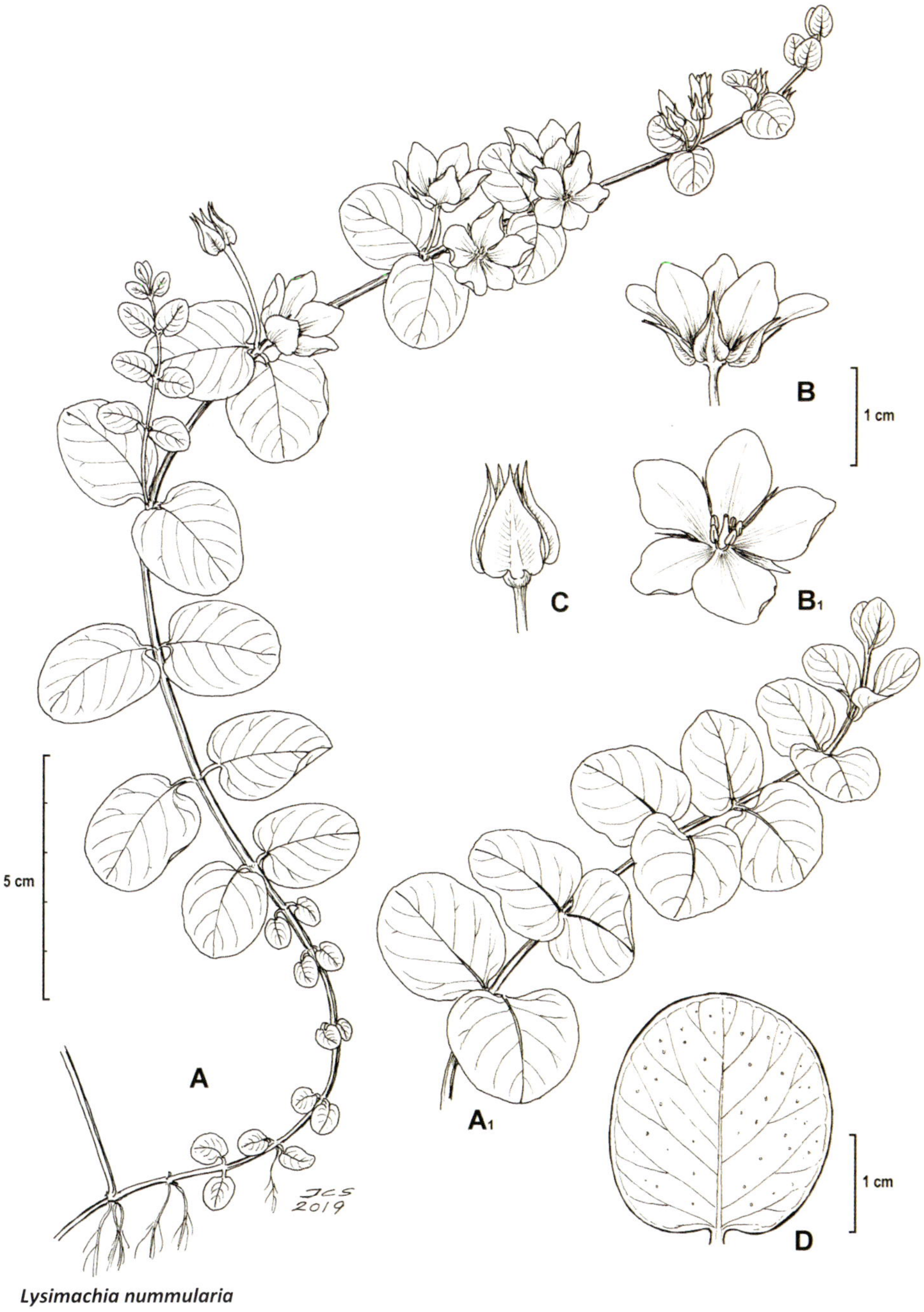

Lysimachia nummularia
A habit, **A1** habit submerged plant, **B**, **B1** flower, **C** calyx, **D** leaf.

Primulaceae

269 *Samolus valerandi* L.

DK: Samel **N:** Strandsamel **S:** Bunge **FIN:** Valkosuolapunka **IS:** -
GB: Brookweed **NL:** Waterpunge **F:** Samole de Valerand
D: Salzbunge **CZ:** Solenka Valerandova **PL:** Jarnik solankowy
EST: Liht-randpung **LV:** - **LT:** Baltažiedis samuolis **Ru:** Самолус Валеранд

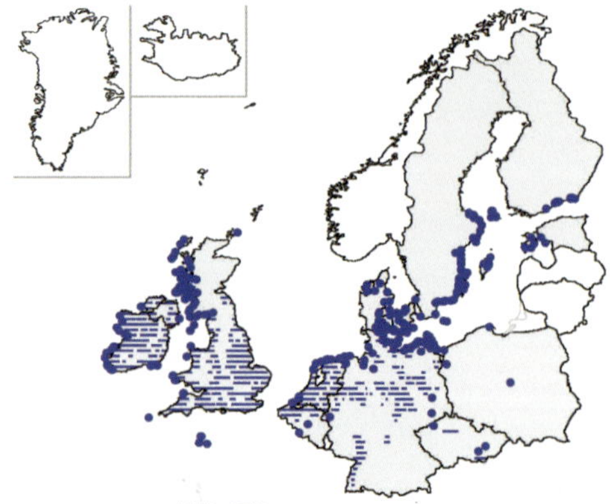

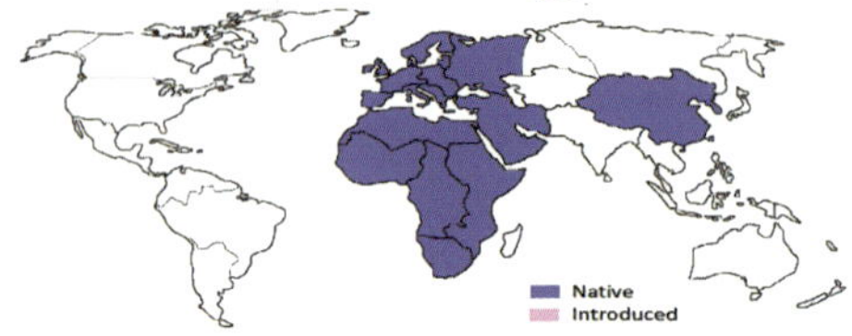

Samolus valerandi in the margin of a swamp close to the sea. Lolland, Denmark. Photo JCS.

Annual or short-lived perennial herbaceous helophyte or hydrophyte with short thin rhizome. Roots fibrous, white or greenish.

Stem: (5–)15–50(–90) cm long, erect, terete, glabrous, green, simple or branched from the basal rosette. Submerged plants may remain as rosettes or develop stems.

Leaves: Alternate and entire. A basal rosette of 7–14 spathulate, fleshy, glabrous, shiny leaves, (15–)20–45(–100) mm long and 8–25(–35) mm wide, yellowish green to green, petiole up to 3 cm long. Stem leaves similar to the rosette leaves but becoming gradually smaller and more or less sessile towards the apex, pale to dark green. Submerged leaves more upright, flaccid, commonly light green.

Inflorescence: Terminal or axillary many-flowered raceme, up to 25 cm long.

Flowers: 5-merous, bisexual. Pedicels 1–25 mm long, normally abruptly angled near the middle and with a small bract on the angle. Calyx cup shaped, the tube c. 1.5 mm long and partly fused with the ovary, sepal lobes triangular and about half as long as the calyx tube. Corolla bell-shaped, the tube c. 2 mm long with 5 spathulate, white lobes c. 2 mm long. Stamens 5, attached to the base of the corolla tube, c. 1.1 mm long, the anthers yellow, not extending beyond the throat of the tube, alternating with 5 linear staminodes (stamens without anthers). Style 0.5 mm long with a rounded stigma.

Fruits: Capsules globular to ovate, c. 3 mm in diameter, shorter than calyx, valves strongly reflexed after dehiscence. Seeds angular, c. 0.5 mm long, minutely granular, red brown.

Flowering: June-September.

Biology: Self-pollinated and self-fertile. Seeds numerous, dispersed by water and animals, germinating readily. Sometimes sterile, stemless and remaining as rosettes when growing submerged, but sometimes forming leafy stems. Permanently submerged plants may persist in deep water, while temporarily submerged plants occur during periods of high water level and flooding. Tolerant of exposure to saltwater. *S. valerandii* is a popular ornamental plant in garden ponds and especially in aquariums.

Habitats: In neutral to alkaline, oligotrophic to eutrophic water bodies on sandy, loamy and clayey substrate. In the marginal zone of ponds, lakes, ditches, canals and streams. Also in meadows, freshwater marshes, flushes and the upper parts of salt marshes. Mainly in open and light-exposed biotopes.

Distribution within the region:
Native. Widespread and abundant throughout much of Britain and Ireland but local in Scotland, east through Belgium, the Netherlands and Denmark, southern Sweden and Finland to Estonia, local in Germany, Poland and the Czech Republic.

Samolus valerandi. Inflorescence. Schadbruch, Kempen/Lower Rhine, Germany. Photo KvdW.

Samolus valerandi. Submerged plants. Sand pit, Beinheim, France, depth approximately 4 metres. Photo KvdW.

Characteristics and similar species: Flowering *S. valerandi* differs from other aquatic species by the basal rosettes of spathulate, fleshy leaves and tall stems bearing small white 5-lobed flowers.

Submerged *Samolus* rosettes may be confused with **241 *Epilobium*** spp. and **275-280 *Myosotis*** spp., from which it is differentiated by the spathulate, glabrous leaves.

The basal leaves resemble those of **343 *Shinnersia rivularis***, which has lobed leaves. *Bellis perennis* (not in this book) can resemble *S. valerandii* when not flowering but differs in the leaves, which are more spathulate and toothed.

Submerged *S. valerandi*. Bourne, Lincolnshire, Britain. Photo RVL.

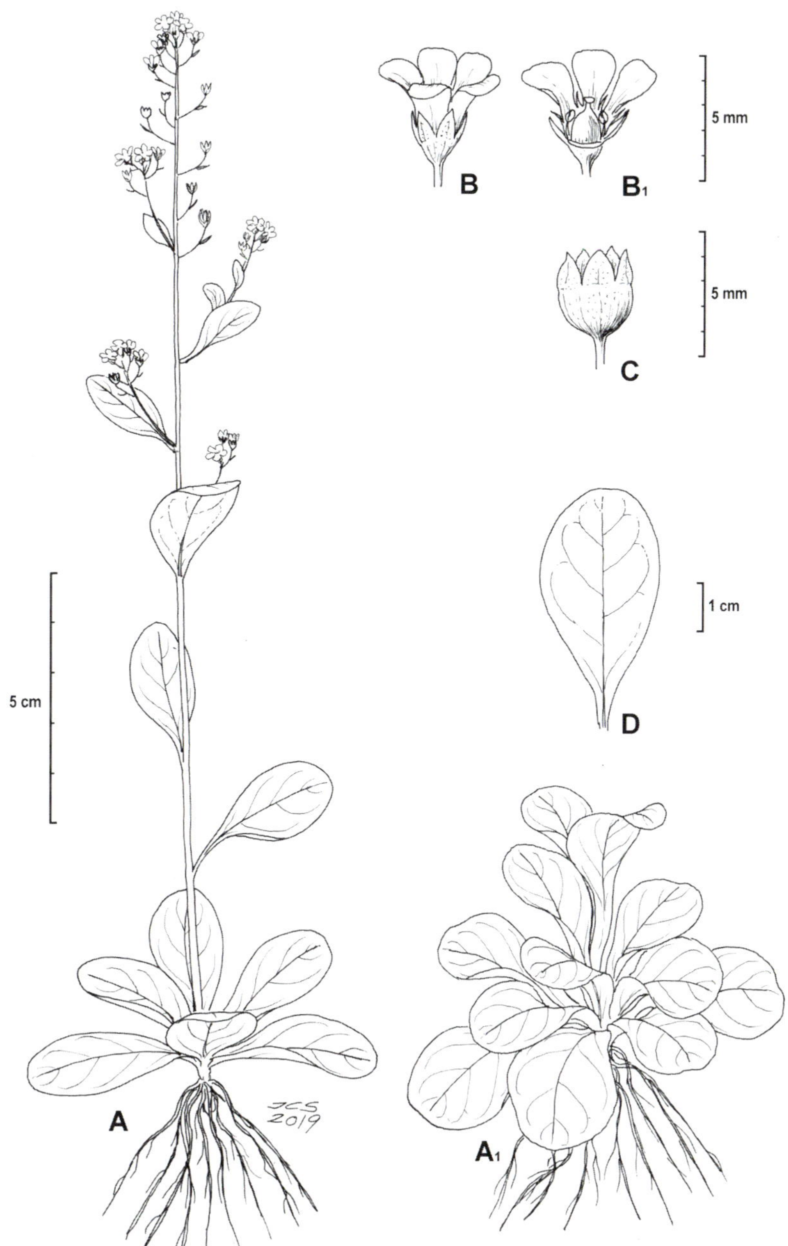

Samolus valerandi
A habit, **A1** habit submerged plant, **B** flower, **B1** flower with some petals removed, **C** calyx, **D** leaf.

Rubiaceae

270 *Galium palustre* L.

DK: Kær-Snerre **N:** Myrmaure **S:** Vattenmåra **FIN:** Rantamatara **IS:** Mýramaðra
GB: Marsh-bedstraw **NL:** Moeraswalstro **F:** Gaillet des marais
D: Sumpf-Labkraut **CZ:** Svízel bahenní **PL:** Przytulia błotna
EST: Soomadar **LV:** Purva madara **LT:** Pelkinis lipikas **Ru:** Подмаренник болотный

Galium palustre growing submerged in late spring on flooded ground in a pond in Stavtrup near Aarhus, Denmark. Photo BM.

Perennial, rarely annual helophyte or temporary hydrophyte without rhizome. Roots white.

Stem: 10–60(–80) cm long, 0.5–3.0 mm diameter, ascending from a prostrate base, weak and often leaning on neighbouring plants, 4-angled, glabrous or scabrous due to few backward-directed prickles on the angles, sparsely branched, green, turning black when dried.

Leaves: Sessile or very shortly petiolate, in whorls of (3–)4(–6). Aerial leaf size variable in individual whorls, becoming smaller towards the tip of the shoot. 5–29 mm long and 1–3 mm wide, linear or narrowly elliptic to narrowly obovate, obtuse or subacute, glabrous or slightly scabrous due to retrorse pricklets on margins and on mid-vein on the lower side of the leaves, 1-veined, green. Submerged plants mostly with whorls of 4 leaves of more or less uniform size, broader, obovate, yellow green to green, mostly glabrous.

Inflorescence: Small, axillary cymes forming an open panicle. Cymes 2–5 from the same axis, each with 3–7 flowers, usually aggregated into terminal panicles of up to 20 flowers. Peduncles c. 2 cm long.

Flowers: 4-merous, bisexual. Pedicels 1–4 mm long, longer and strongly divaricate as fruit ripen. Calyx absent. Corolla 2.0–3.0(–3.5) mm in diameter with a short tube and 4 acute-ovate to triangular, white petal lobes. Stamens 4, anthers purplish. Style 1, stigma bifid.

Fruits: 2 globular, nut-like mericarps, 0.8–1.2 mm long, smooth.

Flowering: June-August.

Biology: Flowers protandrous. Pollinated by insects attracted to nectar pits at the base of the corolla tube. Seeds dispersed by animals and water. Plants mostly rooted above the water line. Submerged plants occur from late autumn to early summer when plants become inundated. Permanently submerged plants mostly grow in small streams and ditches.

Habitats: On moderately acid to alkaline organic or mineral, oligotrophic to eutrophic substrate in the marginal zone of ponds, lakes, ditches, canals and streams. Also in mires, wet meadows and freshwater marshes.

Distribution within the region: Common throughout the region, but very rare in Iceland. Absent from Greenland.

Characteristics and similar species:
Submerged *Galium* spp. may be mistaken for **281-292 *Callitriche*** spp. or *Stellaria alsine* (not in this book), all of which have opposite leaves.

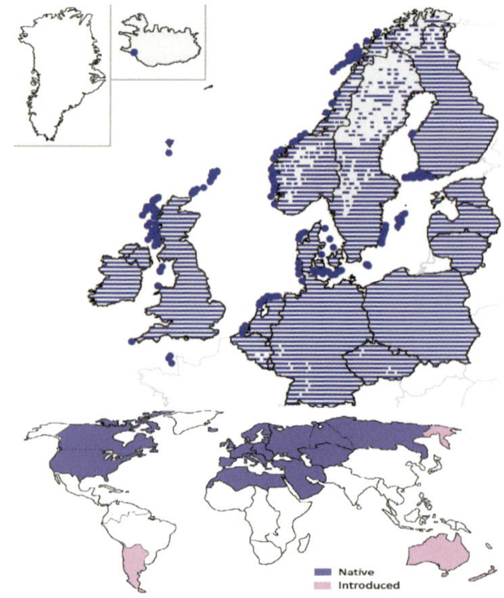

Galium palustre is treated as two or three subspecies by different authors, here we recognise two:

270a *G. palustre* subsp. *palustre*

Leaves 5–14(–17) mm long, stem slender with green angles. Flowers 2.5–4 mm across. Inflorescence more or less cylindrical. Fruit c. 1 mm.

271 *G. palustre* subsp. *elongatum* (C.Presl) Arcang.

Leaves (16–)20–35 mm long, stem robust with whitish angles. Flowers (3–)4–5.5 mm across. Inflorescence more or less conical. Fruit c. 1.9 mm.

Flowering *Galium palustre* in a wet meadow. Farsø, Jutland, Denmark. Photo JCS.

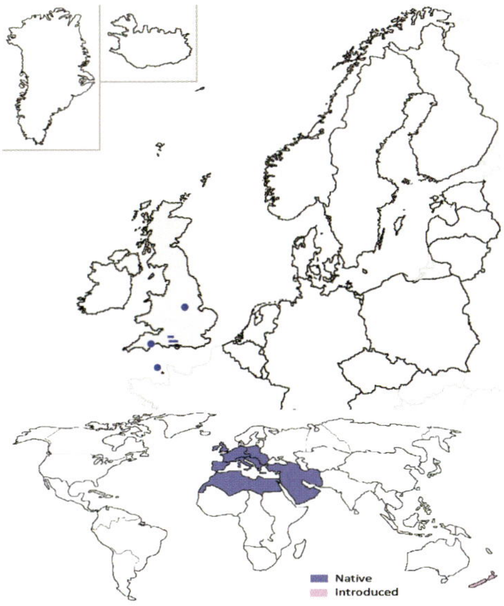

272 *Galium constrictum* Chaub.

Syn.: *G. debile* Desv. non Hoffm. & Link
GB: Slender Bedstraw

Similar in habit to **270** ***G. palustre*** but smaller and more slender. Stem 4-angled, smooth or slightly rough. Leaves 5–10 mm long, up to 1 mm wide, subacute, margins with forward-directed prickles, in whorls of 4–6. Submerged shoots with 1–2(–3) cm long, very narrow, flaccid leaves.

Distribution within the region:
Rare and declining, local in southwestern England and the Channel Islands.

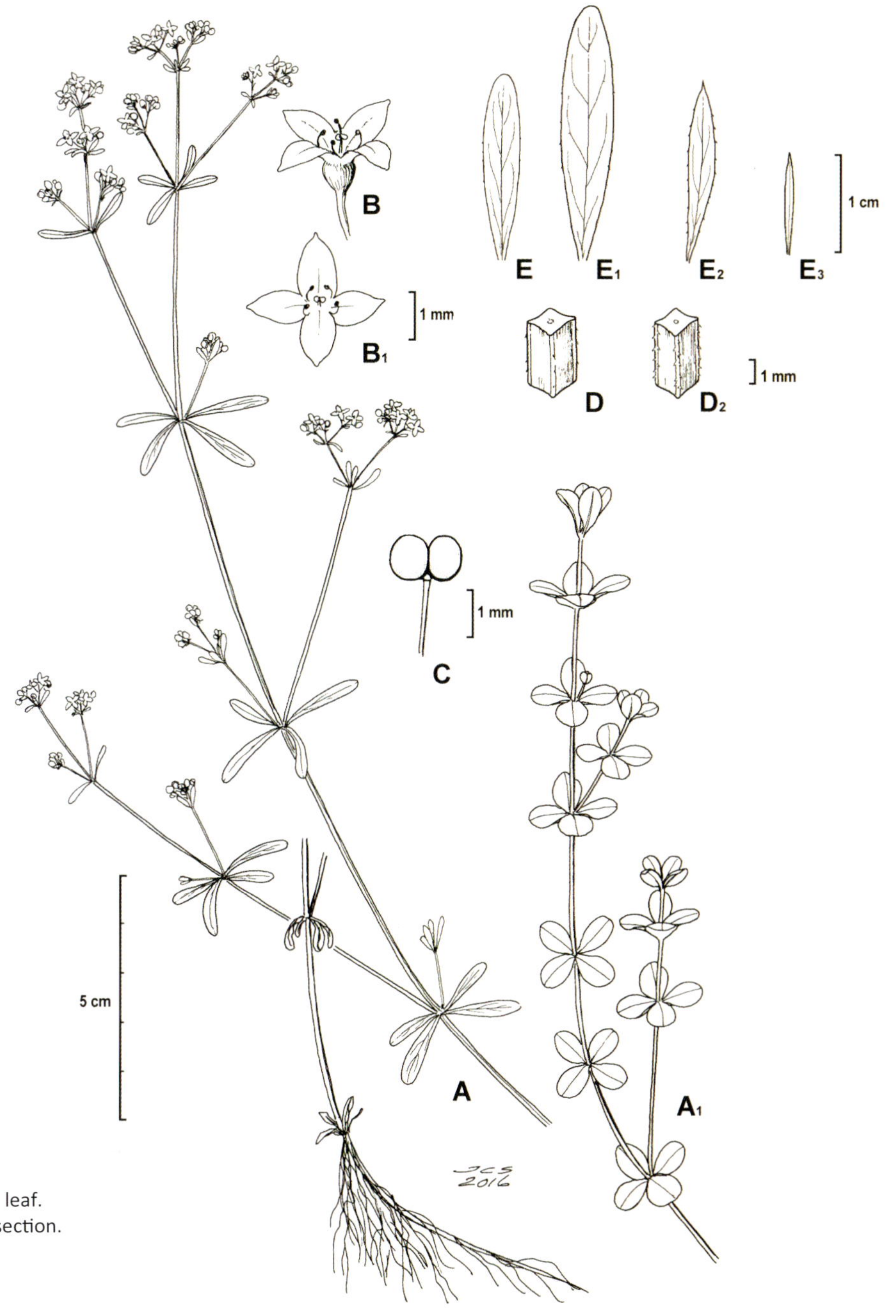

Galium palustre subsp. ***palustre***
A habit, **A1** habit submerged plant, **B**, **B1** flower, **C** fruit, **E** leaf.
Galium palustre subsp. ***elongatum*** **E1** leaf, **D** stem cross section.
Galium uliginosum **E2** leaf, **D2** stem cross section.
Galium constrictum **E3** leaf.

Rubiaceae

273 *Galium uliginosum* L.

DK: Sump-Snerre **N:** Sumpmaure **S:** Sumpmåra **FIN:** Luhtamatara **IS:** Laugamaðra
GB: Fen Bedstraw **NL:** Ruw walstro **F:** Gaillet aquatique
D: Mohr-Labkraut **CZ:** Svízel slatinný **PL:** Przytulia bagienna
EST: Lodumadar **LV:** Dūkstu madara **LT:** Liūninis lipikas **Ru:** Подмаренник топяной

Flowering *Galium uliginosum*.
Hammershøj, Jutland, Denmark. Photo JCS.

Similar in habit to **271** ***G. palustre*** but more slender. Stem angles very rough with backward-directed prickles. Leaves 5–10(–15) mm long, 1–3 mm wide, strongly mucronate at apex, margins with backward-directed prickles, in whorls of (4–)5–8. Submerged plants much less frequent than the aquatic form of *G. palustre*.

Common throughout the region, but rare in Iceland and introduced in Greenland.

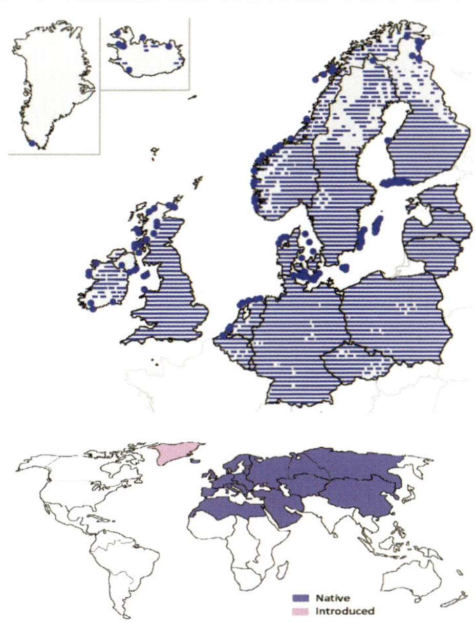

Solanaceae

274 *Solanum dulcamara* L.

DK: Bittersød Natskygge **N:** Slyngsøtvier **S:** Besksöta **FIN:** Punakoiso **IS:** -
GB: Bittersweet **NL:** Bitterzoet **F:** Douce-amère
D: Bittersüßer Nachtschatten **CZ:** Lilek potměchuť **PL:** Psianka słodkogórz
EST: Harilik maavits **LV:** Bebrukārkliņš **LT:** Karklavijas **Ru:** Паслён сладкогорький

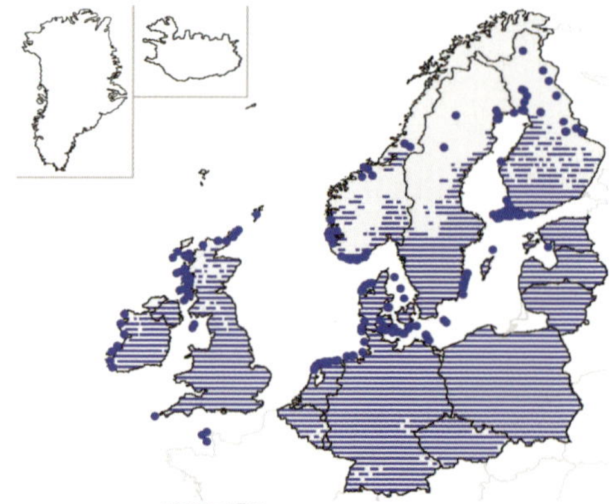

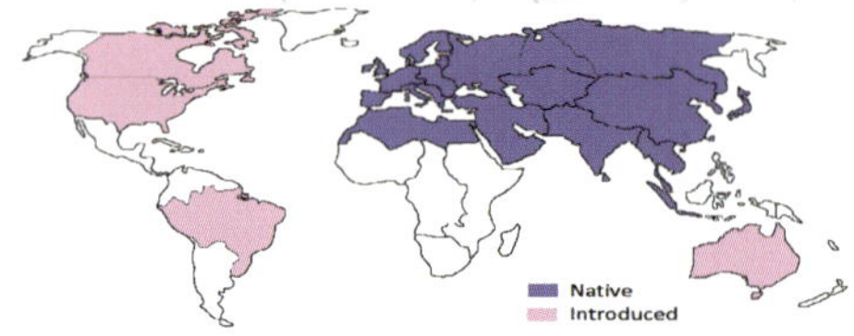

Perennial vine or subshrub with underground creeping rhizome, numerous long runners and brownish roots.

Stem: Up to 3 metres long and 1 cm diameter at the base, branched, erect to ascending or climbing, the upper parts soft and brittle, herbaceous, initially purple, eventually green, the older lower parts hard and woody with buds from which next year's shoots develop, green to grey green. Young plants with hairy stems.

Leaves: Alternate and petiolate, lamina 2.5–8.0 cm long and 1.5–5.0 cm wide, acuminate-obovate with cordate base. Young leaves entire, older leaves usually deeply lobed near the base, dividing the lamina into two, rarely four, lower lobes and one large terminal lobe, all more or less the same shape, green, often with a purple tinge.

Inflorescence: A complex cyme with 5–25 flowers, peduncles up to 4 cm long, arising from the upper leaf axils.

Flowers: 5-merous, bisexual, 10–15 mm in diameter. Sepals fused to a cone-shaped calyx tube with short, rounded lobes, green and usually finely hairy. Corolla of 5 lanceolate and acute tepals fused at the base, 5–9 mm long, purplish violet, with white-edged green dots at the base, with fine hairs dorsally and along the edges, spreading at first but soon strongly recurved. Anthers 4–4.5 mm long, oblong, deep yellow and fused into a tube around the thin style that protrudes beyond the anthers.

Fruits: Obovate berries, 8–10 mm long, glossy, initially dark green, eventually deep red when ripe. Seeds numerous, disc shaped, 2.0–2.5 mm in diameter, beige to light brown.

Flowering: June-September.

Biology: Insect pollinated. Insects are attracted to the large stamens and white-edged green pseudonectaries. Stamens release pollen through an apical pore, while the protruding style to some extent prevents self-fertilisation. Unable to grow permanently submerged, but stems and branches often extend far into water from rooting points at or above the water line, forming large floating or semi-submerged stands. Dense stands of plants with long stems often occur sprawling over bushes and trees. All parts of the plants are poisonous.

Habitats: On moderately acidic to alkaline, mesotrophic to eutrophic, peaty, muddy and mineral substrate. On the margins of a wide range of waters and wetlands, ponds, lakes, streams, rivers,

From a rooting point at the edge of a reed swamp *S. dulcamara* stretches its shoots out onto the water surface and into the water, where the submerged stems support the aerial shoots and eventually allow them to flower and develop berries. Lille Vildmose, Jutland, Denmark. Photo JCS.

Solanum dulcamara. The beautiful violet petals contrast with the fused, bright yellow anthers. Skjellerup, Hobro, Denmark. Photo JCS.

Solanum dulcamara. The egg-shaped berries change from green to red colour as they ripen and may persist long after the leaves have withered away late in the season. Lake Navn Sø, Jutland, Denmark. Photo JCS.

ditches and canals, inland and coastal marshes. Also in drier, but moist habitats. In open, sunlit places as well as more shaded habitats.

Distribution within the region: Widespread and abundant throughout most of the region, but rare in the mountain areas of Norway and Sweden. Absent from Iceland and Greenland.

Characteristics and similar species: *Solanum dulcamara* differs from other aquatic plants by the climbing vine-like growth form, the shape of the leaves, purple-violet flowers with recurved petals and prominent yellow stamens, and green to red berries.

Due to the toxicity of all parts, it is important to get to know the characteristics of this species and avoid confusion with other species.

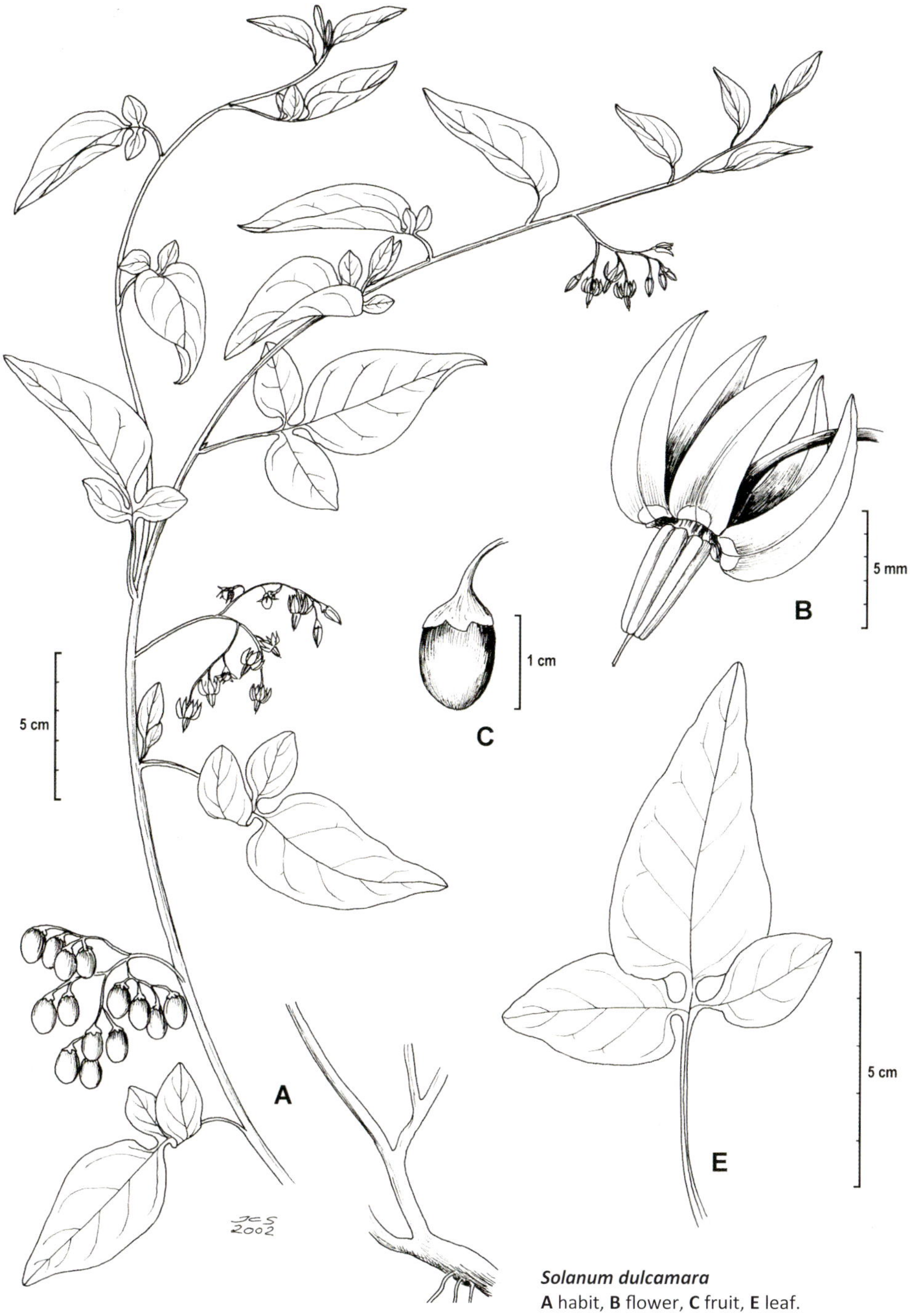

Solanum dulcamara
A habit, **B** flower, **C** fruit, **E** leaf.

Boraginaceae

Key to *Myosotis*

Reliable identification of *Myosotis* depends on flowering or fruiting material above the water line. Submerged *Myosotis* species can be separated into two groups - those with or those without stolons. Vegetative characters, such as whether the stem is angular or terete, and density or direction of hairs are of limited use on submerged plants.

1 Calyx divided less than halfway to base (***1a***). Sepals broad, more or less equilateral triangular — 2

1 Calyx divided halfway or more to base (***1b***). Sepals narrow, more or less isosceles triangular — 4

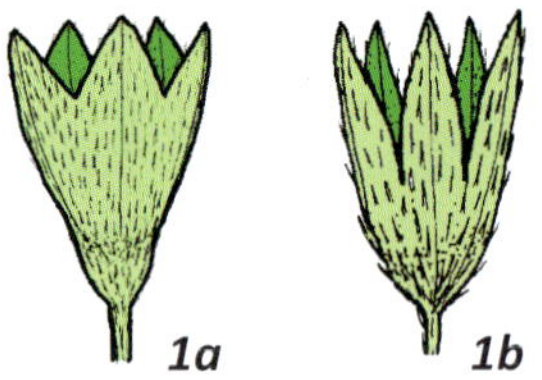

1a ***1b***

2 Stem usually more than 15 cm long in fruit. Not mat forming. Leaves up to 10 cm long and 2 cm wide — 3

2 Stem usually less than 10 cm long in fruit. Stoloniferous, mat forming. Leaves up to 3 cm long and 1 cm wide. Only in the Alps
276 *M. rehsteineri*

3 Perennial with rhizome. Lower part of stem glabrous or with mostly upward-directed hairs, rather dull, bluntly angular. Lower leaves with appressed, antrorse hairs below
277 *M. scorpioides*

3 Biennial without rhizome. Lower part of stem glabrous or with mostly downward-directed hairs, shining, distinctly angular. Lower leaves with patent to retrorse hairs below
278 *M. nemorosa*

4 Lower part of stem with patent hairs
279 *M. secunda*

4 Lower part of stem with appressed hairs — 5

5 Stolons absent. Leaves more than 3 times as long as wide
275 *M. laxa* subsp. ***caespitosa***

5 Stolons present. Leaves 1-2 cm long. Leaves less than 3 times as long as wide
280 *M. stolonifera*

275 *Myosotis laxa* Lehm. subsp. *caespitosa* (Schultz) Hyl. ex Nordh.

Syn.: *M. caespitosa* Schultz ; *M. lingulata* Lehm

DK: Sump-Forglemmigej **N:** Dikeminneblom **S:** Vanlig sumpförgätmigej **FIN:** Rantalemmikki **IS:** -
GB: Tufted Forget-me-not **NL:** Zompvergeet-mij-nietje **F:** Myosotis cespiteux
D: Schlaffes Vergißmeinnicht **CZ:** Pomněnka trsnatá **PL:** Niezapominajka darniowa
EST: Muru-lõosilm **LV:** Ciņu neaizmirstule **LT:** Velėninė neužmirštuolė **Ru:** Незабудка дернистая

Flowering *M. laxa* subsp. *caespitosa*. Hobro, Denmark. Photo JCS.

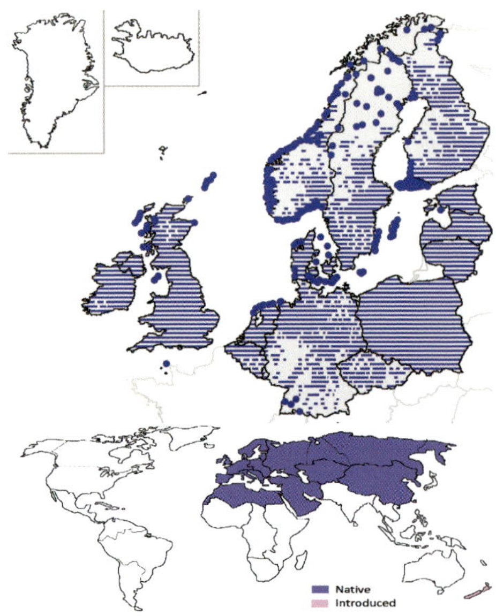

Biennial or annual helophyte or hydrophyte without rhizome or runners. Roots fibrous, white.

Stem: Ascending to erect, 20–50 cm long and 2–3 mm thick, terete or slightly angular, branched from lower half of the stem, with short, forward-directed, compressed hairs.

Leaves: Alternate, all with appressed hairs on both sides, one distinct mid-vein and two less distinct marginal veins, the central and marginal veins connected by secondary veins. Aerial shoots with entire, lanceolate leaves with rounded apices, 1–8 cm long and 1 cm wide, green. Submerged leaves more oblong with more or less parallel margins, softer and more flaccid, light green.

Inflorescence: Monecious. Pedunculate racemes at the tips of branching stems, initially curled with flowers opening in succession as the racemes unfold and elongate.

Flowers: 5-merous. Pedicels up to 10 mm long. Calyx up to 5 mm long, with 5 narrow isosceles-triangular teeth at least as long as the calyx tube, green, with appressed hairs. Corolla 3–6 mm in diameter, tubular with 5 spreading lobes, oval to ovate, initially reddish, finally sky blue. Lobes with swollen appendages forming a yellow collar around the throat. Stamens 5, with short filaments, anthers 0.6–0.9 mm long situated just below the throat of the corolla. Style 1, 1.0–1.5 mm long, shorter than the calyx tube and never longer than the calyx teeth.

Fruits: Capsules hidden inside the persistent calyx, each with 4 obovate, glossy, smooth and brown-black, 1.0–1.5 mm long and c. 1 mm wide nutlets.

Flowering: June–August.

Biology: Insect- or self-pollinated. Ripe capsules remain closed within the calyces, which serve as dispersal units. Released nutlets are dispersed by water. Plants may be temporarily submerged during periods of high water levels, but permanently submerged plants are also frequent. Terrestrial plants die back in the winter, but submerged plants often remain green.

Habitats: Moderately acid to alkaline, moderately nutrient-poor to nutrient-rich, muddy as well as mineral and peaty substrate in the wet or damp margins of ponds, lakes, ditches, canals and streams as well as springs, open freshwater marshes, moors and meadows. Typically in open, sunny areas.

Distribution within the region:
Native. Widespread and abundant throughout most of the region, absent from Iceland and Greenland.

Characteristics and similar species:
Myosotis species differ from other aquatic plants by the sky blue flowers with yellow centre and by the alternate leaves with appressed hairs on both sides.

Submerged *Myosotis* may resemble submerged **297-300** ***Veronica*** spp., which have opposite, glabrous leaves, and submerged **241** ***Epilobium*** spp., which have glabrous leaves with different venation.

Completely submerged *Myosotis* plants can be split into two groups based on the presence or absence of runners and shape of the stem cross section, but identification to species level requires flowers or fruits (see key opposite).

Submerged *M. laxa* subsp. *caespitosa*. Thy, Denmark. Photo JCS.

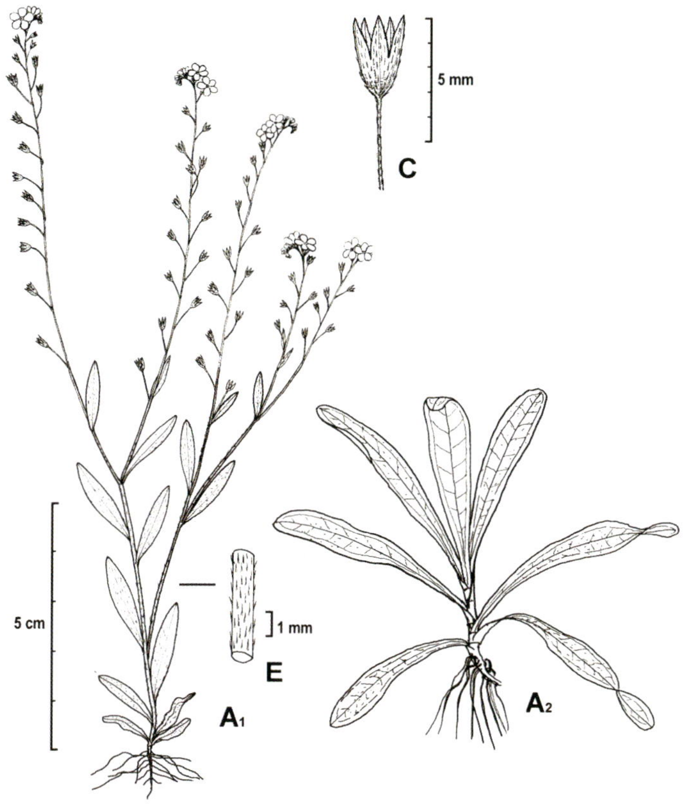

Myosotis laxa subsp. ***caespitosa***
A1 habit, **A2** habit submerged plant, **C** calyx, **E** stem section.

276 *Myosotis rehsteineri* Wartm.

D: Bodensee-Vergißmeinnicht

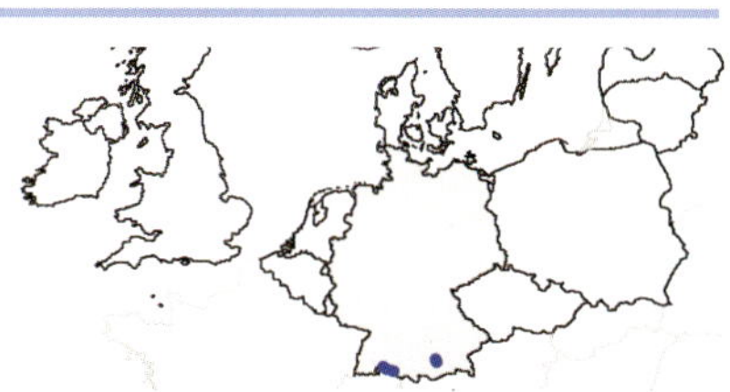

Perennial helophyte or hydrophyte with creeping rhizome and very short stolons, forming small mats. Submerged plants up to 12 cm high with short, robust stolons. Stem 4–8(–13) cm long, ascending, terete, glabrous or with upward-directed hairs. Leaves 2–4 cm long, 4–12 mm wide, sessile to short petiolate. Corolla of bisexual flowers 6–12 mm diameter, those of female flowers 4–6 mm. Pedicels rarely longer than calyx when fruiting. Flowering April-May.

In oligotrophic to oligo-mesotrophic water by periodically flooded, gravelly banks of pre-alpine lakes including in- and out-flowing rivers.

Distribution within the region:
Native. Lake Constance (Bodensee) and outflow (Hochrhein) at c. 400 m altitude. Probably recently introduced at Starnberger See (584 m). Endemic to the pre-alpine region (Germany, Switzerland, Austria, Italy).

Myosotis rehsteineri. Bodensee, Germany. Photo Irene Strang.

Boraginaceae

277 *Myosotis scorpioides* L.

Syn.: *M. palustris* (L.) Hill

DK: Eng-Forglemmigej **N:** Engminneblom **S:** Bäckförgätmigej **FIN:** Meriluhtalemmikki **IS:** Engjamunablóm
GB: Water Forget-me-not **NL:** Moerasvergeet-mij-nietje **F:** Myosotis des marais
D: Sumpf-Vergißmeinnicht **CZ:** Pomněnka bahenní **PL:** Niezapominajka błotna
EST: Soo-lõosilm **LV:** Purva neaizmirstule **LT:** Pelkinė neužmirštuolė **Ru:** Незабудка болотная

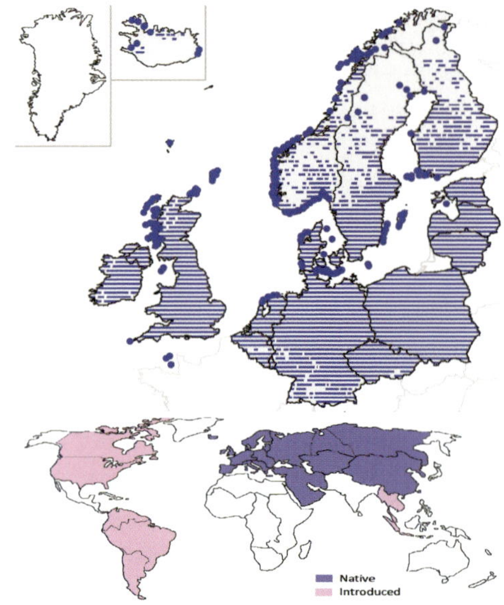

Submerged *M. scorpioides*. Bislicher Insel, Germany. Photo KvdW.

Perennial helophyte or hydrophyte with horizontal rhizome and underground stolons. Roots fibrous, white.

Stem: Ascending to erect, 20–80(–100) cm long and 2–3 mm diameter, bluntly angular, glabrous or with upward-directed hairs, rather dull.

Leaves: Alternate, 2–10 cm long and 0.5–2.0 cm wide, oblong lanceolate, entire, apex rounded, sessile to shortly petiolate, green, with appressed, antrorse hairs on both sides. Submerged leaves softer and more flaccid, light green.

Inflorescence: Monecious. Pedunculate racemes at tips of branching stems, initially curled with flowers opening in succession as the racemes unfold and elongate.

Flowers: 5-merous. Pedicels up to 10 mm long. Calyx up to 5 mm long, with 5 narrow equilateral-triangular teeth at least as long as the calyx tube, green, with appressed hairs. Corolla 7–9 mm in diameter, tubular with 5 spreading, oval to ovate lobes, initially reddish, finally sky blue. Lobes with swollen appendages forming a yellow collar around the throat. Stamens 5, with short filaments, anthers 0.6–0.9 mm long situated just below the throat of the corolla. Style 1, 1.6–2.8 mm long, longer than the calyx tube and often longer than the calyx teeth.

Fruits: Capsules, hidden inside the persistent calyx, each with 4 obovate, glossy, smooth and brown-black nutlets, 1.3–1.8 mm long and c. 1.2 mm wide.

Flowering: June–August.

Biology: Insect- or self-pollinated. Ripe capsules remain closed within the calyces, which serve as dispersal units. Released nutlets are dispersed by water. Plants may be temporarily submerged during periods of high water levels, but permanently submerged plants are also frequent. Terrestrial plants die back in the winter, but submerged plants often remain green. Often forms large, dense stands due to the presence of stolons.

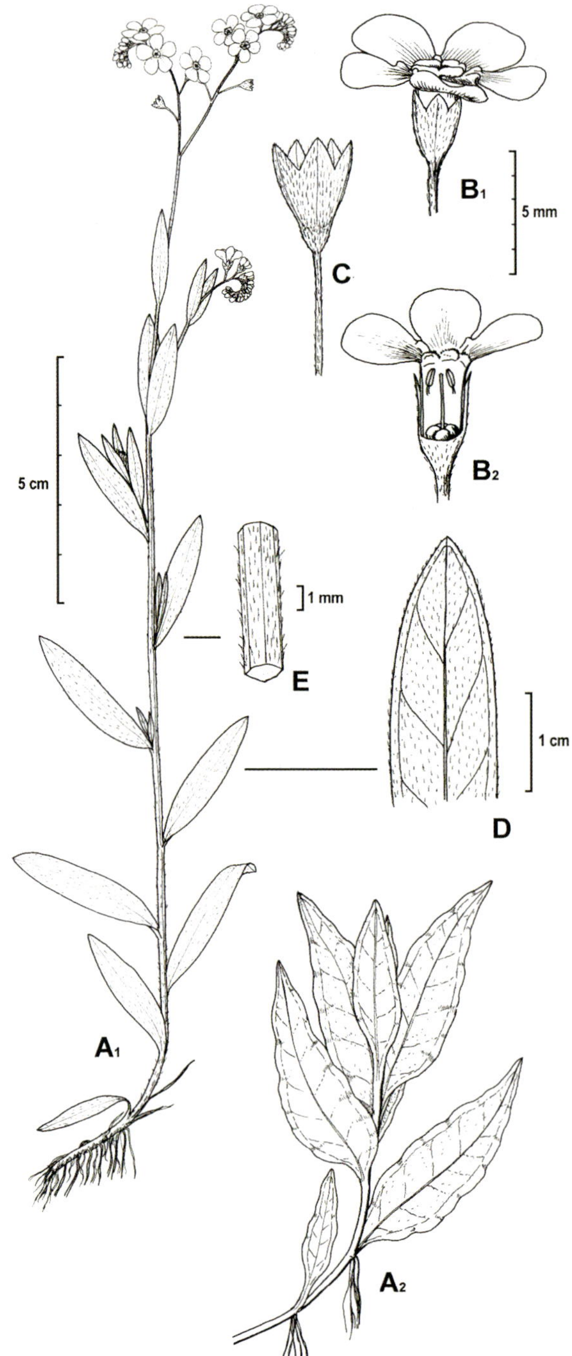

Myosotis scorpioides
A1 habit, **A2** habit submerged plant , **B1** flower, **B2** flower opened, **C** calyx, **D** part of leaf, **E** stem section.

Habitats: Moderately acid to alkaline, moderately nutrient-poor to nutrient-rich, muddy, mineral and peaty soils on the margins of ponds, lakes, ditches, canals and streams as well as in springs. Also in open freshwater marshes, mires and meadows. Typically in open, sunny areas.

Distribution within the region:
Native. Widespread and abundant throughout most of the region, less frequent inland in the north. Absent from Greenland.

Characteristics and similar species:
See **275 *M. laxa*** subsp. ***caespitosa*** and characters given in the key (p. 596).

Hybrids:
Myosotis ×suzae Domin
(*M. scorpioides × laxa*)

Intermediate between the parents and partially fertile. Corolla 5–7 mm across. Style more or less as long as calyx tube. Typically develops elongate inflorescences, particularly late in the year, and combines the stolons of *M. scorpioides* with the narrow calyx teeth of *M. laxa*.

Occasional, typically in areas where both parents occur, often where there is some disturbance of substrate.

M. scorpioides. Flowers. Randers, Denmark. Photo JCS.

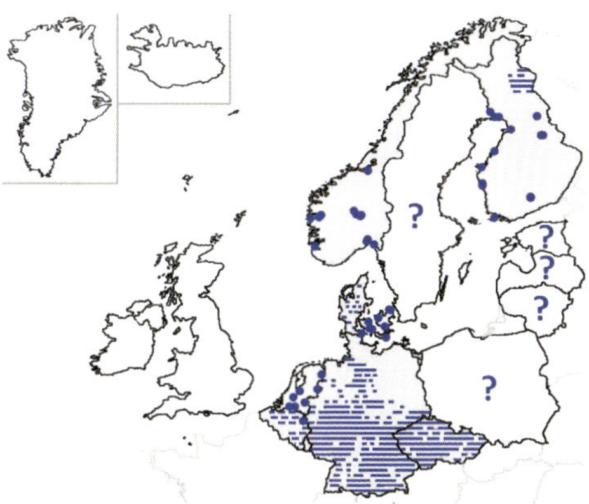

278 *Myosotis nemorosa* Besser

Syn.: *M. strigulosa* Rchb.
DK: - **N:** Lundminneblom **S:** Kantförgätmigej **FIN:** Viitalemmikki **NL:** Weidevergeet-mij-nietj **F:** Myosotis à poils réfractés **D:** Hain-Vergißmeinnicht **CZ:** Pomněnka hajní **PL:** Niezapominajka górsk **Ru:** Незабудка дубравная

Biennial helophyte or hydrophyte habitually like **277 *M. scorpioides***, but without stolons. The lower part of stem is shiny with mostly downward-directed hairs or almost glabrous, distinctly angular. Lower leaves with patent to retrorse hairs below. Corolla 4–6 mm across.

Distribution within the region:
Native. Known from most of the region, but not mapped in all countries.

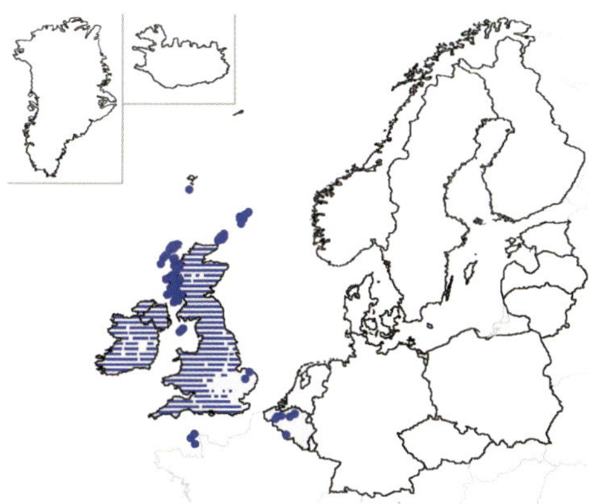

279 *Myosotis secunda* Al.Murray

S: Engelsk förgätmigej **GB:** Creeping Forget-me-not **F:** Myosotis rampant

Annual to perennial helophyte or hydrophyte resembling **275 *M. laxa,*** but stem with dense patent hairs. Corolla up to 6 mm diameter, typically very variable in size. Pedicels 2.5–5 times as long as fruiting calyx. On wet, acidic, peaty substrates by streams, flushes and springs, in pools and bogs.

Distribution within the region:
Native. Widespread and abundant in the British Isles, scarce in Belgium.

Myosotis secunda. Stem with dense clothing of patent hairs. Photo RVL.

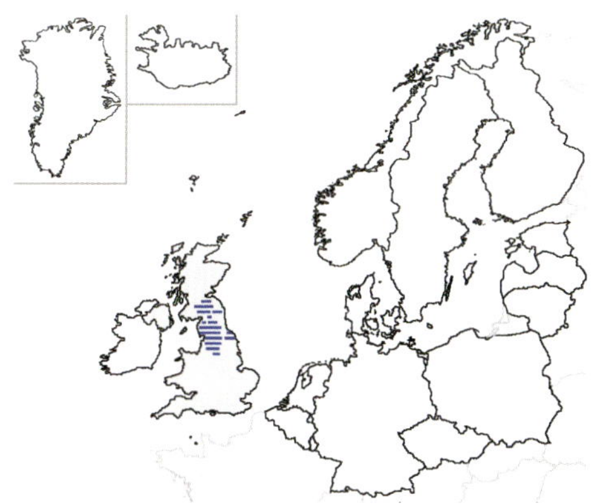

280 *Myosotis stolonifera* (DC.) J.Gay ex Leresche & Levier

Syn.: *M. brevifolia* C.E.Salmon
GB: Pale Forget-me-not

Annual to perennial helophyte or hydrophyte with stolons emerging from lower nodes, mat forming. Leaves 1–2 cm long. Corolla up to 5 mm diameter. Montane. In or by springs, flushes and streams.

Distribution within the region:
Native. Northern England. Endemic to Europe - also Portugal and Spain.

Myosotis stolonifera. Nenthead, Britain. Photo RVL.

Plantaginaceae

Key to *Callitriche*

Callitriche plants lack a rhizome, have unbranched white to brownish roots and opposite leaves. The flowers develop in the leaf axils and lack perianths (sepals and petals). The fruits are composed of four mericarps, which may be fused along one side (and thus parallel) or fused only on one side (and thus divergent).

Fruits seen from above - parallel and divergent.

The main division within the genus is between *C. hermaphroditica* and *C. truncata* and the other species. *C. hermaphroditica* and C. *truncata* never develop floating rosettes, have uniform, somewhat translucent leaves with a single central vein, which are not joined across the stem, lack stem scales and have fruit with divergent mericarps.

All of the other species are capable of developing floating rosettes, have leaves in which the shape and venation can vary widely from simple to complex, are joined across the stem and bear stem scales. The presence of stem scales can be confirmed in the field with a hand lens by examining the stem for tiny raised bumps, particularly just below the nodes. However *Callitriche* populations can only reliably be identified when fertile, ideally with mature anthers and fruit. Non-flowering *C. hermaphroditica* and *C. truncata* populations can be distinguished from the other taxa based on the lack of stem scales and translucent leaves, and outside the range of *C. truncata* this may enable recognition of *C. hermaphroditica* with some confidence. However, *C. truncata* is spreading and occurs far inland in parts of its range. It is therefore difficult to be confident of the limits of its distribution. With familiarity, some other taxa can be recognised when not flowering, however records which are not confirmed based on flowering material are of limited value.

The *Callitriche* taxa which have yellow pollen and aerial pollination typically develop anthers and stigmas which are exserted above the floating rosettes. Examination of rosettes in situ can often enable fertile shoots to be selected and in the case of *C. cophocarpa* can indicate the separation of male and female shoots. If aquatic populations are located but no fruit can be found, it is worth searching muddy areas associated with the same water body, as in some species such as *C. brutia* and *C. palustris* terrestrial plants are typically much more fertile than those growing in water.

The leaf venation is a useful character within the genus, although apart from *C. hermaphroditica* and *C. truncate,* which have only a single vein, the number of veins is strongly related to the width of the leaf. All taxa are capable of developing leaves with only a single central nerve, although this is rare in *C. stagnalis*. Expanded leaf venation involves a single central vein from which two veins arise on opposite sides and rejoin it near the leaf apex. Wider leaves may then have additional loops or short, variously curved veins which anastomose near the leaf margin.

The most important character in identification of *Callitriche* is the wing of the fruit. In general terms, the wing is only fully formed when the fruit starts to change colour as it matures. Identifications based on green, immature fruit must be made with caution.

Callitriche platycarpa floating rosette. Photo JCS.

Callitriche stagnalis fruits. Photo JCS.

Callitriche brutia. Photo JCS.

1 Fruit unwinged or winged only at the apex (***1***) 2
1 Fruit winged all around (***2***, ***3***) 5

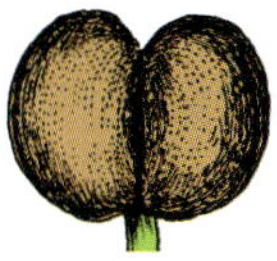

1

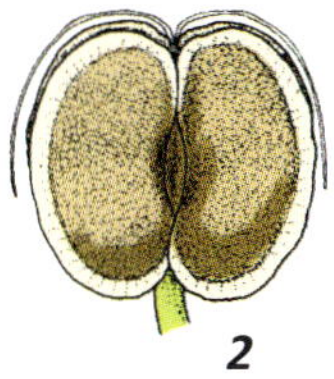

2

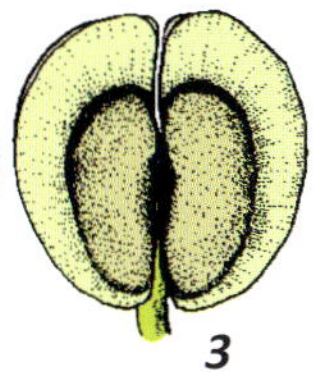

3

2 Fruit black when mature, less than 1 mm long or longer than wide 3
2 Fruit pale brown when mature, more than 1 mm long 4

3 Fruit more or less as wide as long
285 *C. heterophylla*
3 Fruit longer than wide
284 *C. palustris*

4 Bracts present. Mericarps parallel. Fruit longer than wide, sessile. Stem scales present. Pollen yellow. Leaves opaque, joined across stem
288 *C. obtusangula*
4 Bracts lacking. Mericarps divergent. Fruit wider than long, pedunculate. Stem scales lacking. Pollen colourless. Leaves translucent, not joined across stem
283 *C. truncata* subsp. ***occidentalis***

5 Wing very wide, wider than ovule (***3***). Mericarps divergent. Bracts lacking. Stem scales lacking. Leaves translucent, not joined across stem
281 *C. hermaphroditica* 6
5 Wing narrow to wide, narrower than ovule. Mericarps parallel. Bracts present, at least on youngest flowers. Stem scales present. Leaves opaque, joined across stem 7

6 Fruits 1.2–1.7 x 1.2–1.7 mm*
281a *C. hermaphroditica* subsp. ***hermaphroditica***
6 Fruits 1.5–2.4 x 1.6–2.8 mm*
282 *C. hermaphroditica* subsp. ***macrocarpa***

7 Style strongly reflexed (***4***), in young flowers extending across to opposite leaf axil, ultimately appressed to side of fruit. Pollen colourless
286 *C. brutia* 8
7 Style erect to spreading or recurved (***5***, ***6***), not extending to opposite leaf axil, not appressed to side of fruit. Pollen yellow 9

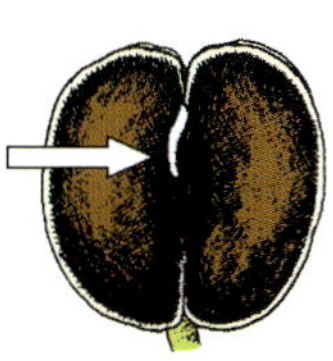

4

5

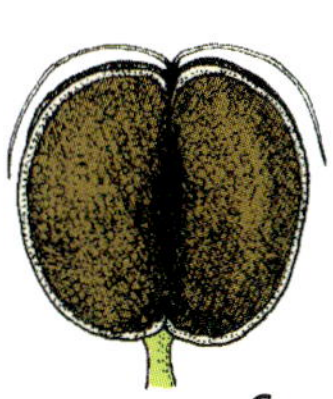

6

8 Fruit pedunculate (***7***). Apices of lingulate leaves not broadly widened to form a pincer-like shape**
286a *C. brutia* subsp. ***brutia***
8. Fruit sessile (***4***). Apices of lingulate leaves broadly widened to form a pincer-like shape**
287 *C. brutia* subsp**. *hamulata***

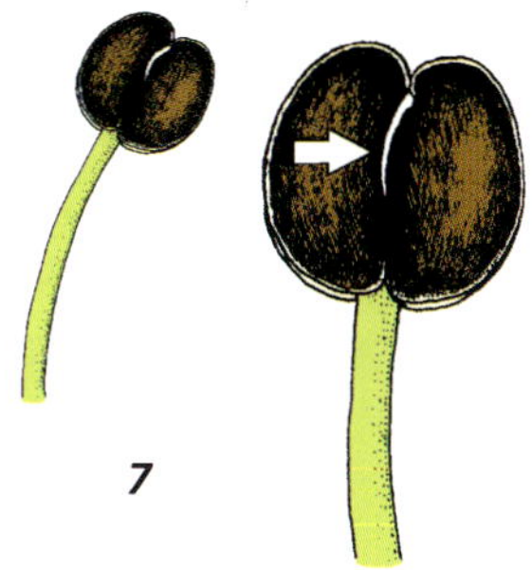

7

9 Fruit longer than wide (***8***). Male and female flowers together in at least one of most leaf axils
284 *C. palustris*
9 Fruit more or less as long as wide. Flowers solitary 10

8

10 Male and female flowers mainly on separate shoots, plants usually with one male shoot and numerous female shoots. Fruit < 1.1 mm wide and long, narrowly winged. Rosette leaves typically narrow***
289 *C. cophocarpa*
10 Male and female flowers mixed on all shoots. Fruit > 1.1 mm wide and long, broadly winged 11

11 Rosette leaves greyish-green. Fruit very broadly winged (***2***). Pollen ellipsoid
292 C. *stagnalis*
11 Rosette leaves bright green. Fruit broadly winged (***6***). Pollen tetrahedroid***
291 *C. platycarpa*

* Intermediate plants cannot be assigned to subspecies.

** Plants lacking either of these characters cannot be assigned to subspecies.

*** **290 *C. ×vigens*** could key out here.

Plantaginaceae

281 *Callitriche hermaphroditica* L.

Syn.: *C. autumnalis* L.

DK: Høst-Vandstjerne **N:** Haustvasshår **S:** Höstlånke **FIN:** Uposvesitähti **IS:** Lækjabrúða
GB: Autumnal Water-starwort **NL:** Herfststerrenkroos **F:** -
D: Herbst-Wasserstern **CZ:** Hvězdoš podzimní **PL:** Rzęśl jesienna
EST: Sügis-vesitäht **LV:** Rudens ūdenīte **LT:** Rudeninė praujenė **Ru:** Болотник обоеполый

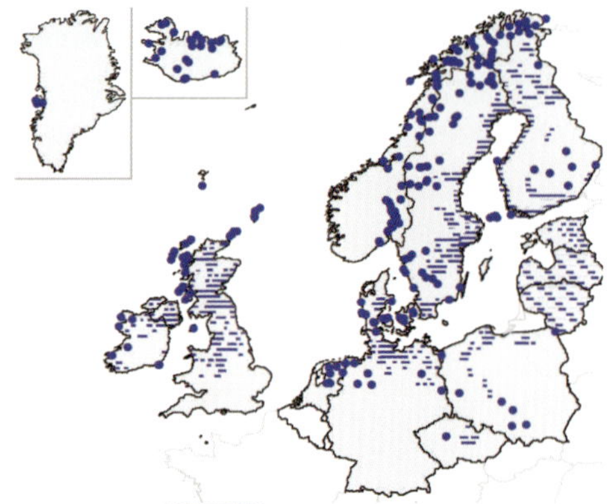

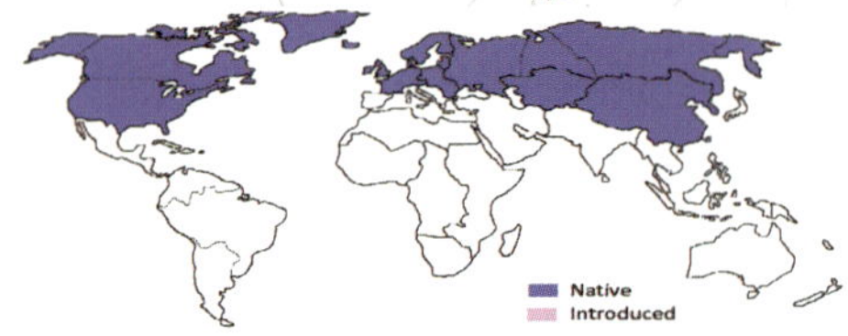

Fruiting *C. hermaphroditica*. Schaalsee, Germany. Photo KvdW.

Perennial or possibly annual. Always submerged, terrestrial form unknown.

Stem: Sparsely to densely branched, slender, light green to white, without stem scales. Internodes short, gradually decreasing in length to shoot tip.

Leaves: More or less uniform, all submerged. Light to dark green, somewhat translucent, sessile, not joined across stem. Lingulate leaves tapering slightly from base, 5.5–15.3 mm long and 0.6–2.2 mm wide, 1-veined. Expanded leaves and rosettes never produced.

Inflorescence: Monoecious. Male and female flowers solitary in leaf axils.

Flowers: Without bracts. Filament erect, 0.2–0.9 mm long, pollen colourless, ellipsoid. Styles, initially erect, later recurved and eventually remains emerge below top of fruit, up to 5 mm long.

Fruits: Sessile, mericarps divergent, as wide as or wider than long, brownish when mature, 1.2–2.4 mm long and 1.2–3.0 mm wide, wings very broad all around, 0.2–0.8 mm wide.

Flowering: June-September.

Biology: Pollination submerged, but probably typically self-fertilised. The only known plant species in which the pollen tube can grow within the tissue to contact the opposite ovary.

Habitats: *C. hermaphroditica* occurs in lakes, gravel pits, canals and slow-flowing reaches of large rivers, as well as in sheltered bays in the northern Gulf of Bothnia. Typically in lowland, naturally mesotrophic waters, it can tolerate some eutrophication.

Distribution within the region:
Native. Scattered or locally frequent throughout the region, absent from or rare in the south.

Characteristics and similar species:
Callitriche hermaphroditica can be distinguished from all other *Callitriche* species in the region by the broadly winged, sessile fruit with divergent mericarps. *C. truncata* subsp. *occidentalis* also has translucent, uniform leaves contrasting in colour with the stem, but the fruits are pedunculate and unwinged.

Compare **283 *C. truncata***.

Callitriche hermaphroditica. Schaalsee, Germany. Photo KvdW.

Callitriche hermaphroditica. The fruits are very broadly winged, consisting of mostly 4 divergent mericarps only fused at the middle. Lake Tange, Ans, Jutland, Denmark. Photo JCS.

Callitriche hermaphroditica. Immature fruits are light green and become dark brown at maturity. Lake Tange, Ans, Jutland, Denmark. Photo JCS.

281a *C. hermaphroditica* L. subsp. *hermaphroditica*

Differs from subsp. *macrocarpa* in smaller fruits 1.2–1.7 x 1.2–1.7 mm, with wings 0.1–0.3 mm wide. It is known from more southern parts of the range of the species, from Ireland northeast to Sweden and Estonia, and south to the Czech Republic.

282 C. *hermaphroditica* L. subsp. *macrocarpa* (Hegelm.) Lansdown

Differs from subsp. *hermaphroditica* in larger fruits 1.5–2.4 x 1.6–2.8 mm, with wings 0.2–0.8 mm wide. It is known from northern Scandinavia, Britain and Ireland.

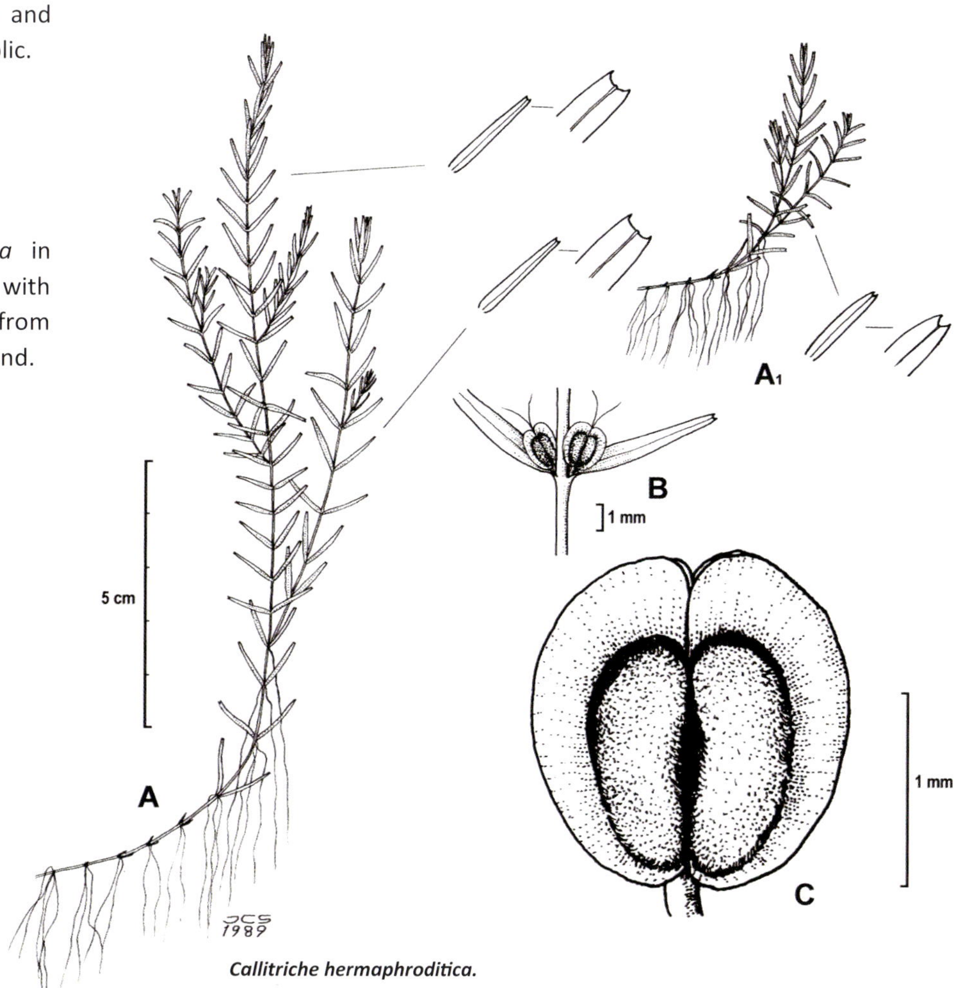

Callitriche hermaphroditica.
A,**A1** habit, **B** flower, **C** fruit.

Plantaginaceae

283 *Callitriche truncata* Guss. subsp. *occidentalis* (Rouy) Schotsman

Syn.: *C. occidentalis* (Rouy) Prain; *C. truncata* var. *occidentalis* (Rouy) Schotsman
DK: - **N:** - **S:** - **FIN:** - **IS:** -
GB: Short-leaved Water-starwort **NL:** Doorschijnend sterrenkroos **F:** Callitriche tronqué
D: Gestutzter Wasserstern **CZ:** - **PL:** -
EST: - **LV:** - **LT:** - **Ru:** -

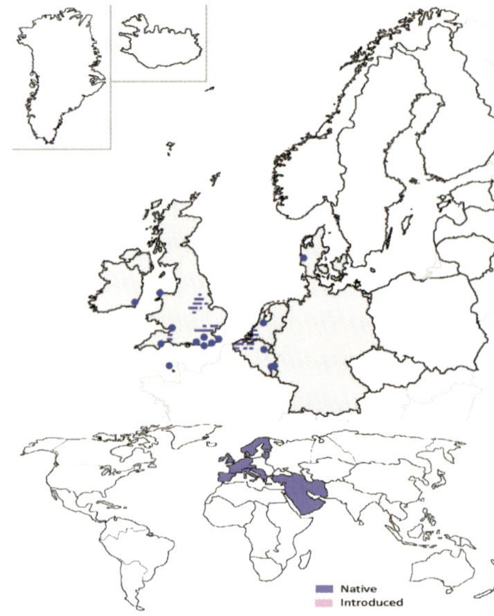

Callitriche truncata subsp. *occidentalis.* Skjern, Denmark. Photo JCS.

Annual, aquatic or occasionally terrestrial.

Stem: Typically densely branched, whitish, without stem scales. Internodes short, abruptly decreasing in length to shoot tip.

Leaves: More or less uniform, all submerged. Dark green, somewhat translucent, sessile, not joined across stem. Lingulate leaves more or less parallel-sided, 4–11 mm long and 0.3–1.1 mm wide, 1-veined. Expanded leaves and rosettes never produced. Terrestrial leaf blades dark green, narrowly elliptic, 1.7–4.7 mm long and 0.5–1.2 mm wide.

Inflorescence: Monoecious, male and female flowers solitary in leaf axils.

Flowers: Without bracts. Filament erect, 0.9–1.7 mm long, pollen colourless, ellipsoid. Styles initially erect, ultimately spreading or reflexed, 0.4–6.1 mm long.

Fruits: On peduncle 0.2–0.9 mm long, mericarps divergent, wider than long, brown when mature, 1–1.4 mm long and 1.2–1.9 mm wide, unwinged.

Flowering: June-August.

Biology: Pollination submerged; fruits typically break up and disperse as solitary mericarps.

Habitats: Occurs in mesotrophic to eutrophic coastal lagoons, lakes, ponds, gravel pits, canals and rivers, occasionally inland, including seasonal water bodies.

Distribution within the region:
Native. Britain, Ireland, the Netherlands, Belgium and Denmark.

Characteristics and similar species:
Callitriche truncata subsp. *occidentalis* differs from all other *Callitriche* species in the region by the unwinged fruits, which are pedunculate, wider than long and with divergent mericarps.

Compare **281 *C. hermaphroditica*.**

C. truncata subsp. *occidentalis* on wet substrate. Llyn Trafwll, Anglesey, Wales. Photo RVL.

Callitriche truncata subsp. *occidentalis.* Fruit. Photos KvdW.

Callitriche truncata subsp. *occidentalis.*
Leaf tips. Photo JCS.

Callitriche truncata subsp. *occidentalis.*
Submerged plant. Photo KvdW.

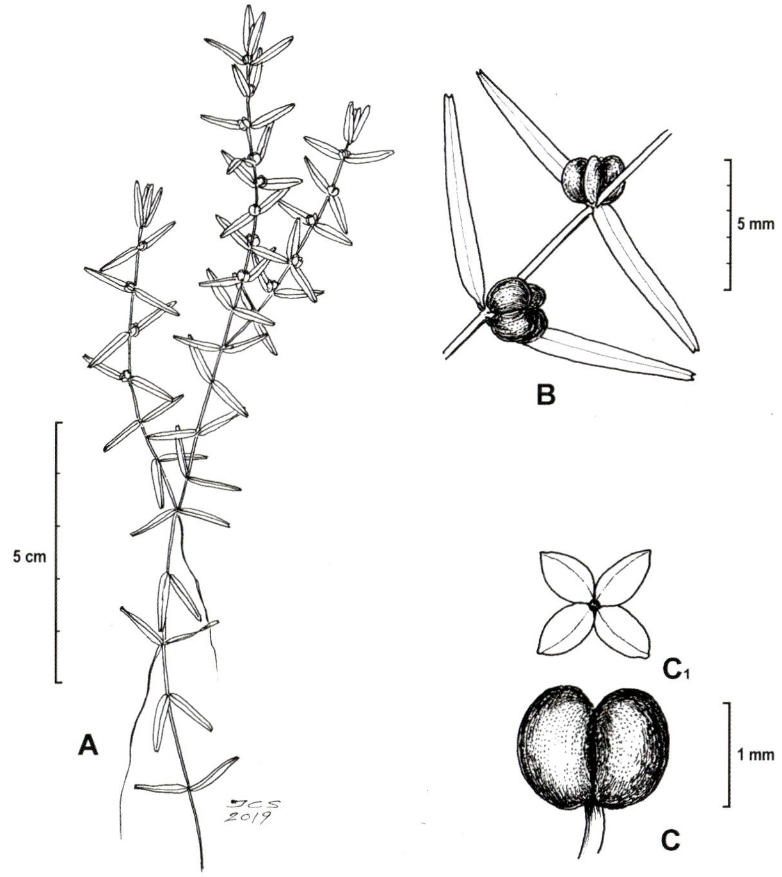

Callitriche truncata subsp. ***occidentalis***
A habit, **B** part of stem with fruits, **C** fruit, **C1** cross section of fruit.

Plantaginaceae

284 *Callitriche palustris* L.

Syn.: *C. verna* L.

DK: Småfrugtet Vandstjerne **N:** Småvasshår **S:** Smålånke **FIN:** Pikkuvesitähti **IS:** Vorbrúða
GB: Narrow-fruited Water-starwort **NL:** Klein sterrenkroos **F:** Callitriche des marais
D: Sumpf-Wasserstern **CZ:** Hvězdoš jarní **PL:** Rzęśl wiosenna
EST: Soo-vesitäht **LV:** Pavasara ūdenīte **LT:** Pavasarinė praujenė **Ru:** Болотник болотный

Terrestrial form of *C. palustris* from moist substrate. Snowshoe Lake, Alaska, USA. Photo RVL.

Perennial but capable of behaving as an annual. Aquatic and terrestrial, probably more often terrestrial.

Stem: Sparsely to densely branched, whitish or yellowish green, with stem scales. Internode length decreases gradually to shoot tips.

Leaves: Light green, joined across stem. Lingulate leaves linear to narrowly expanded, 5.6–9.7 mm long and 0.5–1.2 mm wide. Expanded submerged or floating leaves elliptic to more or less orbicular, 3.6–10.3 mm long and 1.2–4.5 mm wide, with 3 main veins and often additional veins arising from these. Aquatic plants develop floating rosettes of 7–15 leaves. Leaves of terrestrial plants lanceolate to oval, typically very small and slightly fleshy but grade into the size of expanded leaves.

Inflorescence: Typically a male and female in one of a pair of axils and the same or a solitary female in the opposed axil.

Flowers: Bracts whitish and translucent, 0.5–1.2 mm long. Filaments erect, 0.2–2.7 mm long, pollen yellow. Styles erect or slightly spreading, often poorly developed, 0.2–1.7 mm long.

Fruits: Sessile, longer than wide, widest above the middle, 0.9–1.4 mm long and 0.8–1.1 mm wide, winged mainly or only at the top, black and shiny when mature.

Flowering: June-September.

Biology: Pollination aerial, but probably mainly self-pollinated by contact within a leaf axil. Typically abundantly fertile in both aquatic and terrestrial forms. Terrestrial plants continue to grow and flower, often forming extensive mats over the substrate.

Habitats: Occurs in a wide range of habitats, including wet ruts on woodland rides, turloughs, lakes, river bars and margins and reservoirs. A characteristic plant of oligotrophic waters but can tolerate anthropic eutrophication. Populations are most often encountered on wet mud in the drawdown zones of lakes and rivers or in seasonal pools, where they may be abundant over large areas.

Distribution within the region:
Native. Through much of the region, more frequent in the north and east, very local in Britain and Ireland.

Characteristics and similar species:
Callitriche palustris can be distinguished from all other *Callitriche* species in the region by the fruits, which are black and shiny when mature and longer than wide with the wing developed only or mainly at the apex.

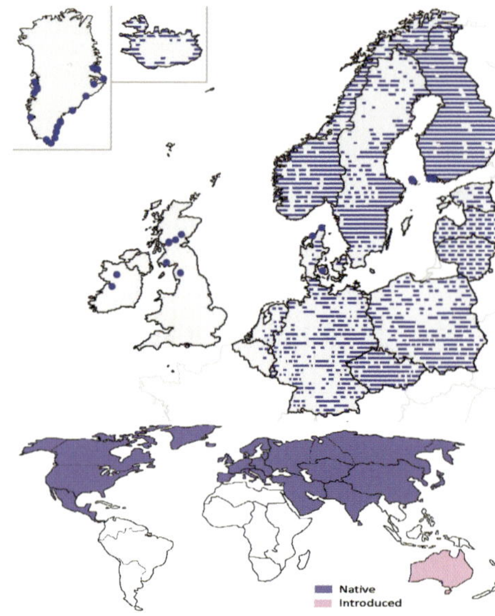

285 *C. heterophylla* and *C. anceps* are taxonomically very close to *C. palustris,* but both have fruits which are as wide as long.

286 *C. brutia,* which can also have very dark fruits, can be distinguished by the reflexed style.

C. brevistyla Lansdown typically has both male and female flowers in leaf axils; the fruits may appear similar and are as wide as or usually wider than long. It has only once been recorded

Callitriche palustris. Submerged plants with floating leaf rosettes. Hausdülmener fishponds, Germany. Photo KvdW.

The fruits of *C. palustris* are longer than wide, obovate to elliptic and winged mainly towards the apex. Voldsted, Jutland, Denmark. Photo BM.

Callitriche palustris. Aquatic forms develop floating rosettes of spathulate to elliptical leaves. Voldsted, Jutland, Denmark. Photo JCS.

within the region, as a weed of green houses in Sweden, where it persisted for many years and may still occur.

Non-native *Callitriche* species, including *C. japonica* Engelm. ex Hegelm., *C. peploides* Nutt. and *C. terrestris* Raf., have been recorded as contaminants of imported plants in the region and could occur in the wild. All three species only occur as terrestrial plants growing on damp mud.

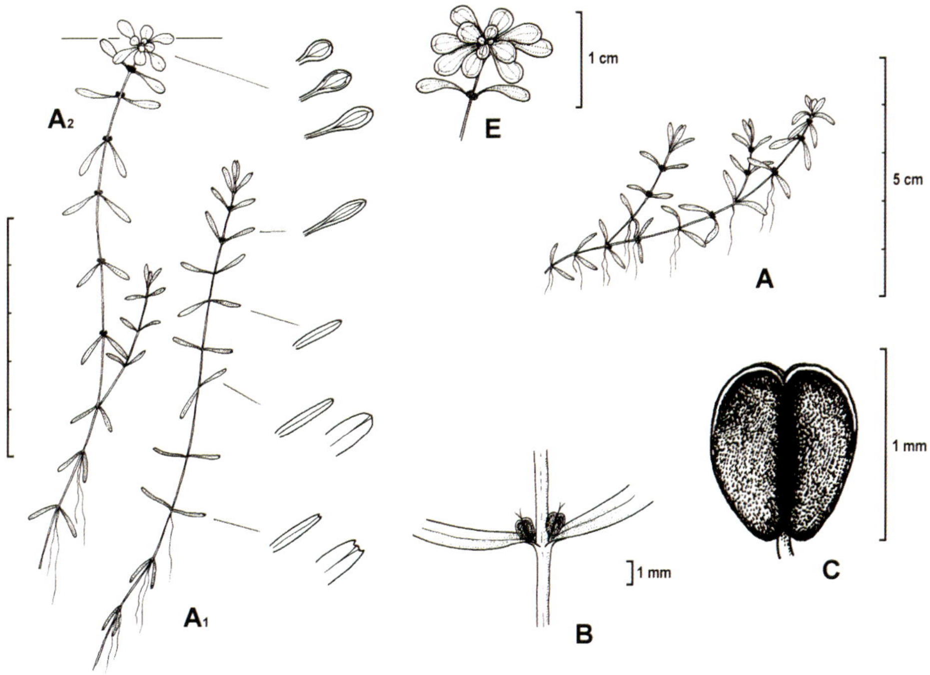

Callitriche palustris
A habit of terrestrial plant**, A1** habit of aquatic plant, **A2** habit of aquatic plant with floating leaf rosette, **B** immature fruits, **C** fruit, **E** floating leaf rosette.

285 *Callitriche heterophylla* Pursh

Syn.: *C. anceps* Fern.
DK: Tveægget Vandstjerne

Within the region, *C. heterophylla* is known only from Greenland. It produces both lingulate and expanded leaves, forming a floating rosette at the shoot tips, often with very distinctive leaf shape. It has fruits which are more or less as wide as long, unwinged or narrowly winged with the wing wider at the top. Plants with very small unwinged fruit have been distinguished as *C. anceps* Fernald, however it may be that this taxon is more closely related to the northern Russian *C. subanceps* Petrov. The taxonomic relationship between *C. anceps, C. heterophylla, C. palustris* and *C. subanceps* is still not yet resolved.

Distribution within the region:
Native to Greenland, where many populations conform to the concept of *C. anceps*.

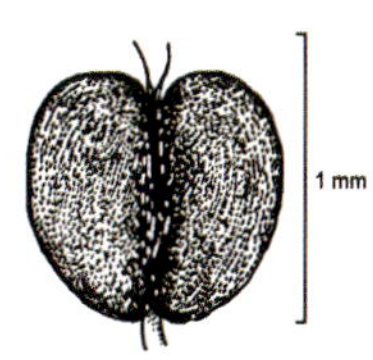

C. heterophylla fruit

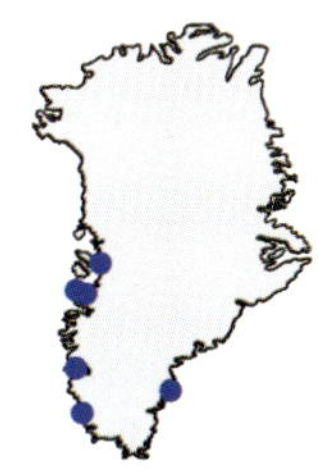

Callitriche heterophylla. Lynchburg, Missouri, USA. Photo RVL.

Plantaginaceae

286 *Callitriche brutia* Petagna

DK: - **N:** - **S:** - **FIN:** - **IS:** -
GB: Intermediate Water-starwort **NL:** - **F:** -
D: - **CZ:** - **PL:** -
EST: - **LV:** - **LT:** - **Ru:** -

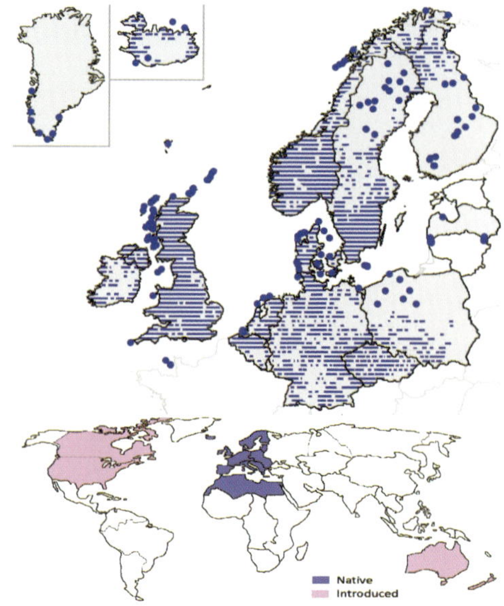

Aquatic plants of both subspecies of *C. brutia* normally produce sessile or very short-stalked fruits. Værnenge, W-Jutland, Denmark. Photo JCS.

Perennial but often behaving as an annual in seasonal wetlands. Aquatic or terrestrial.

Stem: Typically densely branched, pale green, not contrasting strongly with leaf colour, with stem scales. Aquatic plants with long internodes, decreasing in length gradually to shoot tips.

Leaves: Lingulate or expanded, submerged or floating, pale green, opaque, joined across stem. Lingulate leaves typically very long and narrow, 4–16 mm long and 0.1–1.7 mm wide, apex deeply and often broadly notched. Expanded leaves narrowly elliptic, 1.8–18 mm long and 0.8–3.6 mm wide, petiole up to 10 mm long, venation simple, apical leaves form a floating rosette of 7–14 leaves. Leaves of terrestrial plants shortly linear, slightly fleshy, 2.4–5.5 mm long and 0.3–0.8 mm wide.

Inflorescence: Monoecious, male and female flowers solitary in leaf axils, rarely more than one flower in an axil.

Flowers: Bracts small and soon falling, 0.2–1.8 mm long. Filament erect, 0.1–1.2 mm long, pollen colourless, ellipsoid. Style tightly reflexed and appressed to side of fruit.

Fruits: Sessile or subsessile on aquatic plants and sessile or on peduncle up to 12 mm long on terrestrial plants, mericarps parallel, more or less as wide as or wider than long, blackish when mature, 0.8–1.6 mm long and 0.7–1.8 mm wide, wing narrow all around, 0.02–0.3 mm wide.

Flowering: June-September.

Biology: Pollination submerged, but typically self-fertilised by stigmas growing to contact the anther in the opposite leaf axil. Plants are typically abundantly fertile. The fruits of subsp. *brutia* are typically buried in the substrate on long peduncles and are usually white.

Distribution within the region: Native throughout Europe, from Iceland and the Faroe Islands through Scandinavia to Poland and the Czech Republic.

Characteristics and similar species: *Callitriche brutia* can be distinguished from all other *Callitriche* species by the styles so strongly reflexed that they are appressed to the sides of the fruit, which are narrowly winged throughout, combined with the presence of scales on the stem.

Two subspecies of *C. brutia* occur in the region: subsp. *brutia* can be recognised when pedunculate fruits are present, and subsp. *hamulata* can be recognised by the characteristic shape of the lingulate leaf apices. Plants lacking these characters cannot reliably be identified to subspecies, although terrestrial plants with only sessile fruit are likely to be subsp. *hamulata*.

Callitriche brutia. Rosette of floating leaves. Værnenge, W-Jutland, Denmark. Photo JCS.

Plantaginaceae

286a *Callitriche brutia* Petagna subsp. *brutia*

Syn.: *C. pedunculata* DC.; *C. intermedia* Hoffm. subsp. *pedunculata* (DC.) Clapham
DK: Stilkfrugtet Vandstjerne **N:** Stilkvasshår **S:** Skaftlånke **FIN:** - **IS:** Lækjabrúða
GB: Pedunculate Water-starwort **NL:** Gesteeld sterrenkroos **F:** Callitriche pédonculé
D: Stielfrüchtiger Wasserstern **CZ:** - **PL:** -
EST: - **LV:** - **LT:** - **Ru:** -

C. brutia subsp. *brutia* differs from subsp. *hamulata* mainly by the long pedunculate fruits, but these are only present on terrestrial plants. Værnenge, W-Jutland, Denmark. Photos JCS.

Subsp. *brutia* can only be distinguished reliably from subsp. *hamulata* when fruits are pedunculate, with the peduncle more than 2 mm long. Aquatic plants only very rarely develop pedunculate fruit, but most fruit on terrestrial plants will be pedunculate, and so terrestrial plants with sessile fruit are unlikely to be this subspecies.

Habitats: Subsp. *brutia* is a characteristic plant of temporary wetlands, such as seasonal pools and the drawdown zones of lakes and reservoirs, as well as poached pools in gateways and on tracks.

Distribution within the region: Native. Throughout much of the range of the species but more frequent in the south with only scattered records in Scandinavia.

Particularly broad-leaved *C. brutia* subsp. *brutia*. Hatchett Pond, New Forest, England. Photo RVL.

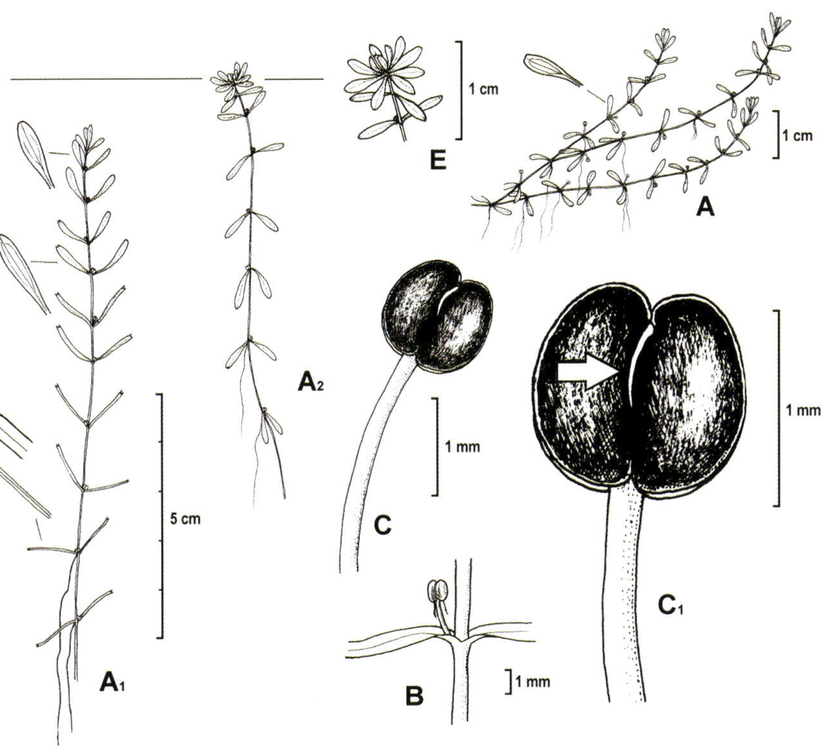

Callitriche brutia subsp. ***brutia***
A habit of terrestrial plant, **A1, A2** habit of aquatic plant, **B** immature fruit, **C, C1** fruit, **E** rosette of floating leaves.

Plantaginaceae

287 *Callitriche brutia* Petagna subsp. *hamulata* (Kütz. ex W.D.J.Koch) O.Bolòs & Vigo

Syn.: *C. hamulata* Kütz. ex W.D.J.Koch; *C. intermedia* Hoffm. subsp. *hamulata* (Kütz. ex W.D.J.Koch) Clapham
DK: Smalbladet Vandstjerne **N:** Klovasshår **S:** Klolånke **FIN:** Lapinvesitähti **IS:** Síkjabrúða
GB: Intermediate Water-starwort **NL:** Haaksterrenkroos **F:** Callitriche en crochet
D: Haken-Wasserstern **CZ:** Hvězdoš háčkatý **PL:** Rzęśl hakowata
EST: Haakjas vesitäht **LV:** - **LT:** Kabliuotoji praujenė **Ru:** Болотник крючковатый

Callitriche brutia subsp. *hamulata* in standing water with both floating rosettes and linear submerged leaves. Skagen, Denmark. Photo JCS.

Callitriche brutia subsp. *hamulata* showing the characteristic broadly expanded leaf apices. River Blackwater, County Cork, Ireland. Photo RVL.

Subsp. *hamulata* can only be distinguished reliably from subsp. *brutia* when the characteristic long, fine submerged leaves with finely pointed, pincer-like apices are developed, and these are usually only found on plants in deep or fast-flowing water. However, terrestrial plants with all fruits sessile or subsessile are also likely to be this subspecies.

Habitats: Subsp. *hamulata* typically occurs in oligotrophic and acid waters, although it will occur in more mesotrophic and rarely basic conditions. Found in a wide range of water bodies from ponds, lakes, ditches and canals to large and very fast-flowing rivers.

Distribution within the region:
Native. Widespread and abundant in northern Europe east to the Czech Republic and south to southern France and northern Italy; it is absent from the Mediterranean.

Callitriche brutia subsp. *hamulata*. Narrow submerged leaves typically develop a distinct emargination, which on very narrow leaves may have the shape of a pincer and are wider than the adjacent part of the leaves. Photo KvdW.

Characteristic appearance of stands of *C. brutia* subsp. *hamulata* in flowing water. River Vyrnwy, Powys, Wales. Photo RVL.

Callitriche brutia subsp. *hamulata*. From left to right, increasingly mature fruit with the remains of the style clearly visible at all stages. Photos KvdW, JCS, JCS.

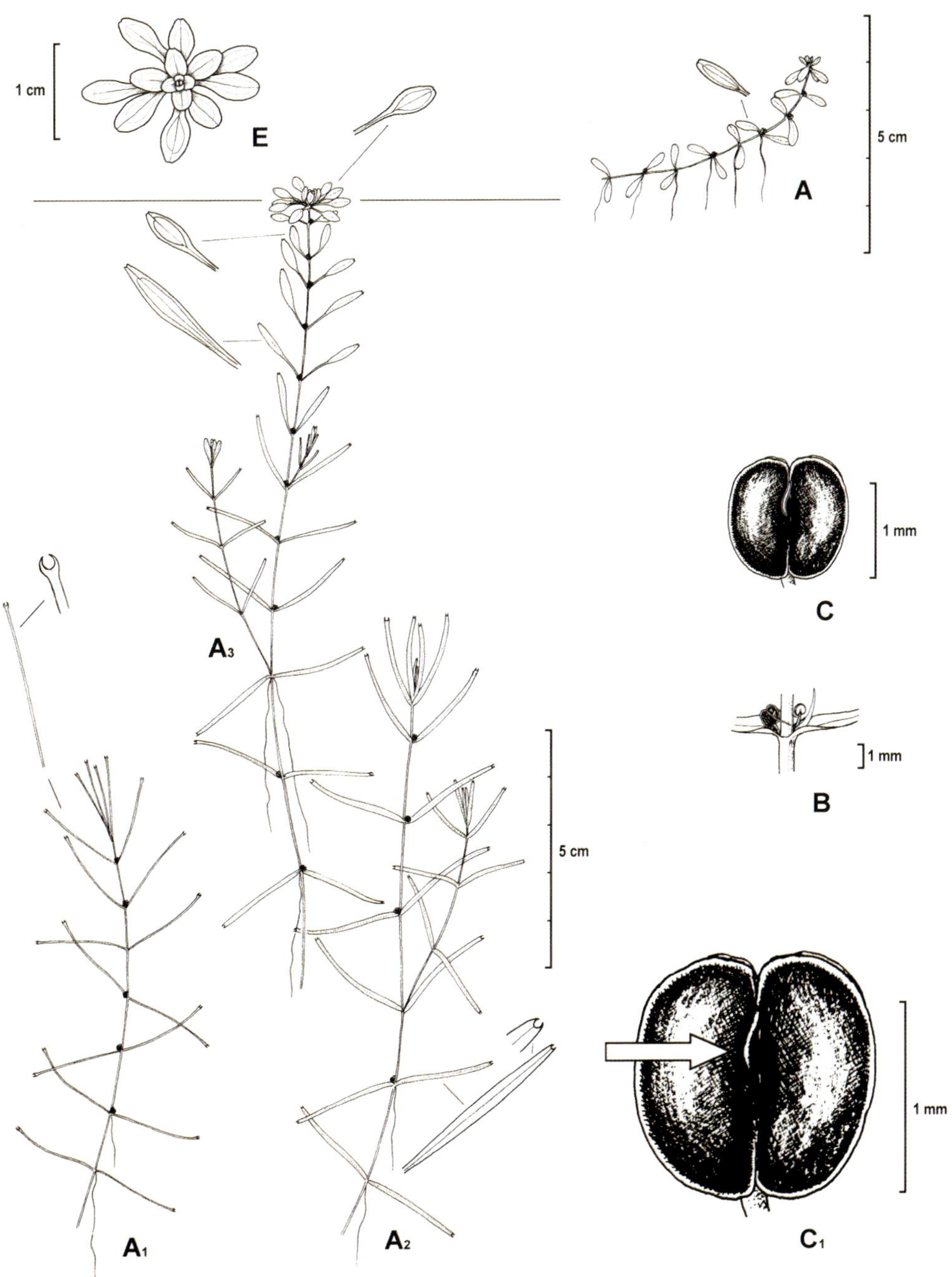

Callitriche brutia subsp. ***hamulata***
A habit terrestrial form, **A1**, **A2** habit submerged plant, **A3** habit plant with floating leaf rosette, **B** flowers, **C** fruit, **C1** fruit with stylus residue, **E** rosette of floating leaves.

Plantaginaceae

288 *Callitriche obtusangula* Le Gall

DK: Butfrugtet Vandstjerne **N:** - **S:** Trublånke **FIN:** - **IS:** -
GB: Blunt-fruited Water-starwort **NL:** Stomphoekig sterrenkroos **F:** Callitriche à angles obtus
D: Nussfrüchtiger Wasserstern **CZ:** - **PL:** -
EST: - **LV:** - **LT:** - **Ru:**

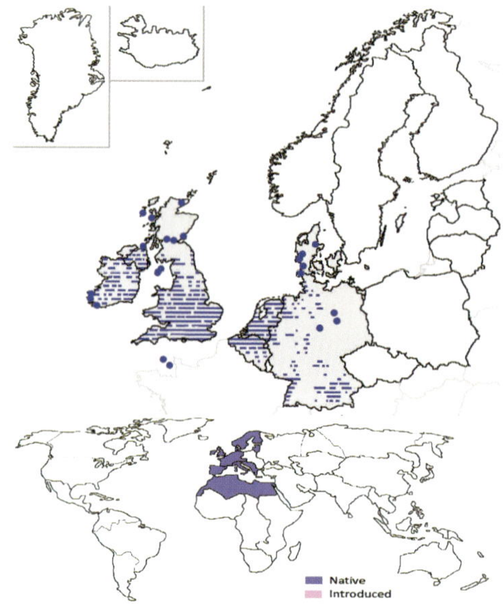

Callitriche obtusangula in still water with floating rosettes, the centres often raised above the water surface. Varming, Ribe, Denmark. Photo JCS.

Perennial, aquatic or less often terrestrial.

Stem: Sparsely to densely branched, robust, pale green, with stem scales. Internodes often long, gradually decreasing in length to shoot tip.

Leaves: Lingulate or expanded, submerged or floating, pale green, opaque, joined across stem. Lingulate leaves fine to very fine, 9.3–32.2 mm long and 0.2–2.2 mm wide. Expanded leaves spathulate, 14.9–29.2 mm long, 1.8–7.0 mm wide, petiole 3.0–11.1 mm long, venation mainly simple with a few short additional veins, apical leaves form a floating or emergent rosette of 13–24 grey-green, rhombic leaves, typically with 3 prominent veins. Leaves of terrestrial plants narrowly spathulate, 4.7–10.2 mm long and 1–3 mm wide.

Inflorescence: Monoecious, flowers solitary in leaf axils.

Flowers: Bracts persistent, 1.1–3.6 mm long. Filaments initially erect, ultimately strongly recurved, 4–12.3 mm long, pollen yellow, elongate ellipsoid and curved. Styles initially erect, becoming strongly recurved, persistent, 1.7–8.3 mm long.

Fruits: Sessile, mericarps parallel, slightly longer than wide, pale brown when mature, 1.1–1.8 mm long and 1.1–1.7 mm wide, unwinged.

Flowering: May to early August.

Biology: Pollination aerial. Aquatic plants are typically fertile, but terrestrial plants rarely so. This is the first *Callitriche* species to stop flowering, with few flowers found after late July.

Habitats: Lowland, often in calcareous, naturally mesotrophic to eutrophic waters, but it can occur in mildly acid and oligotrophic systems. In ponds, lakes, ditches, canals and rivers, extending into the upper limits of brackish water in coastal areas.

Distribution within the region:
Native. Widespread and locally abundant from Ireland north to Denmark, east to Slovakia.

Characteristics and similar species:
C. obtusangula can be distinguished from all other *Callitriche* species by the unwinged fruit with parallel mericarps, as well as by the elongate, curved pollen. The fruits are very distinctive and

Callitriche obtusangula, Large and well-developed floating rosette with long-stalked, rhombic leaves with the characteristic prominent veins. Varming, Ribe, Denmark. Photo JCS.

Callitriche obtusangula differs from all other *Callitriche* species by the unwinged fruits, which are longer than wide and initially green, then brownish. Varming, Ribe, Denmark. Photos JCS.

even young fruits are easily recognised. The prominently ridged grey-green, rhombic rosette leaves, which are often emergent at the centre, are also a good field character, however occasional populations of **291** ***C. platycarpa*** can have prominently ridged rosette leaves and identification should always be confirmed based on the fruit.

289 ***C. cophocarpa*** can be similar to *C. obtusangula,* but the rosette leaves typically have at most one prominent vein, while the fruit is always at least narrowly winged.

Elongate and slightly curved pollen distinguishes *C. obtusangula* from all other *Callitriche* species. Photo JCS.

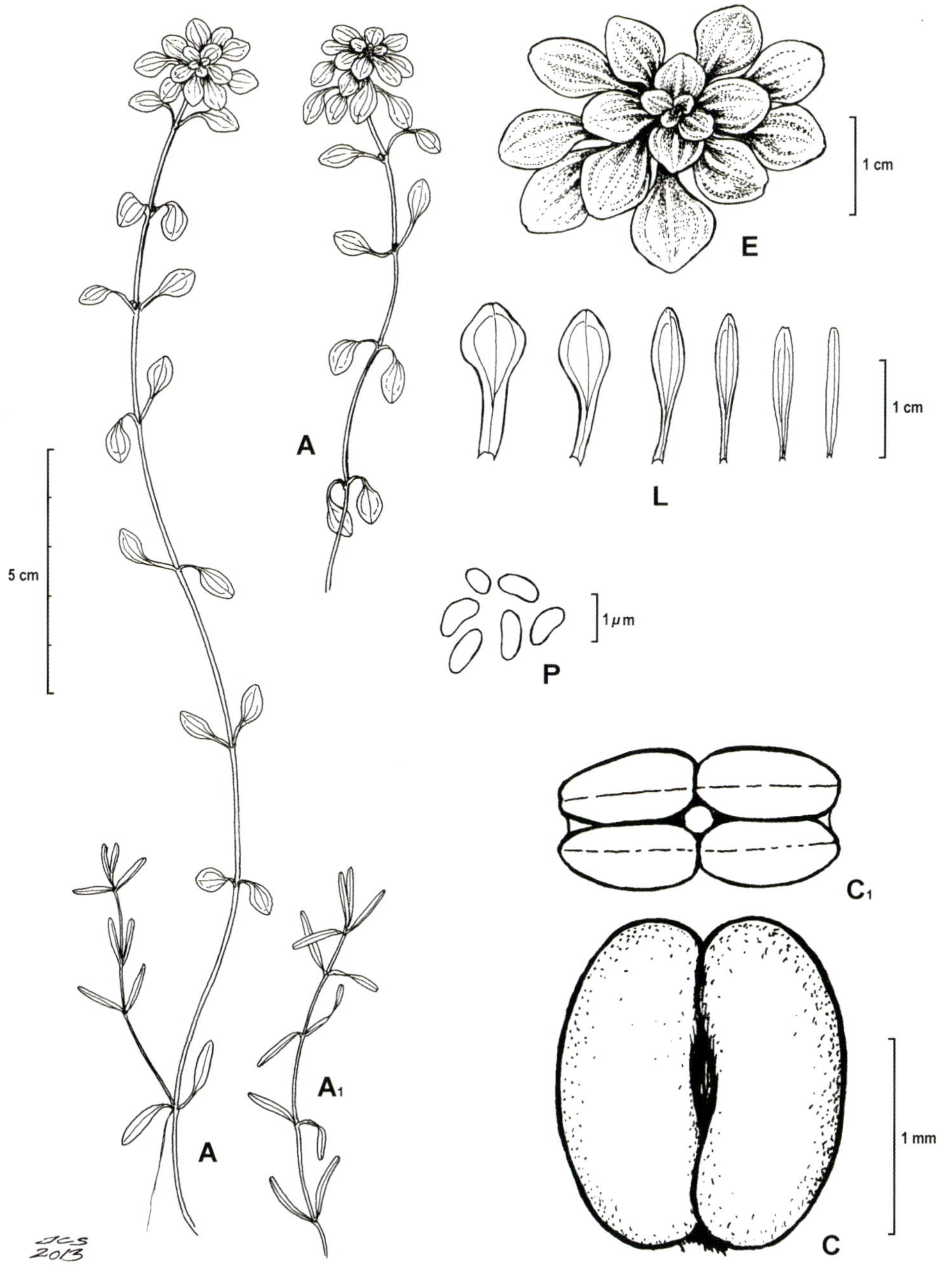

Callitriche obtusangula
A habit plant with floating leaf rosette, **A1** habit submerged plant, **C** fruit, **C1** cross section of fruit, **E** rosette of floating leaves, **L** leaves, **P** pollen.

Plantaginaceae

289 *Callitriche cophocarpa* Sendtn.

Syn.: *C. polymorpha* Lönnr.

DK: Roset-Vandstjerne **N:** Sprikjevasshår **S:** Sommarlånke **FIN:** Isovesitähti **IS:** -
GB: Longstyled Water-starwort **NL:** Gekield sterrenkroos **F:** Callitriche à fruits obtus
D: Stumpfkantiger Wasserstern **CZ:** Hvězdoš mnohotvary **PL:** Rzęśl długoszyjkowa
EST: Harilik vesitäht **LV:** Daudzziedu ūdenīte **LT:** Įvairioji praujenė **Ru:** Болотник коротkoплодный

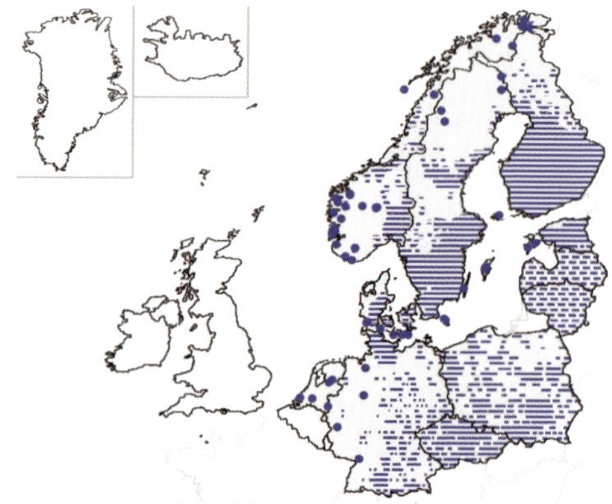

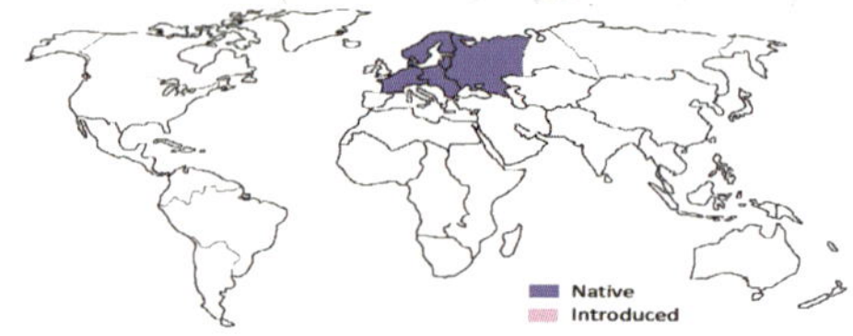

Callitriche cophocarpa with floating leaf rosettes. Randers, Denmark. Photo JCS.

Perennial or possibly occasionally annual; aquatic or occasionally terrestrial.

Stem: Sparsely to densely branched, whitish to pale green, with stem scales. Internodes often long, gradually decreasing in length to shoot tips.

Leaves: Lingulate or expanded, submerged or floating, pale green, opaque, joined across stem. Lingulate leaves fine, 11–18.9 mm long and 0.7–1.8 mm wide. Expanded leaves narrowly angular to elliptic, 4.5–18.4 mm long, 1.3–5.2 mm wide, petiole 0.8–8.7 mm long, venation simple but there may be short additional veins, apical leaves form a floating rosette of 10–25 leaves. Leaves of terrestrial plants shortly petiolate, elliptic, 5.5–9.7 mm long and 2.0–3.8 mm wide.

Inflorescence: Monoecious (probably best termed paroecious), most flowers on a shoot either predominantly male or predominantly female, a single plant typically having one more or less central male shoot and abundant female shoots.

Flowers: Bracts persistent, 0.7–2.2 mm long. Filaments erect but becoming recurved, 2.8–4.7 mm long, pollen yellow, ellipsoid. Styles erect, becoming slightly curved, persistent, 1.2–3.9 mm long.

Fruits: Sessile, mericarps parallel, more or less as wide as long, pale brown when mature, 0.9–1.1 mm long and 1.1–1.2 mm wide, wing very narrow, 0.02–0.06 mm wide.

Flowering: June-September.

Biology: Pollination aerial and possibly occasionally submerged.

Habitats: Occurs in a wide range of water bodies, including neutral, oligotrophic to mesotrophic waters of ponds, lakes, ditches and canals, as well as occasionally in streams and rivers.

Distribution within the region: Native. More or less throughout except Britain, Ireland, Iceland and Greenland. Scarce or local in northern Scandinavia. *C. cophocarpa* appears to be declining and undergoing a range retraction; it may be extinct in Belgium and the Netherlands.

Callitriche cophocarpa. The shape of the rosette leaves is very variable and often narrower than displayed in the photo. Randers, Denmark. Photo JCS.

Characteristics and similar species:
Callitriche cophocarpa is distinguished from all other *Callitriche* species in the region by the small fruits which are more or less as long as wide, combined with the erect to recurved styles, yellow pollen, stem scales and unique arrangement of the male and female flowers. The unique distribution of the flowers can often be seen in the floating rosettes of each plant, where anthers will be visible only in one rosette and stigmas in all others.

The fruits of *Callitriche cophocarpa* are small, almost circular, without or, as shown in the photo, with very narrow wings. Styles remain erect or slightly curved until the fruits are mature. Randers, Denmark. Photo JCS.

290 *Callitriche ×vigens* Martinsson

(*C. cophocarpa x platycarpa*)

Some populations of *C. ×vigens* can be distinguished by very low fertility, combined with a degree of difference in the distribution of male and female flowers on plants. However, some populations appear to be fully fertile and may be indistinguishable from the parents. Populations may persist in the absence of one or both parents. The chromosome number is 15 = the average between the parent species.

Habitats: In similar habitats to its parents, in lowland ditches, canals, streams, slow-flowing reaches and backwaters of rivers, oxbows and field ponds.

Distribution within the region:
C. ×vigens occurs in a zone where the ranges of the parent species overlap or have done so in the past. It has been confirmed from southern Sweden, Denmark, the Netherlands, the Czech Republic and western Switzerland. It may now be extinct in the Netherlands.

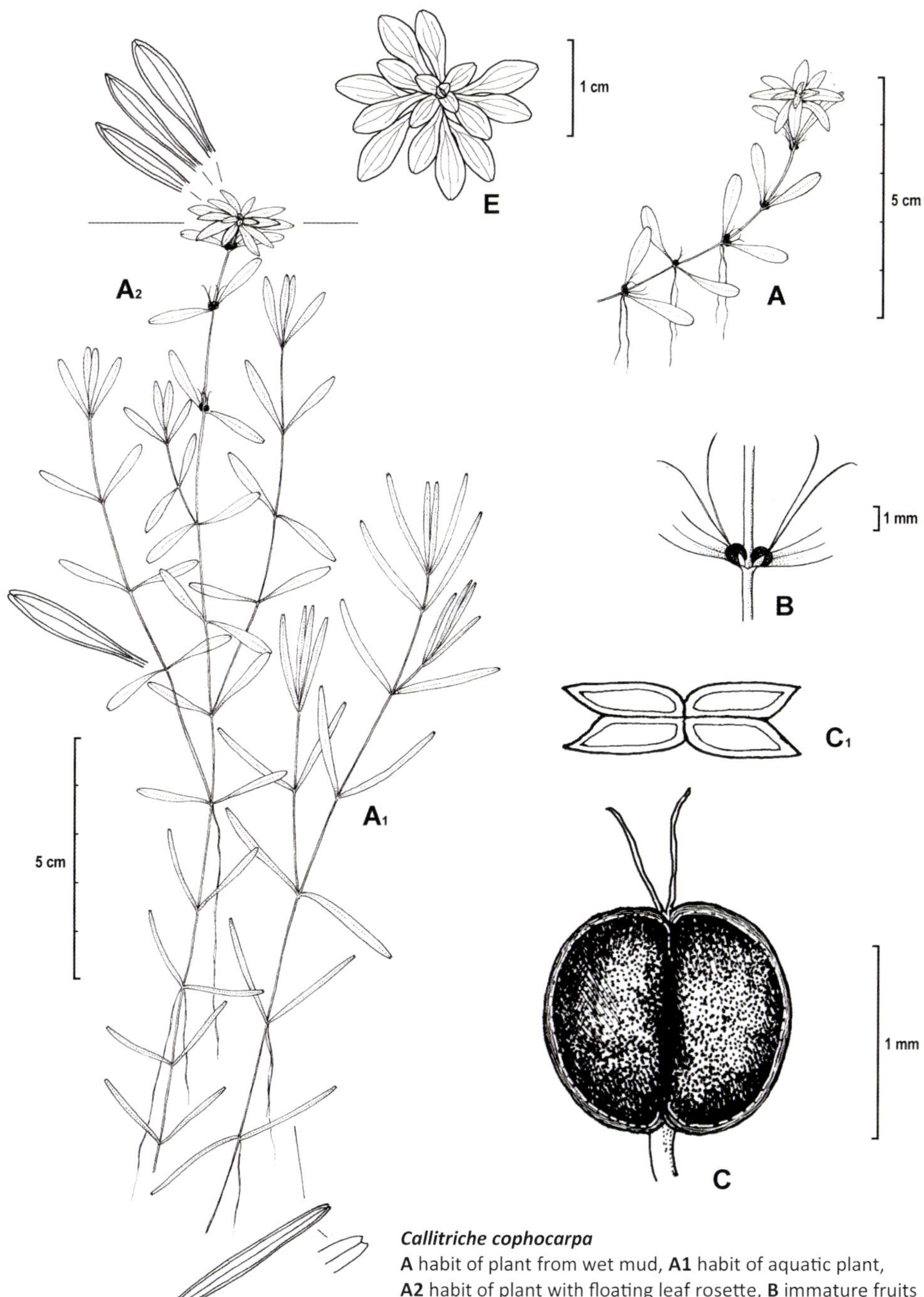

Callitriche cophocarpa
A habit of plant from wet mud, **A1** habit of aquatic plant, **A2** habit of plant with floating leaf rosette, **B** immature fruits (note the very long styles), **C** fruit, **C1** cross section of fruit, **E** rosette of floating leaves.

Plantaginaceae

291 *Callitriche platycarpa* Kütz. in Rchb.

DK: Fladfrugtet Vandstjerne **N:** Mørkvasshår **S:** Plattlånke **FIN:** - **IS:** -
GB: Various-leaved Water-starwort **NL:** Gewoon sterrenkroos **F:** Étoile d'eau à fruits plats
D: Breitfrüchtiger Wasserstern **CZ:** Hvězdoš hranoplodý **PL:** -
EST: - **LV:** - **LT:** - **Ru:** -

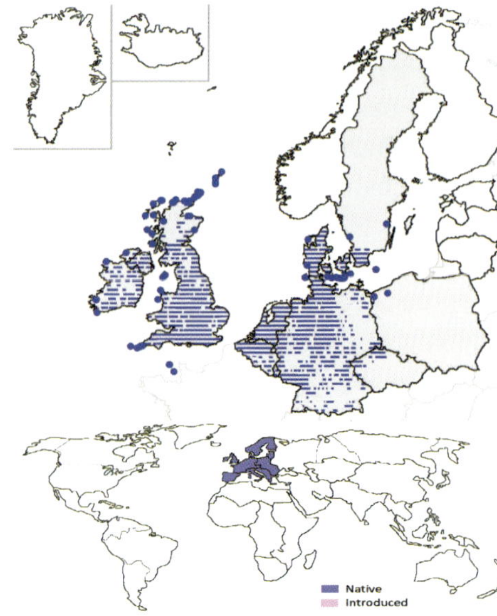

Well-developed floating and flowering rosettes of *C. platycarpa*. Sønderup Å, Jutland, Denmark. Photo JCS.

Perennial, aquatic or terrestrial.

Stem: Sparsely to densely branched, robust, white to greenish, with stem scales. Internodes often long, gradually decreasing in length to shoot tips.

Leaves: Lingulate or expanded, submerged or floating, pale green, opaque, joined across stem. Lingulate leaves fine to broad, often sinuous, 11.1–28.5 mm long and 0.1–2.0 mm wide. Expanded leaves spathulate to elliptic, 11.1–15.9 mm long and 2.0–3.8 mm wide, petiole 0.5–5.5 mm long, venation mainly simple, with a few short additional veins, apical leaves forming a floating rosette of 7–18 greyish-green, elliptic leaves which may occasionally have prominent veins. Leaves of terrestrial plants narrowly elliptic, 1.5–3.8 mm long and 0.5–1.8 mm wide.

Inflorescence: Monoecious. Flowers solitary in leaf axils.

Flowers: Bracts persistent, 0.3–2.6 mm long. Filaments initially erect, ultimately recurved, 1.5–5.1 mm long, pollen yellow, very variable shape, rounded triangular, ellipsoid or rarely tetrahedroid. Styles erect and slightly curved, 0.5–6.5 mm long.

Fruits: Sessile, mericarps parallel, more or less as wide as long, pale brown when mature, 1.3–1.7 mm long and 1.4–1.8 mm wide, narrowly winged throughout, wing 0.05–0.25 mm wide.

Flowering: June-August.

Biology: Pollination aerial. Some populations of *Callitriche platycarpa* may be abundantly fertile but many populations will rarely if ever fruit, while terrestrial populations typically lack fruit. *C. platycarpa* probably arose from hybridisation between **289 *C. cophocarpa*** and **292 *C. stagnalis*** through chromosome doubling and is, in some ways, intermediate between these species.

Habitats: A typical plant of lowland, mesotrophic to eutrophic ditches, canals, streams, ponds and slow-flowing rivers in calcareous to slightly acid waters. It generally occurs in more permanent water bodies than *C. stagnalis* and less often in fast-flowing streams and rivers than *C. obtusangula*. It is apparently able to tolerate quite high anthropogenic eutrophication and appears to be increasing throughout its range.

Submerged *C. platycarpa* showing the distinctive sinuous leaf shape. O'Hara's Lough, County Clare, Ireland. Photo RVL.

Distribution within the region:
Native. *C. platycarpa* is endemic to Western Europe from Britain and Ireland north to southern Sweden and east throughout Germany and south to southern France. It appears to be spreading, with recent records from the Czech Republic and the Mediterranean region.

Characteristics and similar species:
Callitriche platycarpa can be distinguished from all other *Callitriche* in the region (except **292 *C. stagnalis*** and **290 *C. ×vigens***) by the combination of fruits, which are more or less as wide as long, more than 1.1 mm long, winged throughout, erect styles and yellow pollen. It can only be differentiated reliably from *C. stagnalis* by the shape of the pollen, although the leaves of *C. platycarpa* are typically more grey green than those of *C. stagnalis* and the fruit wing narrower. When *C. platycarpa* and *C. stagnalis* grow together, the difference in colour of the rosette leaves can be striking (see photo p. 618). Some populations of *C. ×vigens* can be distinguished by their low fertility, but others are likely to be indistinguishable in the field and can only be distinguished by molecular methods or chromosome counts.

289 *C. cophocarpa* can be very similar to *C. platycarpa,* but the fruits are smaller and the rosette leaves narrower in the former, while the unique floristics of *C. cophocarpa* are usually sufficient to distinguish it from *C. platycarpa.*

Callitriche platycarpa fruits. Photos JCS, BM.

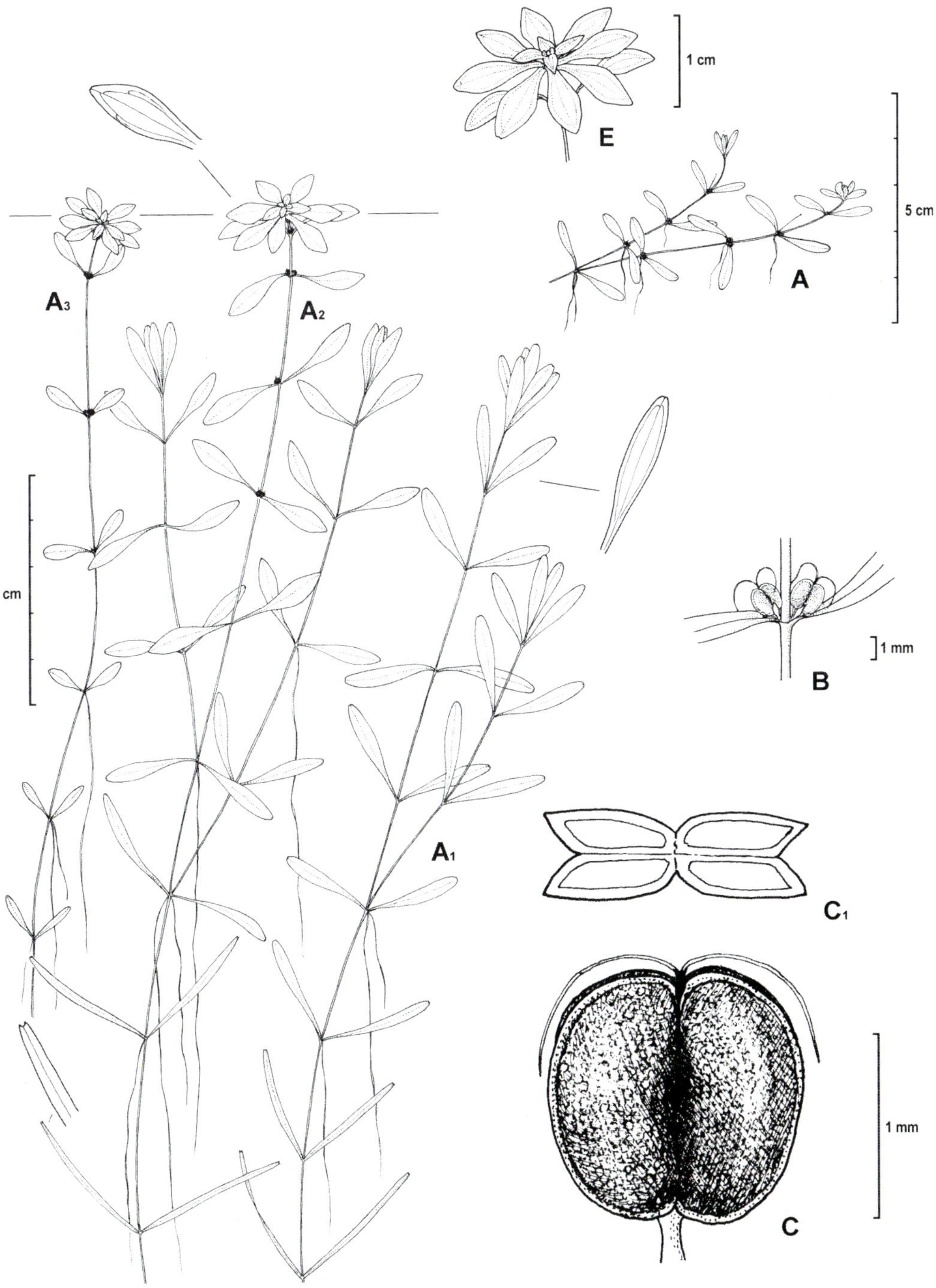

Callitriche platycarpa
A habit plant from moist bottom, **A1** habit submerged plant, **A2-A3** habit plant with floating leaf rosette, **B** immature fruits, **C** fruit, **C1** cross section of fruit, **E** rosette of floating leaves.

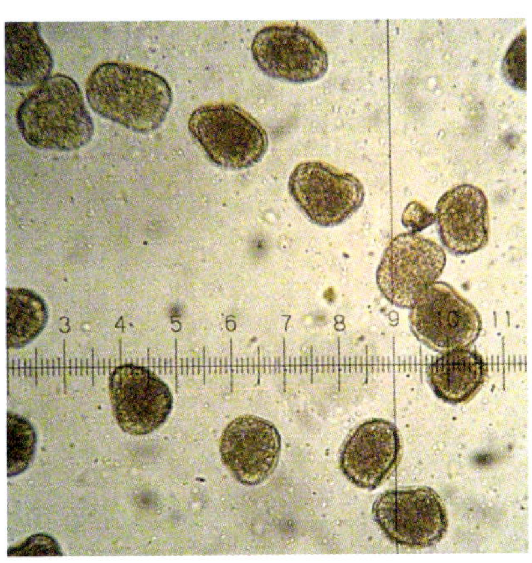

The pollen of *C. platycarpa* is rounded triangular or ellipsoid. Photo RVL.

Plantaginaceae

292 *Callitriche stagnalis* Scop.

DK: Storfrugtet Vandstjerne **N:** Dikevasshår **S:** Dikeslånke **FIN:** Ojasvesitähti **IS:** Laugabrúða
GB: Common Water-starwort **NL:** Gevleugelde sterrenkroos **F:** Callitriche des étangs
D: Teich-Wasserstern **CZ:** Hvězdoš kalužní **PL:** Rzęśl wielkoowockowa
EST: Tiik-vesitäht **LV:** Dumbru ūdenīte **LT:** Tvenkininė praujenė **Ru:** Болотник прудовой

Flowering rosettes of *C. stagnalis*. River Boschbeek, Elmpt, Lower Rhine, Germany. Photo KvdW.

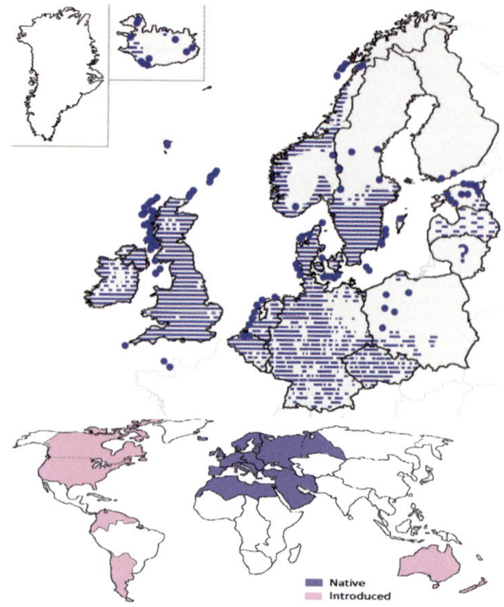

Perennial but annual in seasonal wetlands, aquatic or terrestrial.

Stem: Sparingly to densely branched, delicate or robust, white to greenish, with stem scales. Internodes occasionally long in aquatic forms but typically short, decreasing gradually in length to shoot tips.

Leaves: Expanded, submerged or floating, bright green, opaque, joined across stem. Lingulate leaves unknown in the region. Expanded leaves with ovate-spathulate to almost circular blades, 2.7–21.4 mm long and 1.4–8.3 mm wide, petiole 0.7–6.5 mm long, venation simple to very complex with numerous short additional veins, apical leaves form a floating rosette of 6–12 bright green leaves. Leaves of terrestrial plants elliptic, 2.6–4.4 mm long and 1.6–3.3 mm wide.

Inflorescence: Monoecious, flowers solitary in leaf axils.

Flowers: Bracts persistent, 0.7–2.6 mm long. Filaments erect, becoming recurved, 8.5–10.3 mm long, pollen yellow, ellipsoid. Styles erect, becoming recurved, 5.8–8.6 mm long.

Fruits: Sessile, mericarps parallel, more or less as wide as long, grey to pale greyish brown when mature, 1.3–1.8 mm long and 1.3–2.1 mm wide, wing broad throughout, 0.1–0.4 mm wide.

Flowering: May-November.

Biology: Pollination aerial. Typically abundantly fertile in terrestrial populations, less so in aquatic populations.

Habitats: *C. stagnalis* is the characteristic *Callitriche* species of poached ephemeral pools in gateways, on woodland rides, in pasture, seepages and in flushes, as well as occurring on lake and river margins and in ditches. It is much less often truly aquatic and rarely grows in deep water.

Distribution within the region:
Native. Widespread and abundant from Ireland east to Germany and Denmark, as well as in southern Scandinavia, Estonia, Latvia and southwestern Poland. Scattered or local in most of Poland and Latvia, coastal in Norway and Iceland.

Characteristics and similar species:
Callitriche stagnalis can be distinguished from other *Callitriche* species in the region, apart from **291 *C. platycarpa,*** by the combination of broadly expanded leaves and broadly winged fruits. Although the bright green leaves and greyish (rather than brownish) mature fruit with very broad wings will normally distinguish it from *C. platycar-*

Callitriche stagnalis (top) and *C. platycarpa* (bottom) growing together. The difference in colour is rather striking. Valley, Anglesey, Wales. Photo RVL.

Immature fruits are greenish, turning greyish at maturity. The wide winged and grey to pale, greyish brown fruits are the most important diagnostic character of *C. stagnalis*. Photo left JCS, right KvdW.

Callitriche stagnalis rosette. The pale to bright green, broadly-spathulate inner leaves and almost circular outer leaves are characteristic. Photo JCS.

Callitriche stagnalis. Flowering rosette with whitish styles up to 2 mm long. Jutland, Denmark. Photo JCS.

pa, final confirmation is dependent on examination of pollen.

In addition to confusion with other *Callitriche* species, the broad, almost circular leaf blades of *C. stagnalis* may lead to confusion with **239 *Lythrum portula***, which has completely different leaf veins and flowers as well as commonly red or violet colouration of stems and petioles.

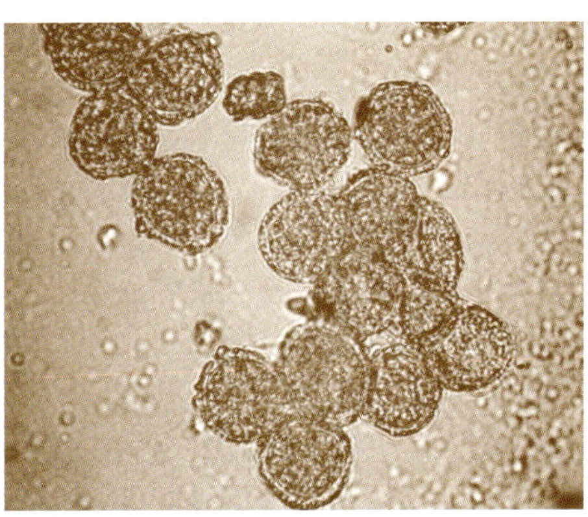

The pollen of C. stagnalis is ellipsoid. Photo KvdW.

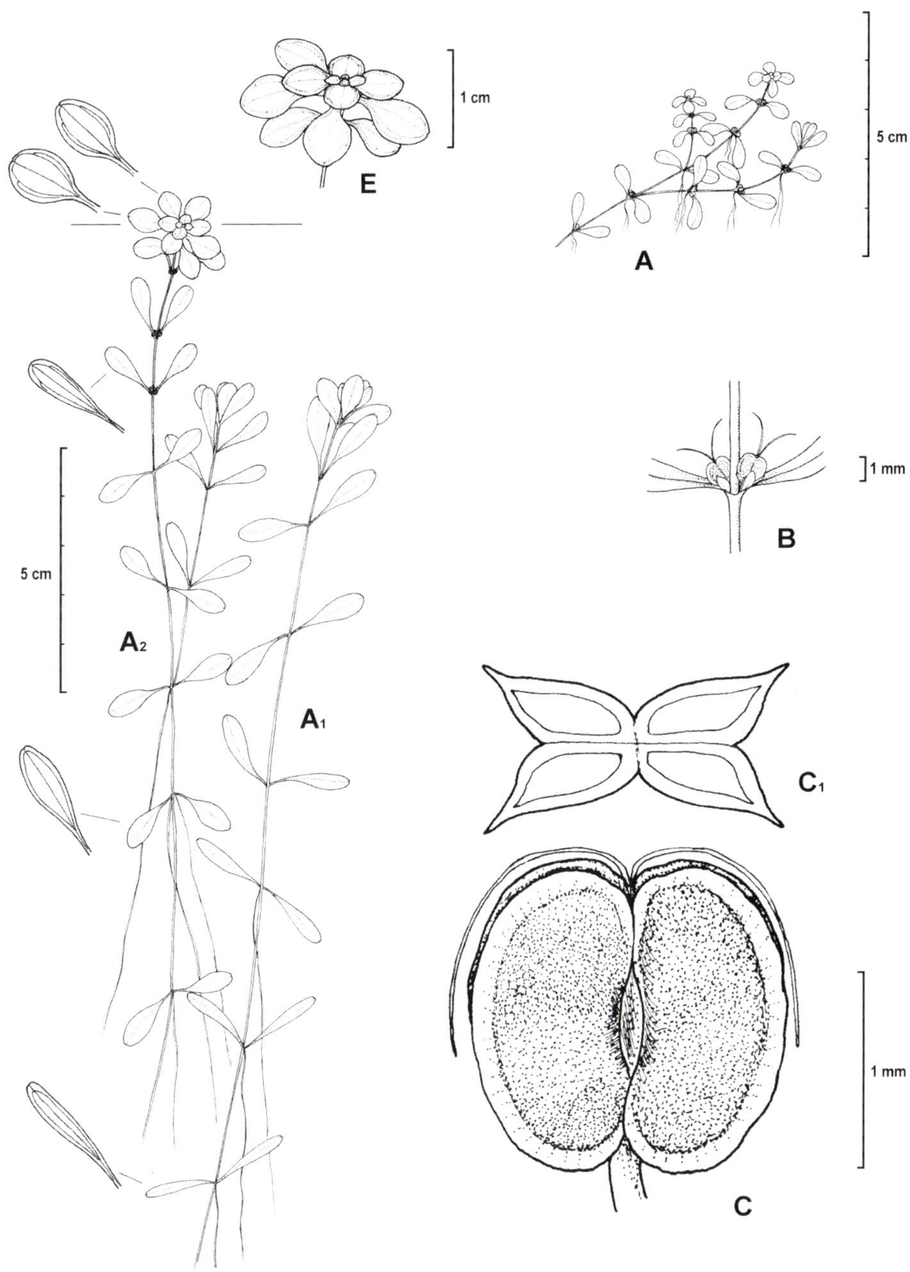

Callitriche stagnalis
A habit plant from moist substrate, **A1** habit submerged plant, **A2** habit plant with floating rosette, **B** immature fruits, **C** fruit, **C1** cross section of fruit, **E** rosette of floating leaves.

Plantaginaceae

293 *Hippuris vulgaris* L.

DK: Almindelig Hestehale **N:** Hesterumpe **S:** Hästsvans **FIN:** Lamparevesikuusi **IS:** Lófótur
GB: Mare's-tail **NL:** Lidsteng **F:** Pesse d'eau
D: Gemeiner Tannenwedel **CZ:** Prustka obecná **PL:** Przęstka pospolita
EST: Harilik kuuskhein **LV:** Parastā skujene **LT:** Paprastoji oodeguonė **Ru:** Хвостник обыкновенный

Flowering *H. vulgaris*. Helgenæs, Jutland, Denmark. Photo JCS.

Perennial hydrophyte or helophyte with richly branched, rigid, horizontal, brownish underground rhizome and unbranched, whitish roots from the nodes.

Stem: 3–6 mm in diameter, hollow or solid, unbranched, light green, the lower part commonly with a brownish to purple tinge. Terrestrial plants 30–80 cm long, ascending to erect, in submerged plants up to 2 metres long, ascending to erect or bent.

Leaves: Leaves sessile, in dense whorls, the upper with 8–12(–16) and the lower commonly with 4–8 leaves. Submerged plants with flaccid, linear leaves, 5–8 cm long and 1.5–2.5 mm wide, acute with entire margin, pale green to grey green or brownish, often with a purple tinge. Terrestrial plants and aerial shoots with rather stiff, narrowly lanceolate to linear, gradually tapering, almost needle-like leaves, 5–20 mm long and 0.5–3.0 mm wide, dark green. Intermediate leaves absent.

Inflorescence: Monoecious. Flowers solitary in leaf axils of aerial shoots, submerged shoots sterile.

Flowers: Perianth reduced to an almost invisible collar at the base of the flower. Flowers bisexual or unisexual, green to purplish, ovoid, c. 1 mm long. Stamen 1, anther purplish, filament usually longer than the mature anther. Style 1, with a single white, thread-like stigma.

Fruits: Nuts 1-seeded, ovoid and glabrous, 2–3 mm long. Seeds are ovoid elliptic and smooth.

Flowering: June-August.

Biology: Wind pollination. Bisexual flowers with stamens mature after the stigmas die back. Mature fruits detach and sink to the bottom, where the seeds eventually become exposed during winter as the surrounding tissue decays. Fruit set is commonly sparse, whereas vegetative propagation by prolific growth and branching of the rhizome is effective, commonly leading to formation of large, dense stands. Vegetative dispersal over distances by detached and readily rooting stems and rhizome fragments. The trigger for development of emergent shoots from submerged plants appears to be high light intensity in combination with temperatures above 10°C. Permanently submerged plants occur in flowing water, deep springs, gravel pits, etc. In springs the species may be wintergreen, whereas plants in other biotopes overwinter as rhizomes. Terrestrial stands occur on seasonally dry lake and pond sediments as well as permanently wet ground in meadows and marshes.

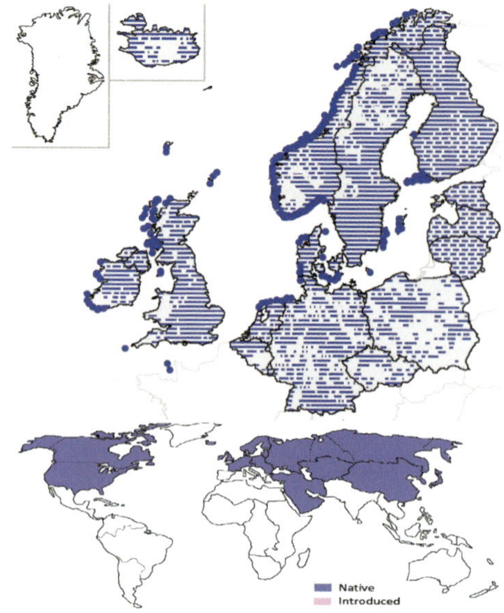

Habitats: Oligotrophic to eutrophic and neutral to alkaline waters – springs, streams, rivers, ditches and canals as well as permanent and temporary lakes and ponds, on mineral as well as soft, muddy substrates. Also in brackish water with low salinity and in meadows and marshes, inland as well as coastal.

Distribution within the region: Native. Widespread and abundant more or less throughout, but absent from Greenland.

Characteristics and similar species: With the dense whorls of 8–12(–16) linear and acute, entire leaves *Hippuris vulgaris* can hardly be confused with other aquatic plants, except for the two *Hippuris* taxa treated on the following pages.

294 *Hippuris tetraphylla* differs by (3–) 4(–5–8) obovate to oblanceolate, obtuse and slightly fleshy leaves in each of

Hippuris vulgaris. Whorls of leaves and flowers. Female flowers with white stigmas and dark red, not yet mature anthers. Photo JCS.

Hippuris vulgaris. Emergent plants close to the shore and submerged plants further out in the stream. River Vorgod Å, Jutland, Denmark. Photo BM.

Hippuris vulgaris. Submerged plants. Budberg, Germany. Foto KvdW.

the wide-spaced whorls and red to purple stems.

295 *Hippuris lanceolata* has (5–)6(–7) mostly linear, obtuse to subacute leaves in each whorl. Both species also have stamens with shorter filaments than in *H. vulgaris*.

Submerged leaves of *H. vulgaris* differ very little from submerged leaves of *H. lanceolata*, but are very different from the submerged leaves of *H. tetraphylla*. All three taxa display significant morphological plasticity, and since *H. lanceolata* appears able to backcross with its parent species, the number of leaves in the whorls as well as the leaf morphology and morphometry should be used with care for identification.

Submerged specimens may be confused with **226-231 *Myriophyllum*** spp., which have whorls of pinnate leaves, and **197-199 *Ceratophyllum*** spp., which have whorls of bi- or trifurcate leaves.

233 *Elatine alsinastrum* has different flowers and fruits.

57 *Hydrilla verticillata* and **53 *Egeria densa*** differ by toothed leaf margins.

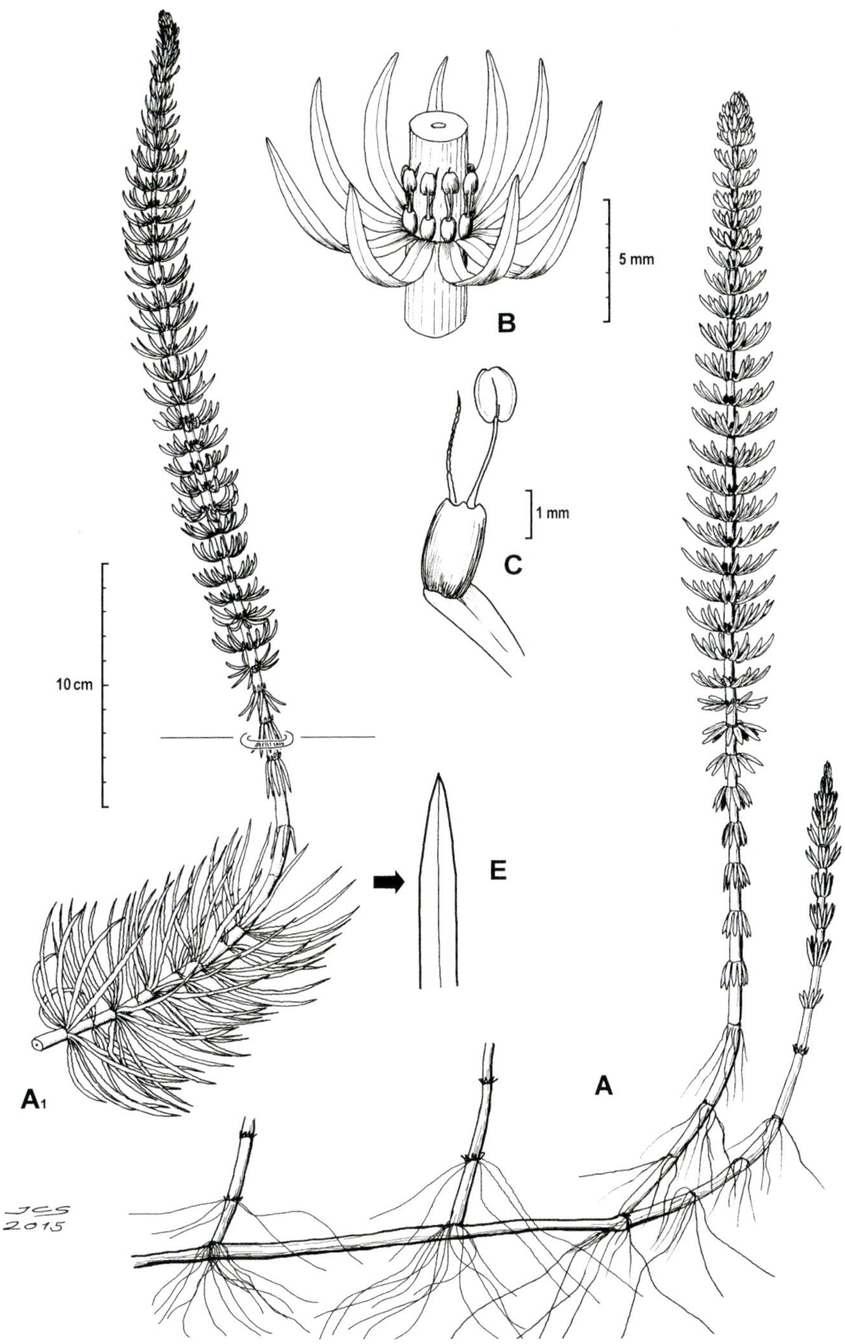

Hippuris vulgaris
A habit emerged plant, **A1** habit partly submerged plant, **B** whorl of leaves with flowers, **C** flower in female state - the anther immature and the filament not yet stretched, **E** leaf tip.

Plantaginaceae

294 *Hippuris tetraphylla* L.f.

DK: Bredbladet Hestehale **N:** Krosshesterumpe **S:** Ishavshästsvans **FIN:** Nelilehtivesikuusi
IS: Flæðalófótur
GB: - **NL:** - **F:** - **D:** - **CZ:** - **PL:** -
EST: -**LV:** - **LT:** - **Ru:** Хвостник четырёхлистный

Hippuris tetraphylla on dried-up substrate. Tauvonniemi, Siikajoki, Finland. Photo Antje Neumann, Biopix.com.

Perennial hydrophyte or helophyte with richly branched, rigid, horizontal rhizome on or below ground. Roots arising from the nodes, whitish, unbranched.

Stem: Ascending or more commonly erect stems, 10–40 cm long and 1.5–3.0 (–5.0) mm in diameter, unbranched, usually distinctly red to purple.

Leaves: Leaves sessile, obovate to broadly lanceolate, in dense to widely spaced whorls of 4, occasionally 3 or 5, or rarely up to 8, subacute with entire margin, 10–15 mm long and 3–5 mm wide, green to dark green, slightly fleshy. Submerged leaves flaccid, linear to lanceolate, blunt, 20–30 mm long, 2–3 mm wide.

Inflorescence: Monoecious. Flowers solitary in leaf axils of aerial shoots, submerged shoots sterile.

Flowers: Perianth reduced to an almost invisible collar or rim at the base of the flower. Flowers dark purple, oval, mostly bisexual. The elongate and slender style as long as or slightly longer than the anther. Stamen 1, anther purplish, filament usually shorter than anther. Style 1, with a single thread-like, white stigma.

Fruits: Nuts 1-seeded, oblong, glabrous, c. 2 mm long. Seeds ovoid elliptic and smooth.

Flowering: June-August.

Biology: Wind pollinated. Seed dispersal by water and birds. Vegetative propagation by growth and branching of the rhizome. Vegetative dispersal occurs over longer distances by detached, rooting stems and rhizome fragments. Growing most commonly in shallow water, producing erect aerial shoots from permanently submerged rhizomes.

Habitats: Coastal brackish waters, tidal flats and salt marshes with shallow water and commonly soft, more or less muddy substrate. Also tidal flats in river mouths.

Distribution within the region: Native. Scattered in Iceland, northern Norway and Sweden, frequent along the coast of Finland.

Characteristics and similar species: *Hippuris tetraphylla* is characterised by 3–5 mm diameter, terete, normally reddish to purplish, emergent stems with whorls of typically 4, rarely 3 or 5 or more, obovate, fleshy leaves and can hardly be confused with other aquatic plants, except the two other *Hippuris* taxa.

Emergent **293** ***H. vulgaris*** has 8–12(–20) linear, acute, 0.5–3.0 mm wide leaves in dense whorls, while emergent **295** ***H. lanceolata*** has whorls of commonly 6, rarely 5 or 7, lanceolate, subacute leaves, 1–2 mm wide.

Hippuris tetraphylla. Whorls of leaves and flowers. Hällängesviken, East of Ostnäs, Sweden. Photo RVL.

Submerged *H. tetraphylla*. Note the blunt leaves in whorls of 3 or 4. Hailuoto, Finland. Photos KvdW.

The relative length of the anther and filament can be useful to differentiate *H. tetraphylla* from *H. vulgaris*.

Hippuris tetraphylla is an almost strictly brackish water species, while *H. vulgaris* occur mainly in freshwater and less commonly in brackish water with low salinity.

Hippuris tetraphylla in 0-5 cm water over thin mud. Hällängesviken, East of Ostnäs, Sweden. Photo RVL.

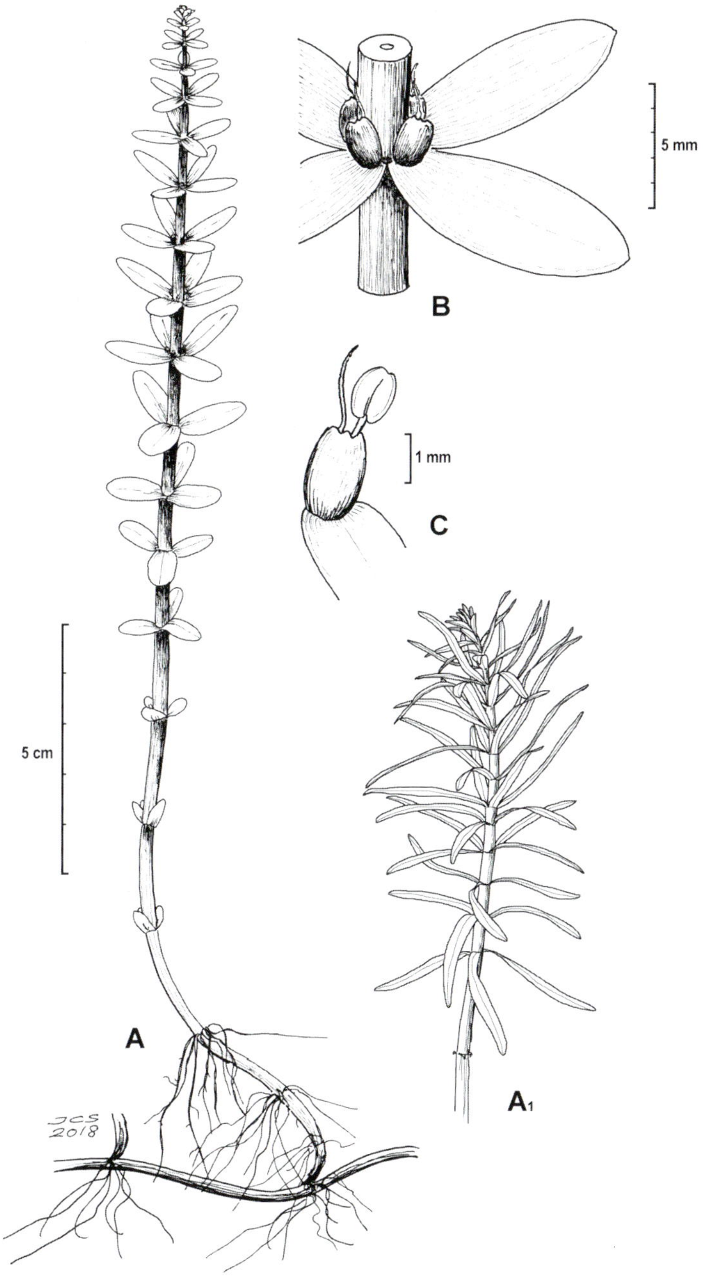

Hippuris tetraphylla
A habit partly emerged plant, **A1** habit submerged plant,
B whorl of leaves with flowers, **C** flower with mature anther.

Plantaginaceae

295 *Hippuris lanceolata* Retz.

DK: Lancetbladet Hestehale **N:** Brakkhesterumpe **S:** Mellanhästsvans **FIN:** Rannikkovesikuusi
IS: Lensulófótur
GB: - **NL:** - **F:** - **D:** - **CZ:** - **PL:** -
EST: Süstlehine kuuskhein **LV:** - **LT:** - **Ru:** Хвостник ланцетный

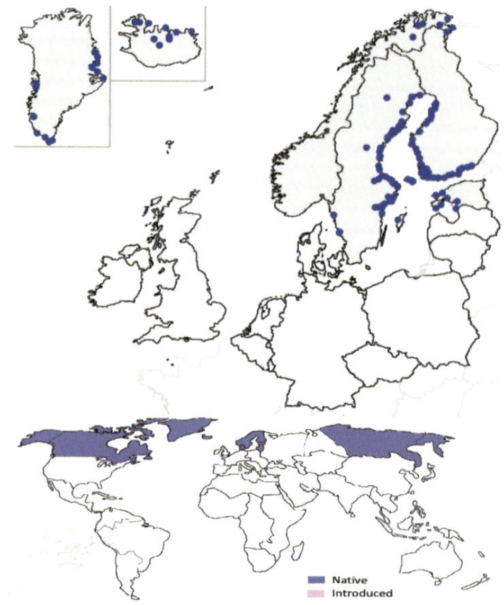

Hippuris lanceolata. Note the submerged leaves. Store Saltsö, Kangerlussuaq, Greenland. Photo Jan-Thomas Johansson.

Perennial hydrophyte or helophyte with richly branched, rigid, horizontal, brownish rhizome on or below ground. Roots unbranched, whitish from the nodes.

Stem: Ascending to erect, 5–50 cm long, hollow or solid, unbranched, light green, often with a brownish or reddish tinge.

Leaves: Sessile, in dense to fairly widely spaced whorls with (5–)6(–7) leaves. Submerged leaves flaccid, linear, (0.5–) 2–8 cm long and 0.6–1.3(–4.0) mm wide, subacute with entire margin, pale green, grey green or brownish, often with a purple tinge. Terrestrial plants and aerial shoots with rather stiff, narrowly lanceolate to linear, gradually tapering, subacute leaves, broadest above the middle, 0.5–1.0 cm long and 1–2 mm wide, green to dark green. Intermediate leaves absent.

Inflorescence: Monoecious. Flowers solitary in leaf axils of aerial shoots – submerged shoots sterile.

Flowers: Perianth reduced to an almost invisible collar or rim at the base of the flower. Flowers bisexual, dark purple, ovoid. Stamen 1, anther purplish, filament usually shorter than anther. Style 1, with a single thread-like, white stigma.

Fruits: 1-seeded oblong and glabrous achenes (nuts), 1.8–2.0 mm long and 1.0–1.2 mm wide. Seeds are oval elliptic and smooth.

Flowering: June-August.

Biology: Wind pollinated. Seeds float and are dispersed by water and birds. Vegetative propagation by prolific growth and branching of the rhizome is effective, commonly leading to formation of large, dense stands. Vegetative dispersal occurs over long distances by detached and readily rooting stem and rhizome fragments.

Habitats: Oligotrophic to eutrophic, slightly acid to alkaline waters, shallow lakes, ponds, slow-flowing streams, brackish water with low salinity, mainly soft and muddy substrates. Growth is typically better when the rhizomes are permanently submerged. In southern part of the region it mainly occurs in brackish waters.

Distribution within the region: Native. Scattered to frequent along the coasts of the Gulf of Bothnia, Estonia, northern Norway, Iceland and Greenland.

Characteristics and similar species: With the fairly dense whorls of usually 6, rarely 5 or 7, lanceolate to linear, subacute, entire leaves *Hippuris lanceolata* can hardly be confused with other species of aquatic plants, except the two other *Hippuris* taxa. **293 *Hippuris vulgaris*** has 8–12(–20) linear, acute, almost needle-like leaves in each of the dense whorls. **294 *Hippuris tetraphylla*** usually has 4, rarely 3 or 5(–8), obovate to oblanceolate, obtuse, broad, slightly fleshy leaves in each of the widely spaced whorls. Submerged leaves of *H. lanceolata* differ little from those of

Submerged *H. lanceolata*. Sandviken, Kalix, Sweden. Photo KvdW.

Hippuris lanceolata. Fruits. Sandviken, Kalix, Sweden. Photo RVL.

Hippuris lanceolata. Emerged shoots. Sandviken, Kalix, Sweden. Photo RVL.

H. vulgaris, but are very different from those of *H. tetraphylla*.

All three taxa display significant morphological plasticity and therefore the number of leaves in whorls, leaf morphology and leaf size should be used with care for identification. The relative length of the anthers and filaments can be of some use to distinguish *H. lanceolata* from *H. vulgaris*.

H. tetraphylla may be confused with **226-231 *Myriophyllum*** spp., which have whorls of pinnate laves, and **197-199 *Ceratophyllum*** spp., which have whorls of forked or trifurcate leaves.

Note: *H. lanceolata* is often treated as a hybrid between *H. vulgaris* and *H. tetraphylla*. However, because of its wide range including, large areas where neither of the parent species occurs, and the rich fruit set, we consider it reasonable to treat it here as a species.

Hippuris lanceolata. Leaf whorls with flowers. Store Saltsö, Kangerlussuaq, Greenland. Photo Jan-Thomas Johansson.

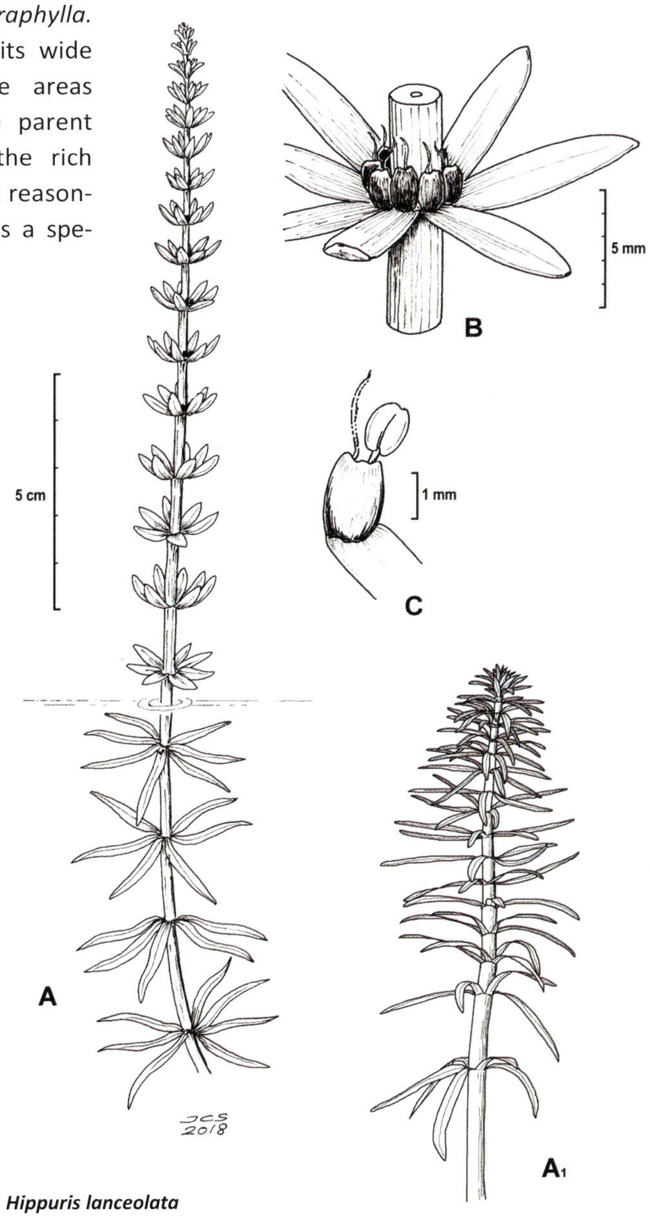

Hippuris lanceolata
A habit partly emerged plant, **A1** habit submerged plant, **B** whorl of leaves with flowers, **C** flower.

Plantaginaceae

296 *Littorella uniflora* (L.) Asch.

Syn.: *Plantago uniflora* L.
DK: Strandbo **N:** Tjønngras **S:** Strandpryl **FIN:** Raani **IS:** Tjarnalaukur
GB: Shoreweed **NL:** Oeverkruid **F:** Littorelle uniflore
D: Strandling **CZ:** Pobřežnice jednokvětá **PL:** Brzeżyca jednokwiatowa
EST: Silmjärvikas **LV:** Vienzieda krastene **LT:** Ežerinė kraštūnė **Ru:** Прибрежница одноцветковая

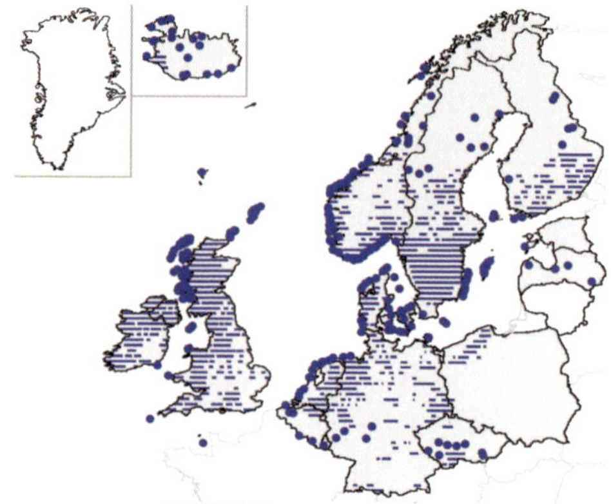

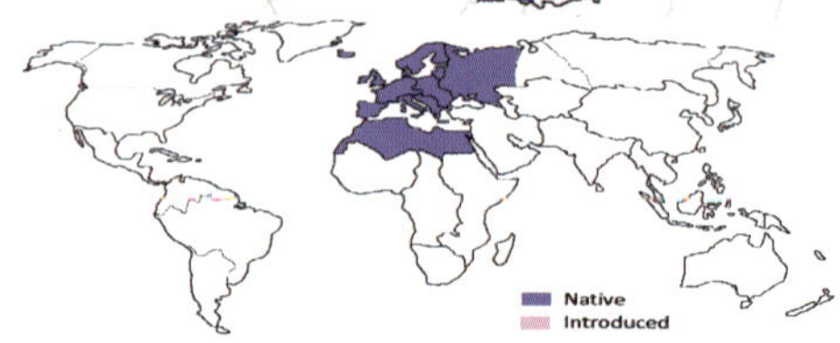

Flowering *L. uniflora,* drawdown zone of a lake. Etang de Bellebouche, Parc Naturel Régional de la Brenne, Indre, France. Photo RVL.

Perennial hydrophyte or temporary helophyte, with 1–6 cm long, thin, creeping stolons. Rhizome absent. Roots numerous, unbranched, white.

Stem: Without erect stem, but with a short, 4–5 mm in diameter, compressed, rhizome-like vertical structure.

Leaves: Leaves in rosettes on top of the rhizome-like structure. Submerged plants with straight, rigid, terete, fleshy, acuminate leaves with numerous lacunal channels, 4–12 cm long and 3–5 mm in diameter, glabrous, without stomata, green to dark green. Emergent plants with spreading, semi-cylindrical leaves with upward-curved, acute tips, the upper side flat with a longitudinal furrow, 2.5–4.0(–10) cm long, 2–3 mm wide, margins hairy, dark green with a brown or purplish tinge.

Inflorescence: Monoecious. Flowers in 1–8, few-flowered spikes from leaf axils. Each spike with 1 male flower and typically 2, rarely up to 5 female flowers.

Flowers: 4-merous, unisexual. Male flower on a pedicel 1–5(–9) cm long, with an acute, ovate bract shortly below the middle. Sepals 4, fused at the base, 4–5 mm long, keeled, green with a brownish tinge. Corolla whitish, membranous and translucent with a 4 mm long tube and 4 obtuse, 2 mm long lobes. Stamens 4, exserted on filaments up to 2 cm long, anthers yellow, 3 mm long. Female flowers sessile, mostly 2 together at base of the spike, each with a thin, lanceolate, 1-veined bract. Sepals 2–4, 3 mm long and similar to the bract. Corolla whitish, tubular, 4-lobed. Style up to 1.2 cm long, filamentous and with numerous papillae.

Fruits: Nuts, 1.5–2.5 mm long, hard, yellow brown.

Flowering: June-September.

Biology: Only terrestrial and partly submerged plants in very shallow water produce flowers. Protogynous. Each plant may produce up to 20 nuts in a growing season, and these remain viable for up to 30 years. Dispersal by water and birds. Submerged plants have prolific vegetative propagation by stolons, often forming very dense, monospecific stands. It grows from above the summer water line to a depth of up to 3 m. The ability to develop a terrestrial form enables it to withstand even prolonged air exposure during drought and low water levels. When submerged plants become exposed, the submerged leaves eventually die back, while terrestrial leaves are formed, and vice versa when terrestrial plants be-

Submerged *L. uniflora* with leaves terete, up to 5 mm in diameter, each with numerous lacunal channels (see insert). Lake Madum Sø, Himmerland, Denmark. Photo JCS.

Recently exposed *L. uniflora* in the littoral zone of a small pond in a dune landscape near Vangså, Thy, Denmark. The erect, thick, terete submerged leaves without stomata are dying back while linear aerial leaves with epidermis and stomata are emerging, enabling the plant to withstand dry conditions. Photo JCS.

come submerged. Submerged plants overwinter with green leaves. The numerous roots provide uptake of CO_2 from the sediment while the lacunal channels conduct the CO_2 to the photosynthetic tissues in the leaves. *L. uniflora* has the capacity to carry out some form of C4 or CAM photosynthesis, which allows it to survive in waters where inorganic carbon is the limiting factor.

Habitats: Mostly in soft, oligotrophic to mesotrophic, clear, moderately acidic or alkaline water. On sandy and gravelly substrate, rarely with organic matter. Lakes, reservoirs, ponds and temporary water bodies in dunes, rarely in rivers.

Distribution within the region: Native. Rather rare to rare in Netherlands, Belgium, Luxembourg, Germany, Czech Republic, Poland and the Baltic states, scattered to common in Scandinavia, Finland, Iceland and British Isles. Declining in lowland areas.

Characteristics and similar species: Submerged plants differ from other isoetids in oligotrophic waters by the thin, rosette-forming runners. The numerous lacunal channels distinguish *L. uniflora* from **359-360** ***Isoetes*** spp., which have leaves with 4 lacunal channels and macrospores at the leaf bases. **334** ***Lobelia dortmanna*** has leaves with two large lacunal channels. **254** ***Subularia aquatica*** differs in having long, tapering, linear leaves, submerged flowers and no runners.

Emergent *L. uniflora* can be recognised by the rosettes of spreading semi-cylindrical leaves with furrowed upper side and hairy edges.

Flowering *L. uniflora* on dry lake sediment in Råbjerg Mile, Denmark. Note the 4 stamens with very long filaments protruding from the male flower as well as the whitish, filamentous stigma (white arrow). Note also the hairy edge of the aerial leaf to the left of the flower. Photo JCS.

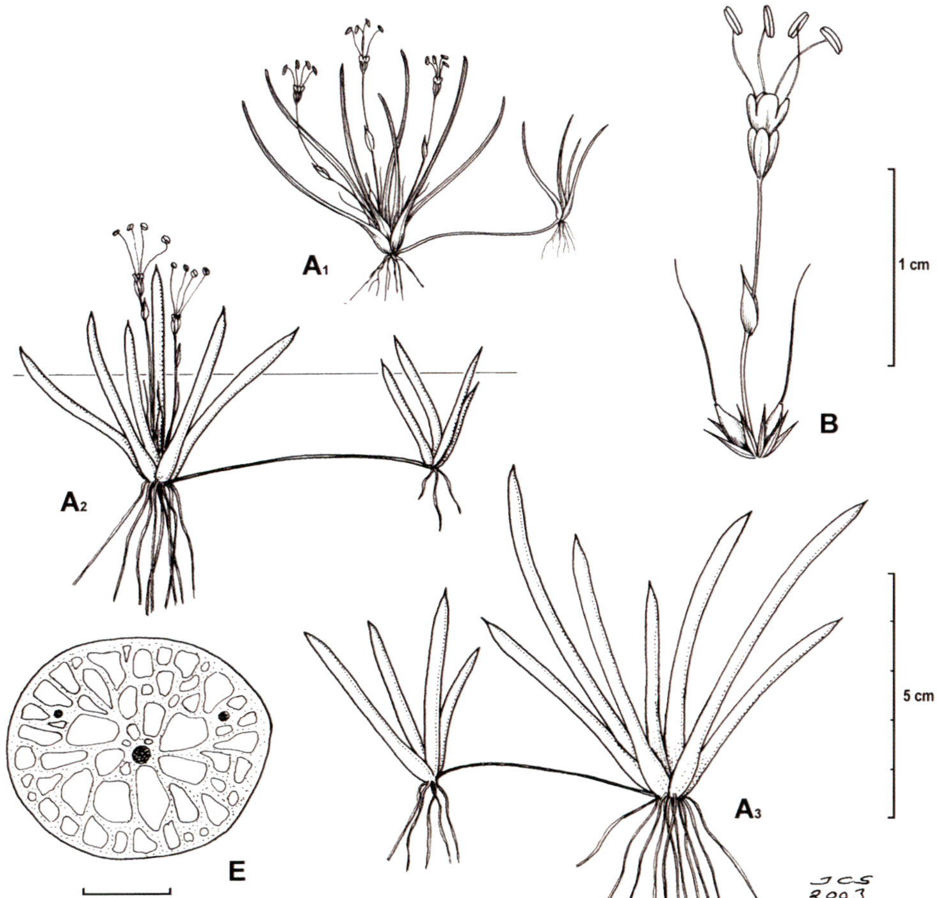

Littorella uniflora
A1 habit emerged plant, **A2** habit partly submerged plant, **A3** habit submerged plant, **B** spike with 1 male and 2 female flowers, **E** leaf cross section of submerged plant.

Plantaginaceae

297 *Veronica anagallis-aquatica* L.

Syn.: *V. anagallis* auct.; *V. aquatica* auct.
DK: Lancetbladet Ærenpris **N:** Vassveronika **S:** Vattenveronika **FIN:** Konnantädyke **IS:** Laugadepla
GB: Blue Water-speedwell **NL:** Blauwe waterereprijs **F:** Véronique mouron d'eau
D: Gauchheil-Ehrenpreis **CZ:** Rozrazil drchničkovitý **PL:** Przetacznik bobownik
EST: Allikmailane **LV:** Upmalas veronika **LT:** Šaltininė veronika **Ru:** Вероника ключевая

Veronica anagallis-aquatica in a shallow stream. Ireviken, Gotland, Sweden. Photo JCS.

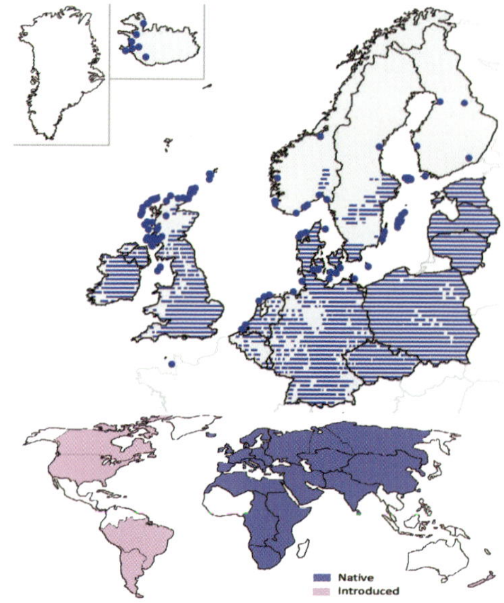

Annual or perennial helophyte or hydrophyte with vertical rhizome rooting from the nodes.

Stem: Bluntly angular, glabrous or rarely basal parts densely pubescent, (10–)30–60(–100) cm long, 3–15 mm in diameter, pale green to green, erect to ascending, prostrate to ascending in submerged plants. With prostrate, stolon-like branches from the lower stem nodes, early in the growing season.

Leaves: Decussate, basal leaves often shortly petiolate, middle and upper leaves sessile, clasping, 5–12(–15) cm long and 1.5–3(–5) cm wide, widest at or above the middle, oblong to lanceolate, acute, serrulate in terrestrial plants. Broadly lanceolate, acute to obtuse with crenate margins, thin, soft and flaccid, pale green to green in submerged plants, often with a brownish tinge.

Inflorescence: Monoecious. Racemes with 10–40 flowers, glabrous or sometimes with glandular hairs, emerging mainly from the upper leaf axils, eventually elongating during flowering. Bracts oblong lanceolate, acute, c. 5 mm long, shorter than pedicel during flowering. Pedicel 4–7 mm long, often slightly curved, protruding from rachis at an acute angle in fruit.

Flowers: 4-merous, bisexual. Calyx divided into 4, ovate-oblong, acute lobes. Corolla (5–)6–7(–8) mm in diameter, with 4 widely spreading lobes, pale violet to blue or rarely white or pinkish, with darker, reddish-violet stripes. Stamens 2, exserted. Style 1, 1.5–2.0 mm long.

Fruits: Capsules ovate, 2.5–4.0 mm long and 2–3 mm wide, distinctly shorter than sepals. Seeds numerous, 0.6–0.7 mm long, oval.

Flowering: June-August.

Biology: Insect pollinated; submerged flowers remain closed and self-fertilise. Seed dispersed by water and animals; vegetative dispersal by detached stem fragments. Aerial shoots usually die back after flowering, while submerged parts and permanently submerged plants remain green and continue growth during the cold months.

Habitats: In mesotrophic to eutrophic, slightly acid to alkaline, mainly flowing water. In springs, streams, rivers, canals and ditches, less often in standing water.

Distribution within the region:
Native. Widespread and abundant throughout most of the region but scarce north of southernmost Sweden.

Veronica anagallis-aquatica. Many-flowered racemes with light blue flowers and fruits with pedicels in an acute angle to the rachis are characteristic. Horsens, Denmark. Photo JCS.

Submerged *V. anagallis-aquatica*. Daugbjerg, Denmark. Photo BM.

Veronica anagallis-aquatica. Young, partly submerged leaves on short shoots in shallow water, early season in a spring. Ravnkilde, Denmark. Photo JCS.

Characteristics and similar species:
Submerged *Veronica anagallis-aquatica* and **300 *Veronica catenata*** cannot be distinguished from each other with certainty, but the venation of the leaf base appears different - see note under *V. catenata* and fig. E1. Even when emergent with flowers and fruits, determination of the two species can be difficult. *V. catenata* only has sessile leaves and does not develop basal stolon-like branches. Fertile specimens of *V. catenata* have white or pink flowers with purple stripes. When fruiting the pedicels are held at a right or obtuse angle to the rachis. The length of mature fruits exceeds the length of the sepals.

The hybrid between these two species, **298 *V. ×lackschewitzii***, can be very abundant and difficult to distinguish from the parents.

Young specimens of *V. anagallis-aquatica* may resemble **241 *Epilobium*** spp., which also has opposite leaves, but with different-shaped laminas. Non-flowering **275-280 *Myosotis*** spp. differ in alternate leaves.

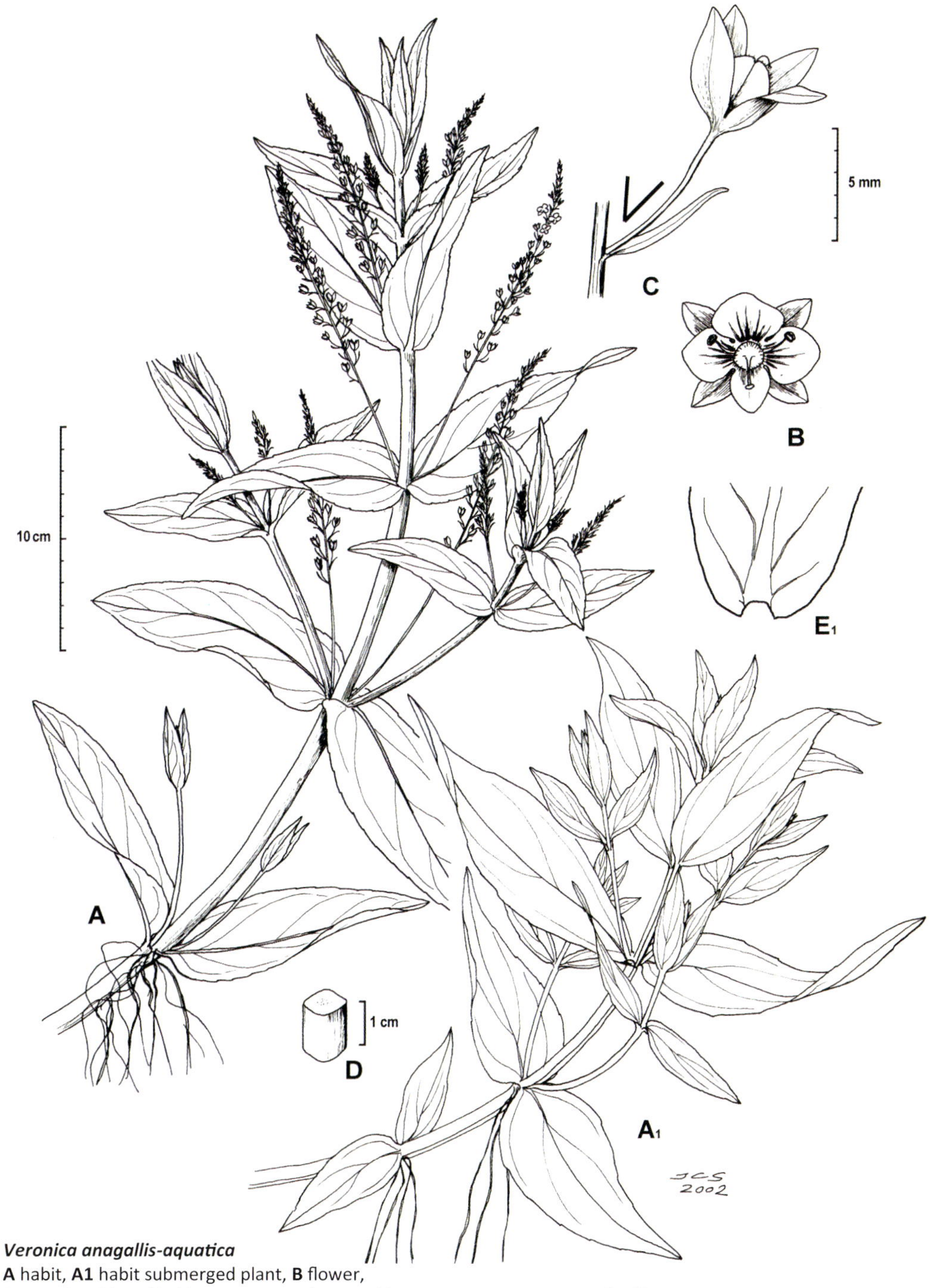

Veronica anagallis-aquatica
A habit, **A1** habit submerged plant, **B** flower,
C capsule with pedicel in an acute angle to rachis, **D** stem cross section, **E1** leaf base.

Plantaginaceae

298 ***Veronica ×lackschewitzii*** J.B.Keller

(*V. anagallis-aquatica × catenata*)

Veronica ×lackschewitzii. River Itchen, Hampshire, Britain. Photo KvdW.

Veronica ×lackschewitzii. River Avon, Hampshire, Britain. Photo RVL.

The hybrid is typically very similar to **297 *V. anagallis-aquatica***; some populations are almost completely sterile (95%) and can be recognised by the many undeveloped capsules. However, other populations, particularly in southern Britain appear to be highly fertile; they are most easily recognised by characters such as the variable flower colour and angle of the pedicels to the rachis.

The plants are very vigorous and continue to grow until late autumn, reaching a stem length of up to 300 cm.

Non-flowering, submerged plants cannot be distinguished from the parental species.

The abundance of this hybrid varies between regions. In some countries it seems to be rare or is maybe overlooked. In some British river systems it is very frequent and may almost completely replace the parent species.

Submerged *V. ×lackschewitzii.* River Itchen, Hampshire, Britain. Photo KvdW.

Plantaginaceae

299 *Veronica anagalloides* L.

DK: - **N:** - **S:** - **FIN:** - **IS:** -
GB: Faded Speedwell **NL:** - **F:** Véronique faux Mouron
D: Schlammehrenpreis **CZ:** Rozrazil bažinný **PL:** Przetacznik mułowy
EST: -**LV:** - **LT:** - **Ru:** Вероника ложноключевая

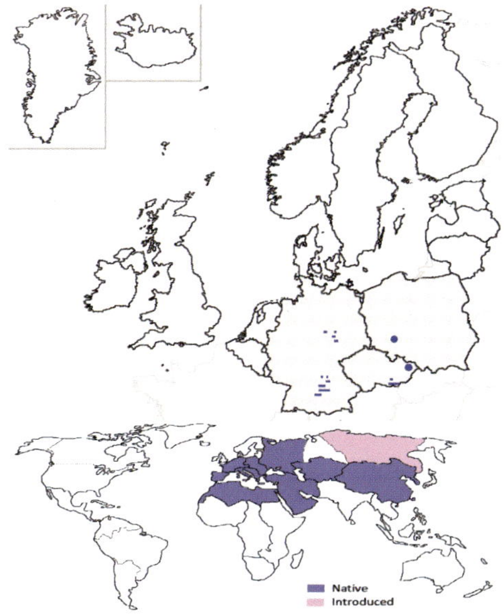

Flowering *V. anagalloides*. Floodplain of River Saale, Halle, Germany. Photo KvdW.

Annual helophyte.

Stem: Slender, angular, erect, 10–30 (–60) cm long.

Leaves: Decussate or sometimes in whorls of 3–4, sessile and clasping, linear to lanceolate, 4–10 mm wide, about 4–5 times as long as wide.

Inflorescence: Monoecious. Petiolate racemes with 10–30 flowers, covered in glandular hairs, emerging mainly from the upper leaf axils. Bracts linear to lanceolate, as long as or shorter than pedicels during flowering.

Flowers: 4-merous, bisexual. Calyx divided into 4 oblong oval lobes. Corolla 3–4 mm diameter, whitish, bluish violet or with differently coloured lobes. Stamens 2, exserted. Style 1, (0.6–)0.8–1.1 (–1.5) mm long. Pedicels 3–4 mm long, straight, glandular hairy, when fruiting held more or less at a right angle to the rachis.

Fruits: Capsules obovoid, c. 3 mm long and 2 mm wide, at maturity distinctly longer than the sepals, glandular hairy. Seeds oval, c. 0.5 mm long.

Flowering: May-October.

Biology: Insect pollinated.

Habitats: In mesotrophic to eutrophic, slightly acidic to alkaline, mainly standing water. On wet or temporarily flooded, muddy substrate, in ditches, riverbanks, wet meadows and fields.

Distribution within the region: Native. Rare in Germany, Poland and Czech Republic.

Characteristics and similar species: *V. anagalloides* differs from similar species by the small capsules which are distinctly longer than wide.

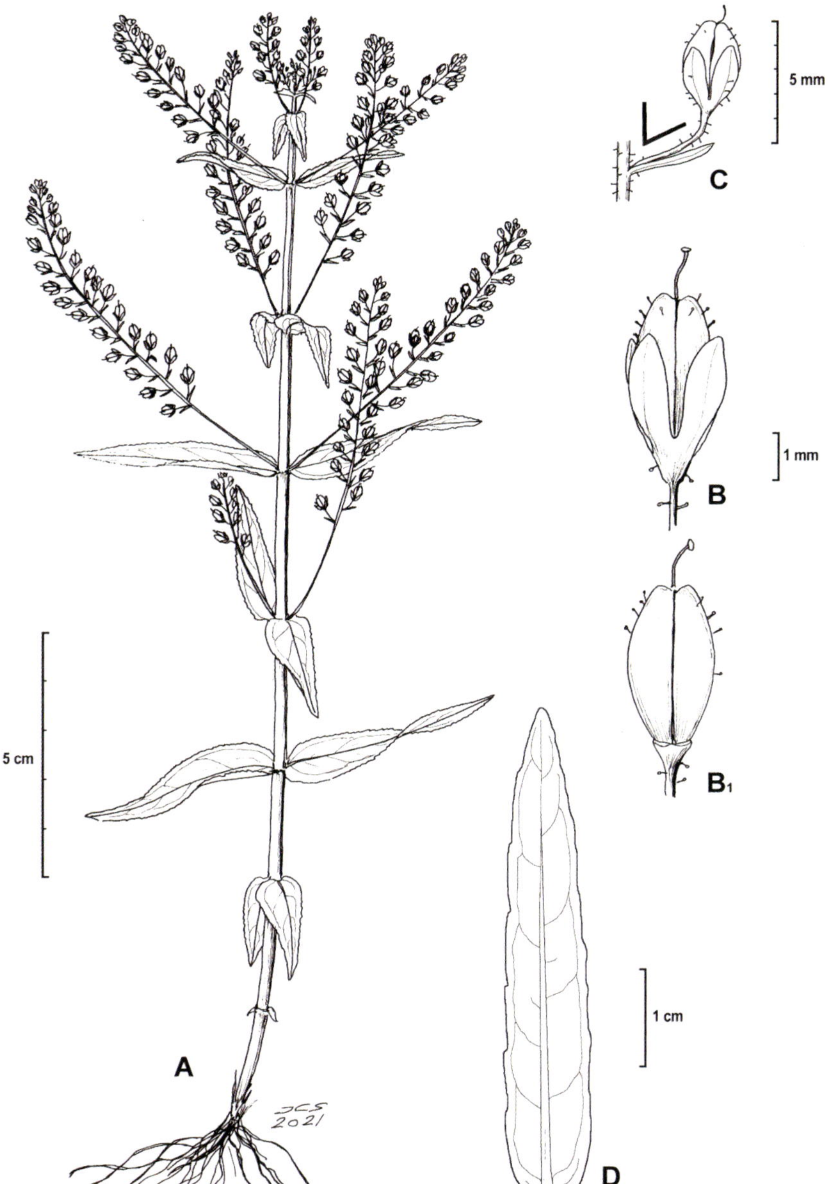

Veronica anagalloides
A habit, **B** capsule, **B1** capsule - calyx removed, **C** capsule with pedicel at a right angle to rachis, **D** leaf.

Plantaginaceae

300 *Veronica catenata* Pennell

Syn.: *V. aquatica* Bernh.; *V. comosa* auct.
DK: Vand-Ærenpris **N:** Dikeveronika **S:** Dikesveronika **FIN:** Vesitädyke **IS:** -
GB: Pink Water-speedwell **NL:** Rode waterereprijs **F:** Véronique aquatique
D: Blasser Gauchheil-Ehrenpreis **CZ:** Rozrazil pobřežní **PL:** Przetacznik wodny
EST: Vesimailane **LV:** Ūdeņu veronika **LT:** Vandeninė veronika **Ru:** Вероника водная

Partly submerged, flowering *V. catenata*. Middelharnis, Netherlands. Photo KvdW.

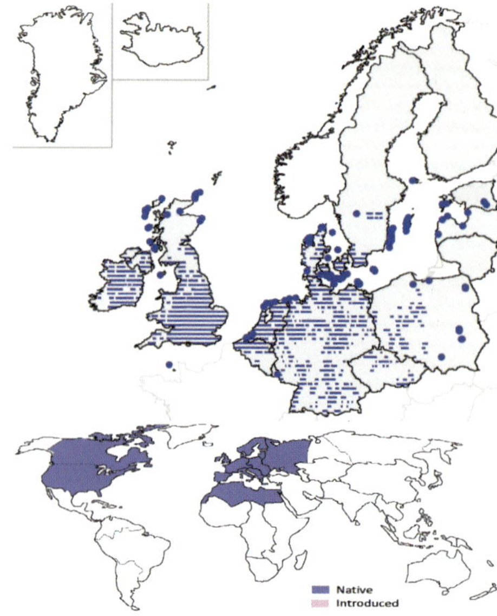

Annual or sometimes perennial, helophyte or hydrophyte with vertical rhizome and numerous adventitious roots from the stem nodes. Without stolon-like branches from the lower stem nodes.

Stem: Usually glabrous, rarely hairy, almost terete or bluntly angular. Terrestrial plants on wet soil and aerial shoots from submerged plants with erect or ascending stem, (10–)30–60(–100) cm long, 4–15 mm in diameter, green, often with a brownish or reddish tinge. Submerged plants with stem 10–30(–50) cm long, ascending to erect and pale green to green.

Leaves: Decussate, sessile and clasping. Leaves of terrestrial plants 2–9 cm long, 0.5–1.5 cm wide, widest below the middle, lanceolate, acute, remotely serrulate, glabrous, slightly glossy, green or with reddish tinge, especially on the underside. Leaves of submerged plants 5–10 cm long, up to 2 cm wide, lanceolate, acute to obtuse, margins crenate, thin, soft and limb pale green to yellow green, often with a light-brownish tinge.

Inflorescence: Monoecious. Petiolate racemes with (10–)20(–30) flowers, often more or less covered with glandular hairs, emerging mainly from the upper leaf axils, eventually elongating during flowering. Bracts oblong lanceolate, obtuse to acute, c. 5 mm long, longer than pedicels during flowering. Pedicels 3–6 mm long, straight, held more or less at a right angle to the rachis when fruiting.

Flowers: 4-merous, bisexual. Calyx divided into 4, oblong oval, acute lobes. Corolla 3.5–5.5 mm in diameter, with 4 widely spread lobes, white, pinkish or pale violet with reddish or purplish stripes. Stamens 2, exserted. Style 1, 1.3–1.5(–2.0) mm long.

Fruits: Capsules more or less circular, 3.0–3.5 mm long and 3–4 mm wide, often emarginate at apex, distinctly longer than sepals at maturity. Seeds numerous, 0.6–0.7 mm long, ovoid.

Flowering: June-August.

Biology: Insect pollinated, submerged flowers self-fertilising. Seed dispersal by water and animals. The seeds germinate on wet substrate in spring, while plants typically die back late in the growing season. Vegetative dispersal by detached stem fragments. Submerged plants often develop flowering aerial shoots which die back after flowering, while the submerged parts remain green.

Habitats: In oligotrophic to eutrophic, slightly acidic to alkaline, mainly still water. In and by ponds and lakes, commonly on muddy soil, on the margins of slow-flowing streams, rivers, canals and ditches as well as wet meadows and marshes. Permanently submerged to a depth of 4 m in lakes.

Distribution within the region: Native. Widespread and abundant in the west, from Ireland east into western Poland, scattered or absent in the northern and eastern part of the region.

Characteristics and similar species: Submerged *Veronica catenata* is very similar to and cannot reliably be distin-

Veronica catenata. Capsules. Selbjerg Vejle, Thy, Denmark. Photo JCS.

Presumed *V. catenata* permanently submerged. Sand pit, Krefeld, Germany. Photo KvdW.

Submerged *V. catenata*. Stavtrup, Aarhus, Denmark. Photo BM.

guished from submerged plants of **297 *V. anagallis-aquatica***. Even when emergent with flowers and fruits, identification of the two species can cause problems. *V. anagallis-aquatica* has oval capsules which are longer than wide, but shorter than the sepals, its pedicels held at an acute angle to the rachis, and it rarely has pink flowers. It usually forms vegetative stolon-like branches from the lower leaf axils. The lower leaves are usually short pedicellate.

Young terrestrial *V. catenata* plants may be confused with species of **241 *Epilobium*** spp., which also have opposite leaves but with laminas of a different shape. Non-flowering **275-280 *Myosotis*** spp. differ in leaves alternate.

Note: An additional vegetative character is given by Foerster (1967) but should be used in conjunction with other characters - especially when referring to submerged plants.

***V. catenata*,** fig. E1: Mid-vein in leaves not fused with the lower side veins toward the base, therefore only slightly widened and much narrower than the stem.

V. anagallis aquatica, fig. E1, p. 629: Mid-vein in leaves fused with the lower side veins toward base, therefore much widened and almost as wide as stem.

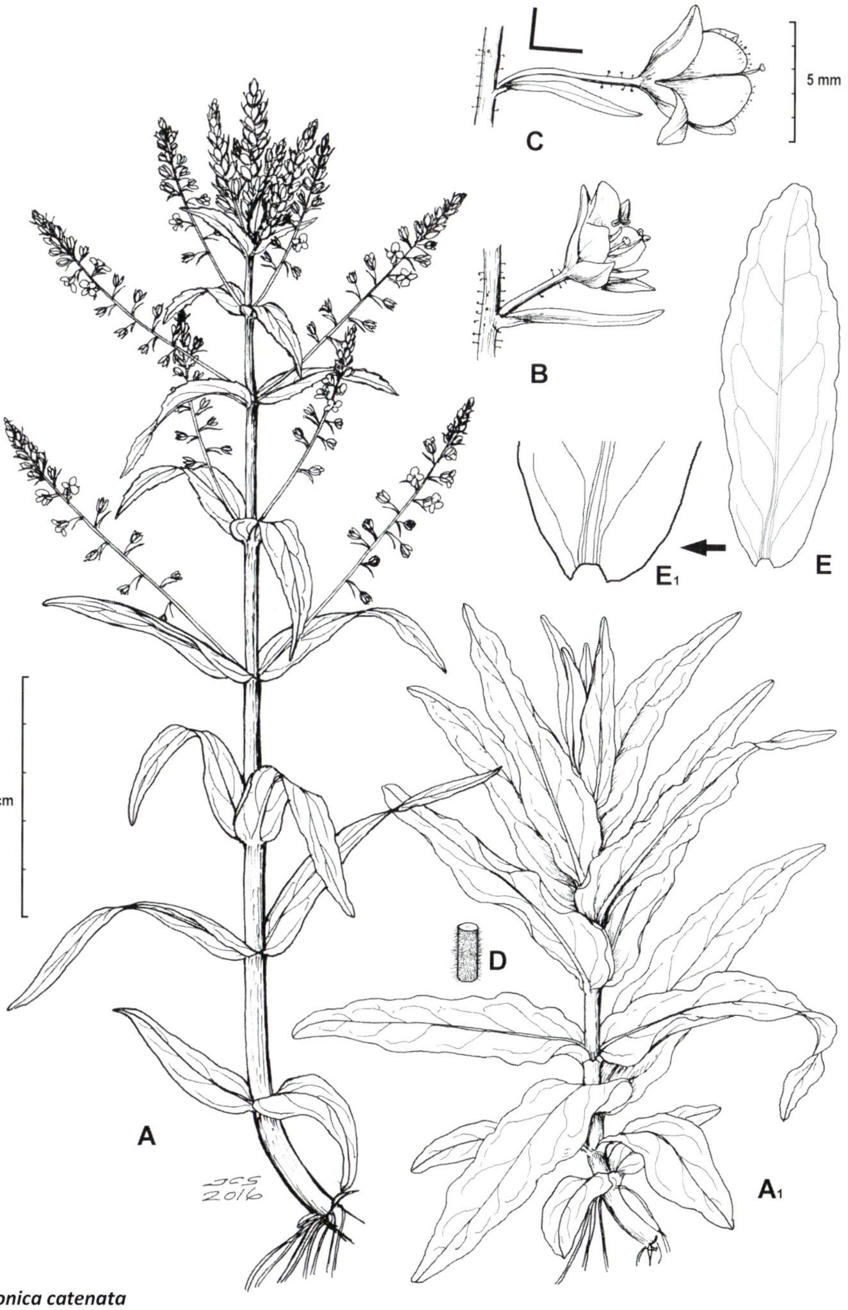

Veronica catenata
A habit, **A1** habit submerged plant, **B** flower, **C** capsule with pedicel at a right angle to rachis, **D** stem - variation with numerous hairs, **E** submerged leaf, **E1** leaf base.

Plantaginaceae

301 *Veronica beccabunga* L.

DK: Tykbladet Ærenpris **N:** Bekkeveronika **S:** Bäckveronika **FIN:** Ojatädyke **IS:** -
GB: Brooklime **NL:** Beekpunge **F:** Cresson de cheval
D: Bachbungen-Ehrenpreis **CZ:** Rozrazil potoční **PL:** Przetacznik bobowniczek
EST: Ojamailane **LV:** Avotu veronika **LT:** Upelinė veronika **Ru:** Вероника поточная

Even without flowers or fruits *V. beccabunga* is easily recognised by the fleshy, ovate-oblong leaves. Ditch, Nettetal, Germany. Photo KvdW.

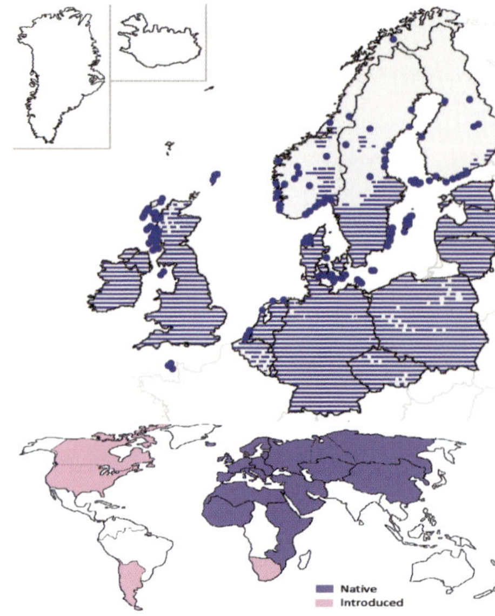

Perennial helophyte with stolons above and below the ground. Rhizome vertical, rooting from the nodes.

Stem: Prostrate or ascending from the stolons, rooting at nodes, 15–50(–100) cm long, 2–5 mm in diameter, terete or slightly compressed, smooth, glabrous, green, often with a brown tinge.

Leaves: Decussate. Lamina 3–4(–6) cm long and 1.5–2.0(–3.5) cm wide, ovate oblong, obtuse, fleshy, with glossy surface, margins entire or remotely serrate, base rounded to tapering. Petioles 2–4 mm long.

Inflorescence: Monoecious. Racemes stalked, with 7–20(–35) flowers, smooth, glabrous, emerging mainly from the upper leaf axils. Bracts linear, acute, more or less as long as the pedicels. Pedicels smooth, shorter than or as long as the bracts during flowering, held more or less at a right angle to the rachis when fruiting.

Flowers: 4-merous, bisexual. Calyx divided into 4 oblong-oval to spathulate lobes. Corolla (5–)7–8(–9) mm in diameter, with 4 widely spread lobes of unequal size, dark to bright blue, rarely white or pinkish, all with darker stripes and white corolla tube. Stamens 2, exserted. Style 1, 1.3–2.2 mm long.

Fruits: Capsules inflated, more or less spherical, 3–4 mm long and wide, apex minutely emarginate, shorter than or slightly longer than sepals at maturity. Seeds numerous, 0.6 mm long, oval.

Flowering: June-August.

Biology: Protogynous, anthers do not open until the stigmas have withered. Insect pollinated, cleistogamous flowers self-fertilising. Abundantly fertile. Seeds sticky after contact with water and dispersed by animals. Vegetative dispersal by detached stem fragments and uprooted plants that readily develop adventitious roots from the stem nodes. Often developing large monospecific stands, commonly rooted above or at the water line and sending floating stems and shoots out over the surface. May survive temporary sub-

The flowers of *V. beccabunga* are typically deep blue, although light blue and more rarely white or pink flowers do occur. Lille Skovgårds Hage, Mariager Fjord, Denmark. Photo JCS.

Veronica beccabunga. The four corolla lobes are of unequal size, the upper larger one formed by fusion of two lobes, often with a minutely emarginate apex. Photo JCS.

Rooting in the bank above the water line *V. beccabunga* sends its prostrate and ascending shoots and stems out over the surface of the water in a small stream, however without developing an aquatic form. Skibstedbro, Jutland, Denmark. Photo BM.

mergence but does not develop an aquatic form. Wintergreen.

Habitats: In mesotrophic to eutrophic, slightly acidic to alkaline water. On the margins of ponds, lakes, streams and rivers, in canals and ditches as well as in springs, flushes and wet meadows.

Distribution within the region: Native. Widespread and abundant throughout, except for the far north.

Characteristics and similar species: Both flowering and non-flowering *V. beccabunga* differ from all other European aquatic plants by the glossy and succulent leaves with ovate, obtuse laminas with entire or minutely serrate margins.

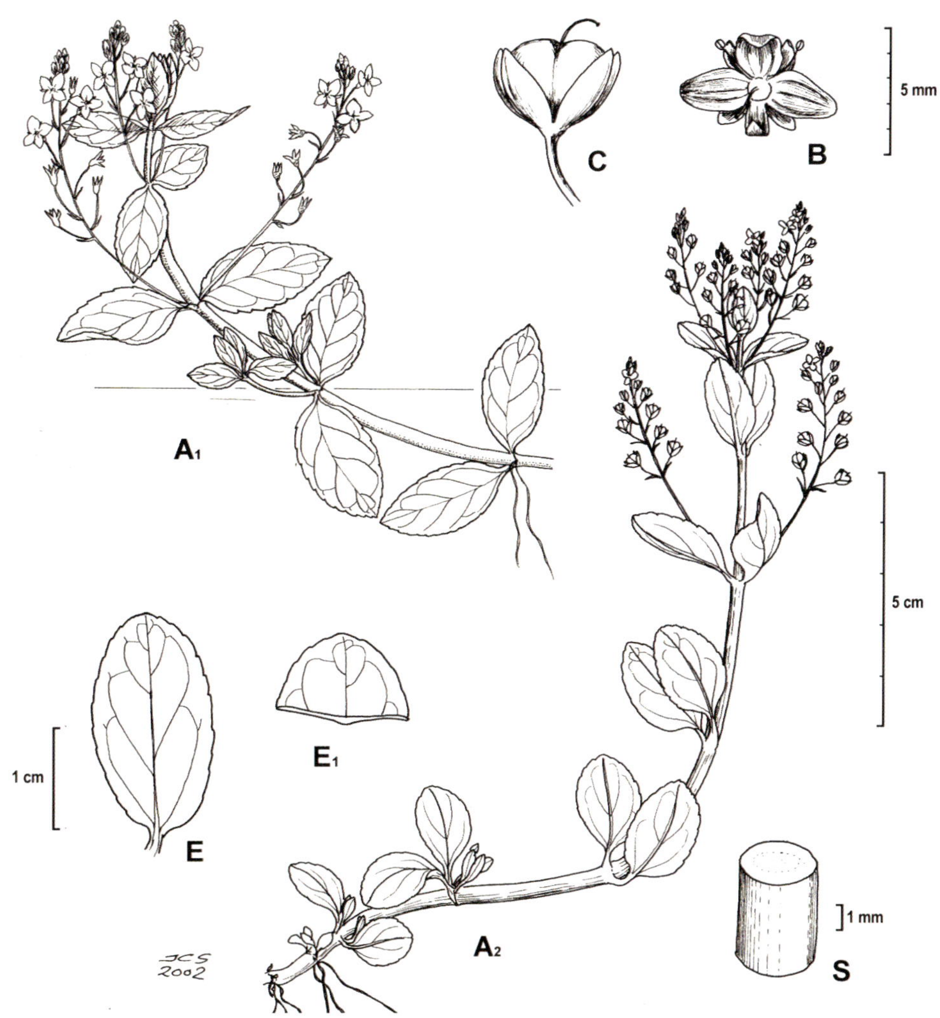

Veronica beccabunga
A1,**A2** habit, **B** flower, **C** capsule, **E** leaf, **E1** cross section of leaf, **S** stem section.

Plantaginaceae

302 *Veronica scutellata* L.

DK: Smalbladet Ærenpris **N:** Veikveronika **S:** Dyveronika **FIN:** Luhtatädyke **IS:** Skriðdepla **GB:** Marsh Speedwell **NL:** Schildereprijs **F:** Véronique à écusson **D:** Schild-Ehrenpreis **CZ:** Rozrazil štítkovitý **PL:** Przetacznik błotny **EST:** Kännasmailane **LV:** Vairogu veronika **LT:** Pelkinė veronika **Ru:** Вероника щитковая

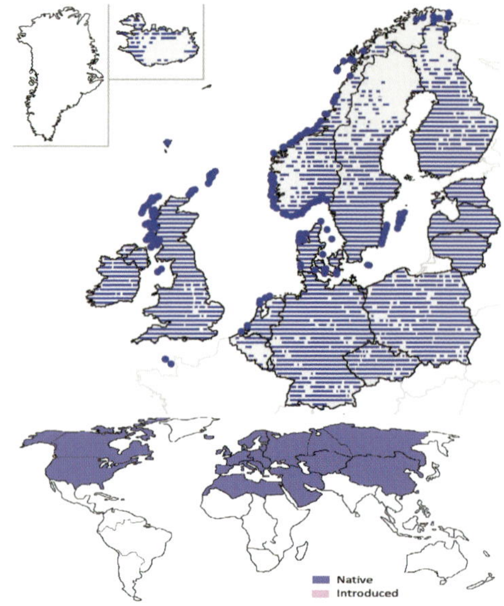

Veronica scutellata. Voldsted, Himmerland, Denmark. Photo JCS.

Perennial helophyte with stolons above and below the ground, with white roots from the nodes.

Stem: Ascending to erect from stolons, (8–)10–30(–100) cm long and 0.6–2.0 mm in diameter, the lower part bluntly angular, the upper part terete, usually smooth and glabrous, sometimes hairy, green, often with a brown or reddish tinge.

Leaves: Decussate and sessile, (1.6–) 2.5–4.5(–5.5) cm long and 2–7(–10) mm wide, lanceolate, acute, the margin with very small, widely spaced, perpendicular teeth. Smooth and glabrous, sometimes sparsely hairy or rarely distinctly hairy on the lower side, green, often with a brown tinge.

Inflorescence: Monoecious. Petiolate racemes with (4–)8–14(–17) flowers, smooth, glabrous or sometimes hairy, emerging mainly from upper leaf axils. Bracts linear, acute, distinctly shorter than pedicels. Pedicels smooth, several times longer than the bracts during flowering. Fruiting pedicels straight, held more or less at a right angle to the rachis.

Flowers: 4-merous, bisexual. Calyx divided into 4, oblong-oval to lanceolate, acute lobes. Corolla 4–6 mm in diameter, with 4 widely spread obovate lobes of unequal size, bright blue, rarely whitish or pinkish, all with darker stripes, corolla tube greenish. Stamens 2, exserted. Anthers white. Style 1.

Fruits: Capsules heart shaped, 2–4 mm long and 4–6 mm wide, distinctly emarginate at apex, longer than the sepals at maturity and usually oriented more or less at a right angle to the pedicels. Seeds up to 20, flattened and elliptic.

Flowering:
June-September.

Biology: Insect pollinated, with only a few flowers open at a time. Abundantly fertile. Vegetative dispersal by detached stem fragments, which readily develop adventitious roots from the stem nodes. May survive temporary submergence but does not develop distinct aquatic forms.

Habitats: In oligotrophic to mesotrophic or even eutrophic, mainly slightly acidic to neutral or alkaline, mineral as well as humic water. In ponds and on lake margins as well as reed swamps and temporary ponds, in mires and wet meadows.

Veronica scutellata var. *villosa* Schum. with numerous hairs on stem, leaves and inflorescences. Öland, Sweden. Photo JCS.

Submerged *V. scutellata.* Pool, Lüsekamp, Germany. Photo KvdW.

Veronica scutellata. Capsule. Photo KvdW.

Distribution within the region: Native. Common throughout most of the region, but scattered in the far north.

Characteristics and similar species: *Veronica scutellata* differs from other aquatic *Veronica* species by the narrowly lanceolate leaves with very small, perpendicular teeth. Flowering plants differ by the very long pedicels and the wide, flattened fruits with distinctly emarginate apex.

220 *Ranunculus flammula* has alternate leaves and yellow flowers. Confusion with other aquatic plants is unlikely.

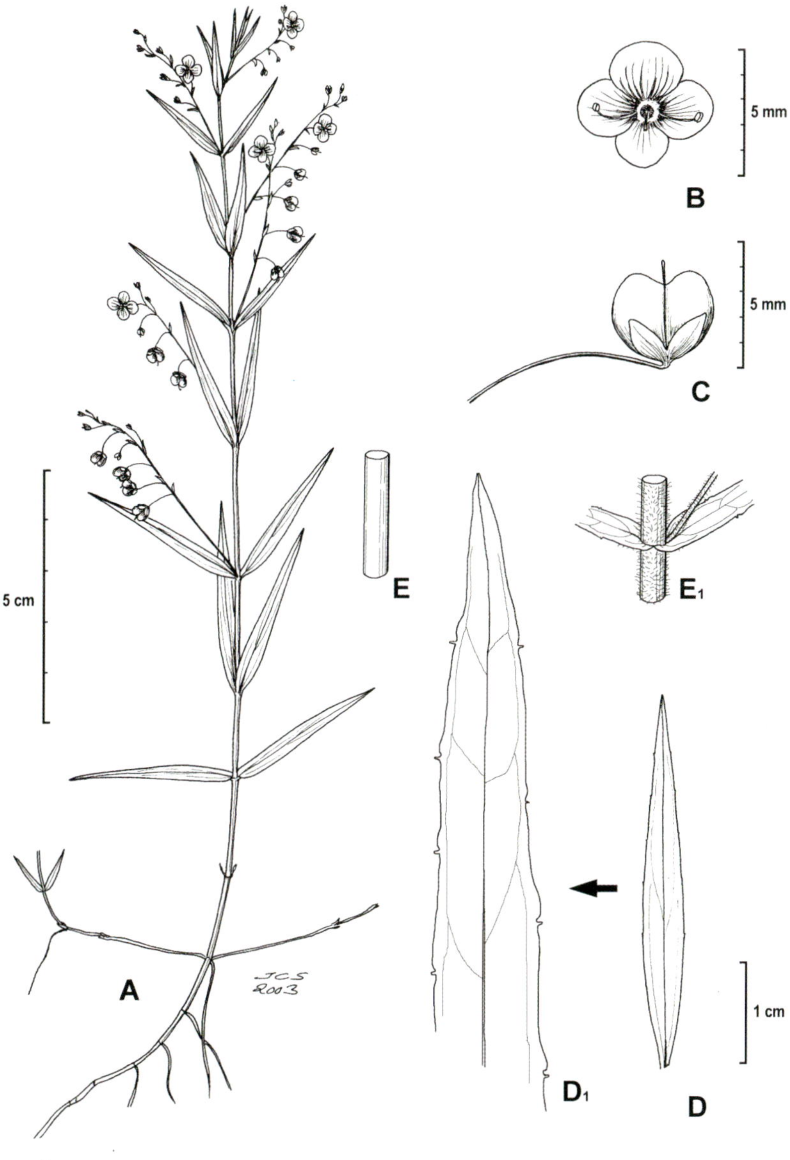

Veronica scutellata
A habit, **B** flower, **C** capsule, **D**,**D1** leaf, **E** stem section, **E1** part of stem and leaves var. *villosa* Schum.

Scrophulariaceae

303 *Limosella aquatica* L.

DK: Dyndurt **N:** Evjebrodd **S:** Ävjebrodd **FIN:** Mutayrtti **IS:** Efjugras
GB: Mudwort **NL:** Slijkgroen **F:** Limoselle
D: Gewöhnliches Schlammkraut **CZ:** Blatěnka vodní **PL:** Namulnik brzegowy
EST: Harilik nõgilillik **LV:** Ūdeņu dūņene **LT:** Vandeninė dumblenė **Ru:** Лужница водяная

Limosella aquatica on open, muddy substrate in a pond. Voldsted, Himmerland, Denmark. Photo JCS.

Annual helophyte or hydrophyte, with subterranean or surface stolons from leaf axils up to 5 cm long. Roots white, numerous.

Stem: Erect stem absent, but with a short, rhizome-like vertical structure.

Leaves: In rosettes, radially spreading, horizontal or curved upwards. The first leaves are subulate, withering shortly after germination. Subsequent leaves have petioles 4–10(–18) cm long, obovate to broadly lanceolate, slightly fleshy, laminas 0.5–2.5 cm long and 2.5–10 mm wide, yellowish- to dark green, glabrous with minute, sessile glands. Permanently submerged plants only develop linear to subulate leaves. Upon submergence terrestrial plants develop long-petiolate leaves with floating blades.

Inflorescence: Monoecious. Flowers solitary, pedicels 1–5 cm long, smooth from leaf axils, 2–10(–18) flowers per rosette.

Flowers: 5-merous. Bisexual. Calyx 5-veined, yellowish green with reddish stripes between the veins, fused and bell shaped at base, with 5 free, triangular lobes often curved outwards. Corolla 2–3 mm in diameter, with 5 ovate, white to pinkish or rarely bright violet lobes. Corolla tube yellowish, slightly longer than the lobes. Stamens 4, with reddish-violet anthers. Style 1, 0.5–0.7 mm long.

Fruits: Oval to spherical capsules, 3–5 mm long and 2–3 mm wide, distinctly shorter than the sepals. Each capsule containing 80–120(–190) cylindrical seeds, 0.5–0.7 mm long and 0.3 mm in diameter.

Flowering: July-September.

Biology: *L. aquatica* populations are markedly sporadic in occurrence, with

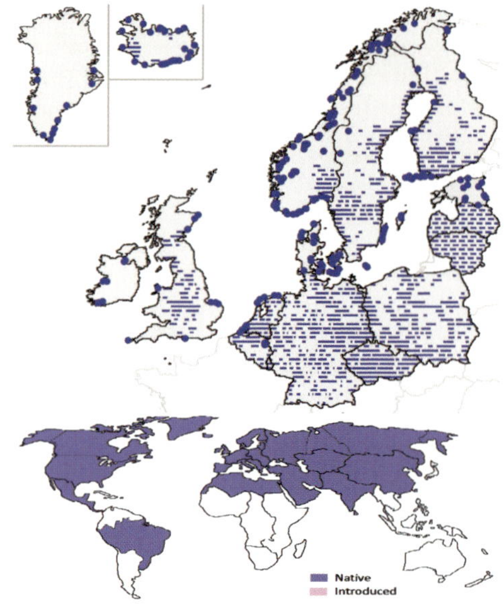

populations often briefly becoming established only as fluctuating water levels expose suitable substrate to varying degrees between years. New plants grow rapidly and with vegetative reproduction by stolons forming new rosettes. Submerged flowers do not open and self-fertilisation is probably frequent within buds, possibly within a small air bubble formed between the petals and the sexual organs. Flowers on emergent and terrestrial plants open, enabling cross-fertilisation. Flowering and fruiting are fast and prolific. Some seeds germinate after only 3 days, while most seeds form a persistent seed bank which remains viable for several years. Seed dispersal by animals and water. Plants die back late in the season.

Habitats: In mesotrophic to hypertrophic, slightly acidic to alkaline water.

The tiny flowers of *Limosella aquatica* have white or pinkish petals. To the left a fully developed green capsule, in the middle an open flower and below this an open capsule with cylindrical brown seeds. Tønder, Denmark. Photo JCS.

Submerged *L. aquatica* in a sand pit at a depth of 4.5 m. Fischeln, Germany. Photo KvdW.

Limosella aquatica. Voldsted, Himmerland, Denmark. Photo JCS.

Typically on muddy substrate in temporary ponds and seasonally inundated hollows, as well as small, shallow lakes with widely fluctuating water levels.

Distribution within the region: Native. Widespread but scattered and declining in most countries.

Characteristics and similar species: *L. aquatica* is characterised by rosettes of spathulate leaves, pedicellate, white to pinkish flowers, spherical capsules with numerous seeds and abundant formation of stolons establishing new rosettes.

221 *Ranunculus reptans* and small specimens of **220 *Ranunculus flammula*** differ in their horizontally creeping stem, as well as sexual characters.

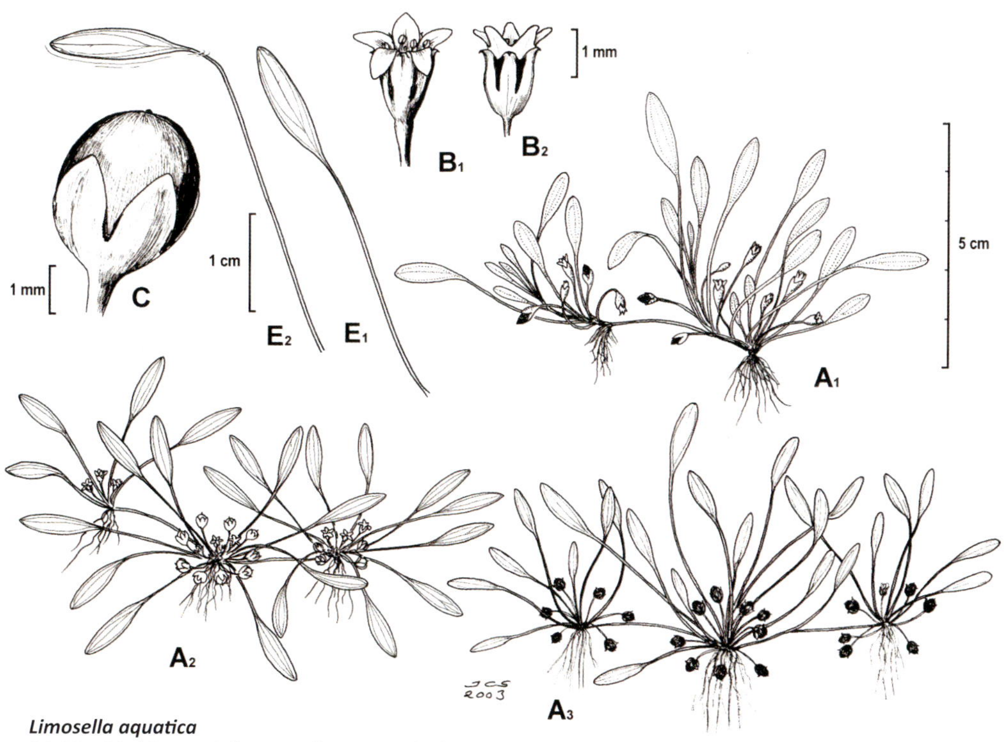

Limosella aquatica
A1,**A2**,**A3** habit, **B1**,**B2** flower, **C** fruit, **E1**,**E2** leaf.

304 *Limosella australis* R.Br.

GB: Welsh Mudwort

Leaves subulate to narrowly spathulate, without distinct lamina and petiole. Flowers 3.5–4 mm diameter, white.

Distribution within the region: Native. In Europe only known from Merionedd and Caernarvonshire in West Wales; formerly more widespread in Wales. The hybrid *L. aquatica × australis* was formerly reported among the parents in Caernarvonshire.

Note: The genus *Limosella* is badly in need of thorough global revision. In particular the taxonomy of plants currently generally referred to in the literature as *L. australis* from both Northern and Southern Hemisphere populations requires clarification. Further research is likely to show that the plants in Britain currently referred to as *L. australis* involve a different taxon, but available information is inadequate to resolve nomenclatural issues. We have therefore retained this name for the purposes of this book.

Limosella australis. Porthmadog, Wales. Photo RVL.

Lamiaceae

Key to *Mentha*

Mentha species are characterised by a distinct odour, which sets them apart from other aquatic species with decussate leaves. The odour is related to the essential oil composition of the leaves. In our flora, two main chemotypes are found. The most frequent is the menthol (including menthone and menthofurane) chemotype, referred to as "peppermint". The carvone chemotype, referred to as "spearmint", is mainly found in cultivated taxa. The amount of essential oils within a species can vary considerably. Odour is furthermore modified by additional components, such as pulegone (often characterised as "pungent") and linalool derivates, mostly regarded as "pleasant" in cultivated varieties.

With some experience, odour can be helpful for identification, even though exact identification of *Mentha* species requires flowering plants. Difficulties can arise from phenotypic plasticity, geographical variation, and frequent hybridisation, which involves all native species except *M. pulegium*. *Mentha* taxa have been cultivated as ornamental and medicinal plants for centuries. Many cultivars and varieties have escaped from gardens and nurseries, surviving as sterile clones or contributing to local hybridisation circles. Together with their hybrids and backcrosses, *M. spicata*, *M. longifolia* and *M. suaveolens* (not treated here) form a very difficult group. Several *Mentha* taxa grow in wet places and close to water bodies, but only *M. aquatica* is regularly found fully submerged. The best way to identify submerged plants is to look for flowering individuals on the margins.

1 Calyx with hairs in the throat, upper 3 calyx teeth wider and shorter than the lower 2 (***1***). Leaves less than 2 cm long, margin subentire or with a few small teeth
305 *M. pulegium*

1 Calyx without hairs in the throat, calyx teeth more or less equal (***2***). Leaves mostly more than 2 cm long, toothed — 2

2 Flowers contracted into 1–2 terminal rounded heads (***3***), sometimes also in a few whorls below
306 *M. aquatica*

2 Flowers in axillary whorls (***4***) and/or in terminal cylindrical to pyramidal spikes (***5***) — 3

3 Flowers in axillary whorls. Bracts leaf-like, usually toothed (***4***) — 4

3 Flowers in a more or less cylindrical to pyramidal head. Bracts scale-like, entire (***5***) — 5

4 Calyx with 5 indistinct veins. Calyx teeth equilateral, < 0.5 mm long (***6***). Flowers in (8–)10–13 axillary whorls, terminal leaves without flower whorls. Stamens exserted
307 *M. arvensis*

4 Calyx with 5–10 distinct veins. Calyx teeth isosceles triangular, 0.5–1.5 mm long (***7***). Flowers in up to 7(–8) whorls, terminal leaves with flower whorls. Stamens usually included
308 *M. ×verticillata*

5 At least lower leaves petiolate, smooth. Plant sterile
309 *M. ×piperita*, 310 *M. x dumetorum*

5 All leaves sessile, often rugose. Plants at least partly fertile
311 *M. spicata*, 312 *M. longifolia*

1

2

3

4

5

6

7

Lamiaceae

305 *Mentha pulegium* L.

DK: Polej-Mynte **N:** - **S:** Poleijmynta **FIN:** - **IS:** -, **GB:** Pennyroyal **NL:** Poleimunt **F:** Menthe Pouliot **D:** Poleiminze **CZ:** Máta polej **PL:** Mięta polej, **EST:** - **LV:** - **LT:** Taškuotoji mėta **Ru:** Мята болотная

Flowering *M. pulegium*. Dordrecht, Netherlands. Photo KvdW.

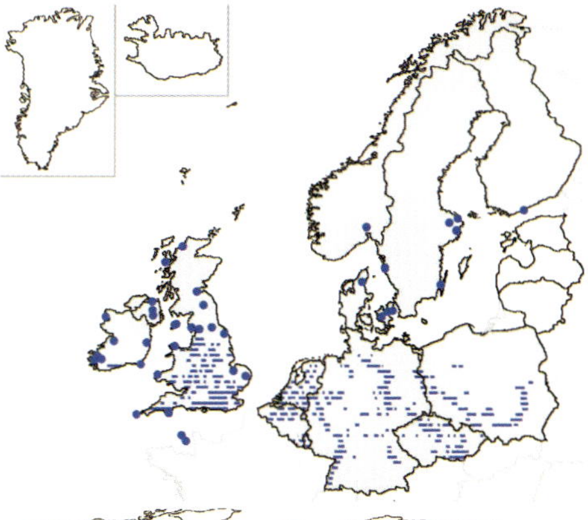

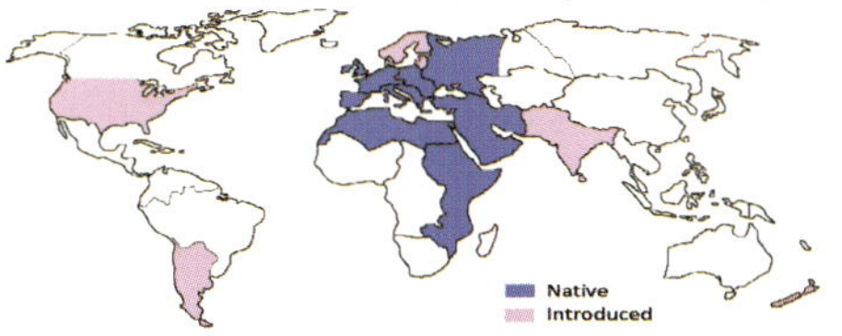

Perennial helophyte sometimes occurring submerged. Leaves with a pungent odour. Stem 10–30 cm long, prostrate to erect, more or less hairy. Leaves 0.8–2.9 cm long, oblong to oval, obscurely crenate to almost entire. Flowers in distant axillary whorls. Bracts leaf-like, longer than flowers; present in young inflorescences, absent from the apices of mature inflorescences. Calyx 2–3 mm long, narrowly bell shaped, hairy on the outside and inside the throat, with 10 distinct veins and 5 teeth - the upper 3 wider and shorter than the lower 2. Corolla lilac.

Distribution within the region: Regarded as native in the British Isles, Netherlands, Belgium, Luxembourg, Germany, Poland and the Czech Republic, introduced in Scandinavia.

M. pulegium was formerly used as a medicinal plant and an insect repellent. It was cultivated in monasteries and farms from where it spread and became established in moist localities. Today it is rarely cultivated and is declining in most of the region.

Characteristics and similar species: *M. pulegium* has been confused with **308 *M. ×verticillata***, from which it differs by the shape of the calyx and the pungent pennyroyal odour.

Hybrids: *M. pulegium* has not been confirmed to hybridise with any other species.

Partly submerged *M. pulegium*. Algarve, Portugal. Photo KvdW.

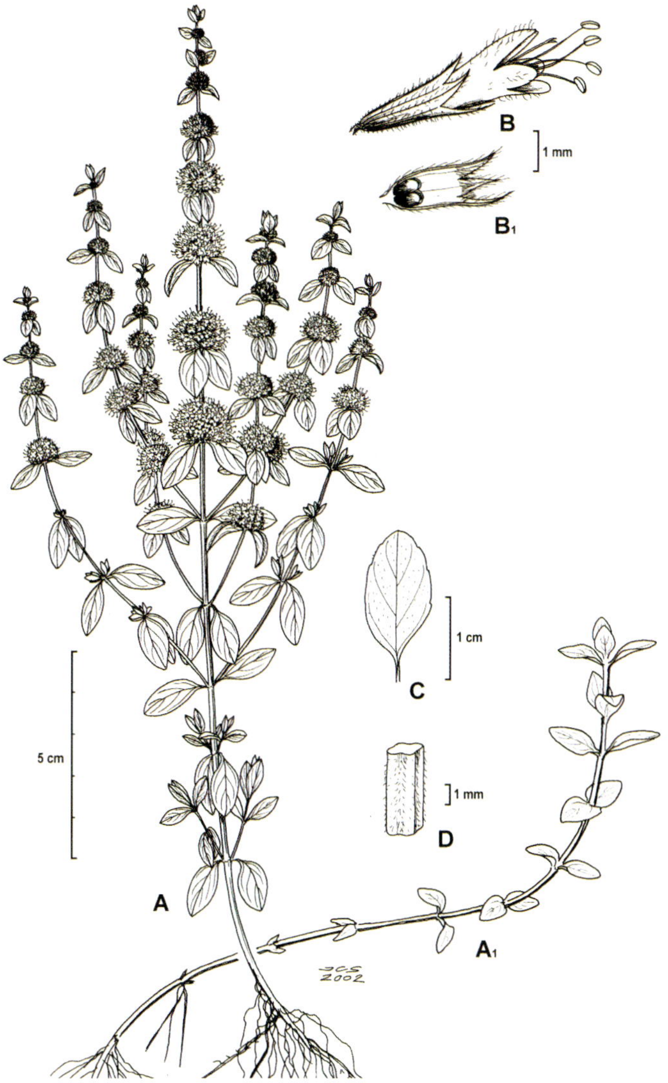

Mentha pulegium
A habit of plant with young inflorescences, **A1** habit of submerged plant, **B** flower side view, **B1** calyx opened, displaying nutlets, **C** leaf, **D** stem section from emergent plant.

Lamiaceae

306 *Mentha aquatica* L.

Syn.: *M. hirsuta* Huds., *M. ×citrata* Ehrh.

DK: Vand-Mynte **N:** Vassmynte **S:** Vattenmynta **FIN:** Vesiminttu **IS:** Vatnamynta
GB: Water Mint **NL:** Watermunt **F:** Menthe aquatique
D: Wasser-Minze **CZ:** Máta vodní **PL:** Mięta nadwodna
EST: Vesimünt **LV:** Ūdeņu mētra **LT:** Vandeninė mėta **Ru:** Мята водная

Flowering M. aquatica. Vramsån, Skåne, Sweden. Photo BM.

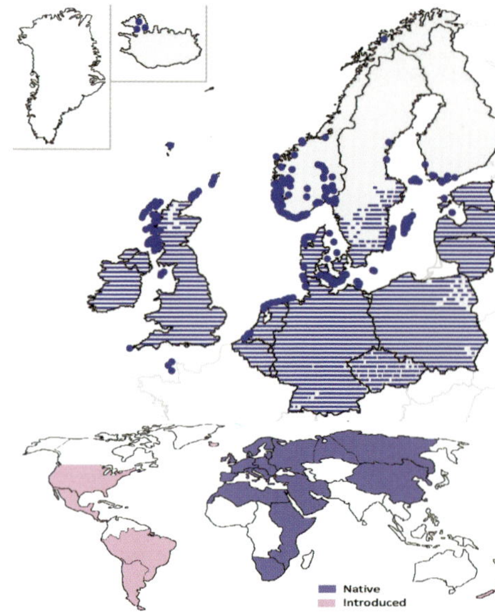

Perennial helophyte or hydrophyte with branched rhizome below or above ground and numerous pale roots.

Stem: Sharply quadrangular, (10–)20–50(–80) cm long, c. 2 mm in diameter. In terrestrial plants erect or ascending, green, often with a purple tinge, hairy to almost glabrous. In flowing water mainly prostrate to ascending, sometimes floating, rooted under or above the water line; in still water commonly ascending to erect, pale green to green. Prostrate stems with adventitious roots from the nodes.

Leaves: Decussate and petiolate. Lamina in terrestrial plants 2–8 cm long and 1–4(–6) cm wide, obovate, bluntly toothed at margins, pinnately veined, more or less hairy. Petioles 3–15 mm long. Aquatic plants with shorter and stiffer, more or less sessile leaves, lamina broadly obovate with entire margin, often with downward-curved apex, commonly light to pale green, glabrous. Aerial as well as submerged leaves with distinct smell of mint when crushed.

Inflorescence: Monoecious. Flowers densely clustered into a rounded to oblong terminal head and sometimes also in a few whorls in the upper leaf axils.

Flowers: 5-merous, bisexual. Calyx approximately 4 mm long, hairy, with 13 veins and 5 equally long, acute teeth, their length about 1/3 the total length of the calyx. Corolla tubular, 5–7 mm long with 4 light purple to purple lobes. Stamens 4, style 1, all exserted.

Fruits: A cluster of 4 one-seeded nutlets. Nutlets with minutely granular surface, yellowish brown.

Flowering: June-September.

Biology: Flowers protandrous with stamens maturing before the stigma. Pollinated by a wide range of insects attracted to nectar. Seeds float and are dispersed by water. Vegetative dispersal by the numerous, long runners, commonly forming large stands. Terrestrial plants overwinter as underground rhizomes, while submerged plants and shoots may remain green during winter.

Habitats: Oligo-, meso- and eutrophic, slightly acidic to alkaline, fresh to brackish water. In and on the banks of a wide range of waters – springs, streams, rivers, canals and ditches, as well as ponds, lakes, seasonally wet depressions and wet meadows.

Submerged *M. aquatica* in a small pond. Åbo Skov, Aarhus, Denmark. Photo BM.

Mentha aquatica. Nees, W-Jutland, Denmark. Photo JCS.

The large, apical cluster of flowers is characteristic for *M. aquatica*. Schadbruch, Kempen/Niederrhein, Germany. Photo KvdW.

Distribution within the region:
Native. Common throughout most of the region, rare in northern Scandinavia and Iceland.

Characteristics and similar species:
Flowering *Mentha aquatica* differs from all other *Mentha* species and hybrids and especially from **308 *M. ×verticillata*** by the distinct apical clusters of flowers and calyx teeth 1/3 the length of the calyx.

307 *Mentha arvensis* L.

DK: Ager-Mynte **N:** Åkermynte **S:** Åkermynta **FIN:** Rantaminttu **GB:** Corn Mint **NL:** Akkermunt **F:** Menthe des champs **D:** Acker-Minze **CZ:** Máta rolní **PL:** Mięta polna **EST:** Põldmünt **LV:** Tīruma mētra **LT:** Dirvinė mėta **Ru:** Мята полевая

Please see the drawing next page.

Perennial helophyte. Stem 10–60 cm long, erect or ascending, more or less branched and hairy. Leaves 2–6 mm long, shallowly serrate, petiolate. Flowers in (8–)10–13 distant axillary whorls. Calyx with 5 indistinct veins and 5 short, triangular teeth.

Distribution within the region:
Native. Throughout the region except Iceland and Greenland.

Characteristics and similar species:
Please see **308 *M. ×verticillata***.

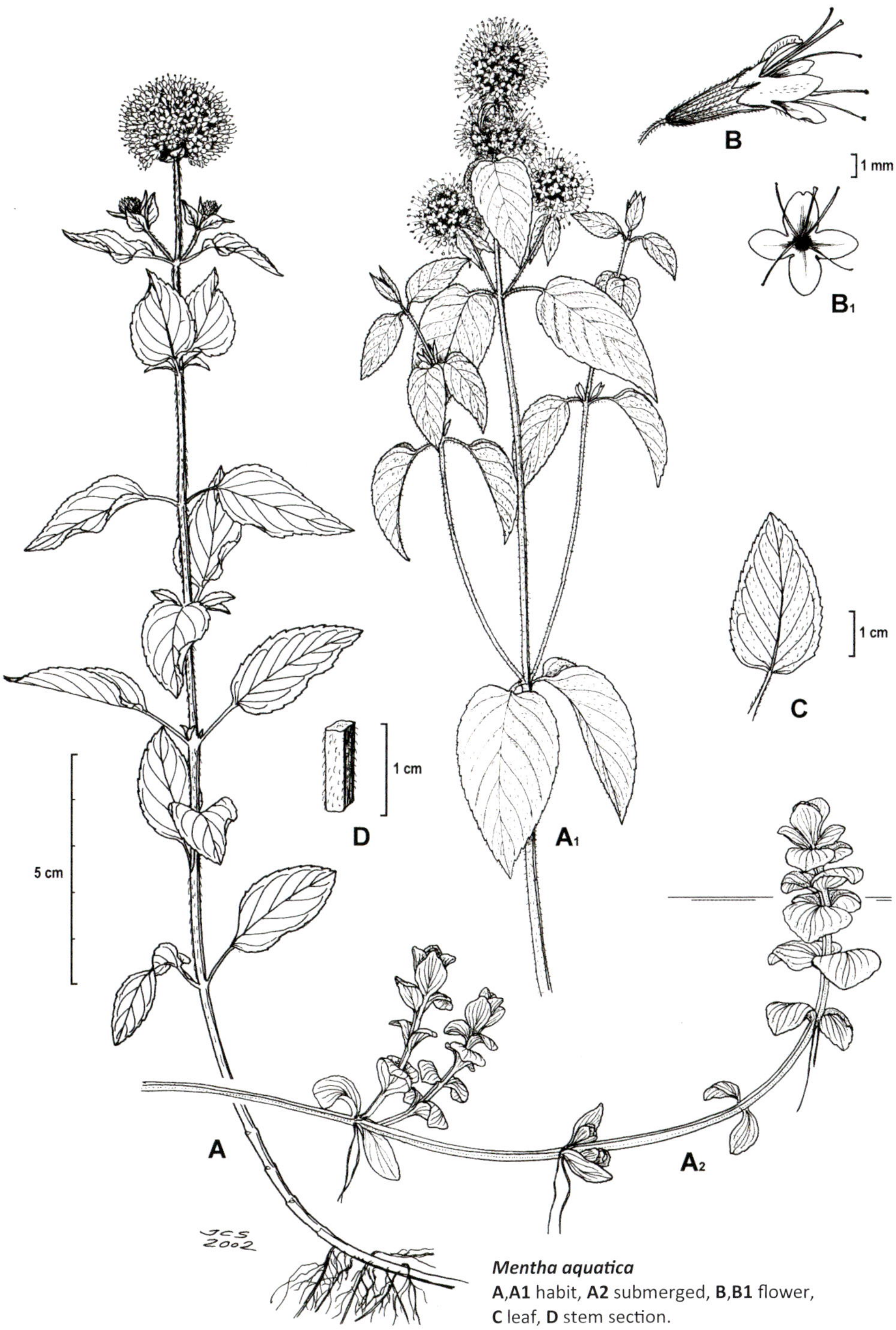

Mentha aquatica
A,A1 habit, **A2** submerged, **B,B1** flower, **C** leaf, **D** stem section.

Lamiaceae

308 *Mentha ×verticillata* L.

(*M. aquatica × arvensis*)

DK: Krans-Mynte **N:** Temynte **S:** Kransmynta **FIN:** Kiehkuraminttu **IS:** -
GB: Whorled Mint **NL:** Kransmunt **F:** Menthe verticillée
D: Quirlminze **CZ:** Máta přeslenitá **PL:** Mięta okręgowa
EST: Männasmünt **LV:** Mieturu mētra **LT:** Pakrantinė mėta
Ru: Мята мутовчатая

A variable hybrid which is often aquatic and usually sterile. Stem up to 90 cm long, more or less hairy. Flowers in axillary whorls. Calyx 2.5–3.5 mm long, with 10 rather distinct veins and 0.5–1.5 mm long, isosceles-triangular calyx teeth.

Distribution within the region: Native. Often occurring in the absence of the parents. Rather frequent in the region, but its actual distribution is poorly known due to confusion with other hybrids.

Characteristics and similar species: Usually similar to **307 *M. arvensis*** (see drawing), but generally more robust. Flowers in up to 7(–8) whorls, terminal leaves with flower whorls. Calyx with 5–10 distinct veins and isolateral teeth. Crushing the leaves releases a faint peppermint odour, while*M. arvensis* is usually poor in menthol (musty odoured). **306 *M. aquatica*** has leaves of equal size up to the inflorescence, while the stem leaves in *M. ×verticillata* are smaller towards the stem apex. *M. aquatica* always smells strongly of menthol. See also **305 *M. pulegium***.

Mentha ×verticillata. Plant morphologically closer to *M. aquatica* than to *M. arvensis*. Hanherred, Jutland, Denmark. Photo JCS.

Mentha ×verticillata. Lake Hampen, Jutland, Denmark. Photo JCS.

M. ×gracilis Sole (*M. arvensis × spicata*) and *M. ×wirtgeniana* F.W.Schulz (*M. arvensis × aquatica × spicata*), which also have the verticillate habit, are occasionally found in and at streams and ditches, but not treated in this work.

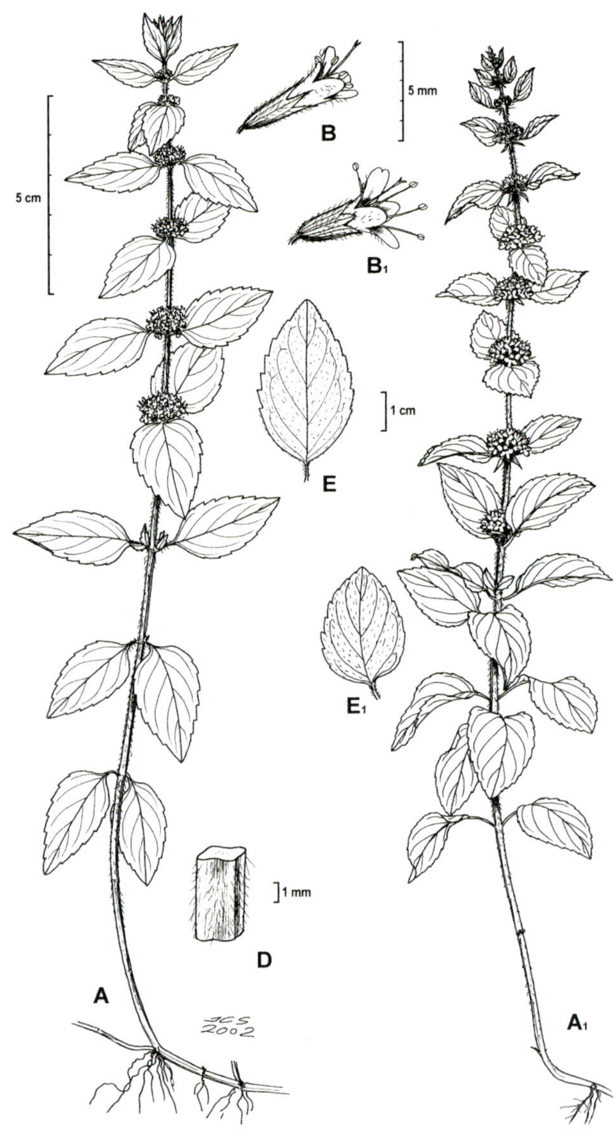

Mentha ×verticillata
A habit, **B** flower side view, **D** stem section, **E** leaf.
Mentha arvensis
A1 habit, **B1** flower side view, **E1** leaf.

309 *Mentha ×piperita* L.

(*M. aquatica × spicata*)

DK: Peber-Mynte **N:** Peppermynte **S:** Pepparmynta **FIN:** Piparminttu **IS:** -
GB: Peppermint **NL:** Pepermunt **F:** Menthe poivrée
D: Pfefferminze **CZ:** Máta peprná **PL:** Mięta pieprzowa
EST: Piparmünt **LV:** Piparmētra **LT:** Pipirmėtė **Ru:** Мята перечная

Perennial helophyte, sometimes growing in water. Crushed leaves usually with a scent of peppermint. Stem 30–80 cm long, reddish to purple, sparsely hairy. Leaves 3.5–8.5 cm long, lanceolate, sharply serrate, almost glabrous, petiolate. Inflorescence a terminal oblong to pyramidal spike-like head, 3.5–6 cm long, composed of several more or less widely spaced, axillary whorls of flowers. Bracts lanceolate, the lower 1–2 pairs leaf-like. Calyx 2.5–4.5 mm long, more or less bell shaped, with subulate teeth. Corolla lilac. Mostly introduced and naturalised from cultivation.

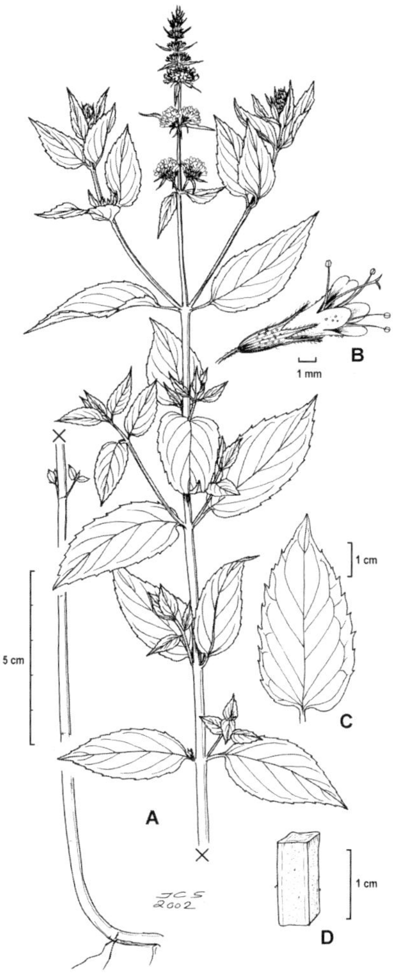

Mentha ×piperita
A habit, **B** flower side view, **C** leaf, **D** stem section.

310 *Mentha×dumetorum* Schult.

(*M. aquatica × longifolia*)

Perennial wetland plant, sometimes a helo- or hydrophyte. Spontaneous hybrid but also cultivated, with peppermint or fruit odour. Differs from **309 *M. ×piperita*** by the densely hairy leaves. Distribution: Germany, Britain, Belgium, Poland, Czech Republic.

311 *Mentha spicata* L.

Syn.: *M. viridis* (L.) L.

DK: Grøn Mynte **N:** Grønnmynte **S:** Grönmynta **FIN:** Viherminttu **IS:** - **GB:** Spearmint **NL:** Groene munt
F: Menthe en épi **D:** Grüne Minze **CZ:** Máta klasnatá **PL:** Mięta zielona **EST:** Rohemünt **LV:** - **LT:** Šaltmėtė
Ru: Мята колосистая

Perennial wetland plant, sometimes a helophyte, generally with intense spearmint odour. Stem 30–90 cm long, glabrous or tomentose. Leaves 4–9 cm long, lanceolate to oblong, sharply serrate, glabrous, sessile or at most with petioles up to 3 mm long. Inflorescence a terminal, spike-like head 3–6 cm long, composed of several more or less widely spaced axillary whorls of flowers. Bracts linear, longer than flowers, entire - the lowest pair sometimes leaf-like. Calyx 1.5–3 mm long, bell shaped, with 5 narrow teeth. Corolla lilac. Considered native in parts of Europe, but mostly escaped and naturalised from cultivation. A variable species derived from *M. longifolia* and *M. suaveolens* (not treated here).

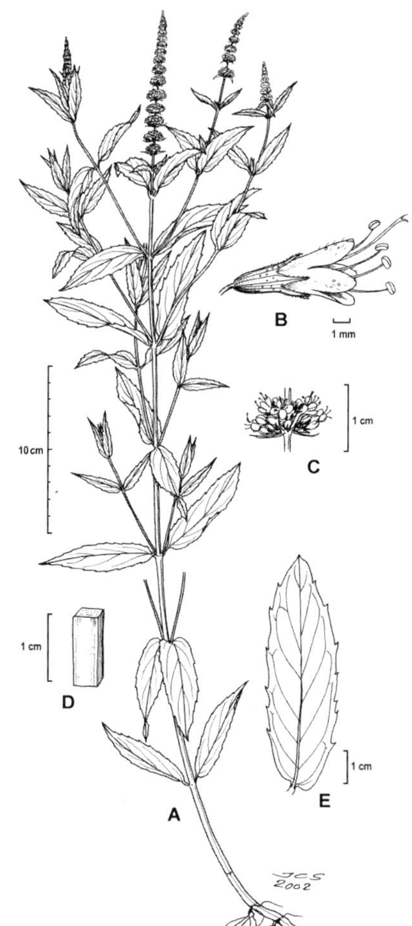

Mentha spicata
A habit, **B** flower side view, **C** whorl of flowers, **D** stem section, **E** leaf.

312 *Mentha longifolia* L.

Syn.: *M. sylvestris* L.

DK: Grå Mynte **N:** Gråmynte **S:** Grå mynta **FIN:** Harmaaminttu **IS:** - **GB:** Horse-mint **NL:** Herzmunt
F: Menthe à longues feuilles **D:** RossMinze **CZ:** Máta dlouholistá **PL:** Mięta długolistna
EST: Pikalehine münt **LV:** Garlapu mētra **LT:** Miškinė mėta **Ru:** Мята длиннолистна

Perennial wetland plant, sometimes a helophyte or hydrophyte, with pungent smell, often unpleasant, with musty, pulegone, and carvone components. Stem 60–120 cm long, more or less hairy. Leaves 3–10 cm long, ovate to oblong, acute at apex, rounded at base, with irregular, patent teeth, green and more or less hairy on the upper side, grey and felted on the lower side, sessile. Inflorescence a terminal, 3–10 cm long, spike-like head composed of several more or less widely spaced axillary whorls of flowers. Bracts narrowly linear elliptic, longer than flowers. Calyx 2.2–2.8 mm long, bell shaped, with 5 narrow teeth. Corolla pale lilac or white. Recorded from most of the region except the British Isles and Iceland, but often misidentified.

Acanthaceae

313 *Hygrophila polysperma* (Roxb.) T.Anderson

DK: - **N:** - **S:** - **FIN:** - **IS:** -
GB: Dwarf Hygrophila **NL:** Belgisch groen **F:** -
D: Indischer Wasserfreund **CZ:** - **PL:** -
EST: - **LV:** - **LT:** - **Ru:** -

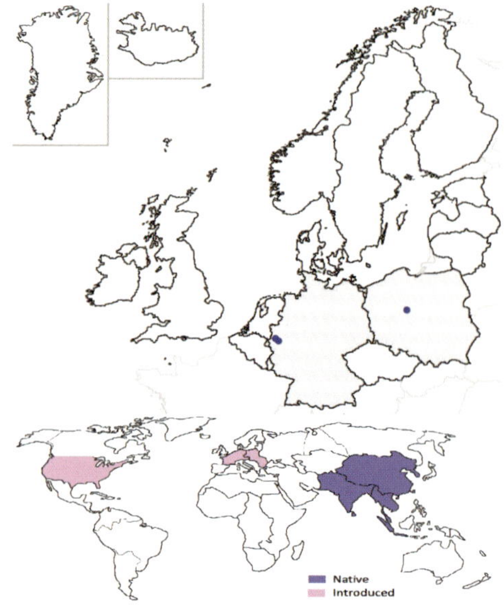

Hygrophila polysperma. River Erft, Germany. Photo KvdW.

Perennial hydrophyte, with a creeping rhizome. Roots whitish to pale brownish.

Stem: Up to 3 m long, mostly submerged, ascending to creeping, very brittle, quadrangular, finely hairy above, glabrous below, rooting from nodes both within the substrate and on floating shoots.

Leaves: Decussate, entire, acute or somewhat obtuse at apex, sessile or very shortly petiolate, fused at base by a narrow flange of tissue with some cilia up to 1.5 mm long on the margin. Emergent leaves 1–3.5 cm long, 0.4–1 cm wide, green to dark green, hairy. Submerged leaves generally larger than emergent, 4–8 cm long and 2 cm wide, elliptic to oblong, pale green or yellowish to reddish in colour, glabrous to sparsely hairy, tapering to a narrow base.

Inflorescence: Monoecious. Flowers solitary in upper, emergent leaf axils, sessile, subtended by two hairy bracts 5–15 mm long.

Flowers: 5-merous. Bisexual. Calyx tubular, 4 mm long with 5 narrow lobes. Corolla 5–9 mm long, ciliate, with a 3 mm long, whitish, tubular base and 5 purplish-white lobes up to 6 mm long, the upper 2 lobes fused for most of their length, the lower 3 free. Stamens 2, fertile with whitish anthers and 2 staminodes. Style 1, about 0.5 mm long.

Fruits: Capsule 6–9 mm long, narrow, hairy, splitting lengthwise, containing 20–30 seeds.

Seeds 0.4–0.6 mm long, 0.3–0.5 mm wide, very flat, obovate to elliptic or circular, narrowly winged, brownish yellow, shiny, sticky.

Flowering: Flowering plants not observed in the region.

Biology: Fragments of the brittle stem root easily, and detached leaves are also capable of rooting and developing new plants. Often forming large monospecific stands. Plants growing at depths of 3 m or even more can reach the surface to flower. Seed dispersal by animals and water, but viable seeds may only develop in the tropics.

Hygrophila polysperma. Channel, Hévíz, Hungary. Photo KvdW.

Hygrophila polysperma. Channel, Hévíz, Hungary. Photo KvdW.

Hygrophila polysperma is classified as a noxious weed in the United States with the potential of being very problematic due to tolerance of herbicides and grass carp.

Habitats: Prefers warm, flowing, eutrophic water. In Europe it is known from a thermal river, a thermal bathing pond, a farm fishpond connected to the cooling system of a power station and a lake.

Distribution within the region:
Alien. Introduced as an aquarium plant but very rarely naturalised in Germany and Poland.

Characteristics and similar species:
Resembles **297-300** ***Veronica*** spp., which differ in the more robust stems, with leaves clasping and without cilia.

242-244 ***Ludwigia*** spp. lack cilia on the fused leaf bases and have small, greenish, 4-merous flowers.

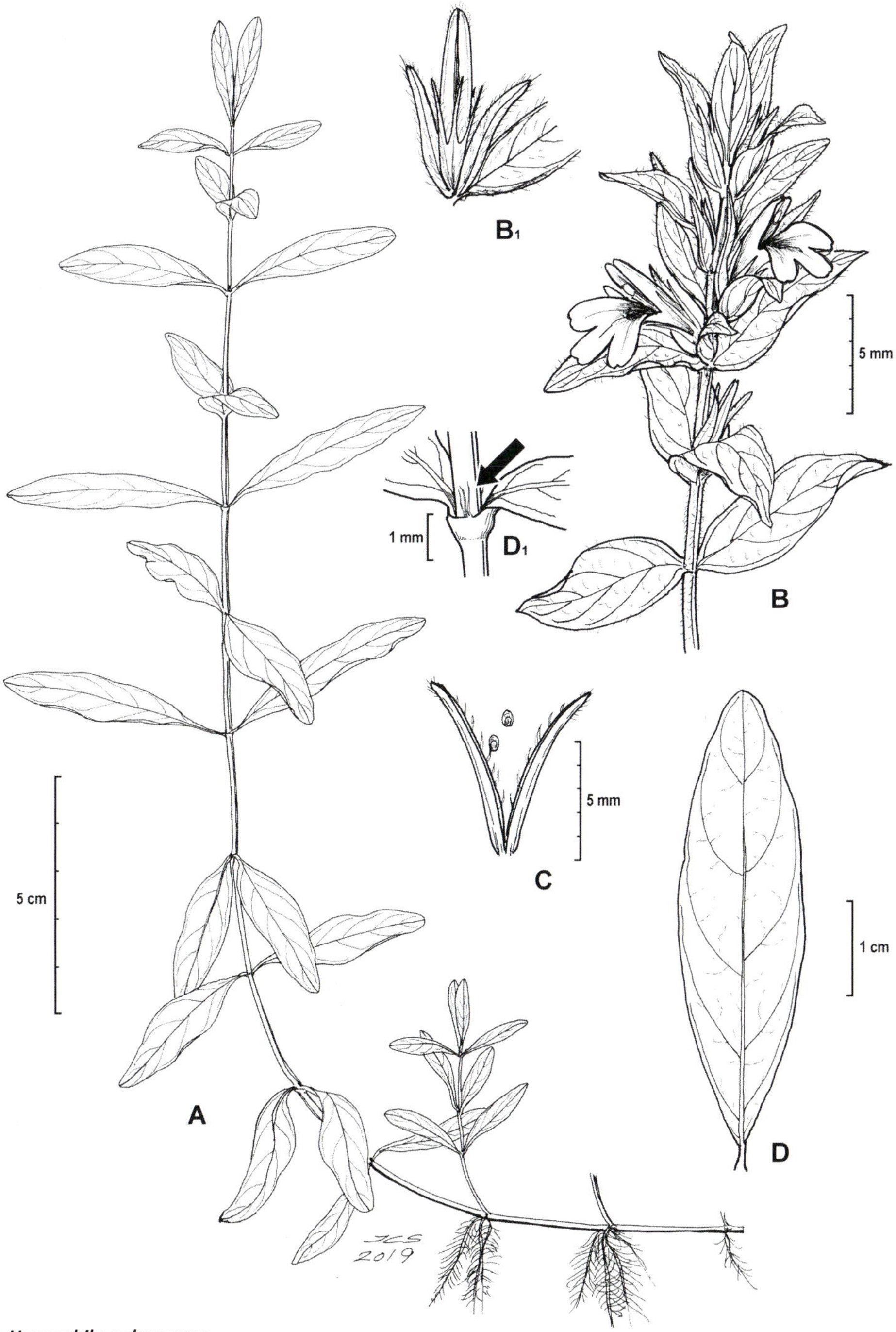

Hygrophila polysperma
A habit submerged plant, **B** inflorescence, **B1** capsule in leaf axil, **C** open capsule, **D** leaf, **D1** fused leaf bases with cilia.

Phrymaceae

Key to *Erythranthe*

Syn.: *Mimulus*

The taxonomy and nomenclature of *Mimulus* has recently been clarified through comprehensive taxonomic and nomenclatural revision. The polyphyletic genus *Mimulus* has recently been split into a number of monophyletic genera (Barker et al. 2012), leaving only seven species within the genus *Mimulus*; all other taxa formerly within *Mimulus* that have been recorded from the region have been reassigned to *Erythranthe* (Barker et al. 2012), although this change has been contested (Lowry et al. 2019, Nesom et al. 2019). This revision included taxonomic changes, such as synonymisation of *M. cupreus* and *M. nummularius* with *E. guttatus,* and *M. variegatus* with *E. lutea* var. *variegata*. The revision has largely been accepted in Europe, however application has not been complete or necessarily consistent; for example, Stace (2019) retained *E. cuprea* as distinct from *E. lutea*, enabling separate recognition of hybrids involving this species (see also Stace et al. 2015), but accepted inclusion of *M. nummularius* and *M. variegatus* within a revised concept of *E. lutea*.

The genera *Mimulus* and *Erythranthe* have their origins in the Americas, but because they have long been popular in horticulture, many species and hybrids have become established in the wild. In addition, two hybrids (*E. ×smithii*, *E. guttata* × *E. lutea* var. *rivularis* × *E. lutea* var. *variegata*) and one species (*E. peregrina*) have arisen in the wild in Britain, involving parents which would not otherwise meet as they are native to different continents; these three may be considered native to Europe. Of the species retained in *Mimulus*, the only species recorded from Europe is *M. ringens* (please see p. 657).

Hybrids of which *E. guttata* is a parent are sterile (although see *E. peregrina*), whereas those between the South American taxa (*E. cuprea*, *E. lutea* var. *rivularis* and *E. lutea* var. *variegata*) are fertile. The occurrence of horticultural cultivars and triple hybrids, as well as the development of an amphidiploid species through chromosome doubling, presumably as a consequence of backcrossing between naturalised populations of the parent and the hybrid, means that the naming of some populations can be very problematic. Hybrids are typically intermediate between the parents, and there is often no precise line to be drawn between the two. However, a parent will often confer a distinctive character on its hybrid offspring; for example, hybrids of which *E. guttata* is one parent have simple white hairs in the inflorescence and the mouth of the corolla closed to varying degrees, while hybrids of which *E. cuprea* is a parent typically have a coppery background colour to the corolla lobes, smaller corolla and small leaves.

The following key is an attempt to enable identification of plants which have been both artificially crossed and selected for particular traits and in some cases have hybridised in nature. In some ways it is unrealistic to attempt to distinguish all populations using field characters, and this key is simply provided to help with naming populations found in the wild. If used in combination with the texts outlining the character of the different taxa, it may be possible to name most plants encountered with some degree of reliability.

---

1 Plant glandular hairy all over, often viscid and occasionally aromatic. Corolla 11–16 mm long. Calyx teeth unequal to subequal, 2–4 mm long
316 *E. moschata*

1 Plant glabrous at base, often glandular hairy above but not viscous or aromatic. Corolla 10–40 mm long. Upper calyx tooth distinctly longer than lower 4 — 2

2 Calyx petaloid, so that there appears to be one corolla inside another on each pedicel, both corolla and calyx coppery in colour. Plant sterile
320 *E. ×burnetii* - petaloid form

2 Calyx not petaloid — 3

3 Throat of corolla closed by two swellings at base of lower lip (***1***). Petal lobes without blotches. Pedicels more or less equal to calyx length. Plant densely glandular pubescent above, particularly on inflorescence and calyx — 4

3 Throat of corolla open or partially open, swellings at base of lower lip small and poorly developed or absent. Pedicels longer than calyx length. Plants may be almost glabrous above or with glandular and simple hairs — 5

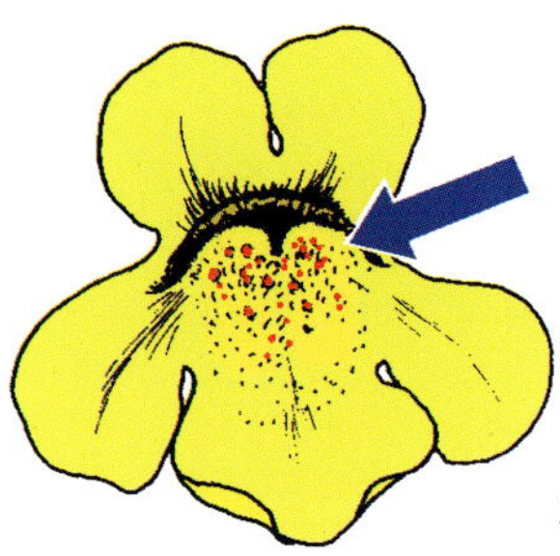

1

4 Highly fertile, most capsules swollen. Lower lobe of corolla of newly opened flowers held almost horizontal. Inflorescences usually short with few flowers
314 *E. guttata*

4 Sterile or low fertility. Capsules generally empty and withering. Lower lobe of corolla drooping. Inflorescences often tall with many flowers
315 *E. ×robertsii*

5 Throat of corolla partially open, swellings at base of lower lip present but small and poorly developed — 6

5 Throat of corolla open without swellings at base of lower lip — 7

6 Highly fertile. Most capsules swollen
315a *E. peregrina*

6 Sterile. Capsules generally empty and withering
315 *E. ×robertsii*

7 Calyx winged on angles (***2***), lobes equal to subequal, more or less straight. Flowers uniformly scarlet
317 *E. cardinalis*

7 Calyx not winged on angles (***3***), upper lobe longer than lower 4, lower teeth of mature capsule folding over the lateral and upper tooth. Flowers variously coloured, uniform, spotted or blotched, rarely uniform scarlet 8

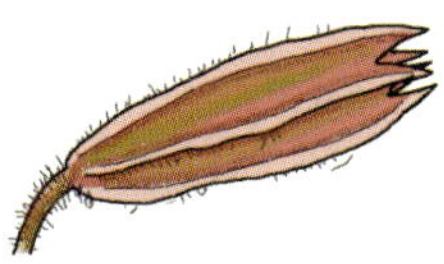

2

3

8 Plant sterile, capsule not developing. Calyx becoming brown and withered containing the undeveloped capsule. Simple white hairs on pedicels and calyces, at least on the angles 9

8 Plant fertile, capsule developing. Calyx not withering until capsule well developed. Simple white hairs lacking 10

9 Corolla, usually bright brownish orange, rarely yellow or apricot, 3.5–5 cm, spotted but not blotched. Plant slender, sprawling or creeping. Leaves usually longer than wide
320 *E. ×burnetii*

9 Corolla with dark blotches and sometimes tinted red. Plant relatively robust. Leaves more or less as wide as long
E. cuprea* × *E. lutea* × *E. guttata or
E. guttata* × *E. lutea var. ***rivularis* ×**
E. lutea var. ***variegata***

10 Calyx short, 0.9–1.2 cm long. Leaves small, largest leaves typically 3-veined. Corolla lobes initially yellow or copper coloured, ultimately copper coloured. Plant sprawling or creeping, stems to 20 cm (plants with corolla bright red are the cultivar "Red Emperor" or "Roter Kaiser")
318 *E. cuprea*

10 Calyx (1–)1.4–2.5 cm long. Leaves larger, largest leaves 5–7-veined. Corolla lobes variously marked but not uniformly copper coloured. Plants robust, more or less upright, often stoloniferous 11

11 Leaves very broad, almost rounded, with broad teeth mostly wider than long. Lower corolla lobe or all five lobes occupied by a single cherry red to purplish-red blotch
323 *E ×smithii*

11 Leaves more or less as long as wide, with teeth coarse, longer than wide and often twisted. Corolla lobes uniform or variously blotched and spotted, not with a single blotch on one or all lobes 12

12 Sprawling to erect plant, with very little lateral growth. Corolla cream or pale yellow (plants with corolla of uniform colour are likely to be of the "Malibu" series)
321 *E. ×hybrida*

12 Plant erect, but typically with abundant lateral growth, mat forming. Corolla bright to golden yellow 13

13 Leaves with even, triangular, flat teeth. Corolla cream or yellow, usually with coppery-orange spots or often confluent blotches, or mainly coppery
322 *E. ×maculosa*

13 Leaves with irregular, oblong, often twisted teeth. Corolla yellow or rarely cream, the lobes unblotched or with 1 or more dark red to purplish-brown blotches 14
319 *E. lutea*

14 Calyx covered with microscopic bristles. Corolla yellow, unblotched
319 *E. lutea* var. ***lutea***

14 Calyx glabrous or in very hot weather may develop a covering of minute glandular hairs. Corolla lobes usually with areas of reddish or purplish blotches, rarely umarked and then pale yellow 15

15 Corolla lobes unmarked or at least the central lobe of the lower lip or commonly all 5 corolla lobes each marked with a round, deep red to red-brown blotch. Pedicels 2.5–5 times the length of the calyx. Leaf margins rather coarsely and irregularly toothed, the teeth often oblong and twisted
319 *E. lutea* var. ***rivularis***

15 Distal areas of corolla lobes carmine to pinkish purple. Pedicels approximately 2 times the length of the calyx. Leaf margins broadly and shallowly toothed
319 *E. lutea* var. ***variegata***

Erythranthe cardinalis. Botanical Garden, Aarhus, Denmark. Photo JCS.

Phrymaceae

314 *Erythranthe guttata* (DC.) G.L.Nesom

Syn.: *Mimulus guttatus* DC.

DK: Gul Abeblomst **N:** Gjøglerblom **S:** Gyckelblomma **FIN:** Täpläapinankukka **IS:** -
GB: Monkey-flower **NL:** Gele maskerbloem **F:** Mimule tachetée
D: Gelbe Gauklerblume **CZ:** Kejklířka skvrnitá **PL:** Kroplik żółty
EST: Kollane pärdiklill **LV:** Lāsainā pērtiķmutīte **LT:** Geltonžiedis puikūnas **Ru:** Губастик крапчатый

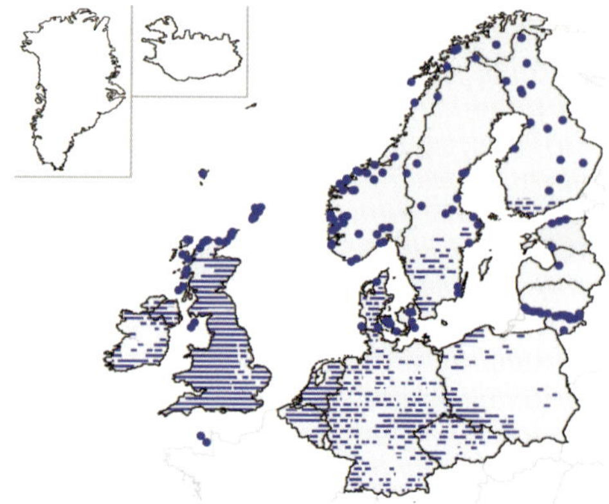

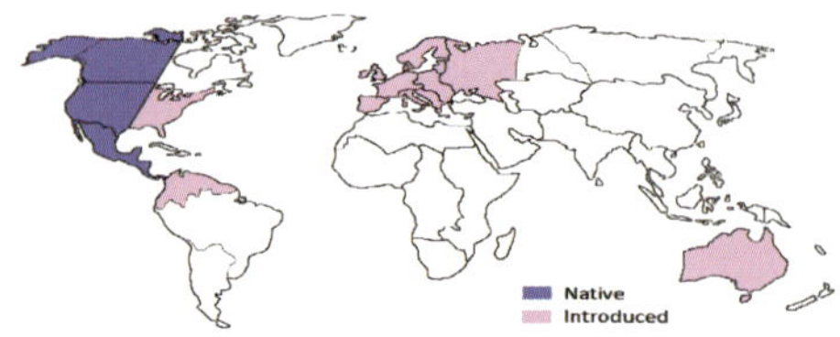

Erythranthe guttata. Villestrup Aa, East-Jutland, Denmark. Photo JCS.

Perennial helophyte. Rhizome creeping above ground, horizontal, stoloniferous and often reddish or purple, sometimes rooting at nodes. Roots fibrous, white or brownish.

Stem: To c. 80 cm high, terete or somewhat 4-angled toward the apex, sometimes hollow, erect or weak and more or less trailing and rooting at the nodes, simple or sparsely branched, commonly glabrous below and puberulent or pubescent above. Internodes generally longer than the leaves.

Leaves: Opposite, glabrous to hairy, membranous, basal leaves on petiole 2–9 cm long, stem leaves sessile, clasping, variable, mostly rounded ovate or ovate oblong, sometimes broadly oblong or oblong lanceolate, occasionally lyrate, apex acute to obtuse, 2.5–15 cm long and 1.5–8 cm wide, 1–2 times longer than wide, pinnately or palmately 5–7-veined from the base, coarsely and irregularly dentate, often with small projections at the base of the blade.

Inflorescence: Mostly racemes of 3–20 flowers, flowers solitary or few together in axils, the pedicels 2–4(–6) cm long, thick and less than twice as long as the calyx, rarely slender and elongate, usually densely glandular hairy, also with eglandular hairs.

Flowers: Calyx campanulate, glabrous or minutely hairy, green dotted or tinged with red, 6–20 mm long, becoming asymmetrically inflated, 10–20 mm long, 8–15 mm in fruit, teeth 5, short, c. 2 mm long, the upper c. 4 mm long, broadly oblong, obtuse, sometimes mucronate, glandular hairy. Corolla 10–30(–40) mm long, yellow without spots or blotches on petals, throat usually finely spotted with red, the ridges densely hairy and nearly closing the throat, strongly 2-lipped, the lips spreading, the upper lip 2-lobed, the lower 3-lobed and typically horizontal. Style glabrous or puberulent. Stamens 4.

Fruits: Capsule broadly oblong, 7–15 mm long, inflated, constricted at the base, seeds longitudinally striate, 0.4–0.5 mm long, moderately to densely hairy with glandular and eglandular hairs.

Flowering: June-September.

Erythranthe guttata. Botanical Garden Aarhus, Denmark. Photo JCS.

Biology: Flowers pollinated by insects, mainly bumblebees. Commonly forming large stands by creeping rhizomes and vigorous seeding. Dispersal of seeds by water. Rooted in gravel or silt, often with basal leaves and stolons submerged throughout much of the year. Submerged leaves stay green during winter.

Habitats: A wide range of habitats, typically growing along the margins of flowing water bodies such as streams and rivers but also along canals, and on the margins of lakes and floodplain pools, marshy ground and in open woodland.

Distribution within the region:
Alien. Widespread throughout much of Europe, apart from the Mediterranean and parts of eastern Europe. Introduced to Britain in 1812 and now very widely established. Elevation: 20–3200(–3700) m. Native in North America.

Characteristics and similar species:
E. guttata can be distinguished from other taxa in the genus by the combination of more or less complete fertility, the throat of the corolla more or less completely closed by two swellings at the base of the lower lip and the presence of simple white hairs on the inflorescence, although these may be hidden among dense, long, glandular hairs and may be restricted to keels and bases of the calyces. These characters also serve to distinguish it from its hybrid with **319 *E. lutea*** (**315 *E.* × *robertsii***) and the recently evolved *E. peregrina*, from both of which additional field characters include the lower lip of the corolla of *E. guttata* held at an almost horizontal angle, at least in newly opened flowers and in its conspicuously inflated fruiting calyces.

The long, slender, small-leaved stolons produced in some abundance by *E. guttata* in the late summer and autumn have not been observed in any clone of its hybrids.

Erythranthe guttata. Marshy ground in riparian woodland, River Itchen, Hampshire, Britain. Photo RVL.

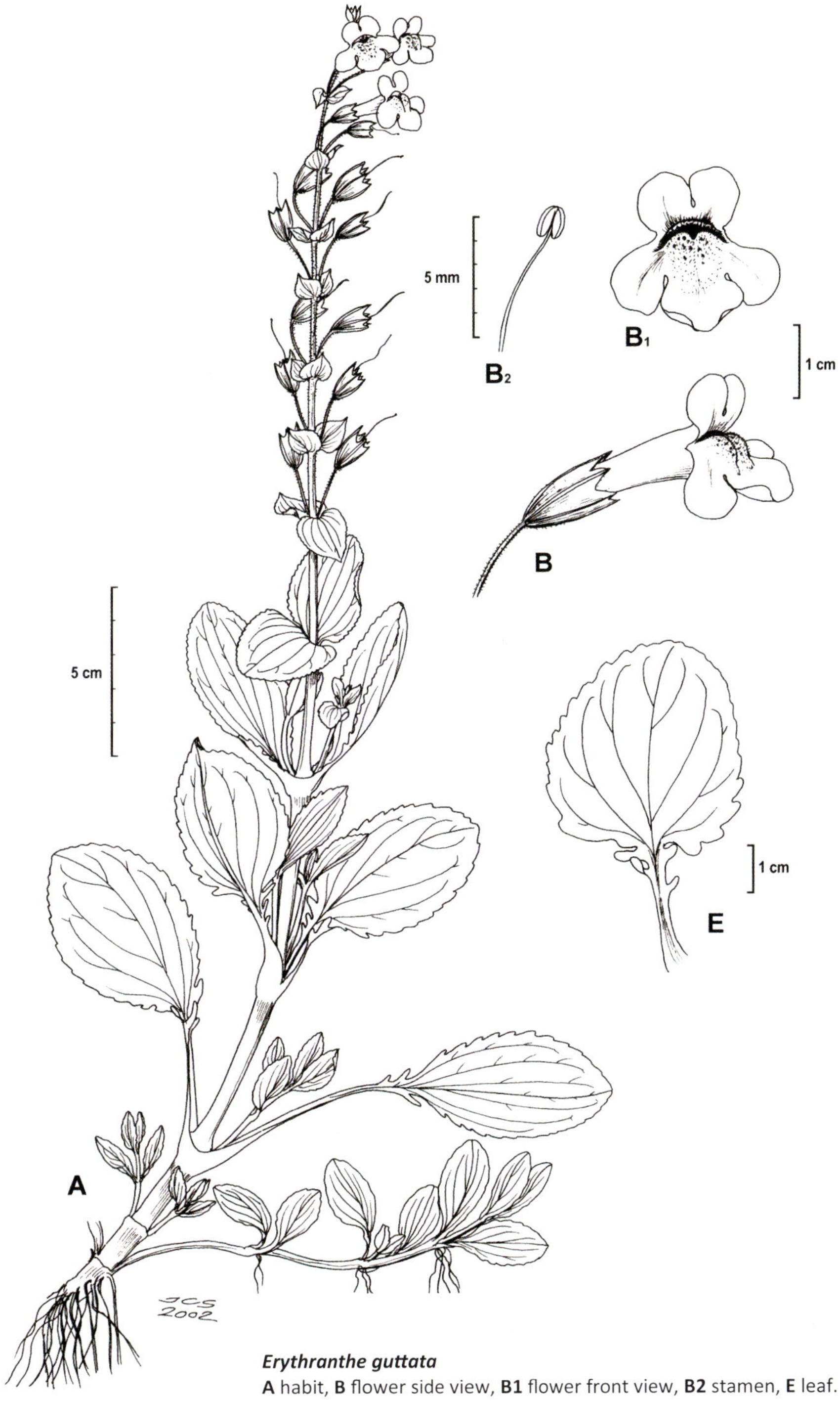

Erythranthe guttata
A habit, **B** flower side view, **B1** flower front view, **B2** stamen, **E** leaf.

Phrymaceae

315 *Erythranthe ×robertsii* (Silverside) G.L.Nesom

(*E. guttata × lutea*)

DK: Broget Abeblomst **N:** - **S:** Brokgyckelblomma **FIN:** - **IS:** -
GB: Hybrid Monkey-flower **NL:** - **F:** -
D: - **CZ:** - **PL:** -
EST: - **LV:** - **LT:** - **Ru:** -

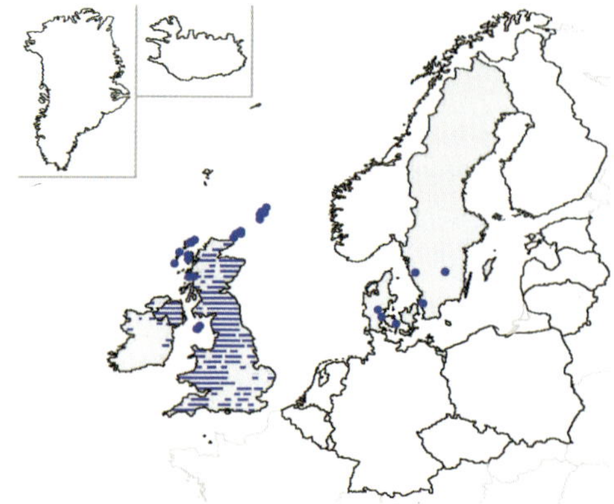

Map incomplete.

Erythranthe ×robertsii. Stream, Randbøldal, Jutland, Denmark. Photo JCS.

Perennial helophyte. Usually shortly rhizomatous or stolons short and thick, with the leaves not notably different in size from other cauline leaves. Roots fibrous, white or brownish.

Stem: Terete or somewhat 4-angled toward the apex, sometimes hollow, erect or weak and more or less trailing and rooting at the nodes, to c. 60 cm high, simple or sparsely branched, commonly glabrous below and variably hairy above, with both glandular and simple white hairs.

Leaves: Opposite, glabrous to hairy, membranous, basal leaves long petiolate, stem leaves sessile, clasping, variable, mostly rounded ovate to rhomboid, apex acute to subacute, 1–2 times longer than wide, pinnately or palmately 5–7-veined from the base, margins toothed, often with acute, more or less forward-directed teeth alternating with shallow, indistinct teeth often longer than wide and twisted.

Inflorescence: Few- to many-flowered leafy racemes, raceme leaves small, broadly ovate, acute, toothed. Flowers solitary in axils, sometimes solitary or few flowered. Lower pedicels up to 3 times the length of corollas at maturity, thick and 1–3 times as long as the calyx, with sparse, minute, simple, bristly hairs (less than 0.1 mm) even when the inflorescence is otherwise hairless.

Flowers: Calyx campanulate, sometimes becoming inflated, teeth 5, the upper c. 4 mm long, the lower four c. 2 mm long, broadly oblong, obtuse, sometimes mucronate, glandular-hairy. Corolla 25–45 mm long, golden yellow. Palate red spotted. Lobes commonly unmarked or else blotched with red on the lower central lobe or on all 5 lobes. Lower lip held angled downwards; throat only partially closed by the ridges of the palate, style glabrous or puberulent, stamens 4.

Fruits: Capsule remaining undeveloped. Sterile.

Flowering: June-September.

Biology: Vegetative propagation by short stolons.

Habitats: Similar to those of *E. guttata*; occurs in a wide range of habitats, typically growing along the margins of flowing water bodies such as streams and rivers but also along canals, and on the margins of lakes and floodplain pools, marshy ground and in open woodland. Rooted in gravel or silt, often with basal leaves and stolons submerged throughout much of the year.

Erythranthe ×robertsii. Leithen Water, Scotland. Photo RVL.

Erythranthe ×robertsii with extremely elongate racemes, with lower calyces withering. Roadside pond, Hampshire, Britain. Photo RVL.

Submerged rosette of *E. ×robertsii*, River Itchen, Hampshire, Britain. Photo RVL.

Distribution within the region:

Alien. Widespread throughout much of Europe, apart from the Mediterranean and parts of Eastern Europe. Elevation: 20–3200(–3700) m.

Characteristics and similar species:

The hybrid can be distinguished from other *Erythranthe* taxa by the combination of simple white hairs, at least on the calyces, the mouth of the corolla almost closed by two swellings at the base of the lower lobe or partially open and its sterility.

315a ***Erythranthe peregrina*** (Vallejo-Marín) G.L.Nesom

A recently established fertile plant, behaving as a species and described as such in 2012, apparently from back-crossing of *E.* × *robertsii* with *E. guttata*. It is understood to be distinguishable from *E.* × *robertsii* (which is sterile) in its high fertility. However, there are reports that plants indistinguishable in morphology from *E.* × *robertsii* sometimes arise from crosses between *E. guttata* and *E. cuprea* and have become naturalised, while plants identifiable as *E.* × *robertsii* may be produced by crosses of *E. guttata* with *E. lutea* var. *rivularis*, *E. lutea* var. *variegatus* or *M.* ×*smithii* Paxton (the latter a hybrid between *E. lutea* var. *rivularis* and *E. lutea* var. *variegata*).

Erythranthe peregrina. Glengonnar Water, Scotland. Photo RVL.

Phrymaceae

316 ***Erythranthe moschata*** (Douglas ex Lindl.) G.L.Nesom

Syn.: *Mimulus moschatus* Douglas ex Lindl.
DK: Moskus-Abeblomst **N:** Moskusgjøglerblom **S:** Myskgyckelblomma **FIN:** Myskiapinankukka **IS:** -
GB: Musk **NL:** Muskusplantje **F:** Mimule musqué
D: Moschus-Gauklerblume **CZ:** Kejklířka pižmová **PL:** Kroplik piżmowy
EST: - **LV:** - **LT:** Muskusinis puikūnas **Ru:** Губастик мускусный

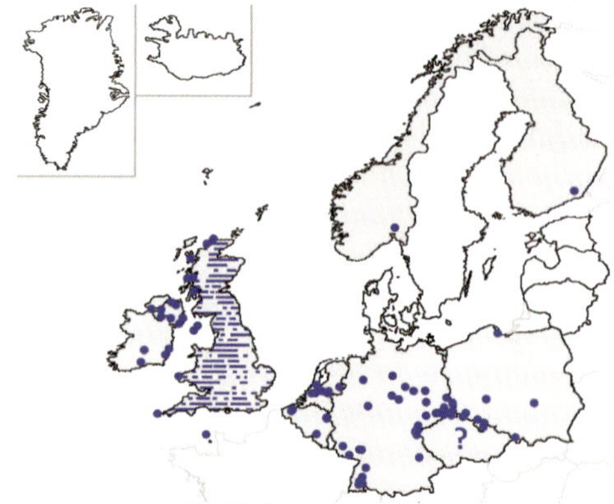

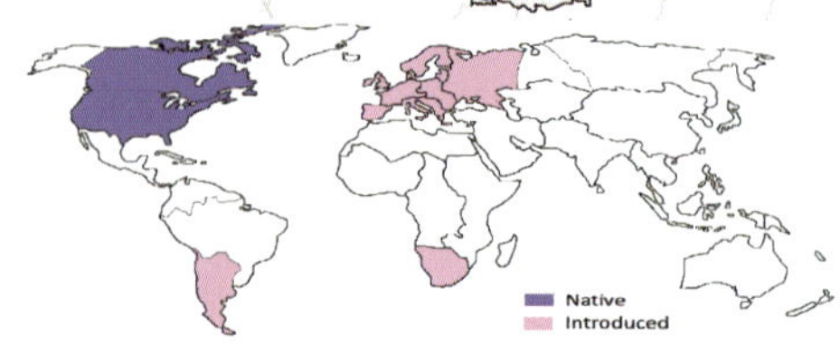

Erythranthe moschata. New Forest, England. Photo RVL.

A more or less slimy, viscid, villous perennial. Rhizomes slender, often moniliform.

Stem: 5–30 cm long, glandular villous, creeping, decumbent or ascending, rooting at the lower nodes. The main branches spreading up to 50 cm, with internodes 2–6.5(–10) cm long.

Leaves:. Ovate, 1–4(–5) cm long and 0.5–2.5 cm wide, acute or obtuse, base subcordate or nearly truncate, pinnately 3-veined from base; not joined at nodes, basal leaves not persistent; petioles (1–)2–5(–10) mm.

Inflorescence: Flowers few, 1 rarely 2 in the axils. Pedicels 10–22 mm long, slender, shorter than the leaves.

Flowers: Calyx campanulate, 6–13 mm long, weakly or not inflated in fruit, villous to glandular villous, teeth lanceolate, acute or often subulate, unequal, 3–4 mm long, frequently nearly one-third the length of the tube. Corolla 15–25 mm long, yellow, tube cylindrical, exserted, throat short, usually striped with red and with a patch of coarse hairs and brown blotches inside the lower side of open mouth, lobes rounded, subequal. Stamens included, filaments glabrous, anthers glabrous or pubescent, coherent nearly to the base. Style glabrous, stigma lips broad, rounded, equal, fimbriate.

Fruits: Capsule ovate, acuminate, 5–6 mm long. Seeds oval, tuberculate.

Flowering: June-September.

Biology: Although commonly described as musk scented, many populations are more or less scentless. It appears that the scented form, once widely cultivated, may have been a "sport" that has not bred true.

Habitats: *Erythranthe moschata* typically occurs in streams, flushes, the margins of ditches and ponds, in wet pasture and on wet woodland rides.

Distribution within the region:
Alien. Widespread but scattered in Britain, mainly in the west and occasional elsewhere in the region, but rare in Scandinavia.

Erythranthe moschata. Botanical Garden Aarhus, Denmark. Photo JCS.

Erythranthe moschata. Botanical Garden Aarhus, Denmark. Photo JCS.

Characteristics and similar species:
Erythranthe moschata differs from similar species by being glandular hairy all over and often viscid.

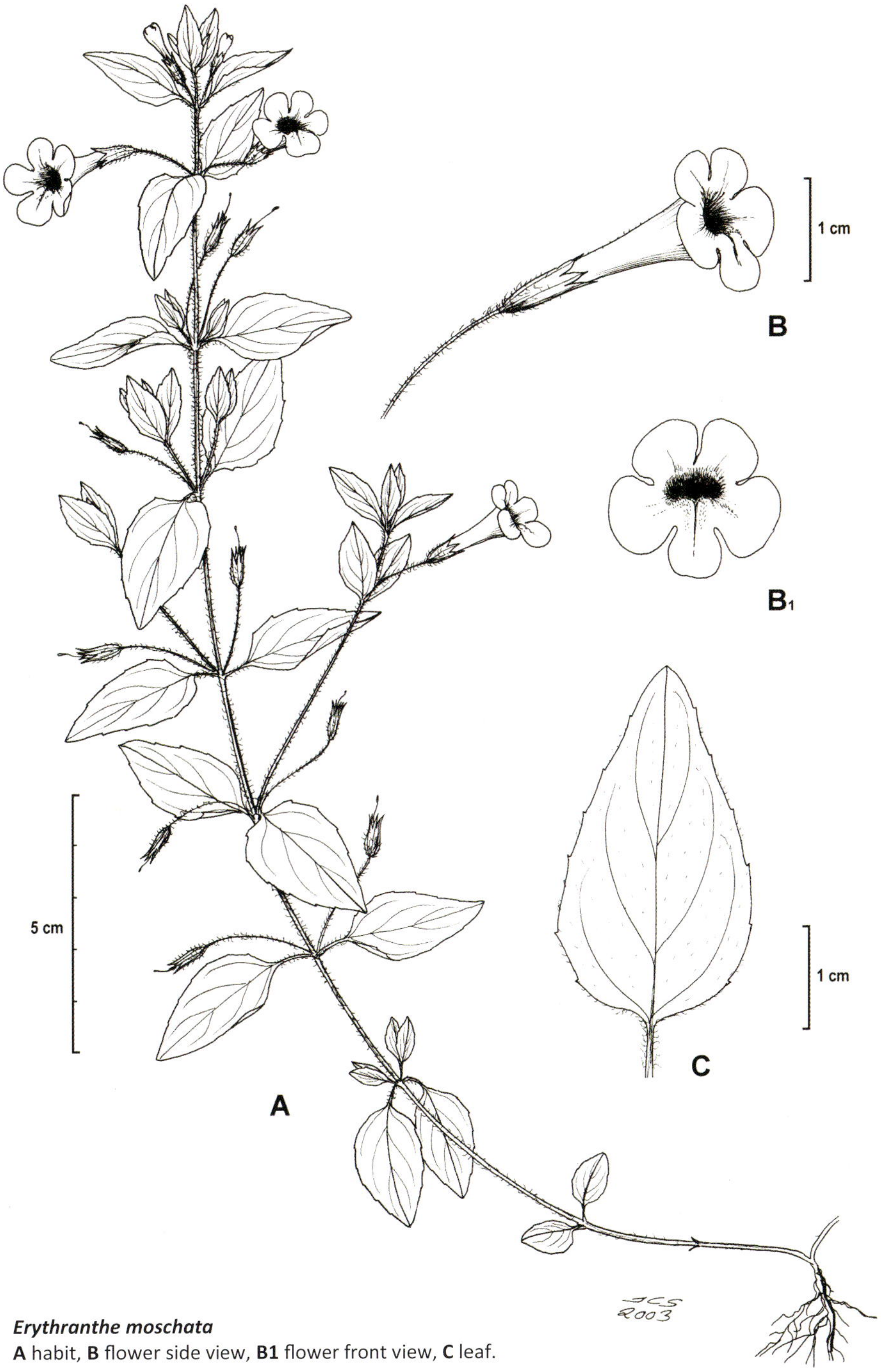

Erythranthe moschata
A habit, **B** flower side view, **B1** flower front view, **C** leaf.

Erythranthe lutea var. *rivularis.* Shingle bar on River Tweed, southern Scotland. Photo RVL.

Erythranthe ×burnetii "Duplex". Shingle bar on River Tweed, southern Scotland. Photo RVL.

317 *E. cardinalis* (Douglas ex Benth.) Spach

GB: Scarlet Monkeyflower

It can be distinguished from other *Erythranthe* species and hybrids in Europe by the calyx lobes, which are equal to subequal in length and more or less straight. Calyx not inflated and with sharp angles. Rarely recorded in Britain, naturalised on waste ground but not near water.

318 *E. cuprea* (Dombrain) G.L.Nesom

Very rarely recorded in the wild in Britain, but a parent of hybrid combinations which are naturalised. It typically has unblotched, coppery-coloured corolla. It can be distinguished from other *Erythranthe* species and hybrids by the high fertility, small corolla (< 12 mm long) and small, typically 3-veined leaves. It is usually a small, compact, decumbent plant. It can be distinguished from its hybrid with *E. guttata*, *E. × burnetiid,* by the lack of simple white hairs on the inflorescence. A cultivar with bright red flowers known as "Roter Kaiser" or "Red Emperor" has been recorded as a garden escape in Britain naturalised on waste ground but not near water.

319 *E. lutea* (L.) G.L.Nesom

GB: Blood-drop-emlets

Erythranthe lutea can be distinguished from other *Erythranthe* species and hybrids by the combination of high fertility, decumbent or prostrate habit and leaves with irregular, oblong, often twisted teeth. Naturalised in northern England and Scotland on the margins of rivers, particularly on bars, as well as locally in wet and marshy areas, however the habitats in which the different varieties occur is obscured by inconsistent identification.

Var. *lutea* has very rarely been recorded, var. *variegatus* is widely reported as naturalised in Britain and there are a small number of records of var. *rivularis*, although it is likely that there is much misidentification of hybrids of these taxa. The varieties can be distinguished as outlined in the key (pp. 648–649), but most naturalised populations appear to involve hybrids.

320 *E. ×burnetii* (S.Arn.) Silverside

(*E. guttata × cuprea*)

GB: Coppery Monkeyflower

Locally abundant on river shingle in northern England and Scotland and includes a petaloid form which is very locally established from Selkirk northwards. The petaloid form "Duplex" is unmistakable, as the calyx resembles the corolla, effectively producing the appearance of a double corolla ("hose in hose"). Typical forms of the hybrid can be distinguished from other *Erythranthe* species and hybrids by its sterility, combined with a coppery-coloured unblotched corolla. Widespread along river margins and on marshy ground, including the "Duplex" form which is more frequent on the Scottish borders.

321 *E. ×hybrida* (Voss) Silverside

(*E. cuprea × lutea* var. *rivularis × variegata)* or

(*E. cuprea × smithii*)

It is difficult to work out the taxonomy of this plant following the taxonomic changes of Nesom and the approach of Stace (2019). This is an extensively cultivated plant, with many horticultural forms showing a very wide range of colours and colour combinations,

Erythranthe ×maculosa. Margin of River Tweed, Berwickshire, Britain. Photo RVL.

Erythranthe ×hybrida. Cultivation, Selkirk, Britain. Photos RVL.

often used in hanging baskets, but due to its sterility appears only rarely to have been recorded in the wild. Is rarely recorded in Britain as a transient escape from cultivation on waste ground but not near water.

322 *E. ×maculosa* (T.Moore) G.L.Nesom

(*E. lutea × cuprea*)

GB: Scottish Monkeyflower

Is at least partly fertile. It can be distinguished from other *Erythranthe* species and hybrids by the combination of coppery tinge to the corollas, combined with evenly toothed, often small leaves. Very locally established in northern England and Scotland on shingle bars on rivers, as well as on river and stream margins and in flushes.

323 *E. ×smithii* (Paxton) Silverside

(*E. guttata × lutea* var. *variegata*)

Similar to *E. lutea* var. *rivularis* from which it differs in being more vigorous, often to 60 cm, with the corolla creamy to bright yellow, and the lower central lobe or all five lobes occupied by single, cherry red to purplish-red blotches; leaves broad, often almost rounded, with teeth mainly broader than long (Stace 2019). Very rarely recorded on river and stream margins and in flushes.

324 *E. guttata × lutea* var. *rivularis × lutea* var. *variegata*

(*E. ×caledonica* Silverside nom. nud.)

It is very similar to *M. × smithii* but sterile, with fruiting calyces rarely inflated, flower stalks and calyces with sparse white hairs, at least on the angles, lower flower stalks about twice the length of the calyx, corolla golden yellow; blotches orange brown, intensity of colour varying within individual blotches. This hybrid seems likely to have arisen naturally among mixed naturalised populations of *E. × smithii* and *E. guttata* in Scotland and northern England, where it has itself become naturalised and been taken into cultivation in cottage gardens. Locally frequent in northern England and southern Scotland on shingle bars on rivers, as well as on river and stream margins and in flushes.

325 *E. guttata × (E. lutea × E. cuprea)*

(*Mimulus × polymaculus* Silverside nom. nud.)

Resembles *E. ×maculosa* from which it differs in being sterile, with less intense spots on the corolla and more often hairy above. Locally naturalised in the Faroe Islands and Britain as a rare, transient garden escape. On wet ground such as stream and river margins, flushes, shingle bars on rivers and marshy ground.

326 *Mimulus ringens* L.

Flowers lavender blue. Cultivated as an ornamental and spread to waste ground and settling ponds; in northern France it is recorded from the River Meuse. It appears to be increasing with 5–6 recent records from the Netherlands (J. Bruinsma personal comment) and single records from Belgium (F. Verloove personal comment), the Czech Republic (Pladias 2021) and Britain, as well as a few sites elsewhere in Europe.

Mimulus ringens. Roeselare, West-Vlaanderen, Belgium. Photo Filip Verloove.

Lentibulariaceae

Key to *Utricularia*

1 Margins of terminal leaf segments without small papillae with bristles (***1***). Spur 1–3 mm long 2
1 Margins of terminal leaf segments with small papillae with bristles (***2***). Spur 3–10 mm long 3

1

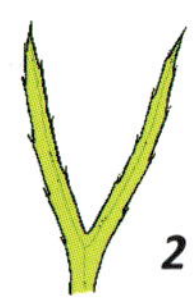

2

2 Lower lip of corolla ovate, 4–7 mm wide and soon with reflexed margins (***3***)
329 *U. minor*
2 Lower lip of corolla almost circular, 8–10(–12) mm wide and remaining flat (***4***)
330 *U. bremii*

3

4

3 Leafy shoots more than 20 cm long. All shoots alike with green leaves and many bladders (***5***) 4
3 Leafy shoots less than 20 cm long. Shoots dimorphic. Green leafy shoots without or with few bladders, colourless shoots with reduced leaves and many bladders (***6***) 5

5

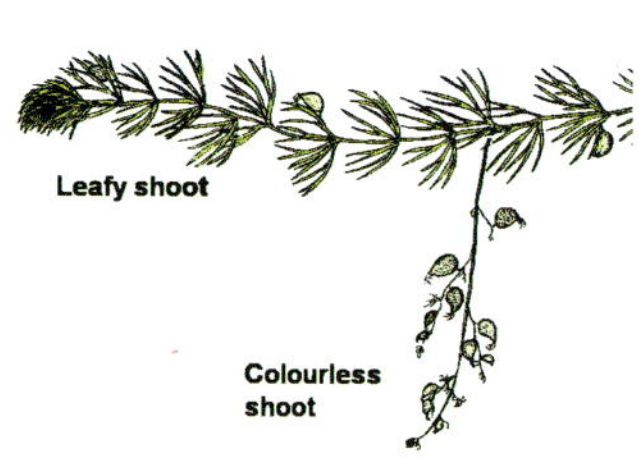

6

4 Lower lip of corolla dark or golden yellow with reflexed margins (saddle shaped) (***7***). Glands only on the abaxial side inside the spur. Pedicels recurved after flowering. Fertile
327 *U. vulgaris*
4 Lower lip of corolla light yellow, flat or with upturned margins (***8***). Glands on both abaxial and adaxial sides inside the spur. Pedicels sinuous after flowering. Sterile
328 *U. australis*

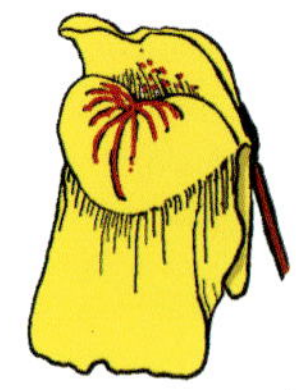

7

8

5 Green, leafy shoots without bladders. Spur pointing forward, appressed to the lower lip and nearly as long (***9***). Quadrifid hairs inside the bladders with arms in more or less parallel pairs (***9***)
331 *U. intermedia*
5 Green, leafy shoots with few bladders. Spur directed downwards at an acute angle, and about half as long as the lower lip (***10***). Quadrifid hairs inside the bladders with divergent arms (***11, 12***) 6

9

10

6 Spur 2–3 mm long. Quadrifid hairs with shorter arms diverging at (117°–)146°–197°(–228°) (***11***)
332 *U. ochroleuca*
6 Spur 4–5 mm long. Quadrifid hairs with shorter arms diverging at (30°–)52°–97°(–140°) (***12***)
333 *U. stygia*

11

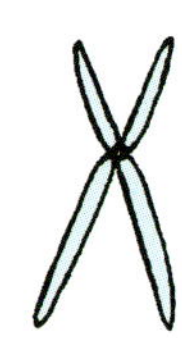

12

Quadrifid hairs

The angles between the short pair of arms and between the longer pair of arms of the quadrifid hairs inside the bladders are of diagnostic value and useful for determination of non-flowering specimens. The bladders need to be cut lengthwise and flattened out before being studied under a microscope, ideally at 100x magnification. Hairs from the middle part should be chosen for the best results, as the quadrifid hairs may vary depending on their position inside the bladder. The quadrifid hairs cannot be used to distinguish *U. vulgaris* from *U. australis*.

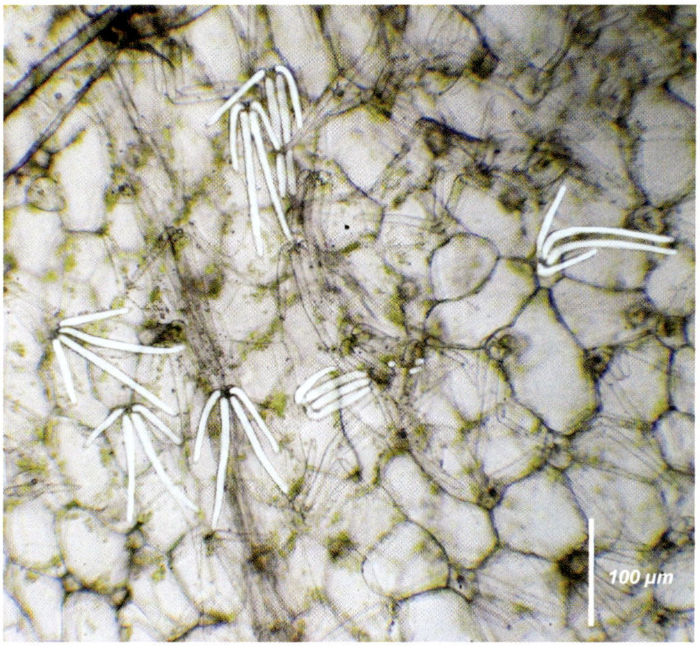

Utricularia minor. Peatbog. Tarm, Jutland, Denmark. Photo JCS.

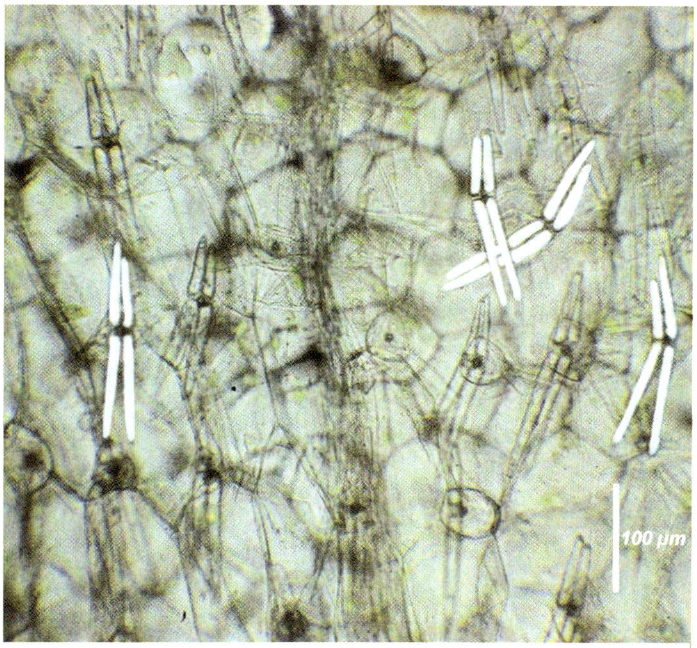

Utricularia intermedia. Lake Navn, Jutland, Denmark. Photo JCS.

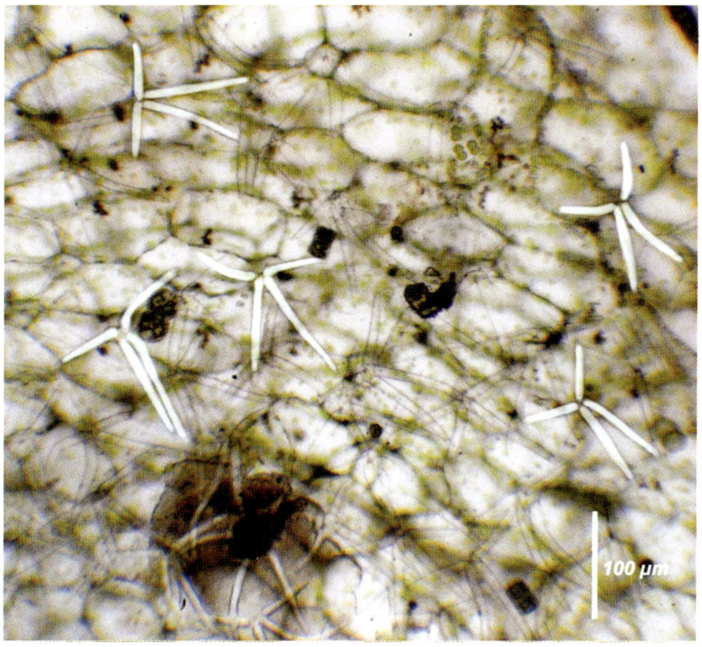

Utricularia ochroleuca. Peatbog. Tarm, Jutland, Denmark. Photo JCS.

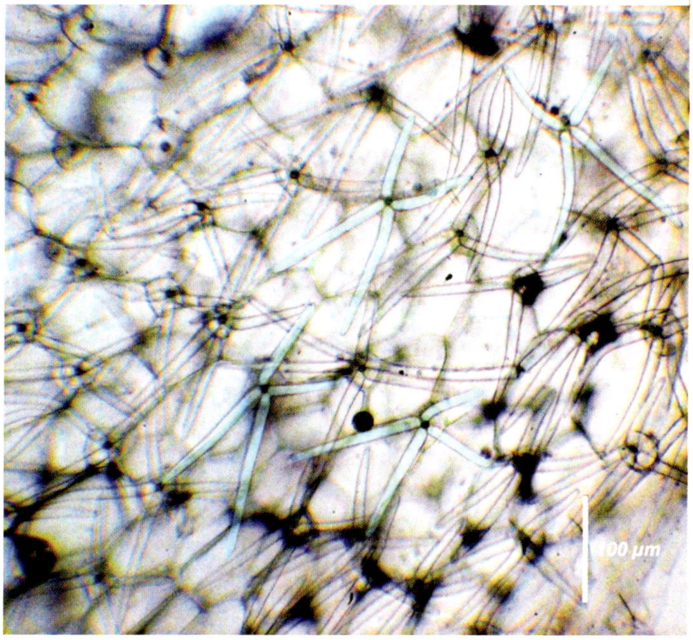

Utricularia stygia. Holmegaard, Sealand, Denmark. Photo JCS.

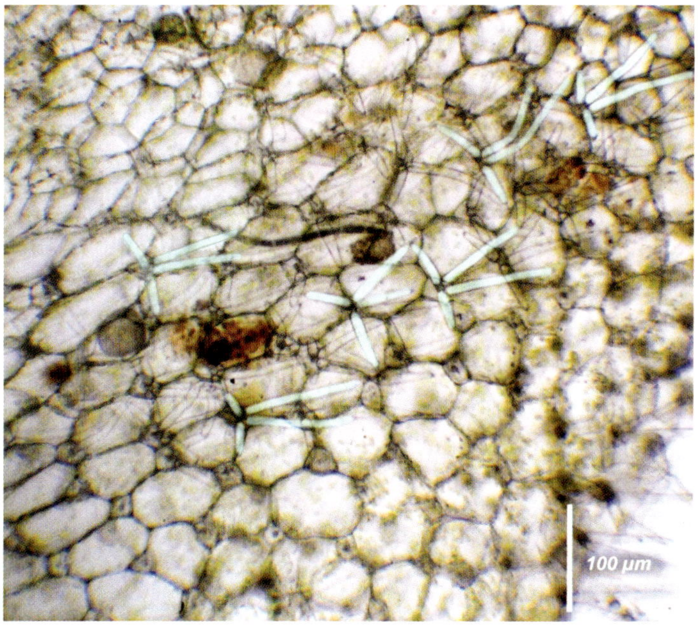

Utricularia australis. Ditch. Lille Vildmose, Jutland, Denmark. Photo JCS.

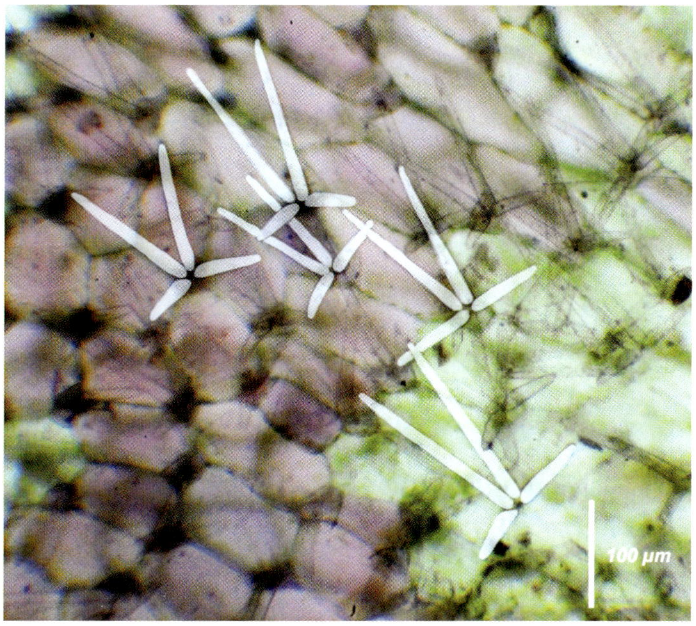

Utricularia vulgaris. Pond. Stavtrup, Aarhus, Denmark. Photo JCS.

Lentibulariaceae

327 *Utricularia vulgaris* L.

DK: Almindelig Blærerod **N:** Storblærerot **S:** Vattenbläddra **FIN:** Isovesiherne **IS:** -
GB: Greater Bladderwort **NL:** Groot blaasjeskruid **F:** Utriculaire commune
D: Gewöhnlicher Wasserschlauch **CZ:** Bublinatka obecná **PL:** Pływacz zwyczajny
EST: Harilik vesihernes **LV:** Parastā pūslene **LT:** Paprastasis skendenis **Ru:** Пузырчатка обыкновенная

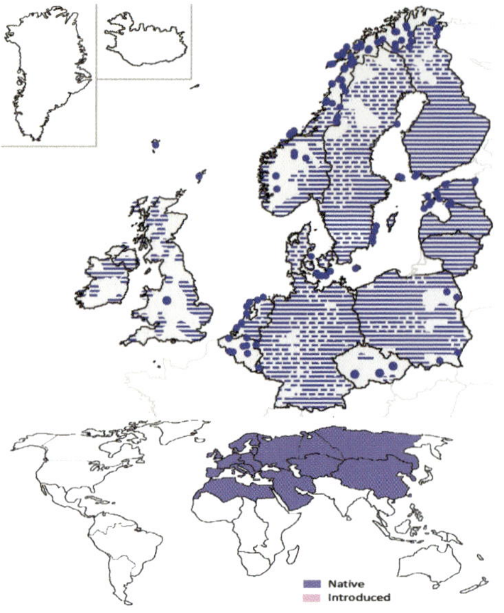

When flowering *U. vulgaris* is easily recognised by the dark yellow corolla with saddle-shaped lower lip. Gråmyr, Gudhjem, Bornholm. Photo JCS.

Perennial, submerged free-floating hydrophyte, without roots.

Stem: Homomorphic. 30–100 cm long, 1.5–3 mm in diameter, branched, yellowish to dark green, often with reddish or brownish tones. Often with branched rhizoids, up to 5, 5–35 mm long, at the base of the peduncle. Rhizoid segments 2–3(–5) mm long, dense.

Leaves: Sessile, alternate, in two rows, 2–8 cm long, green to brownish, 2–3-lobed in outline, each lobe 1–2 times pinnately divided. Terminal segments filiform, without a mid-vein, with 8–15 more or less prominent papillae on the margin, each with 1–5 bristles. Each leaf bearing 10–50 bladders, up to 4 mm long. The inner walls of the bladders are covered with quadrifid hairs c. 0.1 mm long (use a microscope 100x). The 2 longest arms diverge at an acute angle, while the angle between the 2 short arms varies from 60°–180°. Aerial shoots are rare and only occur on flowering plants. They are 2–15 cm long, 1–2 mm in diameter, bearing only a few small bract-like leaves.

Inflorescence: Peduncle emergent, c. 20 cm long, less than 1 mm in diameter, greenish, brownish or reddish with a few scale-like leaves. Racemes with 5–8 flowers. Bracts 3–5 mm long, ovate, blunt, brownish or reddish. Pedicels c. 8–10 mm long, recurved after flowering.

Flowers: Calyx of two weakly-lobed, green to reddish-brown sepals. Corolla 13–20 mm long, two-lipped, dark or golden yellow. Upper lip ovate 9–14 x 8–14 mm, more or less as long as the palate. Lower lip broadly ovate, 13–16 mm wide, with reflexed margins (saddle shaped). Palate swollen with red markings. Spur 6–10 mm long, conical, with glands inside only on the abaxial side. 2 stamens. Stigma 2-lipped.

Fruits: Capsule orbicular, 2–5 mm in diameter.

Flowering: July-August.

Biology: *Utricularia* species are carnivorous, aquatic plants, trapping the prey (mainly crustaceans) in bladders formed from transformed leaves (see

Without flowering stems it is not possible with certainty to distinguish *U. vulgaris* from *U. australis*, although the stems of *U. vulgaris* are, on average, longer. Compared to the other members of the genus the two species are much larger and with only one sort of shoot. Vodskov, Denmark. Photo JCS.

Flowering *U. vulgaris* (left) is recognised by the saddle-shaped lower lip, with reflexed margins, and by the dark yellow or golden yellow corolla. *U. australis* (right) can be indentified by the spreading margins of the lower lip and the lemon yellow or light yellow corolla. Photos JCS.

329 *Utricularia minor*). *Utricularia vulgaris* is pollinated by flies, small wasps and bees seeking nectar and guided towards the mouth by the red markings on the lower lip and palate. The weight of the insect opens the mouth, enabling it to enter past the anthers and stigma to the spur. Plants are moderately fertile. Fruits mature above water, but ripe seeds sink and are dispersed by water movement. Overwinters as dense, orbicular or ovate turions developed from the shoot apices in the late summer.

Habitats: Still or slow-flowing, slightly acidic to moderately alkaline, nutrient-poor to moderately nutrient-rich waters. Lakes in forested or open landscapes, ponds, ditches and canals, peat bogs and swamps.

Distribution within the region:
Native. Throughout the region except Iceland and Greenland but typically scattered or rare.

Characteristics and similar species:
Utricularia vulgaris and **328 *U. australis*** are both large, vigorous plants with only one type of stem unlike all other members of the genus in Europe.

Utricularia australis differs by the light yellow or lemon yellow corolla with the margins of the lower lip flared and spreading (see illustration above). It is almost always sterile, with obliquely upward-pointing pedicels which elongate after flowering. The rhizoids have branched segments (see photo p. 663). Without flowers, fruits or rhizoids it is not possible to distinguish *U. vulgaris* from *U. australis* reliably.

The characters mentioned below are indicative only, they are influenced by individual variation and must be used with caution, they should always be supported by microscopic examination of flowering specimens.

Utricularia australis: Terminal leaf segments with 6–10 prominent papillae - each papilla with 1–2 bristles. Angle between the 2 shorter arms of the quadrifid hairs inside the bladders 120°–180°.

Utricularia vulgaris: Terminal leaf segments with 8–15 more or less prominent papillae - each papilla with 1–5 bristles. Angle between the 2 shorter arms of the quadrifid hairs inside the bladders 60°–180°.

Utricularia vulgaris. Turion. Photo RVL.

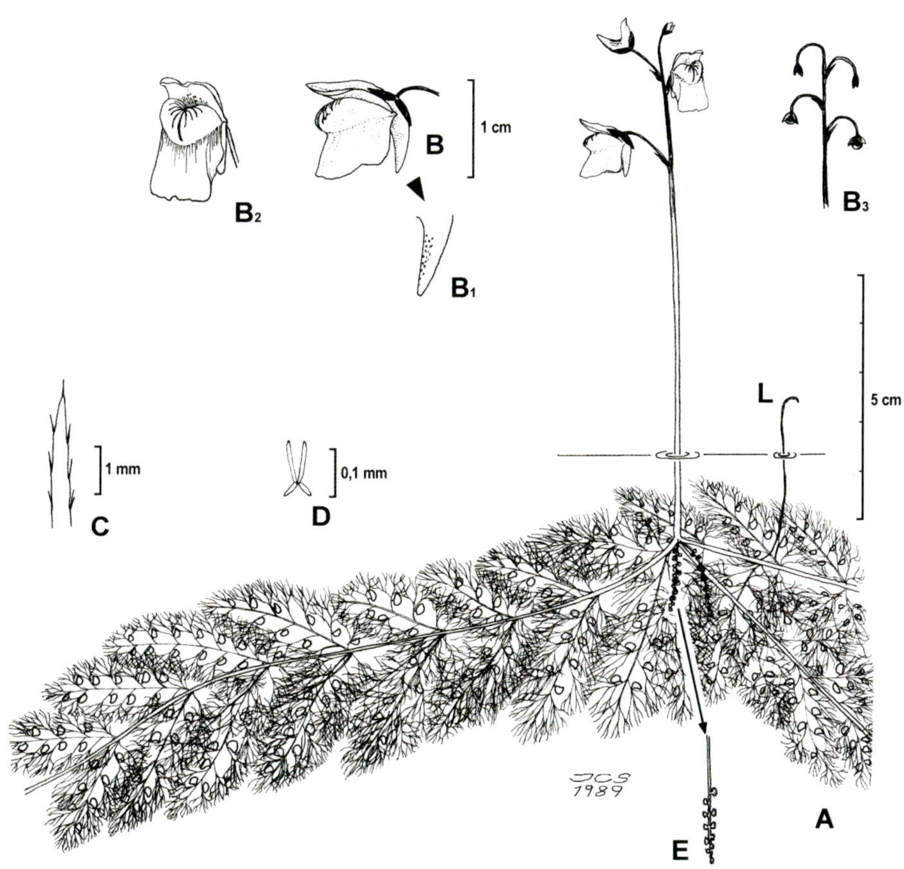

Utricularia vulgaris
A habit, **B** flower - side view, **B1** spur with internal glands, **B2** flower - face view, **B3** fruiting head, **C** terminal leaf segment, **D** quadrifid hair, **E** rhizoid, **L** aerial shoot.

Lentibulariaceae

328 *Utricularia australis* R.Br.

DK: Slank Blærerod **N:** Vrangblærerot **S:** Sydbläddra **FIN:** Lännen vesiherne **IS:** -
GB: Western Bladderwort **NL:** Loos blaasjeskruid **F:** Utriculaire citrine
D: Südlicher Wasserschlauch **CZ:** Bublinatka jižní **PL:** Pływacz zachodni
EST: Lõuna-vesihernes **LV:** Dienvidu pūslene **LT:** Pelkinis skendenis **Ru:** Пузырчатка южная

Utricularia australis in shallow, still water in hollows on the newly formed coastal foreland. Uggerby, North Jutland, Denmark. The light yellow corollas and the flared lower lip are reliable characters to distinguish *U. australis* from *U. vulgaris*. Photo JCS.

Perennial, submerged hydrophyte, free floating without roots.

Stem: Homomorphic. 20–50 cm long, 1.5–3 mm in diameter, branched, light to dark green, often with reddish or brownish tones. Often with branched rhizoids, up to 5, 10–50 mm long, at the base of the peduncle. Rhizoid segments 3–4(–10) mm long, loose.

Leaves: Sessile, alternate, spreading, in two rows, 2–6.5 cm long, green to brownish, 2–3-lobed in outline, each lobe 1–2 times pinnately divided. Terminal segments filiform, without a midvein, with 6–10 rather prominent papillae on the margin, each with 1–2 bristles. Each leaf bearing 10–50 bladders, up to 3.5 mm long. The inner wall of the bladders is covered with quadrifid hairs c. 0.1 mm long (use a microscope 100x). The 2 longest arms diverge in an acute angle, while the angle between the 2 short arms varies from 120°–180°. Aerial shoots are rare and only occur on flowering specimens. They are 2–15 cm long and 1–2 mm in diameter, bearing only a few small bract-like leaves.

Inflorescence: Peduncle emergent, c. 20 cm long, less than 1 mm in diameter, greenish, brownish or reddish with a few scale-like leaves. Raceme with 5–8 flowers. Bracts 3–5 mm long, oval, blunt, brownish or reddish. Pedicels c. 5–15 mm long, more or less straight and elongate but not recurved after flowering. Calyx of two weakly lobed, green to reddish-brown sepals.

Flowers:
Corolla 13–20 mm long, 2-lipped, lemon yellow to yellow. Upper lip oval, 7–10 x 7–9 mm, more or less twice as long as the palate. Lower lip broadly ovate or almost circular, 15–20 mm wide, with spreading margins. Palate swollen, with red markings. Spur 5.5–8 mm long, conical, with glands inside on both the abaxial and adaxial sides. 2 stamens. Stigma 2-lipped. The anthers produce morphologically normal pollen grains (Beretta et al., 2014), even though the species appears to be almost totally sterile.

Fruits: Capsules extremely rare, only known from China and Japan (Taylor, 1989), orbicular, 2–4 mm in diameter.

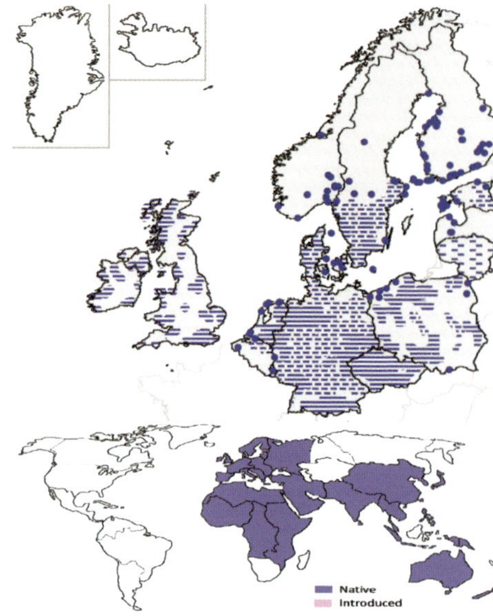

Flowering: July-August.

Biology: Carnivorous aquatic plant (see 329 *Utricularia minor*). *Utricularia australis* is considered to have originated through hybridisation between *U. tenuicaulis* Miki and *U. macrorhiza* Leconte ex Torr (Kameyama et al. 2005). The flowers are visited by insects, but viable seeds are not produced. Overwintering and vegetative reproduction by large, compact, orbicular or ovate turions, 5–15 mm, which develop from stem apices in the late summer.

Habitats: Still or slow- flowing, slightly acidic to slightly alkaline, oligotrophic to mesotrophic or eutrophic waters. Dune depressions, peat bogs, ditches, fish farm ponds, small ponds in sphag-

Utricularia australis inhabits more or less the same habitats as *U. vulgaris*, but appears to be less nutrient demanding. Lille Vildmose, Denmark. Photo JCS.

Utricularia australis (left) and *U. vulgaris* (right). Differentiation using vegetative characters such as the size of the papillae on terminal leaf segments is unreliable, as is using the quadrifid hairs inside the bladders (p. 659). Photos JCS.

num bogs, gravel pits and quarries, ponds in fens, drainage channels and lakes.

Distribution within the region:
Native. Throughout the region except Iceland and Greenland but typically scattered or rare.

Characteristics and similar species:
For distinction from **327 *U. vulgaris*** please see p. 661.

Utricularia australis (left) and *U. vulgaris* (right). It is often possible to find rhizoids at the base of the peduncle of flowering or fruiting specimens, which at least in Northern Europe is a useful character to separate the two species. The rhizoids serve as anchors for some *Utricularia* species, but have no obvious function for the free-floating species. Rhizoids are occasionally produced by *U. intermedia* and rarely by *U. ochroleuca* and *stygia*. *U. minor* and *U. bremii* always lack rhizoids. Photo JCS.

328a *Utricularia inflata*
Walter

Native to the United States, presumably deliberately planted in a bog in Bavaria, Germany. At flowering, it is easily recognised by the stem that rises above the water over a wheel-like float of inflated leaf stalks. In vegetative state it differs from **327 *U. vulgaris*** and **328 *U. australis*** by the zigzag shape of the stem branches.

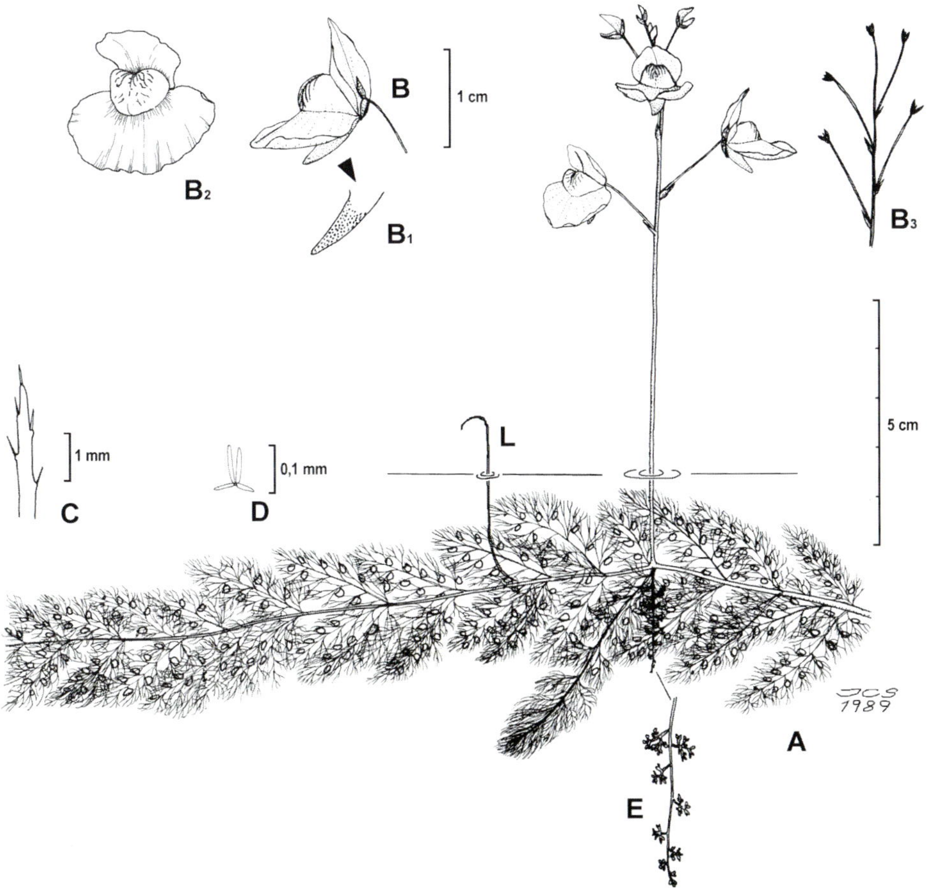

Utricularia australis
A habit, **B** flower - side view, **B1** spur with internal glands, **B2** flower - face view, **B3** fruiting head, **C** terminal leaf segment, **D** quadrifid hair, **E** rhizoid, **L** aerial shoot.

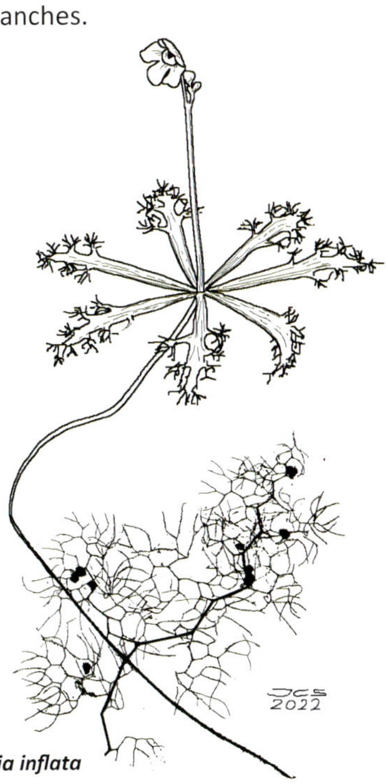

Utricularia inflata

Lentibulariaceae

329 *Utricularia minor* L.

DK: Liden Blærerod **N:** Småblærerot **S:** Dvärgbläddra **FIN:** Pikkuvesiherne **IS:** Blöðrujurt
GB: Lesser Bladderwort **NL:** Klein blaasjeskruid **F:** Petite utriculaire
D: Kleiner Wasserschlauch **CZ:** Bublinatka menší **PL:** Pływacz drobny
EST: Väike vesihernes **LV:** Mazā pūslene **LT:** Mažasis skendenis **Ru:** Пузырчатка малая

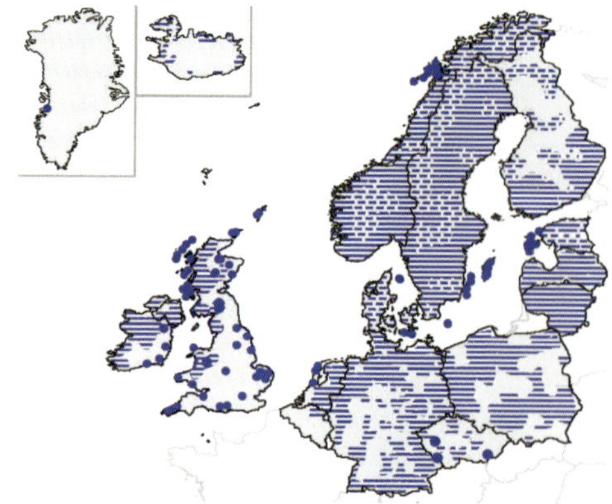

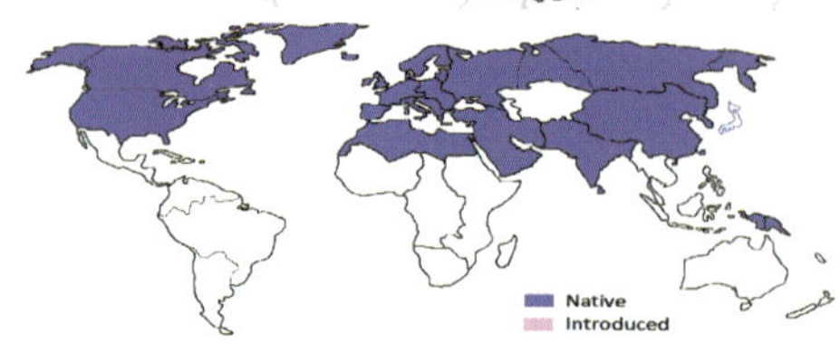

Utricularia minor is often abundant in small acidic, oligotrophic waters. Bog pool. Hammer Bakker, Jutland, Denmark. Photo JCS.

Perennial, submerged hydrophyte, free floating without roots.

Stem: Dimorphic. Leafy stolons 5–40 cm long, up to 1 mm in diameter, branched, green or with purplish tones. In shallow water or on moist sediment, 3–13 cm long; colourless shoots with reduced leaves are often anchored in the substrate. The shoots bear numerous bladders to catch small animals. The base of the peduncle without rhizoids.

Leaves: Submerged shoots appear slender, less than 20 mm wide, with sessile leaves in two alternate rows. The leaves are 3–15 mm long, more or less fan shaped in outline, green or with reddish tones, 2–3(–4) times dichotomously divided into 7–22 filiform, terminal segments. The segments lack a mid-vein and marginal papillae, but have 1 or 2 apical bristles. Most leaves have 1–6 bladders, 1–2.5 mm long. The inner wall of the bladders has numerous quadrifid hairs c. 0.1 mm long (use a microscope 100x), which differ in shape from those of all other European species (except *U. bremii*) in having the 4 arms directed forward like fingers on a hand. The 2 long arms are slightly divergent, while the 2 short arms are directed forward in an acute angle to the longer ones.

Inflorescence: Peduncle emergent, 5–10(–17) cm long, less than 1 mm in diameter, greenish, brownish or reddish with few scale-like leaves. Raceme with 2–5 flowers. Bracts 1–2 mm long, ovate, blunt, brownish or reddish. Pedicels c. 5 mm long, curved after flowering.

Flowers: Calyx of two weakly lobed, green to reddish-brown sepals. Corolla 6.5–8 mm long, 2-lipped, pale yellow to faintly greenish yellow. Upper lip ovate 2–3 x 2 mm, more or less as long as the palate. Lower lip ovate, 4–7 mm wide, with reflexed margins. Palate elongate, with red markings. Spur 1–2 mm long, short, blunt and with glands on all sides inside. 2 stamens. Stigma 2-lipped.

Fruits: Capsules orbicular, 2–2.5 mm in diameter.

Utricularia minor. Because of the reflexed margins of the lower lip, the flowers look smaller than they actually are. The spur is very short and often with brownish markings. Hammer Bakker, Jutland, Denmark. Photo JCS.

Shoot of *U. minor* with purplish tones (anthocyanin). Jordtorp, Öland, Sweden. Photo JCS.

The bladder functions as a trap and consists of transformed leaf segments. Long hair forms a funnel, leading small aquatic animals towards the trapdoor. The quadrifid hairs inside the bladder absorb water, which is then secreted by glands on the outside to the surrounding water. The vacuum so produced inside the bladder is finely balanced with the turgor pressure in the trapdoor cells. Small invertebrates attracted by a sugary slime secreted near the trapdoor release the trap by touching the external trigger hair. In just 300–700 µs the door opens and the prey is sucked in and dissolved by enzymes. The resulting substances are absorbed by the plant as a nutrient supplement that, while not necessary for the plant's survival, produces better growth. *U. minor*. Jordtorp, Öland, Sweden. Photo JCS.

Flowering: June-August.

Biology: Carnivorous aquatic plant (see photo caption above right). *Utricularia minor* flowers abundantly even when beached on wet substrate. The flowers are visited by small, flying insects and typically moderately fertile. Sexual reproduction by seeds and vegetative reproduction by large, globose turions 0.5–5 mm in diameter, which develop at the apices of leafy shoots in late summer.

Habitats: Shallow, oligotrophic, acidic or base-rich waters. Peat bogs and sphagnum bogs, water-filled hollows and ditches in moorland and fens, lakes with peaty water, alkaline springs.

Distribution within the region: Native. Scattered or common more or less throughout.

Characteristics and similar species: *Utricularia minor* can be identified by the small, pale yellow flowers with the margins of the lower lip reflexed and by the slender shoots with dichotomously divided leaves bearing bristles only at the apices.

330 *Utricularia bremii* is more robust and bigger than *U. minor* in all parts. The lower lip of the corolla is 8–10 mm wide, more or less circular and flat; the leaves are larger, with 12–25(–50) terminal segments.

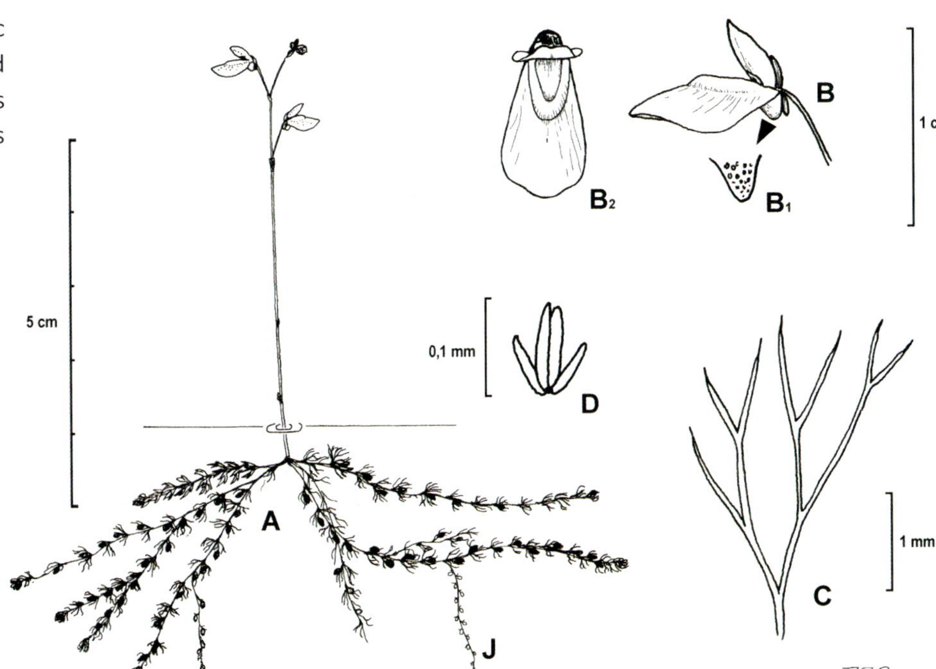

Utricularia minor
A habit, **B** flower - side view, **B1** spur with internal glands, **B2** flower - face view, **C** leaf, **D** quadrifid hair, **J** leafless stolon with bladders.

Lentibulariaceae

330 *Utricularia bremii* Heer

Syn.: *U. minor* R.Br. var. *bremii* (Heer) Franchet; *U. minor* R.Br. subsp. *bremii* (Heer) Bertsch & F.Bertsch
DK: Tallerken-Blærerod **N:** - **S:** - **FIN:** - **IS:** -
GB: New Forest Bladderwort **NL:** - **F:** Utriculaire de Bremi
D: Zierlicher Wasserschlauch **CZ:** Bublinatka vícekvětá **PL:** Pływacz Brema
EST: - **LV:** - **LT:** Liūninis skendenis **Ru:** Пузырчатка Брема

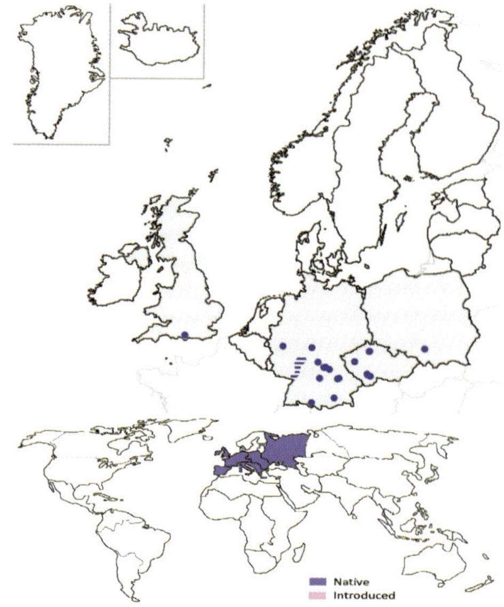

Utricularia bremii. Franken, Bavaria, Germany. Photo Thomas Franke.

Perennial, submerged hydrophyte, free floating without roots.

Stem: Dimorphic. Leafy stolons up to 60 cm long, about 1 mm in diameter, branched, green or with purplish tones. In shallow water or on moist substrate 3–13 cm long, colourless shoots with reduced leaves are often anchored in the substrate. The shoots bear numerous bladders to catch small animals. The base of the peduncle without rhizoids.

Leaves: Submerged shoots appear slender, less than 30 mm wide, with sessile, leaves in two alternate rows. The leaves are 3–20 mm long, 3.5–30 mm wide, more or less fan shaped in outline, green or with reddish tones, often 4–5 times dichotomously divided into 12–25 (–50) filiform, terminal segments. The segments lack a mid-vein and marginal papillae, but have 1 or 2 apical bristles. Most leaves have 1–10 bladders, 1–2.8 mm long. The inner walls of the bladders are covered with quadrifid hairs c. 0.1 mm long (use a microscope 100x), which differ in shape from all other European species (except *U. minor*) by the 4 arms directed forward like fingers on a hand. The 2 long arms are slightly divergent, while the 2 short arms are strongly divergent and directed forward at an acute angle relative to the longer ones.

Inflorescence: Peduncle emergent, 5–40(–60) cm long, c. 1 mm in diameter, greenish, brownish or reddish with 2–5 scale-like leaves. Raceme with 2–14 flowers. Bracts 1–2 mm long, oval, blunt, brownish or reddish. Pedicels c. 10 mm long, curved after flowering.

Flowers: Calyx of two weakly lobed, green to reddish-brown sepals. Corolla 8–10(–12) mm long, 2-lipped, pale- to lime yellow with brownish streaks. Upper lip ovoid, as long as or longer than the palate. Lower lip 8–10 (–12) mm wide, almost circular and flat. Palate elongate with red markings. Spur 2–3(–4) mm long, slightly longer than wide, blunt, with glands inside on all sides. 2 stamens. Stigma 2-lipped.

Fruits: Extremely rare. Capsules orbicular, 2–2.5 mm in diameter.

Flowering: July-September.

Biology: Carnivorous aquatic plant (see **329 *Utricularia minor***). *Utricularia bremii* flowers abundantly, even on dried-up sediment. The flowers are visited by small insects, but this species appears to be male-sterile, almost always producing only malformed and asymmetrical pollen. However, there are a few records of plants producing viable seeds. Propagation is almost completely vegetative and by 0.5–5 mm large, globose turions which develop at the apices of leafy shoots in late summer.

Habitats: Oligotrophic to mesotrophic or dystrophic, slightly acidic to slightly basic waters. Peat bogs, mires, water-filled hollows and ditches in moorland and around calcareous springs.

Distribution within the region:
Native. Scattered and rare in the Czech Republic, Germany and Poland, with a single known population in England. Perhaps overlooked elsewhere.

Utricularia bremii. Franken, Bavaria, Germany. Photo Thomas Franke.

Utricularia bremii. Leaf with bladders. Photo KvdW.

Characteristics and similar species: *Utricularia bremii* is a critical taxon which is very similar to **329 *U. minor***.

Utricularia bremii is more robust and bigger in all its parts than *U. minor* and typically has more bladders. However, because there is an overlap in the size of most parts, particularly vegetative parts, fully developed plants with flowers are needed for reliable identification. The most reliable characters to differentiate the two species are the size and shape of the flower, together with use of a microscope to determine whether the pollen is fully formed (*minor*) or malformed and asymmetrical (*bremii*).

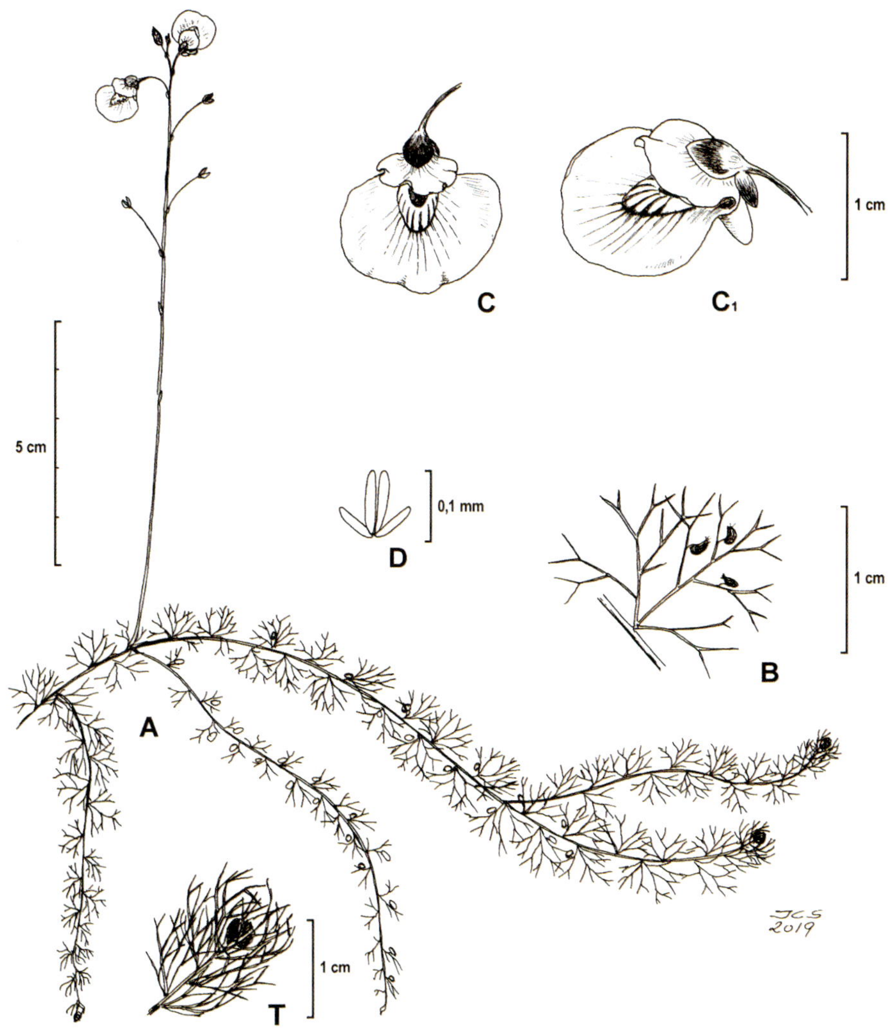

Utricularia bremii
A habit, **B** leaf with bladders, **C** flower - face view, **C1** flower side view, **D** quadrifid hair, **T** turion.

Lentibulariaceae

331 *Utricularia intermedia* Hayne

DK: Storlæbet Blærerod **N:** Gytjeblærerod **S:** Dybläddra **FIN:** Rimpivesiherne **IS:** -
GB: Intermediate Bladderwort **NL:** Plat blaasjeskruid **F:** Utriculaire intermédiaire
D: Mittlerer Wasserschlauch **CZ:** Bublinatka prostřední **PL:** Pływacz średni
EST: Vahelmine vesihernes **LV:** Vidējā pūslene **LT:** Balinis skendenis **Ru:** Пузырчатка средняя

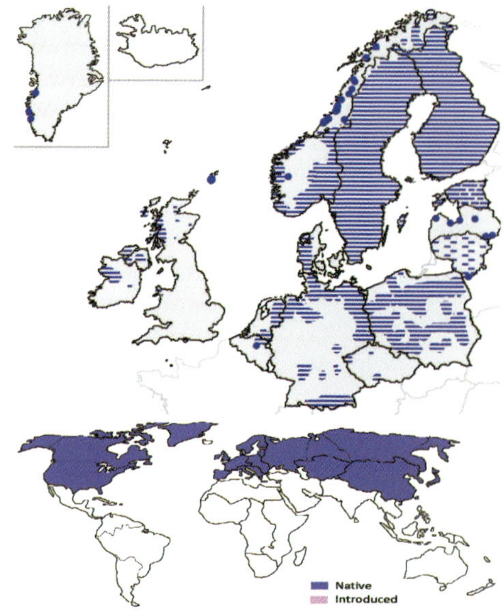

Flowering *U. intermedia* in shallow, acidic, oligotrophic water. Thy, Denmark. Photo JCS.

Perennial, submerged hydrophyte, free floating without roots.

Stem: Shoots represented by two types of stolons. Leafy stolons 5–20(–40) cm, branched, with numerous alternating green leaves and normally with few or no bladders. Leafless stolons 3–20 cm long, colourless, with several bladders and often partly buried in the substrate. Rhizoids at the base of the peduncle rare, 1–4 cm long.

Leaves: 4–20 mm long, sessile and flat. Often tripartite at the base and then dichotomously divided. Terminal segments with 8–15 very small papillae on the margin, each papilla with 1–6 short bristles. Bladders 1.5–4 mm long, on the inside densely covered with quadrifid hairs c. 0.1 mm long (use microscope 100x) with arms in almost parallel pairs. Angle between shorter arms 2°–21°(–37°) and between longer arms 3°–12°(–18°).

Inflorescence: Peduncle emergent, 10–20 cm long, approximately 1 mm in diameter, green to reddish brown, with a few scale-like leaves. Raceme with 2–5 flowers. Bracts 1–2 mm long, ovate, green or brownish. Pedicels 10–15 mm long, straight.

Flowers: Calyx of two weakly lobed, yellowish-green to reddish brown sepals. Corolla 12–14 mm long, 2-lipped, yellow to deep yellow. Upper lip ovate and shorter than the lower lip, which is almost circular, 8–12 mm long, flat or with slightly reflexed margins. Palate swollen, almost closing the mouth and with orange markings. Spur forward-pointing, more or less appressed to and nearly as long as the lower lip. With glands inside on both abaxial and adaxial sides. 2 stamens. Stigma 2-lipped.

Fruits: Capsule globose, 2–2.5 mm wide, with recurved pedicels.

Flowering: June-September.

Biology: Carnivorous aquatic plant (see **329 *Utricularia minor***). Bladders trap small aquatic animals. *Utricularia intermedia* produces few flowers and reproduction is mainly vegetative. Flowers are pollinated by small insects and seed production is moderate. In late autumn, large, globose turions 3–6 mm in diameter are formed at the apex of leafy shoots, while the rest of the plant dies back.

Habitats: Shallow, still, acidic, oligotrophic water in peat bogs, swamps, lake margins, ditches and slow-flowing streams. Also in nutrient-poor, highly calcareous waters.

Distribution within the region:
Native. Rare in the south, from Ireland east through Denmark, Belgium and Germany to Poland. More abundant and locally common in parts of Norway, Sweden, Finland and the Baltic states.

In *U. intermedia* the spur is appressed to the lower side of the lip and almost as long as this. Thy, Denmark. Photo JCS.

Utricularia intermedia as it often appears in shallow water. The green, leafy shoots densely cover the surface, and the bladders on leafless stolons are hidden beneath the mass of leaves. On dried-out substrates the plant might be mistaken for *Lycopodiella inundata*. Navn Sø, Denmark. Photo JCS.

Utricularia intermedia. In deeper water both types of shoots are often visible. The green, leafy shoots normally lack bladders, but occasionally a single bladder can be found. If more bladders are present on the green shoots, it is more likely that the plant is either *U. ochroleuca* or *U. stygia,* and close examination of the quadrifid hairs inside the bladders is needed to confirm its identity. The whitish, leafless stolons bear several bladders and also function as anchors, more or less buried in the substrate. Thy, Denmark. Photo JCS.

Characteristics and similar species: When flowering, this species can be identified by the forward-pointing spur appressed to the underside of the lower lip and more or less of the same length as this. When not flowering, it can be identified by the leafy shoots having no or at most a few bladders and by the quadrifid hairs inside the bladders having almost parallel arms.

332 *Utricularia ochroleuca* and **333 *U. stygia*** have flowers with shorter, downward-pointing spurs, combined with leafy shoots bearing several bladders and the quadrifid hairs inside the bladders with arms forming an X.

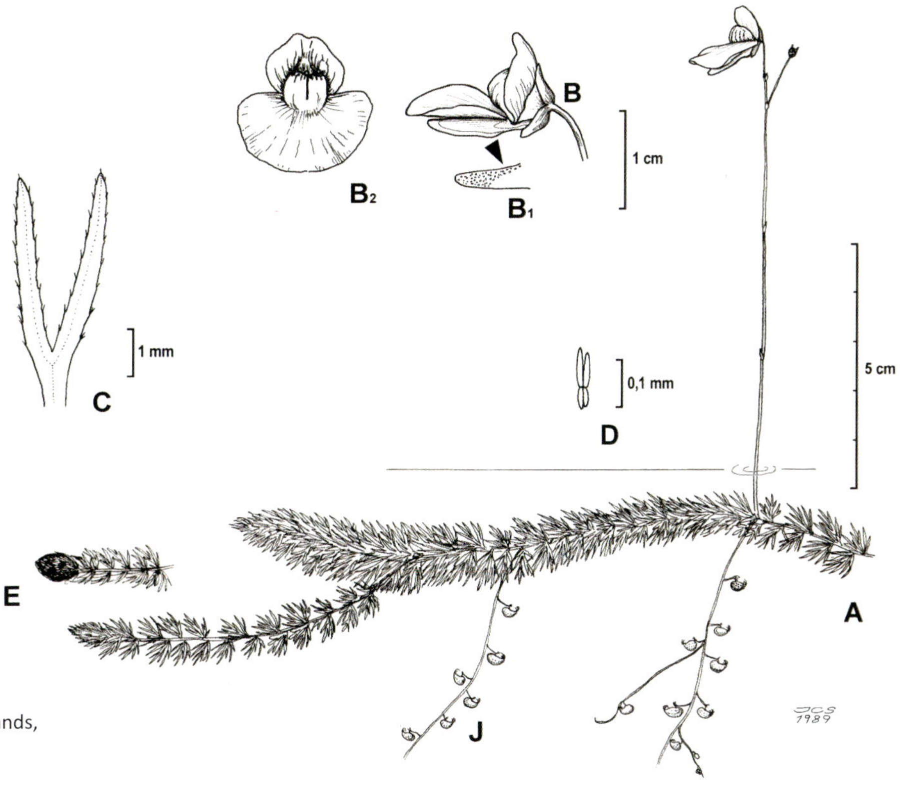

Utricularia intermedia
A habit, **B** flower - side view, **B1** spur with internal glands, **B2** flower - face view, **C** part of leaf, **D** quadrifid hair, **E** turion, **J** leafless stolon with bladders.

Lentibulariaceae

332 *Utricularia ochroleuca* R.W.Hartm.

DK: Kortsporet Blærerod **N:** Mellomblærerot **S:** Blekblāddra **FIN:** Kalvasvesiherne **IS:** -
GB: Pale Bladderwort **NL:** Bleekgeel blaasjeskruid **F:** Utriculaire jaunâtre
D: Blassgelber Wasserschlauch **CZ:** Bublinatka bledožlutá **PL:** Pływacz krótkoostrogowy
EST: Kollakas vesihernes **LV:** Gaišdzeltenā pūslene **LT:** Gelsvasis skendenis **Ru:** Пузырчатка светложёлтая

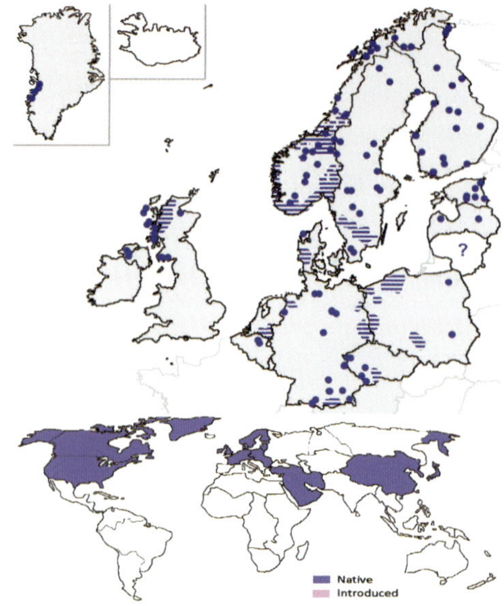

Utricularia ochroleuca growing in an old peat bog near Tarm, Jutland, Denmark. Photo JCS.

Perennial, submerged hydrophyte, free floating without roots.

Stem: Shoots represented by two types of stolons. Leafy stolons 5–20(–25) cm, branched, with numerous alternate green leaves and normally with some bladders. Leafless stolons 3–20 cm long, colourless, with several bladders and often partly buried in the substrate. Rhizoids at the base of the peduncle rare, 2–4 cm long.

Leaves: 6–20 mm long, green to brownish, sessile and flat. Often tripartite at the base and then dichotomously divided. Terminal segments with 8–15 very small papillae on the margin, each papilla with 1–2 short bristles. Bladders 1.5–4 mm long. The inner surfaces densely covered with quadrifid hairs c. 0.1 mm long (use microscope 100x) with spreading arms resembling a butterfly. Angle between shorter arms (117°–)146°–197°(–228°), between longer arms (34°–)60°–93°(–123°).

Inflorescence: Peduncle emergent, 10–20 cm long, not more than 1 mm in diameter, green or greenish brown to reddish brown and with a few scale-like leaves. The raceme with 1–4 flowers. Bracts 1–2 mm long, ovate, blunt, green or brownish. Pedicels 7–15 mm long, straight.

Flowers: Calyx of two weakly lobed, yellow-green to reddish-brown sepals. Corolla 8–10 mm long, 2-lipped, yellow. Upper lip ovate, 7 x 5 mm, shorter than the lower lip. Lower lip almost circular, 8–10 mm long, at first flat, but soon with reflexed margins. Palate swollen, more or less closing the mouth and with orange markings. Spur about 3 mm long, half as long as the lower lip and pointing downwards. With glands inside on both abaxial and adaxial sides. 2 stamens. Stigma 2-lipped.

Fruits: Very rare.

Flowering: June-September.

Biology: Carnivorous aquatic plant (see **329 *Utricularia minor***). *Utricularia ochroleuca* appears to be a vegetative apomict, and is considered to have originated through hybridisation between *U. intermedia* and *U. minor*. It rarely flowers, and although the flowers are visited by small insects, most of the pollen produced is sterile and seed set is extremely rare. Propagation is by division of shoots and by turions, which are 2–4 mm, oval to globose and develop at the apices of leafy shoots. The plant dies back in the late autumn and only the turions overwinter. *U. ochroleuca* is highly morphologically variable because of the predominantly

The spur of *Utricularia ochroleuca* is only half as long as the lower lip and directed downwards in an acute angle. Peat bog near Tarm, Jutland, Denmark. Photo JCS.

Utricularia ochroleuca. The leafy shoots of *U. ochroleuca* are generally less dense than those of *U. intermedia* with more slender leaf branches. Note the scattered bladders and the globose turions at the shoot apices. Peat bog near Tarm, Jutland, Denmark. Photo JCS.

vegetative propagation and clonal growth.

Habitats: Shallow, stagnant, acidic, oligotrophic water in peat bogs, swamps and ditches, also in slightly eutrophic and semi-rich habitats.

Distribution within the region: Native. Rare throughout.

Characteristics and similar species: *Utricularia ochroleuca* resembles **333** ***U. stygia*** and **331** ***U. intermedia***, with which it shares habitat preferences, and the three species can be found growing together in mixed stands. The shoots are easily torn and care must be taken to collect suitable material for determination. All three species flower rather rarely and in the absence of flowers, identification must rely on vegetative characters, which in the field are restricted to the presence of bladders on the leafy shoots. Thorough examination of the quadrifid hairs inside the bladders is needed for reliable identification.

Utricularia stygia differs from *U. ochroleuca* by the spur, which is 4–5 mm long and points downwards, as well as by the angle between the short arms of the quadrifid hairs inside the bladders.

Utricularia intermedia is easily recognised when in flower by the forward-pointing spur almost appressed to and nearly as long as the lower lip, as well as by the quadrifid hairs with the arms in parallel pairs.

Utricularia ochroleuca and *U. stygia* normally have bladders on both leafy and colourless shoots, whereas the leafy shoots of *U. intermedia* normally have at most one or two bladders.

The quadrifid hairs inside the bladders are good characters for differentiating the three species, whereas the degree to which the leaf tip tapers, size of papillae and number of bristles on leaf segments are highly variable and so unreliable for identification.

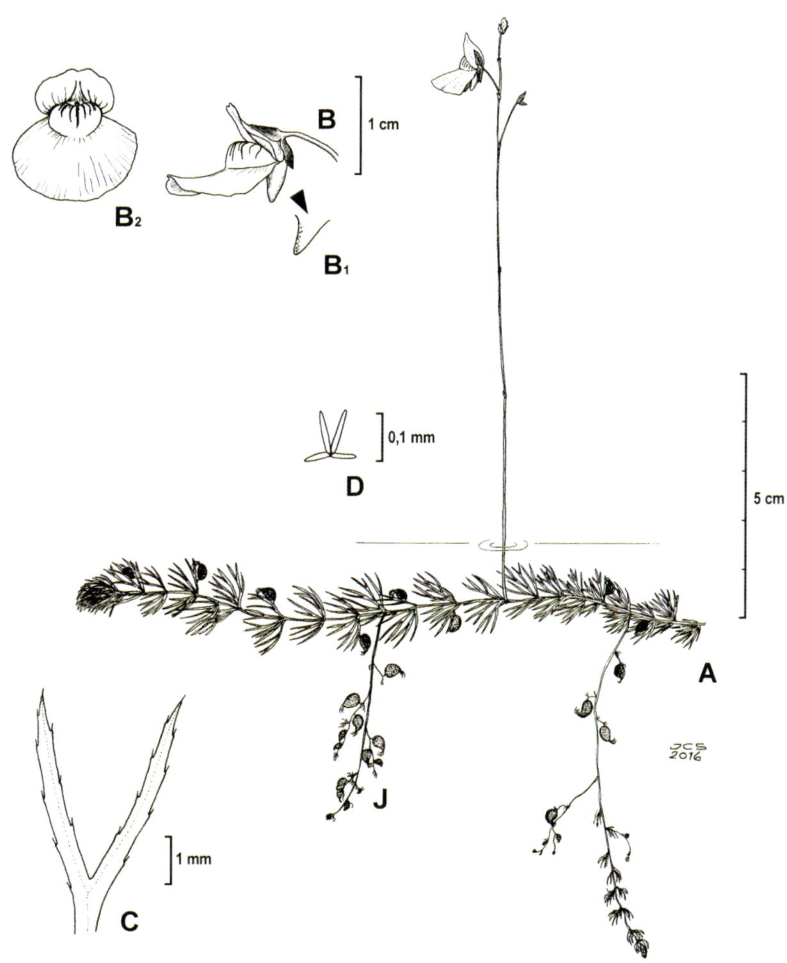

Utricularia ochroleuca
A habit, **B** flower - side view, **B1** spur with internal glands, **B2** flower - face view, **C** part of leaf, **D** quadrifid hair, **J** leafless stolon with bladders.

Lentibulariaceae

333 ***Utricularia stygia*** G.Thor

Syn.: *Utricularia intermedia* Hayne var. *kochiana* Celak

DK: Thors Blærerod **N:** Sumpblærerot **S:** Sumpblåddra **FIN:** Piilovesiherne **IS:** -
GB: Nordic Bladderwort **NL:** - **F:** Utriculaire du Styx
D: Dunkelgelber Wasserschlauch **CZ:** Bublinatka tmavá **PL:** -
EST: Kollakas vesihernes **LV:** - **LT:** - **Ru:** Пузырчатка стигийская

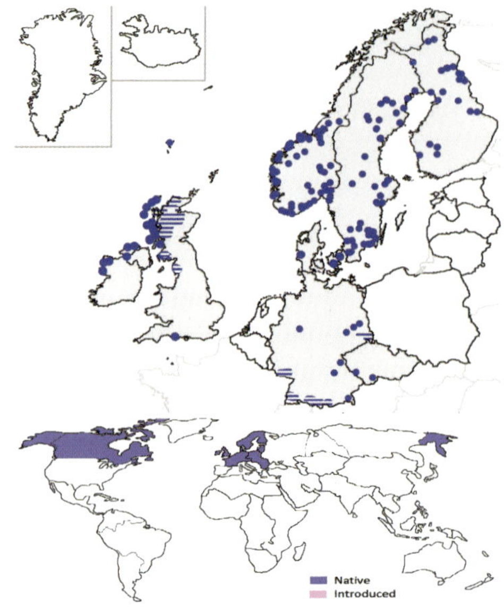

Utricularia stygia. Acid bog pool, Purbeck, Dorset, Britain. Photo RVL.

Perennial, submerged hydrophyte, free floating without roots.

Stem: Shoots represented by two types of stolons. Leafy stolons 5–20 cm long, branched, with numerous alternate green leaves and normally with some bladders. Leafless stolons 3–20 cm long, colourless, with several bladders and often partly buried in the substrate. Rhizoids at the base of the peduncle none or few, 1–4 cm long, and with 2 mm long, divided leaves.

Leaves: 6–20 mm long, green to brownish, sessile and flat. Often tripartite at the base and then dichotomously divided. Terminal segments with 2–7 very small papillae on the margin, each papilla with 1–2 short bristles. Bladders 1.5–4 mm long, on the inside densely covered with quadrifid hairs c. 0.1 mm long (use microscope 100x) with the arms forming an X. Angle between shorter arms (30°–)52°–97°(–140°).

Inflorescence: Peduncle emergent, 10–20 cm long, not more than 1 mm in diameter, green to greenish brown or reddish brown and with only a few scale-like leaves. The raceme 2–3 flowered. Bracts 1–2 mm long, ovate, blunt, green or brownish. Pedicel 7–15 mm long, straight.

Flowers: Calyx of two weakly lobed, yellowish-green to reddish-brown sepals. Corolla 8–10 mm long, 2-lipped, yellow to dark yellow with a reddish tone. Upper lip ovate, approximately 8 x 6 mm, shorter than the lower lip. Lower lip broadly ovate, 9–11 x 12–15 mm, initially flat or eventually with the margins slightly curved upwards. Palate swollen, almost closing the mouth and with orange markings. Spur c. 4–5 mm long, more than half as long as the lower lip and pointing downwards, with glands inside on both abaxial and adaxial sides. 2 stamens. Stigma 2-lipped.

Fruits: Extremely rare.

Flowering: June-September.

Biology: Carnivorous aquatic plant (see **329 *Utricularia minor***). *Utricularia stygia* appears to be a vegetative apomict of uncertain origin, but probably a result of clonal reproduction in *Utricularia ochroleuca*. It is very seldom flowering. The flowers are visited by small

Utricularia stygia. Acid bog pool, Purbeck, Dorest, Britain. Photo RVL.

insects, but the pollen produced is sterile and seed set is extremely rare. Propagation is by shoot division and by turions which are 2–4 mm long, ovoid to globose and develop at the apices of the leafy shoots. The plant dies back in late autumn and only the turions overwinter.

Habitats: Shallow, still, acidic, oligotrophic waters in peat bogs, swamps and ditches, but also in slightly eutrophic habitats.

Distribution within the region:
Native. Rare in Ireland, Britain, Denmark, Finland, Germany and the Czech Republic. Absent or overlooked elsewhere.

Characteristics and similar species:
In terms of habit similar to **332** ***U. ochroleuca*** and **331** ***U. intermedia.*** For differences see these.

Utricularia ochroleuca (left), *U. stygia* (middle) and *U. intermedia* (right) can be distinguished by the size and direction of the spur. Both *U. ochroleuca* and *U. stygia* have the spur directed downwards at an acute angle. It is only about 3 mm long in *U. ochroleuca* but 4–5 mm in *U. stygia*. In *U. intermedia* the spur is appressed to and nearly as long as the lower side of the lip. Del. JCS.

Utricularia stygia. Without close examination of the quadrifid hairs inside the bladders it is not possible to distinguish non-flowering *U. stygia* from *U. intermedia* and *U. ochroleuca*. Syrkhultasjön, Skåne, Sweden. Photo JCS.

Utricularia stygia. Normally some bladders can be found on the green leaves. Holmegaards Mose, Sjælland, Denmark. Photo JCS.

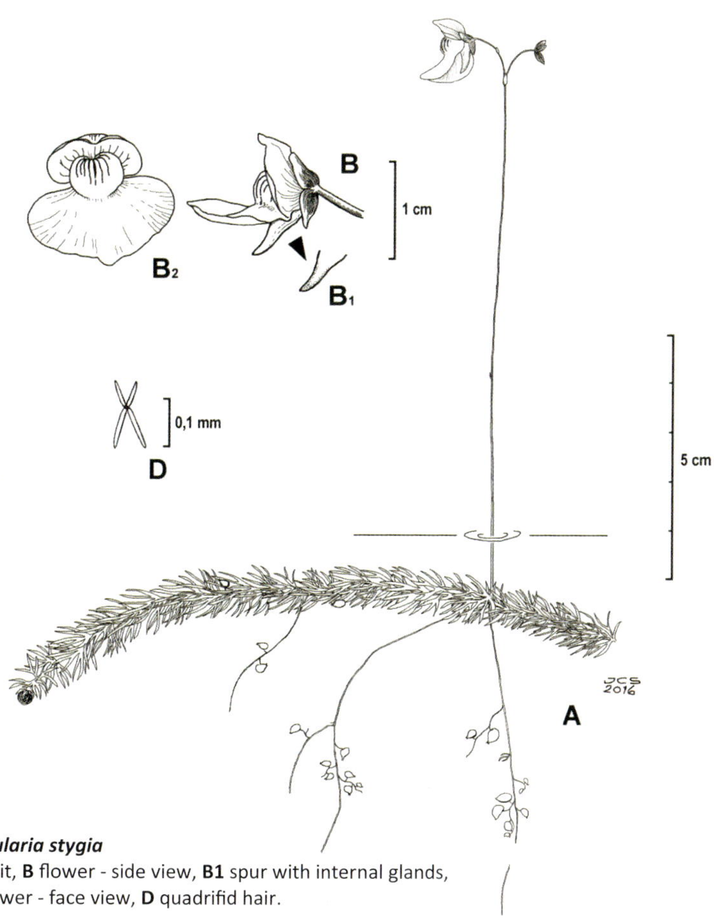

Utricularia stygia
A habit, **B** flower - side view, **B1** spur with internal glands, **B2** flower - face view, **D** quadrifid hair.

Lobeliaceae

334 *Lobelia dortmanna* L.

DK: Tvepibet Lobelie **N:** Botnegras **S:** Notblomster **FIN:** Nuottaruoho **IS:** -
GB: Water Lobelia **NL:** Waterlobelia **F:** Lobélie de Dortmann
D: Wasser-Lobelie **CZ:** - **PL:** Lobelia jeziorna
EST: Vesilobeelia **LV:** Dortmaņa lobēlija **LT:** Ežerinė lobelija **Ru:** Лобелияи Дортмана

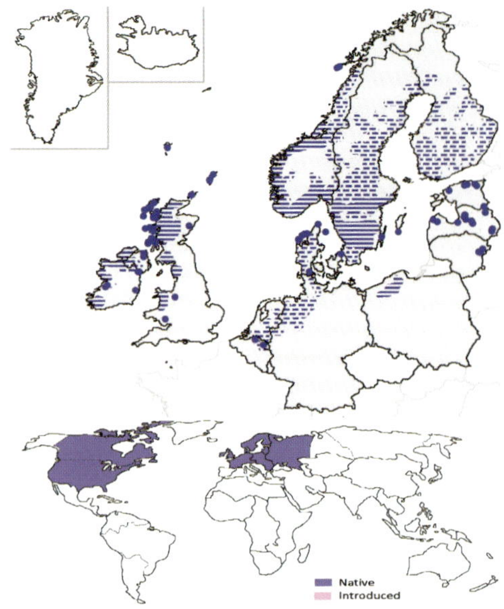

Lobelia dortmanna. The rosettes are easily recognised by the flattened, cylindrical, blunt leaves. Lake Navn Sø, Jutland, Denmark. Photo JCS.

Perennial hydrophyte or helophyte with a short vertical rhizome and numerous long, white, fibrous roots.

Stem: 15–80(–110) cm long, 2–3(–4) mm in diameter, raising the inflorescence above the water surface, unbranched, leafless but with a few bracts, slender, terete, hollow, glabrous, pale green, sometimes with a reddish tinge.

Leaves: Submerged, 8–27 in a rosette, spirally arranged from the upper part of the short vertical rhizome, 3–7(–10) cm long, 2–4(–6) mm wide, linear, flattened, cylindrical, hollow with 2 large, longitudinal air channels, yellowish green to dark green, with rounded, usually slightly recurved apex.

Inflorescence: A lax, elongate raceme of 3–8(–12) flowers. Bracts shorter than pedicels.

Flowers: Pendent. Pedicels 5–10 mm long. Calyx 4.5–5.5 mm long, with 5 short, triangular lobes. Corolla zygomorphic, white or pale lilac, tubular part with deep dorsal split. Upper lip narrowly 2-lobed, lobes lanceolate, recurved. Lower lip with 3 lanceolate, oblong-ovate lobes, longer than those of the upper lip. Stamens 5, fused around the style. Anthers brown to blackish. Stigma pale, 2-lobed.

Fruits: Capsule 6–10 mm long, 3–5 mm wide, clavate, 2-locular, opening into 2 apical valves. Seeds numerous, brown.

Flowering: July-September.

Biology: *Lobelia dortmanna* flowers from the margin to a depth of about 60 cm, raising the inflorescence above the water. In deeper water it is sterile, only producing a submerged inflorescence with closed (cleistogamous) flowers. The flowers are self-pollinated before opening, and are only visited by few insects. Submerged flowers set as much fruit as emergent ones. The seeds sink to the bottom immediately after release from the capsule and become part of the seed bank, remaining viable for up to 30 years, requiring a cold period and light for germination. Shoots die back after flowering, but are often replaced by new rosettes from axils of the outer leaves. Plants growing in shallow water are able to survive dry periods. Wintergreen. CO_2 for photosynthesis derives from the sediment and diffuses to the leaf cells in the continuous air channels. In return, oxygen is released from the roots to the sediment where it supports mycorrhizal fungi.

Lobelia dortmanna. The rosette leaves are characteristically curved towards the apex. Inserted: Cross section of a leaf showing the 2 large air channels. Madum Sø, Jutland, Denmark. Photo JCS.

Habitats: Oligotrophic to slightly mesotrophic lakes, often in dune areas, heathland or upland lakes, with clear, acid to neutral, soft water and sandy or gravelly substrate.

Distribution within the region:
Native. Temperate to arctic areas in Europe and North America. Declining throughout much of its range due to degradation and loss of habitat as well as eutrophication and acidification.

Characteristics and similar species: *Lobelia dortmanna* is easily recognised when flowering. When not flowering it differs from all similar species, such as **296 *Littorella uniflora*** and **359-360 *Isoëtes*** spp., by the 2 large, longitudinal air channels in the leaves - easily seen in cross section.

Lobelia dortmanna occurs as two forms: one almost white-flowered form and a pale lilac form. Only plants with lilac flowers contain anthocyanin and become reddish on the stem and leaves when dried; those with white flowers remain green. Left: Lake Navn Sø. Right: Lake Skån Sø. Both Jutland, Denmark. Photos JCS.

Lobelia dortmanna in shallow water - leaf rosette submerged and inflorescence raised high above the water. Lake Navn, Jutland, Denmark. Photo JCS.

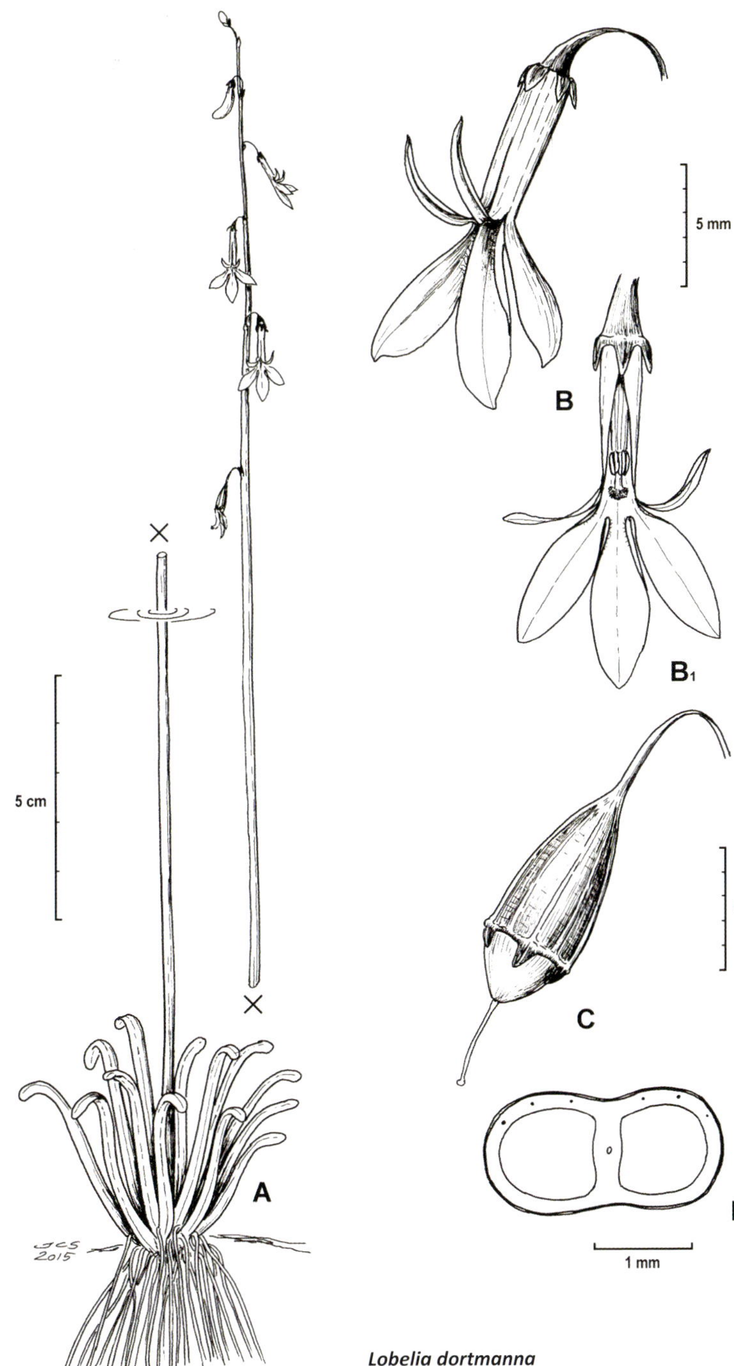

Lobelia dortmanna
A habit, **B**, **B1** flower, **C** fruit, **E1** leaf cross section.

Menyanthaceae

335 *Menyanthes trifoliata* L.

DK: Bukkeblad **N:** Bukkeblad **S:** Vattenklöver **FIN:** Raate **IS:** Horblaðka
GB: Bogbean **NL:** Waterdrieblad **F:** Ményanthe trifolié
D: Fieberklee **CZ:** Vachta trojlistá **PL:** Bobrek trójlistkowy
EST: Ubaleht **LV:** Trejlapu puplaksis **LT:** Trilapis puplaiškis **Ru:** Вахта трёхлистная

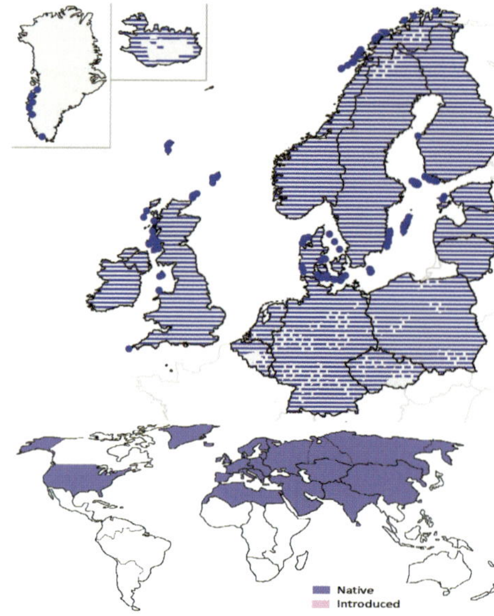

Rooted in the shore *M. trifoliata* is able to cover small water bodies and contribute to establishing floating mats of vegetation. Rødsø, Salten Langsø, Jutland, Denmark. Photo JCS.

Perennial hydrophyte or helophyte. Rhizome 30–130 cm long, 6–12 mm in diameter, creeping or floating, hollow, green to reddish brown, branched, rooting at all nodes. Roots pale green to brownish.

Stem: 10–30 cm long, creeping or ascending, more or less branched.

Leaves: Alternate. Petiole 5–20(–30) cm long, robust, with more or less hyaline stipules up to 5 cm long clasping the stem or rhizome at the base. Lamina emergent, divided into three leaflets resembling a clover leaf. Leaflets obovate to oval, 4–7 cm long and 3–5 cm wide, obtuse or rounded at apex.

Inflorescence: The inflorescence arises from the youngest node of the rhizome, but growth continues by side shoots producing new leaves. Peduncle 10–30 (–40) cm long, leafless, raceme 6–14 cm long, carrying 10–30 flowers. Bracts short, ovate.

Flowers: Pedicel 5–10 mm long, yellowish green. Calyx 4–5 mm long with 5 ovate lobes, obtuse at apex. Corolla tubular, white to pale pink on the outside, white on the inner side, divided to the middle into 5 oblong-ovate, somewhat recurved lobes. Inner side with numerous long, white hairs. Stamens 5, style 1, stigma capitate. The flowers are usually heterostylous, with a long-styled and a short-styled form. A homostylous form with the pistil and stamens of the same length is reported from West Greenland.

Fruits: Capsule unilocular, about 10 mm long, containing numerous seeds and opening by two valves. Seeds 2–2.5 mm long, somewhat flattened, glossy, orange brown with a hard shell.

Flowering: May-June.

Biology: Pollination mainly by bumblebees. The flowers are self-sterile and fruiting is usually sparse. The seeds are buoyant and are dispersed by water and birds. They only germinate after decomposition of the hard seed shell. The species propagates vegetatively by runners and rapidly forms large clones. The runners are able to float and root in water. Wintergreen, overwintering as stem and leaves.

Habitats: Acid to alkaline, oligotrophic to moderately eutrophic habitats, wet depressions in meadows, as well as in bogs, marshes, lakes, ponds and streams.

Menyanthes trifoliata. The clover-like leaves have a bitter taste. Lough Bridget, Ireland. Photo KvdW.

Menyanthes trifoliata flowers have long, white hairs which may prevent small insects from stealing nectar from the base of the ovary. The hairs are no obstacle to bumblebees (*Bombus* sp.), which are very effective pollinators and whose long tongue can reach the throat. Photos left and in the middle show long-styled flowers, the photo to the right a short-styled flower. Hobro, Jutland, Denmark. Photos JCS.

Distribution within the region:

Native. Widespread and generally abundant throughout. In some areas, particularly in southeast England, increasingly occurring as an escape from cultivation.

Characteristics and similar species:

Even without flowers *Menyanthes trifoliata* is easily recognised by the large, trifid leaves arising from a thick, creeping rhizome.

Menyanthes trifoliata
A habit, **B** flower, **C** fruit with long style, **C1** fruit with short style.

Menyanthaceae

336 ***Nymphoides peltata*** (S.G.Gmel.) Kuntze

DK: Søblad **N:** Sjøgull **S:** Sjögull **FIN:** Lammikki **IS:** -
GB: Fringed Water-lily **NL:** Watergentiaan **F:** Faux nénuphar
D: Seekanne **CZ:** Plavín lekninovitý **PL:** Grzybieńczyk wodny
EST: -**LV:** Vairogu palēpe **LT:** Vandeninė plaumuonė **Ru:** Болотоцветник щитолистный

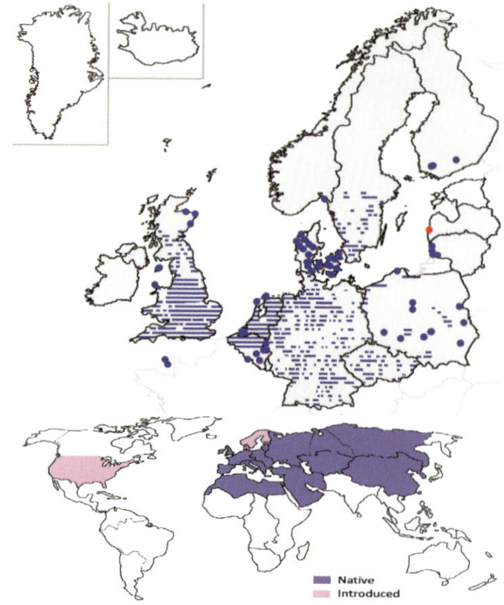

Nymphoides peltata is easily recognised when flowering. Xantener Altrhein (oxbow lake), Germany. Photo KvdW.

Perennial aquatic. Rhizome 50–200 cm long, 2–4 mm in diameter, branched, creeping or partly floating with white to pale brown roots anchored to the bed. Short shoots with very short internodes and numerous roots; long shoots with 1 leaf and 2–7 roots per node.

Stem: Creeping or ascending, more or less branched.

Leaves: Alternate, but opposite below the inflorescence. Pedicels up to 2 metres long, terete and flaccid, attached to the rhizome. Lamina floating, 3–12 cm long, circular to ovate with a deep basal incision and wavy, scalloped margins. Upper side of lamina yellowish green to dark green with veins forming a reticulate pattern. Lower side of lamina green or often purple, with a dense cover of wart-like papillae.

Inflorescence: Flowers 4–8 in 1–3 whorls borne on peduncles arising from the rhizome. Each whorl of flowers subtended by 2 opposite, floating leaves of the same shape as other leaves, but smaller.

Flowers: Pedicels 5–10 cm long, rising above the water surface. Calyx with 5 oblong-triangular lobes, fused at base and somewhat recurved at apex. Corolla funnel shaped, 2.5–4 cm across, golden yellow, divided almost to the base into 5 obovate lobes, with fringed margins. Stamens 5, style 1, stigma capitate. The flowers usually heterostylous, with a long-styled and a short-styled form. Long-styled flowers with style up to 1 cm long, stamens with 1–2 mm long filaments and 2–3.5 mm long anthers. Short-styled flowers with a 1–2 mm long style, stamens with 3–4 mm long filaments and 4–6 mm long anthers.

Fruits: Capsule unilocular, 1.7–2.5 cm long, containing numerous seeds and only opening on one side. Seeds 4–5 mm long, 2.7–3.0 mm wide, ovoid, flat, smooth, brown, with stiff marginal hairs.

Flowering: July-August.

Biology: The evolution of heterostylous flowers is assumed to be a protection against self-pollination. Flowers pollinated from a different morph produce more and better seeds than flowers pollinated from a similar flower. Each flower only lasts for one day before withering. Pollination is by insects, guided towards the throat and nectar by UV-reflecting marks. Self-pollination may also occur. The fruit ripens under water as, after pollination, the stalk bends, submerging the fruit. The seeds are buoyant for a long time and are dispersed by water and waterfowl. Vegetative propagation is by runners and detached fragments of the rhizome.

Nymphoides peltata in dense mats. Lake Fräjen north of Växjö, Sweden. Photo JCS.

Nymphoides peltata flowers with long styles. Xantener Altrhein (oxbow lake), Germany. Photo KvdW.

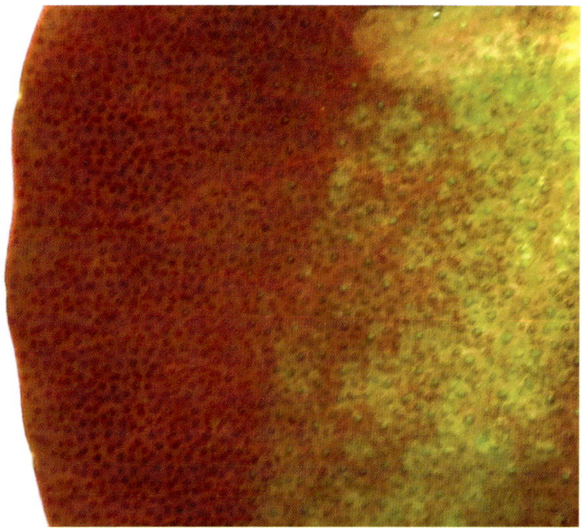

Lower side of leaf with dense covering of wart-like papillae. Photo BM.

Often forming dense mats and thereby excluding other plant species. Overwinters as short, rather thick shoots from the rhizome. It can become invasive in shallow, slow-moving rivers, lakes and ponds.

Habitats: On neutral to alkaline, nutrient-rich substrate in ponds, shallow lakes, canals, ditches, moats and along the banks of creeks.

Distribution within the region: Considered native in parts of England, introduced in Denmark, Sweden and parts of Germany outside the valleys of large rivers and streams.

Characteristics and similar species: Without flowers *Nymphoides peltata* can be mistaken for small plants of **1-3** ***Nymphaea*** spp. or **4-5** ***Nuphar*** spp., but differs from these by the scalloped leaf margin and the lower side of the leaves being densely covered by wart-like papillae.

51 ***Hydrocharis morsus-ranae*** is free-floating, has white flowers and smaller, entire leaves without wart-like papillae on the lower side.

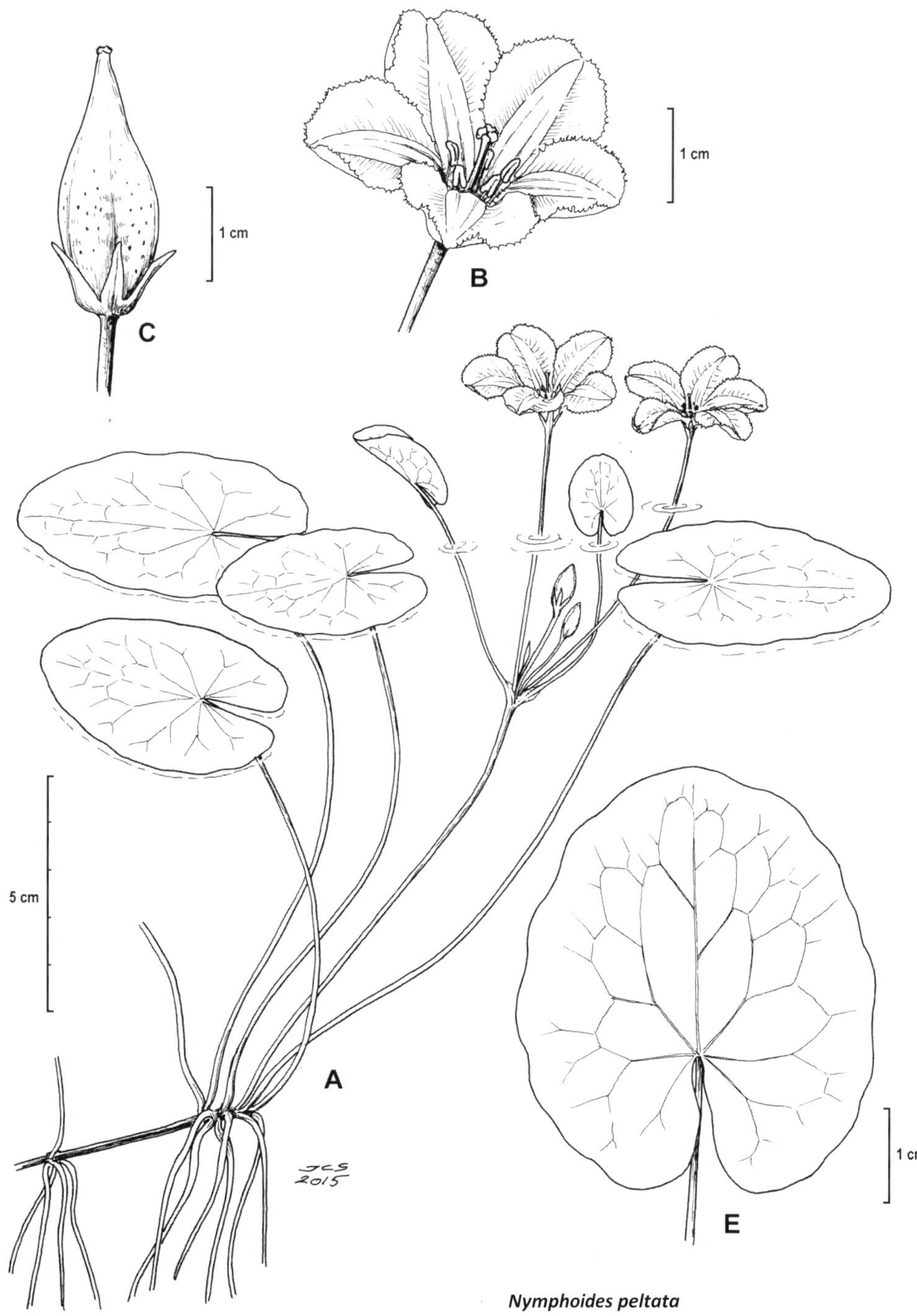

Nymphoides peltata
A habit, **B** flower, **C** capsule, **E** leaf.

Asteraceae

Key to *Bidens, Cotula* and *Shinnersia*

None of the *Bidens* species are true aquatics, they typically occur on the margins of rivers, streams, lakes, ponds and canals, in ditches and in seasonally wet depressions in marshy fields, often on open, poached, muddy, nutrient-rich substrates. The flowers are pollinated by insects and the achenes spread by animals. The ripe achenes and to some extend the leaves offer the best characters for identification.

1 Outer phyllaries leafy (***1***) and much larger than the inner (***2***) — 2

1 Outer and inner phyllaries of equal length (***3***) — 7

2 Leaves simple, unlobed or lobed less than halfway to the midrib (***4,5,6***) — 3

2 Some or all leaves pinnate or lobed almost to the midrib (***7,8,9***) — 5

3 Leaves sessile, unlobed but distinctly serrate (***4***). Yellow ray florets sometimes present. Capitula nodding after flowering
337 *Bidens cernua*

3 Leaves tapering to a winged petiole (***7***), unlobed or lobed. Yellow ray florets never present. Capitula erect after flowering — 4

4 Achenes flattened, more than twice as wide as thick, smooth on faces, awns with downward-pointed barbs (***10***)
341 *Bidens tripartita*

4 Achenes rhombic in cross section, less than twice as wide as thick, warty on faces, awns with upward-pointed barbs (***11***)
338 *Bidens connata*

5 Petioles winged on all leaves (***7,8***). Achene smooth on faces (***10***) — 6

5 Petioles unwinged at least on lower leaves. Achene rugose on faces (***12***)
339 *Bidens frondosa*

6 Capitula with 5–8(–10) outer phyllaries (***2***). Achenes 5–8.5(–11) mm long (excluding awns)
341 *Bidens tripartita*

6 Capitula with 9–12(–15) outer phyllaries (***2***). Achenes 3–4(–5.5) mm long (excluding awns)
340 *Bidens radiata*

7 Leaves alternate, deeply pinnatifid to irregularly toothed or entire and linear
342 *Cotula coronopifolia*

7 Leaves opposite, broadly oblanceolate or 3-lobed, coarsely and bluntly toothed or sinuate
343 *Shinnersia rivularis*

Asteraceae

337 *Bidens cernua* L.

DK: Nikkende Brøndsel **N:** Nikkebrønsle **S:** Nickskära **FIN:** Nuokkurusokki **IS:** -
GB: Nodding Bur-marigold **NL:** Knikkend tandzaad **F:** Bident penché
D: Nickender Zweizahn **CZ:** Dvouzubec nicí **PL:** Uczep zwisły
EST: Longus ruse **LV:** Nokarenais sunītis **LT:** Pelkinis lakišius
Ru: Череда поникшая

Bidens cernua var. *cernua*. Cors Ddyga Nature Reserve, Anglesey, Wales. Photo RVL.

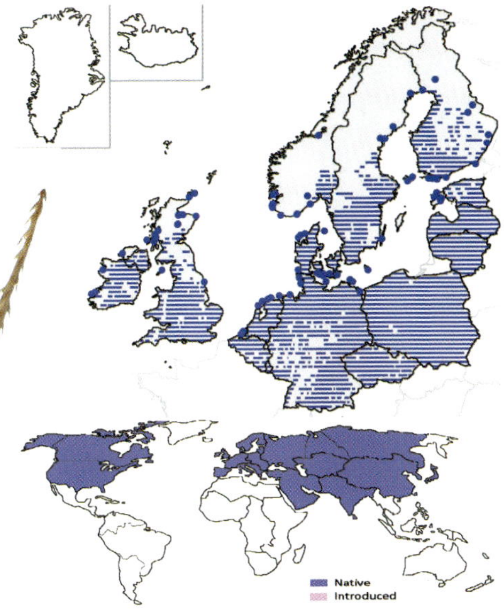

Bidens cernua. Achene. Sjælland, Denmark. Photo JCS.

Annual. Stem 15–75(–130) cm long, erect or ascending, more or less branched, glabrous or sparsely hairy. Leaves 4–15 cm long, simple, coarsely serrate, sessile, those in pairs often somewhat connate. Capitula discoid, 15–25 mm across (excluding ray florets if present), nodding before and after flowering, long stalked. Outer phyllaries 5–8, lanceolate. Achenes 5–6 mm long (excluding the 3–4 awns), strongly compressed, barbs on achene and awns downward pointing.

Flowering: June-September.

Distribution within the region: Widespread and fairly abundant throughout except in northern Scandinavia and Scotland where it is scarce. Absent from Iceland and Greenland.

Characteristics and similar species: Recognised by the sessile, unlobed, but coarsely toothed leaves. Var. *radiata* DC. differs in having yellow, lingulate ray-florets.

Bidens cernua var. *radiata* - with yellow ray flowers. Filsø, SW-Jutland, Denmark. Photo JCS.

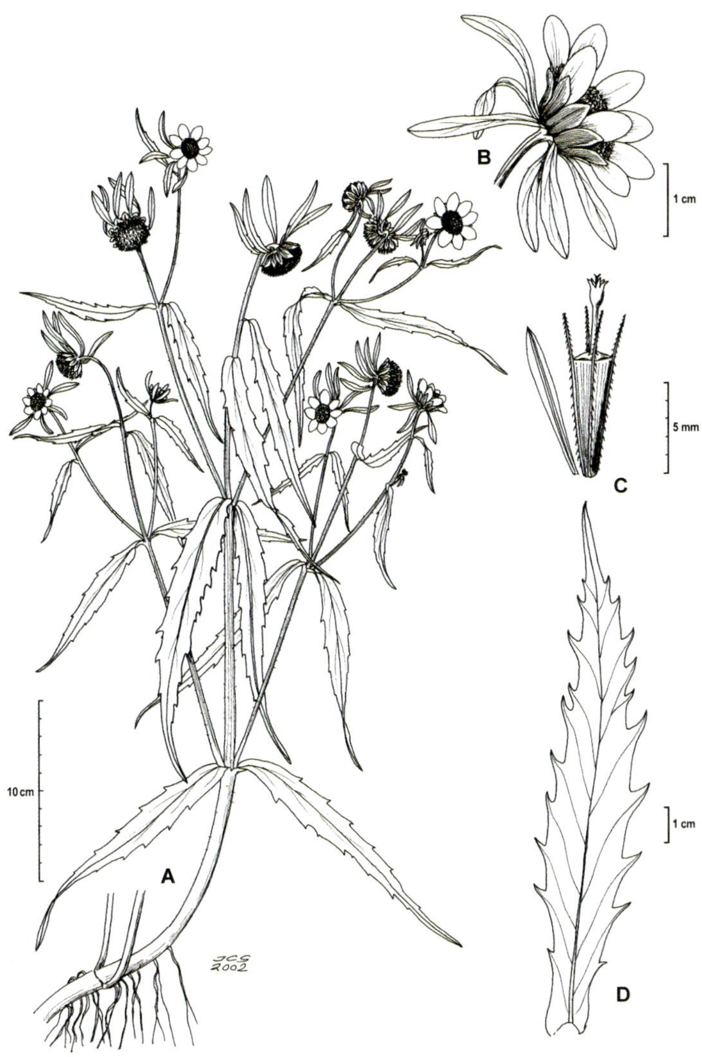

Bidens cernua
A habit, **B** capitulum, **C** achene with receptacular bract, **D** leaf.

Asteraceae

338 *Bidens connata* Muhl. ex Willd.

DK: - **N:** - **S:** - **FIN:** - **IS:** -
GB: London Bur-marigold **NL:** Smal tandzaad **F:** Bident à feuilles connées
D: Verwachsenblättriger Zweizahn **CZ:** Dvouzubec hnědožlutý **PL:** Uczep zwodniczy
EST: - **LV:** - **LT:** - **Ru:** Череда сростнолопастная

Bidens connata. Middlewich, Britain. Photo RVL.

Annual. Stem 15–100(–130) cm long, erect or ascending, branched, light green to purple, glabrous. Leaves 6–15 cm long, lanceolate to oblong lanceolate, coarsely toothed, not lobed or sometimes with 1–2 lobes close to the base, petioles distinct, unwinged or slightly winged. Capitula discoid, 8–10(–15) mm across, stalk 1–6 cm long. Outer phyllaries 3–6, lanceolate. Achenes 5–7(–9) mm long (excluding the 3–4 awns), slightly compressed, warty and somewhat barbed on faces, barbs on achene and awns upward- or rarely downward pointed.

Bidens connata. Middlewich, Britain. Photo RVL.

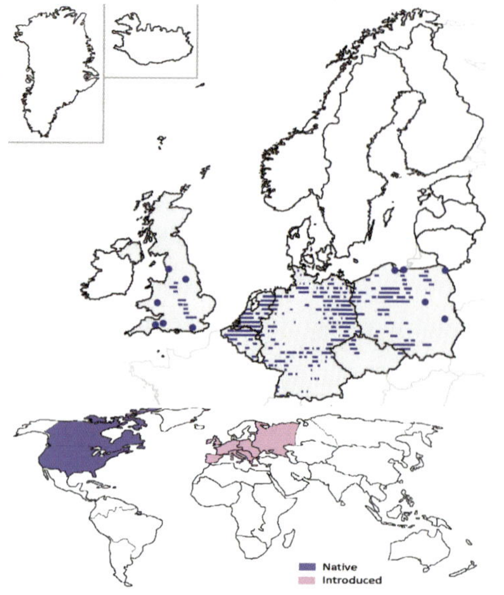

Bidens connata. Achene. Germany. Photo JCS.

Flowering: July-September.

Distribution within the region: Alien. Scattered to locally frequent in the Netherlands, Belgium, Germany, Poland, rare in Czech Republic and Britain.

Characteristics and similar species: Similar to forms of **341 *B. tripartita*** with unlobed leaves, but differs in achene characters - faces warty and somewhat barbed, awns with upward-pointing barbs.

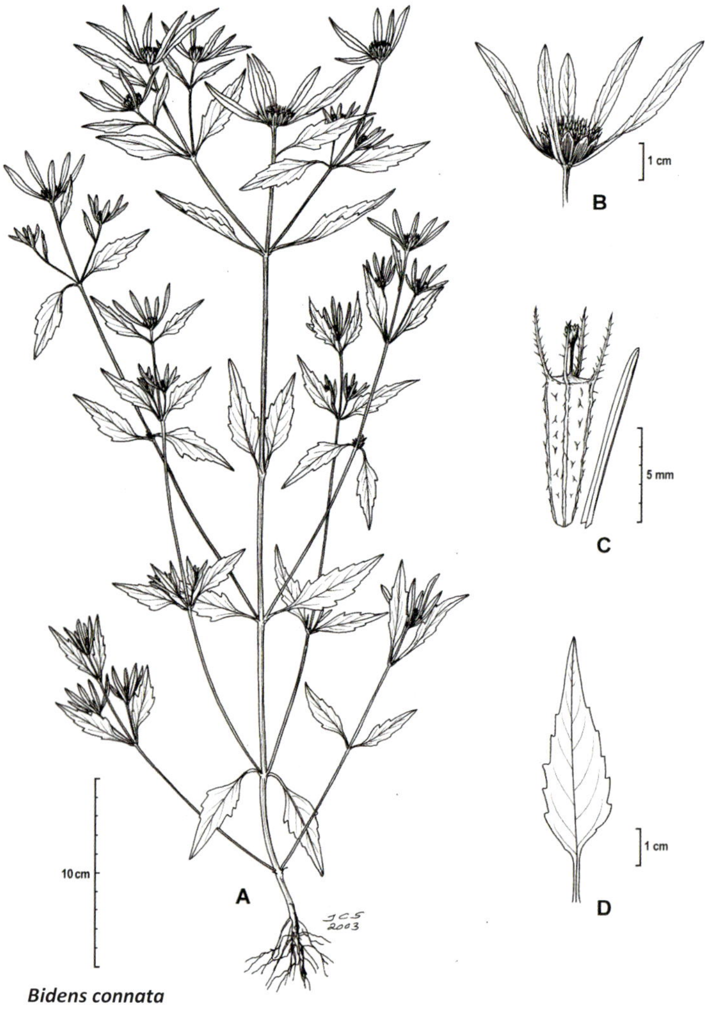

Bidens connata
A habit, **B** capitulum, **C** achene with receptacular bract, **D** leaf.

339 *Bidens frondosa* L.

DK: Sortfrugtet Brøndsel **N:** Hyllebrønsle **S:** Fläderskära **FIN:** Seljarusokki **IS:** -
GB: Beggar's-ticks **NL:** Zwart tandzaad **F:** Bident à fruits noirs
D: Schwarzfrüchtiger Zweizahn **CZ:** Dvouzubec černoplodý
PL: Uczep amerykański **EST:** Ameerika ruse **LV:** Lapainais sunītis
LT: Ilgakotis lakišius **Ru:** Череда олиственная

Bidens frondosa. Birmingham, Britain. Photo RVL.

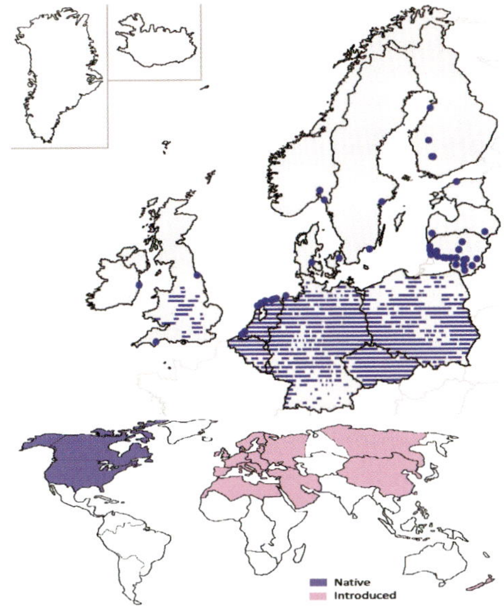

Bidens frondosa. Achene.
Blekinge, Sweden.
Photo Jan-Thomas Johansson.

Hybrids: *B.* × *decipiens* (*B. frondosa* × *cernua*) shares characters of the parents, including both paired and simple multicellular hairs and quadrilateral, rugose achenes, while the capitula resemble those of *B. frondosa* but the leaves resemble those of *B. cernua*.

Annual. Stem 30–100(–180) cm long, erect, branched above or from base, slightly quadrangular, glabrous, often purplish. Leaves 3–8–15 cm long (excluding petiole), mostly pinnate with 1–2 pairs of distinctly stalked leaflets, coarsely serrate, with forward-pointing teeth, green to dark green, petioles 1–4(–6) cm long, scarcely winged. Capitula discoid, 6–10 mm across, erect, peduncles 1–4(–8) cm long. Outer phyllaries 5–8(–10), lanceolate. Achenes 7–9 mm long (excluding the 2 awns), strongly compressed, rugose on faces, barbs on both achenes and awns pointing upwards or those on the barbs pointing downwards.

Flowering: August-September.

Distribution within the region: Alien. Common in the Netherlands, most of Belgium, Germany, the Czech Republic and Poland. Occasional in parts of Britain, casual in Scandinavia and the Baltic states.

Bidens frondosa has characteristic leaves. Merkine, Lithuania. Photo Jan-Thomas Johansson.

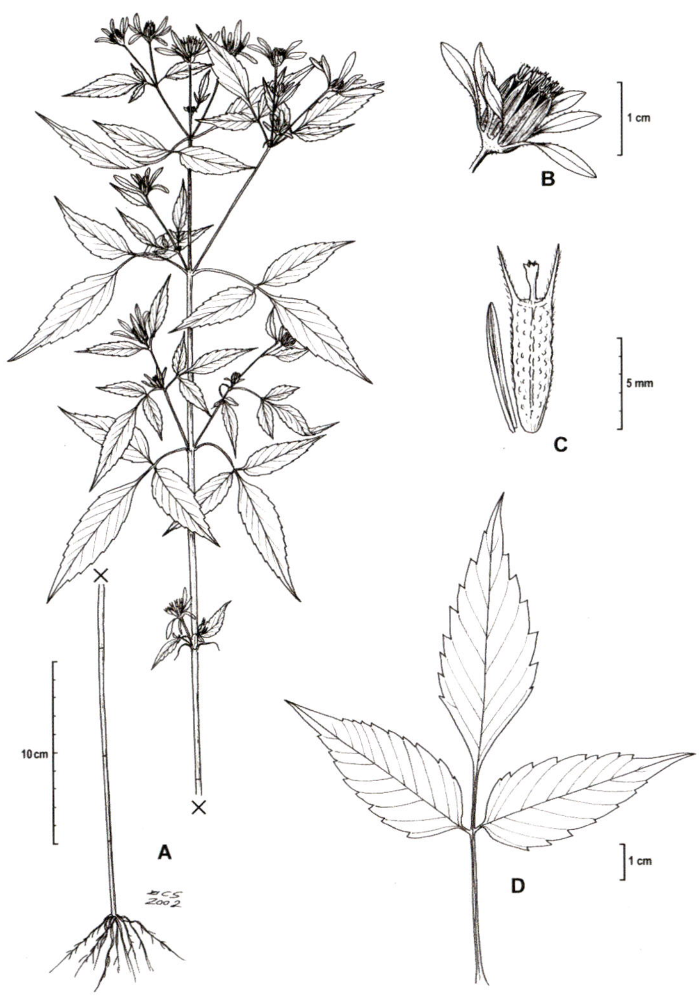

Bidens frondosa
A habit, **B** capitulum, **C** achene with receptacular bract, **D** leaf.

Asteraceae

340 *Bidens radiata* Thuill.

Syn.: *Bidens platycephala* Oerst.

DK: Fladhoved Brøndsel **N:** Grønnbrønsle **S:** Grönskära **FIN:** Säderusokki **IS:** - **GB:** Greater Bur-marigold **NL:** Riviertandzaad **F:** Bident radié **D:** Strahlender Zweizahn **CZ:** Dvouzubec paprsčitý **PL:** Uczep śląski **EST:** Kiirjas ruse **LV:** - **LT:** Krantinis lakišius **Ru:** Череда лучистая

Bidens radiata. Valkiakoski, Hattula, Finland. Photo Jan-Thomas Johansson.

Bidens radiata. Achene. Tavasteburg, Finland. Photo JCS.

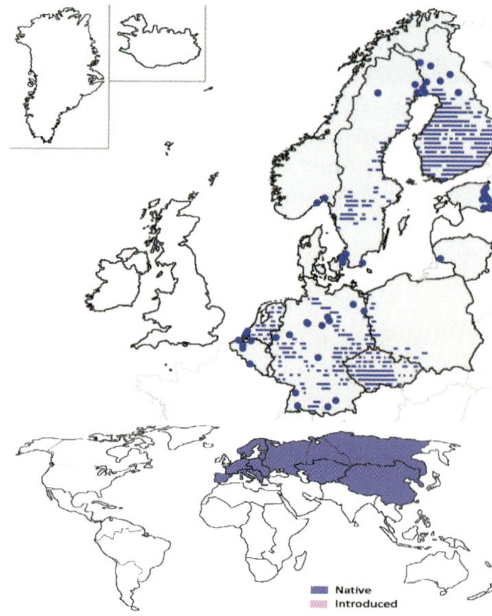

Annual. Stem (5–)15–80(–150) cm long, erect, unbranched or somewhat branched at base, terete, glabrous, yellowish sometimes with a reddish tinge. Leaves opposite, 3–15 cm long, usually with 1–2 pairs of lateral lobes or simple, yellowish green, coarsely serrate with backward- or forward-pointed teeth, petioles winged. Capitula discoid, 10–15(–20) mm across, erect and wider than long after flowering, long stalked. Outer phyllaries 9–12(–15), lanceolate, sometimes leaf-like. Achenes 3–5.5 mm long (excluding the 2 awns), compressed or angular, smooth on faces, barbs on achenes and awns pointing downwards.

Flowering: July-September.

Distribution within the region: Native. Rare or scattered in most of the region, absent from the British Isles and Iceland. Not recorded from Latvia. May suddenly occur in large numbers in places where conditions are optimal, e.g., because of a warm summer, but often absent or sporadic the following year.

Characteristics and similar species: Often confused with **341** ***B. tripartita*** but recognised by smaller achenes, 9–12(–15) outer phyllaries and the yellow-green colour of the leaves.

Hybrids: *Bidens ×polakii* Velen. (*B. radiata x tripartita*) has been reported and is assumed to share characters of the parent species.

Bidens radiata. Valkiakoski, Hattula, Finland. Photo Jan-Thomas Johansson.

Bidens radiata

A habit, **B** capitulum, **C** achenes with receptacular bracts, **D** leaf.

Asteraceae

341 *Bidens tripartita* L.

DK: Fliget Brøndsel **N:** Flikbrønsle **S:** Brunskära **FIN:** Tummarusokki **IS:** -
GB: Trifid Bur-marigold **NL:** Veerdelig tandzaad **F:** Bident trifolié
D: Dreiteiliger Zweizahn **CZ:** Dvouzubec trojdílný **PL:** Uczep trójlistkowy
EST: Kolmisruse **LV:** Trejdaivu sunītis **LT:** Triskiautis lakišius
Ru: Череда трехраздельная

Bidens tripartita at the margin of a small lake. Vitskøl, Jutland, Denmark. Photo JCS.

Bidens tripartita. Achene. Lyngby, Denmark. Photo JCS.

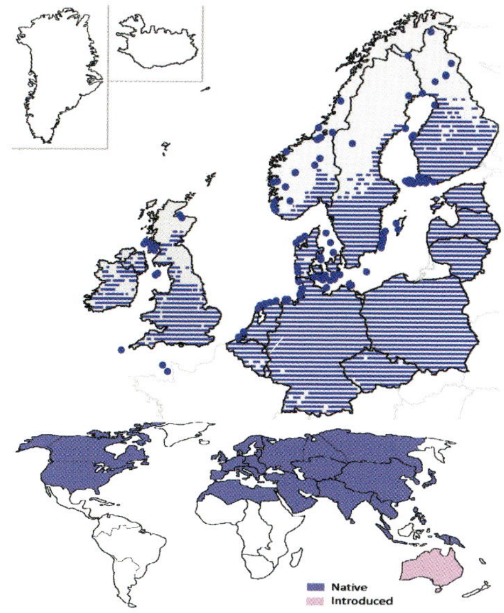

Annual. Stem 15–75(–150) cm long, erect or ascending, often richly branched, slightly quadrangular, glabrous or sparsely hairy. Leaves opposite or the upper alternate, 5–15 cm long, usually with 1–2 pairs of lateral lobes or sometimes simple, dark green, coarsely serrate with straight or outward-pointed teeth, petioles winged. Capitula discoid, 10–20 mm across, erect after flowering, long stalked. Outer phyllaries 5–8(–10), lanceolate. Achenes 5–8.5(–11) mm long (excluding the 2–4 awns), strongly compressed, smooth on faces, barbs on achenes and awns downward pointing.

Flowering: July-October.

Distribution within the region: Common in most of the region, rare in northern Scandinavia and Scotland. Absent from Iceland and Greenland.

Hybrids: *Bidens ×polakii* Velen. (*B. radiata x tripartita*).

Bidens tripartita. Leaf base. Cors Ddyga Nature Reserve, Anglesey, Wales. Photo RVL.

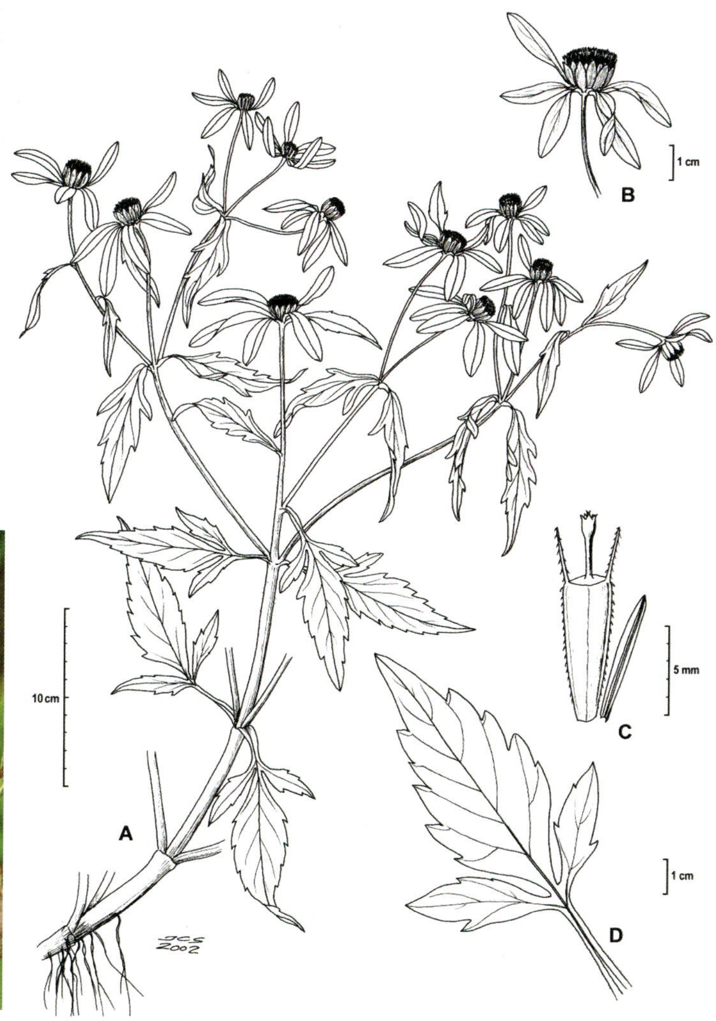

Bidens tripartita
A habit, **B** capitulum, **C** achene with receptacular bract, **D** leaf.

Asteraceae

342 *Cotula coronopifolia* L.

DK: Firkløft **N:** Fjæreknapp **S:** Kotula **FIN:** Nappikotula **IS:** -
GB: Buttonweed **NL:** Goudknopje **F:** Cotule Pied-de-corbeau
D: Krähenfußlaugenblume **CZ:** - **PL:** -
EST: - **LV:** - **LT:** - **Ru:** Котула коронопилистная

Cotula coronopifolia. Razza di Junco, N Portisco, Sardinia, Italy. Photo KvdW.

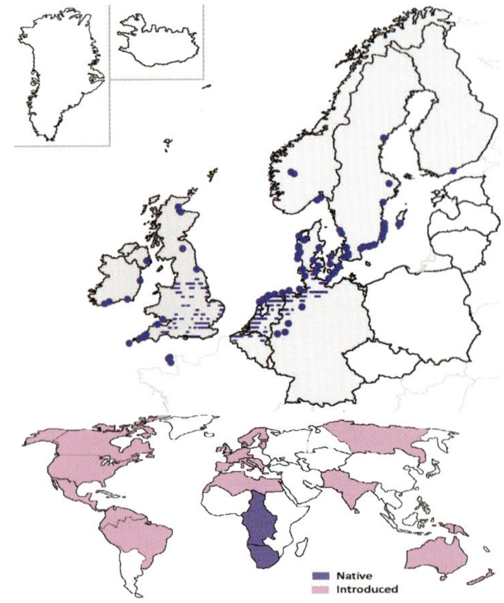

Annual or sometimes perennial helophyte or hydrophyte. Stem ascending to erect, 5–30 cm long, glabrous, branched. Leaves 2–5(–10) cm long, fleshy, lanceolate in outline, deeply pinnatifid to irregularly toothed or, especially in submerged plants, entire and linear. Leaf base sheathing, whitish green. Capitula discoid, 6–10 mm across, long pedunculate, terminal on stem and branches, nodding before and after flowering. Corolla bright yellow, 1–1.5 mm long. Cypselae 0.8–1.8 mm long, more or less winged.

Flowering: July-August.

Biology: The seeds are buoyant and mainly dispersed by water. Salt tolerant.

Habitats: On sunny, nitrogen-rich, brackish to saline or fresh, wet, muddy substrate in salt marshes, marshes, estuaries, along streams and shallow lakes.

Submerged *C. coronopifolia*. Razza di Junco, N Portisco, Sardinia, Italy. Photo KvdW.

Distribution within the region: Alien. Scattered to common in the British Isles, Netherlands, Belgium, northwestern Germany, Denmark, Sweden, rare in Norway and Finland. Apparently increasing.

Characteristics and similar species: When flowering not likely to be confused with other species. Submerged plants are recognised by the alternate, rather fleshy, linear leaves with a characteristic sheathing base.

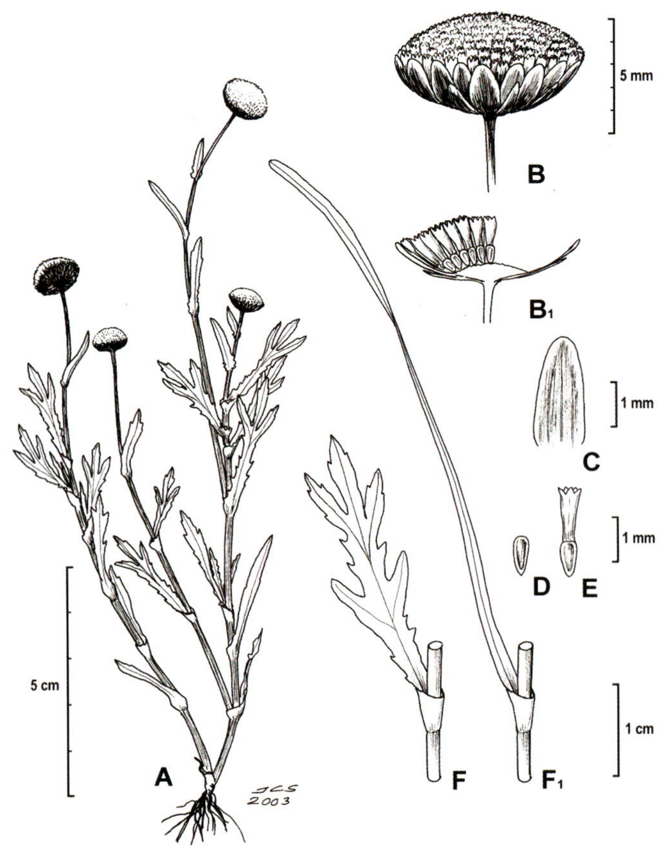

Cotula coronopifolia
A habit, **B** capitulum, **B1** cross section of capitulum, **C** phyllary, **D** cypsela, **E** flower, **F** leaf, **F1** leaf of submerged plant.

Asteraceae

343 ***Shinnersia rivularis*** (A.Gray) R.M.King & H.Rob.

Syn.: *Trichocoronis rivularis* A.Gray
DK: - **N:** - **S:** - **FIN:** - **IS:** -
GB: Mexican Oakleaf Plant **NL:** - **F:** -
D: Mexikanisches Eichenblatt **CZ:** - **PL:** -
EST: - **LV:** - **LT:** - **Ru:** -

Shinnersia rivularis. Balaton region, Hungary. Photo KvdW.

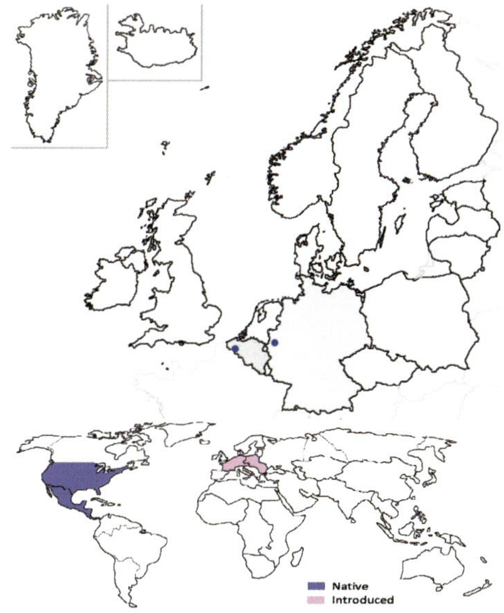

Perennial hydrophyte. Stem 10–30 cm long, ascending, more or less branched, rooting from lower nodes. Leaves 2–5 cm long, opposite, broadly lanceolate to 3-lobed, coarsely toothed or sinuate - resembling oak leaves, sessile, stipulate or widened at base, glabrous or slightly hairy, with glandular dots. Capitula discoid, 6–9 mm across, single, long pedunculate, axillary on emergent parts. Involucre of 20–30 phyllaries in 2–3 rows. Corolla c. 2 mm long, white, campanulate, with 5 lobes. Cypselae c. 2 mm long, prismatic, 4–5 ribbed, sparsely hairy and sparsely glandular.

Flowering: July-October.

Habitats: In streams, rivers and lakes.

Distribution within the region: Alien. In Europe it is cultivated as an aquarium plant and naturalised in Belgium, Germany, Austria, Slovakia and Hungary.

Characteristics and similar species: When flowering unlikely to be confused with other aquatic species.

Non-flowering or submerged plants may resemble **297-302 *Veronica*** spp. and **313 *Hygrophila polysperma*** but differ in leaf shape.

Shinnersia rivularis. Balaton region, Hungary. Photo KvdW.

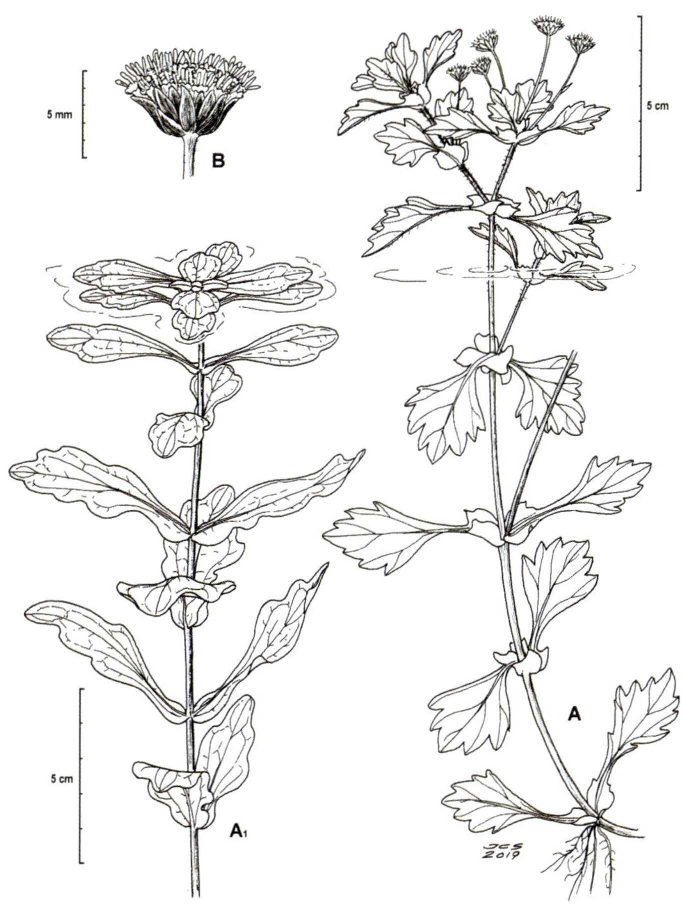

Shinnersia rivularis
A habit flowering, partly submerged plant**, A1** habit submerged plant, **B** capitulum.

Araliceae

344 *Hydrocotyle vulgaris* L.

DK: Vandnavle **N:** Skjoldblad **S:** Spikblad **FIN:** Viitasammakonputki **IS:** Vatnsnafli
GB: Marsh Pennywort **NL:** Gewone waternavel **F:** Écuelle d'eau
D: Gewöhnlicher Wassernabel **CZ:** Pupečník obecný **PL:** Wąkrota zwyczajna
EST: Loim-vesipaunikas **LV:** Parastā vairoglape **LT:** Pelkinė raistenė **Ru:** Щитолистник обыкновенный

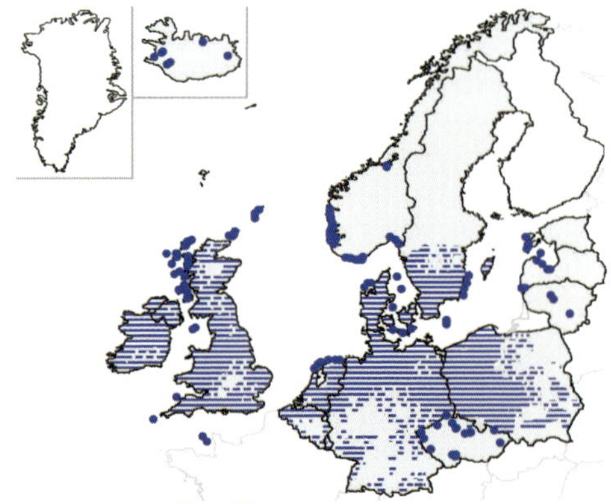

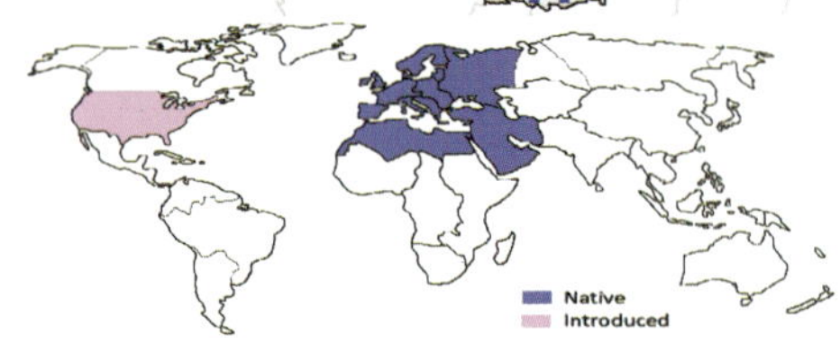

Hydrocotyle vulgaris in association with *Eleocharis palustris* at 25-30 cm depth. Submerged plants do not flower. Lake Madum, Jutland, Denmark. Photo JCS.

Aquatic perennial. Rooting at all nodes, roots pale.

Stem: 10–40(–100) cm long, creeping, slender, terete, glabrous, pale green or with reddish tones.

Leaves: Alternate, 1 from each stem node, but sometimes 2 very close together appearing subopposite. Petioles terete, with 2 triangular, membranous, often somewhat laciniate stipules at the base and with a few patent hairs beneath the lamina. The length of the petioles depends on the biotope - on moist substrate it may be only 2–10 cm but on submerged plants and those growing in dense vegetation up to 30 cm. Lamina peltate, orbicular, 1.5–4 cm diameter, with 5–9 main veins radiating from the petiole, shiny waxy, yellowish- to dark green, glabrous or with a few hairs below.

Inflorescence: 1 or 2 together from the stem nodes, shorter than the adjacent leaf. Flowers 3–6 together in axillary umbels, often with additional flowers in whorls below. Bracts c. 1 mm long, ovoid, membranous.

Flowers: Bisexual, more or less sessile. Sepals lacking or much reduced. Petals 5, pinkish white to purplish green, 0.5–1 mm long, ovoid, apex acute, spreading. 5 stamens with white filaments and pinkish to yellowish anthers. 2 styles not thickened at base.

Fruits: Dry 2-celled schizocarps. Orbicular, 1.5–2.3 mm wide and 1.5–1.8 mm long, strongly laterally compressed, covered with brown resinous warts, ribs rather prominent.

Flowering: July-September.

Biology: Often in dense carpets on moist or seasonally flooded substrate. Emergent plants flower abundantly; submerged are sterile. The flowers are probably mostly self-pollinated, but small insects visit the flowers and may act as pollinators. The fruits are dispersed by water movement or by animals. The fruits can be carried around in the hooves of grazing animals and survive passage through the intestinal system, sprouting from dung. Vegetative propagation by stolons. Overwinters with stem and green leaves.

Habitats: On acidic to neutral or slightly alkaline, nutrient-poor, moist, permanently or seasonally wet substrates. Dune slacks, bogs, fens, marshes, lakeshores.

Flowering *H. vulgaris* on dried-up, sandy bottom. Raabjerg Mile, Denmark. Photo JCS.

Flowering *H. vulgaris* with *Pilularia globulifera* growing on moist sand in a dune slack. Note the few hairs on the petiole. Råbjerg Mile, Jutland, Denmark. Photo JCS.

Hydrocotyle vulgaris. Inflorescences with pinkish flowers and compressed fruits. The small insects may act as pollinators, but self-pollination is probably more common. Råbjerg Mile, Jutland, Denmark. Photo JCS.

Distribution within the region:
Native. Widespread and fairly abundant from Ireland east to western Poland and north into southern Sweden, rare in Norway, the Baltic states and the Czech Republic.

Characteristics and similar species:
H. vulgaris can be identified by the peltate leaves with only 5–9 main veins radiating from the petiole, which in the upper part has few patent hairs. For distinction from other members of the genera see the key (right).

51 *Hydrocharis morsus-ranae* differs in the larger flowers and suborbicular cordate leaves with a deep sinus, mid-vein and broadly arching lateral veins.

H. moschata G.Forster and ***H. novae-zeelandiae*** DC. have occurred occasionally as casuals in Britain. Both have hairy stems and leaves, with fruits longer than wide; they are terrestrial and were recorded in turf and lawns.

Key to *Hydrocotyle* species from wet habitats:

1 Leaves peltate (without a deep basal sinus) — 2
1 Leaves with deep basal sinus (never peltate) — 3

2 Leaves with 5–9 main veins. Petiole with few patent hairs in the upper part. Inflorescence shorter than the adjacent leaf
344 *H. vulgaris*
2 Leaves with 9–13 main veins. Petiole without hairs. Inflorescence as long as or longer than the adjacent leaf
346 *H. verticillata*

3 Leaves fleshy, 2.5–8(–10) cm wide. Fruit with a short but distinct stalk
345 *H. ranunculoides*
3 Leaves thin, 0.3–1.2(–2) cm wide. Fruit sessile
347 *H. sibthorpoides*

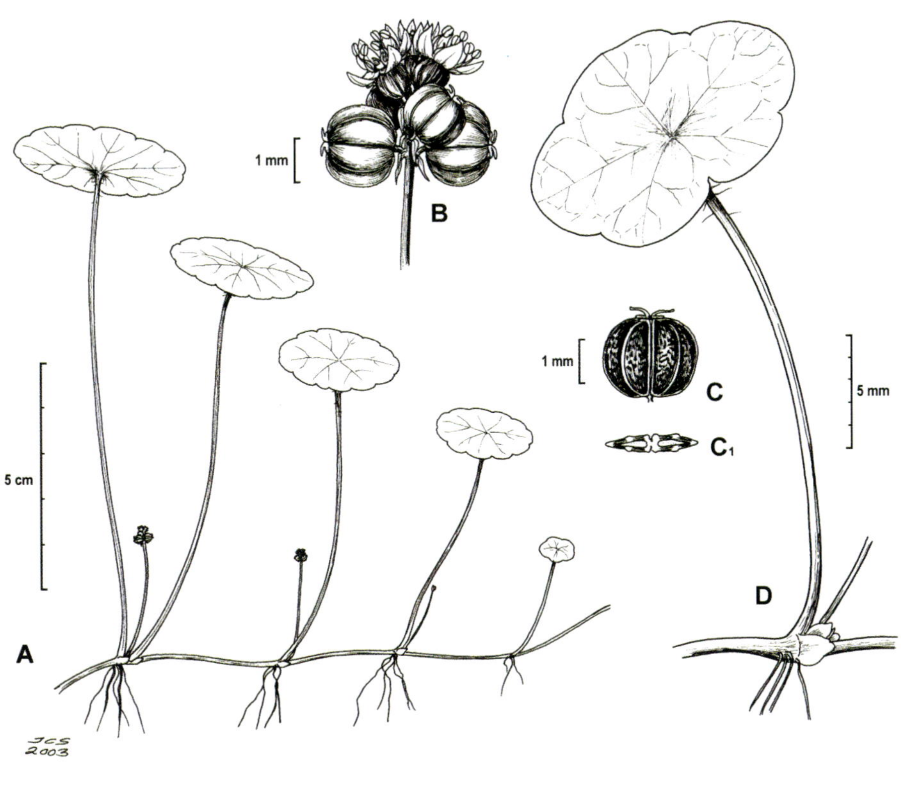

Hydrocotyle vulgaris
A habit, **B** inflorescence, **C** fruit, **C1** section of fruit, **D** part of stem with roots and leaf.

Araliceae

345 *Hydrocotyle ranunculoides* L.f.

DK: Flydende Vandnavle **N:** - **S:** - **FIN:** sumasammakonputki **IS:** -
GB: Floating Pennywort **NL:** Grote waternavel **F:** Hydrocotyle fausse renoncule
D: Grosser Wassernabel **CZ:** - **PL:** Wąkrotka jaskrowata
EST: -**LV:** - **LT:** - **Ru:** -

Hydrocotyle ranunculoides. Pool, Wesel, Lower Rhine, Germany. Photo KvdW.

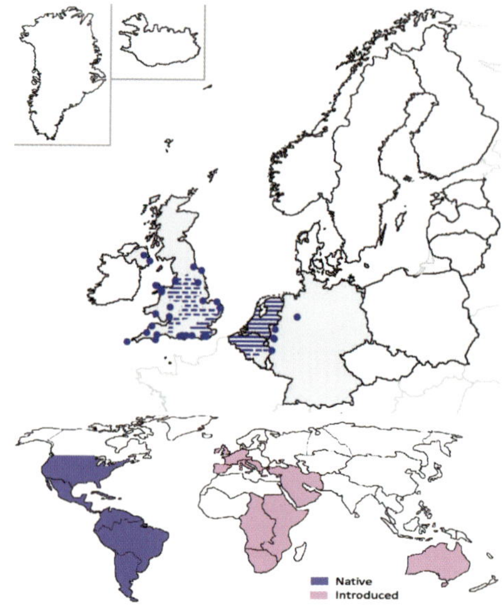

Aquatic perennial. Rooting at all nodes, roots pale.

Stem: Creeping or floating, fleshy, terete, brownish white.

Leaves: Petioles up to 40 cm long. Lamina emergent or floating, reniform, shiny, waxy, yellowish- to dark green, fleshy, 2.5–8(–10) cm in diameter, usually broader than long, shallowly or deeply incised into 3–7 rounded lobes, with a cordate base.

Inflorescence: Peduncle 1–6 cm long, shorter than the adjacent leaf and with 5–10 flowers in a small umbel at the top. Pedicles c. 0.5 mm long.

Flowers: Bisexual. Sepals absent. Petals 5. Stamens 5. Styles 2, not thickened at base.

Fruits: Dry 2-celled schizocarps, orbicular, 2.5–3 mm wide and 2–2.5 mm long, strongly laterally compressed, yellowish green or brownish.

Flowering: July-August.

Biology: The vigorous growth of the branched and fragile stems can reach 20 cm a day. The species can spread rapidly and totally obstruct water bodies.

Habitats: In still and slow-flowing highly eutrophic, base-rich water in rivers, canals and ponds, but also in more nutrient-poor, acid water.

Distribution within the region:
Alien. Widely established in Britain, the Netherlands and Belgium, less abundant in Ireland, Denmark and Germany.

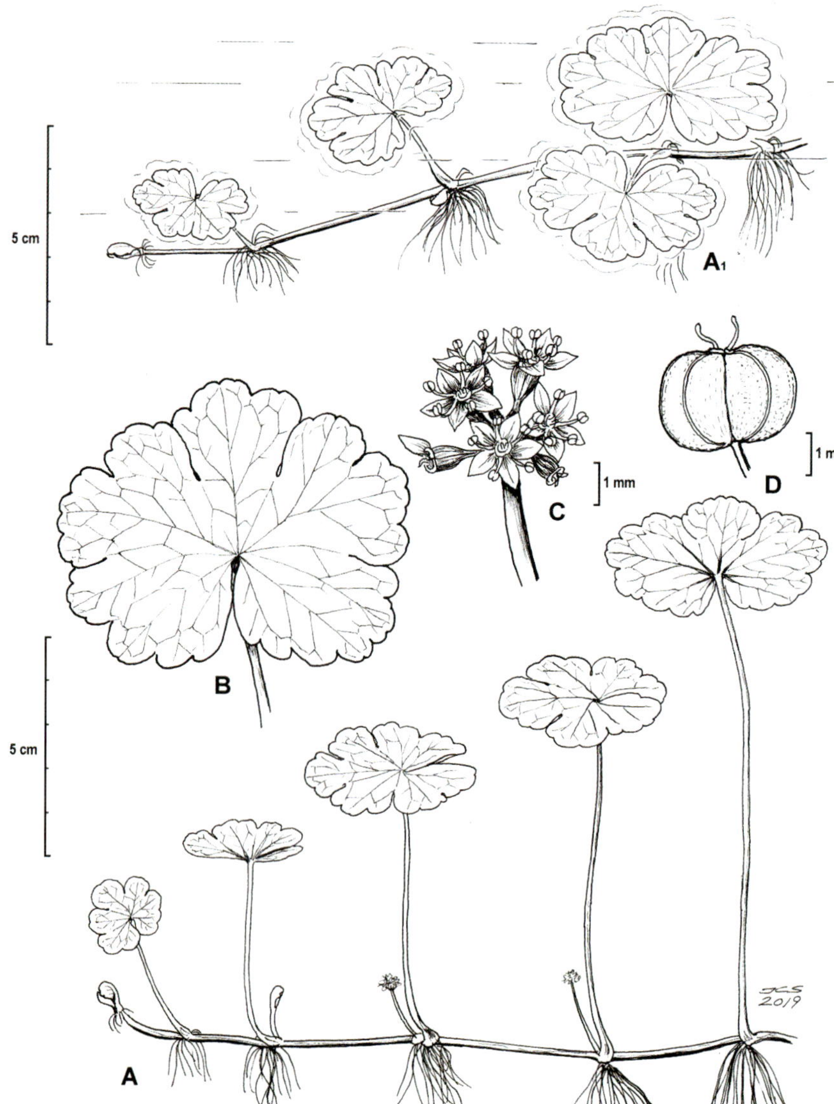

Hydrocotyle ranunculoides
A habit terrestrial plant, **A1** floating stem, **B** leaf, **C** inflorescence, **D** fruit.

Characteristics and similar species:
Similar to **344 *H. vulgaris*** but leaves with a deep basal sinus, generally larger and entirely glabrous. Compare **218 *Ranunculus sceleratus*** and **202-213 *Ranunculus*** sect. ***Batrachium***.

Flowering *H. ranunculoides*. Britain. Photo RVL.

Hydrocotyle ranunculoides. Pool, Wesel, Lower Rhine, Germany. Photo KvdW.

Araliceae

346 *Hydrocotyle verticillata* Thunb.

D: Quirlblättriger Wassernabel

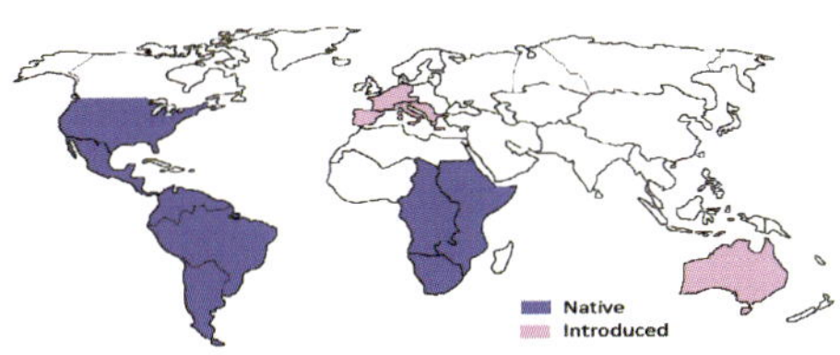

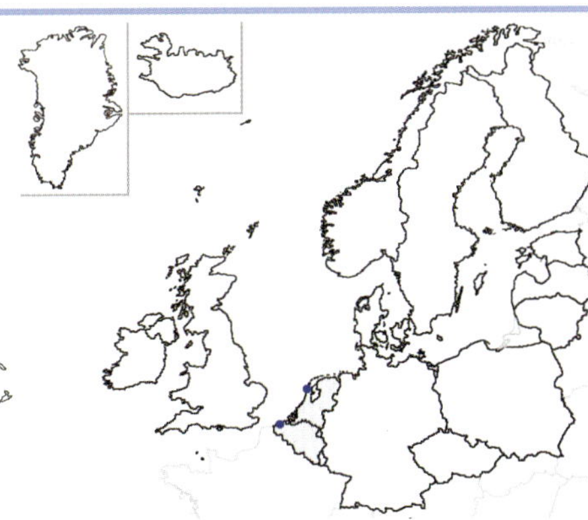

Shape of the peltate leaves very similar to **344 *H. vulgaris,*** but lamina with 9–13 main veins radiating from the entirely glabrous petiole. The characteristic inflorescence is an interrupted spike with flowers in whorls of 2–7. It is equal in length to or longer than the adjacent leaf and sometimes branched into 2 or more spikes.

Distribution within the region:
Alien. Only known from single sites in Belgium and in the Netherlands. May be overlooked in other countries.

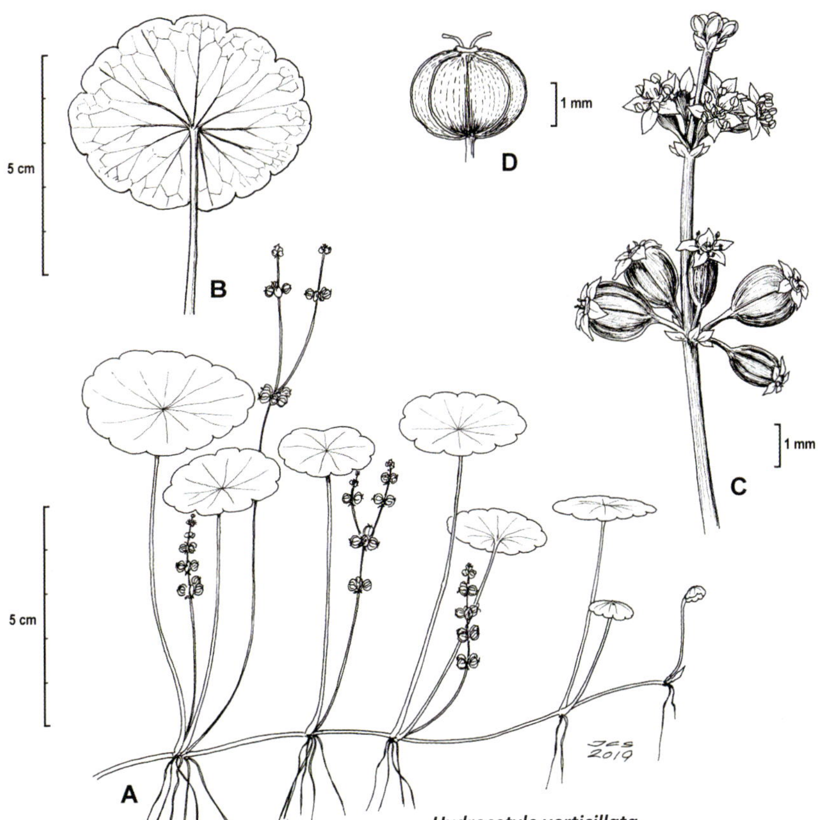

Hydrocotyle verticillata
A habit, **B** leaf, **C** top of inflorescence, **D** fruit.

347 *Hydrocotyle sibthorpioides* Lam.

Similar in habit to a small **344 *H. vulgaris***, but leaves thin, 3–12(–20) mm wide, dentate with a deep basal sinus, petioles 3–10(–30) mm, patent-hairy.

Distribution within the region:
Alien, rarely established in Belgium, Netherlands, Germany, casual in Britain.

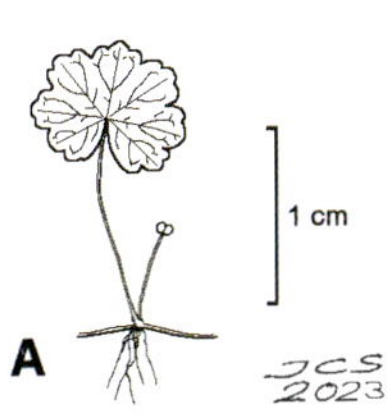

Hydrocotyle sibthorpioides
A habit.

Key to Apiaceae

The vegetative parts of the individual Apiaceae species vary greatly when the species grow in water, making it difficult to make a dichotomous key. It is therefore particularly important to compare the result obtained with the descriptions.

---

1 Stem angular, distinctly ridged to furrowed (***1***)
358 *Sium latifolium*
(Leaves of submerged plants (***17***) very different from aerial leaves (***18***))

1 Stem terete, grooved or slightly furrowed (***2***) 2

2 All leaves 1-pinnate (***3, 4***) 3

2 At least some leaves 2–4-pinnate 7

3 Leaves with 3–7 pairs of short, narrowly linear leaflets (***4***)
356 *Oenanthe fistulosa*

3 Leaves with 1–10 pairs of ovate to almost circular (***3, 8, 9***) or 3-lobed, cuneate leaflets (***6***) 4

4 Leaves with 1–4 pairs of leaflets. These lanceolate to ovate or cuneate, usually 3-lobed. Styles shorter than stylopodium in fruit (***5***)
350 *Helosciadium inundatum*

4 Leaves with 2–10 pairs of leaflets. These ovate to lanceolate or obovate, serrate sometimes with a deep indentation. Styles longer than stylopodium in fruit (***7***) 5

5 Lowest pair of leaflets very small or absent, leaving only a white ring mark at its position (***3a***)
348 *Berula erecta*
(Submerged plants often produce extensive swards of leaves in rivers and canals)

5 Lowest pair of leaflets not very small and no white ring mark is visible 6

6 Leaflets lanceolate to ovate, more or less regularly dentate (***8***). Primary umbels with 0–2(–3) bracts. Peduncle longer than rays
351 *Helosciadium nodiflorum*
(The sterile, intergeneric hybrid between *Berula erecta* and *Helosciadium nodiflorum*, **348a ×*Beruladium***, may key out here, and so will the sterile hybrid **352 *Helosciadium ×moorei*** (*H. nodiflorum × inundatum*))

6 Leaflets obovate to almost circular, irregularly and shallowly serrate often with one deep indentation (***9***). Primary umbels with 3–5(–7) bracts. Peduncle shorter than rays
353 *Helosciadium repens*

7 Leaflets distinctly serrate (***10***)
349 *Cicuta virosa*

7 Leaflets either finely dissected or at most indistinctly serrate 8

8 Submerged leaves present - these with linear to lanceolate segments (***11***) 9

8 Submerged leaves present or not. If present these without linear to lanceolate segments 11

1

3

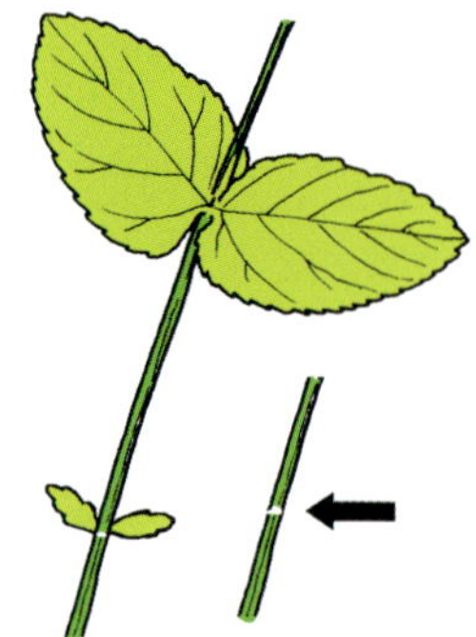
3a

2

6

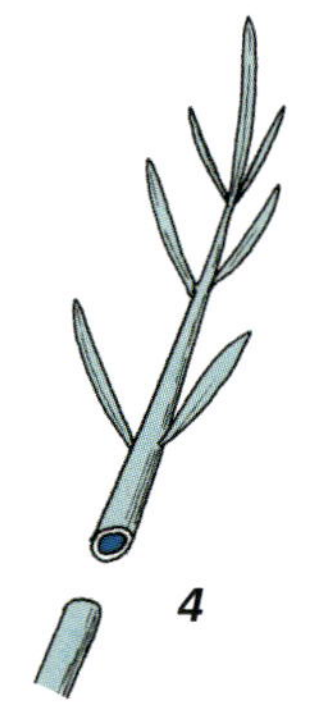
4

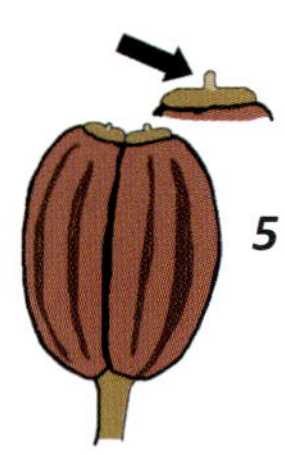
5

7

8

9

10

11

9 Submerged leaves very fine, with 0.1–0.2 mm wide segments. Stem floating, rarely 0.3 cm in diameter at base. Primary umbel with 1–2(–3) rays
350 *Helosciadium inundatum*

9 Submerged leaves coarser, with more than 0.2 mm wide segments. Stem erect or floating, usually more than 2 cm in diameter at base. Primary umbel with 5–15(–17) rays 10

10 Submerged leaves with capillary segments less than 0.3 mm wide. Fruits c. 4 times as long as styles (***12***)
354 *Oenanthe aquatica*

10 Submerged leaves with lanceolate segment more than 0.5 mm wide. Fruits 5–8 times as long as styles (***13***)
357 *Oenanthe fluviatilis*

11 All umbels terminal (***14, 15***). Peduncles longer than rays 12

11 Some umbels leaf opposed (***16***). Peduncles shorter than rays 13

12 Leaves with petioles longer than divided part, lobes not terminated by a hair point.
Bracts absent
356 *Oenanthe fistulosa*

12 Leaves with petioles shorter than divided part, lobes terminated by a small hair point.
Bracts present, 4–8, persistent
355 *Oenanthe crocata*

13 Fruits c. 4 times as long as styles (***12***)
354 *Oenanthe aquatica*

13 Fruits 5–8 times as long as styles (***13***)
357 *Oenanthe fluviatilis*

12

13

14

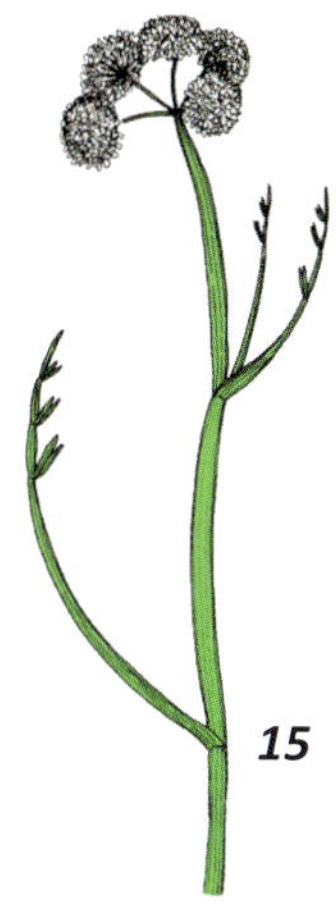
15

16

17

18

Apiaceae

348 *Berula erecta* (Huds.) Coville

Syn.: *Sium erectum* Huds.

DK: Sideskærm **N:** Vasskjeks **S:** Bäckmärke **FIN:** Hanhenputki **IS:** -
GB: Lesser Water-parsnip **NL:** Kleine watereppe **F:** Berle dressée
D: Aufrechter Merk **CZ:** Potočník vzpřímený **PL:** Potocznik wąskolistny
EST: Oja-haneputk **LV:** Stāvā berula **LT:** Siauralapė drėgnuolė **Ru:** Бéрула прямая

Berula erecta richly flowering in a ditch. Uggerby, North Jutland, Denmark. Photo JCS.

Perennial hydrophyte or helophyte with horizontally creeping rhizome and underground or above-ground runners. Roots numerous, commonly also from lower stem nodes.

Stem: Up to 1 m long and more than 1 cm thick, terete, grooved, glabrous, hollow, often richly branched, green. Submerged plants with prostrate to ascending, straight to curved stems, in flowing water totally submerged or with emergent shoots and leaves. Plants in shallow water with more upright stems.

Leaves: 1-pinnate, alternate. Basal and submerged leaves with petioles up to 30 cm long. Sheaths broad, not inflated. Leaves yellowish- to dark green, glabrous, with 5–10 pairs of broad, ovate leaflets with rounded, more or less inconspicuous teeth, the apical leaflet often 3-lobed, the lowest rudimentary or present only as a whitish ring, further from the other leaflets than they are from each other. Stem leaves gradually smaller up the stem, with more sharply, irregularly toothed and narrower, lanceolate leaflets. Submerged leaves flaccid, aerial leaves firm.

Inflorescence: Monoecious. Umbels compound, 3–6 cm in diameter, flat to convex, leaf opposed, peduncles 0.5–3.0 cm long. Primary umbels with 4–8 linear to lanceolate, linear or commonly trifid bracts with membranous margin and 7–16, 1–2.5 cm long rays. Secondary umbels with 14–22 flowers, bracteoles lanceolate, sometimes trifid.

Flowers: 5-merous. Mainly bisexual and all alike. Sepals 0.1–0.5 mm long, very narrow, greenish. Petals white, 0.8–1.5 mm long, 0.8–1.3 mm wide, obovate, moderately emarginate with incurved tips. 5 stamens with 0.9–1.7 mm long filaments and c. 0.3 mm long, pale or sometimes pinkish anthers. 2 styles, 0.5–1 mm long, recurved.

Fruits: Schizocarps, 1.4–1.8 mm long, 0.8–1.1 mm wide, broadly obovate to globular, slightly compressed, dark grey-brown with inconspicuous ridges, sepal teeth inconspicuous, styles curved outwards.

Flowering: June-August.

Biology: Insect pollinated. Fruit dispersal by water and birds. Vegetative dispersal as detached stem fragments.

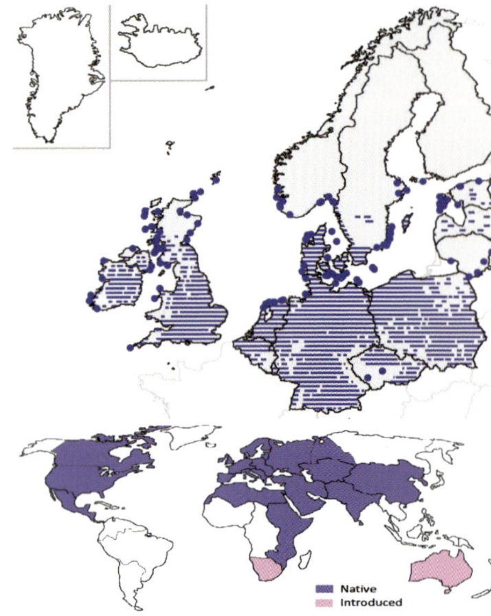

Submerged plants commonly form large stands, propagating by rooting stems and runners. Aerial shoots and leaves die back in autumn while submerged leaves remain green throughout the winter. Plants on wet ground overwinter as leafless rhizomes and runners. Crushed leaves have a smell similar to that of carrot leaves. Plants slightly poisonous.

Habitats: In neutral to alkaline, mesotrophic water. On mineral or humic substrate in fast-flowing brooks and streams, slow-flowing ditches and canals, wet meadows, springs and on the margins of standing waters.

Distribution within the region: Widespread and abundant from Ireland east to eastern Poland, including Denmark and southern Sweden. Scattered or rare further north and in the Baltic states.

Submerged *B. erecta* and *Sparganium emersum* in a stream. Hobro, Denmark. Photo JCS.

Sparsely flowering, emergent *B. erecta*. Rebild, Jutland, Denmark. Photo JCS.

Rudimentary leaflets and "the white ring" positioned below the lowest pair of normal leaflets. Photo KvdW.

Characteristics and similar species:

Often confused with **351 *Helosciadium nodiflorum*** but differs in the umbels with 4–8 bracts and pale ring, often bearing a pair of rudimentary leaflets below and further from the other leaflets than they are from each other.

358 *Sium latifolium* is taller and more robust with an erect and distinctly ridged or grooved stem. Aerial leaves 1-pinnate, finely toothed; submerged leaves 2–3-pinnate with lobes varying from almost linear to broad.

348a ×*Beruladium procurrens* A.C.Leslie

(*Berula erecta* × *H. nodiflorum*)

Sterile, intergeneric hybrid confirmed only from two sites in England. It resembles a sprawling *Berula erecta* with sharply toothed leaves, but these lack the pale ring and distant vesitigial leaflets. The umbels are intermediate, with 1–9 rays subtended by (1–)2–5 bracts, and the stylopodium varies from broadly cushion-like to slightly conical.

×*Beruladium procurrens*. Chippenham Fen, Cambs, England. Photo RVL.

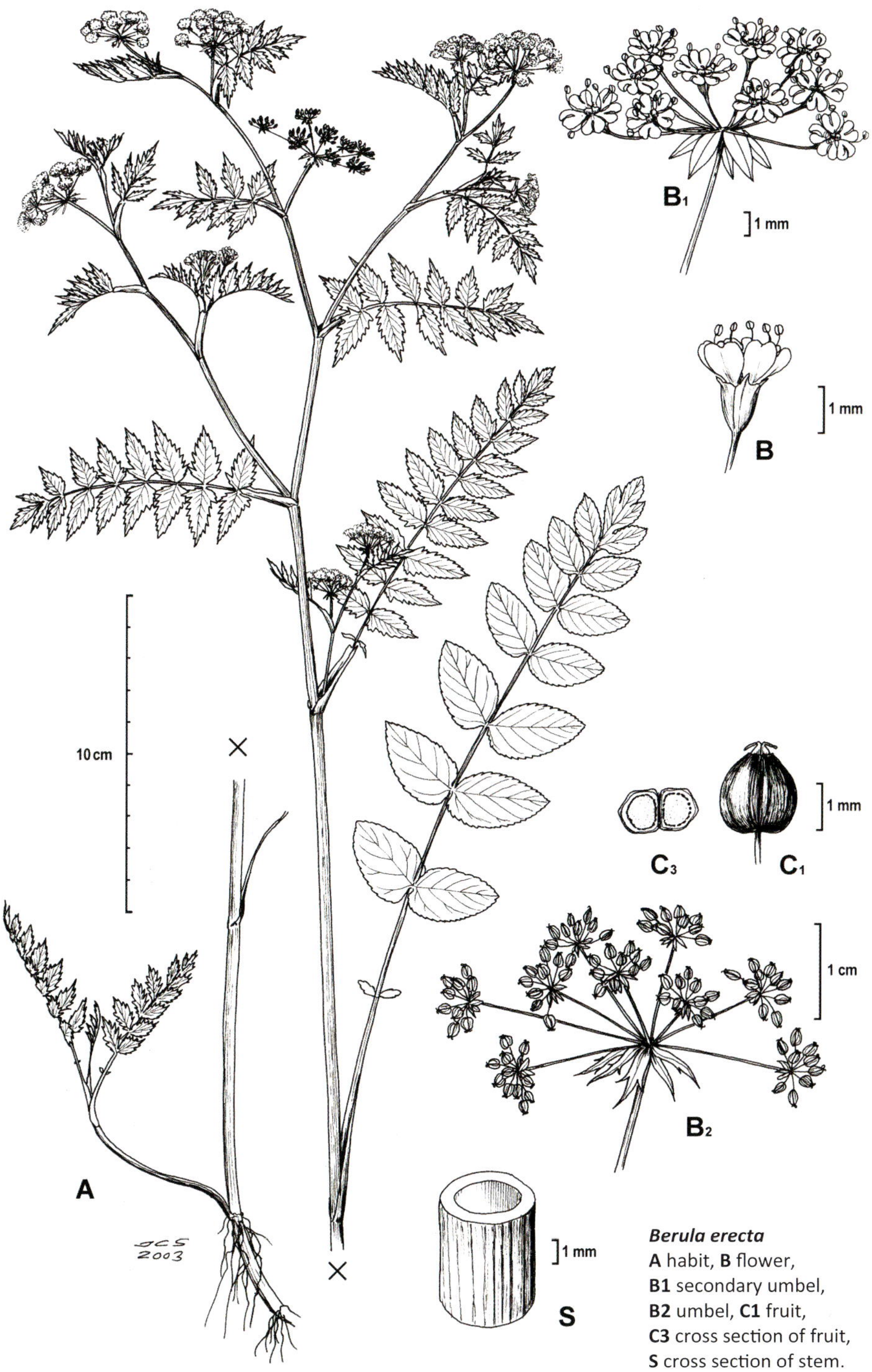

Berula erecta
A habit, **B** flower,
B1 secondary umbel,
B2 umbel, **C1** fruit,
C3 cross section of fruit,
S cross section of stem.

Apiaceae

349 *Cicuta virosa* L.

DK: Gifttyde **N:** Selsnepe **S:** Vanlig sprängört **FIN:** Myrkkykeiso **IS:** -
GB: Cowbane **NL:** Waterscheerling **F:** Cigüe aquatiqu
D: Wasserschierling **CZ:** Rozpuk jizlivý **PL:** Szalej jadowity
EST: Harilik mürkputk **LV:** Indīgais velnarutks **LT:** Nuodingoji nuokana **Ru:** цикута ядовитая

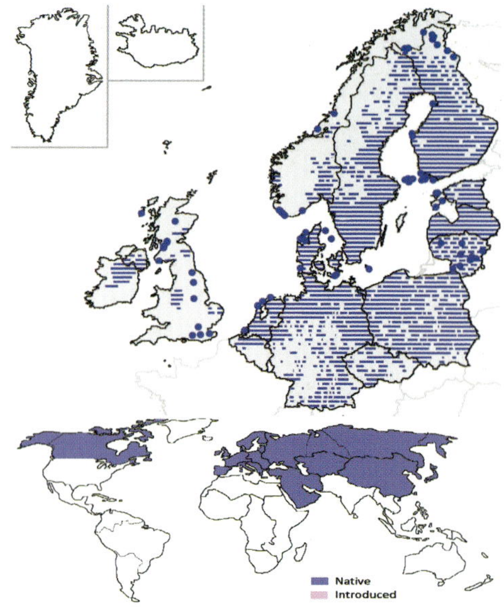

Cicuta virosa in a ditch. Randers, Denmark. Photo JCS.

Perennial helophyte with short horizontal to vertical rhizome, 15–35 mm in diameter, divided into numerous chambers by thin walls. Roots tuberous, grey to light brown.

Stem: Up to 150 cm long, 5–12(–18) mm thick, erect, branched, terete, hollow, glabrous, striate and green, basal parts often with a violet tinge.

Leaves: Basal leaves 2(–3)-pinnate, upper leaves 1-pinnate, lamina 8–25(–33) cm long and 4–13(–22) cm wide, the upper leaves smaller than the basal leaves, green when fresh, but yellow and later violet when they die back, petioles 10–31(–39) cm long with a broad purplish sheath. Leaf lobes lanceolate, sharply dentate. Early leaves of plants growing in water are smaller and narrower.

Inflorescence: Monoecious. Umbels compound, 8–15 cm in diameter, distinctly convex, terminal or opposite leaves on stem and branches, peduncles 4–17 cm long. Primary umbels without bracts or rarely with 1–2 persistent linear bracts and 10–30 rays. Secondary umbels with 20–64 flowers and 6–10 persistent 2.5–7.0(–11) mm long, linear, green bracts.

Flowers: 5-merous. Bisexual. Sepals triangular and acute, 0.3–0.6 mm long, greenish. Petals white, sometimes purplish, 1.1–1.8 mm long and 0.9–1.4 mm wide, uniformly obovate, with incurved, shallowly emarginate tips. 5 stamens with 1.5–2.8 mm long filaments and 0.5–0.6 mm long, whitish-purplish anthers. 2 styles, 1.1–1.8 mm long, divergent.

Fruits: Schizocarps, subglobose, 1.5–2.3 mm long and 2.2–3.0 mm wide, with prominent ridges, brown. Sepals very short, style remains divergent.

Flowering: June-September.

Biology: Insect pollinated. Fruit ribs with air tissue, dispersal by water and birds. Vegetative dispersal by uprooted floating rhizomes and rhizome fragments. The air-filled rhizomes enable plants to form floating or semi-floating stands partly or not rooted in the sediment. Plants in nutrient-rich sites are typically much larger than those in oligotrophic sites. All parts of the plant, and particularly the rhizome, contain the poisonous toxins cicutoxin and cicutol - both of which are lethal to humans and livestock, even in small quantities. Handle with care and avoid physical contact.

Cicuta virosa growing in shallow water. Tange Sø, Ans, Jutland, Denmark. Photo JCS.

Cicuta virosa. Upper stem leaf. Fladbro, Randers, Denmark. Photo JCS.

Cicuta virosa. Umbel. Fladbro, Randers, Denmark. Photo JCS.

Habitats: In moderately acidic to alkaline, mesotrophic to eutrophic water bodies. On muddy or clay substrate in the wet or damp littoral zone of lakes and streams, in open freshwater marshes, moorland and meadows, rarely in brackish marshes.

Distribution within the region: More or less throughout, except Iceland and Greenland, locally abundant or scattered.

Characteristics and similar species: *Cicuta virosa* can be recognised by the pinnate leaves with distinctly serrated and 3-lobed, skewed lanceolate leaflets, almost globular fruits and the distinctly chambered rhizome (warning - very poisonous).

Cicuta virosa is most likely confused with **358 *Sium latifolium***, which has a thicker and distinctly sulcate stem, 1-pinnate leaves with uniform and evenly serrated leaflets, oblong and obovate fruits and a more erect habit compared to the bushier appearance of *Cicuta virosa*.

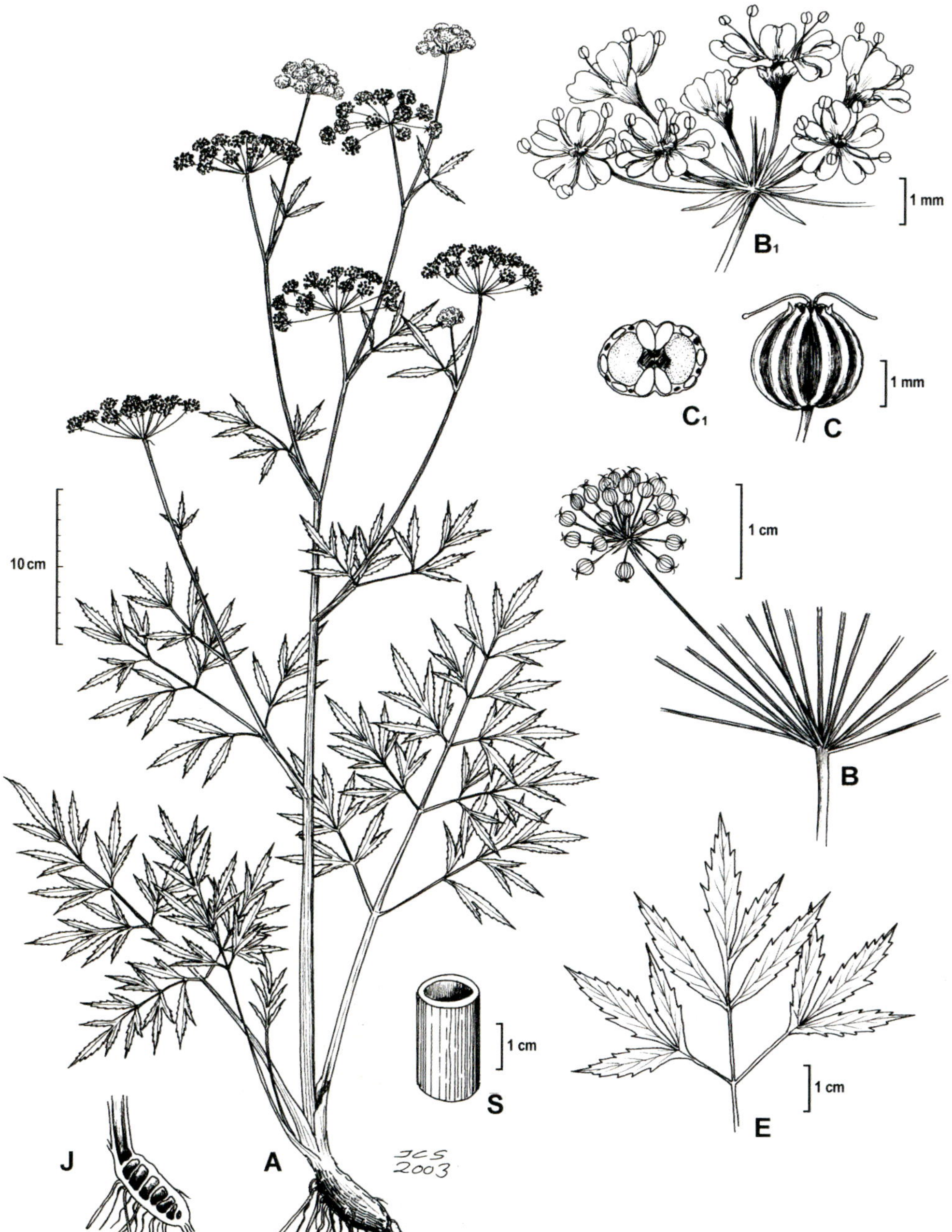

Cicuta virosa
A habit, **B** umbel with a fruiting secondary umbel, **B1** secondary umbel, **C** fruit, **C1** cross section of fruit, **E** part of leaf, **J** cross section of rhizome, **S** stem section.

Apiaceae

350 *Helosciadium inundatum* (L.) W.D.J.Koch

Syn.: *Apium inundatum* (L.) Rchb. f.

DK: Svømmende Sumpskærm **N:** Krypselleri **S:** Krypfloka **FIN:** Tulvaselleri **IS:** -
GB: Lesser Marshwort **NL:** Ondergedoken moerasscherm **F:** Ache inondée
D: Flutender Sumpfsellerie **CZ:** - **PL:** Pęczyna wodne
EST: -**LV:** - **LT:** - **Ru:** -

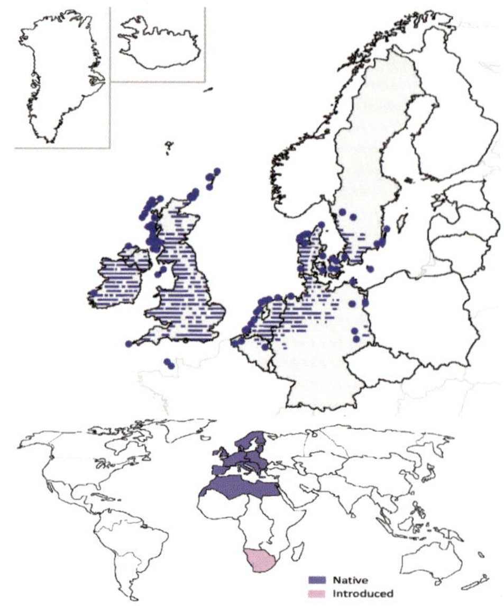

Submerged, flowering *H. inundatum*. Lough Bridget, County Clare, Ireland. Photo RVL.

Perennial hydrophyte or helophyte. Rhizome absent. Roots white to greenish, attached to the substrate or hanging free in the water.

Stem: Up to 1 m long, 1.5–3.0 mm in diameter, terete, smooth, glabrous, hollow, easily compressed, moderately branched, green to yellow green, often with a reddish tinge, rooting at the nodes. Submerged plants with creeping to prostrate or floating stems, with long internodes. Stems of emerged plants ascending with short internodes.

Leaves: Alternate. Aerial leaves variable, mostly 1-pinnate, with 1–4 pairs of usually 3-lobed, cuneate, green to dark green, more or less broad leaflets. Apical leaflet 5–13 mm long and 6–13 mm wide. Petioles up to 3.5 cm long, broadly sheathing with membranous margins. Submerged leaves 3–4-pinnate, with 7–9 pairs of very fine filamentous to narrowly linear, slightly compressed leaflets. Lamina 2–6 cm long and 0.9–2.8 cm wide, yellow green to green, petioles short with the sheath reaching the basal part of the lamina.

Inflorescence: Monoecious. Umbels compound, few cm in diameter, opposite to the leaves, peduncles 0.5–1.8 cm long. Primary umbels ebracteate with 1–2(–3) rays up to c. 1 cm long. Secondary umbels 3–6 flowered. Bracteoles 2–4(–6), lanceolate, 3-veined.

Flowers: 5-merous. Bisexual. Sepals absent or rudimentary. Petals white, 0.6–1.0 mm long and 0.5–0.7 mm wide, entire with short incurved tips. 5 stamens with 0.3–0.5 mm long filaments and 0.2–0.3 mm long, reddish-violet anthers. 2 styles, 0.2–0.3 mm long, divergent.

Fruits: Schizocarps, 2.5–3.5 mm long, 0.7–1.1 mm wide, elliptic to oblong. Carpophore absent. Mericarps light brown or often tinged purplish, with 5 thick, prominent ridges. Stylopodium flat.

Flowering: June-August.

Biology: Self-pollinated. After fertilisation the peduncles bend and the fruits ripen under water or close to the substrate. The fruits have ribs containing air tissue and are dispersed by water and birds. Plants may grow permanently submerged, particularly in fast-flowing water but also occasionally in standing waters. Submerged plants may remain green during the winter, while terrestrial plants on wet ground overwinter as more or less leafless stems or as seeds.

Floating *H. inundatum* with typical 1-pinnate aerial leaves with 3-lobed leaflets. Oligotrophic lake, Per Madsens Kær, Jutland, Denmark. Photo JCS.

Helosciadium inundatum. Typical fruits are light brown, but exposed plants often develop distinctly purple fruits. Temporary pond in dunes near Vangså, Jutland. Photo JCS.

Flowering *H. inundatum* with aerial and submerged leaves. Seasonal pond in pasture, Bragança, Portugal. Photo RVL.

Habitats: In slightly acidic to slightly alkaline, oligotrophic to mesotrophic water. On mineral or organic substrate in or by ponds, lakes, seasonal dune slacks and wet meadows. Less commonly in flowing waters, rivers, streams, ditches and canals.

Distribution within the region:
Native. Widespread but typically scattered and declining in Ireland, Britain, the Netherlands, northern Germany, Denmark and southern Sweden.

Characteristics and similar species:
Flowering *Helosciadium inundatum* is characterised by the small primary umbels without bracts and with only 1–3 secondary umbels.

Sterile plants may be confused with tiny individuals of **206-217 *Ranunculus*** spp. (*Batrachium*), which differs in palmate or 3–8 times tri- and bifurcately branched leaves in contrast to the pinnate leaves of *H. inundatum*. The finely divided submerged leaves may also lead to confusion with other aquatic umbellifers, with filamentous submerged leaves, but these are generally much larger and robust with well-developed rhizomes.

226-231 *Myriophyllum* spp. and **199 *Ceratophyllum submersum*** differ in leaves in whorls of 3–5.

Hybrids: 352 *Helosciadium ×moorei* - see this.

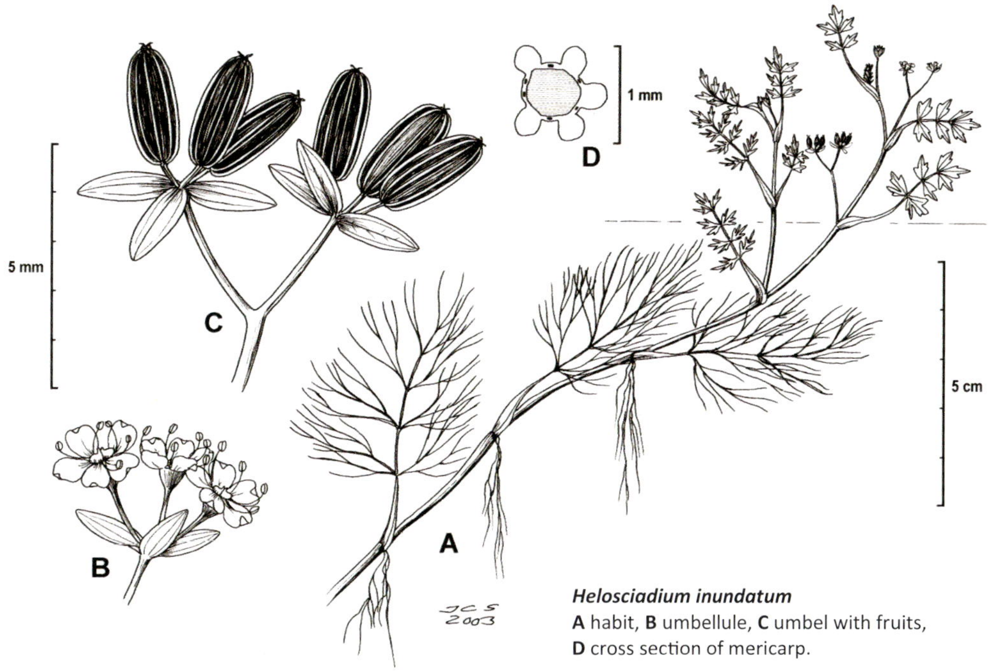

Helosciadium inundatum
A habit, **B** umbellule, **C** umbel with fruits, **D** cross section of mericarp.

Apiaceae

351 *Helosciadium nodiflorum* (L.) W.D.J.Koch

Syn.: *Apium nodiflorum* (L.) Lag.
DK: - **N:** Nikkeselleri **S:** Fräneflоka **FIN:** Koiranselleri **IS:** -
GB: Fool's Watercress **NL:** Groot moerasscherm **F:** Ache
D: Knotenblütiger Sumpfsellerie **CZ:** - **PL:** Pęczyna węzłobaldachowa
EST: -**LV:** - **LT:** - **Ru:** Болотнозонтичник узлоцветковый

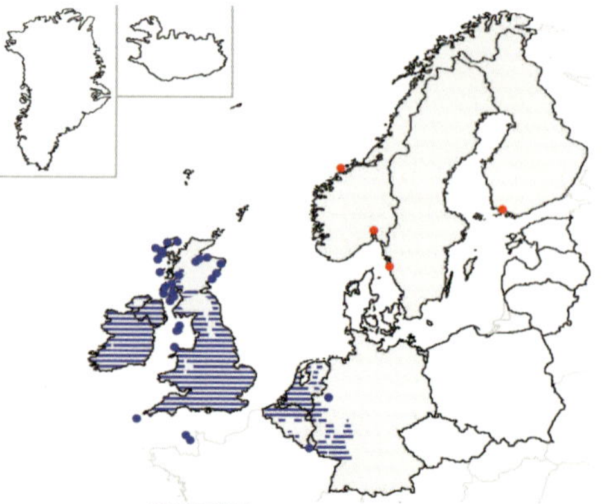

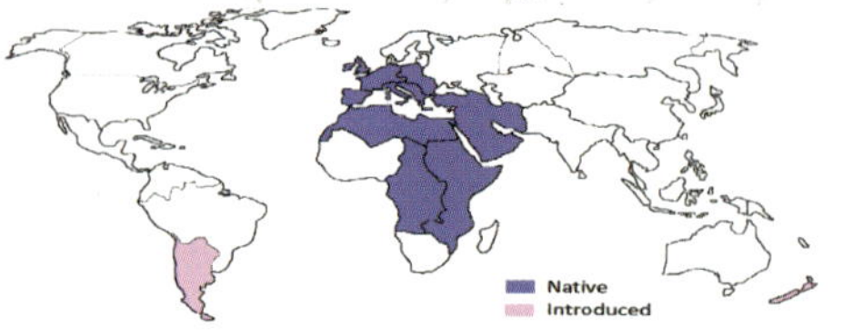

Submerged *H. nodiflorum.* River Itchen, Hampshire, England. Photo RVL.

Perennial hydrophyte or helophyte. Roots in clusters from the lower stem nodes.

Stem: 30–100 cm long, 3–12 mm in diameter, creeping horizontally or ascending, terete, finely grooved, glabrous, hollow, branched or unbranched, green, often with a purple tinge.

Leaves: 1-pinnate, homophyllous, bright green, slightly glossy. Upper leaves sessile, rather short with few leaflets. Lower and middle stem leaves ovate to oblong in outline, 4–40 cm long with 2–6 pairs of sessile, lanceolate to ovate leaflets, these (5–)9–55(–75) mm long, (3.5–)5–20(–40) mm wide, more or less regularly dentate, sometimes lobed. Petioles 1.5–12 cm long with 15–35(–50) mm long, narrowly oblong, striate sheaths with a membranous margin.

Inflorescence: Monoecious. Umbels compound, 2–4 cm in diameter, opposite leaves, sessile or with short peduncles c. 4 mm long. Primary umbels with narrowly lanceolate bracts 0–2(–3), c. 3 mm long and 5–10, 8–30 mm long rays of unequal length. Secondary umbels with 12–20 flowers and 5–7 lanceolate, acute bracteoles 3–6 mm long with pale membranous margins.

Flowers: 5-merous. Mainly bisexual and all alike. Sepals green with rudimentary teeth. Petals white, c. 0.75 mm long, broadly lanceolate, acute with slightly incurved tips. 5 stamens with filaments more than 1 mm long and c. 0.2 mm long, pale to light brown anthers. 2 styles, recurved.

Fruits: Schizocarps, 1.25–2.0 mm long, longer than wide, ovoid, laterally compressed. Carpophore absent. Mericarps brown to purple, with 5 distinctly prominent, pale ribs. Styles slightly longer than the flat stylopodium.

Flowering: June-August.

Biology: Insect pollinated. Fruit dispersal by water and birds. Vegetative dispersal by detached stem fragments. Ascending aerial shoots mainly develop in shallow water or on wet soil. In running water, particularly fast-flowing alkaline streams, often permanently submerged. Favoured by disturbance of the substrate and it tends to decline in stable conditions. Overwinters as leafless stems unless submerged, where plants may be green all year.

Habitats: In shallow, mainly nutrient rich, neutral to alkaline water. In streams, ditches, swamps and marshes, the drawdown zones on the margins of ponds, lakes, rivers and canals, sometimes growing into nearby terrestrial vegetation from wetter habitats.

Distribution within the region:
Native. Widespread and abundant in Ireland, Britain, the Netherlands, parts of Belgium and western Germany, with scattered non-native populations in Scandinavia.

Characteristics and similar species:
Flowering *H. nodiflorum* differs from other aquatic umbellifers by the more or less sessile umbels and by distinctly ridged fruits with no visible sepal teeth.

348 *Berula erecta* has similar leaves, but the lowest pair of leaflets are usually rudimentary or absent with a white ring at its position. Flowering *Berula* may be recognised by the 4–8 bracts, and fruits with inconspicuous ribs.

Sterile plants may also be confused with **248-250 *Nasturtium*** spp., which differ in the leaves with subopposite, more rounded and less serrate leaflets.

Hybrids:
348a ×*Beruladium procurrens* can be recognised by a combination of: the absence of a ring-mark on the petiole, lower leaves with up to 5 pairs of leaflets, peduncles varying from shorter than to longer than rays, umbels

Helosciadium nodiflorum. Flowering plant. Dilborn, Germany. Photo KvdW.

Helosciadium nodiflorum. Umbel. River Itchen, Hampshire, England. Photo RVL.

with several usually untoothed bracts, a relatively long stigma and style and the absence of ripe fruits.

352 *Helosciadium ×moorei* (Syme) Druce

(*H. nodiflorum × inundatum*)

Sterile; distinguished from *H. nodiflorum* by the more strongly divided leaflets, umbel peduncles and rays subterete and smooth walled, rays 2–4 and bracteoles wholly herbaceous (green), or with a narrow hyaline margin basally. Common in parts of Ireland, rare or under-recorded elsewhere.

Submerged *H. ×moorei* . Near Ennis, Co. Clare, Ireland. Photo RVL.

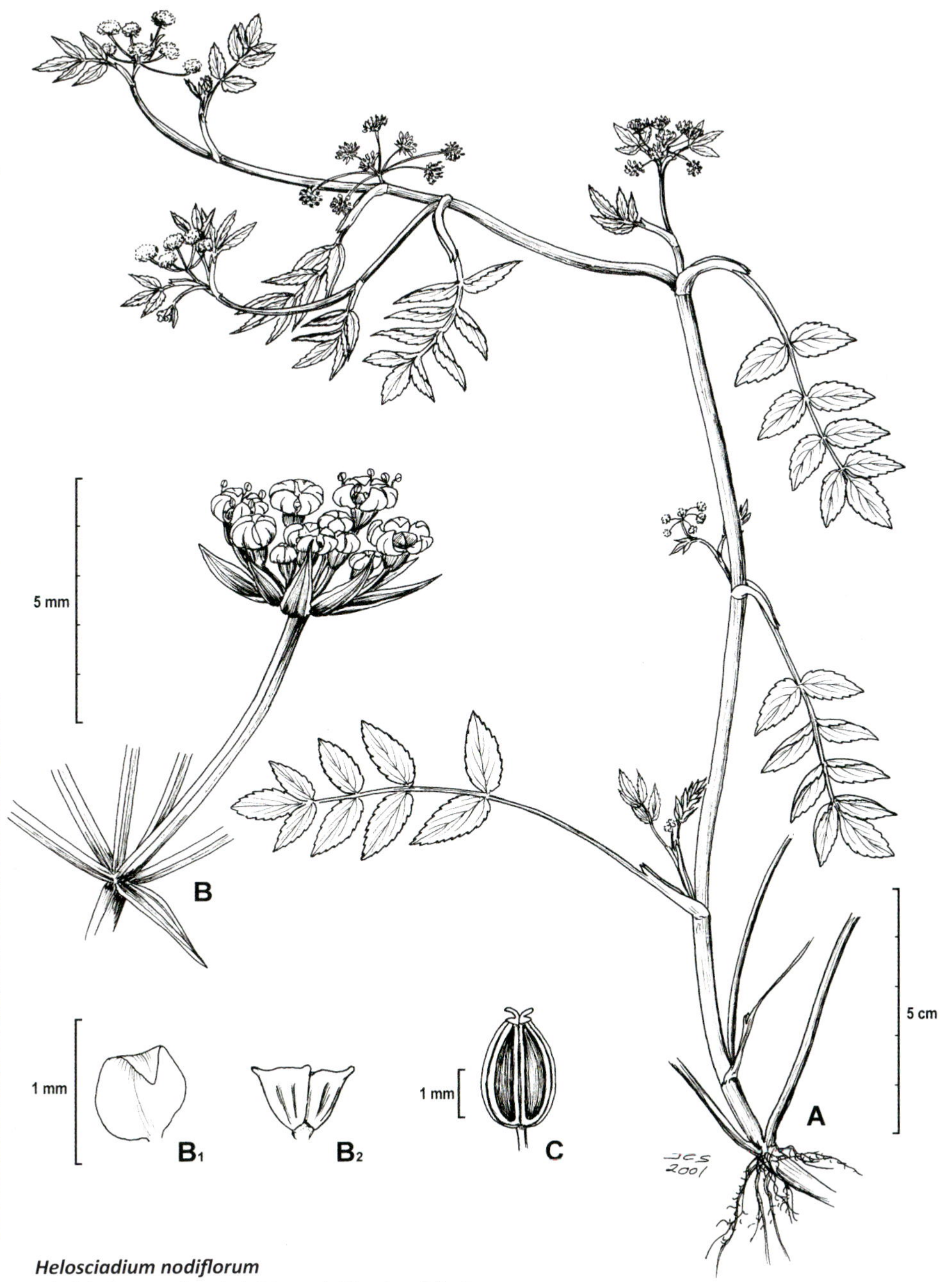

Helosciadium nodiflorum
A habit, **B** part of umbel, **B1** petal, **B2** calyx, **C** fruit.

Apiaceae

353 *Helosciadium repens* (Jacq.) W.D.J.Koch

Syn.: *Apium repens* (Jacq.) Lag.
DK: Krybende Sumpskærm **N:** - **S:** - **FIN:** Juurtoselleri **IS:** -
GB: Creeping Marshwort **NL:** Kruipend moerasscherm **F:** Ache rampante
D: Kriechender Sumpfsellerie **CZ:** Miřík celer **PL:** Pęczyna bloltna
EST: - **LV:** - **LT:** - **Ru:** Сельдерей ползучий

Helosciadium repens partly submerged. Navarredonda de Gredos, Sierra de Gredos, Spain. Photo RVL.

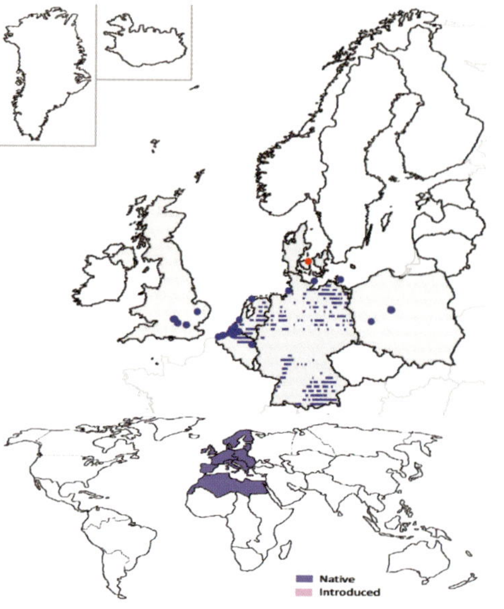

Perennial helophyte or rarely hydrophyte. Rhizomes short, vertical. Roots pale.

Stem: 2.5–9(–13) cm long - up to 30 cm when stem is submerged, 0.8–1.6 mm in diameter, creeping and rooting from all nodes, terete, smooth or sulcate with prominent veins, hollow, pale green to violet.

Leaves: Alternate, cauline, 1–2(–5) together at the nodes, 1-pinnate with 3–5 pairs of leaflets. Lamina 1.6–4.5(–9.0) cm long, green to reddish violet. Petiole 3.4–9.5(–14) cm long, with a sheath 0.4–1.3 cm long, with membranous margins and not usually reddish violet. Leaflets sessile, obovate to almost circular, 3.2–10 mm long and 3.8–13 mm wide, margin irregularly and shallowly serrate often with one deep indentation. Terminal leaflet with 2–3 lobes.

Inflorescence: Monoecious. Umbels compound, opposite to the leaves, peduncles 1.5–7.0 cm long, shorter than the leaves. Primary umbel with 3–5(–7) persistent, obtuse bracts with membranous margins and 4–6 glabrous rays up to 2 cm long. Secondary umbels 13–22 flowered with 5–6 persistent, obtuse, 3-veined bracteoles.

Flowers: 5-merous. Bisexual. Pedicels 1.8–2.6 mm long. Sepals indistinct. Petals white, 0.6–0.8 mm long and 0.5–0.7 mm wide, unevenly obovate, with incurved tips. 5 stamens with filaments 0.5–0.8 mm long and reddish-violet anthers c. 0.3 mm long. 2 styles, 0.4–0.5 mm long, recurved.

Fruits: Schizocarps, 0.9–1.1 mm long, 0.8–1.2 mm wide, slightly wider than long, ovoid to suborbicular. Carpophore absent. Mericarps with 5 slender, prominent ribs, greenish brown. Style approximately twice as long as stylopodium.

Flowering: June-September.

Biology: Mainly self-pollinated; insect pollination may occur. Fruit dispersal by water and animals but probably limited. Prolific vegetative propagation by the creeping stems and runners. Commonly forms large, dense stands. Plants on moist or wet ground have short-petiolate leaves and flowers. Plants growing in water develop leaves with long petioles, but do not flower.

Habitats: In neutral to slightly alkaline, predominantly oligotrophic water, on muddy or mineral substrate in the wet or damp littoral zone of lakes and streams and in temporary, mainly winter-flooded water bodies; also in open freshwater marshes, moors and meadows, rarely in slightly brackish marshes. Predominantly exposed habitats with little competition from tall, shading vegetation.

Helosciadium repens. Leaf. Westphalen, Germany. Photo KvdW.

Helosciadium repens on moist substrate in a ditch at Erp, Nord-Brabant, Netherlands. Photo KvdW.

Helosciadium repens. Secondary umbel. Photo KvdW.

Distribution within the region:
Native. Locally frequent in parts of Germany and the Netherlands, scarce or rare elsewhere, absent from the north.

Characteristics and similar species:
Helosciadium repens is a relatively small plant with creeping stems rooting at all nodes. The leaflets are sessile, shallowly serrate with one deep incision. The fruits are almost globular.

Often confused with **351 *Helosciadium nodiflorum***, from which it can only be distinguished with certainty when flowering or fruiting. *H. repens* has umbels with 3–5(–7) bracts, more or less globular fruits and often bifid leaflets, while *H. nodiflorum* has umbels with 0–2(–3) bracts, more oblong fruits and dentate but not bifid leaflets.

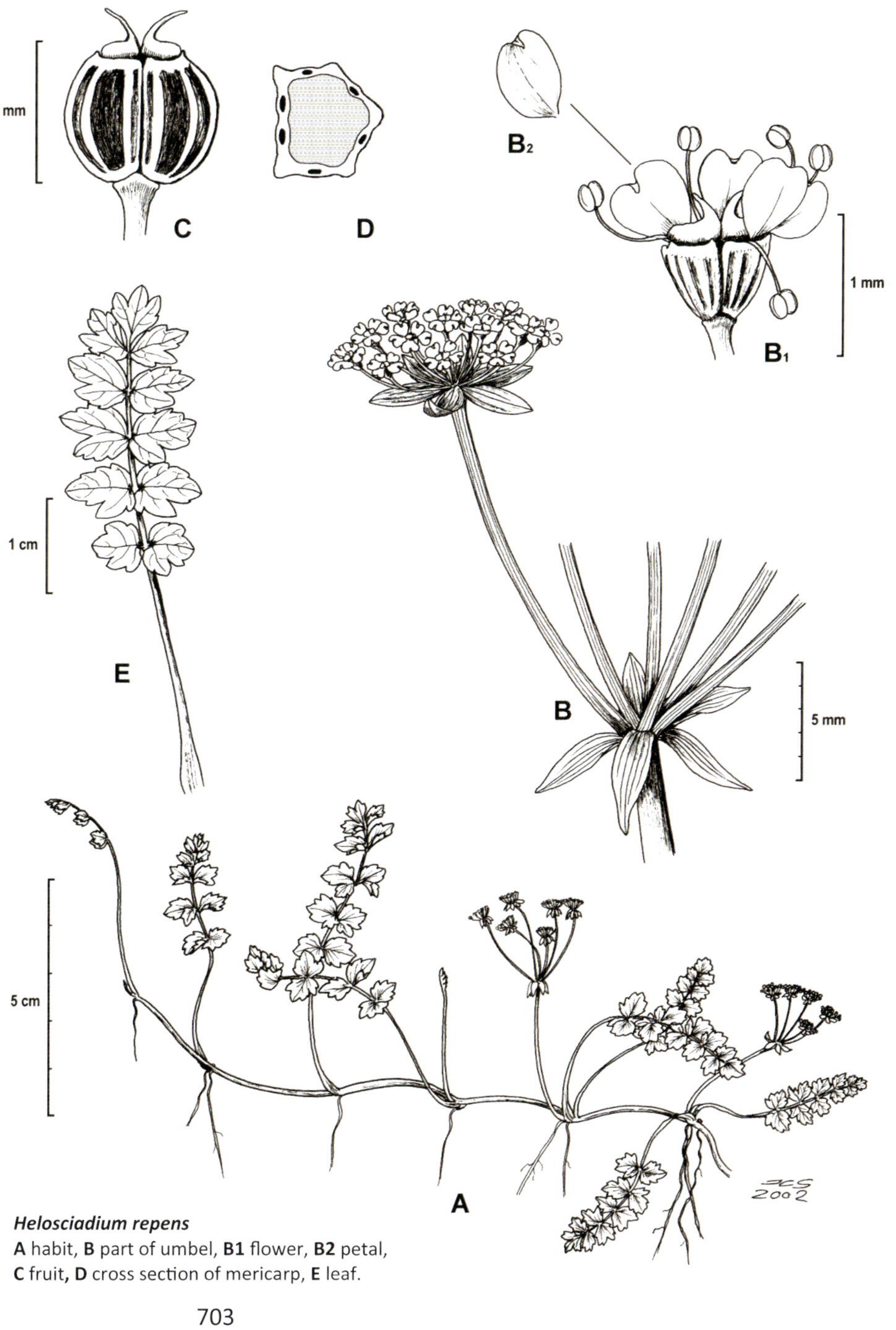

Helosciadium repens
A habit, **B** part of umbel, **B1** flower, **B2** petal, **C** fruit, **D** cross section of mericarp, **E** leaf.

353a *H. ×longipedunculatum* (F.W.Schultz) Desjardins

(*H. nodiflorum × repens*)

is fertile with leaflets ovate to broadly ovate, coarsely serrate and occasionally lobed, peduncles typically longer than umbel rays and an involucre of 1–3 bracts always present. Known from scattered locations in Scotland, England, Germany and France.

Apiaceae

354 *Oenanthe aquatica* (L.) Poir.

DK: Billebo-Klaseskærm **N:** Hestekjørvel **S:** Vattenstäkra **FIN:** Liejupahaputki **IS:** - **GB:** Fine-leaved Water-dropwort **NL:** Watertorkruid **F:** Oenanthe aquatique **D:** Großer Wasserfenchel **CZ:** Halucha vodní **PL:** Kropidło wodne **EST:** Harilik vesiputk **LV:** Ūdenu padille **LT:** Vandeninė išnė **Ru:** Омежник водяной

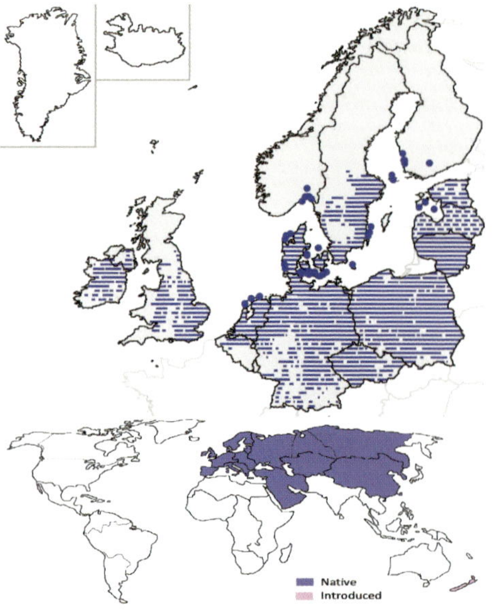

With the richly branched stem and finely divided leaves *O. aquatica* often has a messy appearance. Randers, Denmark. Photo JCS.

Annual or biennial helophyte or more rarely hydrophyte with short vertical rhizome and numerous branched roots. Young plants with long, swollen roots.

Stem: 30–150(–180) cm long, 2.0–2.2 (–2.8) cm diameter at the base, terete, ribbed, glabrous, hollow, richly branched, distinctly inflated towards the base, green, often with purple tinge. Plants in shallow water or on moist ground typically have upright or ascending stems. Permanently submerged plants occasionally have very long stems.

Leaves: 3–5-pinnate. Aerial leaves up to 20 cm long, 3–4-pinnate. Leaflets green, petiolate with fine, lobed, acute filaments. Petiole with inflated, 1–4 cm long basal sheath. Submerged leaves up to 60 cm long, very finely divided, obovate to lanceolate in outline, 4(–5)-pinnate. Leaflets mostly with linear, capillary filaments.

Inflorescence: Monoecious. Umbels compound, pedunculate, slightly convex, terminal on main stem and opposite to leaves on branches. Primary umbels ebracteate with 7–15(–17) scabrid rays 1–4 cm long. Secondary umbels densely flowered, with 5–10 linear to lanceolate bracteoles with membranous margins.

Flowers: 5-merous, most are bisexual, some are male. Sepals 0.4–1.2 mm, greenish, with triangular tips. Petals white, (1.1–)1.5–2.4 mm long, 2.0–1.8 mm wide, deeply emarginate with incurved tips, those of central flowers equal, those of marginal flowers subequal. 5 stamens with long filaments and purple anthers , 0.3–0.5 mm long. 2 styles.

Fruits: Schizocarps 3.5–5.0 mm long, cylindrical or ovoid, greenish brown with pale ribs, c. 4 times longer than styles, with 5 small, apical sepal teeth of unequal length.

Flowering: June-August.

Biology: Insect pollinated. Fruit with air tissue in ribs, dispersed by water and birds. Vegetative dispersal by detached, rooting stem fragments. Submerged plants most commonly occur early in the growing season when water levels are high. Permanently submerged plants are rare.

Habitats: In slightly acid to alkaline, mesotrophic to eutrophic water. In temporary lakes and ponds which are

Oenanthe aquatica. Fruits about 4 times as long as styles. Paderup, Randers, Denmark. Photo JCS.

Oenanthe aquatica. Young plant with finely divided leaves in rosette. Randers, Denmark. Photo JCS.

Oenanthe aquatica. Submerged leaf with linear, capillary filaments. Bislicher Insel, Germany. Photo KvdW.

seasonal or with variable water levels. On the margins of lakes and rivers and in wet meadows and marshes. Rarely in flowing waters.

Distribution within the region: Native. Widespread and abundant throughout most of the region, absent from Scotland and northern Scandinavia.

Characteristics and similar species: The very similar **357** ***Oenanthe fluviatilis*** differs in fruits 5–8 times as long as the styles and in submerged leaves with flat filaments with a cuneate base.

Submerged leaves of *O. aquatica* are very similar to those of **358** ***Sium latifolium,*** which differs in distinctly ribbed and non-inflated stem. **349** ***Cicuta virosa*** differs from *O. aquatica* by serrate leaflets and globular fruits.

Oenanthe aquatica. Part of submerged leaf. Photo KvdW.

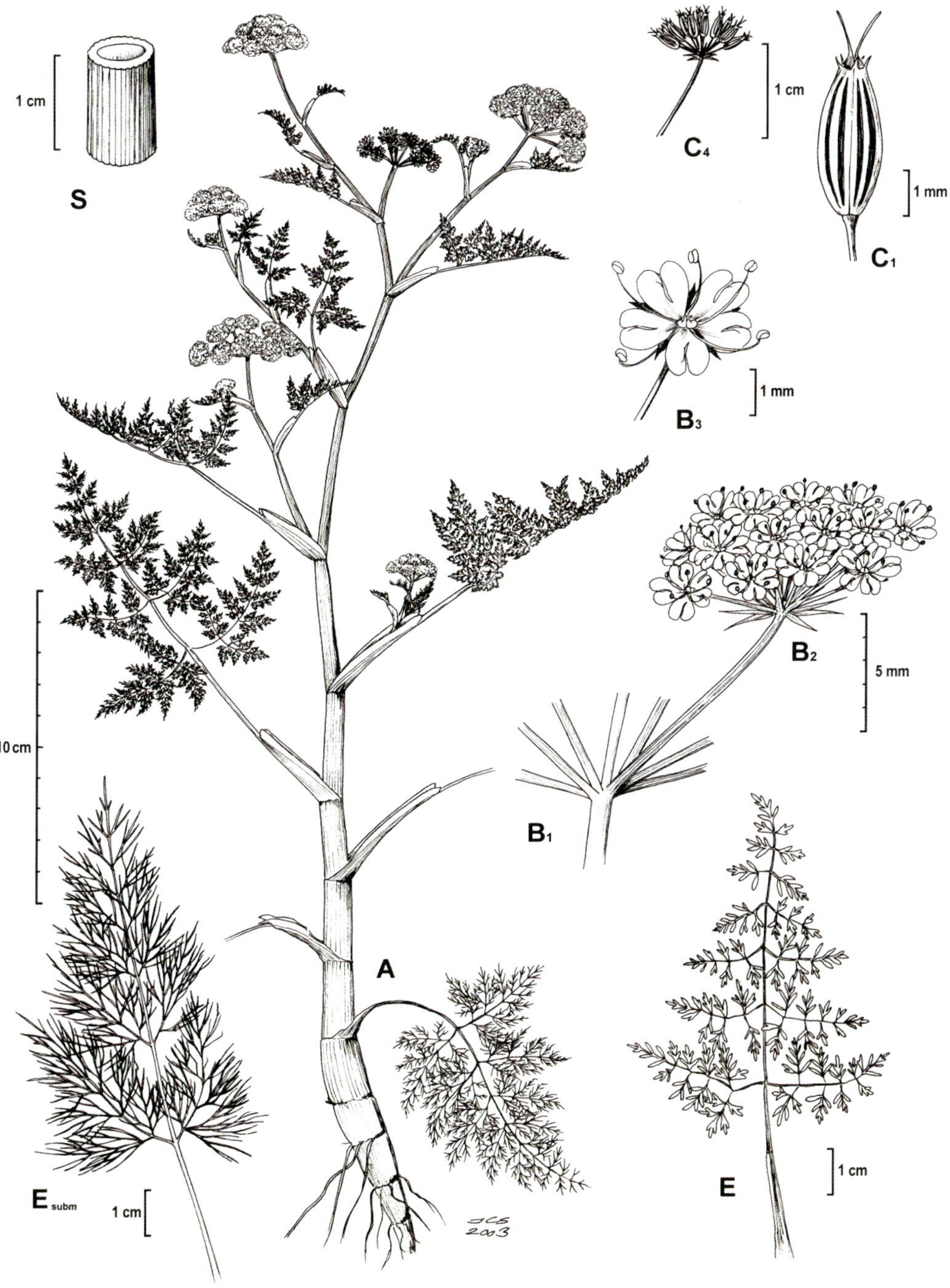

Oenanthe aquatica
A habit, **B1** part of primary umbel, **B2** secondary umbel, **B3** flower, **C1** fruit, **C4** fruiting secondary umbel, **E** aerial leaf, **Esubm** submerged leaf, **S** stem section.

Apiaceae

355 *Oenanthe crocata* L.

DK: Gift-Klaseskærm **N:** -**S:** - **FIN:** - **IS:** -
GB: Hemlock Water-dropwort **NL:** Dodemansvingers **F:** Oenanthe safranée
D: Giftiger Wasserfenchel **CZ:** - **PL:** -
EST: - **LV:** - **LT:** - **Ru:** Кропило шафранное

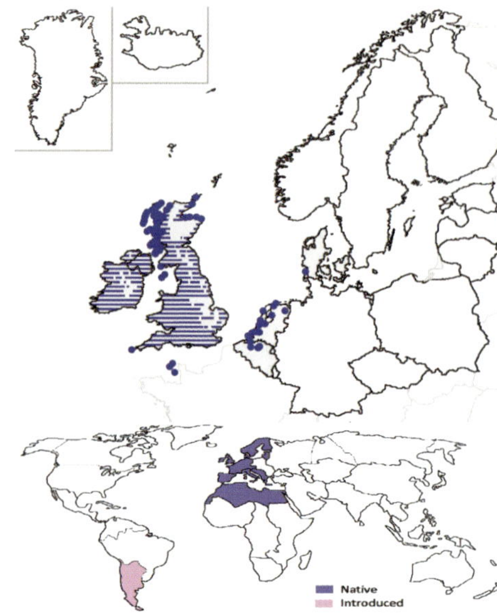

Oenanthe crocata. Stroudwater Canal, Gloucestershire, Britain. Photo RVL.

Perennial helophyte. Rhizome vertical with numerous fine roots and few tuber-like fleshy roots up to 6 cm long and 1(–3) cm in diameter.

Stem: 0.5–1.5 m long, terete, striate, glabrous, hollow, green, richly branched.

Leaves: Petiolate, with a clasping basal sheath. Green to dark green, glabrous and slightly glossy. Basal leaves up to 30 cm long, 3–4-pinnate, broadly triangular in outline. Leaflets with broadly oval to rounded, apiculate lobes tapered at the base. Stem leaves 2–3-pinnate, shortly petiolate, similar to the basal leaves, but much smaller and leaflets with narrower lobes. Submerged leaves are not morphologically different.

Inflorescence: Monoecious. Umbels compound, convex, terminal on main stem and branches, long pedunculate. Primary umbel with 4–8 linear to lanceolate, rarely trifid, persistent bracts and 12–30 smooth, 3–6 cm long rays. Secondary umbels densely flowered, with 6–10 narrowly linear to lanceolate bracteoles.

Flowers: 5-merous. Mainly bisexual. Sepals short, narrowly triangular and green. Petals white, unequal, up to 1.5 mm long, much longer than the sepals, deeply emarginate with incurved tips. 5 stamens with long filaments and 0.3–0.5 mm long purple anthers. 2 styles, 2–3 mm long.

Fruits: Schizocarps, 4–6 mm long, cylindrical or ovoid, green brown with shallow ribs, c. 2 times longer than the styles, with 5 small apical sepal teeth.

Flowering: May-August.

Biology: Insect pollinated. Fruit dispersal by water and birds. Dense stands are commonly formed by seeds on muddy ground above the water line on the margins of flowing as well as still waters. Seeds may germinate under water. In western oceanic areas, plants often grow fully submerged in fast-flowing streams and rivers; they generally do not flower, but some plants will develop an emergent flowering stem in such conditions. All parts, especially the roots, are highly poisonous, and the species is considered to be one of the most poisonous aquatic plants.

Habitats: In slightly acid to alkaline, mesotrophic to eutrophic water. In and above the water line of ponds, lakes, streams, rivers, canals and ditches, in wet meadows, alder or willow carrs and marshes.

Distribution within the region:
Native. Western, widespread and very abundant in Ireland and Britain, local in the Netherlands and Denmark.

Characteristics and similar species:
Typical *Oenanthe crocata* with 2–4-pinnate leaves with broad, rounded

Oenanthe crocata. Submerged plants. Pineda Geres, Portugal. Photo KvdW.

Flowering *O. crocata*. Roadside verge, Berkeley, Gloucestershire, England. Photo RVL.

Oenanthe crocata. Fruiting umbellule. Portugal. Photo KvdW.

lobes, homophyllous basal and stem leaves and umbels with widely separated secondary umbels is unmistakable. The leaves resemble those of the non-aquatic species Celery (*Apium graveolens* L.), which has led to fatal mistakes, but differ in the apical hair-point.

Oenanthe crocata. Leaflet lobes apiculate. Photo RVL.

Oenanthe crocata. Swollen tuber-like roots. River Itchen, Hampshire, England. Photo RVL.

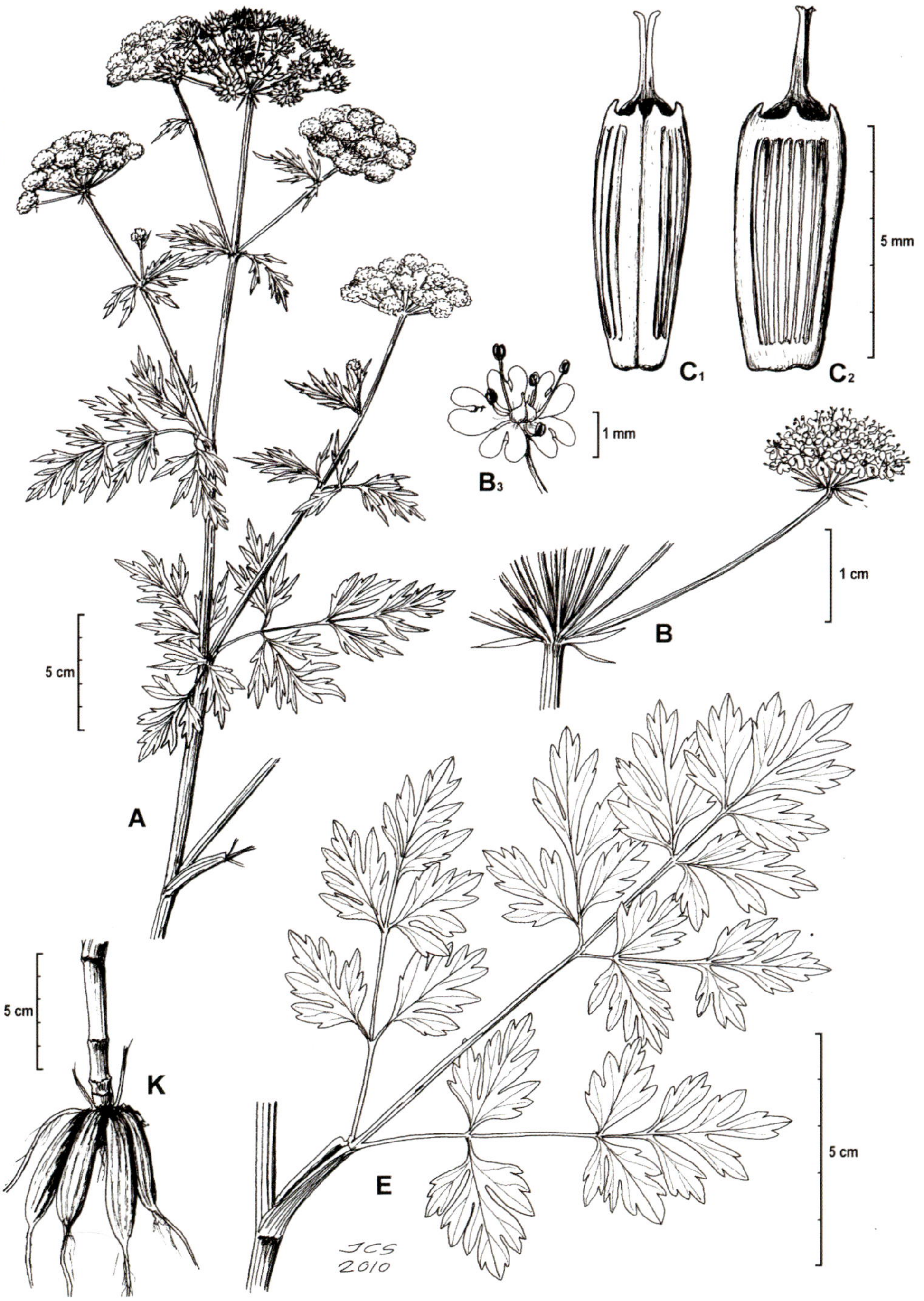

Oenanthe crocata
A habit, **B** part of primary umbel, **B3** flower, **C1**,**C2** fruit, **E** aerial leaf, **K** swollen roots.

Apiaceae

356 *Oenanthe fistulosa* L.

DK: Vand-Klaseskærm **N:** Pipekjørvel **S:** Pipstäkra **FIN:** Pulleapahaputki **IS:** -
GB: Tubular Water-dropwort **NL:** Pijptorkruid **F:** Oenanthe fistuleuse
D: Röhriger Wasserfenchel **CZ:** Halucha dutá **PL:** Kropidło piszczałkowate
EST: - **LV:** - **LT:** - **Ru:** Омежник дудчатый

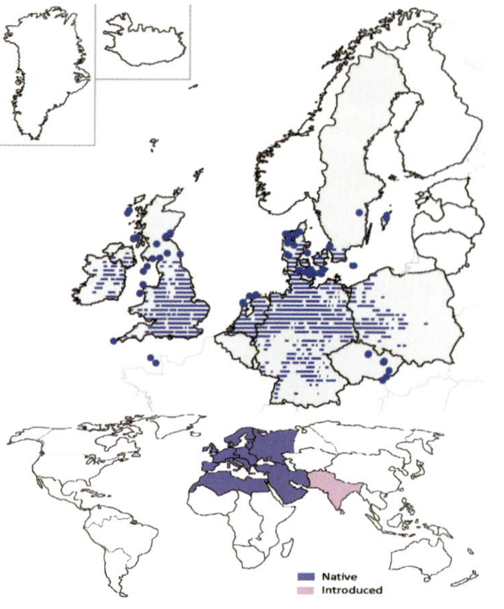

Oenanthe fistulosa. Cantley Marshes, Norfolk, England. Photo RVL.

Perennial helophyte or more rarely hydrophyte with short vertical rhizome and numerous swollen roots.

Stem: 30–80 cm long, 3–8 mm thick, terete, fistular, easily compressible, striate, green to bluish green, sparsely branched, but producing long aboveground stolons from the lower nodes.

Leaves: Stem leaves green, 5–20 cm long, 1-pinnate with inflated petioles and 3–7 pairs of short, narrowly linear, filamentous leaflets, sheaths not inflated. Basal leaves 1–2(–3) pinnate with 4–7 pairs of narrowly lobed leaflets distinctly different from stem leaves, typically dying back early. Submerged leaves similar to basal leaves.

Inflorescence: Monoecious. Umbels compound, terminal on main stem and branches, some opposite to leaves, peduncle 2–13 cm long. Primary umbel ebracteate with 2–6(–7) robust, glabrous rays 1–2 cm long. Secondary umbels with 24–50 flowers, slightly convex when flowering, becoming almost globular when fruiting. Bracteoles 6–12, linear to lanceolate, persistent with membranous margins.

Flowers: 5-merous. Sepals 1.0–1.2 mm, greenish. Petals white to pink, 2.8–4.5 mm long and 1.5–2.8 mm wide, very deeply emarginate with incurved, narrow tips. Central flowers bisexual, symmetrical, more or less sessile. Outer flowers male, enlarged, distinctly zygomorphic and pedicellate. 5 stamens with long filaments and 0.4–0.5 mm long purple anthers. 2 styles.

Fruits: Schizocarps, 3.0–4.5 mm long, conical to obovate, yellow brown, as long as the divergent styles, with 5 prominent apical sepal teeth.

Flowering: June-August.

Biology: The large outer petals with a scent of vanilla attract pollinating insects. Fruit ribs contain air tissue and dispersal is by water and birds. Vegetative dispersal and formation of stands occurs from above-ground runners. Submerged plants most commonly occur early in the growing season, although they may persist in deep water. Submerged leaves typically die back later in the season.

Habitats: In neutral to alkaline, mesotrophic to eutrophic water. In wet meadows, marshes and grazing marsh ditch systems and on the margins of ponds, lakes, streams and rivers, typically on seasonally inundated areas. Favoured by intermediate disturbance of the ground. Less frequent in running waters, where plants are more likely to be permanently submerged.

Oenanthe fistulosa. Submerged leaves. Cantley Marshes, Norfolk, England. Photo RVL.

Oenanthe fistulosa. Inflorescence - note the zygomorphic outer flowers. Randers, Denmark. Photo JCS.

Oenanthe fistulosa. Part of fruiting head. Randers, Denmark. Photo JCS.

Oenanthe fistulosa. Leaf of young plant. Photo BM.

Distribution within the region:
Native. Widespread in Ireland, Britain, the Netherlands, Denmark and northern Germany, scattered or rare elsewhere, absent from the north and east.

Characteristics and similar species:
The fistular stem, 1-pinnate stem leaves with few pairs of linear leaflets and the above-ground runners should exclude confusion with other aquatic plants. The basal leaves which develop early and similarly looking submerged leaves may lead to confusion with other pinnate-leaved plants.

Oenanthe fistulosa. The stem leaves are characteristic. Photo JCS.

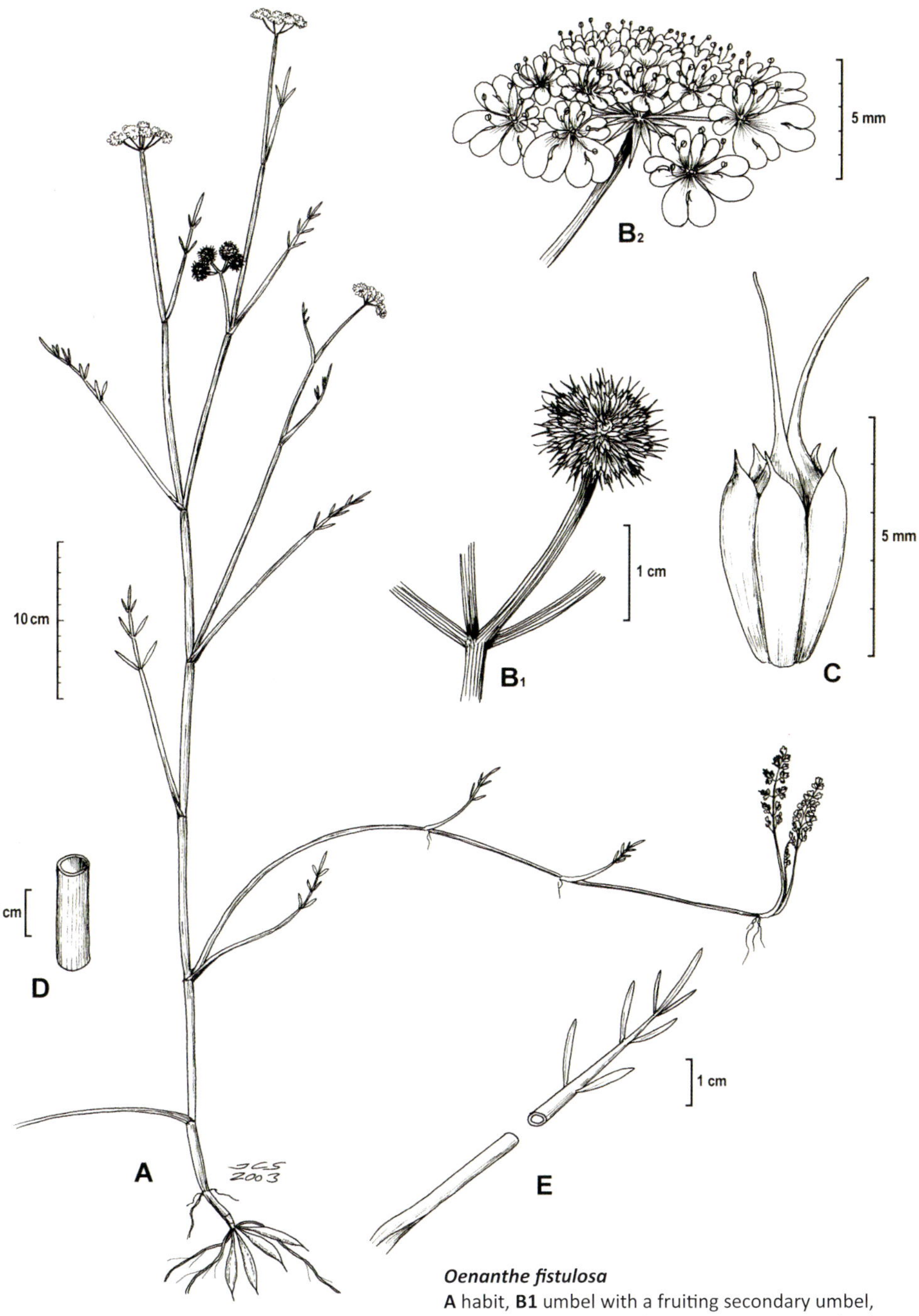

Oenanthe fistulosa
A habit, **B1** umbel with a fruiting secondary umbel, **B2** secondary umbel, **C** fruit, **D** stem section, **E** leaf.

Apiaceae

357 *Oenanthe fluviatilis* (Bab.) Coleman

DK: Flod-Klaseskærm **N:** - **S:** Flodstäkra **FIN:** Jokipahaputki **IS:** -
GB: River Water-dropwort **NL:** Riviertorkruid **F:** Oenanthe des rivières
D: Flütender Wasserfenchel **CZ:** - **PL:** -
EST: -**LV:** - **LT:** - **Ru:** -

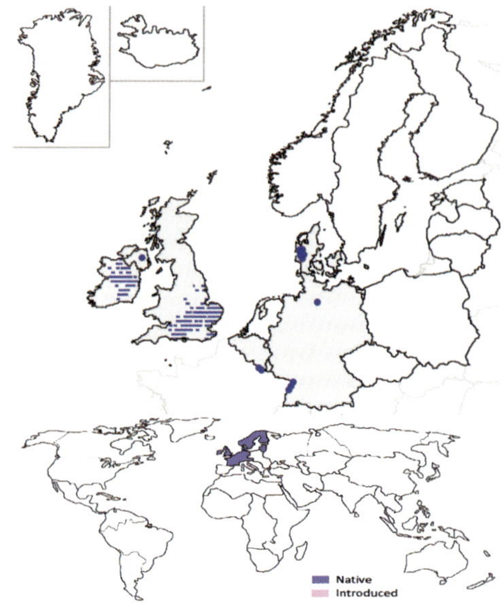

Oenanthe fluviatilis in swiftly flowing water. River Skjern Å, Jutland, Denmark. Photo JCS.

Perennial hydrophyte or more rarely helophyte with horizontally creeping or short vertical rhizome and numerous branched roots.

Stem: Up to 2 m long and 0.4–1.3 cm diameter, terete, ribbed, glabrous, hollow, richly branched. Submerged plants with prostrate to ascending, straight to curved stems, totally submerged in flowing water or with emergent shoots. Plants in still, shallow water with erect stems distinctly inflated at the base.

Leaves: 2-3-pinnate, deeply incised, morphologically variable. Submerged leaves up to 50 cm long, occasionally longer, the lamina finely divided with obovate to lanceolate outline. Leaflets with linear to oblong-lanceolate, flat filaments, with an acute apex and cuneate base, > 0.5 mm wide. Petiole shorter than lamina, basally sheathing.

Inflorescence: Monoecious. Umbels compound, pedunculate, more or less flat, terminal on main stem, opposite to leaves on branches. Primary umbel ebracteate with 5–10(–15) rays. Secondary umbels densely flowered, with 5–10 linear to lanceolate bracteoles with membranous margins.

Flowers: 5-merous. Mainly bisexual, partly male. Sepals 0.4–0.9 mm long with triangular tips, greenish. Petals 1.5–2.5 mm long and 0.9–1.5 mm wide, unevenly obovate, deeply emarginate with incurved tips. 5 stamens with long filaments and 0.4–0.6 mm long purple anthers. 2 styles.

Fruits: Schizocarps, 5.0–6.5 mm long, cylindrical to oval, ribbed, green brown, 5–8 times longer than the styles, with 5 small apical sepal teeth.

Flowering: June-September.

Biology: Insect pollinated. Fruit with air tissue in ribs, dispersed by water and birds. Vegetative dispersal as detached stem fragments. Submerged plants often form large stands, propagating by rooting stems; remains green throughout winter.

Habitats: In slightly acid to alkaline, mesotrophic to eutrophic water. Typically in clear, flowing water of rivers and streams at depths of 0.5–2(–4) m, as well as in backwaters and less frequently in standing waters such as grazing marsh ditches.

Distribution within the region:
Native. Widespread and locally abundant in central Ireland, on the chalk in England, rare in Northern Ireland, Denmark, Belgium and Germany. Rare and endemic to northwestern Europe.

Characteristics and similar species:
Submerged *O. fluviatilis* in flowing water with finely divided leaves and flat,

Oenanthe fluviatilis. Submerged leaves. River Skjern Å, Jutland, Denmark. Photo JCS.

In slow-flowing water *O. fluviatilis* can resemble *O. aquatica*. River Vorgod Å, Denmark. Photo JCS.

Oenanthe fluviatilis. Fruits about 5-8 times as long as styles. River Skjern Å, Denmark. Photo JCS.

> 0.5 mm wide filaments with the base cuneate is unmistakable.

The very similar **354 *Oenanthe aquatica*** differs in fruits c. 4 times as long as the styles and in submerged leaves with capillary filaments less than 0.3 mm wide.

Oenanthe fluviatilis. Part of submerged leaf. Photo KvdW.

Oenanthe fluviatilis
A habit, **B2** umbellule, **C1** fruit, **C4** umbel with a fruiting umbellule, **E** part of submerged leaf.

Apiaceae

358 *Sium latifolium* L.

DK: Bredbladet Mærke **N:** Stor vasskjeks **S:** Vattenmärke **FIN:** Sorsanputki **IS:** -
GB: Greater Water-parsnip **NL:** Grote watereppe **F:** Berle à larges feuilles
D: Großer Merk **CZ:** Sevlák potoční **PL:** Marek szerokolistny
EST: Harilik jõgiputk **LV:** Platlapu cemere **LT:** Plačialapė drėgmenė **Ru:** поручейник широколистный

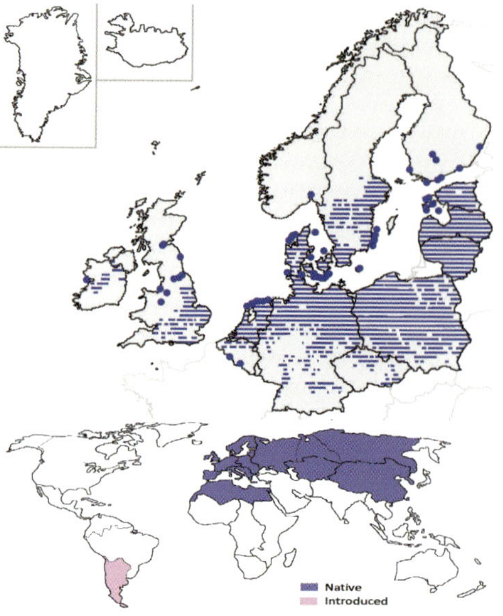

Sium latifolium on the edge of a reed-fringed ditch, Cantley Marshes, Norfolk, England. Photo RVL.

Perennial helophyte with short, horizontally creeping to vertical rhizome 5–11 mm diameter. Roots numerous, white.

Stem: 50–130(–200) cm long, base up to 16 mm diameter, terete, distinctly ridged to furrowed, glabrous, hollow, the upper part sparsely branched with upright branches, dark green.

Leaves: Alternate. Aerial leaves 1-pinnate with lamina up to 30 cm long and 20 cm wide and petioles up to 35 cm long with a broad but not inflated sheath. Leaves with 3–8 pairs of lanceolate-elliptic leaflets with uniformly and finely serrate margin, apical leaflet petiolate, lateral sessile, dark to yellowish green, glabrous. Submerged leaves 2–5-pinnate, up to 30 cm long and 20 cm wide, petioles robust, the leaflets with linear to broad, green to brownish filaments, often appearing bushy. Intermediate forms common when aerial leaves begin to develop.

Inflorescence: Monoecious. Umbels compound, 9–15 cm in diameter, flat to convex, terminal on main stem and branches, with 5–14 cm long peduncles. Primary umbels with 5–12 lanceolate bracts, sometimes large and leaf-like, with membranous margin and 20–30(–35) rays up to c. 6 cm long. Secondary umbels densely flowered with 3–10 lanceolate bracteoles with membranous margin.

Flowers: 5-merous. Mainly bisexual in the first developed secondary umbels, subsequently mostly male. Sepals 0.4–1.0 mm long, lanceolate, greenish. Petals white, 1.2–1.8 mm long, 1.0–1.5 mm wide, moderately emarginate with inwardly curved tips. 5 stamens with 1.6–2.5 mm long filaments and c. 0.5 mm long, violet anthers. 2 styles,

1.0–1.8 mm long, divergent.

Fruits: Schizocarps, 3–4 mm long and 1.0–1.5 mm wide, elliptic to obovate, light brown with distinct whitish ridges, sepal teeth acute with narrow membranous margin, styles divergent. Crushed unripe fruits with a distinct smell like parsley.

Flowering: July-August.

Biology: Insect pollinated. Fruit ribs with air tissue, fruit dispersal by water and birds. Vegetative propagation by shoots from the rhizome and buds in the axils of the lower stem leaves. Submerged leaves develop early in the growth season and die back as the aerial shoots develop. Fully developed flowering plants usually lack submerged leaves. The first aerial leaves may be poly-pinnate with narrow filaments, and grade into submerged leaves. Plants slightly poisonous, especially the roots and the rhizome.

Habitats: In neutral to alkaline water. In the littoral zone of lakes and streams, ponds, ditches and canals, wet meadows and alder swamps. Rare in fast-flowing water.

Sium latifolium. Submerged leaf in deep water of grazing marsh ditch, Cantley Marshes, Norfolk, England. Photo RVL.

Aerial leaf of *Sium latifolium*. Ditch near Byrum, Læsø, Denmark. Photo JCS.

Sium latifolium with poly-pinnate submerged leaves with very narrow filaments and poly-pinnate aerial leaves preceding the typical 1-pinnate aerial leaves early in the growth season. Byrum, Læsø, Denmark. Photo JCS.

Sium latifolium. Stem base and roots. Sjælland, Denmark. Photo Niels Faurholdt.

Distribution within the region: Native. Widespread and locally abundant from central Ireland east to the Baltic states, including Denmark and southern Sweden; scarce or absent in southern and northern areas. Scarce and declining in some areas, such as parts of England.

Characteristics and similar species: Flowering *Sium latifolium* differs from other aquatic umbellifers by the erect appearance with a thick and ridged stem, uniform aerial leaves and elliptic fruits with distinct sepal teeth.

348 *Berula erecta* has a much less ridged stem and leaflets with more unevenly serrated margin, globular fruits with inconspicuous sepal teeth.

349 *Cicuta virosa* has 2-pinnate leaves, primary umbels without bracts and globular fruits.

Submerged *Sium latifolium* may be confused with submerged plants of **354 *Oenanthe aquatica***, less likely with **357 *Oenanthe fluviatilis***, but differs from both mainly by its distinctly ridged stem.

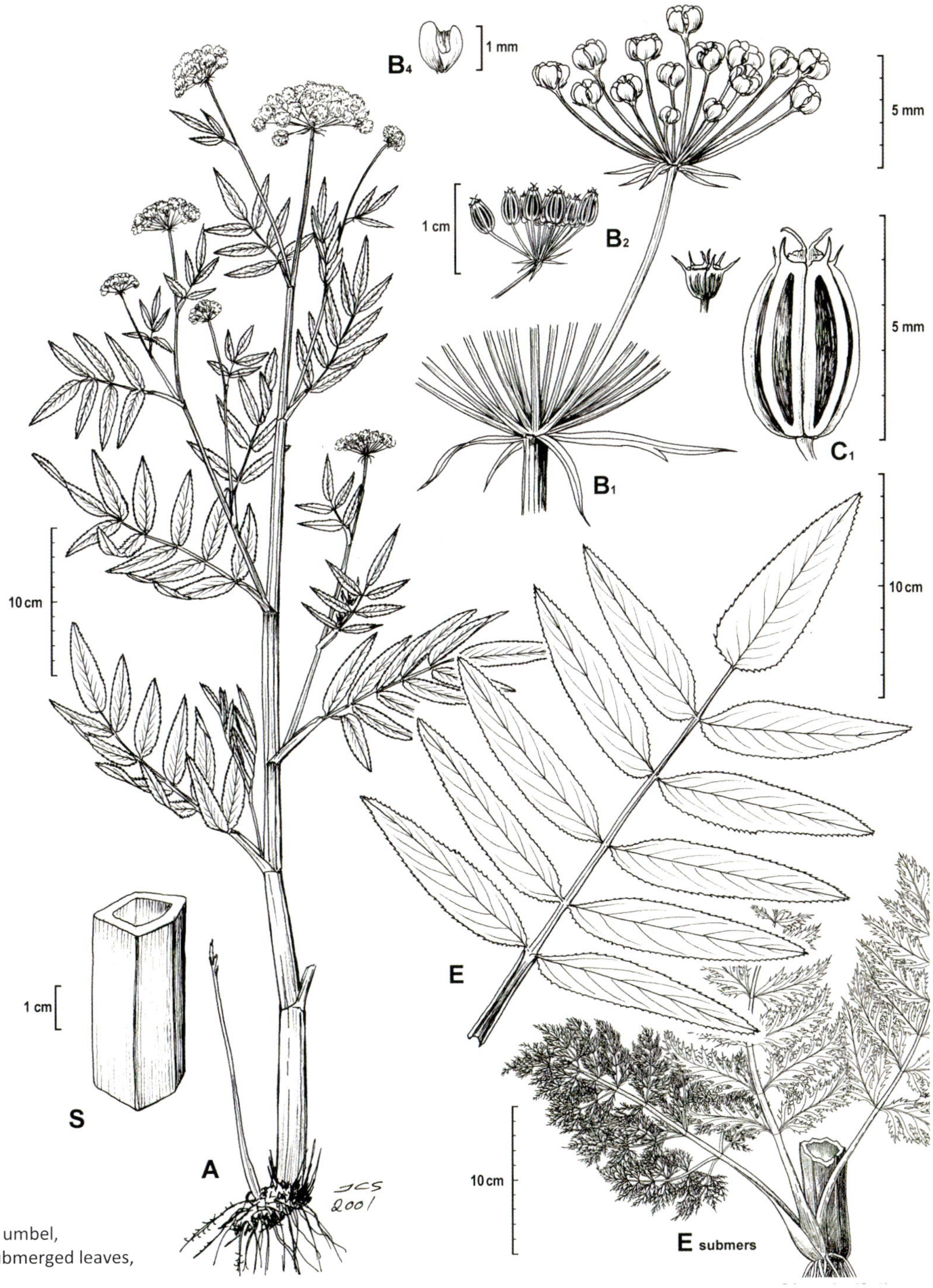

Sium latifolium
A habit, **B1** umbel, **B2** flowering secundary umbel, **B4** petal, **C1** fruit, **E** aerial leaf, **E submers** submerged leaves, **S** stem section.

Isoëtaceae

359 *Isoetes echinospora* Durieu

Syn.: *I. setacea* auct. non Lam.

DK: Gulgrøn Brasenføde **N:** Mjukt brasmegras **S:** Vekt braxengräs **FIN:** Vaalealahnanruoho **IS:** Álftalaukur **GB:** Spring Quillwort **NL:** Kleine biesvaren **F:** Isoète à spores spinuleuses **D:** Igelsporiges Brachsenkraut **CZ:** Šídlatka ostnovýtrusá **PL:** Poryblin kolczasty **EST:** Muda-lahnarohi **LV:** Dzeloņsporu ezerene **LT:** - **Ru:** Полушник колючеспоровый

Isoetes echinospora. Nationalpark Thy, Denmark. Photo KvdW.

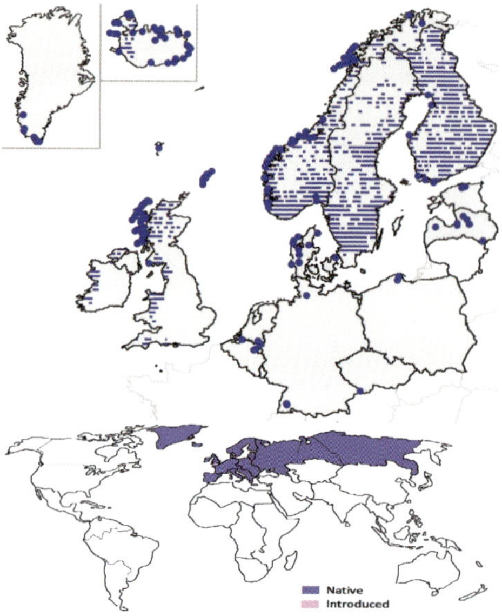

Perennial hydrophyte or very rarely amphibious helophyte. Rhizome short and vertical with brown, dichotomously branched, hollow roots.

Stem: A short, thick, rhizome-like vertical structure.

Leaves: In a dense rosette at the top of the rhizome, spreading, 5–15 cm long and 1–2(–3) mm in diameter, linear to subulate, gradually tapering into a pointed tip, rather soft and flaccid, clinging together when removed from the water, like wet hairs on a soft brush. Leaves widely expanded at the base, with membranous margins and a small tongue-like ligule on the adaxial side, the above-ground parts yellowish green, basal parts white. Each leaf with 4 longitudinal air cavities divided into chambers by thin membranous walls.

Sporangia: Solitary, embedded in cavities on the adaxial side of the leaf bases, covered by a thin membrane. Outer leaves in the rosettes with megaspores (analogous to egg cells), the middle leaves with microspores (analogous to sperm cells) and the inner leaves without sporangia.

Spores: Megaspores spherical, 0.4–0.5 mm in diameter, white (when dry), with conspicuous ridges and covered in fine acute or obtuse spikes. Microspores oblong, 30 µm long and 15–20 µm wide, pale.

Spore maturity: July-September.

Biology: Megaspores and microspores develop late in the summer and are released into the water when leaves become detached from the plants or decay. They fuse to gametophytes (analogous to seeds) in the water. Detached leaves with spores may disperse over long distances and contribute to large-scale dispersal. *I. echinospora* almost always grows submerged but can survive short periods of exposure. It typically grows below the summer minimum water level, forming dense monospecific stands.

Habitats: In mainly moderately acid to neutral, oligotrophic to mesotrophic, clear water. In lakes and ponds, primarily on mineral substrate, but also on peaty soil, in the lower littoral zone at depths of 0.1–1.0 metres (at mean summer water level), to a maximum of c. 3 metres and here often rooting in muddier sediments. Occasionally also in rivers and in slightly saline waters.

Distribution within the region:

Native. Locally frequent to scattered in Scandinavia, parts of Britain and Ice-

Isoetes echinospora. Södra Vixen, Sweden. Photo JCS.

Isoetes echinospora. Megaspores with acute spines. Photo JCS.

Isoetes echinospora and *Lobelia dortmanna* growing in a partly dried out oligotrophic lake. Thy, Denmark. Photo JCS.

land. Rare to very rare elsewhere; absent from Lithuania and Luxembourg.

Characteristics and similar species:
360 *I. lacustris* differs in the megaspores grey (when dry) with wart- or crest-like ornamentation and rigid, dark green leaves with abruptly tapering apices.

Isoetes echinospora and I. lacustris may be confused with other rosette-forming species occurring in oligotrophic lakes and ponds (see below), but differ from all of these by having 4 longitudinal air cavities in the leaves and by presence of sporangia. **254 *Subularia aquatica*, 296 *Littorella uniflora*, 137 *Juncus bulbosus*, 27-28 *Baldellia*** spp., **33 *Alisma plantago-aquatica*** and **36 *Alisma wahlenbergii*.**

334 *Lobelia dortmanna* has 2 large longitudinal air cavities.

133 Eriocaulon aquaticum has flat leaves and white roots, appearing articulated because of brownish striations.

Hybrids:
I. ×jermyi D.F.Brunton & F.J.Rumsey
(*I. lacustris × echinospora*)
Typically occurs in mixed populations of both parents, it is usually more robust and sterile, with aborted megaspores. It is known from scattered locations in Wales and the Western Isles of Scotland.

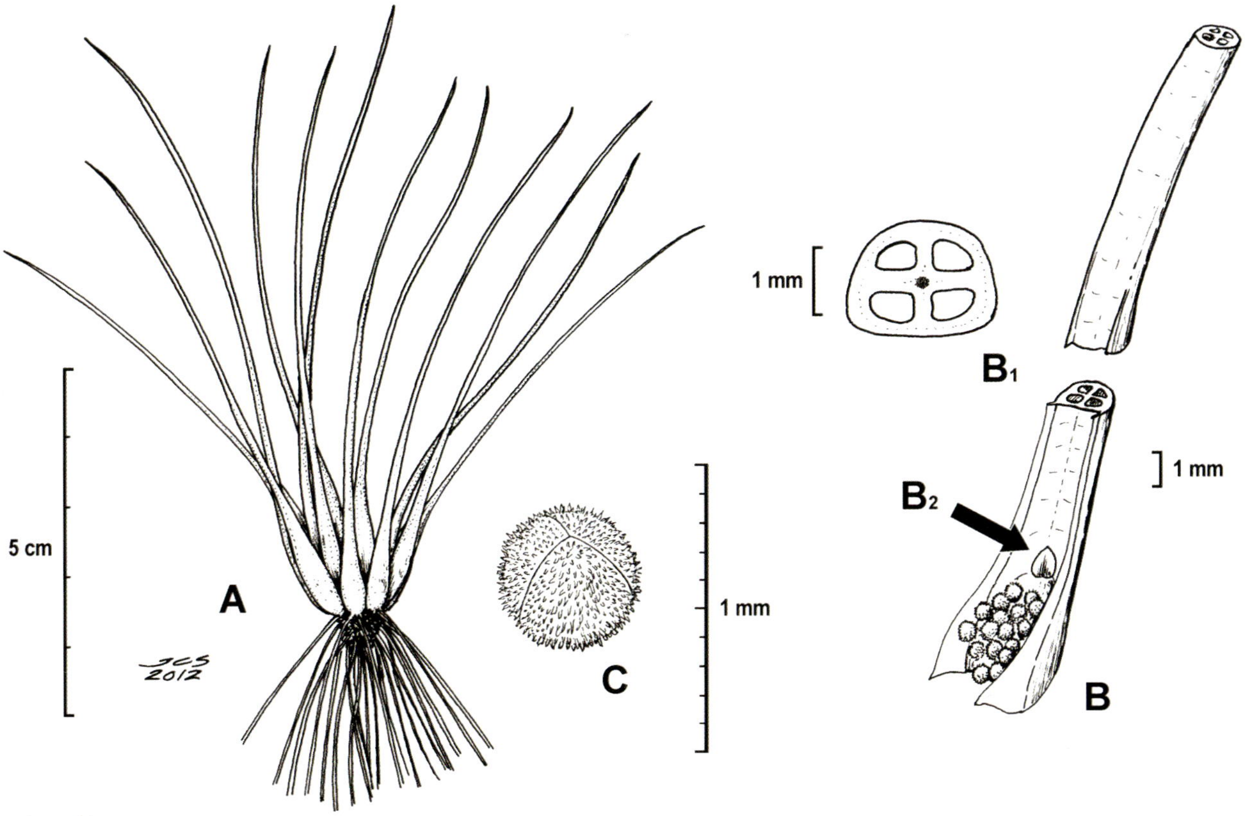

Isoetes echinospora
A habit, **B** lower part of leaf with exposed megaspores, **B1** cross section of leaf, **B2** ligule, **C** megaspore.

Isoëtaceae

360 *Isoetes lacustris* L.

DK: Sortgrøn Brasenføde **N:** Stivt brasmegras **S:** Styvt braxengräs **FIN:** Tummalahnanruoho **IS:** vatnalaukur **GB:** Quillwort **NL:** Grote biesvaren **F:** Isoète des lacs **D:** Gewöhnliches Brachsenkraut **CZ:** Šídlatka jezerní **PL:** Poryblin jeziorny **EST:** Järv-lahnarohi **LV:** Gludsporu ezerene **LT:** Ežerinė slepišerė **Ru:** Полушник озёрный

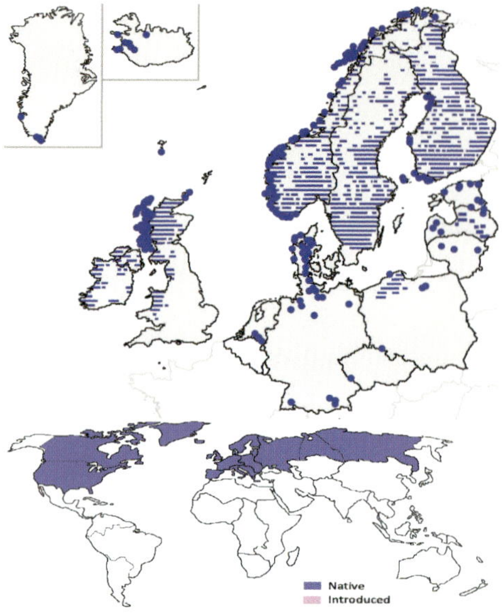

Perennial hydrophyte or very rarely amphibious helophyte. Rhizome short, vertical with brown, dichotomously branched, hollow roots.

Stem: A short, thick, rhizome-like vertical structure.

Leaves: In a dense rosette at the top of the rhizome, spreading, 5–15 cm long and 1.5–3 mm diameter, linear to subulate, with abruptly tapering tips, rigid and spreading when removed from the water. Leaves widely expanded at base, with membranous margins and a small tongue-like ligule on the adaxial side, the above-ground parts dark to almost blackish green, the basal parts lighter. Each leaf with 4 longitudinal air cavities divided into chambers by thin membranous walls.

Plants in deep water have long, more or less straight leaves, while plants in shallow water have shorter, distinctly recurved leaves.

Sporangia: Solitary, embedded in cavities on the adaxial side of the leaf bases, covered by a thin membrane. Outer leaves in the rosettes with megaspores (analogous to egg cells), the middle leaves with microspores (analogous to sperm cells), the inner leaves without sporangia.

Spores: Megaspores spherical, 0.5–0.7 mm in diameter, greyish, the surface covered with wart- or crest-like formations and with 3 more or less conspicuous ridges. Microspores oblong, 40–50 µm long and 20–30 µm wide, pale.

Spore maturity: July-September.

Biology: See **359 *I. echinospora***.

Habitats: Mainly in moderately acid to neutral, nutrient-poor, clear water. In lakes and ponds, primarily on mineral substrates in the lower littoral zone at depths of 0.5–2 metres (at mean summer water level). May grow to a maximum depth of c. 8 metres and here often rooting in muddier sediments. In the far north also in rivers and rarely in slightly saline waters.

Distribution within the region:
Native. Locally frequent in Scandinavia, western and northern Britain and parts of Ireland. Rare to very rare elsewhere, absent from Belgium and Luxembourg.

Isoetes lacustris. Wastwater, Wasdale, Britain. Photo KvdW.

Isoetes lacustris. Megaspores with wart- or crest-like formations. Photo JCS.

Washed up *I. lacustris* lifted out of the water. Note the thick rhizomes and recurved leaves which do not stick together. Also note the brickwork-like appearence of the leaves - a handy characteristic for distinction from other isoetids. Lake Skån Sø, Jutland, Denmark. Photo JCS.

Characteristics and similar species:
Isoetes echinospora differs in megaspores, white (when dry), covered with fine acute or obtuse spikes, and yellowish-green, gradually tapering, rather flaccid leaves, which stick together like wet hairs on a soft brush when removed from the water.

For distinguishing from other similar species see *I. echinospora.*

Hybrids:
I. ×jermyi D.F.Brunton & F.J.Rumsey. See *I. echinospora.*

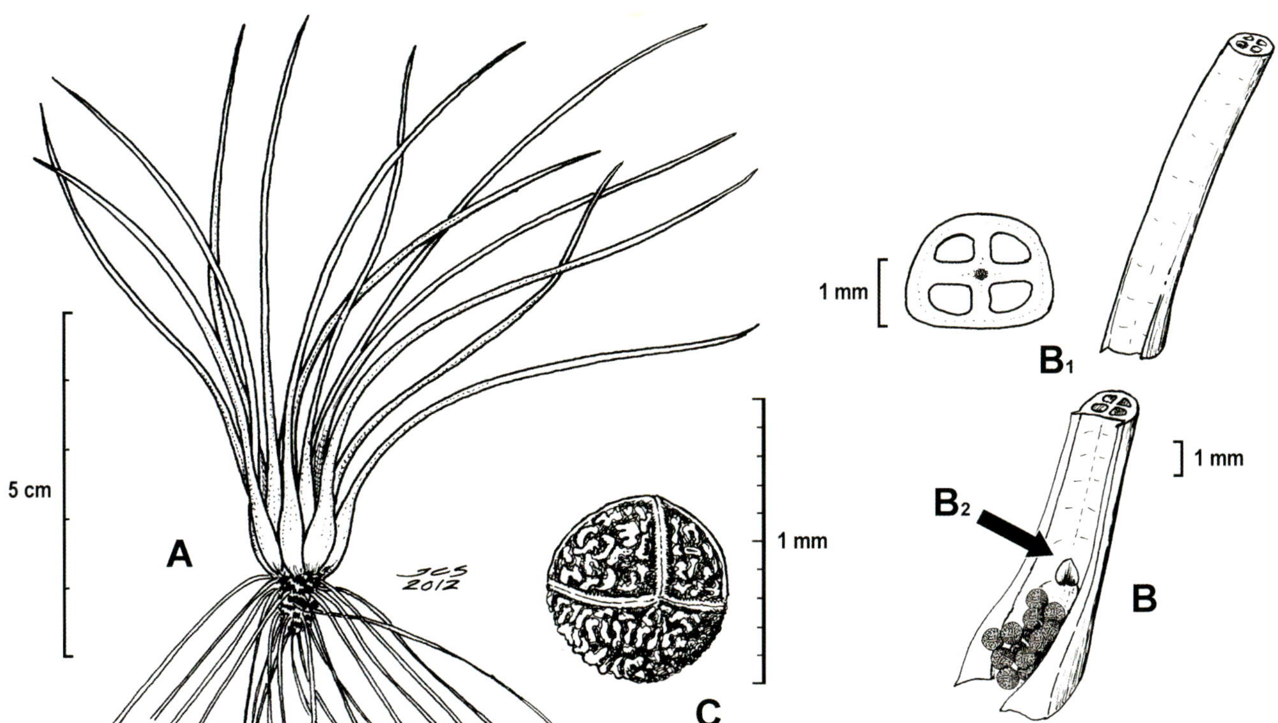

Isoetes lacustris
A habit, **B** lower part of leaf with exposed megaspores, **B1** cross section of leaf, **B2** ligule, **C** megaspore.

Marsileaceae

361 *Pilularia globulifera* L.

DK: Pilledrager **N:** Trådbregne **S:** Klotgräsv **FIN:** Ormio **IS:** -
GB: Pillwort **NL:** Pilvaren **F:** Pilulaire à globules
D: Pillenfarn **CZ:** Míčovka kulkonosná **PL:** Gałuczka kulecznica
EST: - **LV:** Lodaugļu pilulārija **LT:** - **Ru:** пилюльница шариконосная

Dense grass-like mats of *P. globulifera*, which from the damp shore sediment extends into the water. Associated with *Hydrocotyle vulgaris* and *Mentha aquatica*. Lake Navn Sø, Jutland, Denmark. Photo JCS.

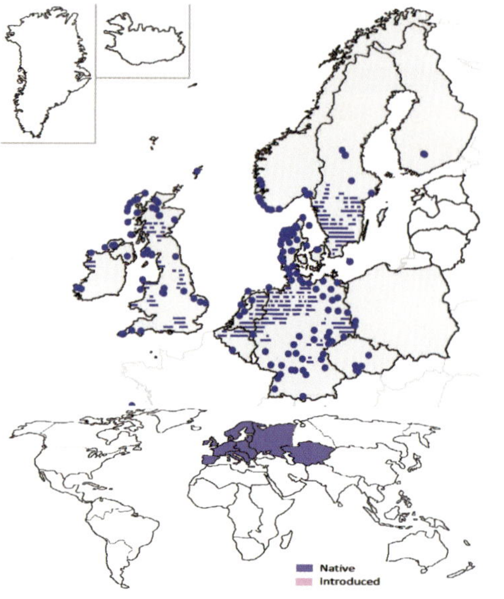

Perennial hydrophyte or amphibious helophyte with above-ground rhizome-like stem. Roots brownish, arising from the lower side of nodes.

Stem: Up to 50 cm long, 1.0–1.5 mm diameter, creeping, richly branched, terete, hairy, green, often with an orange or brown tinge with age and on damp substrates.

Leaves: Solitary or few close together from the upper side of the stem, initially coiled like croziers, later unrolling to become straight and upright. Sterile leaves 1.0–1.5 mm diameter, terete and filamentous, with one longitudinal air channel, 3–10 cm long in emergent plants, green, often becoming orange or brownish; in submerged leaves up to 30 cm or longer, mostly green, occasionally black. Fertile leaves only develop on emergent or partly submerged plants. They are short stalked, globose and enclose the sporangia.

Sporocarps: Spherical, 4-chambered, 2–3 mm in diameter, very shortly stalked, initially greyish green and densely hairy, later as the spores ripen becoming dark brown and often less hairy.

Spores: Each chamber of the sporocarp contains a gel-like tissue with two different types of sporangia - microsporangia and megasporangia - containing microspores and megaspores (analogous to sperm and egg cells).

Spore maturity: July-September.

Biology: Late in the growth season the 4-chambered sporocarps split open with 4 flaps, each flap containing micro- and megasporangia, the latter below the microsporangia. In contact with water the gel-like tissue swells and is squeezed out, releasing the mega- and microspores, which fuse and form gametophytes - analogous to seeds and dispersed by water and water birds. Vegetative dispersal occurs by detached stem fragments. *P. globulifera* is a weak competitor and is favoured by moderate physical disturbance and temporary

Pilularia globulifera on damp substrate. Råbjerg Mile, Jutland, Denmark. Photo JCS.

Pilularia globulifera with sporocarps and young leaves with coiled tips. Lake Navn Sø, Jutland, Denmark. Photo JCS.

Submerged *P. globulifera*. Wankumer Heidesee (left) and a pool in Elmpter Schwalmbruch (right). Lower Rhine region, Germany. Photos KvdW.

drying of the sediment. Under favourable conditions it often forms large, dense stands. Rarely forming floating mats, occasionally grows permanently submerged. Wintering as more or less leafless stems and gametophytes.

Habitats: In moderately acidic to alkaline, oligotrophic to mesotrophic water, on muddy or mineral substrates in the littoral zone of lakes, ponds, ditches and streams as well as in temporary waters, predominantly on areas. which dry out during the growing season.

Distribution within the region:
Native. Locally frequent but declining in the British Isles. Scattered to rare in the Netherlands, Belgium, Germany, Denmark and southern Scandinavia. Very rare or absent in Poland and Finland. Absent from the Baltic states, Iceland and Greenland.

Characteristics and similar species:
Vegetative *Pilularia globulifera* is mainly characterised by the coiled, crozier-like tips of the young leaves. If present, the spherical sporocarps distinguish the species from all other water plants. *Pilularia* may be confused with other species with filamentous leaves.

160 *Eleocharis acicularis* differs by the sheath-like bracts (leaves) at the base of the quadrangular stems and straight leaf tips.

137 *Juncus bulbosus* has nodes at the base of the stem and forms rosettes of filamentous leaves without coiled tips.

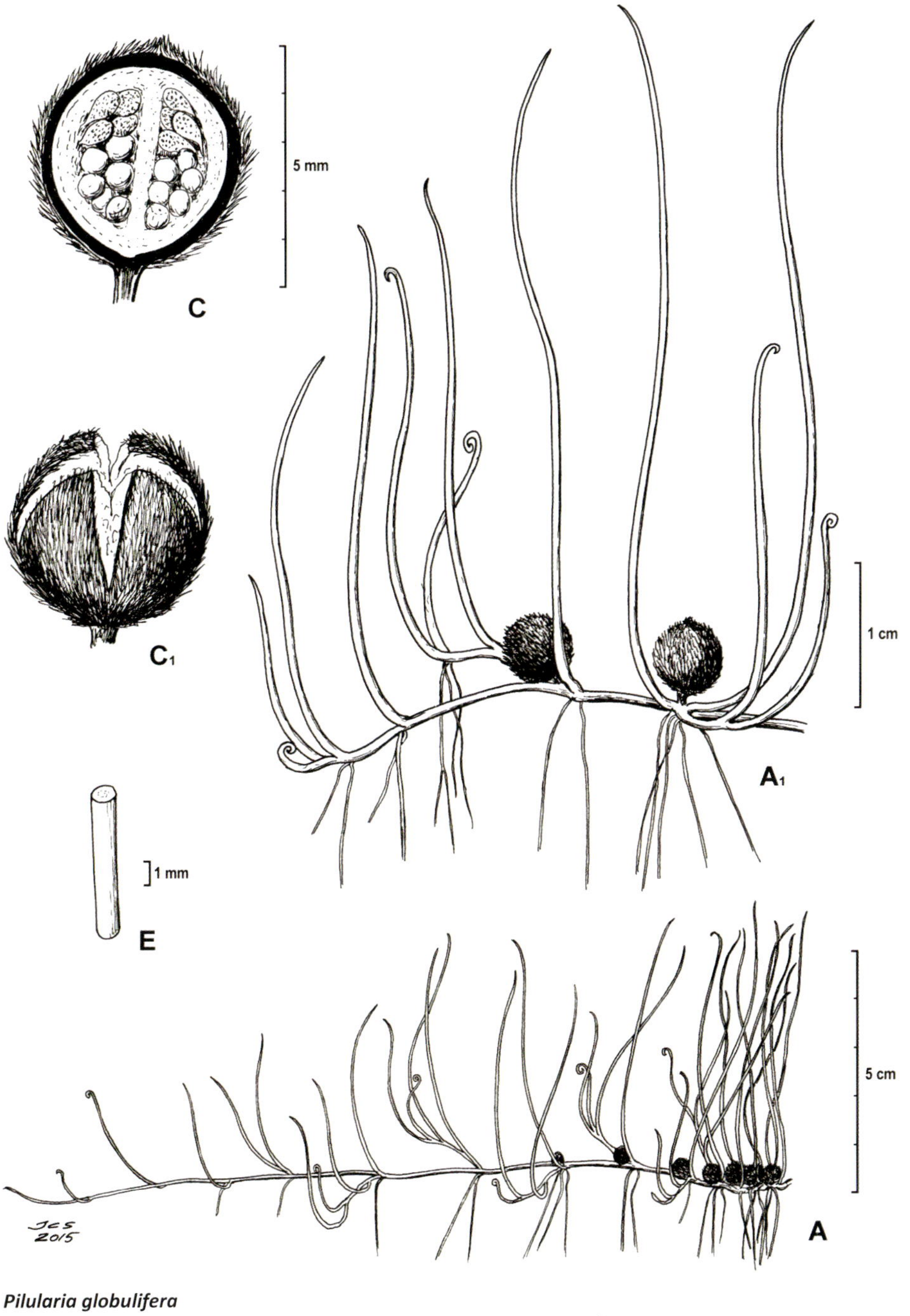

Pilularia globulifera
A, **A1** habit, **C** cross section of sporocarp, **C1** sporocarp opening, **E** leaf section.

Marsileaceae

362 *Marsilea quadrifolia* L.

DK: Kløverbregne **N:** - **S:** Klöverbräken **FIN:** - **IS:** -
GB: Water clover **NL:** Drijvblad **F:** Marsiléa à quatre feuilles
D: Vierblättriger Kleefarn **CZ:** Marsilka čtyřlistá **PL:** Marsylia czterolistna
EST: - **LV:** - **LT:** - **Ru:** Марсилия четырёхлистная

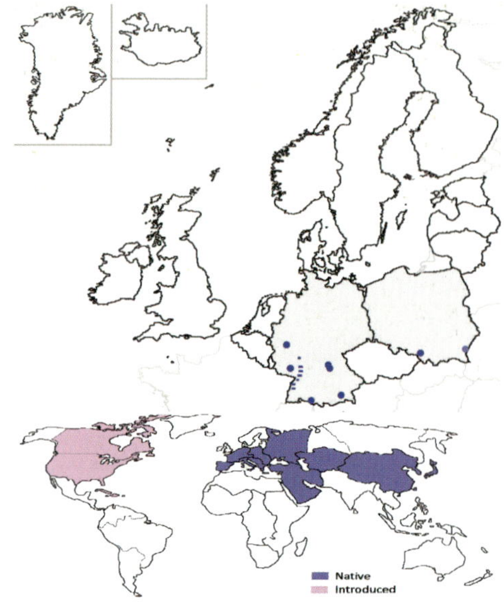

Marsilea quadrifolia. In shallow water on the margin of a large lake. Parc Naturel Regional de la Brenne, Indre, France. Photo RVL.

Perennial hydrophyte or amphibious helophyte with above-ground rhizome-like stem. Roots whitish to brownish, from the lower side of nodes. Primary roots short-lived and soon replaced by adventitious roots.

Stem: Up to 50 cm long or in aquatic forms up to 1 m, 1.0–1.5 mm diameter, creeping, with 0.2–12 cm long internodes, branched, terete, hairy when young, usually brownish.

Leaves: Leaves alternate in two rows from the upper side of the stem. Petioles initially coiled, later unrolling to become upright or sinuous, 5–30 cm long depending on water-depth, slender, terete, more or less hairy. Lamina floating or raised above the water-surface, cruciform 4-foliolate, composed of 2 contiguous pairs of opposite leaflets. Leaflets sessile, 0.6–3.0 cm long and 0.5–2.8 cm wide, cuneate at base, rounded or shallowly 3-lobed distally, dull to brownish green, with flabellate venation.

Sporocarps: On a short stalk from the basal part of the petiole, presumably a modified entire leaf. The sporocarps are long pedicellate and typically fused in groups of 2, 3–6 mm long, 2-chambered, ellipsoid, felted hairy, initially green, later as the spores ripen turning blackish brown.

Spores: The sporocarps contain a mucilaginous tissue and two different types of sporangia, microsporangia and megasporangia containing microspores and megaspores respectively (analogous to sperm and egg cells).

Spore maturity: September-October.

Biology: The sporocarp is very hard and impermeable, taking 2 or 3 years underwater to decay sufficiently for moisture to penetrate. The mucilaginous tissue inside the sporocarp swells and splits the sporocarp open, forming a long extrusion with the sori attached in pairs - each with both mega- and microsporangia. When released the mega- and microspores germinate immediately, fuse and form gametophytes with roots and leaf appearing within 2–3 days. The development from gametophyte to mature plants bearing sporocarps may take only 3 months.

Vegetative dispersal by branching and fragmentation of the stem. Under favourable conditions often forming large stands. Overwintering as more or less leafless stems.

Habitats: In acidic, oligotrophic to mesotrophic, nutrient- and humus-poor, fresh or slightly to mildly saline water. In natural as well as man-made water bodies with still or slow-flowing, shallow water which dries out in summer - oxbow lakes, flood channels, fishponds, ditches, clay-pits and artificial lakes.

Marsilea quadrifolia. In shallow water on the margin of a large lake. Parc Naturel Regional de la Brenne, Indre, France. Photo RVL.

Marsilea quadrifolia growing in a ditch. Meerlo, Netherlands. Photo KvdW.

Marsilea quadrifolia. Cultivated. Aarhus, Denmark. Photo JCS.

Distribution within the region:
Native. Rare in Germany and mostly in the Upper Rhine valley between Kehl and Mannheim. Reintroduced in Poland. Frequently sold as an aquarium plant, so it may occur as an escape in suitable habitats.

Characteristics and similar species:
With the clover-like leaves it is not likely to be mistaken for other aquatic plants.

362a *Marsilea hirsuta* R.Br.

has been naturalised as an escape from cultivation at a single site in England for a number of years.

It can be distinguished from *M. quadrifolia* by the presence of two teeth on the dorsal side of the sporocarp, which are lacking or represented by a single small tooth in *M. quadrifolia*.

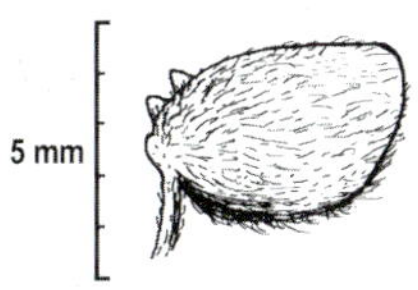

Marsilea hirsuta sporocarp.

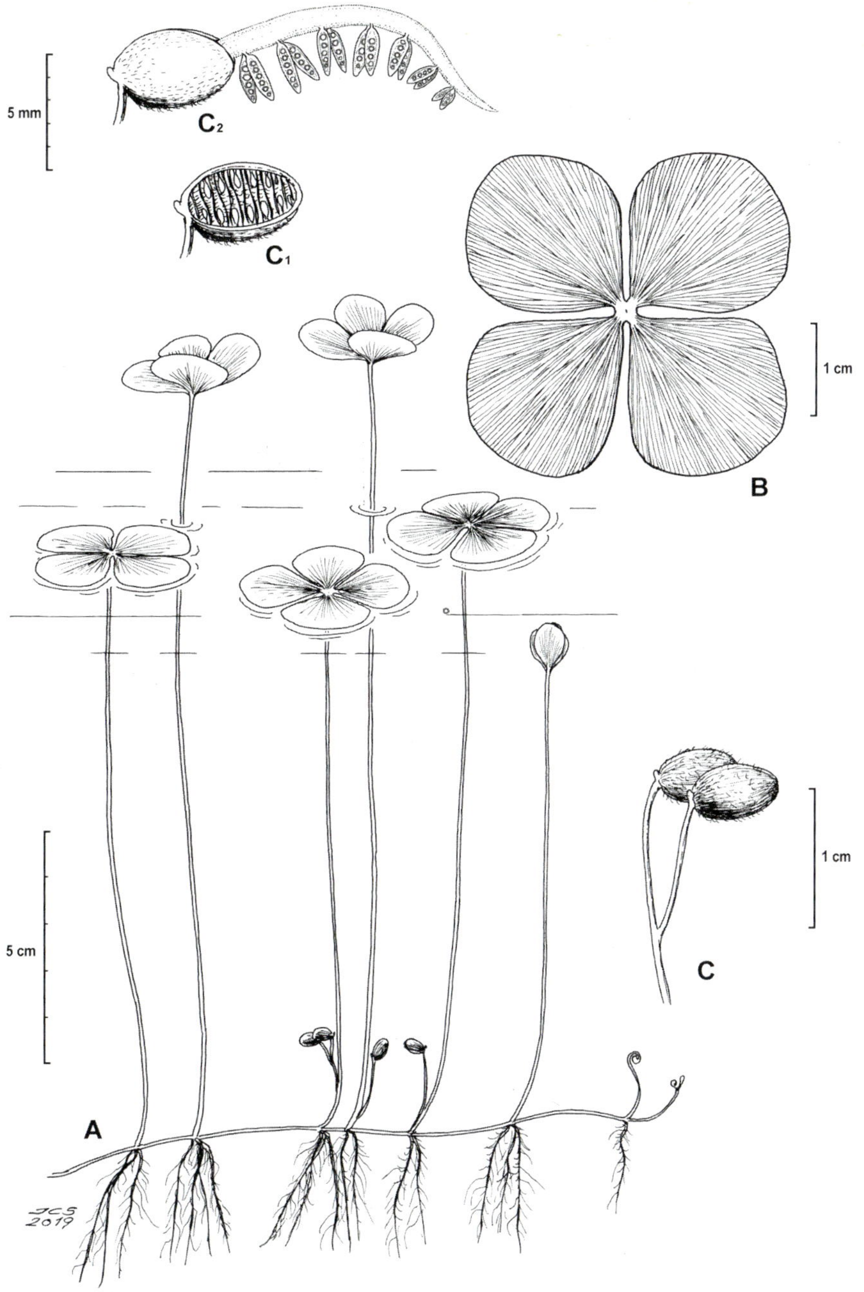

Marsilea quadrifolia
A habit, **B** lamina, **C** sporocarps, **C1** cross section of sporocarp, **C2** sporocarp opened with sori in pairs on a long mucilaginous extrusion.

Salviniaceae

363 *Salvinia natans* (L.) All.

DK: Flydebregne **N:** - **S:** Simbräken **FIN:** Parikellussaniainen **IS:** -
GB: Floating fern **NL:** Kleine vlotvaren **F:** Salvinia nageante
D: Gewöhnlicher Schwimmfarn **CZ:** Nepukalka vzplývající **PL:** Salwinia pływająca
EST: - **LV:** - **LT:** Plūduriuojančioji plūstis **Ru:** САЛЬВИНИЯ ПЛАВАЮЩАЯ

Salvinia natans. Berghäuser Althrein, Germany. Photo KvdW.

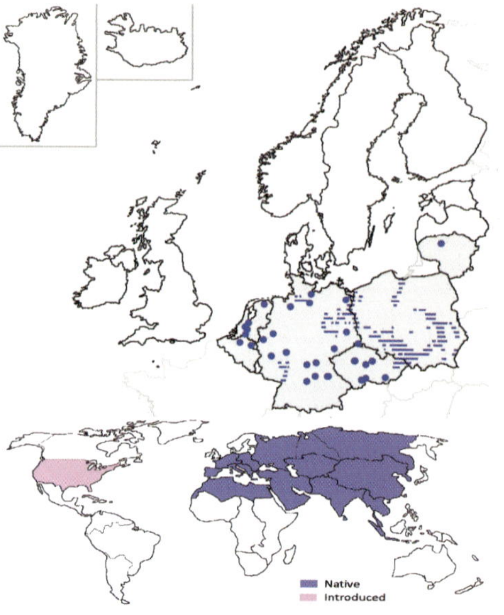

Annual, free-floating plant, without roots, but modified leaves resemble roots.

Stem: Up to 20 cm long, horizontal, thin, densely hairy, more or less branched.

Leaves: In whorls of 3, 2 floating leaves and 1 finely divided, submerged leaf acting as a root. Floating leaves 0.8–1.5 cm long and 0.5–1.0 cm wide, oblong elliptic, sessile or very shortly petiolate, folded when young, with 15–20 lateral veins on each side of the mid-vein. Upper side of the leaf with numerous evenly dispersed papillae, which at the top are divided into 4 multi-cellular, spreading hairs. Submerged leaves divided into 9–14 very fine, hanging leaves covered in hairs.

Sporocarps: 4–8 together at base of submerged leaves. Each sporocarp about 3 mm in diameter, shortly stalked, with a globular, two-layered indusium covering the sorus.

Spores: Each sporocarp contains either microsporangia or megasporangia. The microsporangia produce 64 microspores, while the macrosporangia produce only one surviving spore.

Spore maturity: April-September.

Biology: When released from the sporocarps the microspores develop into male gametophytes and the macrospores into female gametophytes. Fertilisation between the female and male gametophytes takes place on the water surface, and the resulting zygote develops into a new plant. Vegetative dispersal by detached stem fragments. Under optimal conditions capable of rapidly forming large floating mats.

Habitats: In eutrophic, still or slow-flowing water. In sheltered bays and the drawdown zones of large rivers, lakes, oxbows, ponds and ditches.

Distribution within the region:
Native. Scattered to locally frequent in parts of the Netherlands, Belgium, Germany, Poland and the Czech Republic. Extinct in Lithuania.

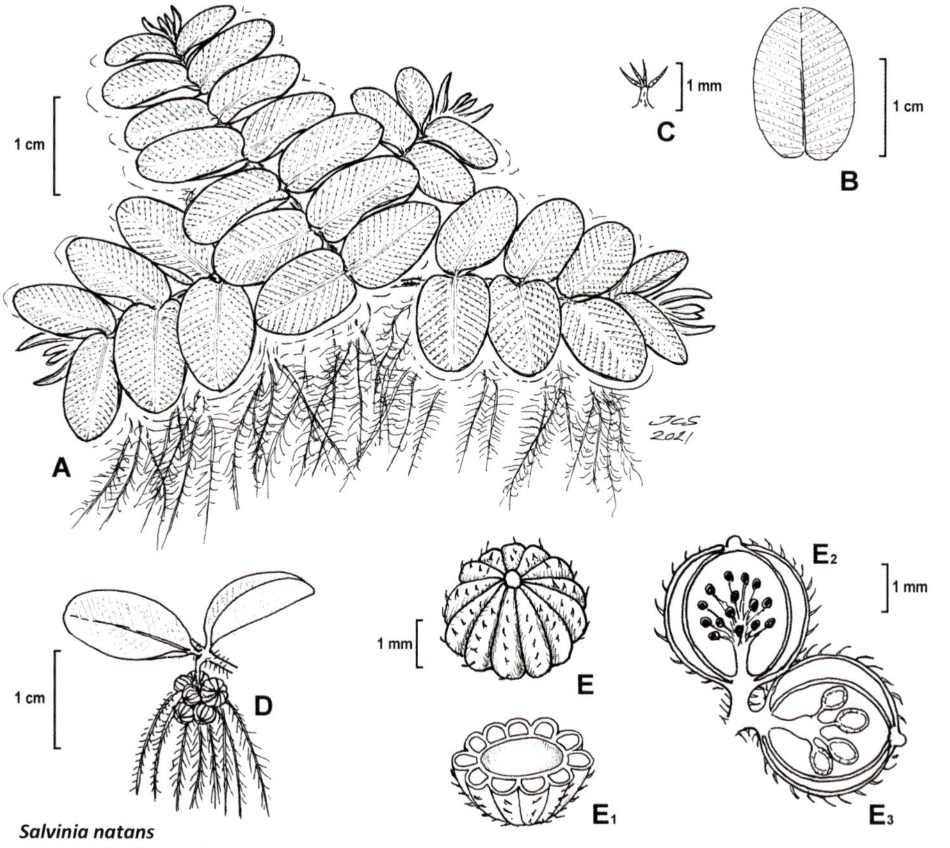

Salvinia natans
A habit, **B** leaf, **C** papilla with 4 multi-cellular hairs spreading distally, **D** plant with sporocarps, **E-E3** sporocarp, **E1** cross section, **E2** with microsporangia, **E3** with macrosporangia.

Characteristics and similar species: *Salvinia* species are unlikely to be confused with other genera, but species identification is often problematic.

Several Salvinia species are cultivated as ornamental plants for garden ponds and by accident or on purpose spread to natural habitats. In warmer climates listed as one of the most noxious aquatic weeds, but in our region it rarely persists.

Salvinia natans. Note the papillae on the enlargement to the right. Berghäuser Althrein, Germany. Photo KvdW.

363a *Salvinia minima* Baker

is very similar to *S. natans*, but can be distinguished from this by having 10 veins on the leaves, rather than 18–20. The 2-4 multi-cellular hairs at the top of the papillae are spreading. Native to South and Central America, introduced in Sweden.

Salviniaceae

364 *Salvinia auriculata* aggr.

S. *auriculata* Aubl., *S. biloba* Raddi, *S. herzogii* De la Sota, *S. molesta* D.S.Mitch.

Salvinia auriculata aggr. Note the papillae on the enlargement to the right. Swimming pond, Algarve, Portugal. Photo KvdW.

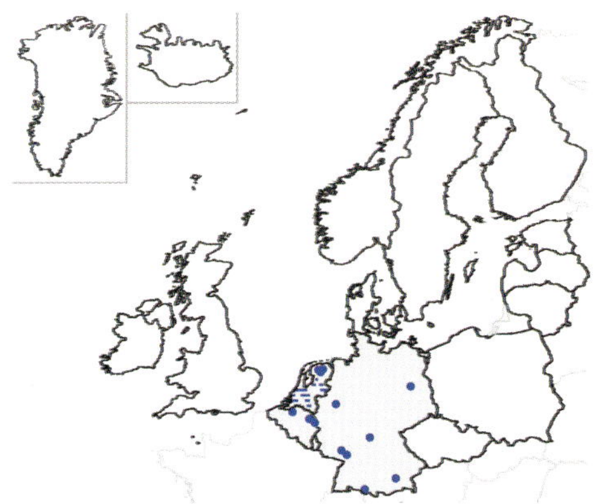

In this book the remaining taxa recorded from the region are collectively treated as **364 *Salvinia auriculata*** aggr.

Salvinia molesta D.S.Mitch. (syn.: *S. adnata* Desv.) appears to be the most common species in culture. It has been recorded from the Netherlands, Belgium and Germany. The leaves are generally larger and more rounded than those of *S. natans*, but the best distinction between the two is the 4 multicellular hairs on the papillae on the upper side of the leaves. In *S. molesta* and the other members of the *S. auriculata* aggr. they are distally united.

For determination of the species within the aggregate please consult specialised floras.

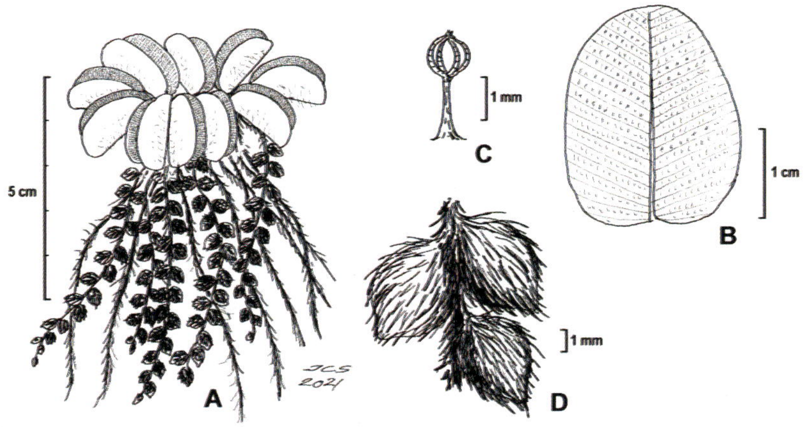

Salvinia molesta
A habit - plant with sporocarps, **B** leaf, **C** papilla with 4 multi-cellular hairs united distally - "egg-beater" hair, **D** sporocarps.

Salviniaceae

365 *Azolla filiculoides* Lam.

Syn.: *Azolla caroliniana* Willd.
DK: Stor Andemadsbregne **N:** Andematbregne **S:** Mossbräken **FIN:** - **IS:** -
GB: Water Fern **NL:** Grote kroosvaren **F:** Azolla flottante
D: Grosse Algenfarn **CZ:** Azola americká **PL:** Azolla karolińska
EST: - **LV:** - **LT:** - **Ru:** Азолла папоротниковая

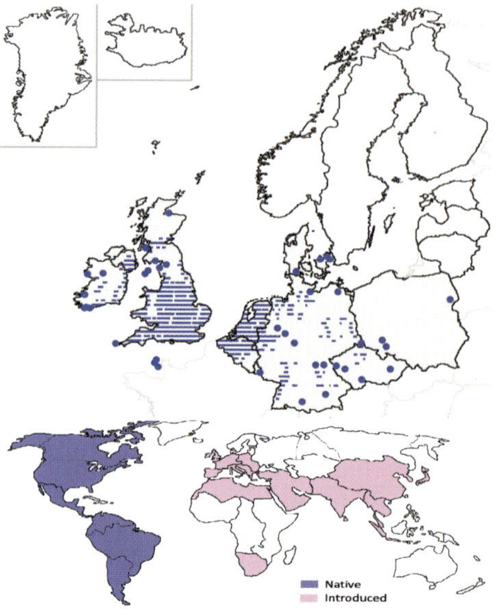

Azolla filiculoides and a few *Lemna minor*. Cultivated. Hobro, Denmark. Photo JCS.

Annual or perennial, free-floating plant. Roots unbranched, pale green to pale brown.

Stem: 1–5 cm long, slender and repeatedly forked.

Leaves: Distichously arranged, overlapping. 2-lobed, green or often pink to dark red, especially in late summer. The upper lobe 1.5–2.5 mm long, blunt, with an open cover of unicellular hairs and a membranous rim. The lower lobe is slightly larger and membranous.

Sporocarps: 2 or 4, greenish to reddish, formed on the lower side of the branches and of two kinds.

Spores: The male sporocarps are about 2 mm long, ovoid, containing 64 microsporangia, each with numerous microspores. The female sporocarps are 0.5 mm long, oblong, containing one macrosporangium which produces only one surviving warty macrospore.

Spore maturity: September-October.

Biology: The plant reproduces mainly vegetatively and by virtue of its coexistence with a blue-green algae (*Anabaena azollae*) is able to fix nitrogen from the air to form nitrate. It is extremely productive and can double in biomass in a matter of days. Plants and spores are dispersed by water and sometimes by birds and other animals. Overwintering as fragments or spores. It is used in tropical areas as a natural fertiliser in rice fields. The species is considered invasive in parts of Western Europe including Britain and Ireland. It rapidly forms large floating carpets that block all light from submerged plants.

Habitats: In acidic to alkaline, nutrient-rich, fresh or slightly brackish, still or slow-flowing water or beached on wet mud. In ponds, ditches and canals. It can tolerate and may benefit from some pollution.

Distribution within the region: Widespread and abundant in southern Britain, the Netherlands and Belgium, scattered in Ireland, northern Britain, Germany, the Czech Republic and Poland, casual in Denmark.

Characteristics and similar species: *Azolla* species are unlikely to be confused with aquatic plants of other genera. Very similar to **366 *Azolla mexicana*** - see this.

Azolla filiculoides. Close-up of a male and two female sporocarps. Cultivated. Hobro, Denmark. Photo JCS.

Azolla filiculoides covering the water surface in a ditch. Grefrath, Lower Rhine, Germany. Photo KvdW.

Azolla filiculoides with sporocarps. Reservoir near Sacramento, California. Photo RVL.

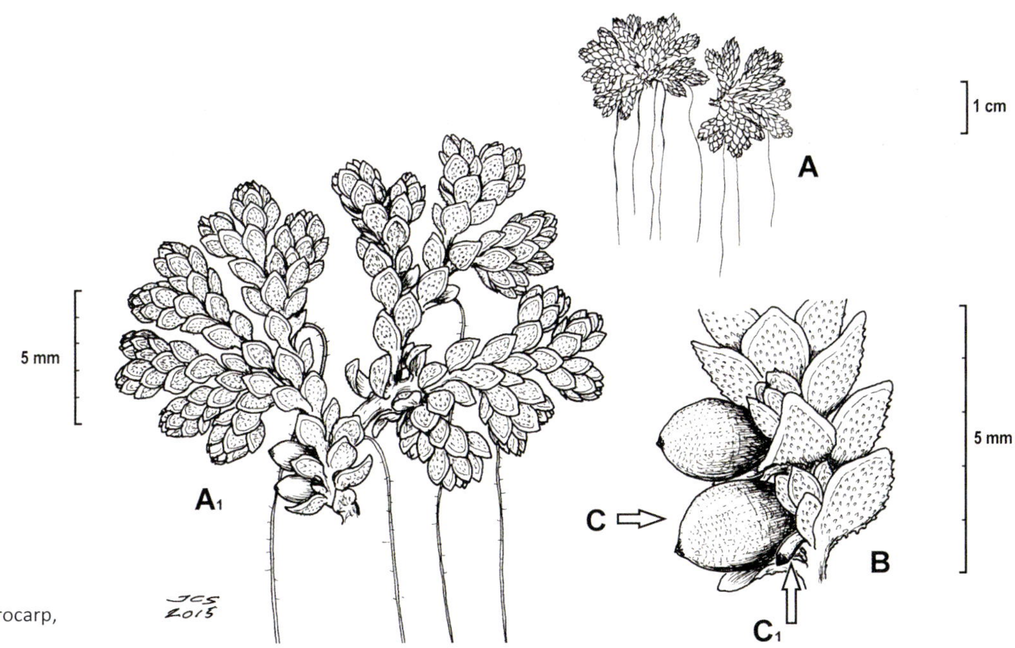

Azolla filiculoides
A og **A1** habit, **B** fertile branch, **C** male sporocarp, **C1** female sporocarp.

Salviniaceae

366 *Azolla mexicana* C.Presl

Syn.: *Azolla caroliniana* auct.
DK: Lille Andemadsbregne **N:** - **S:** Liten Mossbräken **FIN:** - **IS:** -
GB: Lesser Water Fern **NL:** Kleine kroosvaren **F:** -
D: Kleiner Algenfarn **CZ:** - **PL:** Azolla meksykańska
EST: - **LV:** - **LT:** - **Ru:** -

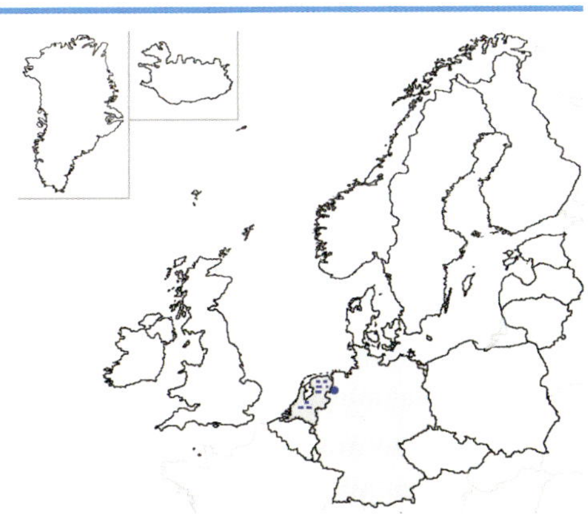

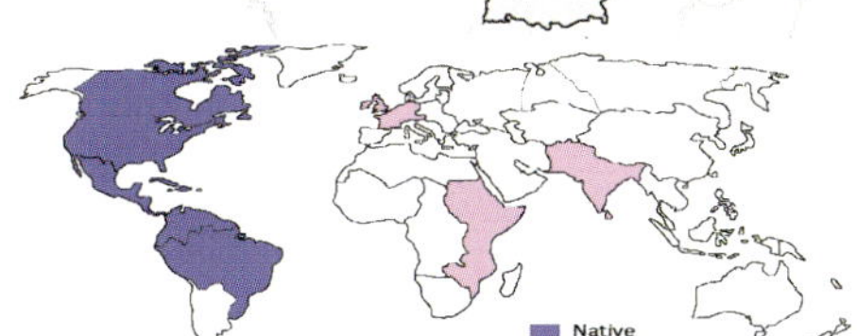

Azolla filiculoides (left) and *A. mexicana*. Reservoir near Sacramento, California. Photo RVL.

Differs from **365 *Azolla filiculoides*** by its smaller leaves - upper lobe 1–1.5 mm long with multi-cellular hairs, as well as by the macrospores without warts.

Distribution within the region: Reported from the Netherlands and Germany.

Equisetaceae

367 *Equisetum fluviatile* L.

DK: Dynd-Padderok **N:** Elvesnelle **S:** Sjöfräken **FIN:** Järvikorte **IS:** Fergin
GB: Water Horsetail **NL:** Holpijp **F:** Prêle des fleuves
D: Teich-Schachtelhalm **CZ:** Přeslička poříční **PL:** Skrzyp bagienny
EST: Konnaosi **LV:** Upes kosa **LT:** Balinis asiūklis **Ru:** Хвощ приречный

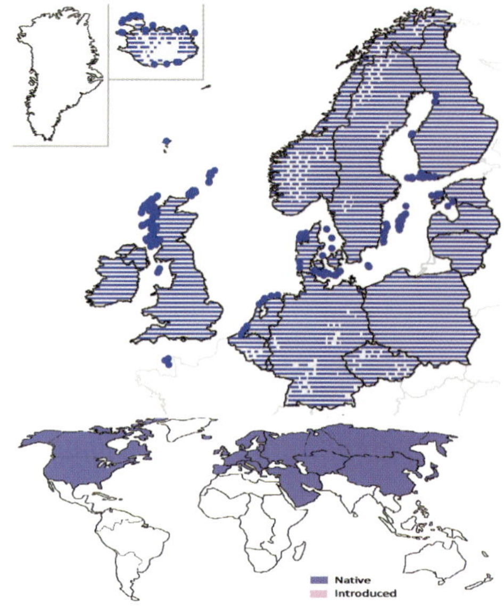

From a depth of 1 m *E. fluviatile* extends its shoots above the water surface. Vigsø, Thy, Denmark. Photo JCS.

Perennial helophyte. Rhizome underground, richly branched and far creeping, glossy brown, thick and with large central cavity.

Stem: 30–150 cm long, 3–9 mm in diameter, indistinctly furrowed with 10–30 shallow ridges, green to brownish black, articulated, with a large central hollow (> 75% of stem diameter), and easily collapsing when squeezed. Sheaths with 10–20(–30) teeth. With or without arcuate ascending side branches.

Cone: 1–2 cm long, terminal, blackish green, becoming brownish at maturity.

Spores: Approximately 50 µm in diameter, with 4 spirally wound filaments (elaters) to improve dispersal.

Spore maturity: June-July (-August).

Biology: Often forming large stands. Cone production is most abundant when growing in fully exposed habitats. The spores are spread by wind. Vegetative propagation by rooting of floating fragments. Overwinters as a rhizome.

Habitats: In oligotrophic to eutrophic, acidic to alkaline water to a depth of 1.5 m. Reed swamps, swamps, wet marshes, meadows, ponds, ditches, at the banks of streams and lakes, as well as in brackish water with a salinity of 2‰-5 ‰ in the Gulf of Bothnia.

Distribution within the region: Native. Widespread and common throughout the region except Greenland.

Characteristics and similar species: *E. fluviatile* can be confused with other species of horsetail occurring in and near water, but may be recognised by the easily collapsible stem with a large central hollow.

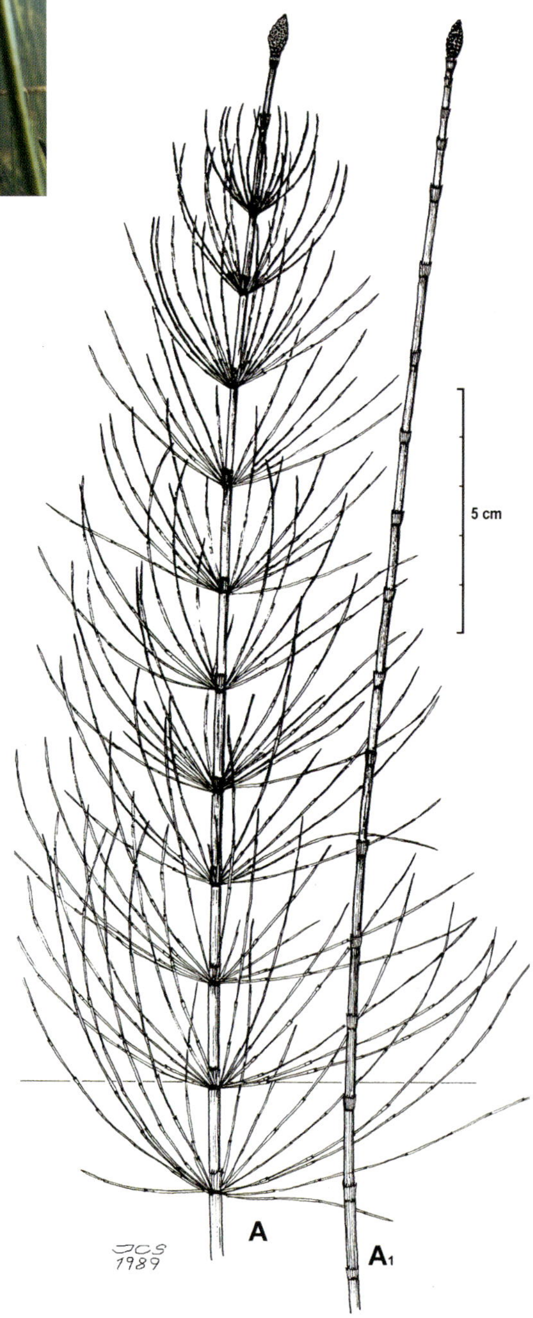

A, A1 habit with and without branches.

Equisetum fluviatile with cones. Paderup, Randers, Denmark. Photos JCS.

Other relevant *Equisetum* species and hybrids

Stems sheaths and cross-sections. Drawings JCS.

E. arvense L.

Cones on short-lived, unbranched brown stems with whitish or dark brown parts, which emerge earlier than the green, branched, sterile stems. **Stem** very firm, with 8–20 ridges, central hollow less than 1/2 diameter. **Sheath teeth** without scarious margin. **Lowest branch** internode much longer than sheath, with (3–)4 subulate, greenish teeth.

Common throughout the region.

E. ×litorale Kuhlew. ex Rupr. (*E. arvense × fluviatile*).

Cones rare, if present remaining closed. **Stem** very firm, with c. 9 ridges, central hollow 1/3– 1/2 diameter. **Sheath teeth** with a very narrow scarious margin. **Lowest branch** internode more or less as long as sheath, with 4–5 greenish teeth.

Rather common, but probably overlooked.

E. fluviatile L.

Cones normally present and open in summer. **Stem** with 10–30 indistinct ridges, easily collapsing when squeezed. **Sheath teeth** without or with a very narrow scarious margin. **Lowest branch** internode shorter or as long as sheath, with 5 greenish teeth.

Common throughout the region, absent from Greenland.

E. × dycei C.N.Page (*E. fluviatile × palustre*)

Cones rare - if present remaining closed. **Stem** with 9–12 ridges, central hollow approximately 1/4 diameter. **Sheath teeth** with a very narrow scarious margin. **Lowest branch** internode shorter or about as long as sheath, with 5, usually blackish teeth.

Scattered, but probably overlooked.

E. palustre L.

Cones normally present and open in summer. **Stem** very firm, with 5–9 ridges, central hollow less than 1/4 diameter. **Sheath teeth** with a rather wide, scarious margin. **Lowest branch** internode much shorter than sheath, with c. 5 short, blackish teeth.

Common throughout the region, absent from Greenland.

Glossary

Abaxial - facing away from the stem of a plant, in the case of leaves normally the lower side

Achene - a dry, indehiscent, usually 1-seeded fruit, more or less hard, with papery to leathery skin. A fruiting head is often composed of many achenes

Acicular - having the shape of a needle

Actinomorphic - a radially symmetrical flower with multiple axes

Acuminate - gradually tapering to a point - see figure p. 731

Acute - with the point < 90° - see figure p. 731

Adaxial - facing towards the stem of a plant, in the case of leaves normally the upper side

Adherent - joined or fused to

Aerial - above ground or above water

Agamospermy - asexual reproduction in which seeds are produced from unfertilised ovules

Alien - not native, deliberately or accidentally introduced to an area by humans

Allopolyploid - having two or more sets of chromosomes derived from parents of different species

Alternate - lateral organs on an axis 1 per node, successive ones on opposite sides - see figure p. 732

Amphiatlantic - occurring on both sides of the Atlantic Ocean

Amphiphyte - a plant able to live either rooted in damp soil above the water level or completely submerged

Amphistomous - having stomata on both surfaces of the leaf

Amplexicaul - clasping; the leaf base completely encloses the stem - see figure p. 732

Annual - completing its life cycle in 12 months or less (but often not within 1 calendar year)

Anther - fertile pollen-bearing part of a stamen, usually terminal on a stalk or filament

Anthesis - the period during which a flower is fully open and functional

Apex - uppermost part of a structure

Apiculate - ending abruptly in a small distinct point, e.g., an apiculate leaf - see figure p. 731

Apomictic; **apomixis** - producing seed wholly female in origin, without fertilisation

Appressed - lying flat against another organ

Areole - the area between the veins on a leaf

Ascending - sloping or curving upwards

Auricle - small ear-like extension, often at the base of a leaf

Auriculate - having auricles

Autopolyploid - having more than two haploid sets of chromosomes derived from the same ancestral species

Axil - point at which the leaf or stalk joins the stem

Beak - a narrow, usually apical projection

Bifid - divided into two, usually deeply, at apex

Biotype - a group of organisms having the same genotype

Bisexual - of a plant or flower, bearing both sexes

Blade - main part of a flat organ, e.g., part of a leaf or petal

Bloom - delicate, waxy, easily removed coating on fruit, leaves, etc.

Bract - small, scale- or leaf-like structure, often subtending (beneath) a flower or inflorescence

Bracteole - a supplementary bract present on the pedicel

Bulb - swollen underground storage organ consisting of condensed stem and fleshy leaves

Calcareous - sediment or soil that is formed from or contains a high proportion of calcium

Calyx - all the sepals of the flower; the outer whorl(s) of the perianth, if different from the inner

Capillary - hair-like

Capitate - head-like, such as a tight inflorescence on a stalk, a knob-like stigma on a style, or a stalked gland

Capsule - a dry, many-seeded dehiscent fruit formed from more than 1 carpel

Carpel - female organ of the flower (reproductive unit of angiosperms), comprising stigma, style and ovary

Carpophore - A thin stalk that raises the pistil above the stamens in Apiaceae

Casual - an alien plant not naturalised, persisting only a short time and not established in the wild

Cauline leaves - all leaves above the base of the stem

Cell - of an ovary, the chambers into which it may be divided (often each one responding to a carpel)

Chasmogamous - flowers open at maturity, usually cross-pollinated

Clavate - club shaped, slender and distally thickened

Claw - proximal, narrow part of a flat organ such as a petal, bearing the blade distally

Cleistogamous - flowers not opening, self-pollinated in the bud

Colliculate - covered with small, rounded bumps

Column - a stout stalk formed by fusion of various floral parts

Compound - not simple, of a leaf divided into leaflets right to the rachis

Compressed - flattened

Connate - fusion of organs of the same type, like petals to one another to form a tubular corolla

Connective - part of anther connecting its 2 halves

Contiguous - touching at the edges with no gap

Converging - lines that tend to meet at a point

Cordate - see figure p. 731

Coriaceous - of leathery texture

Corolla - all the petals of a flower, which may form a tube, the inner whorl(s) of the perianth

Corymbose - raceme in which the lower flowers have longer pedicels, producing a more or less flat-topped inflorescence

Crenate - having the margin notched so as to form rounded teeth

Cuneate - refers to the base of a flat organ, used here to describe the leaf base of vascular plants in which the two sides of the blade become gradually narrower to taper into the petiole - see figure p. 731

Cuspidate - abruptly narrowed to a point

Cyme - a flower cluster with a central stem bearing a single terminal flower that develops first, the other flowers in the cluster developing as terminal buds of lateral stems

Cytotype - an individual of a species that has a different chromosome number, chromosome structure or genomes from another

Deciduous - not persistent, e.g., leaves falling in autumn or petals falling after anthesis

Decurrent - a lateral organ, having its base prolonged down the main axis - see figure p. 732

Decussate - having opposite leaves arranged alternately at right angles

Dehiscent - opening naturally, splitting

Dentate - with a row of more or less patent teeth; cf. serrate

Denticulate - minutely or finely dentate

Dichotomous - divided or dividing into two parts

Dimorphic - of two forms

Dioecious - having the 2 sexes on different plants

Diploid - having 2 matching sets of chromosomes

Disc - anything disc-shaped, e.g., the nectar-secreting ring inside the flower at the base of *Nuphar* fruit

Dissected - deeply divided into segments

Distal - at the end away from the point of attachment; cf. proximal

Distichous - leaves arranged in two vertical rows on opposite sides of an axis

Divergent - of 2 or more organs with apices farther apart than their bases

Drupes - a succulent or spongy fruit, the seeds usually 1 and with a stony coat

Dry - not succulent

Dyads - pollen grains attached in pairs

Ellipsoid - a solid shape, elliptic in side view

Elliptic - a flat shape, widest in the middle and 1.2-3x as long as wide - see figure p. 731

Emarginate - with a pronounced, angled notch at the apex; cf. retuse - see figure p. 731

Endemic - confined to one particular area or region

Entire - whole, of the margin of a flat organ, not toothed or lobed

Epicarp - the outermost layer of a fruit

Epidermis - the outermost cell layer of an organ, such as a stem or leaf

Epipetalous - borne on or attached to the petals or corolla

Epiphyte - a plant growing on another plant

Epistomous - having stomata only on the upper surface of the leaf

Erect - upright

Erecto-patent - between erect and patent

Erose - with an irregular, uneven margin

Escape - a plant growing in the wild but having spread vegetatively or by seed from cultivation

Exceeding - longer than

Extravaginal - developing outside the sheath

Fibrous roots - a root system in which there is no main axis

Filament - the stalk of a stamen

Filiform - thread- or wire-like

Fistular - resembling a hollow cylinder or a pipe

Flexuous - of a stem or hair, wavy

Follicle - a dry, usually many- seeded fruit, usually dehiscent along 1 side, formed from 1 carpel

Free - separate, not fused to another organ or to one another except at point of origin

Fruit - the ripe, fertilised ovary, containing 1 or more seeds

Fusiform - tapering at both ends

Geophyte - a plant that survives the dry summer as a dormant underground bulb or tuber

Glabrous - hairless

Gland - a secreting structure, usually round or more or less so, on the surface of an organ, below the surface, or raised on a stalk

Glaucous - of a dull greyish-green or bluish colour

Glume - empty bract(s) at the base of a spikelet

Grain - cultivated cereal used as food

Hastate - triangular with two spreading lobes at the base - see figure p. 731

Helophyte - a perennial marsh plant having its overwintering buds under water

Hemicryptophyte - a perennial plant having its overwintering buds situated at the soil surface

Herb - a plant dying down to ground level each year

Herbaceous - not woody, dying down each year; leaf-like as opposed to woody, hyaline, scarious or spongy

Heterophyllous - having leaves of 2 or more distinct forms

Homomorphic - A similar appearance of two unrelated organisms or structures

Homophyllous - having the same kinds of leaves on a plant or stem

Hyaline - thin more or less transparent

Hybrid - a cross between plants of 2 different taxa, often 2 species

Hydatode - a structure (modified pore) that discharges water from the interior of the leaf to its surface

Hydromorphic - adapted to growth wholly or partially in water

Hydrophyte - a plant that grows partly or entirely in water, whether rooted in the mud, attached to a solid submerged substrate or free floating

Hypanthium - extension of the receptacle above the base of the ovary

Hypogynous - of a flower with a superior ovary, the calyx, corolla and often stamens being inserted at the base of the ovary

Hypostomous - having stomata on the lower surface of the leaf only

Included - not exserted

Inflated - air filled

Inflorescence - a group of flowers with their branching system and associated bracts and bracteoles, etc.

Infructescence - the fruiting stage of the inflorescence

Internode - the stem between adjacent nodes

Introduced - a plant that is not native and was brought into the area (intentionally or not) by humans

Isolateral - having similar upper and lower (or front and back) surfaces

Isosceles triangular - a triangular shape in which two sides have the same length

Iterative branching - a sequence of branching which is repeated

Keel - a longitudinal ridge on an organ, resembling the keel of a boat

Lacunae - translucent (pale) cells in rows along the midrib in species of *Potamogeton*

Laminar - in the form of a flat leaf

Lanceolate - see figure p. 731

Lax - loose or diffuse, not dense

Leaflet - a division of a compound leaf

Leaf sheath - a structure at the base of a leaf petiole that partly surrounds or protects the stem or another organ that it subtends

Lemma - the lower of the pair of bracts that subtends the flower

Lenticular - lens shaped; can vary from biconvex to biconcave

Ligule - a small membranous projection or ring of hairs at the junction of the leaf sheath and stem

Linear - long and narrow with more or less parallel margins, such as the leaf of a grass

Lip - part of the distal region of a calyx or corolla sharply differentiated from the rest due to fusion or close association of its parts

Lobe - a substantial division of a leaf, calyx or corolla

Membranous - paper-like or resembling a membrane in texture

-merous - divided into or composed of a particular number of parts

Midrib - the central, main vein

Monoecious - having separate male and female flowers on the same plant

Morphology - the appearance, form or structure

Mucronate - having a very short bristle-like tip; **mucro** - the tip itself - see figure p. 731

Naked - not enclosed

Native - opposite of alien, a taxon that colonised an area by natural means, from other native areas

Naturalised - not native but well established

Nectar - a sugary fluid secreted within flowers to encourage pollination by insects and other animals

Nodding - bent over and pendent at the tip

Node - the position on a stem where leaves, flowers or lateral stems arise

Nut - a dry, indehiscent 1-seeded fruit with a hard woody wall, often large

Ob- - the other way up from normal, usually flattened or widened at the distal rather than proximal end, e.g., obovoid, obtrullate

Obovate - see figure p. 731

Oblong - see figure p. 731

Obtuse - with a point > 90°

Opposite - see figure p. 732

Orbicular - a flat shape, circular in outline

Orthogonally - forming an angle of 90°

Ovary - the basal part of the gynoecium containing the ovules

Ovate - a flat shape, widest nearer the base and 1.2-3x as long as wide, egg shaped - see figure p. 731

Ovoid - a solid shape. ovate in side view

Ovule - the organ containing the egg which develops into the seed after fertilisation

Palate - the projecting part of the lower lip that closes the opening of the corolla (in some two-lipped corollas)

Palea - the upper of the pair of bracts that subtends the flower

Panicle - a compound or densely branched inflorescence

Papilla - a small nipple-like projection

Paroecious - having the male and female reproductive organs at different levels on the same stem

Patent - projecting more or less at right angles

Pectinate - having narrow parallel projections or divisions

Pedicel - the stalk of an individual flower in an inflorescence or the stalk of a grass spikelet

Peduncle - stalk of a group of 2 or more flowers

Peltate - of a flat shape with its stalk arising from the plane surface, not the edge

Pendent - hanging down

Penicillate - having, forming, or resembling a small tuft or tufts of hair

Perennial - living 2 or more years

Perianth - the outer non-sexual covering layers of the flower (the calyx and corolla together); usually used when the calyx and corolla are not or are little differentiated

Pericarp - the wall of a fruit, originally the ovary wall

Persistent - remaining intact for most of the lifetime of the plant

Petal - one of the segments of the inner whorl(s) of the perianth

Petiole - the stalk of a leaf

Pistil - the female reproductive part of a flower

Placenta - the point of origin of ovules in an ovary

Pollen sac - the microsporangium of a gymnosperm or angiosperm; one of the chambers in an anther in which the pollen is formed

Polymorphic - occurring as 2 or more forms

Procumbent - trailing along the ground without rooting

Proliferating - with inflorescences bearing plantlets instead of flowers or fruits

Prostrate - lying close to and along the ground

Protandrous - the male reproductive organs mature before the female reproductive organs

Protogynous - the female reproductive organs mature before the male reproductive organs

Proximal - at the end near the point of attachment; cf. distal

Pseudostem - a sterile stem-like shoot formed by several overlapping leaf sheaths

Punctate - marked with dots or transparent spots

Raceme - a simple unbranched inflorescence with stalked flowers on a single axis, the youngest flowers at the top

Racemose - a flowering shoot in which the growing region at the tip of the flower stalk continues to produce new flower buds during growth

Rachilla - in grasses, the short stem of a spikelet that bears the inflorescence

Ray - anything that radiates outwards, e.g., branches of an umbel and stigma ridges

Receptacle - the usually expanded, often capitate, cup-shaped or tubular, apical part of a pedicel on which the flower parts are inserted

Recurved - curved down or back

Reticulate - forming or covered with a network

Retuse - with a shallow blunt notch at the apex; cf. emarginate

Leaf tips

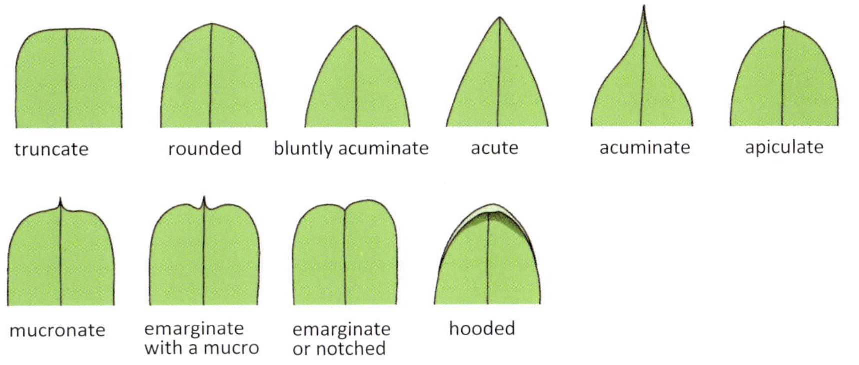

Leaf bases

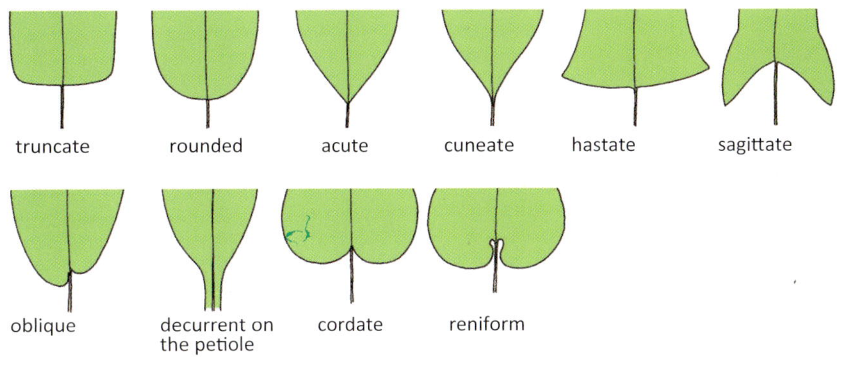

Leaf shapes

Rhizome - a stem which grows underground or at ground level, usually growing horizontally or downwards, often more or less swollen; cf. stolon

Rhombic - a flat shape, widest and more or less angled in the middle, 1.2-3x as long as wide

Rigid - stiff, not flexible

Rounded - without a point or angle

Scabrid - rough to the touch, with minute prickles or bristly hairs

Scale - a modified leaf, often small and membranous

Scalloped - leaf margin with edges having large rounded indentations

Scape - a flowering stem of a plant in which all the leaves are basal

Scarious - of thin, papery texture and not green

Schizocarp - a dry fruit that splits into two or more one-seeded portions at maturity

Sclerenchyma - supportive plant tissue that consists of thick-walled, usually lignified cells

Seed - a fertilised ovule

Self-pollination - fertilising itself

Semi-amplexicaul - leaf bases half-clasping the stem, the leaf margins not touching each other on the side of the stem opposite the leaf blade

Sepal - one of the segments of the outer whorl(s) of the perianth

Septate - with cross-walls dividing a cavity or hollow in smaller parts

Serrate - with a row of more or less apically directed teeth, cf. dentate

Sessile - not stalked - see figure p. 732

Silicula - a short siliqua, less than three times as long as wide

Siliqua - a dry fruit in which the seeds are contained in a cylindrical structure more than three times as long as wide

Simple - not compound

Sinuous - wavy, either of a hair, a stalk, or the margin of a leaf; cf. undulate

Solitary - born singly

Spadix - a floral spike with a fleshy or succulent axis usually enclosed in a spathe

Leaf arrangement

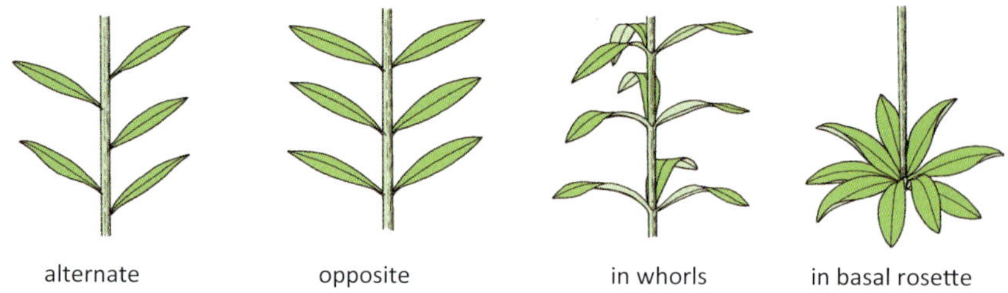

Leaf attachment

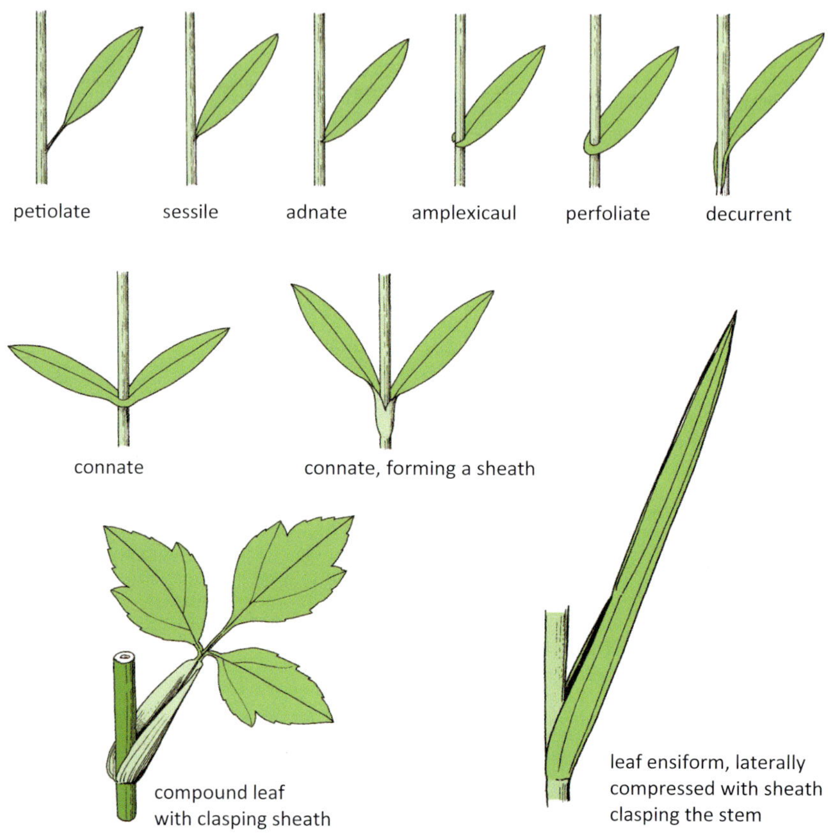

Cross section of stems or petioles

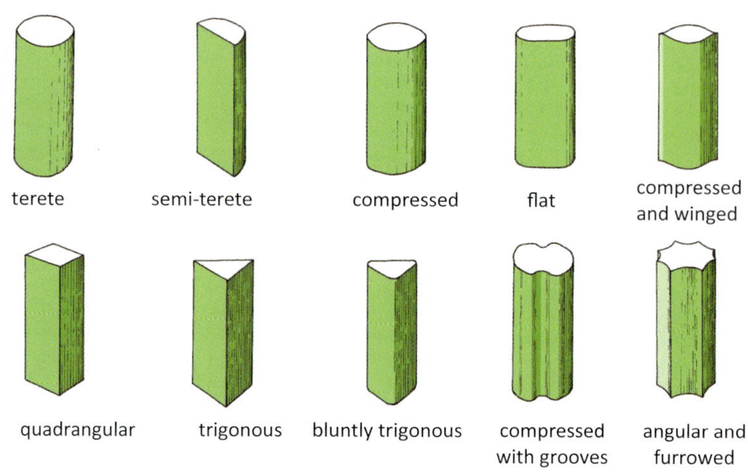

Spathe - large, leafy bract

Spathulate - paddle or spoon shaped

Spike - a racemose inflorescence in which the flowers are sessile

Spikelet - a unit of a grass flower usually comprising two outer bracts (glumes) and one or more flowers each borne between a lemma and a palea

Spine - a narrow, sharp, stiff, straight outgrowth

Spreading - growing out divergently, not straight or erect

Stamen - the basic male reproductive unit of the flower, consisting of a filament and anther

Staminode - a sterile stamen, sometimes modified to perform some other function, e.g., that of a petal or nectary

Stigma - the receptive tip of a style (simple to densely branched) that receives pollen and upon which the pollen germinates

Stipule - small, leaf-like organ at the base of some leaf petioles, often in pairs, either simple or lobed

Stolon - an aerial or procumbent stem, usually not swollen, rooting at the tip and/or along the length; cf. rhizome

Stomata - very small openings in membranes in plants, through which water and gas pass

Style - the stalk of any ovary bearing stigma(s), sometimes absent

Sub- - almost, as in subacute, subentire, etc.

Subtend - to occur beneath or close to

Subulate - tapering more or less constantly from a narrow base to a fine point

Taxon - any taxonomic grouping, such as a genus or species

Tepal - one of the segments of the perianth, used when sepals and petals are not differentiated

Terete - rounded in cross section - see figure p. 732

Terminal - at the very apex

Tetrahedroid - in shape like a triangular pyramid

Throat - the opening where the tube joins either the limb or the lobes of a corolla or calyx

Tomentose - a very dense, often more or less matted hair covering

Tooth - a shallow division of a leaf, calyx or corolla, or the apex of a capsule

Torulose - of a cylindrical or ellipsoid body alternately swollen and pinched along its length

Transverse - lying cross ways

Trifid - partly or entirely split into three divisions or lobes

Truncate - ending abruptly as if cut off across the base or tip - see figure p. 731

Tube - the narrow, cylindrical, proximal part of cthe alyx or corolla, as distinct from the limb, lobes or throat

Tuber - swollen root or subterranean stem

Tubercle - a small more or less spherical or ellipsoid swelling

Tubular - in the form of a hollow cylinder or tube

Tufted - of elongated organs or stems that are clustered together

Turion - a type of bud that is capable of dormancy

Xeromorphic - designed to prevent water loss in a dry atmosphere.

Umbel - an inflorescence in which all the pedicels arise from one point; compound umbel, an umbel of umbels

Undulate - wavy at the margin; cf. sinuous

Unisexual - of a plant or flower, bearing only 1 sex

Vascular - pertaining to the veins or wood of an organ (i.e., conducting tissue)

Vascular bundle - anatomically discrete strands of vascular tissue

Vegetative - without reproductive parts

Vein - a strand of vascular tissue consisting of 1 or more vascular bundles

Venation - the pattern of veins

Vesitigial - having a form that is small or imperfectly developed and not able to function

Whorl - more than two leaves or flowers arising from a node - see figure p. 732

Wing - any membranous or foliaceous extension of an organ, e.g., a seed or fruit

Literature

Important floras etc.

Arber A, 1920. Water Plants: A Study of Aquatic Angiosperms. Cambridge University Press, Cambridge, UK.

Casper SJ, Krausch HD, 1980. Pteridophyta und Anthophyta, in Ettl, H, Gerloff, J, Heynig H (eds.), Süßwasserflora von Mitteleuropa, teil 1, bd. 23. Gustav Fischer Verlag, Stuttgart.

Casper SJ, Krausch HD, 1981. Pteridophyta und Anthophyta, in Ettl H, Gerloff J, Heynig H (eds.), Süßwasserflora von Mitteleuropa, teil 2, bd. 24 Gustav Fischer Verlag, Stuttgart.

Fedorov AA (ed.), 1999. Flora of Russia vol, 1. A. A. Balkema. Moscow.

Gessner F, 1955, 1959. Hydrobotanik I and II. VEB Deutscher Verlag der Wissenschaften, Berlin.

Glück H, 1905, 1908, 1911. Biologische und morphologische Untersuchungen über Wasser und Sumpggewäsche I, II and III. Gustav Fischer, Jena, Germany.

Glück H, 1936: Pteridophyten und Phanerogamen. A Pascher, Die SüsswasserFlora Mitteleuropas, Heft 15. Jena.

Hylander N, 1953-1966. Nordisk Kärlväxtflora, vol. 1-2. Almqvist & Wiksells. Stockholm.

Jäger E (hrsg.), 2011. Rothmaler - Exkursionsflora von Deutschland. Gefässpflanzen: Grundband, 20th ed. Spektrum Akademischer Verlag, Heidelberg, Germany

Jonsell B (ed.), 2000. Flora Nordica 1. Bergius Foundation, Royal Swedish Academy of Sciences, Stockholm.

Jonsell B (ed.), 2001. Flora Nordica 2. Bergius Foundation, Royal Swedish Academy of Sciences, Stockholm.

Jonsell B (ed.), 2004. Flora Nordica, general volume. Bergius Foundation, Royal Swedish Academy of Sciences, Stockholm.

Jonsell B, Karlsson T (eds.), 2010. Flora Nordica 6. Swedish Museum of Natural History, Stockholm.

Karlsson T, 2012. Novelties in the flora of Norden. I. Lycopodiaceae - Annonaceae (in Swedish). Svensk Botanisk Tidskrift. 106: 138.

Karlsson T, Agestam M, 2014. Checklista över Nordens kärlväxter. http://www.euphrasia.nu/checklista/

Lansdown RV, 2009. A Field Guide to the Riverine Plants of Britain and Ireland. Self-published, Stroud, UK.

Lid J, Lid DT, Elven R (eds.), 2005. Norsk Flora 7. Det Norske Samlaget, Oslo.

Mjelde M, Rørslett B, Langangen A, 2022. Fotoflora for Norske vannplanter, versjon 1. Norsk Institutt for Vannforskning. https://www.niva.no/en/projectweb/fotoflora-for-norske-vannplanter.

Moeslund B, Løjtnant B, Mathiesen H, Mathiesen, L, Pedersen A, Thyssen N, Schou JC, 1990. Danske Vandplanter. Miljøministeriet, Miljøstyrelsen, Danmark.

Øllgaard B, 1993. Scandinavian Ferns. Rhodos, Copenhagen.

Poland J, Clement EJ, 2009. The Vegetative Key to the British Flora. Botanical Society of the British Isles, London.

Preston CD, Croft JM, 1995. Database and Atlas of Aquatic Vascular Plants in the British Isles. Part I: Species Accounts. National Rivers Authority, Bristol, UK.

Preston CD, Croft JM, 2014. Aquatic Plants in Britain and Ireland. Brill, Leiden, Netherlands.

Raunkiær C, 1890. De danske Blomsterplanters Naturhistorie I: Enkimbladede. Gyldendal, Copenhagen.

Raunkiær C, 1950. Dansk Ekskursions-Flora. 7th. ed. Gyldendalske Boghandel - Nordisk Forlag. Copenhagen.

Rostrup E, Jørgensen CA, 1973. Den danske flora. 20th ed. Gyldendal. Copenhagen.

Schou JC, Moeslund B, Baastrup-Spohr L, Sand-Jensen K, 2017. Danmarks Vandplanter. BFN's Forlag, Klitmøller, Denmark.

Sculthorpe CD, 1967. The Biology of Aquatic Vascular Plants. Edward Arnold, London.

Sell P, Murrell G, 2007. Flora of Great Britain and Ireland, vol. 5: Butomaceae - Orchidaceae. Cambridge University Press, Cambridge, UK.

Stace C, 2019. New Flora of the British Isles. 4th. ed. C & M Floristics, Cambridge, UK.

Stace CA, Preston CD, Pearman DA, 2015. Hybrid flora of the British Isles. Botanical Society of the British Isles, London.

Weyer K van de, Schmidt C, 2018. Bestimmungsschlüssel für die aquatischen Makrophyten (Gefässpflanzen, Armleuchteralgen und Mosse) in Deutschland. Band 1: Bestimmungsschlüssel, heft 119. Ministerium für Umwelt, Gesundheit und Verbraucherschutz, Brandenburg, Germany.

Weyer K van de, Schmidt C, 2011. Bestimmungsschlüssel für die aquatischen Makrophyten (Gefässpflanzen, Armleuchteralgen und Mosse) in Deutschland. Band 1: Abbildungen, heft 120. Ministerium für Umwelt, Gesundheit und Verbraucherschutz, Brandenburg, Germany.

Distribution

Belgium. https://observations.be.

Britain and Ireland. https://bsbi.org/maps.

Chappuis E, Ballesteros E, Gacia E, 2012. Distribution and richness of aquatic plants across Europe and Mediterranean countries: mutual driving factors. Journal of Vegetation Science 23: 988-97.

Denmark. https://www.naturbasen.dk/ and https://arter.dk/landing-page.

Estonia https://ottluuk.github.io/atlas/archive.html .

Finland. https://laji.fi/en.

Germany. https://www.floraweb.de.

Global Biodiversity Information Facility, 2022. https://www.gbif.org.

Hartvig P, 2015. Atlas Flora Danica. Gyldendal, Copenhagen.

Hultén E, 1971 : Atlas över växternas utbredning i Norden 2. ed. Generalstabens Litografiske Anstalts Förlag, Stockholm.

Hultén E, Fries M, 1986. Atlas of North European Vascular Plants: North of the Tropic of Cancer, vols. 1-3. Koeltz, Königstein, Germany.

Iceland. https://www.ni.is/search/biota/ranunculus.

International Union for Conservation of Nature's Red List of Threatened Species. https://www.iucnredlist.org.

Latvia. www.latvijasdaba.lv/augi/.

Murphy K et al., 2019. World distribution, diversity and endemism of aquatic macrophytes. Aquatic Botany 158: 103127.

Netherlands. NDFF Verspreidingsatlas Vaatplanten. https://www.verspreidingsatlas.nl/vaatplanten.

Norway. https://artskart.artsdatabanken.no.

Pladias – Database of the Czech Flora and Vegetation. www.pladias.cz

Poland. https://www.atlas-roslin.pl/index.html.

POWO, 2022. Plants of the world online. Royal Botanic Gardens, Kew, UK. http://www.plantsoftheworldonline.org/

Samuelsson G, 1934. Die Verbreitung der höheren Wasserpflanzen in Nordeuropa (Fennoskandien und Dänemark). Acta Phytogeogr Suec 6. Almqvist & Wiksells, Uppsala, Sweden.

Sweden. https://artfakta.se/artbestamning.

World Flora online. http://www.worldfloraonline.org.

Environment etc.

Baagøe J, Ravn FK, 1896. Excursion til de jydske Søer og vandløb i Sommeren 1895. Botanisk Tidsskrift 20: 288-326.

Baastrup-Spohr L, Kragh T, Petersen K, Moeslund B, Schou JC, Sand-Jensen K, 2016. Remarkable richness of aquatic macrophytes in 3-years old reestablished Lake Fil, Denmark. Ecological Engineering 95: 375-83.

Baastrup-Spohr L, Sand-Jensen K, Olesen SCH, Bruun HH, 2017. Recovery of lake vegetation following reduced eutrophication and acidification. Freshwater Biology. https://doi.ord/10.1111/fwb 13000.

Bakker ES, Sarneel JM, Gulari RD, Liu ZW, van Donk E, 2013. Restoring macrophyte diversity in shallow temperate lakes: Biotic versus abiotic constraints. Hydrobiologia 710: 23-31.

Demars BOL, Wiegleb G, Harper DM, Bröring U, Brux H, Herr W, 2014. Aquatic plant dynamics in lowland river networks: Connectivity, management and climate change. Water 6: 868-911.

Hutchinson GE, 1975. A Treatise on Limnology III: Limnological Botany. Wiley, New York.

Iversen J, 1929. Studien über die pH-Verhältnisse dänischer gewässer und ihren Einfluss auf die Hydrophyten-Vegetation. Botanisk Tidsskrift 40: 277-333.

Iversen J, Olsen S, 1943. Die Verbreitung der Wasserpflanzen in Relation zur Chemie des Wassers. Botanisk Tidsskrift 46: 136-45.

Iversen, LL et al. 2019. Catchment properties and the photosynthetic trait composition of freshwater plants. Science 366,6467: 878-81.

Kohler A, Vollrath H, Beisl E, 1971. Zur Verbreitung, Vergesellschaftung und Ökologie der Gefäß-Makrophyten im Fließwassersystem Moosach (Münchener Ebene). Arch Hydrobiol 69: 333-65.

Lohammer G, 1965. The vegetation of Swedish lakes. Acta Phytogeogr. Suec. 50: 28-48.

Murphy F, Schmieder K, Baastrup-Spohr L, Pedersen O, Sand-Jensen K, 2018. Five decades of dramatic changes in submerged vegetation in Lake Constance. Aquatic Botany 144: 31-37.

Newbold C, Palmer M, 1979. Trophic adaptations of aquatic plants. CST note no. 18, Nature Conservancy Council, England.

Pietsch W, 1972. Ausgewählte Beispiele für Indikatoreigenschaften höherer Wasserpflanzen. Arch Naturschutz Landschaftsforsch 12,2: 121-51.

Riis T, Sand-Jensen K, 2001. Historical changes in species composition and richness accompanying perturbation and eutrophication of Danish lowland streams. Freshwater Biology 46: 269-80.

Roelofs JGM, 1983. Impact of acidification and eutrophication on macrophyte communities in soft water lakes in the Netherlands. Aquatic Botany 17: 139-55.

Rørslett B, 1991. Principal determinants of aquatic macrophyte richness in Northern European lakes. Aquatic Botany 31: 173-93.

Sand-Jensen K, Bruun HH, Baastrup-Spohr L, 2017. Decade-long time delays in nutrient and plant species dynamics during eutrophication and reoligotrophication of Lake Fure 1900-2015. Journal of Ecology 105: 690-700.

Sand-Jensen, K, Friberg N, Murphy J (eds.), 2006. Running Waters. Historical Development and Restoration of Lowland Danish Streams. National Environmental Research Institute, Denmark.

Sand-Jensen K, Riis T, Vestergaard O, Larsen SE, 2000. Macrophyte decline in Danish lakes and streams over the past 100 years. Journal of Ecology 88: 1030-40.

Spence DHN, 1967. Factors controlling the distribution of freshwater macrophytes with particular reference to the lochs of Scotland. Journal of Ecology 55: 147-70.

Thunmark S, 1931. Die See Fiolen und Seine Vegetation. Acta Phytogeogr. Suec, 2: 1-198.

Vestergaard O, Sand-Jensen K, 2000. Alkalinity and trophic state regulate aquatic plant distribution in Danish lakes. Aquatic Botany 17: 139-55.

Vestergaard O, Sand-Jensen K, 2000. Aquatic macrophyte richness in Danish lakes in relation to alkalinity, transparency, and lake area. Canadian Journal of Fishery and Aquatic Science 57: 2022-31.

Weber-Oldecop DW, 1969. Wasserpflanzengesellschaften im östlichen Niedersachsen. PhD Diss. TU Hannover.

Wesenberg-Lund C, Sand ML, Boye-Petersen J, Raunkiær AS, Stenberg CM, 1917. Furesøstudier en bathymetrisk botanisk zoologisk undersøgelse af Mølleaaens Søer. A. F. Høst & søn. Copenhagen.

Wiegleb G, Bröring U, Filetti M, Brux H, Herr W, 2014. Long-term dynamics of macrophyte dominance and growth form types in two northwest German lowland streams. Freshwater Biology 59: 1012-23.

Wiegleb G, Herr W, Zander B, Bröring U, Brux H, Weyer K van de, 2015. Natural variation of macrophyte vegetation in lowland streams at the regional level. Limnologica 51: 53-62.

Willby NJ, Abernethy VJ, Demars BOL, 2000. Attribute-associated classification of European hydrophytes and its relationship to habitat utllization. Freshwater Biology 43: 43-71.

Species identification and biology

Acanthaceae

Nault ME, Mikulyuk A, 2009. East Indian Hygrophila (*Hygrophila polysperma*): A Technical Review of Distribution, Ecology, Impacts, and Management. Wisconsin Department of Natural Resources Bureau of Science Services, PUB-SS-1049 2009, Madison.

Alismataceae

Jacobson A, 2005. Åtgärdsprogram för Bevarande av Småsvalting (*Alisma wahlenbergii*). Naturvårdsvärket Rapport 5499. Sweden.

Kozlowski G, Jones RA, Nicholls-Vuille F, 2008. Biological flora of Central Europe: *Baldellia ranunculoides* (Alismataceae). Perspectives in Plant Ecology, Evolution and Systematics 10: 109-42.

Lindblad R, Ståhl B, 1990. Revsvalting, *Baldellia repens*, i Norden. Svensk Botanisk Tidskrift 84: 253-58.

Mesterhazy A, 2014. *Caldesia parnassifolia* (L.) Parl. új előfordulása Magyarországon. Kitaibelia 19,1: 50-55.

Moeslund B, Schou, JC, 1994. Noter om danske vandplanter, 2. Vandranke (*Luronium natans*). URT 1994: 1.

Preston CD, Uotila P, 2009. *Sagittaria ×lunata*, a binomial for the widespread North European hybrid between *S. natans* and *S. sagittifolia* (Alismataceae). Annales Botanici Fennici 46,3: 215-30.

Samuelsson G, 1920. De nordiska *Sagittaria*-arterna. Svensk Botanisk Tidskrift 14,1: 21-38.

Smith JG, 1895. A Revision of the North American Species of *Sagittaria* and *Lophotocarpus*. Missouri Botanical Garden Annual Report Vol. 1895: 27-64.

Apiaceae

Desjardins SD, Leslie AC, Stace CA, Schwarzacher T, Bailey JP, 2015. Intergeneric hybridisation between *Berula erecta* and *Helosciadium nodiflorum* (Apiaceae). Taxon 64,4: 784–94.

Faurholdt N, Schou JC, 2012. Danmarks Skærmplanter. BFN´s forlag; Klitmøller, Denmark.

Ronse AC, Popper ZA, Preston JC, Watson MF, 2010. Taxonomic revision of European *Apium* L. s.l.: *Helosciadium* W.D.J.Koch restored. Plant Systematic and Evolution 287: 1–17.

Araceae

Bog M, Sowjanya SK, Fuchs J, Hoang Phuong TN, Schubert I, Kuever J, Rabenstein A, Paolacci S, Jansen MAK, Appenroth KJ, 2020. A taxonomic revision of *Lemna* sect. Uninerves (Lemnaceae). Taxon 69,1: 56-66.

Garve E, Kelm H, Fischer C, Thiel H, Schmitz U, 2017. Die Kolumbianische Zwergwasserlinse (*Wolffia columbiana* H. Karst.) – eine neue Wasserpflanze in Niedersachsen. Tuexenia 37: 355-62.

Iberite M, Duilio L, Silverio A, Giovanna A, 2011. *Lemna valdiviana* Phil. (Araceae) as a potential invasive species in Italy and Europe: Taxonomic study and first observations on its ecology and distribution. Plant Biosystems 145: 751-57.

Landolt E, 1980. Key to the determination of taxa within the family of Lemnaceae. Ver. Geobot. Inst. ETH, Stiftung Rubel, Zürich 70: 13-21.

Landolt E, 1992. Lemnaceae duckweed family. Journal of the Arizona-Nevada Academy of Science 26,1: 10-14.

Landolt E, 1994. Taxonomy and Ecology of the section Wolffia of the genus *Wolffia* (Lemnaceae). Ver. Geobot. Inst. ETH, Stiftung Rubel, Zürich 60: 137-51.

Pereira SF, Pott VJ, Temponi LG, 2016. Lemnoideae (Araceae) no estado do Paraná, Brasil. Rodriguésia 67,3: 839–48.

Schou JC, 2014. Rød Andemad - *Lemna turionifera* Landolt fundet i Danmark. URT 38,3: 84-87.

Ward DB, 2011. *Spirodela oligorrhiza* (Lemnaceae) is the correct name for the lesser greater duckweed. Journal of Biogeography 5,1: 197-201.

Araliaceae

Webb CJ, 1990. New Zealand species of *Hydrocotyle* (Apiaceae) naturalised in Britain and Ireland. Watsonia 18: 93-95.

Asteraceae

Elias P, Hájek M, Hájková P, 2009. A European warm water neophyte *Shinnersia rivularis* - new alien species to the Slovak flora. Biologia 64,4: 684-86.

Brassicaceae

Jonsell B, 2017. Crucifers in the Nordic flora: 1. *Nasturtium* and *Rorippa*. Svensk Botanisk Tidskrift 111,3-4: 192-210.

Moeslund B, Schou JC, 1994. Noter om danske vandplanter: 3. Tykskulpet Brøndkarse (*Nasturtium officinale*). URT 1994: 2.

Cabombaceae

Hussner A, Haese U, Weyer K van der, 2010. *Cabomba caroliniana* (Cabombaceae) - neu für Deutschland. Florist. Rundbriefe 43: 17-23.

Campanulaceae

Makavičiūtė J, Sinkevičienė Z, 2010. Initial data on populations of water lobelia (*Lobelia dortmanna*) in Lithuania. Botanica Lithuanica 16: 13-20.

Ceratophyllaceae

Ganie AH, Tali BA, Khuroo AA, Reshi ZA, Les DH, 2015. *Ceratophyllum platyacanthum* subsp. *oryzetorum* (Kom.) Les (Ceratophyllaceae): An addition to the flora of India from Kashmir Himalaya. Check List 11,3: 1661.

Les DH, 1988. The evolution of achene morphology in *Ceratophyllum* (Ceratophyllaceae), III. Relationships in the "Facially-spined" group. Systematic Botany 13,4: 509-18.

Cyperaceae

Browning J, Gordon-Gray KD, Smith SG, Staden JV, 1996. *Bolboschoenus yagara* (Cyperaceae) newly reported for Europe. Annales Botanici Fennici 33: 129-36.

Foerster E, 2009. Seggen, Binsen, Simsen und andere Scheingräser des Grünlandes. LANUV-Arbeitsblatt 5. Landesamt für Natur, Umwelt und Verbraucherschutz, Nordrhein-Westfalen.

Gregor T, 2003. *Eleocharis mamillata* - Distribution and infraspecific differentiation. Folia Geobotanica 38,1: 49-64.

Hroudová Z, Gregor T, Zákravský P, 2009. Die Verbreitung von *Bolboschoenus* - Arten in Deutschland. Kochia 4: 1-22.

Hroudová Z, Zákravský P, Ducháček M, Marhold K, 2007. Taxonomy, distribution and ecology of *Bolboschoenus* in Europe. Annales Botanici Fennici 44: 81–102.

Hroudová Z, Zákravský P, Frantík T, 1999. Ecological differentiation of Central European *Bolboschoenus* taxa and their relationship to plant communities. Folia Geobot 34, 77–96.

Hroudová Z, Zákravský P, Jarolimova V, 1999. *Bolboschoenus glaucus:* A new species in the Czech Republic. Preslia, 71: 27-32.

Hroudová Z, Zákravský P, Wojcicki JJ, Marhold K, 2005. The genus *Bolboschoenus* (Cyperaceae) in Poland. Polish Botanical Journal 50,2: 117-37.

Lansdown R, Rumsey FJ, 2020. *Schoenoplectus ×flevensis* (D.Bakker) Lansdown & Rumsey (*S. lacustris × S. tabernaemontani*, Cyperaceae) in Britain. British & Irish Botany 2,3: 190-201.

Lindberg H, 1902. Die nordeuropäischen Formen von Scirpus (Heleocharis) paluster. Acta Soc. Pro Fauna et Flora Fennica 23,7: 3–16.

Marhold K, Hroudová Z, Duchácek M, Zákravsky P, 2004. The *Bolboschoenus maritimus* group (Cyperaceae) in Central Europe, including *B. laticarpus*, spec. nova. Phyton (Horn.) 44: 1-21.

Schmidt D, Videki R, Tiborcz V, Mesterhazy A, 2019. Hidden on the shore for half a century: Actual distribution of *Schoenoplectus pungens* (Cyperaceae) in Hungary. Studia botanica hungarica. 50,1: 107-20.

Schou JC, 2006. Danmarks halvgræsser. BFN´s Forlag. Klitmøller, Danmark.

Simons E, Gonggrijp S, 2019. *Bolboschoenus planiculmis* (Cyperaceae), a third species of *Bolboschoenus* for the Netherlands. Gorteria 41: 30-34.

Simons E, Wieringa JJ, Gonggrijp S, 2020. *Eleocharis ovata* and its alien allies in the Netherlands. Gorteria 42: 28-35.

Verlove, F, 2015. *Eleocharis engelmannii* and *E. obtusa* (Cyperaceae), two recent acquisitions from series Ovatae in Belgium. Dumortiera 107: 25-30.

Vuksanović S, Bubanja N, Berg C, 2019. *Eleocharis carniolica* W.D.J.Koch, new species in flora of Montenegro. Hacquetia 18,1: 129-35.

Droseraceae

Adamec L, 2018. Biological flora of Central Europe: *Aldrovanda vesiculosa* L. Perspectives in Plant Ecology, Evolution and Systematics 35: 8-21.

Adamec L, Tichý M, 1997. Flowering of *Aldrovanda vesiculosa* in outdoor culture in the Czech Republic and isozyme variability of its European populations. Carnivorous Plant Newsletter 26: 99-103.

Elatinaceae

Popiela A, Łysko A, Białecka B, Bihun MM, Sramkó G, Staroń W, Wieczorek A, Molnár V A, 2017. Seed morphometric characteristics of European species of Elatine Elatinaceae) PeerJ 5: e3399. https://doi.org/10.7717/peerj.3399

Thiery F, Begue P, Ferrez Y, Hennequin C, Tison JM, 2016. *Elatine orthosperma* Duben, une nouvelle espece pour la flore de France. Les Nouvelles Archives de la Flore jurassienne et du nord-est de la France 14.

Uotila P, 2010. Elatinaceae. In Jonsell B, Karlsson T (eds.). The Swedish Museum of Natural History, Flora Nordica 6: 62-69. Stockholm.

Eriocaulaceae

Löve A, Löve D, 1958. The American element in the flora of the British Isles. Botaniska Notiser III: 376-88.

Haloragaceae

Aiken SG, 1981. A conspectus of *Myriophyllum* (Haloragaceae) in North America. Brittonia 33,1: 57-69.

Ceska O, Ceska A, Warrington PD, 1986. *Myriophyllum quitense* and *Myriophyllum ussuriense* (Haloragaceae) in British Columbia, Canada. Brittonia 38,1: 73-81.

Moody M, Les D, 2010. Systematics of the Aquatic Angiosperm Genus *Myriophyllum* (Haloragaceae). Systematic Botany 35: 121-139.

Preston CD, 1998. *Myriophyllum*. In: Rich TCG, Jermy AC, (eds.) 1998. Plant Crib 1998. Botanical Society of Britain and Ireland. https://bsbi.org/wp-content/uploads/dlm_uploads/Myriophyllum_Crib.pdf

Reznicek AA, Voss EG, Walters BS, 2011. Myriophyllum. Michigan Flora Online. University of Michigan. https://michiganflora.net/genus.aspx?id=Myriophyllum.

Scribailo RW, Alix MS, 2014. Haloragaceae. Flora of North America North of Mexico. Flora of North America Association (May 28, 2014).

Wang D, 2002. Taxonomic revision of the genus *Myriophyllum* (Haloragaceae) in China. Rhodora 104,920: 396-421.

Hydrocharitaceae

Bowmer KH, Jacobs SWL, Sainty GR, 1995. Identification, biology and management of *Elodea canadensis*, Hydrocharitaceae. Journal of Aquatic Plant Management. 33: 13-19.

Bräuchler C, 2015. Towards a better understanding of the Najas marina complex: Notes on the correct application and typification of the names N. intermedia, N. major, and N. marina. Taxon 64,5: 1028-30.

Forbes RS, 2000. Assessing the status of *Stratiotes aloides* L. (Water-soldier) in Co. Fermanagh, Northern Ireland (v.c. H33). Watsonia 23: 179-96.

Gettys LA, Enloe SF, 2016. *Hydrilla*: Florida's Worst Submersed Weed. University of Florida, SS-AGR-400.

Issakainen J, Kemppainen E, Mäkelä K, Hakalisto S, Koistinen M, 2011. Hentonäkinruoho (*Najas tenuissima*) ja notkeanäkinruoho (*Najas flexilis*): Suomen uhanalaisia lajeja. Suomen ympäristö 13: 204-7.

Les DH, Jacobs SWL, Tippery NP, Chen L, Moody ML, Wilstermann-Hildebrand M, 2008. Systematics of *Vallisneria* (Hydrocharitaceae). Botany 33,1: 49–65.

Matthews J et al., 2014.Knowledge document for risk analysis of the non-native Brazilian waterweed (*Egeria densa*) in the Netherlands. Dep. of Envir. Sc., Inst. for Water and Wetland Research, Radboud University Nijmegen, Netherlands.

Meriläinen J, 1968. *Najas minor* All. in North America. Rhodora 70,782: 161-66.

Sinkevičienė Z, 2001. *Najas flexilis* ir *Najas minor* Lietuvoje. Botanica Lithuanica 7,2: 203-8.

Triest L, 1988. A Revision of the Genus *Najas* L. (Najadaceae) in the Old World. Academie Royale des Sciences d'Outre-Mer, Brussels.

Vargot EV, Yana B, Uotila P, 2016. Current distribution and conservation of *Najas tenuissima* (Hydrocharitaceae). Nature Conservation Research 1: 2-10.

Voss EG, Reznicek AA, 2012. Field Manual of Michigan Flora. University of Michigan Press, Ann Arbor.

Wasowicz P, Przedpelska-Wasowicz EM, Guðmundsdóttir L, Tamayo M, 2014. *Vallisneria spiralis* and *Egeria densa* (Hydrocharitaceae) in arctic and subarctic Iceland. New Journal of Botany 4,2: 85–89.

Juncaceae

Schou JC, Wind P, Lægaard S, 2010. Danmarks Siv og Frytler. BFN´s Forlag. Klitmøller, Denmark.

Lentibulariaceae

Beretta M, Rodondi G, Adamec L, Andreis C, 2014. Pollen morphology of European bladderworts (Utricularia L., Lentibulariaceae). Review of Palaeobotany and Palynology 205: 22-30.

Christiansen SG, Schou JC, 2016. Thors Blærerod (*Utricularia stygia*) fundet i Holmegårds Mose. URT 40: 3.

Kameyama Y, Toyama M, Ohara M, 2005. Hybrid origins and F1 dominance in the free-floating, sterile bladderwort, Utricularia australis f. australis (Lentibulariaceae). American Journal of Botany 92,3 (Mar): 469-76.

Taylor P, 1989. The genus Utricularia - a taxonomic monograph. Kew Bulletin Additional Series XIV: London.

Thor G, 1979. *Utricularia* i Sverige, specielt de förbisedda arterna *U. australis* och *U. ochroleuca*. Svensk Botanisk Tidskrift. 73: 381-95.

Thor G, 1987. Sumpblāddra, *Utricularia stygia*, en ny svensk art. Svensk Botanisk Tidskrift. 81:273-80.

Nymphaeaceae

Hansen A, 1960. Hybriden *Nuphar luteum x pumilum* fundet i Danmark. Botansik Tidsskrift. 55: 324.

Miller SG, Stanley PC, 1912. The North American species of *Nymphaea*. US National Herbarium, 16,3: 63-108.

Padgett DJ, 2007. A monograph of *Nuphar* (Nymphaceae). Rhodora 109: 1–95.

Tyler T, 2017. Nya och gamla näckrosor i svenska vatten. Botaniska Notiser 150,2: 15-21.

Uotila P, 2001. Nymphaeaceae. In Jonsell B (ed.). Bergius Foundation, Royal Swedish Academy of Sciences, Flora Nordica 2: 216–227. Stockholm.

Onagraceae

Clement EJ, 2000. *Ludwigia ×kentiana* E. J. Clement: A new hybrid aquatic. Watsonia 23: 167-172.

Phrymaceae

Barker W, Nesom G, Beardsley P, Fraga N, 2012. A taxonomic conspectus of Phrymaceae: a narrowed circumscription for Mimulus, new and resurrected genera, and new names and combinations. Phytoneuron. 39.

Lowry DB et al., 2019. The case for the continued use of the genus name Mimulus for all monkeyflowers. Taxon 68,4: 617-623.

Nesom GL, Fraga NS, Barker WR, Beardsley PM, Tank DC, Baldwin BG, Olmstead RG, 2019. Response to "The case for the continued use of the genus name Mimulus for all monkeyflowers". Taxon 68: 624–627.

Plantaginaceae

Aronsson M, Jonsell L, 1996. *Hippuris tetraphylla* ishavshästsvans. Artfaktablad (Species Fact Sheet), Swedish Species Information Centre. Revised by M. Edqvist 2006.

Elven R, Murray DF, Solstad H, 2012. *Hippuris* (Plantaginaceae) Flora of North America Association. http://beta.floranorthamerica.org/Hippuris.

Ericsson S, Bader P, 1999. Ishavshästsvans *Hippuris tetraphylla* åter i Sveriges flora. Svensk Botanisk Tidskrift 93,5-6: 233-47.

Lansdown R, 2008. Water-Starworts (*Callitriche*) of Europe. BSBI Handbook, vol.11. Botanical Society of Britain and Ireland, London.

Lansdown R, 2009. Nomenclatural Notes on *Callitriche* (Callitrichaceae) in North America. Novon 19: 364-69.

Prancl J, Schou JC, 2014. Butfrugtet Vandstjerne *Callitriche obtusangula* Le Gall - ny art for Danmark. URT 38: 29-33.

Poaceae

Richert E, Achtziger R, Dajdok Z, Günther A, Heilmeier H, Hübner A, John H, Šumberová K, 2016. Rare wetland grass *Coleanthus subtilis* in Central and Western Europe - current distribution, habitat types, and threats. Acta Soc Bot Pol. 85,3: 3511. http://dx.doi. org/10.5586/asbp.3511.

Schou JC , Wind P, Lægaard S, 2014. Danmarks græsser. BFN´s forlag. Klitmøller, Denmark.

Potamogetonaceae

Bobrov AA, 2007. *Potamogeton ×fennnicus* (*P. filiformis × P. vaginatus*, Potamogetonaceae) in East Europe. Komarovia 5,1: 1-23.

Bruinsma J, Weyer K van de, 2010. *Potamogeton ×schreberi* G.Fisch. (*Potamogeton natans × P. nodosus*) nieuw voor Nederland. Gorteria 34: 97 -105.

Fischer G, 1907. Die bayerischen Potamogetonen und Zannichellien. In Berichte der Bayerischen Botanischen Gesellschaft, bd. 11. München, 1891-1914.

Gavrilova, Ģ. 2011. Vascular flora of Latvia: 11. Pondweed - *Potamogeton (*Potamogetonaceae). Institute of Biology, University of Latvia.

Hagström JO, 1916. Critical Researches on the *Potamogetons*. Almqvist & Wiksells, Stockholm.

Kaplan Z, 2008. A taxonomic revision of *Stuckenia* (Potamogetonaceae) in Asia, with notes on the diversity and variation of the genus on a worldwide scale. Folia Geobotanica 43: 159 -234.

Kaplan Z, 2010. Hybridization of *Potamogeton* species in the Czech Republic: diversity, distribution, temporal trends and habitat preferences. Preslia 82: 261-87.

Kaplan Z, Fehrer J, 2004. Evidence for the hybrid origin of *Potamogeton ×cooperi* (Potamogetonaceae): Traditional morphology-based taxonomy and molecular techniques in concert. Folia Geobotanica 39: 431-53.

Kaplan Z, Fehrer J, 2013. Molecular identification of hybrids from a former hot spot of *Potamogeton* hybrid diversity. Aquatic Botany 105: 34-40.

Kaplan Z, Jarolímová V, Fehrer J, 2013. Revision of chromosome numbers of Potamogetonaceae: a new basis for taxonomic and evolutionary implications. Preslia 85: 421-82.

Moeslund B, Schou JC, 1994. Noter om danske vandplanter: 4. Danske Vandaks-hybrider. URT 1995: 3.

Preston CD, 1995. Pondweeds of Great Britain and Ireland. BSBI. Handbook vol. 8. Botanical Society of Britain and Ireland. London.

Preston CD, Hollingsworth PM, Gornall RJ, 1999. The distribution and habitat of *Potamogeton × suecicus* K.Richt. (*P. filiformis* Pers. × *P. pectinatus* L.) in the British Isles. Watsonia 22: 329-42.

Smith SJ, Wolfe-Murphy SA, Preston CD, 1991. Irish pondweeds I: A recent record of *Potamogeton x cooperi* (Fryer) Fryer from Co. Antrim. Irish Naturalists' Journal, 23,11: 457-58.

Talavera S, Murillo PG, Smit H, 1986. Sobre el genero *Zannichellia* L. (Zannichelliaceae). Lagascalia 14: 241–71.

Wiegleb G, 1990. The importance of stem anatomical characters for the systematics of the genus *Potamogeton* L. Flora 184: 197-208.

Wiegleb G, Brux H, 1991. Comparison of life-history characters of broad-leaved species of the genus *Potamogeton* L.: I. General characterization of morphology and reproductive strategies. Aquatic Botany 39: 131-46.

Wiegleb G, Kaplan Z, 1998. An account of the species of *Potamogeton* L. (Potamogetonaceae). Folia Geobotanica 33: 241-316.

Wiegleb G, Weyer K van de, Bolbrinker P, Wolff P, 2008. *Potamogeton* -Hybriden in Deutschland. Feddes Repertorium 119,5–6: 433–48.

Zalewska-Gałosza J, 2002. Occurrence and distribution of *Potamogeton* hybrid (Potamogetonaceae) in Poland. Feddes Repertorium 113,5-6: 380 -93.

Zalewska-Gałosza J, Ronikier M, 2010. Are linear-leaved *Potamogeton* hybrids really so rare? Molecular evidence for multiple hybridizations between *P. acutifolius* and *P. compressus* in central Europe. Nordic Journal of Botany 28: 257-61.

Ranunculaceae

Bobrov A, Zalewska-Gałosz J, Jopek M, Movergoz EA, 2015. *Ranunculus schmalhausenii* (section *Batrachium*, Ranunculaceae), a neglected water crowfoot endemic to Fennoscandia - a case of rapid hybrid speciation in postglacial environment of North Europe. Phytotaxa 233: 101-38.

Cook CDK, 1966. A monographic study of *Ranunculus* subgenus *Batrachium* (DC.) A.Gray. Mitt Bot Staatssamml München 6: 46–237.

Dahlgren G , 1993. *Ranunculus penicillatus* in Norden. Nordic Journal of Botany 13: 593–605.

Gornall RJ, 1987. Notes on a hybrid spearwort, *Ranunculus flammula* L. x *R. reptans* L. Watsonia 16: 383-88.

Hiern WP, 1871. On the forms and distribution over the world of the Batrachium section of Ranunculus. Journal of Botany 9: 43–49, 65–69, 97–107.

Holmes NTH, 1979. A guide to identification of *Batrachium Ranunculus* species of Britain. Nature Conservancy Council, Chief Scientist's Team Notes 14.

Hong DY, 1991. A biosystematic study on *Ranunculus* subgenus *Batrachium* in S Sweden. Nordic Journal of Botany 11: 41–59.

Kaplan Z, Danihelka J, Chrtek Jr J, Zázvorka J, Koutecký P, Ekrt L, Řepka R, Štěpánková J, Jelinek B, Grulich V, Wild J, 2019. Distributions of vascular plants in the Czech Republic: Part 8. Preslia 91: 257–368.

Lansdown RV, 2015. *Ranunculus* subgenus *Batrachium* (DC.) A.Gray. In Stace CA, Preston CD, Pearman DA (eds.), Hybrid flora of the British Isles 36–41. Botanical Society of Britain and Ireland, Bristol, UK.

Padmore PA, 1957. The varieties of *Ranunculus flammula* L. and the status of *R. scoticus* E. S. Marshall and of *R. reptans* L. Watsonia 4: 19-27.

Pizarro J, 1995. Contributión al studio taxonómico de Ranunculus L. subgen. Batrachium (DC.) A.Gray (Ranunculaceae). Lazaroa 15: 21–113.

Prančl J, Koutecký P, Trávníček P, Jarolímová V, Lučanová M, Koutecká E, Kaplan Z, 2018. Cytotype, variation, cryptic diversity and hybridization in *Ranunculus* sect. *Batrachium* revealed by flow cytometry and chromosome number. Preslia 90: 195–223.

Sell P, Murrell G (eds.), 2018. Subgenus 3. *Batrachium* (DC.) A.Gray. In Flora of the British Isles and Ireland, vol. 1, 212-18. Cambridge University Press, Cambridge, UK.

Vollrath H, Kohler A, 1972. *Batrachium*-Fundorte aus bayerischen Naturräumen. Ber. Bayer. Bot. Ges. 43: 63–75.

Webster SD, 1988. *Ranunculus penicillatus* (Dumort.) Bab. in Great Britain and Ireland. Watsonia 17: 1–22.

Webster SD. 1988. *Ranunculus* subgenus *Batrachium*. Botanical Society of Britain and Ireland. https://bsbi.org/wp-content/uploads/dlm_uploads/Ranunculus_Batrachium_Crib.pdf.

Wiegleb G, 2020. *Ranunculus* sektion *Batrachium*. Jäger EJ, Kadereit, JW (eds.), Die Ranunculaceae der Flora von Zentraleuropa, Gesellschaft zur Erforschung der Flora Deutschlands, Berlin. http://www.flora-deutschlands.de/ranunculaceae.html.

Wiegleb G, Bobrov AA, Zalewska-Gałosz J, 2017. A taxonomic account of *Ranunculus* section *Batrachium* (Ranunculaceae). Phytotaxa 319: 1–55.

Zając A, Zając M (eds.), 2019. Distribution atlas of vascular plants in Poland: Appendix. Institute of Botany, Jagiellonian University, Kraków.

Zalewska-Gałosz J, Jopik M, Ilnicki T, 2015. Hybridization in *Batrachium* group: Controversial delimitation between heterophyllous *Ranunculus penicillatus* and the hybrid *Ranunculus fluitans × R. peltatus*. Aquatic Botany 120: 160-68.

Scrophulariaceae

Foerster E, 1967. Die wasserehrenpreis-gruppe. Gött. Flor. Rundbr. 1,3: 10–11.

Typhaceae

Bernhardt KG, Gregor T, 2019. Vielfalt aus aller Welt - neophytische Typha-Arten in Mitteleuropa. Kochia 12: 99-113.

Grace JB, Harrison JS, 1986. The biology of Canadian weeds. 73. Typha latifolia L., Typha angustifolia L. and Typha x glauca Godr. Can. J. Pl. Sci. 66: 361-379.

Nikel A, Bróż E, Czylok A, Michalewska A, Baryła J, Nobis M, Piwowarczyk R, Poloczek A, 2005. *Typha laxmannii* Lepech. The new, expansive kenophyte in Poland: Distribution and taxonomy. Acta Soc. Bot. Pol. 74: 25-28.

Nowinska R, Gawronska B, Czarna A, Wyrzykiewicz-Raszewska M, 2014. *Typha glauca* Godron and its parental plants in Poland: Taxonomic characteristics. Hydrobiologia 737:163-81.

Weyer K van de, 1996. Typha x glauca Godr. (Typha angustifolia L. x T. latifolia L.) an den Hausdülmener Fischteichen (Westfalen). Flor. Rundbr. 39,2: 91-93.

Index of latin names

Accepted names in bold. Synonyms, hybrids, varieties and taxa only shortly described or mentioned in smaller font size.